PRINCIPLES OF COMMUNICATIONS

Systems, Modulation, and Noise

SIXTH EDITION

RODGER E. ZIEMER
University of Colorado at Colorado Springs

WILLIAM H. TRANTER
Virginia Polytechnic Institute and State University

John Wiley & Sons, Inc.

VICE PRESIDENT AND EXECUTIVE PUBLISHER	Donald Fowley
ASSOCIATE PUBLISHER	Daniel Sayre
PRODUCTION SERVICES MANAGER	Dorothy Sinclair
PRODUCTION EDITOR	Janet Foxman
MARKETING MANAGER	Christopher Ruel
CREATIVE DIRECTOR	Harry Nolan
SENIOR DESIGNER	Kevin Murphy
EDITORIAL ASSISTANT	Carolyn Weisman
MEDIA EDITOR	Lauren Sapira
PRODUCTION SERVICES	Sumit Shridhar/Thomson Digital
COVER DESIGN	David Levy

This book was set in 10/12 Times New Roman by Thomson Digital and printed and bound by RRD Crawfordsville. The cover was printed by RRD Crawfordsville.

This book is printed on acid-free paper.

To order books or for customer service, please call 1-800-CALL WILEY (225-5945).

Library of Congress Cataloging in Publication Data:

Ziemer, Rodger E.

Principles of communications : systems, modulation, and noise / R.E. Ziemer, W.H. Tranter.—6th ed.

p. cm.

Includes bibliographical references and index.

ISBN 978-0-470-25254-3 (cloth)

1. Telecommunication. 2. Signal theory (Telecommunication) I. Tranter, William H. II. Title.

TK5105.Z54 2009

621.382'2—dc22

2008042932

Printed in the United States of America

10 9 8 7 6 5 4 3 2 1

To our families.

Rodger Ziemer and Bill Tranter

PREFACE

As in previous editions, the objective of this book is to provide, in a single volume, a thorough treatment of the principles of communication systems, both analog and digital, at the physical layer. As with the previous five editions of this book, the sixth edition targets both senior-level and beginning graduate students in electrical and computer engineering. Although a previous course on signal and system theory would be useful to students using this book, an overview of this fundamental background material is included early in the book (Chapter 2). A significant change in the sixth edition is the addition of a new chapter (Chapter 4) covering the principles of baseband data transmission. Included in this new chapter are line codes, pulse shaping and intersymbol interference, zero-forcing equalization, eye diagrams, and basic ideas on symbol synchronization without the complicating factor of noise. Following overview chapters on probability and random processes (Chapters 5 and 6), the book turns to the central theme of characterizing the performance of both analog (Chapter 7) and digital (Chapters 8–11) communication systems in the presence of noise. Significant additions to the book include an expanded treatment of phase-locked loops, including steady-state tracking errors of first-order, second-order, and third-order loops, the derivation and comparative performances of M-ary digital modulation systems, an expanded treatment of equalization, and the relative bit error rate performance of BCH, Reed-Solomon, Golay, and convolutional codes. Each chapter contains a number of worked examples as well as several computer examples, a summary delineating the important points of the chapter, references, homework problems, and computer problems.

Enabled by rapid and continuing advances in microelectronics, the field of communications has seen many innovations since the first edition of this book was published in 1976. The cellular telephone is a ubiquitous example. Other examples include wireless networks, satellite communications including commercial telephone, television and radio, digital radio and television, and GPS systems, to name only a few. While there is always a strong desire to include a variety of new applications and technologies in a new edition of a book, we continue to believe that a first course in communications serves the student best if the emphasis is placed on fundamentals. We feel that application examples and specific technologies, which often have short lifetimes, are best treated in subsequent courses after students have mastered the basic theory and analysis techniques. We have, however, been sensitive to new techniques that are fundamental in nature and have added material as appropriate. As examples, sections on currently important areas such as spread spectrum techniques, cellular communications, and orthogonal frequency-division multiplexing are provided. Reactions to previous editions have shown that emphasizing fundamentals, as opposed to specific technologies, serve the user well while keeping the length of the book reasonable. This strategy appears to have worked well for advanced undergraduates, for new graduate students who may have forgotten some of the fundamentals, and for the working engineer who may use the book as a reference or who may be taking a course after-hours.

A feature of the previous edition of *Principles of Communications* was the inclusion of several computer examples within each chapter. (MATLAB was chosen for these examples because of its widespread use in both academic and industrial settings, as well as for MATLAB's rich graphics library.) These computer examples, which range from programs for computing performance curves to simulation programs for certain types of communication systems and algorithms, allow the student to observe the behavior of more complex systems without the need for extensive computations. These examples also expose the student to modern computational tools for analysis and simulation in the context of communication systems. Even though we have limited the amount of this material in order to ensure that the character of the book is not changed, the number of computer examples has been increased for the sixth edition. In addition to the in-chapter computer examples, a number of "computer exercises" are included at the end of each chapter. The number of these has also been increased in the sixth edition. These exercises follow the end-of-chapter problems and are designed to make use of the computer in order to illustrate basic principles and to provide the student with additional insight. A number of new problems are included at the end of each chapter in addition to a number of problems that were revised from the previous edition.

The publisher maintains a web site from which the source code for all in-chapter computer examples may be downloaded. The URL is www.wiley.com/college/ziemer. We recommend that, although MATLAB code is included in the text, students download MATLAB code of interest from the publisher website. The code in the text is subject to printing and other types of errors and is included to give the student insight into the computational techniques used for the illustrative examples. In addition, the MATLAB code on the publisher website is periodically updated as need justifies. This web site also contains complete solutions for the end-of-chapter problems and computer exercises. (The solutions manual is password protected and is intended only for course instructors.)

In order to compare the sixth edition of this book with the previous edition, we briefly consider the changes chapter by chapter.

In Chapter 1, the tables have been updated. In particular Table 1.1, which identifies major developments in communications, includes advances since the last edition of this book was published. The role of the ITU and the FCC for allocating spectrum has been reworked. References to turbo codes and to LDPC codes are now included.

Chapter 2, which is essentially a review of signal and system theory, remains basically unchanged. However, several examples have been changed and two new examples have been added. The material on complex envelopes has been clarified.

Chapter 3, which is devoted to basic modulation techniques, makes use of complex envelope notation in the presentation of frequency modulation in order to build upon the ideas presented in Chapter 2. In addition, Chapter 3 has been expanded to include significantly more material on phase-locked loops operating in both the acquisition and tracking modes. The phase-locked loop is a key building block of many communication system components including frequency and phase demodulators, digital demodulators, and carrier and symbol synchronizers.

Chapter 4, which is a new chapter for the sixth edition, covers basic digital transmission techniques including line codes, pulse shaping and filtering, intersymbol interference, equalization, eye diagrams, and basic synchronization techniques. Covering this material early in the book allows the student to appreciate the differences between analog and digital transmission techniques. This material is also presented without considering the complicating effects of noise.

Chapters 5 and 6, which deal with basic probability theory and random processes, have not been significantly changed from the previous edition. Some of the material has been rearranged to increase clarity and readability.

Chapter 7 treats the noise performance of various analog modulation schemes and also contains a brief discussion of pulse-code modulation. The introduction to this chapter has been expanded to reflect the importance of noise and the sources of noise. This also serves to better place Appendix A in context. In addition, this material has been reorganized so that it flows better and is easier for the student to follow.

Binary digital data transmission in the presence of noise is the subject of Chapter 8. A section on the noise performance of M-ary PAM systems has been added. The material dealing with the noise performance of zero-ISI systems has been expanded as well as the material on equalization. An example has been added which compares various digital transmission schemes.

Chapter 9 treats more advanced topics in data communication systems including M-ary systems, synchronization, spread-spectrum systems, multicarrier modulation and OFDM, satellite links, and cellular radio communications. Derivations are now provided for the error probability of M-ary QAM and NCFSK. A figure comparing PSK, DPSK, and QAM has been added as well as a figure comparing CFSK and NCFSK. The derivation of the power density for quadrature modulation schemes has been expanded as well as the material on synchronization. The treatment of multicarrier modulation has also been expanded and information on 3G cellular has been added.

Chapter 10, which deals with optimum receivers and signal-space concepts, is little changed from the previous edition.

Chapter 11 provides the student with a brief introduction to the subjects of information theory and coding. Our goal at the level of this book is not to provide an in-depth treatment of information and coding but to give the student an appreciation of how the concepts of information theory can be used to evaluate the performance of systems and how the concepts of coding theory can be used to mitigate the degrading effects of noise in communication systems. To this end we have expanded the computer examples to illustrate the performance of BCH codes, the Golay code, and convolutional codes in the presence of noise.

We have used this text for various types of courses for a number of years. This book was originally developed for a two-semester course sequence, with the first course covering basic background material on linear systems and noiseless modulation (Chapters 1–4) and the second covering noise effects on analog and digital modulation systems (Chapters 7–11). With a previous background by the students in linear systems and probability theory, we know of several instances where the book has been used for a one-semester course on analog and digital communication system analysis in noise. While probably challenging for all but the best students, this nevertheless gives an option that will get students exposed to modulation system performance in noise in one semester. In short, we feel that it is presumptuous for us to tell instructors using the book what material to cover and in what order. Suffice it to say we feel that there is more than enough material included in the book to satisfy almost any course design at the senior or beginning graduate levels.

We wish to thank the many persons who have contributed to the development of this textbook and who have suggested improvements for the sixth edition. We especially thank our colleagues and students at the University of Colorado at Colorado Springs, the Missouri

University of Science and Technology, and Virginia Tech for their comments and suggestions. The help of Dr. William Ebel at St. Louis University is especially acknowledged. We also express our thanks to the many colleagues who have offered suggestions to us by correspondence or verbally. The industries and agencies that have supported our research deserve special mention since, by working with them on various projects, we have expanded our knowledge and insight significantly. These include the National Aeronautics and Space Administration, the Office of Naval Research, the National Science Foundation, GE Aerospace, Motorola Inc., Emerson Electric Company, Battelle Memorial Institute, DARPA, Raytheon, and the LGIC Corporation. The expert support of Cyndy Graham, who worked through many of the LaTeX-related problems and who contributed significantly to the development of the solutions manual is gratefully acknowledged.

We also thank the reviewers of this and all previous editions of this book. The reviewers for the sixth edition deserve special thanks for their help and guidance. They were:

Larry Milstein, University of California – San Diego

Behnam Kamali, Mercer University

Yao Ma, Iowa State University

Michael Honig, Northwestern University

Emad Ebbini, University of Minnesota

All reviewers, past and present, contributed significantly to this book. They caught many errors and made many valuable suggestions. The authors accept full responsibility for any remaining errors or shortcomings.

Finally, our families deserve much more than a simple thanks for the patience and support that they have given us throughout more than thirty years of seemingly endless writing projects. It is to them that this book is dedicated.

Rodger E. Ziemer
William H. Tranter

CONTENTS

CHAPTER 10
OPTIMUM RECEIVERS AND SIGNAL SPACE CONCEPTS 554

CHAPTER 1

INTRODUCTION

We are said to live in an era called the intangible economy, driven not by the physical flow of material goods but rather by the flow of information. If we are thinking about making a major purchase, for example, chances are we will gather information about the product by an Internet search. Such information gathering is made feasible by virtually instantaneous access to a myriad of facts about the product, thereby making our selection of a particular brand more informed. When one considers the technological developments that make such instantaneous information access possible, two main ingredients surface: a reliable, fast means of communication and a means of storing the information for ready access, sometimes referred to as the *convergence* of communications and computing.

This book is concerned with the theory of systems for the conveyance of information. A *system* is a combination of circuits and/or devices that is assembled to accomplish a desired task, such as the transmission of intelligence from one point to another. Many means for the transmission of information have been used down through the ages ranging from the use of sunlight reflected from mirrors by the Romans to our modern era of electrical communications that began with the invention of the telegraph in the 1800s. It almost goes without saying that we are concerned about the theory of systems for *electrical* communications in this book.

A characteristic of electrical communication systems is the presence of uncertainty. This uncertainty is due in part to the inevitable presence in any system of unwanted signal perturbations, broadly referred to as *noise*, and in part to the unpredictable nature of information itself. Systems analysis in the presence of such uncertainty requires the use of probabilistic techniques.

Noise has been an ever-present problem since the early days of electrical communication, but it was not until the 1940s that probabilistic systems analysis procedures were used to analyze and optimize communication systems operating in its presence (Wiener, 1949; Rice 1944, 1945).[1] It is also somewhat surprising that the unpredictable nature of information was not widely recognized until the publication of Claude Shannon's mathematical theory of communications (Shannon, 1948) in the late 1940s. This work was the beginning of the science of information theory, a topic that will be considered in some detail later.

Major historical facts related to the development of electrical communications are given in Table 1.1.

[1]Refer to Historical References in the Bibliography.

Table 1.1 Major Events and Inventions in the Development of Electrical Communications

Year	Event
1791	Alessandro Volta invents the galvanic cell, or battery.
1826	Georg Simon Ohm establishes a law on the voltage–current relationship in resistors.
1838	Samuel F. B. Morse demonstrates the telegraph.
1864	James C. Maxwell predicts electromagnetic radiation.
1876	Alexander Graham Bell patents the telephone.
1887	Heinrich Hertz verifies Maxwell's theory.
1897	Guglielmo Marconi patents a complete wireless telegraph system.
1904	John Fleming patents the thermionic diode.
1905	Reginald Fessenden transmits speech signals via radio.
1906	Lee De Forest invents the triode amplifier.
1915	The Bell System completes a U.S. transcontinental telephone line.
1918	B. H. Armstrong perfects the superheterodyne radio receiver.
1920	J. R. Carson applies sampling to communications.
1925–1927	First television broadcasts in England and the United States.
1931	Teletypwriter service is initialized.
1933	Edwin Armstrong invents frequency modulation.
1936	Regular television broadcasting begun by the British Broadcasting Corporation.
1937	Alec Reeves conceives pulse-code modulation (PCM).
WWII	Radar and microwave systems are developed. Statistical methods are applied to signal extraction problems.
1944	Computers put into public service (government owned).
1948	The transister is invented by W. Brattain, J. Bardeen, and W. Shockley.
1948	Claude Shannon's *A Mathematical Theory of Communications* is published.
1950	Time-division multiplexing is applied to telephoney.
1956	First successful transoceanic telephone cable.
1959	Jack Kilby patents the "Solid Circuit"—precurser to the integrated circuit.
1960	First working laser demonstrated by T. H. Maiman of Hughes Research Labs. (Patent awarded to G. Gould after a 20 year dispute with Bell Labs.)
1962	First communications satellite, Telstar I, launched.
1966	First successful facsimile (FAX) machine.
1967	U.S. Supreme Court Carterfone decision opens the door for modem development.
1969	Live television coverage of the manned moon exploration (Apollo 11).
1969	First Internet started—ARPANET.
1970	Low-loss optic fiber developed.
1971	Microprocessor invented.
1975	Ethernet patent filed.
1976	Apple I home computer invented.
1977	Live telephone traffic carried by a fiber-optic cable system.
1977	Interplanetary grand tour launched: Jupiter, Saturn, Uranus, and Neptune.
1979	First cellular telephone network started in Japan.
1981	IBM personal computer developed and sold to public.
1981	Hayes Smartmodem marketed (automatic dial-up allowing computer control).
1982	Compact disc (CD) audio based on 16-bit PCM developed.
1983	First 16-bit programmable digital signal processors sold.
1984	Divestiture of AT&T's local operations into seven Regional Bell Operating Companies.
1985	Desktop publishing programs first sold. Ethernet developed.
1988	First commercially available flash memory (later applied in cellular phones, etc.).

1988	Asymmetric digital subscriber lines (ADSL) developed.
1990s	Very small aperture satellites (VSATs) become popular.
1991	Application of echo cancellation results in low-cost 14,400-bps modems.
1993	Invention of turbo coding allows approach to Shannon limit.
mid-1990s	Second generation (2G) cellular systems fielded.
1995	Global Positioning System (GPS) reaches full operational capability.
1996	All-digital phone systems result in modems with 56 kbps download speeds.
late 1990s	Widespread personal and commercial applications of the Internet. High definition TV becomes mainstream.
2001	Apple iPoD first sold (October); 100 million sold by April 2007. Fielding of 3G cellular telephone systems begins. WiFi and WiMAX allow wireless access to the Internet and electronic devices wherever mobility is desired.
2000s	Wireless sensor networks, originally conceived for military applications, find civilian applications such as environment monitoring, healthcare applications, home automation, and traffic control as well.

It is an interesting fact that the first electrical communication system, the telegraph, was digital—that is, it conveyed information from point to point by means of a digital code consisting of words composed of dots and dashes.[2] The subsequent invention of the telephone 38 years after the telegraph, wherein voice waves are conveyed by an analog current, swung the pendulum in favor of this more convenient means of word communication for about 75 years [see Oliver et al. (1948)].

One may rightly ask, in view of this history, why the almost complete domination by digital formatting in today's world? There are several reasons among which are

1. Media integrity: A digital format suffers much less deterioration in reproduction than does an analog record.
2. Media integration: Whether a sound, picture, or naturally digital data such as a word file, all are treated the same when in digital format.
3. Flexible interaction: The digital domain is much more convenient for supporting anything from one-on-one to many-to-many interactions.
4. Editing: Whether text, sound, images, or video, all are conveniently and easily edited when in digital format.

With this brief introduction and history, we now look in more detail at the various components that make up a typical communication system.

1.1 BLOCK DIAGRAM OF A COMMUNICATION SYSTEM

Figure 1.1 shows a commonly used model for a single-link communication system. Although it suggests a system for communication between two remotely located points, this block diagram is also applicable to remote sensing systems, such as radar or sonar, in which the system input and output may be located at the same site. Regardless of the particular application and configuration, all information transmission systems invariably involve three major subsystems—a transmitter, the channel, and a receiver. In this book we will usually be thinking in terms of

[2]In the actual physical telegraph system, a dot was conveyed by a short double click by closing and opening of the circuit with the telegrapher's key (a switch), while a dash was conveyed by a longer double click by an extended closing of the circuit by means of the telegrapher's key.

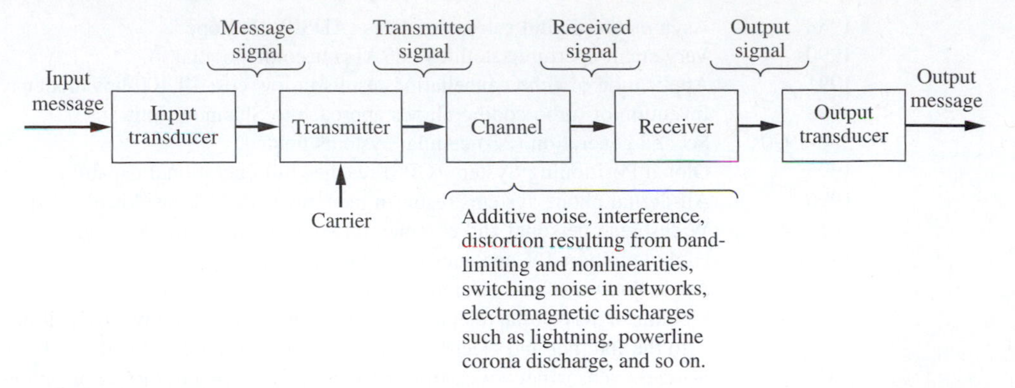

Figure 1.1
The Block Diagram of a Communication System.

systems for transfer of information between remotely located points. It is emphasized, however, that the techniques of systems analysis developed are not limited to such systems.[3]

We will now discuss in more detail each functional element shown in Figure 1.1.

Input Transducer The wide variety of possible sources of information results in many different forms for messages. Regardless of their exact form, however, messages may be categorized as *analog* or *digital.* The former may be modeled as functions of a continuous-time variable (for example, pressure, temperature, speech, music), whereas the latter consist of discrete symbols (for example, written text). Almost invariably, the message produced by a source must be converted by a transducer to a form suitable for the particular type of communication system employed. For example, in electrical communications, speech waves are converted by a microphone to voltage variations. Such a converted message is referred to as the *message signal.* In this book, therefore, a *signal* can be interpreted as the variation of a quantity, often a voltage or current, with time.

Transmitter The purpose of the transmitter is to couple the message to the channel. Although it is not uncommon to find the input transducer directly coupled to the transmission medium, as, for example, in some intercom systems, it is often necessary to *modulate* a carrier wave with the signal from the input transducer. *Modulation* is the systematic variation of some attribute of the carrier, such as amplitude, phase, or frequency, in accordance with a function of the message signal. There are several reasons for using a carrier and modulating it. Important ones are (1) for ease of radiation, (2) to reduce noise and interference, (3) for channel assignment, (4) for multiplexing or transmission of several messages over a single channel, and (5) to overcome equipment limitations. Several of these reasons are self-explanatory; others, such as the second, will become more meaningful later.

[3]More complex communications systems are the rule rather than the norm: a broadcast system, such as television or commercial rado, is a one-to-many type of situation which is composed of several sinks receiving the same information from a single source; a multiple-access communication system is where many users share the same channel and is typified by satellite communications systems; a many-to-many type of communications scenario is the most complex and is illustrated by examples such as the telephone system and the Internet, both of which allow communication between any pair out of a multitude of users. For the most part, we consider only the simplest situation in this book of a single sender to a single receiver, although means for sharing a communication resource will be dealt with under the topics of multiplexing and multiple access.

In addition to modulation, other primary functions performed by the transmitter are filtering, amplification, and coupling the modulated signal to the channel (for example, through an antenna or other appropriate device).

Channel The channel can have many different forms; the most familiar, perhaps, is the channel that exists between the transmitting antenna of a commercial radio station and the receiving antenna of a radio. In this channel, the transmitted signal propagates through the atmosphere, or free space, to the receiving antenna. However, it is not uncommon to find the transmitter hardwired to the receiver, as in most local telephone systems. This channel is vastly different from the radio example. However, all channels have one thing in common: the signal undergoes degradation from transmitter to receiver. Although this degradation may occur at any point of the communication system block diagram, it is customarily associated with the channel alone. This degradation often results from noise and other undesired signals or interference but also may include other distortion effects as well, such as fading signal levels, multiple transmission paths, and filtering. More about these unwanted perturbations will be presented shortly.

Receiver The receiver's function is to extract the desired message from the received signal at the channel output and to convert it to a form suitable for the output transducer. Although amplification may be one of the first operations performed by the receiver, especially in radio communications, where the received signal may be extremely weak, the main function of the receiver is to *demodulate* the received signal. Often it is desired that the receiver output be a scaled, possibly delayed, version of the message signal at the modulator input, although in some cases a more general function of the input message is desired. However, as a result of the presence of noise and distortion, this operation is less than ideal. Ways of approaching the ideal case of perfect recovery will be discussed as we proceed.

Output Transducer The output transducer completes the communication system. This device converts the electric signal at its input into the form desired by the system user. Perhaps the most common output transducer is a loudspeaker. However, there are many other examples, such as tape recorders, personal computers, meters, and cathode ray tubes, to name only a few.

1.2 CHANNEL CHARACTERISTICS

1.2.1 Noise Sources

Noise in a communication system can be classified into two broad categories, depending on its source. Noise generated by components within a communication system, such as resistors, electron tubes, and solid-state active devices is referred to as *internal noise*. The second category, *external noise*, results from sources outside a communication system, including atmospheric, man-made, and extraterrestrial sources.

Atmospheric noise results primarily from spurious radio waves generated by the natural electrical discharges within the atmosphere associated with thunderstorms. It is commonly referred to as *static* or *spherics*. Below about 100 MHz, the field strength of such radio waves is inversely proportional to frequency. Atmospheric noise is characterized in the time domain by large-amplitude, short-duration bursts and is one of the prime examples of noise referred to as *impulsive*. Because of its inverse dependence on frequency, atmospheric noise affects commercial amplitude modulation (AM) broadcast radio, which occupies the frequency range from 540 kHz to 1.6 MHz, more than it affects television and frequency modulation (FM) radio, which operate in frequency bands above 50 MHz.

Man-made noise sources include high-voltage powerline corona discharge, commutator-generated noise in electrical motors, automobile and aircraft ignition noise, and switching-gear noise. Ignition noise and switching noise, like atmospheric noise, are impulsive in character. Impulse noise is the predominant type of noise in switched wireline channels, such as telephone channels. For applications such as voice transmission, impulse noise is only an irritation factor; however, it can be a serious source of error in applications involving transmission of digital data.

Yet another important source of man-made noise is radio-frequency transmitters other than the one of interest. Noise due to interfering transmitters is commonly referred to as *radio-frequency interference* (RFI). Radio-frequency interference is particularly troublesome in situations in which a receiving antenna is subject to a high-density transmitter environment, as in mobile communications in a large city.

Extraterrestrial noise sources include our sun and other hot heavenly bodies, such as stars. Owing to its high temperature (6000°C) and relatively close proximity to the earth, the sun is an intense, but fortunately localized source of radio energy that extends over a broad frequency spectrum. Similarly, the stars are sources of wideband radio energy. Although much more distant and hence less intense than the sun, nevertheless they are collectively an important source of noise because of their vast numbers. Radio stars such as quasars and pulsars are also intense sources of radio energy. Considered a signal source by radio astronomers, such stars are viewed as another noise source by communications engineers. The frequency range of solar and cosmic noise extends from a few megahertz to a few gigahertz.

Another source of interference in communication systems is multiple transmission paths. These can result from reflection off buildings, the earth, airplanes, and ships or from refraction by stratifications in the transmission medium. If the scattering mechanism results in numerous reflected components, the received multipath signal is noiselike and is termed *diffuse*. If the multipath signal component is composed of only one or two strong reflected rays, it is termed *specular*. Finally, signal degradation in a communication system can occur because of random changes in attenuation within the transmission medium. Such signal perturbations are referred to as *fading*, although it should be noted that specular multipath also results in fading due to the constructive and destructive interference of the received multiple signals.

Internal noise results from the random motion of charge carriers in electronic components. It can be of three general types: the first, referred to as *thermal noise*, is caused by the random motion of free electrons in a conductor or semiconductor excited by thermal agitation; the second, called *shot noise*, is caused by the random arrival of discrete charge carriers in such devices as thermionic tubes or semiconductor junction devices; the third, known as *flicker noise*, is produced in semiconductors by a mechanism not well understood and is more severe the lower the frequency. The first type of noise source, *thermal noise*, is modeled analytically in Appendix A, and examples of system characterization using this model are given there.

1.2.2 Types of Transmission Channels

There are many types of transmission channels. We will discuss the characteristics, advantages, and disadvantages of three common types: electromagnetic wave propagation channels, guided electromagnetic wave channels, and optical channels. The characteristics of all three may be explained on the basis of electromagnetic wave propagation phenomena. However, the characteristics and applications of each are different enough to warrant considering them separately.

Electromagnetic Wave Propagation Channels

The possibility of the propagation of electromagnetic waves was predicted in 1864 by James Clerk Maxwell (1831–1879), a Scottish mathematician who based his theory on the experimental work of Michael Faraday. Heinrich Hertz (1857–1894), a German physicist, carried out experiments between 1886 and 1888 using a rapidly oscillating spark to produce electromagnetic waves, thereby experimentally proving Maxwell's predictions. Therefore, by the latter part of the nineteenth century, the physical basis for many modern inventions utilizing electromagnetic wave propagation—such as radio, television, and radar—was already established.

The basic physical principle involved is the coupling of electromagnetic energy into a propagation medium, which can be free space or the atmosphere, by means of a radiation element referred to as an *antenna*. Many different propagation modes are possible, depending on the physical configuration of the antenna and the characteristics of the propagation medium. The simplest case—which never occurs in practice—is propagation from a point source in a medium that is infinite in extent. The propagating wave fronts (surfaces of constant phase) in this case would be concentric spheres. Such a model might be used for the propagation of electromagnetic energy from a distant spacecraft to earth. Another idealized model, which approximates the propagation of radio waves from a commercial broadcast antenna, is that of a conducting line perpendicular to an infinite conducting plane. These and other idealized cases have been analyzed in books on electromagnetic theory. Our purpose is not to summarize all the idealized models but to point out basic aspects of propagation phenomena in practical channels.

Except for the case of propagation between two spacecraft in outer space, the intermediate medium between transmitter and receiver is never well approximated by free space. Depending on the distance involved and the frequency of the radiated waveform, a terrestrial communication link may depend on line-of-sight, ground-wave, or ionospheric skip-wave propagation (see Figure 1.2). Table 1.2 lists frequency bands from 3 kHz to 3×10^6 GHz, along with letter designations for microwave bands used in radar among other applications (WWII and current). Note that the frequency bands are given in decades; the VHF band has 10 times as much frequency space as the HF band. Table 1.3 shows some bands of particular interest.[4]

General spectrum allocations are arrived at by international agreement. The present system of frequency allocations is administered by the International Telecommunications Union (ITU), which is responsible for the periodic convening of Administrative Radio Conferences on a regional or a worldwide basis (WARC before 1995; WRC 1995 and after, standing for World Radiocommunication Conference).[5] The responsibility of the WRC is the

[4]Bennet Z. Kobb, *Spectrum Guide*, 3rd ed., New Signals Press, Falls Church, VA, 1996. Bennet Z. Kobb, *Wireless Spectrum Finder*, McGraw-Hill, New York, 2001.

[5]See A. F. Inglis, *Electronic Communications Handbook*, McGraw-Hill, New York, 1988, Chapter 3. WARC-79, WARC-84, and WARC-92, all held in Geneva, Switzerland, have been the last three held under the WARC designation; WRC-95, WRC-97, WRC-2000 (Istanbul), WRC-03, and WRC-07 are those held under the WRC designation.

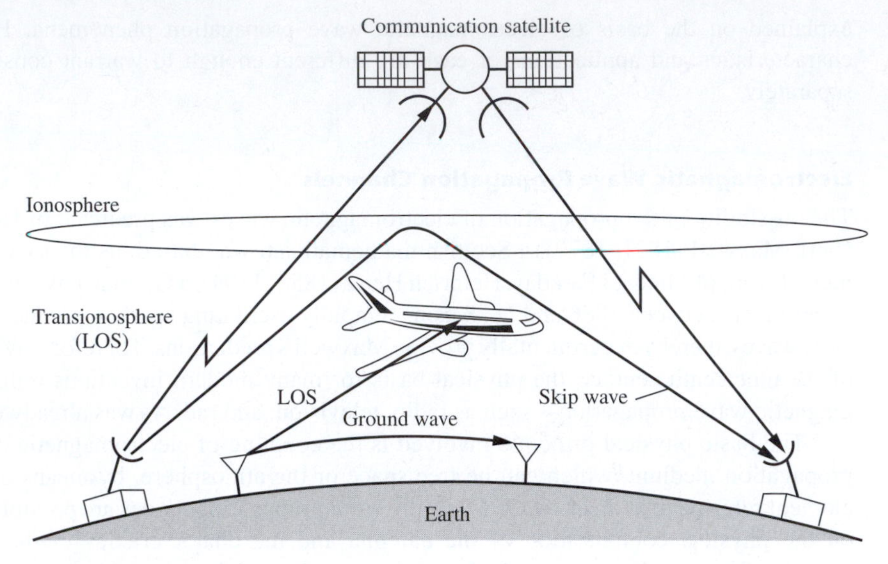

Figure 1.2
The various propagation modes for electromagnetic waves.
(LOS stands for line of sight)

Table 1.2 Frequency Bands with Designations

Frequency band	Name	Microwave band (GHz)	Letter designation Old	Letter designation Current
3–30 kHz	Very low frequency (VLF)	0.5–1.0		C
30–300 kHz	Low frequency (LF)	1.0–2.0	L	D
300–3000 kHz	Medium frequency (MF)	2.0–3.0	S	E
3–30 MHz	High frequency (HF)	3.0–4.0	S	F
30–300 MHz	Very high frequency (VHF)	4.0–6.0	C	G
0.3–3 GHz	Ultrahigh frequency (UHF)	6.0–8.0	C	H
3–30 GHz	Superhigh frequency (SHF)	8.0–10.0	X	I
30–300 GHz	Extremely high frequency (EHF)	10.0–12.4	X	J
43–430 THz	Infrared (0.7–7 μm)	12.4–18.0	Ku	J
430–750 THz	Visible light (0.4–0.7 μm)	18.0–20.0	K	J
750–3000 THz	Ultraviolet (0.1–0.4 μm)	20.0–26.5	K	K
		26.5–40.0	Ka	K

Note: kHz = kilohertz = hertz × 10^3; MHz = megahertz = hertz × 10^6; GHz = gigahertz = hertz × 10^9; THz = terahertz = hertz × 10^{12}; μm = micrometers = × 10^{-6} meters.

drafting, revision, and adoption of the *Radio Regulations* which is an instrument for the international management of the radio spectrum.[6]

[6]Available on the Radio Regulations website: http://www.itu.int/pub/R-REG-RR-2004/en.

Table 1.3 Selected Frequency Bands for Public Use and Military Communications

Use		Frequency
Omega navigation		10–14 kHz
Worldwide submarine communication		30 kHz
Loran C navigation		100 kHz
Standard (AM) broadcast		540–1600 kHz
ISM band	Industrial heaters; welders	40.66–40.7 MHz
Television:	Channels 2–4	54–72 MHz
	Channels 5–6	76–88 MHz
FM broadcast		88–108 MHz
Television	Channels 7–13	174–216 MHz
	Channels 14–83	420–890 MHz
	(In the United States, channels 2–36 and 38–51 will be used for digital TV broadcast; others will be reallocated.)	
Cellular mobile radio (plus other bands in the vacinity of 900 MHz)	Mobile to base station	824–849 MHz
	Base station to mobile	869–894 MHz
ISM band	Microwave ovens; medical	902–928 MHz
Global Positioning System		1227.6, 1575.4 MHz
Point-to-point microwave		2.11–2.13 GHz
Personal communication services	CDMA cellular in North America	1.8–2.0 GHz
Point-to-point microwave	Interconnecting base stations	2.16–2.18 GHz
ISM band	Microwave ovens; unlicensed spread spectrum; medical	2.4–2.4835 GHz
		23.6–24 GHz
		122–123 GHz
		244–246 GHz

In the United States, the Federal Communications Commission (FCC) awards specific applications within a band as well as licenses for their use. The FCC is directed by five commissioners appointed to five-year terms by the President and confirmed by the Senate. One commissioner is appointed as chairperson by the President.[7]

At lower frequencies, or long wavelengths, propagating radio waves tend to follow the earth's surface. At higher frequencies, or short wavelengths, radio waves propagate in straight lines. Another phenomenon that occurs at lower frequencies is reflection (or refraction) of radio waves by the ionosphere (a series of layers of charged particles at altitudes between 30 and 250 mi above the earth's surface). Thus, for frequencies below about 100 MHz, it is possible to have skip-wave propagation. At night, when lower ionospheric layers disappear due to less ionization from the sun (the E, F_1, and F_2 layers coalesce into one layer—the F layer), longer skip-wave propagation occurs as a result of reflection from the higher, single reflecting layer of the ionosphere.

[7]http://www.fcc.gov/.

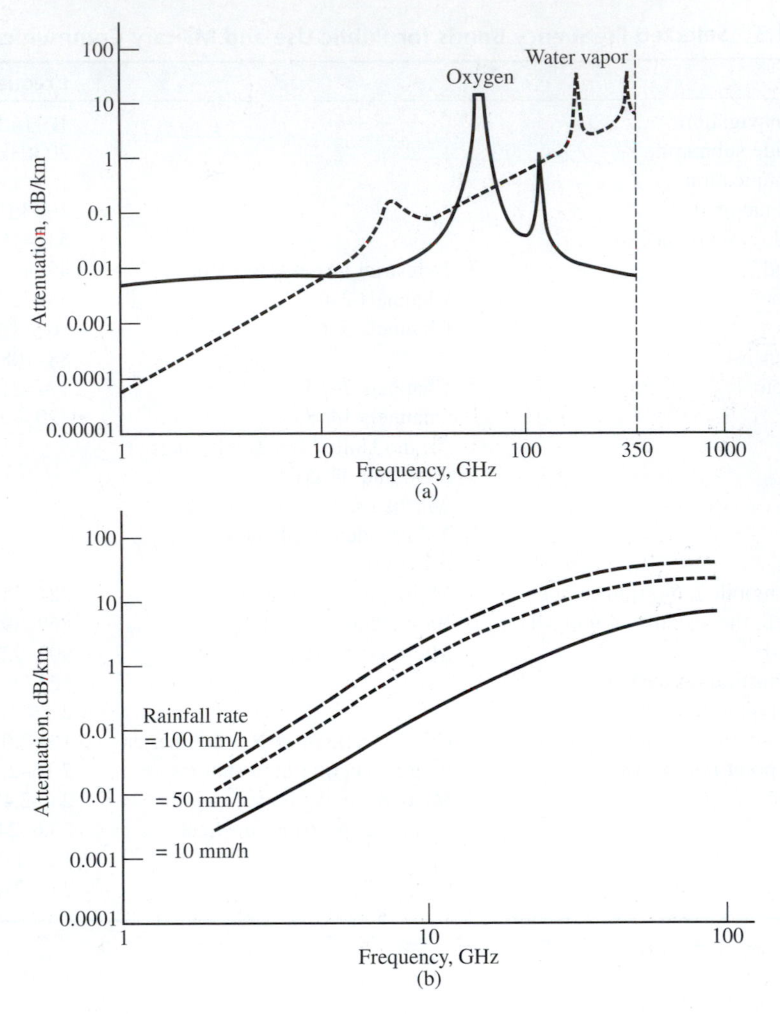

Figure 1.3
Specific attenuation for atmospheric gases and rain. (a) Specific attenuation due to oxygen and water vapor (concentration of 7.5 g/m^3). (b) Specific attenuation due to rainfall at rates of 10, 50, and 100 mm/h.

Above about 300 MHz, propagation of radio waves is by line of sight, because the ionosphere will not bend radio waves in this frequency region sufficiently to reflect them back to the earth. At still higher frequencies, say above 1 or 2 GHz, atmospheric gases (mainly oxygen), water vapor, and precipitation absorb and scatter radio waves. This phenomenon manifests itself as attenuation of the received signal, with the attenuation generally being more severe the higher the frequency (there are resonance regions for absorption by gases that peak at certain frequencies). Figure 1.3 shows specific attenuation curves as a function of frequency[8] for oxygen, water vapor and rain [recall that 1 decibel (dB) is 10 times the logarithm to the base

[8]Data from Louis J. Ippolito, Jr., *Radiowave Propagation in Satellite Communications*, Van Nostrand Reinhold, New York, 1986, Chapters 3 and 4.

10 of a power ratio]. One must account for the possible attenuation by such atmospheric constituents in the design of microwave links, which are used, for example, in transcontinental telephone links and ground-to-satellite communications links.

At about 23 GHz, the first absorption resonance due to water vapor occurs, and at about 62 GHz a second one occurs due to oxygen absorption. These frequencies should be avoided in transmission of desired signals through the atmosphere, or undue power will be expended (one might, for example, use 62 GHz as a signal for cross-linking between two satellites, where atmospheric absorption is no problem, and thereby prevent an enemy on the ground from listening in). Another absorption frequency for oxygen occurs at 120 GHz, and two other absorption frequencies for water vapor occur at 180 and 350 GHz.

Communication at millimeter-wave frequencies (that is, at 30 GHz and higher) is becoming more important now that there is so much congestion at lower frequencies (the Advanced Technology Satellite, launched in the mid-1990s, employs an uplink frequency band around 20 GHz and a downlink frequency band at about 30 GHz). Communication at millimeter-wave frequencies is becoming more feasible because of technological advances in components and systems. Two bands at 30 and 60 GHz, the Local Multipoint Distribution System (LMDS) and Multichannel Multipoint Distribution System (MMDS) bands, have been identified for terrestrial transmission of wideband signals. Great care must be taken to design systems using these bands because of the high atmospheric and rain absorption as well as blockage of objects such as trees and buildings.

Somewhere above 1 THz (1000 GHz), the propagation of radio waves becomes optical in character. At a wavelength of 10 μm (0.00001 m), the carbon dioxide laser provides a source of coherent radiation, and visible light lasers (for example, helium–neon) radiate in the wavelength region of 1 μm and shorter. Terrestrial communications systems employing such frequencies experience considerable attenuation on cloudy days, and laser communications over terrestrial links are restricted to optical fibers for the most part. Analyses have been carried out for the employment of laser communications cross-links between satellites, but there are as yet no optical satellite communications links actually flying.

Guided Electromagnetic Wave Channels

Up until the last part of the 20th century, the most extensive example of guided electromagnetic wave channels is the part of the long-distance telephone network that uses wire lines, but this has almost exclusively been replaced by optical fiber.[9] Communication between persons a continent apart was first achieved by means of voice-frequency transmission (below 10,000 Hz) over open wire. Quality of transmission was rather poor. By 1952, use of the types of modulation known as *double sideband* and *single sideband* on high-frequency carriers was established. Communication over predominantly multipair and coaxial cable lines produced transmission of much better quality. With the completion of the first transatlantic cable in 1956, intercontinental telephone communication was no longer dependent on high-frequency radio, and the quality of intercontinental telephone service improved significantly.

Bandwidths on coaxial cable links are a few megahertz. The need for greater bandwidth initiated the development of millimeter-wave waveguide transmission systems. However, with the development of low-loss optical fibers, efforts to improve millimeter-wave systems to

[9]For a summary of guided transmission systems as applied to telephone systems, see F. T. Andrews, Jr., Communications Technology: 25 Years in Retrospect. Part III, Guided Transmission Systems: 1952–1973, *IEEE Communications Society Magazine*, **16:** 4–10, Jan. 1978.

achieve greater bandwidth ceased. The development of optical fibers, in fact, has made the concept of a wired city—wherein digital data and video can be piped to any residence or business within a city—nearly a reality.[10] Modern coaxial cable systems can carry only 13,000 voice channels per cable, but optical links are capable of carrying several times this number (the limiting factor being the current driver for the light source).[11]

Optical Links The use of optical links was, until recently, limited to short and intermediate distances. With the installation of transpacific and transatlantic optical cables in 1988 and early 1989, this is no longer true.[12] The technological breakthroughs that preceeded the widespread use of light waves for communication were the development of small coherent light sources (semiconductor lasers), low-loss optical fibers or waveguides, and low-noise detectors.[13]

A typical fiber-optic communication system has a light source, which may be either a light-emitting diode or a semiconductor laser, in which the intensity of the light is varied by the message source. The output of this modulator is the input to a light-conducting fiber. The receiver, or light sensor, typically consists of a photodiode. In a photodiode, an average current flows that is proportional to the optical power of the incident light. However, the exact number of charge carriers (that is, electrons) is random. The output of the detector is the sum of the average current which is proportional to the modulation and a noise component. This noise component differs from the thermal noise generated by the receiver electronics in that it is "bursty" in character. It is referred to as *shot noise*, in analogy to the noise made by shot hitting a metal plate. Another source of degradation is the dispersion of the optical fiber itself. For example, pulse-type signals sent into the fiber are observed as "smeared out" at the receiver. Losses also occur as a result of the connections between cable pieces and between cable and system components.

Finally, it should be mentioned that optical communications can take place through free space.[14]

---

[10]The limiting factor here is expense—stringing anything under city streets is a very expensive proposition although there are many potential customers to bear the expense. Providing access to the home in the country is relatively easy from the standpoint of stringing cables or optical fiber, but the number of potential users is small so that the cost per customer goes up. As for cable versus fiber, the "*last mile*" is in favor of cable again because of expense. Many solutions have been proposed for this *last mile problem*, as it is sometimes referred, including special modulation schemes to give higher data rates over telephone lines (see ADSL in Table 1.1), making cable TV access two way (plenty of bandwidth but attenuation a problem), satellite (in remote locations), optical fiber (for those who want wideband and are willing and / or able to pay for it), and wireless or radio access (see the earlier comment about LMDS and MMDS). A universal solution for all situations is most likely not possible. For more on this intriguing topic, see *The IEEE Spectrum*, The Networked House, Dec. 1999.

[11]Wavelength division multiplexing (WDM) is the lastest development in the relatively short existence of optical fiber delivery of information. The idea here is that different wavelength bands ("colors"), provided by different laser light sources, are sent in parallel through an optical fiber to vastly increase the bandwidth—several gigahertz of bandwidth is possible. See, for example, *The IEEE Communcations Magazine*, Feb. 1999 (issue on "Optical Networks, Communication Systems, and Devices"), Oct. 1999 (issue on "Broadband Technologies and Trial's), Feb. 2000 (issue on "Optical Networks Come of Age"), and June, 2000 ("Intelligent Networks for the New Millennium").

[12]See Inglis, op. cit., Chapter 8.

[13]For an overview on the use of signal-processing methods to improve optical communications, see J. H. Winters, R. D. Gitlin, and S. Kasturia, Reducing the Effects of Transmission Impairments in Digital Fiber Optic Systems, *IEEE Communications Magazine*, **31:** 68–76, June 1993.

[14]See *IEEE Communications Magazine*, **38:** 124–139, Aug. 2000 (section on free space laser communications).

1.3 SUMMARY OF SYSTEMS ANALYSIS TECHNIQUES

Having identified and discussed the main subsystems in a communication system and certain characteristics of transmission media, let us now look at the techniques at our disposal for systems analysis and design.

1.3.1 Time-Domain and Frequency-Domain Analyses

From circuits courses or prior courses in linear systems analysis, you are well aware that the electrical engineer lives in the two worlds, so to speak, of time and frequency. Also, you should recall that dual time–frequency analysis techniques are especially valuable for linear systems for which the principle of superposition holds. Although many of the subsystems and operations encountered in communication systems are for the most part linear, many are not. Nevertheless, frequency-domain analysis is an extremely valuable tool to the communications engineer, more so perhaps than to other systems analysts. Since the communications engineer is concerned primarily with signal bandwidths and signal locations in the frequency domain, rather than with transient analysis, the essentially steady-state approach of the Fourier series and transforms is used rather than the Laplace transform. Accordingly, we provide an overview of the Fourier series, the Fourier integral, and their role in systems analysis in Chapter 2.

1.3.2 Modulation and Communication Theories

Modulation theory employs time- and frequency-domain analyses to analyze and design systems for modulation and demodulation of information-bearing signals. To be specific consider the message signal $m(t)$, which is to be transmitted through a channel using the method of double-sideband modulation. The modulated carrier for double-sideband modulation is of the form $x_c(t)=A_c m(t)\cos(\omega_c t)$, where ω_c is the carrier frequency in radians per second and A_c is the carrier amplitude. Not only must a modulator be built that can multiply two signals, but amplifiers are required to provide the proper power level of the transmitted signal. The exact design of such amplifiers is not of concern in a systems approach. However, the frequency content of the modulated carrier, for example, is important to their design and therefore must be specified. The dual time–frequency analysis approach is especially helpful in providing such information.

At the other end of the channel, there must be a receiver configuration capable of extracting a replica of $m(t)$ from the modulated signal, and one can again apply time- and frequency-domain techniques to good effect.

The analysis of the effect of interfering signals on system performance and the subsequent modifications in design to improve performance in the face of such interfering signals are part of *communication theory*, which, in turn, makes use of modulation theory.

This discussion, although mentioning interfering signals, has not explicitly emphasized the uncertainty aspect of the information-transfer problem. Indeed, much can be done without applying probabilistic methods. However, as pointed out previously, the application of probabilistic methods, coupled with optimization procedures, has been one of the key ingredients of the modern communications era and led to the development during the latter half of the twentieth century of new techniques and systems totally different in concept from those which existed before World War II.

We will now survey several approaches to statistical optimization of communication systems.

1.4 PROBABILISTIC APPROACHES TO SYSTEM OPTIMIZATION

The works of Wiener and Shannon, previously cited, were the beginning of modern statistical communication theory. Both these investigators applied probabilistic methods to the problem of extracting information-bearing signals from noisy backgrounds, but they worked from different standpoints. In this section we briefly examine these two approaches to optimum system design.

1.4.1 Statistical Signal Detection and Estimation Theory

Wiener considered the problem of optimally filtering signals from noise, where *optimum* is used in the sense of minimizing the average squared error between the desired output and the actual output. The resulting filter structure is referred to as the *Wiener filter*. This type of approach is most appropriate for analog communication systems in which the demodulated output of the receiver is to be a faithful replica of the message input to the transmitter.

Wiener's approach is reasonable for analog communications. However, in the early 1940s, (North, 1943) provided a more fruitful approach to the digital communications problem, in which the receiver must distinguish between a number of discrete signals in background noise. Actually, North was concerned with radar, which requires only the detection of the presence or absence of a pulse. Since fidelity of the detected signal at the receiver is of no consequence in such signal-detection problems, North sought the filter that would maximize the peak-signal-to-root-mean-square (rms) noise ratio at its output. The resulting optimum filter is called the *matched filter*, for reasons that will become apparent in Chapter 8, where we consider digital data transmission. Later adaptations of the Wiener and matched-filter ideas to time-varying backgrounds resulted in *adaptive filters*. We will consider a subclass of such filters in Chapter 8 when *equalization* of digital data signals is discussed.

The signal-extraction approaches of Wiener and North, formalized in the language of statistics in the early 1950s by several researchers [see Middleton (1960), p. 832, for several references], were the beginnings of what is today called *statistical signal detection* and *estimation theory*. In considering the design of receivers utilizing *all* the information available at the channel output, Woodward and Davies (1952) determined that this so-called ideal receiver computes the probabilities of the received waveform given the possible transmitted messages. These computed probabilities are known as *a posteriori* probabilities. The ideal receiver then makes the decision that the transmitted message was the one corresponding to the largest *a posteriori* probability. Although perhaps somewhat vague at this point, this *maximum a posteriori* (MAP) principle, as it is called, is one of the cornerstones of detection and estimation theory. Another development that had far-reaching consequences in the development of detection theory was the application of generalized vector space ideas (Kotel'nikov, 1959; Wozencraft and Jacobs, 1965). We will examine these ideas in more detail in Chapters 8 through 10.

1.4.2 Information Theory and Coding

The basic problem that Shannon considered is, "Given a message source, how shall the messages produced be represented so as to maximize the information conveyed through a given channel?" Although Shannon formulated his theory for both discrete and analog sources, we will think here in terms of discrete systems. Clearly, a basic consideration in this theory is a measure of information. Once a suitable measure has been defined (and we will do so in Chapter 11), the next step is to define the information carrying capacity, or simply capacity, of a channel as the maximum rate at which information can be conveyed through it. The obvious question that now arises is, "Given a channel, how closely can we approach the capacity of the channel, and what is the quality of the received message?" A most surprising, and the singularly most important, result of Shannon's theory is that by suitably restructuring the transmitted signal, we can transmit information through a channel *at any rate less than the channel capacity with arbitrarily small error*, despite the presence of noise, provided we have an arbitrarily long time available for transmission. This is the gist of Shannon's *second theorem*. Limiting our discussion at this point to binary discrete sources, a proof of Shannon's second theorem proceeds by selecting code words at random from the set of 2^n possible binary sequences n digits long at the channel input. The probability of error in receiving a given n-digit sequence, when averaged over all possible code selections, becomes arbitrarily small as n becomes arbitrarily large. Thus many suitable codes exist, *but we are not told how to find these codes*. Indeed, this has been the dilemma of information theory since its inception and is an area of active research. In recent years, great strides have been made in finding good coding and decoding techniques that are implementable with a reasonable amount of hardware and require only a reasonable amount of time to decode. Several basic coding techniques will be discussed in Chapter 11.[15] Perhaps the most astounding development in the recent history of coding was the invention of turbo coding and subsequent publication by French researchers in 1993.[16] Their results, which were subsequently verified by several researchers, showed performance to within a fraction of a decibel of the Shannon limit.[17]

1.4.3 Recent Advances

There have been great strides made in communications theory and its practical implementation in the past few decades. Some of these will be pointed out later in the book. To capture the gist of these advances at this point would delay the coverage of basic concepts of communications theory, which is the underlying intent of this book. For those wanting additional reading at this point, two recent issues of the *IEEE Proceedings* will provide information in two areas:

---

[15]For a good survey on *Shannon theory*, as it is known, see S. Verdu, Fifty Years of Shannon Theory, *IEEE Trans. Infor. Theory*, **44**: pp. 2057–2078, Oct., 1998.

[16]C. Berrou, A. Glavieux, and P. Thitimajshima, Near Shannon Limit Error-Correcting Coding and Decoding: Turbo Codes, *Proc. 1993 Int. Conf. Commun.*, Geneva, Switzerland, 1064–1070, May 1993. See also D. J. Costello and G. D. Forney, Channel Coding: The Road to Channel Capacity, *Proc. IEEE*, **95:** 1150–1177, June 2007 for an excellent tutorial article on the history of coding theory.

[17]Actually low-density parity-check codes, invented and published by Robert Gallager in 1963, were the first codes to allow data transmission rates close to the theoretical limit (Gallager, 1963). However, they were impractical to implement in 1963, so were forgotten about until the past 10 to 20 years whence practical advances in their theory and substantially advanced processors have spurred a resurgence of interest in them.

turbo-information processing (used in decoding turbo codes among other applications)[18], and multiple-input multiple-output (MIMO) communications theory, which is expected to have far-reaching impact on wireless local- and wide-area network development.[19] An appreciation for the broad sweep of developments from the beginnings of modern communications theory to recent times can be gained from a collection of papers put together in a single volume, spanning roughly 50 years, that were judged to be worthy of note by experts in the field.[20]

1.5 PREVIEW OF THIS BOOK

From the previous discussion, the importance of probability and noise characterization in analysis of communication systems should be apparent. Accordingly, after presenting basic signal, system, and noiseless modulation theory and basic elements of digital data transmission in Chapters 2, 3, and 4, we briefly discuss probability and noise theory in Chapters 5 and 6. Following this, we apply these tools to the noise analysis of analog communications schemes in Chapter 7. In Chapters 8 and 9, we use probabilistic techniques to find optimum receivers when we consider digital data transmission. Various types of digital modulation schemes are analyzed in terms of error probability. In Chapter 10, we approach optimum signal detection and estimation techniques on a generalized basis and use signal-space techniques to provide insight as to why systems that have been analyzed previously perform as they do. As already mentioned, information theory and coding are the subjects of Chapter 11. This provides us with a means of comparing actual communication systems with the ideal. Such comparisons are then considered in Chapter 11 to provide a basis for selection of systems.

In closing, we must note that large areas of communications technology, such as optical, computer, and military communications, are not touched on in this book. However, one can apply the principles developed in this text in those areas as well.

Further Reading

The references for this chapter were chosen to indicate the historical development of modern communications theory and by and large are not easy reading. They are found in the Historical References section of the Bibliography. You also may consult the introductory chapters of the books listed in the Further Reading sections of Chapters 2 and 3. These books appear in the main portion of the Bibliography.

[18] *Proceedings of the IEEE*, **95:** (6), June 2007 (special issue on turbo-information processing).

[19] *Proceedings of the IEEE*, **95:** (7), July 2007 (special issue on multiuser MIMO-OFDM for next-generation wireless).

[20] W. H. Tranter, D. P. Taylor, R. E. Ziemer, N. F. Maxemchuk, and J. W. Mark (eds.), 2007. *The Best of the Best: Fifty Years of Communications and Networking Research*, John Wiley and IEEE Press.

CHAPTER 2

SIGNAL AND LINEAR SYSTEM ANALYSIS

The study of information transmission systems is inherently concerned with the transmission of signals through systems. Recall that in Chapter 1 a *signal* was defined as the time history of some quantity, usually a voltage or current. A *system* is a combination of devices and networks (subsystems) chosen to perform a desired function. Because of the sophistication of modern communication systems, a great deal of analysis and experimentation with trial subsystems occurs before actual building of the desired system. Thus the communications engineer's tools are mathematical models for signals and systems.

In this chapter, we review techniques useful for modeling and analysis of signals and systems used in communications engineering.[1] Of primary concern will be the dual time–frequency viewpoint for signal representation, and models for linear, time-invariant, two-port systems. It is important to always keep in mind that a model is not the signal or the system but a mathematical idealization of certain characteristics of it that are most relevant to the problem at hand.

With this brief introduction, we now consider signal classifications and various methods for modeling signals and systems. These include frequency-domain representations for signals via the complex exponential Fourier series and the Fourier transform, followed by linear system models and techniques for analyzing the effects of such systems on signals.

2.1 SIGNAL MODELS

2.1.1 Deterministic and Random Signals

In this book we are concerned with two broad classes of signals, referred to as *deterministic* and *random*. *Deterministic signals* can be modeled as completely specified functions of time. For example, the signal

$$x(t) = A\cos(\omega_0 t), \quad -\infty < t < \infty \tag{2.1}$$

where A and ω_0 are constants, is a familiar example of a deterministic signal. Another example of a deterministic signal is the unit rectangular pulse, denoted as $\Pi(t)$ and defined as

$$\Pi(t) = \begin{cases} 1, & |t| \le \frac{1}{2} \\ 0, & \text{otherwise} \end{cases} \tag{2.2}$$

[1] More complete treatments of these subjects can be found in texts on linear system theory. See the Bibliography for suggestions.

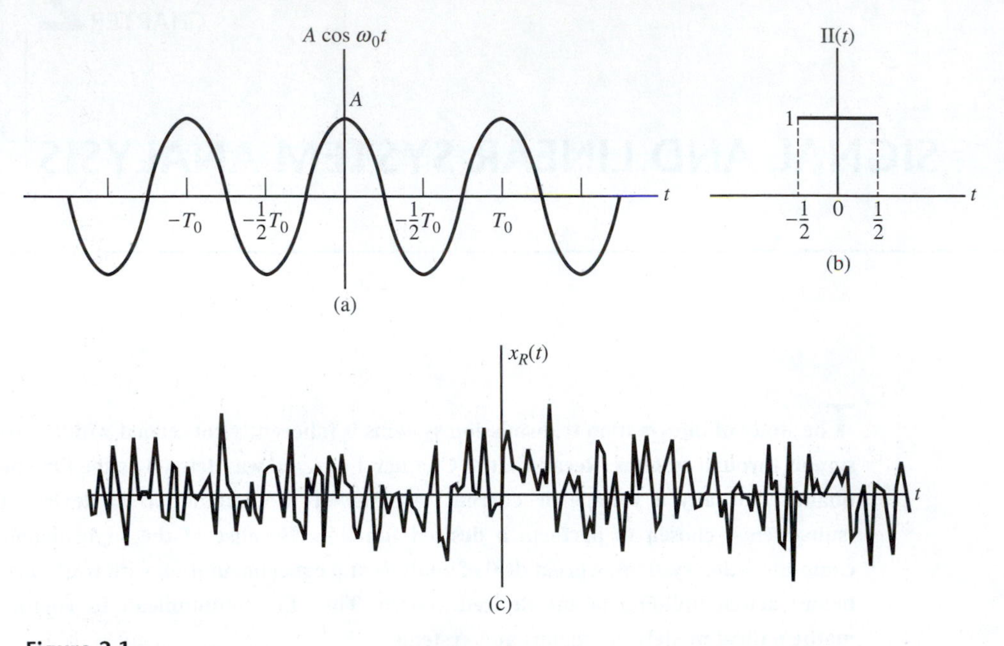

Figure 2.1
Examples of various types of signals. (a) Deterministic (sinusoidal) signal. (b) Unit rectangular pulse signal. (c) Random signal.

Random signals are signals that take on random values at any given time instant and must be modeled probabilistically. They will be considered in Chapters 5 and 6. Figure 2.1 illustrates the various types of signals just discussed.

2.1.2 Periodic and Aperiodic Signals

The signal defined by (2.1) is an example of a *periodic signal*. A signal $x(t)$ is periodic if and only if

$$x(t+T_0) = x(t), \quad -\infty < t < \infty \tag{2.3}$$

where the constant T_0 is the period. The smallest such number satisfying (2.3) is referred to as the *fundamental period* (the modifier *fundamental* is often excluded). Any signal not satisfying (2.3) is called *aperiodic*.

2.1.3 Phasor Signals and Spectra

A useful periodic signal in system analysis is the signal

$$\tilde{x}(t) = Ae^{j(\omega_0 t+\theta)}, \quad -\infty < t < \infty \tag{2.4}$$

which is characterized by three parameters: amplitude A, phase θ in radians, and frequency ω_0 in radians per second or $f_0 = \omega_0/2\pi$ Hz. We will refer to $\tilde{x}(t)$ as a *rotating phasor* to distinguish it from the phasor $Ae^{j\theta}$, for which $e^{j\omega_0 t}$ is implicit. Using Euler's theorem,[2] we may readily

[2]Recall that Euler's theorem is $e^{\pm ju} = \cos u \pm j \sin u$. Also recall that $e^{j2\pi} = 1$.

show that $\tilde{x}(t) = \tilde{x}(t + T_0)$, where $T_0 = 2\pi/\omega_0$. Thus $\tilde{x}(t)$ is a periodic signal with period $2\pi/\omega_0$.

The rotating phasor $Ae^{j(\omega_0 t + \theta)}$ can be related to a real, sinusoidal signal $A\cos(\omega_0 t + \theta)$ in two ways. The first is by taking its real part,

$$\begin{aligned} x(t) = A\cos(\omega_0 t + \theta) &= \text{Re}\,(\tilde{x}(t)) \\ &= \text{Re}\,(Ae^{j(\omega_0 t + \theta)}) \end{aligned} \tag{2.5}$$

and the second is by taking one-half of the sum of $\tilde{x}(t)$ and its complex conjugate,

$$\begin{aligned} A\cos(\omega_0 t + \theta) &= \frac{1}{2}\tilde{x}(t) + \frac{1}{2}\tilde{x}^*(t) \\ &= \frac{1}{2}Ae^{j(\omega_0 t + \theta)} + \frac{1}{2}Ae^{-j(\omega_0 t + \theta)} \end{aligned} \tag{2.6}$$

Figure 2.2 illustrates these two procedures graphically.

Equations (2.5) and (2.6), which give alternative representations of the sinusoidal signal $x(t) = A\cos(\omega_0 t + \theta)$ in terms of the rotating phasor $\tilde{x}(t) = A\exp[j(\omega_0 t + \theta)]$, are time-domain representations for $x(t)$. Two equivalent representations of $x(t)$ in the frequency domain may be obtained by noting that the rotating phasor signal is completely specified if the parameters A and θ are given for a particular f_0. Thus plots of the magnitude and angle of $Ae^{j\theta}$ versus frequency give sufficient information to characterize $x(t)$ completely. Because $\tilde{x}(t)$ exists only at the single frequency f_0, for this case of a single sinusoidal signal, the resulting plots consist of discrete lines and are known as *line spectra*. The resulting plots are referred to as the *amplitude line spectrum* and the *phase line spectrum* for $x(t)$, and are shown in Figure 2.3 (a). These are *frequency-domain* representations not only of $\tilde{x}(t)$ but of $x(t)$ as well, by virtue of (2.5). In addition, the plots of Figure 2.3(a) are referred to as the *single-sided amplitude* and *phase spectra* of $x(t)$ because they exist only for positive frequencies. For a signal consisting of a sum of sinusoids of differing frequencies, the single-sided spectrum consists of a multiplicity of lines, with one line for each sinusoidal component of the sum.

By plotting the amplitude and phase of the complex conjugate phasors of (2.6) versus frequency, one obtains another frequency-domain representation for $x(t)$, referred to as the *double-sided amplitude* and *phase spectra*. This representation is shown in Figure 2.3(b). Two

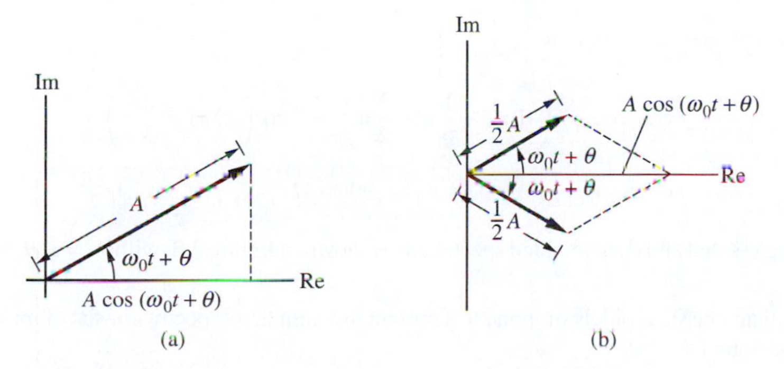

Figure 2.2
Two ways of relating a phasor signal to a sinusoidal signal. (a) Projection of a rotating phasor onto the real axis. (b) Addition of complex conjugate rotating phasors.

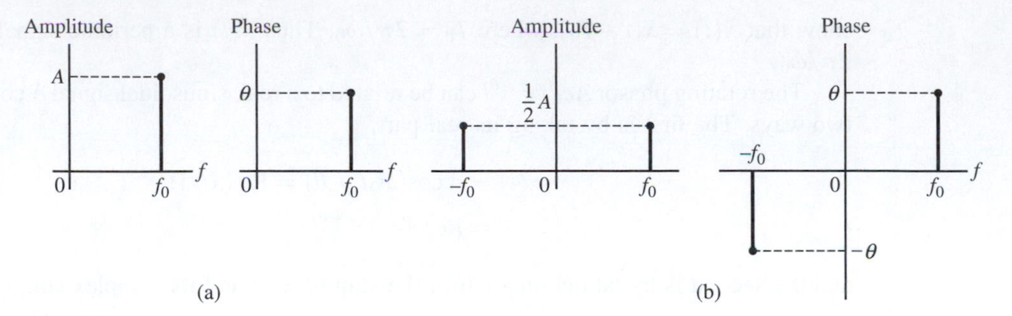

Figure 2.3
Amplitude and phase spectra for the signal $A\cos(\omega_0 t + \theta)$. (a) Single sided. (b) Double sided.

important observations may be made from Figure 2.3(b). First, the lines at the *negative* frequency $f = -f_0$ exist precisely because it is necessary to add complex conjugate (or oppositely rotating) phasor signals to obtain the real signal $A\cos(\omega_0 t + \theta)$. Second, we note that the amplitude spectrum has *even* symmetry and that the phase spectrum has *odd* symmetry about $f = 0$. This symmetry is again a consequence of $x(t)$ being a real signal. As in the single-sided case, the two-sided spectrum for a sum of sinusoids consists of a multiplicity of lines, with one pair of lines for each sinusoidal component.

Figure 2.3(a) and (b) is therefore equivalent spectral representations for the signal $A\cos(\omega_0 t + \theta)$, consisting of lines at the frequency $f = f_0$ (and its negative). For this simple case, the use of spectral plots seems to be an unnecessary complication, but we will find shortly how the Fourier series and Fourier transform lead to spectral representations for more complex signals.

EXAMPLE 2.1

(a) To sketch the single-sided and double-sided spectra of

$$x(t) = 2\sin\left(10\pi t - \frac{1}{6}\pi\right) \tag{2.7}$$

we note that $x(t)$ can be written as

$$\begin{aligned} x(t) &= 2\cos\left(10\pi t - \frac{1}{6}\pi - \frac{1}{2}\pi\right) = 2\cos\left(10\pi t - \frac{2}{3}\pi\right) \\ &= \text{Re}\left(2e^{j(10\pi t - 2\pi/3)}\right) = e^{j(10\pi t - 2\pi/3)} + e^{-j(10\pi t - 2\pi/3)} \end{aligned} \tag{2.8}$$

Thus the single-sided and double-sided spectra are as shown in Figure 2.3, with $A = 2$, $\theta = -\frac{2}{3}\pi$ rad, and $f_0 = 5$ Hz.

(b) If more than one sinusoidal component is present in a signal, its spectra consist of multiple lines. For example, the signal

$$y(t) = 2\sin\left(10\pi t - \frac{1}{6}\pi\right) + \cos(20\pi t) \tag{2.9}$$

can be rewritten as

$$
\begin{aligned}
y(t) &= 2\cos\left(10\pi t - \frac{2}{3}\pi\right) + \cos(20\pi t) \\
&= \mathrm{Re}\left(2e^{j(10\pi t - 2\pi/3)} + e^{j20\pi t}\right) \\
&= e^{j(10\pi t - 2\pi/3)} + e^{-j(10\pi t - 2\pi/3)} + \frac{1}{2}e^{j20\pi t} + \frac{1}{2}e^{-j20\pi t}
\end{aligned}
\tag{2.10}
$$

Its single-sided amplitude spectrum consists of a line of amplitude 2 at $f = 5$ Hz and a line of amplitude 1 at $f = 10$ Hz. Its single-sided phase spectrum consists of a single line of amplitude $-2\pi/3$ at $f = 5$ Hz. To get the double-sided amplitude spectrum, one simply *halves* the amplitude of the lines in the single-sided amplitude spectrum and takes the mirror image of this result about $f = 0$ (amplitude lines at $f = 0$ remain the same). The double-sided phase spectrum is obtained by taking the mirror image of the single-sided phase spectrum about $f = 0$ and inverting the left-hand (negative frequency) portion.

■

2.1.4 Singularity Functions

An important subclass of aperiodic signals is the singularity functions. In this book we will be concerned with only two: the *unit impulse function* $\delta(t)$ (or *delta function*) and the *unit step function* $u(t)$. The unit impulse function is defined in terms of the integral

$$\int_{-\infty}^{\infty} x(t)\delta(t)\,dt = x(0) \tag{2.11}$$

where $x(t)$ is any test function that is continuous at $t = 0$. A change of variables and redefinition of $x(t)$ results in the *sifting property*

$$\int_{-\infty}^{\infty} x(t)\delta(t - t_0)\,dt = x(t_0) \tag{2.12}$$

where $x(t)$ is continuous at $t = t_0$. We will make considerable use of the sifting property in systems analysis. By considering the special case $x(t) = 1$ for $t_1 \leq t \leq t_2$ and $x(t) = 0$ for $t < t_1$ and $t > t_2$, the two properties

$$\int_{t_1}^{t_2} \delta(t - t_0)\,dt = 1, \quad t_1 < t_0 < t_2 \tag{2.13}$$

and

$$\delta(t - t_0) = 0, \quad t \neq t_0 \tag{2.14}$$

are obtained that provide an alternative definition of the unit impulse. Equation (2.14) allows the integrand in (2.12) to be replaced by $x(t_0)\delta(t - t_0)$, and the sifting property then follows from (2.13).

Other properties of the unit impulse function that can be proved from the definition (2.11) are the following:

1. $\delta(at) = (1/|a|)\delta(t)$, a is a constant.
2. $\delta(-t) = \delta(t)$.

3. A generalization of the sifting property, $\int_{t_1}^{t_2} x(t)\delta(t-t_0)dt = \begin{cases} x(t_0), & t_1 < t_0 < t_2 \\ 0, & \text{otherwise} \\ \text{undefined}, & t_0 = t_1 \text{ or } t_2 \end{cases}$
4. $x(t)\delta(t-t_0) = x(t_0)\delta(t-t_0)$, where $x(t)$ is continuous at $t = t_0$.
5. $\int_{t_1}^{t_2} x(t)\delta^{(n)}(t-t_0)\,dt = (-1)^n x^{(n)}(t_0)$, $t_1 < t_0 < t_2$. [In this equation, the superscript n denotes the nth derivative; $x(t)$ and its first n derivatives are assumed continuous at $t = t_0$.]
6. If $f(t) = g(t)$, where $f(t) = a_0\delta(t) + a_1\delta^{(1)}(t) + \cdots + a_n\delta^{(n)}(t)$ and $g(t) = b_0\delta(t) + b_1\delta^{(1)}(t) + \cdots + b_n\delta^{(n)}(t)$, this implies that $a_0 = b_0, a_1 = b_1, \ldots, a_n = b_n$.

It is reassuring to note that (2.13) and (2.14) correspond to the intuitive notion of a unit impulse function as the limit of a suitably chosen conventional function having unity area in an infinitesimally small width. An example is the signal

$$\delta_\epsilon(t) = \frac{1}{2\epsilon}\Pi\left(\frac{t}{2\epsilon}\right) = \begin{cases} \dfrac{1}{2\epsilon}, & |t| < \epsilon \\ 0, & \text{otherwise} \end{cases} \tag{2.15}$$

which is shown in Figure 2.4(a) for $\epsilon = 1/4$ and $\epsilon = 1/2$. It seems apparent that any signal having unity area and zero width in the limit as some parameter approaches zero is a suitable representation for $\delta(t)$, for example, the signal

$$\delta_{1\epsilon}(t) = \epsilon\left(\frac{1}{\pi t}\sin\frac{\pi t}{\epsilon}\right)^2 \tag{2.16}$$

which is sketched in Figure 2.4(b).

Other singularity functions may be defined as integrals or derivatives of unit impulses. We will need only the unit step $u(t)$, defined to be the integral of the unit impulse. Thus

$$u(t) \triangleq \int_{-\infty}^{t} \delta(\lambda)\,d\lambda = \begin{cases} 0, & t < 0 \\ 1, & t > 0 \\ \text{undefined}, & t = 0 \end{cases} \tag{2.17}$$

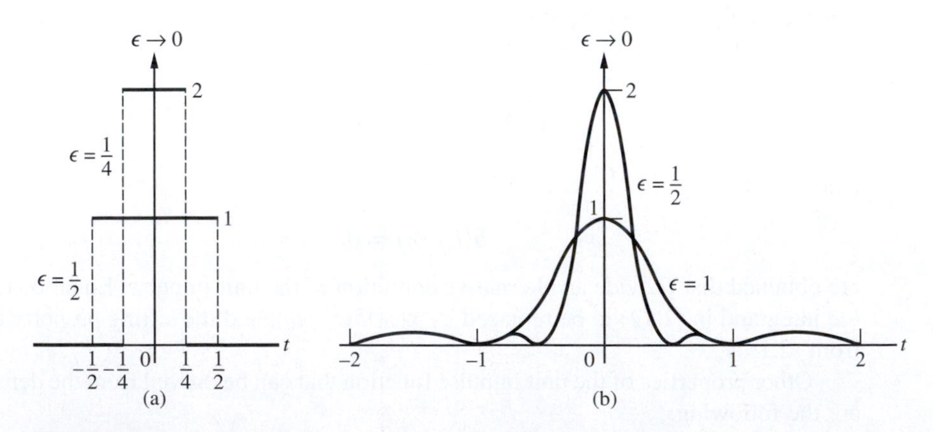

Figure 2.4
Two representations for the unit impulse function in the limit as $\epsilon \to 0$. (a) $(1/2\epsilon)\Pi(t/2\epsilon)$. (b) $\epsilon[(1/\pi t)\sin(\pi t/\epsilon)]^2$.

or

$$\delta(t) = \frac{du(t)}{dt} \tag{2.18}$$

For consistency with the unit pulse function definition, we will define $u(0) = 1$. You are no doubt familiar with the usefulness of the unit step for "turning on" signals of doubly infinite duration and for representing signals of the staircase type. For example, the unit rectangular pulse function defined by (2.2) can be written in terms of unit steps as

$$\Pi(t) = u\left(t + \frac{1}{2}\right) - u\left(t - \frac{1}{2}\right) \tag{2.19}$$

We are now ready to consider power and energy signal classifications.

2.2 SIGNAL CLASSIFICATIONS

Because the particular representation used for a signal depends on the type of signal involved, it is useful to pause at this point and introduce signal classifications. In this chapter we will be considering two signal classes, those with finite energy and those with finite power. As a specific example, suppose $e(t)$ is the voltage across a resistance R producing a current $i(t)$. The instantaneous power per ohm is $p(t) = e(t)i(t)/R = i^2(t)$. Integrating over the interval $|t| \leq T$, the total energy and the average power on a per-ohm basis are obtained as the limits

$$E = \lim_{T \to \infty} \int_{-T}^{T} i^2(t)\, dt \tag{2.20}$$

and

$$P = \lim_{T \to \infty} \frac{1}{2T} \int_{-T}^{T} i^2(t)\, dt \tag{2.21}$$

respectively.

For an arbitrary signal $x(t)$, which may, in general, be complex, we define total (normalized) energy as

$$E \triangleq \lim_{T \to \infty} \int_{-T}^{T} |x(t)|^2\, dt = \int_{-\infty}^{\infty} |x(t)|^2\, dt \tag{2.22}$$

and (normalized) power as

$$P \triangleq \lim_{T \to \infty} \frac{1}{2T} \int_{-T}^{T} |x(t)|^2 dt \tag{2.23}$$

Based on the definitions (2.22) and (2.23), we can define two distinct classes of signals:

1. We say $x(t)$ is an *energy signal* if and only if $0 < E < \infty$, so that $P = 0$.
2. We classify $x(t)$ as a *power signal* if and only if $0 < P < \infty$, thus implying that $E = \infty$.[3]

[3]Signals that are neither energy nor power signals are easily found. For example, $x(t) = t^{-1/4}$, $t \geq t_0 > 0$, and zero otherwise.

EXAMPLE 2.2

As an example of determining the classification of a signal, consider

$$x_1(t) = Ae^{-\alpha t}u(t), \quad \alpha > 0 \tag{2.24}$$

where A and α are positive constants. Using (2.22), we may readily verify that $x_1(t)$ is an *energy signal* since $E = A^2/2\alpha$ by applying (2.22). Letting $\alpha \to 0$, we obtain the signal $x_2(t) = Au(t)$, which has infinite energy. Applying (2.23), we find that $P = \frac{1}{2}A^2$ for $Au(t)$, thus verifying that $x_2(t)$ is a *power signal.*

■

EXAMPLE 2.3

Consider the rotating phasor signal given by (2.4). We may verify that $\tilde{x}(t)$ is a power signal since

$$P = \lim_{T\to\infty} \frac{1}{2T}\int_{-T}^{T} |\tilde{x}(t)|^2\, dt = \lim_{T\to\infty} \frac{1}{2T}\int_{-T}^{T} A^2\, dt = A^2 \tag{2.25}$$

is finite.

■

We note that there is no need to carry out the limiting operation to find P for a periodic signal, since an average carried out over a single period gives the same result as (2.23); that is, for a periodic signal $x_p(t)$,

$$P = \frac{1}{T_0}\int_{t_0}^{t_0+T_0} |x_p(t)|^2\, dt \tag{2.26}$$

where T_0 is the period and t_0 is an arbitrary starting time (chosen for convenience). The proof of (2.26) is left to the problems.

EXAMPLE 2.4

The sinusoidal signal

$$x_p(t) = A\cos(\omega_0 t + \theta) \tag{2.27}$$

has average power

$$\begin{aligned} P &= \frac{1}{T_0}\int_{t_0}^{t_0+T_0} A^2\cos^2(\omega_0 t + \theta)\, dt \\ &= \frac{\omega_0}{2\pi}\int_{t_0}^{t_0+(2\pi/\omega_0)} \frac{A^2}{2}\, dt + \frac{\omega_0}{2\pi}\int_{t_0}^{t_0+(2\pi/\omega_0)} \frac{A^2}{2}\cos[2(\omega_0 t + \theta)]\, dt \\ &= \frac{A^2}{2} \end{aligned} \tag{2.28}$$

where the identity $\cos^2 u = \frac{1}{2} + \frac{1}{2}\cos(2u)$ has been used[4] and the second integral is zero because the integration is over two complete periods of the integrand.

■

[4]See Appendix G.2 for trigonometric identities.

2.3 GENERALIZED FOURIER SERIES

Our discussion of the phasor signal given by (2.4) illustrated the dual time–frequency nature of such signals. Fourier series and transform representations for signals are the key to generalizing this dual nature, since they amount to expressing signals as superpositions of complex exponential functions of the form $e^{j\omega t}$.

In anticipation of signal space concepts, to be introduced and applied to communication systems analysis in Chapters 9 and 10, the discussion in this section is concerned with the representation of signals as a series of orthogonal functions or, as referred to here, a generalized Fourier series. Such generalized Fourier series representations allow signals to be represented as points in a generalized vector space, referred to as *signal space*, thereby allowing information transmission to be viewed in a geometrical context. In the following section, the generalized Fourier series will be specialized to the complex exponential form of the Fourier series.

To begin our consideration of the generalized Fourier series, we recall from vector analysis that any vector **A** in a three-dimensional space can be expressed in terms of any three vectors **a**, **b**, and **c** that do not all lie in the same plane and are not collinear:

$$\mathbf{A} = A_1\mathbf{a} + A_2\mathbf{b} + A_3\mathbf{c} \tag{2.29}$$

where A_1, A_2, and A_3 are appropriately chosen constants. The vectors **a**, **b**, and **c** are said to be *linearly independent*, for no one of them can be expressed as a linear combination of the other two. For example, it is impossible to write $\mathbf{a} = \alpha\mathbf{b} + \beta\mathbf{c}$, no matter what choice is made for the constants α and β.

Such a set of linearly independent vectors **a**, **b**, and **c** is said to form a *basis set* for a three-dimensional vector space. Such vectors *span* a three-dimensional vector space in the sense that *any* vector **A** can be expressed as a linear combination of them.

We may, in an analogous fashion, consider the problem of representing a time function, or signal, $x(t)$ on a T-s interval $(t_0,\ t_0 + T)$, as a similar expansion. Thus we consider a set of time functions $\phi_1(t), \phi_2(t), \ldots, \phi_N(t)$, which are specified independently of $x(t)$, and seek a series expansion of the form

$$x_a(t) = \sum_{n=0}^{N} X_n\phi_n(t), \quad t_0 \leq t \leq t_0 + T \tag{2.30}$$

in which the N coefficients X_n are independent of time and the subscript a indicates that (2.30) is considered an approximation.

We assume that the $\phi_n(t)$s in (2.30) are *linearly independent*; that is, no one of them can be expressed as a weighted sum of the other $N-1$. A set of linearly independent $\phi_n(t)$ will be called a *basis function set*.

We now wish to examine the error in the approximation of $x(t)$ by $x_a(t)$. As in the case of ordinary vectors, the expansion (2.30) is easiest to use if the $\phi_n(t)$ are orthogonal on the interval $(t_0, t_0 + T)$. That is,

$$\int_{t_0}^{t_0+T} \phi_m(t)\phi_n^*(t)\,dt = c_n\delta_{mn} \triangleq \begin{cases} c_n, & n = m \\ 0, & n \neq m \end{cases} \quad \text{(all } m \text{ and } n\text{)} \tag{2.31}$$

where if $c_n = 1$ for all n, the $\phi_n(t)$s are said to be *normalized*. A normalized orthogonal set of functions is called an *orthonormal basis set*. The asterisk in (2.31) denotes complex conjugate, since we wish to allow the possibility of complex-valued $\phi_n(t)$. The symbol δ_{mn}, called the *Kronecker delta function*, is defined as unity if $m = n$ and zero otherwise. The error in the approximation of $x(t)$ by the series of (2.30) will be measured in the integral-squared sense:

$$\text{Error} = \epsilon_N = \int_T |x(t) - x_a(t)|^2 dt \tag{2.32}$$

where $\int_T (\,)\, dt$ denotes integration over t from t_0 to $t_0 + T$. The *integral-squared error* (ISE) is an applicable measure of error only when $x(t)$ is an energy signal or a power signal. If $x(t)$ is an energy signal of infinite duration, the limit as $T \to \infty$ is taken. We now find the set of coefficients X_n that minimizes the ISE. Substituting (2.30) into (2.32), expressing the magnitude squared of the integrand as the integrand times its complex conjugate, and expanding, we obtain

$$\epsilon_N = \int_T |x(t)|^2 dt - \sum_{n=0}^{N} \left[X_n^* \int_T x(t)\phi_n^*(t)\, dt + X_n \int_T x^*(t)\phi_n(t)\, dt \right] + \sum_{n=0}^{N} c_n |X_n|^2 \tag{2.33}$$

in which the orthogonality of the $\phi_n(t)$ has been used after interchanging the orders of summation and integration. To find the X_n that minimize ϵ_N, we add and subtract the quantity

$$\sum_{n=0}^{N} \frac{1}{c_n} \left| \int_T x(t)\phi_n^*(t)\, dt \right|^2$$

which yields, after rearrangement of terms, the following result for ϵ_N:

$$\epsilon_N = \int_T |x(t)|^2\, dt - \sum_{n=0}^{N} \frac{1}{c_n} \left| \int_T x(t)\phi_n^*(t)\, dt \right|^2 + \sum_{n=0}^{N} c_n \left| X_n - \frac{1}{c_n} \int_T x(t)\phi_n^*(t)\, dt \right|^2 \tag{2.34}$$

The first two terms on the right-hand side of (2.34) are independent of the coefficients X_n. Since the last sum on the right-hand side is nonnegative, we will minimize ϵ_N if we choose each X_n such that the corresponding term in the sum is zero. Thus, since $c_n > 0$, the choice of

$$X_n = \frac{1}{c_n} \int_T x(t)\phi_n^*(t)\, dt \tag{2.35}$$

for X_n minimizes the ISE. The resulting minimum-error coefficients will be referred to as the *Fourier coefficients*. The minimum value for ϵ_N, from (2.34), is obviously

$$\begin{aligned} (\epsilon_N)_{\min} &= \int_T |x(t)|^2\, dt - \sum_{n=0}^{N} \frac{1}{c_n} \left| \int^T x(t)\phi_n^*(t)\, dt \right|^2 \\ &= \int_T |x(t)|^2\, dt - \sum_{n=0}^{N} c_n |X_n|^2 \end{aligned} \tag{2.36}$$

If we can find an infinite set of orthonormal functions such that

$$\lim_{N \to \infty} (\epsilon_N)_{\min} = 0 \tag{2.37}$$

for any signal that is integrable square,

$$\int_T |x(t)|^2\, dt < \infty \tag{2.38}$$

we say that the $\phi_n(t)$ are *complete*. In the sense that the ISE is zero, we may then write

$$x(t) = \sum_{n=0}^{\infty} X_n \phi_n(t) \quad (\text{ISE} = 0) \tag{2.39}$$

although there may be a number of isolated points of discontinuity where actual equality does not hold. For almost all points in the interval $(t_0, t_0 + T)$, Equation (2.39) requires that $x(t)$ be equal to $x_a(t)$ as $N \to \infty$.

Assuming a complete orthogonal set of functions, we obtain from (2.36) the relation

$$\int_T |x(t)|^2\, dt = \sum_{n=0}^{\infty} c_n |X_n|^2 \tag{2.40}$$

This equation is known as *Parseval's theorem.*

EXAMPLE 2.5

Consider the set of two orthonormal functions shown in Figure 2.5(a). The signal

$$x(t) = \begin{cases} \sin(\pi t), & 0 \le t \le 2 \\ 0, & \text{otherwise} \end{cases} \tag{2.41}$$

is to be approximated by a two-term generalized Fourier series of the form given by (2.30). The Fourier coefficients, from (2.35), are given by

$$X_1 = \int_0^2 \phi_1(t) \sin(\pi t)\, dt = \int_0^1 \sin(\pi t)\, dt = \frac{2}{\pi} \tag{2.42}$$

and

$$X_2 = \int_0^2 \phi_2(t) \sin(\pi t)\, dt = \int_1^2 \sin(\pi t)\, dt = -\frac{2}{\pi} \tag{2.43}$$

Thus the generalized two-term Fourier series approximation for this signal is

$$x_a(t) = \frac{2}{\pi}\phi_1(t) - \frac{2}{\pi}\phi_2(t) = \frac{2}{\pi}\left[\Pi\left(t - \frac{1}{2}\right) - \Pi\left(t - \frac{3}{2}\right)\right] \tag{2.44}$$

where $\Pi(t)$ is the unit rectangular pulse defined by (2.1). The signal $x(t)$ and the approximation $x_a(t)$ are compared in Figure 2.5(b). Figure 2.5(c) emphasizes the signal space interpretation of $x_a(t)$ by representing it as the point $(2/\pi, -2/\pi)$ in the two-dimensional space spanned by the orthonormal functions $\phi_1(t)$ and $\phi_2(t)$. Representation of an arbitrary $x(t)$ exactly ($\epsilon_N = 0$) would require an infinite set of properly chosen orthogonal functions (that is, a complete set).

The minimum ISE, from (2.36), is

$$(\epsilon_N)_{\min} = \int_0^2 \sin^2(\pi t)\, dt - 2\left(\frac{2}{\pi}\right)^2 = 1 - \frac{8}{\pi^2} \cong 0.189 \tag{2.45}$$

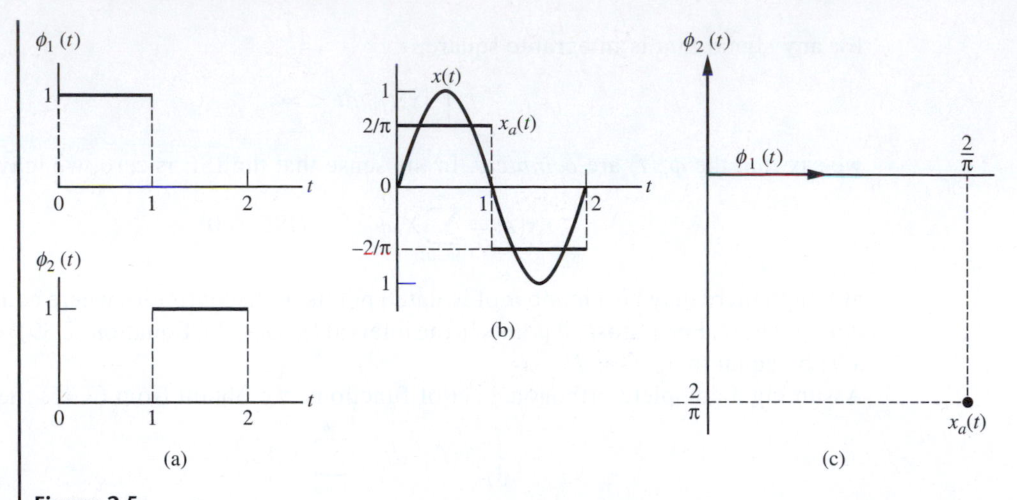

Figure 2.5
Approximation of a sine wave pulse with a generalized Fourier series. (a) Orthonormal functions. (b) Sine wave and approximation. (c) Signal space representation.

■

■ 2.4 FOURIER SERIES

2.4.1 Complex Exponential Fourier Series

Given a signal $x(t)$ defined over the interval $(t_0, t_0 + T_0)$ with the definition

$$\omega_0 = 2\pi f_0 = \frac{2\pi}{T_0}$$

we define the *complex exponential Fourier series* as

$$x(t) = \sum_{n=-\infty}^{\infty} X_n e^{jn\omega_0 t}, \quad t_0 \le t < t_0 + T_0 \tag{2.46}$$

where

$$X_n = \frac{1}{T_0} \int_{t_0}^{t_0 + T_0} x(t) e^{-jn\omega_0 t}\, dt \tag{2.47}$$

It can be shown to represent the signal $x(t)$ exactly in the interval $(t_0, t_0 + T_0)$, except at a point of jump discontinuity where it converges to the arithmetic mean of the left-hand and right-hand limits.[5] Outside the interval $(t_0, t_0 + T_0)$, of course, nothing is guaranteed. However, we note that the right-hand side of (2.46) is periodic with period T_0, since it is the sum of periodic rotating phasors with harmonic frequencies. Thus, if $x(t)$ is periodic with period T_0, the Fourier

[5] Dirichlet's conditions state that sufficient conditions for convergence are that $x(t)$ be defined and bounded on the range $(t_0, t_0 + T_0)$ and have only a finite number of maxima and minima and a finite number of discontinuities on this range.

series of (2.46) is an accurate representation for $x(t)$ for *all t* (except at points of discontinuity). The integration of (2.47) can then be taken over any period.

A useful observation about a complete orthonormal-series expansion of a signal is that the series is unique. For example, if we somehow find a Fourier expansion for a signal $x(t)$, we know that no other Fourier expansion for that $x(t)$ exists, since $\{e^{jn\omega_0 t}\}$ forms a complete set. The usefulness of this observation is illustrated with the following example.

EXAMPLE 2.6

Consider the signal

$$x(t) = \cos(\omega_0 t) + \sin^2(2\omega_0 t) \tag{2.48}$$

where $\omega_0 = 2\pi/T_0$. Find the complex exponential Fourier series.

Solution

We could compute the Fourier coefficients using (2.47), but by using appropriate trigonometric identities and Euler's theorem, we obtain

$$\begin{aligned} x(t) &= \cos(\omega_0 t) + \frac{1}{2} - \frac{1}{2}\cos(4\omega_0 t) \\ &= \frac{1}{2}e^{j\omega_0 t} + \frac{1}{2}e^{-j\omega_0 t} + \frac{1}{2} - \frac{1}{4}e^{j4\omega_0 t} - \frac{1}{4}e^{-j4\omega_0 t} \end{aligned} \tag{2.49}$$

Invoking uniqueness and equating the second line term by term with $\sum_{n=-\infty}^{\infty} X_n e^{jn\omega_0 t}$, we find that

$$\begin{aligned} X_0 &= \frac{1}{2} \\ X_1 &= \frac{1}{2} = X_{-1} \\ X_4 &= -\frac{1}{4} = X_{-4} \end{aligned} \tag{2.50}$$

with all other X_n equal to zero. Thus considerable labor is saved by noting that the Fourier series of a signal is unique. ∎

2.4.2 Symmetry Properties of the Fourier Coefficients

Assuming $x(t)$ is real, it follows from (2.47) that

$$X_n^* = X_{-n} \tag{2.51}$$

by taking the complex conjugate inside the integral and noting that the same result is obtained by replacing n by $-n$. Writing X_n as

$$X_n = |X_n| e^{j\underline{/X_n}} \tag{2.52}$$

we obtain

$$|X_n| = |X_{-n}| \quad \text{and} \quad \underline{/X_n} = -\underline{/X_{-n}} \tag{2.53}$$

Thus, for real signals, the magnitude of the Fourier coefficients is an even function of n, and the argument is odd.

Several symmetry properties can be derived for the Fourier coefficients, depending on the symmetry of $x(t)$. For example, suppose $x(t)$ is even; that is, $x(t) = x(-t)$. Then, using Euler's theorem to write the expression for the Fourier coefficients as (choose $t_0 = -T_0/2$)

$$X_n = \frac{1}{T_0}\int_{-T_0/2}^{T_0/2} x(t)\cos(n\omega_0 t)\,dt - \frac{j}{T_0}\int_{-T_0/2}^{T_0/2} x(t)\sin(n\omega_0 t)\,dt \tag{2.54}$$

we see that the second term is zero, since $x(t)\sin(n\omega_0 t)$ is an odd function. Thus X_n is purely real, and furthermore, X_n is an even function of n since $\cos(n\omega_0 t)$ is an even function of n. These consequences of $x(t)$ being even are illustrated by Example 2.6.

On the other hand, if $x(t) = -x(-t)$ [that is, $x(t)$ is odd], it readily follows that X_n is purely imaginary, since the first term in (2.54) is zero by virtue of $x(t)\cos(n\omega_0 t)$ being odd. In addition, X_n is an odd function of n, since $\sin(n\omega_0 t)$ is an odd function of n.

Another type of symmetry is (*odd*) *half-wave symmetry*, defined as

$$x\left(t \pm \frac{1}{2}T_0\right) = -x(t) \tag{2.55}$$

where T_0 is the period of $x(t)$. For signals with odd half-wave symmetry,

$$X_n = 0, \quad n = 0, \pm 2, \pm 4, \ldots \tag{2.56}$$

which states that the Fourier series for such a signal consists only of odd-indexed terms. The proof of this is left to the problems.

2.4.3 Trigonometric Form of the Fourier Series

Using (2.53) and assuming $x(t)$ real, we can regroup the complex exponential Fourier series by pairs of terms of the form

$$\begin{aligned} X_n e^{jn\omega_0 t} + X_{-n} e^{-jn\omega_0 t} &= |X_n| e^{j\left(n\omega_0 t + \underline{/X_n}\right)} + |X_n| e^{-j\left(n\omega_0 t + \underline{/X_n}\right)} \\ &= 2|X_n|\cos\left(n\omega_0 t + \underline{/X_n}\right) \end{aligned} \tag{2.57}$$

where the facts that $|X_n| = |X_{-n}|$ and $\underline{/X_n} = -\underline{/X_{-n}}$ have been used. Hence, (2.46) can be written in the equivalent trigonometric form:

$$x(t) = X_0 + \sum_{n=1}^{\infty} 2|X_n|\cos\left(n\omega_0 t + \underline{/X_n}\right) \tag{2.58}$$

Expanding the cosine in (2.58), we obtain still another equivalent series of the form

$$x(t) = X_0 + \sum_{n=1}^{\infty} A_n \cos(n\omega_0 t) + \sum_{n=1}^{\infty} B_n \sin(n\omega_0 t) \tag{2.59}$$

where

$$\begin{aligned} A_n &= 2|X_n|\cos \underline{/X_n} \\ &= \frac{2}{T_0}\int_{t_0}^{t_0+T_0} x(t)\cos(n\omega_0 t)\,dt \end{aligned} \tag{2.60}$$

and

$$B_n = -2|X_n| \sin \underline{/X_n}$$
$$= \frac{2}{T_0} \int_{t_0}^{t_0+T_0} x(t) \sin(n\omega_0 t)\, dt \qquad (2.61)$$

In either the trigonometric or the exponential forms of the Fourier series, X_0 represents the average or DC component of $x(t)$. The term for $n = 1$ is called the *fundamental*, the term for $n = 2$ is called the *second harmonic*, and so on.

2.4.4 Parseval's Theorem

Using (2.26) for average power of a periodic signal, substituting (2.46) for $x(t)$, and interchanging the order of integration and summation, we find Parseval's theorem to be

$$P = \frac{1}{T_0} \int_{T_0} |x(t)|^2\, dt = \sum_{n=-\infty}^{\infty} |X_n|^2 \qquad (2.62)$$

$$= X_0^2 + \sum_{n=1}^{\infty} 2|X_n|^2 \qquad (2.63)$$

which is a specialization of (2.40). In words, (2.62) simply states that the average power of a periodic signal $x(t)$ is the sum of the powers in the phasor components of its Fourier series, or (2.63) states that its average power is the sum of the powers in its DC component plus that in its AC components [from (2.58) the power in each cosine component is its amplitude squared divided by 2, or $(2|X_n|)^2/2 = 2|X_n|^2$. Note that powers of the Fourier components can be added because they are orthogonal.

2.4.5 Examples of Fourier Series

Table 2.1 gives Fourier series for several commonly occurring periodic waveforms. The left-hand column specifies the signal over one period. The definition of periodicity,

$$x(t) = x(t + T_0)$$

specifies it for all t. The derivation of the Fourier coefficients given in the right-hand column of Table 2.1 is left to the problems. Note that the full-rectified sine wave actually has the period $\frac{1}{2}T_0$.

For the periodic pulse train, it is convenient to express the coefficients in terms of the *sinc function*, defined as

$$\text{sinc}\, z = \frac{\sin(\pi z)}{\pi z} \qquad (2.64)$$

The sinc function is an even damped oscillatory function with zero crossings at integer values of its argument.

Table 2.1 Fourier Series for Several Periodic Signals

Signal (one period)	Coefficients for exponential Fourier series
1. Asymmetrical pulse train; period $= T_0$:	
$x(t) = A\Pi\left(\frac{t-t_0}{\tau}\right), \tau < T_0$	$X_n = \frac{A\tau}{T_0}\text{sinc}(nf_0\tau)e^{-j2\pi nf_0t_0}$
$x(t) = x(t+T_0)$, all t	$n = 0, \pm1, \pm2, \ldots$
2. Half-rectified sine wave; period $= T_0 = 2\pi/\omega_0$:	
$x(t) = \begin{cases} A\sin(\omega_0 t), & 0 \le t \le T_0/2 \\ 0, & -T_0/2 \le t \le 0 \end{cases}$ $x(t) = x(t+T_0)$ all t	$X_n = \begin{cases} \frac{A}{\pi(1-n^2)}, & n = 0, \pm 2, \pm 4, \cdots \\ 0, & n = \pm3, \pm 5, \cdots \\ -\frac{1}{4}jnA, & n = \pm1 \end{cases}$
3. Full-rectified sine wave; period $= T_0 = \pi/\omega_0$:	
$x(t) = A\lvert\sin(\omega_0 t)\rvert$	$X_n = \frac{2A}{\pi(1-4n^2)}, \quad n = 0, \pm1, \pm2, \ldots$
4. Triangular wave:	
$x(t) = \begin{cases} -\frac{4A}{T_0}t + A, & 0 \le t \le T_0/2 \\ \frac{4A}{T_0}t + A, & -T_0/2 \le t \le 0 \end{cases}$ $x(t) = x(t+T_0)$, all t	$X_n = \begin{cases} \frac{4A}{\pi^2 n^2}, & n \text{ odd} \\ 0, & n \text{ even} \end{cases}$

EXAMPLE 2.7

Specialize the results for the pulse train (item 1) of Table 2.1 to the complex exponential and trigonometric Fourier series of a square wave with even symmetry and amplitudes zero and A.

Solution

The solution proceeds by letting $t_0 = 0$ and $\tau = \frac{1}{2}T_0$ in item 1 of Table 2.1. Thus

$$X_n = \frac{1}{2}A\,\text{sinc}\left(\frac{1}{2}n\right) \tag{2.65}$$

But

$$\text{sinc}\left(\frac{n}{2}\right) = \frac{\sin(n\pi/2)}{n\pi/2}$$

$$= \begin{cases} 1, & n = 0 \\ 0, & n = \text{even} \\ \left|\frac{2}{n\pi}\right|, & n = \pm1, \pm5, \pm9, \ldots \\ -\left|\frac{2}{n\pi}\right|, & n = \pm3, \pm7, \ldots \end{cases}$$

Thus

$$
\begin{aligned}
x(t) &= \cdots + \frac{A}{5\pi}e^{-j5\omega_0 t} - \frac{A}{3\pi}e^{-j3\omega_0 t} + \frac{A}{\pi}e^{-j\omega_0 t} \\
&\quad + \frac{A}{2} + \frac{A}{\pi}e^{j\omega_0 t} - \frac{A}{3\pi}e^{j3\omega_0 t} + \frac{A}{5\pi}e^{j5\omega_0 t} - \cdots \\
&= \frac{A}{2} + \frac{2A}{\pi}\left[\cos(\omega_0 t) - \frac{1}{3}\cos(3\omega_0 t) + \frac{1}{5}\cos(5\omega_0 t) - \cdots\right]
\end{aligned} \tag{2.66}
$$

The first equation is the complex exponential form of the Fourier series and the second equation is the trigonometric form. The DC component of this squarewave is $X_0 = \frac{1}{2}A$. Setting this term to zero in the preceding Fourier series, we have the Fourier series of a square wave of amplitudes $\pm\frac{1}{2}A$. Such a square wave has half-wave symmetry, and this is precisely the reason that no even harmonics are present in its Fourier series. ■

2.4.6 Line Spectra

The complex exponential Fourier series (2.46) of a signal is simply a summation of phasors. In Section 2.1 we showed how a phasor could be characterized in the frequency domain by two plots: one showing its amplitude versus frequency and one showing its phase. Similarly, a periodic signal can be characterized in the frequency domain by making two plots: one showing amplitudes of the separate phasor components versus frequency and the other showing their phases versus frequency. The resulting plots are called the *two-sided amplitude*[6] and *phase spectra*, respectively, of the signal. From (2.53) it follows that for a real signal, the amplitude spectrum is even and the phase spectrum is odd, which is simply a result of the addition of complex conjugate phasors to get a real sinusoidal signal.

Figure 2.6(a) shows the double-sided spectrum for a half-rectified sine wave as plotted from the results given in Table 2.1. For $n = 2, 4, \ldots, X_n$ is represented as follows:

$$X_n = -\left|\frac{A}{\pi(1-n^2)}\right| = \frac{A}{\pi(n^2-1)}e^{-j\pi} \tag{2.67}$$

For $n = -2, -4, \ldots,$ it is represented as

$$X_n = -\left|\frac{A}{\pi(1-n^2)}\right| = \frac{A}{\pi(n^2-1)}e^{j\pi} \tag{2.68}$$

to ensure that the phase is odd, as it must be (note that $e^{\pm j\pi} = -1$). Thus, putting this together with $X_{\pm 1} = \mp jA/4$, we get

$$|X_n| = \begin{cases} \frac{1}{4}A, & n = \pm 1 \\ \left|\frac{A}{\pi(1-n^2)}\right|, & \text{all even } n \end{cases} \tag{2.69}$$

$$\underline{/X_n} = \begin{cases} -\pi, & n = 2, 4, \ldots \\ -\frac{1}{2}\pi & n = 1 \\ 0, & n = 0 \\ \frac{1}{2}\pi, & n = -1 \\ \pi, & n = -2, -4, \ldots \end{cases} \tag{2.70}$$

[6] *Magnitude spectrum* would be a more accurate term, although *amplitude spectrum* is the customary term.

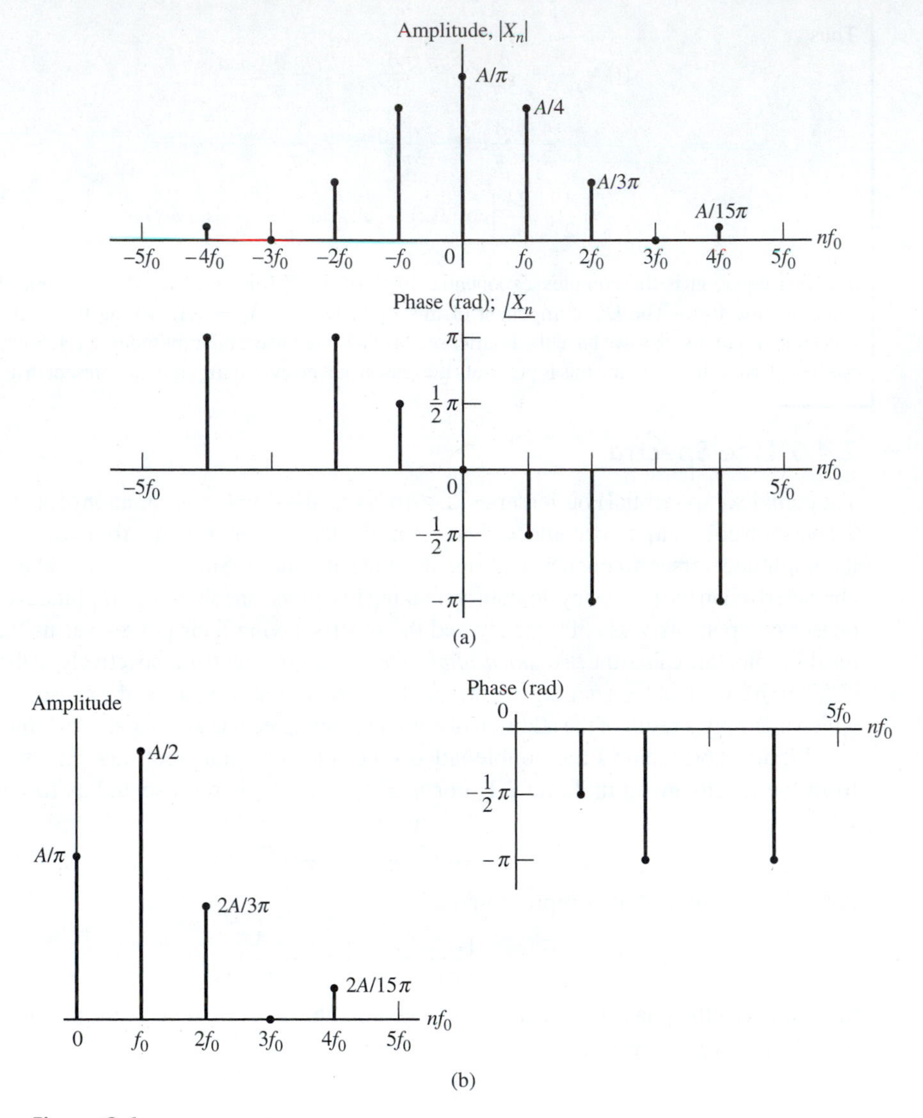

Figure 2.6
Line spectra for half-rectified sinewave. (a) Double sided. (b) Single sided.

The single-sided line spectra are obtained by plotting the amplitudes and phase angles of the terms in the trigonometric Fourier series (2.58) versus nf_0. Because the series (2.58) has only nonnegative frequency terms, the single-sided spectra exist only for $nf_0 \geq 0$. From (2.58) it is readily apparent that the single-sided phase spectrum of a periodic signal is identical to its double-sided phase spectrum for $nf_0 \geq 0$ and zero for $nf_0 < 0$. The single-sided amplitude spectrum is obtained from the double-sided amplitude spectrum by doubling the amplitude of all lines for $nf_0 > 0$. The line at $nf_0 = 0$ stays the same. The single-sided spectra for the half-rectified sinewave are shown in Figure 2.6(b).

As a second example, consider the pulse train

$$x(t) = \sum_{n=-\infty}^{\infty} A\Pi\left(\frac{t - nT_0 - \frac{1}{2}\tau}{\tau}\right) \tag{2.71}$$

From Table 2.1 with $t_0 = \frac{1}{2}\tau$ substituted in item 1, the Fourier coefficients are

$$X_n = \frac{A\tau}{T_0}\operatorname{sinc}(nf_0\tau)e^{-j\pi nf_0\tau} \tag{2.72}$$

The Fourier coefficients can be put in the form $|X_n|\exp(j\,\underline{/X_n})$, where

$$|X_n| = \frac{A_\tau}{T_0}|\operatorname{sinc}(nf_0\tau)| \tag{2.73}$$

and

$$\underline{/X_n} = \begin{cases} -\pi nf_0\tau & \text{if} \quad \operatorname{sinc}(nf_0\tau) > 0 \\ -\pi nf_0\tau + \pi & \text{if } nf_0 > 0 \text{ and } \operatorname{sinc}(nf_0\tau) < 0 \\ -\pi nf_0\tau - \pi & \text{if } nf_0 < 0 \text{ and } \operatorname{sinc}(nf_0\tau) < 0 \end{cases} \tag{2.74}$$

The $\pm\pi$ on the right-hand side of (2.74) on the second and third lines accounts for $|\operatorname{sinc}(nf_0\tau)| = -\operatorname{sinc}(nf_0\tau)$ whenever $\operatorname{sinc}(nf_0\tau) < 0$. Since the phase spectrum must have odd symmetry if $x(t)$ is real, π is subtracted if $nf_0 < 0$ and added if $nf_0 > 0$. The reverse could have been done—the choice is arbitrary. With these considerations, the double-sided amplitude and phase spectra can now be plotted. They are shown in Figure 2.7 for several choices of τ and T_0. Note that appropriate multiples of 2π are added or subtracted from the lines in the phase spectrum ($e^{\pm j2\pi} = 1$).

Comparing Figure 2.7(a) and (b), we note that the zeros of the envelope of the amplitude spectrum, which occur at multiples of $1/\tau$ Hz, move out along the frequency axis as the pulse

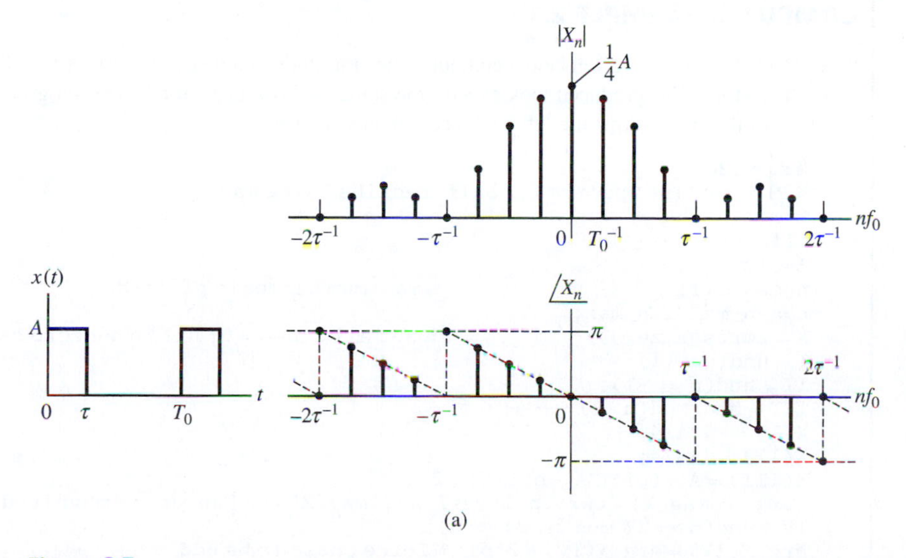

Figure 2.7

Spectra for a periodic pulse train signal. (a) $\tau = \frac{1}{4}T_0$. (b) $\tau = \frac{1}{8}T_0$; T_0 same as in (a). (c) $\tau = \frac{1}{8}T_0$; τ same as in (a).

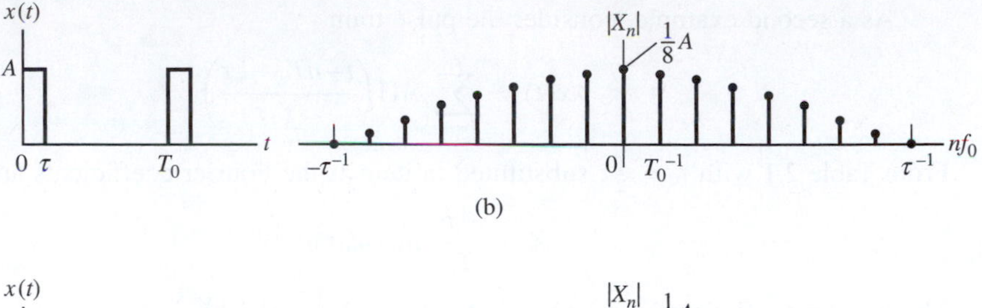

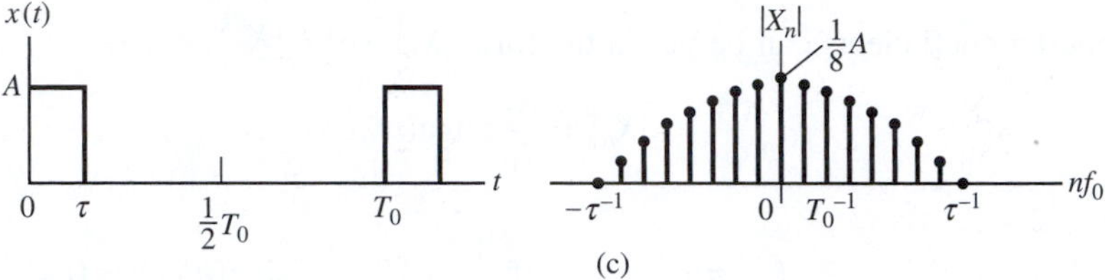

Figure 2.7
Continued.

width decreases. That is, *the time duration of a signal and its spectral width are inversely proportional*, a property that will be shown to be true in general later. Second, comparing Figure 2.7(a) and (c), we note that the separation between lines in the spectra is $1/T_0$. Thus the density of the spectral lines with frequency increases as the period of $x(t)$ increases.

COMPUTER EXAMPLE 2.1

The MATLAB program given below computes the amplitude and phase spectra for a half-rectified sine wave. The stem plots produced look exactly the same as those in Figure 2.6(a). Programs for plotting spectra of other waveforms are left to the computer exercises.

```
% file c2ce1
% Plot of line spectra for half-rectified sine wave
%
clf
A = 1;
n_max = 11;                    % maximum harmonic plotted
n = -n_max:1:n_max;
X = zeros(size(n));            % set all lines = 0; fill in nonzero ones
I = find(n == 1);
II = find(n == -1);
III = find(mod(n, 2) == 0);
X(I) = -j*A/4;
X(II) = j*A/4;
X(III) = A./(pi*(1. - n(III).^2));
[arg_X, mag_X] = cart2pol(real(X),imag(X)); % Convert to magnitude and phase
IV = find(n >= 2 & mod(n, 2) == 0);
arg_X(IV) = arg_X(IV) - 2*pi; % force phase to be odd
subplot(2,1,1), stem(n, mag_X), ylabel('X_n')
subplot(2,1,2), stem(n, arg_X), xlabel('nf_0'), ylabel('angle(X_n)')
```

■

2.5 THE FOURIER TRANSFORM

To generalize the Fourier series representation (2.46) to a representation valid for aperiodic signals, we consider the two basic relationships (2.46) and (2.47). Suppose that $x(t)$ is nonperiodic but is an energy signal, so that it is integrable square in the interval $(-\infty, \infty)$.[7] In the interval $|t| < \frac{1}{2}T_0$, we can represent $x(t)$ as the Fourier series

$$x(t) = \sum_{n=-\infty}^{\infty} \left[\frac{1}{T_0} \int_{-T_0/2}^{T_0/2} x(\lambda) e^{-j2\pi n f_0 \lambda} \, d\lambda \right] e^{j2\pi n f_0 t}, \quad |t| < \frac{T_0}{2} \tag{2.75}$$

where $f_0 = 1/T_0$. To represent $x(t)$ for all time, we simply let $T_0 \rightarrow \infty$ such that $nf_0 = n/T_0$ becomes the continuous variable f, $1/T_0$ becomes the differential df, and the summation becomes an integral. Thus

$$x(t) = \int_{-\infty}^{\infty} \left[\int_{-\infty}^{\infty} x(\lambda) e^{-j2\pi f \lambda} \, d\lambda \right] e^{j2\pi f t} \, df \tag{2.76}$$

Defining the inside integral as

$$X(f) = \int_{-\infty}^{\infty} x(\lambda) e^{-j2\pi f \lambda} \, d\lambda \tag{2.77}$$

we can write (2.76) as

$$x(t) = \int_{-\infty}^{\infty} X(f) e^{j2\pi f t} \, df \tag{2.78}$$

The existence of these integrals is assured, since $x(t)$ is an energy signal. We note that

$$X(f) = \lim_{T_0 \rightarrow \infty} T_0 X_n \tag{2.79}$$

which avoids the problem that $|X_n| \rightarrow 0$ as $T_0 \rightarrow \infty$.

The frequency-domain description of $x(t)$ provided by (2.77) is referred to as the *Fourier transform* of $x(t)$, written symbolically as $X(f) = \Im[x(t)]$. Conversion back to the time domain is achieved via the *inverse Fourier transform* (2.78), written symbolically as $x(t) = \Im^{-1}[X(f)]$.

Expressing (2.77) and (2.78) in terms of $f = \omega/2\pi$ results in easily remembered symmetrical expressions. Integrating (2.78) with respect to the variable ω requires a factor of $(2\pi)^{-1}$.

2.5.1 Amplitude and Phase Spectra

Writing $X(f)$ in terms of magnitude and phase as

$$X(f) = |X(f)| e^{j\theta(f)}, \quad \theta(f) = \underline{/X(f)} \tag{2.80}$$

we can show that for real $x(t)$,

$$|X(f)| = |X(-f)| \quad \text{and} \quad \theta(f) = -\theta(-f) \tag{2.81}$$

[7]This means that $x(t)$ should be an energy signal. Dirichlet's conditions give sufficient conditions for a signal to have a Fourier transform. These condition are that $x(t)$ be (1) single-valued with a finite number of maxima and minima and a finite number of discontinuities in any finite time interval and (2) absolutely integrable, that is, $\int_{-\infty}^{\infty} |x(t)| \, dt < \infty$. These conditions include all energy signals.

just as for the Fourier series. This is done by using Euler's theorem to write

$$R = \mathrm{Re}(X(f)) = \int_{-\infty}^{\infty} x(t)\cos(2\pi ft)\,dt \tag{2.82}$$

and

$$I = \mathrm{Im}(X(f)) = -\int_{-\infty}^{\infty} x(t)\sin(2\pi ft)\,dt \tag{2.83}$$

Thus the real part of $X(f)$ is even and the imaginary part is odd if $x(t)$ is a real signal. Since $|X(f)|^2 = R^2 + I^2$ and $\tan\theta(f) = I/R$, the symmetry properties (2.81) follow. A plot of $|X(f)|$ versus f is referred to as the *amplitude spectrum*[8] of $x(t)$, and a plot of $\underline{/X(f)} = \theta(f)$ versus f is known as the *phase spectrum*.

2.5.2 Symmetry Properties

If $x(t) = x(-t)$, that is, if $x(t)$ is even, then $x(t)\sin(2\pi ft)$ is odd in (2.83) and $\mathrm{Im}\,X(f) = 0$. Furthermore, $\mathrm{Re}(X(f))$ is an even function of f because cosine is an even function. Thus the Fourier transform of a real, even function is real and even.

On the other hand, if $x(t)$ is odd, $x(t)\cos(2\pi ft)$ is odd in (2.82) and $\mathrm{Re}(X(f)) = 0$. Thus the Fourier transform of a real, odd function is imaginary. In addition, $\mathrm{Im}(X(f))$ is an odd function of frequency because $\sin(2\pi ft)$ is an odd function.

EXAMPLE 2.8

Consider the pulse

$$x(t) = A\Pi\left(\frac{t-t_0}{\tau}\right) \tag{2.84}$$

The Fourier transform is

$$\begin{aligned} X(f) &= \int_{-\infty}^{\infty} A\Pi\left(\frac{t-t_0}{\tau}\right)e^{j2\pi ft}\,dt \\ &= A\int_{t_0-\tau/2}^{t_0+\tau/2} e^{-j2\pi ft}\,dt = A\tau\,\mathrm{sinc}(f\tau)e^{-j2\pi ft_0} \end{aligned} \tag{2.85}$$

The amplitude spectrum of $x(t)$ is

$$|X(f)| = A\tau|\mathrm{sinc}(f\tau)| \tag{2.86}$$

and the phase spectrum is

$$\theta(f) = \begin{cases} -2\pi t_0 f & \text{if } \mathrm{sinc}(f\tau) > 0 \\ -2\pi t_0 f \pm \pi & \text{if } \mathrm{sinc}(f\tau) < 0 \end{cases} \tag{2.87}$$

The term $\pm\pi$ is used to account for $\mathrm{sinc}(f\tau)$ being negative, and if $+\pi$ is used for $f > 0$, $-\pi$ is used for $f < 0$, or vice versa, to ensure that $\theta(f)$ is odd. When $|\theta(f)|$ exceeds 2π, an appropriate multiple of 2π may be added or subtracted from $\theta(f)$. Figure 2.8 shows the amplitude and phase spectra for the signal (2.84). The similarity to Figure 2.7 is to be noted, especially the inverse relationship between spectral width and pulse duration.

[8] *Amplitude density spectrum* would be more correct, since its dimensions are (amplitude units)(time) = (amplitude units)/frequency, but we will use the term *amplitude spectrum* for simplicity.

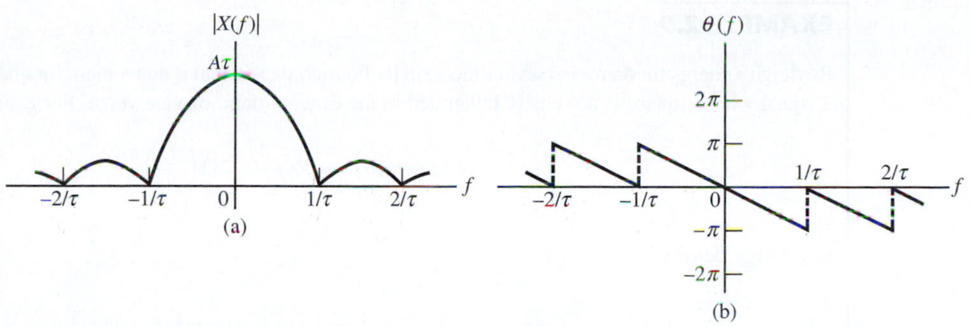

Figure 2.8
Amplitude and phase spectra for a pulse signal. (a) Amplitude spectrum. (b) Phase spectrum ($t_0 = \frac{1}{2}\tau$ is assumed).

2.5.3 Energy Spectral Density

The energy of a signal, defined by (2.22), can be expressed in the frequency domain as follows:

$$
\begin{aligned}
E &\triangleq \int_{-\infty}^{\infty} |x(t)|^2 \, dt \\
&= \int_{-\infty}^{\infty} x^*(t) \left[\int_{-\infty}^{\infty} X(f) e^{j2\pi ft} \, df \right] dt
\end{aligned}
\tag{2.88}
$$

where $x(t)$ has been written in terms of its Fourier transform. Reversing the order of integration, we obtain

$$
\begin{aligned}
E &= \int_{-\infty}^{\infty} X(f) \left[\int_{-\infty}^{\infty} x^*(t) e^{j2\pi ft} \, dt \right] df \\
&= \int_{-\infty}^{\infty} X(f) \left[\int_{-\infty}^{\infty} x(t) e^{-j2\pi ft} \, dt \right]^* df \\
&= \int_{-\infty}^{\infty} X(f) X^*(f) \, df
\end{aligned}
$$

or

$$
E = \int_{-\infty}^{\infty} |x(t)|^2 \, dt = \int_{-\infty}^{\infty} |X(f)|^2 \, df \tag{2.89}
$$

This is referred to as *Rayleigh's energy theorem* or Parseval's theorem for Fourier transforms.

Examining $|X(f)|^2$ and recalling the definition of $X(f)$ given by (2.77), we note that the former has the units of volts-seconds or, since we are considering power on a per-ohm basis, watts−seconds/hertz = joules/hertz. Thus we see that $|X(f)|^2$ has the units of energy density, and we define the energy spectral density of a signal as

$$
G(f) \triangleq |X(f)|^2 \tag{2.90}
$$

By integrating $G(f)$ over all frequency, we obtain the signal's total energy.

EXAMPLE 2.9

Rayleigh's energy theorem (Parseval's theorem for Fourier transforms) is convenient for finding the energy in a signal whose square is not easily integrated in the time domain, or vice versa. For example, the signal

$$x(t) = 40\,\text{sinc}(20t) \longleftrightarrow X(f) = 2\Pi\left(\frac{f}{20}\right) \tag{2.91}$$

has energy density

$$G(f) = |X(f)|^2 = \left[2\Pi\left(\frac{f}{20}\right)\right]^2 = 4\Pi\left(\frac{f}{20}\right) \tag{2.92}$$

where $\Pi(f/20)$ need not be squared because it has unity amplitude. Using Rayleigh's energy theorem, we find that the energy in $x(t)$ is

$$E = \int_{-\infty}^{\infty} G(f)\,df = \int_{-10}^{10} 4\,df = 80\,\text{J} \tag{2.93}$$

This checks with the result that is obtained by integrating $x^2(t)$ over all t using the definite integral $\int_{-\infty}^{\infty} \text{sinc}^2 u\,du = 1$.

The energy contained in the frequency interval $(0, W)$ can be found from the integral

$$E_W = \int_{-W}^{W} G(f)\,df = 2\int_0^W \left[2\Pi\left(\frac{f}{20}\right)\right]^2 df$$

$$= \begin{cases} 8W, & W \le 10 \\ 80, & W > 10 \end{cases} \tag{2.94}$$

which follows because $\Pi(f/20) = 0,\ |f| > 10$.

■

2.5.4 Convolution

We digress somewhat from our consideration of the Fourier transform to define the convolution operation and illustrate it by example.

The convolution of two signals, $x_1(t)$ and $x_2(t)$, is a new function of time, $x(t)$, written symbolically in terms of x_1 and x_2 as

$$x(t) = x_1(t) * x_2(t) = \int_{-\infty}^{\infty} x_1(\lambda)x_2(t-\lambda)\,d\lambda \tag{2.95}$$

Note that t is a parameter as far as the integration is concerned. The integrand is formed from x_1 and x_2 by three operations: (1) time reversal to obtain $x_2(-\lambda)$, (2) time shifting to obtain $x_2(t-\lambda)$, and (3) multiplication of $x_1(\lambda)$ and $x_2(t-\lambda)$ to form the integrand. An example will illustrate the implementation of these operations to form $x_1 * x_2$. Note that the dependence on time is often suppressed.

EXAMPLE 2.10

Find the convolution of the two signals

$$x_1(t) = e^{-\alpha t}u(t) \quad \text{and} \quad x_2(t) = e^{-\beta t}u(t), \quad \alpha > \beta > 0 \tag{2.96}$$

Solution

The steps involved in the convolution are illustrated in Figure 2.9 for $\alpha = 4$ and $\beta = 2$. Mathematically, we can form the integrand by direct substitution:

$$x(t) = x_1(t) * x_2(t) = \int_{-\infty}^{\infty} e^{-\alpha\lambda}u(\lambda)e^{-\beta(t-\lambda)}u(t-\lambda)\,d\lambda \tag{2.97}$$

However,

$$u(\lambda)u(t-\lambda) = \begin{cases} 0, & \lambda < 0 \\ 1, & 0 < \lambda < t \\ 0, & \lambda > t \end{cases} \tag{2.98}$$

Thus,

$$x(t) = \begin{cases} 0, & t < 0 \\ \displaystyle\int_0^t e^{-\beta t}e^{-(\alpha-\beta)\lambda}\,d\lambda = \frac{1}{\alpha-\beta}\left(e^{-\beta t} - e^{-\alpha t}\right), & t \geq 0 \end{cases} \tag{2.99}$$

This result for $x(t)$ is also shown in Figure 2.9.

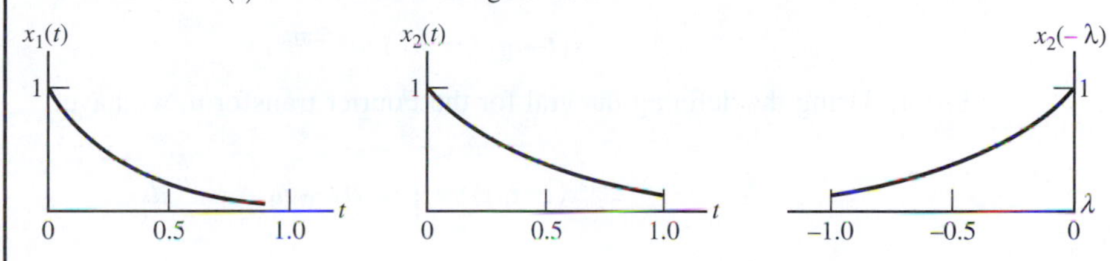

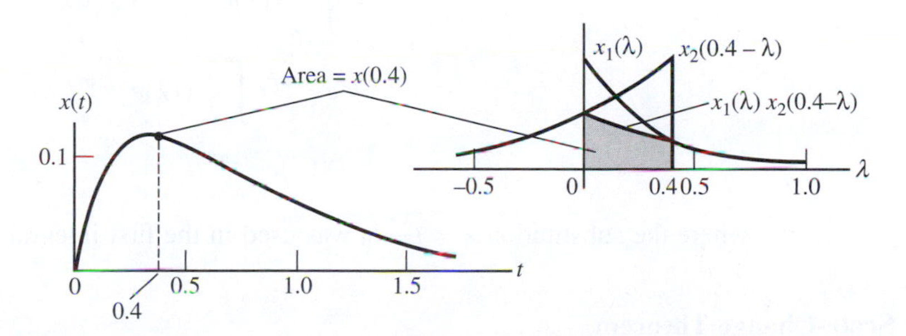

Figure 2.9
The operations involved in the convolution of two exponentially decaying signals.

■

2.5.5 Transform Theorems: Proofs and Applications

Several useful theorems[9] involving Fourier transforms can be proved. These are useful for deriving Fourier transform pairs as well as deducing general frequency-domain relationships. The notation $x(t) \longleftrightarrow X(f)$ will be used to denote a Fourier transform pair.

[9]See Tables G.5 and G.6 in Appendix G for a listing of Fourier transform pairs and theorems.

Each theorem will be stated along with a proof in most cases. Several examples giving applications will be given after the statements of all the theorems. In the statements of the theorems, $x(t)$, $x_1(t)$, and $x_2(t)$ denote signals with $X(f)$, $X_1(f)$, and $X_2(f)$ denoting their respective Fourier transforms. Constants are denoted by a, a_1, a_2, t_0, and f_0.

Superposition Theorem

$$a_1x_1(t) + a_2x_2(t) \longleftrightarrow a_1X_1(f) + a_2X_2(f) \tag{2.100}$$

Proof: By the defining integral for the Fourier transform,

$$\begin{aligned}\Im\{a_1x_1(t) + a_2x_2(t)\} &= \int_{-\infty}^{\infty} [a_1x_1(t) + a_2x_2(t)]e^{-j2\pi ft}\,dt \\ &= a_1\int_{-\infty}^{\infty} x_1(t)e^{-j2\pi ft}dt + a_2\int_{-\infty}^{\infty} x_2(t)e^{-j2\pi ft}\,dt \\ &= a_1X_1(f) + a_2X_2(f)\end{aligned} \tag{2.101}$$

Time-Delay Theorem

$$x(t-t_0) \longleftrightarrow X(f)e^{-j2\pi ft_0} \tag{2.102}$$

Proof: Using the defining integral for the Fourier transform, we have

$$\begin{aligned}\Im\{x(t-t_0)\} &= \int_{-\infty}^{\infty} x(t-t_0)e^{-j2\pi ft}\,dt \\ &= \int_{-\infty}^{\infty} x(\lambda)e^{-j2\pi f(\lambda+t_0)}\,d\lambda \\ &= e^{-j2\pi ft_0}\int_{-\infty}^{\infty} x(\lambda)e^{-j2\pi f\lambda}\,d\lambda \\ &= X(f)e^{-j2\pi ft_0}\end{aligned} \tag{2.103}$$

where the substitution $\lambda = t - t_0$ was used in the first integral.

Scale-Change Theorem

$$x(at) \longleftrightarrow \frac{1}{|a|}X\left(\frac{f}{a}\right) \tag{2.104}$$

Proof: First, assume that $a > 0$. Then

$$\begin{aligned}\Im\{x(at)\} &= \int_{-\infty}^{\infty} x(at)e^{-j2\pi ft}\,dt \\ &= \int_{-\infty}^{\infty} x(\lambda)e^{-j2\pi f\lambda/a}\frac{d\lambda}{a} = \frac{1}{a}X\left(\frac{f}{a}\right)\end{aligned} \tag{2.105}$$

where the substitution $\lambda = at$ has been used. Next, considering $a < 0$, we write

$$\begin{aligned}\Im\{x(at)\} &= \int_{-\infty}^{\infty} x(-|a|t)e^{-j2\pi ft}\,dt = \int_{-\infty}^{\infty} x(\lambda)e^{+j2\pi f\lambda/|a|}\frac{d\lambda}{|a|} \\ &= \frac{1}{|a|}X\left(-\frac{f}{|a|}\right) = \frac{1}{|a|}X\left(\frac{f}{a}\right)\end{aligned} \tag{2.106}$$

where use has been made of the relation $-|a| = a$ if $a < 0$.

Duality Theorem

$$X(t) \longleftrightarrow x(-f) \tag{2.107}$$

That is, if the Fourier transform of $x(t)$ is $X(f)$, then the Fourier transform of $X(f)$ with f replaced by t is the original time-domain signal with t replaced by $-f$.

Proof: The proof of this theorem follows by virtue of the fact that the only difference between the Fourier transform integral and the inverse Fourier transform integral is a minus sign in the exponent of the integrand.

Frequency Translation Theorem

$$x(t)e^{j2\pi f_0 t} \longleftrightarrow X(f - f_0) \tag{2.108}$$

Proof: To prove the frequency translation theorem, note that

$$\int_{-\infty}^{\infty} x(t)e^{j2\pi f_0 t}e^{-j2\pi ft}\,dt = \int_{-\infty}^{\infty} x(t)e^{-j2\pi(f-f_0)t}\,dt = X(f - f_0) \tag{2.109}$$

Modulation Theorem

$$x(t)\cos(2\pi f_0 t) \longleftrightarrow \frac{1}{2}X(f - f_0) + \frac{1}{2}X(f + f_0) \tag{2.110}$$

Proof: The proof of this theorem follows by writing $\cos(2\pi f_0 t)$ in exponential form as $\frac{1}{2}\left(e^{j2\pi f_0 t} + e^{-j2\pi f_0 t}\right)$ and applying the superposition and frequency translation theorems.

Differentiation Theorem

$$\frac{d^n x(t)}{dt^n} \longleftrightarrow (j2\pi f)^n X(f) \tag{2.111}$$

Proof: We prove the theorem for $n = 1$ by using integration by parts on the defining Fourier transform integral as follows:

$$\begin{aligned}\Im\left\{\frac{dx}{dt}\right\} &= \int_{-\infty}^{\infty} \frac{dx(t)}{dt}e^{-j2\pi ft}dt \\ &= x(t)e^{-j2\pi ft}\Big|_{-\infty}^{\infty} + j2\pi f\int_{-\infty}^{\infty} x(t)e^{-j2\pi ft}\,dt \\ &= j2\pi f\,X(f)\end{aligned} \tag{2.112}$$

where $u = e^{-j2\pi ft}$ and $dv = (dx/dt)\,dt$ have been used in the integration-by-parts formula, and the first term of the middle equation vanishes at each end point by virtue of $x(t)$ being an energy signal. The proof for values of $n > 1$ follows by induction.

Integration Theorem

$$\int_{-\infty}^{t} x(\lambda)\,d\lambda \longleftrightarrow (j2\pi f)^{-1}X(f) + \frac{1}{2}X(0)\delta(f) \tag{2.113}$$

Proof: If $X(0) = 0$ the proof of the integration theorem can be carried out by using integration by parts as in the case of the differentiation theorem. We obtain

$$\Im\left\{\int_{-\infty}^{t} x(\lambda)\,d(\lambda)\right\} = \left\{\int_{-\infty}^{t} x(\lambda)\,d(\lambda)\right\}\left(-\frac{1}{j2\pi f}e^{-j2\pi ft}\right)\Big|_{-\infty}^{\infty} + \frac{1}{j2\pi f}\int_{-\infty}^{\infty} x(t)e^{-j2\pi ft}\,dt \tag{2.114}$$

The first term vanishes if $X(0) = \int_{-\infty}^{\infty} x(t)\,dt = 0$, and the second term is just $X(f)/(j2\pi f)$. For $X(0) \neq 0$, a limiting argument must be used to account for the Fourier transform of the nonzero average value of $x(t)$.

Convolution Theorem

$$\int_{-\infty}^{\infty} x_1(\lambda)x_2(t-\lambda)\,d\lambda \triangleq \int_{-\infty}^{\infty} x_1(t-\lambda)x_2(\lambda)d\lambda \longleftrightarrow X_1(f)X_2(f) \tag{2.115}$$

Proof: To prove the convolution theorem of Fourier transforms, we represent $x_2(t-\lambda)$ in terms of the inverse Fourier transform integral as

$$x_2(t-\lambda) = \int_{-\infty}^{\infty} X_2(f)e^{j2\pi f(t-\lambda)}\,df \tag{2.116}$$

Denoting the convolution operation as $x_1(t)*x_2(t)$, we have

$$\begin{aligned} x_1(t) * x_2(t) &= \int_{-\infty}^{\infty} x_1(\lambda)\left[\int_{-\infty}^{\infty} X_2(f)e^{j2\pi f(t-\lambda)}\,df\right]d\lambda \\ &= \int_{-\infty}^{\infty} X_2(f)\left[\int_{-\infty}^{\infty} x_1(\lambda)e^{-j2\pi f\lambda}\,d\lambda\right]e^{j2\pi ft}\,df \end{aligned} \tag{2.117}$$

where the last step results from reversing the orders of integration. The bracketed term inside the integral is $X_1(f)$, the Fourier transform of $x_1(t)$. Thus

$$x_1 * x_2 = \int_{-\infty}^{\infty} X_1(f)X_2(f)e^{j2\pi ft}\,df \tag{2.118}$$

which is the inverse Fourier transform of $X_1(f)X_2(f)$. Taking the Fourier transform of this result yields the desired transform pair.

Multiplication Theorem

$$x_1(t)x_2(t) \longleftrightarrow X_1(f) * X_2(f) = \int_{-\infty}^{\infty} X_1(\lambda)X_2(f-\lambda)\,d\lambda \tag{2.119}$$

Proof: The proof of the multiplication theorem proceeds in a manner analogous to the proof of the convolution theorem.

EXAMPLE 2.11

Use the duality theorem to show that

$$2AW\,\mathrm{sinc}(2Wt) \longleftrightarrow A\Pi\left(\frac{f}{2W}\right) \tag{2.120}$$

Solution

From Example 2.8, we know that

$$x(t) = A\Pi\left(\frac{t}{\tau}\right) \longleftrightarrow A\tau\,\mathrm{sinc}(f\tau) = X(f) \tag{2.121}$$

Considering $X(t)$, and using the duality theorem, we obtain

$$X(t) = A\tau\,\mathrm{sinc}(\tau t) \longleftrightarrow A\Pi\left(-\frac{f}{\tau}\right) = X(-t) \tag{2.122}$$

where τ is a parameter with dimension s^{-1}, which may be somewhat confusing at first sight! By letting $\tau = 2W$ and noting that $\Pi(u)$ is even, the given relationship follows.

■

EXAMPLE 2.12

Obtain the following Fourier transform pairs:

1. $A\delta(t) \longleftrightarrow A$
2. $A\delta(t-t_0) \longleftrightarrow Ae^{-j2\pi f t_0}$
3. $A \longleftrightarrow A\delta(f)$
4. $Ae^{j2\pi f_0 t} \longleftrightarrow A\delta(f-f_0)$

Solution

Even though these signals are not energy signals, we can formally derive the Fourier transform of each by obtaining the Fourier transform of a "proper" energy signal that approaches the given signal in the limit as some parameter approaches zero or infinity. For example, formally,

$$\Im[A\delta(t)] = \Im\left[\lim_{\tau\to 0}\left(\frac{A}{\tau}\right)\Pi\left(\frac{t}{\tau}\right)\right] = \lim_{\tau\to 0} A\,\mathrm{sinc}(f\tau) = A \tag{2.123}$$

We can use a formal procedure such as this to define Fourier transforms for the other three signals as well. It is easier, however, to use the sifting property of the delta function and the appropriate Fourier transform theorems. The same results are obtained. For example, we obtain the first transform pair directly by writing down the Fourier transform integral with $x(t) = \delta(t)$ and invoking the sifting property:

$$\Im[A\delta(t)] = A\int_{-\infty}^{\infty} \delta(t)e^{-j2\pi ft}\,dt = A \tag{2.124}$$

Transform pair 2 follows by application of the time-delay theorem to pair 1.

Transform pair 3 can be obtained by using the inverse-transform relationship or the first transform pair and the duality theorem. Using the latter, we obtain

$$X(t) = A \longleftrightarrow A\delta(-f) = A\delta(f) = x(-f) \tag{2.125}$$

where the eveness property of the impulse function is used.

Transform pair 4 follows by applying the frequency-translation theorem to pair 3. The Fourier transform pairs of Example 2.12 will be used often in the discussion of modulation.

■

EXAMPLE 2.13

Use the differentiation theorem to obtain the Fourier transform of the triangular signal, defined as

$$\Lambda\left(\frac{t}{\tau}\right) \triangleq \begin{cases} 1-|t|/\tau, & |t| < \tau \\ 0, & \text{otherwise} \end{cases} \tag{2.126}$$

Solution

Differentiating $\Lambda(t/\tau)$ twice, we obtain, as shown in Figure 2.10,

$$\frac{d^2\Lambda(t/\tau)}{dt^2} = \frac{1}{\tau}\delta(t+\tau) - \frac{2}{\tau}\delta(t) + \frac{1}{\tau}\delta(t-\tau) \tag{2.127}$$

Using the differentiation, superposition, and time-shift theorems and the result of Example 2.12, we obtain

$$\begin{aligned} \Im\left[\frac{d^2\Lambda(t/\tau)}{dt^2}\right] &= (j2\pi f)^2 \Im\left[\Lambda\left(\frac{t}{\tau}\right)\right] \\ &= \frac{1}{\tau}\left(e^{j2\pi f\tau} - 2 + e^{-j2\pi f\tau}\right) \end{aligned} \tag{2.128}$$

or, solving for $\Im\left[\Lambda\left(\frac{t}{\tau}\right)\right]$ and simplifying, we get

$$\Im\left[\Lambda\left(\frac{t}{\tau}\right)\right] = \frac{2\cos(2\pi f\tau) - 2}{\tau(j2\pi f)^2} = \tau\frac{\sin^2(\pi f\tau)}{(\pi f\tau)^2} \tag{2.129}$$

where the identity $\frac{1}{2}[1 - \cos(2\pi ft)] = \sin^2(\pi ft)$ has been used. Summarizing, we have shown that

$$\Lambda\left(\frac{t}{\tau}\right) \longleftrightarrow \tau\,\text{sinc}^2(f\tau) \tag{2.130}$$

where $\sin(\pi f\tau)/(\pi f\tau)$ has been replaced by $\text{sinc}(f\tau)$.

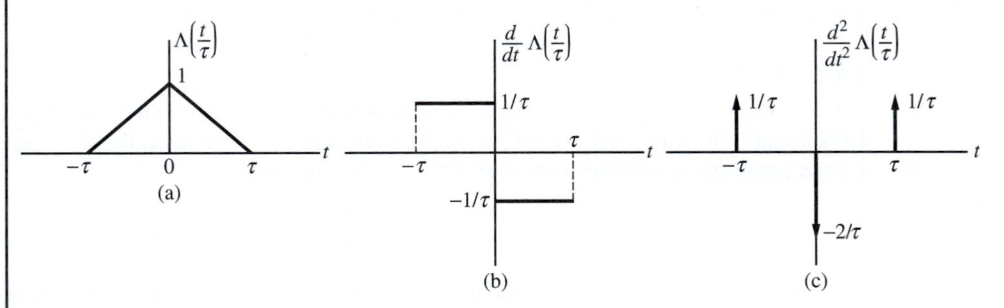

Figure 2.10
Triangular signal and its first two derivatives. (a) Triangular signal. (b) First derivative of the triangular signal. (c) Second derivative of the triangular signal.

■

EXAMPLE 2.14

As another example of obtaining Fourier transforms of signals involving impulses, let us consider the signal

$$y_s(t) = \sum_{m=-\infty}^{\infty} \delta(t - mT_s) \tag{2.131}$$

It is a periodic waveform referred to as the ideal sampling waveform and consists of a doubly infinite sequence of impulses spaced by T_s s.

Solution

To obtain the Fourier transform of $y_s(t)$, we note that it is periodic and, in a formal sense, therefore can be represented by a Fourier series. Thus,

$$y_s(t) = \sum_{m=-\infty}^{\infty} \delta(t - mT_s) = \sum_{n=-\infty}^{\infty} Y_n e^{jn2\pi f_s t}, \quad f_s = \frac{1}{T_s} \tag{2.132}$$

where

$$Y_n = \frac{1}{T_s} \int_{T_s} \delta(t) e^{-jn2\pi f_s t}\, dt = f_s \tag{2.133}$$

by the sifting property of the impulse function. Therefore,

$$y_s(t) = f_s \sum_{n=-\infty}^{\infty} e^{jn2\pi f_s t} \tag{2.134}$$

Fourier transforming term by term, we obtain

$$Y_s(f) = f_s \sum_{n=-\infty}^{\infty} \Im\left[1 \cdot e^{j2\pi n f_s t}\right] = f_s \sum_{n=-\infty}^{\infty} \delta(f - nf_s) \tag{2.135}$$

where we have used the results of Example 2.12. Summarizing, we have shown that

$$\sum_{m=-\infty}^{\infty} \delta(t - mT_s) \longleftrightarrow f_s \sum_{n=-\infty}^{\infty} \delta(f - nf_s) \tag{2.136}$$

The transform pair (2.136) is useful in spectral representations of periodic signals by the Fourier transform, which will be considered shortly.

A useful expression can be derived from (2.136). Taking the Fourier transform of the left-hand side of (2.136) yields

$$\begin{aligned} \Im\left[\sum_{m=-\infty}^{\infty} \delta(t - mT_s)\right] &= \int_{-\infty}^{\infty} \left[\sum_{m=-\infty}^{\infty} \delta(t - mT_s)\right] e^{-j2\pi f t} dt \\ &= \sum_{m=-\infty}^{\infty} \int_{-\infty}^{\infty} \delta(t - mT_s) e^{-j2\pi f t} dt \\ &= \sum_{m=-\infty}^{\infty} e^{-j2\pi m T_s f} \end{aligned} \tag{2.137}$$

where we interchanged the orders of integration and summation and used the sifting property of the impulse function to perform the integration. Replacing m by $-m$ and equating the result to the right-hand side of (2.136) gives

$$\sum\nolimits_{m=-\infty}^{\infty} e^{j2\pi m T_s f} = f_s \sum_{n=-\infty}^{\infty} \delta(f - nf_s) \tag{2.138}$$

This result will be used in Chapter 6.

■

EXAMPLE 2.15

The convolution theorem can be used to obtain the Fourier transform of the triangle $\Lambda(t/\tau)$ defined by (2.126).

Solution

We proceed by first showing that the convolution of two rectangular pulses is a triangle. The steps in computing

$$y(t) = \int_{-\infty}^{\infty} \Pi\left(\frac{t - \lambda}{\tau}\right) \Pi\left(\frac{\lambda}{\tau}\right) d\lambda \tag{2.139}$$

are carried out in Table 2.2. Summarizing the results, we have

Table 2.2 Computation of $\Pi(t/\tau) * \Pi(t/\tau)$

Range	Integrand	Limits	Area
$-\infty < t < -\tau$	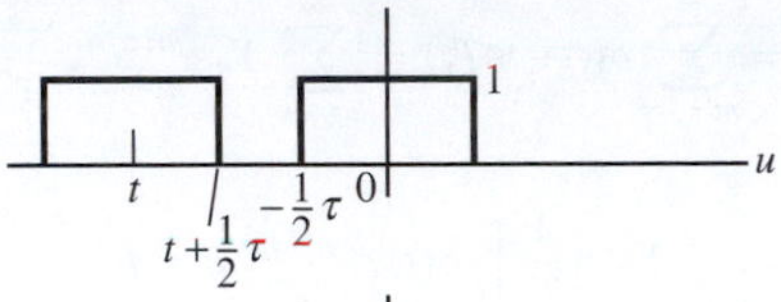		0
$-\tau < t < 0$	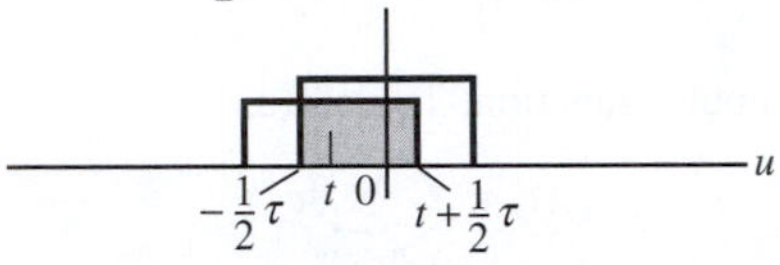	$-\frac{1}{2}\tau$ to $t+\frac{1}{2}\tau$	$\tau + t$
$0 < t < \tau$	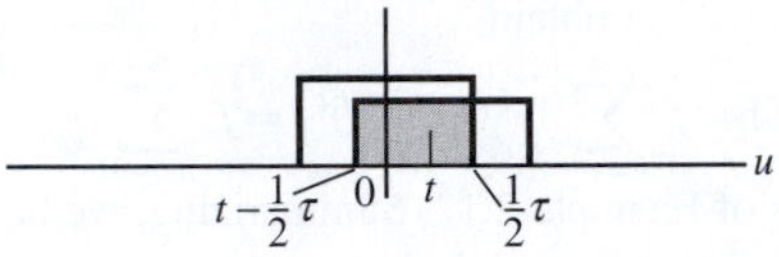	$t-\frac{1}{2}\tau$ to $\frac{1}{2}\tau$	$\tau - t$
$\tau < t < \infty$	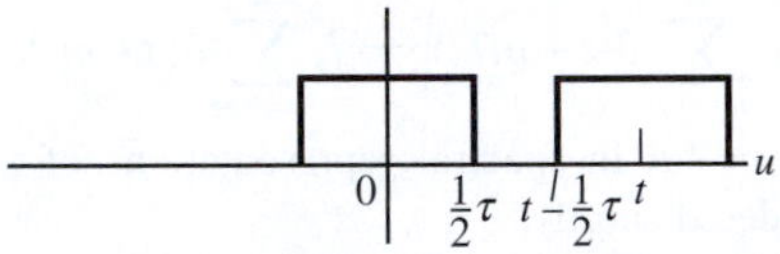		0

$$\tau\Lambda\left(\frac{t}{\tau}\right) = \Pi\left(\frac{t}{\tau}\right) * \Pi\left(\frac{t}{\tau}\right) = \begin{cases} 0, & t < -\tau \\ \tau - |t|, & |t| \le \tau \\ 0, & t > \tau \end{cases} \tag{2.140}$$

$$\text{or} \quad \Lambda\left(\frac{t}{\tau}\right) = \frac{1}{\tau}\Pi\left(\frac{t}{\tau}\right) * \Pi\left(\frac{t}{\tau}\right) \tag{2.141}$$

Using the transform pair

$$\Pi\left(\frac{t}{\tau}\right) \longleftrightarrow \tau \operatorname{sinc} f t \tag{2.142}$$

and the convolution theorem of Fourier transforms (2.115), we obtain the transform pair

$$\Lambda\left(\frac{t}{\tau}\right) \longleftrightarrow \tau \operatorname{sinc}^2 f\tau \tag{2.143}$$

as in Example 2.13 by applying the differentiation theorem. ■

A useful result is the convolution of an impulse $\delta(t - t_0)$ with a signal $x(t)$, where $x(t)$ is assumed continuous at $t = t_0$. Carrying out the operation, we obtain

$$\delta(t - t_0) * x(t) = \int_{-\infty}^{\infty} \delta(\lambda - t_0) x(t - \lambda)\, d\lambda = x(t - t_0) \tag{2.144}$$

by the sifting property of the delta function. That is, convolution of $x(t)$ with an impulse occurring at time t_0 simply shifts $x(t)$ to t_0.

EXAMPLE 2.16

Consider the Fourier transform of the cosinusoidal pulse

$$x(t) = A\Pi\left(\frac{t}{\tau}\right)\cos(\omega_0 t), \quad \omega_0 = 2\pi f_0 \tag{2.145}$$

Using the transform pair (see Example 2.12, item 4)

$$e^{\pm j2\pi f_0 t} \longleftrightarrow \delta(f \mp f_0) \tag{2.146}$$

obtained earlier and Euler's theorem, we find that

$$\cos(2\pi f_0 t) \longleftrightarrow \frac{1}{2}\delta(f - f_0) + \frac{1}{2}\delta(f + f_0) \tag{2.147}$$

We have also shown that

$$A\Pi\left(\frac{t}{\tau}\right) \longleftrightarrow A\tau\,\mathrm{sinc}(f\tau)$$

Therefore, using the multiplication theorem of Fourier transforms (2.118), we obtain

$$\begin{aligned} X(f) &= \Im\left[A\Pi\left(\frac{t}{\tau}\right)\cos(\omega_0 t)\right] = [A\tau\,\mathrm{sinc}(f\tau)] * \left\{\frac{1}{2}[\delta(f - f_0) + \delta(t + f_0)]\right\} \\ &= \frac{1}{2}A\tau\{\mathrm{sinc}[(f - f_0)\tau] + \mathrm{sinc}[(f + f_0)\tau]\} \end{aligned} \tag{2.148}$$

where $\delta(f - f_0) * Z(f) = Z(f - f_0)$ for $Z(f)$ continuous at $f = f_0$ has been used. Figure 2.11(c) shows $X(f)$. The same result can be obtained via the modulation theorem.

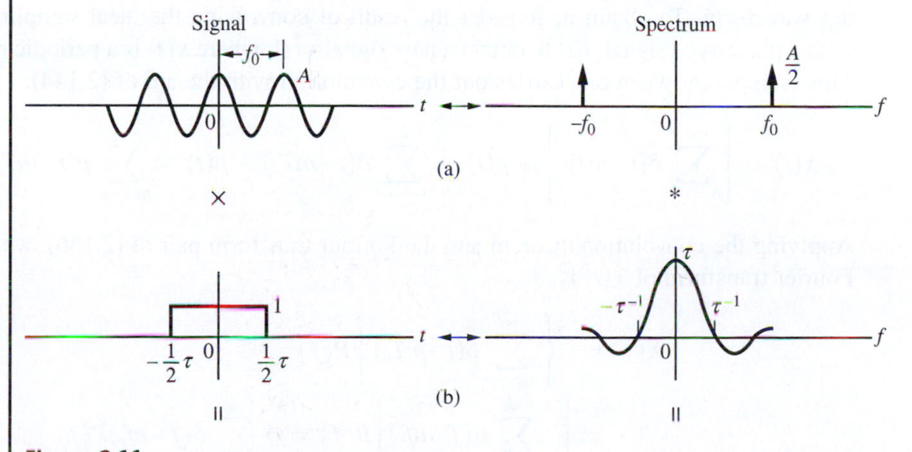

Figure 2.11
(a)–(c) Application of the multiplication theorem. (c)–(e) Application of the convolution theorem. Note: $\times$ denotes multiplication; $*$ denotes convolution, $\longleftrightarrow$ denotes transform pairs.

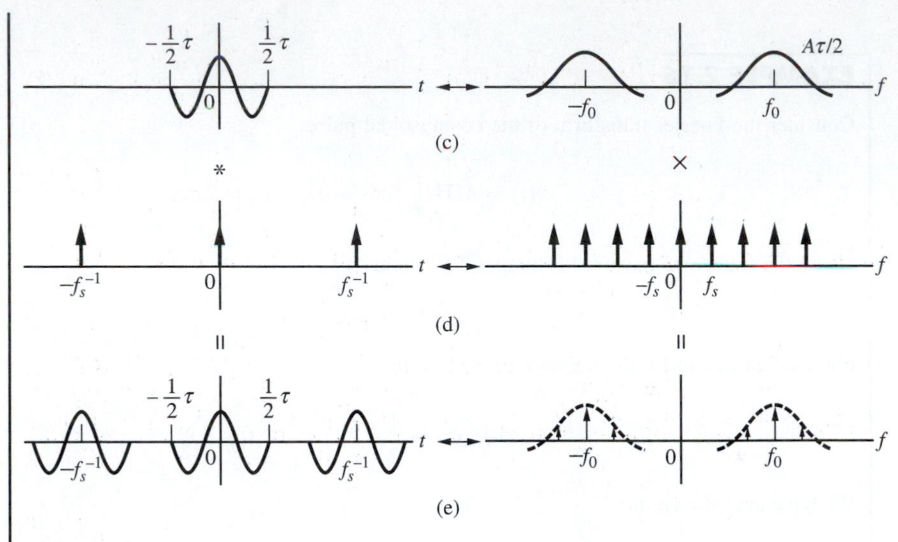

Figure 2.11
Continued.

■

2.5.6 Fourier Transforms of Periodic Signals

The Fourier transform of a periodic signal, in a strict mathematical sense, does not exist, since periodic signals are not energy signals. However, using the transform pairs derived in Example 2.12 for a constant and a phasor signal, we could, in a formal sense, write down the Fourier transform of a periodic signal by Fourier transforming its complex Fourier series term by term.

A somewhat more useful form for the Fourier transform of a periodic signal is obtained by applying the convolution theorem and the transform pair (2.136) for the ideal sampling waveform. To obtain it, consider the result of convolving the ideal sampling waveform with a pulse-type signal $p(t)$ to obtain a new signal $x(t)$, where $x(t)$ is a periodic power signal. This is apparent when one carries out the convolution with the aid of (2.144):

$$x(t) = \left[\sum_{m=-\infty}^{\infty} \delta(t-mT_s)\right] * p(t) = \sum_{m=-\infty}^{\infty} \delta(t-mT_s) * p(t) = \sum_{m=-\infty}^{\infty} p(t-mT_s) \quad (2.149)$$

Applying the convolution theorem and the Fourier transform pair of (2.136), we find that the Fourier transform of $x(t)$ is

$$\begin{aligned} X(f) &= \Im\left\{\sum_{m=-\infty}^{\infty} \delta(t-mT_s)\right\}P(f) \\ &= \left[f_s \sum_{n=-\infty}^{\infty} \delta(f-nf_s)\right]P(f) = f_s \sum_{n=-\infty}^{\infty} \delta(f-nf_s)P(f) \\ &= \sum_{n=-\infty}^{\infty} f_s P(nf_s)\delta(f-nf_s) \end{aligned} \quad (2.150)$$

where $P(f) = \Im[p(t)]$ and the fact that $P(f)\delta(f-nf_s) = P(nf_s)\delta(f-nf_s)$ has been used. Summarizing, we have obtained the Fourier transform pair

$$\sum_{m=-\infty}^{\infty} p(t-mT_s) \leftrightarrow \sum_{n=-\infty}^{\infty} f_s P(nf_s)\delta(f-nf_s) \tag{2.151}$$

The usefulness of (2.151) is illustrated with an example.

EXAMPLE 2.17

The Fourier transform of a single cosinusoidal pulse was found in Example 2.16 and is shown in Figure 2.11(c). The Fourier transform of a periodic cosinusoidal pulse train, which could represent the output of a radar transmitter, for example, is obtained by writing it as

$$\begin{aligned} y(t) &= \left[\sum_{n=-\infty}^{\infty} \delta(t-mT_s)\right] * \Pi\left(\frac{t}{\tau}\right)\cos(2\pi f_0 t), \quad f_0 \gg 1/\tau \\ &= \sum_{m=-\infty}^{\infty} \Pi\left(\frac{t-mT_s}{\tau}\right)\cos[2\pi f_0(t-mT_s)], \quad f_s \leq \tau^{-1} \end{aligned} \tag{2.152}$$

This signal is illustrated in Figure 2.11(e). Identifying $p(t) = \Pi(t/\tau)\cos(2\pi f_0 t)$, we get, by the modulation theorem, that $P(f) = (A\tau/2)[\mathrm{sinc}(f-f_0)\tau + \mathrm{sinc}(f+f_0)\tau]$. Applying (2.151), the Fourier transform of $y(t)$ is

$$Y(f) = \sum_{n=-\infty}^{\infty} \frac{Af_s\tau}{2}[\mathrm{sinc}(nf_s-f_0)\tau + \mathrm{sinc}(nf_s+f_0)\tau]\delta(f-nf_s) \tag{2.153}$$

The spectrum is illustrated on the right-hand side of Figure 2.11(e). ■

2.5.7 Poisson Sum Formula

We can develop the *Poisson sum formula* by taking the inverse Fourier transform of the right-hand side of (2.151). When we use the transform pair $\exp(-j2\pi nf_s t) \leftrightarrow \delta(f-nf_s)$ (see Example 2.12), it follows that

$$\Im^{-1}\left\{\sum_{n=-\infty}^{\infty} f_s P(nf_s)\delta(f-f_s)\right\} = f_s \sum_{n=-\infty}^{\infty} P(nf_s)e^{j2\pi nf_s t} \tag{2.154}$$

Equating this to the left-hand side of (2.151), we obtain the Poisson sum formula:

$$\sum_{m=-\infty}^{\infty} p(t-mT_s) = f_s \sum_{n=-\infty}^{\infty} P(nf_s)e^{j2\pi nf_s t} \tag{2.155}$$

The Poisson sum formula is useful when one goes from the Fourier transform to sampled approximations of it. For example, Equation (2.155) says that the sample values $P(nf_s)$ of $P(f) = \Im\{p(t)\}$ are the Fourier series coefficients of the periodic function $T_s \sum_{n=-\infty}^{\infty} p(t-mT_s)$.

■ 2.6 POWER SPECTRAL DENSITY AND CORRELATION

Recalling the definition of energy spectral density (2.90), we see that it is of use only for energy signals for which the integral of $G(f)$ over all frequencies gives total energy, a finite quantity. For power signals, it is meaningful to speak in terms of *power spectral density*. Analogous to

$G(f)$, we define the power spectral density $S(f)$ of a signal $x(t)$ as a real, even, nonnegative function of frequency, which gives total average power per ohm when integrated; that is,

$$P = \int_{-\infty}^{\infty} S(f)\,df = \langle x^2(t) \rangle \tag{2.156}$$

where $\langle x^2(t) \rangle = \lim_{T \to \infty}(1/2T)\int_{-T}^{T} x^2(t)\,dt$ denotes the time average of $x^2(t)$. Since $S(f)$ is a function that gives the variation of density of power with frequency, we conclude that it must consist of a series of impulses for the periodic power signals that we have so far considered. Later, in Chapter 6, we will consider power spectra of random signals.

EXAMPLE 2.18

Considering the cosinusoidal signal

$$x(t) = A\cos(2\pi f_0 t + \theta) \tag{2.157}$$

we note that its average power per ohm, $\frac{1}{2}A^2$, is concentrated at the single frequency f_0 Hz. However, since the power spectral density must be an even function of frequency, we split this power equally between $+f_0$ and $-f_0$ Hz. Thus the power spectral density of $x(t)$ is, from intuition, given by

$$S(f) = \frac{1}{4}A^2\delta(f - f_0) + \frac{1}{4}A^2\delta(f + f_0) \tag{2.158}$$

Checking this by using (2.156), we see that integration over all frequencies results in the average power per ohm of $\frac{1}{2}A^2$.

■

2.6.1 The Time-Average Autocorrelation Function

To introduce the time-average autocorrelation function, we return to the energy spectral density of an energy signal (2.90). Without any apparent reason, suppose we take the inverse Fourier transform of $G(f)$, letting the independent variable be τ:

$$\begin{aligned} \phi(\tau) &\triangleq \Im^{-1}[G(f)] = \Im^{-1}[X(f)X^*(f)] \\ &= \Im^{-1}[X(f)] * \Im^{-1}[X^*(f)] \end{aligned} \tag{2.159}$$

The last step follows by application of the convolution theorem. Applying the time-reversal theorem (item 3b in Table G.6 in Appendix G) to write $\Im^{-1}[X^*(f)] = x(-\tau)$ and then the convolution theorem, we obtain

$$\begin{aligned} \phi(\tau) &= x(\tau) * x(-\tau) = \int_{-\infty}^{\infty} x(\lambda)x(\lambda + \tau)\,d\lambda \\ &= \lim_{T \to \infty} \int_{-T}^{T} x(\lambda)x(\lambda + \tau)\,d\lambda \text{ (energy signal)} \end{aligned} \tag{2.160}$$

Equation (2.160) will be referred to as the *time-average autocorrelation function* for energy signals. We see that it gives a measure of the similarity, or coherence, between a signal and a delayed version of the signal. Note that $\phi(0) = E$, the signal energy. Also note the similarity of the correlation operation to convolution. The major point of (2.159) is that the autocorrelation function and energy spectral density are Fourier transform pairs. We forgo further discussion of the time-average autocorrelation function for energy signals in favor of analogous results for power signals.

The time-average autocorrelation function $R(\tau)$ of a power signal $x(t)$ is defined as the time average

$$R(\tau) = \langle x(t)x(t+\tau)\rangle$$
$$\triangleq \lim_{T\to\infty} \frac{1}{2T}\int_{-T}^{T} x(t)x(t+\tau)\,dt \quad \text{(power signal)} \tag{2.161}$$

If $x(t)$ is periodic with period T_0, the integrand of (2.161) is periodic, and the time average can be taken over a single period:

$$R(\tau) = \frac{1}{T_0}\int_{T_0} x(t)x(t+\tau)\,dt \quad [x(t) \text{ periodic}]$$

Just like $\phi(\tau)$, $R(\tau)$ gives a measure of the similarity between a power signal at time t and at time $t+\tau$; it is a function of the delay variable τ, since time t is the variable of integration. In addition to being a measure of the similarity between a signal and its time displacement, we note that the total average power of the signal is

$$R(0) = \langle x^2(t)\rangle \triangleq \int_{-\infty}^{\infty} S(f)\,df \tag{2.162}$$

Thus we suspect that the time-average autocorrelation function and power spectral density of a power signal are closely related, just as they are for energy signals. This relationship is stated formally by the *Wiener–Khinchine theorem*, which says that the time-average autocorrelation function of a signal and its power spectral density are Fourier transform pairs:

$$S(f) = \Im[R(\tau)] = \int_{-\infty}^{\infty} R(\tau)e^{-j2\pi f\tau}\,d\tau \tag{2.163}$$

and

$$R(\tau) = \Im^{-1}[S(f)] = \int_{-\infty}^{\infty} S(f)e^{j2\pi f\tau}\,df \tag{2.164}$$

A formal proof of the Wiener–Khinchine theorem will be given in Chapter 6. We simply take (2.163) as the definition of power spectral density at this point. We note that (2.162) follows immediately from (2.164) by setting $\tau = 0$.

2.6.2 Properties of $R(\tau)$

The time-average autocorrelation function has several useful properties, which are listed below:

1. $R(0) = \langle x^2(t)\rangle \geq |R(\tau)|$, for all τ; that is, a relative maximum of $R(\tau)$ exists at $\tau = 0$.
2. $R(-\tau) = \langle x(t)x(t-\tau)\rangle = R(\tau)$; that is, $R(\tau)$ is even.
3. $\lim_{|\tau|\to\infty} R(\tau) = \langle x(t)\rangle^2$ if $x(t)$ does not contain periodic components.
4. If $x(t)$ is periodic in t with period T_0, then $R(\tau)$ is periodic in τ with period T_0.
5. The time-average autocorrelation function of any power signal has a Fourier transform that is nonnegative.

Property 5 results by virtue of the fact that normalized power is a nonnegative quantity. These properties will be proved in Chapter 6.

The autocorrelation function and power spectral density are important tools for systems analysis involving random signals.

EXAMPLE 2.19

We desire the autocorrelation function and power spectral density of the signal $x(t) = \text{Re}(2+3\exp(j10\pi t)+4j\exp(j10\pi t))$ or $x(t) = 2+3\cos(10\pi t)-4\sin(10\pi t)$. The first step is to write the signal as a constant plus a single sinusoid. To do so, we note that

$$x(t) = \text{Re}\left(2+\sqrt{3^2+4^2}\exp\left[j\tan^{-1}\left(\frac{4}{3}\right)\right]\exp(j\,10\pi t)\right) = 2+5\cos\left[10\pi t+\tan^{-1}\left(\frac{4}{3}\right)\right]$$

We may proceed in one of two ways. The first is to find the autocorrelation function of $x(t)$ and Fourier transform it to get the power spectral density. The second is to write down the power spectral density and inverse Fourier transform it to get the autocorrelation function.

Following the first method, we find the autocorrelation function:

$$\begin{aligned}
R(\tau) &= \frac{1}{T_0}\int_{T_0} x(t)x(t+\tau)\,dt \\
&= \frac{1}{0.2}\int_0^{0.2}\left\{2+5\cos\left[10\pi t+\tan^{-1}\left(\frac{4}{3}\right)\right]\right\}\left\{2+5\cos\left[10\pi\left(t+\tau\right)+\tan^{-1}\left(\frac{4}{3}\right)\right]\right\}dt \\
&= 5\int_0^{0.2}\left\{4+10\cos\left[10\pi t+\tan^{-1}\left(\frac{4}{3}\right)\right]+10\cos\left[10\pi(t+\tau)+\tan^{-1}\left(\frac{4}{3}\right)\right]\right. \\
&\quad \left.+25\cos\left[10\pi t+\tan^{-1}\left(\frac{4}{3}\right)\right]\cos\left[10\pi(t+\tau)+\tan^{-1}\left(\frac{4}{3}\right)\right]\right\}dt \\
&= 5\int_0^{0.2}4dt+50\int_0^{0.2}\cos\left[10\pi t+\tan^{-1}\left(\frac{4}{3}\right)\right]dt \\
&\quad +50\int_0^{0.2}\cos\left[10\pi(t+\tau)+\tan^{-1}\left(\frac{4}{3}\right)\right]dt \\
&\quad +\frac{125}{2}\int_0^{0.2}\cos(10\pi\tau)dt+\frac{125}{2}\int_0^{0.2}\cos\left[20\pi t+10\pi\tau+2\tan^{-1}\left(\frac{4}{3}\right)\right]dt \\
&= 5\int_0^{0.2}4dt+0+0+\frac{125}{2}\int_0^{0.2}\cos(10\pi\tau)dt \\
&\quad +\frac{125}{2}\int_0^{0.2}\cos\left[20\pi t+10\pi\tau+2\tan^{-1}\left(\frac{4}{3}\right)\right]dt \\
&= 4+\frac{25}{2}\cos(10\pi\tau)
\end{aligned} \tag{2.165}$$

where integrals involving cosines of t are zero by virtue of integrating a cosine over an integer number of periods, and the trigonometric relationship $\cos x\cos y = \frac{1}{2}\cos(x+y)+\frac{1}{2}\cos(x-y)$ has been used. The power spectral density is the Fourier transform of the autocorrelation function, or

$$\begin{aligned}
S(f) &= \Im\left[4+\frac{25}{2}\cos(10\pi\tau)\right] \\
&= 4\Im[1]+\frac{25}{2}\Im[\cos(10\pi\tau)] \\
&= 4\delta(f)+\frac{25}{4}\delta(f-5)+\frac{25}{4}\delta(f+5)
\end{aligned} \tag{2.166}$$

Note that integration of this over all f gives $P = 4 + \frac{25}{2} = 16.5\ \mathrm{W}/\Omega$, which is the DC power plus the AC power (the latter is split between 5 and -5 Hz). We could have proceeded by writing down the power spectral density first, using power arguments, and inverse Fourier transforming it to get the autocorrelation function. Note that all properties of the autocorrelation function are satisfied except the third which does not apply. ■

EXAMPLE 2.20

The sequence 1110010 is an example of a *pseudo noise* or m-sequence; they are important in the implementation of digital communication systems and will be discussed further in Chapter 9. For now, we use this m-sequence as another illustration for computing autocorrelation functions and power spectra. Consider Figure 2.12 (a), which shows the waveform equivalent of this m-sequence obtained by replacing each 0 by -1, multiplying each sequence member by a square pulse function $\Pi((t-t_0)/\Delta)$, summing, and assuming the resulting waveform is repeated forever thereby making it periodic. To compute the autocorrelation function, we apply

$$R(\tau) = \frac{1}{T_0}\int_{T_0} x(t)x(t+\tau)\,dt$$

since a periodic repetition of the waveform is assumed. Consider the waveform $x(t)$ multiplied by $x(t+n\Delta)$ [shown in Figure 2.12 (b) for $n = 2$]. The product is shown in Figure 2.12 (c), where it is seen that the net area under the product $x(t)x(t+n\Delta)$ is $-\Delta$ which gives $R(2\Delta) = -\Delta/7\Delta = -\frac{1}{7}$ for this case. In fact, this answer results for any τ equal to a nonzero integer multiple of Δ. For $\tau = 0$, the net area under the product $x(t)x(t+0)$ is 7Δ, which gives $R(0) = 7\Delta/7\Delta = 1$. These correlation results are shown in Figure 2.12(d) by the open circles where it is noted that they repeat each $\tau = 7\Delta$. For a given noninteger delay

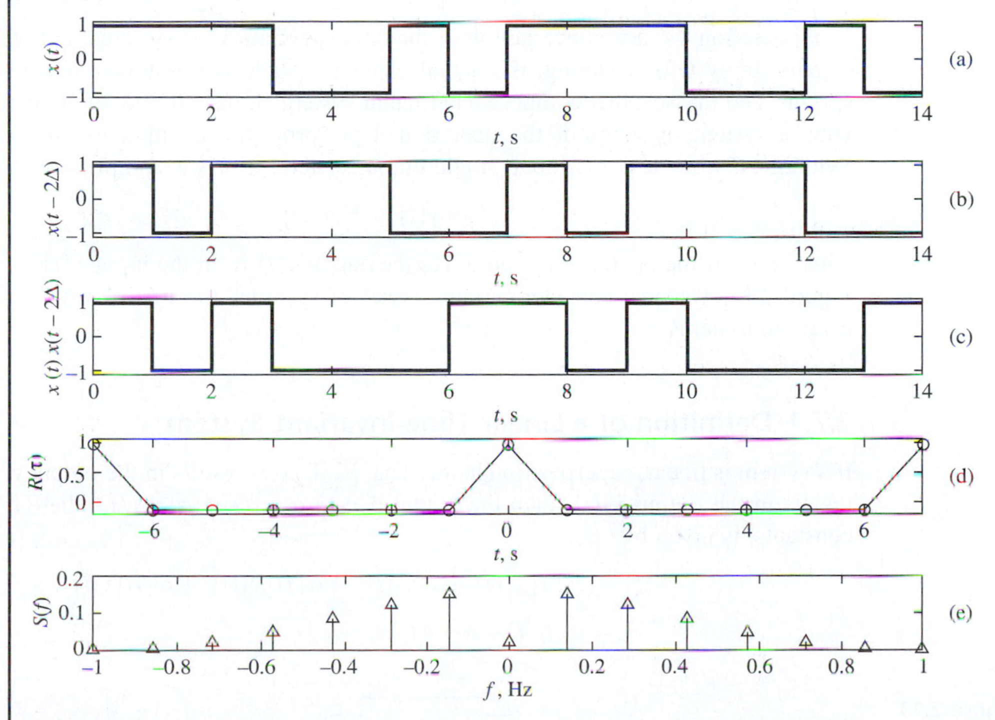

Figure 2.12
Waveforms pertinent to computing the autocorrelation function and power spectrum of an m-sequence of length 7.

value, the autocorrelation function is obtained as the linear interpolation of the autocorrelation function values for the integer delays bracketing the desired delay value. One can see that this is the case by considering the integral $\int_{T_0} x(t)x(t+\tau)\,dt$ and noting that the area under the product $x(t)x(t+\tau)$ must be a linear function of τ due to $x(t)$ being composed of square pulses. Thus the autocorrelation function is as shown in Figure 2.12(d) by the solid line. For one period, it can be expressed as

$$R(\tau) = \frac{8}{7}\Lambda\left(\frac{\tau}{\Delta}\right) - \frac{1}{7}, \quad |\tau| \leq \frac{T_0}{2}$$

The power spectral density is the Fourier transform of the autocorrelation function which can be obtained by applying (2.149). The detailed derivation of it is left to the problems. The result is

$$S(f) = \frac{8}{49}\sum_{n=-\infty}^{\infty} \operatorname{sinc}^2\left(\frac{n}{7\Delta}\right)\delta\left(f - \frac{n}{7\Delta}\right) - \frac{1}{7}\delta(f)$$

and is shown in Figure 2.12(e). Note that near $f = 0$, $S(f) = \left(\frac{8}{49} - \frac{1}{7}\right)\delta(f) = \frac{1}{49}\delta(f)$, which says that the DC power is $\frac{1}{49} = 1/7^2$ W. The student should think about why this is the correct result. (*Hint:* What is the DC value of $x(t)$ and to what power does this correspond?) ■

The autocorrelation function and power spectral density are important tools for systems analysis involving random signals.

■ 2.7 SIGNALS AND LINEAR SYSTEMS

In this section we are concerned with the characterization of systems and their effects on signals. In system modeling, the actual elements, such as resistors, capacitors, inductors, springs, and masses, that compose a particular system are usually not of concern. Rather, we view a system in terms of the operation it performs on an input to produce an output. Symbolically, for a single-input, single-output system, this is accomplished by writing

$$y(t) = \mathcal{H}[x(t)] \tag{2.167}$$

where $\mathcal{H}[\cdot]$ is the operator that produces the output $y(t)$ from the input $x(t)$, as illustrated in Figure 2.13. We now consider certain classes of systems, the first of which is linear time-invariant systems.

2.7.1 Definition of a Linear Time-Invariant System

If a system is linear, superposition holds. That is, if $x_1(t)$ results in the output $y_1(t)$ and $x_2(t)$ results in the output $y_2(t)$, then the output due to $\alpha_1 x_1(t) + \alpha_2 x_2(t)$, where α_1 and α_2 are constants, is given by

$$\begin{aligned} y(t) &= \mathcal{H}[\alpha_1 x_1(t) + \alpha_2 x_2(t)] = \alpha_1 \mathcal{H}[x_1(t)] + \alpha_2 \mathcal{H}[x_2(t)] \\ &= \alpha_1 y_1(t) + \alpha_2 y_2(t) \end{aligned} \tag{2.168}$$

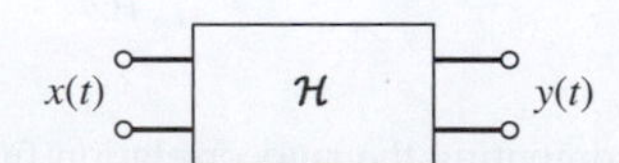

Figure 2.13
Operator representation of a linear system.

If the system is *time invariant*, or *fixed*, the delayed input $x(t-t_0)$ gives the delayed output $y(t-t_0)$; that is,

$$y(t-t_0) = \mathcal{H}[x(t-t_0)] \tag{2.169}$$

With these properties explicitly stated, we are now ready to obtain more concrete descriptions of linear time-invariant (LTI) systems.

2.7.2 Impulse Response and the Superposition Integral

The *impulse response* $h(t)$ of an LTI system is defined to be the response of the system to an impulse applied at $t = 0$, that is

$$h(t) \triangleq \mathcal{H}[\delta(t)] \tag{2.170}$$

By the time-invariant property of the system, the response to an impulse applied at any time t_0 is $h(t-t_0)$, and the response to the linear combination of impulses $\alpha_1\delta(t-t_1)+\alpha_2\delta(t-t_2)$ is $\alpha_1 h(t-t_1)+\alpha_2 h(t-t_2)$ by the superposition property and time invariance. Through induction, we may therefore show that the response to the input

$$x(t) = \sum_{n=1}^{N} \alpha_n \delta(t-t_n) \tag{2.171}$$

is

$$y(t) = \sum_{n=1}^{N} \alpha_n h(t-t_n) \tag{2.172}$$

We will use (2.172) to obtain the superposition integral, which expresses the response of an LTI system to an arbitrary input (with suitable restrictions) in terms of the impulse response of the system. Considering the arbitrary input signal $x(t)$ of Figure 2.14(a), we can represent it as

$$x(t) = \int_{-\infty}^{\infty} x(\lambda)\delta(t-\lambda)\,d\lambda \tag{2.173}$$

by the sifting property of the unit impulse. Approximating the integral of (2.173) as a sum, we obtain

$$x(t) \cong \sum_{n=N_1}^{N_2} x(n\,\Delta t)\,\delta(t-n\Delta t)\,\Delta t, \quad \Delta t \ll 1 \tag{2.174}$$

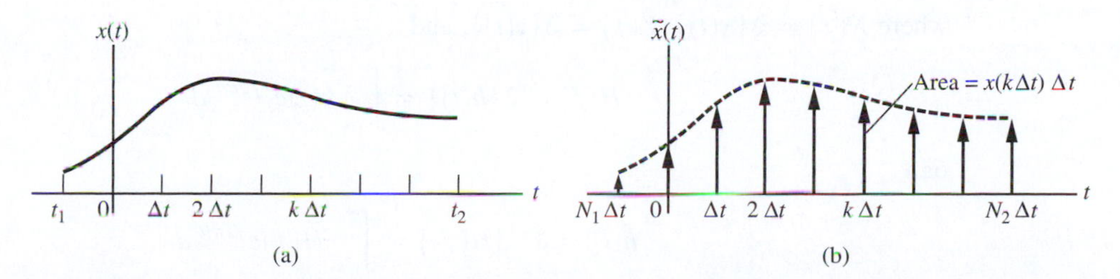

Figure 2.14
A signal and an approximate representation. (a) Signal. (b) Approximation with a sequence of impulses.

where $t_1 = N_1 \Delta t$ is the starting time of the signal and $t_2 = N_2 \Delta t$ is the ending time. The output, using (2.172) with $\alpha_n = x(n\Delta t)\Delta t$ and $t_n = n\Delta t$, is

$$\tilde{y}(t) = \sum_{n=N_1}^{N_2} x(n\,\Delta t)h(t - n\,\Delta t)\,\Delta t \tag{2.175}$$

where the tilde denotes the output resulting from the approximation to the input given by (2.174). In the limit as Δt approaches $d\lambda$ and $n\,\Delta t$ approaches the continuous variable λ, the sum becomes an integral, and we obtain

$$y(t) = \int_{-\infty}^{\infty} x(\lambda)h(t-\lambda)\,d\lambda \tag{2.176}$$

where the limits have been changed to $\pm\infty$ to allow arbitrary starting and ending times for $x(t)$. Making the substitution $\sigma = t - \lambda$, we obtain the equivalent result

$$y(t) = \int_{-\infty}^{\infty} x(t-\sigma)h(\sigma)\,d\sigma \tag{2.177}$$

Because these equations were obtained by superposition of a number of elementary responses due to each individual impulse, they are referred to as *superposition integrals.* A simplification results if the system under consideration is causal, that is, is a system that does not respond before an input is applied. For a causal system, $h(t-\lambda) = 0$ for $t < \lambda$, and the upper limit on (2.176) can be set equal to t. Furthermore, if $x(t) = 0$ for $t < 0$, the lower limit becomes zero.

2.7.3 Stability

A fixed, linear system is bounded-input, bounded-output (BIBO) stable if every bounded input results in a bounded output. It can be shown[10] that a system is BIBO stable if and only if

$$\int_{-\infty}^{\infty} |h(t)|\,dt < \infty \tag{2.178}$$

2.7.4 Transfer (Frequency-Response) Function

Applying the convolution theorem of Fourier transforms (item 8 of Table G.6 in Appendix G) to either (2.176) or (2.177), we obtain

$$Y(f) = H(f)X(f) \tag{2.179}$$

where $X(f) = \Im\{x(t)\}$, $Y(f) = \Im\{y(t)\}$, and

$$H(f) = \Im\{h(t)\} = \int_{-\infty}^{\infty} h(t)e^{-j2\pi ft}\,dt \tag{2.180}$$

or

$$h(t) = \Im^{-1}\{H(f)\} = \int_{-\infty}^{\infty} H(f)e^{j2\pi ft}\,df \tag{2.181}$$

[10]See Ziemer et al. (1998), Chapter 2.

$H(f)$ is referred to as the *transfer (frequency-response) function* of the system. We see that either $h(t)$ or $H(f)$ is an equally good characterization of the system. By an inverse Fourier transform on (2.179), the output becomes

$$y(t) = \int_{-\infty}^{\infty} X(f)H(f)e^{j2\pi ft}\,df \tag{2.182}$$

2.7.5 Causality

A system is causal if it does not anticipate the input. In terms of the impulse response, it follows that for a time-invariant causal system,

$$h(t) = 0, \ t < 0 \tag{2.183}$$

When causality is viewed from the standpoint of the frequency-response function of the system, a celebrated theorem by Wiener and Paley[11] states that if

$$\int_{-\infty}^{\infty} |h(t)|^2\,dt = \int_{-\infty}^{\infty} |H(f)|^2\,df < \infty \tag{2.184}$$

with $h(t) \equiv 0$ for $t < 0$, it is then necessary that

$$\int_{-\infty}^{\infty} \frac{|\ln|H(f)||}{1+f^2}\,df < \infty \tag{2.185}$$

Conversely, if $|H(f)|$ is square integrable and if the integral in (2.185) is unbounded, then we cannot make $h(t) \equiv 0, t < 0$ no matter what we choose for $\underline{/H(f)}$. Consequences of (2.185) are that no filter can have $|H(f)| \equiv 0$ over a finite band of frequencies (i.e., a filter cannot perfectly reject any band of frequencies). In fact, the Paley–Wiener criterion restricts the rate at which $|H(f)|$ for a linear causal time-invariant system can vanish. For example,

$$|H(f)| = e^{-k_1|f|} \Rightarrow |\ln|H(f)|| = k_1|f| \tag{2.186}$$

and

$$|H(f)| = e^{-k_2f^2} \Rightarrow |\ln|H(f)|| = k_2f^2 \tag{2.187}$$

where k_1 and k_2 are positive constants, are not allowable amplitude responses for causal filters because (2.185) does not give a finite result in either case.

The sufficiency statement of the Paley–Wiener criterion is stated as follows: Given any square-integrable function $|H(f)|$ for which (2.185) is satisfied, there exists an $\underline{/H(f)}$ such that $H(f) = |H(f)|\exp\left[j\underline{/H(f)}\right]$ is the Fourier transform of $h(t)$ for a causal filter.

2.7.6 Symmetry Properties of *H(f)*

The frequency response function of an LTI system $H(f)$ is, in general, a complex quantity. We therefore write it in terms of magnitude and argument as

$$H(f) = |H(f)|\exp\left[j\underline{/H(f)}\right] \tag{2.188}$$

[11]See William Siebert, *Circuits, Signals, and Systems*, McGraw-Hill, New York, 1986, p. 476.

where $|H(f)|$ is called the *amplitude- (magnitude-) response function* and $\underline{/H(f)}$ is called the *phase-response function* of the LTI system. Also, $H(f)$ is the Fourier transform of a real-time function $h(t)$. Therefore, it follows that

$$|H(f)| = |H(-f)| \tag{2.189}$$

and

$$\underline{/H(f)} = -\underline{/H(-f)} \tag{2.190}$$

That is, the amplitude response of a system with real-valued impulse response is an even function of frequency and its phase response is an odd function of frequency.

EXAMPLE 2.21

Consider the lowpass RC filter shown in Figure 2.15. We may find its frequency-response function by a number of methods. First, we may write down the governing differential equation (integral-differential equations, in general) as

$$RC\frac{dy}{dt} + y(t) = x(t) \tag{2.191}$$

and Fourier transform it, obtaining

$$(j2\pi fRC + 1)Y(f) = X(f)$$

or

$$\begin{aligned} H(f) &= \frac{Y(f)}{X(f)} = \frac{1}{1 + j(f/f_3)} \\ &= \frac{1}{\sqrt{1 + (f/f_3)^2}} e^{-j\tan^{-1}(f/f_3)} \end{aligned} \tag{2.192}$$

where $f_3 = 1/(2\pi RC)$ is the 3-dB frequency, or half-power frequency. Second, we can use Laplace transform theory with s replaced by $j2\pi f$. Third, we can use AC sinusoidal steady-state analysis. The amplitude and phase responses of this system are illustrated in Figure 2.16(a) and (b), respectively.

Using the Fourier transform pair

$$\alpha e^{-\alpha t}u(t) \longleftrightarrow \frac{\alpha}{\alpha + j2\pi f} \tag{2.193}$$

we find the impulse response of the filter to be

$$h(t) = \frac{1}{RC}e^{-t/RC}u(t) \tag{2.194}$$

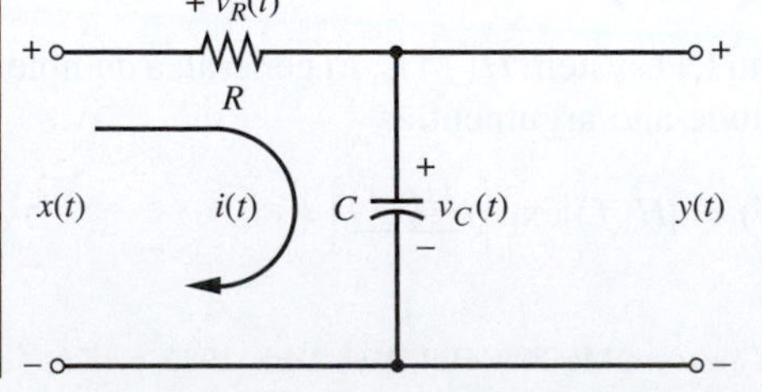

Figure 2.15
An RC lowpass filter.

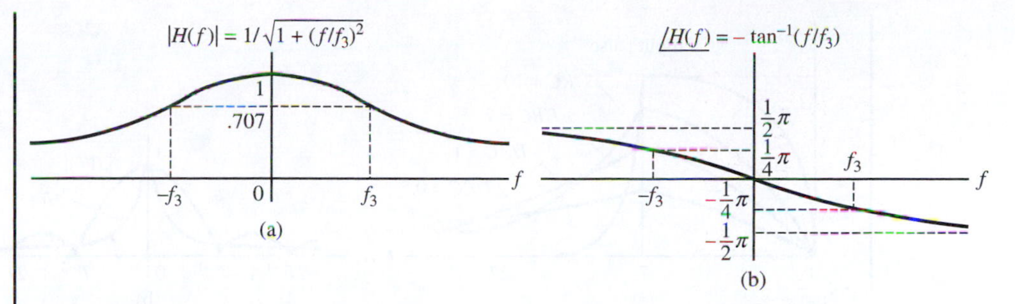

Figure 2.16
Amplitude and phase responses of the lowpass RC filter. (a) Amplitude response. (b) Phase response.

Finally, we consider the response of the filter to the pulse

$$x(t) = A\Pi\left(\frac{t - \frac{1}{2}T}{T}\right) \tag{2.195}$$

Using appropriate Fourier transform pairs, we can readily find $Y(f)$, but its inverse Fourier transformation requires some effort. Thus it appears that the superposition integral is the best approach in this case. Choosing the form

$$y(t) = \int_{-\infty}^{\infty} h(t-\sigma)x(\sigma)\,d\sigma \tag{2.196}$$

we find, by direct substitution in $h(t)$, that

$$h(t-\sigma) = \frac{1}{RC}e^{-(t-\sigma)/RC}u(t-\sigma) = \begin{cases} \dfrac{1}{RC}e^{-(t-\sigma)/RC}, & \sigma < t \\ 0, & \sigma > t \end{cases} \tag{2.197}$$

Since $x(\sigma)$ is zero for $\sigma < 0$ and $\sigma > T$, we find that

$$y(t) = \begin{cases} 0, & t < 0 \\ \displaystyle\int_0^t \frac{A}{RC}e^{-(t-\sigma)/RC}\,d\sigma, & 0 \le t \le T \\ \displaystyle\int_0^T \frac{A}{RC}e^{-(t-\sigma)/RC}\,d\sigma, & t > T \end{cases} \tag{2.198}$$

Carrying out the integrations, we obtain

$$y(t) = \begin{cases} 0, & t < 0 \\ A\left(1 - e^{-t/RC}\right), & 0 < t < T \\ A\left(e^{-(t-T)/RC} - e^{-t/RC}\right), & t > T \end{cases} \tag{2.199}$$

This result is plotted in Figure 2.17 for several values of T/RC. Also shown are $|X(f)|$ and $|H(f)|$. Note that $T/RC = 2\pi f_3/T^{-1}$ is proportional to the ratio of the 3-dB frequency of the filter to the spectral width (T^{-1}) of the pulse. When this ratio is large, the spectrum of the input pulse is essentially passed undistorted by the system, and the output looks like the input. On the other hand, for $2\pi f_3/T^{-1} \ll 1$, the

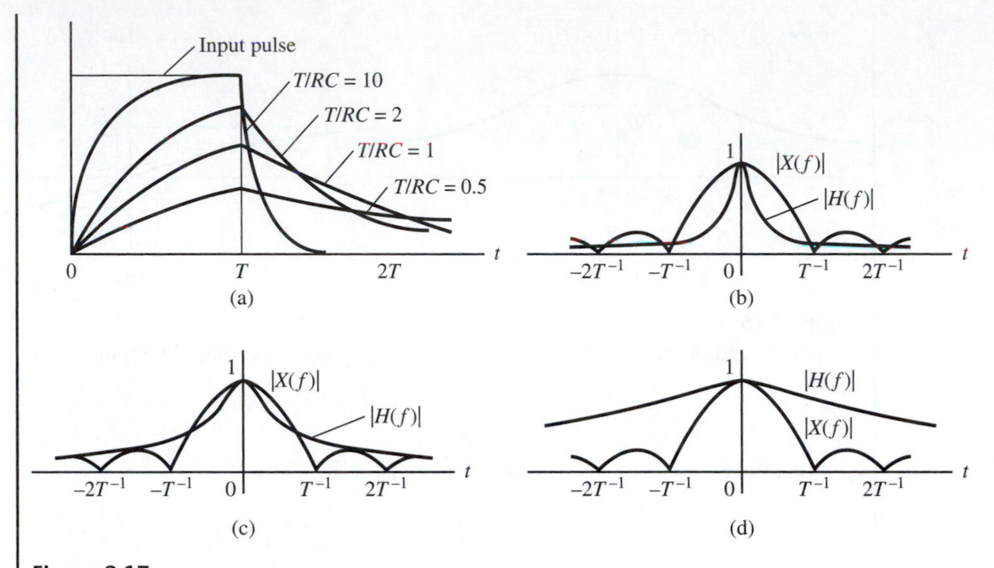

Figure 2.17
(a) Waveforms and (b)–(d) spectra for a lowpass RC filter with pulse input. (a) Input and output signals. (b) $T/RC = 0.5$. (c) $T/RC = 2$. (d) $T/RC = 10$.

system distorts the input signal spectrum, and $y(t)$ looks nothing like the input. These ideas will be put on a firmer basis when signal distortion is discussed.

■

2.7.7 Input–Output Relationships for Spectral Densities

Consider a fixed linear two-port system with frequency-response function $H(f)$, input $x(t)$, and output $y(t)$. If $x(t)$ and $y(t)$ are energy signals, their energy spectral densities are $G_x(f) = |X(f)|^2$ and $G_y(f) = |Y(f)|^2$, respectively. Since $Y(f) = H(f)X(f)$, it follows that

$$G_y(f) = |H(f)|^2 G_x(f) \tag{2.200}$$

A similar relationship holds for power signals and spectra:

$$S_y(f) = |H(f)|^2 S_x(f) \tag{2.201}$$

This will be proved in Chapter 6.

2.7.8 Response to Periodic Inputs

Consider the steady-state response of a fixed linear system to the complex exponential input signal $Ae^{j2\pi f_0 t}$. Using the superposition integral, we obtain

$$\begin{aligned} y_{ss}(t) &= \int_{-\infty}^{\infty} h(\lambda) A e^{j2\pi f_0 (t-\lambda)} d\lambda \\ &= A e^{j2\pi f_0 t} \int_{-\infty}^{\infty} h(\lambda) e^{-j2\pi f_0 \lambda} d\lambda \\ &= H(f_0) A e^{j2\pi f_0 t} \end{aligned} \tag{2.202}$$

That is, the output is a complex exponential signal of the same frequency but with amplitude scaled by $|H(f_0)|$ and phase-shifted by $\underline{/H(f_0)}$ relative to the amplitude and phase of the input. Using superposition, we conclude that the steady-state output due to an arbitrary periodic input is represented by the complex exponential Fourier series

$$y(t) = \sum_{n=-\infty}^{\infty} X_n H(nf_0) e^{jn2\pi f_0 t} \tag{2.203}$$

or

$$y(t) = \sum_{n=-\infty}^{\infty} |X_n||H(nf_0)| \exp\left\{ j \left[n2\pi f_0 t + \underline{/X_n} + \underline{/H(nf_0)} \right] \right\} \tag{2.204}$$

Thus, for a periodic input, the magnitude of each spectral component of the input is attenuated (or amplified) by the amplitude-response function *at the frequency of the particular spectral component*, and the phase of each spectral component is shifted by the value of the phase-shift function of the system at the *frequency of the particular spectral component*.

EXAMPLE 2.22

Consider the response of a filter having the frequency-response function

$$H(f) = 2\Pi\left(\frac{f}{42}\right) e^{-j\pi f/10} \tag{2.205}$$

to a unit-amplitude triangular signal with period 0.1 s. From Table 2.1 and (2.46), the exponential Fourier series of the input signal is

$$\begin{aligned} x(t) &= \cdots \frac{4}{25\pi^2} e^{-j100\pi t} + \frac{4}{9\pi^2} e^{-j60\pi t} + \frac{4}{\pi^2} e^{-j20\pi t} \\ &\quad + \frac{4}{\pi^2} e^{j20\pi t} + \frac{4}{9\pi^2} e^{j60\pi t} + \frac{4}{25\pi^2} e^{j100\pi t} + \cdots \\ &= \frac{8}{\pi^2} \left[\cos(20\pi t) + \frac{1}{9} \cos(60\pi t) + \frac{1}{25} \cos(100\pi t) + \cdots \right] \end{aligned} \tag{2.206}$$

The filter eliminates all harmonics above 21 Hz and passes all those below 21 Hz, imposing an amplitude scale factor of 2 and a phase shift of $-\pi f/10$ rad. The only harmonic of the triangular wave to be passed by the filter is the fundamental, which has a frequency of 10 Hz, giving a phase shift of $-\pi(10)/10 = -\pi$ rad. The output is therefore

$$y(t) = \frac{16}{\pi^2} \cos\left[20\pi \left(t - \frac{1}{20} \right) \right] \tag{2.207}$$

where the phase shift is seen to be equivalent to a delay of $\frac{1}{20}$ s. ■

2.7.9 Distortionless Transmission

Equation (2.204) shows that both the amplitudes and phases of the spectral components of a periodic input signal will, in general, be altered as the signal is sent through a two-port LTI system. This modification may be desirable in signal processing applications, but it amounts to distortion in signal *transmission* applications. While it may appear at first that ideal signal transmission results only if there is *no* attenuation and phase shift of the spectral components of the input, this requirement is too stringent. A system will be classified as distortionless if it introduces the same attenuation and time delay to all spectral components of the input, for then the output looks like the input. In particular, if the output of a system is given in terms of the input as

$$y(t) = H_0 x(t - t_0) \tag{2.208}$$

where H_0 and t_0 are constants, the output is a scaled, delayed replica of the input ($t_0 > 0$ for causality). Employing the time-delay theorem to Fourier transform (2.208) and using the definition $H(f) = Y(f)/X(f)$, we obtain

$$H(f) = H_0 e^{-j2\pi f t_0} \tag{2.209}$$

as the frequency-response function of a distortionless system; that is, the amplitude response of a distortionless system is constant, and the phase shift is linear with frequency. Of course, these restrictions are necessary only within the frequency ranges where the input has significant spectral content. Figure 2.18 and Example 2.23, considered shortly, will illustrate these comments.

In general, we can isolate three major types of distortion. First, if the system is linear but the amplitude response is not constant with frequency, the system is said to introduce *amplitude distortion*. Second, if the system is linear but the phase shift is not a linear function of frequency, the system introduces *phase*, or *delay, distortion*. Third, if the system is not linear, we have *nonlinear distortion*. Of course, these three types of distortion may occur in combination with one another.

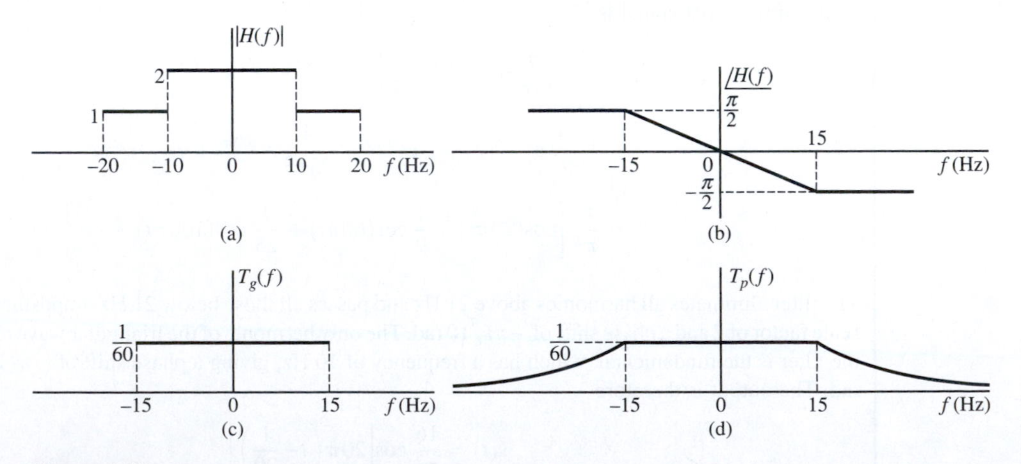

Figure 2.18
Amplitude and phase response and group and phase delays of the filter for Example 2.23. (a) Amplitude response. (b) Phase response. (c) Group delay. (d) Phase delay.

2.7.10 Group and Phase Delay

One can often identify phase distortion in a linear system by considering the derivative of phase with respect to frequency. A distortionless system exhibits a phase response in which phase is directly proportional to frequency. Thus the derivative of phase-response function with respect to frequency of a distortionless system is a constant. The negative of this constant is called the *group delay* of the LTI system. In other words, the group delay is defined by the equation

$$T_g(f) = -\frac{1}{2\pi}\frac{d\theta(f)}{df} \tag{2.210}$$

in which $\theta(f)$ is the phase response of the system. For a distortionless system, the phase-response function is given by (2.209) as

$$\theta(f) = -2\pi f t_0 \tag{2.211}$$

This yields a group delay of

$$T_g(f) = -\frac{1}{2\pi}\frac{d}{df}(-2\pi f t_0)$$

or

$$T_g(f) = t_0 \tag{2.212}$$

This confirms the preceding observation that the group delay of a distortionless LTI system is a constant.

Group delay is the delay that a group of two or more frequency components undergo in passing through a linear system. If a linear system has a single-frequency component as the input, the system is always distortionless, since the output can be written as an amplitude-scaled and phase-shifted (time-delayed) version of the input. As an example, assume that the input to a linear system is given by

$$x(t) = A\cos(2\pi f_1 t) \tag{2.213}$$

It follows from (2.204) that the output can be written as

$$y(t) = A|H(f_1)|\cos[2\pi f_1 t + \theta(f_1)] \tag{2.214}$$

where $\theta(f_1)$ is the phase response of the system evaluated at $f = f_1$. Equation (2.214) can be written as

$$y(t) = A|H(f_1)|\cos\left[2\pi f_1\left(t + \frac{\theta(f_1)}{2\pi f_1}\right)\right] \tag{2.215}$$

The delay of the single component is defined as the phase delay:

$$T_p(f) = -\frac{\theta(f)}{2\pi f} \tag{2.216}$$

Thus (2.215) can be written as

$$y(t) = A|H(f_1)|\cos\{2\pi f_1[t - T_p(f_1)]\} \tag{2.217}$$

Use of (2.211) shows that for a distortionless system, the phase delay is given by

$$T_p(f) = -\frac{1}{2\pi f}(-2\pi f t_0) = t_0 \tag{2.218}$$

The following example should clarify the preceding definitions.

EXAMPLE 2.23

Consider a system with amplitude response and phase shift as shown in Figure 2.18 and the following four inputs:

1. $x_1(t) = \cos(10\pi t) + \cos(12\pi t)$.
2. $x_2(t) = \cos(10\pi t) + \cos(26\pi t)$.
3. $x_3(t) = \cos(26\pi t) + \cos(34\pi t)$.
4. $x_4(t) = \cos(32\pi t) + \cos(34\pi t)$.

Although this system is somewhat unrealistic from a practical standpoint, we can use it to illustrate various combinations of amplitude and phase distortion. Using (2.204) and superposition, we obtain the following corresponding outputs:

1.
$$y_1(t) = 2\cos\left(10\pi t - \frac{1}{6}\pi\right) + 2\cos\left(12\pi t - \frac{1}{5}\pi\right)$$
$$= 2\cos\left[10\pi\left(t - \frac{1}{60}\right)\right] + 2\cos\left[12\pi\left(t - \frac{1}{60}\right)\right]$$

2.
$$y_2(t) = 2\cos\left(10\pi t - \frac{1}{6}\pi\right) + \cos\left(26\pi t - \frac{13}{30}\pi\right)$$
$$= 2\cos\left[10\pi\left(t - \frac{1}{60}\right)\right] + \cos\left[26\pi\left(t - \frac{1}{60}\right)\right]$$

3.
$$y_3(t) = \cos\left(26\pi t - \frac{13}{30}\pi\right) + \cos\left(34\pi t - \frac{1}{2}\pi\right)$$
$$= \cos\left[26\pi\left(t - \frac{1}{60}\right)\right] + \cos\left[34\pi\left(t - \frac{1}{68}\right)\right]$$

4.
$$y_4(t) = \cos\left(32\pi t - \frac{1}{2}\pi\right) + \cos\left(34\pi t - \frac{1}{2}\pi\right)$$
$$= \cos\left[32\pi\left(t - \frac{1}{64}\right)\right] + \cos\left[34\pi\left(t - \frac{1}{68}\right)\right]$$

Checking these results with (2.208), we see that only the input $x_1(t)$ is passed without distortion by the system. For $x_2(t)$, amplitude distortion results, and for $x_3(t)$ and $x_4(t)$, phase (delay) distortion is introduced.

The group delay and phase delay are also illustrated in Figure 2.18. It can be seen that for $|f| \leq 15$ Hz, the group and phase delays are both equal to $\frac{1}{60}$ s. For $|f| > 15$ Hz, the group delay is zero, and the phase delay is

$$T_p(f) = \frac{1}{4|f|}, \quad |f| > 15\,\text{Hz} \tag{2.219}$$

■

2.7.11 Nonlinear Distortion

To illustrate the idea of nonlinear distortion, let us consider a zero memory nonlinear system with the input–output characteristic

$$y(t) = a_1 x(t) + a_2 x^2(t) \tag{2.220}$$

where a_1 and a_2 are constants, and with the input

$$x(t) = A_1 \cos(\omega_1 t) + A_2 \cos(\omega_2 t) \tag{2.221}$$

The output is therefore

$$y(t) = a_1[A_1 \cos(\omega_1 t) + A_2 \cos(\omega_2 t)] + a_2[A_1 \cos(\omega_1 t) + A_2 \cos(\omega_2 t)]^2 \tag{2.222}$$

Using trigonometric identities, we can write the output as

$$\begin{aligned} y(t) &= a_1[A_1 \cos(\omega_1 t) + A_2 \cos(\omega_2 t)] \\ &+ \frac{1}{2} a_2 (A_1^2 + A_2^2) + \frac{1}{2} a_2 [A_1^2 \cos(2\omega_1 t) + A_2^2 \cos(2\omega_2 t)] \\ &+ a_2 A_1 A_2 \{\cos[(\omega_1 + \omega_2)t] + \cos[(\omega_1 - \omega_2)t]\} \end{aligned} \tag{2.223}$$

As can be seen from (2.223) and as shown in Figure 2.19, the system has produced frequencies in the output other than the frequencies of the input. In addition to the first term in (2.223), which may be considered the desired output, there are distortion terms at harmonics of the input frequencies (in this case, second) as well as distortion terms involving sums and differences of the harmonics (in this case, first) of the input frequencies. The former are referred to as *harmonic distortion terms*, and the latter are referred to as *intermodulation distortion terms*. Note that a second-order nonlinearity could be used as a device to *double* the frequency of an input sinusoid. Third-order nonlinearities can be used as *triplers*, and so forth.

A general input signal can be handled by applying the multiplication theorem given in Table G.6 in Appendix G. Thus, for the nonlinear system with the transfer characteristic given by (2.220), the output spectrum is

$$Y(f) = a_1 X(f) + a_2 X(f) * X(f) \tag{2.224}$$

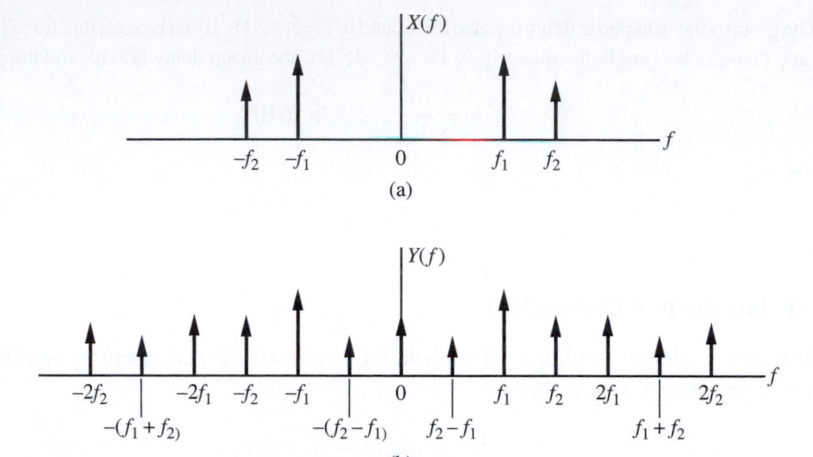

Figure 2.19
Input and output spectra for a nonlinear system with discrete frequency input. (a) Input spectrum. (b) Output spectrum.

The second term is considered distortion and is seen to give interference at all frequencies occupied by the desired output (the first term). It is usually not possible to isolate harmonic and intermodulation distortion components as before. For example, if

$$X(f) = A\Pi\left(\frac{f}{2W}\right) \tag{2.225}$$

Then the distortion term is

$$a_2 X(f) * X(f) = 2a_2 WA^2 \Lambda\left(\frac{f}{2W}\right) \tag{2.226}$$

The input and output spectra are shown in Figure 2.20. Note that the spectral width of the distortion term is *double* that of the input.

2.7.12 Ideal Filters

It is often convenient to work with filters having idealized transfer functions with rectangular amplitude-response functions that are constant within the passband and zero elsewhere. We

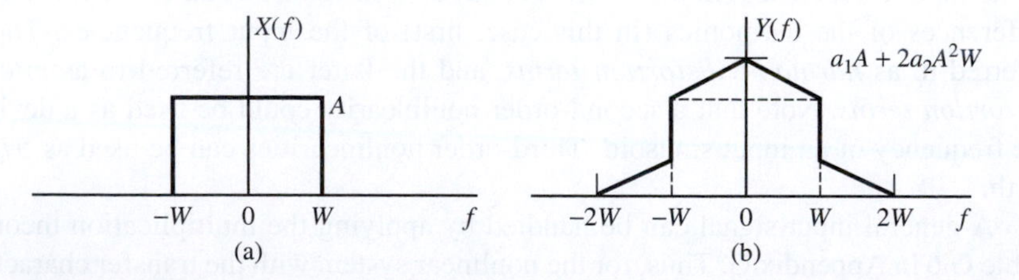

Figure 2.20
Input and output spectra for a nonlinear system with continuous frequency input. (a) Input spectrum. (b) Output spectrum.

will consider three general types of ideal filters: lowpass, highpass, and bandpass. Within the passband, a linear phase-response characteristic is assumed. Thus, if B is the single-sided bandwidth (*width of the stopband*[12] *for the highpass filter*) of the filter in question, the transfer functions of ideal lowpass, highpass and bandpass filters are easily written.

1. For the ideal lowpass filter

$$H_{\mathrm{LP}}(f) = H_0 \Pi(f/2B) e^{-j2\pi f t_0} \tag{2.227}$$

2. For the ideal highpass filter

$$H_{\mathrm{HP}}(f) = H_0[1 - \Pi(f/2B)] e^{-j2\pi f t_0} \tag{2.228}$$

3. Finally, for the ideal bandpass filter

$$H_{\mathrm{BP}}(f) = [H_1(f - f_0) + H_1(f + f_0)] e^{-j2\pi f t_0} \tag{2.229}$$

where $H_1(f) = H_0 \Pi(f/B)$.

The amplitude-response and phase-response functions for these filters are shown in Figure 2.21.

The corresponding impulse responses are obtained by inverse Fourier transformation of the respective frequency-response function. For example, the impulse response of an ideal lowpass filter is, from Example 2.11 and the time-delay theorem, given by

$$h_{\mathrm{LP}}(t) = 2BH_0 \operatorname{sinc}[2B(t - t_0)] \tag{2.230}$$

Since $h_{\mathrm{LP}}(t)$ is not zero for $t<0$, we see that an ideal lowpass filter is noncausal. Nevertheless, ideal filters are useful concepts because they simplify calculations and can give satisfactory results for spectral considerations.

Turning to the ideal bandpass filter, we may use the modulation theorem to write its impulse response as

$$h_{\mathrm{BP}}(t) = 2h_1(t - t_0) \cos[2\pi f_0 (t - t_0)] \tag{2.231}$$

where

$$h_1(t) = \Im^{-1}[H_1(f)] = H_0 B \operatorname{sinc}(Bt) \tag{2.232}$$

Thus the impulse response of an ideal bandpass filter is the oscillatory signal

$$h_{\mathrm{BP}}(t) = 2H_0 B \operatorname{sinc}[B(t - t_0)] \cos[2\pi f_0 (t - t_0)] \tag{2.233}$$

Figure 2.22 illustrates $h_{\mathrm{LP}}(t)$ and $h_{\mathrm{BP}}(t)$. If $f_0 \gg B$, it is convenient to view $h_{\mathrm{BP}}(t)$ as the slowly varying envelope $2H_0 \operatorname{sinc}(Bt)$ *modulating* the high-frequency oscillatory signal $\cos(2\pi f_0 t)$ and shifted to the right by t_0 s.

Derivation of the impulse response of an ideal high pass filter is left to the problems (Problem 2.63).

[12]The *stopband* of a filter will be defined here as the frequency range(s) for which $|H(f)|$ is below 3 dB of its maximum value.

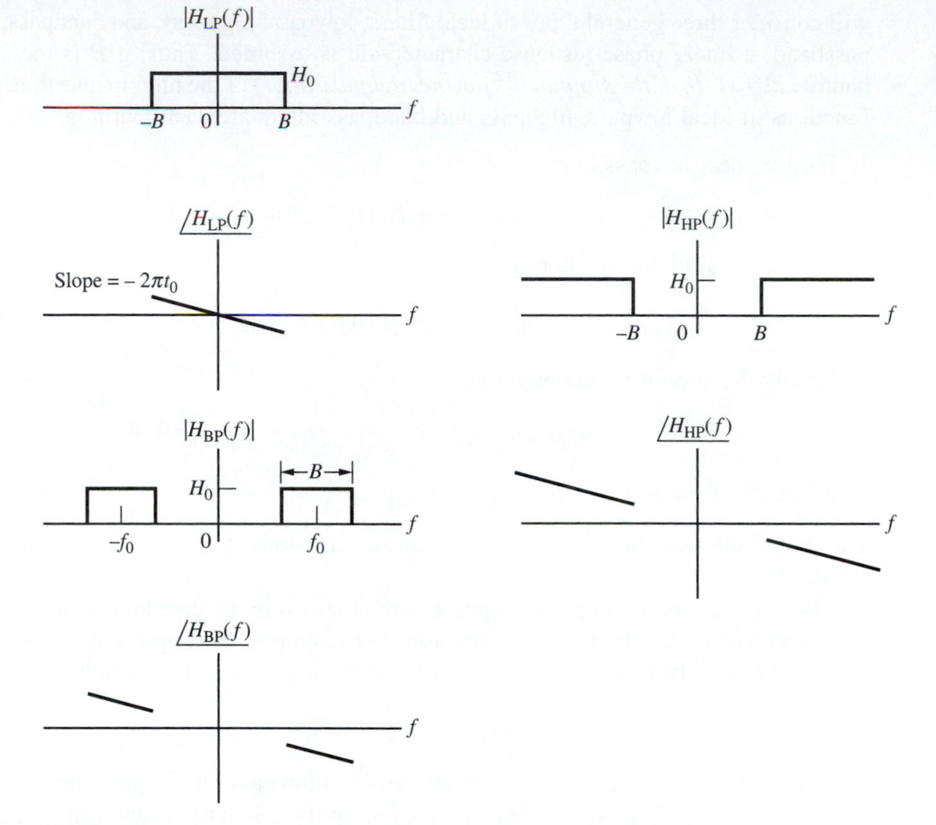

Figure 2.21
Amplitude-response and phase-response functions for ideal filters.

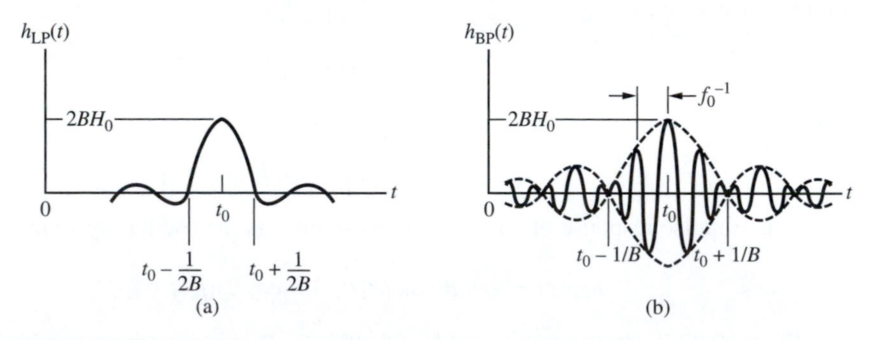

Figure 2.22
Impulse responses for ideal lowpass and bandpass filters. (a) $h_{\mathrm{LP}}(t)$. (b) $h_{\mathrm{BP}}(t)$.

2.7.13 Approximation of Ideal Lowpass Filters by Realizable Filters

Although ideal filters are noncausal and therefore unrealizable devices,[13] there are several practical filter types that may be designed to approximate ideal filter characteristics as closely

[13]See Williams and Taylor (1988), Chapter 2, for a detailed discussion of classical filter designs.

as desired. In this section we consider three such approximations for the lowpass case. Bandpass and highpass approximations may be obtained through suitable frequency transformation. The three filter types to be considered are (1) Butterworth, (2) Chebyshev, and (3) Bessel.

The Butterworth filter is a filter design chosen to maintain a constant amplitude response in the passband at the cost of less stopband attenuation. An nth-order Butterworth filter is characterized by a transfer function, in terms of the complex frequency s, of the form

$$H_{\mathrm{BW}}(s)=\frac{\omega_3^n}{(s-s_1)(s-s_2)\cdots(s-s_n)} \tag{2.234}$$

where the poles $s_1, s_2, \ldots, s_n$ are symmetrical with respect to the real axis and equally spaced about a semicircle of radius ω_3 in the left half s plane and $f_3=\omega_3/2\pi$ is the 3-dB cutoff frequency.[14] Typical pole locations are shown in Figure 2.23(a). For example, the system

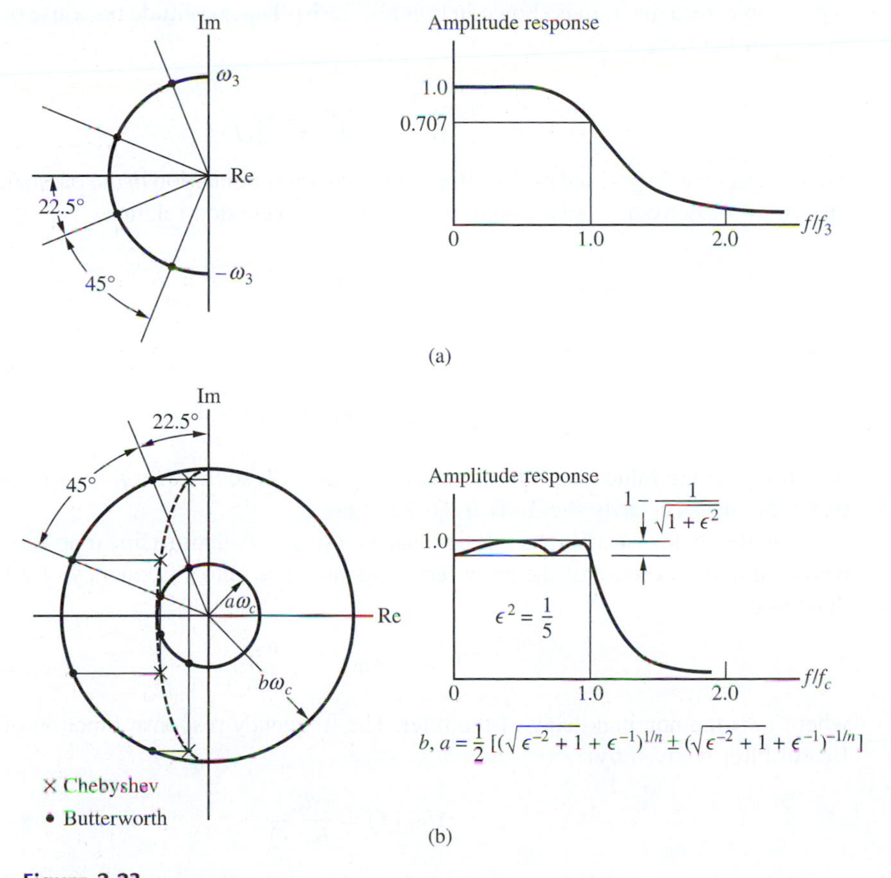

Figure 2.23
Pole locations and amplitude responses for fourth-order Butterworth and Chebyshev filters. (a) Butterworth filter. (b) Chebyshev filter.

[14]From basic circuit theory courses you will recall that the poles and zeros of a rational function of s, $H(s)=N(s)/D(s)$, are those values of complex frequency $s \triangleq \sigma+j\omega$ for which $D(s)=0$ and $N(s)=0$, respectively.

function of a second-order Butterworth filter is

$$H_{\text{2nd-order BW}}(s) = \frac{\omega_3^2}{\left(s+\left[(1+j)/\sqrt{2}\right]\omega_3\right)\left(s+\left[(1-j)/\sqrt{2}\right]\omega_3\right)} = \frac{\omega_3^2}{s^2+\sqrt{2}\omega_3 s+\omega_3^2} \tag{2.235}$$

where $f_3 = \omega_3/2\pi$ is the 3-dB cutoff frequency in hertz. The amplitude response for an nth-order Butterworth filter is of the form

$$|H_{\text{BU}}(f)| = \frac{1}{\sqrt{1+(f/f_3)^{2n}}} \tag{2.236}$$

Note that as n approaches infinity, $|H_{\text{BU}}(f)|$ approaches an ideal lowpass filter characteristic. However, the filter delay also approaches infinity.

The Chebyshev lowpass filter has an amplitude response chosen to maintain a minimum allowable attenuation in the passband while maximizing the attenuation in the stopband. A typical pole-zero diagram is shown in Figure 2.23(b). The amplitude response of a Chebyshev filter is of the form

$$|H_{\text{C}}(f)| = \frac{1}{\sqrt{1+\epsilon^2 C_n^2(f)}} \tag{2.237}$$

The parameter ϵ is specified by the minimum allowable attenuation in the passband, and $C_n(f)$, known as a Chebyshev polynomial, is given by the recursion relation

$$C_n(f) = 2\left(\frac{f}{f_c}\right)C_{n-1}(f) - C_{n-2}(f), \quad n = 2, 3, \ldots \tag{2.238}$$

where

$$C_1(f) = \frac{f}{f_c} \quad \text{and} \quad C_0(f) = 1 \tag{2.239}$$

Regardless of the value of n, it turns out that $C_n(f_c) = 1$, so that $H_{\text{C}}(f_c) = (1+\epsilon^2)^{-1/2}$. (Note that f_c is not necessarily the 3-dB frequency here.)

The Bessel lowpass filter is a design that attempts to maintain a linear phase response in the passband at the expense of the amplitude response. The cutoff frequency of a Bessel filter is defined by

$$f_c = (2\pi t_0)^{-1} = \frac{\omega_c}{2\pi} \tag{2.240}$$

where t_0 is the nominal delay of the filter. The frequency response function of an nth-order Bessel filter is given by

$$H_{\text{BE}}(f) = \frac{K_n}{B_n(f)} \tag{2.241}$$

where K_n is a constant chosen to yield $H(0) = 1$, and $B_n(f)$ is a Bessel polynomial of order n defined by

$$B_n(f) = (2n-1)B_{n-1}(f) - \left(\frac{f}{f_c}\right)^2 B_{n-2}(f) \tag{2.242}$$

where

$$B_0(f) = 1 \quad \text{and} \quad B_1(f) = 1 + j\left(\frac{f}{f_c}\right) \tag{2.243}$$

Figure 2.24 illustrates the amplitude-response and group-delay characteristics of third-order Butterworth, Bessel, and Chebyshev filters. All four filters are normalized to have 3-dB amplitude attenuation at a frequency of $f_c = 1$ Hz. The amplitude responses show that the Chebyshev filters have more attenuation than the Butterworth and Bessel filters do for frequencies exceeding the 3-dB frequency. Increasing the passband $(f < f_c)$ ripple of a Chebyshev filter increases the stopband $(f > f_c)$ attenuation.

The group-delay characteristics shown in Figure 2.24(b) illustrate, as expected, that the Bessel filter has the most constant group delay. Comparison of the Butterworth and the 0.1-dB ripple Chebyshev group delays shows that although the group delay of the Chebyshev filter has a higher peak, it has a more constant group delay for frequencies less than about $0.3f_c$.

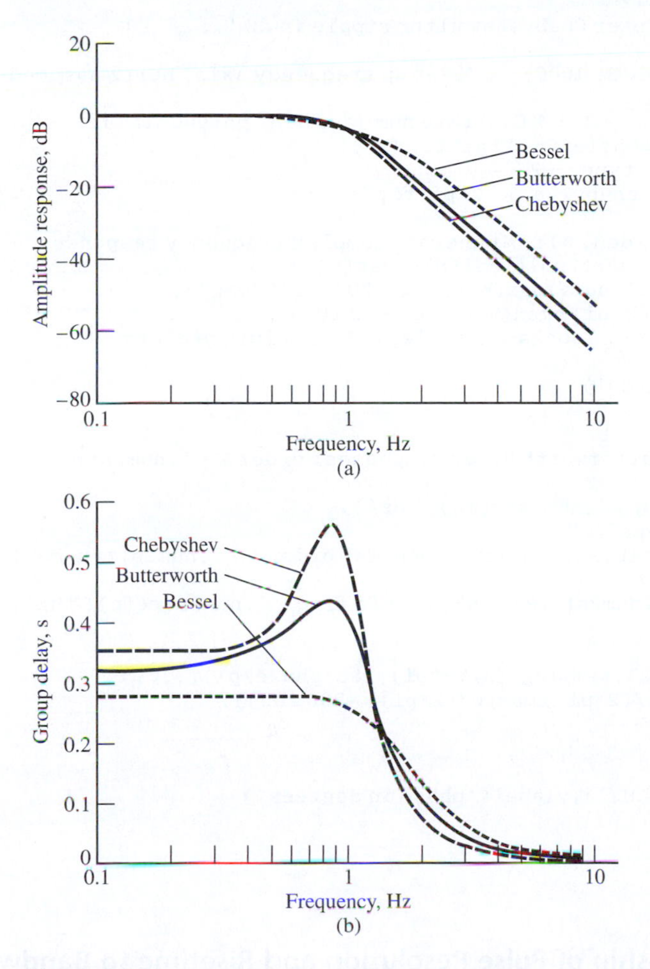

Figure 2.24
Comparison of third-order Butterworth, Chebyshev (0.1-dB ripple), and Bessel filters. (a) Amplitude response. (b) Group delay. All filters are designed to have a 1-Hz, 3-dB bandwidth.

COMPUTER EXAMPLE 2.2

The MATLAB program given below can be used to plot the amplitude and phase responses of Butterworth and Chebyshev filters of any order and any cutoff frequency (3-dB frequency for Butterworth). The ripple is also an input for the Chebyshev filter. Several MATLAB subprograms are used, such as logspace, butter, cheby1, freqs, and cart2pol. It is suggested that the student use the help feature of MATLAB to find out how these are used. For example, a line `freqs(num, den, W)` in the command window automatically plots amplitude and phase responses. However, we have used semilogx here to plot the amplitude response in decibel versus frequency in hertz on a logarithmic scale.

```
% file: c2ce2
% Frequency response for Butterworth and Chebyshev 1 filters
%
clf
filt_type = input('Enter filter type; 1 = Butterworth; 2 = Chebyshev type 1 ');
n_max = input('Enter maximum order of filter ');
fc = input('Enter cutoff frequency (3-dB for Butterworth) in Hz ');
if filt_type == 2
  R = input('Enter Chebyshev filter ripple in dB ');
end
W = logspace(0, 3, 1000);    % Set up frequency axis; hertz assumed
for n = 1:n_max
    if filt_type == 1    % Generate num. and den. polynomials
    [num,den]=butter(n, 2*pi*fc, 's');
    elseif filt_type == 2
    [num,den]=cheby1(n, R, 2*pi*fc, 's');
end
H = freqs(num, den, W);  % Generate complex frequency response
[phase, mag] = cart2pol(real(H),imag(H));
subplot(2,1,1),semilogx(W/(2*pi),20*log10(mag)),...
axis([min(W/(2*pi)) max(W/(2*pi)) -20 0]),...
if n == 1  % Put on labels and title; hold for future plots
  grid on
  ylabel('H in dB')
  hold on
  if filt_type == 1
    title(['Butterworth filter responses: order 1 - ',num2str
    (n_max),'; ...
    cutoff freq = ',num2str(fc),' Hz'])
elseif filt_type == 2
    title(['Chebyshev filter responses: order 1 - ',num2str(n_max),';
    ...
    ripple = ',num2str(R),' dB; cutoff freq = ',num2str(fc),' Hz'])
  end
end
subplot(2,1,2),semilogx(W/(2*pi),180*phase/pi),...
  axis([min(W/(2*pi)) max(W/(2*pi)) -200 200]),...
if n == 1
    grid on
    hold on
    xlabel('f, Hz'),ylabel('phase in degrees')
  end
end
```

■

2.7.14 Relationship of Pulse Resolution and Risetime to Bandwidth

In our consideration of signal distortion, we assumed bandlimited signal spectra. We found that the input signal to a filter is merely delayed and attenuated if the filter has constant amplitude response and linear phase response throughout the passband of the signal. But suppose the input signal is not bandlimited. What rule of thumb can we use to estimate the required bandwidth? This is a particularly important problem in pulse transmission, where the detection and resolution of pulses at a filter output are of interest.

A satisfactory definition for pulse duration and bandwidth, and the relationship between them, is obtained by consulting Figure 2.25. In Figure 2.25(a), a pulse with a single maximum, taken at $t = 0$ for convenience, is shown with a rectangular approximation of height $x(0)$ and duration T. It is required that the approximating pulse and $|x(t)|$ have equal areas. Thus

$$Tx(0) = \int_{-\infty}^{\infty} |x(t)|\, dt \geq \int_{-\infty}^{\infty} x(t)\, dt = X(0) \tag{2.244}$$

where we have used the relationship

$$X(0) = \Im[x(t)]|_{f=0} = \int_{-\infty}^{\infty} x(t) e^{-j2\pi t \cdot 0} dt \tag{2.245}$$

Turning to Figure 2.25(b), we obtain a similar inequality for the rectangular approximation to the pulse spectrum. Specifically, we may write

$$2W\, X(0) = \int_{-\infty}^{\infty} |X(f)|\, df \geq \int_{-\infty}^{\infty} X(f)\, df = x(0) \tag{2.246}$$

where we have used the relationship

$$x(0) = \Im^{-1}[X(f)]|_{t=0} = \int_{-\infty}^{\infty} X(f) e^{j2\pi f \cdot 0} df \tag{2.247}$$

Thus we have the pair of inequalities

$$\frac{x(0)}{X(0)} \geq \frac{1}{T} \quad \text{and} \quad 2W \geq \frac{x(0)}{X(0)} \tag{2.248}$$

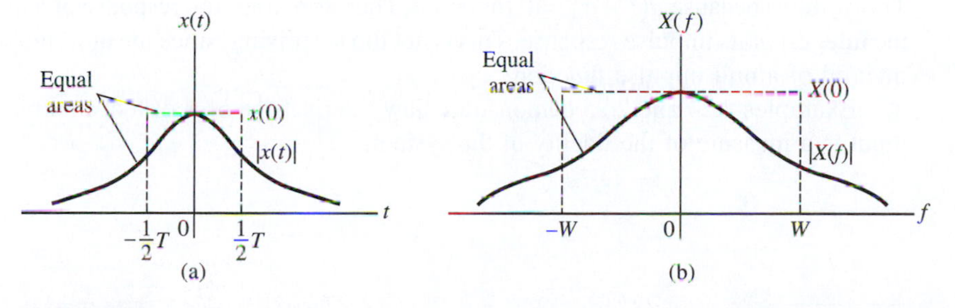

Figure 2.25
Arbitrary pulse signal and spectrum. (a) Pulse and rectangular approximation. (b) Amplitude spectrum and rectangular approximation.

which, when combined, result in the relationship of pulse duration and bandwidth

$$2W \geq \frac{1}{T} \tag{2.249}$$

or

$$W \geq \frac{1}{2T}\text{ Hz} \tag{2.250}$$

Other definitions of pulse duration and bandwidth could have been used, but a relationship similar to (2.249) and (2.250) would have resulted.

This inverse relationship between pulse duration and bandwidth has been illustrated by all the examples involving pulse spectra that we have considered so far (such as Examples 2.8, 2.11, and 2.13).

If pulses with bandpass spectra are considered, the relationship is

$$W \geq \frac{1}{T}\text{ Hz} \tag{2.251}$$

This is illustrated by Example 2.16.

A result similar to (2.249) and (2.250) also holds between the risetime T_R and bandwidth of a pulse. A suitable definition of *risetime* is the time required for a pulse's leading edge to go from 10% to 90% of its final value. For the bandpass case, (2.251) holds with T replaced by T_R, where T_R is the risetime of the *envelope* of the pulse.

Risetime can be used as a measure of a system's distortion. To see how this is accomplished, we will express the step response of a filter in terms of its impulse response. From the superposition integral of (2.177), with $x(t-\sigma) = u(t-\sigma)$, the step response of a filter with impulse response $h(t)$ is

$$\begin{aligned} y_s(t) &= \int_{-\infty}^{\infty} h(\sigma)u(t-\sigma)\,d\sigma \\ &= \int_{-\infty}^{t} h(\sigma)\,d\sigma \end{aligned} \tag{2.252}$$

This follows because $u(t-\sigma) = 0$ for $\sigma > t$. Therefore, the step response of a linear system is the integral of its impulse response. This is not too surprising, since the unit step function is the integral of a unit impulse function.[15]

Examples 2.24 and 2.25 demonstrate how the risetime of a system's output due to a step input is a measure of the fidelity of the system.

[15]This result is a special case of a more general result for an LTI system: If the response of a system to a given input is known and that input is modified through a linear operation, such as integration, then the output to the modified input is obtained by performing the same linear operation on the output due to the original input.

EXAMPLE 2.24

The impulse response of a lowpass RC filter is given by

$$h(t) = \frac{1}{RC} e^{-t/RC} u(t) \tag{2.253}$$

for which the step response is found to be

$$y_s(t) = \left(1 - e^{-2\pi f_3 t}\right) u(t) \tag{2.254}$$

where the 3-dB bandwidth of the filter, defined following (2.192), has been used. The step response is plotted in Figure 2.26(a), where it is seen that the 10% to 90% risetime is approximately

$$T_R = \frac{0.35}{f_3} = 2.2RC \tag{2.255}$$

which demonstrates the inverse relationship between bandwidth and risetime.

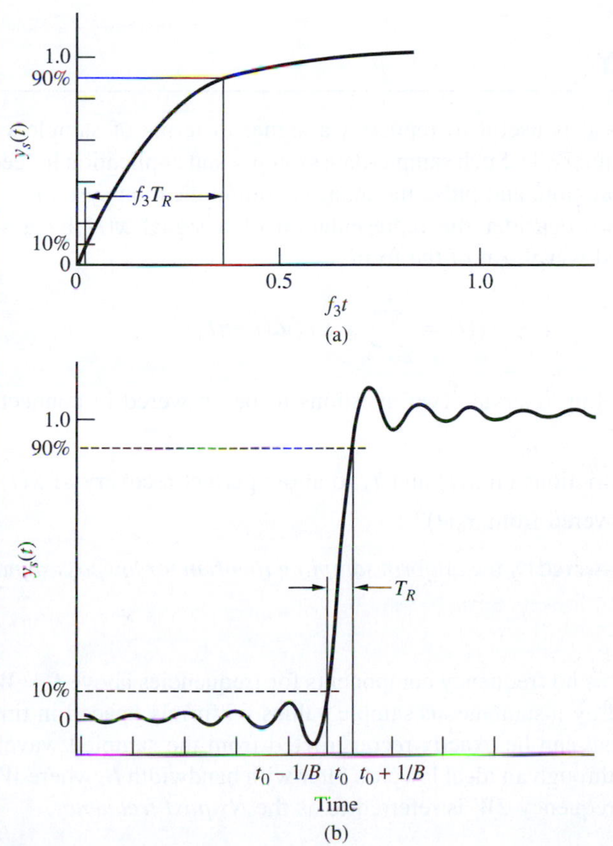

Figure 2.26
Step response of (a) a lowpass RC filter and (b) an ideal lowpass filter, illustrating 10% to 90% risetime of each.

EXAMPLE 2.25

Using (2.230) with $H_0 = 1$, the step response of an ideal lowpass, filter is

$$\begin{aligned} y_s(t) &= \int_{-\infty}^{t} 2B\,\mathrm{sinc}[2B(\sigma - t_0)]d\sigma \\ &= \int_{-\infty}^{t} 2B\frac{\sin[2\pi B(\sigma - t_0)]}{2\pi B(\sigma - t_0)}d\sigma \end{aligned} \tag{2.256}$$

By changing variables in the integrand to $u = 2\pi B(\sigma - t_0)$, the step response becomes

$$y_s(t) = \frac{1}{2\pi}\int_{-\infty}^{2\pi B(t-t_0)} \frac{\sin u}{u}du = \frac{1}{2} + \frac{1}{\pi}\mathrm{Si}[2\pi B(t - t_0)] \tag{2.257}$$

where $\mathrm{Si}(x) = \int_0^x (\sin u/u)\,du = -\mathrm{Si}(-x)$ is the sine-integral function.[16] A plot of $y_s(t)$ for an ideal lowpass filter, such as is shown in Figure 2.26(b), reveals that the 10% to 90% risetime is approximately

$$T_R \cong \frac{0.44}{B} \tag{2.258}$$

Again, the inverse relationship between bandwidth and risetime is demonstrated.

■

■ 2.8 SAMPLING THEORY

In many applications it is useful to represent a signal in terms of sample values taken at appropriately spaced intervals. Such sample-data systems find application in feedback control, digital computer simulation, and pulse-modulation communication systems.

In this section we consider the representation of a signal $x(t)$ by a so-called ideal instantaneous sampled waveform of the form

$$x_\delta(t) = \sum_{n=-\infty}^{\infty} x(nT_s)\delta(t-nT_s) \tag{2.259}$$

where T_s is the sampling interval. Two questions to be answered in connection with such sampling are

What are the restrictions on $x(t)$ and T_s to allow perfect recovery of $x(t)$ from $x_\delta(t)$?

How is $x(t)$ recovered from $x_\delta(t)$?

Both questions are answered by the *uniform sampling theorem for lowpass signals,* which may be stated as follows:

Theorem

If a signal $x(t)$ contains no frequency components for frequencies above $f = W$ Hz, then it is completely described by instantaneous sample values uniformly spaced in time with period $T_s < 1/2W$. The signal can be exactly reconstructed from the sampled waveform given by (2.259) by passing it through an ideal lowpass filter with bandwidth B, where $W < B < f_s - W$ with $f_s = T_s^{-1}$. The frequency $2W$ is referred to as the *Nyquist frequency*.

[16]See M. Abramowitz and I. Stegun (1972), pp. 238ff.

To prove the sampling theorem, we find the spectrum of (2.259). Since $\delta(t-nT_s)$ is zero everywhere except at $t=nT_s$, (2.259) can be written as

$$x_\delta(t) = \sum_{n=-\infty}^{\infty} x(t)\delta(t-nT_s) = x(t)\sum_{n=-\infty}^{\infty}\delta(t-nT_s) \tag{2.260}$$

Applying the multiplication theorem of Fourier transforms (2.119), the Fourier transform of (2.260) is

$$X_\delta(f) = X(f) * \left[f_s \sum_{n=-\infty}^{\infty}\delta(f-nf_s)\right] \tag{2.261}$$

where the transform pair (2.136) has been used. Interchanging the orders of summation and convolution and noting that

$$X(f) * \delta(f-nf_s) = \int_{-\infty}^{\infty} X(u)\delta(f-u-nf_s)\,du = X(f-nf_s) \tag{2.262}$$

by the sifting property of the delta function, we obtain

$$X_\delta(f) = f_s \sum_{n=-\infty}^{\infty} X(f-nf_s) \tag{2.263}$$

Thus, assuming that the spectrum of $x(t)$ is bandlimited to W Hz and that $f_s > 2W$ as stated in the sampling theorem, we may readily sketch $X_\delta(f)$. Figure 2.27 shows a typical choice for $X(f)$ and the corresponding $X_\delta(f)$. We note that sampling simply results in a periodic repetition of $X(f)$ in the frequency domain with a spacing f_s. If $f_s < 2W$, the separate terms in (2.263) overlap, and there is no apparent way to recover $x(t)$ from $x_\delta(t)$ without distortion. On the other hand, if $f_s > 2W$, the term in (2.263) for $n=0$ is easily separated from the rest by ideal lowpass filtering. Assuming an ideal lowpass filter with the frequency-response function

$$H(f) = H_0\Pi\left(\frac{f}{2B}\right)e^{-j2\pi f t_0}, \quad W \le B \le f_s - W \tag{2.264}$$

the output spectrum, with $x_\delta(t)$ at the input, is

$$Y(f) = f_s H_0 X(f) e^{-j2\pi f t_0} \tag{2.265}$$

and by the time-delay theorem, the output waveform is

$$y(t) = f_s H_0 x(t-t_0) \tag{2.266}$$

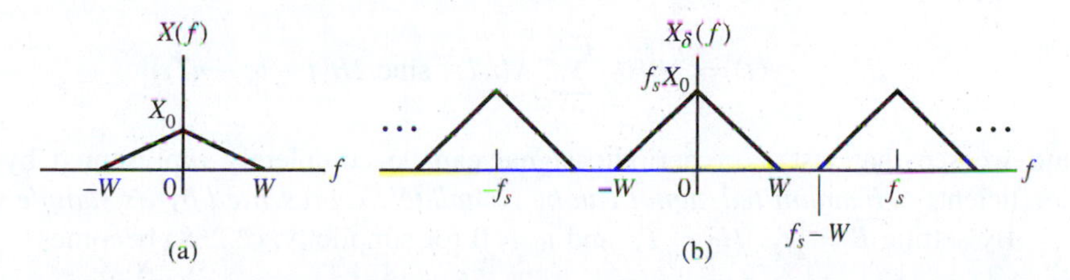

Figure 2.27
Signal spectra for lowpass sampling. (a) Assumed spectrum for $x(t)$. (b) Spectrum of the sampled signal.

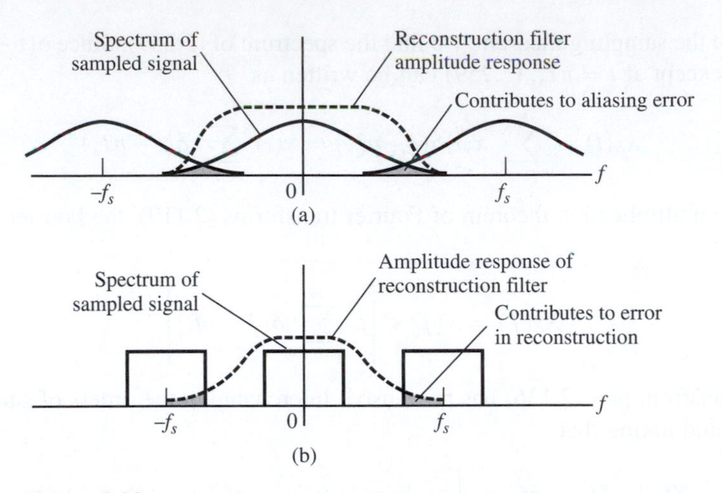

Figure 2.28
Spectra illustrating two types of errors encountered in reconstruction of sampled signals. (a) Illustration of aliasing error in the reconstruction of sampled signals. (b) Illustration of error due to nonideal reconstruction filter.

Thus, if the conditions of the sampling theorem are satisfied, we see that distortionless recovery of $x(t)$ from $x_\delta(t)$ is possible. Conversely, if the conditions of the sampling theorem are not satisfied, either because $x(t)$ is not bandlimited or because $f_s < 2W$, we see that distortion at the output of the reconstruction filter is inevitable. Such distortion, referred to as *aliasing*, is illustrated in Figure 2.28(a). It can be combated by filtering the signal before sampling or by increasing the sampling rate. A second type of error, illustrated in Figure 2.28 (b), occurs in the reconstruction process and is due to the nonideal frequency response characteristics of practical filters. This type of error can be minimized by choosing reconstruction filters with sharper roll-off characteristics or by increasing the sampling rate. Note that the error due to aliasing and the error due to imperfect reconstruction filters are both *proportional to signal level*. Thus increasing the signal amplitude does not improve the signal-to-error ratio.

An alternative expression for the reconstructed output from the ideal lowpass filter can be obtained by noting that when (2.259) is passed through a filter with impulse response $h(t)$, the output is

$$y(t) = \sum_{n=-\infty}^{\infty} x(nT_s)h(t - nT_s) \tag{2.267}$$

but $h(t)$ corresponding to (2.264) is given by (2.230). Thus

$$y(t) = 2BH_0 \sum_{n=-\infty}^{\infty} x(nT_s)\,\text{sinc}[2B(t - t_0 - nT_s)] \tag{2.268}$$

and we see that just as a periodic signal can be completely represented by its Fourier coefficients, a *bandlimited signal can be completely represented by its sample values.*

By setting $B = \frac{1}{2}f_s$, $H_0 = T_s$, and $t_0 = 0$ for simplicity, (2.268) becomes

$$y(t) = \sum_{n} x(nT_s)\,\text{sinc}(f_s t - n) \tag{2.269}$$

This expansion is equivalent to a generalized Fourier series of the form given by (2.39), for we may show that

$$\int_{-\infty}^{\infty} \operatorname{sinc}(f_s t - n)\, \operatorname{sinc}(f_s t - m)\, dt = \delta_{nm} \tag{2.270}$$

where $\delta_{nm} = 1,\ n = m$, and is 0 otherwise.

Turning next to bandpass spectra, for which the upper limit on frequency f_u is much larger than the single-sided bandwidth W, one may naturally inquire as to the feasibility of sampling at rates less than $f_s > 2f_u$. The *uniform sampling theorem for bandpass signals* gives the conditions for which this is possible.

Theorem

If a signal has a spectrum of bandwidth W Hz and upper frequency limit f_u, then a rate f_s at which the signal can be sampled is $2f_u/m$, where m is the largest integer not exceeding f_u/W. All higher sampling rates are not necessarily usable unless they exceed $2f_u$.

EXAMPLE 2.26

Consider the bandpass signal $x(t)$ with the spectrum shown in Figure 2.29. According to the bandpass sampling theorem, it is possible to reconstruct $x(t)$ from sample values taken at a rate of

$$f_s = \frac{2f_u}{m} = \frac{2(3)}{2} = 3 \text{ samples per second} \tag{2.271}$$

whereas the lowpass sampling theorem requires 6 samples per second.

To show that this is possible, we sketch the spectrum of the sampled signal. According to (2.263), which holds in general,

$$X_\delta(f) = 3 \sum_{-\infty}^{\infty} X(f - 3n) \tag{2.272}$$

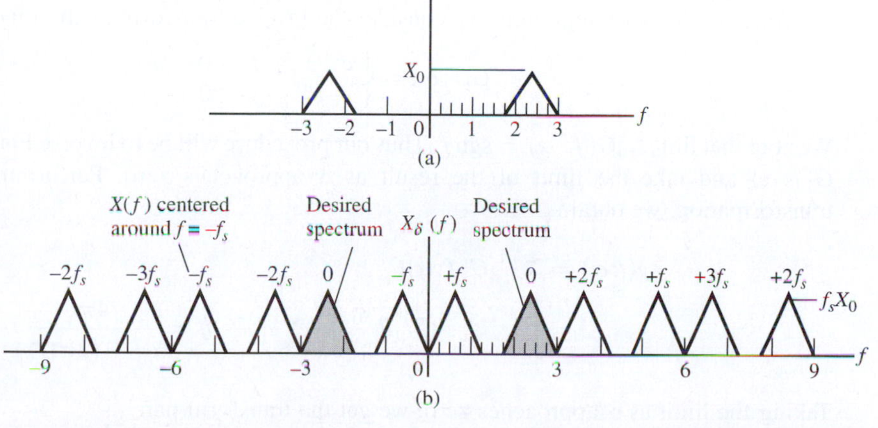

Figure 2.29
Signal spectra for bandpass sampling. (a) Assumed bandpass signal spectrum. (b) Spectrum of the sampled signal.

The resulting spectrum is shown in Figure 2.29(b), and we see that it is theoretically possible to recover $x(t)$ from $x_\delta(t)$ by bandpass filtering.

Another way of sampling a bandpass signal of bandwidth W is to resolve it into two lowpass quadrature signals of bandwidth $\frac{1}{2}W$. Both of these may then be sampled at a minimum rate of $2(\frac{1}{2}W) = W$ samples per second, thus resulting in an overall minimum sampling rate of $2W$ samples per second.

■

■ 2.9 THE HILBERT TRANSFORM

(It may be advantageous to postpone this section until consideration of single-sideband systems in Chapter 3.)

2.9.1 Definition

Consider a filter that simply phase shifts all frequency components of its input by $-\frac{1}{2}\pi$ rad; that is, its frequency-response function is

$$H(f) = -j\,\mathrm{sgn}f \tag{2.273}$$

where the sgn function (read "signum f") is defined as

$$\mathrm{sgn}f = \begin{cases} 1, & f > 0 \\ 0, & f = 0 \\ -1, & f < 0 \end{cases} \tag{2.274}$$

We note that $|H(f)| = 1$ and $\underline{/H(f)}$ is odd, as it must be. If $X(f)$ is the input spectrum to the filter, the output spectrum is $-j(\mathrm{sgn}f)X(f)$, and the corresponding time function is

$$\begin{aligned} \hat{x}(t) &= \Im^{-1}[-j(\mathrm{sgn}f)X(f)] \\ &= h(t) * x(t) \end{aligned} \tag{2.275}$$

where $h(t) = -j\Im^{-1}[\mathrm{sgn}f]$ is the impulse response of the filter. To obtain $\Im^{-1}[\mathrm{sgn}f]$ without resorting to contour integration, we consider the inverse transform of the function

$$G(f;\alpha) = \begin{cases} e^{-\alpha f}, & f > 0 \\ -e^{\alpha f}, & f < 0 \end{cases} \tag{2.276}$$

We note that $\lim_{\alpha \to 0} G(f;\,\alpha) = \mathrm{sgn}f$. Thus our procedure will be to inverse Fourier transform $G(f;\alpha)$ and take the limit of the result as α approaches zero. Performing the inverse transformation, we obtain

$$\begin{aligned} g(t;\alpha) &= \Im^{-1}[G(f;\alpha)] \\ &= \int_0^\infty e^{-\alpha f} e^{j2\pi f t}\,df - \int_\infty^0 e^{\alpha f} e^{j2\pi f t}\,df = \frac{j4\pi t}{\alpha^2 + (2\pi t)^2} \end{aligned} \tag{2.277}$$

Taking the limit as α approaches zero, we get the transform pair

$$\frac{j}{\pi t} \longleftrightarrow \mathrm{sgn}f \tag{2.278}$$

Using this result in (2.275), we obtain the output of the filter:

$$\widehat{x}(t) = \int_{-\infty}^{\infty} \frac{x(\lambda)}{\pi(t-\lambda)}\, d\lambda = \int_{-\infty}^{\infty} \frac{x(t-\eta)}{\pi\eta}\, d\eta \tag{2.279}$$

The signal $\widehat{x}(t)$ is defined as the *Hilbert transform* of $x(t)$. Since the Hilbert transform corresponds to a phase shift of $-\frac{1}{2}\pi$, we note that the Hilbert transform of $\widehat{x}(t)$ corresponds to the frequency-response function $(-j\,\mathrm{sgn} f)^2 = -1$, or a phase shift of π rad. Thus

$$\widehat{\widehat{x}}(t) = -x(t) \tag{2.280}$$

EXAMPLE 2.27

For an input to a Hilbert transform filter of

$$x(t) = \cos(2\pi f_0 t) \tag{2.281}$$

which has a spectrum given by

$$X(f) = \frac{1}{2}\delta(f - f_0) + \frac{1}{2}\delta(f + f_0) \tag{2.282}$$

we obtain an output spectrum from the Hilbert transformer of

$$\widehat{X}(f) = \frac{1}{2}\delta(f - f_0)e^{-j\pi/2} + \frac{1}{2}\delta(f + f_0)e^{j\pi/2} \tag{2.283}$$

Taking the inverse Fourier transform of (2.283), we find the output signal to be

$$\begin{aligned} \widehat{x}(f) &= \frac{1}{2}e^{j2\pi f_0 t}e^{-j\pi/2} + \frac{1}{2}e^{-j2\pi f_0 t}e^{j\pi/2} \\ &= \cos\left(2\pi f_0 t - \frac{\pi}{2}\right) \end{aligned} \tag{2.284}$$

or

$$\widehat{\cos(2\pi f_0 t)} = \sin(2\pi f_0 t)$$

Of course, the Hilbert transform could have been found by inspection in this case by adding $-\frac{1}{2}\pi$ to the argument of the cosine. Doing this for the signal $\sin\omega_0 t$, we find that

$$\widehat{\sin(2\pi f_0 t)} = \sin\left(2\pi f_0 t - \frac{1}{2}\pi\right) = -\cos(2\pi f_0 t) \tag{2.285}$$

We may use the two results obtained to show that

$$\widehat{e^{j2\pi f_0 t}} = -j\,\mathrm{sgn}(2\pi f_0)e^{j2\pi f_0 t} \tag{2.286}$$

This is done by considering the two cases $f_0 > 0$ and $f_0 < 0$ and using Euler's theorem in conjunction with the results of (2.284) and (2.285). The result (2.286) also follows directly by considering the response of a Hilbert transform filter with frequency response $H_{\mathrm{HT}}(f) = -j\,\mathrm{sgn}(2\pi f)$ to the input $x(t) = e^{j2\pi f_0 t}$. ■

2.9.2 Properties

The Hilbert transform has several useful properties that will be illustrated later. Three of these properties will be proved here:

1. The energy (or power) in a signal $x(t)$ and its Hilbert transform $\widehat{x}(t)$ are equal. To show this, we consider the energy spectral densities at the input and output of a Hilbert transform filter. Since $H(f) = -j\,\mathrm{sgn}\,f$, these densities are related by

$$|\widehat{X}(f)|^2 \triangleq |\Im[\widehat{x}(t)]|^2 = |-j\,\mathrm{sgn}\,f|^2 |X(f)|^2 = |X(f)|^2 \tag{2.287}$$

where $\widehat{X}(f) = \Im[\widehat{x}(t)] = -j(\mathrm{sgn}\,f)X(f)$. Thus, since the energy spectral densities at input and output are equal, so are the total energies. A similar proof holds for power signals.

2. A signal and its Hilbert transform are orthogonal; that is,

$$\int_{-\infty}^{\infty} x(t)\widehat{x}(t)\,dt = 0 \quad \text{(energy signals)} \tag{2.288}$$

or

$$\lim_{T\to\infty} \frac{1}{2T}\int_{-T}^{T} x(t)\widehat{x}(t)\,dt = 0 \quad \text{(power signals)} \tag{2.289}$$

Considering (2.288), we note that the left-hand side can be written as

$$\int_{-\infty}^{\infty} x(t)\widehat{x}(t)\,dt = \int_{-\infty}^{\infty} X(f)\widehat{X}^*(f)\,df \tag{2.290}$$

by Parseval's theorem generalized, where $\widehat{X}(f) = \Im[\widehat{x}(t)] = -j(\mathrm{sgn} f)\,X(f)$. It therefore follows that

$$\int_{-\infty}^{\infty} x(t)\widehat{x}(t)\,dt = \int_{-\infty}^{\infty} (+j\,\mathrm{sgn}\,f)|X(f)|^2\,df \tag{2.291}$$

However, the integrand of the right-hand side of (2.291) is odd, being the product of the even function $|X(f)|^2$ and the odd function $j\,\mathrm{sgn}\,f$. Therefore, the integral is zero, and (2.288) is proved. A similar proof holds for (2.289).

3. If $c(t)$ and $m(t)$ are signals with nonoverlapping spectra, where $m(t)$ is lowpass and $c(t)$ is highpass, then

$$\widehat{m(t)c(t)} = m(t)\hat{c}(t) \tag{2.292}$$

To prove this relationship, we use the Fourier integral to represent $m(t)$ and $c(t)$ in terms of their spectra $M(f)$ and $C(f)$, respectively. Thus

$$m(t)c(t) = \int_{-\infty}^{\infty}\int_{-\infty}^{\infty} M(f)C(f')\exp[j2\pi(f+f')t]\,df\,df' \tag{2.293}$$

where we assume $M(f) = 0$ for $|f| > W$ and $C(f') = 0$ for $|f'| < W$. The Hilbert transform of (2.293) is

$$\begin{aligned}\widehat{m(t)c(t)} &= \int_{-\infty}^{\infty}\int_{-\infty}^{\infty} M(f)C(f')\widehat{\exp[j2\pi(f+f')t]}\,df\,df' \\ &= \int_{-\infty}^{\infty}\int_{-\infty}^{\infty} M(f)C(f')[-j\,\mathrm{sgn}(f+f')]\exp[j2\pi(f+f')t]\,df\,df'\end{aligned} \tag{2.294}$$

where (2.286) has been used. However, the product $M(f)C(f')$ is nonvanishing only for $|f| < W$ and $|f'| > W$, and we may replace $\operatorname{sgn}(f+f')$ by $\operatorname{sgn} f'$ in this case. Thus

$$\widehat{m(t)c(t)} = \int_{-\infty}^{\infty} M(f)\exp(j2\pi ft)\,df \int_{-\infty}^{\infty} C(f')[-j(\operatorname{sgn} f')\exp(j2\pi f't)]\,df' \tag{2.295}$$

However, the first integral on the right-hand side is just $m(t)$, and the second integral is $\hat{c}(t)$, since

$$c(t) = \int_{-\infty}^{\infty} C(f')\exp(j2\pi f't)\,df'$$

and

$$\begin{aligned}\widehat{c}(t) &= \int_{-\infty}^{\infty} C(f')\widehat{\exp(j2\pi f't)}\,df' \\ &= \int_{-\infty}^{\infty} C(f')[-j \operatorname{sgn} f' \exp(j2\pi f't)]\,df'\end{aligned} \tag{2.296}$$

Hence (2.295) is equivalent to (2.292), which was the relationship to be proved.

EXAMPLE 2.28

Given that $m(t)$is a lowpass signal with $M(f) = 0$ for $|f| > W$, we may directly apply (2.292) in conjunction with (2.291) and (2.285) to show that

$$\widehat{m(t)\cos(\omega_0 t)} = m(t)\sin(\omega_0 t) \tag{2.297}$$

and

$$\widehat{m(t)\sin(\omega_0 t)} = -m(t)\cos(\omega_0 t) \tag{2.298}$$

if $f_0 = \omega_0/2\pi > W$. ■

2.9.3 Analytic Signals

An *analytic signal* $x_p(t)$, corresponding to the real signal $x(t)$, is defined as

$$x_p(t) = x(t) + j\widehat{x}(t) \tag{2.299}$$

where $\widehat{x}(t)$ is the Hilbert transform of $x(t)$. We now consider several properties of an analytic signal.

We used the term *envelope* in connection with the ideal bandpass filter. The *envelope* of a signal is defined mathematically as the magnitude of the analytic signal $x_p(t)$. The concept of an envelope will acquire more importance when we discuss modulation in Chapter 3.

EXAMPLE 2.29

In Section 2.7.12, (2.233), we showed that the impulse response of an ideal bandpass filter with bandwidth B, delay t_0, and center frequency f_0 is given by

$$h_{BP}(t) = 2H_0B\,\operatorname{sinc}[B(t-t_0)]\cos[\omega_0(t-t_0)] \tag{2.300}$$

Assuming that $B < f_0$, we can use the result of Example 2.28 to determine the Hilbert transform of $h_{BP}(t)$. The result is

$$\widehat{h}_{BP}(t) = 2H_0B\ \text{sinc}[B(t-t_0)]\sin[\omega_0(t-t_0)] \tag{2.301}$$

The envelope is

$$\begin{aligned} |h_{BP}(t)| &= |x(t)+j\widehat{x}(t)| \\ &= \sqrt{[x(t)]^2+[\widehat{x}(t)]^2} \\ &= \sqrt{\{2H_0B\ \ \text{sinc}[B(t-t_0)]\}^2\{\cos^2[\omega_0(t-t_0)]+\sin^2[\omega_0(t-t_0)]\}} \end{aligned} \tag{2.302}$$

or

$$|h_{BP}(t)| = 2H_0B|\text{sinc}[B(t-t_0)]| \tag{2.303}$$

as shown in Figure 2.22(b) by the dashed lines. The envelope is obviously easy to identify if the signal is composed of a lowpass signal multiplied by a high-frequency sinusoid. Note, however, that the envelope is mathematically defined for any signal.

■

The spectrum of the analytic signal is also of interest. We will use it to advantage in Chapter 3 when we investigate single-sideband modulation. Since the analytic signal, from (2.299), is defined as

$$x_p(t) = x(t)+j\widehat{x}(t)$$

it follows that the Fourier transform of $x_p(t)$ is

$$X_p(f) = X(f)+j[-j(\text{sgn}f)X(f)] \tag{2.304}$$

where the term in brackets is the Fourier transform of $\widehat{x}(t)$. Thus

$$X_p(f) = X(f)[1+\text{sgn}f] \tag{2.305}$$

or

$$X_p(f) = \begin{cases} 2X(f), & f>0 \\ 0, & f<0 \end{cases} \tag{2.306}$$

The subscript p is used to denote that the spectrum is nonzero only for positive frequencies.

Similarly, we can show that the signal

$$x_n(t) = x(t)-j\widehat{x}(t) \tag{2.307}$$

is nonzero only for negative frequencies. Replacing $\widehat{x}(t)$ by $-\widehat{x}(t)$ in the preceding discussion results in

$$X_n(f) = X(f)(1-\text{sgn}f) \tag{2.308}$$

or

$$X_n(f) = \begin{cases} 0, & f>0 \\ 2X(f), & f<0 \end{cases} \tag{2.309}$$

These spectra are illustrated in Figure 2.30.

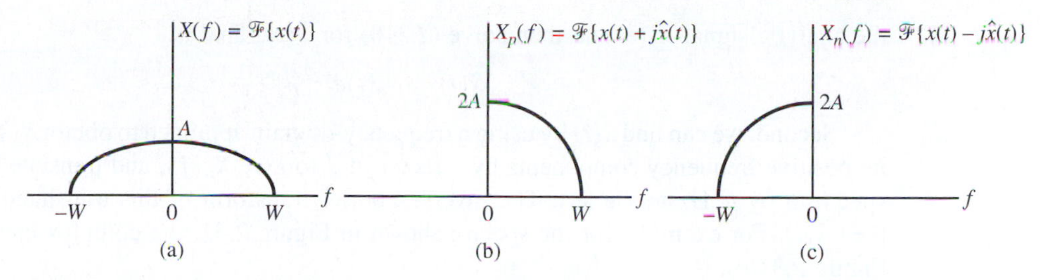

Figure 2.30
Specta of analytic signals. (a) Spectrum of $x(t)$. (b) Spectrum of $x(t)+j\widehat{x}(t)$. (c) Spectrum of $x(t)-j\widehat{x}(t)$.

Two observations may be made at this point. First, if $X(f)$ is nonzero at $f=0$, then $X_p(f)$ and $X_n(f)$ will be discontinuous at $f=0$. Also, we should not be confused that $|X_n(f)|$ and $|X_p(f)|$ are not even, since the corresponding time-domain signals are not real.

2.9.4 Complex Envelope Representation of Bandpass Signals

If $X(f)$ in (2.304) corresponds to a signal with a bandpass spectrum, as shown in Fig. 2.31(a), it then follows by (2.306) that $X_p(f)$ is just twice the positive frequency portion of $X(f)=\Im\{x(t)\}$, as shown in Fig. 2.31(b). By the frequency-translation theorem, it follows that $x_p(t)$ can be written as

$$x_p(t) = \tilde{x}(t)e^{j2\pi f_0 t} \tag{2.310}$$

where $\tilde{x}(t)$ is a complex-valued lowpass signal (hereafter referred to as the *complex envelope*) and f_0 is a reference frequency chosen for convenience.[17] The spectrum (assumed to be real for ease of plotting) of $\tilde{x}(t)$ is shown in Figure 2.31(c).

To find $\tilde{x}(t)$, we may proceed along one of two paths [note that simply taking the magnitude of (2.310) gives only $|\tilde{x}(t)|$ but not its arguement]. First, using (2.299), we can find

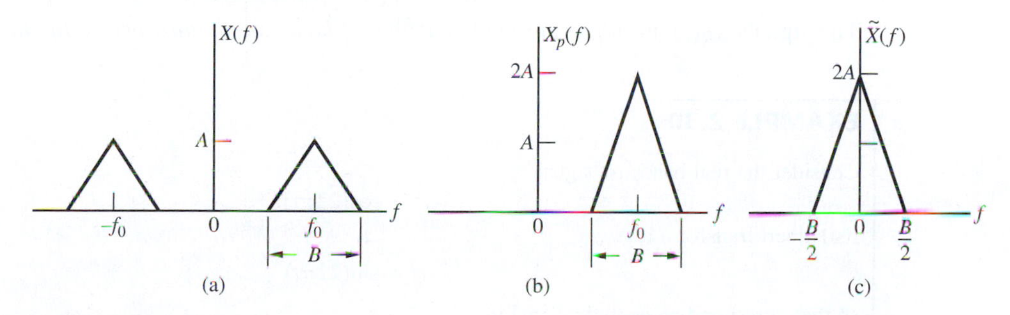

Figure 2.31
Spectra pertaining to the formation of a complex envelope of a signal $x(t)$. (a) A bandpass signal spectrum. (b) Twice the positive-frequency portion of $X(f)$ corresponding to $\Im[x(t)+j\widehat{x}(t)]$. (c) Spectrum of $\tilde{x}(t)$.

[17]If the spectrum of $x_p(t)$ has a center of symmetry, a natural choice for f_0 would be this point of symmetry, but it need not be.

the analytic signal $x_p(t)$ and then solve (2.310) for $\tilde{x}(t)$. That is,

$$\tilde{x}(t) = x_p(t)e^{-j2\pi f_0 t} \tag{2.311}$$

Second, we can find $\tilde{x}(t)$ by using a frequency-domain approach to obtain $X(f)$, then scale its positive frequency components by a factor of 2 to give $X_p(f)$, and translate the resultant spectrum by f_0 Hz to the left. The inverse Fourier transform of this translated spectrum is then $\tilde{x}(t)$. For example, for the spectra shown in Figure 2.31, the complex envelope, using Figure 2.31(c), is

$$\tilde{x}(t) = \Im^{-1}\left[2A\Lambda\left(\frac{2f}{B}\right)\right] = AB\,\text{sinc}^2(Bt/2) \tag{2.312}$$

The complex envelope is real in this case because the spectrum $X(f)$ is symmetrical around $f = f_0$.

Since $x_p(t) = x(t) + j\widehat{x}(t)$, where $x(t)$ and $\widehat{x}(t)$ are the real and imaginary parts, respectively, of $x_p(t)$, it follows from (2.310) that

$$x_p(t) = \tilde{x}(t)e^{j2\pi f_0 t} \triangleq x(t) + j\widehat{x}(t) \tag{2.313}$$

or

$$x(t) = \text{Re}\left(\tilde{x}(t)e^{j2\pi f_0 t}\right) \tag{2.314}$$

and

$$\widehat{x}(t) = \text{Im}\left(\tilde{x}(t)e^{j2\pi f_0 t}\right) \tag{2.315}$$

Thus, from (2.314), the real signal $x(t)$ can be expressed in terms of its complex envelope as

$$\begin{aligned} x(t) &= \text{Re}\left(\tilde{x}(t)e^{j2\pi f_0 t}\right) \\ &= \text{Re}(\tilde{x}(t))\cos(2\pi f_0 t) - \text{Im}(\tilde{x}(t))\sin(2\pi f_0 t) \\ &= x_R(t)\cos(2\pi f_0 t) - x_I(t)\sin(2\pi f_0 t) \end{aligned} \tag{2.316}$$

where

$$\tilde{x}(t) \triangleq x_R(t) + jx_I(t) \tag{2.317}$$

The signals $x_R(t)$ and $x_I(t)$ are known as the *inphase* and *quadrature components* of $x(t)$.

EXAMPLE 2.30

Consider the real bandpass signal

$$x(t) = \cos(22\pi t) \tag{2.318}$$

Its Hilbert transform is

$$\widehat{x}(t) = \sin(22\pi t) \tag{2.319}$$

so the corresponding analytic signal is

$$\begin{aligned} x_p(t) &= x(t) + j\widehat{x}(t) \\ &= \cos(22\pi t) + j\sin(22\pi t) \\ &= e^{j22\pi t} \end{aligned} \tag{2.320}$$

In order to find the corresponding complex envelope, we need to specify f_0, which for the purposes of this example, we take as $f_0 = 10$ Hz. Thus, from (2.311), we have

$$\begin{aligned}\tilde{x}(t) &= x_p(t)e^{-j2\pi f_0 t} \\ &= e^{j22\pi t}e^{-j20\pi t} \\ &= e^{j2\pi t} \\ &= \cos(2\pi t) + j\sin(2\pi t)\end{aligned} \tag{2.321}$$

so that, from (2.317), we obtain

$$x_R(t) = \cos(2\pi t) \quad \text{and} \quad x_I(t) = \sin(2\pi t) \tag{2.322}$$

Putting these into (2.316), we get

$$\begin{aligned}x(t) &= x_R(t)\cos(2\pi f_0 t) - x_I(t)\sin(2\pi f_0 t) \\ &= \cos(2\pi t)\cos(20\pi t) - \sin(2\pi t)\sin(20\pi t) \\ &= \cos(22\pi t)\end{aligned} \tag{2.323}$$

which is, not surprisingly, what we began with in (2.318). ■

2.9.5 Complex Envelope Representation of Bandpass Systems

Consider a bandpass system with impulse response $h(t)$ that is represented in terms of a complex envelope $\tilde{h}(t)$ as

$$h(t) = \text{Re}\left(\tilde{h}(t)e^{j2\pi f_0 t}\right) \tag{2.324}$$

where $\tilde{h}(t) = h_R(t) + jh_I(t)$. Assume that the input is also bandpass with representation (2.314). The output, by the superposition integral, is

$$y(t) = x(t) * h(t) = \int_{-\infty}^{\infty} h(\lambda)x(t-\lambda)\,d\lambda \tag{2.325}$$

By Euler's theorem, we can represent $h(t)$ and $x(t)$ as

$$h(t) = \frac{1}{2}\tilde{h}(t)e^{j2\pi f_0 t} + \text{c.c.} \tag{2.326}$$

and

$$x(t) = \frac{1}{2}\tilde{x}(t)e^{j2\pi f_0 t} + \text{c.c.} \tag{2.327}$$

respectively, where c.c. stands for the complex conjugate of the immediately preceding term. Using these in (2.325), the output can be expressed as

$$\begin{aligned}y(t) &= \int_{-\infty}^{\infty}\left[\frac{1}{2}\tilde{h}(t)e^{j2\pi f_0 \lambda} + \text{c.c.}\right]\left[\frac{1}{2}\tilde{x}(t-\lambda)e^{j2\pi f_0(t-\lambda)} + \text{c.c.}\right]d\lambda \\ &= \frac{1}{4}\int_{-\infty}^{\infty}\tilde{h}(\lambda)\tilde{x}(t-\lambda)\,d\lambda\, e^{j2\pi f_0 t} + \text{c.c.} \\ &\quad + \frac{1}{4}\int_{-\infty}^{\infty}\tilde{h}(\lambda)\tilde{x}^*(t-\lambda)e^{j4\pi f_0 \lambda}\,d\lambda\, e^{-j2\pi f_0 t} + \text{c.c.}\end{aligned} \tag{2.328}$$

The second pair of terms, $\frac{1}{4}\int_{-\infty}^{\infty}\tilde{h}(\lambda)\tilde{x}^*(t-\lambda)e^{j4\pi f_0\lambda}\,d\lambda\, e^{-j2\pi f_0 t}+\text{c.c}$, is approximately zero by virtue of the factor $e^{j4\pi f_0\lambda}=\cos(4\pi f_0\lambda)+j\sin(4\pi f_0\lambda)$ in the integrand ($\tilde{h}$ and $\tilde{x}$ are slowly varying with respect to this complex exponential, and therefore, the integrand cancels to zero, half-cycle by half-cycle). Thus

$$\begin{aligned} y(t) &\cong \frac{1}{4}\int_{-\infty}^{\infty}\tilde{h}(\lambda)\tilde{x}(t-\lambda)\,d\lambda\, e^{j2\pi f_0 t}+\text{c.c.} \\ &= \frac{1}{2}\text{Re}\left(\left[\tilde{h}(t) * \tilde{x}(t)\right]e^{j2\pi f_0 t}\right) \triangleq \frac{1}{2}\text{Re}\left(\tilde{y}(t)e^{j2\pi f_0 t}\right) \end{aligned} \tag{2.329}$$

where

$$\tilde{y}(t)=\tilde{h}(t) * \tilde{x}(t)=\Im^{-1}\left[\tilde{H}(f)\tilde{X}(f)\right] \tag{2.330}$$

in which $\tilde{H}(f)$ and $\tilde{X}(f)$ are the respective Fourier transforms of $\tilde{h}(t)$ and $\tilde{x}(t)$.

EXAMPLE 2.31

As an example of the application of (2.329), consider the input

$$x(t)=\Pi\left(\frac{t}{\tau}\right)\cos(2\pi f_0 t) \tag{2.331}$$

to a filter with impulse response

$$h(t)=\alpha e^{-\alpha t}u(t)\cos(2\pi f_0 t) \tag{2.332}$$

Using the complex envelope analysis just developed with $\tilde{x}(t)=\Pi(t/\tau)$ and $\tilde{h}(t)=\alpha e^{-\alpha t}u(t)$, we have as the complex envelope of the filter output

$$\begin{aligned} \tilde{y}(t) &= \Pi(t/\tau) * \alpha e^{-\alpha t}u(t) \\ &= \left[1-e^{-\alpha(t+\tau/2)}\right]u\left(t+\frac{\tau}{2}\right)-\left[1-e^{-(t-\tau/2)}\right]u\left(t-\frac{\tau}{2}\right) \end{aligned} \tag{2.333}$$

Multiplying this by $\frac{1}{2}e^{j2\pi f_0 t}$ and taking the real part results in the output of the filter in accordance with (2.329). The result is

$$y(t)=\frac{1}{2}\left[\left(1-e^{-\alpha(t+\tau/2)}\right)u(t+\tau/2)-\left(1-e^{-(t-\tau/2)}\right)u(t-\tau/2)\right]\cos(2\pi f_0 t) \tag{2.334}$$

To check this result, we convolve (2.331) and (2.332) directly. The superposition integral becomes

$$\begin{aligned} y(t) &= x(t) * h(t) \\ &= \int_{-\infty}^{\infty}\Pi(\lambda/\tau)\cos(2\pi f_0\lambda)\alpha e^{-\alpha(t-\lambda)}u(t-\lambda)\cos[2\pi f_0(t-\lambda)]\,d\lambda \end{aligned} \tag{2.335}$$

However,

$$\cos(2\pi f_0\lambda)\cos[2\pi f_0(t-\lambda)]=\frac{1}{2}\cos(2\pi f_0 t)+\frac{1}{2}\cos[2\pi f_0(t-2\lambda)] \tag{2.336}$$

so that the superposition integral becomes

$$\begin{aligned} y(t) &= \frac{1}{2}\int_{-\infty}^{\infty}\Pi(\lambda/\tau)\alpha e^{-\alpha(t-\lambda)}u(t-\lambda)\,d\lambda\cos(2\pi f_0 t) \\ &\quad +\frac{1}{2}\int_{-\infty}^{\infty}\Pi(\lambda/\tau)\alpha e^{-\alpha(t-\lambda)}u(t-\lambda)\cos[2\pi f_0(t-2\lambda)]\,d\lambda \end{aligned} \tag{2.337}$$

If $f_0^{-1} \ll \tau$ and $f_0^{-1} \ll \alpha^{-1}$, the second integral is approximately zero, so that we have only the first integral, which is $\Pi(t/\tau)$ convolved with $\alpha e^{-\alpha t}u(t)$ and the result multiplied by $\frac{1}{2}\cos(2\pi f_0 t)$, which is the same as (2.334).

■

■ 2.10 DISCRETE FOURIER TRANSFORM AND FAST FOURIER TRANSFORM

In order to compute the Fourier spectrum of a signal by means of a digital computer, the time-domain signal must be represented by sample values, and the spectrum must be computed at a discrete number of frequencies. It can be shown that the following sum gives an approximation to the Fourier spectrum of a signal at frequencies $k/(NT_s), k = 0, 1, \ldots, N-1$:

$$X_k = \sum_{n=0}^{N-1} x_n e^{-j2\pi nk/N}, \quad k = 0, 1, \ldots, N-1 \tag{2.338}$$

where $x_0, x_1, x_2, \ldots, x_{N-1}$ are N sample values of the signal taken at T_s-s intervals for which the Fourier spectrum is desired. The sum (2.338) is called the *discrete Fourier transform* (DFT) of the sequence $\{x_n\}$. According to the sampling theorem, if the samples are spaced by T_s, the spectrum repeats every $f_s = T_s^{-1}$ Hz. Since there are N frequency samples in this interval, it follows that the frequency resolution of (2.338) is $f_s/N = 1/(NT_s) \triangleq 1/T$. To obtain the sample sequence $\{x_n\}$ from the DFT sequence $\{X_k\}$, the sum

$$x_n = \frac{1}{N}\sum_{k=0}^{N-1} X_k e^{j2\pi nk/N}, \quad k = 0, 1, 2, \ldots, N-1 \tag{2.339}$$

is used. That (2.338) and (2.339) form a transform pair can be shown by substituting (2.338) into (2.339) and using the sum formula for a geometric series:

$$S_N \equiv \sum_{k=0}^{N-1} x^k = \begin{cases} \dfrac{1-x^N}{1-x}, & x \neq 1 \\ N, & x = 1 \end{cases} \tag{2.340}$$

As indicated above, the DFT and inverse DFT are approximations to the true Fourier spectrum of a signal $x(t)$ at the discrete set of frequencies $\{0, 1/T, 2/T, \ldots, (N-1)/T\}$. The error can be small if the DFT and its inverse are applied properly to a signal. To indicate the approximations involved, we must visualize the spectrum of a sampled signal that is truncated to a finite number of sample values and whose spectrum is then sampled at a discrete number N of points. To see the approximations involved, we use the following Fourier transform theorems:

1. The Fourier transform of an ideal sampling waveform (Example 2.14):

$$y_s(t) = \sum_{m=-\infty}^{\infty} \delta(t - mT_s) \longleftrightarrow f_s^{-1} \sum_{n=-\infty}^{\infty} \delta(f - nf_s), \quad f_s = T_s^{-1}$$

2. The Fourier transform of a rectangular window function:

$$\Pi(t/T) \longleftrightarrow T\,\mathrm{sinc}(fT)$$

3. The convolution theorem of Fourier transforms:

$$x_1(t) * x_2(t) \longleftrightarrow X_1(f)X_2(f)$$

4. The multiplication theorem of Fourier transforms:

$$x_1(t)x_2(t) \longleftrightarrow X_1(f) * X_2(f)$$

The approximations involved are illustrated by the following example.

EXAMPLE 2.32

An exponential signal is to be sampled, the samples truncated to a finite number, and the result represented by a finite number of samples of the Fourier spectrum of the sampled truncated signal. The continuous-time signal and its Fourier transform are

$$x(t) = e^{-|t|/\tau} \longleftrightarrow X(f) = \frac{2\tau}{1 + (2\pi f \tau)^2} \tag{2.341}$$

This signal and its spectrum are shown in Figure 2.32(a). However, we are representing the signal by sample values spaced by T_s s, which entails multiplying the original signal by the ideal sampling waveform $y_s(t)$, given by (2.131). The resulting spectrum of this sampled signal is the convolution of $X(f)$ with the Fourier transform of $y_s(t)$, given by (2.136), which is $Y_s(f) = f_s \sum_{n=-\infty}^{\infty} \delta(f - nf_s)$. The result of this convolution in the frequency domain is

$$X_s(f) = f_s \sum_{n=-\infty}^{\infty} \frac{2\tau}{1 + [2\pi\tau(f - f_s)]^2} \tag{2.342}$$

The resulting sampled signal and its spectrum are shown in Figure 2.32(b).

In calculating the DFT, only a T-s segment of $x(t)$ can be used (N samples spaced by $T_s = T/N$). This means that the sampled time-domain signal is effectively multiplied by a window function $\Pi(t/T)$. In the frequency domain, this corresponds to convolution with the Fourier transform of the rectangular window function, which is use $T\,\mathrm{sinc}(fT)$. The resulting windowed, sampled signal and its spectrum are sketched in Figure 2.32(c). Finally, the spectrum is available only at N discrete frequencies separated by the reciprocal of the window duration $1/T$. This corresponds to convolution in the time domain with a sequence of delta functions. The resulting signal and spectrum are shown in Figure 2.32(d). It can be seen that unless one is careful, there is indeed a considerable likelihood that the DFT spectrum will look nothing like the spectrum of the original continuous-time signal. Means for minimizing these errors are discussed in several references on the subject.[18]

A little thought will indicate that to compute the complete DFT spectrum of a signal, approximately N^2 complex multiplications are required in addition to a number of complex additions. It is possible to find algorithms that allow the computation of the DFT spectrum of a signal using only approximately $N \log_2 N$ complex multiplications, which gives significant computational savings for N large. Such algorithms are referred to as *fast Fourier transform* (FFT) algorithms. Two main types of FFT algorithms are those based on *decimation in time* (DIT) and those based on *decimation in frequency* (DIF).

Fortunately, FFT algorithms are included in most computer mathematics packages such as MATLAB, so we do not have to go to the trouble of writing our own FFT programs, although it is an instructive exercise to do so. The following computer example computes the FFT of a sampled double-sided exponential pulse and compares spectra of the continuous-time and sampled pulses.

---

[18]Ziemer et al. (1998), Chapter 10.

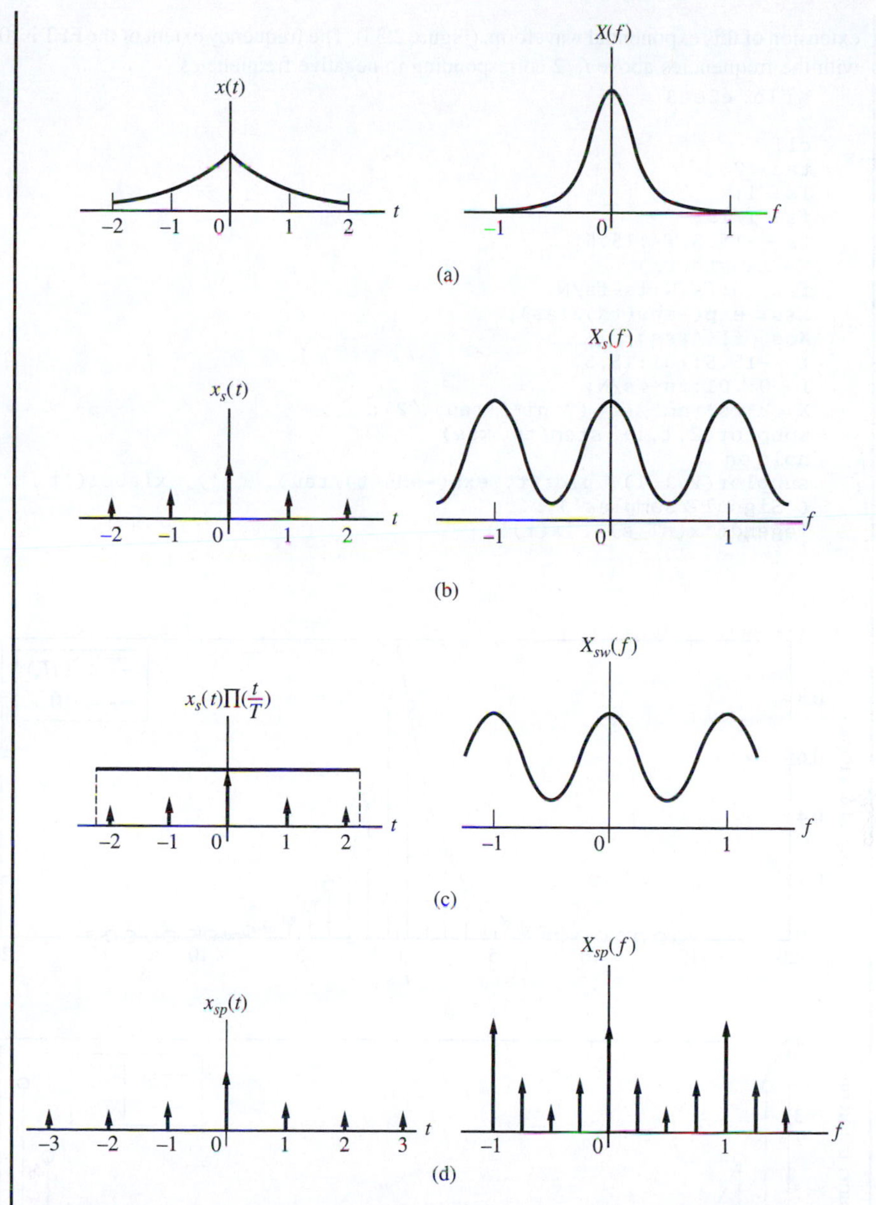

Figure 2.32
Signals and spectra illustrating the computation of the DFT. (a) Signal to be sampled and its spectrum ($\tau = 1$ s). (b) Sampled signal and its spectrum ($f_s = 1$ Hz). (c) Windowed, sampled signal and its spectrum $T \geq 4$ s). (d) Sampled signal spectrum and corresponding periodic repetition of the sampled, windowed signal. ■

COMPUTER EXAMPLE 2.3

The MATLAB program given below computes the fast Fourier transform (FFT) of a double-sided exponentially decaying signal truncated to $-15.5 \leq t \leq 15.5$ sampled each $T_s = 1$ s. The periodicity property of the FFT means that the resulting FFT coefficients correspond to a waveform that is the periodic extension of this exponential waveform. (Figure 2.33). The frequency extent of the FFT is $[0, f_s(1-1/N)]$ with the frequencies above $f_s/2$ corresponding to negative frequencies.

```
% file: c2ce3
%
clf
tau = 2;
Ts = 1;
fs = 1/Ts;
ts = -15.5:Ts:15.5;
N = length(ts);
fss = 0:fs/N:fs-fs/N;
xss = exp(-abs(ts)/tau);
Xss = fft(xss);
t = -15.5:.01:15.5;
f = 0:.01:fs-fs/N;
X = 2*fs*tau./(1+(2*pi*f*tau).^2);
subplot(2,1,1), stem(ts, xss)
hold on
subplot(2,1,1), plot(t, exp(-abs(t)/tau), '-'), xlabel('t, s'), ylabel
('Signal & samples'),...
legend('x(nT_s)', 'x(t)')
```

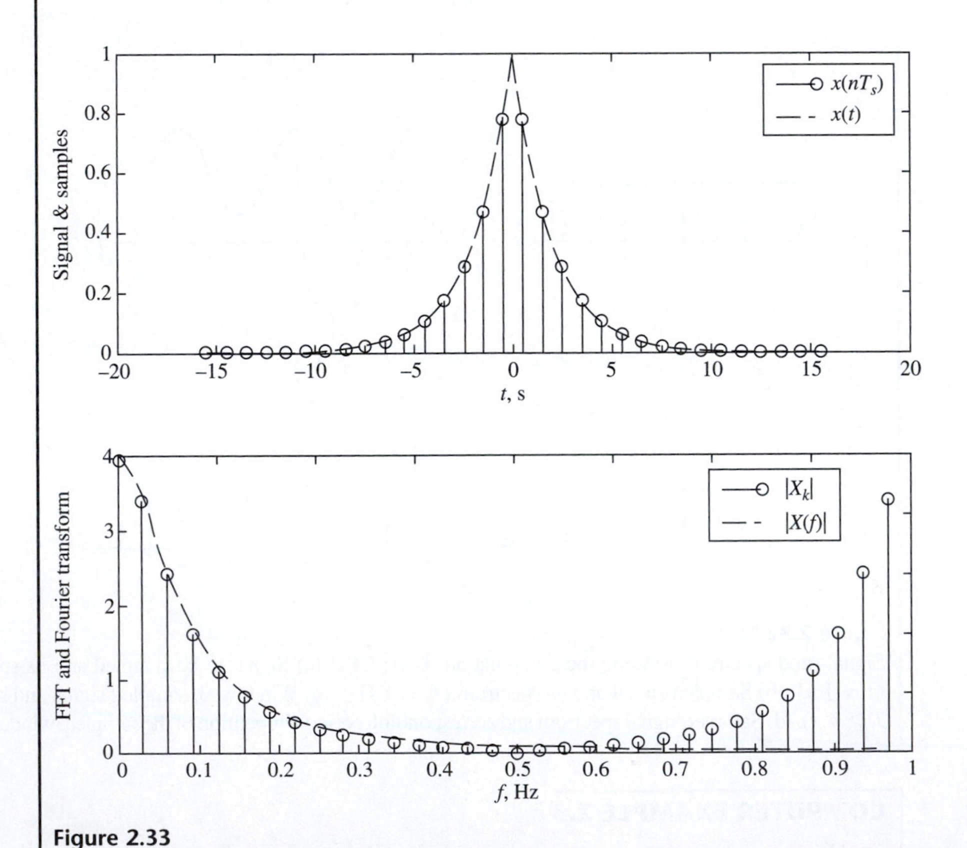

Figure 2.33
(a) $x(t) = \exp(-|t|/\tau)$ and samples taken each $T_s = 1$ s for $\tau = 2$ s. (b) Magnitude of the 32-point FFT of the sampled signal compared with the Fourier transform of $x(t)$. The spectral plots deviate from each other around $f_s/2$ due to aliasing.

```
subplot(2,1,2), stem(fss, abs(Xss))
hold on
subplot(2,1,2), plot(f, X, '-'), xlabel('f, Hz'), ylabel('FFT and Fourier
transform')
legend('|tX_k|'. '|X(f)|')
```

■

Summary

1. Two general classes of signals are deterministic and random. The former can be written as a completely known function of time, whereas the amplitudes of random signals must be described probabilistically.
2. A periodic signal of period T_0 is one for which $x(t) = x(t + T_0)$ for all t.
3. A single-sided spectrum for a rotating phasor $\tilde{x}(t) = Ae^{j(2\pi f_0 t + \theta)}$ shows A (amplitude) and θ (phase) versus f (frequency). The real, sinusoidal signal corresponding to this phasor is obtained by taking the real part of $\tilde{x}(t)$. A double-sided spectrum results if we think of forming $x(t) = \frac{1}{2}\tilde{x}(t) + \frac{1}{2}\tilde{x}^*(t)$. Graphs of amplitude and phase (two plots) of this rotating phasor sum versus f are known as two-sided amplitude and phase spectra, respectively. Such spectral plots are referred to as *frequency-domain representations* of the signal $A\cos(2\pi f_0 t + \theta)$.
4. The unit impulse function, $\delta(t)$, can be thought of as a zero-width, infinite-height pulse with unity area. The sifting property, $\int_{-\infty}^{\infty} x(\lambda)\delta(\lambda - t_0)\, d\lambda = x(t_0)$, where $x(t)$ is continuous at $t = t_0$, is a generalization of the defining relation for a unit impulse. The unit step function $u(t)$ is the integral of a unit impulse.
5. A signal $x(t)$ for which $E = \int_{-\infty}^{\infty} |x(t)|^2\, dt$ is finite is called an *energy signal.* If $x(t)$ is such that $P = \lim_{T \to \infty}(1/2T)\int_{-T}^{T} |x(t)|^2 dt$ is finite, the signal is known as a *power signal.* Example signals may be either or neither.
6. A set of orthogonal functions, $\phi_1(t), \phi_2(t), \ldots, \phi_N(t)$, can be used as a series approximation of the form

$$x_a(t) = \sum_{n=0}^{N} X_n \phi_n(t)$$

for a signal $x(t)$, which has finite energy in the interval $(t_0, t_0 + T)$. The ISE between $x_a(t)$ and $x(t)$ is minimized if the coefficients are chosen as

$$X_n = \frac{1}{c_n}\int_{t_0}^{t_0+T} x(t)\phi_n^*(t)\, dt$$

where

$$\int_{t_0}^{t_0+T_0} \phi_n(t)\phi_m^*(t)\, dt = c_n \delta_{nm}, \quad c_n = \text{real constant}$$

For a complete set of $\phi_n(t)$s, the ISE approaches zero as N approaches infinity, and Parseval's theorem then holds:

$$\int_{t_0}^{t_0+T} |x(t)|^2\, dt = \sum_{n=0}^{\infty} c_n |X_n|^2$$

7. If $\phi_n(t) = e^{jn\omega_0 t}$, $n = 0, \pm 1, \pm 2, \ldots$, where $\omega_0 = 2\pi/T_0$ and T_0 is the expansion interval, is used in an orthogonal function series, the result is the complex exponential Fourier series. If $x(t)$ is periodic with period T_0, the exponential Fourier series represents $x(t)$ exactly for all t, except at points of discontinuity.

8. For exponential Fourier series of real signals, the Fourier coefficients obey $X_n = X_{-n}^*$, which implies that $|X_n| = |X_{-n}|$ and $\underline{/X_n} = -\underline{/X_{-n}}$. Plots of $|X_n|$ and $\underline{/X_n}$ versus nf_0 are referred to as the *discrete, double-sided amplitude* and *phase spectra*, respectively, of $x(t)$. If $x(t)$ is real, the amplitude spectrum is even and the phase spectrum is odd as functions of nf_0.

9. Parseval's theorem for periodic signals is

$$\frac{1}{T_0}\int_{T_0} |x(t)|^2\, dt = \sum_{n=-\infty}^{\infty} |X_n|^2$$

10. The Fourier transform of a signal $x(t)$ is

$$X(f) = \int_{-\infty}^{\infty} x(t) e^{-j2\pi ft}\, dt$$

and the inverse Fourier transform is

$$x(t) = \int_{-\infty}^{\infty} X(f) e^{j2\pi ft}\, df$$

For real signals, $|X(f)| = |X(-f)|$ and $\underline{/X(f)} = -\underline{/X(-f)}$.

11. Plots of $|X(f)|$ and $\underline{/X(f)}$ versus f are referred to as the *double-sided amplitude* and *phase spectra*, respectively, of $x(t)$. As functions of frequency, the amplitude spectrum of a real signal is even and its phase spectrum is odd.

12. The energy of a signal is

$$\int_{-\infty}^{\infty} |x(t)|^2\, dt = \int_{-\infty}^{\infty} |X(f)|^2 df$$

This is known as *Rayleigh's energy theorem*. The energy spectral density of a signal is $G(f) = |X(f)|^2$. It is the density of energy with frequency of the signal.

13. The convolution of two signals, $x_1(t)$ and $x_2(t)$, is

$$x(t) = x_1 * x_2 = \int_{-\infty}^{\infty} x_1(\lambda) x_2(t-\lambda)\, d\lambda = \int_{-\infty}^{\infty} x_1(t-\lambda) x_2(\lambda)\, d\lambda$$

The convolution theorem of Fourier transforms states that $X(f) = X_1(f)X_2(f)$, where $X(f)$, $X_1(f)$, and $X_2(f)$ are the Fourier transforms of $x(t)$, $x_1(t)$, and $x_2(t)$, respectively.

14. The Fourier transform of a periodic signal can be obtained formally by Fourier transforming its exponential Fourier series term by term using $Ae^{j2\pi f_0 t} \longleftrightarrow A\delta(f - f_0)$, even though, mathematically speaking, Fourier transforms of power signals do not exist.

15. The power spectrum $S(f)$ of a power signal $x(t)$ is a real, even, nonnegative function that integrates to give total average power: $\langle x^2(t)\rangle = \int_{-\infty}^{\infty} S(f)\,df$, where $\langle w(t)\rangle \triangleq \lim_{T\to\infty}(1/2T)\int_{-T}^{T} w(t)dt$. The time-average autocorrelation function of a power signal is defined as $R(\tau) = \langle x(t)x(t+\tau)\rangle$. The Wiener–Khinchine theorem states that $S(f)$ and $R(\tau)$ are Fourier transform pairs.

16. A linear system, denoted operationally as $\mathcal{H}()$, is one for which superposition holds; that is, if $y_1 = \mathcal{H}(x_1)$ and $y_2 = \mathcal{H}(x_2)$, then $\mathcal{H}(\alpha_1 x_1 + \alpha_2 x_2) = \alpha_1 y_1 + \alpha_2 y_2$, where x_1 and x_2 are inputs and y_1 and y_2 are outputs (the time variable t is suppressed for simplicity). A system is fixed, or time invariant, if, given $y(t) = \mathcal{H}[x(t)]$, the input $x(t-t_0)$ results in the output $y(t-t_0)$.

17. The impulse response $h(t)$ of a linear, time-invariant (LTI) system is its response to an impulse applied at $t = 0$: $h(t) = \mathcal{H}[\delta(t)]$. The output of an LTI system to an input $x(t)$ is given by $y(t) = h(t) * x(t) = \int_{-\infty}^{\infty} h(\tau)x(t-\tau)d\tau$.

18. A *causal system* is one which does not anticipate its input. For such an LTI system, $h(t) = 0$ for $t < 0$. A *stable system* is one for which every bounded input results in a bounded output. An LTI system is stable if and only if $\int_{-\infty}^{\infty} |h(t)|dt < \infty$.

19. The frequency-response function $H(f)$ of an LTI system is the Fourier transform of $h(t)$. The Fourier transform of the system output $y(t)$ due to an input $x(t)$ is $Y(f) = H(f)X(f)$, where $X(f)$ is the Fourier transform of the input. $|H(f)| = |H(-f)|$ is called the *amplitude response* of the system, and $\underline{/H(f)} = -\underline{/H(-f)}$ is called the *phase response*.

20. For a fixed linear system with a periodic input, the Fourier coefficients of the output are given by $Y_n = H(nf_0)X_n$, where X_n represents the Fourier coefficients of the input.

21. Input and output spectral densities for a fixed linear system are related by

$$G_y(f) = |H(f)|^2 G_x(f) \quad \text{(energy signals)}$$
$$S_y(f) = |H(f)|^2 S_x(f) \quad \text{(power signals)}$$

22. A system is distortionless if its output looks like its input except for a time delay and amplitude scaling: $y(t) = H_0 x(t-t_0)$. The frequency response function of a distortionless system is $H(f) = H_0 e^{-j2\pi f t_0}$. Such a system's amplitude response is $|H(f)| = H_0$ and its phase response is $\underline{/H(f)} = -2\pi t_0 f$ over the band of frequencies occupied by the input. Three types of distortion that a system may introduce are amplitude, phase (or delay), and nonlinear, depending on whether $|H(f)| \neq$ constant, $\underline{/H(f)} \neq$ $-$ constant $\times f$, or the system is nonlinear, respectively. Two other important

properties of a linear system are the group and phase delays. These are defined by

$$T_g(f) = -\frac{1}{2\pi}\frac{d\theta(f)}{df} \quad \text{and} \quad T_p(f) = -\frac{\theta(f)}{2\pi f}$$

respectively, in which $\theta(f)$ is the phase response of the LTI system. Phase distortionless systems have equal group and phase delays (constant).

23. Ideal filters are convenient in communication system analysis, even though they are unrealizable. Three types of ideal filters are lowpass, bandpass, and highpass. Throughout their passband, ideal filters have constant amplitude response and linear phase response. Outside their passbands, ideal filters perfectly reject all spectral components of the input. In the stop band the phase response is arbitrary.

24. Approximations to ideal filters are Butterworth, Chebyshev, and Bessel filters. The first two are attempts at approximating the amplitude response of an ideal filter, and the latter is an attempt to approximate the linear phase response of an ideal filter.

25. An inequality relating the duration T of a pulse and its single-sided bandwidth W is $W \geq 1/2T$. Pulse risetime T_R and signal bandwidth are related approximately by $W = 1/2T_R$. These relationships hold for the lowpass case. For bandpass filters and signals, the required bandwidth is doubled, and the risetime is that of the envelope of the signal.

26. The sampling theorem for lowpass signals of bandwidth W states that a signal can be perfectly recovered by lowpass filtering from sample values taken at a rate of $f_s > 2W$ samples per second. The spectrum of an impulse-sampled signal is

$$X_\delta(f) = f_s \sum_{n=-\infty}^{\infty} X(f - nf_s)$$

where $X(f)$ is the spectrum of the original signal. For bandpass signals, lower sampling rates than specified by the lowpass sampling theorem may be possible.

27. The Hilbert transform $\widehat{x}(t)$ of a signal $x(t)$ corresponds to a $-90°$ phase shift of all the signal's positive-frequency components. Mathematically,

$$\widehat{x}(t) = \int_{-\infty}^{\infty} \frac{x(\lambda)}{\pi(t-\lambda)}\, d\lambda$$

In the frequency domain, $\widehat{X}(f) = -j(\text{sgn} f)X(f)$, where sgn f is the signum function, $X(f) = \Im[x(t)]$, and $\hat{X}(f) = \Im[\widehat{x}(t)]$. The Hilbert transform of $\cos(\omega_0 t)$ is $\sin(\omega_0 t)$, and the Hilbert transform of $\sin(\omega_0 t)$ is $-\cos(\omega_0 t)$. The power (or energy) in a signal and its Hilbert transform are equal. A signal and its Hilbert transform are orthogonal in the range $(-\infty, \infty)$. If $m(t)$ is a lowpass signal and $c(t)$ is a highpass signal with nonoverlapping spectra,

$$\widehat{m(t)c(t)} = m(t)\widehat{c}(t)$$

The Hilbert transform can be used to define the analytic signal

$$z(t) = x(t) \pm j\widehat{x}(t)$$

The magnitude of the analytic signal, $|z(t)|$, is the real envelope of the signal. The Fourier transform of an analytic signal, $Z(f)$, is identically zero for $f<0$ or $f>0$, respectively, depending on whether the $+$ sign or $-$ sign is chosen for the imaginary part of $z(t)$.

28. The complex envelope $\tilde{x}(t)$ of a bandpass signal is defined by

$$x(t) + j\widehat{x}(t) = \tilde{x}(t)e^{j2\pi f_0 t}$$

where f_0 is the reference frequency for the signal. Similarly, the complex envelope $\tilde{h}(t)$ of the impulse response of a bandpass system is defined by

$$h(t) + j\widehat{h}(t) = \tilde{h}(t)e^{j2\pi f_0 t}$$

The complex envelope of the bandpass system output is conveniently obtained in terms of the complex envelope of the output which can be found from either of the operations

$$\tilde{y}(t) = \tilde{h}(t) * \tilde{x}(t)$$

or

$$\tilde{y}(t) = \Im^{-1}\left[\tilde{H}(f)\tilde{X}(f)\right]$$

where $\tilde{H}(f)$ and $\tilde{X}(f)$ are the Fourier transforms of $\tilde{h}(t)$ and $\tilde{x}(t)$, respectively. The actual (real) output is then given by

$$y(t) = \frac{1}{2}\mathrm{Re}\left[\tilde{y}(t)e^{j2\pi f_0 t}\right]$$

29. The DFT of a signal sequence $\{x_n\}$ is defined as

$$X_k = \sum_{n=0}^{N-1} x_n e^{j2\pi nk/N} = \mathrm{DFT}\left[\{x_n\}\right], \quad k = 0, 1, \ldots, N-1$$

and the inverse DFT can be found from

$$x_n = \frac{1}{N}\mathrm{DFT}\left[\{X_k^*\}\right]^*, \quad k = 0, 1, \ldots, N-1$$

The DFT can be used to digitally compute spectra of sampled signals and to approximate operations carried out by the normal Fourier transform, for example, filtering.

Further Reading

Bracewell (1986) is a text concerned exclusively with Fourier theory and applications. Ziemer et al. (1998) and Kamen and Heck (2007) are devoted to continuous and discrete signal and system theory and provide background for this chapter.

Problems

Section 2.1

2.1. Sketch the single-sided and double-sided amplitude and phase spectra of the following signals:

a. $x_a(t) = 10\cos(4\pi t + \pi/8) + 6\sin(8\pi t + 3\pi/4)$.

b. $x_b(t) = 8\cos(2\pi t + \pi/3) + 4\cos(6\pi t + \pi/4)$.

c. $x_c(t) = 2\sin(4\pi t + \pi/8) + 12\sin(10\pi t)$.

2.2. A signal has the double-sided amplitude and phase spectra shown in Figure 2.34. Write a time-domain expression for the signal.

2.3. The sum of two or more sinusoids may or may not be periodic depending on the relationship of their separate frequencies. For the sum of two sinusoids, let the frequencies of the individual terms be f_1 and f_2, respectively. For the sum to be periodic, f_1 and f_2 must be commensurable; i.e., there must be a number f_0 contained in each an integral number of times. Thus, if f_0 is the largest such number,

$$f_1 = n_1 f_0 \quad \text{and} \quad f_2 = n_2 f_0$$

where n_1 and n_2 are integers; f_0 is the fundamental frequency. Which of the signals given below are periodic? Find the periods of those that are periodic.

a. $x_1(t) = 2\cos(2t) + 4\sin(6\pi t)$.

b. $x_2(t) = \cos(6\pi t) + 7\cos(30\pi t)$.

c. $x_3(t) = \cos(4\pi t) + 9\sin(21\pi t)$.

d. $x_4(t) = 2\cos(4\pi t) + 5\cos(6\pi t) + 6\sin(17\pi t)$.

2.4. Sketch the single-sided and double-sided amplitude and phase spectra of

a. $x_a(t) = 5\cos(12\pi t - \pi/6)$.

b. $x_b(t) = 3\sin(12\pi t) + 4\cos(16\pi t)$.

c. $x_c(t) = 4\cos(8\pi t)\cos(12\pi t)$.

(*Hint:* use an appropriate trigonometric identity to write as the sum of cosines.)

d. $x_d(t) = 8\sin(2\pi t)\cos^2(5\pi t)$.

(*Hint:* use appropriate trigonometric identities.)

2.5.

a. Show that the function $\delta_\epsilon(t)$ sketched in Figure 2.4 (b) has unity area.

b. Show that

$$\delta_\epsilon(t) = \epsilon^{-1} e^{-t/\epsilon} u(t)$$

has unity area. Sketch this function for $\epsilon = 1, \frac{1}{2}$, and $\frac{1}{4}$. Comment on its suitability as an approximation for the unit impulse function.

c. Show that a suitable approximation for the unit impulse function as $\epsilon \to 0$ is given by

$$\delta_\epsilon(t) = \begin{cases} \epsilon^{-1}\left(1 - \frac{|t|}{\epsilon}\right), & |t| \le \epsilon \\ 0, & \text{otherwise} \end{cases}$$

2.6. Use the properties of the unit impulse function given after (2.14) to evaluate the following relations.

a. $\int_{-\infty}^{\infty} [t^2 + \sin(2\pi t)]\delta(2t - 5)\,dt$.

b. $\int_{-10^-}^{10+} (t^2 + 1)\left[\sum_{n=-\infty}^{\infty} \delta(t - 5n)\right] dt$.

(*Note:* 10^+ means just to the right of 10; -10^- means just to the left of -10.)

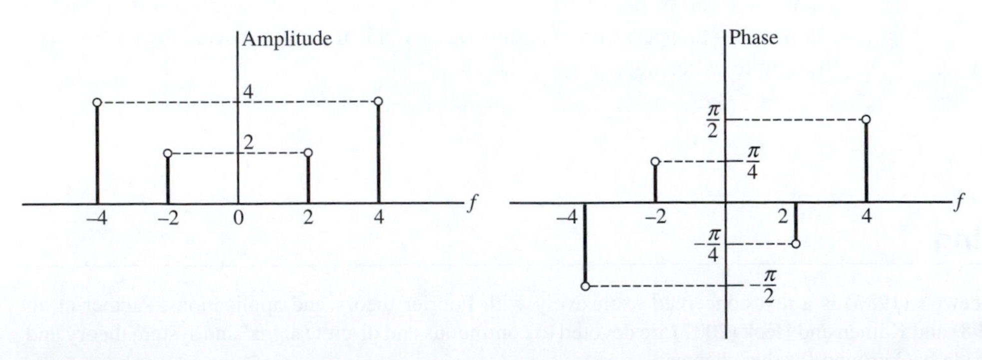

Figure 2.34

c. $10\delta(t) + A\, d\delta(t)/dt = B\delta(t) + 5\, d\delta(t)/dt$; find A and B.

d. $\int_2^{11} [e^{-4\pi t} + \tan(10\pi t)]\delta(3t+6)\, dt$.

e. $\int_{-\infty}^{\infty} [\cos(8\pi t) + e^{-2t}][d^2\delta(t-2)/dt^2]\, dt$.

2.7. Which of the following signals are periodic and which are aperiodic? Find the periods of those which are periodic. Sketch all signals.

a. $x_1(t) = \cos(5\pi t) + \sin(7\pi t)$.

b. $x_2(t) = \sum_{n=0}^{\infty} \Lambda(t-2n)$.

c. $x_3(t) = \sum_{n=-\infty}^{\infty} \Lambda(t-2n)$.

d. $x_4(t) = \sin(3t) + \cos(2\pi t)$.

e. $x_5(t) = \sum_{n=-\infty}^{\infty} \Pi(t-3n)$.

f. $x_6(t) = \sum_{n=0}^{\infty} \Pi(t-3n)$.

2.8. Write the signal $x(t) = \sin(6\pi t) + 2\cos(10\pi t)$ as

a. The real part of a sum of rotating phasors.

b. A sum of rotating phasors plus their complex conjugates.

c. From your results in parts (a) and (b), sketch the single-sided and double-sided amplitude and phase spectra of $x(t)$.

Section 2.2

2.9. Find the normalized power for each signal below that is a power signal and the normalized energy for each signal that is an energy signal. If a signal is neither a power signal nor an energy signal, so designate it. Sketch each signal (α is a positive constant).

a. $x_1(t) = 2\cos(4\pi t + 2\pi/3)$.

b. $x_2(t) = e^{-\alpha t}u(t)$.

c. $x_3(t) = e^{\alpha t}u(-t)$.

d. $x_4(t) = (\alpha^2 + t^2)^{-1/2}$.

e. $x_5(t) = e^{-\alpha|t|}$.

f. $x_6 = e^{-\alpha t}u(t) - e^{-\alpha(t-1)}u(t-1)$.

2.10. Classify each of the following signals as an energy signal or a power signal by calculating the energy E or the power P (A, θ, ω, and τ are positive constants).

a. $A|\sin(\omega t + \theta)|$.

b. $A\tau/\sqrt{\tau + jt}$, $j = \sqrt{-1}$.

c. $Ate^{-t/\tau}u(t)$.

d. $\Pi(t/\tau) + \Pi(t/2\tau)$.

2.11. Sketch each of the following periodic waveforms and compute their average powers.

a. $x_1(t) = \sum_{n=-\infty}^{\infty} \Pi[(t-6n)/3]$.

b. $x_2(t) = \sum_{n=-\infty}^{\infty} \Lambda[(t-5n)/2]$.

c. $x_3(t) = \sum_{n=-\infty}^{\infty} \Lambda[(t-3n)/2]u(t-3n)$.

d. $x_4(t) = 2\sin(5\pi t)\cos(5\pi t)$.

(*Hint:* use an appropriate trigonometric identy to simplify.

2.12. For each of the following signals, determine both the normalized energy and power. (Note: 0 and ∞ are possible answers.)

a. $x_1(t) = 6e^{(-3+j4\pi)t}u(t)$.

b. $x_2(t) = \Pi[(t-3)/2] + \Pi[(t-3)/6]$.

c. $x_3(t) = 7e^{j6\pi t}u(t)$.

d. $x_4(t) = 2\cos(4\pi t)$.

2.13. Show that the following are energy signals. Sketch each signal

a. $x_1(t) = \Pi(t/12)\cos(6\pi t)$

b. $x_2(t) = e^{-|t|/3}$

c. $x_3(t) = 2u(t) - 2u(t-8)$

d. $x_4(t) = \int_{-\infty}^{t} u(\lambda)\, d\lambda - 2\int_{-\infty}^{t-10} u(\lambda)\, d\lambda + \int_{-\infty}^{t-20} u(\lambda)\, d\lambda$

(*Hint:* Consider the integral of a step function.)

Section 2.3

2.14.

a. Fill in the steps for obtaining (2.33) from (2.32).

b. Obtain (2.34) from (2.33).

c. Given the set of orthogonal functions

$$\phi_n(t) = \Pi\left(\frac{4[t-(2n-1)T/8]}{T}\right), \quad n = 1, 2, 3, 4$$

sketch and dimension accurately these functions.

d. Approximate the ramp signal

$$x(t) = \frac{t}{T}\Pi\left(\frac{t-T/2}{T}\right)$$

by a generalized Fourier series using this set.

e. Do the same for the set

$$\phi_n(t) = \Pi\left\{\frac{2[t-(2n-1)T/4]}{T}\right\}, \quad n = 1, 2$$

f. Compute the integral-squared error for both part (b) and part (c). What do you conclude about the dependence of ϵ_N on N?

Section 2.4

2.15. Using the uniqueness property of the Fourier series, find exponential Fourier series for the following signals (f_0 is an arbitrary frequency):

a. $x_1(t) = \sin^2(2\pi f_0 t)$.

b. $x_2(t) = \cos(2\pi f_0 t) + \sin(4\pi f_0 t)$.

c. $x_3(t) = \sin(4\pi f_0 t)\cos(4\pi f_0 t)$.

d. $x_4(t) = \cos^3(2\pi f_0 t)$.

Hint: Use appropriate trigonometric identities and Euler's theorem.

2.16. Expand the signal $x(t) = 2t^2$ in a complex exponential Fourier series over the interval $|t| \le 2$. Sketch the signal to which the Fourier series converges for all t.

2.17. If $X_n = |X_n| \exp[j\underline{/X_n}]$ are the Fourier coefficients of a real signal, $x(t)$, fill in all the steps to show that:

a. $|X_n| = |X_{-n}|$ and $\underline{/X_n} = -\underline{/X_{-n}}$.

b. X_n is a real, even function of n for $x(t)$ even.

c. X_n is imaginary and an odd function of n for $x(t)$ odd.

d. $x(t) = -x(t + T_0/2)$ (half wave odd symmetry) implies that $X_n = 0$, n even.

2.18. Obtain the complex exponential Fourier series coefficients for the (a) pulse train, (b) half-rectified sinewave, (c) full-rectified sine wave, and (d) triangular waveform as given in Table 2.1.

2.19. Find the ratio of the power contained in a pulse train for $|nf_0| \le \tau^{-1}$ to the total power for each of the following cases:

a. $\tau/T_0 = \frac{1}{2}$.

b. $\tau/T_0 = \frac{1}{5}$

c. $\tau/T_0 = \frac{1}{10}$.

d. $\tau/T_0 = \frac{1}{20}$.

Hint: You can save work by noting the spectra are even about $f = 0$.

2.20.

a. If $x(t)$ has the Fourier series

$$x(t) = \sum_{n=-\infty}^{\infty} X_n e^{j2\pi n f_0 t}$$

and $y(t) = x(t - t_0)$, show that

$$Y_n = X_n e^{-j2\pi n f_0 t_0}$$

where the Y_n are the Fourier coefficients for $y(t)$.

b. Verify the theorem proved in part (a) by examining the Fourier coefficients for $x(t) = \cos(\omega_0 t)$ and $y(t) = \sin(\omega_0 t)$.

Hint: What delay, t_0, will convert a cosine into a sine. Use the uniqueness property to write down the corresponding Fourier series.

2.21. Use the Fourier series expansions of periodic square wave and triangular wave signals to find the sum of the following series:

a. $1 - \frac{1}{3} + \frac{1}{5} - \frac{1}{7} + \cdots$.

b. $1 + \frac{1}{9} + \frac{1}{25} + \frac{1}{49} + \cdots$.

Hint: Write down the Fourier series in each case and evaluate it for a particular, appropriately chosen value of t.

2.22. Using the results given in Table 2.1 for the Fourier coefficients of a pulse train, plot the double-sided amplitude and phase spectra for the waveforms shown in Figure 2.35.
Hint: Note that $x_b(t) = -x_a(t) + A$. How is a sign change and DC level shift manifested in the spectrum of the waveform?

2.23.

a. Plot the single-sided and double-sided amplitude and phase spectra of the square wave shown in Figure 2.36(a).

b. Obtain an expression relating the complex exponential Fourier series coefficients of the triangular

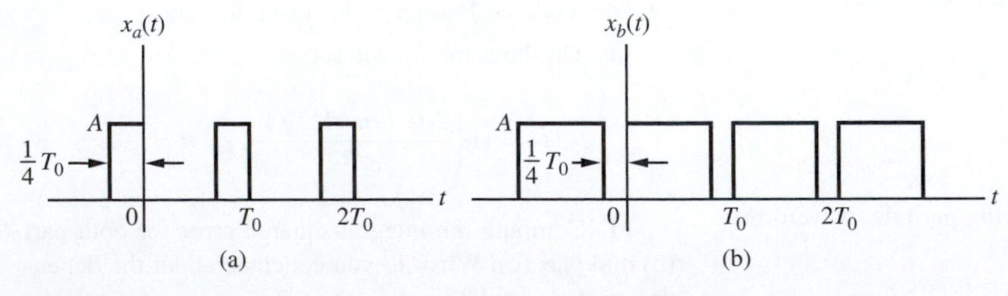

Figure 2.35

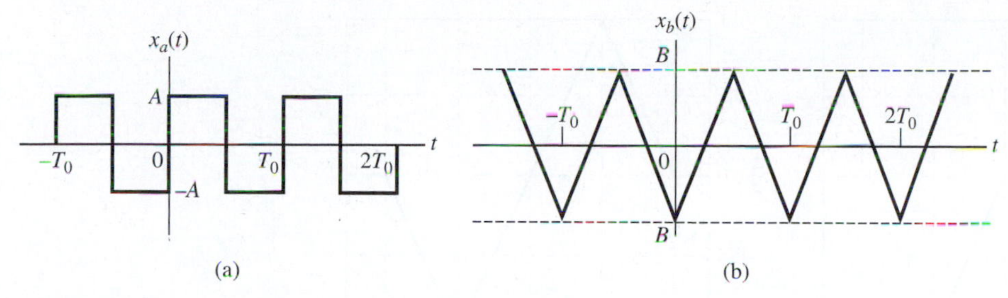

(a)

(b)

Figure 2.36

waveform shown in Figure 2.36(b) and those of $x_a(t)$ shown in Figure 2.35(a).

Hint: Note that $x_a(t) = K[dx_b(t)/dt]$, where K is an appropriate scale change.

c. Plot the double-sided amplitude and phase spectra for $x_b(t)$.

Section 2.5

2.24. Sketch each signal given below and find its Fourier transform. Plot the amplitude and phase spectra of each signal (A and τ are positive constants).

a. $x_1(t) = A\exp(-t/\tau)u(t)$.

b. $x_2(t) = A\exp(t/\tau)u(-t)$.

c. $x_3(t) = x_1(t) - x_2(t)$.

d. $x_4(t) = x_1(t) + x_2(t)$. Does it check with the answer found using Fourier transform tables?

2.25.

a. Use the Fourier transform of

$$x(t) = \exp(-\alpha t)u(t) - \exp(\alpha t)u(-t)$$

where $\alpha > 0$ to find the Fourier transform of the signum function defined as

$$\text{sgn}\, t = \begin{cases} 1, & t > 0 \\ -1, & t < 0 \end{cases}$$

(*Hint:* Take the limit as $\alpha \to 0$ of the Fourier transform found.)

b. Use the result above and the relation $u(t) = \frac{1}{2}[\text{sgn}\, t + 1]$ to find the Fourier transform of the unit step.

c. Use the integration theorem and the Fourier transform of the unit impulse function to find the Fourier transform of the unit step. Compare the result with part (b).

2.26. Using only the Fourier transform of the unit impulse function and the differentiation theorem, find the Fourier transforms of the signals shown in Figure 2.37.

2.27.

a. Write the signals of Figure 2.37 as the linear combination of two delayed triangular functions. That is, write $x_a(t) = a_1\Lambda((t-t_1)/T_1) + a_2\Lambda((t-t_2)/T_2)$ by finding appropriate values for a_1, a_2, t_1, t_2, T_1, and T_2. Do similar expressions for all four signals shown in Figure 2.37.

b. Given the Fourier transform pair $\Lambda(t) \longleftrightarrow \text{sinc}^2 f$, find their Fourier transforms using the superposition, scale change, and time delay theorems. Compare your results with the answers obtained in Problem 2.26.

2.28.

a. Given $\Pi(t) \longleftrightarrow \text{sinc}\, f$, find the Fourier transforms of the following signals using the frequency translation followed by the time delay theorem.

i. $x_1(t) = \Pi(t-1)\exp[j4\pi(t-1)]$.

ii. $x_2(t) = \Pi(t+1)\exp[j4\pi(t+1)]$.

b. Repeat the above, but now applying the time delay followed by the frequency translation theorem.

2.29. By applying appropriate theorems and using the signals defined in Problem 2.28, find Fourier transforms of the following signals:

a. $x_a(t) = \frac{1}{2}x_1(t) + \frac{1}{2}x_1(-t)$.

b. $x_b(t) = \frac{1}{2}x_2(t) + \frac{1}{2}x_2(-t)$.

2.30. Use the scale change and time delay theorems along with the transform pairs $\Pi(t) \longleftrightarrow \text{sinc}\, f$, $\text{sinc}\, t \longleftrightarrow \Pi(f)$, $\Lambda(t) \longleftrightarrow \text{sinc}^2 f$, and $\text{sinc}^2 t \longleftrightarrow \Lambda(f)$ to find Fourier transforms of the following:

a. $x_a(t) = \Pi[(t-1)/2]$.

b. $x_b(t) = 2\,\text{sinc}[2(t-1)]$.

c. $x_c(t) = \Lambda[(t-2)/8]$.

d. $x_d(t) = \text{sinc}^2[(t-3)/4]$.

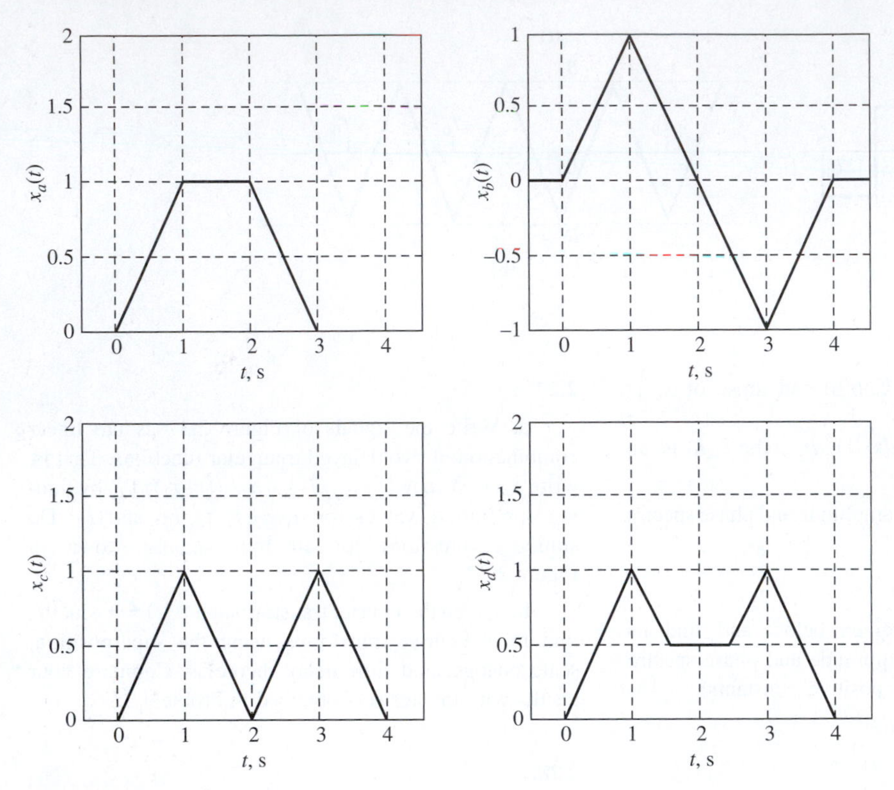

Figure 2.37

2.31. Without actually computing them, but using appropriate sketches, tell if the Fourier transforms of the signals given below are real, imaginary, or neither; even, odd, or neither. Give your reasoning in each case.

a. $x_1(t) = \Pi(t+1/2) - \Pi(t-1/2)$.

b. $x_2(t) = \Pi(t/2) + \Pi(t)$.

c. $x_3(t) = \sin(2\pi t)\Pi(t)$.

d. $x_4(t) = \sin(2\pi t + \pi/4)\Pi(t)$.

e. $x_5(t) = \cos(2\pi t)\Pi(t)$.

f. $x_6(t) = 1/[1+(t/5)^4]$.

2.32. Using the sifting property of the delta function, find the Fourier transforms of the signals given below. Discuss how any symmetry properties a given signal may have affect its Fourier transform in terms of being real or purely imaginary.

a. $x_1(t) = \delta(t+4) + 3\delta(t) + \delta(t-4)$.

b. $x_2(t) = 2\delta(t+8) - 2\delta(t-8)$.

c. $x_3(t) = \sum_{n=0}^{4} (n^2+1)\delta(t-2n)$.

(*Hint:* Write out the terms for this signal.)

d. $x_4(t) = \sum_{n=-2}^{2} n^2\delta(t-2n)$

(*Hint:* Write out the terms for this signal.)

2.33. Find and plot the energy spectral densities of the following signals. Dimension your plots fully. Use appropriate Fourier transform pairs and theorems.

a. $x_1(t) = 2e^{-3|t|}$.

b. $x_2(t) = 20\,\mathrm{sinc}(30t)$.

c. $x_3(t) = 4\Pi(5t)$.

d. $x_4(t) = 4\Pi(5t)\cos(40\pi t)$.

2.34. Evaluate the following integrals using Rayleigh's energy theorem (Parseval's theorem for Fourier transforms).

a. $I_1 = \int_{-\infty}^{\infty} \frac{df}{[\alpha^2+(2\pi f)^2]}$.

(*Hint:* Consider the Fourier transform of $\exp(-\alpha t)u(t)$).

b. $I_2 = \int_{-\infty}^{\infty} \mathrm{sinc}^2(\tau f)\,df$.

c. $I_3 = \int_{-\infty}^{\infty} \frac{df}{[\alpha^2+(2\pi f)^2]^2}$.

d. $I_4 = \int_{-\infty}^{\infty} \mathrm{sinc}^4(\tau f)\,df$.

2.35. Obtain and sketch the convolutions of the following signals.

a. $y_1(t) = e^{-\alpha t}u(t) * \Pi(t-\tau)$,
α and τ positive constants.

b. $y_2(t) = [\Pi(t/2) + \Pi(t)] * \Pi(t)$.

c. $y_3(t) = e^{-\alpha|t|} * \Pi(t), \quad \alpha > 0$.

d. $y_4(t) = x(t) * u(t)$, where $x(t)$ is any energy signal [you will have to assume a particular form for $x(t)$ to sketch this one, but obtain the general result before doing so].

2.36. Obtain the Fourier transforms of the signals $y_1(t)$, $y_2(t)$, and $y_3(t)$ in Problem 2.35 using the convolution theorem of Fourier transforms.

2.37. Given the following signals, suppose that all energy spectral components outside the bandwidth $|f| \leq W$ are removed by an ideal filter, while all energy spectral components within this bandwidth are kept. Find the ratio of output energy to total energy in each case. (α, β, and τ are positive constants.)

a. $x_1(t) = e^{-\alpha t}u(t)$.

b. $x_2(t) = \Pi(t/\tau)$ (requires numerical integration).

c. $x_3(t) = e^{-\alpha t}u(t) - e^{-\beta t}u(t)$ ($\beta = 2\alpha$).

2.38.

a. Find the Fourier transform of the cosine pulse

$$x(t) = A\Pi\left(\frac{2t}{T_0}\right)\cos(\omega_0 t)$$

where $\omega_0 = \frac{2\pi}{T_0}$. Express your answer in terms of a sum of sinc functions. Provide MATLAB plots of $x(t)$ and $X(f)$ [note that $X(f)$ is real].

b. Obtain the Fourier transform of the raised cosine pulse

$$y(t) = \frac{1}{2}A\Pi\left(\frac{2t}{T_0}\right)[1 + \cos(2\omega_0 t)]$$

Provide MATLAB plots of $y(t)$ and $Y(f)$ [note that $Y(f)$ is real]. Compare with part (a).

c. Use (2.151) with the result of part (a) to find the Fourier transform of the half-rectified cosine wave.

2.39. Provide plots of the following functions of time and find their Fourier transforms. Tell which Fourier transforms should be real and even functions of f and which ones should be imaginary and odd functions of f. Do your results bear this out?

a. $x_1(t) = \Lambda\left(\frac{t}{2}\right) + \Pi\left(\frac{t}{2}\right)$.

b. $x_2(t) = \Pi(t/2) - \Lambda(t)$.

c. $x_3(t) = \Pi\left(t + \frac{1}{2}\right) - \Pi\left(t - \frac{1}{2}\right)$.

d. $x_4(t) = \Lambda(t-1) - \Lambda(t+1)$.

Section 2.6

2.40.

a. Obtain the time-average autocorrelation function of $x(t) = 3 + 6\cos(20\pi t) + 3\sin(20\pi t)$.
(*Hint:* Combine the cosine and sine terms into a single cosine with a phase angle.)

b. Obtain the power spectral density of the signal of part (a). What is its total average power?

2.41. Find the power spectral densities and average powers of the following signals.

a. $x_1(t) = 2\cos(20\pi t + \pi/3)$.

b. $x_2(t) = 3\sin(30\pi t)$.

c. $x_3(t) = 5\sin(10\pi t - \pi/6)$.

d. $x_4(t) = 3\sin(30\pi t) + 5\sin(10\pi t - \pi/6)$.

2.42. Find the autocorrelation functions of the signals having the following power spectral densities. Also give their average powers.

a. $S_1(f) = 4\delta(f-15) + 4\delta(f+15)$.

b. $S_2(f) = 9\delta(f-20) + 9\delta(f+20)$.

c. $S_3(f) = 16\delta(f-5) + 16\delta(f+5)$.

d. $S_4(f) = 9\delta(f-20) + 9\delta(f+20) + 16\delta(f-5) + 16\delta(f+5)$.

2.43. By applying the properties of the autocorrelation function, determine whether the following are acceptable for autocorrelation functions. In each case, tell why or why not.

a. $R_1(\tau) = 2\cos(10\pi\tau) + \cos(30\pi\tau)$.

b. $R_2(\tau) = 1 + 3\cos(30\pi\tau)$.

c. $R_3(\tau) = 3\cos(20\pi\tau + \pi/3)$.

d. $R_4(\tau) = 4\Lambda(\tau/2)$.

e. $R_5(\tau) = 3\Pi(\tau/6)$.

f. $R_6(\tau) = 2\sin(10\pi\tau)$.

2.44. Find the autocorrelation functions corresponding to the following signals:

a. $x_1(t) = 2\cos(10\pi t + \pi/3)$.

b. $x_2(t) = 2\sin(10\pi t + \pi/3)$.

c. $x_3(t) = \text{Re}(3\exp(j10\pi t) + 4j\exp(j10\pi t))$.

d. $x_4(t) = x_1(t) + x_2(t)$.

2.45. Show that the $R(\tau)$ of Example 2.20 has the Fourier transform given there. Plot the power spectral density.

Section 2.7

2.46. A system is governed by the differential equation (a, b, and c are nonnegative constants)

$$\frac{dy}{dt} + ay = b\frac{dx}{dt} + cx$$

a. Find $H(f)$.

b. Find and plot $|H(f)|$ and $\underline{/H(f)}$ for $c = 0$.

c. Find and plot $|H(f)|$ and $\underline{/H(f)}$ for $b = 0$.

2.47. For each of the following transfer functions, determine the unit impulse response of the system.

a. $H_1(f) = \frac{1}{(5+j2\pi f)}$

b. $H_2(f) = \frac{j2\pi f}{(5+j2\pi f)}$

(*Hint:* Use long division first.)

c. $H_3(f) = \frac{e^{-j6\pi f}}{(5+j2\pi f)}$.

d. $H_4(f) = \frac{1-e^{-j6\pi f}}{(5+j2\pi f)}$.

2.48. A filter has frequency-response function $H(f) = \Pi(f/2B)$ and input $x(t) = 2W\,\text{sinc}(2Wt)$.

a. Find the output $y(t)$ for $W < B$.

b. Find the output $y(t)$ for $W > B$.

c. In which case does the output suffer distortion? What influenced your answer?

2.49. A second-order active bandpass filter (BPF), known as a bandpass Sallen–Key circuit, is shown in Figure 2.38.

a. Show that the frequency-response function of this filter is given by

$$H(j\omega) = \frac{(K\omega_0/\sqrt{2})(j\omega)}{-\omega^2 + (\omega_0/Q)(j\omega) + \omega_0^2}, \quad \omega = 2\pi f$$

where

$$\omega_0 = \sqrt{2}(RC)^{-1}$$

$$Q = \frac{\sqrt{2}}{4-K}$$

$$K = 1 + \frac{R_a}{R_b}$$

b. Plot $|H(f)|$.

c. Show that the 3-dB bandwidth in hertz of the filter can be expressed as $B = f_0/Q$, where $f_0 = \omega_0/2\pi$.

d. Design a BPF using this circuit with center frequency $f_0 = 1000$ Hz and 3-dB bandwidth of 300 Hz. Find values of $R_a, R_b, R,$ and C that will give these desired specifications.

2.50. For the two circuits shown in Figure 2.39, determine $H(f)$ and $h(t)$. Sketch accurately the amplitude and phase responses. Plot the amplitude response in decibels. Use a logarithmic frequency axis.

2.51. Using the Paley–Wiener criterion, show that

$$|H(f)| = \exp(-\beta f^2)$$

is not a suitable amplitude response for a causal, linear time-invariant filter.

2.52. Determine whether the filters with impulse responses given below are BIBO stable.

a. $h_1(t) = \exp(-\alpha t)\cos(2\pi f_0 t)u(t)$.

b. $h_2(t) = \cos(2\pi f_0 t)u(t)$.

c. $h_3(t) = t^{-1}u(t-1)$.

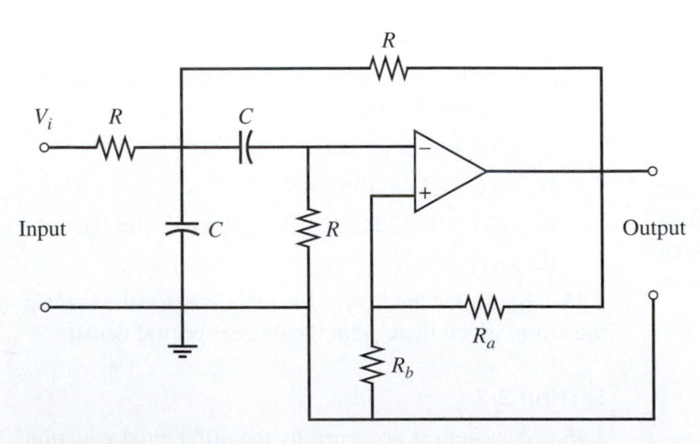

Figure 2.38

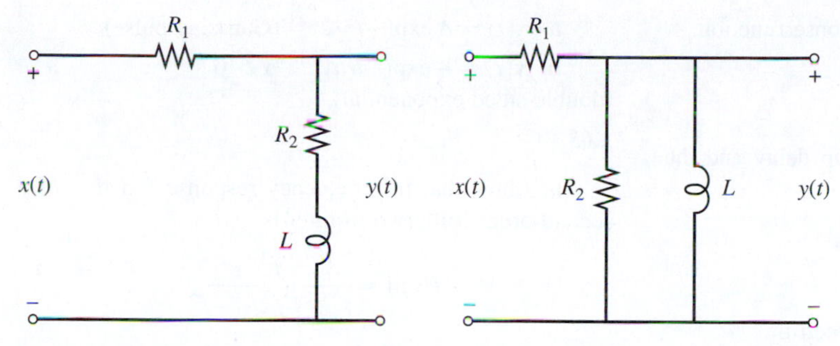

Figure 2.39

2.53. Given a filter with frequency-response function

$$H(f) = \frac{5}{4 + j(2\pi f)}$$

and input $x(t) = e^{-3t}u(t)$, obtain and plot accurately the energy spectral densities of the input and output.

2.54. A filter with frequency-response function

$$H(f) = 3\Pi\left(\frac{f}{26}\right)$$

has, as an input, a half-rectified cosine waveform of fundamental frequency 10 Hz. Determine the output of the filter.

2.55. Another definition of bandwidth for a signal is the 90% energy containment bandwidth. For a signal with energy spectral density $G(f) = |X(f)|^2$, it is given by B_{90} in the relation

$$\begin{aligned} 0.9E_{\text{Total}} &= \int_{-B_{90}}^{B_{90}} G(f)\, df = 2\int_0^{B_{90}} G(f)\, df \\ E_{\text{Total}} &= \int_{-\infty}^{\infty} G(f)\, df = 2\int_0^{\infty} G(f)\, df \end{aligned}$$

Obtain B_{90} for the following signals if it is defined. If it is not defined for a particular signal, state why it is not.

a. $x_1(t) = e^{-\alpha t}u(t)$, where α is a positive constant.

b. $x_2(t) = 2W \operatorname{sinc}(2Wt)$.

c. $x_3(t) = \Pi(t/\tau)$ (requires numerical integration).

2.56. An ideal quadrature phase shifter has

$$H(f) = \begin{cases} e^{-j\pi/2}, & f > 0 \\ e^{+j\pi/2}, & f < 0 \end{cases}$$

Find the outputs for the following inputs:

a. $x_1(t) = \exp(j100\pi t)$.

b. $x_2(t) = \cos(100\pi t)$.

c. $x_3(t) = \sin(100\pi t)$.

d. $x_4(t) = \Pi(t/2)$.

2.57. A filter has amplitude response and phase shift shown in Figure 2.40. Find the output for each of the inputs given below. For which cases is the transmission distortionless? Tell what type of distortion is imposed for the others.

a. $x_1(t) = \cos(48\pi t) + 5\cos(126\pi t)$.

b. $x_2(t) = \cos(126\pi t) + 0.5\cos(170\pi t)$.

c. $x_3(t) = \cos(126\pi t) + 3\cos(144\pi t)$.

d. $x_4(t) = \cos(10\pi t) + 4\cos(50\pi t)$.

2.58. Determine and accurately plot, on the same set of axes, the group delay and the phase delay for the systems with unit impulse responses:

a. $h_1(t) = 3e^{-5t}u(t)$.

b. $h_2(t) = 5e^{-3t}u(t) - 2e^{-5t}u(t)$.

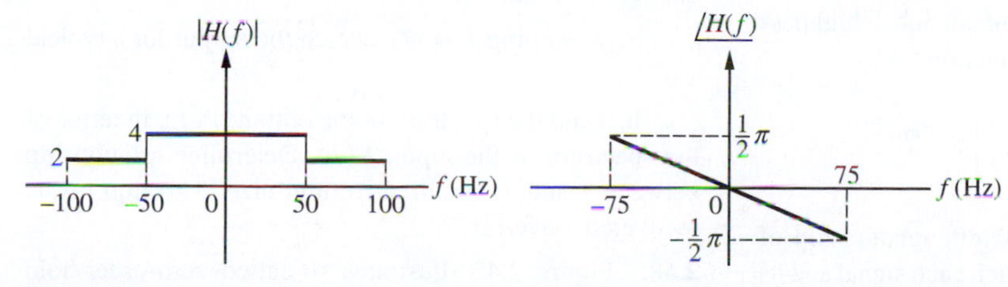

Figure 2.40

2.59. A system has the frequency-response function

$$H(f) = \frac{4\pi + j2\pi f}{8\pi + j2\pi f}$$

Determine and accurately plot the group delay and the phase delay.

2.60. The nonlinear system defined by

$$y(t) = x(t) + 0.1x^2(t)$$

has an input signal with the bandpass spectrum

$$X(f) = 4\Pi\left(\frac{f-20}{6}\right) + 4\Pi\left(\frac{f+20}{6}\right)$$

Sketch the spectrum of the output, labeling all important frequencies and amplitudes.

2.61.

a. Consider a nonlinear device with the transfer characteristic $y(t) = x(t) + 0.1x^3(t)$. The frequency of the input signal $x(t) = \cos(2000\pi t)$ is to be tripled by passing the signal through the nonlinearity and then through a second-order BPF with a frequency response function approximated by

$$H(f) = \frac{1}{1 + j2Q(f-3000)} + \frac{1}{1 + j2Q(f+3000)}$$

Neglecting negative frequency contributions, compute, in terms of the parameter Q, the *total harmonic distortion* (THD) at the tripler output, defined as

$$\text{THD} = \frac{\text{total power in all output distortion terms}}{\text{power in desired output component}} \times 100\%$$

Note that the desired output component in this case is the third harmonic of the input frequency.

b. Find the minimum value of Q that will result in $\text{THD} \leq 0.001\%$.

2.62. A nonlinear device has $y(t) = a_0 + a_1 x(t) + a_2 x^2(t) + a_3 x^3(t)$. If $x(t) = \cos(\omega_1 t) + \cos(\omega_2)t$, list all the frequency components present in $y(t)$. Discuss the use of this device as a frequency multiplier.

2.63. Find the impulse response of an ideal highpass filter with the frequency response function

$$H_{\text{HP}}(f) = H_0\left[1 - \Pi\left(\frac{f}{2W}\right)\right]e^{-j2\pi f t_0}$$

2.64. Verify the pulsewidth–bandwidth relationship of (2.250) for the following signals. Sketch each signal and its spectrum.

a. $x(t) = A\exp(-t^2/2\tau^2)$ (Gaussian pulse)

b. $x(t) = A\exp(-\alpha|t|), \quad \alpha > 0$ (double-sided exponential).

2.65.

a. Show that the frequency response function of a second-order Butterworth filter is

$$H(f) = \frac{f_3^2}{f_3^2 + j\sqrt{2}f_3 f - f^2}$$

where f_3 is the 3-dB frequency in hertz.

b. Find an expression for the group delay of this filter. Plot the group delay as a function of f/f_3.

c. Given that the step response for a second-order Butterworth filter is

$$y_s(t) = \left\{1 - \exp\left(-\frac{2\pi f_3 t}{\sqrt{2}}\right)\left[\cos\left(\frac{2\pi f_3 t}{\sqrt{2}}\right) + \sin\left(\frac{2\pi f_3 t}{\sqrt{2}}\right)\right]\right\}u(t)$$

where $u(t)$ is the unit step function, find the 10% to 90% risetime in terms of f_3.

Section 2.8

2.66. A sinusoidal signal of frequency 1 Hz is to be sampled periodically.

a. Find the maximum allowable time interval between samples.

b. Samples are taken at $\frac{1}{3}$-s intervals (i.e., at a rate of $f_s = 3$ sps). Construct a plot of the sampled signal spectrum that illustrates that this is an acceptable sampling rate to allow recovery of the original sinusoid.

c. The samples are spaced $\frac{2}{3}$ s apart. Construct a plot of the sampled signal spectrum that shows what the recovered signal will be if the samples are passed through a lowpass filter such that only the lowest frequency spectral lines are passed.

2.67. A flat-top sampler can be represented as the block diagram of Figure 2.41.

a. Assuming $T \ll T_s$, sketch the output for a typical $x(t)$.

b. Find the spectrum of the output, $Y(f)$, in terms of the spectrum of the input, $X(f)$. Determine relationship between τ and T_s required to minimize distortion in the recovered waveform?

2.68. Figure 2.42 illustrates so-called zero-order-hold reconstruction.

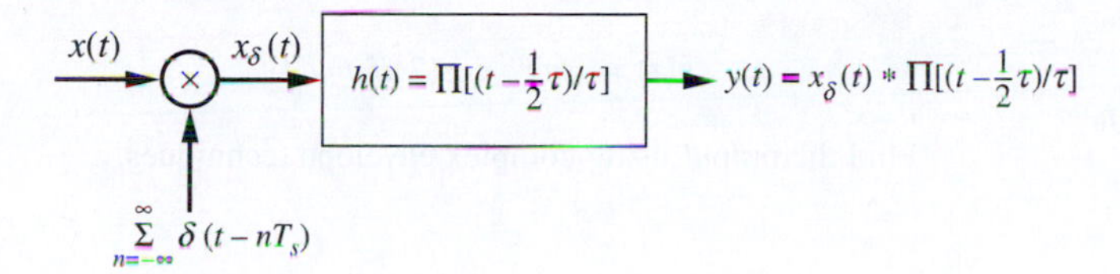

Figure 2.41

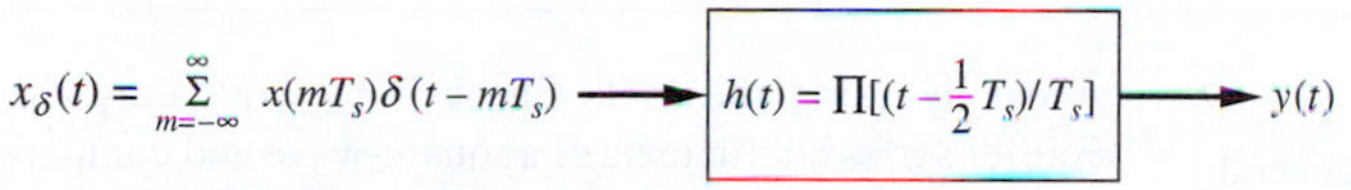

$$x_\delta(t) = \sum_{m=-\infty}^{\infty} x(mT_s)\delta(t - mT_s) \longrightarrow \boxed{h(t) = \Pi[(t - \tfrac{1}{2}T_s)/T_s]} \longrightarrow y(t)$$

Figure 2.42

a. Sketch $y(t)$ for a typical $x(t)$. Under what conditions is $y(t)$ a good approximation to $x(t)$?

b. Find the spectrum of $y(t)$ in terms of the spectrum of $x(t)$. Discuss the approximation of $y(t)$ to $x(t)$ in terms of frequency-domain arguments.

2.69. Determine the range of permissible cutoff frequencies for the ideal lowpass filter used to reconstruct the signal

$$x(t) = 10\cos(600\pi t)\ \cos^2(2400\pi t)$$

which is sampled at 6000 samples per second. Sketch $X(f)$ and $X_\delta(f)$. Find the minimum allowable sampling frequency.

2.70. Given the bandpass signal spectrum shown in Figure 2.43, sketch spectra for the following sampling rates f_s and indicate which ones are suitable: (a) $2B$, (b) $2.5B$, (c) $3B$, (d) $4B$, (e) $5B$, (f) $6B$.

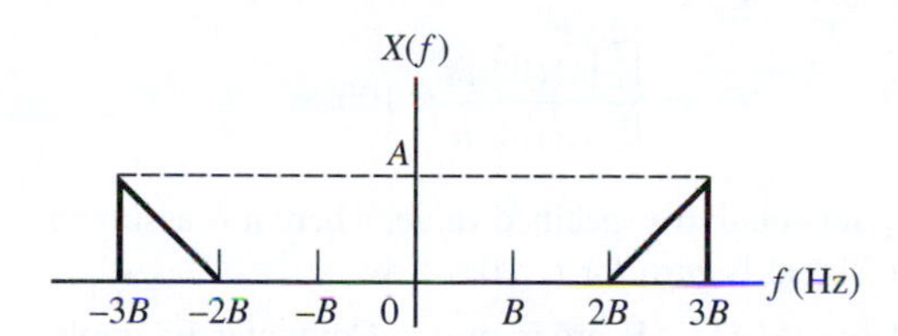

Figure 2.43

Section 2.9

2.71. Using appropriate Fourier transform theorems and pairs, express the spectrum $Y(f)$ of

$$y(t) = x(t)\cos(\omega_0 t) + \widehat{x}(t)\sin(\omega_0 t)$$

in terms of the spectrum $X(f)$ of $x(t)$, where $X(f)$ is lowpass with bandwidth

$$B < f_0 = \frac{\omega_0}{2\pi}$$

Sketch $Y(f)$ for a typical $X(f)$.

2.72. Show that $x(t)$ and $\widehat{x}(t)$ are orthogonal for the following signals ($\omega_0 > 0$):

a. $x_a(t) = \sin(\omega_0 t)$

b. $x_b(t) = 2\cos(\omega_0 t) + \sin(\omega_0 t)\cos(2\omega_0 t)$

c. $x_c(t) = A\exp(j\omega_0 t)$

2.73. Assume that the Fourier transform of $x(t)$ is real and has the shape shown in Figure 2.44. Determine and plot the spectrum of each of the following signals:

a. $x_1(t) = \frac{2}{3}x(t) + \frac{1}{3}j\widehat{x}(t)$.

b. $x_2(t) = \left[\frac{3}{4}x(t) + \frac{3}{4}j\widehat{x}(t)\right]e^{j2\pi f_0 t}, \quad f_0 \gg W$.

c. $x_3(t) = \left[\frac{2}{3}x(t) + \frac{1}{3}j\widehat{x}(t)\right]e^{j2\pi W t}$.

d. $x_4(t) = \left[\frac{2}{3}x(t) - \frac{1}{3}j\widehat{x}(t)\right]e^{j\pi W t}$.

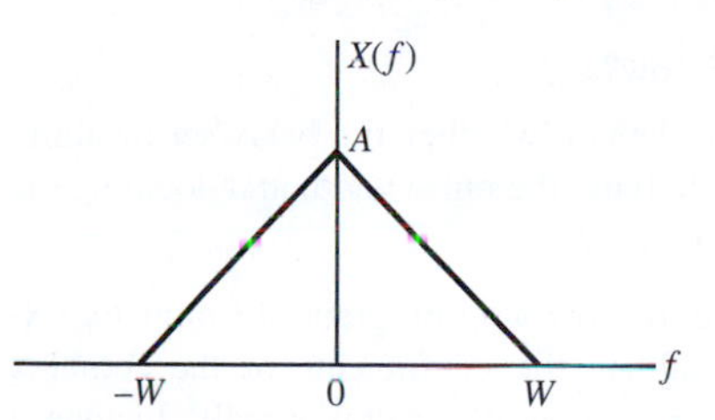

Figure 2.44

2.74. Consider the signal

$$x(t) = 2W\,\text{sinc}(2Wt)\cos(2\pi f_0 t), \quad f_0 > W$$

a. Obtain and sketch the spectrum of $x_p(t) = x(t) + j\widehat{x}(t)$.

b. Obtain and sketch the spectrum of $x_n(t) = x(t) - j\widehat{x}(t)$.

c. Obtain and sketch the spectrum of the complex envelope $\tilde{x}(t)$, where the complex envelope is defined by (2.310).

d. Find the complex envelope $\tilde{x}(t)$.

2.75. Consider the input

$$x(t) = \Pi(t/\tau)\cos[2\pi(f_0+\Delta f)t], \quad \Delta f \ll f_0$$

to a filter with impulse response

$$h(t) = \alpha e^{-\alpha t}\cos(2\pi f_0 t)u(t)$$

Find the output using complex envelope techniques.

Computer Exercises[19]

2.1.

a. Write a computer program to obtain the generalized Fourier series for an energy signal using the orthonormal basis set

$$\Phi_n(t) = \Pi(t-0.5-n), \quad n = 0, 1, 2, \cdots, T-1, \quad T \text{ integer}$$

where the signal extent is $(0, T)$ with T assumed to be integer valued. Your program should compute the generalized Fourier coefficients and the integral-squared error and should make a plot of the signal being approximated and the approximating waveform. Test your program with the signal $e^{-2t}u(t)$, $0 \le t \le 5$.

b. Repeat part (a) with the orthonormal basis set

$$\Phi_n(t) = \sqrt{2}\Pi\left(\frac{t-0.5-n}{0.5}\right), \quad n = 0, 1, 2, \cdots, 2T-1, \quad T \text{ integer}$$

What is the ISE now?

c. Can you deduce whether the basis set resulting from repeatedly halving the pulse width and doubling the amplitude is complete?

2.2. Generalize the computer program of Computer Example 2.1 to evaluate the coefficients of the complex exponential Fourier series of several signals. Include a plot of the amplitude and phase spectrum of the signal for which the Fourier series coefficients are evaluated. Check by evaluating the Fourier series coefficients of a square wave. Plot the square-wave approximation by summing the series through the seventh harmonic.

2.3. Write a computer program to evaluate the coefficients of the complex exponential Fourier series of a signal by using the FFT. Check it by evaluating the Fourier series coefficients of a square-wave and comparing your results with Computer Exercise 2.2.

2.4. How would you use the same approach as in Computer Exercise 2.3 to evaluate the Fourier transform of a pulse-type signal. How do the two outputs differ? Compute an approximation to the Fourier transform of a square pulse signal 1 unit wide and compare with the theoretical result.

2.5. Write a computer program to find the bandwidth of a lowpass energy signal that contains a certain specified percentage of its total energy, for example, 95%. In other words, write a program to find W in the equation

$$E_W = \frac{\int_0^W G_x(f)\,df}{\int_0^\infty G_x(f)\,df} \times 100\%$$

with E_W set equal to a specified value, where $G_X(f)$ is the energy spectral density of the signal.

2.6. Write a computer program to find the time duration of a lowpass energy signal that contains a certain specified percentage of its total energy, for example, 95%. In other words, write a program to find T in the equation

$$E_T = \frac{\int_0^T |x(t)|^2\,dt}{\int_0^\infty |x(t)|^2\,dt} \times 100\%$$

with E_T set equal to a specified value, where it is assumed that the signal is zero for $t < 0$.

2.7. Use a MATLAB program like Computer Example 2.2 to investigate the frequency response of the Sallen–Key circuit for various Q-values.

[19]When doing these computer exercises, we suggest that the student make use of a mathematics package such as MATLAB. Considerable time will be saved in being able to use the plotting capability of MATLAB. You should strive to use the vector capability of MATLAB as well.

CHAPTER 3

BASIC MODULATION TECHNIQUES

Before an information-bearing signal is transmitted through a communication channel, some type of modulation process is typically utilized to produce a signal that can easily be accommodated by the channel. In this chapter we will discuss various types of modulation techniques. The modulation process commonly translates an information-bearing signal, usually referred to as the *message signal*, to a new spectral location depending upon the intended frequency for transmission. For example, if the signal is to be transmitted through the atmosphere or free space, frequency translation is necessary to raise the signal spectrum to a frequency that can be radiated efficiently with antennas of reasonable size. If more than one signal utilizes a channel, modulation allows translation of different signals to different spectral locations, thus allowing the receiver to select the desired signal. Multiplexing allows two or more message signals to be transmitted by a single transmitter and received by a single receiver simultaneously. The logical choice of a modulation technique for a specific application is influenced by the characteristics of the message signal, the characteristics of the channel, the performance desired from the overall communication system, the use to be made of the transmitted data, and the economic factors that are always important in practical applications.

The two basic types of analog modulation are *continuous-wave modulation* and *pulse modulation*. In *continuous-wave modulation*, a parameter of a high-frequency carrier is varied proportionally to the message signal such that a one-to-one correspondence exists between the parameter and the message signal. The carrier is usually assumed to be sinusoidal, but as will be illustrated, this is not a necessary restriction. For a sinusoidal carrier, a general modulated carrier can be represented mathematically as

$$x_c(t) = A(t)\cos[2\pi f_c t + \phi(t)] \tag{3.1}$$

where f_c is the *carrier frequency*. Since a sinusoid is completely specified by its amplitude, $A(t)$, and instantaneous phase, $2\pi f_c + \phi(t)$, it follows that once the carrier frequency is specified, only two parameters are candidates to be varied in the modulation process: the instantaneous amplitude $A(t)$ and the phase deviation $\phi(t)$. When the amplitude $A(t)$ is linearly related to the modulating signal, the result is *linear modulation*. Letting $\phi(t)$ or the time derivative of $\phi(t)$ be linearly related to the modulating signal yields phase or frequency modulation, respectively. Collectively, phase and frequency modulation are referred to as *angle modulation*, since the instantaneous phase angle of the modulated carrier conveys the information.

In *analog pulse modulation*, the message waveform is sampled at discrete time intervals, and the amplitude, width, or position of a pulse is varied in one-to-one correspondence with the values of the samples. Since the samples are taken at discrete times, the periods between the samples are available for other uses, such as insertion of samples from other message signals. This is referred to as *time-division multiplexing*. If the value of each sample is quantized and encoded, *pulse-code modulation* results. We also briefly consider *delta modulation.* Pulse-code modulation and delta modulation are digital rather than analog modulation techniques, but they are considered in this chapter for completeness and as an introduction to the digital systems that are to be considered in following chapters of this book.

3.1 LINEAR MODULATION

A general linearly modulated carrier is represented by setting the instantaneous phase deviation $\phi(t)$ in (3.1) equal to zero. Thus, a linearly modulated carrier is represented by

$$x_c(t) = A(t)\cos(2\pi f_c t) \tag{3.2}$$

in which the carrier amplitude $A(t)$ varies in one-to-one correspondence with the message signal. We next discuss several different types of linear modulation as well as techniques that can be used for demodulation.

3.1.1 Double-Sideband Modulation

Double-sideband (DSB) modulation results when $A(t)$ is proportional to the message signal $m(t)$. Thus the output of a DSB modulator can be represented as

$$x_c(t) = A_c m(t)\cos(2\pi f_c t) \tag{3.3}$$

which illustrates that DSB modulation is simply the multiplication of a carrier, $A_c\cos(2\pi f_c t)$, by the message signal. It follows from the modulation theorem for Fourier transforms that the spectrum of a DSB signal is given by

$$X_c(f) = \frac{1}{2}A_c M(f+f_c) + \frac{1}{2}A_c M(f-f_c) \tag{3.4}$$

The process of DSB modulation is described in Figure 3.1. Figure 3.1(a) illustrates a DSB system and shows that a DSB signal is demodulated by multiplying the received signal, denoted by $x_r(t)$, by the demodulation carrier $2\cos(2\pi f_c t)$ and lowpass filtering. For the idealized system that we are considering here, the received signal $x_r(t)$ is identical to the transmitted signal $x_c(t)$. The output of the multiplier is

$$d(t) = 2A_c[m(t)\cos(2\pi f_c t)]\cos(2\pi f_c t) \tag{3.5}$$

or

$$d(t) = A_c m(t) + A_c m(t)\cos(4\pi f_c t) \tag{3.6}$$

where we have used the trigonometric identity $2\cos^2 x = 1 + \cos 2x$.

The time-domain signals are shown in Figure 3.1(b) for an assumed $m(t)$. The message signal $m(t)$ forms the envelope, or instantaneous magnitude, of $x_c(t)$. The waveform for $d(t)$ can be best understood by realizing that since $\cos^2(2\pi f_c t)$ is nonnegative for all t, $d(t)$ is

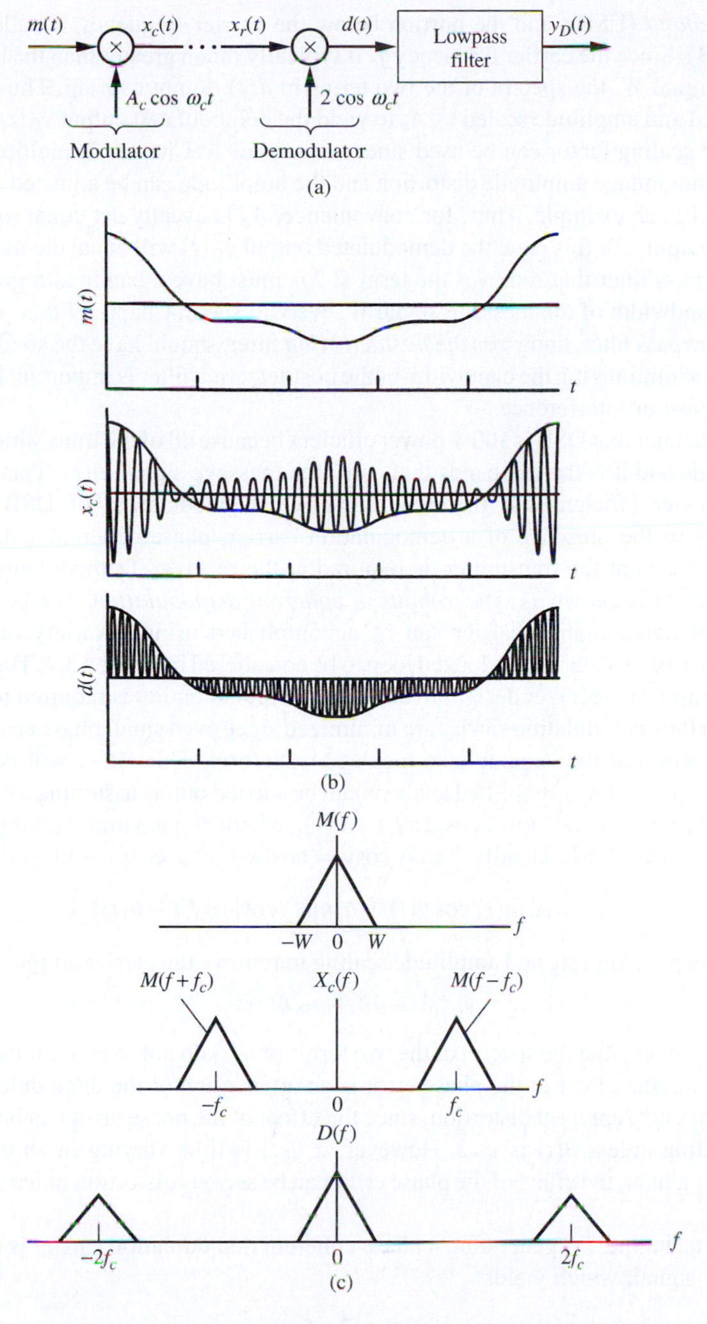

Figure 3.1 Double-sideband modulation. (a) System. (b) Waveforms. (c) Spectra.

positive if $m(t)$ is positive and $d(t)$ is negative if $m(t)$ is negative. Also note that $m(t)$ (appropriately scaled) forms the envelope of $d(t)$ and that the frequency of the sinusoid under the envelope is $2f_c$ rather than f_c.

The spectra of the signals $m(t)$, $x_c(t)$, and $d(t)$, are shown in Figure 3.1(c) for an assumed $M(f)$ having a bandwidth W. The spectra $M(f+f_c)$ and $M(f-f_c)$ are simply the message spectrum translated to $f = \pm f_c$. The portion of $M(f-f_c)$ above the carrier frequency is called the *upper sideband* (USB), and the portion below the carrier frequency is called the *lower sideband* (LSB). Since the carrier frequency f_c is typically much greater than the bandwidth of the message signal W, the spectra of the two terms in $d(t)$ do not overlap. Thus $d(t)$ can be lowpass filtered and amplitude scaled by A_c to yield the demodulated output $y_D(t)$. In practice, any amplitude scaling factor can be used since, as we saw in Chapter 2, multiplication by a constant does not induce amplitude distortion and the amplitude can be adjusted as desired. A volume control is an example. Thus, for convenience, A_c is usually set equal to unity at the demodulator output. For this case, the demodulated output $y_D(t)$ will equal the message signal $m(t)$. The lowpass filter that removes the term at $2f_c$ must have a bandwidth greater than or equal to the bandwidth of the message signal W. We will see in Chapter 7 that when noise is present, this lowpass filter, known as the *postdetection* filter, should have the smallest possible bandwidth since minimizing the bandwidth of the postdetection filter is important for removing out-of-band noise or interference.

We will see later that DSB is 100% power efficient because all of the transmitted power lies in the sidebands and it is the sidebands that carry the message signal $m(t)$. This makes DSB modulation power efficient and therefore attractive. Demodulation of DSB is difficult, however, because the presence of a demodulation carrier, phase coherent with the carrier used for modulation at the transmitter, is required at the receiver. Demodulation utilizing a coherent reference is known as *synchronous* or *coherent demodulation.* The generation of a phase coherent demodulation carrier can be accomplished using a variety of techniques, including the use of a Costas phase-locked loop to be considered in Section 3.4. The use of these techniques complicate receiver design. In addition, careful attention is required to ensure that phase errors in the demodulation carrier are minimized since even small phase errors can result in serious distortion of the demodulated message waveform. This effect will be thoroughly analyzed in Chapter 7, but a simplified analysis can be carried out by assuming a demodulation carrier in Figure 3.1(a) of the form $2\cos[2\pi f_c t+\theta(t)]$, where $\theta(t)$ is a time-varying phase error. Applying the trigonometric identity $2\cos x\cos y=\cos(x+y)+\cos(x-y)$, yields

$$d(t)=A_c m(t)\cos\theta(t)+A_c m(t)\cos[4\pi f_c t+\theta(t)] \tag{3.7}$$

which, after lowpass filtering and amplitude scaling to remove the carrier amplitude, becomes

$$y_D(t)=m(t)\cos\theta(t) \tag{3.8}$$

assuming, once again, that the spectra of the two terms of $d(t)$ do not overlap. If the phase error $\theta(t)$ is a constant, the effect of the phase error is an attenuation of the demodulated message signal. This does not represent distortion, since the effect of the phase error can be removed by amplitude scaling unless $\theta(t)$ is $\pi/2$. However, if $\theta(t)$ is time varying in an unknown and unpredictable manner, the effect of the phase error can be serious distortion of the demodulated output.

A simple technique for generating a phase coherent demodulation carrier is to square the received DSB signal, which yields

$$\begin{aligned} x_r^2(t) &= A_c^2 m^2(t)\cos^2(2\pi f_c t) \\ &= \frac{1}{2}A_c^2 m^2(t)+\frac{1}{2}A_c^2 m^2(t)\cos(4\pi f_c t) \end{aligned} \tag{3.9}$$

If $m(t)$ is a power signal, $m^2(t)$ has a nonzero DC value. Thus, by the modulation theorem, $x_r^2(t)$ has a discrete frequency component at $2f_c$, which can be extracted from the spectrum of $x_r^2(t)$

using a narrowband bandpass filter. The frequency of this component can be divided by 2 to yield the desired demodulation carrier. Later we will discuss a convenient technique for implementing the required frequency divider.

The analysis of DSB illustrates that the spectrum of a DSB signal does not contain a discrete spectral component at the carrier frequency unless $m(t)$ has a DC component. For this reason, DSB systems with no carrier frequency component present are often referred to as *suppressed carrier systems*. However, if a carrier component is transmitted along with the DSB signal, demodulation can be simplified. The received carrier component can be extracted using a narrowband bandpass filter and can be used as the demodulation carrier. If the carrier amplitude is sufficiently large, the need for generating a demodulation carrier can be completely avoided. This naturally leads to the subject of amplitude modulation.

3.1.2 Amplitude Modulation

Amplitude modulation results when a DC bias A is added to $m(t)$ prior to the modulation process. The result of the DC bias is that a carrier component is present in the transmitted signal. For AM, the transmitted signal is typically defined as

$$x_c(t) = A_c[1 + am_n(t)] \cos(2\pi f_c t) \tag{3.10}$$

in which A_c is the amplitude of the unmodulated carrier $A_c \cos(2\pi f_c t)$, $m_n(t)$ is the normalized message signal to be discussed in the following paragraph, and the parameter $a \leq 1$ is known as the *modulation index*.[1] We shall assume that $m(t)$ has zero DC value so that the carrier component in the transmitted signal arises entirely from the bias. The time-domain representation of AM is illustrated in Figure 3.2(a) and (b), and the block diagram of the modulator for producing AM is shown in Figure 3.2(c).

An AM signal can be demodulated using the same coherent demodulation technique that was used for DSB (see Problem 3.2). However, the use of coherent demodulation negates the advantage of AM. The advantage of AM over DSB is that a very simple technique, known as envelope detection or envelope demodulation, can be used. An envelope demodulator is implemented as shown in Figure 3.3(a). It can be seen from Figure 3.3(b) that as the carrier frequency is increased, the envelope, defined as $A_c[1 + am_n(t)]$, becomes easier to observe. More importantly, it also follows from observation of Figure 3.3(b) that, if the envelope of the AM signal $A_c[1 + am_n(t)]$ goes negative, distortion will result in the demodulated signal assuming that envelope demodulation is used. The normalized message signal is defined so that this distortion is prevented. Thus, for $a = 1$, the minimum value of $1 + am_n(t)$ is zero. In order to ensure that the envelope is nonnegative for all t we require that $1 + m_n(t) \geq 0$ or, equivalently, $m_n(t) \geq -1$ for all t. The normalized message signal $m_n(t)$ is therefore found by dividing $m(t)$ by a positive constant so that the condition $m_n(t) \geq -1$ is satisfied. This normalizing constant is $|\min m(t)|$. In many cases of practical interest, such as speech or music signals, the maximum and minimum values of the message signal are equal. We will see why this is true when we study probability and random signals in Chapters 5 and 6.

In order for the envelope detection process to operate properly, the RC time constant of the detector, shown in Figure 3.3(a), must be chosen carefully. The appropriate value for the time constant is related to the carrier frequency and to the bandwidth of $m(t)$. In practice,

[1]The parameter a as used here is sometimes called the *negative modulation factor*. Also, the quantity $a \times 100\%$ is often referred to as the *percent modulation*.

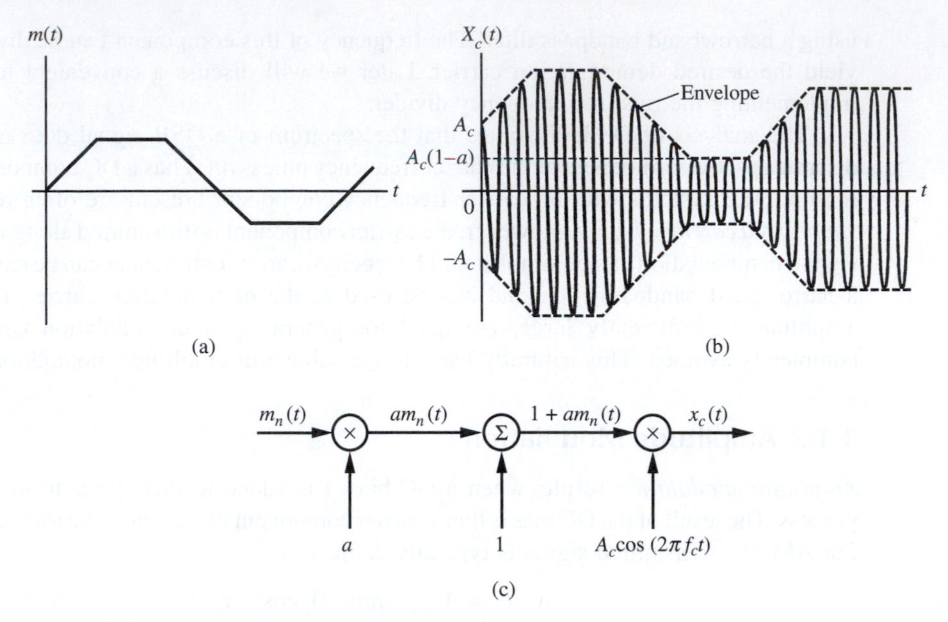

Figure 3.2
Amplitude modulation. (a) Message signal. (b) Modulator output for $a < 1$. (c) Modulator.

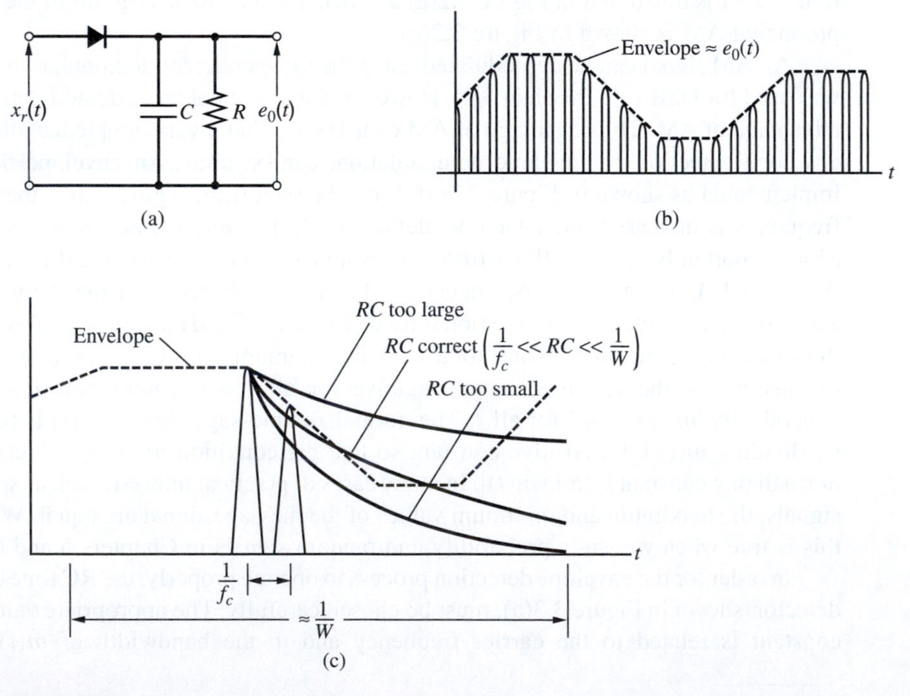

Figure 3.3
Envelope detection. (a) Circuit. (b) Waveforms. (c) Effect of RC time constant.

satisfactory operation requires a carrier frequency of at least 10 times the bandwidth of $m(t)$, W. Also, the cutoff frequency of the RC circuit must lie between f_c and W and must be well separated from both. This is illustrated in Figure 3.3(c).

All information in the modulator output is contained in the sidebands. Thus, the carrier component of (3.10), $A_c \cos\omega_c t$, is wasted power as far as information transfer is concerned. This fact can be of considerable importance in an environment where power is limited and can completely preclude the use of AM as a modulation technique in power-limited applications.

From (3.10) we see that the total power contained in the AM modulator output is

$$\langle x_c^2(t)\rangle = \langle A_c^2[1+am_n(t)]^2 \cos^2(2\pi f_c t)\rangle \tag{3.11}$$

where $\langle\cdot\rangle$ denotes the time average value. If $m_n(t)$ is *slowly* varying with respect to the carrier

$$\begin{aligned}\langle x_c^2(t)\rangle &= \left\langle A_c^2[1+am_n(t)]^2\left[\frac{1}{2}+\frac{1}{2}\cos(4\pi f_c t)\right]\right\rangle \\ &= \left\langle \frac{1}{2}A_c^2\left[1+2am_n(t)+a^2m_n^2(t)\right]\right\rangle\end{aligned} \tag{3.12}$$

Assuming $m_n(t)$ to have zero average value and taking the time average term-by-term gives

$$\langle x_c^2(t)\rangle = \frac{1}{2}A_c^2 + \frac{1}{2}A_c^2a^2\langle m_n^2(t)\rangle \tag{3.13}$$

The first term in the preceding expression represents the carrier power, and the second term represents the sideband (information) power. The efficiency of the modulation process is defined as the ratio of the power in the information-bearing signal (the sideband power) to the total power in the transmitted signal. This is

$$E_{ff} = \frac{a^2\langle m_n^2(t)\rangle}{1+a^2\langle m_n^2(t)\rangle} \tag{3.14}$$

The efficiency is typically multiplied by 100 so that efficiency can be expressed as a percent.

If the message signal has symmetrical maximum and minimum values, such that $|\min m(t)|$ and $|\max m(t)|$ are equal, then $\langle m_n^2(t)\rangle \leq 1$. It follows that for $a \leq 1$, the maximum efficiency is 50% and is achieved for square-wave-type message signals. If $m(t)$ is a sine wave, $\langle m_n^2(t)\rangle = \frac{1}{2}$ and the efficiency is 33.3% for $a = 1$. Note that if we allow the modulation index to exceed 1, efficiency can exceed 50% and that $E_{ff} \to 100\%$ as $a \to \infty$. Values of a greater than 1, as we have seen, preclude the use of envelope detection. Efficiency obviously declines rapidly as the index is reduced below unity. If the message signal does not have symmetrical maximum and minimum values, then higher values of efficiency can be achieved (see Problem 3.6).

The main advantage of AM is that since a coherent reference is not needed for demodulation as long as $a \leq 1$, the demodulator becomes simple and inexpensive. In many applications, such as commercial radio, this fact alone is sufficient to justify its use.

The AM modulator output $x_c(t)$ is shown in Figure 3.4 for three values of the modulation index: $a = 0.5, a = 1.0$, and $a = 1.5$. The message signal $m(t)$ is assumed to be a unity amplitude sinusoid with a frequency of 1 Hz. A unity amplitude carrier is also assumed. The envelope detector output $e_o(t)$, as identified in Figure 3.3, is also shown for each value of the

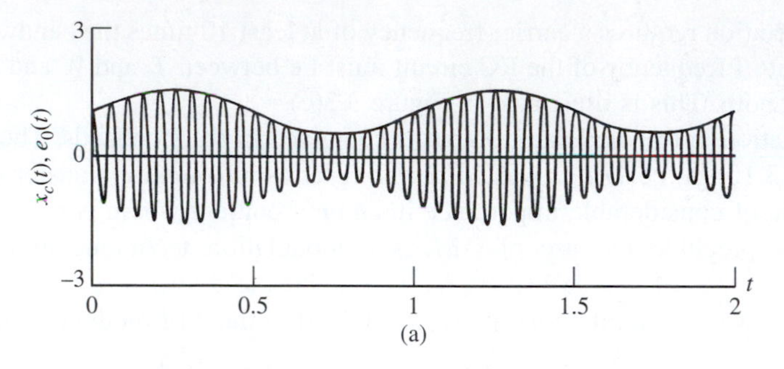

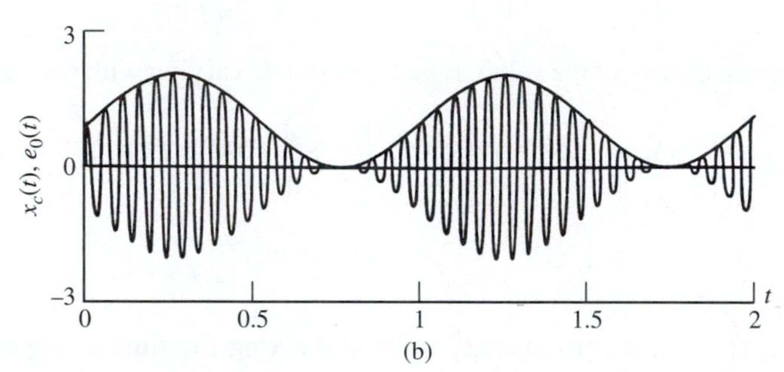

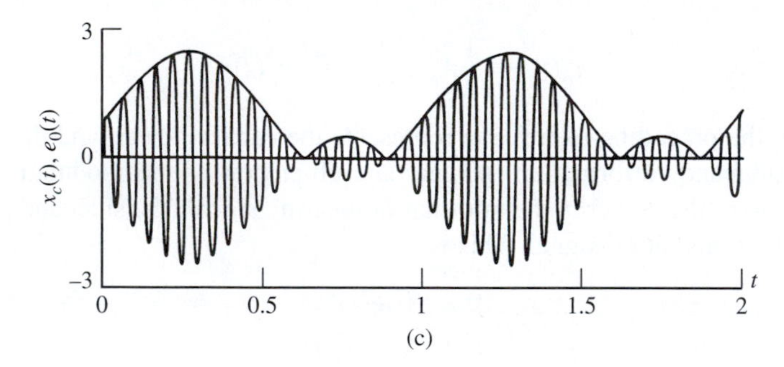

Figure 3.4 Modulated carrier and envelope detector outputs for various values of the modulation index. (a) $a = 0.5$. (b) $a = 1.0$. (c) $a = 1.5$.

modulation index. Note that for $a = 0.5$ the envelope is always positive. For $a = 1.0$ the minimum value of the envelope is exactly zero. Thus, envelope detection can be used for both of these cases. For $a = 1.5$ the envelope goes negative and $e_o(t)$, which is the absolute value of the envelope, is a badly distorted version of the message signal.

EXAMPLE 3.1

In this example we determine the efficiency and the output spectrum for an AM modulator operating with a modulation index of 0.5. The carrier power is 50 W, and the message signal is

$$m(t) = 4\cos\left(2\pi f_m t - \frac{\pi}{9}\right) + 2\sin(4\pi f_m t) \tag{3.15}$$

The first step is to determine the minimum value of $m(t)$. There are a number of ways to accomplish this. Perhaps the easiest way is to simply plot $m(t)$ and pick off the minimum value. MATLAB is very useful for this purpose as shown in the following program.

```
% File:  c3ex1.m
fmt = 0:0.0001:1;
m = 4*cos(2*pi*fmt-pi/9) + 2*sin(4*pi*fmt);
[minmessage,index] = min(m);
plot(fmt,m,'k'),
grid, xlabel('Normalized Time'), ylabel('Amplitude')
minmessage, mintime = 0.0001*(index-1)
% End of script file.
```

Executing the program yields the plot of the message signal, the minimum value of $m(t)$, and the occurrence time for the minimum value as follows:

```
  c3ex1
minmessage = -4.3642
mintime = 0.4352
```

The message signal as generated by the MATLAB program is shown in Figure 3.5(a). Note that the time axis is normalized by dividing by f_m. As shown, the minimum value of $m(t)$ is -4.364 and occurs at $f_m t = 0.435$, as shown. The normalized message signal is therefore given by

$$m_n(t) = \frac{1}{4.364}\left[4\cos\left(2\pi f_m t - \frac{\pi}{9}\right) + 2\sin(4\pi f_m t)\right] \tag{3.16}$$

or

$$m_n(t) = 0.9166\cos\left(2\pi f_m t - \frac{\pi}{9}\right) + 0.4583\sin(4\pi f_m t) \tag{3.17}$$

The mean-square value of $m_n(t)$ is

$$\langle m_n^2(t)\rangle = \frac{1}{2}(0.9166)^2 + \frac{1}{2}(0.4583)^2 = 0.5251 \tag{3.18}$$

Thus, the efficiency is

$$E_{ff} = \frac{(0.25)(0.5251)}{1 + (0.25)(0.5251)} = 0.116 \tag{3.19}$$

or 11.6%.

Since the carrier power is 50 W, we have

$$\frac{1}{2}(A_c)^2 = 50 \tag{3.20}$$

from which

$$A_c = 10 \tag{3.21}$$

Also, since $\sin x = \cos(x - \pi/2)$, we can write $x_c(t)$ as

$$x_c(t) = 10\left\{1 + 0.5\left[0.9166\cos\left(2\pi f_m t - \frac{\pi}{9}\right) + 0.4583\cos\left(4\pi f_m t - \frac{\pi}{2}\right)\right]\right\}\cos(2\pi f_c t) \tag{3.22}$$

In order to plot the spectrum of $x_c(t)$, we write the preceding equation as

$$\begin{aligned} x_c(t) = {} & 10\cos(2\pi f_c t) \\ & + 2.292\left\{\cos\left[2\pi(f_c + f_m)t - \frac{\pi}{9}\right] + \cos\left[2\pi(f_c + f_m)t + \frac{\pi}{9}\right]\right\} \\ & + 1.146\left\{\cos\left[2\pi(f_c + 2f_m)t - \frac{\pi}{2}\right] + \cos\left[2\pi(f_c + 2f_m)t + \frac{\pi}{2}\right]\right\} \end{aligned} \tag{3.23}$$

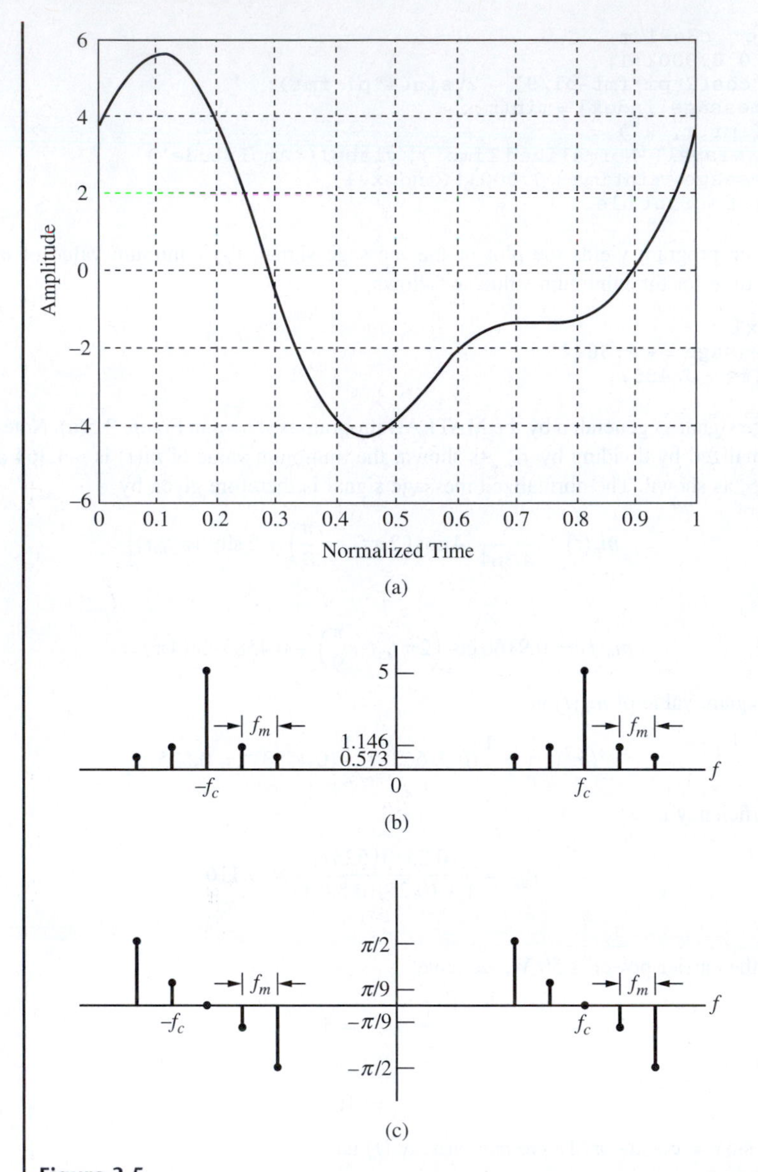

Figure 3.5
Waveform and spectra for Example 3.1. (a) Message signal. (b) Amplitude spectrum of modulator output. (c) Phase spectrum of modulator output.

Figure 3.5 (b) and (c) shows the amplitude and phase spectra of $x_c(t)$. Note that the amplitude spectrum has even symmetry about the carrier frequency and that the phase spectrum has odd symmetry about the carrier frequency. Of course, since $x_c(t)$ is a real signal, the overall amplitude spectrum is also even about $f = 0$, and the overall phase spectrum is odd about $f = 0$.

■

3.1.3 Single-Sideband Modulation

In our development of DSB, we saw that the USB and LSB have even amplitude and odd phase symmetry about the carrier frequency. Thus transmission of both sidebands is not necessary, since either sideband contains sufficient information to reconstruct the message signal $m(t)$. Elimination of one of the sidebands prior to transmission results in single sideband (SSB), which reduces the bandwidth of the modulator output from $2W$ to W, where W is the bandwidth of $m(t)$. However, this bandwidth savings is accompanied by a considerable increase in complexity.

On the following pages, two different methods are used to derive the time-domain expression for the signal at the output of an SSB modulator. Although the two methods are equivalent, they do present different viewpoints. In the first method, the transfer function of the filter used to generate an SSB signal from a DSB signal is derived using the Hilbert transform. The second method derives the SSB signal directly from $m(t)$ using the results illustrated in Figure 2.30 and the frequency-translation theorem.

The generation of an SSB signal by sideband filtering is illustrated in Figure 3.6. First, a DSB signal, $x_{\text{DSB}}(t)$, is formed. Sideband filtering of the DSB signal then yields an upper-sideband or a lower-sideband SSB signal, depending on the filter passband selected.

The filtering process that yields lower-sideband SSB is illustrated in detail in Figure 3.7. A lower-sideband SSB signal can be generated by passing a DSB signal through an ideal filter that passes the LSB and rejects the USB. It follows from Figure 3.7(b) that the transfer function of this filter is

$$H_L(f) = \frac{1}{2}[\text{sgn}(f + f_c) - \text{sgn}(f - f_c)] \tag{3.24}$$

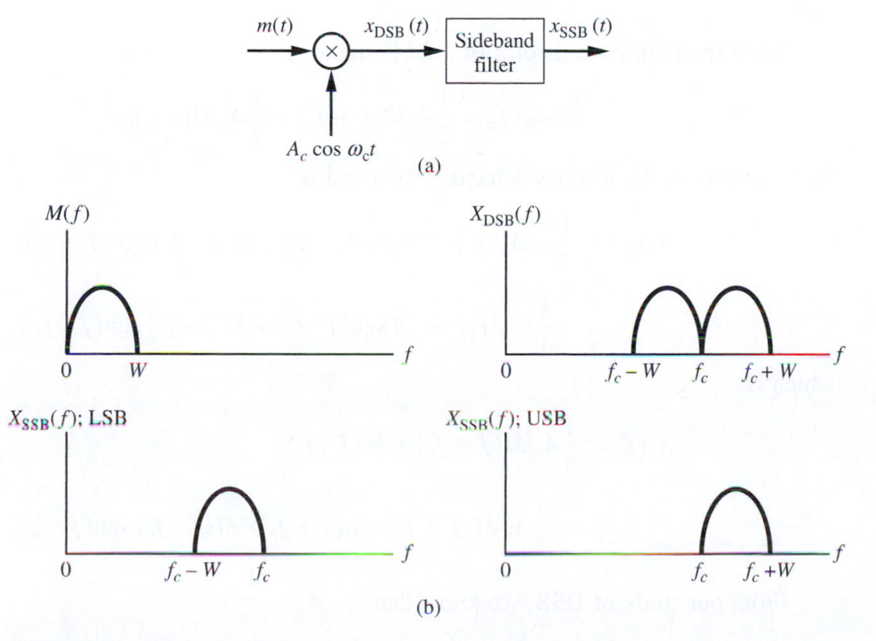

Figure 3.6
Generation of SSB by sideband filtering. (a) SSB modulator. (b) Spectra (single sided).

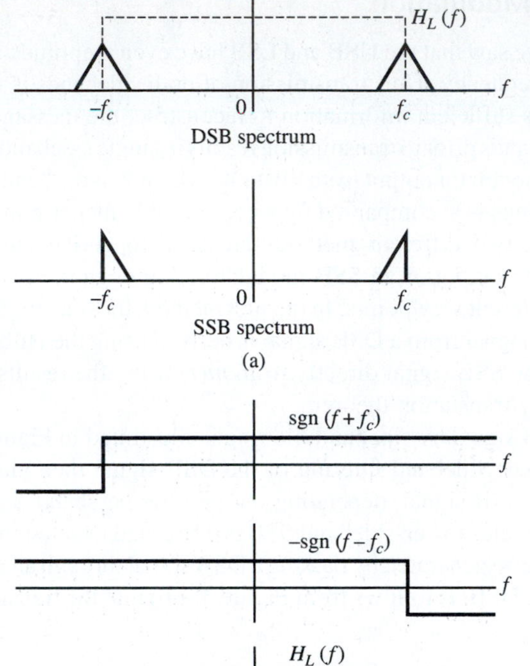

Figure 3.7
Generation of lower-sideband SSB. (a) Sideband filtering process. (b) Generation of lower-sideband filter.

Since the Fourier transform of a DSB signal is

$$X_{\text{DSB}}(f) = \frac{1}{2}A_c M(f+f_c) + \frac{1}{2}A_C M(f-f_c) \tag{3.25}$$

the transform of the lower-sideband SSB signal is

$$\begin{aligned} X_c(f) = &\frac{1}{4}A_c[M(f+f_c)\,\text{sgn}(f+f_c) + M(f-f_c)\,\text{sgn}(f+f_c)] \\ &-\frac{1}{4}A_c[M(f+f_c)\,\text{sgn}(f-f_c) + M(f-f_c)\,\text{sgn}(f-f_c)] \end{aligned} \tag{3.26}$$

which is

$$\begin{aligned} X_c(f) = &\frac{1}{4}A_c[M(f+f_c) + M(f-f_c)] \\ &+\frac{1}{4}A_c[M(f+f_c)\,\text{sgn}(f+f_c) - M(f-f_c)\,\text{sgn}(f-f_c)] \end{aligned} \tag{3.27}$$

From our study of DSB, we know that

$$\frac{1}{2}A_c m(t)\cos(2\pi f_c t) \leftrightarrow \frac{1}{4}A_c[M(f+f_c) + M(f-f_c)] \tag{3.28}$$

and from our study of Hilbert transforms in Chapter 2, we recall that

$$\widehat{m}(t) \leftrightarrow -j(\operatorname{sgn} f)M(f)$$

By the frequency-translation theorem, we have

$$m(t)e^{\pm j2\pi f_c t} \leftrightarrow M(f \mp f_c) \tag{3.29}$$

Replacing $m(t)$ by $\widehat{m}(t)$ in the previous equation yields

$$\widehat{m}(t)e^{\pm j2\pi f_c t} \leftrightarrow -jM(f \mp f_c)\operatorname{sgn}(f \mp f_c) \tag{3.30}$$

Thus

$$\Im^{-1}\left\{\frac{1}{4}A_c[M(f+f_c)\operatorname{sgn}(f+f_c)-M(f-f_c)\operatorname{sgn}(f-f_c)]\right\}$$

$$= -A_c\frac{1}{4j}\widehat{m}(t)e^{-j2\pi f_c t} + A_c\frac{1}{4j}\widehat{m}(t)e^{+j2\pi f_c t} = \frac{1}{2}A_c\widehat{m}(t)\sin(2\pi f_c t) \tag{3.31}$$

Combining (3.28) and (3.31), we get the general form of a lower-sideband SSB signal:)

$$x_c(t) = \frac{1}{2}A_c m(t)\cos(2\pi f_c t) + \frac{1}{2}A_c\widehat{m}(t)\sin(2\pi f_c t) \tag{3.32}$$

A similar development can be carried out for upper-sideband SSB. The result is

$$x_c(t) = \frac{1}{2}A_c m(t)\cos(2\pi f_c t) - \frac{1}{2}A_c\widehat{m}(t)\sin(2\pi f_c t) \tag{3.33}$$

which shows that LSB and USB modulators have the same defining equations except for the sign of the term representing the Hilbert transform of the modulation. Observation of the spectrum of an SSB signal illustrates that SSB systems do not have DC response.

The generation of SSB by the method of sideband filtering the output of DSB modulators requires the use of filters that are very nearly ideal if low-frequency information is contained in $m(t)$. Another method for generating an SSB signal, known as *phase-shift modulation*, is illustrated in Figure 3.8. This system is a term-by-term realization of (3.32) or (3.33). Like the ideal filters required for sideband filtering, the ideal wideband phase shifter, which performs the Hilbert transforming operation, is impossible to implement exactly. However, since the

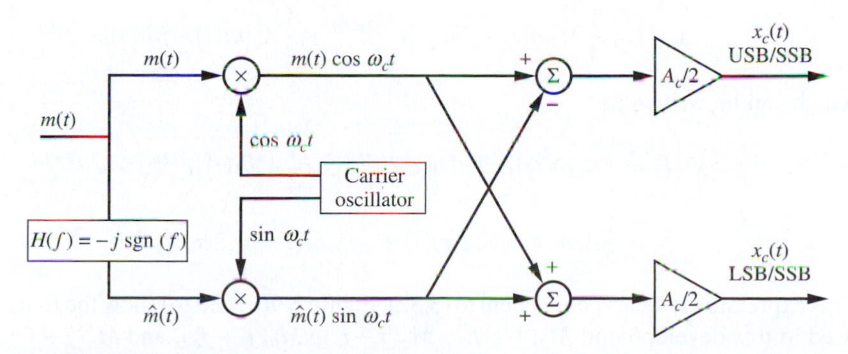

Figure 3.8
Phase-shift modulator.

frequency at which the discontinuity occurs is $f = 0$ instead of $f = f_c$, ideal phase shift devices can be closely approximated.

An alternative derivation of $x_c(t)$ for an SSB signal is based on the concept of the analytic signal. The positive-frequency portion of $M(f)$ is given by

$$M_p(f) = \frac{1}{2}\Im\{m(t) + j\widehat{m}(t)\} \tag{3.34}$$

and the negative-frequency portion of $M(f)$ is given by

$$M_n(f) = \frac{1}{2}\Im\{m(t) - j\widehat{m}(t)\} \tag{3.35}$$

By definition, an upper-sideband SSB signal is given in the frequency domain by

$$X_c(f) = \frac{1}{2}A_c M_p(f - f_c) + \frac{1}{2}A_c M_n(f + f_c) \tag{3.36}$$

Inverse Fourier-transforming yields

$$x_c(t) = \frac{1}{4}A_c[m(t) + j\widehat{m}(t)]e^{j2\pi f_c t} + \frac{1}{4}A_c[m(t) - j\widehat{m}(t)]e^{-j2\pi f_c t} \tag{3.37}$$

which is

$$\begin{aligned} x_c(t) &= \frac{1}{4}A_c m(t)\left[e^{j2\pi f_c t} + e^{-j2\pi f_c t}\right] + j\frac{1}{4}A_c\widehat{m}(t)\left[e^{j2\pi f_c t} - e^{-j2\pi f_c t}\right] \\ &= \frac{1}{2}A_c m(t)\cos(2\pi f_c t) - \frac{1}{2}A_c\widehat{m}(t)\sin(2\pi f_c t) \end{aligned} \tag{3.38}$$

The preceding expression is clearly equivalent to (3.33).

The lower-sideband SSB signal is derived in a similar manner. By definition, for a lower-sideband SSB signal,

$$X_c(f) = \frac{1}{2}A_c M_p(f + f_c) + \frac{1}{2}A_c M_n(f - f_c) \tag{3.39}$$

This becomes, after inverse Fourier-transforming,

$$x_c(t) = \frac{1}{4}A_c[m(t) + j\widehat{m}(t)]e^{-j2\pi f_c t} + \frac{1}{4}A_c[m(t) - j\widehat{m}(t)]e^{j2\pi f_c t} \tag{3.40}$$

which can be written as

$$\begin{aligned} x_c(t) &= \frac{1}{4}A_c m(t)\left[e^{j2\pi f_c t} + e^{-j2\pi f_c t}\right] - j\frac{1}{4}A_c\widehat{m}(t)\left[e^{j2\pi f_c t} - e^{-j2\pi f_c t}\right] \\ &= \frac{1}{2}A_c m(t)\cos(2\pi f_c t) + \frac{1}{2}A_c\widehat{m}(t)\sin(2\pi f_c t) \end{aligned} \tag{3.41}$$

This expression is clearly equivalent to (3.32). Figure 3.9(b) and (c) show the four signal spectra used in this development: $M_p(f + f_c)$, $M_p(f - f_c)$, $M_n(f + f_c)$, and $M_n(f - f_c)$.

There are several methods that can be employed to demodulate SSB. The simplest technique is to multiply $x_c(t)$ by a demodulation carrier and lowpass filter the result, as

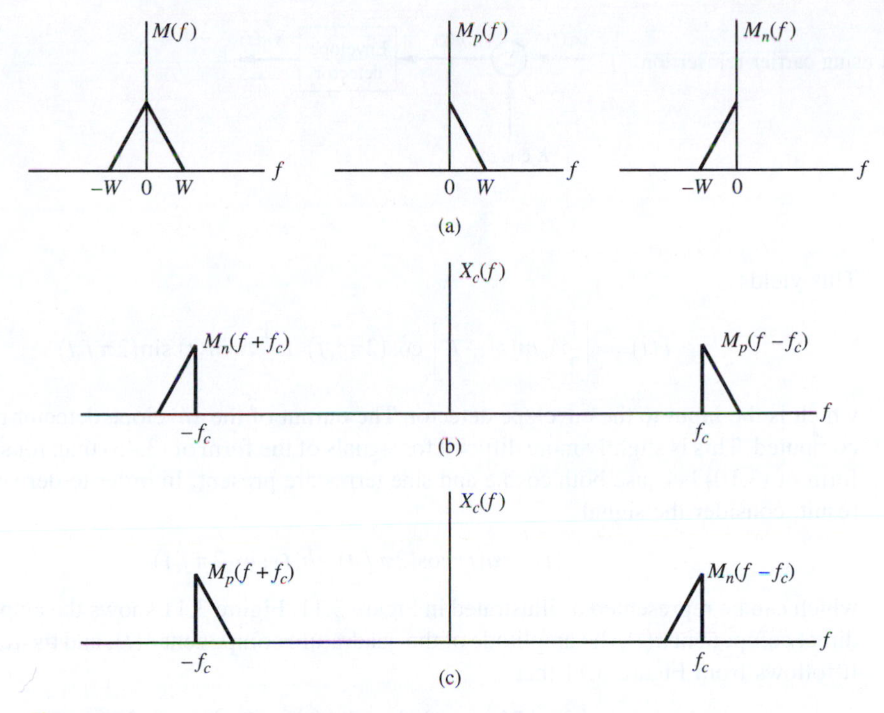

Figure 3.9
Alternative derivation of SSB signals. (a) $M(f), M_p(f)$, and $M_n(f)$. (b) Upper-sideband SSB signal. (c) Lower-sideband SSB signal.

illustrated in Figure 3.1(a). We assume a demodulation carrier having a phase error $\theta(t)$ that yields

$$d(t) = \left[\frac{1}{2}A_c m(t)\cos(2\pi f_c t) \pm \frac{1}{2}A_c \widehat{m}(t)\sin(2\pi f_c t)\right]\{4\cos[2\pi f_c t + \theta(t)]\} \qquad (3.42)$$

where the factor of 4 is chosen for mathematical convenience. The preceding expression can be written as

$$\begin{aligned} d(t) \;\; = \;\; & A_c m(t)\cos\theta(t) + A_c m(t)\cos[4\pi f_c t + \theta(t)] \\ & \mp A_c \widehat{m}(t)\sin\theta(t) \pm A_c \widehat{m}(t)\sin[4\pi f_c t + \theta(t)] \end{aligned} \qquad (3.43)$$

Lowpass filtering and amplitude scaling yield

$$y_D(t) = m(t)\cos\theta(t) \mp \widehat{m}(t)\sin\theta(t) \qquad (3.44)$$

for the demodulated output. Observation of (3.44) illustrates that for $\theta(t)$ equal to zero, the demodulated output is the desired message signal. However, if $\theta(t)$ is nonzero, the output consists of the sum of two terms. The first term is a time-varying attenuation of the message signal and is the output present in a DSB system operating in a similar manner. The second term is a crosstalk term and can represent serious distortion if $\theta(t)$ is not small.

Another useful technique for demodulating an SSB signal is carrier reinsertion, which is illustrated in Figure 3.10. The output of a local oscillator is added to the received signal $x_r(t)$.

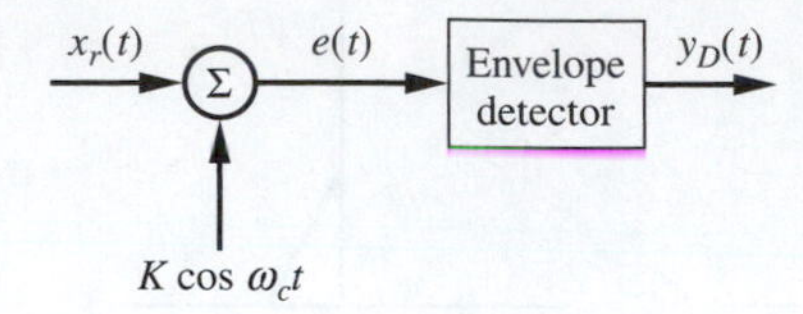

Figure 3.10
Demodulation using carrier reinsertion.

This yields

$$e(t) = \left[\frac{1}{2}A_c m(t) + K\right] \cos(2\pi f_c t) \pm \frac{1}{2}A_c \widehat{m}(t) \sin(2\pi f_c t) \tag{3.45}$$

which is the input to the envelope detector. The output of the envelope detector must next be computed. This is slightly more difficult for signals of the form of (3.45) than for signals of the form of (3.10) because both cosine and sine terms are present. In order to derive the desired result, consider the signal

$$x(t) = a(t) \cos(2\pi f_c t) - b(t) \sin(2\pi f_c t) \tag{3.46}$$

which can be represented as illustrated in Figure 3.11. Figure 3.11 shows the amplitude of the direct component $a(t)$, the amplitude of the quadrature component $b(t)$, and the resultant $R(t)$. It follows from Figure 3.11 that

$$a(t) = R(t) \cos \theta(t) \text{ and } b(t) = R(t) \sin \theta(t)$$

This yields

$$x(t) = R(t)[\cos \theta(t) \cos(2\pi f_c t) - \sin \theta(t) \sin(2\pi f_c t)] \tag{3.47}$$

which is

$$x(t) = R(t) \cos[2\pi f_c t + \theta(t)] \tag{3.48}$$

where

$$\theta(t) = \tan^{-1}\left(\frac{b(t)}{a(t)}\right) \tag{3.49}$$

The instantaneous amplitude $R(t)$, which is the envelope of the signal, is given by

$$R(t) = \sqrt{a^2(t) + b^2(t)} \tag{3.50}$$

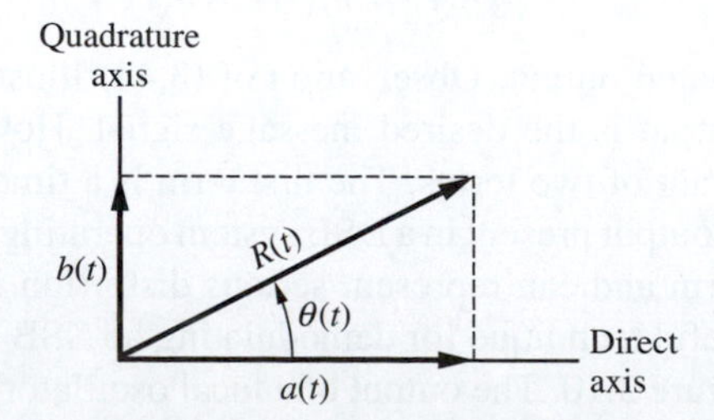

Figure 3.11
Direct-quadrature signal representation.

and will be the output of an envelope detector with $x(t)$ on the input if $a(t)$ and $b(t)$ are slowly varying with respect to $\cos\omega_c t$.

A comparison of (3.45) and (3.50) illustrates that the envelope of an SSB signal, after carrier reinsertion, is given by

$$y_D(t) = \sqrt{\left[\frac{1}{2}A_c m(t) + K\right]^2 + \left[\frac{1}{2}A_c \widehat{m}(t)\right]^2} \tag{3.51}$$

which is the demodulated output $y_D(t)$ in Figure 3.10. If K is chosen large enough such that

$$\left[\frac{1}{2}A_c m(t) + K\right]^2 \gg \left[\frac{1}{2}A_c \widehat{m}(t)\right]^2$$

the output of the envelope detector becomes

$$y_D(t) \cong \frac{1}{2}A_c m(t) + K \tag{3.52}$$

from which the message signal can easily be extracted. The development shows that carrier reinsertion requires that the locally generated carrier must be phase coherent with the original modulation carrier. This is easily accomplished in speech-transmission systems. The frequency and phase of the demodulation carrier can be manually adjusted until intelligibility of the speech is obtained.

EXAMPLE 3.2

As we saw in the preceding analysis, the concept of single sideband is probably best understood by using frequency-domain analysis. However, the SSB time-domain waveforms are also interesting and are the subject of this example. Assume that the message signal is given by

$$m(t) = \cos(2\pi f_1 t) - 0.4\cos(4\pi f_1 t) + 0.9\cos(6\pi f_1 t) \tag{3.53}$$

The Hilbert transform of $m(t)$ is

$$\widehat{m}(t) = \sin(2\pi f_1 t) - 0.4\sin(4\pi f_1 t) + 0.9\sin(6\pi f_1 t) \tag{3.54}$$

These two waveforms are shown in Figures 3.12(a) and (b).

As we have seen, the SSB signal is given by

$$x_c(t) = \frac{A_c}{2}[m(t)\cos(2\pi f_c t) \pm \widehat{m}(t)\sin(2\pi f_c t)] \tag{3.55}$$

with the choice of sign depending upon the sideband to be used for transmission. Using (3.46) to (3.50), we can place $x_c(t)$ in the standard form of (3.1). This gives

$$x_c(t) = R(t)\cos[2\pi f_c t + \theta(t)] \tag{3.56}$$

where the envelope $R(t)$ is

$$R(t) = \frac{A_c}{2}\sqrt{m^2(t) + \widehat{m}^2(t)} \tag{3.57}$$

and $\theta(t)$, which is the phase deviation of $x_c(t)$, is given by

$$\theta(t) = \pm\tan^{-1}\left(\frac{\widehat{m}(t)}{m(t)}\right) \tag{3.58}$$

The instantaneous frequency of $\theta(t)$ is therefore

$$\frac{d}{dt}[2\pi f_c t + \theta(t)] = 2\pi f_c \pm \frac{d}{dt}\left[\tan^{-1}\left(\frac{\widehat{m}(t)}{m(t)}\right)\right] \tag{3.59}$$

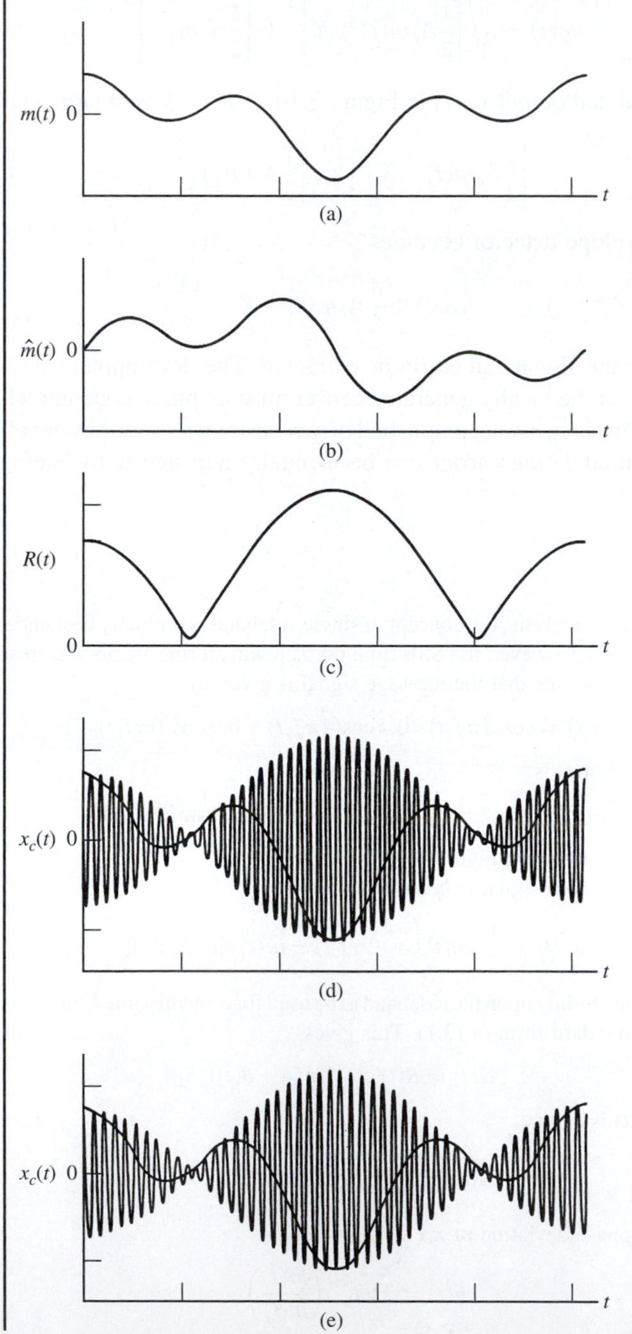

Figure 3.12
Time-domain signals for SSB system. (a) Message signal. (b) Hilbert transform of message signal. (c) Envelope of SSB signal. (d) Upper-sideband SSB signal with message signal. (e) Lower-sideband SSB signal with message signal.

From (3.57) we see that the envelope of the SSB signal is independent of the choice of the sideband. The instantaneous frequency, however, is a rather complicated function of the message signal and also depends upon the choice of sideband. We therefore see that the message signal $m(t)$ affects both the envelope and phase of the modulated carrier $x_c(t)$. In DSB and AM the message signal affected only the envelope of $x_c(t)$.

The envelope of the SSB signal, $R(t)$, is shown in Figure 3.12(c). The upper-sideband SSB signal is illustrated in Figure 3.12(d) and the lower-sideband SSB signal is shown in Figure 3.12(e). It is easily seen that both the upper-sideband and lower-sideband SSB signals have the envelope shown in Figure 3.12(c). The message signal $m(t)$ is also shown in Figure 3.12 (d) and (e).

■

3.1.4 Vestigial-Sideband Modulation

Vestigial-sideband (VSB) modulation overcomes two of the difficulties present in SSB modulation. By allowing a small amount, or vestige, of the unwanted sideband to appear at the output of an SSB modulator, the design of the sideband filter is simplified, since the need for sharp cutoff at the carrier frequency is eliminated. In addition, a VSB system has improved low-frequency response compared to SSB and can even have DC response. A simple example will illustrate the technique.

EXAMPLE 3.3

For simplicity, let the message signal be the sum of two sinusoids:

$$m(t) = A\cos(2\pi f_1 t) + B\cos(2\pi f_2 t) \tag{3.60}$$

This message signal is then multiplied by a carrier, $\cos(2\pi f_c t)$, to form a DSB signal

$$\begin{aligned} e_{\text{DSB}}(t) &= \frac{1}{2}A\cos[2\pi(f_c - f_1)t] + \frac{1}{2}A\cos[2\pi(f_c + f_1)t] \\ &\quad + \frac{1}{2}B\cos[2\pi(f_c - f_2)t] + \frac{1}{2}B\cos[2\pi(f_c + f_2)t] \end{aligned} \tag{3.61}$$

Figure 3.13(a) shows the single-sided spectrum of this signal. Prior to transmission a VSB filter is used to generate the VSB signal. Figure 3.13(b) shows the assumed amplitude response of the VSB filter. The phase response will be the subject of the next example. The skirt of the VSB filter must have the symmetry about the carrier frequency as shown. Figure 3.13(c) shows the single-sided spectrum of the VSB filter output. The spectrum shown in Figure 3.13(c) corresponds to the VSB signal

$$\begin{aligned} x_c(t) &= \frac{1}{2}A\epsilon\cos[2\pi(f_c - f_1)t] \\ &\quad + \frac{1}{2}A(1-\epsilon)\cos[2\pi(f_c + f_1)t] + \frac{1}{2}B\cos[2\pi(f_c + f_2)t] \end{aligned} \tag{3.62}$$

This signal can be demodulated by multiplying by $4\cos(2\pi f_c t)$ and lowpass filtering to remove the terms about $2f_c$. The result is

$$e(t) = A\epsilon\cos(2\pi f_1 t) + A(1-\epsilon)\cos(2\pi f_1 t) + B\cos(2\pi f_2 t) \tag{3.63}$$

or, combining the first two terms in the preceding expression,

$$e(t) = A\cos(2\pi f_1 t) + B\cos(2\pi f_2 t) \tag{3.64}$$

which is the assumed message signal.

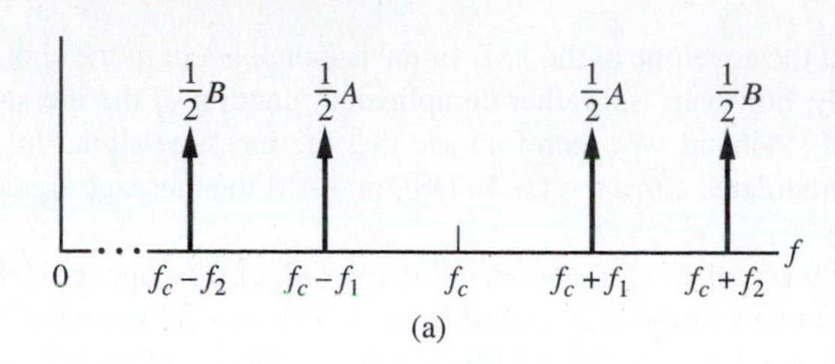

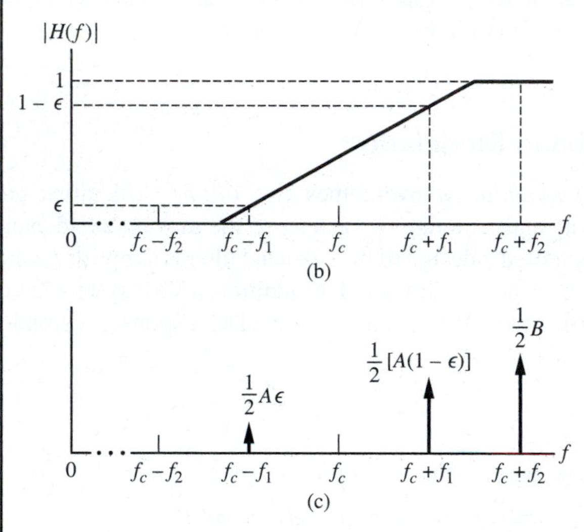

Figure 3.13 Generation of vestigial sideband. (a) DSB spectrum (single-sided). (b) VSB filter characteristic near f_c. (c) VSB spectrum.

■

EXAMPLE 3.4

The preceding example demonstrated the required amplitude response of the VSB filter. We now consider the phase response. Assume that the VSB filter has the following amplitude and phase responses for $f > 0$:

$$H(f_c - f_1) = \epsilon e^{-j\theta_a} \qquad H(f_c + f_1) = (1-\epsilon)e^{-j\theta_b}, \qquad H(f_c + f_2) = 1e^{-j\theta_c} \tag{3.65}$$

The VSB filter input is the DSB signal that, in complex envelope form, can be expressed as

$$x_{\text{DSB}}(t) = \text{Re}\left[\left(\frac{A}{2}e^{-j2\pi f_1 t} + \frac{A}{2}e^{j2\pi f_1 t} + \frac{B}{2}e^{-j2\pi f_2 t} + \frac{B}{2}e^{j2\pi f_2 t}\right)e^{j2\pi f_c t}\right] \tag{3.66}$$

Using the amplitude and phase characteristics of the VSB filter yields the VSB signal

$$x_c(t) = \text{Re}\left\{\left[\frac{A}{2}\epsilon e^{-j(2\pi f_1 t + \theta_a)} + \frac{A}{2}(1-\epsilon)e^{j(2\pi f_1 t - \theta_b)} + \frac{B}{2}e^{j(2\pi f_2 t - \theta_c)}\right]e^{j2\pi f_c t}\right\} \tag{3.67}$$

Demodulation is accomplished by multiplying by $2e^{-j2\pi f_c t}$ and taking the real part. This gives

$$e(t) = A\epsilon\cos(2\pi f_1 t + \theta_a) + A(1-\epsilon)\cos(2\pi f_1 t - \theta_b) + B\cos(2\pi f_2 t - \theta_c) \tag{3.68}$$

In order for the first two terms to combine as in (3.64), we must satisfy

$$\theta_a = -\theta_b \tag{3.69}$$

which shows that the phase response must have odd symmetry about f_c and, in addition, since $e(t)$ is real, the phase response of the VSB filter must also have odd phase response about $f = 0$. With $\theta_a = -\theta_b$

we have

$$e(t) = A\cos(2\pi f_1 t - \theta_b) + B\cos(2\pi f_2 t - \theta_c) \tag{3.70}$$

We still must determine the relationship between θ_c and θ_b.

As we saw in Chapter 2, in order for the demodulated signal $e(t)$ to be an undistorted (no amplitude or phase distortion) version of the original message signal $m(t)$, $e(t)$ must be an amplitude scaled and time-delayed version of $m(t)$. In other words

$$e(t) = Km(t-\tau) \tag{3.71}$$

Clearly the amplitude scaling $K = 1$. With time delay τ, $e(t)$ is

$$e(t) = A\cos[2\pi f_1(t-\tau)] + B\cos[2\pi f_2(t-\tau)] \tag{3.72}$$

Comparing (3.70) and (3.72) shows that

$$\theta_b = 2\pi f_1 \tau \tag{3.73}$$

and

$$\theta_c = 2\pi f_2 \tau \tag{3.74}$$

In order to have no phase distortion, the time delay must be the same for both components of $e(t)$. This gives

$$\theta_c = \frac{f_2}{f_1}\theta_b \tag{3.75}$$

We therefore see that the phase response of the VSB filter must be linear over the bandwidth of the input signal, which was to be expected from our discussion of distortionless systems in Chapter 2. ■

The slight increase in bandwidth required for VSB over that required for SSB is often more than offset by the resulting electronic simplifications. As a matter of fact, if a carrier component is added to a VSB signal, envelope detection can be used. The development of this technique is similar to the development of envelope detection of SSB with carrier reinsertion and is relegated to the problems. The process, however, is demonstrated in the following example.

EXAMPLE 3.5

In this example we consider the time-domain waveforms corresponding to VSB modulation and consider demodulation using envelope detection or carrier reinsertion. We assume the same message signal as was assumed Example 3.4. In other words,

$$m(t) = \cos(2\pi f_1 t) - 0.4\cos(4\pi f_1 t) + 0.9\cos(6\pi f_1 t) \tag{3.76}$$

The message signal $m(t)$ is shown in Figure 3.14(a). The VSB signal can be expressed as

$$\begin{aligned} x_c(t) = A_c[&\epsilon_1\cos[2\pi(f_c - f_1)t] + (1-\epsilon_1)\cos[2\pi(f_c - f_1)t]] \\ &-0.4\epsilon_2\cos[2\pi(f_c - 2f_1)t] - 0.4(1-\epsilon_2)\cos[2\pi(f_c - 2f_1)t \\ &+0.9\epsilon_3\cos[2\pi(f_c - 3f_1)t] + 0.9(1-\epsilon_3)\cos 2\pi(f_c - 3f_1)t] \end{aligned} \tag{3.77}$$

The modulated carrier, along with the message signal, is shown in Figure 3.14(b) for $\epsilon_1 = 0.64$, $\epsilon_2 = 0.78$, and $\epsilon_3 = 0.92$. The result of carrier reinsertion and envelope detection is shown in Figure 3.14(c). The message signal, biased by the amplitude of the carrier component, is clearly shown and will be the output of an envelope detector.

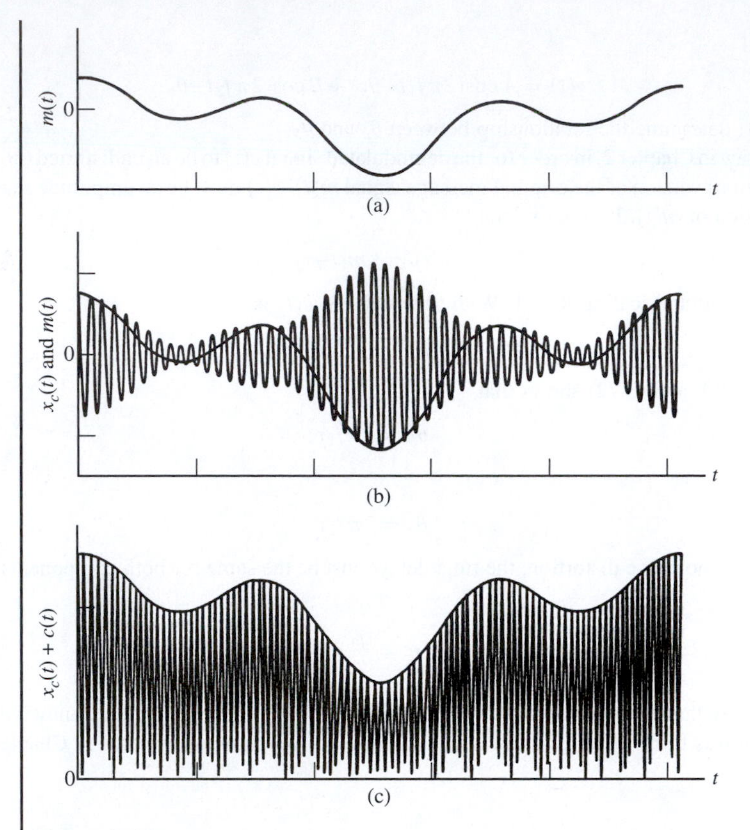

Figure 3.14
Time-domain signals for VSB system. (a) Message signal. (b) VSB signal and message signal. (c) Sum of VSB signal and carrier signal.

■

Vestigial sideband is currently used (at least until March 2009, when TV transmission becomes digital) in the United States for transmission of the video signal in commercial analog television broadcasting. However, exact shaping of the vestigial sideband is not carried out at the transmitter, but at the receiver, where signal levels are low. The filter in the transmitter simply bandlimits the video signal, as shown in Figure 3.15(a). The video carrier frequency is denoted f_v, and, as can be seen, the bandwidth of the video signal is approximately 5.25 MHz. The spectrum of the audio signal is centered about the audio carrier, which is 4.5 MHz above the video carrier. The modulation method used for audio transmission is FM. When we study FM in the following sections, you will understand the shape of the audio spectrum. Since the spectrum centered on the audio carrier is a line spectrum in Figure 3.15(a), a periodic audio signal is implied. This was done for clarity, and in practice the audio signal will have a continuous spectrum. Figure 3.15(b) shows the amplitude response of the receiver VSB filter.

Also shown in Figure 3.15(a), at a frequency 3.58 MHz above the video carrier, is the color carrier. Quadrature multiplexing, which we shall study in Section 3.6, is used with the color subcarrier so that two different signals are transmitted with the color carrier. These two signals, commonly referred to as the *I-channel* and *Q-channel* signals, carry luminance and chrominance (color) information necessary to reconstruct the image at the receiver. One

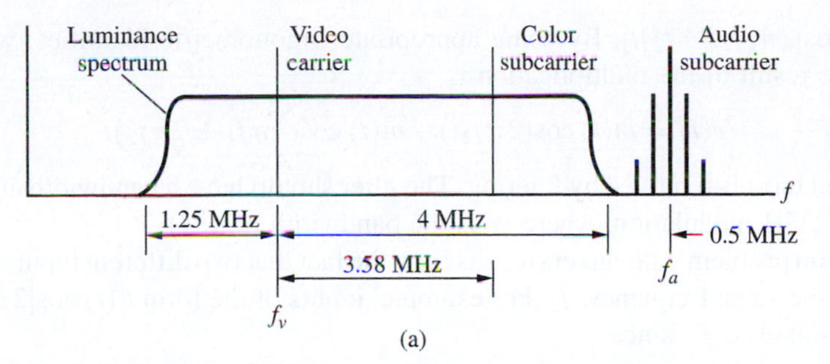

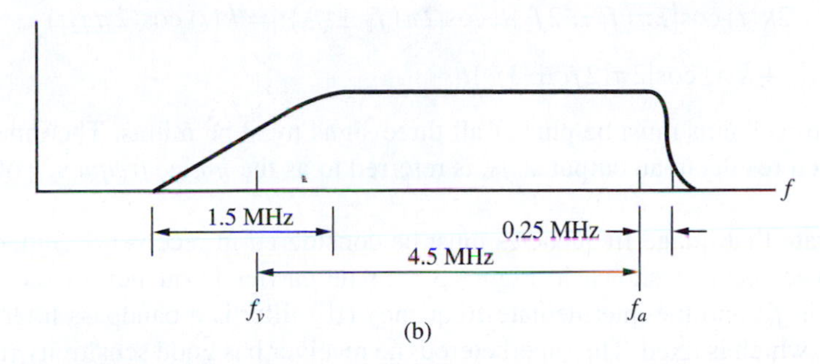

Figure 3.15 Transmitted spectrum and VSB filtering for television. (a) Spectrum of transmitted signal. (b) VSB filter in receiver.

problem faced by the designers of a system for the transmission of a color TV signal was that the transmitted signal was required to be compatible with existing black-and-white television receivers. Such a consideration is a significant constraint in the design process. A similar problem was faced by the designers of stereophonic radio receivers. We shall study this system in Section 3.7, and since the audio system is simpler than a color TV system, we can see how the compatibility problem was solved.

3.1.5 Frequency Translation and Mixing

It is often desirable to translate a bandpass signal to a new center frequency. Frequency translation is used in the implementation of communications receivers as well as in a number of other applications. The process of frequency translation can be accomplished by multiplication of the bandpass signal by a periodic signal and is referred to as *mixing*. A block diagram of a mixer is given in Figure 3.16. As an example, the bandpass signal $m(t)\cos(2\pi f_1 t)$ can be translated from f_1 to a new carrier frequency f_2 by multiplying it by a local oscillator signal of

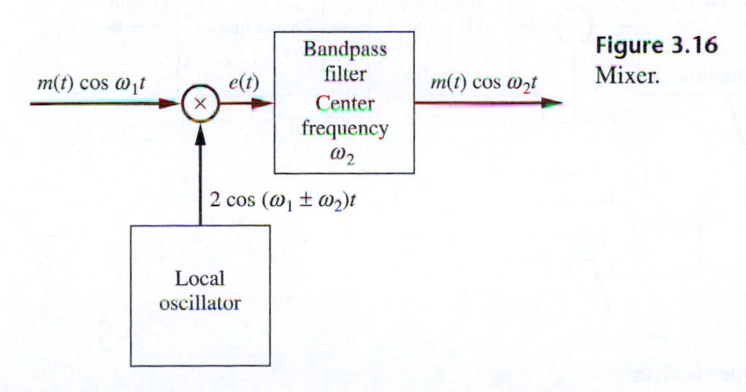

Figure 3.16 Mixer.

the form $2\cos[2\pi(f_1 \pm f_2)t]$. By using appropriate trigonometric identities, we can easily show that the result of the multiplication is

$$e(t) = m(t)\cos(2\pi f_2 t) + m(t)\cos(4\pi f_1 \pm 2\pi f_2)t \tag{3.78}$$

The undesired term is removed by filtering. The filter should have a bandwidth at least $2W$ for the assumed DSB modulation, where W is the bandwidth of $m(t)$.

A common problem with mixers results from the fact that two different input signals can be translated to the same frequency, f_2. For example, inputs of the form $k(t)\cos[2\pi(f_1 \pm 2f_2)t]$ are also translated to f_2, since

$$\begin{aligned} 2k(t)\cos[2\pi(f_1 \pm 2f_2)t]\cos[2\pi(f_1 \pm f_2)t] &= k(t)\cos(2\pi f_2 t) \\ &+ k(t)\cos[2\pi(2f_1 \pm 3f_2)t] \end{aligned} \tag{3.79}$$

In (3.79), all three signs must be plus or all three signs must be minus. The input frequency $f_1 \pm 2f_2$, which results in an output at f_2, is referred to as the *image frequency* of the desired frequency f_1.

To illustrate that image frequencies must be considered in receiver design, consider the superheterodyne receiver shown in Figure 3.17. The carrier frequency of the signal to be demodulated is f_c, and the intermediate-frequency (IF) filter is a bandpass filter with center frequency f_{IF}, which is fixed. The superheterodyne receiver has good sensitivity (the ability to detect weak signals) and selectivity (the ability to separate closely spaced signals). This results because the IF filter, which provides most of the predetection filtering, need not be tunable. Thus it can be a rather complex filter. Tuning of the receiver is accomplished by varying the frequency of the local oscillator. The superheterodyne receiver of Figure 3.17 is the mixer of Figure 3.16 with $f_c = f_1$ and $f_{\text{IF}} = f_2$. The mixer translates the input frequency f_c to the IF frequency f_{IF}. As shown previously, the image frequency $f_c \pm 2f_{\text{IF}}$, where the sign depends on the choice of local oscillator frequency, also will appear at the IF output. This means that if we are attempting to receive a signal having carrier frequency f_c, we can also receive a signal at $f_c + 2f_{\text{IF}}$ if the local oscillator frequency is $f_c + f_{\text{IF}}$ or a signal at $f_c - 2f_{\text{IF}}$ if the local oscillator frequency is $f_c - f_{\text{IF}}$. There is only one image frequency, and it is always

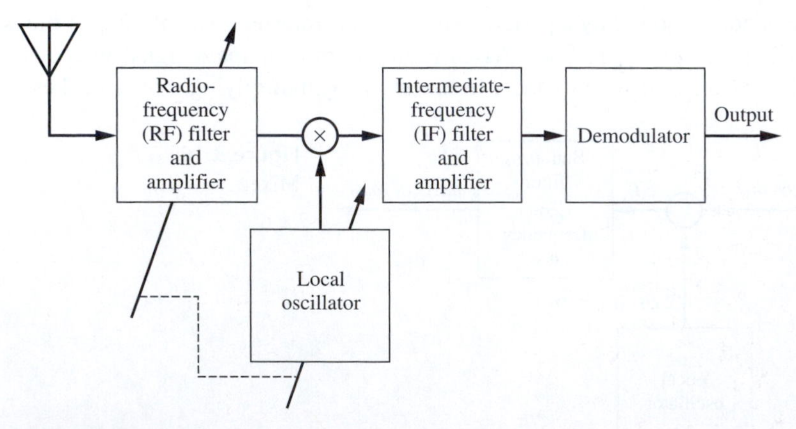

Figure 3.17
Superheterodyne receiver.

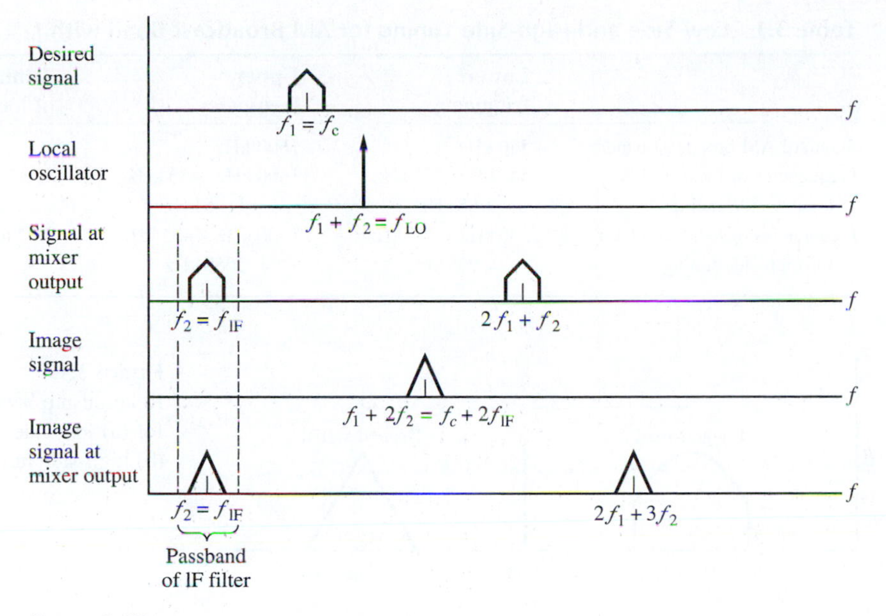

Figure 3.18
Illustration of image frequency (high-side tuning).

separated from the desired frequency by $2f_{IF}$. Figure 3.18 shows the desired signal and image signal for a local oscillator having the frequency

$$f_{LO} = f_c + f_{IF} \tag{3.80}$$

The image frequency can be eliminated by the radio-frequency (RF) filter. A standard IF frequency for AM radio is 455 kHz. Thus the image frequency is separated from the desired signal by almost 1 MHz. This shows that the RF filter need not be narrowband. Furthermore, since the AM broadcast band occupies the frequency range 540 kHz to 1.6 MHz, it is apparent that a tunable RF filter is not required, provided that stations at the high end of the band are not located geographically near stations at the low end of the band. Some inexpensive receivers take advantage of this fact. Additionally, if the RF filter is made tunable, it need be tunable only over a narrow range of frequencies.

One decision to be made when designing a superheterodyne receiver is whether the frequency of the local oscillator is to be below the frequency of the input carrier (*low-side tuning*) or above the frequency of the input carrier (*high-side tuning*). A simple example based on the standard AM broadcast band illustrates one major consideration. The standard AM broadcast band extends from 540 kHz to 1600 kHz. For this example, let us choose a common intermediate frequency, 455 kHz. As shown in Table 3.1, for low-side tuning, the frequency of the local oscillator must be variable from 85 to 1600 kHz, which represents a frequency range in excess of 13 to 1. If high-side tuning is used, the frequency of the local oscillator must be variable from 995 to 2055 kHz, which represents a frequency range slightly in excess of 2 to 1. Oscillators whose frequency must vary over a large ratio are much more difficult to implement than are those whose frequency varies over a small ratio.

Table 3.1 Low-Side and High-Side Tuning for AM Broadcast Band with f_{IF} = 455 kHz

	Lower frequency	Upper frequency	Tuning range of local oscillator
Standard AM broadcast band	540 kHz	1600 kHz	
Frequencies of local oscillator for low-side tuning	540 kHz−455 kHz = 85 kHz	1600 kHz−455 kHz = 1145 kHz	13.47 to 1
Frequencies of local oscillator for high-side tuning	540 kHz + 455 kHz = 995 kHz	1600 kHz + 455 kHz = 2055 kHz	2.07 to 1

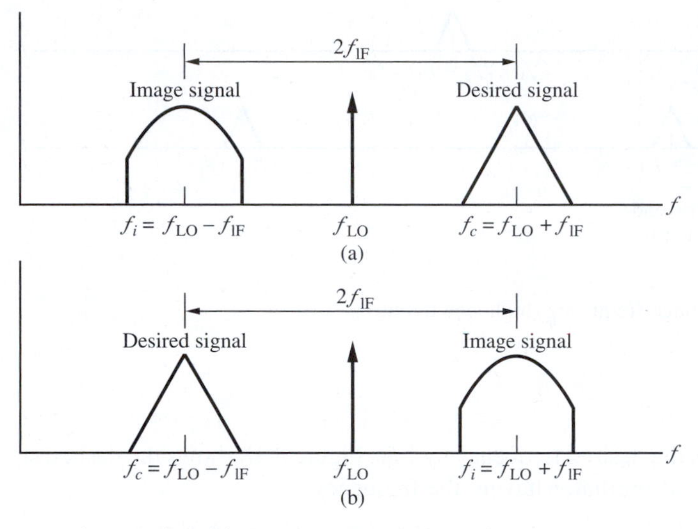

Figure 3.19
Relationship between f_c and f_i for (a) low-side tuning and (b) high-side tuning.

The relationship between the desired signal to be demodulated and the image signal is summarized in Figure 3.19 for low-side and high-side tuning. The desired signal to be demodulated has a carrier frequency of f_c and the image signal has a carrier frequency of f_i.

■ 3.2 ANGLE MODULATION

To generate angle modulation, the amplitude of the modulated carrier is held constant and either the phase or the time derivative of the phase of the carrier is varied linearly with the message signal $m(t)$. Thus the general angle-modulated signal is given by

$$x_c(t) = A_c \cos[2\pi f_c t + \phi(t)] \tag{3.81}$$

The instantaneous phase of $x_c(t)$ is defined as

$$\theta_i(t) = 2\pi f_c t + \phi(t) \tag{3.82}$$

and the instantaneous frequency, in hertz, is defined as

$$f_i(t) = \frac{1}{2\pi}\frac{d\theta_i}{dt} = f_c + \frac{1}{2\pi}\frac{d\phi}{dt} \tag{3.83}$$

The functions $\phi(t)$ and $d\phi/dt$ are known as the *phase deviation* and *frequency deviation* (in radians per second), respectively.

The two basic types of angle modulation are *phase modulation* (PM) and *frequency modulation* (FM). Phase modulation implies that the phase deviation of the carrier is proportional to the message signal. Thus, for phase modulation,

$$\phi(t) = k_p m(t) \tag{3.84}$$

where k_p is the *deviation constant* in radians per unit of $m(t)$. Similarly, FM implies that the frequency deviation of the carrier is proportional to the modulating signal. This yields

$$\frac{d\phi}{dt} = k_f m(t) \tag{3.85}$$

The phase deviation of a frequency-modulated carrier is given by

$$\phi(t) = k_f \int_{t_0}^{t} m(\alpha)\, d\alpha + \phi_0 \tag{3.86}$$

in which ϕ_0 is the phase deviation at $t = t_0$. It follows from (3.85) that k_f is the frequency-deviation constant, expressed in radians per second per unit of $m(t)$. Since it is often more convenient to measure frequency deviation in hertz, we define

$$k_f = 2\pi f_d \tag{3.87}$$

where f_d is known as the *frequency-deviation constant* of the modulator and is expressed in hertz per unit of $m(t)$.

With these definitions, the phase modulator output is

$$x_c(t) = A_c \cos\left[2\pi f_c t + k_p m(t)\right] \tag{3.88}$$

and the frequency modulator output is

$$x_c(t) = A_c \cos\left[2\pi f_c t + 2\pi f_d \int^{t} m(\alpha)\, d\alpha\right] \tag{3.89}$$

The lower limit of the integral is typically not specified, since to do so would require the inclusion of an initial condition as shown in (3.86).

Figures 3.20 and 3.21 illustrate the outputs of PM and FM modulators. With a unit step message signal, the instantaneous frequency of the PM modulator output is f_c for both $t < t_0$ and $t > t_0$. The phase of the unmodulated carrier is advanced by $k_p = \pi/2$ radians for $t > t_0$ giving rise to a signal that is discontinuous at $t = t_0$. The frequency of the output of the FM modulator is f_c for $t < t_0$, and the frequency is $f_c + f_d$ for $t > t_0$. The modulator output phase is, however, continuous at $t = t_0$.

With a sinusoidal message signal, the phase deviation of the PM modulator output is proportional to $m(t)$. The frequency deviation is proportional to the derivative of the phase deviation. Thus the instantaneous frequency of the output of the PM modulator is maximum when the *slope* of $m(t)$ is maximum and minimum when the *slope* of $m(t)$ is minimum. The frequency deviation of the FM modulator output is proportional to $m(t)$. Thus the instantaneous frequency of the FM modulator output is maximum when $m(t)$ is maximum and minimum when $m(t)$ is minimum. It should be noted that if $m(t)$ were not shown along with the modulator outputs, it would not be possible to distinguish the PM and FM modulator

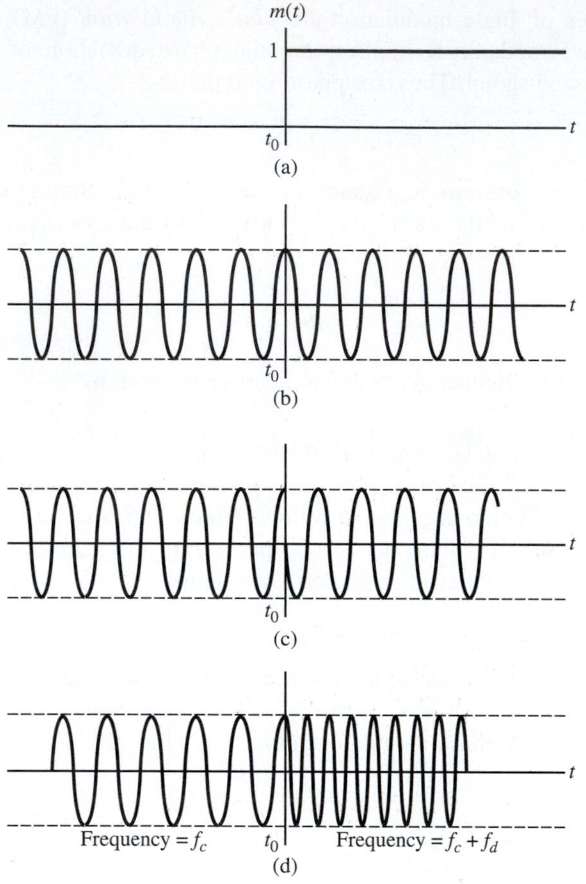

Figure 3.20
Comparison of PM and FM modulator outputs for a unit-step input. (a) Message signal. (b) Unmodulated carrier. (c) Phase modulator output $\left(k_p = \frac{1}{2}\pi\right)$. (d) Frequency modulator output.

outputs. In the following sections we will devote considerable attention to the case in which $m(t)$ is sinusoidal.

3.2.1 Narrowband Angle Modulation

An angle-modulated carrier can be represented in exponential form by writing (3.85) as

$$x_c(t) = \text{Re}\left(A_c e^{j\phi(t)} e^{j2\pi f_c t}\right) \tag{3.90}$$

where Re(·)implies that the real part of the argument is to be taken. Expanding $e^{j\phi(t)}$ in a power series yields

$$x_c(t) = \text{Re}\left\{A_c\left[1 + j\phi(t) - \frac{\phi^2(t)}{2!} - \cdots\right] e^{j2\pi f_c t}\right\} \tag{3.91}$$

If the maximum value of $|\phi(t)|$ is much less than unity, the modulated carrier can be approximated as

$$x_c(t) \cong \text{Re}\left[A_c e^{j2\pi f_c t} + A_c \phi(t) j e^{j2\pi f_c t}\right]$$

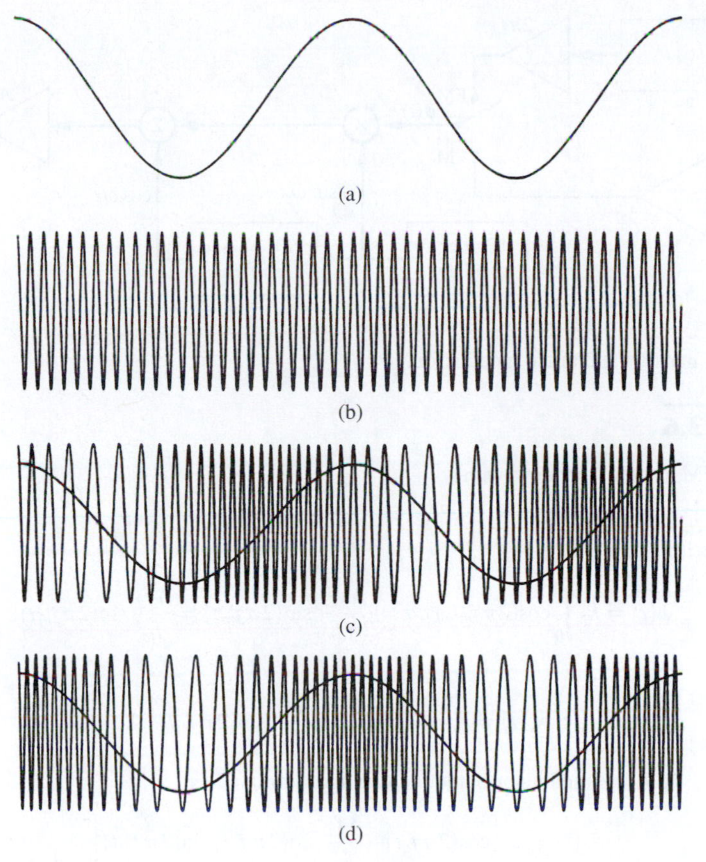

Figure 3.21
Angle modulation with sinusoidal message signal. (a) Message signal $m(t)$. (b) Unmodulated carrier $A_c \cos(2\pi f_c t)$. (c) Output of phase modulator with $m(t)$. (d) Output of frequency modulator with $m(t)$.

Taking the real part yields

$$x_c(t) \cong A_c \cos(2\pi f_c t) - A_c \phi(t) \sin(2\pi f_c t) \tag{3.92}$$

The form of (3.92) is reminiscent of AM. The modulator output contains a carrier component and a term in which a function of $m(t)$ multiplies a 90° phase-shifted carrier. This multiplication generates a pair of sidebands. Thus, if $\phi(t)$ has a bandwidth W, the bandwidth of a narrowband angle modulator output is $2W$. It is important to note, however, that the carrier and the resultant of the sidebands for narrowband angle modulation with sinusoidal modulation are in phase quadrature, whereas for AM they are not. This will be illustrated in Example 3.6.

The generation of narrowband angle modulation is easily accomplished using the method shown in Figure 3.22. The switch allows for the generation of either narrowband FM or narrowband PM. We will show later that narrowband angle modulation is useful for the generation of angle-modulated signals that are not necessarily narrowband, through a process called *narrowband-to-wideband conversion*.

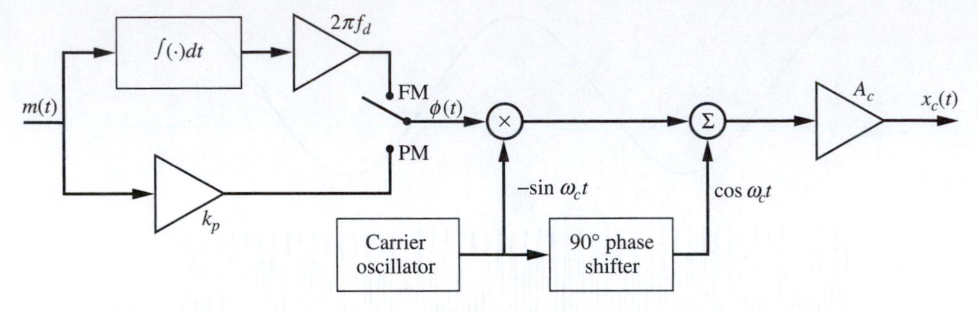

Figure 3.22
Generation of narrowband angle modulation.

EXAMPLE 3.6

Consider an FM system operating with

$$m(t) = A\cos(2\pi f_m t) \tag{3.93}$$

From (3.86), with t_0 equal to zero,

$$\phi(t) = k_f \int_0^t \cos(2\pi f_m \alpha)\, da = \frac{Ak_f}{2\pi f_m}\sin(2\pi f_m t) = \frac{Af_d}{f_m}\sin(2\pi f_m t) \tag{3.94}$$

so that

$$x_c(t) = A_c \cos\left[2\pi f_c t + \frac{Af_d}{f_m}\sin(2\pi f_m t)\right] \tag{3.95}$$

If $Af_d/f_m \ll 1$, the modulator output can be approximated as

$$x_c(t) = A_c\left[\cos(2\pi f_c t) - \frac{Af_d}{f_m}\sin(2\pi f_c t)\sin(2\pi f_m t)\right] \tag{3.96}$$

which is

$$x_c(t) = A_c\cos(2\pi f_c t) + \frac{A_c}{2}\frac{Af_d}{f_m}\{\cos[2\pi(f_c+f_m)t] - \cos[2\pi(f_c-f_m)t]\} \tag{3.97}$$

Thus, $x_c(t)$ can be written as

$$x_c(t) = A_c \operatorname{Re}\left\{\left[1 + \frac{Af_d}{2f_m}\left(e^{j2\pi f_m t} - e^{-j2\pi f_m t}\right)\right]e^{j2\pi f_c t}\right\} \tag{3.98}$$

It is interesting to compare this result with the equivalent result for an AM signal. Since sinusoidal modulation is assumed, the AM signal can be written as

$$x_c(t) = A_c[1 + a\cos(2\pi f_m t)]\cos(2\pi f_c t) \tag{3.99}$$

where $a = Af_d/f_m$ is the modulation index. Combining the two cosine terms yields

$$x_c(t) = A_c\cos(2\pi f_c t) + \frac{A_c a}{2}[\cos 2\pi(f_c+f_m)t + \cos 2\pi(f_c-f_m)t] \tag{3.100}$$

This can be written in exponential form as

$$x_c(t) = A_c \operatorname{Re}\left\{\left[1 + \frac{a}{2}\left(e^{j2\pi f_m t} + e^{-j2\pi f_m t}\right)\right]e^{j2\pi f_c t}\right\} \tag{3.101}$$

Comparing (3.98) and (3.101) illustrates the similarity between the two signals. The first, and most important, difference is the sign of the term at frequency $f_c - f_m$, which represents the lower sideband. The other difference is that the index a in the AM signal is replaced by Af_d/f_m in the narrowband FM signal.

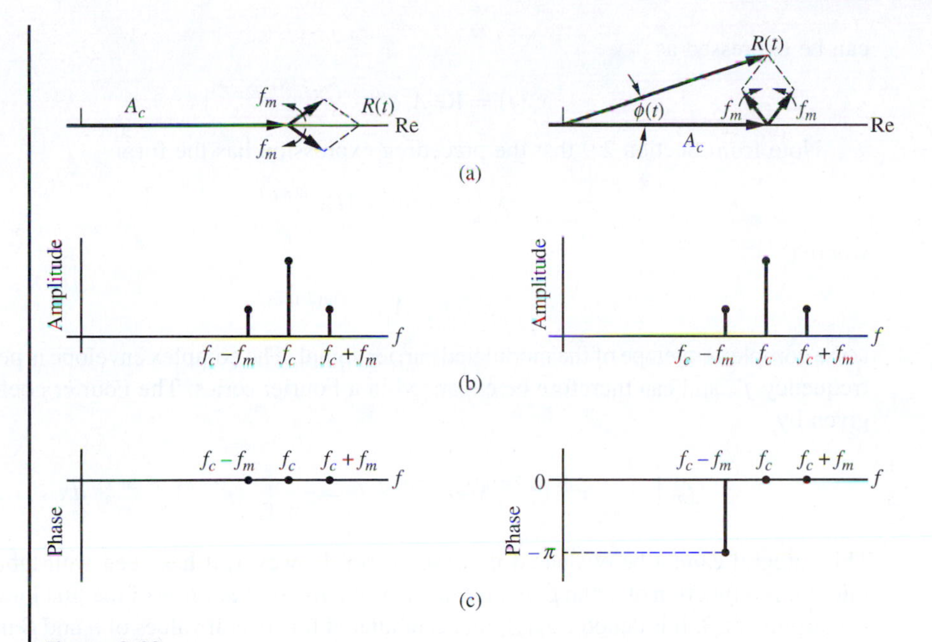

Figure 3.23
Comparison of AM and narrowband angle modulation. (a) Phasor diagrams. (b) Single-sided amplitude spectra. (c) Single-sided phase spectra.

We will see in the following section that $A f_d / f_m$ determines the modulation index for an FM signal. Thus these two parameters are in a sense equivalent since each defines the modulation index.

Additional insight is gained by sketching the phasor diagrams and the amplitude and phase spectra for both signals. These are given in Figure 3.23. The phasor diagrams are drawn using the carrier phase as a reference. The difference between AM and narrowband angle modulation with a sinusoidal message signal lies in the fact that the phasor resulting from the LSB and USB phasors adds to the carrier for AM but is in phase quadrature with the carrier for angle modulation. This difference results from the minus sign in the LSB component and is also clearly seen in the phase spectra of the two signals. The amplitude spectra are equivalent.

■

3.2.2 Spectrum of an Angle-Modulated Signal

The derivation of the spectrum of an angle-modulated signal is typically a very difficult task. However, if the message signal is sinusoidal, the instantaneous phase deviation of the modulated carrier is sinusoidal for both FM and PM, and the spectrum can be obtained with ease. This is the case we will consider. Even though we are restricting our attention to a very special case, the results provide much insight into the frequency-domain behavior of angle modulation. In order to compute the spectrum of an angle-modulated signal with a sinusoidal message signal, we assume that

$$\phi(t) = \beta \sin(2\pi f_m t) \tag{3.102}$$

The parameter β is known as the *modulation index* and is the maximum value of phase deviation for both FM and PM. The signal

$$x_c(t) = A_c \cos[2\pi f_c t + \beta \sin(2\pi f_m t)] \tag{3.103}$$

can be expressed as

$$x_c(t) = \text{Re}[A_c e^{j\beta \sin(2\pi f_m t)} e^{j2\pi f_c t}] \tag{3.104}$$

Note from Section 2.9 that the preceding expression has the form

$$x_c(t) = \text{Re}\left[\tilde{x}_c(t) e^{j2\pi f_c t}\right] \tag{3.105}$$

where

$$\tilde{x}_c(t) = A_c e^{j\beta \sin(2\pi f_m t)} \tag{3.106}$$

is the complex envelope of the modulated carrier signal. The complex envelope is periodic with frequency f_m and can therefore be expanded in a Fourier series. The Fourier coefficients are given by

$$f_m \int_{-1/2f_m}^{1/2f_m} e^{j\beta \sin(2\pi f_m t)} e^{-j2\pi n f_m t}\, dt = \frac{1}{2\pi} \int_{-\pi}^{\pi} e^{-[jnx - \beta \sin(x)]}\, dx \tag{3.107}$$

This integral cannot be evaluated in closed form. However, it has been well tabulated. The integral is a function of n and β and is known as the *Bessel Function* of the first kind of order n and argument β. It is denoted $J_n(\beta)$ and is tabulated for several values of n and β in Table 3.2. The significance of the underlining of various values in the table will be explained later.

Thus, with the aid of Bessel functions, the Fourier series for the complex envelope can be written as

$$e^{j\beta \sin(2\pi f_m t)} = \sum_{n=-\infty}^{\infty} J_n(\beta) e^{j2\pi n f_m t} \tag{3.108}$$

Table 3.2 Bessel Functions

n	$\beta=0.05$	$\beta=0.1$	$\beta=0.2$	$\beta=0.3$	$\beta=0.5$	$\beta=0.7$	$\beta=1.0$	$\beta=2.0$	$\beta=3.0$	$\beta=5.0$	$\beta=7.0$	$\beta=8.0$	$\beta=10.0$
0	0.999	0.998	0.990	0.978	0.938	0.881	0.765	0.224	−0.260	−0.178	0.300	0.172	−0.246
1	0.025	0.050	0.100	0.148	0.242	0.329	0.440	0.577	0.339	−0.328	−0.005	0.235	0.043
2		0.001	0.005	0.011	0.031	0.059	0.115	0.353	0.486	0.047	−0.301	−0.113	0.255
3				0.001	0.003	0.007	0.020	0.129	0.309	0.365	−0.168	−0.291	0.058
4						0.001	0.002	0.034	0.132	0.391	0.158	−0.105	−0.220
5								0.007	0.043	0.261	0.348	0.186	−0.234
6								0.001	0.011	0.131	0.339	0.338	−0.014
7									0.003	0.053	0.234	0.321	0.217
8										0.018	0.128	0.223	0.318
9										0.006	0.059	0.126	0.292
10										0.001	0.024	0.061	0.207
11											0.008	0.026	0.123
12											0.003	0.010	0.063
13											0.001	0.003	0.029
14												0.001	0.012
15													0.005
16													0.002
17													0.001

which allows the modulated carrier to be written as

$$x_c(t) = \text{Re}\left[\left(A_c \sum_{n=-\infty}^{\infty} J_n(\beta) e^{j2\pi n f_m t}\right) e^{j2\pi f_c t}\right] \tag{3.109}$$

Taking the real part yields

$$x_c(t) = A_c \sum_{n=-\infty}^{\infty} J_n(\beta) \cos[2\pi(f_c + n f_m)t] \tag{3.110}$$

from which the spectrum of $x_c(t)$ can be determined by inspection. The spectrum has components at the carrier frequency and has an infinite number of sidebands separated from the carrier frequency by integer multiples of the modulation frequency f_m. The amplitude of each spectral component can be determined from a table of values of the Bessel function. Such tables typically give $J_n(\beta)$ only for positive values of n. However, from the definition of $J_n(\beta)$ it can be determined that

$$J_{-n}(\beta) = J_n(\beta), \qquad n \text{ even} \tag{3.111}$$

and

$$J_{-n}(\beta) = -J_n(\beta), \qquad n \text{ odd} \tag{3.112}$$

These relationships allow us to plot the spectrum of (3.110), which is shown in Figure 3.24. The single-sided spectrum is shown for convenience.

A useful relationship between values of $J_n(\beta)$ for various values of n is the recursion formula

$$J_{n+1}(\beta) = \frac{2n}{\beta} J_n(\beta) + J_{n-1}(\beta) \tag{3.113}$$

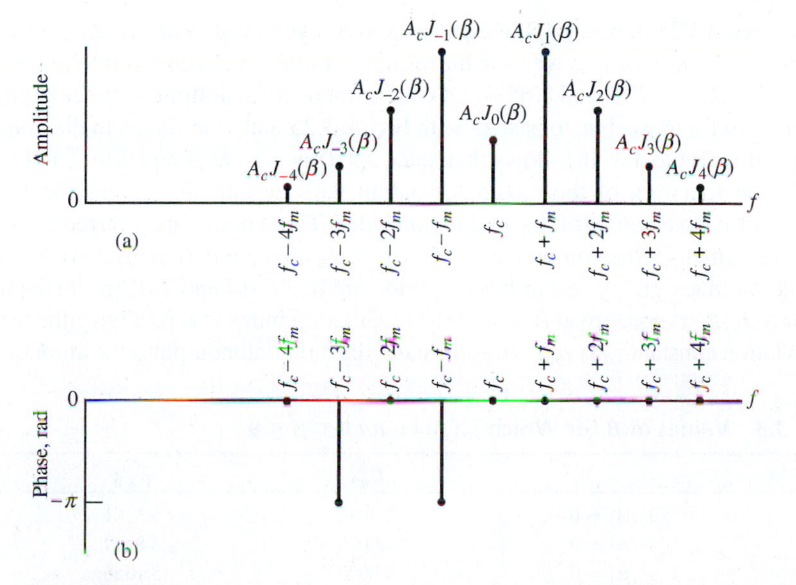

Figure 3.24
Spectra of an angle-modulated signal. (a) Single-sided amplitude spectrum. (b) Single-sided phase spectrum.

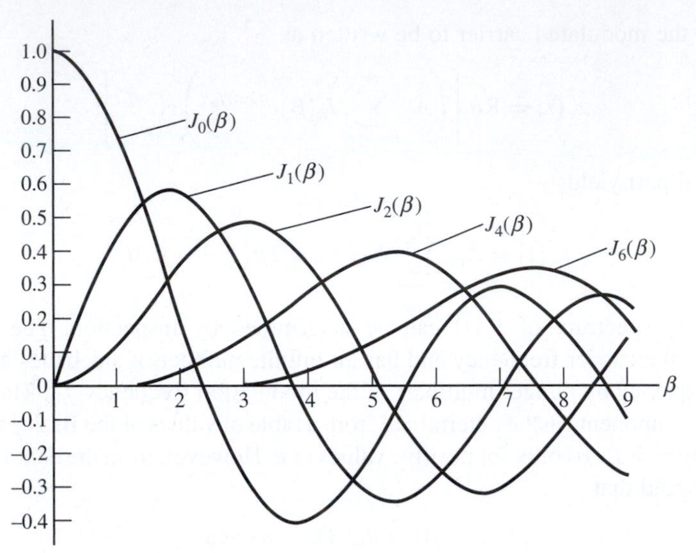

Figure 3.25
$J_n(\beta)$ as a function of β.

Thus, $J_{n+1}(\beta)$ can be determined from knowledge of $J_n(\beta)$ and $J_{n-1}(\beta)$. This enables us to compute a table of values of the Bessel function, as shown in Table 3.2, for any value of n from $J_0(\beta)$ and $J_1(\beta)$.

Figure 3.25 illustrates the behavior of the Fourier Bessel coefficients $J_n(\beta)$, for $n = 0, 1, 2, 4$, and 6 with $0 \leq \beta \leq 9$. Several interesting observations can be made. First, for $\beta \ll 1$, it is clear that $J_0(\beta)$ predominates, giving rise to narrowband angle modulation. It also can be seen that $J_n(\beta)$ oscillates for increasing β but that the amplitude of oscillation decreases with increasing β. Also of interest is the fact that the maximum value of $J_n(\beta)$ decreases with increasing n.

As Figure 3.25 shows, $J_n(\beta)$ is equal to zero at several values of β. Denoting these values of β by β_{nk}, where $k = 0, 1, 2$, we have the results in Table 3.3. As an example, $J_0(\beta)$ is zero for β equal to 2.4048, 5.5201, and 8.6537. Of course, there are an infinite number of points at which $J_n(\beta)$ is zero for any n, but consistent with Figure 3.25, only the values in the range $0 \leq \beta \leq 9$ are shown in Table 3.3. It follows that since $J_0(\beta)$ is zero at β equal to 2.4048, 5.5201, and 8.6537, the spectrum of the modulator output will not contain a component at the carrier frequency for these values of the modulation index. These points are referred to as *carrier nulls*. In a similar manner, the components at $f = f_c \pm f_m$ are zero if $J_1(\beta)$ is zero. The values of the modulation index giving rise to this condition are 0, 3.8317 and 7.0156. It should be obvious why only $J_0(\beta)$ is nonzero at $\beta = 0$. If the modulation index is zero, then either $m(t)$ is zero or the deviation constant f_d is zero. In either case, the modulator output is the unmodulated carrier,

Table 3.3 Values of β for Which $J_n(\beta) = 0$ for $0 \leq \beta \leq 9$

n		β_{n0}	β_{n1}	β_{n2}
0	$J_0(\beta) = 0$	2.4048	5.5201	8.6537
1	$J_1(\beta) = 0$	0.0000	3.8317	7.0156
2	$J_2(\beta) = 0$	0.0000	5.1356	8.4172
4	$J_4(\beta) = 0$	0.0000	7.5883	—
6	$J_6(\beta) = 0$	0.0000	—	—

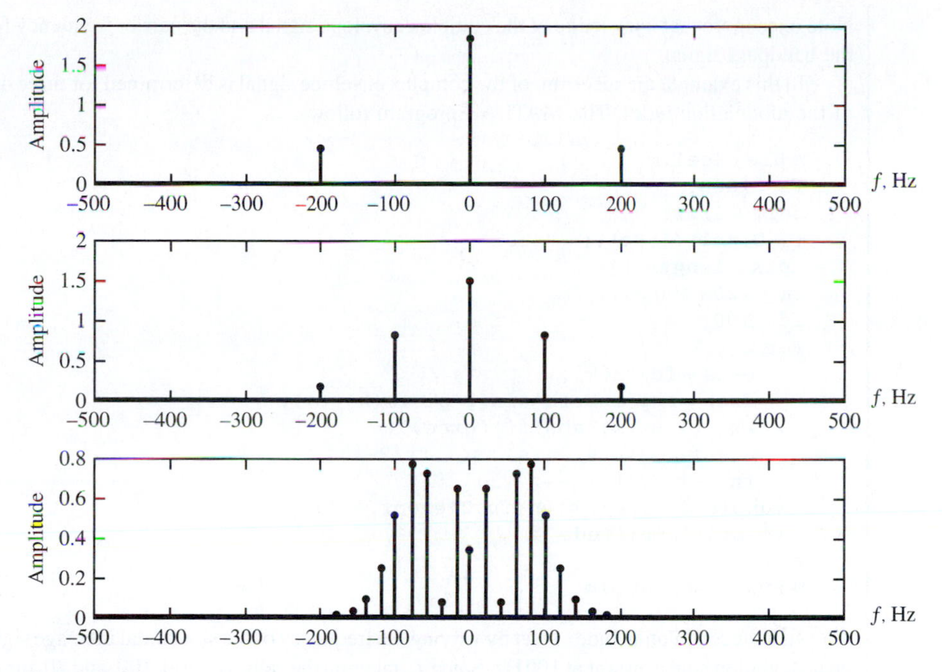

Figure 3.26
Amplitude spectrum of an FM complex envelope signal for increasing β and decreasing f_m.

which has frequency components only at the carrier frequency. In computing the spectrum of the modulator output, our starting point was the assumption that

$$\phi(t) = \beta \sin(2\pi f_m t) \tag{3.114}$$

Note that in deriving the spectrum of the angle modulated signal defined by (3.110), the modulator type (FM or PM) was not specified. The assumed $\phi(t)$, defined by (3.114), could represent either the phase deviation of a PM modulator with $m(t) = A\sin(\omega_m t)$ and an index $\beta = k_p A$, or an FM modulator with $m(t) = A\cos(\omega_m t)$ with index

$$\beta = \frac{2\pi f_d A}{\omega_m} = \frac{f_d A}{f_m} \tag{3.115}$$

Equation (3.115) shows that the modulation index for FM is a function of the modulation frequency. This is not the case for PM. The behavior of the spectrum of an FM signal is illustrated in Figure 3.26, as f_m is decreased while holding Af_d constant. For large values of f_m, the signal is narrowband FM, since only two sidebands are significant. For small values of f_m, many sidebands have significant value. Figure 3.26 is derived in the following computer example.

COMPUTER EXAMPLE 3.1

In this computer example we determine the spectrum of the complex envelope signal given by (3.106). In the next computer example we will determine and plot the two-sided spectrum which is determined from the complex envelope by writing the real bandpass signal as

$$x_c(t) = \frac{1}{2}\tilde{x}(t)e^{j2\pi f_c t} + \frac{1}{2}\tilde{x}_c^*(t)e^{-j2\pi f_c t}$$

Note once more that knowledge of the complex envelope signal and the carrier frequency fully determine the bandpass signal.

In this example the spectrum of the complex envelope signal is determined for three different values of the modulation index. The MATLAB program follows.

```
% file c3ce1.m
fs = 1000;
delt = 1/fs;
t = 0:delt:1-delt;
npts = length(t);
fm = [200 100 20];
fd = 100;
for k=1:3
    beta = fd/fm(k);
    cxce = exp(i*beta*sin(2*pi*fm(k)*t));
    as = (1/npts)*abs(fft(cxce));
    evenf = [as(fs/2:fs) as(1:fs/2-1)];
    fn = -fs/2:fs/2-1;
  subplot(3,1,k); stem(fn,2*evenf,'.')
  ylabel('Amplitude')
end
% End of script file.
```

Note that the modulation index is set by varying the frequency of the sinusoidal message signal f_m with the peak deviation held constant at 100 Hz. Since f_m takes on the values of 200, 100, and 20, the corresponding values of the modulation index are 0.5, 1, and 5, respectively. The corresponding spectra of the complex envelope signal are illustrated as a function of frequency in Figure 3.26.

■

COMPUTER EXAMPLE 3.2

We now consider the calculation of the two-sided amplitude spectrum of an FM (or PM) signal using the FFT algorithm. As can be seen from the MATLAB code, a modulation index of 3 is assumed. Note the manner in which the amplitude spectrum is divided into positive frequency and negative frequency segments (line nine in the following program). The student should verify that the various spectral components fall at the correct frequencies and that the amplitudes are consistent with Bessel function values given in Table 3.2. The output of the MATLAB program are illustrated in Figure 3.27.

```
% File: c3ce2.m
fs = 1000; % sampling frequency
delt = 1/fs; % sampling increment
t = 0:delt:1-delt; % time vector
npts = length(t); % number of points
fn = (0:npts)-(fs/2); % frequency vector for plot
m = 3*cos(2*pi*25*t); % modulation
xc = sin(2*pi*200*t+m); % modulated carrier
asxc = (1/npts)*abs(fft(xc)); % amplitude spectrum
evenf = [asxc((npts/2):npts) asxc(1:npts/2)]; % even amplitude spectrum
stem(fn,evenf,'.');
xlabel('Frequency - Hz')
ylabel('Amplitude')
% End of script.file.
```

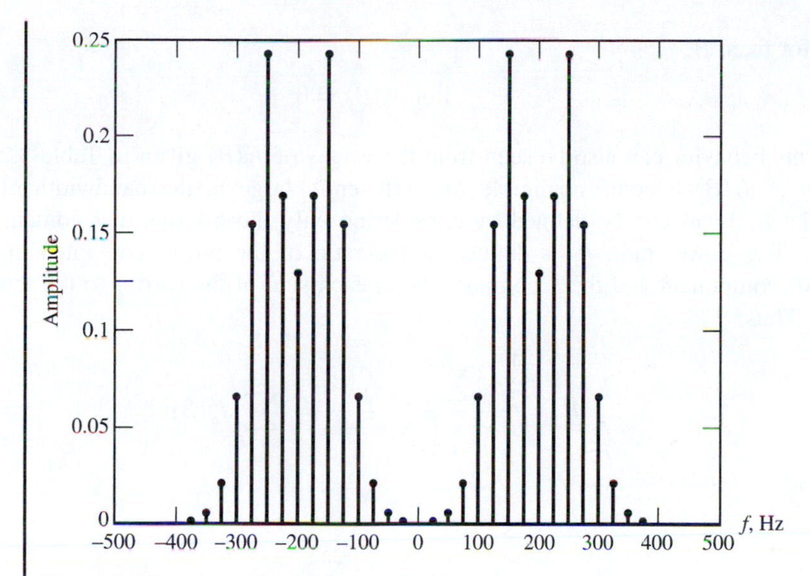

Figure 3.27
Two-sided amplitude spectrum computed using the FFT algorithm.

■

3.2.3 Power in an Angle-Modulated Signal

The power in an angle-modulated signal is easily computed from (3.81). Squaring (3.81) and taking the time-average value yields

$$\langle x_c^2(t)\rangle = A_c^2\langle\cos^2[\omega_c t + \phi(t)]\rangle \tag{3.116}$$

which can be written as

$$\langle x_c^2(t)\rangle = \frac{1}{2}A_c^2 + \frac{1}{2}A_c^2\langle\cos\{2[\omega_c t + \phi(t)]\}\rangle \tag{3.117}$$

If the carrier frequency is large so that $x_c(t)$ has negligible frequency content in the region of DC, the second term in (3.117) is negligible and

$$\langle x_c^2(t)\rangle = \frac{1}{2}A_c^2 \tag{3.118}$$

Thus the power contained in the output of an angle modulator is independent of the message signal. Constant transmitter power, independent of the message signal, is one important difference between angle modulation and linear modulation.

3.2.4 Bandwidth of Angle-Modulated Signals

Strictly speaking, the bandwidth of an angle-modulated signal is infinite, since angle modulation of a carrier results in the generation of an infinite number of sidebands. However, it can be seen from the series expansion of $J_n(\beta)$ (Appendix G, Table G.3) that for large n

$$J_n(\beta) \approx \frac{\beta^n}{2^n n!} \tag{3.119}$$

Thus for fixed β,

$$\lim_{n \to \infty} J_n(\beta) = 0 \tag{3.120}$$

This behavior can also be seen from the values of $J_n(\beta)$ given in Table 3.2. Since the values of $J_n(\beta)$ become negligible for sufficiently large n, the bandwidth of an angle-modulated signal can be defined by considering only those terms that contain significant power. The power ratio P_r is defined as the ratio of the power contained in the carrier $(n = 0)$ component and the k components on each side of the carrier to the total power in $x_c(t)$. Thus

$$P_r = \frac{\frac{1}{2}A_c^2 \sum_{n=-k}^{k} J_n^2(\beta)}{\frac{1}{2}A_c^2} = \sum_{n=-k}^{k} J_n^2(\beta) \tag{3.121}$$

or simply

$$P_r = J_0^2(\beta) + 2\sum_{n=1}^{k} J_n^2(\beta) \tag{3.122}$$

Bandwidth for a particular application is often determined by defining an acceptable power ratio, solving for the required value of k using a table of Bessel functions, and then recognizing that the resulting bandwidth is

$$B = 2kf_m \tag{3.123}$$

The acceptable value of the power ratio is dictated by the particular application of the system. Two power ratios are depicted in Table 3.2: $P_r \geq 0.7$ and $P_r \geq 0.98$. The value of n corresponding to k for $P_r \geq 0.7$ is indicated by a single underscore, and the value of n corresponding to k for $P_r \geq 0.98$ is indicated by a double underscore. For $P_r \geq 0.98$ it is noted that n is equal to the integer part of $1 + \beta$, so that

$$B \cong 2(\beta + 1)f_m \tag{3.124}$$

which will take on greater significance when Carson's rule is discussed in the following paragraph.

The preceding expression assumes sinusoidal modulation, since the modulation index β is defined only for sinusoidal modulation. For arbitrary $m(t)$, a generally accepted expression for bandwidth results if the deviation ratio D is defined as

$$D = \frac{\text{peak frequency deviation}}{\text{bandwidth of } m(t)} \tag{3.125}$$

which is

$$D = \frac{f_d}{W}(\max|m(t)|) \tag{3.126}$$

The deviation ratio plays the same role for nonsinusoidal modulation as the modulation index plays for sinusoidal systems. Replacing β by D and replacing f_m by W in (3.124), we obtain

$$B = 2(D + 1)W \tag{3.127}$$

This expression for bandwidth is generally referred to as *Carson's rule*. If $D \ll 1$, the bandwidth is approximately $2W$, and the signal is known as a *narrowband angle-modulated signal*. Conversely, if $D \gg 1$, the bandwidth is approximately $2DW = 2f_d\ (\max|m(t)|)$, which is twice the peak frequency deviation. Such a signal is known as a *wideband angle-modulated signal*.

EXAMPLE 3.7

In this example we consider an FM modulator with output

$$x_c(t) = 100\cos[2\pi(1000)t + \phi(t)] \tag{3.128}$$

The modulator operates with $f_d = 8$ and has the input message signal

$$m(t) = 5\cos 2\pi(8)t \tag{3.129}$$

The modulator is followed by a bandpass filter with a center frequency of 1000 Hz and a bandwidth of 56 Hz, as shown in Figure 3.28(a). Our problem is to determine the power at the filter output.

The peak deviation is $5f_d$ or 40 Hz, and $f_m = 8$ Hz. Thus, the modulation index is $40/5 = 8$. This yields the single-sided amplitude spectrum shown in Figure 3.28(b). Figure 3.28(c) shows the passband of

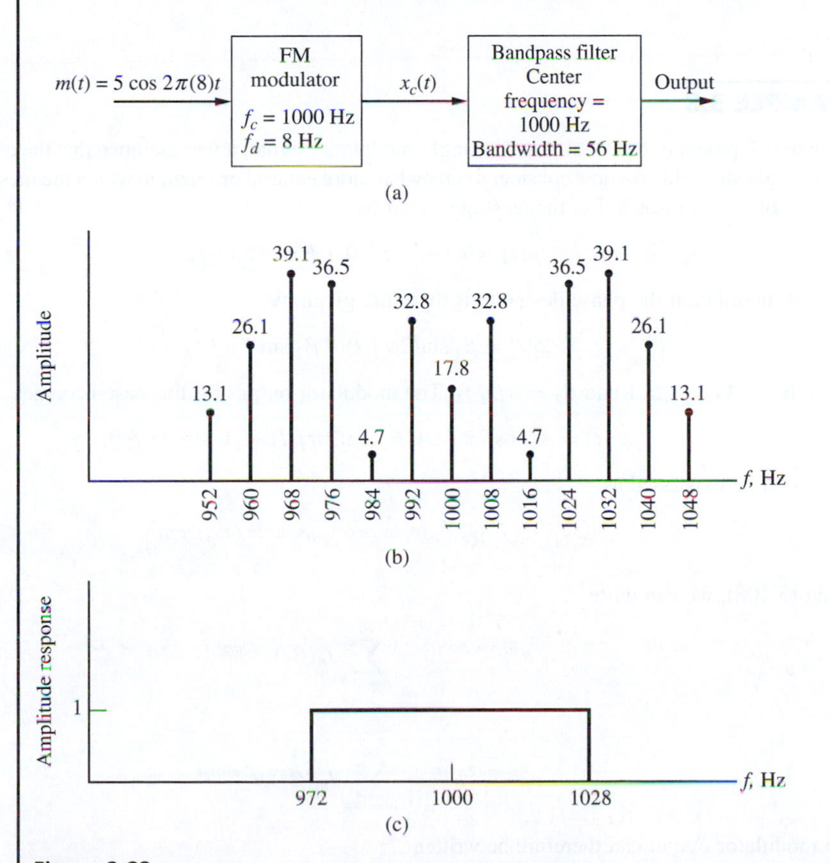

Figure 3.28
System and spectra for Example 3.5. (a) FM system. (b) Single-sided spectrum of modulator output. (c) Amplitude response of bandpass filter.

the bandpass filter. The filter passes the component at the carrier frequency and three components on each side of the carrier. Thus the power ratio is

$$P_r = J_0^2(5) + 2\left[J_1^2(5) + J_2^2(5) + J_3^2(5)\right] \tag{3.130}$$

which is

$$P_r = (0.178)^2 + 2\left[(0.328)^2 + (0.047)^2 + (0.365)^2\right] \tag{3.131}$$

This yields

$$P_r = 0.518 \tag{3.132}$$

The power at the output of the modulator is

$$\overline{x_c^2} = \frac{1}{2}A_c^2 = \frac{1}{2}(100)^2 = 5000 \text{ W} \tag{3.133}$$

The power at the filter output is the power of the modulator output multiplied by the power ratio. Thus the power at the filter output is

$$P_r\overline{x_c^2} = 2589 \text{ W} \tag{3.134}$$

■

EXAMPLE 3.8

In the development of the spectrum of an angle-modulated signal, it was assumed that the message signal was a single sinusoid. We now consider a somewhat more general problem in which the message signal is the sum of two sinusoids. Let the message signal be

$$m(t) = A\cos(2\pi f_1 t) + B\cos(2\pi f_2 t) \tag{3.135}$$

For FM modulation the phase deviation is therefore given by

$$\phi(t) = \beta_1 \sin(2\pi f_1 t) + \beta_2 \sin(2\pi f_2 t) \tag{3.136}$$

where $\beta_1 = Af_d/f_1 > 1$ and $\beta_2 = Bf_d/f_2$. The modulator output for this case becomes

$$x_c(t) = A_c \cos[2\pi f_c t + \beta_1 \sin(2\pi f_1 t) + \beta_2 \sin(2\pi f_2 t)] \tag{3.137}$$

which can be expressed as

$$x_c(t) = A_c \operatorname{Re}\left(e^{j\beta_1 \sin(2\pi f_1 t)} e^{j\beta_2 \sin(2\pi f_2 t)} e^{j2\pi f_c t}\right) \tag{3.138}$$

Using (3.108), we can write

$$e^{j\beta_1 \sin(2\pi f_1 t)} = \sum_{n=-\infty}^{\infty} J_n(\beta_1) e^{j2\pi n f_1 t} \tag{3.139}$$

and

$$e^{j\beta_2 \sin(2\pi f_2 t)} = \sum_{m=-\infty}^{\infty} J_m(\beta_2) e^{j2\pi m f_2 t} \tag{3.140}$$

The modulator output can therefore be written

$$x_c(t) = A_c \operatorname{Re}\left\{\left[\sum_{n=-\infty}^{\infty} J_n(\beta_1) e^{j2\pi n f_1 t} \sum_{m=-\infty}^{\infty} J_{m}, (\beta_2) e^{j2\pi m f_2 t}\right] e^{j2\pi f_c t}\right\} \tag{3.141}$$

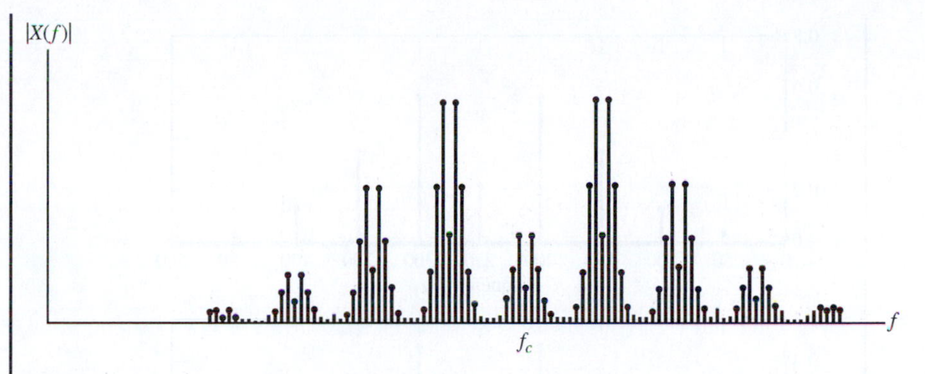

Figure 3.29
Amplitude spectrum for (3.142) with $\beta_1 = \beta_2$ and $f_2 = 12f_1$.

which, upon taking the real part, can be expressed

$$x_c(t) = A_c \sum_{n=-\infty}^{\infty} \sum_{m=-\infty}^{\infty} J_n(\beta_1) J_m(\beta_2) \cos[2\pi(f_c + n f_1 + m f_2)t] \tag{3.142}$$

Examination of the signal $x_c(t)$ shows that it not only contains frequency components at $f_c + nf_1$ and $f_c + mf_2$ but also contains frequency components at $f_c + nf_1 + mf_2$ for all combinations of n and m. Therefore, the spectrum of the modulator output due to a message signal consisting of the sum of two sinusoids contains additional components over the spectrum formed by the superposition of the two spectra resulting from the individual message components. This example therefore illustrates the nonlinear nature of angle modulation. The spectrum resulting from a message signal consisting of the sum of two sinusoids is shown in Figure 3.29 for the case in which $\beta_1 = \beta_2$ and $f_2 = 12f_1$.

■

COMPUTER EXAMPLE 3.3

In this computer example we consider a MATLAB program for computing the amplitude spectrum of an FM (or PM) signal having a message signal consisting of a pair of sinusoids. The single-sided amplitude spectrum is calculated (Note the multiplication by 2 in lines 10 and 11 in the following computer program.) The single sided spectrum is determined by using only the positive portion of the spectrum represented by the first $N/2$ points generated by the FFT program. In the following program N is represented by the variable `npts`.

Two plots are generated for the output. Figure 3.30(a) illustrates the spectrum with a single sinusoid for the message signal. The frequency of this sinusoidal component (50 Hz) is evident. Figure 3.30(b) illustrates the amplitude spectrum of the modulator output when a second component, having a frequency of 5 Hz, is added to the message signal. For this exercise the modulation index associated with each component of the message signal was carefully chosen to insure that the spectra were essentially constrained to lie within the bandwidth defined by the carrier frequency (250 Hz).

```
% File: c3ce3.m
fs = 1000; % sampling frequency
delt = 1/fs; % sampling increment
t = 0:delt:1-delt; % time vector
npts = length(t); % number of points
fn = (0:(npts/2))*(fs/npts); % frequency vector for plot
m1 = 2*cos(2*pi*50*t); % modulation signal 1
m2 = 2*cos(2*pi*50*t)+1*cos(2*pi*5*t); % modulation signal 2
```

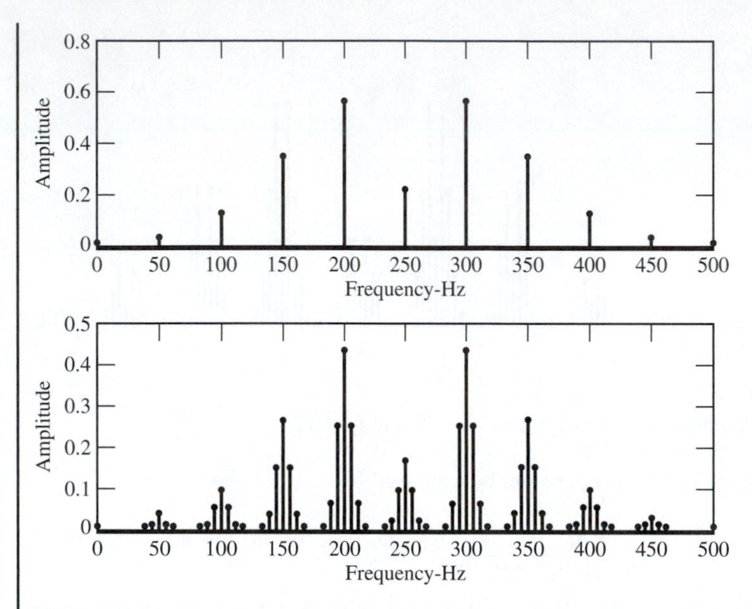

Figure 3.30
Frequency modulation spectra. (a) Single-tone modulating signal. (b) Two-tone modulating signal.

```
xc1 = sin(2*pi*250*t+m1); % modulated carrier 1
xc2 = sin(2*pi*250*t+m2); % modulated carrier 2
asxc1 = (2/npts)*abs(fft(xc1)); % amplitude spectrum 1
asxc2 = (2/npts)*abs(fft(xc2)); % amplitude spectrum 2
ampspec1 = asxc1(1:((npts/2)+1)); % positive frequency portion 1
ampspec2 = asxc2(1:((npts/2)+1)); % positive frequency portion 2
subplot(211)
stem(fn,ampspec1,'.k');
xlabel('Frequency - Hz')
ylabel('Amplitude')
subplot(212)
stem(fn,ampspec2,'.k');
xlabel('Frequency - Hz')
ylabel('Amplitude')
subplot(111)
% End of script file.
```

■

3.2.5 Narrowband-to-Wideband Conversion

One technique for generating wideband FM is illustrated in Figure 3.31. The carrier frequency of the narrowband frequency modulator is $f_{c1,}$ and the peak frequency deviation is f_{d1}. The frequency multiplier multiplies the argument of the input sinusoid by n. In other words, if the input of a frequency multiplier is

$$x(t) = A_c \cos[2\pi f_0 t + \phi(t)] \tag{3.143}$$

the output of the frequency multiplier is

$$y(t) = A_c \cos[2\pi n f_0 t + n\phi(t)] \tag{3.144}$$

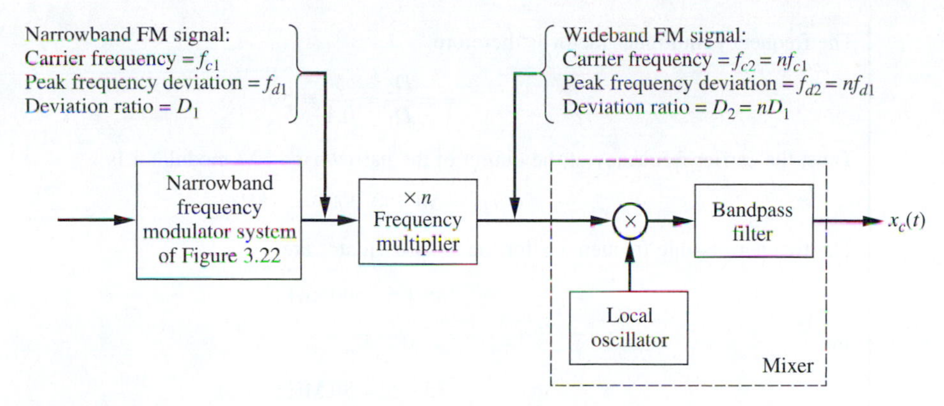

Figure 3.31
Frequency modulation utilizing narrowband-to-wideband conversion.

Assuming that the output of the local oscillator is

$$e_{\text{LO}}(t) = 2\cos(2\pi f_{\text{LO}}t) \tag{3.145}$$

results in

$$\begin{aligned} e(t) &= A_c \cos[2\pi(nf_0 + f_{\text{LO}})t + n\phi(t)] \\ &\quad + A_c \cos[2\pi(nf_0 - f_{\text{LO}})t + n\phi(t)] \end{aligned} \tag{3.146}$$

for the multiplier output. This signal is then filtered, using a bandpass filter having center frequency f_c, given by

$$f_c = nf_0 + f_{\text{LO}} \quad \text{or} \quad f_c = nf_0 - f_{\text{LO}}$$

This yields the output

$$x_c(t) = A_c \cos[2\pi f_c t + n\phi(t)] \tag{3.147}$$

The bandwidth of the bandpass filter is chosen in order to pass the desired term in (3.146). One can use Carson's rule to determine the bandwidth of the bandpass filter if the transmitted signal is to contain 98% of the power in $x_c(t)$.

The central idea in narrowband-to-wideband conversion is that the frequency multiplier changes both the carrier frequency and the deviation ratio by a factor of n, whereas the mixer changes the effective carrier frequency but does not affect the deviation ratio. This technique of implementing wideband frequency modulation is known as *indirect frequency modulation*.

EXAMPLE 3.9

A narrowband-to-wideband converter is implemented as shown in Figure 3.31. The output of the narrowband frequency modulator is given by (3.143) with $f_0 = 100,000$ Hz. The peak frequency deviation of $\phi(t)$ is 50 Hz and the bandwidth of $\phi(t)$ is 500 Hz. The wideband output $x_c(t)$ is to have a carrier frequency of 85 MHz and a deviation ratio of 5. Determine the frequency multiplier factor, n. Also determine two possible local oscillator frequencies. Finally, determine the center frequency and the bandwidth of the bandpass filter.

Solution

The deviation ratio at the output of the narrowband FM modulator is

$$D = \frac{f_{d1}}{W} = \frac{50}{500} = 0.1 \tag{3.148}$$

The frequency multiplier factor is therefore

$$n = \frac{D_2}{D_1} = \frac{5}{0.1} = 50 \tag{3.149}$$

Thus, the carrier frequency at the output of the narrowband FM modulator is

$$nf_0 = 50(100,000) = 5\text{ MHz} \tag{3.150}$$

The two permissible frequencies for the local oscillator are

$$85 + 5 = 90\text{ MHz} \tag{3.151}$$

and

$$85 - 5 = 80\text{ MHz} \tag{3.152}$$

The center frequency of the bandpass filter must be equal to the desired carrier frequency of the wideband output. Thus the center frequency of the bandpass filter is 85 MHz. The bandwidth of the bandpass filter is established using Carson's rule. From (3.127) we have

$$B = 2(D+1)W = 2(5+1)(500) \tag{3.153}$$

Thus

$$B = 6000\text{ Hz} \tag{3.154}$$

■

3.2.6 Demodulation of Angle-Modulated Signals

The demodulation of an FM signal requires a circuit that yields an output proportional to the frequency deviation of the input. Such circuits are known as *frequency discriminators*. If the input to an ideal discriminator is the angle modulated signal

$$x_r(t) = A_c \cos[2\pi f_c t + \phi(t)] \tag{3.155}$$

the output of the ideal discriminator is

$$y_D(t) = \frac{1}{2\pi} K_D \frac{d\phi}{dt} \tag{3.156}$$

For FM, $\phi(t)$ is given by

$$\phi(t) = 2\pi f_d \int^t m(\alpha)\, d\alpha \tag{3.157}$$

so that (3.156) becomes

$$y_D(t) = K_D f_d m(t) \tag{3.158}$$

The constant K_D is known as the *discriminator constant* and has units of volts per hertz. Since an ideal discriminator yields an output signal proportional to the frequency deviation from a carrier, it has a linear frequency-to-voltage transfer function, which passes through zero at $f = f_c$. This is illustrated in Figure 3.32.

The system characterized by Figure 3.32 can also be used to demodulate PM signals. Since $\phi(t)$ is proportional to $m(t)$ for PM, $y_D(t)$ given by (3.156) is proportional to the time derivative

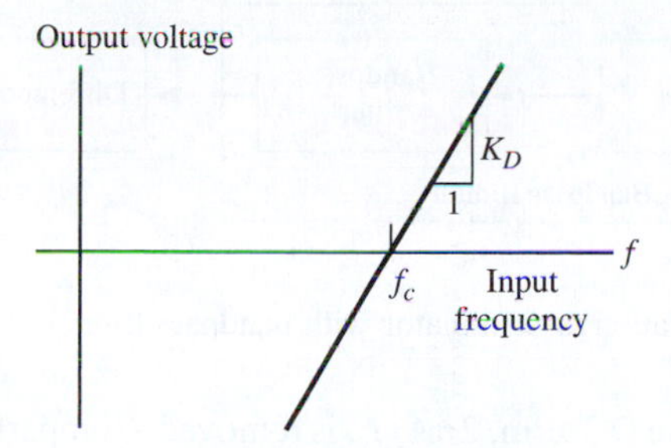

Figure 3.32
Ideal discriminator characteristic.

of $m(t)$ for PM inputs. Integration of the discriminator output yields a signal proportional to $m(t)$. Thus a demodulator for PM can be implemented as an FM discriminator followed by an integrator. We define the output of a PM discriminator as

$$y_D(t) = K_D k_p m(t) \tag{3.159}$$

It will be clear from the context whether $y_D(t)$ and K_D refer to an FM or a PM system.

An approximation to the characteristic illustrated in Figure 3.32 can be obtained by the use of a differentiator followed by an envelope detector, as shown in Figure 3.33. If the input to the differentiator is

$$x_r(t) = A_c \cos[2\pi f_c t + \phi(t)] \tag{3.160}$$

the output of the differentiator is

$$e(t) = -A_c\left(2\pi f_c + \frac{d\phi}{dt}\right)\sin[2\pi f_c t + \phi(t)] \tag{3.161}$$

This is exactly the same form as an AM signal, except for the phase deviation $\phi(t)$. Thus, after differentiation, envelope detection can be used to recover the message signal. The envelope of $e(t)$ is

$$y(t) = A_c\left(2\pi f_c + \frac{d\phi}{dt}\right) \tag{3.162}$$

and is always positive if

$$f_c > -\frac{1}{2\pi}\frac{d\phi}{dt} \quad \text{for all } t$$

which is usually satisfied since f_c is typically significantly greater than the bandwidth of the message signal. Thus, the output of the envelope detector is

$$y_D(t) = A_c\frac{d\phi}{dt} = 2\pi A_c f_d m(t) \tag{3.163}$$

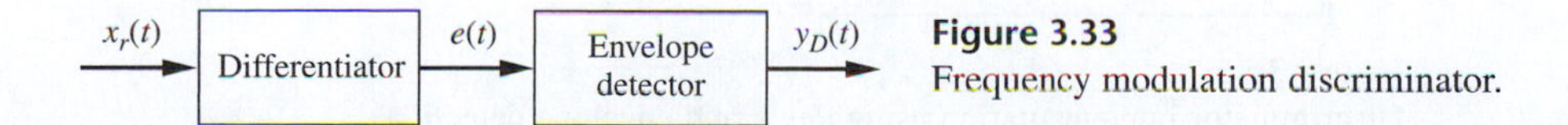

Figure 3.33
Frequency modulation discriminator.

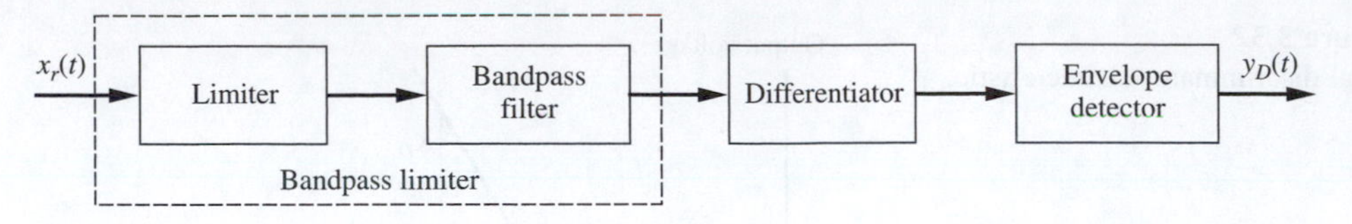

Figure 3.34
Frequency modulation discriminator with bandpass limiter.

assuming that the DC term, $2\pi A_c f_c$, is removed. Comparing (3.163) and (3.158) shows that the discriminator constant for this discriminator is

$$K_D = 2\pi A_c \tag{3.164}$$

We will see later that interference and channel noise perturb the amplitude A_c of $x_r(t)$. In order to ensure that the amplitude at the input to the differentiator is constant, a *limiter* is placed before the differentiator. The output of the limiter is a signal of square-wave type, which is $K \operatorname{sgn}[x_r(t)]$. A bandpass filter having center frequency f_c is then placed after the limiter to convert the signal back to the sinusoidal form required by the differentiator to yield the response defined by (3.161). The cascade combination of a limiter and a bandpass filter is known as a *bandpass limiter*. The complete discriminator is illustrated in Figure 3.34.

The process of differentiation can often be realized using a time-delay implementation, as shown in Figure 3.35. The signal $e(t)$, which is the input to the envelope detector, is given by

$$e(t) = x_r(t) - x_r(t-\tau) \tag{3.165}$$

which can be written

$$\frac{e(t)}{\tau} = \frac{x_r(t) - x_r(t-\tau)}{\tau} \tag{3.166}$$

Since, by definition,

$$\lim_{\tau \to 0} \frac{e(t)}{\tau} = \lim_{\tau \to 0} \frac{x_r(t) - x_r(t-\tau)}{\tau} = \frac{dx_r(t)}{dt} \tag{3.167}$$

it follows that for small τ,

$$e(t) \cong \tau \frac{dx_r(t)}{dt} \tag{3.168}$$

This is, except for the constant factor τ, identical to the envelope detector input shown in Figure 3.33 and defined by (3.161). The resulting discriminator constant K_D is $2\pi A_c \tau$. There are many

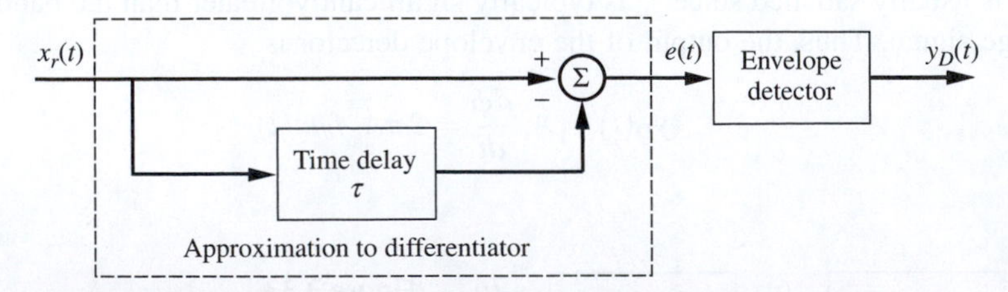

Figure 3.35
Discriminator implementation using delay and envelope detection.

other techniques that can be used to implement a discriminator. In Section 3.4 we will examine the phase-locked loop, which is an especially attractive implementation.

EXAMPLE 3.10

Consider the simple RC network shown in Figure 3.36(a). The transfer function is

$$H(f) = \frac{R}{R + 1/j2\pi fC} = \frac{j2\pi f\, RC}{1 + j2\pi f\, RC} \tag{3.169}$$

The amplitude response is shown in Figure 3.36(b). If all frequencies present in the input are low, so that

$$f \ll \frac{1}{2\pi RC}$$

the transfer function can be approximated by

$$H(f) = j2\pi f\, RC \tag{3.170}$$

Thus, for small f, the RC network has the linear amplitude–frequency characteristic required of an ideal discriminator. Equation (3.170) illustrates that for small f, the RC filter acts as a differentiator with gain RC. Thus, the RC network can be used in place of the differentiator in Figure 3.34 to yield a discriminator with

$$K_D = 2\pi A_c RC \tag{3.171}$$

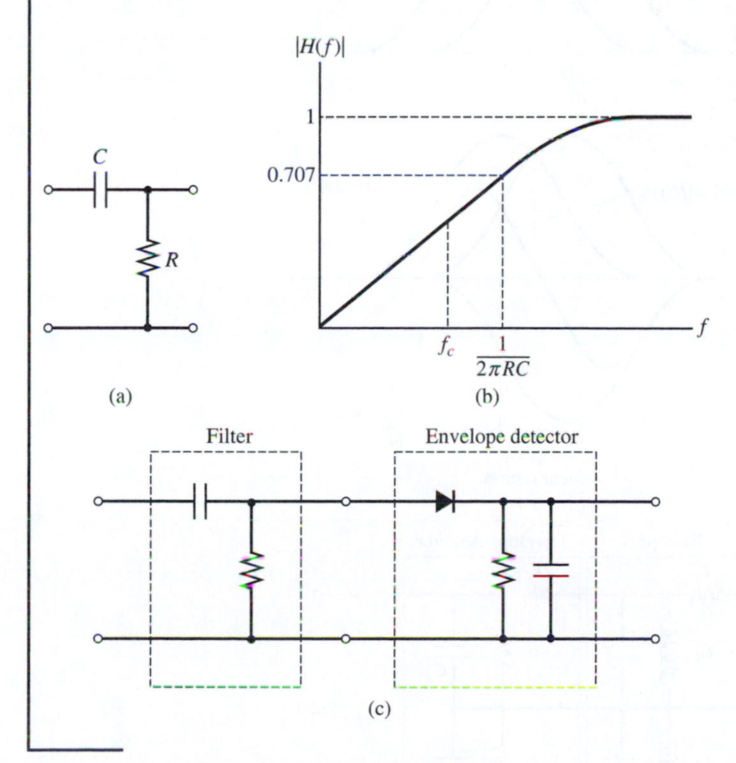

Figure 3.36
Implementation of a simple discriminator. (a) RC network. (b) Transfer function. (c) Simple discriminator.

Example 3.10 illustrates the essential components of a frequency discriminator, a circuit that has an amplitude response linear with frequency and an envelope detector. However, a highpass filter does not in general yield a practical implementation. This can be seen from the expression for K_D. Clearly the 3-dB frequency of the filter, $1/2\pi RC$, must exceed the carrier frequency f_c. In commercial FM broadcasting, the carrier frequency at the discriminator input, i.e., the IF frequency, is on the order of 10 MHz. As a result, the discriminator constant K_D is very small indeed.

A solution to the problem of a very small K_D is to use a bandpass filter, as illustrated in Figure 3.37. However, as shown in Figure 3.37(a), the region of linear operation is often

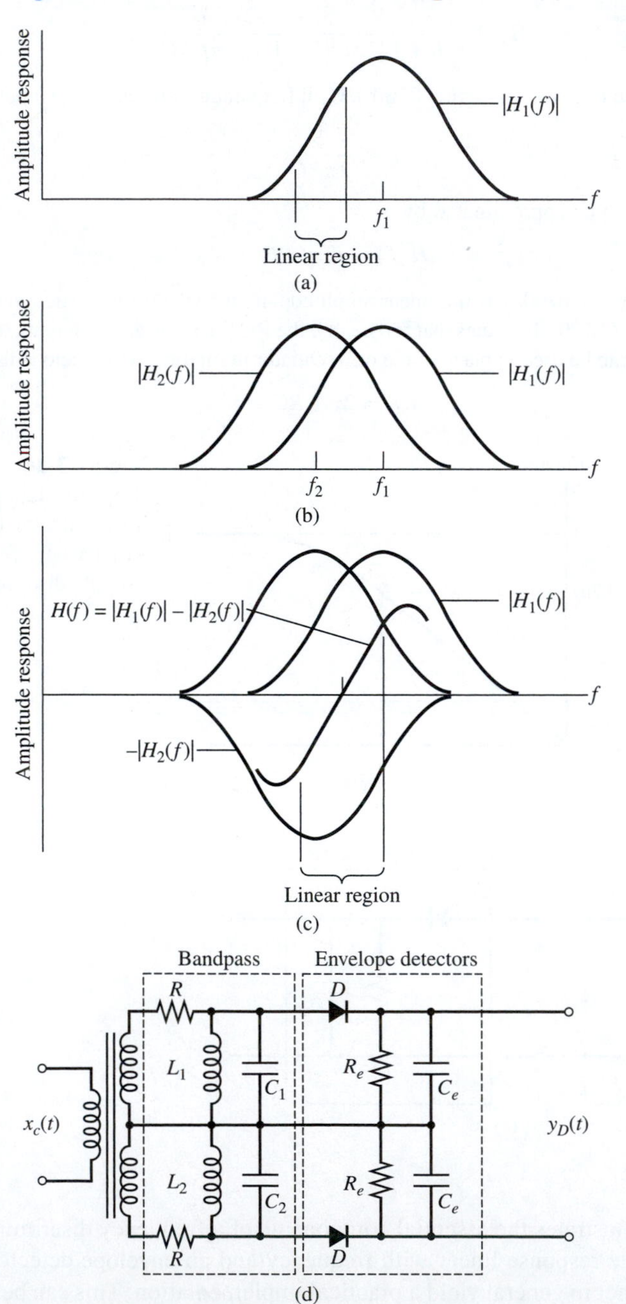

Figure 3.37
Derivation of balanced discriminator. (a) Bandpass filter. (b) Stagger-tuned bandpass filters. (c) Amplitude response $H(f)$ of balanced discriminator. (d) Balanced discriminator.

unacceptably small. In addition, use of a bandpass filter results in a DC bias on the discriminator output. This DC bias could of course be removed by a blocking capacitor, but the blocking capacitor would negate an inherent advantage of FM—namely, that FM has DC response. One can solve these problems by using two filters with staggered center frequencies f_1 and f_2, as shown in Figure 3.37(b). The magnitudes of the envelope detector outputs following the two filters are proportional to $|H_1(f)|$ and $|H_2(f)|$. Subtracting these two outputs yields the overall characteristic

$$H(f) = |H_1(f)| - |H_2(f)| \tag{3.172}$$

as shown in Figure 3.37(c). The combination is linear over a wider frequency range than would be the case for either filter used alone, and it is clearly possible to make $H(f_c) = 0$.

There are several techniques that can be used to combine the outputs of two envelope detectors. A differential amplifier can be used, for example. Another alternative, using a strictly passive circuit, is shown in Figure 3.37(d). A center-tapped transformer supplies the input signal $x_c(t)$ to the inputs of the two bandpass filters. The center frequencies of the two bandpass filters are given by

$$f_i = \frac{1}{2\pi\sqrt{L_i C_i}} \tag{3.173}$$

for $i = 1, 2$. The envelope detectors are formed by the diodes and the resistor–capacitor combinations $R_e C_e$. The output of the upper envelope detector is proportional to $|H_1(f)|$, and the output of the lower envelope detector is proportional to $|H_2(f)|$. The output of the upper envelope detector is the positive portion of its input envelope, and the output of the lower envelope detector is the negative portion of its input envelope. Thus $y_D(t)$ is proportional to $|H_1(f)| - |H_2(f)|$. This system is known as a *balanced discriminator* because the response to the undeviated carrier is balanced so that the net response is zero.

■ 3.3 INTERFERENCE

We now consider the effect of interference in communication systems. In real-world systems interference occurs from various sources, such as RF emissions from transmitters having carrier frequencies close to that of the carrier being demodulated. We also study interference because the analysis of systems in the presence of interference provides us with important insights into the behavior of systems operating in the presence of noise, which is the topic of Chapter 7. In this section we consider both linear modulation and angle modulation. It is important to understand the very different manner in which these two systems behave in the presence of interference.

3.3.1 Interference in Linear Modulation

As a simple case of linear modulation in the presence of interference, we consider the received signal having the spectrum (single sided) shown in Figure 3.38. The received signal consists of three components: a carrier component, a pair of sidebands representing a sinusoidal message

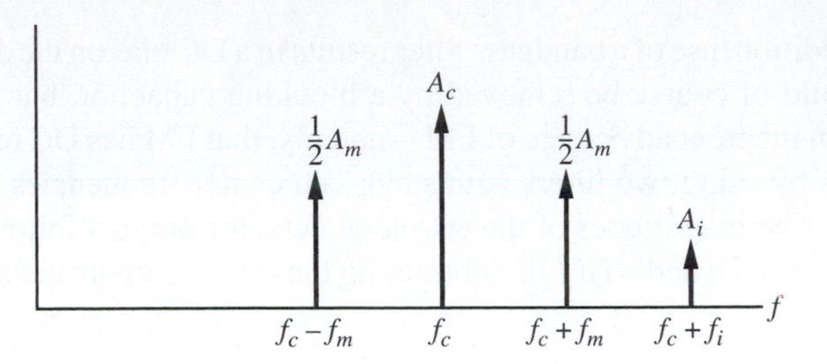

Figure 3.38
Assumed received-signal spectrum.

signal, and an undesired interfering tone of frequency $f_c + f_i$. The input to the demodulator is therefore

$$x_c(t) = A_c \cos(2\pi f_c t) + A_i \cos[2\pi(f_c + f_i)t] + A_m \cos(2\pi f_m t) \cos(2\pi f_c t) \tag{3.174}$$

Multiplying $x_c(t)$ by $2\cos(2\pi f_c t)$ and lowpass filtering (coherent demodulation) yields

$$y_D(t) = A_m \cos(2\pi f_m t) + A_i \cos(2\pi f_i t) \tag{3.175}$$

where we have assumed that the interference component is passed by the filter and that the DC term resulting from the carrier is blocked. From this simple example we see that the signal and interference are additive at the receiver output if the interference is additive at the receiver input. This result was obtained because the coherent demodulator operates as a linear demodulator.

The effect of interference with envelope detection is quite different because of the nonlinear nature of the envelope detector. The analysis with envelope detection is much more difficult than the coherent demodulation case. Some insight can be gained by writing $x_c(t)$ in a form that leads to the phasor diagram. In order to develop the phasor diagram, we write (3.174) in the form

$$x_r(t) = \text{Re}\left[\left(A_c + A_i e^{j2\pi f_i t} + \frac{1}{2}A_m e^{j2\pi f_m t} + \frac{1}{2}A_m e^{-j2\pi f_m t}\right) e^{j2\pi f_c t}\right] \tag{3.176}$$

The phasor diagram is constructed with respect to the carrier by taking the carrier frequency as equal to zero. In other words, we plot the phasor diagram corresponding to the complex envelope signal. The phasor diagrams are illustrated in Figure 3.39, both with and without interference. The output of an ideal envelope detector is $R(t)$ in both cases. The phasor diagrams illustrate that interference induces both an amplitude distortion and a phase deviation.

The effect of interference with envelope detection is determined by writing (3.174) as

$$\begin{aligned} x_r(t) = A_c \cos(2\pi f_c t) + A_m \cos(2\pi f_m t) \cos(2\pi f_c t) \\ + A_i[\cos(2\pi f_c t) \cos(2\pi f_i t) - \sin(2\pi f_c t) \sin(2\pi f_i t)] \end{aligned} \tag{3.177}$$

which is

$$x_r(t) = [A_c + A_m \cos(2\pi f_c t) + A_i \cos(2\pi f_i t)] \cos(2\pi f_c t) - A_i \sin(2\pi f_i t) \sin(2\pi f_c t) \tag{3.178}$$

If $A_c \gg A_i$, which is the usual case of interest, the last term in (3.178) is negligible compared to the first term and the output of the envelope detector is

$$y_D(t) \cong A_m \cos(2\pi f_m t) + A_i \cos(2\pi f_i t) \tag{3.179}$$

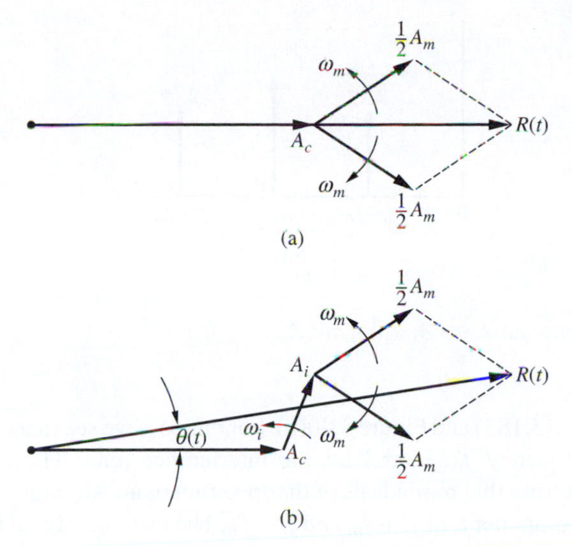

Figure 3.39
Phasor diagrams illustrating interference.
(a) Phasor diagram without interference.
(b) Phasor diagram with interference.

assuming that the DC term is blocked. Thus, for the small interference case, envelope detection and coherent demodulation are essentially equivalent.

If $A_c \ll A_i$, the assumption cannot be made that the last term of (3.178) is negligible, and the output is significantly different. To show this, (3.174) is rewritten as

$$\begin{aligned} x_r(t) &= A_c \cos[2\pi(f_c + f_i - f_i)t] + A_i \cos[2\pi(f_c + f_i)t] \\ &\quad + A_m \cos(2\pi f_m t) \cos[2\pi(f_c + f_i - f_i)t] \end{aligned} \tag{3.180}$$

which, when we use appropriate trigonometric identities, becomes

$$\begin{aligned} x_r(t) = A_c\{&\cos[2\pi(f_c + f_i)t] \cos(2\pi f_i t) + \sin[2\pi(f_c + f_i)t] \sin(2\pi f_i t)\} \\ &+ A_i \cos[2\pi(f_c + f_i)t] + A_m \cos(2\pi f_m t)\{\cos[2\pi(f_c + f_i)t] \cos(2\pi f_i t) \\ &+ \sin[2\pi(f_c + f_i)t] \sin(2\pi f_i t)\} \end{aligned} \tag{3.181}$$

Equation (3.181) can also be written as

$$\begin{aligned} x_r(t) &= [A_i + A_c \cos(2\pi f_i t) + A_m \cos(2\pi f_m t) \cos(2\pi f_i t)] \cos[2\pi(f_c + f_i)t] \\ &\quad + [A_c \sin(2\pi f_i t) + A_m \cos(2\pi f_m t) \sin(2\pi f_i t)] \sin[2\pi(f_c + f_i)t] \end{aligned} \tag{3.182}$$

If $A_i \gg A_c$, the last term in (3.182) is negligible with respect to the first term. It follows that the envelope detector output is approximated by

$$y_D(t) \cong A_c \cos(2\pi f_i t) + A_m \cos(2\pi f_m t) \cos(2\pi f_i t) \tag{3.183}$$

At this point, several observations are in order. In envelope detectors, the largest high-frequency component is treated as the carrier. If $A_c \gg A_i$, the effective demodulation carrier has a frequency f_c, whereas if $A_i \gg A_c$, the *effective* carrier frequency becomes the interference frequency $f_c + f_i$.

The spectra of the envelope detector output are illustrated in Figure 3.40 for $A_c \gg A_i$ and for $A_c \ll A_i$. For $A_c \gg A_i$ the interfering tone simply appears as a sinusoidal component, having frequency f_i at the output of the envelope detector. This illustrates that for $A_c \gg A_i$, the envelope detector performs as a linear demodulator. The situation is much different for

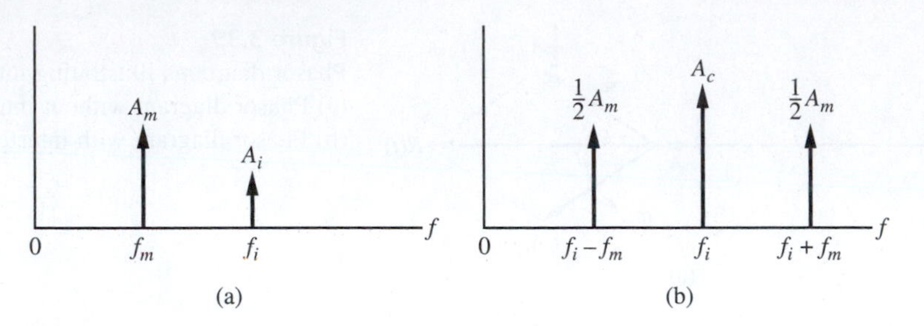

Figure 3.40
Envelope detector output spectra. (a) $A_c \gg A_i$. (b) $A_c \ll A_i$.

$A_c \ll A_i$, as can be seen from (3.183) and Figure 3.40(b). For this case we see that the sinusoidal message signal, having frequency f_m, modulates the interference tone. The output of the envelope detector has a spectrum that reminds us of the spectrum of an AM signal with carrier frequency f_i and sideband components at $f_i + f_m$ and $f_i - f_m$. The message signal is effectively lost. This degradation of the desired signal is called the *threshold effect* and is a consequence of the nonlinear nature of the envelope detector. We shall study the threshold effect in detail in Chapter 7 when we investigate the effect of noise in analog systems.

3.3.2 Interference in Angle Modulation

We now consider the effect of interference in angle modulation. We will see that the effect of interference in angle modulation is quite different from what was observed in linear modulation. Furthermore, we will see that the effect of interference in an FM system can be reduced by placing a lowpass filter at the discriminator output. We will consider this problem in considerable detail since the results will provide significant insight into the behavior of FM discriminators operating in the presence of noise.

Assume that the input to a PM or FM ideal discriminator is an unmodulated carrier plus an interfering tone at frequency $f_c + f_i$. Thus the input to the discriminator is assumed to have the form

$$x_r(t) = A_c\cos(2\pi f_c t) + A_i\cos[2\pi(f_c + f_i)t] \tag{3.184}$$

which can be written as

$$x_r(t) = A_c \cos\omega_c t + A_i \cos(2\pi f_i t)\cos(2\pi f_c t) - A_i \sin(2\pi f_i)\sin(2\pi f_c t) \tag{3.185}$$

Using (3.46) through (3.50), the preceding expression can be written as

$$x_r(t) = R(t)\cos[2\pi f_c t + \psi(t)] \tag{3.186}$$

in which the amplitude $R(t)$ is given by

$$R(t) = \sqrt{[A_c + A_i\cos(2\pi f_i t)]^2 + [A_i \sin(2\pi f_i t)]^2} \tag{3.187}$$

and the phase deviation $\psi(t)$ is given by

$$\psi(t) = \tan^{-1}\left(\frac{A_i \sin(2\pi f_i t)}{A_c + A_i\cos(2\pi f_i t)}\right) \tag{3.188}$$

If $A_c \gg A_i$, Equations (3.187) and (3.188) can be approximated

$$R(t) = A_c + A_i \cos(2\pi f_i t) \tag{3.189}$$

and

$$\psi(t) = \frac{A_i}{A_c} \sin(2\pi f_i t) \tag{3.190}$$

Thus (3.186) is

$$x_r(t) = A_c\left[1 + \frac{A_i}{A_c}\cos(2\pi f_i t)\right]\cos\left[2\pi f_i t + \frac{A_i}{A_c}\sin(2\pi f_i t)\right] \tag{3.191}$$

The instantaneous phase deviation $\psi(t)$ is given by

$$\psi(t) = \frac{A_i}{A_c} \sin(2\pi f_i t) \tag{3.192}$$

Thus, the ideal discriminator output for PM is

$$y_D(t) = K_D \frac{A_i}{A_c} \sin(2\pi f_i t) \tag{3.193}$$

and the output for FM is

$$\begin{aligned} y_D(t) &= \frac{1}{2\pi} K_D \frac{d}{dt} \frac{A_i}{A_c} \sin(2\pi f_i t) \\ &= K_D \frac{A_i}{A_c} f_i \cos(2\pi f_i t) \end{aligned} \tag{3.194}$$

For both cases, the discriminator output is a sinusoid of frequency f_i. The amplitude of the discriminator output, however, is proportional to the frequency f_i for the FM case. It can be seen that for small f_i, the interfering tone has less effect on the FM system than on the PM system and that the opposite is true for large values of f_i. Values of $f_i > W$, the bandwidth of $m(t)$, are of little interest, since they can be removed by a lowpass filter following the discriminator.

For larger values of A_i the assumption that $A_i \ll A_c$ cannot be made and (3.194) no longer can describe the discriminator output. If the condition $A_i \ll A_c$ does not hold, the discriminator is not operating above threshold and the analysis becomes much more difficult. Some insight into this case can be obtained from the phasor diagram, which is obtained by writing (3.184) in the form

$$x_r(t) = \operatorname{Re}\left[\left(A_c + A_i e^{j2\pi f_i t}\right) e^{j2\pi f_c t}\right] \tag{3.195}$$

The term in parentheses defines a phasor, which is the complex envelope signal. The phasor diagram is shown in Figure 3.41(a). The carrier phase is taken as the reference and the interference phase is

$$\theta(t) = 2\pi f_i t \tag{3.196}$$

Approximations to the phase of the resultant $\psi(t)$ can be determined using the phasor diagram.

From Figure 3.41(b) we see that the magnitude of the discriminator output will be small when $\theta(t)$ is near zero. This results because for $\theta(t)$ near zero, a given change in $\theta(t)$ will result

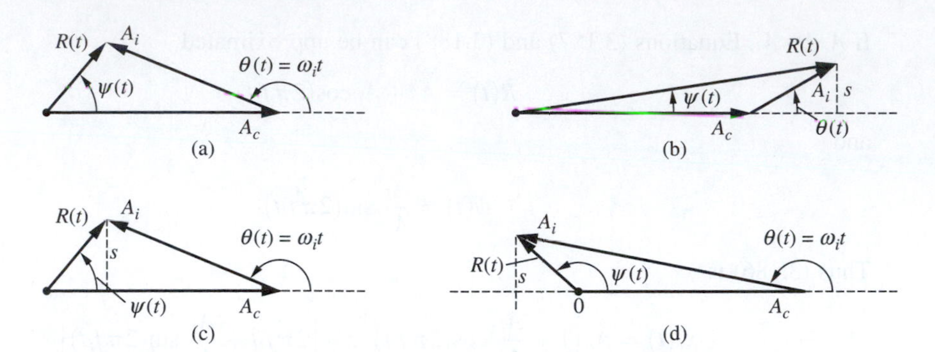

Figure 3.41
Phasor diagram for carrier plus single-tone interference. (a) Phasor diagram for general $\theta(t)$. (b) Phasor diagram for $\theta(t) \approx 0$. (c) Phasor diagram for $\theta(t) \approx \pi$ and $A_i \lesssim A_c$. (d) Phasor diagram for $\theta(t) \approx \pi$ and $A_i \gtrsim A_c$.

in a much smaller change in $\psi(t)$. Using the relationship between arc length s, angle θ and radius r, which is $s = \theta r$, we obtain

$$s = \theta(t)A_i \approx (A_c + A_i)\psi(t), \quad \theta(t) \approx 0 \tag{3.197}$$

Solving for $\psi(t)$ yields

$$\psi(t) \approx \frac{A_i}{A_c + A_i}\omega_i t \tag{3.198}$$

Since the discriminator output is defined by

$$y_D(t) = \frac{K_D}{2\pi}\frac{d\psi}{dt} \tag{3.199}$$

we have

$$y_D(t) = K_D \frac{A_i}{A_c - A_i} f_i, \quad \theta(t) \approx 0 \tag{3.200}$$

This is a positive quantity for $f_i > 0$ and a negative quantity for $f_i < 0$.

If A_i is slightly less than A_c, denoted $A_i \lesssim A_c$, and $\theta(t)$ is near π, a small positive change in $\theta(t)$ will result in a large negative change in $\psi(t)$. The result will be a negative spike appearing at the discriminator output. From Figure 3.41 (c) we can write

$$s = A_i(\pi - \theta(t)) \approx (A_c - A_i)\psi(t), \quad \theta(t) \approx \pi \tag{3.201}$$

which can be expressed

$$\psi(t) \approx \frac{A_i(\pi - 2\pi f_i t)}{A_c - A_i} \tag{3.202}$$

Using (3.199), we see that the discriminator output is

$$y_D(t) = -K_D \frac{A_i}{A_c - A_i} f_i, \quad \theta(t) \approx \pi \tag{3.203}$$

This is a negative quantity for $f_i > 0$ and a positive quantity for $f_i < 0$.

If A_i is slightly greater than A_c, denoted $A_i \gtrsim A_c$, and $\theta(t)$ is near π, a small positive change in $\theta(t)$ will result in a large positive change in $\psi(t)$. The result will be a positive spike appearing at the discriminator output. From Figure 3.41(d) we can write

$$s = A_i[\pi - \theta(t)] \approx (A_i - A_c)[\pi - \psi(t)], \quad \theta(t) \approx \pi \tag{3.204}$$

Solving for $\psi(t)$ and differentiating gives the discriminator output

$$y_D(t) \approx -K_D \frac{A_i}{A_c - A_i} f_i \tag{3.205}$$

Note that this is a positive quantity for $f_i > 0$ and a negative quantity for $f_i < 0$.

The phase deviation and discriminator output waveforms are shown in Figure 3.42 for $A_i = 0.1A_c$, $A_i = 0.9A_c$, and $A_i = 1.1A_c$. Figure 3.42(a) illustrates that for small A_i the phase deviation and the discriminator output are nearly sinusoidal as predicted by the results of the small interference analysis given in (3.192) and (3.194). For $A_i = 0.9A_c$, we see that we have a negative spike at the discriminator output as predicted by (3.203). For $A_c = 1.1A_c$, we have a

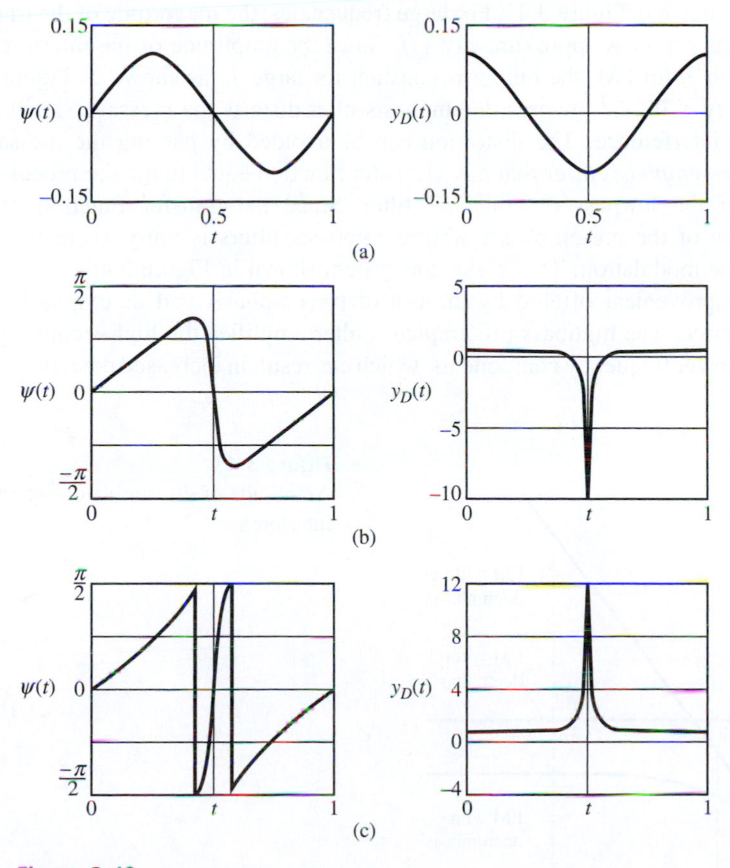

Figure 3.42
Phase deviation and discriminator outputs due to interference. (a) Phase deviation and discriminator output for $A_i = 0.1A_c$. (b) Phase deviation and discriminator output for $A_i = 0.9A_c$. (c) Phase deviation and discriminator output for $A_i = 1.1A_c$.

positive spike at the discriminator output as predicted by (3.205). Note that for $A_i > A_c$, the origin of the phasor diagram is encircled as $\theta(t)$ goes from 0 to 2π. In other words, $\psi(t)$ goes from 0 to 2π as $\theta(t)$ goes from 0 to 2π. The origin is not encircled if $A_i < A_c$. Thus the integral

$$\int_T \left(\frac{d\psi}{dt}\right) dt = \begin{cases} 2\pi, & A_i > A_c \\ 0, & A_i < A_c \end{cases} \tag{3.206}$$

where T is the time required for $\theta(t)$ to go from $\theta(t) = 0$ to $\theta(t) = 2\pi$. In other words, $T = 1/f_i$. Thus the area under the discriminator output curve is 0 for parts (a) and (b) of Figure 3.42 and $2\pi K_D$ for the discriminator output curve in Figure 3.42(c).The origin encirclement phenomenon will be revisited in Chapter 7 when demodulation of FM signals in the presence of noise is examined. An understanding of the interference results presented here will provide valuable insights when noise effects are considered.

For operation above threshold $A_i \ll A_c$, the severe effect of interference on FM for large f_i can be reduced by placing a filter, called a *de-emphasis filter*, at the FM discriminator output. This filter is typically a simple RC lowpass filter with a 3-dB frequency considerably less than the modulation bandwidth W. The de-emphasis filter effectively reduces the interference for large f_i, as shown in Figure 3.43. For large frequencies, the magnitude of the transfer function of a first-order filter is approximately $1/f$. Since the amplitude of the interference increases linearly with f_i for FM, the output is constant for large f_i, as shown in Figure 3.43.

Since $f_3 < W$, the lowpass de-emphasis filter distorts the message signal in addition to combating interference. The distortion can be avoided by passing the message through a highpass *pre-emphasis filter* that has a transfer function equal to the reciprocal of the transfer function of the lowpass de-emphasis filter. Since the transfer function of the cascade combination of the pre-emphasis and de-emphasis filters is unity, there is no detrimental effect on the modulation. This yields the system shown in Figure 3.44.

The improvement offered by the use of pre-emphasis and de-emphasis is not gained without a price. The highpass pre-emphasis filter amplifies the high-frequency components relative to lower frequency components, which can result in increased deviation and bandwidth

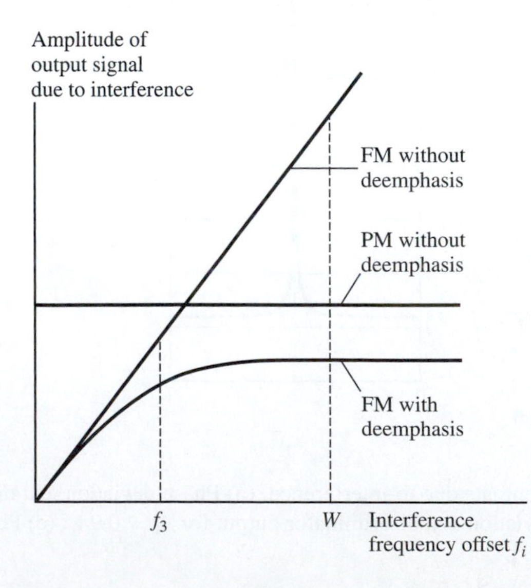

Figure 3.43
Amplitude of discriminator output due to interference.

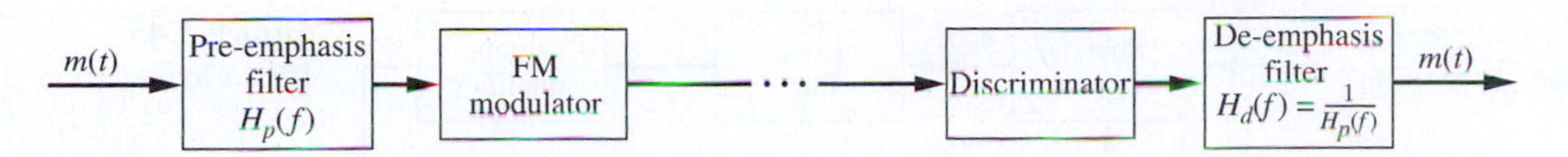

Figure 3.44
Frequency modulation system with pre-emphasis and de-emphasis.

requirements. We shall see in Chapter 7, when the impact of channel noise is studied, that the use of pre-emphasis and de-emphasis often provides significant improvement in system performance with very little added complexity or implementation costs.

The idea of pre-emphasis and/or de-emphasis filtering has found application in a number of areas. For example, signals recorded on long playing (LP) records are, prior to recording, filtered using a highpass pre-emphasis filter. This attenuates the low-frequency content of the signal being recorded. Since the low-frequency components typically have large amplitudes, the distance between the groves on the record must be increased to accommodate these large amplitude signals if pre-emphasis filtering were not used. The impact of more widely spaced record groves is reduced recording time. The playback equipment applies de-emphasis filtering to compensate for the pre-emphasis filtering used in the recording process. In the early days of LP recording, several different pre-emphasis filter designs were used among different record manufacturers. The playback equipment was consequently required to provide for all of the different pre-emphasis filter designs in common use. This later became standardized. With modern digital recording techniques this is no longer an issue.

3.4 FEEDBACK DEMODULATORS: THE PHASE-LOCKED LOOP

We have previously studied the technique of FM to AM conversion for demodulating an angle-modulated signal. We shall see in Chapter 7 that improved performance in the presence of noise can be gained by utilizing a feedback demodulator. The subject of this section is the phase-locked loop (PLL), which is a basic form of the feedback demodulator. Phase-locked loops are widely used in today's communication systems, not only for demodulation of angle modulated signals but also for carrier and symbol synchronization, for frequency synthesis, and as the basic building block for a variety of digital demodulators. Phase-locked loops are flexible in that they can be used in a wide variety of applications, are easily implemented, and PLLs give superior performance to many other techniques. It is therefore not surprising that they are ubiquitous in modern communications systems. Therefore, a detailed look at the PLL is justified.

3.4.1 Phase-Locked Loops for FM and PM Demodulation

A block diagram of a PLL is shown in Figure 3.45. The basic PLL contains four basic elements. These are

1. Phase detector
2. Loop filter
3. Loop amplifier (assume $\mu = 1$)
4. Voltage-controlled oscillator (VCO).

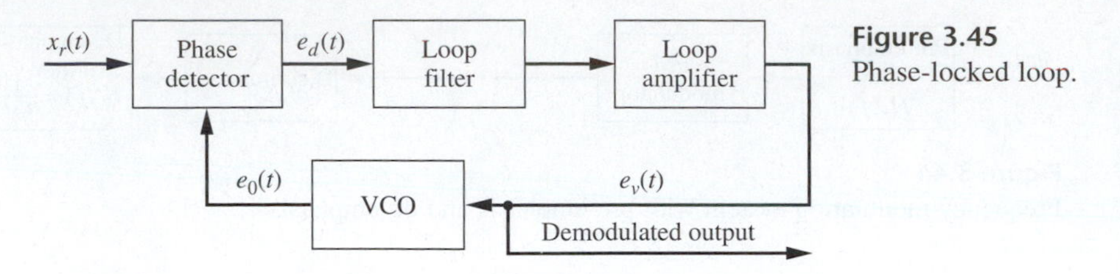

Figure 3.45
Phase-locked loop.

In order to understand the operation of the PLL, assume that the input signal is given by

$$x_r(t) = A_c \cos[2\pi f_c t + \phi(t)] \tag{3.207}$$

and that the VCO output signal is given by

$$e_0(t) = A_v \sin[2\pi f_c t + \theta(t)] \tag{3.208}$$

There are many different types of phase detectors, all having different operating properties. For our application, we assume that the phase detector is a multiplier followed by a lowpass filter to remove the second harmonic of the carrier. We also assume that an inverter is present to remove the minus sign resulting from the multiplication. With these assumptions, the output of the phase detector becomes

$$e_d(t) = \frac{1}{2} A_c A_v K_d \sin[\phi(t) - \theta(t)] = \frac{1}{2} A_c A_v K_d \sin[\psi(t)] \tag{3.209}$$

where K_d is the phase detector constant and $\psi(t) = \phi(t) - \theta(t)$ is the phase error. Note that for small phase error the two inputs to the multiplier are approximately orthogonal so that the result of the multiplication is an odd function of the phase error $\phi(t) - \theta(t)$. This is a necessary requirement so that the phase detector can distinguish between positive and negative phase errors.

The output of the phase detector is filtered, amplified, and applied to the VCO. A VCO is essentially a frequency modulator in which the frequency deviation of the output, $d\theta/dt$, is proportional to the VCO input signal. In other words,

$$\frac{d\theta}{dt} = K_v e_v(t) \text{ rad/s} \tag{3.210}$$

which yields

$$\theta(t) = K_v \int^t e_v(\alpha)\, d\alpha \tag{3.211}$$

The parameter K_v is known as the *VCO constant* and is measured in radians per second per unit of input.

From the block diagram of the PLL it is clear that

$$E_v(s) = F(s) E_d(s) \tag{3.212}$$

where $F(s)$ is the transfer function of the loop filter. In the time domain the preceding expression is

$$e_v(\alpha) = \int^t e_d(\lambda) f(\alpha - \lambda)\, d\lambda \tag{3.213}$$

which follows by simply recognizing that multiplication in the frequency domain is convolution in the time domain. Substitution of (3.209) into (3.213) and this result into (3.211) gives

$$\theta(t) = K_t \int^t \int^\alpha \sin[\phi(\lambda) - \theta(\lambda)] f(\alpha - \lambda)\, d\lambda\, d\alpha \tag{3.214}$$

where K_t is the total loop gain defined by

$$K_t = \frac{1}{2} A_v A_c K_d K_v \tag{3.215}$$

Equation (3.214) is the general expression relating the VCO phase $\theta(t)$ to the input phase $\phi(t)$. The system designer must select the loop filter transfer function $F(s)$, thereby defining the filter impulse response $f(t)$, and the loop gain K_t. We see from (3.215) that the loop gain is a function of the input signal amplitude A_v. Thus PLL design requires knowledge of the input signal level, which is often unknown and time varying. This dependency on the input signal level is typically removed by placing a hard limiter at the loop input. If a limiter is used, the loop gain K_t is selected by appropriately choosing A_v, K_d, and K_v, which are all parameters of the PLL. The individual values of these parameters are arbitrary so long as their product gives the desired loop gain. However, hardware considerations typically place constraints on these parameters.

Equation (3.214) defines the nonlinear model of the PLL, which is illustrated in Figure 3.46. Since (3.214) is nonlinear, analysis of the PLL using (3.214) is difficult and often involves a number of approximations. In practice, we typically have interest in PLL operation in either the tracking mode or in the acquisition mode. In the acquisition mode the PLL is attempting to acquire a signal by synchronizing the frequency and phase of the VCO with the input signal. In the acquisition mode of operation, the phase errors are typically large, and the nonlinear model is required for analysis.

In the tracking mode, however, the phase error $\phi(t) - \theta(t)$ is often small and (3.214) simplifies to the linear model defined by

$$\theta(t) = K_t \int^t \int^\alpha [\phi(\lambda) - \theta(\lambda)] f(\alpha - \lambda)\, d\lambda\, d\alpha \tag{3.216}$$

Thus, if the phase error is sufficiently small, the sinusoidal nonlinearity can be neglected, and the PLL becomes a linear feedback control system, which is easily analyzed. The linear model that results is illustrated in Figure 3.47. While both the nonlinear and linear models involve $\theta(t)$

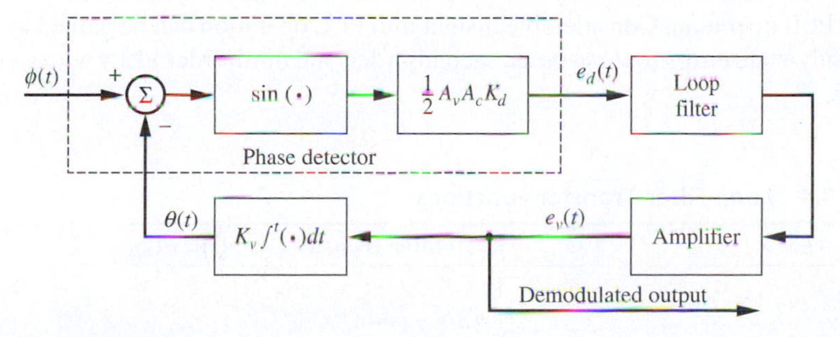

Figure 3.46
Nonlinear PLL model.

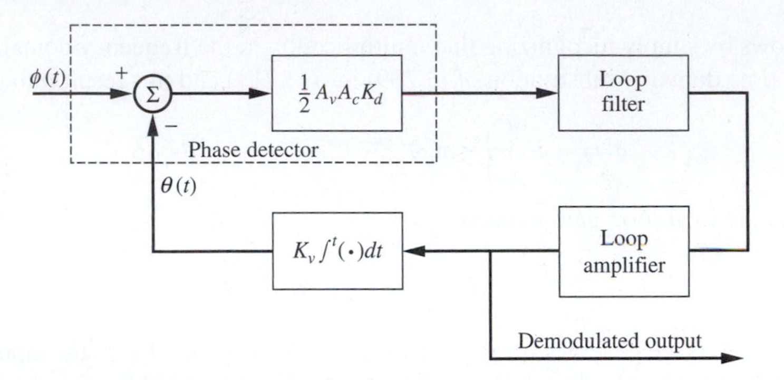

Figure 3.47
Linear PLL model.

and $\phi(t)$ rather than $x_r(t)$ and $e_0(t)$, knowledge of $\theta(t)$ and $\phi(t)$ fully determines $x_r(t)$ and $e_0(t)$, as can be seen from (3.207) and (3.208). If $\theta(t) \cong \phi(t)$, it follows that

$$\frac{d\theta(t)}{dt} \cong \frac{d\phi(t)}{dt} \tag{3.217}$$

and the VCO frequency deviation is a good estimate of the input frequency deviation. For an FM system, the frequency deviation of the PLL input signal is proportional to the message signal $m(t)$. Since the VCO frequency deviation is proportional to the VCO input $e_v(t)$, it follows that $e_v(t)$ is proportional to $m(t)$ if (3.217) is satisfied. Thus $e_v(t)$ is the demodulated output for FM systems.

The form of the loop filter transfer function $F(s)$ has a profound effect on both the tracking and acquisition behavior of the PLL. In the work to follow we will have interest in first-order, second-order, and third-order PLLs. The loop filter transfer functions for these three cases are given in Table 3.4. Note that the order of the PLL exceeds the order of the loop filter by one. The extra integration results from the VCO as we will see in the next section. We now consider the PLL in both the tracking and acquisition mode. Tracking mode operation is considered first since it is more straightforward.

3.4.2 Phase-Locked Loop Operation in the Tracking Mode: The Linear Model

As we have seen, in the tracking mode the phase error is small, and linear analysis can be used to define PLL operation. Considerable insight into PLL operation can be gained by investigating the steady-state errors for first-order, second-order, and third-order PLLs with a variety of input signals.

Table 3.4 Loop Filter Transfer Functions

PLL order	Loop filter transfer function, $F(s)$
1	1
2	$1+a/s=(s+a)/s$
3	$1+a/s+b/s^2=(s^2+as+b)/s^2$

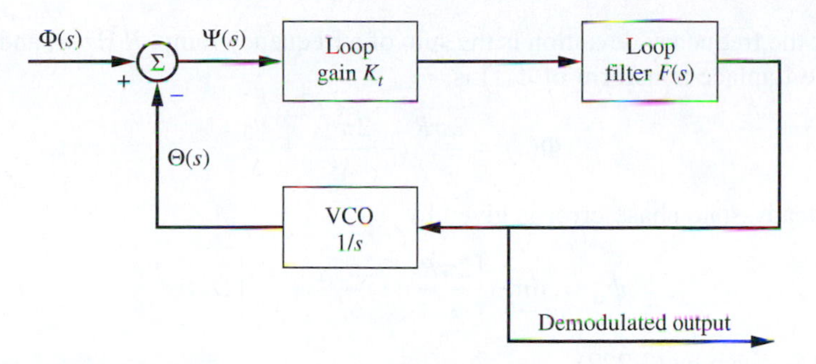

Figure 3.48
Linear PLL model in the frequency domain.

The Loop Transfer Function and Steady-State Errors

The frequency-domain equivalent of Figure 3.47 is illustrated in Figure 3.48. It follows from Figure 3.48 and (3.216) that

$$\Theta(s) = K_t[\Phi(s)-\Theta(s)]\frac{F(s)}{s} \tag{3.218}$$

from which the transfer function relating the VCO phase to the input phase is

$$H(s) = \frac{\Theta(s)}{\Phi(s)} = \frac{K_t F(s)}{s+K_t F(s)} \tag{3.219}$$

immediately follows. Since the Laplace transform of the phase error is

$$\Psi(s) = \Phi(s)-\Theta(s) \tag{3.220}$$

we can write the transfer function relating the phase error to the input phase as

$$G(s) = \frac{\Psi(s)}{\Phi(s)} = \frac{\Phi(s)-\Theta(s)}{\Phi(s)} = 1-H(s) \tag{3.221}$$

so that

$$G(s) = \frac{s}{s+K_t F(s)} \tag{3.222}$$

The steady-state error can be determined through the final value theorem from Laplace transform theory. The final value theorem states that the $\lim_{t\to\infty} a(t)$ is given by $\lim_{s\to 0} sA(s)$, where $a(t)$ and $A(s)$ are a Laplace transform pair.

In order to determine the steady-state errors for various loop orders, we assume that the phase deviation has the somewhat general form

$$\phi(t) = \pi R t^2 + 2\pi f_\Delta t + \theta_0, \quad t > 0 \tag{3.223}$$

The corresponding frequency deviation is

$$\frac{1}{2\pi}\frac{d\phi}{dt} = Rt + f_\Delta, \quad t > 0 \tag{3.224}$$

We see that the frequency deviation is the sum of a frequency ramp, R Hz/s, and a frequency step f_Δ. The Laplace transform of $\phi(t)$ is

$$\Phi(s) = \frac{2\pi R}{s^3} + \frac{2\pi f_\Delta}{s^2} + \frac{\theta_0}{s} \tag{3.225}$$

Thus, the steady-state phase error is given by

$$\psi_{ss} = \lim_{s\to 0} s\left[\frac{2\pi R}{s^3} + \frac{2\pi f_\Delta}{s^2} + \frac{\theta_0}{s}\right]G(s) \tag{3.226}$$

where $G(s)$ is given by (3.222).

In order to generalize, consider the third-order filter transfer function defined in Table 3.4:

$$F(s) = \frac{1}{s^2}(s^2 + as + b) \tag{3.227}$$

If $a = 0$ and $b = 0$, $F(s) = 1$, the loop filter transfer function for a first-order PLL. If $a \neq 0$, and $b = 0$, $F(s) = (s+a)/s$, which defines the loop filter for second-order PLL. With $a \neq 0$ and $b \neq 0$ we have a third-order PLL. We can therefore use $F(s)$, as defined by (3.227) with a and b taking on appropriate values, to analyze first-order, second-order, and third-order PLLs.

Substituting (3.227) into (3.222) yields

$$G(s) = \frac{s^3}{s^3 + K_t s^2 + K_t a s + K_t b} \tag{3.228}$$

Using the expression for $G(s)$ in (3.226) gives the steady-state phase error expression

$$\psi_{ss} = \lim_{s\to 0} \frac{s(\theta_0 s^2 + 2\pi f_\Delta s + 2\pi R)}{s^3 + K_t s^2 + K_t a s + K_t b} \tag{3.229}$$

We now consider the steady-state phase errors for first-order, second-order, and third-order PLLs. For various input signal conditions, defined by θ_0, f_Δ, and R and the loop filter parameters a and b, the steady-state errors given in Table 3.5 can be determined. Note that a first-order PLL can track a phase step with a zero steady-state error. A second-order PLL can track a frequency step with zero steady-state error, and a third-order PLL can track a frequency ramp with zero steady-state error.

Note that for the cases given in Table 3.5 for which the steady-state error is nonzero and finite, the steady-state error can be made as small as desired by increasing the loop gain K_t. However, increasing the loop gain increases the loop bandwidth. When we consider the effects of noise in later chapters, we will see that increasing the loop bandwidth makes the PLL

Table 3.5 Steady-State Errors

PLL order	$\theta_0 \neq 0$, $f_\Delta = 0$, $R = 0$	$\theta_0 \neq 0$, $f_\Delta \neq 0$, $R = 0$	$\theta_0 \neq 0$, $f_\Delta \neq 0$, $R \neq 0$
1 ($a = 0$, $b = 0$)	0	$2\pi f_\Delta / K_t$	∞
2 ($a \neq 0$, $b = 0$)	0	0	$2\pi R / K_t$
3 ($a \neq 0$, $b \neq 0$)	0	0	0

performance more sensitive to the presence of noise. We therefore see a trade-off between steady-state error and loop performance in the presence of noise.

EXAMPLE 3.11

We now consider a first-order PLL, which from (3.222) and (3.227), with $a = 0$ and $b = 0$, has the transfer function

$$H(s) = \frac{\Theta(s)}{\Phi(s)} = \frac{K_t}{s + K_t} \tag{3.230}$$

The loop impulse response is therefore

$$h(t) = K_t e^{-K_t t} u(t) \tag{3.231}$$

The limit of $h(t)$ as the loop gain K_t tends to infinity satisfies all properties of the delta function. Therefore,

$$\lim_{K_t \to \infty} K_t e^{-K_t t} u(t) = \delta(t) \tag{3.232}$$

which illustrates that for large loop gain $\theta(t) \approx \phi(t)$. This also illustrates that the PLL serves as a demodulator for angle-modulated signals. Used as an FM demodulator, the VCO input is the demodulated output since the VCO input signal is proportional to the frequency deviation of the PLL input signal. For PM the VCO input is simply integrated to form the demodulated output, since phase deviation is the integral of frequency deviation.

■

EXAMPLE 3.12

As an extension of the preceding example, assume that the input to an FM modulator is $m(t) = Au(t)$. The resulting modulated carrier

$$x_c(t) = A_c \cos\left[2\pi f_c t + k_f A \int^t u(\alpha)\, d\alpha\right] \tag{3.233}$$

is to be demodulated using a first-order PLL. The demodulated output is to be determined.

This problem will be solved using linear analysis and the Laplace transform. The loop transfer function (3.230) is

$$\frac{\Theta(s)}{\Phi(s)} = \frac{K_t}{s + K_t} \tag{3.234}$$

The phase deviation of the PLL input $\phi(t)$ is

$$\phi(t) = A\, k_f \int^t u(\alpha)\, d\alpha \tag{3.235}$$

The Laplace transform of $\phi(t)$ is

$$\Phi(s) = \frac{Ak_f}{s^2} \tag{3.236}$$

which gives

$$\Theta(s) = \frac{AK_f}{s^2} \frac{K_t}{s + K_t} \tag{3.237}$$

The Laplace transform of the defining equation of the VCO, (3.211), yields

$$E_v(s) = \frac{s}{K_v} \Theta(s) \tag{3.238}$$

so that

$$E_v(s) = \frac{Ak_f}{K_v}\frac{K_t}{s(s+K_t)} \tag{3.239}$$

Partial fraction expansion gives

$$E_v(s) = \frac{Ak_f}{K_v}\left(\frac{1}{s} - \frac{1}{s+K_t}\right) \tag{3.240}$$

Thus the demodulated output is given by

$$e_v(t) = \frac{Ak_f}{K_v}\left(1 - e^{-K_t t}\right) u(t) \tag{3.241}$$

Note that for $t \gg 1/K_t$ and $k_f = K_v$ we have $e_v(t) = Au(t)$ as the demodulated output. The transient time is set by the total loop gain K_t, and k_f/K_v is simply an amplitude scaling of the demodulated output signal.

■

As previously mentioned, very large values of loop gain cannot be used in practical applications without difficulty. However, the use of appropriate loop filters allows good performance to be achieved with reasonable values of loop gain and bandwidth. These filters make the analysis more complicated than our simple example, as we shall soon see.

Even though the first-order PLL can be used for demodulation of angle-modulated signals and for synchronization, the first-order PLL has a number of drawbacks that limit its use for most applications. Among these drawbacks are the limited lock range and the nonzero steady-state phase error to a step-frequency input. Both these problems can be solved by using a second-order PLL, which is obtained by using a loop filter of the form

$$F(s) = \frac{s+a}{s} = 1 + \frac{a}{s} \tag{3.242}$$

This choice of loop filter results in what is generally referred to as a *perfect second-order PLL*. Note that the loop filter defined by (3.242) can be implemented using a single integrator, as will be demonstrated in a computer example to follow.

The Second-Order PLL: Loop Natural Frequency and Damping Factor

With $F(s)$ given by (3.242), the transfer function (3.219) becomes

$$H(s) = \frac{\Theta(s)}{\Phi(s)} = \frac{K_t(s+a)}{s^2 + K_t s + K_t a} \tag{3.243}$$

We also can write the relationship between the phase error $\Psi(s)$ and the input phase $\Phi(s)$. From Figure 3.48 or (3.222), we have

$$G(s) = \frac{\Psi(s)}{\Phi(s)} = \frac{s^2}{s^2 + K_t as + K_t a} \tag{3.244}$$

Since the performance of a linear second-order system is typically parameterized in terms of the natural frequency and damping factor, we now place the transfer function in the standard form for a second-order system. The result is

$$\frac{\Psi(s)}{\Phi(s)} = \frac{s^2}{s^2 + 2\zeta\omega_n s + \omega_n^2} \tag{3.245}$$

in which ζ is the damping factor and ω_n is the natural frequency. It follows from the preceding expression that the natural frequency is

$$\omega_n = \sqrt{K_t a} \tag{3.246}$$

and that the damping factor is

$$\zeta = \frac{1}{2}\sqrt{\frac{K_t}{a}} \tag{3.247}$$

A typical value of the damping factor is $1/\sqrt{2} = 0.707$. Note that this choice of damping factor gives a second-order Butterworth response.

In simulating a second-order PLL, one usually specifies the loop natural frequency and the damping factor and determines loop performance as a function of these two fundamental parameters. The PLL simulation model, however, is a function of the physical parameters K_t and a. Equations (3.246) and (3.247) allow K_t and a to be written in terms of ω_n and ζ. The results are

$$a = \frac{\omega_n}{2\zeta} = \frac{\pi f_n}{\zeta} \tag{3.248}$$

and

$$K_t = 4\pi\zeta f_n \tag{3.249}$$

where $2\pi f_n = \omega_n$. These last two expressions will be used to develop the simulation program for the second-order PLL that is given in Computer Example 3.4.

EXAMPLE 3.13

We now work a simple second-order example. Assume that the input signal to the PLL experiences a small step change in frequency. (The step in frequency must be small to ensure that the linear model is applicable. We will consider the result of large step changes in PLL input frequency when we consider operation in the acquisition mode.) Since instantaneous phase is the integral of instantaneous frequency and integration is equivalent to division by s, the input phase due to a step in frequency of magnitude Δf is

$$\Phi(s) = \frac{2\pi\Delta f}{s^2} \tag{3.250}$$

From (3.245) we see that the Laplace transform of the phase error $\psi(t)$ is

$$\Psi(s) = \frac{\Delta\omega}{s^2 + 2\zeta\omega_n s + \omega_n^2} \tag{3.251}$$

Inverse transforming and replacing ω_n by $2\pi f_n$ yields, for $\zeta < 1$,

$$\psi(t) = \frac{\Delta f}{f_n\sqrt{1-\zeta^2}} e^{-2\pi\zeta f_n t}\left[\sin\left(2\pi f_n\sqrt{1-\zeta^2}\, t\right)\right]u(t) \tag{3.252}$$

and we see that $\psi(t) \to 0$ as $t \to \infty$. Note that the steady-state phase error is zero as we first saw in Table 3.5.

■

3.4.3 Phase-Locked Loop Operation in the Acquisition Mode

In the acquisition mode we must determine that the PLL actually achieves phase lock and the time required for the PLL to achieve phase lock. In order to show that the phase error signal tends to drive the PLL into lock, we will simplify the analysis by assuming a first-order PLL for which the loop filter transfer function $F(s) = 1$ or $f(t) = \delta(t)$. Simulation will be used for higher-order loops. Using the general nonlinear model defined by (3.214) with $h(t) = \delta(t)$ and applying the sifting property of the delta function yields

$$\theta(t) = K_t \int^t \sin[\phi(\alpha) - \theta(\alpha)]\, d\alpha \tag{3.253}$$

Taking the derivative of $\theta(t)$ gives

$$\frac{d\theta}{dt} = K_t \sin[\phi(t) - \theta(t)] \tag{3.254}$$

Assume that the input to the FM modulator is a unit step so that the frequency deviation $d\phi/dt$ is a unit step of magnitude $2\pi\Delta f = \Delta\omega$. Let the phase error $\phi(t) - \theta(t)$ be denoted $\psi(t)$. This yields

$$\frac{d\theta}{dt} = \frac{d\phi}{dt} - \frac{d\psi}{dt} = \Delta\omega - \frac{d\psi}{dt} = K_t \sin\psi(t), \qquad t \geq 0 \tag{3.255}$$

or

$$\frac{d\psi}{dt} + K_t \sin\psi(t) = \Delta\omega \tag{3.256}$$

This equation is sketched in Figure 3.49. It relates the frequency error and the phase error.

A plot of the derivative of a function versus the function is known as a *phase-plane plot* and tells us much about the operation of a nonlinear system. The PLL must operate with a phase error $\psi(t)$ and a frequency error $d\psi/dt$ that are consistent with (3.256). To demonstrate that the PLL achieves lock, assume that the PLL is operating with zero phase and frequency error prior to the application of the frequency step. When the step in frequency is applied, the frequency error becomes $\Delta\omega$. This establishes the initial operating point, point B in Figure 3.49, assuming $\Delta\omega > 0$. In order to determine the trajectory of the operating point, we need only recognize that since dt, a time increment, is always a positive quantity, $d\psi$ must be positive if $d\psi/dt$ is positive. Thus, in the upper half plane ψ increases. In other words, the operating point moves from left-to-right in the upper half plane. In the same manner, the operating point moves from right-to-left in the lower half plane, the region for which $d\psi/dt$ is less than zero. Thus the operating point must move from point B to point A. When the operating point attempts to move

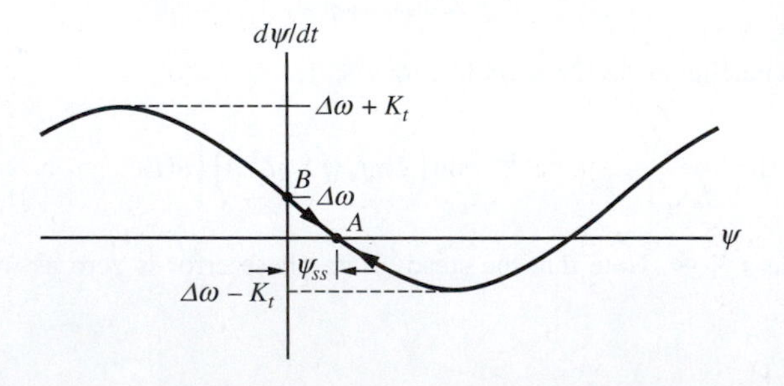

Figure 3.49
Phase-plane plot.

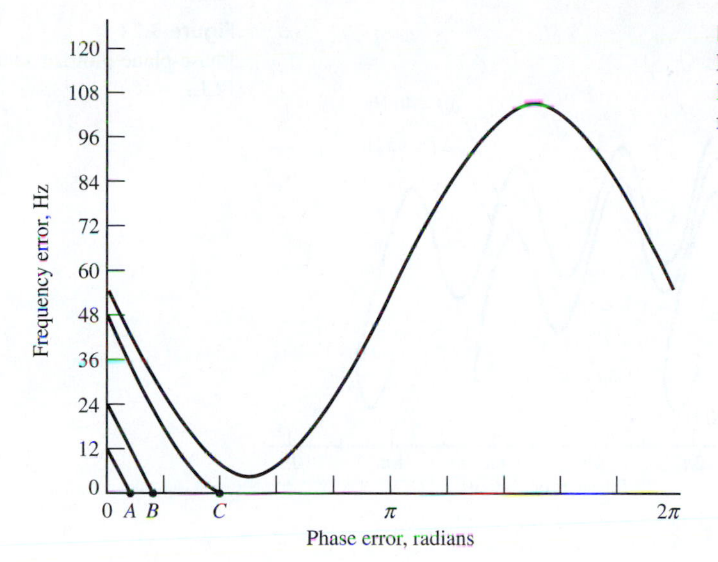

Figure 3.50
Phase-plane plot of first-order PLL for several initial frequency errors.

from point A by a small amount, it is forced back to point A. Thus point A is a stable operating point and is the steady-state operating point of the system. The steady-state phase error is ψ_{ss}, and the steady-state frequency error is zero as shown.

The preceding analysis illustrates that the loop locks only if there is an intersection of the operating curve with the $d\psi/dt = 0$ axis. Thus, if the loop is to lock, $\Delta\omega$ must be less than K_t. For this reason, K_t is known as the *lock range* for the first-order PLL.

The phase-plane plot for a first-order PLL with a frequency-step input is illustrated in Figure 3.50. The loop gain is $2\pi(50)$, and four values for the frequency step are shown: $\Delta f = 12, 24, 48$, and 55 Hz. The steady-state phase errors are indicated by A, B, and C for frequency-step values of 12, 24, and 48 Hz, respectively. For $\Delta f = 55$, the loop does not lock but forever oscillates.

A mathematical development of the phase-plane plot of a second-order PLL is well beyond the level of our treatment here. However, the phase-plane plot is easily obtained, using computer simulation. For illustrative purposes, assume a second-order PLL having a damping factor ζ of 0.707 and a natural frequency f_n of 10 Hz. For these parameters, the loop gain K_t is 88.9, and the filter parameter a is 44.4. The input to the PLL is assumed to be a step change in frequency at time $t = t_0$. Four values were used for the step change in frequency $\Delta\omega = 2\pi(\Delta f)$. These were $\Delta f = 20, 35, 40$, and 45 Hz.

The results are illustrated in Figure 3.51. Note that for $\Delta f = 20$ Hz, the operating point returns to a steady-state value for which the frequency and phase error are both zero, as should be the case from Table 3.5. For $\Delta f = 35$ Hz, the phase plane is somewhat more complicated. The steady-state frequency error is zero, but the steady-state phase error is 2π rad. We say that the PLL has slipped one cycle. Note that the steady-state error is zero mod(2π). The cycle-slipping phenomenon accounts for the nonzero steady-state phase error. The responses for $\Delta f = 40$ and 45 Hz illustrate that three and four cycles are slipped, respectively. The instantaneous VCO frequency is shown in Figure 3.52 for these four cases. The cycle-slipping behavior is clearly shown. The second-order PLL does indeed have an infinite lock range, and cycle slipping occurs until the phase error is within π rad of the steady-state value.

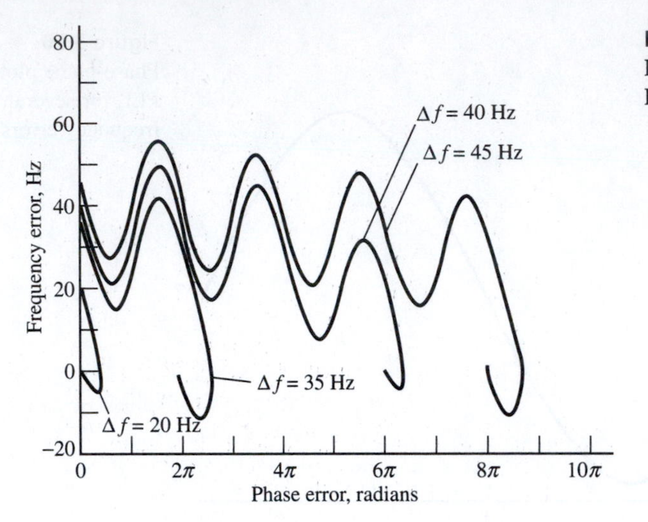

Figure 3.51
Phase-plane plot for second-order PLL.

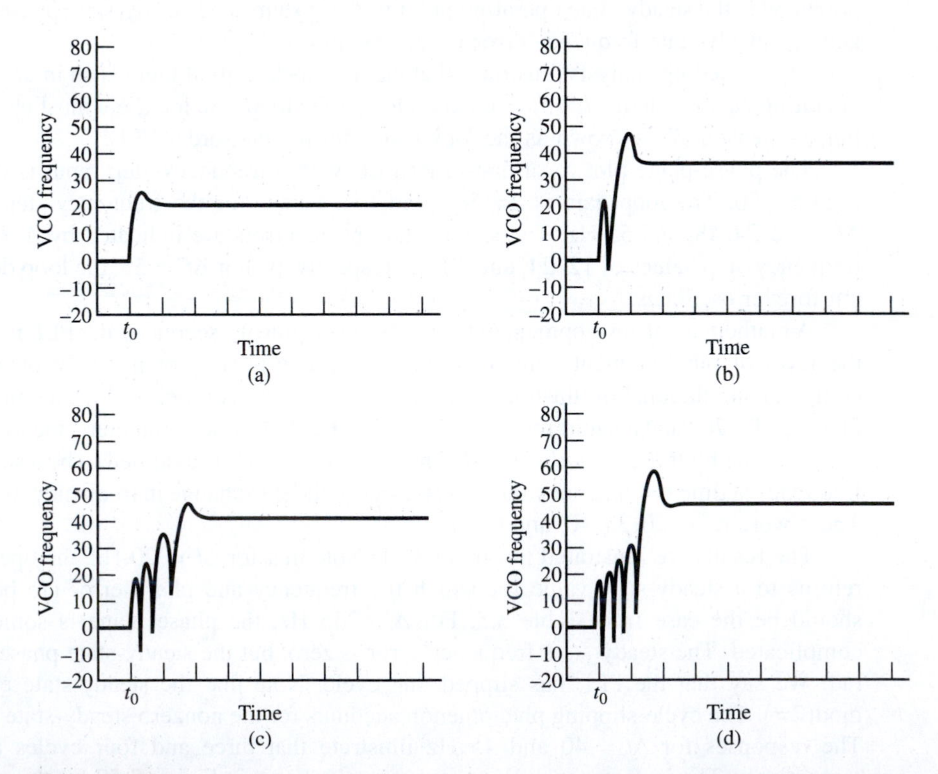

Figure 3.52
voltage-controlled oscillator frequency for four values of input frequency step. (a) VCO frequency for $\Delta f = 20$ Hz. (b) VCO frequency for $\Delta f = 35$ Hz. (c) VCO frequency for $\Delta f = 40$ Hz. (d) VCO frequency for $\Delta f = 45$ Hz.

COMPUTER EXAMPLE 3.4

A simulation program is easily developed for the PLL. Two integration routines are required; one for the loop filter and one for the VCO. The trapezoidal approximation is used for these integration routines. The trapezoidal approximation is

```
y[n] = y[n-1] + (T/2)[x[n] + x[n-1]
```

where `y[n]` represents the current output of the integrator, `y[n-1]` represents the previous integrator output, `x[n]` represents the current integrator input, `x[n-1]` represents the previous integrator input, and T represents the simulation step size, which is the reciprocal of the sampling frequency. The values of `y[n-1]` and `x[n-1]` must be initialized prior to entering the simulation loop. Initializing the integrator inputs and outputs usually result in a transient response. The parameter `settle`, which in the simulation program to follow is set equal to 10% of the simulation run length, allows any initial transients to decay to negligible values prior to applying the loop input.

The following simulation program is divided into three parts. The preprocessor defines the system parameters, the system input, and the parameters necessary for execution of the simulation, such as the sampling frequency. The simulation loop actually performs the simulation. Finally, the postprocessor allows for the data generated by the simulation to be displayed in a manner convenient for interpretation by the simulation user. Note that the postprocessor used here is interactive in that a menu is displayed and the simulation user can execute postprocessor commands without typing them.

The simulation program given here assumes a frequency step on the loop input and can therefore be used to generate Figures 3.51 and 3.52.

```
% File:  c3ce4.m
% beginning of preprocessor
clear all % be safe
fdel = input('Enter frequency step size in Hz > ');
fn = input('Enter the loop natural frequency in Hz > ');
zeta = input('Enter zeta (loop damping factor) > ');
npts = 2000; % default number of simulation points
fs = 2000; % default sampling frequency
T = 1/fs;
t = (0:(npts-1))/fs; % time vector
nsettle = fix(npts/10); % set nsettle time as 0.1*npts

Kt = 4*pi*zeta*fn; % loop gain
a = pi*fn/zeta; % loop filter parameter

filt_in_last = 0; filt_out_last=0;
vco_in_last = 0; vco_out = 0; vco_out_last=0;
% end of preprocessor

% beginning of simulation loop
for i=1:npts
    if i < nsettle
            fin(i) = 0;
            phin = 0;
    else
            fin(i) = fdel;
            phin = 2*pi*fdel*T*(i-nsettle);
    end
      s1=phin - vco_out;
      s2=sin(s1); % sinusoidal phase detector
      s3=Kt*s2;
    filt_in = a*s3;
    filt_out = filt_out_last + (T/2)*(filt_in + filt_in_last);
```

```
        filt_in_last = filt_in;
        filt_out_last = filt_out;
        vco_in = s3 + filt_out;
        vco_out = vco_out_last + (T/2)*(vco_in + vco_in_last);
        vco_in_last = vco_in;
        vco_out_last = vco_out;
        phierror(i)=s1;
        fvco(i)=vco_in/(2*pi);
        freqerror(i) = fin(i)-fvco(i);
end
% end of simulation loop

% beginning of postprocessor
kk = 0;
while kk == 0
    k = menu('Phase Lock Loop Postprocessor',...
    'Input Frequency and VCO Frequency',...
    'Phase Plane Plot',...
    'Exit Program');
    if k == 1
        plot(t,fin,t,fvco)
        title('Input Frequency and VCO Freqeuncy')
        xlabel('Time - Seconds')
        ylabel('Frequency - Hertz')
        pause
    elseif k == 2
        plot(phierror/2/pi,freqerror)
        title('Phase Plane')
        xlabel('Phase Error / pi')
        ylabel('Frequency Error - Hz')
        pause
    elseif k == 3
        kk = 1;
    end
end
% end of postprocessor
```

■

3.4.4 Costas PLLs

We have seen that systems utilizing feedback can be used to demodulate angle-modulated carriers. A feedback system also can be used to generate the coherent demodulation carrier necessary for the demodulation of DSB signals. One system that accomplishes this is the Costas PLL illustrated in Figure 3.53. The input to the loop is the assumed DSB signal

$$x_r(t) = m(t)\cos(2\pi f_c t) \tag{3.257}$$

The signals at the various points within the loop are easily derived from the assumed input and VCO output and are included in Figure 3.53. The lowpass filter preceding the VCO is assumed sufficiently narrow so that the output is $K\sin(2\theta)$, essentially the DC value of the input. This signal drives the VCO such that θ is reduced. For sufficiently small θ, the output of the top lowpass filter is the demodulated output, and the output of

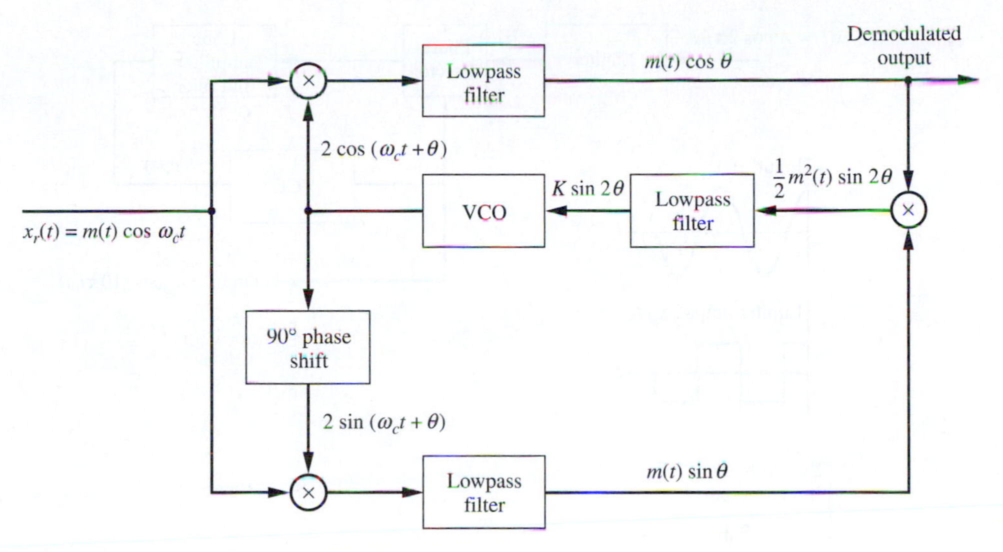

Figure 3.53
Costas PLL.

the lower filter is negligible. We will see in Chapter 8 that the Costas PLL is useful in the implementation of digital receivers.

3.4.5 Frequency Multiplication and Frequency Division

Phase-locked loops also allow for simple implementation of frequency multipliers and dividers. There are two basic schemes. In the first scheme, harmonics of the input are generated, and the VCO tracks one of these harmonics. This scheme is most useful for implementing frequency multipliers. The second scheme is to generate harmonics of the VCO output and to phase lock one of these frequency components to the input. This scheme can be used to implement either frequency multipliers or frequency dividers.

Figure 3.54 illustrates the first technique. The limiter is a nonlinear device and therefore generates harmonics of the input frequency. If the input is sinusoidal, the output of the limiter is a square wave; therefore, odd harmonics are present. In the example illustrated, the VCO quiescent frequency [VCO output frequency f_c with $e_v(t)$ equal to zero] is set equal to $5f_0$. The result is that the VCO phase locks to the fifth harmonic of the input. Thus the system shown multiplies the input frequency by 5.

Figure 3.55 illustrates frequency division by a factor of 2. The VCO quiescent frequency is $f_0/2$ Hz, but the VCO output waveform is a narrow pulse that has the spectrum shown. The component at frequency f_0 phase locks to the input. A bandpass filter can be used to select the component desired from the VCO output spectrum. For the example shown, the center frequency of the bandpass filter should be $f_0/2$. The bandwidth of the bandpass filter must be less than the spacing between the components in the VCO output spectrum; in this case, this spacing is $f_0/2$. It is worth noting that the system shown in Figure 3.55 could be used to multiply the input frequency by 5 by setting the center frequency of the bandpass filter to $5f_0$. Thus this system could also serve as a $\times 5$ frequency multiplier, like the first example. Many variations of these basic techniques are possible.

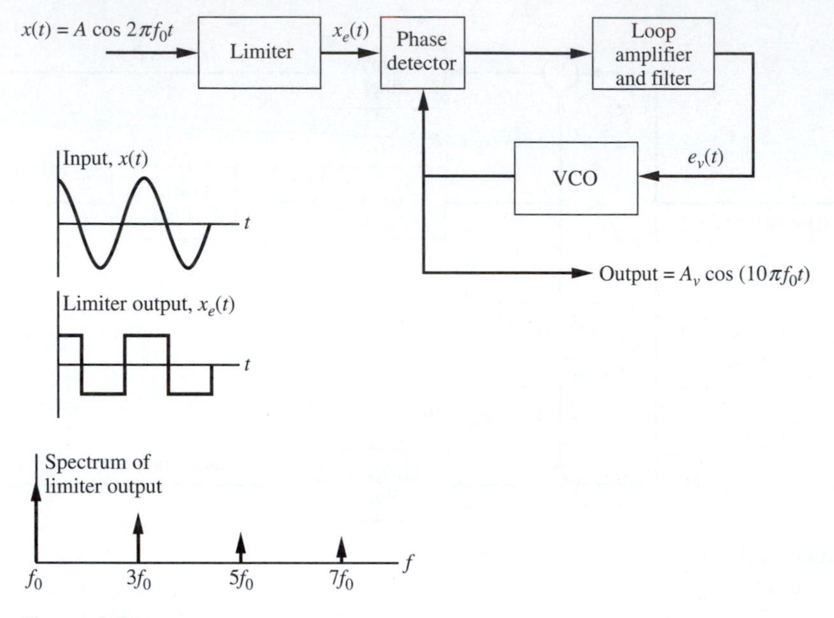

Figure 3.54
Phase-locked loop used as a frequency multiplier.

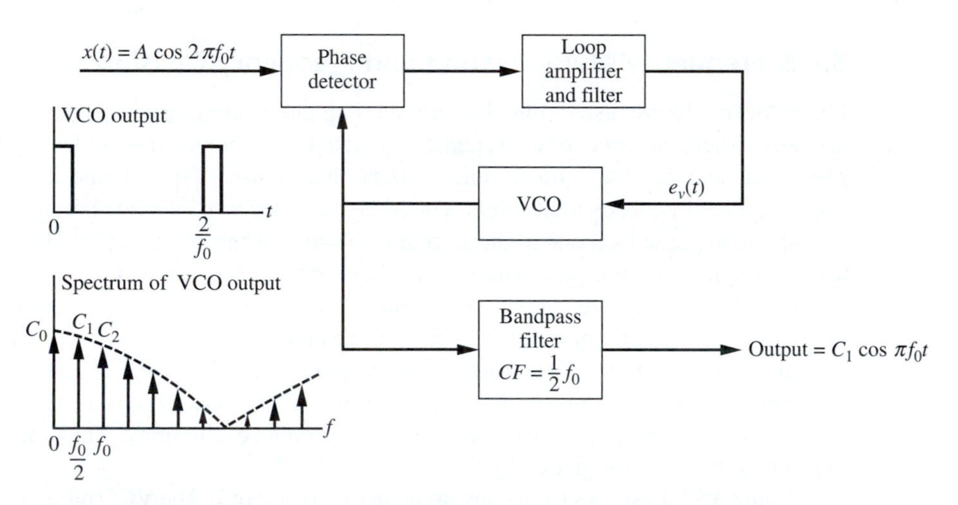

Figure 3.55
Phase-locked loop used as a frequency divider.

3.5 ANALOG PULSE MODULATION

In Section 2.8 we saw that continuous bandlimited signals can be represented by a sequence of discrete samples and that the continuous signal can be reconstructed with negligible error if the sampling rate is sufficiently high. Consideration of sampled signals leads us to the topic of pulse modulation. Pulse modulation can be either analog, in which some attribute of a pulse varies continuously in one-to-one correspondence with a sample value, or digital, in which some

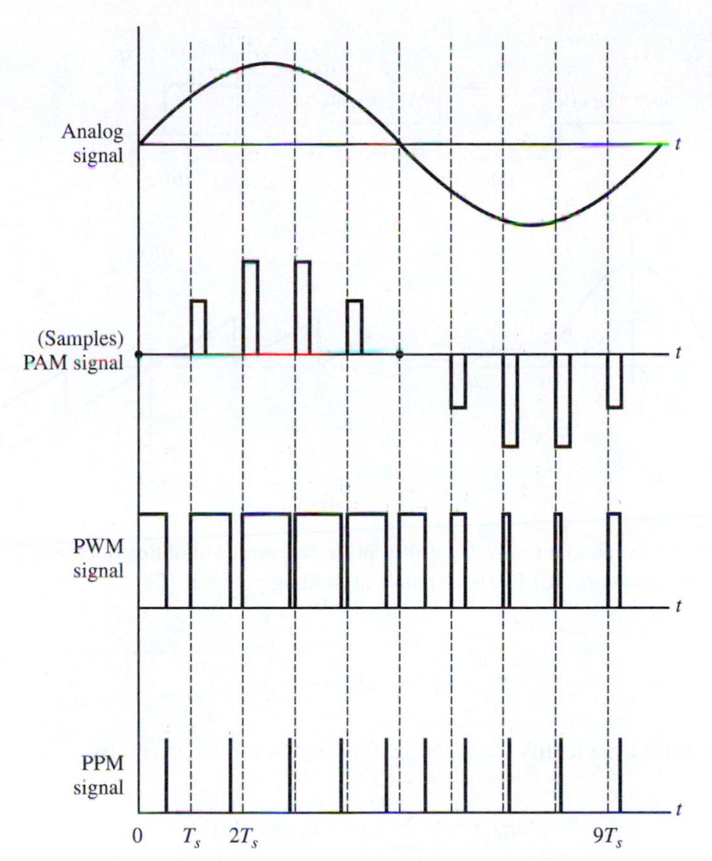

Figure 3.56
Illustration of PAM, PWM, and PPM.

attribute of a pulse can take on a certain value from a set of allowable values. In this section we examine analog pulse modulation. In the following section we examine a couple of examples of digital pulse modulation.

As mentioned, *analog pulse modulation* results when some attribute of a pulse varies continuously in one-to-one correspondence with a sample value. Three attributes can be readily varied: amplitude, width, and position. These lead to pulse amplitude modulation (PAM), pulse-width modulation (PWM), and pulse-position modulation (PPM) as illustrated in Figure 3.56.

3.5.1 Pulse-Amplitude Modulation

A PAM waveform consists of a sequence of flat-topped pulses designating sample values. The amplitude of each pulse corresponds to the value of the message signal $m(t)$ at the leading edge of the pulse. The essential difference between PAM and the sampling operation discussed in the previous chapter is that in PAM we allow the sampling pulse to have finite width. The finite-width pulse can be generated from the impulse-train sampling function by passing the impulse-train samples through a holding circuit as shown in Figure 3.57. The impulse response of the ideal holding circuit is given by

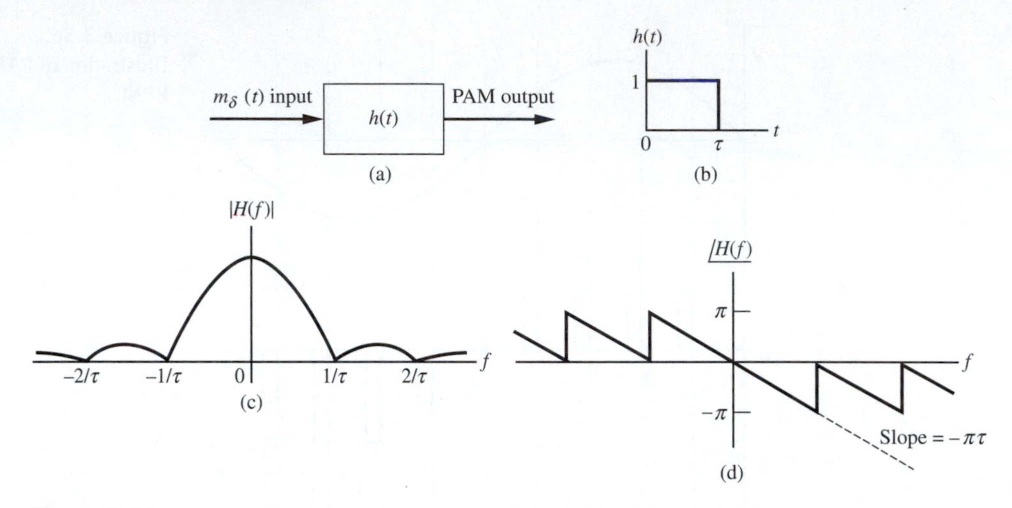

Figure 3.57
Generation of PAM. (a) Holding network. (b) Impulse response of holding network. (c) Amplitude response of holding network. (d) Phase response of holding network.

$$h(t) = \Pi\left(\frac{t - \frac{1}{2}\tau}{\tau}\right) \tag{3.258}$$

The holding circuit transforms the impulse function samples, given by

$$m_\delta(t) = \sum_{n=-\infty}^{\infty} m(nT_s)\delta(t - nT_s) \tag{3.259}$$

to the PAM waveform given by

$$m_c(t) = \sum_{n=-\infty}^{\infty} m(nT_s)\Pi\left(\frac{t - \left(nT_s + \frac{1}{2}\tau\right)}{\tau}\right) \tag{3.260}$$

as illustrated in Figure 3.57. The transfer function of the holding circuit is

$$H(f) = \tau \operatorname{sinc}(f\tau)\, e^{-j\pi f\tau} \tag{3.261}$$

Since the holding network does not have a constant amplitude response over the bandwidth of $m(t)$, amplitude distortion results. This amplitude distortion, which can be significant unless the pulse width τ is very small, can be removed by passing the samples, prior to reconstruction of $m(t)$, through a filter having an amplitude response equal to $1/|H(f)|$, over the bandwidth of $m(t)$. This process is referred to as *equalization* and will be treated in more detail in Chapters 5 and 8. Since the phase response of the holding network is linear, the effect is a time delay and can usually be neglected.

3.5.2 Pulse-Width Modulation (PWM)

A PWM waveform, as illustrated in Figure 3.56, consists of a sequence of pulses with each pulse having a width proportional to the values, and of a message signal at the sampling instants. If the message is 0 at the sampling time, the width of the PWM pulse is typically $\frac{1}{2}T_s$. Thus, pulse widths less than $\frac{1}{2}T_s$ correspond to negative sample values, and pulse widths greater than $\frac{1}{2}T_s$ correspond to positive sample values. The modulation index β is defined so that for $\beta = 1$, the maximum pulse width of the PWM pulses is exactly equal to the sampling period $1/T_s$. Pulse-Width Modulation is seldom used in modern communications systems. Pulse-Width Modulation is used extensively for DC motor control in which motor speed is proportional to the width of the pulses. Since the pulses have equal amplitude, the energy in a given pulse is proportional to the pulse width. Thus, the sample values can be recovered from a PWM waveform by lowpass filtering.

COMPUTER EXAMPLE 3.5

In this computer example we determine the spectrum of a PWM signal. The MATLAB code follows:

```
% File: c3ce5.m
clear all; % be safe
N = 20000; % FFT size
N_samp = 200; % 200 samples per period
f = 1; % frequency
beta = 0.7; % modulation index
period = N/N_samp; % sample period (Ts)
Max_width = beta*N/N_samp; % maximum width
y = zeros(1,N); % initialize
for n=1:N_samp
    x = sin(2*pi*f*(n-1)/N_samp);
    width = (period/2)+round((Max_width/2)*x);
    for k=1:Max_width
        nn = (n-1)*period+k;
        if k<width
            y(nn) = 1; % pulse amplitude
        end
    end
end
ymm = y-mean(y); % remove mean
z = (1/N)*fft(ymm,N); % compute FFT
subplot(211)
stem(0:999,abs(z(1:1000)),'.k')
xlabel('Frequency - Hz.')
ylabel('Amplitude')
subplot(212)
stem(180:220,abs(z(181:221)),'.k')
xlabel('Frequency - Hz.')
ylabel('Amplitude')
% End of script file.
```

In the preceding program the message signal is a sinusoid having a frequency of 1 Hz. The message signal is sampled at 200 samples per period or 200 Hz. The FFT covers 10 periods of the waveform. The spectrum, as determined by the FFT, is illustrated in Figure 3.58(a) and (b). Figure 3.58(a) illustrates the spectrum in the range $0 \leq f \leq 1000$. Since the individual spectral components are spaced 1 Hz apart, corresponding to the 1-Hz sinusoid, they cannot be clearly seen. Figure 3.58(b) illustrates the spectrum in the neighborhood of $f = 200$ Hz. The spectrum in this region reminds us of a Fourier–Bessel spectrum for a sinusoid FM modulated by a pair of sinusoids (see Figure 3.29). We observe that PWM is a nonlinear modulation process.

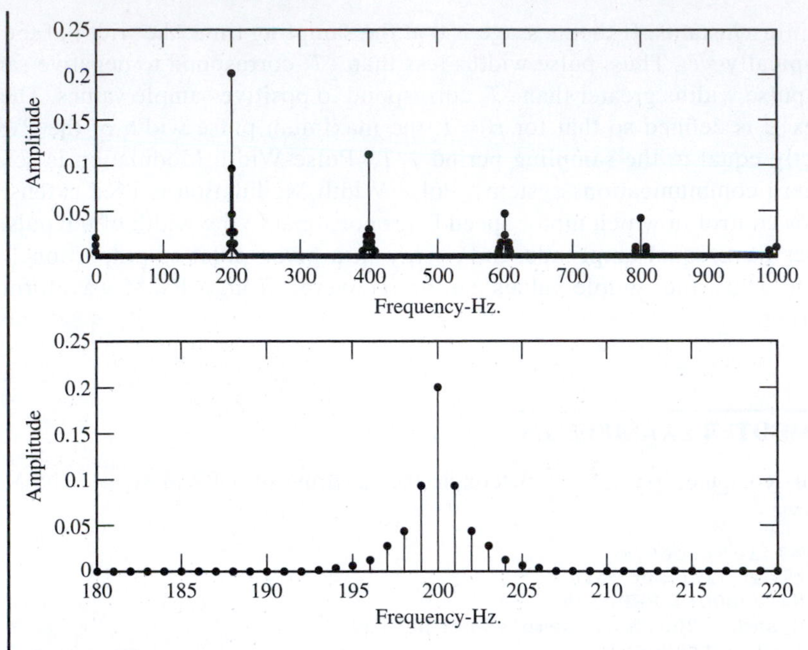

Figure 3.58
Spectrum of a PWM signal. (a) Spectrum for $0 \leq f \leq 1000$ Hz. (b) Spectrum in the neighborhood of $f = 200$ Hz.

■

3.5.3 Pulse-Position Modulation (PPM)

A PPM signal consists of a sequence of pulses in which the pulse displacement from a specified time reference is proportional to the sample values of the information-bearing signal. A PPM signal is illustrated in Figure 3.56 and can be represented by the expression

$$x(t) = \sum_{n=-\infty}^{\infty} g(t-t_n) \tag{3.262}$$

where $g(t)$ represents the shape of the individual pulses, and the occurrence times t_n are related to the values of the message signal $m(t)$ at the sampling instants nT_s, as discussed in the preceding paragraph. The spectrum of a PPM signal is very similar to the spectrum of a PWM signal. (See the computer examples at the end of the chapter.)

If the time axis is slotted so that a given range of sample values is associated with each slot, the pulse positions are quantized, and a pulse is assigned to a given slot depending upon the sample value. Slots are nonoverlaping and are therefore orthogonal. If a given sample value is assigned to one of M slots, the result is M-ary orthogonal communications, which will be studied in detail in Chapter 10. Pulse-Position Modulation is finding a number of applications in the area of ultra-wideband communications.[2]

[2]See, for example, R. A. Scholtz, Multiple Access withTime-Hopping Impulse Modulation, *Proceedings of the IEEE 1993 MILCOM Conference*, 1993, and J. H. Reed (ed.), *An Introduction to Ultra Wideband Communicaion Systems*, Prentice Hall PTR, 2005.

3.6 DELTA MODULATION AND PCM

In analog pulse modulation systems, the amplitude, width, or position of a pulse can vary over a continuous range of values in accordance with the message amplitude at the sampling instant. In systems utilizing digital pulse modulation, the transmitted samples take on only discrete values. We now examine two types of digital pulse modulation: delta modulation and pulse-code modulation (PCM).

3.6.1 Delta Modulation

Delta modulation (DM) is a modulation technique in which the message signal is encoded into a sequence of binary symbols. These binary symbols are represented by the polarity of impulse functions at the modulator output. The electronic circuits to implement both the modulator and the demodulator are extremely simple. This simplicity makes DM an attractive technique for a number of applications.

A block diagram of a delta modulator is illustrated in Figure 3.59(a). The input to the pulse modulator portion of the circuit is

$$d(t) = m(t) - m_s(t) \tag{3.263}$$

where $m(t)$ is the message signal and $m_s(t)$ is a reference waveform. The signal $d(t)$ is hard-limited and multiplied by the pulse-generator output. This yields

$$x_c(t) = \Delta(t) \sum_{n=-\infty}^{\infty} \delta(t - nT_s) \tag{3.264}$$

where $\Delta(t)$ is a hard-limited version of $d(t)$. The preceding expression can be written as

$$x_c(t) = \sum_{n=-\infty}^{\infty} \Delta(nT_s)\delta(t - nT_s) \tag{3.265}$$

Thus the output of the delta modulator is a series of impulses, each having positive or negative polarity depending on the sign of $d(t)$ at the sampling instants. In practical applications, the output of the pulse generator is not, of course, a sequence of impulse functions but rather a sequence of pulses that are narrow with respect to their periods. Impulse functions are assumed here because of the resulting mathematical simplicity. The reference signal $m_s(t)$ is generated by integrating $x_c(t)$. This yields at

$$m_s(t) = \sum_{n=-\infty}^{\infty} \Delta(nT_s) \int^{t} \delta(\alpha - nT_s)\, d\alpha \tag{3.266}$$

which is a stairstep approximation of $m(t)$. The reference signal $m_s(t)$ is shown in Figure 3.59 (b) for an assumed $m(t)$. The transmitted waveform $x_c(t)$ is illustrated in Figure 3.59(c).

Demodulation of DM is accomplished by integrating $x_c(t)$ to form the stairstep approximation $m_s(t)$. This signal can then be lowpass filtered to suppress the discrete jumps in $m_s(t)$. Since a lowpass filter approximates an integrator, it is often possible to eliminate the integrator portion of the demodulator and to demodulate DM simply by lowpass filtering, as was done for PAM and PWM. A difficulty with DM is the problem of slope overload. Slope

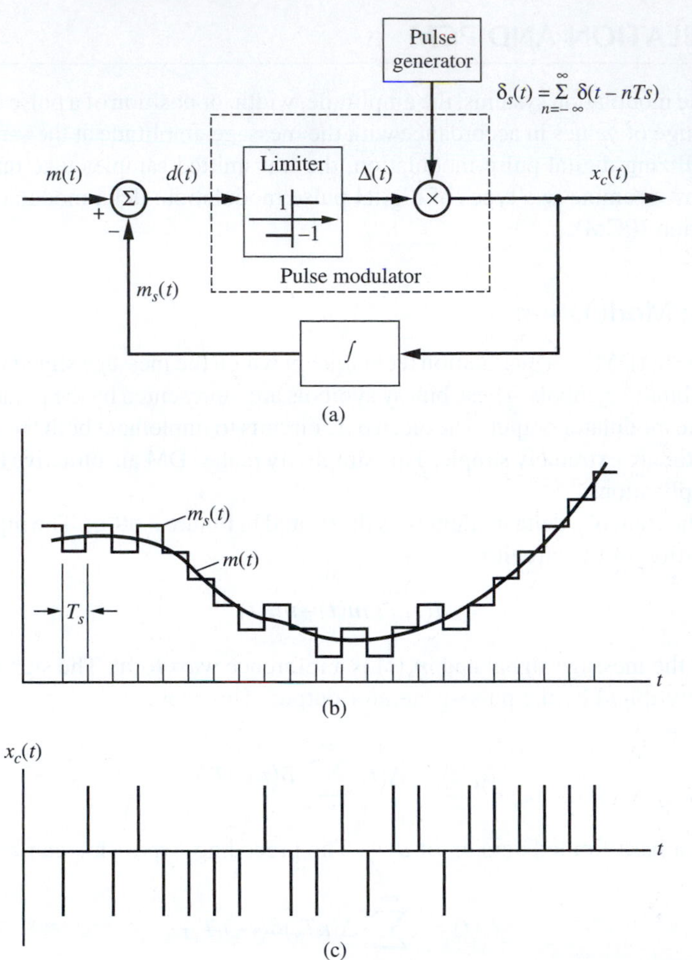

Figure 3.59 Delta modulation. (a) Delta modulator. (b) Modulation waveform and stairstep approximation. (c) Modulator output.

overload occurs when the message signal $m(t)$ has a slope greater than can be followed by the stairstep approximation $m_s(t)$. This effect is illustrated in Figure 3.60(a), which shows a step change in $m(t)$ at time t_0. Assuming that each pulse in $x_c(t)$ has weight δ_0, the maximum slope that can be followed by $m_s(t)$ is δ_0/T_s, as shown. Figure 3.60(b) shows the resulting error signal due to a step change in $m(t)$ at t_0. It can be seen that significant error exists for some time following the step change in $m(t)$. The duration of the error due to slope overload depends on the amplitude of the step, the impulse weights δ_0, and the sampling period T_s.

A simple analysis can be carried out assuming that the message signal $m(t)$ is the sinusoidal signal

$$m(t) = A \sin(2\pi f_1 t) \tag{3.267}$$

The maximum slope that $m_s(t)$ can follow is

$$S_m = \frac{\delta_0}{T_s} \tag{3.268}$$

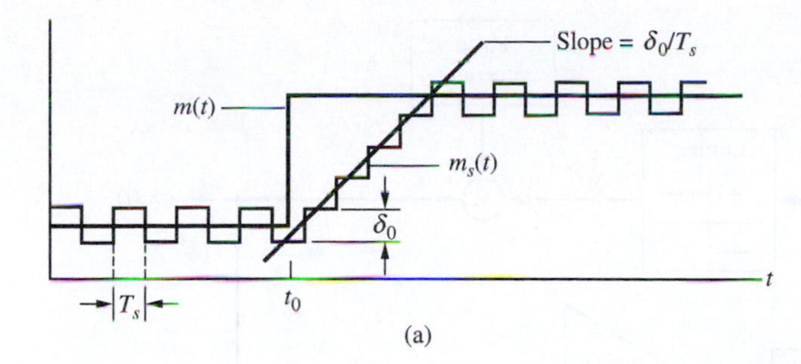

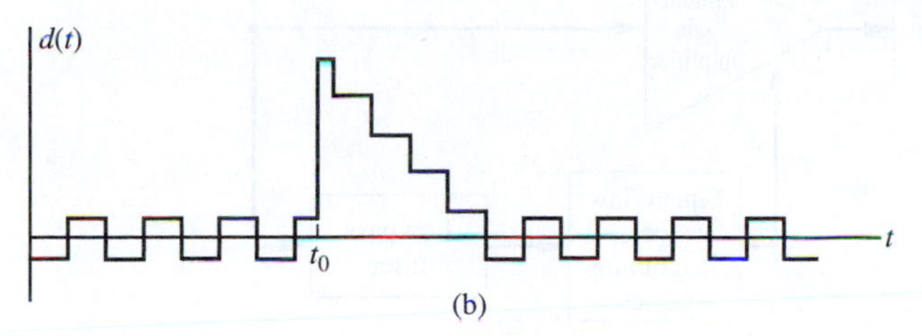

Figure 3.60
Illustration of slope overload. (a) Illustration of $m(t)$ and $m_s(t)$ with step change in $m(t)$. (b) Error between $m(t)$ and $m_s(t)$.

and the derivative of $m(t)$ is

$$\frac{d}{dt}m(t) = 2\pi A f_1 \cos(2\pi f_1 t) \tag{3.269}$$

It follows that $m_s(t)$ can follow $m(t)$ without slope overload if

$$\frac{\delta_0}{T_s} \geq 2\pi A f_1 \tag{3.270}$$

This example illustrates the bandwidth constraint on $m(t)$ if slope overload is to be avoided.

One technique for overcoming the problem of slope overload is to modify the modulator as shown in Figure 3.61. The result is known as *adaptive delta modulation*. The system is explained by recognizing that the weights δ_0, and consequently the step size of $m_s(t)$, can be very small if $m(t)$ is nearly constant. A rapidly changing $m(t)$ requires a larger value of δ_0 if slope overload is to be avoided. A lowpass filter is used as shown, with $x_c(t)$ as input. If $m(t)$ is constant or nearly constant, the pulses constituting $x_c(t)$ will alternate in sign. Thus the DC value, determined over the time constant of the lowpass filter, is nearly zero. This small value controls the gain of the variable-gain amplifier such that it is very small under this condition. Thus, δ_0 is made small at the integrator input. The square-law or magnitude device is used to ensure that the control voltage and amplifier gain $g(t)$ are always positive. If $m(t)$ is increasing or decreasing rapidly, the pulses $x_c(t)$ will have the same polarity over this period. Thus the magnitude of the output of the lowpass filter will be relatively large. The result is an increase in the gain of the variable-gain amplifier and consequently an increase in δ_0. This in turn reduces the time span of significant slope overload. The use of an adaptive delta modulator requires that the receiver be adaptive also, so that the step size at the receiver changes to match the changes in δ_0 at the modulator. This is illustrated in Figure 3.62.

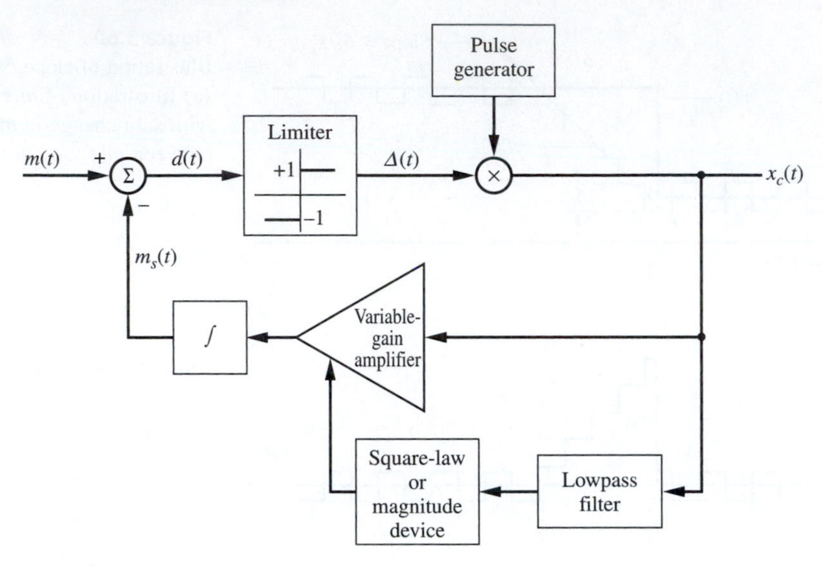

Figure 3.61
Adaptive delta modulator.

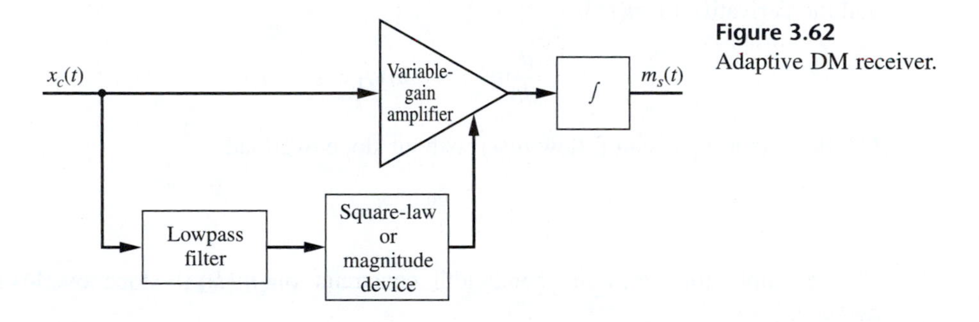

Figure 3.62
Adaptive DM receiver.

3.6.2 Pulse-Code Modulation

The generation of PCM is a three-step process, as illustrated in Figure 3.63(a). The message signal $m(t)$ is first sampled, and the resulting sample values are then quantized. In PCM, the quantizing level of each sample is the transmitted quantity instead of the sample value. Typically, the quantization level is encoded into a binary sequence, as shown in Figure 3.63(b). The modulator output is a pulse representation of the binary sequence, which is shown in Figure 3.63(c). A binary "one" is represented as a pulse, and a binary "zero" is represented as the absence of a pulse. This absence of a pulse is indicated by a dashed line in Figure 3.63(c). The PCM waveform of Figure 3.63(c) shows that a PCM system requires synchronization so that the starting points of the digital words can be determined at the demodulator.

To consider the bandwidth requirements of a PCM system, suppose that q quantization levels are used, satisfying

$$q = 2^n \tag{3.271}$$

where n, the word length, is an integer. For this case, $n = \log_2 q$ binary pulses must be transmitted for each sample of the message signal. If this signal has bandwidth W and the

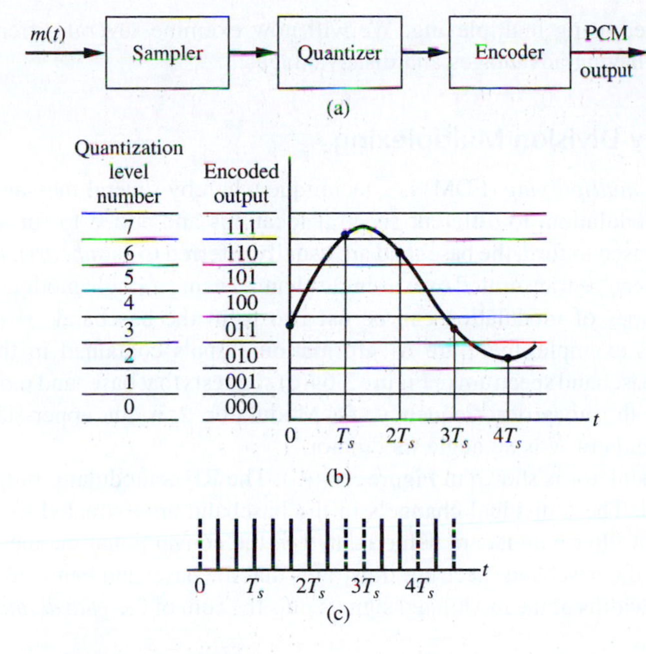

Figure 3.63 Generation of PCM. (a) PCM modulator. (b) Quantization and encoding. (c) Transmitted output.

sampling rate is $2W$, then $2nW$ binary pulses must be transmitted per second. Thus the maximum width of each binary pulse is

$$(\Delta\tau)_{\max} = \frac{1}{2nW} \tag{3.272}$$

We saw in Section 2.7 that the bandwidth required for transmission of a pulse is inversely proportional to the pulse width, so that

$$B = 2knW \tag{3.273}$$

where B is the required bandwidth of the PCM system and k is a constant of proportionality. Note that we have assumed both a minimum sampling rate and a minimum value of bandwidth for transmitting a pulse. Equation (3.273) shows that the PCM signal bandwidth is proportional to the product of the message signal bandwidth W and the wordlength n.

If the major source of error in the system is quantizing error, it follows that a small error requirement dictates large word length resulting in large transmission bandwidth. Thus, in a PCM system, quantizing error can be exchanged for bandwidth. We shall see that this behavior is typical of many nonlinear systems operating in noisy environments. However, before noise effects can be analyzed, we must take a detour and develop the theory of probability and random processes. Knowledge of this area enables one to accurately model realistic and practical communication systems operating in everyday, nonidealized environments.

3.7 MULTIPLEXING

In many applications, a large number of data sources are located at a common point, and it is desirable to transmit these signals simultaneously using a single communication channel.

This is accomplished using multiplexing. We will now examine several different types of multiplexing, each having advantages and disadvantages.

3.7.1 Frequency-Division Multiplexing

Frequency-division multiplexing (FDM) is a technique whereby several message signals are translated, using modulation, to different spectral locations and added to form a baseband signal. The carriers used to form the baseband are usually referred to as *subcarriers*. If desired, the baseband signal can be transmitted over a single channel using a single modulation process. Several different types of modulation can be used to form the baseband, as illustrated in Figure 3.64. In this example, there are N information signals contained in the baseband. Observation of the baseband spectrum in Figure 3.64(c) suggests that baseband modulator 1 is a DSB modulator with subcarrier frequency f_1. Modulator 2 is an upper-sideband SSB modulator, and modulator N is an angle modulator.

An FDM demodulator is shown in Figure 3.64(b). The RF demodulator output is ideally the baseband signal. The individual channels in the baseband are extracted using bandpass filters. The bandpass filter outputs are demodulated in the conventional manner.

Observation of the baseband spectrum illustrates that the baseband bandwidth is equal to the sum of the bandwidths of the modulated signals plus the sum of the *guardbands*, the empty

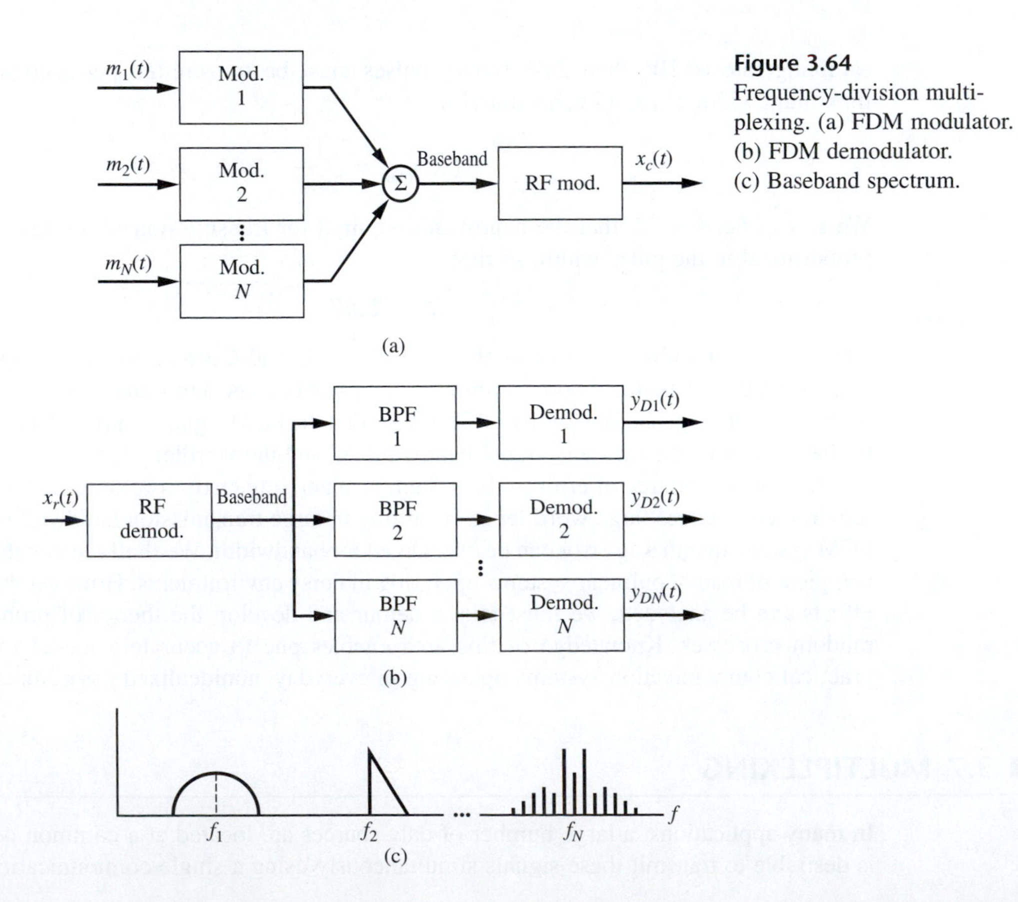

Figure 3.64 Frequency-division multiplexing. (a) FDM modulator. (b) FDM demodulator. (c) Baseband spectrum.

spectral bands between the channels necessary for filtering. This bandwidth is lower bounded by the sum of the bandwidths of the message signals. This bandwidth,

$$B = \sum_{i=1}^{N} W_i \tag{3.274}$$

where W_i is the bandwidth of $m_i(t)$, is achieved when all baseband modulators are SSB and all guardbands have zero width.

3.7.2 Example of FDM: Stereophonic FM Broadcasting

As an example of FDM, we now consider stereophonic FM broadcasting. A necessary condition established in the early development of stereophonic FM is that stereo FM be compatible with monophonic FM receivers. In other words, the output from a monophonic FM receiver must be the composite (left-channel plus right-channel) stereo signal.

The scheme adopted for stereophonic FM broadcasting is shown in Figure 3.65(a). As can be seen, the first step in the generation of a stereo FM signal is to first form the sum and the difference of the left- and right-channel signals, $l(t) \pm r(t)$. The difference signal, $l(t) - r(t)$, is then translated to 38 kHz using DSB modulation with a carrier derived from a 19-kHz oscillator. A frequency doubler is used to generate a 38-kHz carrier from a 19-kHz oscillator. We previously saw that a PLL could be used to implement this frequency doubler.

The baseband signal is formed by adding the sum and difference signals and the 19-kHz pilot tone. The spectrum of the baseband signal is shown in Figure 3.65 for assumed left-channel and right-channel signals. The baseband signal is the input to the FM modulator. It is important to note that if a monophonic FM transmitter, having a message bandwidth of 15 kHz, and a stereophonic FM transmitter, having a message bandwidth of 53 kHz, both have the same constraint on the peak deviation, the deviation ratio D, of the stereophonic FM transmitter is reduced by a factor of $53/15 = 3.53$. The impact of this reduction in the deviation ratio will be seen when we consider noise effects in Chapter 7.

The block diagram of a stereophonic FM receiver is shown in Figure 3.65(c). The output of the FM discriminator is the baseband signal $x_b(t)$ which, under ideal conditions, is identical to the baseband signal at the input to the FM modulator. As can be seen from the spectrum of the baseband signal, the left-plus right-channel signal can be generated by filtering the baseband signal with a lowpass filter having a bandwidth of 15 kHz. Note that this signal constitutes the monophonic output. The left-minus right-channel signal is obtained by coherently demodulating the DSB signal using a 38-kHz demodulation carrier. This coherent demodulation carrier is obtained by recovering the 19-kHz pilot using a bandpass filter and then using a frequency doubler as was done in the modulator. The left-plus right-channel signal and the left-minus right-channel signal are added and subtracted, as shown in Figure 3.65(c) to generate the left-channel signal and the right-channel signal.

3.7.3 Quadrature Multiplexing

Another type of multiplexing is *quadrature multiplexing* (QM), in which quadrature carriers are used for frequency translation. For the system shown in Figure 3.66, the signal

$$x_c(t) = A_c[m_1(t)\cos(2\pi f_c t) + m_2(t)\sin(2\pi f_c t)] \tag{3.275}$$

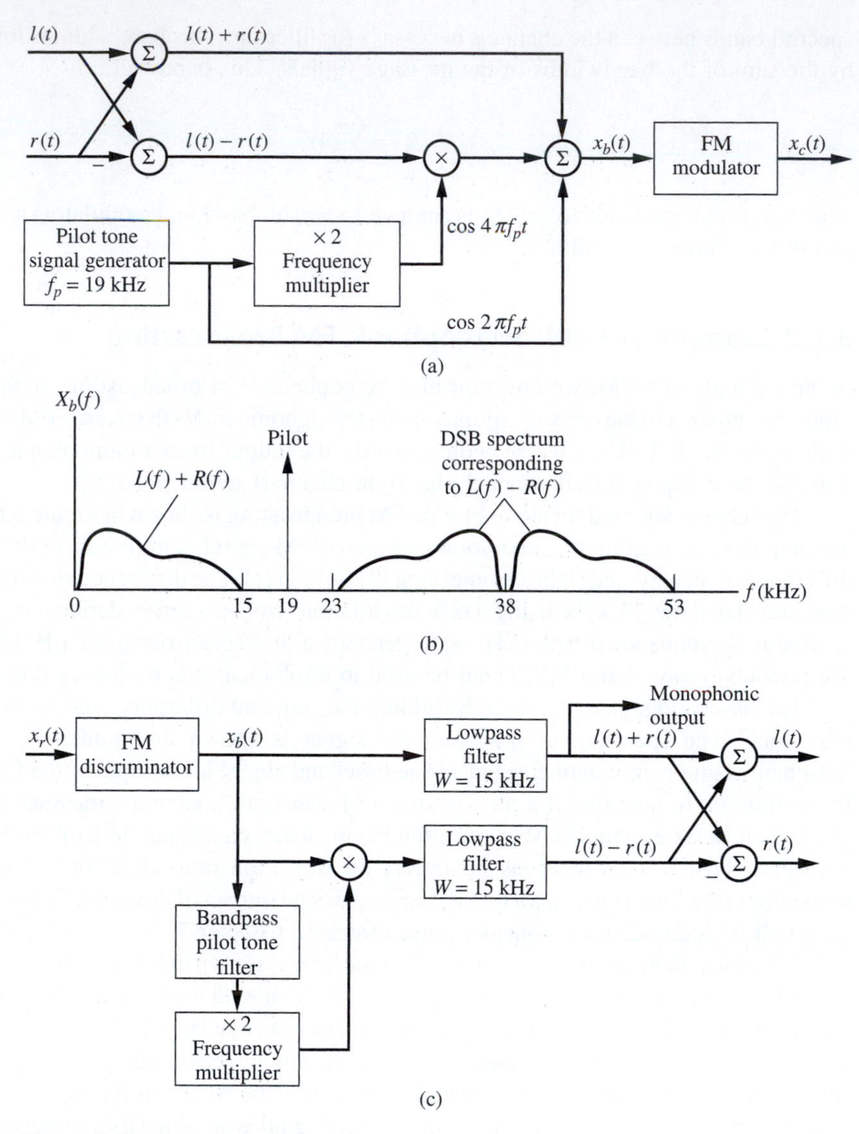

Figure 3.65
Stereophonic FM transmitter and receiver. (a) Stereophonic FM transmitter. (b) Single-sided spectrum of FM baseband signal. (c) Stereophonic FM receiver.

is a quadrature-multiplexed signal. By sketching the spectra of $x_c(t)$ we see that these spectra overlap in frequency if the spectra of $m_1(t)$ and $m_2(t)$ overlap. Even though frequency translation is used in QM, it is not a FDM technique since the two channels do not occupy disjoint spectral locations. Note that SSB is a QM signal with $m_1(t) = m(t)$ and $m_2(t) = \pm\widehat{m}(t)$.

A QM signal is demodulated by using quadrature demodulation carriers. To show this, multiply $x_r(t)$ by $2\cos(2\pi f_c t + \theta)$. This yields

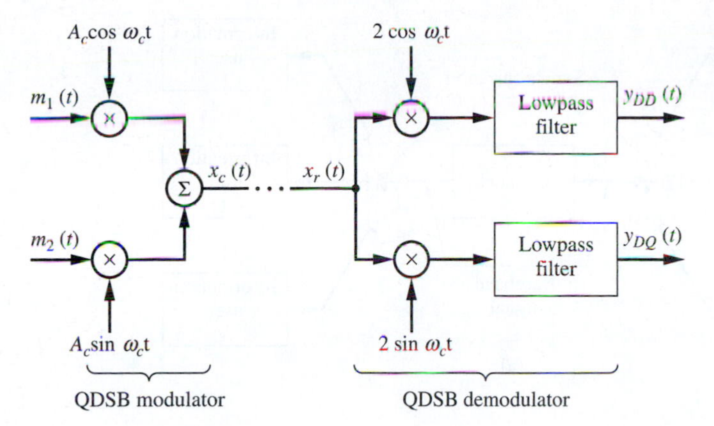

Figure 3.66 Quadrature multiplexing.

$$2x_r(t)\cos(2\pi f_c t+\theta) = A_c[m_1(t)\cos\theta - m_2(t)\sin\theta] + A_c[m_1(t)\cos(4\pi f_c t+\theta) + m_2(t)\sin(4\pi f_c t+\theta)] \tag{3.276}$$

The terms on the second line of the preceding equation have spectral content about $2f_c$ and can be removed by using a lowpass filter. The output of the lowpass filter is

$$y_{DD}(t) = A_c[m_1(t)\cos\theta - m_2(t)\sin\theta] \tag{3.277}$$

which yields $m_1(t)$, the desired output for $\theta = 0$. The quadrature channel is demodulated using a demodulation carrier of the form $2\sin(2\pi f_c t)$.

The preceding result illustrates the effect of a demodulation phase error on QM. The result of this phase error is both an attenuation, which can be time varying, of the desired signal and crosstalk from the quadrature channel. It should be noted that QM can be used to represent both DSB and SSB with appropriate definitions of $m_1(t)$ and $m_2(t)$. We will take advantage of this observation when we consider the combined effect of noise and demodulation phase errors in Chapter 7.

Frequency-division multiplexing can be used with QM by translating pairs of signals, using quadrature carriers, to each subcarrier frequency. Each channel has bandwidth $2W$ and accommodates two message signals, each having bandwidth W. Thus, assuming zero-width guardbands, a baseband of bandwidth NW can accommodate N message signals, each of bandwidth W, and requires $\frac{1}{2}N$ separate subcarrier frequencies.

3.7.4 Time-Division Multiplexing

Time-division multiplexing (TDM) is best understood by considering Figure 3.67(a). The data sources are assumed to have been sampled at the Nyquist rate or higher. The commutator then interlaces the samples to form the baseband signal shown in Figure 3.67(b). At the channel output, the baseband signal is demultiplexed by using a second commutator as illustrated. Proper operation of this system obviously depends on proper synchronization between the two commutators.

If all message signals have equal bandwidth, then the samples are transmitted sequentially, as shown in Figure 3.67(b). If the sampled data signals have unequal bandwidths, more samples must be transmitted per unit time from the wideband channels. This is easily accomplished if

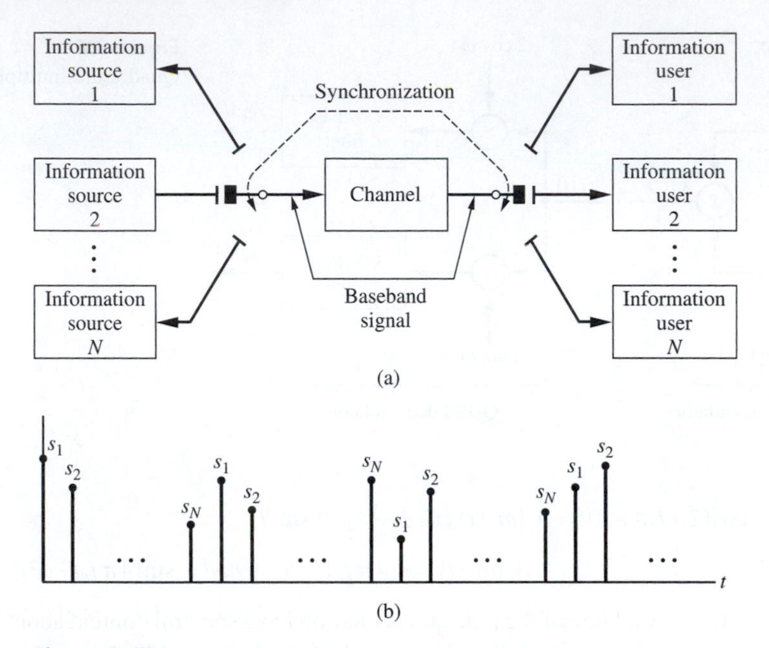

Figure 3.67
Time-division multiplexing. (a) TDM system. (b) Baseband signal.

the bandwidths are harmonically related. For example, assume that a TDM system has four channels of data. Also assume that the bandwidth of the first and second data sources, $s_1(t)$ and $s_2(t)$, is W Hz, the bandwidth of $s_3(t)$ is $2W$ Hz, and the bandwidth of $s_4(t)$ is $4W$ Hz. It is easy to show that a permissible sequence of baseband samples is a periodic sequence, one period of which is $\ldots s_1 s_4 s_3 s_4 s_2 s_4 s_3 s_4 \ldots$

The minimum bandwidth of a TDM baseband is easy to determine using the sampling theorem. Assuming Nyquist rate sampling, the baseband contains $2W_iT$ samples from the ith channel in each T-s interval, where W is the bandwidth of the ith channel. Thus the total number of baseband samples in a T-s interval is

$$n_s = \sum_{i=1}^{N} 2W_iT \tag{3.278}$$

Assuming that the baseband is a lowpass signal of bandwidth B, the required sampling rate is $2B$. In a T-s interval, we then have $2BT$ total samples. Thus

$$n_s = 2BT = \sum_{i=1}^{N} 2W_iT \tag{3.279}$$

or

$$B = \sum_{i=1}^{N} W_i \tag{3.280}$$

which is the same as the minimum required bandwidth obtained for FDM.

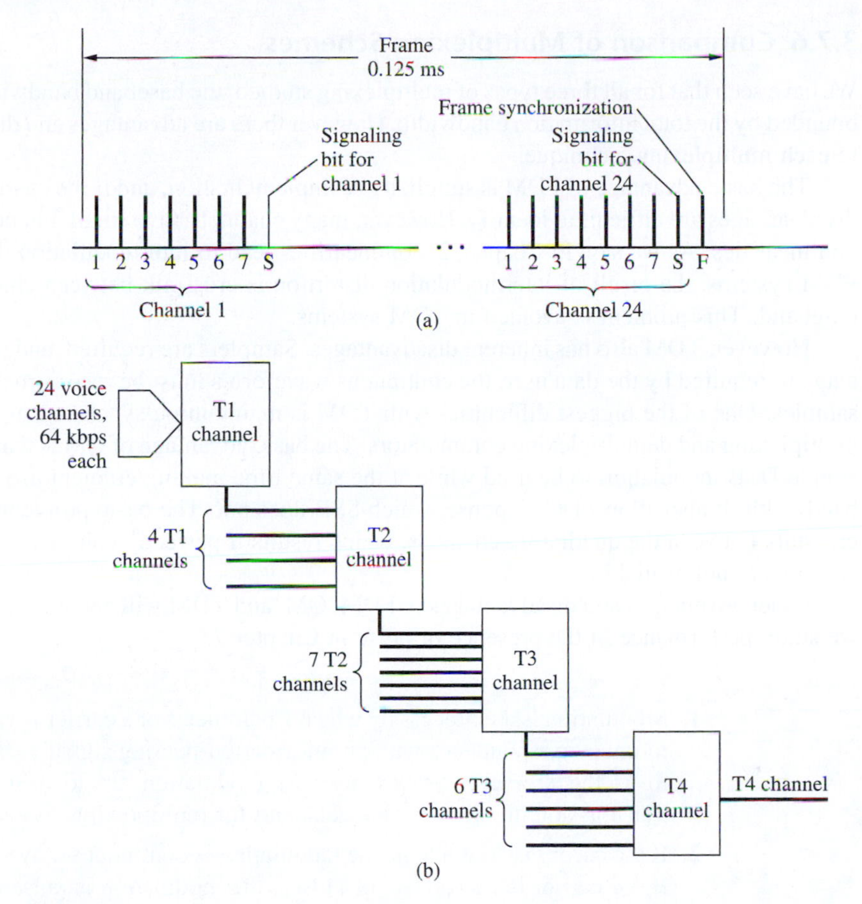

Figure 3.68
Digital multiplexing scheme for digital telephone. (a) T1 frame. (b) Digital multiplexing.

3.7.5 An Example: The Digital Telephone System

As an example of a digital TDM system, we consider a multiplexing scheme common to many telephone systems. The sampling format is illustrated in Figure 3.68(a). A voice signal is sampled at 8000 samples per second, and each sample is quantized into seven binary digits. An additional binary digit, known as a *signaling bit*, is added to the basic seven bits that represent the sample value. The signaling bit is used in establishing calls and for synchronization. Thus eight bits are transmitted for each sample value, yielding a bit rate of 64,000 bit/s (64 kbps). Twenty-four of these 64-kbps voice channels are grouped together to yield a T1 carrier. The T1 frame consists of $24(8)+1=193$ bits. The extra bit is used for frame synchronization. The frame duration is the reciprocal of the fundamental sampling frequency, or 0.125 ms. Since the frame rate is 8000 frames per second, with 193 bits per frame, the T1 data rate is 1.544 Mbps.

As shown in Figure 3.68(b), four T1 carriers can be multiplexed to yield a T2 carrier, which consists of 96 voice channels. Seven T2 carriers yield a T3 carrier, and six T3 carriers yield a T4 carrier. The bit rate of a T4 channel, consisting of 4032 voice channels with signaling bits and framing bits, is 274.176 Mbps. A T1 link is typically used for short transmission distances in areas of heavy usage. T4 and T5 channels are used for long transmission distances.

3.7.6 Comparison of Multiplexing Schemes

We have seen that for all three types of multiplexing studied, the baseband bandwidth is lower-bounded by the total information bandwidth. However there are advantages and disadvantages to each multiplexing technique.

The basic advantage of FDM is simplicity of implementation, and if the channel is linear, disadvantages are difficult to identify. However, many channels have small, but nonnegligible nonlinearities. As we saw in Chapter 2, nonlinearities lead to intermodulation distortion. In FDM systems, the result of intermodulation distortion is crosstalk between channels in the baseband. This problem is avoided in TDM systems.

However, TDM also has inherent disadvantages. Samplers are required, and if continuous data are required by the data user, the continuous waveforms must be reconstructed from the samples. One of the biggest difficulties with TDM is maintaining synchronism between the multiplexing and demultiplexing commutators. The basic advantage of QM is that QM allows simple DSB modulation to be used while at the same time making efficient use of baseband bandwidth. It also allows DC response, which SSB does not. The basic problem with QM is crosstalk between the quadrature channels, which results if perfectly coherent demodulation carriers are not available.

Other advantages and disadvantages of FDM, QM, and TDM will become apparent when we study performance in the presence of noise in Chapter 7.

Summary

1. Modulation is the process by which a parameter of a carrier is varied in one-to-one correspondence with an information-bearing signal usually referred to as the *message*. Several uses of modulation are to achieve efficient transmission, to allocate channels, and for multiplexing.
2. If the carrier is continuous, the modulation is continuous-wave modulation. If the carrier is a sequence of pulses, the modulation is pulse modulation.
3. There are two basic types of continuous-wave modulation: linear modulation and angle modulation.
4. Assume that a general modulated carrier is given by

$$x_c(t) = A(t)\cos[2\pi f_c t + \phi(t)]$$

If $A(t)$ is proportional to the message signal, the result is linear modulation. If $\phi(t)$ is proportional to the message signal, the result is PM. If the time derivative of $\phi(t)$ is proportional to the message signal, the result is FM. Both PM and FM are examples of angle modulation. Angle modulation is a nonlinear process.
5. The simplest example of linear modulation is DSB. Double sideband is implemented as a simple product device, and coherent demodulation must be used, where coherent demodulation means that a local reference at the receiver that is of the same frequency and phase as the incoming carrier is used in demodulation.
6. If a carrier component is added to a DSB signal, the result is AM. This is a useful modulation technique because it allows simple envelope detection to be used.

7. The efficiency of a modulation process is defined as the percentage of total power that conveys information. For AM, this is given by

$$E = \frac{a^2 \langle m_n^2(t) \rangle}{1 + a^2 \langle m_n^2(t) \rangle} (100\%)$$

where the parameter a is known as the *modulation index* and $m_n(t)$ is $m(t)$ normalized so that the peak value is unity. If envelope demodulation is used, the index must be less than unity.

8. A SSB signal is generated by transmitting only one of the sidebands in a DSB signal. Single-sideband signals are generated either by sideband filtering a DSB signal or by using a phase-shift modulator. Single-sideband signals can be written as

$$x_c(t) = \frac{1}{2} A_c m(t) \cos(2\pi f_c t) \pm \frac{1}{2} A_c \widehat{m}(t) \sin(2\pi f_c t)$$

in which the plus sign is used for lower-sideband SSB and the minus sign is used for upper-sideband SSB. These signals can be demodulated either through the use of coherent demodulation or through the use of carrier reinsertion.

9. Vestigial sideband results when a vestige of one sideband appears on an otherwise SSB signal. Vestigial sideband is easier to generate than SSB. Demodulation can be coherent, or carrier reinsertion can be used.

10. Frequency translation is accomplished by multiplying a signal by a carrier and filtering. These systems are known as mixers.

11. The concept of mixing is used in superheterodyne receivers. Mixing results in *image frequencies*, which can be troublesome.

12. The general expression for an angle-modulated signal is

$$x_c(t) = A_c \cos[2\pi f_c t + \phi(t)]$$

For a PM signal, $\phi(t)$ is given by

$$\phi(t) = k_p m(t)$$

and for an FM signal, it is

$$\phi(t) = 2\pi f_d \int^t m(\alpha)\, d\alpha$$

where k_p and f_d are the phase and frequency deviation constants, respectively.

13. Angle modulation results in an infinite number of sidebands for sinusoidal modulation. If only a single pair of sidebands is significant, the result is narrowband angle modulation. Narrowband angle modulation, with sinusoidal message, has approximately the same spectrum as an AM signal except for a 180° phase shift of the lower sideband.

14. An angle-modulated carrier with a sinusoidal message signal can be expressed as

$$x_c(t) = A_c \sum_{n=-\infty}^{\infty} J_n(\beta) \cos[2\pi(f_c + nf_m)t]$$

The term $J_n(\beta)$ is the Bessel function of the first kind of order n and argument β. The parameter β is known as the *modulation index*. If $m(t) = A \sin \omega_m t$, then $\beta = k_p A$ for PM, and $\beta = f_d A / f_m$ for FM.

15. The power contained in an angle-modulated carrier is $\langle x_c^2(t) \rangle = \frac{1}{2} A_c^2$, if the carrier frequency is large compared to the bandwidth of the modulated carrier.

16. The bandwidth of an angle-modulated signal is, strictly speaking, infinite. However, a measure of the bandwidth can be obtained by defining the power ratio

$$P_r = J_0^2(\beta) + 2 \sum_{n=1}^{k} J_n^2(\beta)$$

which is the ratio of the total power $\frac{1}{2} A_c^2$ to the power in the bandwidth $B = 2kf_m$. A power ratio of 0.98 yields $B = 2(\beta + 1) f_m$.

17. The deviation ratio of an angle-modulated signal is

$$D = \frac{\text{peak frequency deviation}}{\text{bandwith of } m(t)}$$

18. Carson's rule for estimating the bandwidth of an angle-modulated carrier with an arbitrary message signal is $B = 2(D + 1)W$.

19. Narrowband-to-wideband conversion is a technique whereby a wideband FM signal is generated from a narrowband FM signal. The system makes use of a frequency multiplier, which, unlike a mixer, multiplies the deviation as well as the carrier frequency.

20. Demodulation of an angle-modulated signal is accomplished through the use of a frequency discriminator. This device yields an output signal proportional to the frequency deviation of the input signal. Placing an integrator at the discriminator output allows PM signals to be demodulated.

21. An FM discriminator can be implemented as a differentiator followed by an envelope detector. Bandpass limiters are used at the differentiator input to eliminate amplitude variations.

22. *Interference*, the presence of undesired signal components, can be a problem in demodulation. Interference at the input of a demodulator results in undesired components at the demodulator output. If the interference is large and if the demodulator is nonlinear, thresholding can occur. The result of this is a drastic loss of the signal component.

23. Interference is also a problem in angle modulation. In FM systems, the effect of interference is a function of both the amplitude and frequency of the interfering tone. In PM systems, the effect of interference is a function only of the amplitude of the interfering tone. In FM systems interference can be reduced by the use of pre-emphasis and de-emphasis wherein the high-frequency message components are boosted at the transmitter before modulation and the inverse process is done at the receiver after demodulation.

24. A PLL is a simple and practical system for the demodulation of angle-modulated signals. It is a feedback control system and is analyzed as such. Phase-locked loops also provide simple implementations of frequency multipliers and frequency dividers.

25. The Costas PLL, which is a variation of the basic PLL, is a system for the demodulation of DSB signals.

26. Analog pulse modulation results when the message signal is sampled and a pulse train carrier is used. A parameter of each pulse is varied in one-to-one correspondence with the value of each sample.

27. Pulse-amplitude modulation results when the amplitude of each carrier pulse is proportional to the value of the message signal at each sampling instant. Pulse-amplitude modulation is essentially a sample-and-hold operation. Demodulation of PAM is accomplished by lowpass filtering.

28. Pulse-width modulation results when the width of each carrier pulse is proportional to the value of the message signal at each sampling instant. Demodulation of PWM is also accomplished by lowpass filtering.

29. Pulse-position modulation results when the position of each carrier pulse, as measured by the displacement of each pulse from a fixed reference, is proportional to the value of the message signal at each sampling instant.

30. Digital pulse modulation results when the sample values of the message signal are quantized and encoded prior to transmission.

31. Delta modulation is an easily implemented form of digital pulse modulation. In DM, the message signal is encoded into a sequence of binary symbols. The binary symbols are represented by the polarity of impulse functions at the modulator output. Demodulation is ideally accomplished by integration, but lowpass filtering is often a simple and satisfactory substitute.

32. Pulse-code modulation results when the message signal is sampled and quantized, and each quantized sample value is encoded as a sequence of binary symbols. Pulse-code modulation differs from DM in that in PCM each quantized sample value is transmitted but in DM the transmitted quantity is the polarity of the change in the message signal from one sample to the next.

33. Multiplexing is a scheme allowing two or more message signals to be communicated simultaneously using a single system.

34. Frequency-division multiplexing results when simultaneous transmission is accomplished by translating message spectra, using modulation to *non-overlapping* locations in a baseband spectrum. The baseband signal is then transmitted using any carrier modulation method.

35. Quadrature multiplexing results when two message signals are translated, using linear modulation with quadrature carriers, to the same spectral locations. Demodulation is accomplished coherently using quadrature demodulation carriers. A phase error in a demodulation carrier results in serious distortion of the demodulated signal. This distortion has two components: a time-varying attenuation of the desired output signal and crosstalk from the quadrature channel.

36. Time-division multiplexing results when samples from two or more data sources are interlaced, using commutation, to form a baseband signal. Demultiplexing is accomplished by using a second commutator, which must be synchronous with the multiplexing commutator.

Further Reading

One can find basic treatments of modulation theory at about the same technical level of this text in a wide variety of books. Examples are Carlson et al. (2001), Haykin (2000), Lathi (1998), and Couch (2007). Taub and Schilling (1986) have an excellent treatment of PLLs. The performance of the PLL in the absence of noise is discussed by Viterbi (1966, Chapters 2 and 3) and Gardner (1979). The simulation of a PLL is treated by Tranter et al. (2004).

Problems

Section 3.1

3.1. Assume that a DSB signal

$$x_c(t) = A_c m(t)\cos(2\pi f_c t + \phi_0)$$

is demodulated using the demodulation carrier $2\cos[2\pi f_c t + \theta(t)]$. Determine, in general, the demodulated output $y_D(t)$. Let $A_c = 1$ and $\theta(t) = \theta_0$, where θ_0 is a constant, and determine the mean-square error between $m(t)$ and the demodulated output as a function of ϕ_0 and θ_0. Now let $\theta(t) = 2\pi f_0 t$ and compute the mean-square error between $m(t)$ and the demodulated output.

3.2. Show that an AM signal can be demodulated using coherent demodulation by assuming a demodulation carrier of the form

$$2\cos[2\pi f_c t + \theta(t)]$$

where $\theta(t)$ is the demodulation phase error.

3.3. Design an envelope detector that uses a full-wave rectifier rather than the half-wave rectifier shown in Figure 3.3. Sketch the resulting waveforms, as was done in Figure 3.3(b) for a half-wave rectifier. What are the advantages of the full-wave rectifier?

3.4. Three message signals are periodic with period T, as shown in Figure 3.69. Each of the three message signals is applied to an AM modulator. For each message signal, determine the modulation efficiency for $a = 0.2$, $a = 0.4$, $a = 0.7$, and $a = 1$.

3.5. The positive portion of the envelope of the output of an AM modulator is shown in Figure 3.70. The message signal is a waveform having zero DC value. Determine the modulation index, the carrier power, the efficiency, and the power in the sidebands.

3.6. In this problem we examine the efficiency of AM for the case in which the message signal does not have symmetrical maximum and minimum values. Two message signals are shown in Figure 3.71. Each is periodic with period T, and τ is chosen such that the DC value of $m(t)$ is zero. Calculate the efficiency for each $m(t)$ for $a = 1$.

3.7. An AM modulator operates with the message signal

$$m(t) = 9\cos(20\pi t) - 8\cos(60\pi t)$$

The unmodulated carrier is given by $110\cos(200\pi t)$, and the system operates with an index of $\frac{1}{2}$.

a. Write the equation for $m_n(t)$, the normalized signal with a minimum value of -1.

b. Determine $\langle m_n^2(t)\rangle$, the power in $m_n(t)$.

c. Determine the efficiency of the modulator.

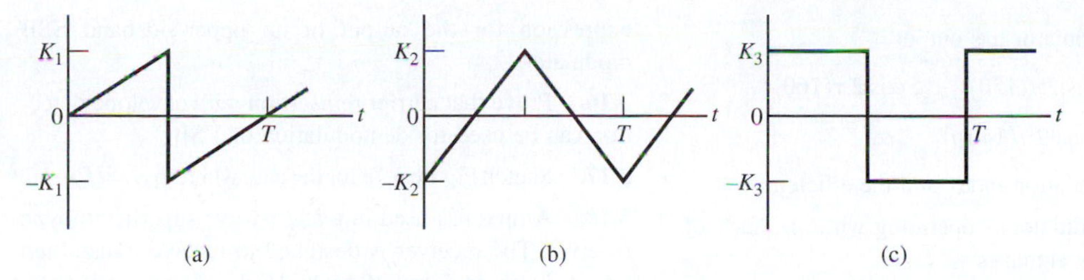

Figure 3.69

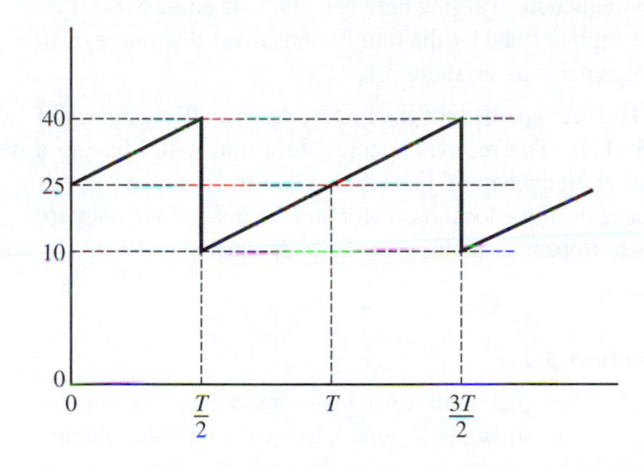

Figure 3.70

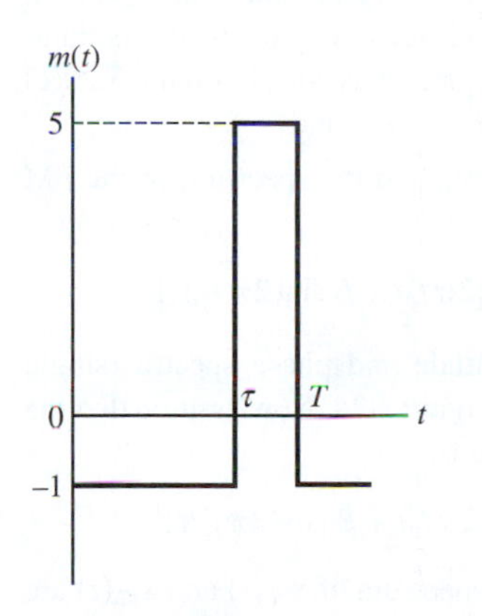

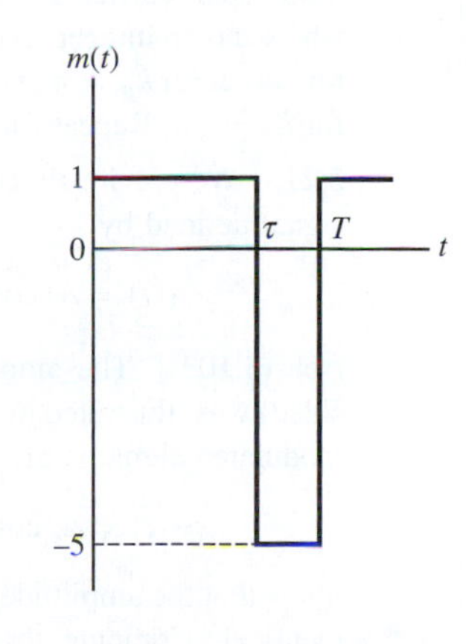

Figure 3.71

d. Sketch the double-sided spectrum of $x_c(t)$, the modulator output, giving the weights and frequencies of all components.

3.8. Rework Problem 3.7 for the message signal

$$m(t) = 9\cos(20\pi t) + 8\cos(60\pi t)$$

3.9. An AM modulator has output

$$x_c(t) = 30\cos[2\pi(200)t] + 4\cos[2\pi(180)t] + 4\cos[2\pi(220)t]$$

Determine the modulation index and the efficiency.

3.10. An AM modulator has output

$$x_c(t) = A\cos[2\pi(200)t] + B\cos[2\pi(180)t] + B\cos[2\pi(220)t]$$

The carrier power is P_0 and the efficiency is E_{ff}. Derive an expression for E_{ff} in terms of P_0, A, and B. Determine A, B, and the modulation index for $P_0 = 100\,W$ and $E_{ff} = 40\,\%$.

3.11. An AM modulator has output

$$x_c(t) = 25\cos[2\pi(150)t] + 5\cos[2\pi(160)t] + 5\cos[2\pi(140)t]$$

Determine the modulation index and the efficiency.

3.12. An AM modulator is operating with an index of 0.7. The modulating signal is

$$m(t) = 2\cos(2\pi f_m t) + \cos(4\pi f_m t) + 2\cos(10\pi f_m t)$$

a. Sketch the spectrum of the modulator output showing the weights of all impulse functions.

b. What is the efficiency of the modulation process?

3.13. Consider the system shown in Figure 3.72. Assume that the average value of $m(t)$ is zero and that the maximum value of $|m(t)|$ is M. Also assume that the square-law device is defined by $y(t) = 4x(t) + 2x^2(t)$.

a. Write the equation for $y(t)$.

b. Describe the filter that yields an AM signal for $g(t)$. Give the necessary filter type and the frequencies of interest.

c. What value of M yields a modulation index of 0.1?

d. What is an advantage of this method of modulation?

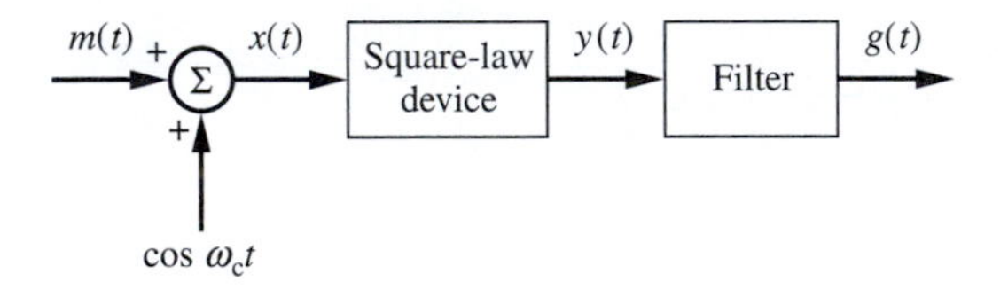

Figure 3.72

3.14. Assume that a message signal is given by

$$m(t) = 2\cos(2\pi f_m t) + \cos(4\pi f_m t)$$

Calculate an expression for

$$x_c(t) = \frac{1}{2}A_c m(t)\cos(2\pi f_c t) \pm \frac{1}{2}A_c \widehat{m}(t)\sin(2\pi f_c t)$$

for $A_c = 4$. Show that the result is upper-sideband or lower-sideband SSB depending upon the choice of the algebraic sign.

3.15. Redraw Figure 3.7 to illustrate the generation of upper-sideband SSB. Give the equation defining the upper-sideband filter. Complete the analysis by deriving the expression for the output of an upper-sideband SSB modulator.

3.16. Prove that carrier reinsertion with envelope detection can be used for demodulation of VSB.

3.17. Sketch Figure 3.18 for the case where $f_{LO} = f_c - f_{IF}$.

3.18. A mixer is used in a short-wave superheterodyne receiver. The receiver is designed to receive transmitted signals between 5 and 25 MHz. High-side tuning is to be used. Determine the tuning range of the local oscillator for IF frequencies varying between 400 kHz and 2 MHz. Plot the ratio defined by the tuning range over this range of IF frequencies as in Table 3.1.

3.19. A superheterodyne receiver uses an IF frequency of 455 kHz. The receiver is tuned to a transmitter having a carrier frequency of 1120 kHz. Give two permissible frequencies of the local oscillator and the image frequency for each. Repeat assuming that the IF frequency is 2500 kHz.

Section 3.2

3.20. Let the input to a phase modulator be $m(t) = u(t - t_0)$, as shown in Figure 3.20(a). Assume that the unmodulated carrier is $A_c\cos(2\pi f_c t)$ and that $f_c t_0 = n$, where n is an integer. Sketch accurately the phase modulator output for $k_p = \pi$ and $\frac{1}{4}\pi$ as was done in Figure 3.20(c) for $k_p = \frac{1}{2}\pi$. Repeat for $k_p = -\pi$ and $-\frac{\pi}{4}$.

3.21. We previously computed the spectrum of the FM signal defined by

$$x_{c1}(t) = A_c\cos[2\pi f_c t + \beta\sin(2\pi f_m t)]$$

[see (3.103)]. The amplitude and phase spectra (single sided) was illustrated in Figure 3.24. Now assume that the modulated signal is given by

$$x_{c2}(t) = A_c\cos[2\pi f_c t + \beta\cos(2\pi f_m t)]$$

Show that the amplitude spectrum of $x_{c1}(t)$ and $x_{c2}(t)$ are identical. Compute the phase spectrum of $x_{c2}(t)$ and compare with the phase spectrum of $x_{c1}(t)$.

3.22. Compute the single-sided amplitude and phase spectra of

$$x_{c3}(t) = A\sin[2\pi f_c t + \beta\sin(2\pi f_m t)]$$

and

$$x_{c4}(t) = A_c\sin[2\pi f_c t + \beta\cos(2\pi f_m t)]$$

Compare the results with Figure 3.24.

3.23. The power of an unmodulated carrier signal is 50 W, and the carrier frequency is $f_c = 50$ Hz. A sinusoidal message signal is used to FM modulate it with index $\beta = 10$. The sinusoidal message signal has a frequency of 5 Hz. Determine the average value of $x_c(t)$. By drawing appropriate spectra, explain this apparent contradiction.

3.24. Given that $J_0(3) = -0.2601$ and that $J_1(3) = 0.3391$, determine $J_4(3)$. Use this result to calculate $J_5(3)$.

3.25. Determine and sketch the spectrum (amplitude and phase) of an angle-modulated signal assuming that the instantaneous phase deviation is $\phi(t) = \beta \sin(2\pi f_m t)$. Also assume $\beta = 10$, $f_m = 20$ Hz, and $f_c = 1000$ Hz.

3.26. A modulated signal is given by

$$x_c(t) = 6\cos[2\pi(70)t] + 6\cos[2\pi(100)t] + 6\cos[2\pi(130)t]$$

Assuming a carrier frequency of 100 Hz, write this signal in the form of (3.1). Give equations for the envelope $R(t)$ and the phase deviation $\phi(t)$.

3.27. A transmitter uses a carrier frequency of 1000 Hz so that the unmodulated carrier is $A_c \cos(2\pi f_c t)$. Determine both the phase and frequency deviation for each of the following transmitter outputs:

a. $x_c(t) = \cos[2\pi(1000)t + 40t^2]$
b. $x_c(t) = \cos[2\pi(500)t^2]$
c. $x_c(t) = \cos[2\pi(1200)t]$
d. $x_c(t) = \cos\left[2\pi(900)t + 10\sqrt{t}\right]$

3.28. An FM modulator has output

$$x_c(t) = 100\cos\left[2\pi f_c t + 2\pi f_d \int^t m(\alpha)\, d\alpha\right]$$

where $f_d = 20$ Hz/V. Assume that $m(t)$ is the rectangular pulse $m(t) = 4\Pi\left[\frac{1}{8}(t-4)\right]$

a. Sketch the phase deviation in radians.
b. Sketch the frequency deviation in hertz.
c. Determine the peak frequency deviation in hertz.
d. Determine the peak phase deviation in radians.
e. Determine the power at the modulator output.

3.29. Repeat the preceding problem assuming that $m(t)$ is the triangular pulse $4\Lambda\left[\frac{1}{3}(t-6)\right]$.

3.30. An FM modulator with $f_d = 10$ Hz/V. Plot the frequency deviation in hertz and the phase deviation in radians for the three message signals shown in Figure 3.73.

3.31. An FM modulator has $f_c = 2000$ Hz and $f_d = 14$ Hz/V. The modulator has input $m(t) = 5\cos 2\pi(10)t$.

a. What is the modulation index?

b. Sketch, approximately to scale, the magnitude spectrum of the modulator output. Show all frequencies of interest.

c. Is this narrowband FM? Why?

d. If the same $m(t)$ is used for a phase modulator, what must k_p be to yield the index given in (a)?

3.32. An audio signal has a bandwidth of 12 kHz. The maximum value of $|m(t)|$ is 6 V. This signal frequency modulates a carrier. Estimate the peak deviation and the bandwidth of the modulator output, assuming that the deviation constant of the modulator is

a. 20 Hz/V
b. 200 Hz/V
c. 2 kHz/V
d. 20 kHz/V.

3.33. By making use of (3.110) and (3.118), show that

$$\sum_{n=-\infty}^{\infty} J_n^2(\beta) = 1$$

3.34. Prove that $J_n(\beta)$ can be expressed as

$$J_n(\beta) = \frac{1}{\pi}\int_0^{\pi} \cos(\beta \sin x - nx)\, dx$$

and use this result to show that

$$J_{-n}(\beta) = (-1)^n J_n(\beta)$$

3.35. An FM modulator is followed by an ideal bandpass filter having a center frequency of 500 Hz and a bandwidth of 70 Hz. The gain of the filter is 1 in the passband. The unmodulated carrier is given by $10\cos(1000\pi t)$, and the message signal is $m(t) = 10\cos(20\pi t)$. The transmitter frequency deviation constant f_d is 8 Hz/V.

a. Determine the peak frequency deviation in hertz.

b. Determine the peak phase deviation in radians.

c. Determine the modulation index.

d. Determine the power at the filter input and the filter output

e. Draw the single-sided spectrum of the signal at the filter input and the filter output. Label the amplitude and frequency of each spectral component.

3.36. A sinusoidal message signal has a frequency of 150 Hz. This signal is the input to an FM modulator with an index of 10. Determine the bandwidth of the modulator

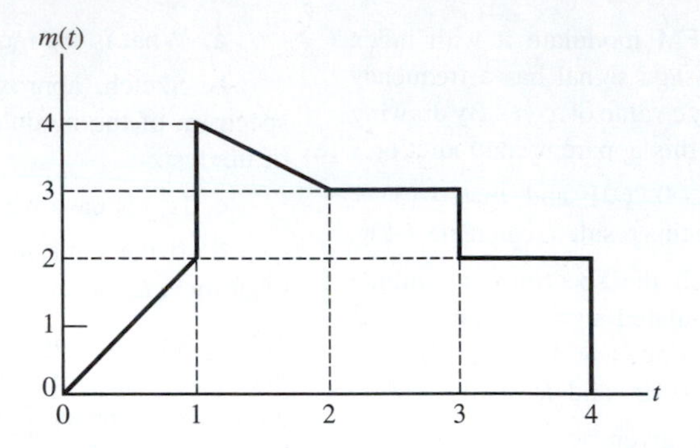

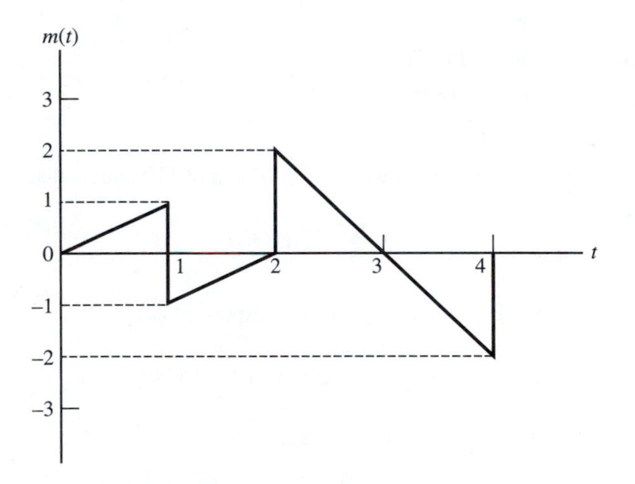

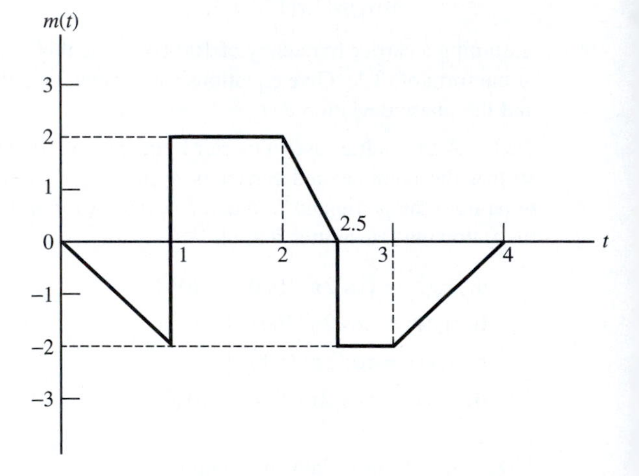

Figure 3.73

output if a power ratio, P_r, of 0.8 is needed. Repeat for a power ratio of 0.9.

3.37. A narrowband FM signal has a carrier frequency of 110 kHz and a deviation ratio of 0.05. The modulation bandwidth is 10 kHz. This signal is used to generate a wideband FM signal with a deviation ratio of 20 and a carrier frequency of 100 MHz. The scheme utilized to accomplish this is illustrated in Figure 3.31. Give the required value of frequency multiplication, n. Also, fully define the mixer by giving *two* permissible frequencies for the local oscillator, and define the required bandpass filter (center frequency and bandwidth).

3.38. Consider the FM discriminator shown in Figure 3.74. The envelope detector can be considered ideal with an infinite input impedance. Plot the magnitude of the transfer function $E(f)/X_r(f)$. From this plot, determine a

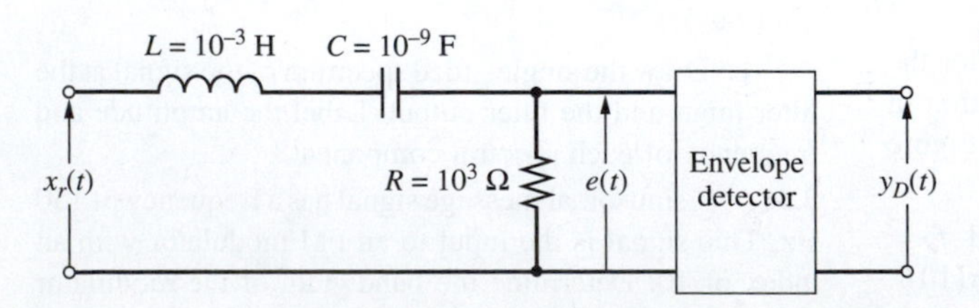

Figure 3.74

suitable carrier frequency and the discriminator constant K_D, and estimate the allowable peak frequency deviation of the input signal.

3.39. By adjusting the values of R, L, and C in Problem 3.38, design a discriminator for a carrier frequency of 100 MHz, assuming that the peak frequency deviation is 4 MHz. What is the discriminator constant K_D for your design?

Section 3.3

3.40. Assume that an FM demodulator operates in the presence of sinusoidal interference. Show that the discriminator output is a nonzero constant for each of the following cases: $A_i = A_c, A_i = -A_c$, and $A_i \gg A_c$. Determine the FM demodulator output for each of these three cases.

Section 3.4

3.41. Starting with (3.229) verify the steady-state errors given in Table 3.5.

3.42. Rework Example 3.12 for $m(t) = A\cos(2\pi f_m t)u(t)$.

3.43. Using $x_r(t) = m(t)\cos(2\pi f_c t)$ and $e_0(t) = 2\cos(2\pi f_c t + \theta)$ for the assumed Costas PLL input and VCO output, respectively, verify that all signals shown at the various points in Figure 3.53 are correct. Assuming that the VCO frequency deviation is defined by $d\theta/dt = -K_v e_v(t)$, where $e_v(t)$ is the VCO input and K_v is a positive constant, derive the phase plane. Using the phase plane, verify that the loop locks.

3.44. Using a single PLL, design a system that has an output frequency equal to $\frac{7}{3}f_0$, where f_0 is the input frequency. Describe fully, by sketching, the output of the VCO for your design. Draw the spectrum at the VCO output and at any other point in the system necessary to explain the operation of your design. Describe any filters used in your design by defining the center frequency and the appropriate bandwidth of each.

3.45. A first-order PLL is operating with zero frequency and phase error when a step in frequency of magnitude $\Delta\omega$ is applied. The loop gain K_t is $2\pi(100)$. Determine the steady-state phase error, in degrees, for $\Delta\omega = 2\pi(30)$, $2\pi(50)$, $2\pi(80)$, and $-2\pi(80)$ rad/s. What happens if $\Delta\omega = 2\pi(120)$ rad/s?

3.46. Verify (3.232) by showing that $K_t e^{-K_t t}u(t)$ satisfies all properties of an impulse function in the limit as $K_t \to \infty$.

3.47. The imperfect second-order PLL is defined as a PLL with the loop filter

$$F(s) = \frac{s+a}{s+\lambda a}$$

in which λ is the offset of the pole from the origin relative to the zero location. In practical implementations λ is small but often cannot be neglected. Use the linear model of the PLL and derive the transfer function for $\Theta(s)/\Phi(s)$. Derive expressions for ω_n and ζ in terms of K_t, a, and λ.

3.48. Assuming the loop filter model for an imperfect second-order PLL described in the preceding problem, derive the steady-state phase errors under the three conditions of θ_0, f_Δ, and R given in Table 3.5.

3.49. A Costas PLL operates with a small phase error so that $\sin\psi \approx \psi$ and $\cos\psi \approx 1$. Assuming that the lowpass filter preceding the VCO is modeled as $a/(s+a)$, where a is an arbitrary constant, determine the response to $m(t) = u(t-t_0)$.

3.50. In this problem we wish to develop a baseband (lowpass equivalent model) for a Costas PLL. We assume that the loop input is the complex envelope signal

$$\tilde{x}(t) = A_c m(t) e^{j\phi(t)}$$

and that the VCO output is $e^{j\theta(t)}$. Derive and sketch the model giving the signals at each point in the model.

Section 3.6

3.51. A continuous data signal is quantized and transmitted using a PCM system. If each data sample at the receiving end of the system must be known to within ± 0.25 % of the peak-to-peak full-scale value, how many binary symbols must each transmitted digital word contain? Assume that the message signal is speech and has a bandwidth of 4 kHz. Estimate the bandwidth of the resulting PCM signal (choose k).

3.52. A delta modulator has the message signal

$$m(t) = 3\sin[2\pi(10)t] + 4\sin[2\pi(20)t]$$

Determine the minimum sampling frequency required to prevent slope overload, assuming that the impulse weights δ_0 are 0.05π.

Section 3.7

3.53. Five messages bandlimited to W, W, $2W$, $4W$, and $4W$ Hz, respectively, are to be time-division multiplexed. Devise a commutator configuration such that each signal is periodically sampled at its own minimum rate and the samples are properly interlaced. What is the minimum transmission bandwidth required for this TDM signal?

3.54. In an FDM communication system, the transmitted baseband signal is

$$x(t) = m_1(t)\cos(2\pi f_1 t) + m_2(t)\cos(2\pi f_2 t)$$

This system has a second-order nonlinearity between transmitter output and receiver input. Thus the received baseband signal $y(t)$ can be expressed as

$$y(t) = a_1 x(t) + a_2 x^2(t)$$

Assuming that the two message signals, $m_1(t)$ and $m_2(t)$, have the spectra

$$M_1(f) = M_2(f) = \Pi\left(\frac{f}{W}\right)$$

sketch the spectrum of $y(t)$. Discuss the difficulties encountered in demodulating the received baseband signal. In many FDM systems, the subcarrier frequencies f_1 and f_2 are harmonically related. Describe any additional problems this presents.

Computer Exercises

3.1. In Example 3.1 we determined the minimum value of $m(t)$ using MATLAB. Write a MATLAB program that provides a complete solution for Example 3.1. Use the FFT for finding the amplitude and phase spectra of the transmitted signal $x_c(t)$.

3.2. The purpose of the exercise is to demonstrate the properties of SSB modulation. Develop a computer program to generate both upper-sideband and lower-sideband SSB signals and display both the time-domain signals and the amplitude spectra of these signals. Assume the message signal

$$m(t) = 2\cos(2\pi f_m t) + \cos(4\pi f_m t)$$

Select both f_m and f_c so that both the time and frequency axes can be easily calibrated. Plot the envelope of the SSB signals, and show that both the upper-sideband and the lower-sideband SSB signals have the same envelope. Use the FFT algorithm to generate the amplitude spectrum for both the upper-sideband and the lower sideband SSB signal.

3.3. Using the same message signal and value for f_m used in the preceding computer exercise, show that carrier reinsertion can be used to demodulate a SSB signal. Illustrate the effect of using a demodulation carrier with insufficient amplitude when using the carrier reinsertion technique.

3.4. In this computer exercise we investigate the properties of VSB modulation. Develop a computer program (using MATLAB) to generate and plot a VSB signal and the corresponding amplitude spectrum. Using the program, show that VSB can be demodulated using carrier reinsertion.

3.5. Using Computer Example 3.1 as a guide, reconstruct Figure 3.26 for the case in which 3 values of the modulation index (0.5, 1, and 5) are achieved by adjusting the peak frequency deviation while holding f_m constant.

3.6. Develop a computer program to generate the amplitude spectrum at the output of an FM modulator assuming a square-wave message signal. Plot the output for various values of the peak deviation. Compare the result with Figure 3.29 and comment on your observations.

3.7. Develop a computer program and use the program to verify the simulation results shown in Figure 3.42.

3.8. Referring to Computer Example 3.4, draw the block diagram of the system represented by the simulation loop, and label the inputs and outputs of the various loop components with the names used in the simulation code. Using this block diagram, verify that the simulation program is correct. What are the sources of error in the simulation program?

3.9. Modify the simulation program given in Computer Example 3.4 to allow the sampling frequency to be entered interactively. Examine the effect of using different sampling frequencies by executing the simulation with a range of sampling frequencies. Be sure that you start with a sampling frequency that is clearly too low and gradually increase the sampling frequency until you reach a sampling frequency that is clearly higher than is required for an accurate simulation result. Comment on the results. How do you know that the sampling frequency is sufficiently high?

3.10. Modify the simulation program given in Computer Example 3.4 so that the phase detector includes a limiter so that the phase detector characteristic is defined by

$$e_d(t) = \begin{cases} \sin[\psi(t)], & -1 < -A \le \sin[\psi(t)] \le A < 1 \\ A, & \sin[\psi(t)] > A \\ -A, & \sin[\psi(t)] < -A \end{cases}$$

where $\psi(t)$ is the phase error $\phi(t) - \theta(t)$ and A is a parameter that can be adjusted by the simulation user. Adjust the value of A and comment on the impact that decreasing A has on the number of cycles slipped and therefore on the time required to achieve phase lock.

3.11. Using Computer Example 3.5 as a guide, develop a simulation program for PAM and PPM

CHAPTER 4

PRINCIPLES OF BASEBAND DIGITAL DATA TRANSMISSION

So far we have dealt primarily with the transmission of analog signals. In this chapter we introduce the idea of transmission of digital data—that is, signals that can assume one of only a finite number of values during each transmission interval. This may be the result of sampling and quantizing an analog signal, as in the case of pulse-code modulation discussed in Chapter 3, or it might be the result of the need to transmit a message that is naturally discrete, such as a data or text file. In this chapter, we will discuss several features of a digital data transmission system. One feature that will not be covered in this chapter is the effect of random noise. This will be dealt with in Chapter 8 and following chapters. Another restriction of our discussion is that modulation onto a carrier signal is not assumed—hence the modifier *baseband*. Thus the types of data transmission systems to be dealt with utilize signals with power concentrated from 0 Hz to a few kilohertz or megahertz, depending on the application. Digital data transmission systems that utilize bandpass signals will be considered in Chapter 8 and following.

4.1 BASEBAND DIGITAL DATA TRANSMISSION SYSTEMS

Figure 4.1 shows a block diagram of a baseband digital data transmission system which includes several possible signal processing operations. Each will be discussed in detail in future sections of the chapter. For now we give only a short description.

As already mentioned, the analog-to-digital converter (ADC) block is present only if the source produces analog messages. It can be thought of as consisting of two operations: sampling and quantization. The quantization operation can be thought of as broken up into rounding the samples to the nearest quantizing level and then converting them to a binary number representation (designated as 0s and 1s, although their actual waveform representation will be determined by the *line code* used, to be discussed shortly). The requirements of sampling in order to minimize errors were discussed in Chapter 2, where it was shown that, in order to avoid aliasing, the source had to be lowpass bandlimited, say to W Hz, and the sampling rate had to satisfy $f_s > 2W$ samples per second (sps). If the signal being sampled is not strictly bandlimited or if the sampling rate is less than $2W$ sps, aliasing results. Error characterization due to quantizing will be dealt with in Chapter 7. If the message is analog, necessitating the use of an ADC at the transmitter, the inverse operation must take place at the receiver output in order to convert the digital signal back to analog form (called *digital-to-analog conversion*,

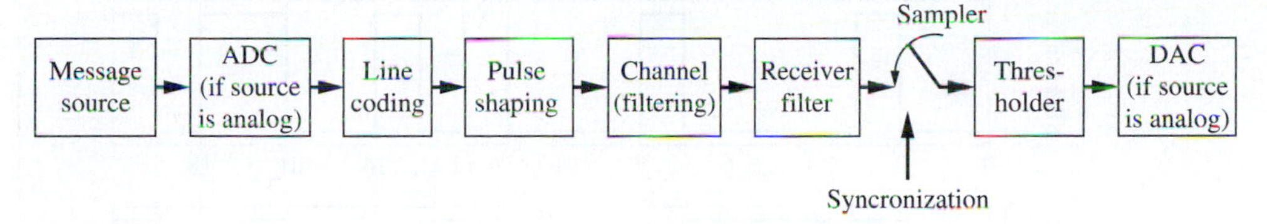

Figure 4.1
Block diagram of a baseband digital data transmission system.

or DAC). As seen in Chapter 2, after converting from binary format to quantized samples, this can be as simple as a lowpass filter or, as analyzed in Problem 2.68, a zero- or higher-order hold operation can be used.

The next block, line coding, will be dealt with in the next section. It is sufficient for now to simply state that the purposes of line coding are varied and include spectral shaping, synchronization considerations, and bandwidth considerations, among other reasons.

Pulse shaping might be used to shape the transmitted signal spectrum in order for it to be better accommodated by the transmission channel available. In fact, we will discuss the effects of filtering and how, if inadequate attention is paid to it, severe degradation can result from transmitted pulses interfering with each other. This is termed *intersymbol interference* (ISI) and can very severely impact overall system performance if steps are not taken to counteract it. On the other hand, we will also see that careful selection of the combination of pulse shaping (transmitter filtering) and receiver filtering (it is assumed that any filtering done by the channel is not open to choice) can completely eliminate ISI.

At the output of the receiver filter, it is necessary to synchronize the sampling times to coincide with the received pulse epochs. The samples of the received pulses are then compared with a threshold in order to make a decision as to whether a 0 or a 1 was sent (depending on the line code used, this may require some additional processing). If the data transmission system is operating reliably, these 1–0 decisions are correct with high probability, and the resulting DAC output is a close replica of the input message waveform.

Although the present discussion is couched in terms of two possible levels, designated as a 0 or 1, being sent, it is found to be advantageous in certain situations to utilize more than two levels. If two levels are used, the data format is referred to as *binary*; if $M > 2$ levels are utilized, the data format is called *M-ary*. If a binary format is used, the 0–1 symbols are called *bits*. If an M-ary format is used, each transmission is called a *symbol*.

4.2 LINE CODES AND THEIR POWER SPECTRA

4.2.1 Description of Line Codes

The spectrum of a digitally modulated signal is influenced both by the particular baseband data format used to represent the digital data and any additional pulse shaping (filtering) used to prepare the signal for transmission. Several commonly used baseband data formats are illustrated in Figure 4.2. Names for the various data formats shown are given on the vertical axis of the respective sketch of a particular waveform, although these are not the only terms applied to certain of these. Briefly, during each signaling interval, the following descriptions apply:

- Nonreturn-to-zero (NRZ) change: A 1 is represented by a positive level, A; a 0 is represented by $-A$.

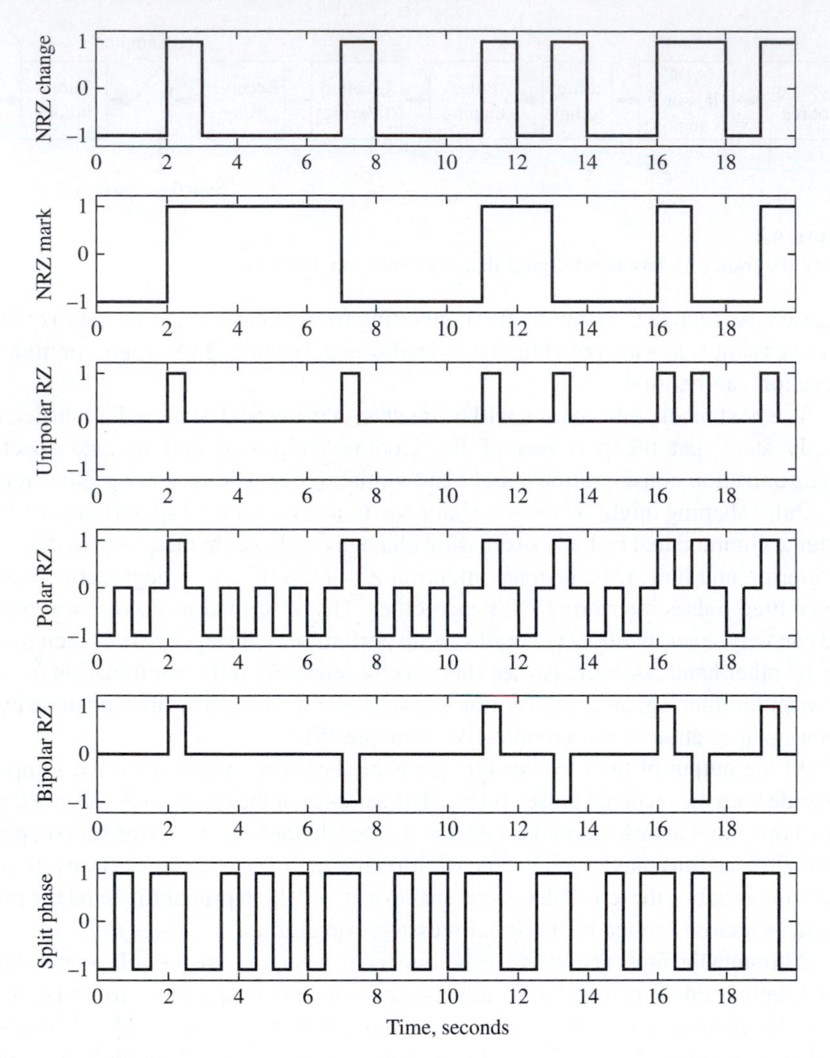

Figure 4.2
Abbreviated list of binary data formats.

Adapted from Holmes 1982.

- NRZ mark: A 1 is represented by a change in level (i.e., if the previous level sent was A, $-A$ is sent to represent a 1 and vice versa); a 0 is represented by no change in level.
- Unipolar return-to-zero (RZ): A 1 is represented by a $\frac{1}{2}$-width pulse (i.e., a pulse that "returns to zero"); a 0 is represented by no pulse.
- Polar RZ: A 1 is represented by a positive RZ pulse; a 0 is represented by a negative RZ pulse.
- Bipolar RZ: A 0 is represented by a 0 level; 1s are represented by RZ pulses that alternate in sign.
- Split phase (Manchester): A 1 is represented by A switching to $-A$ at $\frac{1}{2}$ the symbol period; a 0 is represented by $-A$ switching to A at $\frac{1}{2}$ the symbol period.

Two of the most commonly used formats are NRZ and split phase. Split phase, we note, can be thought of as being obtained from NRZ by multiplication by a square-wave clock waveform with a period equal to the symbol duration.

Several considerations should be taken into account in choosing an appropriate data format for a given application. Among these are

- *Self-synchronization*: Is there sufficient timing information built into the code so that synchronizers can be easily designed to extract a timing clock from the code?
- *Power spectrum suitable for the particular channel available*: For example, if the channel does not pass low frequencies, does the power spectrum of the chosen data format have a null at zero frequency?
- *Transmission bandwidth*: If the available transmission bandwidth is scarce, which it often is, a data format should be conservative in terms of bandwidth requirements. Sometimes conflicting requirements may force difficult choices.
- *Transparency*: Every possible data sequence should be faithfully and transparently received, regardless of whether it is infrequent or not.
- *Error detection capability*: Although the subject of forward error correction deals with the design of codes to provide error correction, inherent data correction capability is an added bonus for a given data format.
- *Good bit error probability performance*: There should be nothing about a data format that makes it difficult to implement minimum error probability receivers.

4.2.2 Power Spectra for Line Coded Data

It is important to know the spectral occupancy of line-coded data in order to predict the bandwidth requirements for the data transmission system (conversely, given a certain system bandwidth specification, the line code used will imply a certain maximum data rate). We now consider the power spectra for line-coded data assuming that the data source produces a random coin-toss sequence of 1s and 0s, with a binary digit being produced each T (recall that each binary digit is referred to as a bit which is a contraction for "binary digit").

To compute the power spectra for line-coded data, we use a result to be derived in Section 6.3.4 for the autocorrelation function of pulse-train-type signals. While it may be pedagogically unsound to use a result yet to be described, the avenue suggested to the student is to simply accept the result of Section 6.3.4 for now and concentrate on the results to be derived and the system implications of these results. In particular, this result is shown in Section 6.3.4 for a pulse train signal of the form

$$X(t) = \sum_{k=-\infty}^{\infty} a_k p(t - kT - \Delta) \tag{4.1}$$

where $\ldots a_{-1},\ a_0,\ a_1, \ldots,\ a_k, \ldots$ is a sequence of random variables with the averages

$$R_m = \langle a_k a_{k+m} \rangle \quad m = 0,\ \pm 1,\ \pm 2,\ \ldots \tag{4.2}$$

The function $p(t)$ is a deterministic pulse-type waveform, where T is the separation between pulses and Δ is a random variable that is independent of the value of a_k and uniformly distributed in the interval $(-T/2, T/2)$. The autocorrelation function of this waveform is

$$R_X(\tau) = \sum_{m=-\infty}^{\infty} R_m r(\tau - mT) \tag{4.3}$$

in which

$$r(\tau) = \frac{1}{T} \int_{-\infty}^{\infty} p(t+\tau) p(t)\, dt \tag{4.4}$$

The power spectral density is the Fourier transform of $R_X(\tau)$, which is

$$\begin{aligned} S_X(f) = \Im\left[R_X(\tau)\right] &= \Im\left[\sum_{m=-\infty}^{\infty} R_m r(\tau - mT)\right] \\ &= \sum_{m=-\infty}^{\infty} R_m \Im[r(\tau - mT)] \\ &= \sum_{m=-\infty}^{\infty} R_m S_r(f) e^{-j2\pi mTf} \\ &= S_r(f) \sum_{m=-\infty}^{\infty} R_m e^{-j2\pi mTf} \end{aligned} \tag{4.5}$$

where $S_r(f) = \Im[r(\tau)]$. Noting that $r(\tau) = 1/T \int_{-\infty}^{\infty} p(t+\tau)p(t)\, dt = 1/Tp(-t) * p(t)$, we obtain

$$S_r(f) = \frac{|P(f)|^2}{T} \tag{4.6}$$

where $P(f) = \Im[p(t)]$.

EXAMPLE 4.1

In this example we apply the above result to find the power spectral density of NRZ. For NRZ, the pulse shape function is $p(t) = \Pi(t/T)$ so that

$$P(f) = T \operatorname{sinc}(Tf) \tag{4.7}$$

and

$$S_r(f) = \frac{1}{T} |T \operatorname{sinc}(Tf)|^2 = T \operatorname{sinc}^2(Tf) \tag{4.8}$$

The time average $R_m = \langle a_k a_{k+m} \rangle$ can be deduced by noting that for a given pulse, the amplitude is $+A$ half the time and $-A$ half the time, while, for a sequence of two pulses with a given sign on the first pulse, the second pulse is $+A$ half the time and $-A$ half the time. Thus

$$R_m = \begin{cases} \frac{1}{2}A^2 + \frac{1}{2}(-A)^2 = A^2, & m = 0 \\ \frac{1}{4}A(A) + \frac{1}{4}A(-A) + \frac{1}{4}(-A)A + \frac{1}{4}(-A)(-A) = 0, & m \neq 0 \end{cases} \tag{4.9}$$

Thus the power spectral density, from (4.5) and (4.6), for NRZ is

$$S_{\mathrm{NRZ}}(f) = A^2 T \operatorname{sinc}^2(Tf) \tag{4.10}$$

This is plotted in Figure 4.3(a) where it is seen that the bandwidth to the first null of the power spectral density is $B_{\text{NRZ}} = 1/T$ Hz. Note that $A = 1$ gives unit power as seen from squaring and averaging the time-domain waveform.

■

EXAMPLE 4.2

The computation of the power spectral density for split phase differs from that for NRZ only in the spectrum of the pulse shape function because the coefficients R_m are the same as for NRZ. The pulse shape function for split phase is given by

$$p(t) = \Pi\left(\frac{t+T/4}{T/2}\right) - \Pi\left(\frac{t-T/4}{T/2}\right) \tag{4.11}$$

By applying the time-delay and superposition theorems of Fourier transforms, we have

$$\begin{aligned} P(f) &= \frac{T}{2}\,\text{sinc}\left(\frac{T}{2}f\right)e^{j2\pi(T/4)f} - \frac{T}{2}\,\text{sinc}\left(\frac{T}{2}f\right)e^{-j2\pi(T/4)f} \\ &= \frac{T}{2}\,\text{sinc}\left(\frac{T}{2}f\right)\left(e^{j\pi Tf/2} - e^{-j\pi Tf/2}\right) \\ &= jT\,\text{sinc}\left(\frac{T}{2}f\right)\sin\left(\frac{\pi T}{2}f\right) \end{aligned} \tag{4.12}$$

Thus

$$\begin{aligned} S_r(f) &= \frac{1}{T}\left|jT\,\text{sinc}\left(\frac{T}{2}f\right)\sin\left(\frac{\pi T}{2}f\right)\right|^2 \\ &= T\,\text{sinc}^2\left(\frac{T}{2}f\right)\sin^2\left(\frac{\pi T}{2}f\right) \end{aligned} \tag{4.13}$$

Hence, for split phase the power spectral density is

$$S_{\text{SP}}(f) = A^2 T\,\text{sinc}^2\left(\frac{T}{2}f\right)\sin^2\left(\frac{\pi T}{2}f\right) \tag{4.14}$$

This is plotted in Figure 4.3(b) where it is seen that the bandwidth to the first null of the power spectral density is $B_{\text{SP}} = 2/T$ Hz. However, unlike NRZ, split phase has a null at $f = 0$, which might have favorable implications if the transmission channel does not pass DC. Note that by squaring the time waveform and averaging the result, it is evident that $A = 1$ gives unit power.

■

EXAMPLE 4.3

In this example, we compute the power spectrum of unipolar RZ, which provides the additional challenge of discrete spectral lines. For unipolar RZ, the data correlation coefficients are

$$R_m = \begin{cases} \frac{1}{2}A^2 + \frac{1}{2}(0)^2 = \frac{1}{2}A^2, & m = 0 \\ \frac{1}{4}(A)(A) + \frac{1}{4}(A)(0) + \frac{1}{4}(0)(A) + \frac{1}{4}(0)(0) = \frac{1}{4}A^2, & m \neq 0 \end{cases} \tag{4.15}$$

The pulse shape function is given by

$$p(t) = \Pi\left(\frac{2t}{T}\right) \tag{4.16}$$

Therefore, we have

$$P(f) = \frac{T}{2}\,\mathrm{sinc}\left(\frac{T}{2}f\right) \tag{4.17}$$

and

$$\begin{aligned} S_r(f) &= \frac{1}{T}\left|\frac{T}{2}\,\mathrm{sinc}\left(\frac{T}{2}f\right)\right|^2 \\ &= \frac{T}{4}\,\mathrm{sinc}^2\left(\frac{T}{2}f\right) \end{aligned} \tag{4.18}$$

For unipolar RZ, we therefore have

$$\begin{aligned} S_{\mathrm{URZ}}(f) &= \frac{T}{4}\mathrm{sinc}^2\left(\frac{T}{2}f\right)\left(\frac{1}{2}A^2 + \frac{1}{4}A^2 \sum_{m=-\infty,\, m\neq 0}^{\infty} e^{-j2\pi mTf}\right) \\ &= \frac{T}{4}\mathrm{sinc}^2\left(\frac{T}{2}f\right)\left(\frac{1}{4}A^2 + \frac{1}{4}A^2 \sum_{m=-\infty}^{\infty} e^{-j2\pi mTf}\right) \end{aligned} \tag{4.19}$$

However, from (2.138) we have

$$\sum_{m=-\infty}^{\infty} e^{-j2\pi mTf} = \sum_{m=-\infty}^{\infty} e^{j2\pi mTf} = \frac{1}{T}\sum_{n=-\infty}^{\infty} \delta\left(f - \frac{n}{T}\right) \tag{4.20}$$

Thus, $S_{\mathrm{URZ}}(f)$ can be written as

$$\begin{aligned} S_{\mathrm{URZ}}(f) &= \frac{T}{4}\mathrm{sinc}^2\left(\frac{T}{2}f\right)\left[\frac{1}{4}A^2 + \frac{1}{4}\frac{A^2}{T}\sum_{n=-\infty}^{\infty}\delta\left(f-\frac{n}{T}\right)\right] \\ &= \frac{A^2T}{16}\mathrm{sinc}^2\left(\frac{T}{2}f\right) + \frac{A^2}{16}\delta(f) + \frac{A^2}{16}\mathrm{sinc}^2\left(\frac{1}{2}\right)\left[\delta\left(f-\frac{1}{T}\right) + \delta\left(f+\frac{1}{T}\right)\right] \\ &\quad + \frac{A^2}{16}\mathrm{sinc}^2\left(\frac{3}{2}\right)\left[\delta\left(f-\frac{3}{T}\right) + \delta\left(f+\frac{3}{T}\right)\right] + \cdots \end{aligned} \tag{4.21}$$

where the fact that $Y(f)\,\delta(f-f_n) = Y(f_n)\,\delta(f-f_n)$ for $Y(f)$ continuous at $f = f_n$ has been used to simplify the $\mathrm{sinc}^2[Tf/2]\,\delta(f-n/T)$ terms.

The power spectrum of unipolar RZ is plotted in Figure 4.3(c) where it is seen that the bandwidth to the first null of the power spectral density is $B_{\mathrm{URZ}} = 2/T$ Hz. The reason for the impulses in the spectrum is because the unipolar nature of this waveform is reflected in finite power at DC and harmonics of $1/T$ Hz. This can be a useful feature for synchronization purposes.

Note that for unit power in unipolar RZ, $A = 2$ because the average of the time-domain waveform squared is

$$\frac{1}{T}\left[\frac{1}{2}\left(A^2\frac{T}{2} + 0^2\frac{T}{2}\right) + \frac{1}{2}0^2T\right] = \frac{A^2}{4}$$

■

EXAMPLE 4.4

The power spectral density of polar RZ is straightforward to compute based on the results for NRZ. The data correlation coeffients are the same as for NRZ. The pulse shape function is $p(t) = \Pi(2t/T_b)$, the same as for unipolar RZ, so $S_r(f) = \frac{T}{4}\text{sinc}^2\left(\frac{T}{2}f\right)$. Thus

$$S_{\text{PRZ}}(f) = \frac{A^2T}{4}\text{sinc}^2\left(\frac{T}{2}f\right) \tag{4.22}$$

The power spectrum of polar RZ is plotted in Figure 4.3(d) where it is seen that the bandwidth to the first null of the power spectral density is $B_{\text{PRZ}} = 2/T$ Hz. Unlike unipolar RZ, there are no discrete spectral lines. Note that by squaring and averaging the time-domain waveform, we get $1/T(A^2T/2 + 0^2T/2) = A^2/2$, so $A = \sqrt{2}$ for unit average power.

■

EXAMPLE 4.5

The final line code for which we will compute the power spectrum is bipolar RZ. For $m = 0$, the possible $a_k a_k$ products are $AA = (-A)(-A) = A^2$, each of which occurs $\frac{1}{4}$ the time, and $(0)(0) = 0$, which occurs $\frac{1}{2}$ the time. For $m = \pm 1$, the possible data sequences are (1, 1), (1, 0), (0, 1), and (0, 0) for which the possible $a_k a_{k+1}$ products are $-A^2$, 0, 0, and 0, respectively, each of which occurs with probability $\frac{1}{4}$. For $m > 1$ the possible products are A^2 and $-A^2$, each of which occurs with probability $\frac{1}{8}$, and $\pm A(0)$, and $(0)(0)$, each of which occur with probability $\frac{1}{4}$. Thus the data correlation coefficients become

$$R_m = \begin{cases} \frac{1}{4}A^2 + \frac{1}{4}(-A)^2 + \frac{1}{2}(0)^2 = \frac{1}{2}A^2, & m = 0 \\ (-A)^2\left(\frac{1}{4}\right) + (A)(0)\left(\frac{1}{4}\right) + (0)(A)\left(\frac{1}{4}\right) + (0)(0)\left(\frac{1}{4}\right) = -\frac{A^2}{4}, & m = \pm 1 \\ A^2\left(\frac{1}{8}\right) + (-A^2)\left(\frac{1}{8}\right) + (A)(0)\left(\frac{1}{4}\right) + (-A)(0)\left(\frac{1}{4}\right) + (0)(0)\left(\frac{1}{4}\right) = 0, & |m| > 1 \end{cases} \tag{4.23}$$

The pulse shape function is

$$p(t) = \Pi\left(\frac{2t}{T}\right) \tag{4.24}$$

Therefore, we have

$$P(f) = \frac{T}{2}\text{sinc}\left(\frac{T}{2}f\right) \tag{4.25}$$

and

$$\begin{aligned} S_r(f) &= \frac{1}{T}\left|\frac{T}{2}\text{sinc}\left(\frac{T}{2}f\right)\right|^2 \\ &= \frac{T}{4}\text{sinc}^2\left(\frac{T}{2}f\right) \end{aligned} \tag{4.26}$$

Therefore, for bipolar RZ we have

$$\begin{aligned} S_{\mathrm{BRZ}}(f) &= S_r(f) \sum_{m=-\infty}^{\infty} R_m e^{-j2\pi mTf} \\ &= \frac{A^2T}{8} \operatorname{sinc}^2\left(\frac{T}{2}f\right)\left(1-\frac{1}{2}e^{j2\pi Tf}-\frac{1}{2}e^{-j2\pi Tf}\right) \\ &= \frac{A^2T}{8} \operatorname{sinc}^2\left(\frac{T}{2}f\right)[1-\cos(2\pi Tf)] \\ &= \frac{A^2T}{4} \operatorname{sinc}^2\left(\frac{T}{2}f\right)\sin^2(\pi Tf) \end{aligned} \tag{4.27}$$

which is shown in Figure 4.3(f).

Note that by squaring the time-domain waveform and accounting for it being 0 for the time when logic 0s are sent and it being 0 half the time when logic 1s are sent, we get for the power

$$\frac{1}{T}\left[\frac{1}{2}\left(\frac{1}{2}A^2\frac{T}{2}+\frac{1}{2}(-A)^2\frac{T}{2}+0^2\frac{T}{2}\right)+\frac{1}{2}0^2T\right]=\frac{A^2}{4} \tag{4.28}$$

so $A = 2$ for unit average power.

■

Typical power spectra are shown in Figure 4.3 for all of the data modulation formats shown in Figure 4.2, assuming a random (coin toss) bit sequence. For data formats lacking power spectra with significant frequency content at multiples of the bit rate $1/T$, nonlinear operations are required to generate power at a frequency of $1/T$ Hz or multiples thereof for symbol synchronization purposes. Note that split phase guarantees at least one zero crossing per bit interval but requires twice the transmission bandwidth of NRZ. Around 0 Hz, NRZ possesses significant power. Generally, no data format possesses all the desired features listed in Section 4.2.1, and the choice of a particular data format will involve trade-offs.

COMPUTER EXAMPLE 4.1

A MATLAB script file for plotting the power spectra of Figure 4.3 is given below.

```
% File:  c4ce1.m
%
clf
ANRZ = 1;
T = 1;
f = -40:.005:40;
SNRZ = ANRZ^2*T*(sinc(T*f)).^2;
areaNRZ = trapz(f, SNRZ) % Area of NRZ spectrum as check
ASP = 1;
SSP = ASP^2*T*(sinc(T*f/2)).^2.*(sin(pi*T*f/2)).^2;
areaSP = trapz(f, SSP) % Area of split phase spectrum as check
AURZ = 2;
SURZc = AURZ^2*T/16*(sinc(T*f/2)).^2;
areaRZc = trapz(f, SURZc)
fdisc = -40:1:40;
SURZd = zeros(size(fdisc));
```

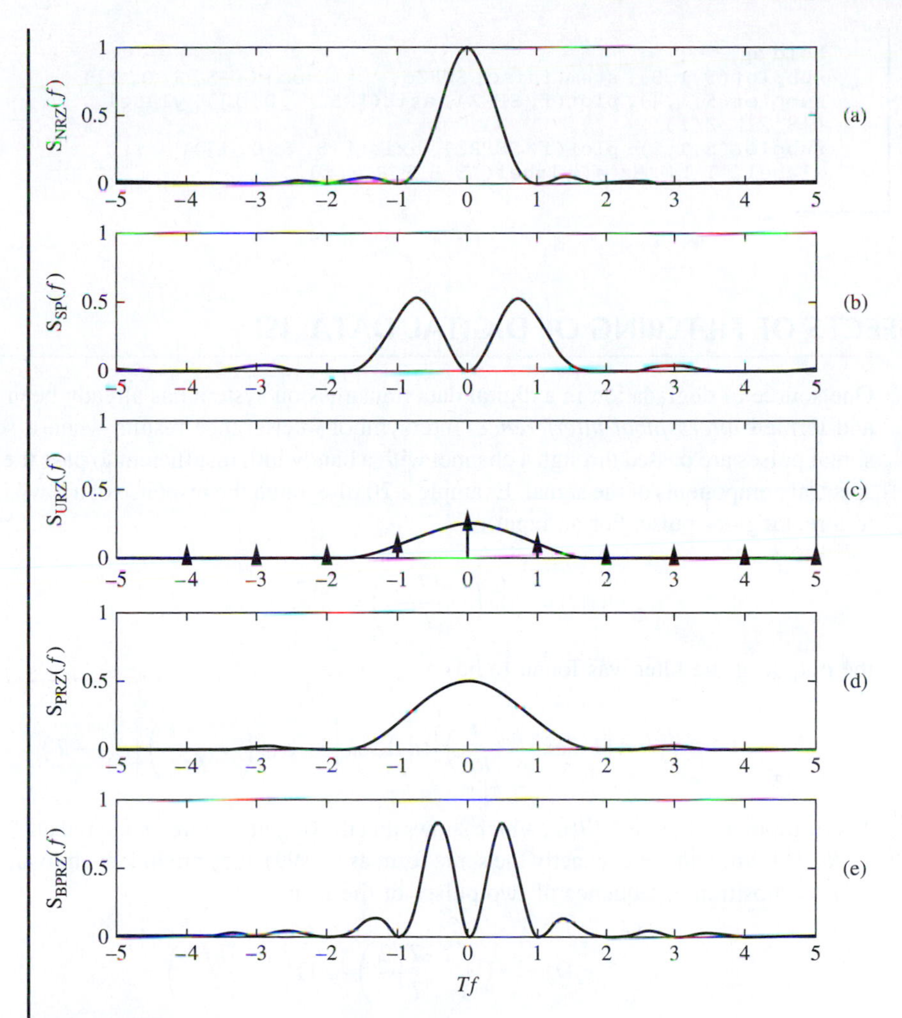

Figure 4.3
Power spectra for line-coded binary data formats.

```
SURZd = AURZ^2/16*(sinc(fdisc/2)).^2;
areaRZ = sum(SURZd)+areaRZc  %Area of unipolar return-to-zero spect as check
APRZ = sqrt(2);
SPRZ = APRZ^2*T/4*(sinc(T*f/2)).^2;
areaSPRZ = trapz(f, SPRZ) % Area of polar return-to-zero spectrum as check
ABPRZ = 2;
SBPRZ = ABPRZ^2*T/4*((sinc(T*f/2)).^2).*(sin(pi*T*f)).^2;
areaBPRZ = trapz(f, SBPRZ)  % Area of bipolar return-to-zero
spectrum as check
subplot(5,1,1), plot(f, SNRZ), axis([-5, 5, 0, 1]), ylabel
('S_N_R_Z(f)')
subplot(5,1,2), plot(f, SSP), axis([-5, 5, 0, 1]), ylabel
('S_S_P(f)')
subplot(5,1,3), plot(f, SURZc), axis([-5, 5, 0, 1]),
ylabel('S_U_R_Z(f)')
```

```
hold on
subplot(5,1,3), stem(fdisc, SURZd, '^'), axis([-5, 5, 0, 1])
subplot(5,1,4), plot(f, SPRZ), axis([-5, 5, 0, 1]), ylabel
('S_P_R_Z(f)')
subplot(5,1,5), plot(f, SBPRZ), axis([-5, 5, 0, 1]),
xlabel('T_bf, Hz'), ylabel('S_B_P_R_Z(f)')
```

■

■ 4.3 EFFECTS OF FILTERING OF DIGITAL DATA: ISI

One source of degradation in a digital data transmission system has already been mentioned and termed *intersymbol interference*. Intersymbol interference results when a sequence of signal pulses are passed through a channel with a bandwidth insufficient to pass the significant spectral components of the signal. Example 2.20 illustrated the response of a lowpass RC filter to a rectangular pulse. For an input of

$$x_1(t) = A\Pi\left(\frac{t - T/2}{T}\right) = A[u(t) - u(t - T)] \tag{4.29}$$

the output of the filter was found to be

$$y_1(t) = A\left[1 - \exp\left(-\frac{t}{RC}\right)\right]u(t) - A\left[1 - \exp\left(-\frac{t - T}{RC}\right)\right]u(t - T) \tag{4.30}$$

This is plotted in Figure 2.17(a), which shows that the output is more "smeared out" the smaller T/RC is [although not in exactly the same form as (2.199), they are in fact equivalent]. In fact, by superposition, a sequence of two pulses of the form

$$\begin{aligned} x_2(t) &= A\Pi\left(\frac{t - T/2}{T}\right) - A\Pi\left(\frac{t - 3T/2}{T}\right) \\ &= A[u(t) - 2u(t - T) + u(t - 2T)] \end{aligned} \tag{4.31}$$

will result in the response

$$\begin{aligned} y_2(t) = {} & A\left[1 - \exp\left(-\frac{t}{RC}\right)\right]u(t) - 2A\left[1 - \exp\left(-\frac{t - T}{RC}\right)\right]u(t - T) \\ & + A\left[1 - \exp\left(-\frac{t - 2T}{RC}\right)\right]u(t - 2T) \end{aligned} \tag{4.32}$$

At a simple level, this illustrates the idea of ISI. If the channel, represented by the lowpass RC filter, has only a single pulse at its input, there is no problem from the transient response of the channel. However, when two or more pulses are input to the channel in time sequence [in the case of the input $x_2(t)$, a positive pulse followed by a negative one], the transient response due to the initial pulse interferes with the responses due to the second and

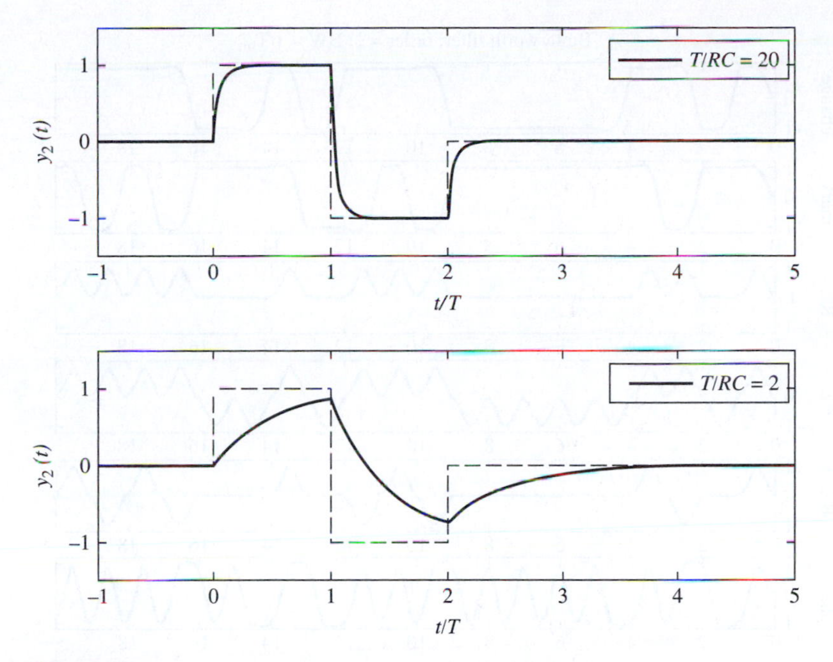

Figure 4.4
Response of a lowpass RC filter to a positive rectangular pulse followed by a negative rectangular pulse to illustrate the concept of ISI. (a) $T/RC = 20$. (b) $T/RC = 2$.

following pulses. This is illustrated in Figure 4.4, where the two-pulse response (4.32) is plotted for two values of T/RC, the first of which results in negligible ISI and the second of which results in significant ISI in addition to distortion of the output pulses. In fact, the smaller T/RC, the more severe the ISI effects are because the time constant, RC, of the filter is large compared with the pulse width, T.

To consider a more realistic example, we reconsider the line codes of Figure 4.2. These waveforms are shown filtered by a lowpass, second-order Butterworth filter in Figure 4.5 for the filter 3-dB frequency equal to $f_3 = 1/T$ and in Figure 4.6 for $f_3 = 0.5/T$. The effects of ISI are evident. In Figure 4.5 the bits are fairly discernable, even for data formats using pulses of width $T/2$ (i.e., all the RZ cases and split phase). In Figure 4.6, the NRZ cases have fairly distinguishable bits, but the RZ and split-phase formats suffer greatly from ISI. Recall that from the plots of Figure 4.3 and the analysis that led to them, the RZ and split-phase formats occupy essentially twice the bandwidth of the NRZ formats for a given data rate.

The question about what can be done about ISI naturally arises. One perhaps surprising solution is that with proper transmitter and receiver filter design (the filter representing the channel is whatever it is) the effects of ISI can be completely eliminated. We investigate this solution in the following section. Another somewhat related solution is the use of special filtering at the receiver called *equalization*. At a very rudimentary level, an equalization filter can be looked upon as the inverse of the channel filter, or a close approximation to it. We consider one form of equalization filtering in Section 4.5.

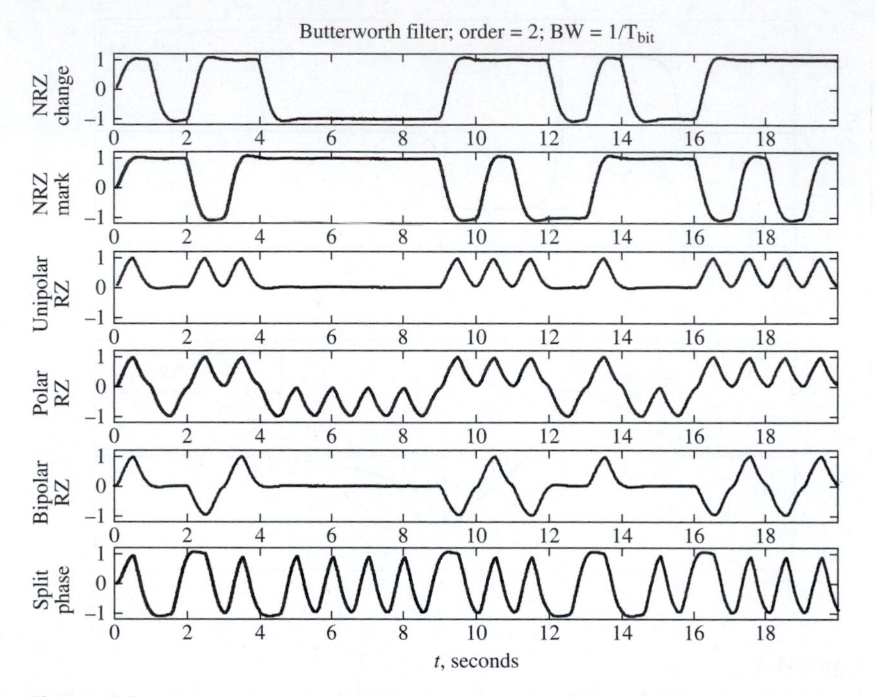

Figure 4.5
Data sequences formatted with various line codes passed through a channel represented by a second-order lowpass Butterworth filter of bandwidth 1 bit rate.

4.4 PULSE SHAPING: NYQUIST'S CRITERION FOR ZERO ISI

In this section we examine designs for the transmitter and receiver filters that shape the overall signal pulse shape function so as to ideally eliminate interference between adjacent pulses. This is formally stated as Nyquist's criterion for zero ISI.

4.4.1 Pulses Having the Zero-ISI Property

To see how one might implement this approach, we recall the sampling theorem, which gives a theoretical minimum spacing between samples to be taken from a signal with an ideal lowpass spectrum in order that the signal can be reconstructed exactly from the sample values. In particular, the transmission of a lowpass signal with bandwidth W Hz can be viewed as sending a minimum of $2W$ independent samples per second. If these $2W$ samples per second represent $2W$ independent pieces of data, this transmission can be viewed as sending $2W$ pulses per second through a channel represented by an ideal lowpass filter of bandwidth W. The transmission of the nth piece of information through the channel at time $t = nT = n/2W$ is accomplished by sending an impulse of amplitude a_n. The output of the channel due to this impulse at the input is

$$y_n(t) = a_n \text{sinc}\left[2W\left(t - \frac{n}{2W}\right)\right] \tag{4.33}$$

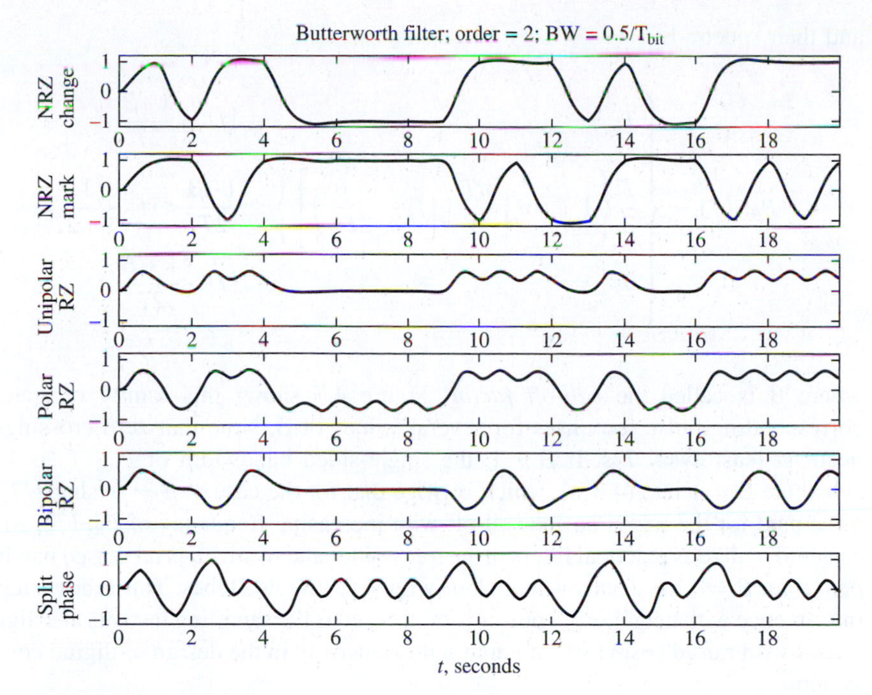

Figure 4.6
Data sequences formatted with various line codes passed through a channel represented by a second-order lowpass Butterworth filter of bandwidth $\frac{1}{2}$ bit rate.

For an input consisting of a train of impulses spaced by $T = 1/2W$ s, the channel output is

$$y(t) = \sum_n y_n(t) = \sum_n a_n \operatorname{sinc}\left[2W\left(t - \frac{n}{2W}\right)\right] \tag{4.34}$$

where $\{a_n\}$ is the sequence of sample values (i.e., the information). If the channel output is sampled at time $t_m = m/2W$, the sample value is a_m because

$$\operatorname{sinc}(m-n) = \begin{cases} 1, & m = n \\ 0, & m \neq n \end{cases} \tag{4.35}$$

which results in all terms in (4.34) except the mth being zero. In other words, the mth sample value at the output is not affected by preceding or succeeding sample values; it represents an independent piece of information.

Note that the bandlimited channel implies that the time response due to the nth impulse at the input is infinite in extent; a waveform cannot be simultaneously bandlimited and time limited. It is of interest to inquire if there are any bandlimited waveforms other than sinc $(2Wt)$ that have the property of (4.35), that is, that their zero crossings are spaced by $T = 1/2W$s. One such family of pulses are those having raised cosine spectra. Their time response is given by

$$p_{RC}(t) = \frac{\cos(\pi\beta t/T)}{1-(2\beta t/T)^2} \operatorname{sinc}\left(\frac{t}{T}\right) \tag{4.36}$$

and their spectra by

$$P_{\mathrm{RC}}(f) = \begin{cases} T, & |f| \le \dfrac{1-\beta}{2T} \\ \dfrac{T}{2}\left\{1+\cos\left[\dfrac{\pi T}{\beta}\left(|f| - \dfrac{1-\beta}{2T}\right)\right]\right\}, & \dfrac{1-\beta}{2T} < |f| \le \dfrac{1+\beta}{2T} \\ 0, & |f| > \dfrac{1+\beta}{2T} \end{cases} \tag{4.37}$$

where β is called the *roll-off factor*. Figure 4.7 shows this family of spectra and the corresponding pulse responses for several values of β. Note that zero crossings for $p_{RC}(t)$ occur at least every T s. If $\beta = 1$, the single-sided bandwidth of $P_{\mathrm{RC}}(f)$ is $1/T$ Hz [just substitute $\beta = 1$ into (4.37)], which is twice that for the case of $\beta = 0$ [$\mathrm{sinc}(t/T)$pulse]. The price paid for the raised cosine roll-off with increasing frequency of $P_{\mathrm{RC}}(f)$, which may be easier to realize as practical filters in the transmitter and receiver, is increased bandwidth. Also, $p_{\mathrm{RC}}(t)$ for $\beta = 1$ has a narrow main lobe with very low side lobes. This is advantageous in that interference with neighboring pulses is minimized if the sampling instants are slightly in error. Pulses with raised cosine spectra are used extensively in the design of digital communication systems.

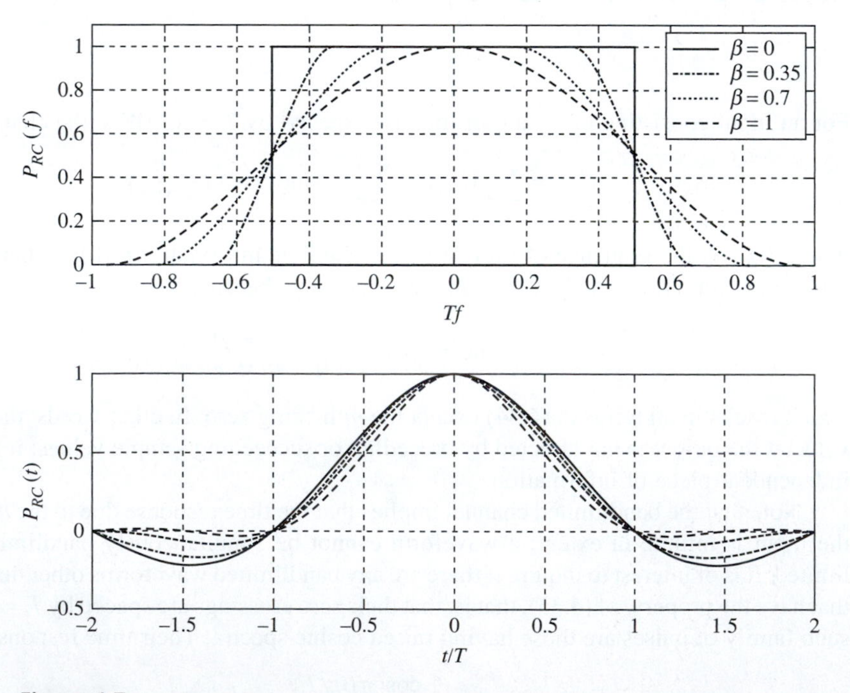

Figure 4.7
(a) Raised cosine spectra and (b) corresponding pulse responses.

4.4.2 Nyquist's Pulse Shaping Criterion

Nyquist's pulse shaping criterion states that a pulse shape function $p(t)$, having a Fourier transform $P(f)$ that satisfies the criterion

$$\sum_{k=-\infty}^{\infty} P\left(f+\frac{k}{T}\right) = T, \quad |f| \leq \frac{1}{2T} \tag{4.38}$$

results in a pulse shape function with sample values

$$p(nT) = \begin{cases} 1, & n=0 \\ 0, & n \neq 0 \end{cases} \tag{4.39}$$

Using this result, we can see that no adjacent pulse interference will be obtained if the received data stream is represented as

$$y(t) = \sum_{n=-\infty}^{\infty} a_n p(t-nT) \tag{4.40}$$

and the sampling at the receiver occurs at integer multiples of T s, at the pulse epochs. For example, to obtain the $n=$ 10th sample, one simply sets $t = 10T$ in (4.40), and the resulting sample is a_{10}, given that the result of Nyquist's pulse shaping criterion of (4.39) holds.

The proof of Nyquist's pulse shaping criterion follows easily by making use of the inverse Fourier representation for $p(t)$, which is

$$p(t) = \int_{-\infty}^{\infty} P(f)\exp(j2\pi ft)\, df \tag{4.41}$$

For the nth sample value, this expression can be written as

$$p(nT) = \sum_{k=-\infty}^{\infty} \int_{-(2k+1)/2T}^{(2k+1)/2T} P(f)\exp(j2\pi fnT)\, df \tag{4.42}$$

where the inverse Fourier transform integral for $p(t)$ has been broken up into contiguous frequency intervals of length $1/T$ Hz. By the change of variables $u = f - k/T$, Equation (4.42) becomes

$$\begin{aligned} p(nT) &= \sum_{k=-\infty}^{\infty} \int_{-1/2T}^{1/2T} P\left(u+\frac{k}{T}\right)\exp(j2\pi nTu)\, du \\ &= \int_{-1/2T}^{1/2T} \sum_{k=-\infty}^{\infty} P\left(u+\frac{k}{T}\right)\exp(j2\pi nTu)\, du \end{aligned} \tag{4.43}$$

where the order of integration and summation has been reversed. By hypothesis

$$\sum_{k=-\infty}^{\infty} P\left(u+\frac{k}{T}\right) = T \tag{4.44}$$

between the limits of integration, so that (4.43) becomes

$$\begin{aligned} p(nT) &= \int_{-1/2T}^{1/2T} T\exp(j2\pi nTu)\, du = \text{sinc}(n) \\ &= \begin{cases} 1, & n=0 \\ 0, & n \neq 0 \end{cases} \end{aligned} \tag{4.45}$$

which completes the proof of Nyquist's pulse shaping criterion.

With the aid of this result, it is now apparent why the raised cosine pulse family is free of ISI, even though the family is by no means unique. Note that what is excluded from the raised cosine spectra for $|f| < 1/T$ Hz is filled by the spectral translate tail for $|f| > 1/T$ Hz. Example 4.6 illustrates this for a simpler, although more impractical, spectrum than the raised cosine spectrum.

EXAMPLE 4.6

Consider the triangular spectrum

$$P_\Delta(f) = T\,\Lambda(Tf) \tag{4.46}$$

It is shown in Figure 4.8(a) and in Figure 4.8(b) $\sum_{k=-\infty}^{\infty} P_\Delta(f + k/T)$ is shown, where it is evident that the sum is a constant. Using the transform pair $\Lambda(t/B) \leftrightarrow B\,\mathrm{sinc}^2(Bf)$ and duality to get the transform pair $p_\Delta(t) = \mathrm{sinc}^2(t/T) \leftrightarrow T\Lambda(T/f) = P_\Delta(f)$, we see that this pulse shape function does indeed have the zero-ISI property because $p_\Delta(nT) = \mathrm{sinc}^2(n) = 0,\ n \neq 0$ integer.

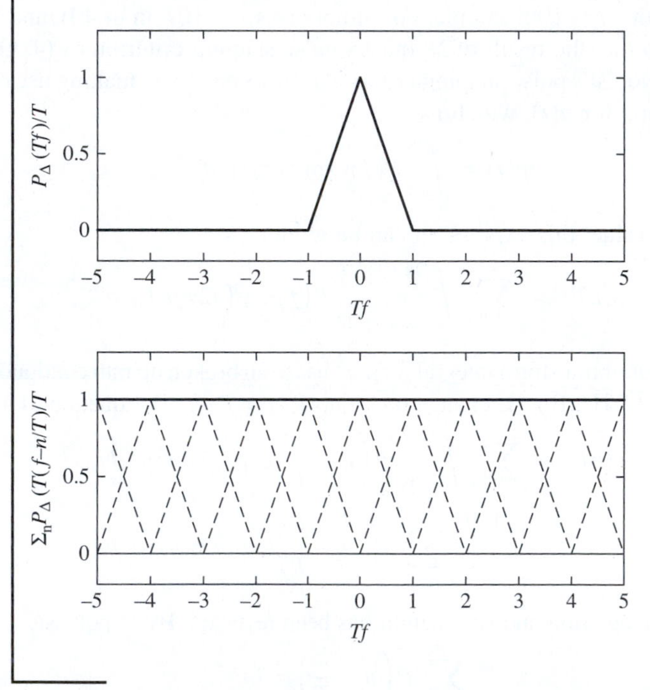

Figure 4.8
Illustration that (a) a triangular spectrum satisfies (b) Nyquist's zero ISI criterion.

■

4.4.3 Transmitter and Receiver Filters for Zero ISI

Consider the simplified pulse transmission system of Figure 4.9. A source produces a sequence of sample values $\{a_n\}$. Note that these are not necessarily quantized or binary digits, but they could be. For example, two bits per sample could be sent with four possible levels, representing 00, 01, 10, and 11. In the simplified transmitter model under consideration here, the kth sample

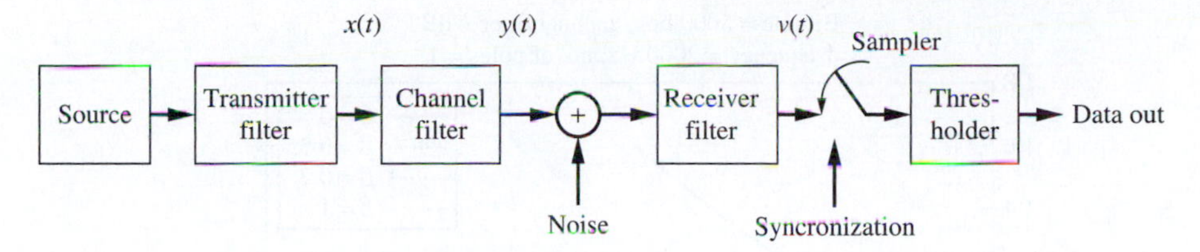

Figure 4.9
Transmitter, channel, and receiver cascade illustrating the implementation of a zero-ISI communication system.

value multiplies a unit impulse occuring at time kT and this weighted impulse train is the input to a transmitter filter with impulse response $h_T(t)$ and corresponding frequency response $H_T(f)$. The noise for now is assumed to be zero (effects of noise will be considered in Chapter 8). Thus, the input signal to the transmission channel, represented by a filter having impulse response $h_C(t)$ and corresponding frequency response $H_C(f)$, for all time is

$$\begin{aligned} x(t) &= \sum_{k=-\infty}^{\infty} a_k \delta(t-kT) * h_T(t) \\ &= \sum_{k=-\infty}^{\infty} a_k h_T(t-kT) \end{aligned} \tag{4.47}$$

The output of the channel is

$$y(t) = x(t) * h_C(t) \tag{4.48}$$

and the output of the receiver filter is

$$v(t) = y(t) * h_R(t) \tag{4.49}$$

We want the output of the receiver filter to have the zero-ISI property, and to be specific, we set

$$v(t) = \sum_{k=-\infty}^{\infty} a_k A p_{\mathrm{RC}}(t - kT - t_d) \tag{4.50}$$

where $p_{\mathrm{RC}}(t)$ is the raised cosine pulse function, t_d represents the delay introduced by the cascade of filters and A represents an amplitude scale factor. Putting this all together, we have

$$A p_{\mathrm{RC}}(t-t_d) = h_T(t) * h_C(t) * h_R(t) \tag{4.51}$$

or, by Fourier transforming both sides, we have

$$A P_{\mathrm{RC}}(f) \exp(-j2\pi f t_d) = H_T(f) H_C(f) H_R(f) \tag{4.52}$$

In terms of amplitude responses this becomes

$$A P_{\mathrm{RC}}(f) = |H_T(f)||H_C(f)||H_R(f)| \tag{4.53}$$

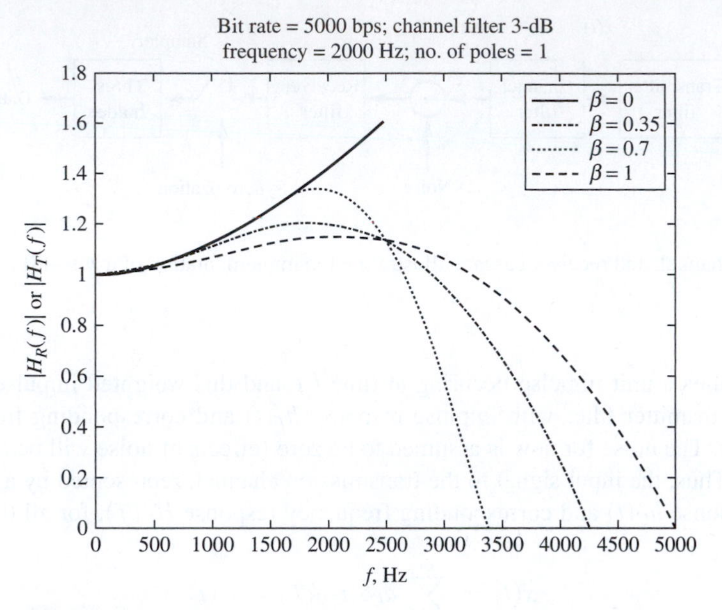

Figure 4.10
Transmitter and receiver filter amplitude responses that implement the zero-ISI condition assuming a first-order Butterworth channel filter and raised cosine pulse shapes.

Now $|H_C(f)|$ is fixed (the channel is whatever it is), and $P_{\text{RC}}(f)$ is specified. Suppose we want the transmitter and receiver filter amplitude responses to be the same. Then, solving (4.53) with $|H_T(f)| = |H_R(f)|$, we have

$$|H_T(f)|^2 = |H_R(f)|^2 = \frac{AP_{\text{RC}}(f)}{|H_C(f)|} \tag{4.54}$$

or

$$|H_T(f)| = |H_R(f)| = \frac{AP_{\text{RC}}^{1/2}(f)}{|H_C(f)|^{1/2}} \tag{4.55}$$

This amplitude response is shown in Figure 4.10 for raised cosine spectra of various roll-off factors and for a channel filter assumed to have a first-order Butterworth amplitude response. We have not accounted for the effects of additive noise. If the noise spectrum is flat, the only change would be another multiplicative constant. The constants are arbitrary since they multiply both signal and noise alike.

■ 4.5 ZERO-FORCING EQUALIZATION

In the previous section, it was shown how to choose transmitter and receiver filter amplitude responses, given a certain channel filter, to provide output pulses satisfying the zero-ISI condition. In this section, we present a procedure for designing a filter that will accept a channel

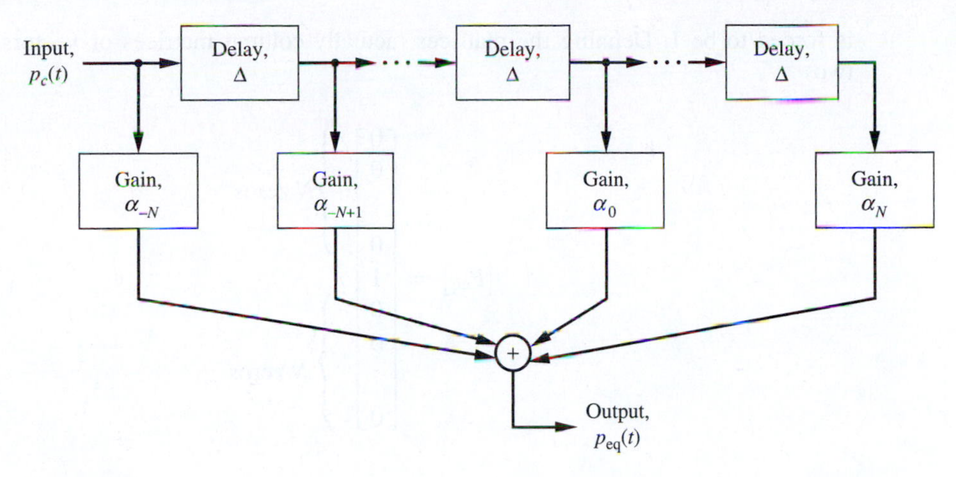

Figure 4.11
A transversal filter implementation for equalization of ISI.

output pulse response not satisfying the zero-ISI condition and produce a pulse at its output that has N zero-valued samples on either side of its maximum sample value taken to be 1 for convenience. This filter will be called a *zero-forcing equalizer*. We specialize our considerations of an equalization filter to a particular form—a transversal or tapped-delay-line filter. Figure 4.11 is a block diagram of such a filter.

There are at least two reasons for considering a transversal structure for the purpose of equalization. First, it is simple to analyze. Second, it is easy to mechanize by electronic means (i.e., transmission line delays and analog multipliers) at high frequencies and by digital signal processors at lower frequencies.

Let the pulse response of the channel output be $p_c(t)$. The output of the equalizer in response to $p_c(t)$ is

$$p_{\text{eq}}(t) = \sum_{n=-N}^{N} \alpha_n p_c(t - n\Delta) \tag{4.56}$$

where Δ is the tap spacing and the total number of transversal filter taps is $2N + 1$. We want $p_{\text{eq}}(t)$ to satisfy Nyquist's pulse shaping criterion, which we will call the *zero-ISI condition*. Since the output of the equalizer is sampled every T s, it is reasonable that the tap spacing be $\Delta = T$. The zero-ISI condition therefore becomes

$$\begin{aligned} p_{\text{eq}}(mT) &= \sum_{n=-N}^{N} \alpha_n p_c[(m-n)T] \\ &= \begin{cases} 1, & m = 0 \\ 0, & m \neq 0 \end{cases} \qquad m = 0, \pm 1, \pm 2, \ldots, \pm N \end{aligned} \tag{4.57}$$

Note that the zero-ISI condition can be satisfied at only $2N$ time instants because there are only $2N + 1$ coefficients to be selected in (4.57) and the output of the filter for $t = 0$

is forced to be 1. Defining the matrices (actually column matrices or vectors for the first two)

$$[P_{\text{eq}}] = \begin{bmatrix} 0 \\ 0 \\ \vdots \\ 0 \\ 1 \\ 0 \\ 0 \\ \vdots \\ 0 \end{bmatrix} \begin{matrix} \left.\begin{matrix} \\ \\ \\ \\ \end{matrix}\right\} N \text{ zeros} \\ \\ \left.\begin{matrix} \\ \\ \\ \\ \end{matrix}\right\} N \text{ zeros} \end{matrix} \tag{4.58}$$

$$[A] = \begin{bmatrix} \alpha_{-N} \\ \alpha_{-N+1} \\ \vdots \\ \alpha_N \end{bmatrix} \tag{4.59}$$

and

$$[P_c] = \begin{bmatrix} p_c(0) & p_c(-T) & \cdots & p_c(-2NT) \\ p_c(T) & p_c(0) & \cdots & p_c(-2N+1)T \\ \vdots & & & \vdots \\ p_c(2NT) & & & p_c(0) \end{bmatrix} \tag{4.60}$$

it follows that (4.57) can be written as the matrix equation

$$[P_{\text{eq}}] = [P_c][A] \tag{4.61}$$

The method of solution of the zero-forcing coefficients is now clear. Since $[P_{\text{eq}}]$ is specified by the zero-ISI condition, all we must do is multiply through by the inverse of $[P_c]$. The desired coefficient matrix $[A]$ is then the middle column of $[P_c]^{-1}$, which follows by multiplying $[P_c]^{-1}$ times $[P_{\text{eq}}]$:

$$A = [P_c]^{-1}[P_{\text{eq}}] = [P_c]^{-1} \begin{bmatrix} 0 \\ 0 \\ \vdots \\ 0 \\ 1 \\ 0 \\ 0 \\ \vdots \\ 0 \end{bmatrix} = \text{middle column of } [P_c]^{-1} \tag{4.62}$$

EXAMPLE 4.7

Consider a channel for which the following sample values of the channel pulse response are obtained:

$$p_c(-3T) = 0.02 \quad p_c(-2T) = -0.05 \quad p_c(-T) = 0.2 \quad p_c(0) = 1.0$$
$$p_c(T) = 0.3 \quad p_c(2T) = -0.07 \quad p_c(3T) = 0.03$$

The matrix $[P_c]$ for $N = 1$ is

$$[P_c] = \begin{bmatrix} 1.0 & 0.2 & -0.05 \\ 0.3 & 1.0 & 0.2 \\ -0.07 & 0.3 & 1.0 \end{bmatrix} \tag{4.63}$$

and the inverse of this matrix is

$$[P_c]^{-1} = \begin{bmatrix} 1.0815 & -0.2474 & 0.1035 \\ -0.3613 & 1.1465 & -0.2474 \\ 0.1841 & -0.3613 & 1.0815 \end{bmatrix} \tag{4.64}$$

Thus, by (4.62)

$$A = \begin{bmatrix} 1.0815 & -0.2474 & 0.1035 \\ -0.3613 & 1.1465 & -0.2474 \\ 0.1841 & -0.3613 & 1.0815 \end{bmatrix} \begin{bmatrix} 0 \\ 1 \\ 0 \end{bmatrix} = \begin{bmatrix} -0.2474 \\ 1.1465 \\ -0.3613 \end{bmatrix} \tag{4.65}$$

Using these coefficients, the equalizer output is

$$p_{\text{eq}}(m) = -0.2474 p_c[(m+1)T] + 1.1465 p_c(mT)$$
$$-0.3613 p_c[(m-1)T], \quad m = \ldots, -1, 0, 1, \ldots$$

Putting values in shows that $p_{\text{eq}}(0) = 1$ and that the single samples on either side of $p_{\text{eq}}(0)$ are zero. Samples more than one away from the center sample are not necessarily zero for this example. Calculation using the extra samples for $p_c(nT)$ gives $p_c(-2T) = -0.1140$ and $p_c(2T) = -0.1961$. Samples for the channel and the equalizer outputs are shown in Figure 4.12.

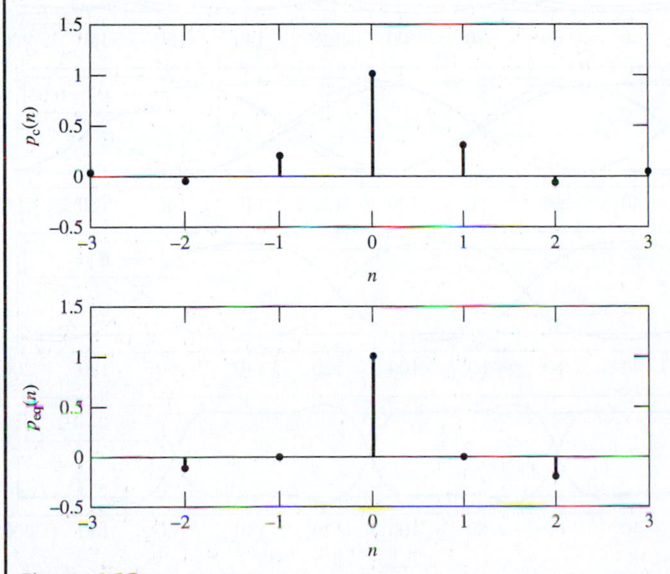

Figure 4.12
Samples for (a) an assumed channel response and for (b) the output of a zero-forcing equalizer of length 3.

■

4.6 EYE DIAGRAMS

We now consider eye diagrams which, although not a quantitative measure of system performance, are simple to construct and give significant insight into system performance. An eye diagram is constructed by plotting overlapping k-symbol segments of a baseband signal. In other words, an eye diagram can be displayed on an oscilloscope by triggering the time sweep of the oscilloscope, as shown in Figure 4.13, at times $t = nkT_s$, where T_s is the symbol period, kT_s is the eye period, and n is an integer. A simple example will demonstrate the process of generating an eye diagram.

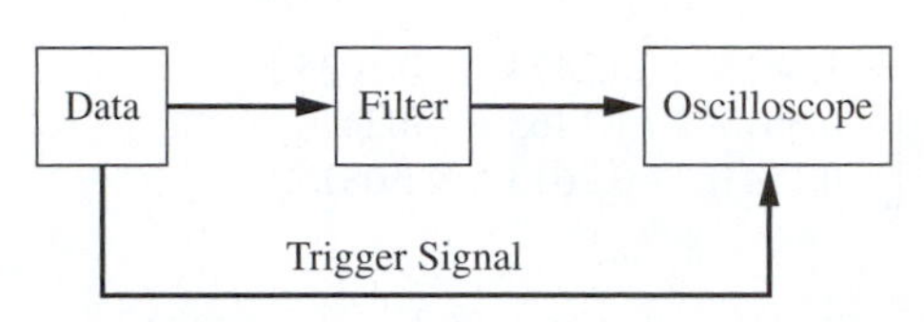

Figure 4.13
Simple technique for generating an eye diagram for a bandlimited signal.

EXAMPLE 4.8

Consider the eye diagram of a bandlimited digital NRZ baseband signal. In this example the signal is generated by passing a NRZ waveform through a third-order Butterworth filter as illustrated in Figure 4.13. The filter bandwidth is normalized to the symbol rate. In other words, if the symbol rate of the NRZ waveform is 1000 symbols per second and the normalized filter bandwidth is $B_N = 0.6$, the filter bandwidth is 600 Hz. The eye diagrams corresponding to the signal at the filter output are those illustrated in Figure 4.14

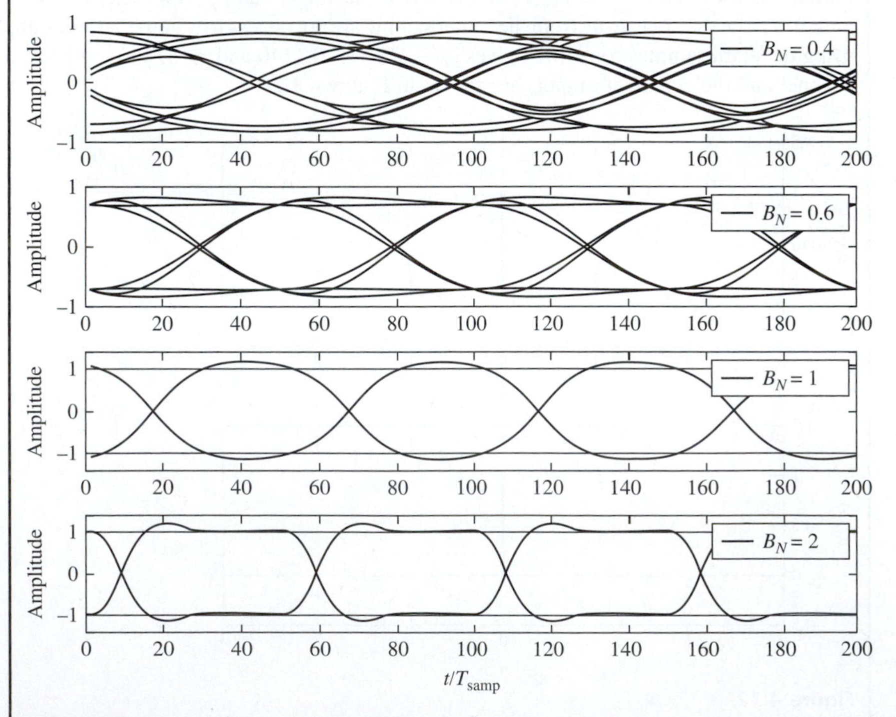

Figure 4.14
Eye diagrams for $B_N = 0.4$, 0.6, 1.0, and 2.0.

for normalized bandwidths, B_N, of 0.4, 0.6, 1.0, and 2.0. Each of the four eye diagrams span $k = 4$ symbols. Sampling is performed at 50 samples per symbol and therefore the sampling index ranges from 1 to 200 as shown. The effect of bandlimiting by the filter, leading to ISI, on the eye diagram is clearly seen. ■

We now look at an eye diagram in more detail. Figure 4.15 shows the top pane of Figure 4.14 ($B_N = 0.4$), in which two symbols are illustrated rather than four. Observation of Figure 4.15 suggests that the eye diagram is composed of two fundamental waveforms, each of which approximates a sine wave. One wave form goes through two periods in the two-symbol eye and the other waveform goes through a single period. A little thought shows that the high-frequency waveform corresponds to the binary sequences 01 or 10, while the low frequency waveform corresponds to the binary sequences 00 or 11.

Also shown in Figure 4.15 is the optimal sampling time, which is when the eye is most open. Note that for significant bandlimiting the eye will be more closed due to ISI. This shrinkage of the eye opening due to ISI is labeled *amplitude jitter*, A_j. Referring back to Figure 4.14, we see that increasing the filter bandwidth decreases the amplitude jitter. When we consider the effects of noise in later chapters of this book, we will see that if the vertical eye opening is reduced, the probability of symbol error increases. Note also that ISI leads to *timing jitter*, denoted T_j in Figure 4.15, which is a perturbation of the zero crossings of the filtered signal. Also note that a large slope of the signal at the zero crossings will result in a more open eye and that increasing this slope is accomplished by increasing the signal bandwidth. If the signal bandwidth is decreased leading to increased ISI, T_j increases and synchronization becomes more difficult. As we will see in later chapters, increasing the bandwith of a channel often results in increased noise levels. This leads to both an increase in timing jitter and amplitude jitter. Thus many trade-offs exist in the design of communication systems, several of which will be explored in later sections of this book.

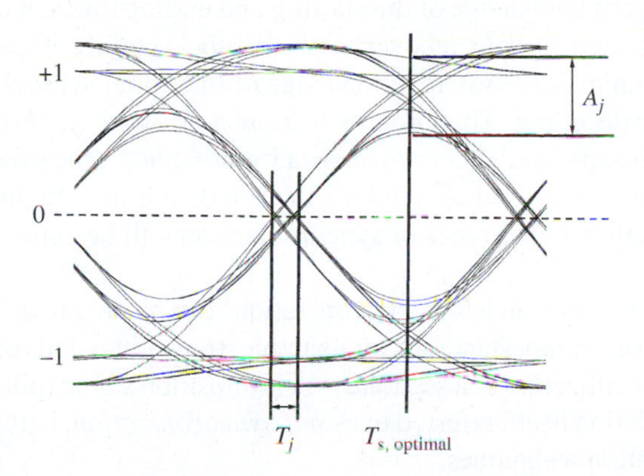

Figure 4.15
Two-symbol eye diagrams for $B_N = 0.4$.

COMPUTER EXAMPLE 4.2

The eye diagrams illustrated in Figure 4.14 were generated using the following MATLAB code:

```
% File: c4ce2.m
clf
nsym = 1000; nsamp = 50; bw = [0.4 0.6 1 2];
```

```
for k = 1:4
    lambda = bw(k);
    [b,a] = butter(3,2*lambda/nsamp);
    l = nsym*nsamp;                          % Total sequence length
    y = zeros(1,l-nsamp+1);                 % Initalize output vector
    x = 2*round(rand(1,nsym))-1;            % Components of x = +1 or -1
    for i = 1:nsym                          % Loop to generate info symbols
        kk = (i-1)*nsamp+1;
        y(kk) = x(i);
    end
    datavector=conv(y,ones(1,nsamp));  % Each symbol is nsamp long
    filtout = filter(b, a, datavector);
    datamatrix = reshape(filtout, 4*nsamp, nsym/4);
    datamatrix1 = datamatrix(:, 6:(nsym/4));
    subplot(4,1,k),plot(datamatrix1, 'k'),ylabel('Amplitude'), ...
    axis([0 200 -1.4 1.4]), legend(['{itB_N} = ', num2str(lambda)])
    if k == 4
        xlabel('{\itt/T}_s_a_m_p')
    end
end
% End of script file.
```

■

■ 4.7 SYNCHRONIZATION

We now briefly look at the important subject of synchronization. There are many different levels of synchronization in a communications system. Coherent demodulation requires carrier synchronization as we discussed in the preceding chapter, where we noted that a Costas PLL could be used to demodulate a DSB signal. In a digital communications system bit or symbol synchronization gives us knowledge of the starting and ending times of discrete-time symbols.This is a necessary step in data recovery. When block coding is used for error correction in a digital communications system, knowledge of the initial symbols in the code words must be identified for decoding. This process is known as *word synchronization*. In addition, symbols are often grouped together to form data frames, and frame synchronization is required to identify the starting and ending symbols in each data frame. In this section we focus on symbol synchronization. Other types of synchronization will be considered later in this book.

Three general methods exist by which bit synchronization[1] can be obtained. These are (1) derivation from a primary or secondary standard (for example, transmitter and receiver slaved to a master timing source), (2) utilization of a separate synchronization signal (pilot clock), and (3) derivation from the modulation itself, referred to as *self-synchronization*. In this section we explore two self-synchronization techniques.

As we saw earlier in this chapter (see Figure 4.2), several binary data formats, such as polar RZ and split phase, guarantee a level transition within every symbol period. For these data formats a discrete spectral component is generated at the symbol frequency. A PLL, such as we studied in the preceding chapter, can then be used to track this component in

[1]See Stiffler (1971), Part II, or Lindsey and Simon (1973), Chapter 9, for a more extensive discussion.

order to recover symbol timing. Symbol synchronization is therefore easy but comes at the cost of increased bandwidth. For data formats that do not have a level transition within each symbol period, a nonlinear operation is performed on the signal in order to generate a spectral component at the symbol frequency. A number of techniques are in common use for accomplishing this. The following examples illustrate two basic techniques, both of which make use of the PLL for timing recovery. Techniques for acquiring symbol synchronization that are similar in form to the Costas loop are also possible but will not be discussed here (see Chapter 9).[2]

COMPUTER EXAMPLE 4.3

To demonstrate the first method, we assume that a data signal is represented by an NRZ signal that has been bandlimited by passing it through a bandlimited channel. If this NRZ signal is squared, a component is generated at the symbol frequency. The component generated at the symbol frequency can then be phase tracked by a PLL in order to generate the symbol synchronization, as illustrated by the following MATLAB simulation:

```
% File: c4ce3.m
nsym = 1000; nsamp = 50; lambda = 0.7;
[b,a] = butter(3,2*lambda/nsamp);
l = nsym*nsamp;                              % Total sequence length
y = zeros(1,l-nsamp+1);                      % Initalize output vector
x =2*round(rand(1,nsym))-1;   % Components of x = +1 or -1
for i = 1:nsym                               % Loop to generate info symbols
    k = (i-1)*nsamp+1;
    y(k) = x(i);
end
datavector1 = conv(y,ones(1,nsamp));  % Each symbol is nsamp long
subplot(3,1,1), plot(datavector1(1,200:799),'k', 'LineWidth', 1.5)
axis([0 600 -1.4 1.4]), ylabel('Amplitude')
filtout = filter(b,a,datavector1);
datavector2 = filtout.*filtout;
subplot(3,1,2), plot(datavector2(1,200:799),'k', 'LineWidth', 1.5)
ylabel('Amplitude')
y = fft(datavector2);
yy = abs(y)/(nsym*nsamp);
subplot(3,1,3), stem(yy(1,1:2*nsym),'k.')
xlabel('FFT Bin'), ylabel('Spectrum')
% End of script file.
```

The results of executing the preceding MATLAB program are illustrated in Figure 4.16. Assume that the 1000 symbols generated by the MATLAB program occur in a time span of 1 s. Thus the symbol rate is 1000 symbols/s, and since the NRZ signal is sampled at 10 samples/symbol, the sampling frequency is 10,000 Hz. Figure 4.16(a) illustrates 600 samples of the NRZ signal. Filtering by a third-order Butterworth filter having a bandwidth of twice the symbol rate and squaring this signal results in the signal shown in Figure 4.16(b). The second-order harmonic created by the squaring operation can clearly be seen by observing a data segment consisting of alternating data symbols. The spectrum, generated using the FFT algorithm, is illustrated in Figure 4.16(c). Two spectral components can clearly be seen; a

[2]Again, see Stiffler (1971) or Lindsey and Simon (1973)

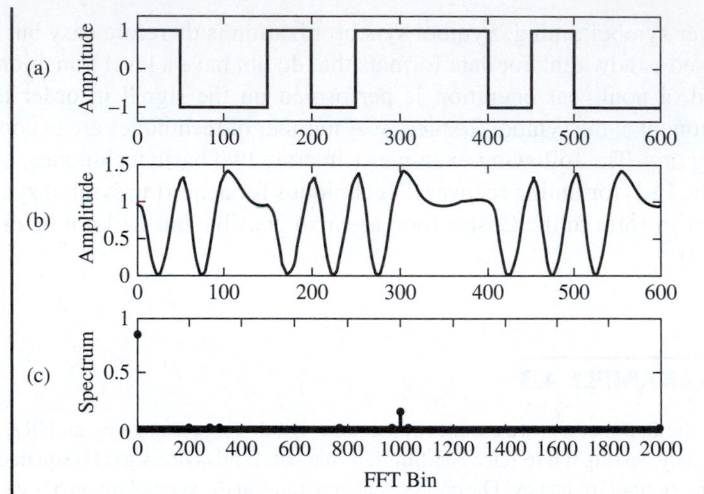

Figure 4.16
Simulation results for Computer Example 4.3. (a) NRZ waveform. (b) NRZ waveform filtered and squared. (c) FFT of squared NRZ waveform.

component at DC (0 Hz), which results from the squaring operation, and a component at 1000 Hz, which represents the component at the symbol rate. This component is tracked by a PLL to establish symbol timing.

It is interesting to note that a sequence of alternating data states, e.g., 101010. . ., will result in an NRZ waveform that is a square wave. If the spectrum of this square wave is determined by forming the Fourier series, the period of the square wave will be twice the symbol period. The frequency of the fundamental will therefore be one-half the symbol rate. The squaring operation doubles the frequency to the symbol rate of 1000 symbols/s.

■

COMPUTER EXAMPLE 4.4

To demonstrate a second self-synchronization method, consider the system illustrated in Figure 4.17. Because of the nonlinear operation provided by the delay-and-multiply operation, power is produced at the symbol frequency. The following MATLAB program simulates the symbol synchronizer:

```
% File:  c4ce4.m
nsym = 1000; nsamp = 50;              % Make nsamp even
m = nsym*nsamp;
y = zeros(1,m-nsamp+1);               % Initalize output vector
x =2*round(rand(1,nsym))-1;           % Components of x = +1 or -1
for i = 1:nsym                        % Loop to generate info symbols
   k = (i-1)*nsamp+1;
   y(k) = x(i);
end
datavector1 = conv(y,ones(1,nsamp));  % Make symbols nsamp samples long
subplot(3,1,1), plot(datavector1(1,200:10000),'k', 'LineWidth', 1.5)
axis([0 600 -1.4 1.4]), ylabel('Amplitude')
datavector2 = [datavector1(1,m-nsamp/2+1:m) datavector1(1,1:
m-nsamp/2)];
```

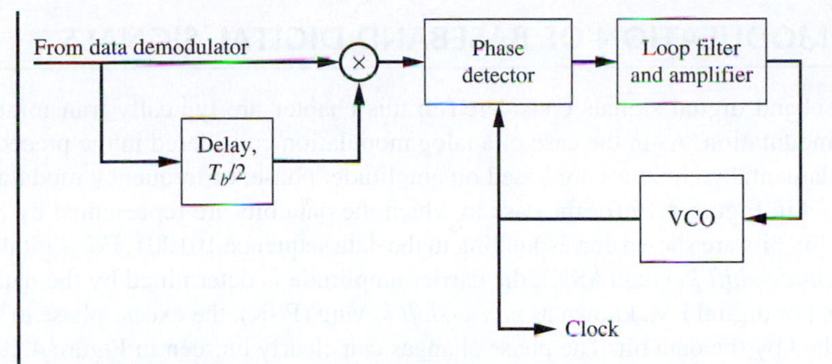

Figure 4.17
System for deriving a symbol clock simulated in Computer Example 4.3.

```
datavector3 = datavector1.*datavector2;
subplot(3,1,2), plot(datavector3(1,200:10000),'k', 'LineWidth', 1.5),
axis([0 600 -1.4 1.4]), ylabel('Amplitude')
y = fft(datavector3);
yy=abs(y)/(nsym*nsamp);
subplot(3,1,3), stem(yy(1,1:4*nsym),'k.')
xlabel('FFT Bin'), ylabel('Spectrum')
% End of script file.
```

The data waveform is shown in Figure 4.18(a), and this waveform multiplied by its delayed version is shown in Figure 4.18(b). The spectral component at 1000 Hz, as seen in Figure 4.18(c), represents the symbol-rate component and is tracked by a PLL for timing recovery.

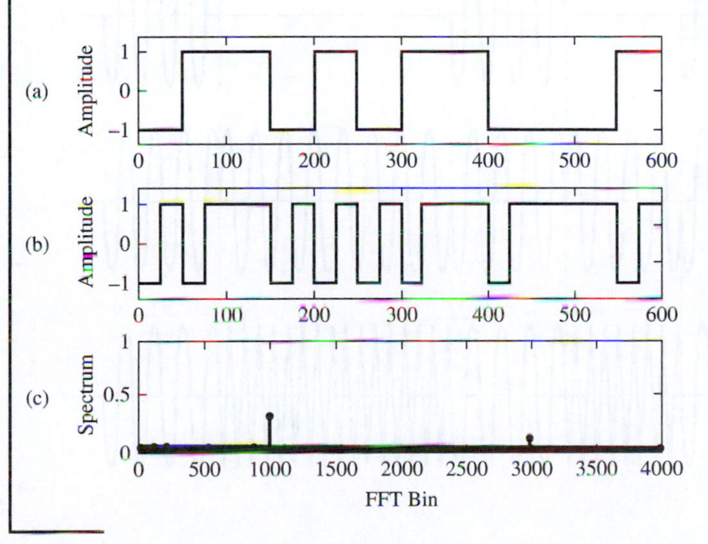

Figure 4.18
Simulation results for Computer Example 4.4. (a) Data waveform. (b) Data waveform multiplied by a half-bit delayed version of itself. (c) FFT spectrum of (b).

■

4.8 CARRIER MODULATION OF BASEBAND DIGITAL SIGNALS

The baseband digital signals considered in this chapter are typically transmitted using RF carrier modulation. As in the case of analog modulation considered in the preceding chapter, the fundamental techniques are based on amplitude, phase, or frequency modulation. This is illustrated in Figure 4.19 for the case in which the data bits are represented by an NRZ data format. Six bits are shown corresponding to the data sequence 101001. For digital AM, known as *amplitude-shift keying* (ASK), the carrier amplitude is determined by the data bit for that interval. For digital PM, known as *phase-shift keying* (PSK), the excess phase of the carrier is established by the data bit. The phase changes can clearly be seen in Figure 4.19. For digital frequency modulation, known as *frequency-shift keying* (FSK), the carrier frequency deviation is established by the data bit.

To illustrate the similarity to the material studied in Chapter 3, note that the ASK RF signal can be represented by

$$x_{\text{ASK}}(t) = A_c[1 + d(t)]\cos(2\pi f_c t) \tag{4.67}$$

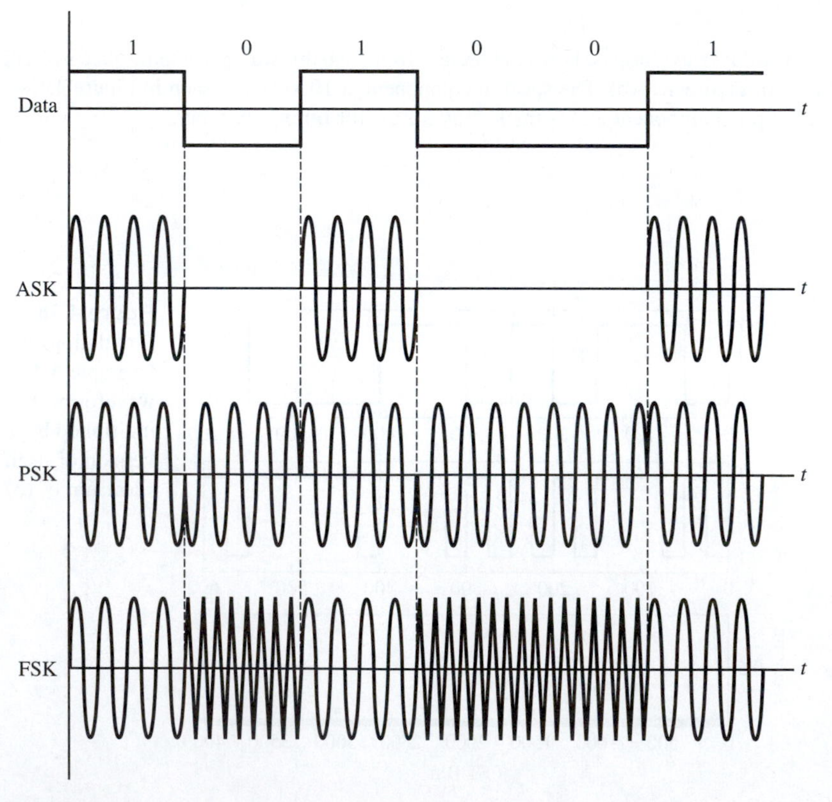

Figure 4.19
Examples of digital modulation schemes.

where $d(t)$ is the NRZ waveform. Note that this is identical to AM with the only essential difference being the definition of the message signal. Phase-shift keying and FSK can be similarly represented by

$$x_{\mathrm{PSK}}(t) = A_c \cos\left[2\pi f_c t + \frac{\pi}{2} d(t)\right] \tag{4.68}$$

and

$$x_{\mathrm{FSK}}(t) = A_c \cos\left[2\pi f_c t + k_f \int^t d(\alpha)\, d\alpha\right] \tag{4.69}$$

respectively. We therefore see that many of the concepts introduced in Chapter 3 carry over to digital data systems. These techniques will be studied in detail in Chapters 8 and 9.

A major concern of both analog and digital communication systems is system performance in the presence of channel noise and other random disturbances. In order to have the tools required to undertake a study of system performance, we interrupt our discussion of communication systems to study random variables and stochastic processes.

Summary

1. The block diagram of the baseband model of a digital communications systems contains several components not present in the analog systems studied in the preceding chapter. The underlying message signal may be analog or digital. If the message signal is analog, an analog-to-digital converter must be used to convert the signal from analog to digital form. In such cases a digital-to-analog converter is usually used at the receiver output to convert the digital data back to analog form. Three operations covered in detail in this chapter were line coding, pulse shaping, and symbol synchronization.
2. Digital data can be represented using a number of formats, generally referred to as *line codes*. The two basic classifications of line codes are those that do not have an amplitude transition within each symbol period and those that do have an amplitude transition within each symbol period. A number of possibilities exist within each of these classifications. Two of the most popular data formats are nonreturn to zero (NRZ), which does not have an amplitude transition within each symbol period and split phase, which does have an amplitude transition within each symbol period. The power spectral density corresponding to various data formats is important because of the impact on transmission bandwidth. Data formats having an amplitude transition within each symbol period have a discrete line at the symbol rate. This simplifies symbol synchronization at the cost of increased bandwidth. Thus, a number of design trade-offs exist.
3. A major source of performance degradation in a digital system is intersymbol interference (ISI). Distortion due to ISI results when the bandwith of a channel is not sufficient to pass all significant components of the channel input signal. Channel equalization is often used to combat the effects of ISI. Equalization,

in its simplest form, can be viewed as filtering the channel output using a filter having a transfer function that is the inverse of the transfer function of the channel.

4. A number of pulse shapes satisfy Nyquist pulse-shaping criterion and result in zero ISI. A simple example is the pulse defined by $p(t) = \text{sinc}(t/T)$, where T is the sampling period. Zero ISI results since $p(t) = 1$ for $t = 0$ and $p(t) = 0$ for $t = nT$, $n \neq 0$.
5. A popular technique for implementing zero-ISI conditions is to use identical filters in both the transmitter and receiver. If the transfer function of the channel is known and the underlying pulse shape is defined, the transfer function of the transmitter–receiver filters can easily be found so that the Nyquist zero-ISI condition is satisfied. This technique is typically used with raised cosine pulses.
6. A zero-forcing equalizer is a digital filter which operates upon a channel output to produce a sequence of samples satisfying the Nyquist zero-ISI condition. The implementation takes the form of an tapped delay line, or transversal, filter. The tap weights are determined by the inverse of the matrix defining the pulse response of the channel. Attributes of the zero-forcing equalizer include ease of implementation and ease of analysis.
7. Eye diagrams are formed by overlaying segments of signals representing k data symbols. The eye diagrams, while not a quantitative measure of system performance, provide a qualitative mesure of system performance. Signals with large vertical eye openings display lower levels of ISI than those with smaller vertical openings. Eyes with small horizontal openings have high levels of timing jitter, which makes symbol synchronization more difficult.
8. Many levels of synchronization are required in digital communication systems, including carrier, symbol, word, and frame synchronization. In this chapter we considered only symbol synchronization. Symbol synchronization is typically accomplished by using a PLL to track a component in the data signal at the symbol frequency. Data signals in which the data format has an amplitude transition in every symbol period have a naturally occuring spectral component at the symbol rate. If the data format does not have an amplitude transition within symbol periods, such as NRZ, a nonlinear operation must be applied to the data signal in order to generate a spectral component at the symbol rate.

Further Reading

Further discussions on the topics of this chapter may be found in to Ziemer and Peterson (2001), Couch (2007), and Proakis and Salehi (2005).

Problems

Section 4.2

4.1. Given the channel features or objectives below. For each part, tell which line code(s) is (are) the best choice(s).

a. The channel frequency response has a null at $f = 0$ Hz.

b. The channel has a passband from 0 to 10 kHz, and it is desired to transmit data through it at 10,000 bps.

c. At least one zero crossing per bit is desired for synchronization purposes.

d. Built in redundancy is desired for error checking purposes.

e. For simplicity of detection, distinct positive pulses are desired for ones and distinct negative pulses are desired for zeros.

f. A discrete spectral line at the bit rate is desired from which to derive a clock at the bit rate.

4.2. For the ± 1-amplitude waveforms of Figure 4.2, show that the average powers are

a. NRZ change – $P_{\text{ave}} = 1$ W.

b. NRZ mark – $P_{\text{ave}} = 1$ W.

c. Unipolar RZ – $P_{\text{ave}} = \frac{1}{4}$ W.

d. Polar RZ – $P_{\text{ave}} = \frac{1}{2}$ W.

e. Bipolar RZ – $P_{\text{ave}} = \frac{1}{4}$ W.

f. Split phase – $P_{\text{ave}} = 1$ W.

4.3.

a. Given the random binary data sequence 0 1 1 0 0 0 1 0 1 1, provide waveform sketches for (i) NRZ change and (ii) split phase.

b. Demonstrate satisfactorily that the split-phase waveform can be obtained from the NRZ waveform by multiplying the NRZ waveform by a ± 1-valued clock signal of period T.

4.4. For the data sequence of Problem 4.3 provide a waveform sketch for NRZ mark.

4.5. For the data sequence of Problem 4.3 provide waveform sketches for

a. Unipolar RZ

b. Polar RZ

c. Bipolar RZ

4.6. A channel of bandwidth 4 kHz is available. Determine the data rate that can be accommodated for the following line codes (assume a bandwidth to the first spectral null):

a. NRZ change

b. Split phase

c. Unipolar RZ and polar RZ

d. Bipolar RZ

Section 4.3

4.7. Using the superposition and time-invariance properties of an RC filter, show that (4.30) is the response of a lowpass RC filter to (4.29) given that the filter's response to a unit step is $[1-\exp(-t/RC)]\,u(t)$.

Section 4.4

4.8. Show that (4.37) is an ideal rectangular spectrum for $\beta = 0$. What is the corresponding pulse shape function?

4.9. Show that (4.36) and (4.37) are Fourier transform pairs.

4.10. Sketch the following spectra and tell which ones satisfy Nyquist's pulse shape criterion. For those that do, find the appropriate sample interval, T, in terms of W. Find the corresponding pulse shape function $p(t)$. (Recall that $\Pi(f/A)$ is a unit-high rectangular pulse from $-A/2$ to $A/2$; $\Lambda(f/B)$ is a unit-high triangle from $-B$ to B.)

a. $P_1(f) = \Pi(f/2W) + \Pi(f/W)$.

b. $P_2(f) = \Lambda(f/2W) + \Pi(f/W)$.

c. $P_3(f) = \Pi(f/4W) - \Lambda(f/W)$.

d. $P_4(f) = \Pi[(f-W)/W] + \Pi[(f+W)/W]$.

e. $P_5(f) = \Lambda(f/2W) - \Lambda(f/W)$.

4.11. If $|H_C(f)| = [1 + (f/5000)^2]^{-1/2}$, provide a plot for $|H_T(f)| = |H_R(f)|$ assuming the pulse spectrum $P_{\text{RC}}(f)$ with $1/T = 5000$ Hz for (a) $\beta = 1$ and (b) $\beta = \frac{1}{2}$.

4.12. It is desired to transmit data at 9 kbps over a channel of bandwidth 7 kHz using raised-cosine pulses. What is the maximum value of the roll-off factor β that can be used?

4.13.

a. Show by a suitable sketch that the triangular spectrum of Figure 4.8(a) satisfies Nyquist's pulse shaping criterion.

b. Find the pulse shape function corresponding to this spectrum.

Section 4.5

4.14. Given the following channel pulse-response samples:

$$P_c(-4T) = -0.001 \quad p_c(-3T) = 0.001 \quad p_c(-2T) = -0.01 \quad p_c(-T) = 0.1 \quad p_c(0) = 1.0$$
$$p_c(T) = 0.2 \quad p_c(2T) = -0.02 \quad p_c(3T) = 0.005 \quad P_c(4T) = -0.003$$

a. Find the tap coefficients for a three-tap zero-forcing equalizer.

b. Find the output samples for $mT = -2T, -T, 0, T$, and $2T$.

4.15. Repeat Problem 4.14 for a five-tap zero-forcing equalizer.

4.16. A simple model for a multipath communications channel is shown in Figure 4.20(a).

a. Find $H_c(f) = Y(f)/X(f)$ for this channel and plot $|H_c(f)|$ for $\beta = 1$ and 0.5.

b. In order to equalize, or undo, the channel-induced distortion, an equalization filter is used. Ideally, its transfer function should be

$$H_{\text{eq}}(f) = \frac{1}{H_c(f)}$$

if the effects of noise are ignored and only distortion caused by the channel is considered. A tapped delay-line or transversal filter, as shown in Figure 4.20(b), is commonly used to approximate $H_{\text{eq}}(f)$. Write down a series expression for $H'_{\text{eq}}(f) = Z(f)/Y(f)$.

c. Using $(1+x)^{-1} = 1-x+x^2-x^3+\ldots, |x|<1$, find a series expression for $1/H_c(f)$. Equating this with $H_{\text{eq}}(f)$ found in part (b), find the values for $\beta_1, \beta_2, \ldots, \beta_N$, assuming $\tau_m = \Delta$.

Section 4.6

4.17. In a certain digital data transmission system the probability of a bit error as a function of timing jitter is given by

$$P_E = \frac{1}{4}\exp(-z) + \frac{1}{4}\exp\left[-z\left(1-2\frac{|\Delta T|}{T}\right)\right]$$

where z is the signal-to-noise ratio, $|\Delta T|$ is the timing jitter, and T is the bit period. From observations of an eye diagram for the system, it is determined that $|\Delta T|/T = 0.05$ (5%).

a. Find the value of signal-to-noise ratio, z_0, that gives a probability of error of 10^{-6} for a timing jitter of 0.

b. With the jitter of 5%, tell what value of signal-to-noise ratio, z_1, is necessary to maintain the probability of error at 10^{-6}. Express the ratio z_1/z_0 in decibels, where

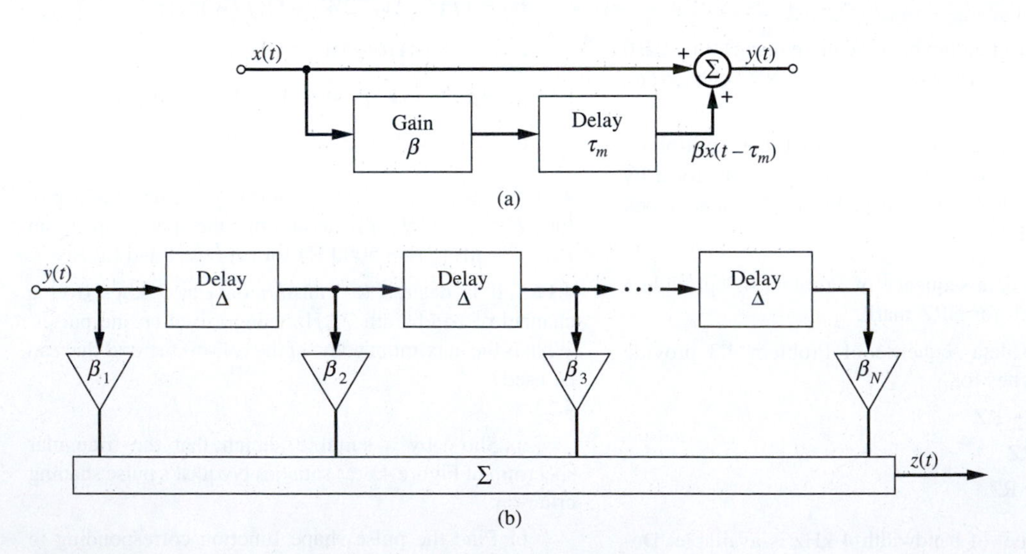

Figure 4.20

$[z_1/z_0]_{\text{dB}} = 10 \log_{10}(z_1/z_0)$. Call this the degradation due to jitter.

c. Recalculate (a) and (b) for a probability of error of 10^{-4}. Is the degradation due to jitter better or worse than for a probability of error of 10^{-6}?

Section 4.7

4.18. Rewrite the MATLAB simulation of Computer Example 4.3 for the case of an absolute-value type of nonlinearity. Is the spectral line at the bit rate stronger or weaker than for the square-law type of nonlinearity?

4.19. Assume that the bit period of Computer Example 4.3 is $T = 1$ s. That means that the sampling rate is $f_s = 50$ samples per second because `nsamp` $= 50$ in the program. Assuming that a $N_{\text{FFT}} = 10,000$-point FFT was used to produce Figure 4.16 and that the 10,000th point corresponds to f_s, justify that the FFT output at bin 1000 corresponds to the bin rate of $1/T = 1$ bps in this case.

Section 4.8

4.20. Referring to (4.68), it is sometimes desirable to leave a residual carrier component in a PSK-modulated waveform for carrier synchronization purposes at the receiver. Thus, instead of (4.68), we would have

$$x_{\text{PSK}}(t) = A_c \cos\left[2\pi f_c t + \alpha \frac{\pi}{2} d(t)\right], \quad 0 < \alpha < 1$$

Find α so that 10% of the power of $x_{\text{PSK}}(t)$ is in the carrier (unmodulated) component.
(*Hint:* Use $\cos(u+v)$ to write $x_{\text{PSK}}(t)$ as two terms, one dependent on $d(t)$ and the other independent of $d(t)$. Make use of the facts that $d(t) = \pm 1$ and cosine is even and sine is odd.)

4.21. Referring to (4.69) and using the fact that $d(t) = \pm 1$ in T-second intervals, find the value of k_f such that the peak frequency deviation of $x_{\text{FSK}}(t)$ is 10,000 Hz if the bit rate is 1,000 bits per second.

Computer Exercises

4.1. Write a MATLAB program that will produce plots like those shown in Figure 4.2, assuming a random binary data sequence. Include as an option a Butterworth channel filter whose number of poles and bandwidth (in terms of bit rate) are inputs.

4.2. Write a MATLAB program that will produce plots like those shown in Figure 4.10. The Butterworth channel filter poles and 3-dB frequency should be inputs as well as the roll-off factor, β.

4.3. Write a MATLAB program that will compute the weights of a transversal-filter zero forcing equalizer for a given input pulse sample sequence.

4.4. A symbol synchronizer uses a fourth-power device instead of a squarer. Modify the MATLAB program of Computer Example 4.3 accordingly, and show that a useful spectral component is generated at the output of the fourth-power device.

CHAPTER 5

OVERVIEW OF PROBABILITY AND RANDOM VARIABLES

The objective of this chapter is to review probability theory in order to provide a background for the mathematical description of random signals. In the analysis and design of communication systems it is necessary to develop mathematical models for random signals and noise, or *random processes*, which will be accomplished in Chapter 6.

5.1 WHAT IS PROBABILITY?

Two intuitive notions of probability may be referred to as the *equally likely outcomes* and *relative-frequency* approaches.

5.1.1 Equally Likely Outcomes

The equally likely outcomes approach defines probability as follows: if there are N possible *equally likely* and *mutually exclusive* outcomes (that is, the occurrence of a given outcome precludes the occurrence of any of the others) to a random, or chance, experiment and if N_A of these outcomes correspond to an *event* A of interest, then the probability of event A , or $P(A)$, is

$$P(A) = \frac{N_A}{N} \tag{5.1}$$

There are practical difficulties with this definition of probability. One must be able to break the chance experiment up into two or more equally likely outcomes, and this is not always possible. The most obvious experiments fitting these conditions are card games, dice, and coin tossing. Philosophically, there is difficulty with this definition in that use of the words *equally likely* really amounts to saying something about being equally probable, which means we are using probability to define probability.

Although there are difficulties with the equally likely definition of probability, it is useful in engineering problems when it is reasonable to list N equally likely, mutually exclusive outcomes. The following example illustrates its usefulness in a situation where it applies.

EXAMPLE 5.1

Given a deck of 52 playing cards, (a) What is the probability of drawing the ace of spades? (b) What is the probability of drawing a spade?

Solution

(a) Using the principle of equal likelihood, we have one favorable outcome in 52 possible outcomes. Therefore, $P(\text{ace of spades}) = \frac{1}{52}$. (b) Again using the principle of equal likelihood, we have 13 favorable outcomes in 52, and $P(\text{spade}) = \frac{13}{52} = \frac{1}{4}$.

■

5.1.2 Relative Frequency

Suppose we wish to assess the probability of an unborn child being a boy. Using the classical definition, we predict a probability of $\frac{1}{2}$, since there are two possible mutually exclusive outcomes, which from outward appearances appear equally probable. However, yearly birth statistics for the United States consistently indicate that the ratio of males to total births is about 0.51. This is an example of the relative-frequency approach to probability.

In the *relative-frequency approach*, we consider a random experiment, enumerate all possible outcomes, repeatedly perform the experiment, and take the ratio of the number of outcomes, N_A, favorable to an event of interest, A, to the total number of trials, N. As an approximation of the probability of A, $P(A)$, we define the limit of N_A/N, called the *relative frequency* of A, as $N \to \infty$, as $P(A)$:

$$P(A) \triangleq \lim_{N \to \infty} \frac{N_A}{N} \tag{5.2}$$

This definition of probability can be used to estimate $P(A)$. However, since the infinite number of experiments implied by (5.2) cannot be performed, only an approximation to $P(A)$ is obtained. Thus the relative-frequency notion of probability is useful for estimating a probability but is not satisfactory as a mathematical basis for probability.

The following example fixes these ideas and will be referred to later in this chapter.

EXAMPLE 5.2

Consider the simultaneous tossing of two fair coins. Thus, on any given trial, we have the possible outcomes HH, HT, TH, and TT, where, for example, HT denotes a head on the first coin and a tail on the second coin. (We imagine that numbers are painted on the coins so we can tell them apart.) What is the probability of two heads on any given trial?

Solution

By distinguishing between the coins, the correct answer, using equal likelihood, is $\frac{1}{4}$. Similarly, it follows that $P(\text{HT}) = P(\text{TH}) = P(\text{TT}) = \frac{1}{4}$.

■

5.1.3 Sample Spaces and the Axioms of Probability

Because of the difficulties mentioned for the preceding two definitions of probability, mathematicians prefer to approach probability on an axiomatic basis. The axiomatic approach,

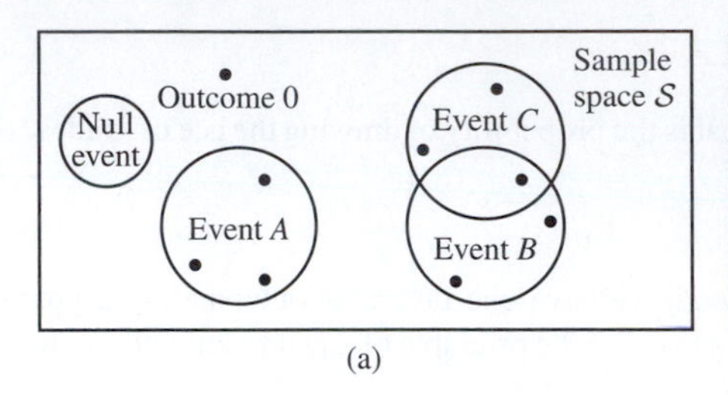

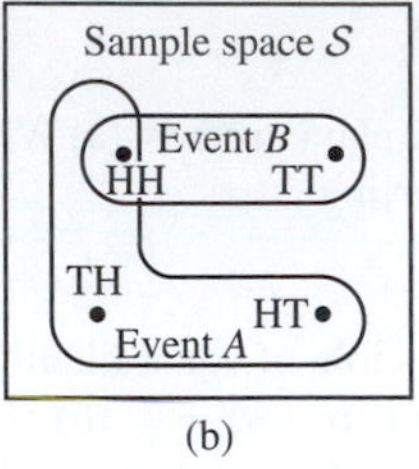

Figure 5.1
Sample spaces.(a) Pictorial representation of an arbitrary sample space. Points show outcomes; circles show events. (b) Sample space representation for the tossing of two coins.

which is general enough to encompass both the equally likely and relative-frequency definitions of probability, will now be briefly described.

A chance experiment can be viewed geometrically by representing its possible outcomes as elements of a space referred to as a *sample space S*. An *event* is defined as a collection of outcomes. An impossible collection of outcomes is referred to as the *null event*, ϕ. Figure 5.1(a) shows a representation of a sample space. Three events of interest, A, B, and C, which do not encompass the entire sample space, are shown.

A specific example of a chance experiment might consist of measuring the direct current (DC) voltage at the output terminals of a power supply. The sample space for this experiment would be the collection of all possible numerical values for this voltage. On the other hand, if the experiment is the tossing of two coins, as in Example 5.2, the sample space would consist of the four outcomes HH, HT, TH, and TT enumerated earlier. A sample-space representation for this experiment is shown in Figure 5.1(b). Two events of interest, A and B, are shown. Event A denotes at least one head, and event B consists of the coins matching. Note that A and B encompass all possible outcomes for this particular example.

Before proceeding further, it is convenient to summarize some useful notation from set theory. The event "A or B or both" will be denoted as $A \cup B$ or sometimes as $A + B$. The event "both A and B" will be denoted either as $A \cap B$ or sometimes as (A, B) or AB (called the *joint event* A and B). The event "not A" will be denoted $\overline{A}$. An event such as $A \cup B$, which is composed of two or more events, will be referred to as a *compound event*. In set theory terminology, mutually exclusive events are referred to as *disjoint sets*; if two events, A and B, are mutually exclusive, then $A \cap B = \phi$ where ϕ is the null set.

In the axiomatic approach, a *measure*, called *probability* is somehow assigned to the events of a sample space[1] such that this measure possesses the properties of probability. The properties or axioms of this probability measure are chosen to yield a satisfactory theory such that results from applying the theory will be consistent with experimentally observed phenomena. A set of satisfactory axioms is the following:

Axiom 1

$P(A) \geq 0$ for all events A in the sample space S.

Axiom 2

The probability of all possible events occurring is unity, $P(S) = 1$.

[1]For example, by the relative-frequency or the equally likely approaches.

Axiom 3

If the occurrence of A precludes the occurrence of B, and vice versa (that is, A and B are mutually exclusive), then $P(A \cup B) = P(A) + P(B)$.[2]

It is emphasized that this approach to probability does not give us the number $P(A)$; it must be obtained by some other means.

5.1.4 Venn Diagrams

It is sometimes convenient to visualize the relationships between various events for a chance experiment in terms of a *Venn diagram*. In such diagrams, the sample space is indicated as a rectangle, with the various events indicated by circles or ellipses. Such a diagram looks exactly as shown in Figure 5.1(a), where it is seen that events B and C are not mutually exclusive, as indicated by the overlap between them, whereas event A is mutually exclusive of events B and C.

5.1.5 Some Useful Probability Relationships

Since it is true that $A \cup \overline{A} = S$ and that A and $\overline{A}$ are mutually exclusive, it follows by Axioms 2 and 3 that $P(A) + P(\overline{A}) = P(S) = 1$, or

$$P(\overline{A}) = 1 - P(A) \tag{5.3}$$

A generalization of Axiom 3 to events that are not mutually exlcusive is obtained by noting that $A \cup B = A \cup (B \cap \overline{A})$, where A and $B \cap \overline{A}$ are disjoint (this is most easily seen by using a Venn diagram). Therefore, Axiom 3 can be applied to give

$$P(A \cup B) = P(A) + P(B \cap \overline{A}) \tag{5.4}$$

Similarly, we note from a Venn diagram that the events $A \cap B$ and $B \cap \overline{A}$ are disjoint and that $(A \cap B) \cup (B \cap \overline{A}) = B$ so that

$$P(A \cap B) + P(B \cap \overline{A}) = P(B) \tag{5.5}$$

Solving for $P\left(B \cap \overline{A}\right)$ from (5.5) and substituting into (5.4) yields the following for $P(A \cup B)$:

$$P(A \cup B) = P(A) + P(B) - P(A \cap B) \tag{5.6}$$

This is the desired generalization of Axiom 3.

Now consider two events A and B, with individual probabilities $P(A) > 0$ and $P(B) > 0$, respectively, and joint event probability $P(A \cap B)$. We define the *conditional probability* of event A given that event B occurred as

$$P(A|B) = \frac{P(A \cap B)}{P(B)} \tag{5.7}$$

---

[2]This can be generalized to $P(A \cup B \cup C) = P(A) + P(B) + P(C)$ for A, B, and C mutually exclusive by considering $B_1 = B \cup C$ to be a composite event in Axiom 3 and applying Axiom 3 twice: i.e., $P(A \cup B_1) = P(A) + P(B_1) = P(A) + P(B) + P(C)$. Clearly, in this way we can generalize this result to any finite number of mutually exclusive events.

Similarly, the conditional probability of event B given that event A has occurred is defined as

$$P(B|A) = \frac{P(A \cap B)}{P(A)} \tag{5.8}$$

Putting (5.7) and (5.8) together, we obtain

$$P(A|B)\,P(B) = P(B|A)\,P(A) \tag{5.9}$$

or

$$P(B|A) = \frac{P(B)\,P(A|B)}{P(A)} \tag{5.10}$$

This is a special case of *Bayes' rule.*

Finally, suppose that the occurrence or nonoccurrence of B in no way influences the occurrence or nonoccurrence of A. If this is true, A and B are said to be *statistically independent*. Thus, if we are given B, this tells us nothing about A, and therefore, $P(A|B) = P(A)$. Similarly, $P(B|A) = P(B)$. From (5.7) or (5.8) it follows that, for such events,

$$P(A \cap B) = P(A)P(B) \tag{5.11}$$

Equation (5.11) will be taken as the definition of statistically independent events.

EXAMPLE 5.3

Referring to Example 5.2, suppose A denotes at least one head and B denotes a match. The sample space is shown in Figure 5.1(b). To find $P(A)$ and $P(B)$, we may proceed in several different ways.

Solution

First, if we use equal likelihood, there are three outcomes favorable to A (that is, HH, HT, and TH) among four possible outcomes, yielding $P(A) = \frac{3}{4}$. For B, there are two favorable outcomes in four possibilities, giving $P(B) = \frac{1}{2}$.

As a second approach, we note that, if the coins do not influence each other when tossed, the outcomes on separate coins are statistically independent with $P(H) = P(T) = \frac{1}{2}$. Also, event A consists of any of the mutually exclusive outcomes HH, TH, and HT, giving

$$P(A) = \frac{1}{2}\left(\frac{1}{2}\right) + \frac{1}{2}\left(\frac{1}{2}\right) + \frac{1}{2}\left(\frac{1}{2}\right) = \frac{3}{4} \tag{5.12}$$

by (5.11) and Axiom 3, generalized. Similarly, since B consists of the mutually exclusive outcomes HH and TT,

$$P(B) = \frac{1}{2}\left(\frac{1}{2}\right) + \frac{1}{2}\left(\frac{1}{2}\right) = \frac{1}{2} \tag{5.13}$$

again through the use of (5.11) and Axiom 3. Also, $P(A \cap B) = P$ (at least one head and a match) $=$ $P(\text{HH}) = \frac{1}{4}$.

Next, consider the probability of at least one head given a match, $P(A|B)$. Using Bayes' rule, we obtain

$$P(A|B) = \frac{P(A \cap B)}{P(B)} = \frac{\frac{1}{4}}{\frac{1}{2}} = \frac{1}{2} \tag{5.14}$$

which is reasonable, since given B, the only outcomes under consideration are HH and TT, only one of which is favorable to event A. Next, finding $P(B|A)$, the probability of a match given at least one head, we obtain

$$P(B|A) = \frac{P(A \cap B)}{P(A)} = \frac{\frac{1}{4}}{\frac{3}{4}} = \frac{1}{3} \tag{5.15}$$

Checking this result using the principle of equal likelihood, we have one favorable event among three candidate events (HH, TH, and HT), which yields a probability of $\frac{1}{3}$. We note that

$$P(A \cap B) \neq P(A)P(B) \tag{5.16}$$

Thus events A and B are not statistically independent, although the events H and T on either coin are independent.

Finally, consider the joint probability $P(A \cup B)$. Using (5.6), we obtain

$$P(A \cup B) = \frac{3}{4} + \frac{1}{2} - \frac{1}{4} = 1 \tag{5.17}$$

Remembering that $P(A \cup B)$ is the probability of at least one head or a match or both, we see that this includes all possible outcomes, thus confirming the result. ■

EXAMPLE 5.4

This example illustrates the reasoning to be applied when trying to determine if two events are independent. A single card is drawn at random from a deck of cards. Which of the following pairs of events are independent? (a) The card is a club, and the card is black. (b) The card is a king, and the card is black.

Solution

We use the relationship $P(A \cap B) = P(A|B)P(B)$ (always valid) and check it against the relation $P(A \cap B) = P(A)P(B)$ (valid only for independent events). For part (a), we let A be the event that the card is a club and B be the event that it is black. Since there are 26 black cards in an ordinary deck of cards, 13 of which are clubs, the conditional probability $P(A|B)$ is $\frac{13}{26}$ (given we are considering only black cards, we have 13 favorable outcomes for the card being a club). The probability that the card is black is $P(B) = \frac{26}{52}$, because half the cards in the 52-card deck are black. The probability of a club (event A), on the other hand, is $P(A) = \frac{13}{52}$ (13 cards in a 52-card deck are clubs). In this case,

$$P(A|B)P(B) = \frac{13}{26}\left(\frac{26}{52}\right) \neq P(A)P(B) = \frac{13}{52}\left(\frac{26}{52}\right) \tag{5.18}$$

so the events are not independent.

For part (b), we let A be the event that a king is drawn and event B be that it is black. In this case, the probability of a king given that the card is black is $P(A|B) = \frac{2}{26}$ (two cards of the 26 black cards are kings). The probability of a king is simply $P(A) = \frac{4}{52}$ (four kings in the 52-card deck) and $P(B) = P(\text{black}) = \frac{26}{52}$. Hence,

$$P(A|B)P(B) = \frac{2}{26}\left(\frac{26}{52}\right) = P(A)P(B) = \frac{4}{52}\left(\frac{26}{52}\right) \tag{5.19}$$

which shows that the events king and black are statistically independent. ■

EXAMPLE 5.5

As an example more closely related to communications, consider the transmission of binary digits through a channel as might occur, for example, in computer networks. As is customary, we denote the two possible symbols as 0 and 1. Let the probability of receiving a zero, given a zero was sent, $P(0r|0s)$, and the probability of receiving a 1, given a 1 was sent, $P(1r|1s)$, be

$$P(0r|0s) = P(1r|1s) = 0.9 \tag{5.20}$$

Thus the probabilities $P(1r|0s)$ and $P(0r|1s)$ must be

$$P(1r|0s) = 1 - P(0r|0s) = 0.1 \tag{5.21}$$

and

$$P(0r|1s) = 1 - P(1r|1s) = 0.1 \tag{5.22}$$

respectively. These probabilities characterize the channel and would be obtained through experimental measurement or analysis. Techniques for calculating them for particular situations will be discussed in Chapters 8 and 9.

In addition to these probabilities, suppose that we have determined through measurement that the probability of sending a 0 is

$$P(0s) = 0.8 \tag{5.23}$$

and therefore the probability of sending a 1 is

$$P(1s) = 1 - P(0s) = 0.2 \tag{5.24}$$

Note that once $P(0r|0s)$, $P(1r|1s)$, and $P(0s)$ are specified, the remaining probabilities are calculated using Axioms 2 and 3.

The next question we ask is, If a 1 was received, what is the probability, $P(1s|1r)$, that a 1 was sent? Applying Bayes' rule, we find that

$$P(1s|1r) = \frac{P(1r|1s)P(1s)}{P(1r)} \tag{5.25}$$

To find $P(1r)$, we note that

$$P(1r, 1s) = P(1r|1s)P(1s) = 0.18 \tag{5.26}$$

and

$$P(1r, 0s) = P(1r|0s)P(0s) = 0.08 \tag{5.27}$$

Thus

$$P(1r) = P(1r, 1s) + P(1r, 0s) = 0.18 + 0.08 = 0.26 \tag{5.28}$$

and

$$P(1s|1r) = \frac{0.9(0.2)}{0.26} = 0.69 \tag{5.29}$$

Similarly, one can calculate $P(0s|1r) = 0.31$, $P(0s|0r) = 0.97$, and $P(1s|0r) = 0.03$. For practice, you should go through the necessary calculations.

■

5.1.6 Tree Diagrams

Another handy device for determining probabilities of compound events is a *tree diagram*, particularly if the compound event can be visualized as happening in time sequence. This device is illustrated by the following example.

EXAMPLE 5.6

Suppose five cards are drawn without replacement from a standard 52-card deck. What is the probability that three of a kind results (e.g., three kings)?

Solution

The tree diagram for this chance experiment is shown in Figure 5.2. On the first draw we focus on a particular card, denoted as *X*, which we either draw or do not. The second draw results in four possible events of interest: a card is drawn that matches the first card with probability $\frac{3}{51}$, or a match is not obtained with probability $\frac{48}{51}$. If some card other than X was drawn on the first draw, then X results with probability $\frac{4}{51}$ on the second draw (lower half of Figure 5.2). At this point, 50 cards are left in the deck. If we follow the upper branch, which corresponds to a match of the first card, two events of interest are again possible: another match that will be referred to as a *triple* with probability of $\frac{2}{50}$ *on that draw* or a card that does not match the first two with probability $\frac{48}{50}$. If a card other than *X* was obtained on the second draw, then *X* occurs with probability $\frac{4}{50}$ if *X* was obtained on the first draw, and probability $\frac{46}{50}$ if it was not. The remaining branches are filled in similarly. Each path through the tree will either result in success or failure, and the probability of drawing the cards along a particular path will be the product of the separate probabilities along each path. Since a particular sequence of draws resulting in success is mutually exclusive of the sequence of draws resulting in any other success, we simply add up all the products of probabilities along all paths that result in success. In addition to these sequences involving card *X*, there are 12 others involving other face values that result in three of a kind. Thus we multiply the result obtained from Figure 5.2 by 13. The probability of drawing three cards of the same value, in any order, is then given by

$$\begin{aligned} P(3 \text{ of a kind}) &= 13\,\frac{10(4)(3)(2)(48)(47)}{52(51)(50)(49)(48)} \\ &= 0.02257 \end{aligned} \tag{5.30}$$

■

EXAMPLE 5.7

Another type of problem very closely related to those amenable to tree-diagram solutions is a reliability problem. Reliability problems can result from considering the overall failure of a system composed of several components each of which may fail with a certain probability *p*. An example is shown in Figure 5.3, where a battery is connected to a load through the series–parallel combination of relay switches each of which may fail to close with probability *p* (or close with probability $q = 1 - p$). The problem is to find the probability that current flows in the load. From the diagram, it is clear that a circuit is completed if S1 or S2 and S3 are closed. Therefore

$$\begin{aligned} P(\text{success}) &= P(\text{Sl or S2 and S3 closed}) \\ &= P(\text{S1 or S2 or both closed})P(\text{S3 closed}) \\ &= [1 - P(\text{both switches open})]P(\text{S3 closed}) \\ &= (1 - p^2)q \end{aligned} \tag{5.31}$$

where it is assumed that the separate switch actions are statistically independent.

■

5.1.7 Some More General Relationships

Some useful formulas for a somewhat more general case than those considered above will now be derived. Consider an experiment that is composed of compound events (A_i, B_j) that are mutually exclusive. The totality of all these compound events, $i = 1, 2, \ldots, M$, $j = 1, 2, \ldots, N$, composes the entire sample space (that is, the events are said to be exhaustive or to form a partition of

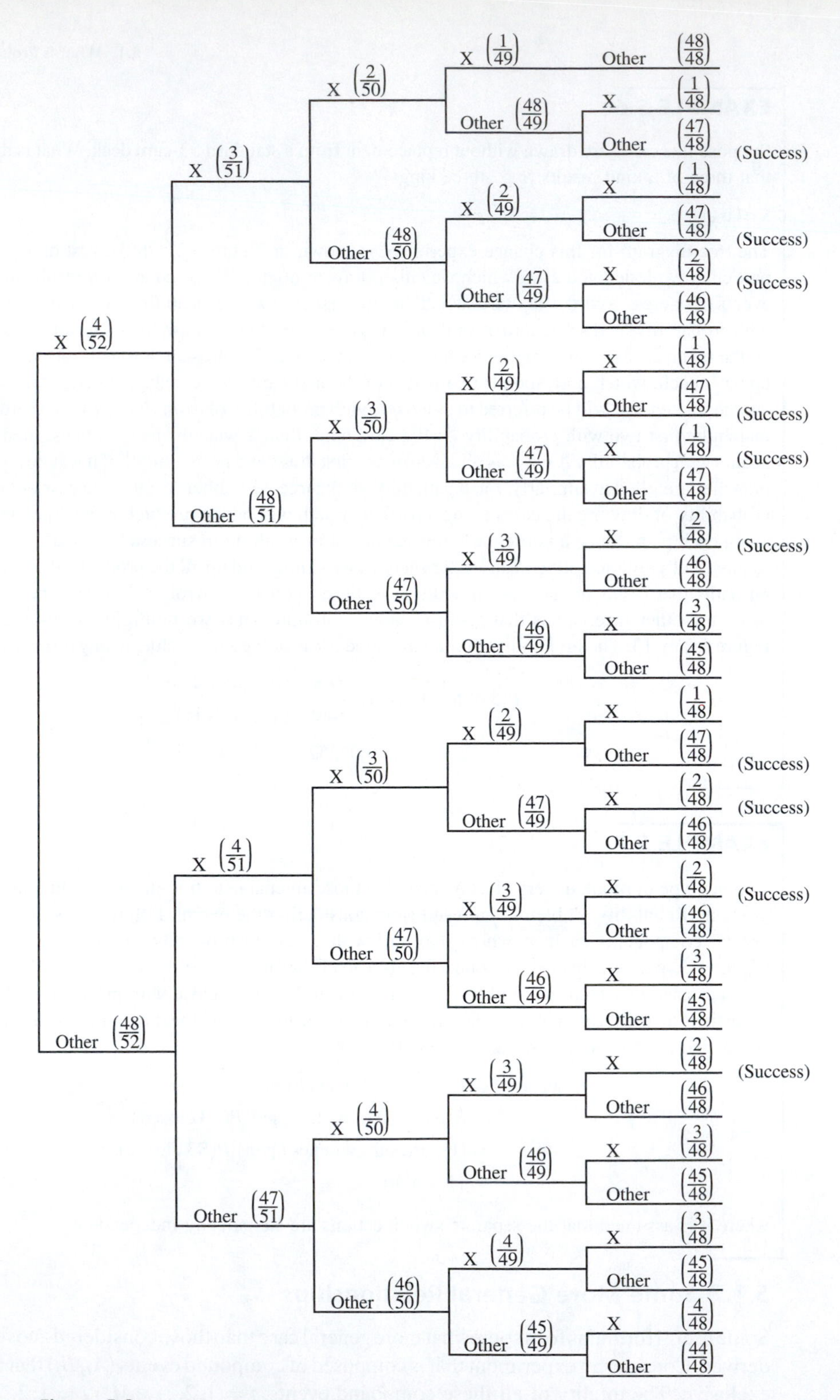

Figure 5.2

A card-drawing problem illustrating the use of a tree diagram.

Figure 5.3
Circuit illustrating the calculation of reliability

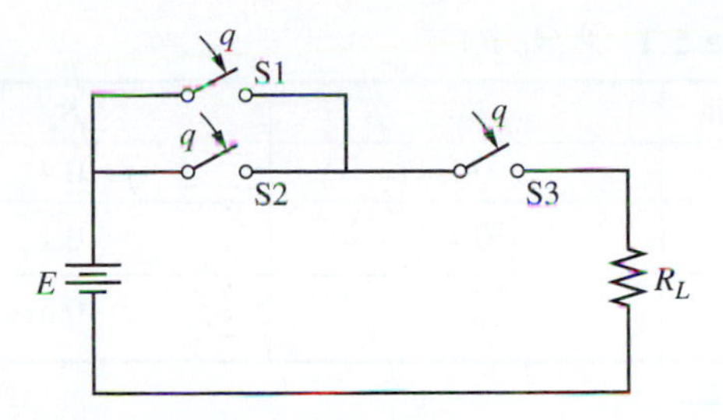

the sample space). For example, the experiment might consist of rolling a pair of dice with (A_i, B_j) = (number of spots showing on die 1, number of spots showing on die 2).

Suppose the probability of the joint event (A_i, B_j) is $P(A_i, B_j)$. Each compound event can be thought of as a simple event, and if the probabilities of all these mutually exclusive, exhaustive events are summed, a probability of 1 will be obtained, since the probabilities of all possible outcomes have been included. That is,

$$\sum_{i=1}^{M} \sum_{j=1}^{N} P(A_i, B_j) = 1 \tag{5.32}$$

Now consider a particular event B_j. Associated with this particular event, we have M possible mutually exclusive, but not exhaustive outcomes $(A_1, B_j), (A_2, B_j), \ldots, (A_M, B_j)$. If we sum over the corresponding probabilities, we obtain the probability of B_j irrespective of the outcome on A. Thus

$$P(B_j) = \sum_{i=1}^{M} P(A_i, B_j) \tag{5.33}$$

Similar reasoning leads to the result

$$P(A_i) = \sum_{j=1}^{N} P(A_i, B_j) \tag{5.34}$$

$P(A_i)$ and $P(B_j)$ are referred to as *marginal probabilities.*

Suppose the conditional probability of B_m given A_n, $P(B_m|A_n)$, is desired. In terms of the joint probabilities $P(A_i, B_j)$, we can write this conditional probability as

$$P(B_m|A_n) = \frac{P(A_n, B_m)}{\sum_{j=1}^{N} P(A_n, B_j)} \tag{5.35}$$

which is a more general form of Bayes' rule than that given by (5.10).

EXAMPLE 5.8

A certain experiment has the joint and marginal probabilities shown in Table 5.1. Find the missing probabilities.

Solution

Using $P(B_1) = P(A_1, B_1) + P(A_2, B_1)$, we obtain $P(B_1) = 0.1 + 0.1 = 0.2$. Also, since $P(B_1) + P(B_2) + P(B_3) = 1$, we have $P(B_3) = 1 - 0.2 - 0.5 = 0.3$. Finally, using $P(A_1, B_3) + P(A_2, B_3) = P(B_3)$, we get $P(A_1, B_3) = 0.3 - 0.1 = 0.2$, and therefore, $P(A_1) = 0.1 + 0.4 + 0.2 = 0.7$.

Table 5.1 $P(A_i, B_j)$

A_i B_j	B_1	B_2	B_3	$P(A_i)$
A_1	0.1	0.4	?	?
A_2	0.1	0.1	0.1	0.3
$P(B_j)$	?	0.5	?	1

■

5.2 RANDOM VARIABLES AND RELATED FUNCTIONS

5.2.1 Random Variables

In the applications of probability it is often more convenient to work in terms of numerical outcomes (for example, the number of errors in a digital data message) rather than nonnumerical outcomes (for example, failure of a component). Because of this, we introduce the idea of a *random variable*, which is defined as a rule that assigns a numerical value to each possible outcome of a chance experiment. (The term *random variable* is a misnomer; a random variable is really a function, since it is a rule that assigns the members of one set to those of another.)

As an example, consider the tossing of a coin. Possible assignments of random variables are given in Table 5.2. These are examples of *discrete random variables* and are illustrated in Figure 5.4(a).

As an example of a *continuous random variable*, consider the spinning of a pointer, such as is typically found in children's games. A possible assignment of a random variable would be the angle Θ_1 in radians, that the pointer makes with the vertical when it stops. Defined in this fashion, Θ_1 has values that continuously increase with rotation of the pointer. A second possible random variable, Θ_2, would be Θ_1 minus integer multiples of 2π rad, such that $0 \leq \Theta_2 < 2\pi$, which is commonly denoted as Θ_1 modulo 2π. These random variables are illustrated in Figure 5.4(b).

At this point, we introduce a convention that will be adhered to, for the most part, throughout this book. Capital letters (X, Θ, and so on) denote random variables, and the corresponding lowercase letters (x, θ, and so on) *denote the values that the random variables take on or running values for them.*

5.2.2 Probability (Cumulative) Distribution Functions

We need some way of probabilistically describing random variables that works equally well for discrete and continuous random variables. One way of accomplishing this is by means of the *cumulative distribution function* (cdf).

Consider a chance experiment with which we have associated a random variable X. The cdf

Table 5.2 Possible Random Variables (RV)

Outcome: S_i	RV No. 1: $X_1(S_i)$	RV No. 2: $X_2(S_i)$
S_1 = heads	$X_1(S_1) = 1$	$X_2(S_1) = \pi$
S_2 = tails	$X_1(S_2) = -1$	$X_2(S_2) = \sqrt{2}$

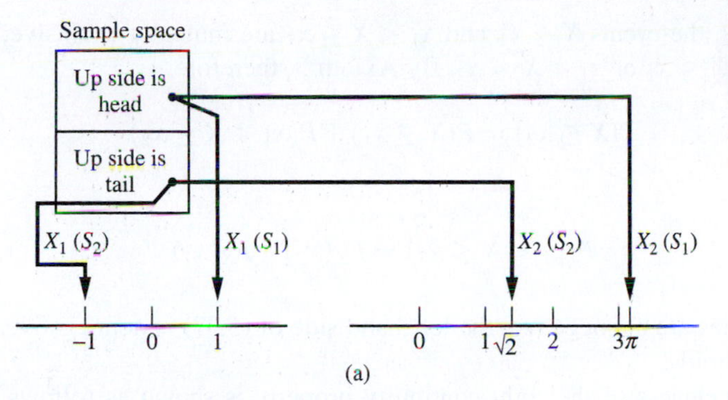

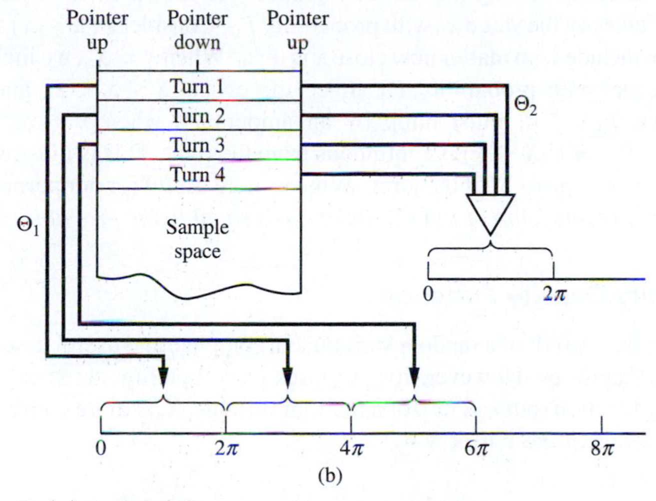

Figure 5.4 Pictorial representation of sample spaces and random variables. (a) Coin-tossing experiment. (b) Pointer-spinning experiment.

$F_X(x)$ is defined as

$$F_X(x) = \text{probability that } X \leq x = P(X \leq x) \tag{5.36}$$

We note that $F_X(x)$ is a function of x, not of the random variable X. However, $F_X(x)$ also depends on the assignment of the random variable X, which accounts for the subscript.

The cdf has the following properties:

Property 1

$0 \leq F_X(x) \leq 1$, with $F_X(-\infty) = 0$ and $F_X(\infty) = 1$.

Property 2

$F_X(x)$ is continuous from the right; that is, $\lim_{x \to x_0+} F_X(x) = F_X(x_0)$.

Property 3

$F_X(x)$ is a nondecreasing function of x; that is, $F_X(x_1) \leq F_X(x_2)$ if $x_1 < x_2$.

The reasonableness of the preceding properties is shown by the following considerations.

Since $F_X(x)$ is a probability, it must, by the previously stated axioms, lie between 0 and 1, inclusive. Since $X = -\infty$ excludes all possible outcomes of the experiment, $F_X(-\infty) = 0$, and since $X = \infty$ includes all possible outcomes, $F_X(\infty) = 1$, which verifies Property 1.

For $x_1 < x_2$, the events $X \le x_1$ and $x_1 < X \le x_2$ are mutually exclusive; furthermore, $X \le x_2$ implies $X \le x_1$ or $x_1 < X \le x_2$. By Axiom 3, therefore,

$$P(X \le x_2) = P(X \le x_1) + P(x_1 < X \le x_2)$$

or

$$P(x_1 < X \le x_2) = F_X(x_2) - F_X(x_1) \tag{5.37}$$

Since probabilities are nonnegative, the left-hand side of (5.37) is nonnegative. Thus we see that Property 3 holds.

The reasonableness of the right-continuity property is shown as follows. Suppose the random variable X takes on the value x_0 with probability P_0. Consider $P(X \le x)$. If $x < x_0$, the event $X = x_0$ is not included, no matter how close x is to x_0. When $x = x_0$, we include the event $X = x_0$, which occurs with probability P_0. Since the events $X \le x < x_0$ and $X = x_0$ are mutually exclusive, $P(X \le x)$ must jump by an amount P_0 when $x = x_0$, as shown in Figure 5.5. Thus $F_X(x) = P(X \le x)$ is continuous from the right. This is illustrated in Fig 5.5 by the dot on the curve to the right of the jump. What is more useful for our purposes, however, is that the *magnitude of any jump of* $F_X(x)$, *say at* x_0, *is equal to the probability that* $X = x_0$.

5.2.3 Probability Density Function

From (5.37) we see that the cdf of a random variable is a complete and useful description for the computation of probabilities. However, for purposes of computing statistical averages, the probability density function (pdf), $f_X(x)$, of a random variable, X, is more convenient. The pdf of X is defined in terms of the cdf of X by

$$f_X(x) = \frac{dF_X(x)}{dx} \tag{5.38}$$

Since the cdf of a discrete random variable is discontinuous, its pdf, mathematically speaking, does not exist at the points of discontinuity. By representing the derivative of a jump-discontinuous function at a point of discontinuity by a delta function of area equal to the magnitude of the jump, we can define pdfs for discrete random variables. In some books, this problem is avoided by defining a *probability mass function* for a discrete random variable, which consists simply of lines equal in magnitude to the probabilities that the random variable takes on at its possible values.

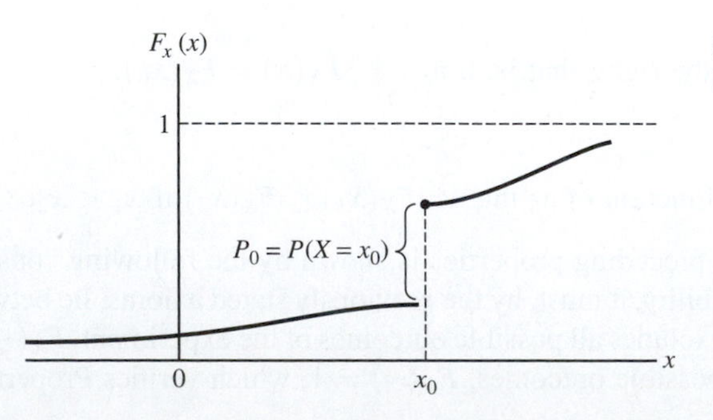

Figure 5.5
Illustration of the jump property of $F_X(x)$.

Recalling that $F_X(-\infty) = 0$, we see from (5.38) that

$$F_X(x) = \int_{-\infty}^{x} f_X(\eta)\, d\eta \tag{5.39}$$

That is, the *area* under the pdf from $-\infty$ to x is the probability that the observed value will be less than or equal to x.

From (5.38), (5.39), and the properties of $F_X(x)$, we see that the pdf has the following properties:

$$f_X(x) = \frac{dF_X(x)}{dx} \geq 0 \tag{5.40}$$

$$\int_{-\infty}^{\infty} f_X(x)\, dx = 1 \tag{5.41}$$

$$P(x_1 < X \leq x_2) = F_X(x_2) - F_X(x_1) = \int_{x_1}^{x_2} f_X(x)\, dx \tag{5.42}$$

To obtain another enlightening and very useful interpretation of $f_X(x)$, we consider (5.42) with $x_1 = x - dx$ and $x_2 = x$. The integral then becomes $f_X(x)\, dx$, so

$$f_X(x)\, dx = P(x - dx < X \leq x) \tag{5.43}$$

That is, the ordinate at any point x on the pdf curve multiplied by dx gives the probability of the random variable X lying in an infinitesimal range around the point x assuming that $f_X(x)$ is continuous at x.

The following two examples illustrate cdfs and pdfs for discrete and continuous cases, respectively.

EXAMPLE 5.9

Suppose two fair coins are tossed and X denotes the number of heads that turn up. The possible outcomes, the corresponding values of X, and the respective probabilities are summarized in Table 5.3. The cdf and pdf for this experiment and random variable definition are shown in Figure 5.6. The properties of the cdf and pdf for discrete random variables are demonstrated by this figure, as a careful examination will reveal. It is emphasized that the cdf and pdf change if the definition of the random variable or the probability assigned is changed.

Table 5.3 Outcomes and Probabilities

Outcome	X	$P(X = x_j)$
TT	$x_1 = 0$	$\frac{1}{4}$
TH, HT	$x_2 = 1$	$\frac{1}{2}$
HH	$x_3 = 2$	$\frac{1}{4}$

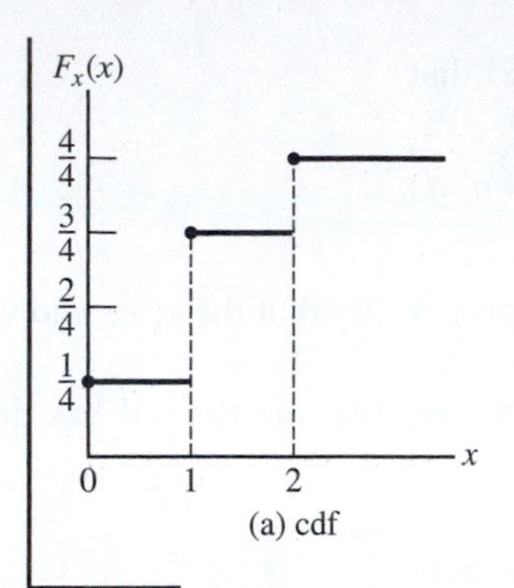

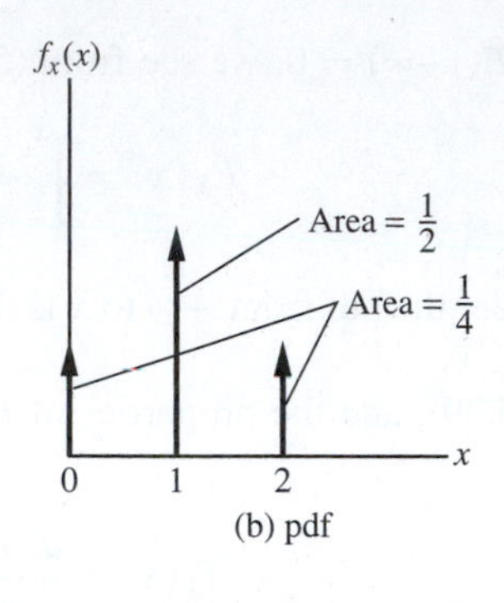

Figure 5.6
The cdf and pdf for a coin-tossing experiment.

■

EXAMPLE 5.10

Consider the pointer-spinning experiment described earlier. We assume that any one stopping point is not favored over any other and that the random variable Θ is defined as the angle that the pointer makes with the vertical, modulo 2π. Thus Θ is limited to the range $[0, 2\pi)$, and for any two angles θ_1 and θ_2 in $[0, 2\pi)$, we have

$$P(\theta_1 - \Delta\theta < \Theta \leq \theta_1) = P(\theta_2 - \Delta\theta < \Theta \leq \theta_2) \tag{5.44}$$

by the assumption that the pointer is equally likely to stop at any angle in $[0, 2\pi)$. In terms of the pdf $f_\Theta(\theta)$, this can be written, using (5.43), as

$$f_\Theta(\theta_1) = f_\Theta(\theta_2), \quad 0 \leq \theta_1, \theta_2 < 2\pi \tag{5.45}$$

Thus, in the interval $[0, 2\pi)$, $f_\Theta(\theta)$ is a constant, and outside $[0, 2\pi)$, $f_\Theta(\theta)$ is zero by the modulo 2π condition (this means that angles less than or equal to 0 or greater than 2π are impossible). By (5.41), it follows that

$$f_\Theta(\theta) = \begin{cases} \dfrac{1}{2\pi}, & 0 \leq \theta < 2\pi \\ 0, & \text{otherwise} \end{cases} \tag{5.46}$$

The pdf $f_\Theta(\theta)$ is shown graphically in Figure 5.7(a). The cdf $F_\Theta(\theta)$ is easily obtained by performing a graphical integration of $f_\Theta(\theta)$ and is shown in Figure 5.7(b).

To illustrate the use of these graphs, suppose we wish to find the probability of the pointer landing anyplace in the interval $\left[\frac{1}{2}\pi, \pi\right]$. The desired probability is given either as the area under the pdf curve from $\frac{1}{2}\pi$ to π, shaded in Figure 5.7(a), or as the value of the ordinate at $\theta = \pi$ minus the value of the ordinate at $\theta = \frac{1}{2}\pi$ on the cdf curve. The probability that the pointer lands exactly at $\frac{1}{2}\pi$, however, is 0.

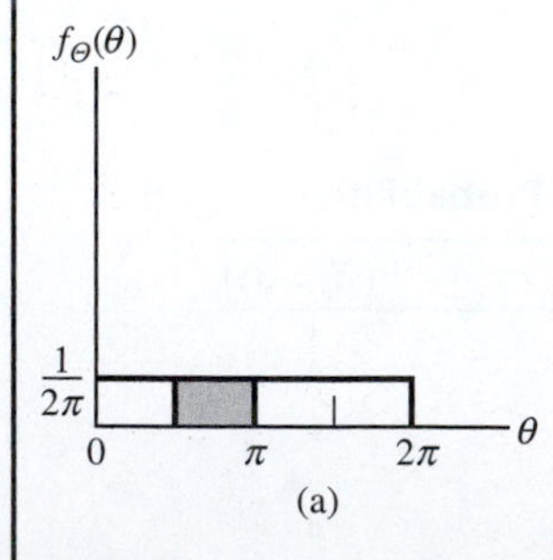

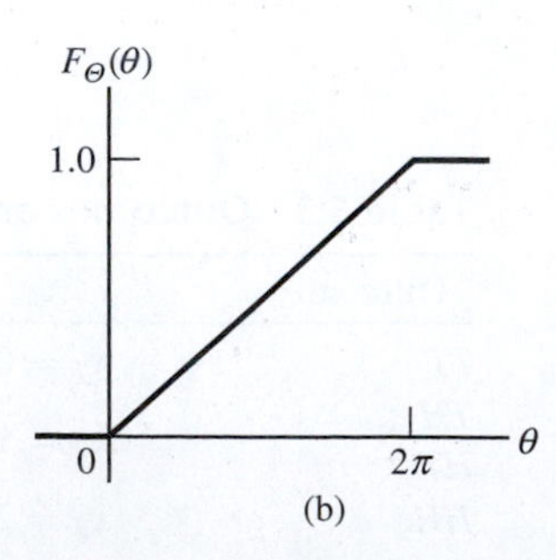

Figure 5.7
The (a) pdf and (b) cdf for a pointer-spinning experiment.

■

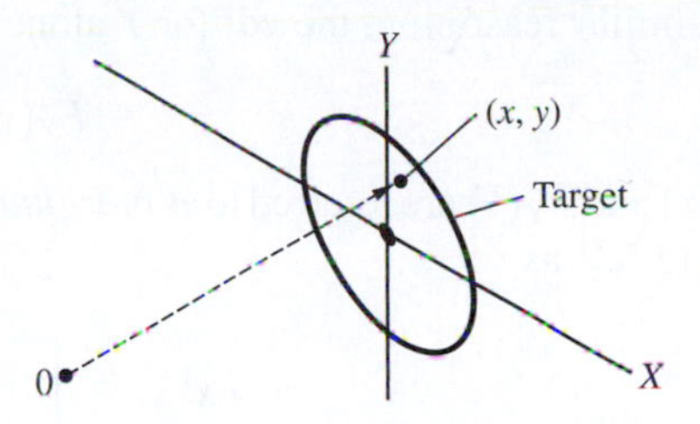

Figure 5.8
The dart-throwing experiment.

5.2.4 Joint cdfs and pdfs

Some chance experiments must be characterized by two or more random variables. The cdf or pdf description is readily extended to such cases. For simplicity, we will consider only the case of two random variables.

To give a specific example, consider the chance experiment in which darts are repeatedly thrown at a target, as shown schematically in Figure 5.8. The point at which the dart lands on the target must be described in terms of two numbers. In this example, we denote the impact point by the two random variables X and Y, whose values are the xy coordinates of the point where the dart sticks, with the origin being fixed at the bull's eye. The joint cdf of X and Y is defined as

$$F_{XY}(x, y) = P(X \leq x, Y \leq y) \tag{5.47}$$

where the comma is interpreted as "and." The *joint pdf* of X and Y is defined as

$$f_{XY}(x, y) = \frac{\partial^2 F_{XY}(x, y)}{\partial x \, \partial y} \tag{5.48}$$

Just as we did in the case of single random variables, we can show that

$$P(x_1 < X \leq x_2, y_1 < Y \leq y_2) = \int_{y_1}^{y_2} \int_{x_1}^{x_2} f_{XY}(x, y) \, dx \, dy \tag{5.49}$$

which is the two-dimensional equivalent of (5.42). Letting $x_1 = y_1 = -\infty$ and $x_2 = y_2 = \infty$, we include the entire sample space. Thus

$$F_{XY}(\infty, \infty) = \int_{-\infty}^{\infty} \int_{-\infty}^{\infty} f_{XY}(x, y) \, dx \, dy = 1 \tag{5.50}$$

Letting $x_1 = x - dx$, $x_2 = x$, $y_1 = y - dy$, and $y_2 = y$, we obtain the following enlightening special case of (5.49):

$$f_{XY}(x, y) \, dx \, dy = P(x - dx < X \leq x, y - dy < Y \leq y) \tag{5.51}$$

Thus the probability of finding X in an infinitesimal interval around x while simultaneously finding Y in an infinitesimal interval around y is $f_{XY}(x, y) \, dx \, dy$ assuming a continuous pdf.

Given a joint cdf or pdf, we can obtain the cdf or pdf of one of the random variables using the following considerations. The cdf for X irrespective of the value Y takes on is simply

$$\begin{aligned} F_X(x) &= P(X \leq x, Y < \infty) \\ &= F_{XY}(x, \infty) \end{aligned} \tag{5.52}$$

By similar reasoning, the cdf for Y alone is

$$F_Y(y) = F_{XY}(\infty, y) \tag{5.53}$$

$F_X(x)$ and $F_Y(y)$ are referred to as *marginal cdfs*. Using (5.49) and (5.50), we can express (5.52) and (5.53) as

$$F_X(x) = \int_{-\infty}^{\infty} \int_{-\infty}^{x} f_{XY}(x', y')\, dx'\, dy' \tag{5.54}$$

and

$$F_Y(y) = \int_{-\infty}^{y} \int_{-\infty}^{\infty} f_{XY}(x', y')\, dx'\, dy' \tag{5.55}$$

respectively. Since

$$f_X(x) = \frac{dF_X(x)}{dx} \quad \text{and} \quad f_Y(y) = \frac{dF_Y(y)}{dy} \tag{5.56}$$

we obtain

$$f_X(x) = \int_{-\infty}^{\infty} f_{XY}(x, y')\, dy' \tag{5.57}$$

and

$$f_Y(y) = \int_{-\infty}^{\infty} f_{XY}(x', y)\, dx' \tag{5.58}$$

from (5.54) and (5.55), respectively. Thus, to obtain the marginal pdfs $f_X(x)$ and $f_Y(y)$ from the joint pdf $f_{XY}(x, y)$, we simply integrate out the undesired variable (or variables for more than two random variables). Hence the joint cdf or pdf contains all the information possible about the joint random variables X and Y. Similar results hold for more than two random variables.

Two random variables are statistically independent (or simply independent) if the values one takes on do not influence the values that the other takes on. Thus, for any x and y, it must be true that

$$P(X \leq x, Y \leq y) = P(X \leq x)P(Y \leq y) \tag{5.59}$$

or, in terms of cdfs,

$$F_{XY}(x, y) = F_X(x)F_Y(y) \tag{5.60}$$

That is, the joint cdf of independent random variables factors into the product of the separate marginal cdfs. Differentiating both sides of (5.59) with respect to first x and then y, and using the definition of the pdf, we obtain

$$f_{XY}(x, y) = f_X(x)\, f_Y(y) \tag{5.61}$$

which shows that the joint pdf of independent random variables also factors. If two random variables are not independent, we can write their joint pdf in terms of conditional pdfs $f_{X|Y}(x|y)$ and $f_{Y|X}(y|x)$ as

$$\begin{aligned} f_{XY}(x, y) &= f_X(x)\, f_{Y|X}(y|x) \\ &= f_Y(y)\, f_{X|Y}(x|y) \end{aligned} \tag{5.62}$$

These relations *define* the conditional pdfs of two random variables. An intuitively satisfying interpretation of $f_{X|Y}(x|y)$ is

$$f_{X|Y}(x|y)\,dx = P(x - dx < X \leq x \quad \text{given } Y = y) \tag{5.63}$$

with a similar interpretation for $f_{Y|X}(y|x)$. Equation (5.62) is reasonable in that if X and Y are dependent, a given value of Y should influence the probability distribution for X. On the other hand, if X and Y are independent, information about one of the random variables tells us nothing about the other. Thus, for independent random variables,

$$f_{X|Y}(x|y) = f_X(x) \qquad \text{and} \qquad f_{Y|X}(y|x) = f_Y(y) \tag{5.64}$$

which could serve as an alternative definition of statistical independence. The following example illustrates the preceding ideas.

EXAMPLE 5.11

Two random variables X and Y have the joint pdf

$$f_{XY}(x,y) = \begin{cases} Ae^{-(2x+y)}, & x, y \geq 0 \\ 0, & \text{otherwise} \end{cases} \tag{5.65}$$

where A is a constant. We evaluate A from

$$\int_{-\infty}^{\infty}\int_{-\infty}^{\infty} f_{XY}(x,y)\,dx\,dy = 1 \tag{5.66}$$

Since

$$\int_0^{\infty}\int_0^{\infty} e^{-(2x+y)}\,dx\,dy = \frac{1}{2} \tag{5.67}$$

$A = 2$. We find the marginal pdfs from (5.57) and (5.58) as follows:

$$\begin{aligned} f_X(x) = \int_{-\infty}^{\infty} f_{XY}(x,y)\,dy &= \begin{cases} \int_0^{\infty} 2e^{-(2x+y)}dy, & x \geq 0 \\ 0, & x < 0 \end{cases}. \\ &= \begin{cases} 2e^{-2x}, & x \geq 0 \\ 0, & x < 0 \end{cases} \end{aligned} \tag{5.68}$$

$$f_Y(y) = \begin{cases} e^{-y}, & y \geq 0 \\ 0, & y < 0 \end{cases} \tag{5.69}$$

These joint and marginal pdfs are shown in Figure 5.9. From these results, we note that X and Y are statistically independent since $f_{XY}(x,y) = f_X(x)f_Y(y)$.

We find the joint cdf by integrating the joint pdf on both variables, using (5.42) and (5.40), which gives

$$\begin{aligned} F_{XY}(x,y) &= \int_{-\infty}^{y}\int_{-\infty}^{x} f_{XY}(x', y')\,dx',dy' \\ &= \begin{cases} (1 - e^{-2x})(1 - e^{-y}), & x, y \geq 0 \\ 0, & \text{otherwise} \end{cases} \end{aligned} \tag{5.70}$$

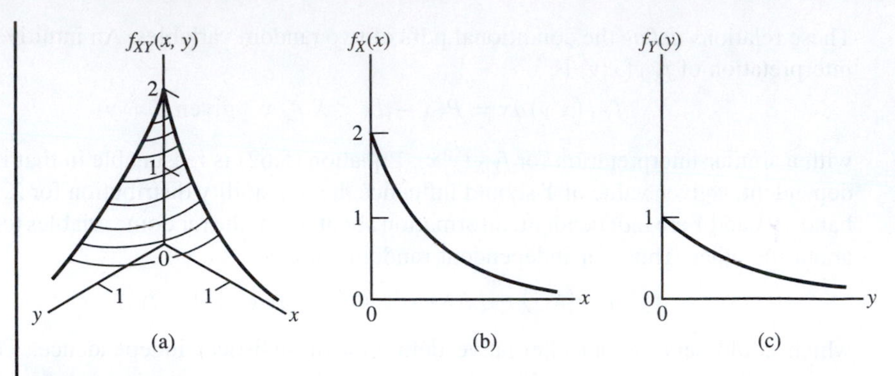

Figure 5.9
Joint and marginal pdfs for two random variables. (a) Joint pdf. (b) Marginal pdf for X. (c) Marginal pdf for Y.

Dummy variables are used in the integration to avoid confusion. Note that $F_{XY}(-\infty, -\infty) = 0$ and $F_{XY}(\infty, \infty) = 1$, as they should, since the first case corresponds to the probability of an impossible event and the latter corresponds to the inclusion of all possible outcomes. We also can use the result for $F_{XY}(x, y)$ to obtain

$$F_X(x) = F_{XY}(x, \infty) = \begin{cases} (1 - e^{-2x}), & x \geq 0 \\ 0, & \text{otherwise} \end{cases} \tag{5.71}$$

and

$$F_Y(y) = F_{XY}(\infty, y) = \begin{cases} (1 - e^{-y}), & y \geq 0 \\ 0, & \text{otherwise} \end{cases} \tag{5.72}$$

Also note that the joint cdf factors into the product of the marginal cdfs, as it should, for statistically independent random variables.

The conditional pdfs are

$$f_{X|Y}(x|y) = \frac{f_{XY}(x, y)}{f_Y(y)} = \begin{cases} 2e^{-2x}, & x \geq 0 \\ 0, & x < 0 \end{cases} \tag{5.73}$$

and

$$f_{Y|X}(y|x) = \frac{f_{XY}(x, y)}{f_X(x)} = \begin{cases} e^{-y}, & y \geq 0 \\ 0, & y < 0 \end{cases} \tag{5.74}$$

They are equal to the respective marginal pdfs, as they should be for independent random variables. ■

EXAMPLE 5.12

To illustrate the processes of normalization of joint pdfs, finding marginal from joint pdfs, and checking for statistical independence of the corresponding random variables, we consider the joint pdf

$$f_{XY}(x, y) = \begin{cases} \beta xy, & 0 \leq x \leq y,\ 0 \leq y \leq 4 \\ 0, & \text{otherwise} \end{cases} \tag{5.75}$$

For independence, the joint pdf should be the product of the marginal pdfs.

Solution

This example is somewhat tricky because of the limits; so a diagram of the pdf is given in Figure 5.10. We find the constant β by normalizing the volume under the pdf to unity by integrating $f_{XY}(x, y)$ over all x and

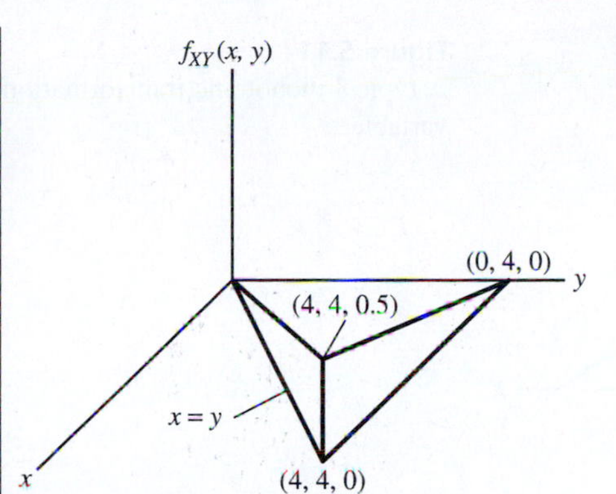

Figure 5.10
Probability density function for Example 5.12.

y. This gives

$$\beta \int_0^4 y \left[\int_0^y x\, dx \right] dy = \beta \int_0^4 y \frac{y^2}{2}\, dy$$
$$= \beta \frac{y^4}{2 \times 4} \bigg|_0^4$$
$$= 32\beta = 1$$

so $\beta = \frac{1}{32}$.

We next proceed to find the marginal pdfs. Integrating over x first and checking Figure 5.10 to obtain the proper limits of integration, we obtain

$$f_Y(y) = \int_0^y \frac{xy}{32} dx, \quad 0 \le y \le 4$$
$$= \begin{cases} \frac{y^3}{64}, & 0 \le y \le 4 \\ 0, & \text{otherwise} \end{cases} \tag{5.76}$$

The pdf on X is similarly obtained as

$$f_X(x) = \int_x^4 \frac{xy}{32} dy, \quad 0 \le y \le 4$$
$$= \begin{cases} \frac{x}{4}\left[1 - \left(\frac{x}{4}\right)^2\right], & 0 \le x \le 4 \\ 0, & \text{otherwise} \end{cases} \tag{5.77}$$

A little work shows that both marginal pdfs integrate to 1, as they should.

It is clear that the product of the marginal pdfs is not equal to the joint pdf so the random variables X and Y are not statistically independent. ∎

5.2.5 Transformation of Random Variables

Situations are often encountered where the pdf (or cdf) of a random variable X is known and we desire the pdf of a second random variable Y defined as a function of X, for example,

$$Y = g(X) \tag{5.78}$$

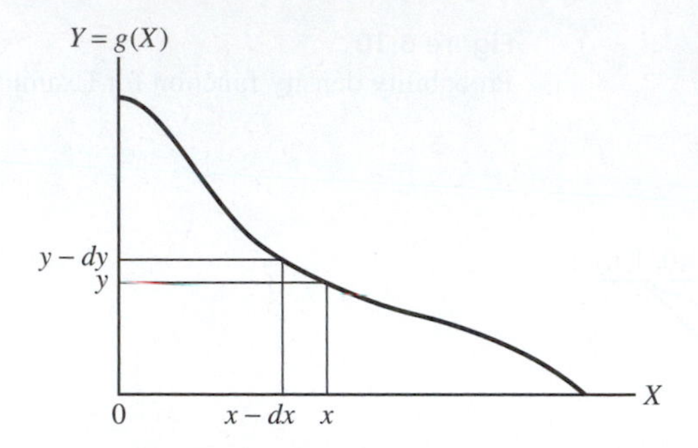

Figure 5.11
A typical monotonic transformation of a random variable.

We initially consider the case where $g(X)$ is a monotonic function of its argument (for example, it is either nondecreasing or nonincreasing as the independent variable ranges from $-\infty$ to ∞), a restriction that will be relaxed shortly.

A typical function is shown in Figure 5.11. The probability that X lies in the range $(x - dx, x)$ is the same as the probability that Y lies in the range $(y - dy, y)$, where $y = g(x)$. Therefore, we obtain

$$f_X(x)\,dx = f_Y(y)\,dy \tag{5.79}$$

if $g(X)$ is monotonically increasing, and

$$f_X(x)\,dx = -f_Y(y)\,dy \tag{5.80}$$

if $g(X)$ is monotonically decreasing, since an *increase* in x results in a *decrease* in y. Both cases are taken into account by writing

$$f_Y(y) = f_X(x)\left|\frac{dx}{dy}\right|_{x=g^{-1}(y)} \tag{5.81}$$

where $x = g^{-1}(y)$ denotes the inversion of (5.78) for x in terms of y.

EXAMPLE 5.13

To illustrate the use of (5.81), let us consider the pdf of Example 5.10, namely

$$f_\Theta(\theta) = \begin{cases} \dfrac{1}{2\pi}, & 0 \leq \theta \leq 2\pi \\ 0, & \text{otherwise} \end{cases} \tag{5.82}$$

Assume that the random variable Θ is transformed to the random variable Y according to

$$Y = -\left(\frac{1}{\pi}\right)\Theta + 1 \tag{5.83}$$

Since, $\theta = -\pi y + \pi$, $\frac{d\theta}{dy} = -\pi$ and the pdf of Y, by (5.81) and (5.83), is

$$f_Y(y) = f_\Theta(\theta = -\pi y + \pi)|-\pi| = \begin{cases} \dfrac{1}{2}, & -1 \leq y \leq 1 \\ 0, & \text{otherwise} \end{cases} \tag{5.84}$$

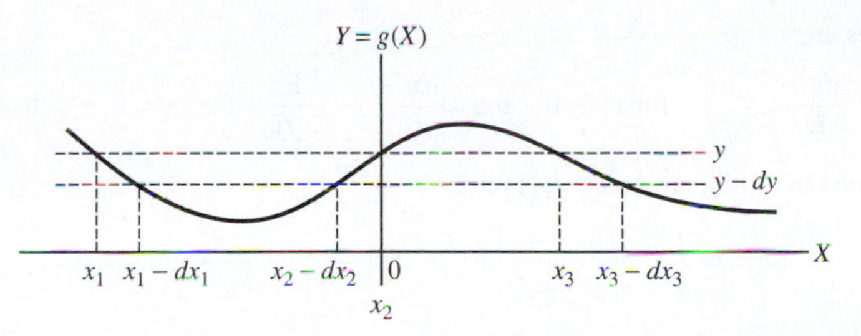

Figure 5.12
A nonmonotonic transformation of a random variable.

Note that from (5.83), $\Theta = 2\pi$ gives $Y = -1$ and $\Theta = 0$ gives $Y = 1$, so we would expect the pdf of Y to be nonzero only in the interval $[-1, 1)$; furthermore, since the transformation is linear, it is not surprising that the pdf of Y is uniform as is the pdf of Θ. ■

Consider next the case of $g(x)$ nonmonotonic as illustrated in Figure 5.12. For the case shown, the infinitesimal interval $(y - dy, y)$ corresponds to three infinitesimal intervals on the x-axis: $(x_1 - dx_1, x_1)$, $(x_2 - dx_2, x_2)$, and $(x_3 - dx_3, x_3)$. The probability that X lies in any one of these intervals is equal to the probability that Y lies in the interval $(y - dy, y)$. This can be generalized to the case of N disjoint intervals where it follows that

$$P(y - dy < Y \leq y) = \sum_{i=1}^{N} P(x_i - dx_i < X \leq x_i) \tag{5.85}$$

where we have generalized to N intervals on the X axis corresponding to the interval $(y - dy, y)$ on the Y axis. Since

$$P(y - dy < Y \leq y) = f_Y(y)|dy| \tag{5.86}$$

and

$$P(x_i - dx_i < X \leq x_i) = f_X(x_i)|dx_i| \tag{5.87}$$

we have

$$f_Y(y) = \sum_{i=1}^{N} f_X(x_i) \left| \frac{dx_i}{dy} \right|_{x_i = g_i^{-1}(y)} \tag{5.88}$$

where the absolute value signs are used because a probability must be positive, and $x_i = g_i^{-1}(y)$ is the ith solution to $g(y) = x$.

EXAMPLE 5.14

Consider the transformation

$$y = x^2 \tag{5.89}$$

If $f_X(x) = 0.5 \exp(-|x|)$, find $f_Y(y)$.

Solution

There are two solutions to $x^2 = y$; these are

$$x_1 = \sqrt{y} \quad \text{for } x_1 \geq 0 \quad \text{and} \quad x_2 = -\sqrt{y} \quad \text{for } x_2 < 0, y \geq 0 \tag{5.90}$$

Their derivatives are

$$\frac{dx_1}{dy} = \frac{1}{2\sqrt{y}} \quad \text{for } x_1 \geq 0 \quad \text{and} \quad \frac{dx_2}{dy} = -\frac{1}{2\sqrt{y}} \quad \text{for } x_2 < 0,\ y > 0 \tag{5.91}$$

Using these results in (5.88), we obtain $f_Y(y)$ to be

$$f_Y(y) = \frac{1}{2}e^{-\sqrt{y}}\left| -\frac{1}{2\sqrt{y}}\right| + \frac{1}{2}e^{-\sqrt{y}}\left|\frac{1}{2\sqrt{y}}\right| = \frac{e^{-\sqrt{y}}}{2\sqrt{y}}, \quad y > 0 \tag{5.92}$$

Since Y cannot be negative, $f_Y(y) = 0,\ y < 0$.

■

For two or more random variables, we consider only one-to-one transformations and the probability of the joint occurrence of random variables lying within infinitesimal areas (or volumes for more than two random variables). Thus, suppose two new random variables U and V are defined in terms of two original joint random variables X and Y by the relations

$$U = g_1(X, Y) \quad \text{and} \quad V = g_2(X, Y) \tag{5.93}$$

The new pdf $f_{UV}(u, v)$ is obtained from the old pdf $f_{XY}(x, y)$ by using (5.51) to write

$$P(u - du < U \leq u, v - dv < V \leq v) = P(x - dx < X \leq x, y - dy < Y \leq y)$$

or

$$f_{UV}(u, v)\, dA_{UV} = f_{XY}(x, y)\, dA_{XY} \tag{5.94}$$

where dA_{UV} is the infinitesimal area in the uv plane corresponding to the infinitesimal area dA_{XY} in the xy plane through the transformation (5.93).

The ratio of elementary area dA_{XY} to dA_{UV} is given by the Jacobian

$$\frac{\partial(x, y)}{\partial(u, v)} = \begin{vmatrix} \dfrac{\partial x}{\partial u} & \dfrac{\partial x}{\partial v} \\ \dfrac{\partial y}{\partial u} & \dfrac{\partial y}{\partial v} \end{vmatrix} \tag{5.95}$$

so that

$$f_{UV}(u, v) = f_{XY}(x, y)\left|\frac{\partial(x, y)}{\partial(u, v)}\right|_{\substack{x = g_1^{-1}(u, v) \\ y = g_2^{-1}(u, v)}} \tag{5.96}$$

where the inverse functions $g_1^{-1}(u, v)$ and $g_2^{-1}(u, v)$ exist because the transformations defined by (5.93) are assumed to be one to one. An example will help clarify this discussion.

EXAMPLE 5.15

Consider the dart-throwing game discussed in connection with joint cdfs and pdfs. We assume that the joint pdf in terms of rectangular coordinates for the impact point is

$$f_{XY}(x, y) = \frac{\exp[-(x^2 + y^2)/2\sigma^2]}{2\pi\sigma^2}, \quad -\infty < x,\ y < \infty \tag{5.97}$$

where σ^2 is a constant. This is a special case of the *joint Gaussian pdf*, which we will discuss in more detail shortly.

Instead of cartesian coordinates, we wish to use polar coordinates R and Θ, defined by

$$R = \sqrt{X^2 + Y^2} \tag{5.98}$$

and

$$\Theta = \tan^{-1}\frac{Y}{X} \tag{5.99}$$

so that

$$X = R\cos\Theta = g_1^{-1}(R, \Theta) \tag{5.100}$$

and

$$Y = R\sin\Theta = g_2^{-1}(R, \Theta) \tag{5.101}$$

where $0 \le \Theta < 2\pi$, and $0 \le R < \infty$, so that the whole plane is covered. Under this transformation, the infinitesimal area $dx\,dy$ in the xy plane transforms to the area $r\,dr\,d\theta$ in the $r\theta$ plane, as determined by the Jacobian, which is

$$\frac{\partial(x, y)}{\partial(r, \theta)} = \begin{vmatrix} \frac{\partial x}{\partial r} & \frac{\partial x}{\partial \theta} \\ \frac{\partial y}{\partial r} & \frac{\partial y}{\partial \theta} \end{vmatrix} = \begin{vmatrix} \cos\theta & -r\sin\theta \\ \sin\theta & r\cos\theta \end{vmatrix} = r \tag{5.102}$$

Thus the joint pdf of R and θ is

$$f_{R\Theta}(r, \theta) = \frac{re^{-r^2/2\sigma^2}}{2\pi\sigma^2}, \quad 0 \le \theta < 2\pi,\ 0 \le r < \infty \tag{5.103}$$

which follows from (5.96), which for this case takes the form

$$f_{R\Theta}(r, \theta) = rf_{XY}(x, y)\big|_{\substack{x = r\cos\theta \\ y = r\sin\theta}} \tag{5.104}$$

If we integrate $f_{R\Theta}(r, \theta)$ over θ to get the pdf for R alone, we obtain

$$f_R(r) = \frac{r}{\sigma^2}e^{-r^2/2\sigma^2}, \quad 0 \le r < \infty \tag{5.105}$$

which is referred to as the *Rayleigh pdf.* The probability that the dart lands in a ring of radius r from the bull's eye and having thickness dr is given by $f_R(r)\,dr$. From the sketch of the Rayleigh pdf given in Figure 5.13, we see that the most probable distance for the dart to land from the bull's eye is $R = \sigma$. By integrating (5.103) over r, it can be shown that the pdf of Θ is uniform in $[0, 2\pi)$.

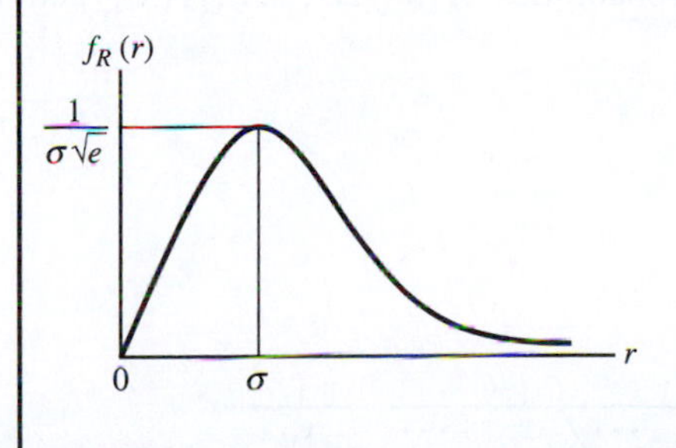

Figure 5.13
The Rayleigh pdf.

■

5.3 STATISTICAL AVERAGES

The probability functions (cdf and pdf) we have just discussed provide us with all the information possible about a random variable or a set of random variables. Often, such complete descriptions as provided by the pdf or cdf are not required, or in many cases, we are not able to obtain the cdf or pdf. A partial description of a random variable or set of random variables is then used and is given in terms of various statistical averages or mean values.

5.3.1 Average of a Discrete Random Variable

The statistical average, or expectation, of a discrete random variable X, which takes on the possible values $x_1, x_2, \ldots, x_M$ with the respective probabilities $P_1, P_2, \ldots, P_M$, is defined as

$$\overline{X} = E[X] = \sum_{j=1}^{M} x_j P_j \tag{5.106}$$

To show the reasonableness of this definition, we look at it in terms of relative frequency. If the underlying chance experiment is repeated a large number of times N, and $X = x_1$ is observed n_1 times and $X = x_2$ is observed n_2 times, etc., the arithmetical average of the observed values is

$$\frac{n_1 x_1 + n_2 x_2 + \cdots + n_M x_M}{N} = \sum_{j=1}^{M} x_j \frac{n_j}{N} \tag{5.107}$$

By the relative-frequency interpretation of probability (5.2), n_j/N approaches P_j, $j = 1, 2, \ldots, M$, the probability of the event $X = x_j$, as N becomes large. Thus, in the limit as $N \to \infty$, (5.107) becomes (5.106).

5.3.2 Average of a Continuous Random Variable

For the case where X is a continuous random variable with the pdf $f_X(x)$, we consider the range of values that X may take on, say x_0 to x_M, to be broken up into a large number of small subintervals of length Δx, as shown in Figure 5.14.

For example, consider a discrete approximation for finding the expectation of a continuous random variable X. The probability that X lies between $x_i - \Delta x$ and x_i is, from (5.43), given by

$$P(x_i - \Delta x < X \le x_i) \cong f_X(x_i)\,\Delta x, \quad i = 1, 2, \ldots, M \tag{5.108}$$

for Δx small. Thus we have approximated X by a discrete random variable that takes on the values $x_0, x_1, \ldots, x_M$ with probabilities $f_X(x_0)\,\Delta x, \ldots, f_X(x_M)\,\Delta x$, respectively.

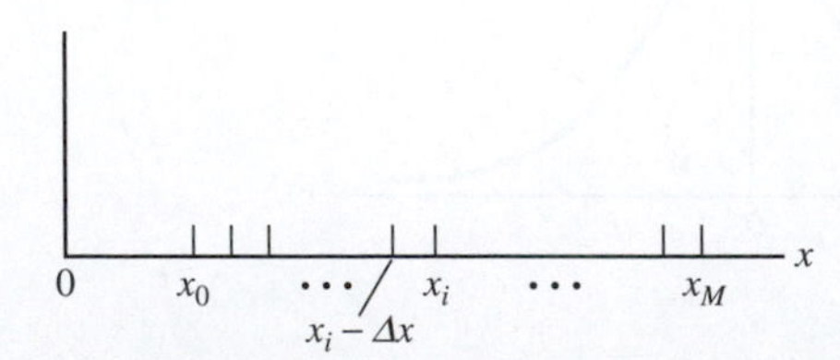

Figure 5.14
A discrete approximation for a continuous random variable X.

Using (5.106), the expectation of this random variable is

$$E[X] \cong \sum_{i=0}^{M} x_i f_X(x_i)\,\Delta x \tag{5.109}$$

As $\Delta x \to 0$, this becomes a better and better approximation for $E[X]$. In the limit, as $\Delta x \to dx$, the sum becomes an integral, giving

$$E[X] = \int_{-\infty}^{\infty} x f_X(x)\,dx \tag{5.110}$$

for the expectation of X.

5.3.3 Average of a Function of a Random Variable

We are interested not only in $E[X]$, which is referred to as the *mean* or *first moment* of X, but also in statistical averages of functions of X. Letting $Y = g(X)$, the statistical average or expectation of the new random variable Y could be obtained as

$$E[Y] = \int_{-\infty}^{\infty} y f_Y(y)\,dy \tag{5.111}$$

where $f_Y(y)$ is the pdf of Y, which can be found from $f_X(x)$ by application of (5.81). However, it is often more convenient simply to find the expectation of the function $g(X)$ as given by

$$\overline{g(X)} \triangleq E[g(X)] = \int_{-\infty}^{\infty} g(x) f_X(x)\,dx \tag{5.112}$$

which is identical to $E[Y]$ as given by (5.111). Two examples follow to illustrate the use of (5.111) and (5.112).

EXAMPLE 5.16

Suppose the random variable Θ has the pdf

$$f_\Theta(\theta) = \begin{cases} \dfrac{1}{2\pi}, & |\theta| \le \pi \\ 0, & \text{otherwise} \end{cases} \tag{5.113}$$

Then $E[\Theta^n]$ is referred to as the n*th moment of* Θ and is given by

$$E[\Theta^n] = \int_{-\infty}^{\infty} \theta^n f_\Theta(\theta)\,d\theta = \int_{-\pi}^{\pi} \theta^n \frac{d\theta}{2\pi} \tag{5.114}$$

Since the integrand is odd if n is odd, $E[\Theta^n] = 0$ for n odd. For n even,

$$E[\Theta^n] = \frac{1}{\pi}\int_0^{\pi} \theta^n\,d\theta = \frac{1}{\pi}\frac{\theta^{n+1}}{n+1}\bigg|_0^{\pi} = \frac{\pi^n}{n+1} \tag{5.115}$$

The first moment or mean of Θ, $E[\Theta]$, is a measure of the location of $f_\Theta(\theta)$ (that is, the " center of mass"). Since $f_\Theta(\theta)$ is symmetrically located about $\theta = 0$, it is not surprising that $E[\Theta] = 0$. ■

EXAMPLE 5.17

Later we shall consider certain random waveforms that can be modeled as sinusoids with random phase angles having uniform pdf in $[-\pi, \pi)$. In this example, we consider a random variable X that is defined in terms of the uniform random variable Θ considered in Example 5.16 by

$$X = \cos \Theta \tag{5.116}$$

The density function of X, $f_X(x)$, is found as follows. First $-1 \le \cos\theta \le 1$; so $f_X(x) = 0$ for $|x| > 1$. Second, the transformation is not one-to-one, there being two values of Θ for each value of X, since $\cos\theta = \cos(-\theta)$. However, we can still apply (5.81) by noting that positive and negative angles have equal probabilities and writing

$$f_X(x) = 2f_\Theta(\theta)\left|\frac{d\theta}{dx}\right|, \quad |x| < 1 \tag{5.117}$$

Now $\theta = \cos^{-1} x$ and $|d\theta/dx| = (1-x^2)^{-1/2}$, which yields

$$f_X(x) = \begin{cases} \dfrac{1}{\pi\sqrt{1-x^2}}, & |x| < 1 \\ 0, & |x| > 1 \end{cases} \tag{5.118}$$

This pdf is illustrated in Figure 5.15. The mean and second moment of X can be calculated using either (5.111) or (5.112). Using (5.111), we obtain

$$\overline{X} = \int_{-1}^{1} \frac{x}{\pi\sqrt{1-x^2}}\, dx = 0 \tag{5.119}$$

because the integrand is odd, and

$$\overline{X^2} = \int_{-1}^{1} \frac{x^2\, dx}{\pi\sqrt{1-x^2}}\, dx = \frac{1}{2} \tag{5.120}$$

by a table of integrals. Using (5.112), we find that

$$\overline{X} = \int_{-\pi}^{\pi} \cos\theta\, \frac{d\theta}{2\pi} = 0 \tag{5.121}$$

and

$$\overline{X^2} = \int_{-\pi}^{\pi} \cos^2\theta\, \frac{d\theta}{2\pi} = \int_{-\pi}^{\pi} \frac{1}{2}(1+\cos 2\theta)\, \frac{d\theta}{2\pi} = \frac{1}{2} \tag{5.122}$$

as obtained by finding $E[X]$ and $E[X^2]$ directly.

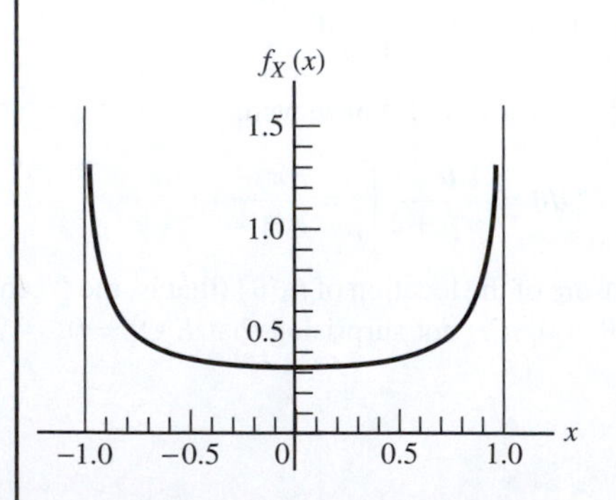

Figure 5.15
Probability density function of a sinusoid with uniform random phase.

■

5.3.4 Average of a Function of More Than One Random Variable

The expectation of a function $g(X, Y)$ of two random variables X and Y is defined in a manner analogous to the case of a single random variable. If $f_{XY}(x, y)$ is the joint pdf of X and Y, the expectation of $g(X, Y)$ is

$$E[g(X, Y)] = \int_{-\infty}^{\infty} \int_{-\infty}^{\infty} g(x, y) f_{XY}(x, y) \, dx \, dy \tag{5.123}$$

The generalization to more than two random variables should be obvious.

Equation (5.123) and its generalization to more than two random variables include the single-random-variable case, for suppose $g(X, Y)$ is replaced by a function of X alone, say $h(X)$. Then using (5.57) we obtain the following from (5.123):

$$\begin{aligned} E[h(X)] &= \int_{-\infty}^{\infty} \int_{-\infty}^{\infty} h(x) f_{XY}(x, y) \, dx \, dy \\ &= \int_{-\infty}^{\infty} h(x) f_X(x) \, dx \end{aligned} \tag{5.124}$$

where the fact that $\int_{-\infty}^{\infty} f_{XY}(x, y) \, dy = f_X(x)$ has been used.

EXAMPLE 5.18

Consider the joint pdf of Example 5.11 and the expectation of $g(X, Y) = XY$. From (5.123), this expectation is

$$\begin{aligned} E[XY] &= \int_{-\infty}^{\infty} \int_{-\infty}^{\infty} xy f_{XY}(x, y) \, dx \, dy \\ &= \int_0^{\infty} \int_0^{\infty} 2xye^{-(2x+y)} \, dx \, dy \\ &= 2 \int_0^{\infty} xe^{-2x} dx \int_0^{\infty} ye^{-y} \, dy = \frac{1}{2} \end{aligned} \tag{5.125}$$

We recall from Example 5.11 that X and Y are statistically independent. From the last line of the preceding equation for $E[XY]$, we see that

$$E[XY] = E[X]E[Y] \tag{5.126}$$

a result that holds in general for statistically independent random variables. In fact, for X and Y statistically independent random variables, it readily follows that

$$E[h(X)g(Y)] = E[h(X)]E[g(Y)] \tag{5.127}$$

where $h(X)$ and $g(Y)$ are two functions of X and Y, respectively. ■

In the special case where $h(X) = X^m$ and $g(Y) = Y^n$ and X and Y are not necessarily statistically independent in general, the expectations $E[X^m Y^n]$ are referred to as the *joint moments* of order $m + n$ of X and Y. According to (5.127), the *joint moments of statistically independent random variables factor into the products of the corresponding marginal moments.*

When finding the expectation of a function of more than one random variable, it may be easier to use the concept of conditional expectation. Consider, for example, a function $g(X, Y)$ of two random variables X and Y, with the joint pdf $f_{XY}(x, y)$. The expectation of $g(X, Y)$ is

$$\begin{aligned} E[g(X,Y)] &= \int_{-\infty}^{\infty}\int_{-\infty}^{\infty} g(x,y)f_{XY}(x,y)\,dx\,dy \\ &= \int_{-\infty}^{\infty}\left[\int_{-\infty}^{\infty} g(x,y)f_{X|Y}(x|y)\,dx\right]f_Y(y)\,dy \\ &= E[E[g(X,Y)|Y]] \end{aligned} \tag{5.128}$$

where $f_{X|Y}(x|y)$ is the conditional pdf of X given Y, and $E[g(X,Y)|Y] = \int_{-\infty}^{\infty} g(x,y)f_{X|Y}(x|y)\,dx$ is called the *conditional expectation of* $g(X, Y)$ *given* $Y = y$.

EXAMPLE 5.19

As a specific application of conditional expectation, consider the firing of projectiles at a target. Projectiles are fired until the target is hit for the first time, after which firing ceases. Assume that the probability of a projectile's hitting the target is p and that the firings are independent of one another. Find the average number of projectiles fired at the target.

Solution

To solve this problem, let N be a random variable denoting the number of projectiles fired at the target. Let the random variable H be 1 if the first projectile hits the target and 0 if it does not. Using the concept of conditional expectation, we find the average value of N is given by

$$\begin{aligned} E[N] &= E[E[N|H]] = pE[N|H=1] + (1-p)E[N|H=0] \\ &= p \times 1 + (1-p)(1+E[N]) \end{aligned} \tag{5.129}$$

where $E[N|H=0] = 1 + E[N]$ because $N \geq 1$ if a miss occurs on the first firing. By solving the last expression for $E[N]$, we obtain

$$E[N] = \frac{1}{p} \tag{5.130}$$

If $E[N]$ is evaluated directly, it is necessary to sum the series:

$$E[N] = 1 \times p + 2 \times (1-p)p + 3 \times (1-p)^2 p + \cdots \tag{5.131}$$

which is not too difficult in this instance.[3] However, the conditional-expectation method clearly makes it easier to keep track of the bookkeeping.

■

5.3.5 Variance of a Random Variable

The statistical average

$$\sigma_x^2 \triangleq E\left[(X - E[X])^2\right] \tag{5.132}$$

[3] Consider $E(N) = p(1 + 2q + 3q^2 + 4q^4 + \cdots)$ where $q = 1 - p$. The sum $S = 1 + q + q^2 + q^3 + \cdots = 1/(1-q)$ can be used to derive the sum of $1 + 2q + 3q^2 + 4q^4 + \cdots$ by differentiation with respect to q: $dS/dq = 1 + 2q + 3q^2 + \cdots = d/dq(1/(1-q)) = 1/(1-q)^2$ so that $E(N) = p[1/(1-q)^2] = 1/p$.

is called the *variance* of the random variable X; σ_x is called the *standard deviation* of X and is a measure of the concentration of the pdf of X, or $f_X(x)$, about the mean. The notation var[X] for σ_x^2 is sometimes used. A useful relation for obtaining σ_x^2 is

$$\sigma_x^2 = E[X^2] - E^2[X] \tag{5.133}$$

which, in words, says that the variance of X is simply its second moment minus its mean, squared. To prove (5.133), let $E[X] = m_x$. Then

$$\begin{aligned}\sigma_x^2 &= \int_{-\infty}^{\infty} (x - m_x)^2 f_X(x)\,dx = \int_{-\infty}^{\infty} (x^2 - 2xm_x + m_x^2) f_X(x)\,dx \\ &= E[X^2] - 2m_x^2 + m_x^2 = E[X^2] - E^2[X]\end{aligned} \tag{5.134}$$

which follows because $\int_{-\infty}^{\infty} x f_X(x)\,dx = m_x$.

EXAMPLE 5.20

Let X have the uniform pdf

$$f_X(x) = \begin{cases} \dfrac{1}{b-a}, & a \le x \le b \\ 0, & \text{otherwise} \end{cases} \tag{5.135}$$

Then

$$E[X] = \int_a^b x \frac{dx}{b-a} = \frac{1}{2}(a+b) \tag{5.136}$$

and

$$E[X^2] = \int_a^b x^2 \frac{dx}{b-a} = \frac{1}{3}(b^2 + ab + a^2) \tag{5.137}$$

which follows after a little work. Thus

$$\sigma_x^2 = \frac{1}{3}(b^2 + ab + a^2) - \frac{1}{4}(a^2 + 2ab + b^2) = \frac{1}{12}(a-b)^2 \tag{5.138}$$

Consider the following special cases:

1. $a = 1$ and $b = 2$, for which $\sigma_x^2 = \frac{1}{12}$.
2. $a = 0$ and $b = 1$, for which $\sigma_x^2 = \frac{1}{12}$.
3. $a = 0$ and $b = 2$, for which $\sigma_x^2 = \frac{1}{3}$.

For cases 1 and 2, the pdf of X has the same width but is centered about different means; the variance is the same for both cases. In case 3, the pdf is wider than it is for cases 1 and 2, which is manifested by the larger variance.

■

5.3.6 Average of a Linear Combination of N Random Variables

It is easily shown that the expected value, or average, of an arbitrary linear combination of random variables is the same as the linear combination of their respective means. That is,

$$E\left[\sum_{i=1}^{N} a_i X_i\right] = \sum_{i=1}^{N} a_i E[X_i] \tag{5.139}$$

where $X_1, X_2, \ldots, X_N$ are random variables and $a_1, a_2, \ldots, a_N$ are arbitrary constants. Equation (5.139) will be demonstrated for the special case $N = 2$; generalization to the case $N > 2$ is not difficult, but results in unwieldy notation (proof by induction can also be used).

Let $f_{X_1X_2}(x_1, x_2)$ be the joint pdf of X_1 and X_2. Then, using the definition of the expectation of a function of two random variables in (5.123), it follows that

$$\begin{aligned} E[a_1 X_1 + a_2 X_2] &\triangleq \int_{-\infty}^{\infty}\int_{-\infty}^{\infty} (a_1 x_1 + a_2 x_2) f_{X_1X_2}(x_1, x_2)\, dx_1\, dx_2 \\ &= a_1 \int_{-\infty}^{\infty}\int_{-\infty}^{\infty} x_1 f_{X_1X_2}(x_1, x_2)\, dx_1\, dx_2 \\ &\quad + a_2 \int_{-\infty}^{\infty}\int_{-\infty}^{\infty} x_2 f_{X_1X_2}(x_1, x_2)\, dx_1\, dx_2 \end{aligned} \tag{5.140}$$

Considering the first double integral and using (5.57) (with $x_1 = x$ and $x_2 = y$) and (5.110), we find that

$$\begin{aligned} \int_{-\infty}^{\infty}\int_{-\infty}^{\infty} x_1 f_{X_1X_2}(x_1, x_2)\, dx_1\, dx_2 &= \int_{-\infty}^{\infty} x_1 \left[\int_{-\infty}^{\infty} f_{X_1X_2}(x_1, x_2)\, dx_2\right] dx_1 \\ &= \int_{-\infty}^{\infty} x_1 f_X(x_1)\, dx_1 \\ &= E[X_1] \end{aligned} \tag{5.141}$$

Similarly, it can be shown that the second double integral reduces to $E[X_2]$. Thus (5.139) has been proved for the case $N = 2$. Note that (5.139) holds regardless of whether the X_i terms are independent. Also, it should be noted that a similar result holds for a linear combination of functions of N random variables.

5.3.7 Variance of a Linear Combination of Independent Random Variables

If $X_1, X_2, \ldots, X_N$ are *statistically independent* random variables, then

$$\text{var}\left[\sum_{i=1}^{N} a_i X_i\right] = \sum_{i=1}^{N} a_i^2 \text{var}[X_i] \tag{5.142}$$

where $a_1, a_2, \ldots, a_N$ are arbitrary constants and $\text{var}[X_i] = E[(X_i - \overline{X}_i)^2]$. This relation will be demonstrated for the case $N = 2$. Let $Z = a_1X_1 + a_2X_2$, and let $f_{X_i}(x_i)$ be the marginal pdf of X_i. Then the joint pdf of X_1 and X_2 is $f_{X_1}(x_1)f_{X_2}(x_2)$ by the assumption of statistical independence. Also, $\overline{Z} = a_1\overline{X}_1 + a_2\overline{X}_2$ by (5.139). Also, $\text{var}[Z] = E[(Z - \overline{Z})^2]$. However, since $Z = a_1X_1 + a_2X_2$, we may write var[Z] as

$$\begin{aligned}\text{var}[Z] &= E\left\{\left[(a_1 X_1 + a_2 X_2) - (a_1 \overline{X}_1 + a_2 \overline{X}_2)\right]^2\right\} \\ &= E\left\{\left[a_1 (X_1 - \overline{X}_1) + a_2 (X_2 - \overline{X}_2)\right]^2\right\} \\ &= a_1^2 E\left[(X_1 - \overline{X}_1)^2\right] + 2a_1 a_2 E\left[(X_1 - \overline{X}_1)(X_2 - \overline{X}_2)\right] \\ &\quad + a_2^2 E\left[(X_2 - \overline{X}_2)^2\right]\end{aligned} \tag{5.143}$$

The first and last terms in the preceding equation are a_1^2 var$[X_1]$ and a_2^2 var$[X_2]$, respectively. The middle term is zero, since

$$\begin{aligned}E\left[(X_1 - \overline{X}_1)(X_2 - \overline{X}_2)\right] &= \int_{-\infty}^{\infty}\int_{-\infty}^{\infty} (x_1 - \overline{X}_1)(x_2 - \overline{X}_2) f_{X_1}(x_1) f_{X_2}(x_2)\, dx_1\, dx_2 \\ &= \int_{-\infty}^{\infty} (x_1 - \overline{X}_1) f_{X_1}(x_1)\, dx_1 \int_{-\infty}^{\infty} (x_2 - \overline{X}_2) f_{X_2}(x_2)\, dx_2 \\ &= (\overline{X}_1 - \overline{X}_1)(\overline{X}_2 - \overline{X}_2) = 0\end{aligned} \tag{5.144}$$

Note that the assumption of *statistical independence* was used to show that the middle term above is zero (it is a sufficient, but not necessary, condition).

5.3.8 Another Special Average: The Characteristic Function

Letting $g(X) = e^{jvX}$ in (5.112), we obtain an average known as the *characteristic function* of X, or $M_X(jv)$, defined as

$$M_X(jv) \triangleq E\left[e^{jvX}\right] = \int_{-\infty}^{\infty} f_X(x) e^{jvx}\, dx \tag{5.145}$$

It is seen that $M_X(jv)$ would be the *Fourier transform* of $f_X(x)$, as we have defined the Fourier transform in Chapter 2, provided a minus sign were used in the exponent instead of a plus sign. Thus, if $j\omega$ is replaced by $-jv$ in Fourier transform tables, they can be used to obtain characteristic functions from pdfs (sometimes it is convenient to use the variable s in place of jv; the resulting function is called the *moment generating function)*.

A pdf is obtained from the corresponding characteristic function by the inverse transform relationship

$$f_X(x) = \frac{1}{2\pi}\int_{-\infty}^{\infty} M_X(jv) e^{-jvx}\, dv \tag{5.146}$$

This illustrates one possible use of the characteristic function. It is sometimes easier to obtain the characteristic function than the pdf, and the latter is then obtained by inverse Fourier transformation, either analytically or numerically.

Another use for the characteristic function is to obtain the moments of a random variable. Consider the differentiation of (5.145) with respect to v. This gives

$$\frac{\partial M_X(jv)}{\partial v} = j\int_{-\infty}^{\infty} x f_X(x) e^{jvx}\, dx \tag{5.147}$$

Setting $v = 0$ after differentiation and dividing by j, we obtain

$$E[X] = (-j)\left.\frac{\partial M_X(jv)}{\partial v}\right|_{v=0} \tag{5.148}$$

For the nth moment, the relation

$$E[X^n] = (-j)^n \left.\frac{\partial^n M_X(jv)}{\partial v^n}\right|_{v=0} \tag{5.149}$$

can be proved by repeated differentiation.

EXAMPLE 5.21

By use a table of Fourier transforms, the one-sided exponential pdf

$$f_X(x) = \exp(-x)u(x) \tag{5.150}$$

is found to have the characteristic function

$$M_X(jv) = \int_0^\infty e^{-x} e^{jvx}\, dx = \frac{1}{1-jv} \tag{5.151}$$

By repeated differentiation or expansion of the characteristic function in a power series in jv, it follows from (5.149) that $E[X^n] = n!$ for this random variable. ■

5.3.9 The pdf of the Sum of Two Independent Random Variables

Given two *statistically independent* random variables X and Y with known pdfs $f_X(x)$ and $f_Y(y)$, respectively, the pdf of their sum $Z = X + Y$ is often of interest. The characteristic function will be used to find the pdf of Z, or $f_Z(z)$, even though we could find the pdf of Z directly.

From the definition of the characteristic function of Z, we write

$$\begin{aligned} M_Z(jv) &= E\left[e^{jvZ}\right] = E\left[e^{jv(X+Y)}\right] \\ &= \int_{-\infty}^{\infty}\int_{-\infty}^{\infty} e^{jv(x+y)} f_X(x) f_Y(y)\, dx\, dy \end{aligned} \tag{5.152}$$

since the joint pdf of X and Y is $f_X(x)f_Y(y)$ by the assumption of statistical independence of X and Y. We can write (5.152) as the product of two integrals since $e^{jv(x+y)} = e^{jvx}e^{jvy}$. This results in

$$\begin{aligned} M_Z(jv) &= \int_{-\infty}^{\infty} f_X(x) e^{jvx}\, dx \int_{-\infty}^{\infty} f_Y(y) e^{jvy}\, dy \\ &= E\left[e^{jvX}\right]E\left[e^{jvY}\right] \end{aligned} \tag{1.153}$$

From the definition of the characteristic function, given by (5.145), we see that

$$M_Z(jv) = M_X(jv)\, M_Y(jv) \tag{5.154}$$

where $M_X(jv)$ and $M_Y(jv)$ are the characteristic functions of X and Y, respectively. Remembering that the characteristic function is the Fourier transform of the corresponding pdf and that a product in the frequency domain corresponds to convolution in the time domain, it follows that

$$f_Z(z) = f_X(x) * f_Y(y) = \int_{-\infty}^{\infty} f_X(z-u) f_Y(u)\, du \tag{5.155}$$

This result generalizes to more than two random variables. The following example illustrates the use of (5.155).

EXAMPLE 5.22

Consider the sum of four identically distributed, independent random variables,

$$Z = X_1 + X_2 + X_3 + X_4 \tag{5.156}$$

where the pdf of each X_i is

$$f_{X_i}(x_i) = \Pi(x_i) = \begin{cases} 1, & |x_i| \leq \frac{1}{2} \\ 0, & \text{otherwise}, i = 1, 2, 3, 4 \end{cases} \tag{5.157}$$

where $\Pi(x_i)$ is the unit rectangular pulse function defined in Chapter 2. We find $f_Z(z)$ by applying (5.155) twice. Thus, let

$$Z_1 = X_1 + X_2 \quad \text{and} \quad Z_2 = X_3 + X_4 \tag{5.158}$$

The pdfs of Z_1 and Z_2 are identical, both being the convolution of a uniform density with itself. From Table 2.2, we can immediately write down the following result:

$$f_{Z_i}(z_i) = \Lambda(z_i) = \begin{cases} 1 - |z_i|, & |z_i| \leq 1 \\ 0, & \text{otherwise} \end{cases} \tag{5.159}$$

where $f_{Z_i}(z_i)$ is the pdf of $Z_i, i = 1, 2$. To find $f_Z(z)$, we simply convolve $f_{Z_i}(z_i)$ with itself. Thus

$$f_Z(z) = \int_{-\infty}^{\infty} f_{Z_i}(z-u) f_{Z_i}(u)\, du \tag{5.160}$$

The factors in the integrand are sketched in Figure 5.16(a). Clearly, $f_Z(z) = 0$ for $z < 2$ or $z > 2$. Since $f_{Z_i}(z_i)$ is even, $f_Z(z)$ is also even. Thus we need not consider $f_Z(z)$ for $z < 0$. From Figure 5.16(a) it follows that for $1 \leq z \leq 2$,

$$f_Z(z) = \int_{z-1}^{1} (1-u)(1+u-z)\, du = \frac{1}{6}(2-z)^3 \tag{5.161}$$

and for $0 \leq z \leq 1$, we obtain

$$\begin{aligned} f_Z(z) &= \int_{z-1}^{0} (1+u)(1+u-z)\, du + \int_{0}^{z} (1-u)(1+u-z)\, du \\ &\quad + \int_{z}^{1} (1-u)(1-u+z)\, du \\ &= (1-z) - \frac{1}{3}(1-z)^3 + \frac{1}{6}z^3 \end{aligned} \tag{5.162}$$

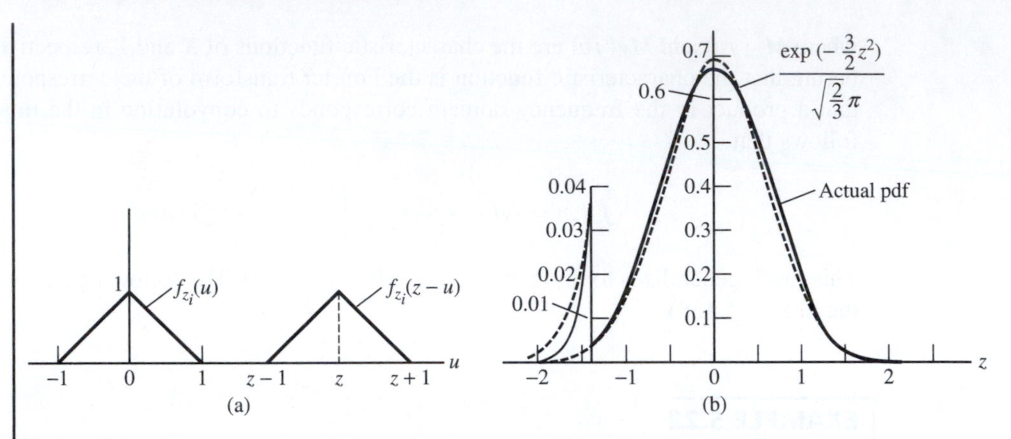

Figure 5.16
The pdf for the sum of four independent uniformly distributed random variables. (a) Convolution of two triangular pdfs. (b) Comparison of actual and Gaussian pdfs.

A graph of $f_Z(z)$ is shown in Figure 5.16(b) along with the graph of the function

$$\frac{\exp\left(-\frac{3}{2}z^2\right)}{\sqrt{\frac{2}{3}\pi}} \tag{5.163}$$

which represents a marginal Gaussian pdf of mean 0 and variance $\frac{1}{3}$, the same variance as $Z = X_1 + X_2 + X_3 + X_4$ [the results of Example 5.20 and (5.142) can be used to obtain the variance of Z]. We will describe the Gaussian pdf more fully later.

The reason for the striking similarity of the two pdfs shown in Figure 5.16(b) will become apparent when the central-limit theorem is discussed in Section 5.4.5.

■

5.3.10 Covariance and the Correlation Coefficient

Two useful joint averages of a pair of random variables X and Y are their covariance μ_{XY}, defined as

$$\mu_{XY} = E\left[\left(X - \overline{X}\right)\left(Y - \overline{Y}\right)\right] = E[XY] - E[X]E[Y] \tag{5.164}$$

and their correlation coefficient ρ_{XY}, which is written in terms of the covariance as

$$\rho_{XY} = \frac{\mu_{XY}}{\sigma_X \sigma_Y} \tag{5.165}$$

From the preceding two expressions we have the relationship

$$E[XY] = \sigma_X \sigma_Y \rho_{XY} + E[X]E[Y] \tag{5.166}$$

Both μ_{XY} and ρ_{XY} are measures of the interdependence of X and Y. The correlation coefficient is more convenient because it is normalized such that $-1 \le \rho_{XY} \le 1$. If $\rho_{XY} = 0$, X and Y are said to be *uncorrelated*. (Note that this does not imply statistical independence.)

It is easily shown that $\rho_{XY} = 0$ for statistically independent random variables. If X and Y are independent, their joint pdf $f_{XY}(x, y)$ is the product of the respective marginal pdfs; that is,

$f_{XY}(x,y) = f_X(x) f_Y(y)$. Thus

$$
\begin{aligned}
\mu_{XY} &= \int_{-\infty}^{\infty}\int_{-\infty}^{\infty} (x-\overline{X})(y-\overline{Y}) f_X(x) f_Y(y)\, dx\, dy \\
&= \int_{-\infty}^{\infty} (x-\overline{X}) f_X(x)\, dx \int_{-\infty}^{\infty} (y-\overline{Y}) f_Y(y)\, dy \\
&= (\overline{X}-\overline{X})(\overline{Y}-\overline{Y}) = 0
\end{aligned}
\tag{5.167}
$$

Considering next the cases $X = \pm\alpha Y$, so that $\overline{X} = \pm\alpha\overline{Y}$, where α is a positive constant, we obtain

$$
\begin{aligned}
\mu_{XY} &= \int_{-\infty}^{\infty}\int_{-\infty}^{\infty} (\pm\alpha y \mp \alpha\overline{Y})(y-\overline{Y}) f_{XY}(x,y)\, dx\, dy \\
&= \pm\alpha \int_{-\infty}^{\infty}\int_{-\infty}^{\infty} (y-\overline{Y})^2 f_{XY}(x,y)\, dx\, dy \\
&= \pm\alpha\sigma_Y^2
\end{aligned}
\tag{5.168}
$$

Using (5.142) with $N = 1$, we can write the variance of X as $\sigma_X^2 = \alpha^2\sigma_Y^2$. Thus the correlation coefficient is

$$
\rho_{XY} = +1 \text{ for } X = +\alpha Y \quad \text{and} \quad \rho_{XY} = -1 \text{ for } X = -\alpha Y
$$

To summarize, the correlation coefficient of two independent random variables is zero. When two random variables are linearly related, their correlation is $+1$ or -1 depending on whether one is a positive or a negative constant times the other.

■ 5.4 SOME USEFUL pdfs

We have already considered several often used probability distributions in the examples.[4] These have included the Rayleigh pdf (Example 5.15), the pdf of a sine wave of random phase (Example 5.17), and the uniform pdf (Example 5.20). Some others, which will be useful in our future considerations, are given below.

5.4.1 Binomial Distribution

One of the most common discrete distributions in the application of probability to systems analysis is the binomial distribution. We consider a chance experiment with two mutually exclusive, exhaustive outcomes A and $\overline{A}$, where $\overline{A}$ denotes the compliment of A, with probabilities $P(A) = p$ and $P(\overline{A}) = q = 1 - p$, respectively. Assigning the discrete random variable K to be numerically equal to the number of times event A occurs in n trials of our chance experiment, we seek the probability that exactly $k \leq n$ occurrences of the event A occur in n repetitions of the experiment. (Thus our actual chance experiment is the replication of the basic experiment n times.) The resulting distribution is called the *binomial distribution*.

Specific examples in which the binomial distribution is the result are the following: In n tosses of a coin, what is the probability of $k \leq n$ heads? In the transmission of n messages

[4]Useful probability distributions are summarized in Table 5.5 at the end of this chapter.

through a channel, what is the probability of $k \leq n$ errors? Note that in all cases we are interested in exactly k occurrences of the event, not, for example, at least k of them, although we may find the latter probability if we have the former.

Although the problem being considered is very general, we solve it by visualizing the coin-tossing experiment. We wish to obtain the probability of k heads in n tosses of the coin if the probability of a head on a single toss is p and the probability of a tail is $1 - p = q$. One of the possible sequences of k heads in n tosses is

$$\underbrace{HH\ldots H}_{k \text{ heads}}\underbrace{T\ldots T}_{n-k \text{ tails}}$$

Since the trials are independent, the probability of this particular sequence is

$$\underbrace{p \cdot p \cdot p \ldots p}_{k \text{ factors}} \cdot \underbrace{q \cdot q \cdot q \ldots q}_{n-k \text{ factors}} = p^k q^{n-k} \tag{5.169}$$

The preceding sequence of k heads in n trials is only one of

$$\binom{n}{k} \triangleq \frac{n!}{k!(n-k)!} \tag{5.170}$$

possible sequences, where $\binom{n}{k}$ is the binomial coefficient. To see this, we consider the number of ways k *identifiable* heads can be arranged in n slots. The first head can fall in any of the n slots, the second in any of $n - 1$ slots (the first head already occupies one slot), the third in any of $n - 2$ slots, and so on for a total of

$$n(n-1)(n-2)\ldots(n-k+1) = \frac{n!}{(n-k)!} \tag{5.171}$$

possible arrangements in which each head is identified. However, we are not concerned about which head occupies which slot. For each possible identifiable arrangement, there are $k!$ arrangements for which the heads can be switched with the same slots occupied. Thus the total number of arrangements, if we do not identify the particular head occupying each slot, is

$$\frac{n(n-1)\ldots(n-k+1)}{k!} = \frac{n!}{k!(n-k)!} = \binom{n}{k} \tag{5.172}$$

Since the occurrence of any of these $\binom{n}{k}$ possible arrangements precludes the occurrence of any other [that is, the $\binom{n}{k}$ outcomes of the experiment are mutually exclusive], and since each occurs with probability $p^k q^{n-k}$, the probability of *exactly* k heads in n trials in *any* order is

$$P(K = k) \triangleq P_n(k) = \binom{n}{k} p^k q^{n-k}, \quad k = 0, 1, \ldots, n \tag{5.173}$$

Equation (5.173), known as the *binomial probability distribution* (note that it is not a pdf or a cdf but rather a *probability distribution*), is plotted in Figure 5.17(a) to (e) for six different values of p and n.

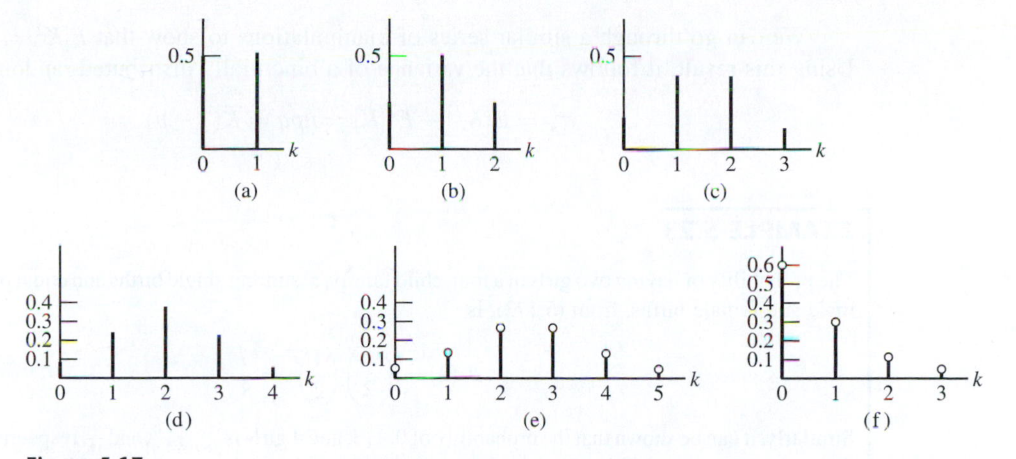

Figure 5.17
The binomial distribution with comparison to Laplace and Poisson approximations. (a) $n = 1, p = 0.5$. (b) $n = 2$, $p = 0.5$.(c) $n = 3, p = 0.5$. (d) $n = 4, p = 0.5$. (e) $n = 5, p = 0.5$. Circles are Laplace approximations. (f) $n = 5$, $p = \frac{1}{10}$. Circles are Poisson approximations.

The mean of a binomially distributed random variable K, by (5.109), is given by

$$E[K] = \sum_{k=0}^{n} k \frac{n!}{k!(n-k)!} p^k q^{n-k} \tag{5.174}$$

Noting that the sum can be started at $k = 1$ since the first term is zero, we can write

$$E[K] = \sum_{k=1}^{n} \frac{n!}{(k-1)!(n-k)!} p^k q^{n-k} \tag{5.175}$$

where the relation $k! = k(k-1)!$ has been used. Letting $m = k - 1$, we get the sum

$$\begin{aligned} E[K] &= \sum_{m=0}^{n-1} \frac{n!}{m!(n-m-1)!} p^{m+1} q^{n-m-1} \\ &= np \sum_{m=0}^{n-1} \frac{(n-1)!}{m!(n-m-1)!} p^m q^{n-m-1} \end{aligned} \tag{5.176}$$

Finally, letting $l = n - 1$ and recalling that, by the binomial theorem,

$$(x+y)^l = \sum_{m=0}^{l} \binom{l}{m} x^m y^{l-m} \tag{5.177}$$

we obtain

$$\overline{K} = E[K] = np(p+q)^l = np \tag{5.178}$$

since $p + q = 1$. The result is reasonable; in a long sequence of n tosses of a fair coin ($p = q = \frac{1}{2}$), we would expect about $np = \frac{1}{2}n$ heads.

We can go through a similar series of manipulations to show that $E[K^2] = np(np+q)$. Using this result, it follows that the variance of a binomially distributed random variable is

$$\sigma_K^2 = E[K^2] - E^2[K] = npq = \overline{K}(1-p) \tag{5.179}$$

EXAMPLE 5.23

The probability of having two girls in a four-child family, assuming single births and equal probabilities of male and female births, from (5.173), is

$$P_4(2) = \binom{4}{2}\left(\frac{1}{2}\right)^4 = \frac{3}{8} \tag{5.180}$$

Similarly, it can be shown that the probability of 0, 1, 3, and 4 girls is $\frac{1}{16}$, $\frac{1}{4}$, $\frac{1}{4}$, and $\frac{1}{16}$, respectively. Note that the sum of the probabilities for 0, 1, 2, 3, and 4 girls (or boys) is 1, as it should be. ■

5.4.2 Laplace Approximation to the Binomial Distribution

When n becomes large, computations using (5.173) become unmanageable. In the limit as $n \to \infty$, it can be shown that for $|k - np| \leq \sqrt{npq}$

$$P_n(k) \cong \frac{1}{\sqrt{2\pi npq}} \exp\left(-\frac{(k-np)^2}{2npq}\right) \tag{5.181}$$

which is called the Laplace approximation to the binomial distribution. A comparison of the Laplace approximation with the actual binomial distribution is given in Figure 5.17(e)

5.4.3 Poisson Distribution and Poisson Approximation to the Binomial Distribution

Consider a chance experiment in which an event whose probability of occurrence in a very small time interval ΔT is $P = \alpha \Delta T$, where α is a constant of proportionality. If successive occurrences are statistically independent, then the probability of k events in time T is

$$P_T(k) = \frac{(\alpha T)^k}{k!} e^{-\alpha T} \tag{5.182}$$

For example, the emission of electrons from a hot metal surface obeys this law, which is called the *Poisson distribution.*

The Poisson distribution can be used to approximate the binomial distribution when the number of trials n is large, the probability of each event p is small, and the product $np \cong npq$. The approximation is

$$P_n(k) \cong \frac{(\overline{K})^k}{k!} e^{-\overline{K}} \tag{5.183}$$

where, as calculated previously, $\overline{K} = E[K] = np$ and $\sigma_k^2 = E[K]q = npq \cong E[K]$ for $q = 1 - p \cong 1$. This approximation is compared with the binomial distribution in Figure 5.17(f)

EXAMPLE 5.24

The probability of error on a single transmission in a digital communication system is $P_E = 10^{-4}$. What is the probability of more than three errors in 1000 transmissions?

Solution

We find the probability of three errors or less from (5.183):

$$P(K \leq 3) = \sum_{k=0}^{3} \frac{(\overline{K})^k}{k!} e^{-\overline{K}} \tag{5.184}$$

where $\overline{K} = (10^{-4})(1000) = 0.1$. Hence

$$P(K \leq 3) = e^{-0.1}\left(\frac{(0.1)^0}{0!} + \frac{(0.1)^1}{1!} + \frac{(0.1)^2}{2!} + \frac{(0.1)^3}{3!}\right) \cong 0.999996 \tag{5.185}$$

Therefore, $P(K > 3) = 1 - P(K \leq 3) \cong 4 \times 10^{-6}$. ■

COMPUTER EXAMPLE 5.1

The MATLAB program given below does a Monte Carlo simulation of the digital communication system described in the above example.

```
% file: c5ce1
% Simulation of errors in a digital communication system
%
N_sim = input('Enter number of trials ');
N = input('Bit block size for simulation ');
N_errors = input('Simulate the probability of more than __ errors
occurring ');
PE = input('Error probability on each bit ');
count = 0;
for n = 1:N_sim
   U = rand(1, N);
   Error = (-sign(U-PE) + 1)/2;    % Error array - elements are 1 where
                                    errors occur
   if sum(Error) > N_errors
      count = count + 1;
   end
end
P_greater = count/N_sim
```

A typical run follows. To cut down on the simulation time, blocks of 1000 bits are simulated with a probability of error on each bit of 10^{-3}. Note that the Poisson approximation does not hold in this case because $\overline{K} = (10^{-3})(1000) = 1$ is not much less than 1. Thus, to check the results analytically, we must use the binomial distribution. Calculation gives $P(0\ \text{errors}) = 0.3677$, $P(1\ \text{error}) = 0.3681$, $P(2\ \text{errors}) = 0.1840$, and $P(3\ \text{errors}) = 0.0613$ so that $P(> 3\ \text{errors}) = 1 - 0.3677 - 0.3681 - 0.1840 - 0.0613 = 0.0189$. This matches with the simulated result if both are rounded to two decimal places.

```
error_sim
Enter number of trials 10000
```

```
Bit block size for simulation 1000
Simulate the probability of more than __ errors occurring 3
Error probability on each bit .001
P_greater = 0.0199
```

■

5.4.4 Geometric Distribution

Suppose we are interested in the probability of the first head in a series of coin tossings or the first error in a long string of digital signal transmissions occurring on the kth trial. The distribution describing such experiments is called the *geometric distribution* and is

$$P(k) = pq^{k-1},\ 1 \le k < \infty \tag{5.186}$$

where p is the probability of the event of interest occurring (i.e., head, error, etc.) and q is the probability of it not occurring.

EXAMPLE 5.25

The probability of the first error occurring at the 1000th transmission in a digital data transmission system where the probability of error is $p = 10^{-6}$ is

$$P(1000) = 10^{-6}\left(1 - 10^{-6}\right)^{999} = 9.99 \times 10^{-7} \cong 10^{-6}$$

■

5.4.5 Gaussian Distribution

In our future considerations, the Gaussian pdf will be used repeatedly. There are at least two reasons for this. One is that the assumption of Gaussian statistics for random phenomena often makes an intractable problem tractable. The other, more fundamental reason, is that because of a remarkable phenomenon summarized by a theorem called the *central-limit theorem*, many naturally occurring random quantities, such as noise or measurement errors, are Gaussianly distributed. The following is a statement of the central-limit theorem.

Theorem 5.1. The Central-Limit Theorem

Let $X_1, X_2, \ldots$ be independent, identically distributed random variables, each with finite mean m and finite variance σ^2. Let Z_n be a sequence of unit-variance, zero-mean random variables, defined as

$$Z_n \triangleq \frac{\sum_{i=1}^{n} X_i - nm}{\sigma\sqrt{n}} \tag{5.187}$$

Then

$$\lim_{n\to\infty} P(Z_n \le z) = \int_{-\infty}^{z} \frac{e^{-t^2/2}}{\sqrt{2\pi}}\, dt \tag{5.188}$$

In other words, the cdf of the normalized sum (5.187) approaches a Gaussian cdf, no matter what the distribution of the component random variables. The only restriction is that they be independent and that their means and variances be finite. In some cases the independence assumption can be relaxed. It is important, however, that no one of the component random variables or a finite combination of them dominate the sum.

We will not prove the central-limit theorem or use it in later work. We state it here simply to give partial justification for our almost exclusive assumption of Gaussian statistics from now on. For example, electrical noise is often the result of a superposition of voltages due to a large number of charge carriers. Turbulent boundary-layer pressure fluctuations on an aircraft skin are the superposition of minute pressures due to numerous eddies. Random errors in experimental measurements are due to many irregular fluctuating causes. In all these cases, the Gaussian approximation for the fluctuating quantity is useful and valid. Example 5.22 illustrates that surprisingly few terms in the sum are required to give a Gaussian-appearing pdf, even where the component pdfs are far from Gaussian.

The generalization of the joint Gaussian pdf first introduced in Example 5.15 is

$$f_{XY}(x,y)=\frac{1}{2\pi\sigma_x\sigma_y\sqrt{1-\rho^2}} \times\exp\left(-\frac{[(x-m_x)/\sigma_x]^2-2\rho[(x-m_x)/\sigma_x][(y-m_y)/\sigma_y]+[(y-m_y)/\sigma_y]^2}{2(1-\rho^2)}\right) \tag{5.189}$$

where, through straightforward but tedious integrations, it can be shown that

$$m_x=E[X] \quad \text{and} \quad m_y=E[Y] \tag{5.190}$$

$$\sigma_x^2=\text{var}[X] \tag{5.191}$$

$$\sigma_y^2=\text{var}[Y] \tag{5.192}$$

and

$$\rho=\frac{E\left[(X-m_x)(Y-m_y)\right]}{\sigma_x\sigma_y} \tag{5.193}$$

The joint pdf for $N>2$ Gaussian random variables may be written in a compact fashion through the use of matrix notation. The general form is given in Appendix B.

Figure 5.18 illustrates the bivariate Gaussian density function, and the associated contour plots, as the five parameters m_x, m_y, σ_x^2, σ_y^2, and ρ are varied. The contour plots provide information on the shape and orientation of the pdf that is not always apparent in a three-dimensional illustration of the pdf from a single viewing point. Figure 5.18(a) illustrates the bivariate Gaussian pdf for which X and Y are zero mean, unit variance, and uncorrelated. Since the variances of X and Y are equal and since X and Y are uncorrelated, the contour plots are circles in the XY plane. We can see why two-dimensional Gaussian noise, in which the two components have equal variance and are uncorrelated, is said to exhibit circular symmetry. Figure 5.18(b) shows the case in which X and Y are uncorrelated but $m_x = 1$, $m_y = -2$, $\sigma_x^2 = 2$, and $\sigma_y^2 = 1$. The means are clear from observation of the contour plot. In addition the spread of the pdf is greater in the X direction than in the Y direction because $\sigma_x^2 > \sigma_y^2$. In Figure 5.18(c) the means of X and Y are both zero, but the correllation coefficient is set equal to 0.9. We see that the contour lines denoting a constant value of the density function are

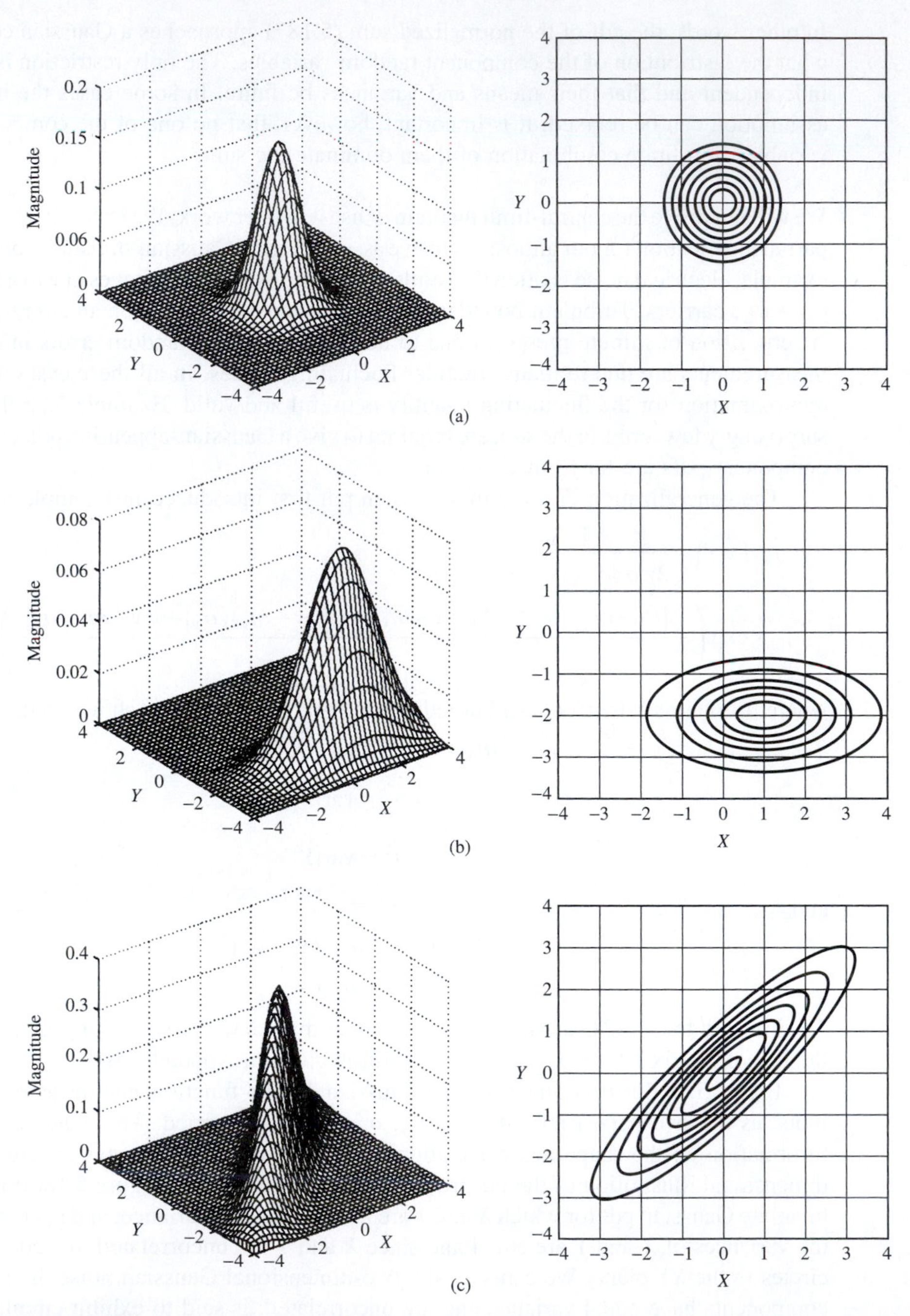

Figure 5.18

Bivariate Gaussian pdfs and corresponding contour plots. (a) $m_x = 0$, $m_y = 0$, $\sigma_x^2 = 1$, $\sigma_y^2 = 1$ and $\rho = 0$. (b) $m_x = 1$, $m_y = -2$, $\sigma_x^2 = 2$, $\sigma_y^2 = 1$, and $\rho = 0$.(c) $m_x = 0$, $m_y = 0$, $\sigma_x^2 = 1$, $\sigma_y^2 = 1$, and $\rho = 0.9$.

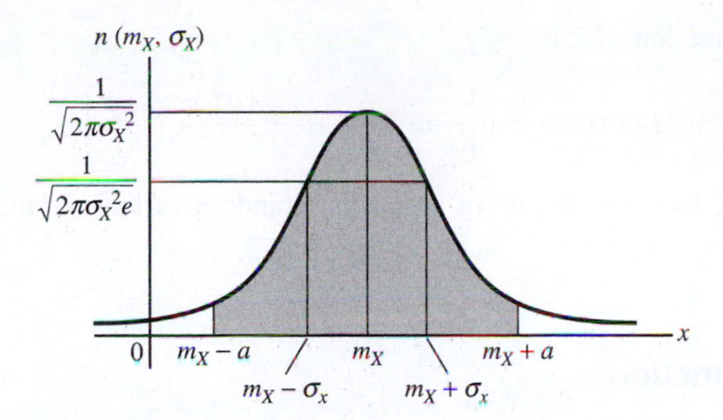

Figure 5.19
The Gaussian pdf with mean m_x and variance σ_x^2.

symmetrical about the line $X = Y$ in the XY plane. This results, of course, because the correlation coefficient is a measure of the linear relationship between X and Y. In addition, note that the pdfs described in Figures 5.18(a) and (b) can be factored into the product of two marginal pdfs since, for these two cases, X and Y are uncorrelated.

The marginal pdf for X (or Y) can be obtained by integrating (5.189) over y (or x). Again, the integration is tedious. The marginal pdf for X is

$$n(m_x, \sigma_x) = \frac{1}{\sqrt{2\pi\sigma_x^2}} \exp\left(\frac{-(x - m_x)^2}{2\sigma_x^2}\right) \tag{5.194}$$

where the notation $n(m_x, \sigma_x)$ has been introduced to denote a Gaussian pdf of mean m_x and standard deviation σ_x. A similar expression holds for the pdf of Y with appropriate parameter changes. This function is shown in Figure 5.19.

We will sometimes assume in the discussions to follow that $m_x = m_y = 0$ in (5.189) and (5.194), for if they are not zero, we can consider new random variables X' and Y' defined as $X' = X - m_x$ and $Y' = Y - m_y$ that do have zero means. Thus no generality is lost in assuming zero means.

For $\rho = 0$, that is, X and Y uncorrelated, the cross term in the exponent of (5.189) is zero, and $f_{XY}(x, y)$, with $m_x = m_y = 0$, can be written as

$$f_{XY}(x, y) = \frac{\exp(-x^2/2\sigma_x^2)}{\sqrt{2\pi\sigma_x^2}} \frac{\exp(-y^2/2\sigma_y^2)}{\sqrt{2\pi\sigma_y^2}} = f_X(x) f_Y(y) \tag{5.195}$$

Thus *uncorrelated Gaussian random variables are also statistically independent*. We emphasize that this does not hold for all pdfs, however.

It can be shown that the sum of any number of Gaussian random variables, independent or not, is Gaussian. The sum of two independent Gaussian random variables is easily shown to be Gaussian. Let $Z = X_1 + X_2$, where the pdf of X_i is $n(m_i, \sigma_i)$. Using a table of Fourier transforms or completing the square and integrating, we find that the characteristic function of X_i is

$$\begin{aligned} M_{X_i}(jv) &= \int_{-\infty}^{\infty} \left(2\pi\sigma_i^2\right)^{-1/2} \exp\left(\frac{-(x_i - m_i)^2}{2\sigma_i^2}\right) \exp(jvx_i)\, dx_i \\ &= \exp\left(jm_i v - \frac{\sigma_i^2 v^2}{2}\right) \end{aligned} \tag{5.196}$$

Thus the characteristic function of Z is

$$M_Z(jv) = M_{X_1}(v)M_{X_2}(jv) = \exp\left[j(m_1+m_2)v - \frac{(\sigma_1^2+\sigma_2^2)v^2}{2}\right] \tag{5.197}$$

which is the characteristic function (5.196) of a Gaussian random variable of mean m_1+m_2 and variance $\sigma_1^2+\sigma_2^2$.

5.4.6 Gaussian Q-Function

As Figure 5.19 shows, $n(m_x, \sigma_x)$ describes a continuous random variable that may take on any value in $(-\infty, \infty)$ but is most likely to be found near $X = m_x$. The even symmetry of $n(m_x, \sigma_x)$ about $x = m_x$ leads to the conclusion that $P(X \le m_x) = P(X \ge m_x) = \frac{1}{2}$.

Suppose we wish to find the probability that X lies in the interval $[m_x - a,\ m_x + a]$. Using (5.42), we can write this probability as

$$P(m_x - a \le X \le m_x + a) = \int_{m_x-a}^{m_x+a} \frac{\exp\left[-(x-m_x)^2/2\sigma_x^2\right]}{\sqrt{2\pi\sigma_x^2}}\,dx \tag{5.198}$$

which is the shaded area in Figure 5.19. With the change of variables $y = (x - m_x)/\sigma_x$, this gives

$$\begin{aligned} P(m_x - a \le X \le m_x + a) &= \int_{-a/\sigma_x}^{a/\sigma_x} \frac{e^{-y^2/2}}{\sqrt{2\pi}}\,dy \\ &= 2\int_0^{a/\sigma_x} \frac{e^{-y^2/2}}{\sqrt{2\pi}}\,dy \end{aligned} \tag{5.199}$$

where the last integral follows by virtue of the integrand being even. Unfortunately, this integral cannot be evaluated in closed form.

The Gaussian Q-function, or simply Q-function, is defined as[5]

$$Q(u) = \int_u^\infty \frac{e^{-y^2/2}}{\sqrt{2\pi}}\,dy \tag{5.200}$$

This function has been evaluated numerically, and rational and asymptotic approximations are available to evaluate it for moderate and large arguments, respectively.[6] Using this transcendental function definition, we may rewrite (5.199) as

$$\begin{aligned} P(m_x - a \le X \le m_x + a) &= 2\left(\frac{1}{2} - \int_{a/\sigma_x}^\infty \frac{e^{-y^2/2}}{\sqrt{2\pi}}\,dy\right) \\ &= 1 - 2Q\left(\frac{a}{\sigma_x}\right) \end{aligned} \tag{5.201}$$

---

[5]An integral representation with finite limits for the Q-function is $Q(x) = \frac{1}{\pi}\int_0^{\pi/2} \exp[-x^2/(2\sin^2\phi)]d\phi$.

[6]These are provided in M. Abramowitz and I. Stegun (eds.), *Handbook of Mathematical Functions with Formulas, Graphs, and Mathematical Tables*. National Bureau of Standards, Applied Mathematics Series No. 55, issued June 1964 (pp. 931ff); also New York: Dover, 1972.

A useful approximation for the Q-function for large arguments is

$$Q(u) \cong \frac{e^{-u^2/2}}{u\sqrt{2\pi}}, \quad u \gg 1 \tag{5.202}$$

Numerical comparison of (5.200) and (5.202) shows that less than a 6% error results for $u \geq 3$ in using this approximation. This, and other results for the Q-function, are given in Appendix G (See Section G.1).

Related integrals are the error function and the complementary error function, defined as

$$\begin{aligned} \operatorname{erf}(u) &= \frac{2}{\sqrt{\pi}} \int_0^u e^{-y^2}\, dy \\ \operatorname{erfc}(u) = 1 - \operatorname{erf}(u) &= \frac{2}{\sqrt{\pi}} \int_u^\infty e^{-y^2}\, dy \end{aligned} \tag{5.203}$$

respectively. The latter can be shown to be related to the Q-function by

$$Q(u) = \frac{1}{2}\operatorname{erfc}\left(\frac{u}{\sqrt{2}}\right) \text{ or } \operatorname{erfc}(v) = 2Q\left(\sqrt{2}v\right) \tag{5.204}$$

MATLAB includes function programs for erf and erfc, and the inverse error and complementary error functions, erfinv and erfcinv, respectively.

5.4.7 Chebyshev's Inequality

The difficulties encountered above in evaluating (5.198) and probabilities like it make an approximation to such probabilities desirable. Chebyshev's inequality gives us a lower bound, regardless of the specific form of the pdf involved, provided its second moment is finite. The probability of finding a random variable X within $\pm k$ standard deviations of its mean is at least $1 - 1/k^2$, according to Chebyshev's inequality. That is,

$$P(|X - m_x| \leq k\sigma_x) \geq 1 - \frac{1}{k^2},\ k > 0 \tag{5.205}$$

Considering $k = 3$, we obtain

$$P(|X - m_x| \leq 3\sigma_x) \geq \frac{8}{9} \cong 0.889 \tag{5.206}$$

Assuming X is Gaussian and using the Q-function, this probability can be computed to be 0.9973. In words, according to Chebyshev's inequality, the probability that a random variable deviates from its mean by more than ± 3 standard deviations is not greater than 0.111, regardless of its pdf. (There is the restriction that its second moment must be finite.) Note that the bound for this example is not very tight.

5.4.8 Collection of Probability Functions and Their Means and Variances

The probability functions (pdfs and probability distributions) discussed above are collected in Table 5.4 along with some additional functions that come up from time to time. Also given are the means and variances of the corresponding random variables.

Table 5.4 Probability Distributions of Some Random Variables with Means and Variances

Probability density or mass function	Mean	Variance
Uniform: $f_X(x) = \begin{cases} \frac{1}{b-a}, & a \le x \le b \\ 0, & \text{otherwise} \end{cases}$	$\frac{1}{2}(a+b)$	$\frac{1}{12}(b-a)^2$
Gaussian: $f_X(x) = \frac{1}{\sqrt{2\pi\sigma^2}}\exp\left(\frac{-(x-m)^2}{2\sigma^2}\right)$	m	σ^2
Rayleigh: $f_R(r) = \frac{r}{\sigma^2}\exp\left(\frac{-r^2}{2\sigma^2}\right), \quad r \ge 0$	$\sqrt{\frac{\pi}{2}}\sigma$	$\frac{1}{2}(4-\pi)\sigma^2$
Laplacian: $f_X(x) = \frac{\alpha}{2}\exp(-\alpha\lvert x\rvert), \quad \alpha > 0$	0	$2/\alpha^2$
One-sided exponential: $f_X(x) = \alpha\exp(-\alpha x)\,u(x)$	$1/\alpha$	$1/\alpha^2$
Hyperbolic: $f_X(x) = \frac{(m-1)h^{m-1}}{2(\lvert x\rvert+h)^m}, \quad m > 3, h > 0$	0	$\frac{2h^2}{(m-3)(m-2)}$
Nakagami-m: $f_X(x) = \frac{2m^m}{\Gamma(m)}x^{2m-1}\exp(-mx^2), \quad x \ge 0$	$\frac{1 \times 3 \times \ldots \times (2m-1)}{2^m\Gamma(m)}$	$\frac{\Gamma(m+1)}{\Gamma(m)\sqrt{m}}$
Central chi-square (n = degrees of freedom):* $f_X(x) = \frac{x^{n/2-1}}{\sigma^n 2^{n/2}\Gamma(n/2)}\exp\left(\frac{-x}{2\sigma^2}\right)$	$n\sigma^2$	$2n\sigma^4$
Lognormal:† $f_X(x) = \frac{1}{x\sqrt{2\pi\sigma_y^2}}\exp\left(\frac{-(\ln x - m_y)^2}{2\sigma_y^2}\right)$	$\exp\left(m_y + 2\sigma_y^2\right)$	$\exp(2m_y + \sigma_y^2)$ $\times[\exp\sigma_y^2 - 1]$
Binomial: $P_n(k) = \binom{n}{k}p^k q^{n-k}, \quad k = 0, 1, 2, \ldots, n; \quad p+q=1$	np	npq
Poisson: $P(k) = \frac{\lambda^k}{k!}\exp(-\lambda), \quad k = 0, 1, 2, \ldots$	λ	λ
Geometric: $P(k) = pq^{k-1}, \quad k = 1, 2, \ldots$	$1/p$	q/p^2

*$\Gamma(m)$ is the gamma function and equals $(m-1)!$ for m an integer.

†The lognormal random variable results from the transformation $Y = \ln X$, where Y is a Gaussian random variable with mean m_y and variance σ_y^2.

Summary

1. The objective of probability theory is to attach real numbers between 0 and 1, called *probabilities*, to the *outcomes* of chance experiments—that is, experiments in which the outcomes are not uniquely determined by the causes but depend on chance—and to interrelate probabilities of events, which are defined to be combinations of outcomes.
2. Two events are *mutually exclusive* if the occurrence of one of them precludes the occurrence of the other. A set of events is said to be *exhaustive* if one of them must occur in the performance of a chance experiment. The null event happens with probability zero, and the certain event happens with probability one in the performance of a chance experiment.
3. The equally likely definition of the probability $P(A)$ of an event A states that if a chance experiment can result in a number N of mutually exclusive, equally likely outcomes, then $P(A)$ is the ratio of the number of outcomes favorable to A, or N_A, to the total number. It is a circular definition in that

probability is used to define probability, but it is nevertheless useful in many situations such as drawing cards from well-shuffled decks.

4. The relative-frequency definition of the probability of an event A assumes that the chance experiment is replicated a large number of times N and

$$P(A) = \lim_{N \to \infty} \frac{N_A}{N}$$

where N_A is the number of replications resulting in the occurrence of A.

5. The axiomatic approach defines the probability $P(A)$ of an event A as a real number satisfying the following axioms:
   a. $P(A) \geq 0$
   b. $P(\text{certain event}) = 1.$
   c. If A and B are mutually exclusive events, $P(A \cup B) = P(A) + P(B)$.
   The axiomatic approach encompasses the equally likely and relative-frequency definitions.

6. Given two events A and B, the compound event "A or B or both" is denoted as $A \cup B$, the compound event "both A and B" is denoted as $(A \cap B)$ or as AB, and the event "not A" is denoted as $\overline{A}$. If A and B are not necessarily mutually exclusive, the axioms of probability may be used to show that $P(A \cup B) = P(A) + P(B) - P(A \cap B)$. Letting $P(A|B)$ denote the probability of A occurring given that B occurred and $P(B|A)$ denote the probability of B given A, these probabilities are defined, respectively, as

$$P(A|B) = \frac{P(AB)}{P(B)} \quad \text{and} \quad P(B|A) = \frac{P(AB)}{P(A)}$$

A special case of Bayes' rule results by putting these two definitions together:

$$P(B|A) = \frac{P(A|B)P(B)}{P(A)}$$

Statistically independent events are events for which $P(AB) = P(A)P(B)$.

7. A random variable is a rule that assigns real numbers to the outcomes of a chance experiments. For example, in flipping a coin, assigning $X = +1$ to the occurrence of a head and $X = -1$ to the occurrence of a tail constitutes the assignment of a discrete-valued random variable.

8. The *cumulative distribution function* (cdf) $F_X(x)$ of a random variable X is defined as the probability that $X \leq x$, where x is a running variable. $F_X(x)$ lies between 0 and 1 with $F_X(-\infty) = 0$ and $F_X(\infty) = 1$, is continuous from the right, and is a nondecreasing function of its argument. Discrete random variables have step-discontinuous cdfs, and continuous random variables have continuous cdfs.

9. The *probability density function* (pdf) $f_X(X)$ of a random variable X is defined to be the derivative of the cdf. Thus

$$F_X(x) = \int_{-\infty}^{x} f_X(\eta)\, d\eta$$

The pdf is nonnegative and integrates over all x to unity. A useful interpretation of the pdf is that $f_X(x)\, dx$ is the probability of the random variable X lying in an infinitesimal range dx about x.

10. The *joint* cdf $F_{XY}(x, y)$ of two random variables X and Y is defined as the probability that $X \leq x$ and $Y \leq y$, where x and y are particular values of X and Y. Their joint pdf $f_{XY}(x, y)$ is the second partial derivative of the cdf first with respect to x and then with respect to y. The cdf of $X(Y)$ alone (that is, the marginal cdf) is found by setting y (x) to infinity in the argument of F_{XY}. The pdf of X (Y) alone (that is, the marginal pdf) is found by integrating f_{XY} over all y (x).

11. Two statistically independent random variables have joint cdfs and pdfs that factor into the respective marginal cdfs or pdfs.

12. The *conditional* pdf of X given Y is defined as

$$f_{X|Y}(x|y) = \frac{f_{XY}(x, y)}{f_Y(y)}$$

with a similar definition for $f_{Y|X}(y|x)$. The expression $f_{X|Y}(x|y)\, dx$ can be interpreted as the probability that $x - dx < X \leq x$ given $Y = y$.

13. Given $Y = g(X)$ where $g(X)$ is a monotonic function,

$$f_Y(y) = f_X(x) \left| \frac{dx}{dy} \right|_{x = g^{-1}(y)}$$

where $g^{-1}(y)$ is the inverse of $y = g(x)$. Joint pdfs of functions of more than one random variable can be transformed similarly.

14. Important probability functions defined in Chapter 5 are the Rayleigh pdf (5.105), the pdf of a random-phased sinusoid (Example 5.17), the uniform pdf [Example 5.20, (5.135)], the binomial probability distribution (5.174), the Laplace and Poisson approximations to the binomial distribution [(5.181) and (5.183), respectively] and the Gaussian pdf (5.189) and (5.194).

15. The statistical average, or expectation, of a function $g(X)$ of a random variable X with pdf $f_X(x)$ is defined as

$$E[g(X)] = \overline{g(X)} = \int_{-\infty}^{\infty} g(x) f_X(x)\, dx$$

The average of $g(X) = X^n$ is called the nth *moment of* X. The first moment is known as the *mean* of X. Averages of functions of more than one random variable are found by integrating the function times the joint pdf over the ranges of its arguments. The averages $\overline{g(X, Y)} = \overline{X^n Y^m} \triangleq E[X^n Y^m]$ are called the *joint moments* of the order $m + n$. The variance of a random variable X is the average $\overline{(X - \overline{X})^2} = \overline{X^2} - \overline{X}^2$.

16. The average $E[\sum a_i X_i]$ is $\sum a_i E[X_i]$; that is, the operations of summing and averaging can be interchanged. The variance of a sum of random variables is the sum of the respective variances *if the random variables are statistically independent*.
17. The characteristic function $M_X(jv)$ of a random variable X that has the pdf $f_X(x)$ is the expectation of $\exp(jvX)$ or, equivalently, the Fourier transform of $f_X(x)$ with a plus sign in the exponential of the Fourier transform integral. Thus the pdf is the inverse Fourier transform (with the sign in the exponent changed from plus to minus) of the characteristic function.
18. The nth moment of X can be found from $M_X(jv)$ by differentiating with respect to v for n times, multiplying by $(-j)^n$, and setting $v = 0$. The characteristic function of $Z = X + Y$, where X and Y are independent, is the product of the respective characteristic functions of X and Y. Thus, by the convolution theorem of Fourier transforms, the pdf of Z is the convolution of the pdfs of X and Y.
19. The covariance μ_{XY} of two random variables X and Y is the average

$$\mu_{XY} = E\left[\left(X - \overline{X}\right)\left(Y - \overline{Y}\right)\right]$$

The correlation coefficient ρ_{XY} is

$$\rho_{XY} = \frac{\mu_{XY}}{\sigma_X \sigma_Y}$$

Both give a measure of the linear interdependence of X and Y, but ρ_{XY} is handier because it is bounded by ± 1. If $\rho_{XY} = 0$, the random variables are said to be uncorrelated.
20. The central-limit theorem states that under suitable restrictions, the sum of a large number N of independent random variables with finite variances (not necessarily with the same pdfs) tends to a Gaussian pdf as N becomes large.
21. The Q-function can be used to compute probabilities of Gaussian random variables being in certain ranges. The Q-function is tabulated in Appendix G.1, and an asymptotic approximation is given for computing it. It can be related to the error function through (5.204).
22. Chebyshev's inequality gives the lower bound of the probability that a random variable is within k standard deviations of its mean as $1 - 1/k^2$, regardless of the pdf of the random variable (its second moment must be finite).
23. Table 5.4 summarizes a number of useful probability distributions with their means and variances.

Further Reading

Several books are available that deal with probability theory for engineers. Among these are Leon-Garcia (1994), Ross (2002), and Walpole, et al. (2007). A good overview with many examples is Ash (1992). Simon (2002) provides a compendium of relations involving the Gaussian distribution.

Problems

Section 5.1

5.1. A circle is divided into 21 equal parts. A pointer is spun until it stops on one of the parts, which are numbered from 1 through 21. Describe the sample space, and assuming equally likely outcomes, find

a. P(an even number)

b. P(the number 21)

c. P(the numbers 4, 5, or 9)

d. P(a number greater than 10)

5.2. If five cards are drawn without replacement from an ordinary deck of cards, what is the probability that

a. Three kings and two aces result.

b. Four of a kind result.

c. All are of the same suit.

d. An ace, king, queen, jack, and 10 of the same suit result.

e. Given that an ace, king, jack, and 10 have been drawn, what is the probability that the next card drawn will be a queen (not all of the same suit)?

5.3. What equations must be satisfied in order for three events A, B, and C to be independent?
(*Hint:* They must be independent by pairs, but this is not sufficient.)

5.4. Two events, A and B, have marginal probabilities $P(A) = 0.2$ and $P(B) = 0.5$, respectively. Their joint probability is $P(A \cap B) = 0.4$.

a. Are they statistically independent? Why or why not?

b. What is the probability of A or B or both occurring?

c. In general, what must be true for two events be both statistically independent and mutually exclusive?

5.5. Figure 5.20 is a graph that represents a communication network, where the nodes are receiver–repeater boxes and the edges (or links) represent communication channels which, if connected, convey the message perfectly. However, there is the probability p that a link will be broken and the probability $q = 1 - p$ that it will be whole. *Hint*: Use a tree diagram like Figure 5.2.

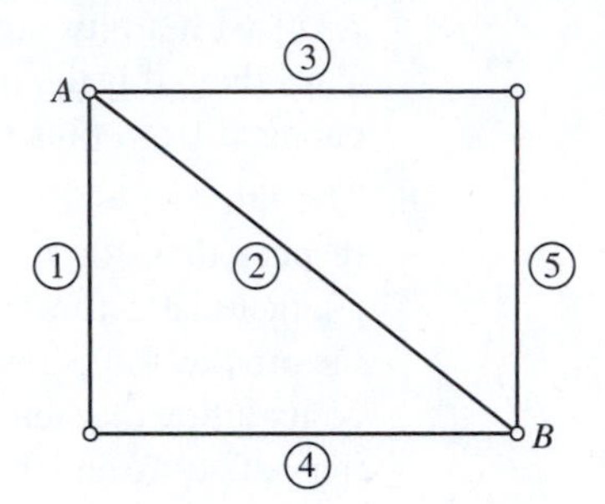

Figure 5.20

a. What is the probability that at least one working path is available between the nodes labeled A and B?

b. Remove link 4. Now what is the probability that at least one working path is available between nodes A and B?

c. Remove link 2. What is the probability that at least one working path is available between nodes A and B?

d. Which is the more serious situation, the removal of link 4 or link 2? Why?

5.6. Given a binary communication channel where $A =$ input and $B =$ output, let $P(A) = 0.45$, $P(B|A) = 0.95$, and $P(\overline{B}|\overline{A}) = 0.65$. Find $P(A|B)$ and $P(A|\overline{B})$.

5.7. Given the table of joint probabilities of Table 5.5.

a. Find the probabilities omitted from Table 5.5.

b. Find the probabilities $P(A_3|B_3)$, $P(B_2|A_1)$, and $P(B_3|A_2)$.

Table 5.5 Probabilities for Problem 5.7

	B_1	B_2	B_3	$P(A_i)$
A_1	0.05		0.45	0.55
A_2		0.15	0.10	
A_3	0.05	0.05		0.15
$P(B_j)$				1.0

Section 5.2

5.8. Two dice are tossed.

a. Let X_1 be a random variable that is numerically equal to the total number of spots on the up faces of the dice. Construct a table that defines this random variable.

b. Let X_2 be a random variable that has the value of 1 if the sum of the number of spots up on both dice is even and the value zero if it is odd. Repeat part (a) for this case.

5.9. Three fair coins are tossed simultaneously such that they don't interact. Define a random variable $X = 1$ if an even number of heads is up and $X = 0$ otherwise. Plot the cumulative distribution function and the probability density function corresponding to this random variable.

5.10. A certain *continuous* random variable has the cumulative distribution function

$$F_X(x) = \begin{cases} 0, & x < 0 \\ Ax^4, & 0 \le x \le 12 \\ B, & x > 12 \end{cases}$$

a. Find the proper values for A and B.

b. Obtain and plot the pdf $f_X(x)$.

c. Compute $P(X > 5)$.

d. Compute $P(4 \le X < 6)$.

5.11. The following functions can be pdfs if constants are chosen properly. Find the proper conditions on the constants [$A, B, C, D, \alpha, \beta, \gamma$, and τ are positive constants, and $u(x)$ is the unit step function.]

a. $f(x) = Ae^{-\alpha x}u(x)$, where $u(x)$ is the unit step.

b. $f(x) = Be^{\beta x}u(-x)$.

c. $f(x) = Ce^{-\gamma x}u(x-1)$.

d. $f(x) = D[u(x) - u(x-\tau)]$.

5.12. Test X and Y for independence if

a. $f_{XY}(x,y) = Ae^{-|x|-2|y|}$.

b. $f_{XY}(x,y) = C(1-x-y), 0 \le x \le 1-y$ and $0 \le y \le 1$.

Prove your answers.

5.13. The joint pdf of two random variables is

$$f_{XY}(x,y) = \begin{cases} C(1+xy), & 0 \le x \le 4, 0 \le y \le 2 \\ 0, & \text{otherwise} \end{cases}$$

Find the following:

a. The constant C.

b. $f_{XY}(1, 1.5)$.

c. $f_{XY}(x, 3)$.

d. $f_{X|Y}(x|1)$.

5.14. The joint pdf of the random variables X and Y is

$$f_{XY}(x,y) = Axye^{-(x+y)}, \quad x \ge 0 \text{ and } y \ge 0$$

a. Find the constant A.

b. Find the marginal pdfs of X and Y, $f_X(x)$ and $f_Y(y)$.

c. Are X and Y statistically independent? Justify your answer.

5.15.

a. For what value of $\alpha > 0$ is the function

$$f(x) = \alpha x^{-2}u(x-\alpha)$$

a probability density function? Use a sketch to illustrate your reasoning and recall that a pdf has to integrate to 1. [$u(x)$ is the unit step function.]

b. Find the corresponding cumulative distribution function.

c. Compute $P(X \ge 10)$.

5.16. Given the Gaussian random variable with the pdf

$$f_X(x) = \frac{e^{-x^2/2\sigma^2}}{\sqrt{2\pi}\sigma}$$

where $\sigma > 0$ is the standard deviation. If $Y = X^2$ find the pdf of Y. *Hint*: Note that $Y = X^2$ is symmetrical about $X = 0$ and that it is impossible for Y to be less than zero.

5.17. A nonlinear system has input X and output Y. The pdf for the input is Gaussian as given in Problem 5.16. Determine the pdf of the output, assuming that the nonlinear system has the following input–output relationship:

a. $Y = \begin{cases} aX, & X \ge 0 \\ 0, & X < 0 \end{cases}$

Hint: When $X < 0$, what is Y? How is this manifested in the pdf for Y?

b. $Y = |X|$.

c. $Y = X - X^3/3$.

Section 5.3

5.18. Let $f_X(x) = A\exp(-bx)u(x-2)$ for all x where A and b are positive constants.

a. Find the relationship between A and b such that this function is a pdf.

b. Calculate $E(X)$ for this random variable.

c. Calculate $E(X^2)$ for this random variable.

d. What is the variance of this random variable?

5.19.

a. Consider a random variable uniformly distributed between 0 and 2. Show that $E(X^2) > E^2(X)$.

b. Consider a random variable uniformly distributed between 0 and 4. Show that $E(X^2) > E^2(X)$.

c. Can you show in general that for any random variable it is true that $E(X^2) > E^2(X)$ unless the random variable is zero almost always?
(*Hint:* Expand $E[(X - E[X])^2] \geq 0$, and note that it is 0 only if $X = 0$ with probability 1.)

5.20. Verify the entries in Table 5.5 for the mean and variance of the following probability distributions:

a. Rayleigh

b. One-sided exponential

c. Hyperbolic

d. Poisson

e. Geometric

5.21. A random variable X has the pdf

$$f_X(x) = Ae^{-bx}[u(x) - u(x - B)]$$

where $u(x)$ is the unit step function and A, B, *and* b are positive constants.

a. Find the proper relationship between the constants A, b, and B. Express b in terms of A and B.

b. Determine and plot the cdf.

c. Compute $E[X]$.

d. Determine $E[X^2]$.

e. What is the variance of X?

5.22. If

$$f_X(x) = (2\pi\sigma^2)^{-1/2}\exp\left(-\frac{x^2}{2\sigma^2}\right)$$

show that

a. $E[X^{2n}] = 1\times3\times5\times\ldots(2n-1)\sigma^{2n}$, for $n = 1, 2, \ldots$

b. $E[X^{2n-1}] = 0$ for $n = 1, 2, \ldots$

5.23. The random variable has pdf

$$f_X(x) = \frac{1}{2}\delta(x-5) + \frac{1}{8}[u(x-4) - u(x-8)]$$

where $u(x)$ is the unit step. Determine the mean and the variance of the random variable thus defined.

5.24. Two random variables X and Y have means and variances given below:

$$m_x = 1,\ \sigma_x^2 = 4,\ m_y = 3,\ \sigma_y^2 = 7$$

A new random variable Z is defined as

$$Z = 3X - 4Y$$

Determine the mean and variance of Z for each of the following cases of correlation between the random variables X and Y:

a. $\rho_{XY} = 0$.

b. $\rho_{XY} = 0.2$.

c. $\rho_{XY} = 0.7$.

d. $\rho_{XY} = 1.0$.

5.25. Two Gaussian random variables X and Y, with zero means and variances σ^2, between which there is a correlation coefficient ρ, have a joint probability density function given by

$$f(x,y) = \frac{1}{2\pi\sigma^2\sqrt{1-\rho^2}}\exp\left(-\frac{x^2-2\rho xy+y^2}{2\sigma^2(1-\rho^2)}\right)$$

The marginal pdf of Y can be shown to be

$$f_Y(y) = \frac{\exp(-y^2/2\sigma^2)}{\sqrt{2\pi\sigma^2}}$$

Find the conditional pdf $f_{X|Y}(x \mid y)$. Simplify.

5.26. Using the definition of a conditional pdf given by (5.62) and the expressions for the marginal and joint Gaussian pdfs, show that for two jointly Gaussian random variables X and Y, the conditional density function of X given Y has the form of a Gaussian density with conditional mean and the conditional variance given by

$$E[X|Y] = m_x + \frac{\rho\sigma_x}{\sigma_y}(Y - m_y)$$

and

$$\text{var}(X|Y) = \sigma_x^2(1-\rho^2)$$

respectively.

5.27. The random variable X has a probability density function uniform in the range $0 \leq x \leq 2$ and zero elsewhere. The independent variable Y has a density uniform in the range $1 \leq y \leq 5$ and zero elsewhere. Find and plot the density of $Z = X + Y$.

5.28. A random variable X is defined by

$$f_X(x) = 4e^{-8|x|}$$

The random variable Y is related to X by $Y = 4 + 5X$.

a. Determine $E[X], E[X^2]$, and σ_x^2.

b. Determine $f_Y(y)$.

c. Determine $E[Y], E[Y^2]$, and σ_y^2. (Hint: The result of part (b) is not necessary to do this part, although it may be used)

5.29. A random variable X has the probability density function

$$f_X(x) = \begin{cases} ae^{-ax}, & x \geq 0 \\ 0, & x < 0 \end{cases}$$

where a is an arbitrary positive constant.

a. Determine the characteristic function $M_x(jv)$.

b. Use the characteristic function to determine $E[X]$ and $E[X^2]$.

c. Check your results by computing

$$\int_{-\infty}^{\infty} x^n f_X(x)\, dx$$

for $n = 1$ and 2.

d. Compute σ_x^2.

Section 5.4

5.30. Compare the binomial, Laplace, and Poisson distributions for

a. $n = 3$ and $p = \frac{1}{5}$.

b. $n = 3$ and $p = \frac{1}{10}$.

c. $n = 10$ and $p = \frac{1}{5}$.

d. $n = 10$ and $p = \frac{1}{10}$.

5.31. An honest coin is flipped 10 times.

a. Determine the probability of the occurrence of either five or six heads.

b. Determine the probability of the first head occurring at toss number 5.

c. Repeat parts (a) and (b) for flipping 100 times and the probability of the occurrence of 50 to 60 heads inclusive and the probability of the first head occurring at toss number 50.

5.32. Passwords in a computer installation take the form $X_1X_2X_3X_4$, where each character X_i is one of the 26 letters of the alphabet. Determine the maximum possible number of different passwords available for assignment for each of the two following conditions:

a. A given letter of the alphabet can be used only once in a password.

b. Letters can be repeated if desired, so that each X_i is completely arbitrary.

c. If selection of letters for a given password is completely random, what is the probability that your competitor could access, on a single try, your computer in part (a)? and part (b)?

5.33. Assume that 20 honest coins are tossed.

a. By applying the binomial distribution, find the probability that there will be fewer than three heads.

b. Do the same computation using the Laplace approximation.

c. Compare the results of parts (a) and (b) by computing the percent error of the Laplace approximation.

5.34. A digital data transmission system has an error probability of 10^{-5} per digit.

a. Find the probability of exactly one error in 10^5 digits.

b. Find the probability of exactly two errors errors in 10^5 digits.

c. Find the probability of more than five errors in 10^5 digits.

5.35. Assume that two random variables X and Y are jointly Gaussian with $m_x = m_y = 1$, $\sigma_x^2 = \sigma_y^2 = 4$, and correlation coeficient $\rho = 0.5$.

a. Making use of (5.194), write down an expression for the margininal pdfs of X and of Y.

b. Write down an expression for the conditional pdf $f_{X|Y}(x|y)$ by using the result of (a) and an expression for $f_{XY}(x, y)$ written down from (5.189). Deduce that $f_{Y|X}(y|x)$ has the same form with y replacing x.

c. Put $f_{X|Y}(x|y)$ into the form of a marginal Gaussian pdf. What is its mean and variance? (The mean will be a function of y.)

5.36. Consider the Cauchy density function

$$f_X(x) = \frac{K}{1 + x^2}, \quad -\infty \leq x \leq \infty$$

a. Find K.

b. Show that var$[X]$ is not finite.

c. Show that the characteristic function of a Cauchy random variable is $M_x(jv) = \pi K e^{-|v|}$.

d. Now consider $Z = X_1 + \ldots + X_N$ where the X_i's are independent Cauchy random variables. Thus their characteristic function is

$$M_Z(jv) = (\pi K)^N \exp(-N|v|)$$

Show that $f_Z(z)$ is Cauchy. (*Comment*: $f_Z(z)$ is not Gaussian as $N \to \infty$ because var$[X_i]$ is not finite and the conditions of the central-limit theorem are therefore violated.)

5.37. (Chi-squared pdf) Consider the random variable $Y = \sum_{i=1}^{N} X_i^2$, where the X_i', are independent Gaussian random variables with pdfs $n(0, \sigma)$.

a. Show that the characteristic function of X_i^2 is

$$M_{X_i^2}(jv) = \left(1 - 2jv\sigma^2\right)^{-1/2}$$

b. Show that the pdf of Y is

$$f_Y(y) = \begin{cases} \dfrac{y^{N/2-1}e^{-y/2\sigma^2}}{2^{N/2}\sigma^N\Gamma(N/2)}, & y \geq 0 \\ 0, & y < 0 \end{cases}$$

where $\Gamma(x)$ is the gamma function, which, for $x = n$ an integer, is $\Gamma(n) = (n-1)!$. This pdf is known as the χ^2 (*chi-squared*) pdf with N degrees of freedom. *Hint*: Use the Fourier transform pair

$$\frac{y^{N/2-1}e^{-y/\alpha}}{\alpha^{N/2}\Gamma(N/2)} \leftrightarrow (1 - j\alpha v)^{-N/2}$$

c. Show that for N large, the χ^2 pdf can be approximated as

$$f_Y(y) \cong \frac{\exp\left\{-\frac{1}{2}\left[\frac{(y - N\sigma^2)}{\sqrt{4N\sigma^4}}\right]^2\right\}}{\sqrt{4N\pi\sigma^4}}, \quad N \gg 1$$

Hint: Use the central-limit theorem. Since the x_i's are independent,

$$\overline{Y} = \sum_{i=1}^{N} \overline{X_i^2} = N\sigma^2$$

and

$$\text{var}[Y] = \sum_{i=1}^{N} \text{var}\left[X_i^2\right] = N\text{var}\left[X_i^2\right]$$

d. Compare the approximation obtained in part (c) with $f_Y(y)$ for $N = 2, 4, 8$.

e. Let $R^2 = Y$. Show that the pdf of R for $N = 2$ is Rayleigh.

5.38. Compare the Q-function and the approximation to it for large arguments given by (5.202) by plotting both expressions on a log–log graph. (Note: MATLAB is handy for this problem.)

5.39. Determine the cdf for a Gaussian random variable of mean m and variance σ^2. Express in terms of the Q-function. Plot the resulting cdf for $m = 0$ and $\sigma = 0.5$, 1, and 2.

5.40. Prove that the Q function may also be represented as

$$Q(x) = \frac{1}{\pi}\int_0^{\pi/2} \exp\left(-\frac{x^2}{2\sin^2\phi}\right)d\phi.$$

5.41. A random variable X has the

$$f_X(x) = \frac{e^{-(x-10)^2/50}}{\sqrt{50\pi}}$$

Express the following probabilities in terms of the Q function and calculate numerical answers for each:

a. $(P(|X| \leq 15)$

b. $P(10 < X \leq 20)$

c. $P(5 < X \leq 25)$

d. $P(20 < X \leq 30)$

5.42.

a. Prove Chebyshev's inequality. *Hint*: Let $Y = (X - m_x)/\sigma_x$, and find a bound for $P(|Y| < k)$ in terms of k.

b. Let X be uniformly distributed over $|x| \leq 1$. Plot $P(|X| \leq k\sigma_x)$ versus k and the corresponding bound given by Chebyshev's inequality.

5.43. If the random variable X is Gaussian with zero mean and variance σ^2, obtain numerical values for the following probabilities:

a. $P(|X| > \sigma)$

b. $P(|X| > 2\sigma)$

c. $P(|X| > 3\sigma)$

5.44. Speech is sometimes idealized as having a Laplacian-amplitude pdf. That is, the amplitude is distributed according to

$$f_X(x) = \frac{a}{2}\exp(-a|x|)$$

a. Express the variance of X, σ^2, in terms of a. Show your derivation; don't just simply copy the result given in Table 5.4.

b. Compute the following probabilities: $P(|X| > \sigma)$; $P(|X| > 2\sigma)$; $P(|X| > 3\sigma)$.

5.45. Two jointly Gaussian zero-mean random variables, X and Y, have respective variances of 3 and 4 and correlation coefficient $\rho_{XY} = -0.4$. A new random variable is defined as $Z = X + 2Y$. Write down an expression for the pdf of Z.

5.46. Two jointly Gaussian random variables, X and Y, have means of 1 and 2, and variances of 3 and 2, respectively. Their correlation coefficient is $\rho_{XY} = 0.2$. A new random variable is defined as $Z = 3X + Y$. Write down an expression for the pdf of Z.

5.47. Two Gaussian random variables, X and Y, are independent. Their respective means are 5 and 3, and their respective variances are 1 and 2.

a. Write down expressions for their marginal pdfs.

b. Write down an expression for their joint pdf.

c. What is the mean of $Z_1 = X + Y$? $Z_2 = X - Y$?

d. What is the variance of $Z_1 = X + Y$? $Z_2 = X - Y$?

e. Write down an expression for the pdf of $Z_1 = X + Y$.

f. Write down an expression for the pdf of $Z_2 = X - Y$.

5.48. Two Gaussian random variables, X and Y, are independent. Their respective means are 4 and 2, and their respective variances are 3 and 5.

a. Write down expressions for their marginal pdfs.

b. Write down an expression for their joint pdf.

c. What is the mean of $Z_1 = 3X + Y$? $Z_2 = 3X - Y$?

d. What is the variance of $Z_1 = 3X + Y$? $Z_2 = 3X - Y$?

e. Write down an expression for the pdf of $Z_1 = 3X + Y$.

f. Write down an expression for the pdf of $Z_2 = 3X - Y$.

5.49. Find the probabilities of the following random variables, with pdfs as given in Table 5.4, exceeding their means. That is, in each case, find the probability that $X \geq m_X$, where X is the respective random variable and m_X is its mean.

a. Uniform

b. Rayleigh

c. One-sided exponential

Computer Exercises

5.1. In this exercise we examine a useful technique for generating a set of samples having a given pdf.

a. First, prove the following theorem: If X is a continuous random variable with cdf $F_X(x)$, the random variable

$$Y = F_X(X)$$

is a uniformly distributed random variable in the interval [0,1).

b. Using this theorem, design a random number generator to generate a sequence of exponentially distributed random variables having the pdf

$$f_X(x) = \alpha e^{-\alpha x} u(x)$$

where $u(x)$ is the unit step. Plot histograms of the random numbers generated to check the validity of the random number generator you designed.

5.2. An algorithm for generating a Gaussian random variable from two independent uniform random variables is easily derived.

a. Let U and V be two statistically independent random numbers uniformly distributed in $[0, 1]$. Show that the following transformation generates two statistically independent Gaussian random numbers with unit variance and zero mean:

$$\begin{aligned} X &= R\cos(2\pi U) \\ Y &= R\sin(2\pi U) \end{aligned}$$

where

$$R = \sqrt{-2\ln V}$$

Hint: First show that R is Rayleigh.

b. Generate 1000 random variable pairs according to the above algorithm. Plot histograms for each set (i.e., X and Y), and compare with Gaussian pdfs after properly scaling the histograms (i.e., divide each cell by the total number of counts times the cell width so that the histogram approximates a probability density function).
Hint: Use the `hist` function of MATLAB.

5.3. Using the results of Problem 5.26 and the Gaussian random number generator designed in Computer Exercise 5.2, design a Gaussian random number generator that will provide a specified correlation between adjacent samples.

Let

$$\rho(\tau) = e^{-\alpha|\tau|}$$

and plot sequences of Gaussian random numbers for various choices of α. Show how stronger correlation between adjacent samples affects the variation from sample to sample. (Note: To get memory over more than adjacent samples, a digital filter should be used with independent Gaussian samples at the input.)

5.4. Check the validity of the central-limit theorem by repeatedly generating n independent uniformly distributed random variables in the interval $(-0.5, 0.5)$, forming the sum given by (5.187), and plotting the histogram. Do this for $N = 5$, 10, and 20. Can you say anything qualitatively and quantitatively about the approach of the sums to Gaussian random numbers? Repeat for exponentially distributed component random variables (do Computer Exercise 5.1 first). Can you think of a drawback to the approach of summing uniformly distributed random variables to generating Gaussian random variables? (*Hint:* Consider the probability of the sum of uniform random variables being greater than $0.5N$ or less than $-0.5N$. What are the same probabilities for a Gaussian random variable?)

CHAPTER 6

RANDOM SIGNALS AND NOISE

The mathematical background reviewed in Chapter 5 on probability theory provides the basis for developing the statistical description of random waveforms. The importance of considering such waveforms, as pointed out in Chapter 1, lies in the fact that noise in communication systems is due to unpredictable phenomena, such as the random motion of charge carriers in conducting materials and other unwanted sources.

In the relative-frequency approach to probability, we imagined repeating the underlying chance experiment many times, the implication being that the replication process was carried out sequentially in time. In the study of random waveforms, however, the outcomes of the underlying chance experiments are mapped into functions of time, or waveforms, rather than numbers, as in the case of random variables. The particular waveform is not predictable in advance of the experiment, just as the particular value of a random variable is not predictable before the chance experiment is performed. We now address the statistical description of chance experiments that result in waveforms as outputs. To visualize how this may be accomplished, we again think in terms of relative frequency.

6.1 A RELATIVE-FREQUENCY DESCRIPTION OF RANDOM PROCESSES

For simplicity, consider a binary digital waveform generator whose output randomly switches between $+1$ and -1 in T_0 intervals as shown in Figure 6.1. Let $X(t, \zeta_i)$ be the random waveform corresponding to the output of the ith generator. Suppose relative frequency is used to estimate $P(X = +1)$ by examining the outputs of all generators at a particular time. Since the outputs are functions of time, we must specify the time when writing down the relative frequency. The following table may be constructed from an examination of the generator outputs in each time interval shown:

Time interval	(0,1)	(1,2)	(2,3)	(3,4)	(4,5)	(5,6)	(6,7)	(7,8)	(8,9)	(9,10)
Relative frequency	$\frac{5}{10}$	$\frac{6}{10}$	$\frac{8}{10}$	$\frac{6}{10}$	$\frac{7}{10}$	$\frac{8}{10}$	$\frac{8}{10}$	$\frac{8}{10}$	$\frac{8}{10}$	$\frac{9}{10}$

From this table it is seen that the relative frequencies change with the time interval. Although this variation in relative frequency could be the result of *statistical irregularity*, we highly suspect that some phenomenon is making $X = +1$ more probable as time increases. To reduce the possibility that statistical irregularity is the culprit, we might repeat the experiment with 100 generators or 1000 generators. This is obviously a mental experiment in that it would be very difficult to obtain a set of identical generators and prepare them all in identical fashions.

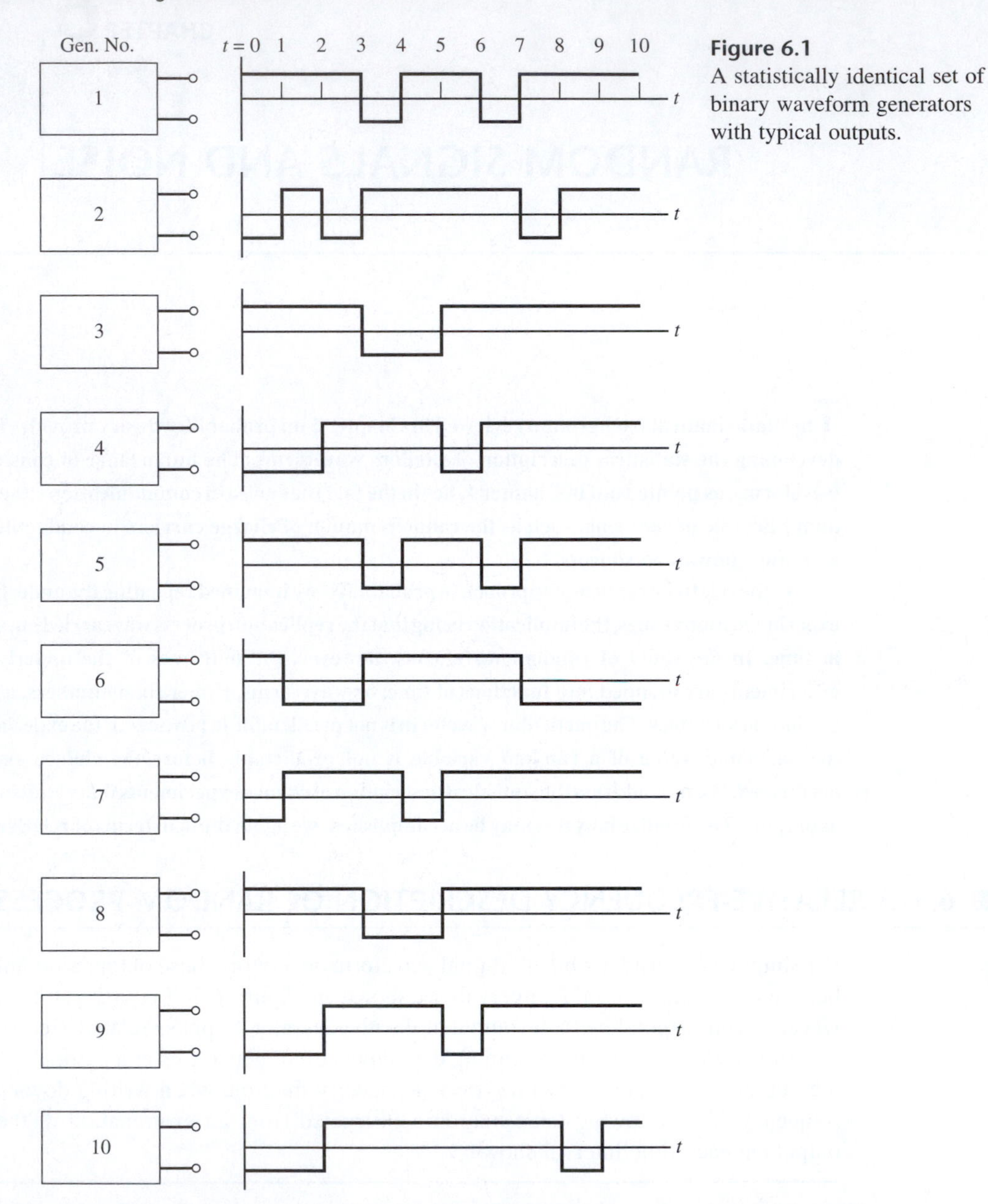

Figure 6.1 A statistically identical set of binary waveform generators with typical outputs.

■ 6.2 SOME TERMINOLOGY OF RANDOM PROCESSES

6.2.1 Sample Functions and Ensembles

In the same fashion as is illustrated in Figure 6.1, we could imagine performing any chance experiment many times simultaneously. If, for example, the random quantity of interest is the voltage at the terminals of a noise generator, the random variable X_1 may be assigned to represent the possible values of this voltage at time t_1 and the random variable X_2 the values at

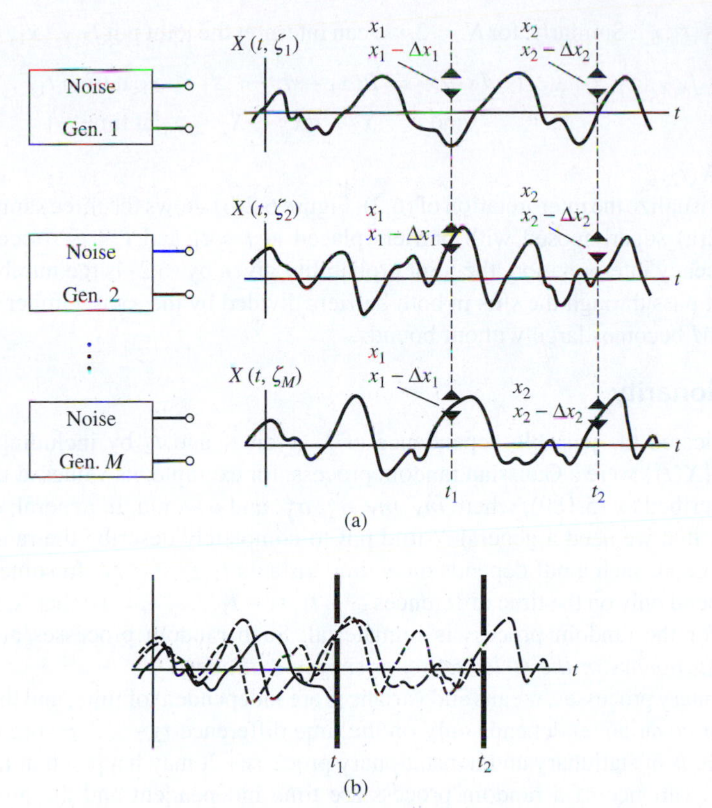

Figure 6.2 Typical sample functions of a random process and illustration of the relative-frequency interpretation of its joint pdf. (a) Ensemble of sample functions. (b) Superposition of the sample functions shown in (a).

time t_2. As in the case of the digital waveform generator, we can imagine many noise generators all constructed in an identical fashion, insofar as we can make them, and run under identical conditions. Figure 6.2(a) shows typical waveforms generated in such an experiment. Each waveform $X(t, \zeta_i)$, is referred to as a *sample function*, where ζ_i is a member of a *sample space S*. The totality of all sample functions is called an *ensemble*. The underlying chance experiment that gives rise to the ensemble of sample functions is called a *random*, or *stochastic*, *process*. Thus, to every outcome ζ we assign, according to a certain rule, a time function $X(t, \zeta)$. For a specific ζ, say ζ_i, $X(t, \zeta_i)$ signifies a single time function. For a specific time t_j, $X(t_j, \zeta)$ denotes a random variable. For fixed $t = t_j$ and fixed $\zeta = \zeta_i$, $X(t_j, \zeta_i)$ is a *number*. In what follows, we often suppress the ζ.

To summarize, the difference between a random variable and a random process is that for a random variable, an outcome in the sample space is mapped into a number, whereas for a random process it is mapped into a function of time.

6.2.2 Description of Random Processes in Terms of Joint pdfs

A complete description of a random process $\{X(t, \zeta)\}$ is given by the N-fold joint pdf that probabilistically describes the possible values assumed by a typical sample function at times $t_N > t_{N-1} > \cdots > t_1$, where N is arbitrary. For $N = 1$, we can interpret this joint pdf $f_{X_1}(x_1, t_1)$ as

$$f_{X_1}(x_1, t_1)\, dx_1 = P(x_1 - dx_1 < X_1 \leq x_1 \text{ at time } t_1) \tag{6.1}$$

where $X_1 = X(t_1, \zeta)$. Similarly, for $N = 2$, we can interpret the joint pdf $f_{X_1X_2}(x_1,\ t_1;\ x_2,\ t_2)$ as

$$f_{X_1X_2}(x_1,\ t_1;\ x_2,\ t_2)dx_1dx_2 = P(x_1 - dx_1 < X_1 \leq x_1 \text{ at time } t_1, \\ \text{and} \quad x_2 - dx_2 < X_2 \leq x_2 \text{ at time } t_2) \tag{6.2}$$

where $X_2 = X(t_2, \zeta)$.

To help visualize the interpretation of (6.2), Figure 6.2(b) shows the three sample functions of Figure 6.2(a) superimposed with barriers placed at $t = t_1$ and $t = t_2$. According to the relative-frequency interpretation, the joint probability given by (6.2) is the number of sample functions that pass through the slits in both barriers divided by the total number M of sample functions as M becomes large without bound.

6.2.3 Stationarity

We have indicated the possible dependence of $f_{X_1X_2}$ on t_1 and t_2 by including them in its argument. If $\{X(t)\}$ were a Gaussian random process, for example, its values at time t_1 and t_2 would be described by (5.189), where $m_X, m_Y, \sigma_X^2, \sigma_Y^2$, and ρ would, in general, depend on t_1 and t_2.[1] Note that we need a general N-fold pdf to completely describe the random process $\{X(t)\}$. In general, such a pdf depends on N time instants $t_1, t_2, \ldots, t_N$. In some cases, these joint pdfs depend only on the time differences $t_2 - t_1, t_3 - t_1, \ldots, t_N - t_1$; that is, the choice of time origin for the random process is immaterial. Such random processes are said to be *statistically stationary in the strict sense*, or simply *stationary*.

For stationary processes, means and variances are independent of time, and the correlation coefficient (or covariance) depends only on the time difference $t_2 - t_1$.[2] Figure 6.3 contrasts sample functions of stationary and nonstationary processes. It may happen that in some cases the mean and variance of a random process are time independent and the covariance is a function only of the time difference, but the N-fold joint pdf depends on the time origin. Such random processes are called *wide-sense stationary processes* to distinguish them from strictly stationary processes (that is, processes whose N-fold pdf is independent of time origin). Strict-sense stationarity implies wide-sense stationarity, but the reverse is not necessarily true. An exception occurs for *Gaussian random processes for which wide-sense stationarity does imply strict-sense stationarity*, since the joint Gaussian pdf is completely specified in terms of the means, variances, and covariances of $X(t_1), X(t_2), \ldots, X(t_N)$.

6.2.4 Partial Description of Random Processes: Ergodicity

As in the case of random variables, we may not always require a complete statistical description of a random process, or we may not be able to obtain the N-fold joint pdf even if desired. In such cases, we work with various moments, either by choice or by necessity. The most important averages are the mean,

$$m_X(t) = E[X(t)] = \overline{X(t)} \tag{6.3}$$

the variance,

---

[1]For a stationary process, all joint moments are independent of time origin. We are interested primarily in the covariance, however.

[2]At N instants of time, if Gaussian, its values would be described by (B.1) of Appendix B.

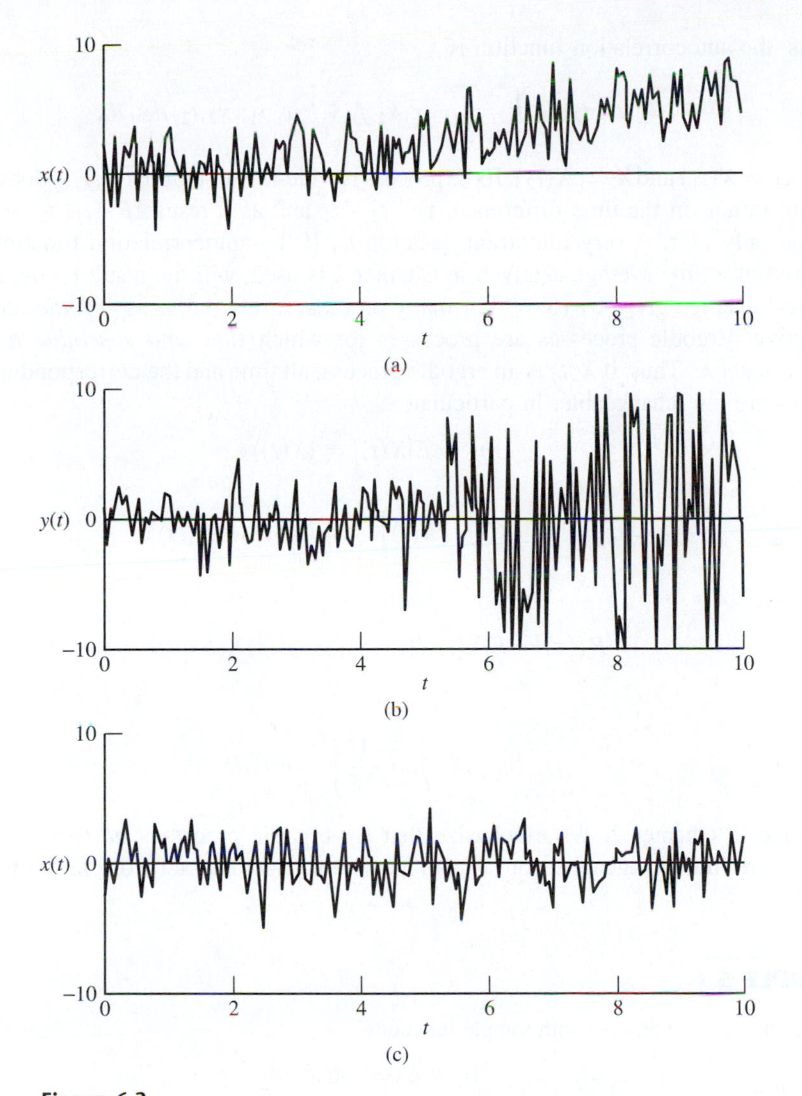

Figure 6.3
Sample functions of nonstationary processes contrasted with a sample function of a stationary process. (a) Time-varying mean. (b) Time-varying variance. (c) Stationary.

$$\sigma_X^2(t) = E\left[\left[X(t) - \overline{X(t)}\right]^2\right] = \overline{X^2(t)} - \left[\overline{X(t)}\right]^2 \tag{6.4}$$

and the covariance,

$$\begin{aligned} \mu_X(t, t+\tau) &= E\left[\left[X(t) - \overline{X(t)}\right]\left[X(t+\tau) - \overline{X(t+\tau)}\right]\right] \\ &= E[X(t)X(t+\tau)] - \overline{X(t)}\;\overline{X(t+\tau)} \end{aligned} \tag{6.5}$$

In (6.5), we let $t = t_1$ and $t + \tau = t_2$. The first term on the right-hand side is the *autocorrelation function* computed as a *statistical*, or *ensemble*, *average* (that is, the average is across the sample functions at times t and $t + \tau$). In terms of the joint pdf of the random process, the autocorrelation function is

$$R_X(t_1, t_2) = \int_{-\infty}^{\infty}\int_{-\infty}^{\infty} x_1 x_2 \, f_{X_1 X_2}(x_1, t_1; x_2, t_2)\, dx_1\, dx_2 \tag{6.6}$$

where $X_1 = X(t_1)$ and $X_2 = X(t_2)$. If the process is wide-sense stationary, $f_{X_1 X_2}$ does not depend on t but rather on the time difference, $\tau = t_2 - t_1$ and as a result, $R_X(t_1, t_2) = R_X(\tau)$ is a function only of τ. A very important question is: If the autocorrelation function using the definition of a time average as given in Chapter 2 is used, will the result be the same as the statistical average given by (6.6)? For many processes, referred to as *ergodic*, the answer is affirmative. Ergodic processes are processes for which *time and ensemble averages are interchangeable*. Thus, if $X(t)$ is an ergodic process, all time and the corresponding ensemble averages are interchangeable. In particular,

$$m_X = E[X(t)] = \langle X(t) \rangle \tag{6.7}$$

$$\sigma_X^2 = E\left[\left[X(t) - \overline{X(t)}\right]^2\right] = \langle [X(t) - \langle X(t) \rangle]^2 \rangle \tag{6.8}$$

and

$$R_X(\tau) = E[X(t)X(t+\tau)] = \langle X(t)X(t+\tau) \rangle \tag{6.9}$$

where

$$\langle v(t) \rangle \triangleq \lim_{T\to\infty} \frac{1}{2T}\int_{-T}^{T} v(t)\, dt \tag{6.10}$$

as defined in Chapter 2. We emphasize that for ergodic processes *all* time and ensemble averages are interchangeable, not just the mean, variance, and autocorrelation function.

EXAMPLE 6.1

Consider the random process with sample functions[3]

$$n(t) = A\cos(2\pi f_0 t + \theta)$$

where f_0 is a constant and Θ is a random variable with the pdf

$$f_\Theta(\theta) = \begin{cases} \dfrac{1}{2\pi}, & |\theta| \leq \pi \\ 0, & \text{otherwise} \end{cases} \tag{6.11}$$

Computed as statistical averages, the first and second moments are

$$\begin{aligned} \overline{n(t)} &= \int_{-\infty}^{\infty} A\cos(2\pi f_0 t + \theta) f_\Theta(\theta)\, d\theta \\ &= \int_{-\pi}^{\pi} A\cos(2\pi f_0 t + \theta)\frac{d\theta}{2\pi} = 0 \end{aligned} \tag{6.12}$$

[3]In this example we violate our earlier established convention that sample functions are denoted by capital letters. This is quite often done if confusion will not result.

and

$$\overline{n^2(t)} = \int_{-\pi}^{\pi} A^2 \cos^2(2\pi f_0 t + \theta) \frac{d\theta}{2\pi} = \frac{A^2}{4\pi} \int_{-\pi}^{\pi} [1 + \cos(4\pi f_0 t + 2\theta)]\, d\theta = \frac{A^2}{2} \tag{6.13}$$

respectively. The variance is equal to the second moment, since the mean is zero.

Computed as time averages, the first and second moments are

$$\langle n(t) \rangle = \lim_{T \to \infty} \frac{1}{2T} \int_{-T}^{T} A \cos(2\pi f_0 t + \theta)\, dt = 0 \tag{6.14}$$

and

$$\langle n^2(t) \rangle = \lim_{T \to \infty} \frac{1}{2T} \int_{-T}^{T} A^2 \cos^2(2\pi f_0 t + \theta)\, dt = \frac{A^2}{2} \tag{6.15}$$

respectively. In general, the time average of some function of an ensemble member of a random process is a random variable. In this example, $\langle n(t) \rangle$ and $\langle n^2(t) \rangle$ are constants! We suspect that this random process is stationary and ergodic, even though the preceding results do not *prove* this. It turns out that this is indeed true.

To continue the example, consider the pdf

$$f_\Theta(\theta) = \begin{cases} \dfrac{2}{\pi}, & |\theta| \leq \dfrac{1}{4}\pi \\ 0, & \text{otherwise} \end{cases} \tag{6.16}$$

For this case, the expected value, or mean, of the random process computed at an arbitrary time t is

$$\begin{aligned} \overline{n^2(t)} &= \int_{-\pi/4}^{\pi/4} A \cos(2\pi f_0 t + \theta) \frac{2}{\pi}\, d\theta \\ &= \frac{2}{\pi} A \sin(2\pi f_0 t + \theta) \Big|_{-\pi/4}^{\pi/4} = \frac{2\sqrt{2}A}{\pi} \cos(2\pi f_0 t) \end{aligned} \tag{6.17}$$

The second moment, computed as a statistical average, is

$$\begin{aligned} \overline{n^2(t)} &= \int_{-\pi/4}^{\pi/4} A^2 \cos^2(2\pi f_0 t + \theta) \frac{2}{\pi}\, d\theta \\ &= \int_{-\pi/4}^{\pi/4} \frac{A^2}{\pi} [1 + \cos(4\pi f_0 t + 2\theta)]\, d\theta \\ &= \frac{A^2}{2} + \frac{A^2}{\pi} \cos(4\pi f_0 t) \end{aligned} \tag{6.18}$$

Since stationarity of a random process implies that all moments are independent of time origin, these results show that this process is not stationary. In order to comprehend the physical reason for this, you should sketch some typical sample functions. In addition, this process cannot be ergodic, since ergodicity requires stationarity. Indeed, the time-average first and second moments are still $\langle n(t) \rangle = 0$ and $\langle n^2(t) \rangle = \frac{1}{2}A^2$, respectively. Thus we have exhibited two time averages that are not equal to the corresponding statistical averages.

■

6.2.5 Meanings of Various Averages for Ergodic Processes

It is useful to pause at this point and summarize the meanings of various averages for an ergodic process:

1. The mean $\overline{X(t)} = \langle X(t) \rangle$ is the DC component.
2. $\overline{X(t)}^2 = \langle X(t) \rangle^2$ is the DC power.
3. $\overline{X^2(t)} = \langle X^2(t) \rangle$ is the total power.
4. $\sigma_X^2 = \overline{X^2(t)} - \overline{X(t)}^2 = \langle X^2(t) \rangle - \langle X(t) \rangle^2$ is the power in the alternating current (AC) (time-varying) component.
5. The total power $\overline{X^2(t)} = \sigma_X^2 + \overline{X(t)}^2$ is the AC power plus the direct current (DC) power.

Thus, in the case of ergodic processes, we see that these moments are measurable quantities in the sense that they can be replaced by the corresponding time averages and that a finite-time approximation to these time averages can be measured in the laboratory.

EXAMPLE 6.2

Consider a random telegraph waveform $X(t)$, as illustrated in Figure 6.4. The sample functions of this random process have the following properties:

1. The values taken on at any time instant t_0 are either $X(t_0) = A$ or $X(t_0) = -A$ with equal probability.
2. The number k of switching instants in any time interval T obeys a Poisson distribution, as defined by (5.182), with the attendant assumptions leading to this distribution. (That is, the probability of more than one switching instant occurring in an infinitesimal time interval dt is zero, with the probability of exactly one switching instant occurring in dt being $\alpha\, dt$, where α is a constant. Furthermore, successive switching occurrences are independent.)

If τ is any positive time increment, the autocorrelation function of the random process defined by the preceding properties can be calculated as

$$\begin{aligned} R_X(\tau) &= E[X(t)X(t+\tau)] \\ &= A^2 P[X(t) \text{ and } X(t+\tau) \text{ have the same sign}] \\ &\quad + (-A^2)P[X(t) \text{ and } X(t+\tau) \text{ have different signs}] \\ &= A^2 P\,[\text{even number of switching instants in } (t,\ t+\tau)] \\ &\quad - A^2 P\,[\text{odd number of switching instants in } (t,\ t+\tau)] \\ &= A^2 \sum_{\substack{k=0 \\ k \text{ even}}}^{\infty} \frac{(\alpha\tau)^k}{k!} \exp(-\alpha\tau) - A^2 \sum_{\substack{k=0 \\ k \text{ odd}}}^{\infty} \frac{(\alpha\tau)^k}{k!} \exp(-\alpha\tau) \\ &= A^2 \exp(-\alpha\tau) \sum_{k=0}^{\infty} \frac{(-\alpha\tau)^k}{k!} \\ &= A^2 \exp(-\alpha\tau)\exp(-\alpha\tau) = A^2 \exp(-2\alpha\tau) \end{aligned} \tag{6.19}$$

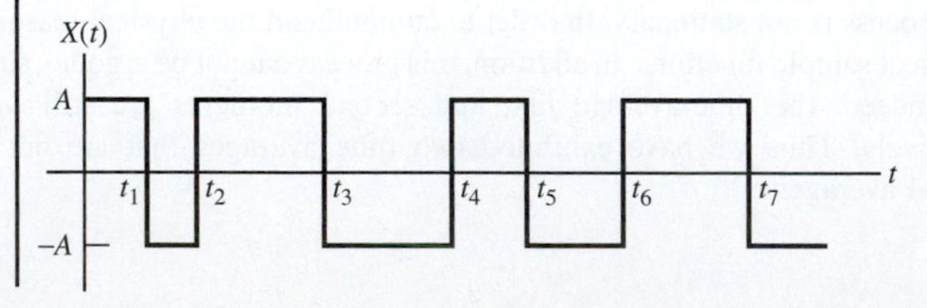

Figure 6.4
Sample function of a random telegraph waveform.

The preceding development was carried out under the assumption that τ was positive. It could have been similarly carried out with τ negative, such that

$$R_X(\tau) = E[X(t)X(t-|\tau|)] = E[X(t-|\tau|)X(t)] = A^2 \exp(-2\alpha|\tau|) \tag{6.20}$$

This is a result that holds for all τ. That is, $R_X(\tau)$ is an even function of τ, which we will show in general shortly.

■

6.3 CORRELATION AND POWER SPECTRAL DENSITY

The autocorrelation function, computed as a statistical average, has been defined by (6.6). If a process is ergodic, the autocorrelation function computed as a time average, as first defined in Chapter 2, is equal to the statistical average of (6.6). In Chapter 2, we defined the power spectral density $S(f)$ as the Fourier transform of the autocorrelation function $R(\tau)$. The *Wiener–Khinchine theorem* is a formal statement of this result for stationary random processes, for which $R(t_1, t_2) = R(t_2 - t_1) = R(\tau)$. For such processes, previously defined as wide-sense stationary, the power spectral density and autocorrelation function are Fourier transform pairs. That is,

$$S(f) \longleftrightarrow R(\tau) \tag{6.21}$$

If the process is ergodic, $R(\tau)$ can be calculated as either a time or an ensemble average.

Since $R_X(0) = \overline{X^2(t)}$ is the average power contained in the process, we have from the inverse Fourier transform of $S_X(f)$ that

$$\text{Average power} = R_X(0) = \int_{-\infty}^{\infty} S_X(f)\, df \tag{6.22}$$

which is reasonable, since the definition of $S_X(f)$ is that it is *power density* with respect to frequency.

6.3.1 Power Spectral Density

An intuitively satisfying, and in some cases computationally useful, expression for the power spectral density of a stationary random process can be obtained by the following approach. Consider a particular sample function $n(t, \zeta_i)$ of a stationary random process. To obtain a function giving power density versus frequency using the Fourier transform, we consider a truncated version, $n_T(t, \zeta_i)$, defined as[4]

$$n_T(t, \zeta_i) = \begin{cases} n(t, \zeta_i) & |t| < \frac{1}{2}T \\ 0, & \text{otherwise} \end{cases} \tag{6.23}$$

Since sample functions of stationary random processes are power signals, the Fourier transform of $n(t, \zeta_i)$ does not exist, which necessitates defining $n_T(t, \zeta_i)$. The Fourier transform of a

[4] Again, we use a lowercase letter to denote a random process for the simple reason that we need to denote the Fourier transform of $n(t)$ by an uppercase letter.

truncated sample function is

$$N_T(f, \zeta_i) = \int_{-T/2}^{T/2} n(t, \zeta_i) e^{-j2\pi ft}\, dt \tag{6.24}$$

and its energy spectral density, according to (2.90), is $|N_T(f, \zeta_i)|^2$. The time-average power density over the interval $\left[-\frac{1}{2}T, \frac{1}{2}T\right]$ for this sample function is $|N_T(f, \zeta_i)|^2/T$. Since this time-average power density depends on the particular sample function chosen, we perform an ensemble average and take the limit as $T \to \infty$ to obtain the distribution of power density with frequency. This is defined as the power spectral density $S_n(f)$ which can be expressed as

$$S_n(f) = \lim_{T \to \infty} \frac{\overline{|N_T(f, \zeta_i)|^2}}{T} \tag{6.25}$$

The operations of taking the limit and taking the ensemble average in (6.25) cannot be interchanged.

EXAMPLE 6.3

Let us find the power spectral density of the random process considered in Example 6.1 using (6.25). In this case,

$$n_T(t, \Theta) = A\Pi\left(\frac{t}{T}\right) \cos\left[2\pi f_0 \left(t + \frac{\Theta}{2\pi f_0}\right)\right] \tag{6.26}$$

By the time-delay theorem of Fourier transforms and using the transform pair

$$\cos(2\pi f_0 t) \longleftrightarrow \frac{1}{2}\delta(f - f_0) + \frac{1}{2}\delta(f + f_0) \tag{6.27}$$

we obtain

$$\Im[\cos(2\pi f_0 t + \Theta)] = \frac{1}{2}\delta(f - f_0)e^{j\Theta} + \frac{1}{2}\delta(f + f_0)e^{-j\Theta} \tag{6.28}$$

We also recall from Chapter 2 (Example 2.8) that $\Pi(t/T) \longleftrightarrow T \operatorname{sinc} Tf$, so by the multiplication theorem of Fourier transforms,

$$\begin{aligned} N_T(f, \Theta) &= (AT \operatorname{sinc} Tf) * \left[\frac{1}{2}\delta(f - f_0)e^{j\Theta} + \frac{1}{2}\delta(f + f_0)e^{-j\Theta}\right] \\ &= \frac{1}{2}AT\left[e^{j\Theta}\operatorname{sinc}(f - f_0)T + e^{-j\Theta}\operatorname{sinc}(f + f_0)T\right] \end{aligned} \tag{6.29}$$

Therefore, the energy spectral density of the sample function is

$$\begin{aligned} |N_T(f, \Theta)|^2 = \left(\frac{1}{2}AT\right)^2 &\left[\operatorname{sinc}^2 T(f - f_0) + e^{2j\Theta}\operatorname{sinc}T(f - f_0)\operatorname{sinc}T(f + f_0)\right. \\ &\left. + e^{-2j\Theta}\operatorname{sinc}T(f - f_0)\operatorname{sinc}T(f + f_0) + \operatorname{sinc}^2 T(f + f_0)\right] \end{aligned} \tag{6.30}$$

In obtaining $\left[\overline{|N_T(f, \Theta)|^2}\right]$, we note that

$$\overline{\exp(\pm j2\Theta)} = \int_{-\pi}^{\pi} e^{\pm j2\Theta} \frac{d\theta}{2\pi} = \int_{-\pi}^{\pi} (\cos 2\theta \pm j \sin 2\theta) \frac{d\theta}{2\pi} = 0 \tag{6.31}$$

Thus we obtain

$$\overline{|N_T(f,\Theta)|^2} = \left(\frac{1}{2}AT\right)^2 \left[\text{sinc}^2 T(f-f_0) + \text{sinc}^2 T(f+f_0)\right] \tag{6.32}$$

and the power spectral density is

$$S_n(f) = \lim_{T\to\infty} \frac{1}{4}A^2\left[T\,\text{sinc}^2 T(f-f_0) + T\,\text{sinc}^2 T(f+f_0)\right] \tag{6.33}$$

However, a representation of the delta function is $\lim_{T\to\infty} T\text{sinc}^2\, Tu = \delta(u)$. [See Figure 2.4(b).] Thus

$$S_n(f) = \frac{1}{4}A^2\delta(f-f_0) + \frac{1}{4}A^2\delta(f+f_0) \tag{6.34}$$

The average power is $\int_{-\infty}^{\infty} S_n(f)\,df = \frac{1}{2}A^2$, the same as obtained in Example 6.1. ■

6.3.2 The Wiener–Khinchine Theorem

The Wiener–Khinchine theorem states that the autocorrelation function and power spectral density of a stationary random process are Fourier transform pairs. It is the purpose of this subsection to provide a formal proof of this statement.

To simplify the notation in the proof of the Wiener–Khinchine theorem, we rewrite (6.25) as

$$S_n(f) = \lim_{T\to\infty} \frac{E\left[|\Im[n_{2T}(t)]|^2\right]}{2T} \tag{6.35}$$

where, for convenience, we have truncated over a $2T$-s interval and dropped ζ in the argument of $n_{2T}(t)$. Note that

$$\begin{aligned} |\Im[n_{2T}(t)]|^2 &= \left|\int_{-T}^{T} n(t)e^{-j\omega t}\,dt\right|^2, \qquad \omega = 2\pi f \\ &= \int_{-T}^{T}\int_{-T}^{T} n(t)n(\sigma)e^{-j\omega(t-\sigma)}\,dt\,d\sigma \end{aligned} \tag{6.36}$$

where the product of two integrals has been written as an iterated integral. Taking the ensemble average and interchanging the orders of averaging and integration, we obtain

$$\begin{aligned} E\left\{|\Im[n_{2T}(t)]|^2\right\} &= \int_{-T}^{T}\int_{-T}^{T} E\{n(t)n(\sigma)\}e^{-j\omega(t-\sigma)}\,dt\,d\sigma \\ &= \int_{-T}^{T}\int_{-T}^{T} R_n(t-\sigma)e^{-j\omega(t-\sigma)}\,dt\,d\sigma \end{aligned} \tag{6.37}$$

by the definition of the autocorrelation function. The change of variables $u = t - \sigma$ and $v = t$ is now made with the aid of Figure 6.5. In the uv plane, we integrate over v first and then over u by breaking the integration over u up into two integrals, one for u negative and one for u

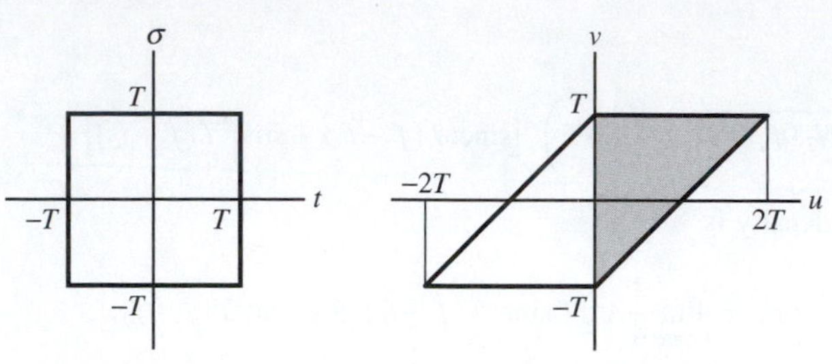

Figure 6.5 Regions of integration for (6.37).

positive. Thus

$$\begin{aligned} E\left\{|\Im[n_{2T}(t)]|^2\right\} &= \int_{u=-2T}^{0} R_n(u)e^{-j\omega u}\left(\int_{-T}^{u+T} dv\right)du + \int_{u=0}^{2T} R_n(u)e^{-j\omega u}\left(\int_{u-T}^{T} dv\right)du \\ &= \int_{-2T}^{0} (2T+u)R_n(u)e^{-j\omega u} + \int_{0}^{2T} (2T-u)R_n(u)e^{-j\omega u}\,du \\ &= 2T\int_{-2T}^{2T}\left(1-\frac{|u|}{2T}\right)R_n(u)e^{-j\omega u}\,du \end{aligned} \tag{6.38}$$

The power spectral density is, by (6.35),

$$S_n(f) = \lim_{T\to\infty}\int_{-2T}^{2T}\left(1-\frac{|u|}{2T}\right)R_n(u)e^{-j\omega u}\,du \tag{6.39}$$

which is the limit as $T \to \infty$ results in (6.21).

EXAMPLE 6.4

Since the power spectral density and the autocorrelation function are Fourier transform pairs, the autocorrelation function of the random process defined in Example 6.1 is, from the result of Example 6.3, given by

$$\begin{aligned} R_n(\tau) &= \Im^{-1}\left[\frac{1}{4}A^2\delta(f-f_0) + \frac{1}{4}A^2\delta(f+f_0)\right] \\ &= \frac{1}{2}A^2\cos(2\pi f_0\tau) \end{aligned} \tag{6.40}$$

Computing $R_n(\tau)$ as an ensemble average, we obtain

$$\begin{aligned} R_n(\tau) &= E[n(t)n(t+\tau)] \\ &= \int_{-\pi}^{\pi} A^2\cos(2\pi f_0 t+\theta)\cos[2\pi f_0(t+\tau)+\theta]\,\frac{d\theta}{2\pi} \\ &= \frac{A^2}{4\pi}\int_{-\pi}^{\pi}\left\{\cos 2\pi f_0\tau + \cos[2\pi f_0(2t+\tau)+2\theta]\right\}d\theta \\ &= \frac{1}{2}A^2\cos(2\pi f_0\tau) \end{aligned} \tag{6.41}$$

which is the same result as that obtained using the Wiener–Khinchine theorem. ■

6.3.3 Properties of the Autocorrelation Function

The properties of the autocorrelation function for a stationary random process $X(t)$ were stated in Chapter 2, at the end of Section 2.6, and all time averages may now be replaced by statistical averages. These properties are now easily proved.

Property 1 states that $|R(\tau)| \leq R(0)$ for all τ. To show this, consider the nonnegative quantity

$$[X(t) \pm X(t+\tau)]^2 \geq 0 \tag{6.42}$$

where $\{X(t)\}$ is a stationary random process. Squaring and averaging term by term, we obtain

$$\overline{X^2(t)} \pm 2\overline{X(t)X(t+\tau)} + \overline{X^2(t+\tau)} \geq 0 \tag{6.43}$$

which reduces to

$$2R(0) \pm 2R(\tau) \geq 0 \quad \text{or} \quad -R(0) \leq R(\tau) \leq R(0) \tag{6.44}$$

because $\overline{X^2(t)} = \overline{X^2(t+\tau)} = R(0)$ by the stationarity of $\{X(t)\}$.

Property 2 states that $R(-\tau) = R(\tau)$. This is easily proved by noting that

$$R(\tau) \triangleq \overline{X(t)X(t+\tau)} = \overline{X(t'-\tau)X(t')} = \overline{X(t')X(t'-\tau)} \triangleq R(-\tau) \tag{6.45}$$

where the change of variables $t' = t+\tau$ has been made.

Property 3 states that $\lim_{|\tau|\to\infty} R(\tau) = \overline{X(t)}^2$ if $\{X(t)\}$ does not contain a periodic component. To show this, we note that

$$\begin{aligned} \lim_{|\tau|\to\infty} R(\tau) &\triangleq \lim_{|\tau|\to\infty} \overline{X(t)X(t+\tau)} \\ &\cong \overline{X(t)}\,\overline{X(t+\tau)}, \quad \text{where } |\tau| \text{ is large} \\ &= \overline{X(t)}^2 \end{aligned} \tag{6.46}$$

where the second step follows intuitively because the interdependence between $X(t)$ and $X(t+\tau)$ becomes smaller as $|\tau| \to \infty$ (if no periodic components are present) and the last step results from the stationarity of $\{X(t)\}$.

Property 4, which states that $R(\tau)$ is periodic if $\{X(t)\}$ is periodic, follows by noting from the time-average definition of the autocorrelation function given by (2.161) that periodicity of the integrand implies periodicity of $R(\tau)$.

Property 5, which says that $\Im[R(\tau)]$ is nonnegative, is a direct consequence of the Wiener–Khinchine theorem (6.21) and (6.25) from which it is seen that the power spectral density is nonnegative.

EXAMPLE 6.5

Processes for which

$$S(f) = \begin{cases} \frac{1}{2}N_0, & |f| \leq B \\ 0, & \text{otherwise} \end{cases} \tag{6.47}$$

where N_0 is constant, are commonly referred to as *bandlimited white noise*, since as $B \to \infty$, all frequencies are present, in which case the process is simply called *white*. N_0 is the single-sided power spectral density of the nonbandlimited process. For a bandlimited white-noise process,

$$
\begin{aligned}
R(\tau) &= \int_{-B}^{B} \frac{1}{2} N_0 \exp(j2\pi f\tau)\, df \\
&= \frac{N_0}{2} \frac{\exp(j2\pi f\tau)}{j2\pi\tau}\Bigg|_{-B}^{B} = BN_0 \frac{\sin(2\pi B\tau)}{2\pi B\tau} \\
&= BN_0 \operatorname{sinc}(2B\tau)
\end{aligned}
\tag{6.48}
$$

As $B \to \infty$, $R(\tau) \to \frac{1}{2}N_0\delta(\tau)$. That is, no matter how close together we sample a white-noise process, the samples have zero correlation. If, in addition, the process is Gaussian, the samples are independent. A white-noise process has infinite power and is therefore a mathematical idealization, but it is nevertheless useful in systems analysis.

■

6.3.4 Autocorrelation Functions for Random Pulse Trains

As another example of calculating autocorrelation functions, consider a random process with sample functions that can be expressed as

$$
X(t) = \sum_{k=-\infty}^{\infty} a_k p(t - kT - \Delta) \tag{6.49}
$$

where $\ldots, a_{-1}, a_0, a_1, \ldots, a_k, \ldots$ is a doubly infinite sequence of random variables with

$$
E[a_k\, a_{k+m}] = R_m \tag{6.50}
$$

The function $p(t)$ is a deterministic pulse-type waveform, where T is the separation between pulses; Δ is a random variable that is independent of the value of a_k and uniformly distributed in the interval $(-T/2,\ T/2)$.[5] The autocorrelation function of this waveform is

$$
\begin{aligned}
R_X(\tau) &= E[X(t)X(t+\tau)] \\
&= E\left[\sum_{k=-\infty}^{\infty}\sum_{m=-\infty}^{\infty} a_k a_{k+m} p(t-kT-\Delta)p[t+\tau-(k+m)T-\Delta]\right]
\end{aligned}
\tag{6.51}
$$

Taking the expectation inside the double sum and making use of the independence of the sequence $\{a_k\, a_{k+m}\}$ and the delay variable Δ, we obtain

$$
\begin{aligned}
R_X(\tau) &= \sum_{k=-\infty}^{\infty}\sum_{m=-\infty}^{\infty} E[a_k a_{k+m}]E[p(t-kT-\Delta)p[t+\tau-(k+m)T-\Delta]] \\
&= \sum_{m=-\infty}^{\infty} R_m \sum_{k=-\infty}^{\infty} \int_{-T/2}^{T/2} p(t-kT-\Delta)p[t+\tau-(k+m)T-\Delta]\frac{d\Delta}{T}
\end{aligned}
\tag{6.52}
$$

The change of variables $u = t - kT - \Delta$ inside the integral results in

[5]Including the random variable Δ in the definition of the sample functions for the process guarantees wide-sense stationarity. If it was not included, $X(t)$ would be what is referred to as a *cyclostationary random process.*

$$\begin{aligned} R_X(\tau) &= \sum_{m=-\infty}^{\infty} R_m \sum_{k=-\infty}^{\infty} \int_{t-(k+1/2)T}^{t-(k-1/2)T} p(u)p(u+\tau-mT)\frac{du}{T} \\ &= \sum_{m=-\infty}^{\infty} R_m \left[\frac{1}{T}\int_{-\infty}^{\infty} p(u+\tau-mT)p(u)\,du\right] \end{aligned} \tag{6.53}$$

Finally, we have

$$R_X(\tau) = \sum_{m=-\infty}^{\infty} R_m\, r(\tau-mT) \tag{6.54}$$

where

$$r(\tau) \triangleq \frac{1}{T}\int_{-\infty}^{\infty} p(t+\tau)p(t)\,dt \tag{6.55}$$

is the pulse-correlation function. We consider the following example as an illustration.

EXAMPLE 6.6

In this example we consider a situation where the sequence $\{a_k\}$ has memory built into it by the relationship

$$a_k = g_0 A_k + g_1 A_{k-1} \tag{6.56}$$

where g_0 and g_1 are constants and the A_k are random variables such that $A_k = \pm A$, where the sign is determined by a random coin toss independently from pulse to pulse for all k (note that if $g_1 = 0$ there is no memory). It can be shown that

$$E[a_k\, a_{k+m}] = \begin{cases} (g_0^2+g_1^2)A^2, & m=0 \\ g_0 g_1 A^2, & m=\pm 1 \\ 0, & \text{otherwise} \end{cases} \tag{6.57}$$

The assumed pulse shape is $p(t) = \Pi(t/\tau)$ so that the pulse-correlation function is

$$\begin{aligned} r(\tau) &= \frac{1}{T}\int_{-\infty}^{\infty} \Pi\left(\frac{t+\tau}{T}\right)\Pi\left(\frac{t}{T}\right)dt \\ &= \frac{1}{T}\int_{-T/2}^{T/2} \Pi\left(\frac{t+\tau}{T}\right)dt = \Lambda\left(\frac{\tau}{T}\right) \end{aligned} \tag{6.58}$$

where, from Chapter 2, $\Lambda(\tau/T)$ is a unit-height triangular pulse symmetrical about $t=0$ of width $2T$. Thus, the autocorrelation function (6.54) becomes

$$R_X(\tau) = A^2\left\{(g_0^2+g_1^2)\Lambda\left(\frac{\tau}{T}\right) + g_0 g_1\left[\Lambda\left(\frac{\tau+T}{T}\right)+\Lambda\left(\frac{\tau-T}{T}\right)\right]\right\} \tag{6.59}$$

Applying the Wiener–Khinchine theorem, the power spectral density of $X(t)$ is found to be

$$S_X(f) = \Im[R_X(\tau)] = A^2 T \operatorname{sinc}^2(fT)\left[g_0^2+g_1^2+2g_0g_1\cos(2\pi fT)\right] \tag{6.60}$$

Figure 6.6 compares the power spectra for the two cases: (1) $g_0 = 1$ and $g_1 = 0$ (i.e., no memory), and (2) $g_0 = g_1 = 1/\sqrt{2}$ (reinforcing memory between adjacent pulses). For case (1), the resulting power

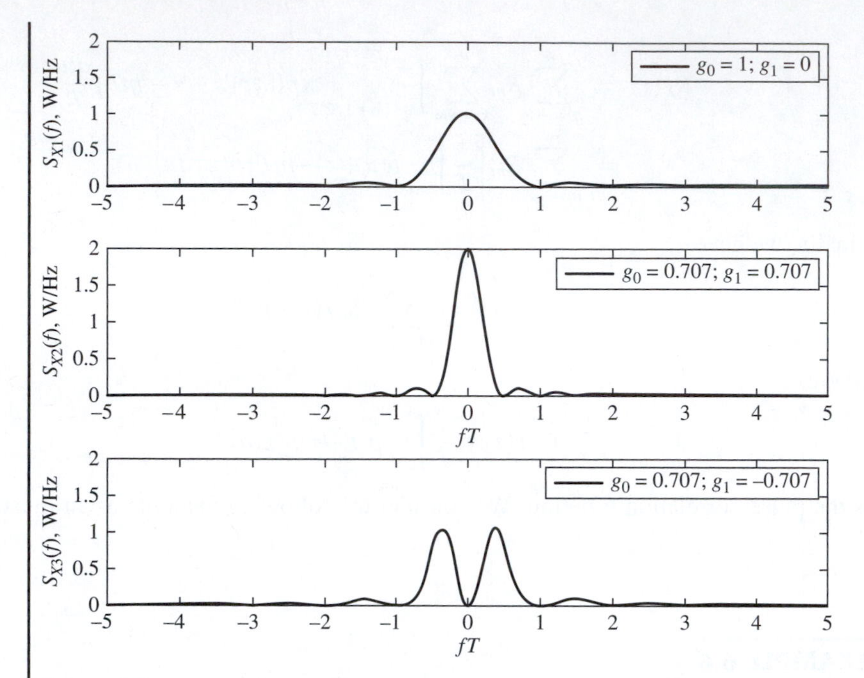

Figure 6.6
Power spectra of binary-valued waveforms. (a) Case in which there is no memory. (b) Case in which there is reinforcing memory between adjacent pulses.(c) Case where the memory between adjacent pulses is antipodal.

spectral density is

$$S_X(f) = A^2 T \operatorname{sinc}^2(fT) \tag{6.61}$$

while for case (2) it is

$$S_X(f) = 2A^2 T \operatorname{sinc}^2(fT) \cos^2(\pi f T) \tag{6.62}$$

In both cases, g_0 and g_1 have been chosen to give a total power of 1 W, which is verified from the plots by numerical integration. Note that in case (2) memory has confined the power sepectrum more than without it. Yet a third case is shown in the bottom plot for which (3) $g_0 = -g_1 = 1/\sqrt{2}$. Now the spectral width is doubled over case (2), but a spectral null appears at $f = 0$.

Other values for g_0 and g_1 can be assumed, and memory between more than just adjacent pulses also can be assumed.

■

6.3.5 Cross-Correlation Function and Cross-Power Spectral Density

Suppose we wish to find the power in the sum of two noise voltages $X(t)$ and $Y(t)$. We might ask if we can simply add their separate powers. The answer is, in general, no. To see why, consider

$$n(t) = X(t) + Y(t) \tag{6.63}$$

where $X(t)$ and $Y(t)$ are two stationary random voltages that may be related (that is, that are not necessarily statistically independent). The power in the sum is

$$\begin{aligned} E\left[n^2(t)\right] &= E\left[\left[X(t)+Y(t)\right]^2\right] \\ &= E\left[X^2(t)\right]+2E[X(t)Y(t)]+E\left[Y^2(t)\right] \\ &= P_X+2P_{XY}+P_Y \end{aligned} \tag{6.64}$$

where P_X and P_Y are the powers of $X(t)$ and $Y(t)$, respectively, and P_{XY} is the cross power. More generally, we define the *cross-correlation function* as

$$R_{XY}(\tau) = E[X(t)Y(t+\tau)] \tag{6.65}$$

In terms of the cross-correlation function, $P_{XY} = R_{XY}(0)$. A *sufficient* condition for P_{XY} to be zero, so that we may simply add powers to obtain total power, is that

$$R_{XY}(\tau) = 0 \quad \text{for all } \tau \tag{6.66}$$

Such processes are said to be *orthogonal*. If processes are statistically independent and at least one of them has zero mean, they are orthogonal. However, orthogonal processes are not necessarily statistically independent.

Cross-correlation functions can be defined for nonstationary processes also, in which case we have a function of two independent variables. We will not need to be this general in our considerations.

A useful symmetry property of the cross-correlation function for jointly stationary processes is

$$R_{XY}(\tau) = R_{YX}(-\tau) \tag{6.67}$$

which can be shown as follows. By definition,

$$R_{XY}(\tau) = E[X(t)Y(t+\tau)] \tag{6.68}$$

Defining $t' = t+\tau$, we obtain

$$R_{XY}(\tau) = E[Y(t')X(t'-\tau)] \triangleq R_{YX}(-\tau) \tag{6.69}$$

since the choice of time origin is immaterial for stationary processes.

The cross-power spectral density of two stationary random processes is defined as the Fourier transform of their cross-correlation function:

$$S_{XY}(f) = \Im[R_{XY}(\tau)] \tag{6.70}$$

It provides, in the frequency domain, the same information about the random processes as does the cross-correlation function.

■ 6.4 LINEAR SYSTEMS AND RANDOM PROCESSES

6.4.1 Input–Output Relationships

In the consideration of the transmission of stationary random waveforms through fixed linear systems, a basic tool is the relationship of the output power spectral density to the input power spectral density, given as

$$S_y(f) = |H(f)|^2 S_x(f) \tag{6.71}$$

The autocorrelation function of the output is the inverse Fourier transform of $S_y(f)$:[6]

$$R_y(\tau) = \Im^{-1}[S_y(f)] = \int_{-\infty}^{\infty} |H(f)|^2 S_x(f) e^{j2\pi f\tau}\, df \tag{6.72}$$

$H(f)$ is the system's frequency-response function, $S_x(f)$ is the power spectral density of the input $x(t)$, $S_y(f)$ is the power spectral density of the output $y(t)$, and $R_y(\tau)$ is the autocorrelation function of the output. The analogous result for energy signals was proved in Chapter 2 (2.200), and the result for power signals was simply stated.

A proof of (6.71) could be carried out by employing (6.25). We will take a somewhat longer route, however, and obtain several useful intermediate results. In addition, the proof provides practice in manipulating convolutions and expectations.

We begin by obtaining the cross-correlation function between input and output, $R_{xy}(\tau)$, defined as

$$R_{xy}(\tau) = E[x(t)y(t+\tau)] \tag{6.73}$$

Using the superposition integral, we have

$$y(t) = \int_{-\infty}^{\infty} h(u)x(t-u)\, du \tag{6.74}$$

where $h(t)$ is the system's impulse response. Equation (6.74) relates each sample function of the input and output processes, so we can write (6.73) as

$$R_{xy}(\tau) = E\left[x(t)\int_{-\infty}^{\infty} h(u)x(t+\tau-u)\, du\right] \tag{6.75}$$

Since the integral does not depend on t, we can take $x(t)$ inside and interchange the operations of expectation and convolution. (Both are simply integrals over different variables.) Since $h(u)$ is not random, (6.75) becomes

$$R_{xy}(\tau) = \int_{-\infty}^{\infty} h(u)E[x(t)x(t+\tau-u)]du \tag{6.76}$$

By definition of the autocorrelation function of $x(t)$,

$$E[x(t)x(t+\tau-u)] = R_x(\tau-u) \tag{6.77}$$

Thus (6.76) can be written as

$$R_{xy}(\tau) = \int_{-\infty}^{\infty} h(u)R_x(\tau-u)\, du \triangleq h(\tau)*R_x(\tau) \tag{6.78}$$

That is, the cross-correlation function of input $x(t)$ with output $y(t)$ is *the autocorrelation function of the input convolved with the system's impulse response*, an easily remembered result. Since (6.78) is a convolution, the Fourier transform of $R_{xy}(\tau)$, the cross-power spectral density of $x(t)$ with $y(t)$, is

$$S_{xy}(f) = H(f)S_x(f) \tag{6.79}$$

[6]For the remainder of this chapter we use lower case x and y to denote input and output random-process signals in keeping with Chapter 2 notation.

From the time-reversal theorem of Table G.6, the cross-power spectral density $S_{yx}(f)$ is

$$S_{yx}(f) = \Im\left[R_{yx}(\tau)\right] = \Im\left[R_{xy}(-\tau)\right] = S_{xy}^{*}(f) \tag{6.80}$$

Employing (6.79) and using the relationships $H^{*}(f) = H(-f)$ and $S_{x}^{*}(f) = S_{x}(f)$ (where $S_{x}(f)$ is real), we obtain

$$S_{yx}(f) = H(-f)S_{x}(f) = H^{*}(f)S_{x}(f) \tag{6.81}$$

where the order of the subscripts is important. Taking the inverse Fourier transform of (6.81) with the aid of the convolution theorem of Fourier transforms in Table G.6, and again using the time-reversal theorem, we obtain

$$R_{yx}(\tau) = h(-\tau)*R_{x}(\tau) \tag{6.82}$$

Let us pause to emphasize what we have obtained. By definition, $R_{xy}(\tau)$ can be written as

$$R_{xy}(\tau) \triangleq E[x(t)\underbrace{[h(t)*x(t+\tau)]}_{y(t+\tau)}] \tag{6.83}$$

Combining this with (6.78), we have

$$E[x(t)[h(t)*x(t+\tau)]] = h(\tau)*R_{x}(\tau) \triangleq h(\tau)*E[x(t)x(t+\tau)] \tag{6.84}$$

Similarly, (6.82) becomes

$$\begin{aligned} R_{yx}(\tau) &\triangleq E[\underbrace{[h(t)*x(t)]}_{y(t)}x(t+\tau)] = h(-\tau)*R_{x}(\tau) \\ &\triangleq h(-\tau)*E[x(t)x(t+\tau)] \end{aligned} \tag{6.85}$$

Thus, bringing the convolution operation outside the expectation gives a convolution of $h(\tau)$ with the autocorrelation function if $h(t)*x(t+\tau)$ is inside the expectation, or a convolution of $h(-\tau)$ with the autocorrelation function if $h(t)*x(t)$ is inside the expectation.

These results are combined to obtain the autocorrelation function of the output of a linear system in terms of the input autocorrelation function as follows:

$$R_{y}(\tau) \triangleq E[y(t)y(t+\tau)] = E[y(t)[h(t)*x(t+\tau)]] \tag{6.86}$$

which follows because $y(t+\tau) = h(t)*x(t+\tau)$. Using (6.84) with $x(t)$ replaced by $y(t)$, we obtain

$$\begin{aligned} R_{y}(\tau) &= h(\tau)*E[y(t)x(t+\tau)] \\ &= h(\tau)*R_{yx}(\tau) \\ &= h(\tau)*[h(-\tau)*R_{x}(\tau)] \end{aligned} \tag{6.87}$$

where the last line follows by substituting from (6.82). Written in terms of integrals, (6.87) is

$$R_{y}(\tau) = \int_{-\infty}^{\infty}\int_{-\infty}^{\infty} h(u)h(v)R_{x}(\tau+v-u)\,dv\,du \tag{6.88}$$

The Fourier transform of (6.87) or (6.88) is the output power spectral density and is easily obtained as follows:

$$\begin{aligned} S_y(f) &\triangleq \Im[R_y(\tau)] = \Im[h(\tau)*R_{yx}(\tau)] \\ &= H(f)S_{yx}(f) \\ &= |H(f)|^2 S_x(f) \end{aligned} \tag{6.89}$$

where (6.81) has been substituted to obtain the last line.

EXAMPLE 6.7

The input to a filter with impulse response $h(t)$ and frequency response function $H(f)$ is a white-noise process with power spectral density,

$$S_x(f) = \frac{1}{2}N_0, \quad -\infty < f < \infty \tag{6.90}$$

The cross-power spectral density between input and output is

$$S_{xy}(f) = \frac{1}{2}N_0 H(f) \tag{6.91}$$

and the cross-correlation function is

$$R_{xy}(\tau) = \frac{1}{2}N_0 h(\tau) \tag{6.92}$$

Hence, we could measure the impulse response of a filter by driving it with white noise and determining the cross-correlation function of input with output. Applications include system identification and channel measurement.

■

6.4.2 Filtered Gaussian Processes

Suppose the input to a linear system is a stationary random process. What can we say about the output statistics? For general inputs and systems, this is usually a difficult question to answer. However, *if the input to a linear system is Gaussian, the output is also Gaussian.*

A nonrigorous demonstration of this is carried out as follows. The sum of two independent Gaussian random variables has already been shown to be Gaussian. By repeated application of this result, we can find that the sum of any number of independent Gaussian random variables is Gaussian.[7] For a fixed linear system, the output $y(t)$ in terms of the input $x(t)$ is given by

$$\begin{aligned} y(t) &= \int_{-\infty}^{\infty} x(\tau)h(t-\tau)dt \\ &= \lim_{\Delta\tau \to 0} \sum_{k=-\infty}^{\infty} x(k\Delta\tau)h(t-k\Delta\tau)\Delta\tau \end{aligned} \tag{6.93}$$

where $h(t)$ is the impulse response. By writing the integral as a sum, we have demonstrated that if $x(t)$ is a white Gaussian process, the output is also Gaussian (but not white) because, at any time t, the right-hand side of (6.93) is simply a linear combination of independent Gaussian

[7]This also follows from Appendix B (B.13).

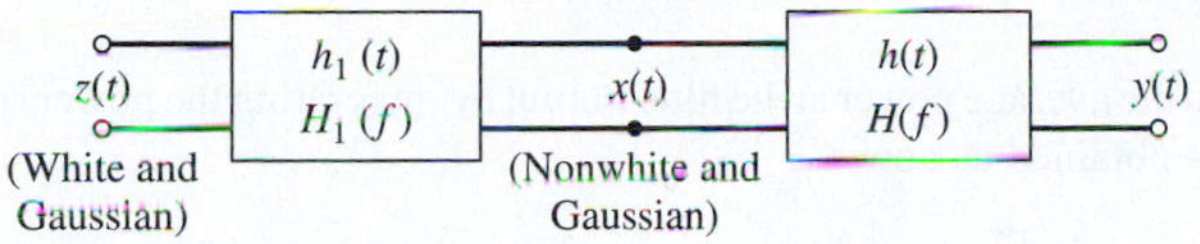

$z(t)$ (White and Gaussian) → $h_1(t)$, $H_1(f)$ → $x(t)$ (Nonwhite and Gaussian) → $h(t)$, $H(f)$ → $y(t)$

Figure 6.7
Cascade of two linear systems with Gaussian input.

random variables. (Recall Example 6.5, where the autocorrelation function of white noise was shown to be a constant times an impulse. Also recall that uncorrelated Gaussian random variables are independent.)

If the input is not white, we can still show that the output is Gaussian by considering the cascade of two linear systems, as shown in Figure 6.7. The system in question is the one with the impulse response $h(t)$. To show that its output is Gaussian, we note that the cascade of $h_1(t)$ with $h(t)$ is a linear system with the impulse response

$$h_2(t) = h_1(t)*h(t) \tag{6.94}$$

This system's input, $z(t)$, is Gaussian and white. Therefore, its output, $y(t)$, is also Gaussian by application of the theorem just proved. However, the output of the system with impulse response $h_1(t)$ is Gaussian by application of the same theorem, but not white. Hence the output of a linear system with nonwhite Gaussian input is Gaussian.

EXAMPLE 6.8

The input to the lowpass RC filter shown in Figure 6.8 is white Gaussian noise with the power spectral density $S_{n_i}(f) = \frac{1}{2}N_0$, $-\infty < f < \infty$. The power spectral density of the output is

$$S_{n_0}(f) = S_{n_i}(f)|H(f)|^2 = \frac{\frac{1}{2}N_0}{1+(f/f_3)^2} \tag{6.95}$$

where $f_3 = (2\pi RC)^{-1}$ is the filter's 3-dB cutoff frequency. Inverse Fourier transforming $S_{n_0}(f)$, we obtain $R_{n_0}(\tau)$, the output autocorrelation function, which is

$$R_{n_0}(\tau) = \frac{\pi f_3 N_0}{2} e^{-2\pi f_3|\tau|} = \frac{N_0}{4RC} e^{-|\tau|/RC}, \quad \frac{1}{RC} = 2\pi f_3 \tag{6.96}$$

The square of the mean of $n_0(t)$ is

$$\overline{n_0(t)}^2 = \lim_{|\tau|\to\infty} R_{n_0}(\tau) = 0 \tag{6.97}$$

and the mean-squared value, which is also equal to the variance since the mean is zero, is

$$\overline{n_0^2(t)} = \sigma_{n_0}^2 = R_{n_0}(0) = \frac{N_0}{4RC} \tag{6.98}$$

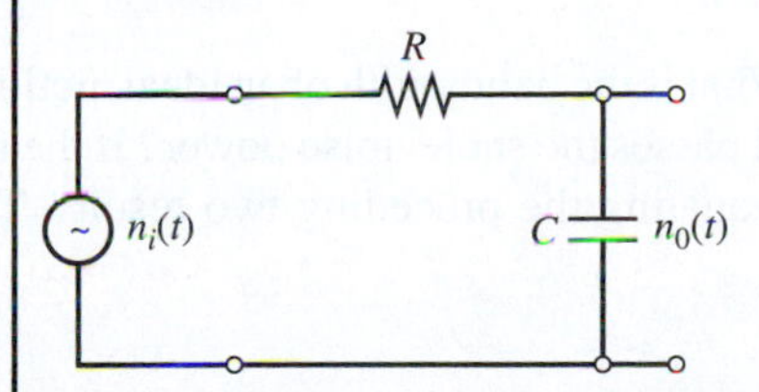

Figure 6.8
A lowpass RC filter with a white-noise input.

Alternatively, we can find the average power at the filter output by integrating the power spectral density of $n_0(t)$. The same result is obtained as above:

$$\overline{n_0^2(t)} = \int_{-\infty}^{\infty} \frac{\frac{1}{2}N_0}{1+(f/f_3)^2}\, df = \frac{N_0}{2\pi RC}\int_0^{\infty} \frac{dx}{1+x^2} = \frac{N_0}{4RC} \tag{6.99}$$

Since the input is Gaussian, the output is Gaussian as well. The first-order pdf is

$$f_{n_0}(y,t) = f_{n_0}(y) = \frac{e^{-2RCy^2/N_0}}{\sqrt{\pi N_0/2RC}} \tag{6.100}$$

by employing (5.194). The second-order pdf at time t and $t+\tau$ is found by substitution into (5.189). Letting X be a random variable that refers to the values the output takes on at time t and Y be a random variable that refers to the values the output takes on at time $t+\tau$, we have, from the preceding results,

$$m_x = m_y = 0 \tag{6.101}$$

$$\sigma_x^2 = \sigma_y^2 = \frac{N_0}{4RC} \tag{6.102}$$

and the correlation coefficient is

$$\rho(\tau) = \frac{R_{n_0}(\tau)}{R_{n_0}(0)} = e^{-|\tau|/RC} \tag{6.103}$$

Referring to Example 6.2, one can see that the random telegraph waveform has the same autocorrelation function as that of the output of the lowpass RC filter of Example 6.8 (with constants appropriately chosen). This demonstrates that processes with drastically different sample functions can have the same second-order averages.

■

6.4.3 Noise-Equivalent Bandwidth

If we pass white noise through a filter that has the frequency-response function $H(f)$, the average power at the output, by (6.72) with $\tau = 0$, is

$$P_{n_0} = \int_{-\infty}^{\infty} \frac{1}{2}N_0|H(f)|^2\, df = N_0\int_0^{\infty} |H(f)|^2\, df \tag{6.104}$$

where $\frac{1}{2}N_0$ is the two-sided power spectral density of the input. If the filter were ideal with bandwidth B_N and midband (maximum) gain[8] H_0, as shown in Figure 6.9, the noise power at the output would be

$$P_{n_0} = H_0^2\left(\frac{1}{2}N_0\right)(2B_N) = N_0 B_N H_0^2 \tag{6.105}$$

The question we now ask is the following: What is the bandwidth of an ideal, fictitious filter that has the same midband gain as $H(f)$ and that passes the same noise power? If the midband gain of $H(f)$ is H_0, the answer is obtained by equating the preceding two results. Thus

[8]Assumed to be finite.

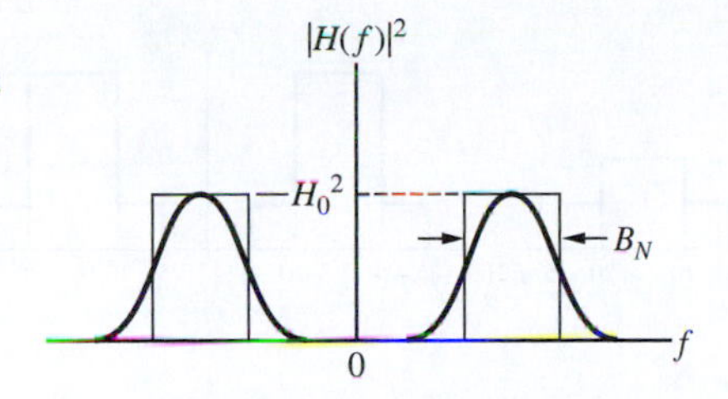

Figure 6.9
Comparison between $|H(f)|^2$ and an idealized approximation.

$$B_N = \frac{1}{H_0^2}\int_0^{\infty} |H(f)|^2\, df \tag{6.106}$$

is the single-side bandwidth of the fictitious filter. B_N is called the *noise-equivalent bandwidth* of $H(f)$.

It is sometimes useful to determine the noise-equivalent bandwidth of a system using time-domain integration. Assume a lowpass system with maximum gain at $f = 0$ for simplicity. By Rayleigh's energy theorem [see (2.89)], we have

$$\int_{-\infty}^{\infty} |H(f)|^2\, df = \int_{-\infty}^{\infty} |h(t)|^2\, dt \tag{6.107}$$

Thus, (6.106) can be written as

$$B_N = \frac{1}{2H_0^2}\int_0^{\infty} |h(t)|^2\, dt = \frac{\int_{-\infty}^{\infty} |h(t)|^2\, dt}{2\left[\int_{-\infty}^{\infty} h(t)\, dt\right]^2} \tag{6.108}$$

where it is noted that

$$H_0 = H(f)|_{f=0} = \int_{-\infty}^{\infty} h(t) e^{-j2\pi f t}\, dt|_{f=0} = \int_{-\infty}^{\infty} h(t)\, dt \tag{6.109}$$

For some systems, (6.108) is easier to evaluate than (6.106).

EXAMPLE 6.9

Assume that a filter has the amplitude-response function illustrated in Figure 6.10(a). Note that assumed filter is not realizable. The purpose of this problem is to provide an illustration of the computation of B_N for a simple filter. The first step is to square $|H(f)|$ to give $|H(f)|^2$, as shown in Figure 6.10(b). By simple geometry, the area under $|H(f)|^2$ for nonnegative frequencies is

$$A = \int_0^{\infty} |H(f)|^2 df = 50 \tag{6.110}$$

Note also that the maximum gain of the actual filter is $H_0 = 2$. For the ideal filter with amplitude response denoted by $H_e(f)$, which is ideal bandpass centered at 15 Hz of single-sided bandwidth B_N and passband gain H_0, we want

$$\int_0^{\infty} |H(f)|^2 df = H_0^2 B_N \tag{6.111}$$

or

$$50 = 2^2 B_N \tag{6.112}$$

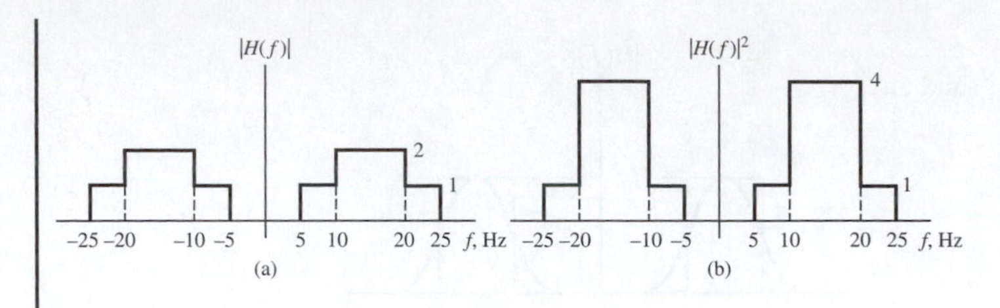

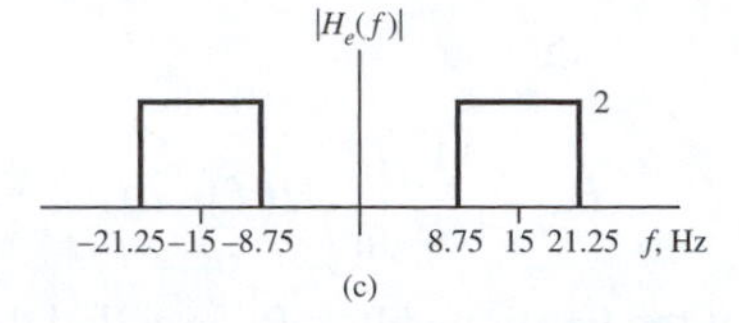

Figure 6.10
Illustrations for Example 6.9.

from which

$$B_N = 12.5 \text{ Hz} \tag{6.113}$$

■

EXAMPLE 6.10

The noise-equivalent bandwidth of an nth-order Butterworth filter for which

$$|H_n(f)|^2 = \frac{1}{1+(f/f_3)^{2n}} \tag{6.114}$$

is

$$\begin{aligned} B_N(n) &= \int_0^\infty \frac{1}{1+(f/f_3)^{2n}}\,df = f_3 \int_0^\infty \frac{1}{1+x^{2n}}\,dx \\ &= \frac{\pi f_3/2n}{\sin(\pi/2n)}, \quad n = 1, 2, \ldots \end{aligned} \tag{6.115}$$

where f_3 is the 3-dB frequency of the filter. For $n = 1$, Equation (6.115) gives the result for a lowpass RC filter, namely, $B_N(1) = \frac{\pi}{2} f_3$. As n approaches infinity, $H_n(f)$ approaches the frequency-response function of an ideal lowpass filter of single-sided bandwidth f_3. The noise-equivalent bandwidth is

$$\lim_{n\to\infty} B_N(n) = f_3 \tag{6.116}$$

as it should be by its definition. As the cutoff of a filter becomes sharper, its noise-equivalent bandwidth approaches its 3-dB bandwidth.

■

EXAMPLE 6.11

To illustrate the application of (6.108), consider the computation of the noise-equivalent bandwidth of a first-order Butterworth filter computed in the time domain. Its impulse response is

$$h(t) = \Im^{-1}\left[\frac{1}{1+jf/f_3}\right] = 2\pi f_3 e^{-2\pi f_3 t} u(t) \tag{6.117}$$

According to (6.108), the noise-equivalent bandwidth of this filter is

$$B_N = \frac{\int_0^\infty (2\pi f_3)^2 e^{-4\pi f_3 t}\, dt}{2\left[\int_0^\infty 2\pi f_3 e^{-2\pi f_3 t}\, dt\right]^2} = \frac{\pi f_3}{2} \frac{\int_0^\infty e^{-v} dv}{2\left(\int_0^\infty e^{-u}\, du\right)^2} = \frac{\pi f_3}{2} \tag{6.118}$$

which checks with (6.115) if $n = 1$ is substituted. ■

■ 6.5 NARROWBAND NOISE

6.5.1 Quadrature-Component and Envelope-Phase Representation

In most communication systems operating at a carrier frequency f_0, the bandwidth of the channel, B, is small compared with f_0. In such situations, it is convenient to represent the noise in terms of quadrature components as

$$n(t) = n_c(t)\cos(2\pi f_0 t + \theta) - n_s(t)\sin(2\pi f_0 t + \theta) \tag{6.119}$$

where θ is an arbitrary phase angle. In terms of envelope and phase components, $n(t)$ can be written as

$$n(t) = R(t)\cos[2\pi f_0 t + \phi(t) + \theta] \tag{6.120}$$

where

$$R(t) = \sqrt{n_c^2 + n_s^2} \tag{6.121}$$

and

$$\phi(t) = \tan^{-1}\left(\frac{n_s(t)}{n_c(t)}\right) \tag{6.122}$$

Actually, any random process can be represented in either of these forms, but if a process is narrowband, $R(t)$ and $\phi(t)$ can be interpreted as the slowly varying envelope and phase, respectively, as sketched in Figure 6.11.

Figure 6.12 shows the block diagram of a system for producing $n_c(t)$ and $n_s(t)$ where θ is, as yet, an arbitrary phase angle. Note that the composite operations used in producing $n_c(t)$ and $n_s(t)$ constitute linear systems (superposition holds from input to output). Thus, if $n(t)$ is a Gaussian process, so are $n_c(t)$ and $n_s(t)$. (The system of Figure 6.12 is to be interpreted as relating input and output processes sample function by sample function.)

We will prove several properties of $n_c(t)$ and $n_s(t)$. Most important, of course, is whether equality really holds in (6.119) and in what sense. It is shown in Appendix C that

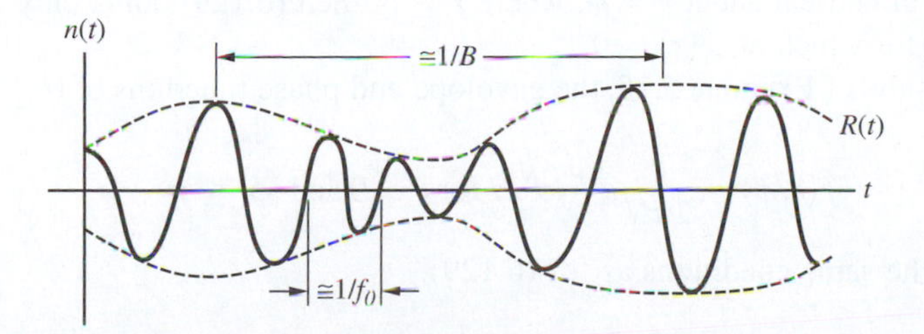

Figure 6.11
A typical narrowband noise waveform.

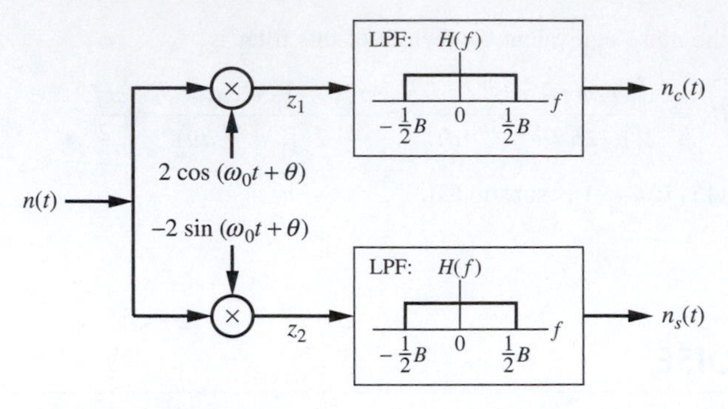

Figure 6.12 The operations involved in producing $n_c(t)$ and $n_s(t)$.

$$E\left[\{n(t) - [n_c(t)\cos(2\pi f_0 t + \theta) - n_s(t)\sin(2\pi f_0 t + \theta)]\}^2\right] = 0 \tag{6.123}$$

That is, the mean-squared error between a sample function of the actual noise process and the right-hand side of (6.119) is zero (averaged over the ensemble of sample functions).

More useful when using the representation in (6.119), however, are the following properties:

Means

$$\overline{n(t)} = \overline{n_c(t)} = \overline{n_s(t)} = 0 \tag{6.124}$$

Variances

$$\overline{n^2(t)} = \overline{n_c^2(t)} = \overline{n_s^2(t)} \triangleq N \tag{6.125}$$

Power spectral densities

$$S_{n_c}(f) = S_{n_s}(f) = \text{Lp}[S_n(f - f_0) + S_n(f + f_0)] \tag{6.126}$$

Cross-power spectral density

$$S_{n_c n_s}(f) = j\text{Lp}[S_n(f - f_0) - S_n(f + f_0)] \tag{6.127}$$

where Lp[ ] denotes the lowpass part of the quantity in brackets; $S_n(f)$, $S_{n_c}(f)$, and $S_{n_s}(f)$ are the power spectral densities of $n(t)$, $n_c(t)$, and $n_s(t)$, respectively; $S_{n_c n_s}(f)$ is the cross-power spectral density of $n_c(t)$ and $n_s(t)$. From (6.127), we see that

$$R_{n_c n_s}(\tau) \equiv 0 \quad \text{for all } \tau \text{ if } \text{Lp}[S_n(f - f_0) - S_n(f + f_0)] = 0 \tag{6.128}$$

This is an especially useful property in that it tells us that $n_c(t)$ and $n_s(t)$ are uncorrelated if the power spectral density of $n(t)$ is symmetrical about $f = f_0$, where $f > 0$. If, in addition, $n(t)$ is Gaussian, $n_c(t)$ and $n_s(t)$ will be independent Gaussian processes because they are uncorrelated, and the joint pdf of $n_c(t)$ and $n_s(t + \tau)$ for *any* delay τ, will simply be of the form

$$f(n_c, t; n_s, t + \tau) = \frac{1}{2\pi N} e^{-\left(n_c^2 + n_s^2\right)/2N} \tag{6.129}$$

If $S_n(f)$ is not symmetrical about $f = f_0$, where $f \geq 0$, then (6.129) holds only for $\tau = 0$ or other values of τ for which $R_{n_c n_s}(\tau) = 0$.

Using the results of Example 5.15, the envelope and phase functions of (6.120) have the joint pdf

$$f(r, \phi) = \frac{r}{2\pi N} e^{-r^2/2N} \quad \text{for } r > 0 \text{ and } |\phi| \leq \pi \tag{6.130}$$

which holds for the same conditions as for (6.129).

6.5.2 The Power Spectral Density Function of $n_c(t)$ and $n_s(t)$

To prove (6.126), we first find the power spectral density of $z_1(t)$, as defined in Figure 6.12, by computing its autocorrelation function and Fourier transforming the result. To simplify the derivation, it is assumed that θ is a uniformly distributed random variable in $[0, 2\pi]$ and is statistically independent of $n(t)$.[9]

The autocorrelation function of $z_1(t) = 2n(t)\cos(\omega_0 t + \theta)$ is

$$\begin{aligned} R_{z_1}(\tau) &= E[4n(t)n(t+\tau)\cos(2\pi f_0 t + \theta)\cos[2\pi f_0(t+\tau)+\theta]] \\ &= 2E[n(t)n(t+\tau)]\cos(2\pi f_0 \tau) \\ &\quad + 2E[n(t)n(t+\tau)\cos(4\pi f_0 t + 2\pi f_0 \tau + 2\theta)] \\ &= 2R_n(\tau)\cos(2\pi f_0 \tau) \end{aligned} \tag{6.131}$$

where $R_n(\tau)$ is the autocorrelation function of $n(t)$ and $\omega_0 = 2\pi f_0$ in Figure 6.12. In obtaining (6.131), we used appropriate trigonometric identities in addition to the independence of $n(t)$ and θ. Thus, by the multiplication theorem of Fourier transforms, the power spectral density of $z_1(t)$ is

$$\begin{aligned} S_{z_1}(f) &= S_n(f)*[\delta(f-f_0)+\delta(f+f_0)] \\ &= S_n(f-f_0)+S_n(f+f_0) \end{aligned} \tag{6.132}$$

of which only the lowpass part is passed by $H(f)$. Thus the result for $S_{n_c}(f)$ expressed by (6.126) follows. A similar proof can be carried out for $S_{n_s}(f)$. Equation (6.125) follows by integrating (6.126) over all f.

Next, let us consider (6.127). To prove it, we need an expression for $R_{z_1 z_2}(\tau)$, the cross-correlation function of $z_1(t)$ and $z_2(t)$. (See Figure 6.12.) By definition, and from Figure 6.12,

$$\begin{aligned} R_{z_1 z_2}(\tau) &= E[z_1(t)z_2(t+\tau)] \\ &= E[4n(t)\,n(t+\tau)\cos(2\pi f_0 t + \theta)\sin[2\pi f_0(t+\tau)+\theta]] \\ &= 2R_n(\tau)\sin(2\pi f_0 \tau) \end{aligned} \tag{6.133}$$

where we again used appropriate trigonometric identities and the independence of $n(t)$ and θ. Letting $h(t)$ be the impulse response of the lowpass filters in Figure 6.12 and employing (6.84) and (6.85), the cross-correlation function of $n_c(t)$ and $n_s(t)$ can be written as

$$\begin{aligned} R_{n_c n_s}(\tau) &\triangleq E[n_c(t)n_s(t+\tau)] = E[[h(t)*z_1(t)]n_s(t+\tau)] \\ &= h(-\tau)*E[z_1(t)n_s(t+\tau)] \\ &= h(-\tau)*E[z_1(t)[h(t)*z_2(t+\tau)]] \\ &= h(-\tau)*h(\tau)*E[z_1(t)z_2(t+\tau)] \\ &= h(-r)*[h(\tau)*R_{z_1 z_2}(\tau)] \end{aligned} \tag{6.134}$$

The Fourier transform of $R_{n_c n_s}(\tau)$ is the cross-power spectral density, $S_{n_c n_s}(f)$, which, from the convolution theorem, is given by

$$\begin{aligned} S_{n_c n_s}(f) &= H(f)\Im[h(-\tau)*R_{z_1 z_2}(\tau)] \\ &= H(f)H*(f)S_{z_1 z_2}(f) \\ &= |H(f)|^2 S_{z_1 z_2}(f) \end{aligned} \tag{6.135}$$

[9]This might be satisfactory for modeling noise where the phase can be viewed as completely random. In other situations, where knowledge of the phase makes this an inappropriate assumption, a cyclostationary model may be more appropriate.

From (6.133) and the frequency-translation theorem, it follows that

$$\begin{aligned} S_{z_1 z_2}(f) &= \Im\left[jR_n(\tau)\left(e^{j2\pi f_0 \tau} - e^{-j2\pi f_0 \tau}\right)\right] \\ &= j[S_n(f-f_0) - S_n(f+f_0)] \end{aligned} \tag{6.136}$$

Thus, from (6.135),

$$\begin{aligned} S_{n_c n_s}(f) &= j|H(f)|^2[S_n(f-f_0) - S_n(f+f_0)] \\ &= j\text{Lp}[S_n(f-f_0) - S_n(f+f_0)] \end{aligned} \tag{6.137}$$

which proves (6.127). Note that since the cross-power spectral density $S_{n_c n_s}(f)$ is imaginary, the cross-correlation function $R_{n_c n_s}(\tau)$ is odd. Thus $R_{n_c n_s}(0)$ is zero if the cross-correlation function is continuous at $\tau = 0$, which is the case for bandlimited signals.

EXAMPLE 6.12

Let us consider a bandpass random process with the power spectral density shown in Figure 6.13(a). Choosing the center frequency of $f_0 = 7$ Hz results in $n_c(t)$ and $n_s(t)$ being uncorrelated. Figure 6.13(b) shows $S_{z_1}(f)$ [or $S_{z_2}(f)$] for $f_0 = 7$ Hz with $S_{n_c}(f)$ [or $S_{n_s}(f)$], that is, the lowpass part of $S_{z_1}(f)$, shaded. The integral of $S_n(f)$ is $2(6)(2) = 24$ W, which is the same result obtained from integrating the shaded portion of Figure 6.13(b).

Now suppose f_0 is chosen as 5 Hz. Then $S_{z_1}(f)$ and $S_{z_2}(t)$ are as shown in Figure 6.12(c), with $S_{n_c}(f)$ shown shaded. From (6.127), it follows that $-jS_{n_c n_s}(f)$ is the shaded portion of Figure 6.12(d). Because of the asymmetry that results from the choice of f_0, $n_c(t)$ and $n_s(t)$ are not uncorrelated. As a matter of interest, we can calculate $R_{n_c n_s}(\tau)$ easily by using the transform pair

$$2AW\,\text{sinc}(2W\tau) \longleftrightarrow A\Pi\left(\frac{f}{2W}\right) \tag{6.138}$$

and the frequency-translation theorem. From Figure 6.12(d), it follows that

$$S_{n_c n_s}(f) = 2j\left\{-\Pi\left[\frac{1}{4}(f-3)\right] + \Pi\left[\frac{1}{4}(f+3)\right]\right\} \tag{6.139}$$

which results in the cross-correlation function

$$\begin{aligned} R_{n_c n_s}(\tau) &= 2j\left[-4\,\text{sinc}(4\tau)e^{j6\pi\tau} + 4\,\text{sinc}(4\tau)e^{-j6\pi\tau}\right] \\ &= 16\,\text{sinc}(4\tau)\sin(6\pi\tau) \end{aligned} \tag{6.140}$$

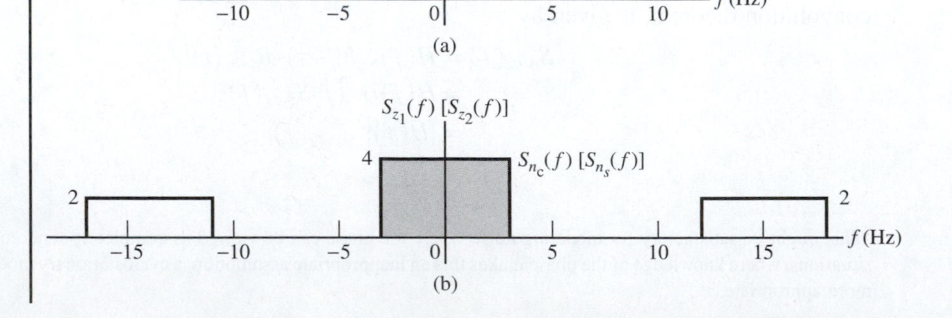

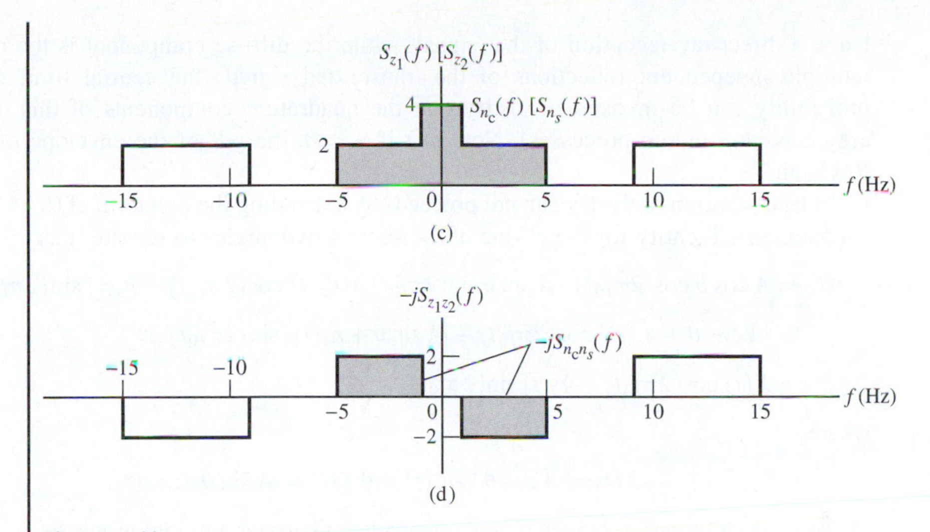

Figure 6.13
Spectra for Example 6.11. (a) Bandpass spectrum. (b) Lowpass spectra for $f_0 = 7$ Hz. (c) Lowpass spectra for $f_0 = 5$ Hz. (d) Cross-spectra for $f_0 = 5$ Hz.

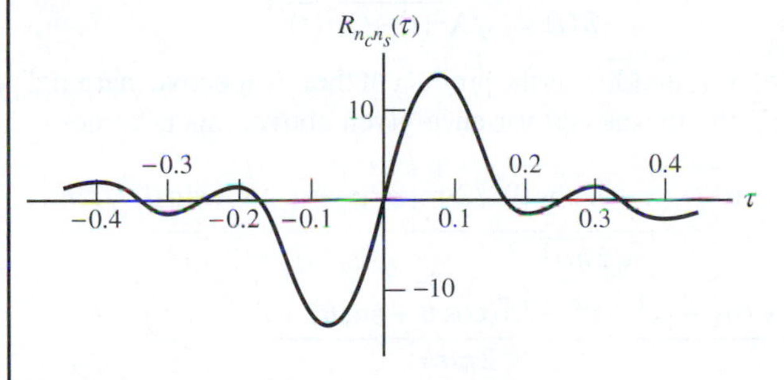

Figure 6.14
Cross-correlation function of $n_c(t)$ and $n_s(t)$ for Example 6.11.

This cross-correlation function is shown in Figure 6.14. Although $n_c(t)$ and $n_s(t)$ are not uncorrelated, we see that τ may be chosen such that $R_{n_cn_s}(\tau) = 0$ for particular values of τ ($\tau = 0, \ \pm\frac{1}{6}, \ \pm\frac{1}{3}, \ \ldots$). ■

6.5.3 Ricean Probability Density Function

A useful random process model for many applications, for example, signal fading, is the sum of a random phased sinusoid and bandlimited Gaussian random noise. Thus, consider a sample function of this process expressed as

$$z(t) = A\cos(2\pi f_0 t + \theta) + n_c(t)\cos(2\pi f_0 t) - n_s(t)\sin(2\pi f_0 t) \tag{6.141}$$

where $n_c(t)$ and $n_s(t)$ are Gaussian quadrature components of the bandlimited, stationary, Gaussian random process $n_c(t)\cos(2\pi f_0 t) - n_s(t)\sin(2\pi f_0 t)$, A is a constant amplitude, and θ is a random variable uniformly distributed in $[0, 2\pi]$. The pdf of the envelope of this stationary random process at any time t is said to be *Ricean* after its originator, S. O. Rice. The first term is often referred to as the *specular* component and the latter two terms make up the *diffuse* component. This is in keeping with the idea that (6.141) results from transmitting an unmodulated sinusoidal signal through a dispersive channel, with the specular component being a direct-ray reception of that signal, while the diffuse component is the resultant of multiple independent reflections of the transmitted signal (the central limit theorem of probability can be invoked to justify that the quadrature components of this diffuse part are Gaussian random processes). Note that if $A = 0$, the pdf of the envelope of (6.141) is Rayleigh.

The derivation of the Ricean pdf proceeds by expanding the first term of (6.141) using the trigonometric identity for the cosine of the sum of two angles to rewrite it as

$$\begin{aligned} z(t) &= A\cos\theta\cos(2\pi f_0 t) - A\sin\theta\sin(2\pi f_0 t) + n_c(t)\cos(2\pi f_0 t) - n_s(t)\sin(2\pi f_0 t) \\ &= [A\cos\theta + n_c(t)]\cos(2\pi f_0 t) - [A\sin\theta + n_s(t)]\sin(2\pi f_0 t) \\ &= X(t)\cos(2\pi f_0 t) - Y(t)\sin(2\pi f_0 t) \end{aligned} \tag{6.142}$$

where

$$X(t) = A\cos\theta + n_c(t) \text{ and } Y(t) = A\sin\theta + n_s(t) \tag{6.143}$$

These random processes, given θ, are independent Gaussian random processes with variance σ^2. Their means are $E[X(t)] = A\cos\theta$ and $E[Y(t)] = A\sin\theta$, respectively. The goal is to find the pdf of

$$R(t) = \sqrt{X^2(t) + Y^2(t)} \tag{6.144}$$

Given θ, the joint pdf of $X(t)$ and $Y(t)$ is the product of their respective marginal pdfs since they are independent. Using the means and variance given above, this becomes

$$\begin{aligned} f_{XY}(x, y) &= \frac{\exp\left[-(x - A\cos\theta)^2/2\sigma^2\right]}{\sqrt{2\pi\sigma^2}} \frac{\exp\left[-(y - A\sin\theta)^2/2\sigma^2\right]}{\sqrt{2\pi\sigma^2}} \\ &= \frac{\exp\{-[x^2 + y^2 - 2A(\cos\theta + \sin\theta) + A^2]/2\sigma^2\}}{2\pi\sigma^2} \end{aligned} \tag{6.145}$$

Now make the change of variables

$$\begin{aligned} x &= r\cos\phi, \\ y &= r\sin\phi, \end{aligned} \quad r \geq 0 \text{ and } 0 \leq \phi < 2\pi \tag{6.146}$$

Recall that transformation of a joint pdf requires multiplication by the Jacobian of the transformation, which in this case is just r. Thus, the joint pdf of the random variables R and Φ is

$$\begin{aligned} f_{R\Phi}(r, \phi) &= \frac{\exp\{-[r^2 + A^2 - 2rA(\cos\theta\cos\phi + \sin\theta\sin\phi)]/2\sigma^2\}}{2\pi\sigma^2} \\ &= \frac{r}{2\pi\sigma^2}\exp\{-[r^2 + A^2 - 2rA\cos(\theta - \phi)]/2\sigma^2\} \end{aligned} \tag{6.147}$$

The pdf over R alone may be obtained by integrating over ϕ with the aid of the definition

$$I_0(u) = \frac{1}{2\pi}\int_0^{2\pi} \exp(u\cos\alpha)\,d\alpha \tag{6.148}$$

where $I_0(u)$ is referred to as the modified Bessel function of order zero. Since the integrand of (6.148) is periodic with period 2π, the integral can be over any 2π range. The result of the integration of (6.147) over ϕ produces

$$f_R(r) = \frac{r}{\sigma^2} \exp\left[-(r^2 + A^2)/2\sigma^2 \right] I_0\left(\frac{Ar}{\sigma^2}\right), \quad r \geq 0 \tag{6.149}$$

Since the result is independent of θ, this is the marginal pdf of R alone. From (6.148), it follows that $I_0(0) = 1$ so that with $A = 0$ (6.149) reduces to the Rayleigh pdf, as it should.

Often, (6.149) is expressed in terms of the parameter $K = A^2/2\sigma^2$, which is the ratio of the powers in the steady component [first term of (6.141)] to the random Gaussian component [second and third terms of (6.141)] When this is done, (6.149) becomes

$$f_R(r) = \frac{r}{\sigma^2} \exp\left[-\left(\frac{r^2}{2\sigma^2} + K\right)\right] I_0\left(\sqrt{2K}\frac{r}{\sigma}\right), \quad r \geq 0 \tag{6.150}$$

As K becomes large, (6.150) approaches a Gaussan pdf. The parameter K is often referred to as the *Ricean K-factor*.

From (6.144) it follows that

$$\begin{aligned} E\left[R^2\right] &= E\left[X^2\right] + E\left[Y^2\right] \\ &= E\left[\left[A\cos\theta + n_c(t)\right]^2 + \left[A\sin\theta + n_s(t)\right]^2\right] \\ &= E\left[A^2\cos^2\theta + A^2\sin^2\theta\right] + 2AE\left[n_c(t)\cos\theta + n_s(t)\sin\theta\right] + E\left[n_c^2(t)\right] + E\left[n_s^2(t)\right] \\ &= A^2 + 2\sigma^2 \\ &= 2\sigma^2(1+K) \end{aligned} \tag{6.151}$$

Other moments for a Ricean random variable must be expressed in terms of confluent hypergeometric functions.[10]

Summary

1. A random process is completely described by the N-fold joint pdf of its amplitudes at the arbitrary times $t_1, t_2, \ldots, t_N$. If this pdf is invariant under a shift of the time origin, the process is said to be *statistically stationary in the strict sense*.
2. The autocorrelation function of a random process, computed as a statistical average, is defined as

$$R(t_1, t_2) = \int_{-\infty}^{\infty}\int_{-\infty}^{\infty} x_1 x_2 f_{X_1X_2}(x_1, t_1;\ x_2, t_2)\, dx_1\, dx_2$$

where $f_{X_1X_2}(x_1, t_1;\ x_2, t_2)$ is the joint amplitude pdf of the process at times t_1 and t_2. *If the process is strict-sense stationary,*

$$R(t_1, t_2) = R(t_2 - t_1) = R(\tau)$$

where $\tau \triangleq t_2 - t_1$.

3. A process whose statistical average mean and variance are time independent and whose autocorrelation function is a function only of $t_2 - t_1 = \tau$ is termed

[10]See, for example, J. Proakis and M. Saleni, *Digital Communications,* 5th ed., New York: McGraw Hill, 2007.

wide-sense stationary. Strict-sense stationary processes are also wide-sense stationary. The converse is true only for special cases; for example, wide-sense stationarity for a Gaussian process guarantees strict-sense stationarity.

4. A process for which statistical averages and time averages are equal is called *ergodic. Ergodicity implies stationarity, but the reverse is not necessarily true.*

5. The Wiener-Khinchine theorem states that the autocorrelation function and the power spectral density of a stationary random process are a Fourier transform pair. An expression for the power spectral density of a random process that is often useful is

$$S_n(f) = \lim_{T \to \infty} \frac{1}{T} E\left[|\Im[n_T(t)]|^2 \right]$$

where $n_T(t)$ is a sample function truncated to T s, centered about $t = 0$.

6. The autocorrelation function of a random process is a real, even function of the delay variable τ with an absolute maximum at $\tau = 0$. It is periodic for periodic random processes, and its Fourier transform is nonnegative for all frequencies. As $\tau \to \pm\infty$, the autocorrelation function approaches the square of the mean of the random process unless the random process is periodic. $R(0)$ gives the total average power in a random process.

7. White noise has a constant power spectral density $\frac{1}{2}N_0$ for all f. Its autocorrelation function is $\frac{1}{2}N_0\delta(\tau)$. For this reason, it is sometimes called *delta-correlated noise.* It has infinite power and is therefore a mathematical idealization. However, it is, nevertheless, a useful approximation in many cases.

8. The cross-correlation function of two stationary random processes $X(t)$ and $Y(t)$ is defined as

$$R_{XY}(\tau) = E[X(t)Y(t+\tau)]$$

Their cross-power spectral density is

$$S_{XY}(f) = \Im[R_{XY}(\tau)]$$

They are said to be *orthogonal* if $R_{XY}(\tau) = 0$ for all τ.

9. Consider a linear system with the impulse response $h(t)$ and the frequency-response function $H(f)$ with random input $x(t)$ and output $y(t)$. Then

$$\begin{aligned}
S_Y(f) &= |H(f)|^2 S_X(f) \\
R_Y(\tau) &= \Im^{-1}[S_Y(f)] = \int_{-\infty}^{\infty} |H(f)|^2 S_X(f) e^{j2\pi f\tau}\, df \\
R_{XY}(\tau) &= h(\tau) * R_X(\tau) \\
S_{XY}(f) &= H(f) S_X(f) \\
R_{YX}(\tau) &= h(-\tau) * R_X(\tau) \\
S_{YX}(f) &= H^*(f) S_X(f)
\end{aligned}$$

where $S(f)$ denotes the spectral density and $R(\tau)$ denotes the autocorrelation function.

10. The output of a linear system with Gaussian input is Gaussian.

11. The noise-equivalent bandwidth of a linear system with a frequency-response function $H(f)$ is defined as

$$B_N = \frac{1}{H_0^2}\int_0^\infty |H(f)|^2\, df$$

where H_0 represents the maximum value of $|H(f)|$. If the input is white noise with the single-sided power spectral density N_0, the output power is

$$P_0 = H_0^2 N_0 B_N$$

An equivalent expression for the noise-equivalent bandwidth written in terms of the impulse response of the filter is

$$B_N = \frac{\int_{-\infty}^{\infty} |h(t)|^2\, dt}{2\left[\int_{-\infty}^{\infty} h(t)\, dt\right]^2}$$

12. The quadrature-component representation of a bandlimited random process $n(t)$ is

$$n(t) = n_c(t)\cos(2\pi f_0 t + \theta) - n_s(t)\sin(2\pi f_0 t + \theta)$$

where θ is an arbitrary phase angle. The envelope-phase representation is

$$n(t) = R(t)\cos[2\pi f_0 t + \phi(t) + \theta]$$

where $R^2(t) = n_c^2(t) + n_s^2(t)$ and $\tan[\phi(t)] = n_s(t)/n_c(t)$. If the process is narrowband, n_c, n_s, R, and ϕ vary slowly with respect to $\cos(2\pi f_0 t)$ and $\sin(2\pi f_0 t)$. If the power spectral density of $n(t)$ is $S_n(f)$, the power spectral densities of $n_c(t)$ and $n_s(t)$ are

$$S_{n_c}(f) = S_{n_s}(f) = \text{Lp}[S_n(f - f_0) + S_n(f + f_0)]$$

where Lp[ ] denotes the low-frequency part of the quantity in the brackets. If $\text{Lp}[S_n(f + f_0) - S_n(f - f_0)] = 0$, then $n_c(t)$ and $n_s(t)$ are orthogonal. The average powers of $n_c(t)$, $n_s(t)$, and $n(t)$ are equal. The processes $n_c(t)$ and $n_s(t)$ are given by

$$n_c(t) = \text{Lp}[2n(t)\cos(2\pi f_0 t + \theta)]$$

and

$$n_s(t) = -\text{Lp}[2n(t)\sin(2\pi f_0 t + \theta)]$$

Since these operations are linear, $n_c(t)$ and $n_s(t)$ will be Gaussian if $n(t)$ is Gaussian. Thus, $n_c(t)$ and $n_s(t)$ are independent if $n(t)$ is zero-mean Gaussian with a power spectral density that is symmetrical about $f = f_0$ for $f > 0$.

13. The Ricean pdf gives the distribution of envelope values assumed by the sum of a sinusoid with phase uniformly distributed in $[0, 2\pi]$ plus bandlimited Gaussian noise. It is convenient in various applications including modeling of fading channels.

Further Reading

Papoulis (1991) is a recommended book for random processes. The references given in Chapter 5 also provide further reading on the subject matter of this chapter.

Problems

Section 6.1

6.1. A fair die is thrown. Depending on the number of spots on the up face, the following random processes are generated. Sketch several examples of sample functions for each case. (A is a constant.)

a. $X(t,\zeta) = \begin{cases} 2A, & \text{1 or 2 spots up} \\ 0, & \text{3 or 4 spots up} \\ -2A, & \text{5 or 6 spots up} \end{cases}$

b. $X(t,\zeta) = \begin{cases} 3A, & \text{1 spot up} \\ 2A, & \text{2 spots up} \\ A, & \text{3 spots up} \\ -A, & \text{4 spots up} \\ -2A, & \text{5 spots up} \\ -3A, & \text{6 spots up} \end{cases}$

c. $X(t,\zeta) = \begin{cases} 4A, & \text{1 spot up} \\ 2A, & \text{2 spots up} \\ At, & \text{3 spots up} \\ -At, & \text{4 spots up} \\ -2A, & \text{5 spots up} \\ -4A, & \text{6 spots up} \end{cases}$

Section 6.2

6.2. Referring to Problem 6.1, what are the following probabilities for each case?

a. $F_X(X \leq 2A, t = 4)$

b. $F_X(X \leq 0, t = 4)$

c. $F_X(X \leq 2A, t = 2)$

6.3. A random process is composed of sample functions that are square waves, each with constant amplitude A, period T_0, and random delay τ as sketched in Figure 6.15. The pdf of τ is

$$f(\tau) = \begin{cases} 1/T_0, & |\tau| \leq T_0/2 \\ 0, & \text{otherwise} \end{cases}$$

a. Sketch several typical sample functions.

b. Write the first-order pdf for this random process at some arbitrary time t_0.
(*Hint*: Because of the random delay τ, the pdf is independent of t_0. Also, it might be easier to deduce the cdf and differentiate it to get the pdf.)

6.4. Let the sample functions of a random process be given by

$$X(t) = A\cos(2\pi f_0 t)$$

where f_0 is fixed and A has the pdf

$$f_A(a) = \frac{e^{-\alpha^2/2\sigma_a^2}}{\sqrt{2\pi}\sigma_a}$$

This random process is passed through an ideal integrator to give a random process $Y(t)$.

a. Find an expression for the sample functions of the output process $Y(t)$.

b. Write down an expression for the pdf of $Y(t)$ at time t_0.
Hint: Note that $\sin 2\pi f_0 t_0$ is just a constant.

c. Is $Y(t)$ stationary? Is it ergodic?

6.5. Consider the random process of Problem 6.3.

a. Find the time-average mean and the autocorrelation function.

b. Find the ensemble-average mean and the autocorrelation function.

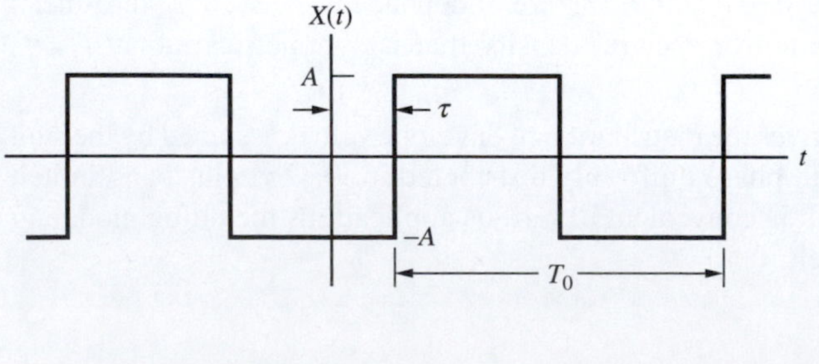

Figure 6.15

c. Is this process wide-sense stationary? Why or why not?

6.6. Consider the random process of Example 6.1 with the pdf of θ given by

$$p(\theta) = \begin{cases} 2/\pi, & \pi/2 \le \theta \le \pi \\ 0, & \text{otherwise} \end{cases}$$

a. Find the statistical-average and time-average mean and variance.

b. Find the statistical-average and time-average autocorrelation functions.

c. Is this process ergodic?

6.7. Consider the random process of Problem 6.4.

a. Find the time-average mean and autocorrelation function.

b. Find the ensemble-average mean and autocorrelation function.

c. Is this process wide-sense stationary? Why or why not?

6.8. The voltage of the output of a noise generator whose statistics are known to be closely Gaussian and stationary is measured with a DC voltmeter and a true root-mean-square (rms) voltmeter that is AC coupled. The DC meter reads 6 V, and the true rms meter reads 7 V. Write down an expression for the first-order pdf of the voltage at any time $t = t_0$. Sketch and dimension the pdf.

Section 6.3

6.9. Which of the following functions are suitable autocorrelation functions? Tell why or why not. (ω_0, τ_0, τ_1, A, B, C, and f_0 are positive constants.)

a. $A\cos(\omega_0 \tau)$.

b. $A\Lambda(\tau/\tau_0)$, where $\Lambda(x)$ is the unit-area triangular function defined in Chapter 2.

c. $A\Pi(\tau/\tau_0)$, where $\Pi(x)$ is the unit-area pulse function defined in Chapter 2.

d. $A\exp(-\tau/\tau_0)u(\tau)$, where $u(x)$ is the unit step function.

e. $A\exp(-|\tau|/\tau_0)$.

f. $A\,\text{sinc}(f_0\tau) = A\sin(\pi f_0 \tau)/\pi f_0 \tau$.

6.10. A bandlimited white-noise process has a double-sided power spectral density of 2×10^{-5} W/Hz in the frequency range $|f| \le 1$ kHz. Find the autocorrelation function of the noise process. Sketch and fully dimension the resulting autocorrelation function.

6.11. Consider a random binary pulse waveform as analyzed in Example 6.6, but with half-cosine pulses given by $p(t) = \cos(2\pi t/2T)\Pi(t/T)$. Obtain and sketch the autocorrelation function for the two cases considered in Example 6.6, namely,

a. $a_k = \pm A$ for all k, where A is a constant, with $R_m = A^2$, $m = 0$, and $R_m = 0$ otherwise.

b. $a_k = A_k + A_{k-1}$ with $A_k = \pm A$ and $E[A_k A_{k+m}] = A^2, m = 0$, and zero otherwise.

c. Find and sketch the power spectral density for each preceding case.

6.12. Two random processes are given by

$$X(t) = n(t) + A\cos(2\pi f_0 t + \theta)$$

and

$$Y(t) = n(t) + A\sin(2\pi f_0 t + \theta)$$

where A and f_0 are constants and θ is a random variable uniformly distributed in the interval $[-\pi, \pi]$. The first term, $n(t)$, represents a stationary random noise process with autocorrelation function $R_n(\tau) = B\Lambda(\tau/\tau_0)$, where B and τ_0 are nonnegative constants.

a. Find and sketch their autocorrelation functions. Assume values for the various constants involved.

b. Find and sketch the cross-correlation function of these two random processes.

6.13. Given two independent, wide-sense stationary random processes $X(t)$ and $Y(t)$ with autocorrelation functions $R_X(\tau)$ and $R_Y(\tau)$, respectively.

a. Show that the autocorrelation function $R_Z(\tau)$ of their product $Z(t) = X(t)Y(t)$ is given by

$$R_Z(\tau) = R_X(\tau)R_Y(\tau)$$

b. Express the power spectral density of $Z(t)$ in terms of the power spectral densities of $X(t)$ and $Y(t)$, denoted as $S_X(f)$ and $S_Y(f)$, respectively.

c. Let $X(t)$ be a bandlimited stationary noise process with power spectral density $S_X(f) = 10\Pi(f/200)$, and let $Y(t)$ be the process defined by sample functions of the form

$$Y(t) = 5\cos(50\pi t + \theta)$$

where θ is a uniformly distributed random variable in the interval $[0, 2\pi)$. Using the results derived in parts (a) and (b), obtain the autocorrelation function and power spectral density of $Z(t) = X(t)Y(t)$.

6.14. A random signal has the autocorrelation function

$$R(\tau) = 9 + 3\Lambda(\tau/5)$$

where $\Lambda(x)$ is the unit-area triangular function defined in Chapter 2. Determine the following:

a. The AC power.

b. The DC power.

c. The total power.

d. The power spectral density. Sketch it and label carefully.

6.15. A random process is defined as $Y(t) = X(t) + X(t-T)$, where $X(t)$ is a wide-sense stationary random process with autocorrelation function $R_X(\tau)$ and power spectral density $S_x(f)$.

a. Show that $R_Y(\tau) = 2R_X(\tau) + R_X(\tau+T) + R_X(\tau-T)$.

b. Show that $S_Y(f) = 4S_X(f)\cos^2(\pi f T)$.

c. If $X(t)$ has autocorrelation function $R_X(\tau) = 5\Lambda(\tau)$, where $\Lambda(\tau)$ is the unit-area triangular function, and $T = 0.5$, find and sketch the power spectral density of $Y(t)$ as defined in the problem statement.

6.16. The power spectral density of a wide-sense stationary random process is given by

$$S_X(f) = 10\delta(f) + 25\,\text{sinc}^2(5f) + 5\delta(f-10) + 5\delta(f+10)$$

a. Sketch and fully dimension this power spectral density function.

b. Find the power in the DC component of the random process.

c. Find the total power.

d. Given that the area under the main lobe of the sinc-squared function is approximately 0.9 of the total area, which is unity if it has unity amplitude, find the fraction of the total power contained in this process for frequencies between 0 and 0.2 Hz.

6.17. Given the following functions of τ,

$$R_{X_1}(\tau) = 4\exp(-\alpha|\tau|)\cos 2\pi\tau$$
$$R_{X_2}(\tau) = 2\exp(-\alpha|\tau|) + 4\cos 2\pi b\tau$$
$$R_{X_3}(f) = 5\exp(-4\tau^2)$$

a. Sketch each function and fully dimension.

b. Find the Fourier transforms of each and sketch. With the information of part (a) and the Fourier transforms justify that each is suitable for an autocorrelation function.

c. Determine the value of the DC power, if any, for each one.

d. Determine the total power for each.

e. Determine the frequency of the periodic component, if any, for each.

Section 6.4

6.18. A stationary random process $n(t)$ has a power spectral density of 10^{-6} W/Hz, $-\infty < f < \infty$. It is passed through an ideal lowpass filter with frequency-response function $H(f) = \Pi(f/500\text{ kHz})$, where $\Pi(x)$ is the unit-area pulse function defined in Chapter 2.

a. Find and sketch the power spectral density of the output?

b. Obtain sketch the autocorrelation function of the output.

c. What is the power of the output process? Find it two different ways.

6.19. An ideal finite-time integrator is characterized by the input-output relationship

$$Y(t) = \frac{1}{T}\int_{t-T}^{t} X(\alpha)\,d\alpha$$

a. Justify that its impulse response is $h(t) = \frac{1}{T}[u(t) - u(t-T)]$.

b. Obtain its frequency response function. Sketch it.

c. The input is white noise with two-sided power spectral density $N_0/2$. Find the power spectral density of the output of the filter.

d. Show that the autocorrelation function of the output is

$$R_0(\tau) = \frac{N_0}{2T}\Lambda(\tau/T)$$

where $\Lambda(x)$ is the unit-area triangular function defined in Chapter 2.

e. What is the equivalent noise bandwidth of the integrator?

f. Show that the result for the output noise power obtained using the equivalent noise bandwidth found in part (e) coincides with the result found from the autocorrelation function of the output found in part (d).

6.20. White noise with two-sided power spectral density $N_0/2$ drives a second-order Butterworth filter with frequency-response function magnitude

$$|H_{2\text{bu}}(f)| = \frac{1}{\sqrt{1+(f/f_3)^4}}$$

where f_3 is its 3-dB cutoff frequency.

a. What is the power spectral density of the filter's output?

b. Show that the autocorrelation function of the output is

$$R_0(r) = \frac{\pi f_3 N_0}{2} \exp\left(-\sqrt{2}\pi f_3|\tau|\right)\cos\left(\sqrt{2}\pi f_3|\tau| - \pi/4\right)$$

Plot as a function of $f_3\tau$. *Hint*: Use the integral given below:

$$\int_0^\infty \frac{\cos(ax)}{b^4+x^4}dx = \frac{\sqrt{2}\pi}{4b^3}\exp\left(-ab/\sqrt{2}\right)\times$$

$$\left[\cos\left(ab/\sqrt{2}\right)+\sin\left(ab/\sqrt{2}\right)\right], \quad a,b>0$$

c. Does the output power obtained by taking $\lim_{\tau\to 0} R_0(\tau)$ check with that calculated using the equivalent noise bandwidth for a Butterworth filter as given by (6.115)?

6.21. A power spectral density given by

$$S_Y(f) = \frac{f^2}{f^4+100}$$

is desired. A white-noise source of two-sided power spectral density 1 W/Hz is available. What is the frequency response function of the filter to be placed at the noise-source output to produce the desired power spectral density?

6.22. Obtain the autocorrelation functions and power spectral densities of the outputs of the following systems with the input autocorrelation functions or power spectral densities given.

a.

Transfer function

$$H(f) = \Pi(f/2B)$$

Autocorrelation function of input

$$R_X(\tau) = \frac{N_0}{2}\delta(\tau)$$

N_0 and B are positive constants.

b.

Impulse response

$$h(t) = A\exp(-\alpha t)u(t)$$

Power spectral density of input :

$$S_X(f) = \frac{B}{1+(2\pi\beta f)^2}$$

A, α, B, and β are positive constants.

6.23. The input to a lowpass filter with impulse response

$$h(t) = \exp(-10t)u(t)$$

is white, Gaussian noise with single-sided power spectral density of 2 W/Hz. Obtain the following:

a. The mean of the output

b. The power spectral density of the output

c. The autocorrelation function of the output

d. The probability density function of the output at an arbitrary time t_1

e. The joint probability density function of the output at times t_1 and $t_1+0.03$ s

6.24. Find the noise-equivalent bandwidths for the following first-and second-order lowpass filters in terms of their 3-dB bandwidths. Refer to Chapter 2 to determine the magnitudes of their transfer functions.

a. Chebyshev

b. Butterworth

6.25. A second-order Butterworth filter, has 3-dB bandwidth of 500 Hz. Determine the unit impulse response of the filter, and use it to compute the noise-equivalent bandwidth of the filter. Check your result against the appropriate special case of Example 6.9.

6.26. Determine the noise-equivalent bandwidths for the filters having transfer functions given below:

a. $H_a(f) = \Pi(f/4)+\Pi(f/2)$.

b. $H_b(f) = 2\Lambda(f/50)$.

c. $H_c(f) = 10/(10+j2\pi f)$.

d. $H_d(f) = \Pi(f/10)+\Lambda(f/5)$.

6.27. A filter has frequency-response function

$$H(f) = H_0(f-500)+H_0(f+500)$$

where

$$H_0(f) = 2\Lambda(f/100)$$

Find the noise-equivalent bandwidth of the filter.

6.28. Determine the noise-equivalent bandwidths of the systems having the following transfer functions. *Hint*: Use the time-domain approach.

a. $H_a(f) = 10/[(j2\pi f + 2)(j2\pi f + 25)]$.

b. $H_b(f) = 100/(j2\pi f + 10)^2$.

Section 6.5

6.29. Noise $n(t)$ has the power spectral density shown in Figure 6.16. We write

$$n(t) = n_c(t)\cos(2\pi f_0 t + \theta) - n_s(t)\sin(2\pi f_0 t + \theta)$$

Make plots of the power spectral densities of $n_c(t)$ and $n_s(t)$ for the following cases:

a. $f_0 = f_1$.

b. $f_0 = f_2$.

c. $f_0 = \frac{1}{2}(f_2 + f_1)$.

d. For which of these cases are $n_c(t)$ and $n_s(t)$ uncorrelated?

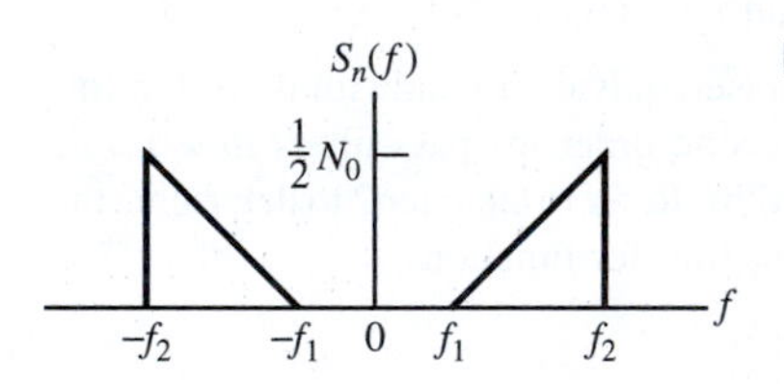

Figure 6.16

6.30.

a. If $S_n(f) = \alpha^2/(\alpha^2 + 4\pi^2 f^2)$, show that $R_n(\tau) = Ke^{-\alpha|\tau|}$. Find K.

b. Find $R_n(\tau)$ if

$$S_n(f) = \frac{\frac{1}{2}\alpha^2}{\alpha^2 + 4\pi^2(f - f_0)^2} + \frac{\frac{1}{2}\alpha^2}{\alpha^2 + 4\pi^2(f + f_0)^2}$$

c. if $n(t) = n_c(t)\cos(2\pi f_0 t + \theta) - n_s(t)\sin(2\pi f_0 t + \theta)$, find $S_{n_c}(f)$, and $S_{n_c n_s}(f)$, where $S_n(f)$ is as given in part (b). Sketch each spectral density.

6.31. The double-sided power spectral density of noise $n(t)$ is shown in Figure 6.17. If $n(t) = n_c(t)\cos(2\pi f_0 t + \theta) - n_s(t)\sin(2\pi f_0 t + \theta)$, find and plot $S_{n_c}(f)$, $S_{n_s}(f)$, and $S_{n_c n_s}(f)$ for the following cases:

a. $f_0 = \frac{1}{2}(f_1 + f_2)$.

b. $f_0 = f_1$.

c. $f_0 = f_2$.

d. Find $R_{n_c n_s}(\tau)$ for each case where $S_{n_c n_s}(f)$ is not zero. Plot.

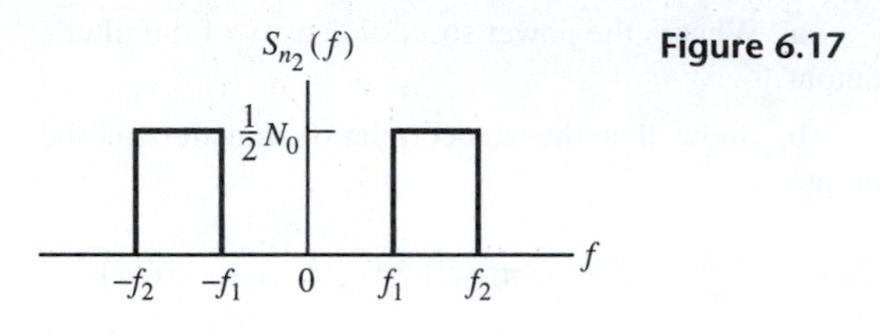

Figure 6.17

6.32. A noise waveform $n_1(t)$ has the bandlimited power spectral density shown in Figure 6.18. Find and plot the power spectral density of $n_2(t) = n_1(t)\cos(\omega_0 t + \theta) - n_1(t)\sin(\omega_0 t + \theta)$, where θ is a uniformly distributed random variable in $[0, 2\pi)$.

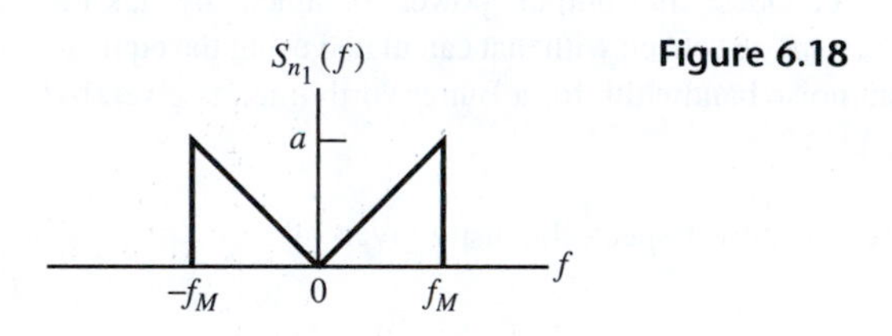

Figure 6.18

Section 6.5
Problems Extending Text Material

6.33. Consider a signal-plus-noise process of the form

$$z(t) = A\cos[2\pi(f_0 + f_d)t] + n(t)$$

with

$$n(t) = n_c(t)\cos(2\pi f_0 t) - n_s(t)\sin(2\pi f_0 t)$$

an ideal bandlimited Gaussian white-noise process with double-sided power spectral density equal to $N_0/2$ for $f_0 - B/2 \le |f| \le f_0 + B/2$ and zero otherwise. Write $z(t)$ as

$$z(t) = A\cos[2\pi(f_0 + f_d)t] + n_c'(t)\cos[2\pi(f_0 + f_d)t] - n_s'(t)\sin[2\pi(f_0 + f_d)t]$$

a. Express $n_c'(t)$ and $n_s'(t)$ in terms of $n_c(t)$ and $n_s(t)$. Using the techniques developed in Section 6.5, find the power spectral densities of $n_c'(t)$ and $n_s'(t)$, $S_{n_c'}(t)$ and $S_{n_s'}(f)$, respectively.

b. Find the cross-spectral density of $n_c'(t)$ and $n_s'(t)$, $S_{n_c' n_s'}(f)$, and the cross-correlation function, $R_{n_c' n_s'}(\tau)$. Are $n_c'(t)$ and $n_s'(t)$ correlated? Are $n_c'(t)$ and $n_s'(t)$ sampled at the same time instant independent?

6.34. A random process is composed of sample functions of the form

$$x(t) = n(t) \sum_{k=-\infty}^{\infty} \delta(t - kT_s) = \sum_{k=-\infty}^{\infty} n_k \delta(t - kT_s)$$

where $n(t)$ is a wide-sense stationary random process with the autocorrelation function $R_n(\tau)$, and $n_k = n(kT_s)$.

a. If T_s is chosen to satisfy

$$R_n(kT_s) = 0, \quad k = 1, 2, \ldots$$

so that the samples $n_k = n(kT_s)$ are orthogonal, use (6.35) to show that the power spectral density of $tx(t)$ is

$$S_x(f) = \frac{R_n(0)}{T_s} = f_s R_n(0) = f_s \overline{n^2(t)}, \quad -\infty < f < \infty$$

b. If $x(t)$ is passed through a filter with impulse response $h(t)$ and frequency-response function $H(f)$, show that the power spectral density of the output random process $y(t)$ is

$$S_y(f) = f_s \overline{n^2(t)} \, |H(f)|^2, \quad -\infty < f < \infty$$

6.35. Consider the system shown in Figure 6.19 as a means of approximately measuring $R_x(\tau)$, where $x(t)$ is stationary.

a. Show that $E[y] = R_x(\tau)$.

b. Find an expression for σ_y^2 if $x(t)$ is Gaussian and has zero mean.

Hint: If x_1, x_2, x_3, and x_4 are Gaussian with zero mean, it can be shown that

$$E[x_1 x_2 x_3 x_4] = E[x_1 x_2]E[x_3 x_4] + E[x_1 x_3]E[x_2 x_4] + E[x_1 x_4]E[x_2 x_3]$$

6.36. A useful average in the consideration of noise in FM demodulation is the cross-correlation

$$R_{y\dot{y}}(\tau) \triangleq E\left[y(t)\frac{dy(t+\tau)}{dt}\right]$$

where $y(t)$ is assumed stationary.

a. Show that

$$R_{y\dot{y}}(\tau) = \frac{dR_y(\tau)}{d\tau}$$

where $R_y(\tau)$ is the autocorrelation function of $y(t)$. (*Hint*: The frequency-response function of a differentiator is $H(f) = j2\pi f$.)

b. If $y(t)$ is Gaussian, write down the joint pdf of

$$Y \triangleq y(t) \quad \text{and} \quad Z \triangleq \frac{dy(t)}{dt}$$

at any time t, assuming the ideal lowpass power spectral density

$$S_y(f) = \frac{1}{2} N_0 \Pi\left(\frac{f}{2B}\right)$$

Express your answer in terms of N_0 and B.

c. Can one obtain a result for the joint pdf of $y(t)$ and $dy(t)/dt$ if $y(t)$ is obtained by passing white noise through a lowpass RC filter? Why or why not?

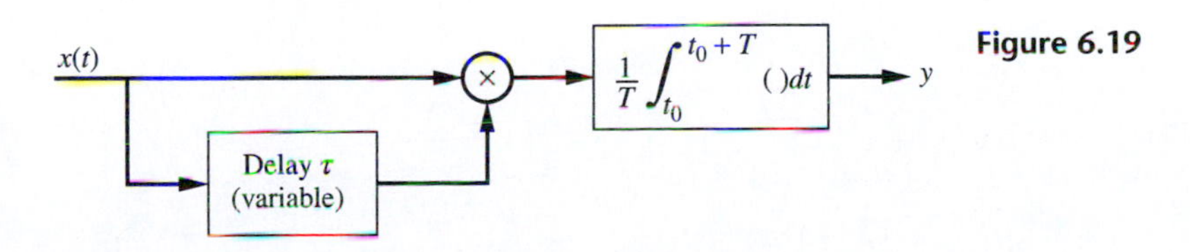

Figure 6.19

Computer Exercises

6.1. In this computer exercise we reexamine Example 6.1. A random process is defined by

$$X(t) = A \cos(2\pi f_0 t + \theta)$$

Using a random number generator program generate 20 values of θ uniformly distributed in the range $0 \leq \theta < 2\pi$. Using these 20 values of θ generate 20 sample functions of the process $X(t)$. Using these 20 sample functions do the following:

a. Plot the sample functions on a single set of axes.

b. Determine $E[X(t)]$ and $E[X^2(t)]$ as time averages.

c. Determine $E[X(t)]$ and $E[X^2(t)]$ as ensemble averages.

d. Compare the results with those obtained in Example 6.1.

6.2. Repeat the previous computer exercise with 20 values of θ uniformly distributed in the range $-\pi/4 \leq \theta < \pi/4$.

6.3. Check the correlation between the random variable X and Y generated by the random number generator of Computer Exercise 5.2 by computing the sample correlation coefficient of 1000 pairs according to the definition

$$\rho(X, Y) = \frac{1}{(N-1)\widehat{\sigma}_1\widehat{\sigma}_2} \sum_{n=1}^{N} (X_n - \widehat{\mu}_X)(Y_n - \widehat{\mu}_Y)$$

where

$$\widehat{\mu}_X = \frac{1}{N}\sum_{n=1}^{N} X_n$$

$$\widehat{\mu}_Y = \frac{1}{N}\sum_{n=1}^{N} Y_n$$

$$\widehat{\sigma}_X^2 = \frac{1}{N-1}\sum_{n=1}^{N}(X_n - \widehat{\mu}_X)^2$$

and

$$\widehat{\sigma}_Y^2 = \frac{1}{N-1}\sum_{n=1}^{N}(Y_n - \widehat{\mu}_X)^2$$

6.4. Write a MATLAB program to plot the Ricean pdf. Use the form (6.150) and plot for $K = 1$, 10, and 100 on the same axes. Use r/σ as the independent variable and plot $\sigma^2 f(r)$.

CHAPTER 7

NOISE IN MODULATION SYSTEMS

In Chapters 5 and 6 the subjects of probability and random processes were studied. These concepts led to a representation for bandlimited noise, which will now be used for the analysis of basic analog communication systems and for introductory considerations of digital systems operating in the presence of noise. The remaining chapters of this book will focus on digital systems in more detail. This chapter is essentially a large number of example problems, most of which focus on different systems and modulation techniques.

Noise is present in varying degrees in all electrical systems. This noise is often low level and can often be neglected in those portions of a system where the signal level is high. However, in many communications applications the receiver input signal level is very small, and the effects of noise significantly degrade system performance. Noise can take several different forms, depending upon the source, but the most common form is due to the random motion of charge carriers. As discussed in more detail in Appendix A, whenever the temperature of a conductor is above 0 K, the random motion of charge carriers results in *thermal noise*. The variance of thermal noise, generated by a resistive element, such as a cable, and measured in a bandwidth B, is given by

$$\sigma_n^2 = 4kTRB \quad \text{V}^2 \tag{7.1}$$

where k is Boltzman's constant (1.38×10^{-23} J/K), T is the temperature of the element in degrees kelvin, and R is the resistance in ohms. Note that the noise variance is directly proportional to temperature, which illustrates the reason for using supercooled amplifiers in low-signal environments, such as for radio astronomy. Note also that the noise variance is independent of frequency, which implies that the noise power spectral density may be treated as constant or white. The range of B over which the thermal noise can be assumed white is a function of temperature. However, for temperatures greater than approximately 3 K, the white noise assumption holds for bandwidths less than approximately 10 GHz. As the temperature increases, the bandwidth over which the white-noise assumption is valid increases. At standard temperature (290 K) the white-noise assumption is valid to bandwidths exceeding 1000 GHz. At very high frequencies other noise sources, such as quantum noise, become significant, and the white-noise assumption is no longer valid. These ideas are discussed in more detail in Appendix A.

We also assume that thermal noise is Gaussian (has a Gaussian amplitude pdf). Since thermal noise results from the random motion of a large number of charge carriers, with each charge carrier making a small contribution to the noise, the Gaussian assumption is justified through the central-limit theorem. Thus, if we assume that the noise of interest is thermal noise, and the bandwidth is

smaller than 10 to 1000 GHz (depending on temperature), the additive white Gaussian noise (AWGN) model is a valid and useful noise model. We will make this assumption throughout this chapter.

As pointed out in Chapter 1, system noise results from sources external to the system as well as from sources internal to the system. Since noise is unavoidable in any practical system, techniques for minimizing the impact of noise on system performance must often be used if high-performance communications are desired. In the present chapter, appropriate performance criteria for system performance evaluation will be developed. After this, a number of systems will be analyzed to determine the impact of noise on system operation. It is especially important to note the differences between linear and nonlinear systems. We will find that the use of nonlinear modulation, such as FM, allows *improved performance* to be obtained at the expense of *increased transmission bandwidth*. Such trade-offs do not exist when linear modulation is used.

■ 7.1 SIGNAL-TO-NOISE RATIOS

In Chapter 3, systems that involve the operations of modulation and demodulation were studied. In this section we extend that study to the performance of linear demodulators in the presence of noise. We concentrate our efforts on the calculation of signal-to-noise ratios since the signal-to-noise ratio is often a useful and easily determined figure of merit of system performance.

7.1.1 Baseband Systems

In order to have a basis for comparing system performance, we determine the signal-to-noise ratio at the output of a baseband system. Recall that a baseband system involves no modulation or demodulation. Consider Figure 7.1(a). Assume that the signal power is finite at P_T W and that the additive noise has the double-sided power spectral density $\frac{1}{2}N_0$ W/Hz over a bandwidth B,

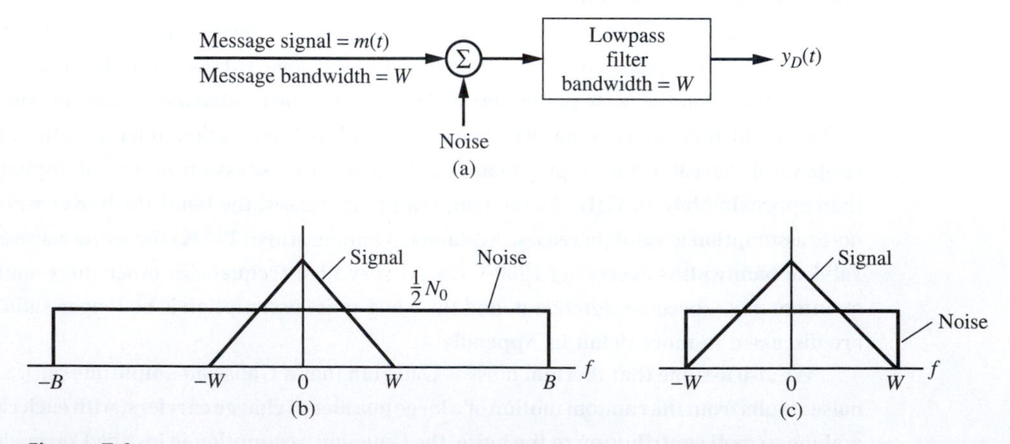

Figure 7.1
Baseband system. (a) Diagram. (b) Spectra at filter input. (c) Spectra at filter output.

which is assumed to exceed W, as illustrated in Figure 7.1(b). The total noise power in the bandwidth B is

$$\int_{-B}^{B} \frac{1}{2} N_0 \, df = N_0 B \tag{7.2}$$

and, therefore, the signal-to-noise ratio (SNR) at the filter input is

$$(\text{SNR})_i = \frac{P_T}{N_0 B} \tag{7.3}$$

Since the message signal $m(t)$ is assumed to be bandlimited with bandwidth $W < B$, a simple lowpass filter can be used to enhance the SNR. This filter is assumed to pass the signal component without distortion but removes the out-of-band noise as illustrated in Figure 7.1(c). Assuming an ideal filter with bandwidth W, the signal is passed without distortion. Thus the signal power at the lowpass filter output is P_T, which is the signal power at the filter input. The noise at the filter output is

$$\int_{-W}^{W} \frac{1}{2} N_0 \, df = N_0 W \tag{7.4}$$

which is less than $N_0 B$ since $W < B$. Thus the SNR at the filter output is

$$(\text{SNR})_o = \frac{P_T}{N_0 W} \tag{7.5}$$

The filter therefore enhances the SNR by the factor

$$\frac{(\text{SNR})_o}{(\text{SNR})_i} = \frac{P_T}{N_0 W} \frac{N_0 B}{P_T} = \frac{B}{W} \tag{7.6}$$

Since (7.5) describes the SNR achieved with a simple baseband system in which all out-of-band noise is removed by filtering, it is a reasonable standard for making comparisons of system performance. This reference, $P_T/N_0 W$, will be used extensively in the work to follow, in which the output SNR is determined for a variety of basic systems.

7.1.2 Double-Sideband Systems

As a first example, we compute the noise performance of the coherent DSB demodulator first considered in Chapter 3. Consider the block diagram in Figure 7.2, which illustrates a coherent demodulator preceded by a predetection filter. Typically, the predetection filter is the IF filter as discussed in Chapter 3. The input to this filter is the modulated signal plus white Gaussian noise of double-sided power spectral density $\frac{1}{2}N_0$ W/Hz. Since the

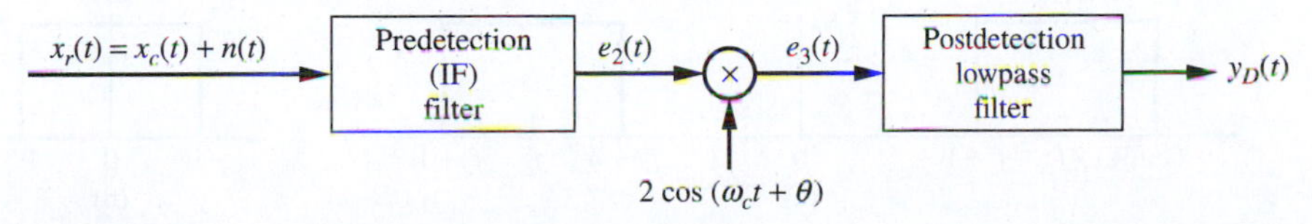

Figure 7.2
Double-sideband demodulator.

transmitted signal $x_c(t)$ is assumed to be a DSB signal, the received signal $x_r(t)$ can be written as

$$x_r(t) = A_c m(t)\cos(2\pi f_c t + \theta) + n(t) \tag{7.7}$$

where $m(t)$ is the message and θ is used to denote our uncertainty of the carrier phase or, equivalently, the time origin. Note that, using this model, the SNR at the input to the predetection filter is zero since the power in white noise is infinite. If the predetection filter bandwidth is (ideally) $2W$, the DSB signal is completely passed by the filter. Using the technique developed in Chapter 6, the noise at the predetection filter output can be expanded into its direct and quadrature components. This gives

$$\begin{aligned} e_2(t) &= A_c m(t)\cos(2\pi f_c t + \theta) \\ &\quad + n_c(t)\cos(2\pi f_c t + \theta) - n_s(t)\sin(2\pi f_c t + \theta) \end{aligned} \tag{7.8}$$

where the total noise power is $\overline{n_0^2(t)} = \frac{1}{2}\overline{n_c^2(t)} + \frac{1}{2}\overline{n_s^2(t)}$ and is equal to $2N_0 W$.

The predetection SNR, measured at the input to the multiplier, is easily determined. The signal power is $\frac{1}{2}A_c^2\overline{m^2}$, where m is understood to be a function of t and the noise power is $2N_0 W$ as shown in Figure 7.3(a). This yields the predetection SNR,

$$(\mathrm{SNR})_T = \frac{A_c^2\overline{m^2}}{4WN_0} \tag{7.9}$$

In order to compute the postdetection SNR, $e_3(t)$ is first computed. This gives

$$\begin{aligned} e_3(t) &= A_c m(t) + n_c(t) + A_c m(t)\cos[2(2\pi f_c t + \theta)] \\ &\quad + n_c(t)\cos[2(2\pi f_c t + \theta)] - n_s(t)\sin[2(2\pi f_c t + \theta)] \end{aligned} \tag{7.10}$$

The double-frequency terms about $2f_c$ are removed by the postdetection filter to produce the baseband (demodulated) signal

$$y_D(t) = A_c m(t) + n_c(t) \tag{7.11}$$

Note that additive noise on the demodulator input gives rise to additive noise at the demodulator output. This is a property of linearity.

The postdetection signal power is $A_c^2\overline{m^2}$, and the postdetection noise power is $\overline{n_c^2}$ or $2N_0 W$, as shown on Figure 7.3(b). This gives the postdetection SNR as

$$(\mathrm{SNR})_D = \frac{A_c^2\overline{m^2}}{2N_0 W} \tag{7.12}$$

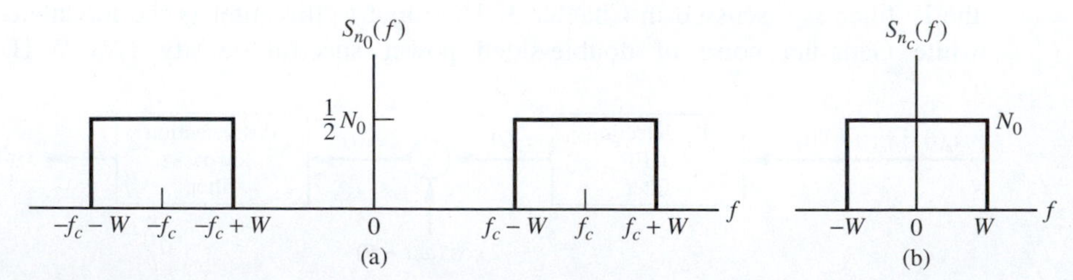

Figure 7.3
(a) Predetection and (b) postdetection filter output noise spectra for DSB demodulation.

Since the signal power is $\frac{1}{2}A_c^2\overline{m^2} = P_T$, we can write the postdetection SNR as

$$(\text{SNR})_D = \frac{P_T}{N_0 W} \tag{7.13}$$

which is equivalent to the ideal baseband system.

The ratio of $(\text{SNR})_D$ to $(\text{SNR})_T$ is referred to as *detection gain* and is often used as a figure of merit for a demodulator. Thus, for the coherent DSB demodulator, the detection gain is

$$\frac{(\text{SNR})_D}{(\text{SNR})_T} = \frac{A_c^2\overline{m^2}}{2N_0 W}\frac{4N_0 W}{A_c^2\overline{m^2}} = 2 \tag{7.14}$$

At first sight, this result is somewhat misleading, for it appears that we have gained 3 dB. This is true for the demodulator because it suppresses the quadrature noise component. However, a comparison with the baseband system reveals that nothing is gained, insofar as the SNR at the system output is concerned. The predetection filter bandwidth must be $2W$ if DSB modulation is used. This results in double the noise bandwidth at the output of the predetection filter and, consequently, double the noise power. The 3 dB detection gain is exactly sufficient to overcome this effect and give an overall performance equivalent to the baseband reference given by (7.5). Note that this ideal performance is only achieved if all out-of-band noise is removed and if the demodulation carrier is perfectly phase coherent with the original carrier used for modulation.

In practice PLLs, as we studied in Chapter 3, are used to establish carrier recovery at the demodulator. If noise is present in the loop bandwidth, phase jitter will result. We will consider the effect on performance resulting from a combination of additive noise and demodulation phase errors in a later section.

7.1.3 Single-Sideband Systems

Similar calculations are easily carried out for SSB systems. For SSB, the predetection filter input can be written as

$$x_r(t) = A_c[m(t)\cos(2\pi f_c t + \theta) \pm \widehat{m}(t)\sin(2\pi f_c t + \theta)] + n(t) \tag{7.15}$$

where $\widehat{m}(t)$ denotes the Hilbert transform of $m(t)$. Recall from Chapter 3 that the plus sign is used for LSB SSB and the minus sign is used for USB SSB. Since the minimum bandwidth of the predetection bandpass filter is W for SSB, the center frequency of the predetection filter is $f_x = f_c \pm \frac{1}{2}W$, where the sign depends on the choice of sideband.

We could expand the noise about the center frequency $f_x = f_c \pm \frac{1}{2}W$ since, as we saw in Chapter 6, we are free to expand the noise about any frequency we choose. It is slightly more convenient, however, to expand the noise about the carrier frequency f_c. For this case, the predetection filter output can be written as

$$\begin{aligned} e_2(t) &= A_c[m(t)\cos(2\pi f_c t + \theta) \pm \widehat{m}(t)\sin(2\pi f_c t + \theta)] \\ &\quad + n_c(t)\cos(2\pi f_c t + \theta) - n_s(t)\sin(2\pi f_c t + \theta) \end{aligned} \tag{7.16}$$

where, as can be seen from Figure 7.4(a),

$$N_T = \overline{n^2} = \overline{n_c^2} = \overline{n_s^2} = N_0 W \tag{7.17}$$

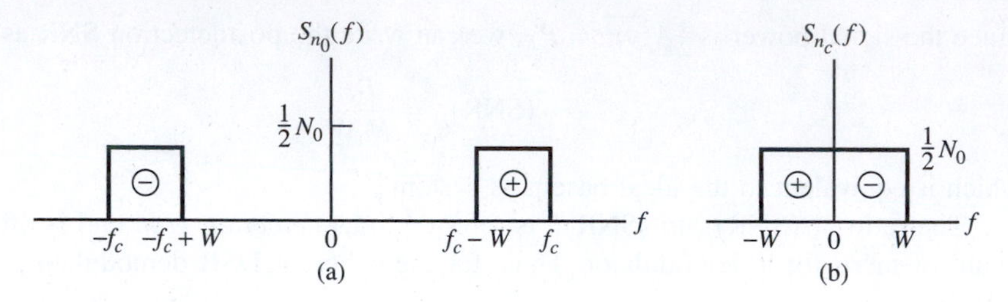

Figure 7.4
(a) Predetection and (b) postdetection filter output spectra for lower-sideband SSB (+ and − signs denote spectral translation of positive and negative portions of $S_{n_0}(f)$ due to demodulation, respectively).

Equation (7.16) can be written

$$\begin{aligned} e_2(t) = & \; [A_c m(t) + n_c(t)] \cos(2\pi f_c t + \theta) \\ & + [A_c \widehat{m}(t) \mp n_s(t)] \sin(2\pi f_c t + \theta) \end{aligned} \tag{7.18}$$

As discussed in Chapter 3, demodulation is accomplished by multiplying $e_2(t)$ by the demodulation carrier $2\cos(2\pi f_c t + \theta)$ and lowpass filtering. Thus the coherent demodulator illustrated in Figure 7.2 also accomplishes demodulation of SSB. It follows that

$$y_D(t) = A_c m(t) + n_c(t) \tag{7.19}$$

We see that coherent demodulation removes $\widehat{m}(t)$ as well as the quadrature noise component $n_s(t)$. The power spectral density of $n_c(t)$ is illustrated in Figure 7.4(b) for the case of LSB SSB. Since the postdetection filter passes only $n_c(t)$, the postdetection noise power is

$$N_D = \overline{n_c^2} = N_0 W \tag{7.20}$$

From (7.19) it follows that the postdetection signal power is

$$S_D = A_c^2 \overline{m^2} \tag{7.21}$$

We now turn our attention to the predetection terms.

The predetection signal power is

$$S_T = \overline{\{A_c[m(t)\cos(2\pi f_c t + \theta) \pm \widehat{m}(t)\sin(2\pi f_c t + \theta)]\}^2} \tag{7.22}$$

In Chapter 2 we pointed out that a function and its Hilbert transform are orthogonal. If $\overline{m(t)} = 0$, it follows that $\overline{m(t)\widehat{m}(t)} = E\{m(t)\}E\{\widehat{m}(t)\} = 0$. Thus the preceding expression becomes

$$S_T = A_c^2 \left\{ \left[\frac{1}{2}\overline{m^2(t)} \right] + \left[\frac{1}{2}\overline{\widehat{m}^2(t)} \right] \right\} \tag{7.23}$$

It was also shown in Chapter 2 that a function and its Hilbert transform have equal power. Applying this to (7.23) yields

$$S_T = A_c^2 \overline{m^2} \tag{7.24}$$

Since both the predetection and postdetection bandwidths are W, it follows that they have equal power. Therefore,

$$N_T = N_D = N_0 W \tag{7.25}$$

and the detection gain is

$$\frac{(\text{SNR})_D}{(\text{SNR})_T} = \frac{A_c^2\overline{m^2}}{N_0 W}\frac{N_0 W}{A_c^2\overline{m^2}} = 1 \tag{7.26}$$

The SSB system lacks the 3-dB detection gain of the DSB system. However, the predetection noise power of the SSB system is 3 dB less than that for the DSB system if the predetection filters have minimum bandwidth. This results in equal performance, given by

$$(\text{SNR})_D = \frac{A_c^2\overline{m^2}}{N_0 W} = \frac{P_T}{N_0 W} \tag{7.27}$$

Thus coherent demodulation of both DSB and SSB results in performance equivalent to baseband.

7.1.4 Amplitude Modulation Systems

The main reason for using AM is that simple envelope demodulation (or detection) can be used at the receiver. In many applications the receiver simplicity more than makes up for the loss in efficiency that we observed in Chapter 3. Therefore, coherent demodulation is not often used in AM. Despite this fact, we consider coherent demodulation briefly since it provides a useful insight into performance in the presence of noise.

Coherent Demodulation of AM Signals

We saw in Chapter 3 that an AM signal is defined by

$$x_c(t) = A_c[1 + am_n(t)]\cos(2\pi f_c t + \theta) \tag{7.28}$$

where $m_n(t)$ is the modulation signal normalized so that the maximum value of $|m_n(t)|$ is unity (assuming $m(t)$ has a symmetrical pdf about zero) and a is the modulation index. Assuming coherent demodulation, it is easily shown, by using a development parallel to that used for DSB systems, that the demodulated output in the presence of noise is

$$y_D(t) = A_c am_n(t) + n_c(t) \tag{7.29}$$

The DC term resulting from multiplication of $x_c(t)$ by the demodulation carrier is not included in (7.29) for two reasons. First, this term is not considered part of the signal since it contains no information. [Recall that we have assumed $\overline{m(t)} = 0$.] Second, most practical AM demodulators are not DC-coupled, so a DC term does not appear on the output of a practical system. In addition, the DC term is frequently used for automatic gain control (AGC) and is therefore held constant at the transmitter.

From (7.29) it follows that the signal power in $y_D(t)$ is

$$S_D = A_c^2 a^2 \overline{m_n^2} \tag{7.30}$$

and, since the bandwidth of the transmitted signal is $2W$, the noise power is

$$N_D = \overline{n_c^2} = 2N_0W \tag{7.31}$$

For the predetection case, the signal power is

$$S_T = P_T = \frac{1}{2}A_c^2\left(1 + a^2\overline{m_n^2}\right) \tag{7.32}$$

and the predetection noise power is

$$N_T = 2N_0W \tag{7.33}$$

Thus the detection gain is

$$\frac{(\text{SNR})_D}{(\text{SNR})_T} = \frac{A_c^2 a^2 \overline{m_n^2}/2N_0W}{\left(A_c^2 + A_c^2 a^2 \overline{m_n^2}\right)/4N_0W} = \frac{2a^2\overline{m_n^2}}{1 + a^2\overline{m_n^2}} \tag{7.34}$$

which is dependent on the modulation index.

Recall that when we studied AM in Chapter 3 the efficiency of an AM transmission system was defined as the ratio of sideband power to total power in the transmitted signal $x_c(t)$. This resulted in the efficiency E_{ff} being expressed as

$$E_{ff} = \frac{a^2\overline{m_n^2}}{1 + a^2\overline{m_n^2}} \tag{7.35}$$

where the overbar, denoting a statistical average, has been substituted for the time-average notation $\langle \cdot \rangle$ used in Chapter 3. It follows from (7.34) and (7.35) that the detection gain can be expressed as

$$\frac{(\text{SNR})_D}{(\text{SNR})_T} = 2E_{ff} \tag{7.36}$$

Since the predetection SNR can be written as

$$(\text{SNR})_T = \frac{S_T}{2N_0W} = \frac{P_T}{2N_0W} \tag{7.37}$$

it follows that the SNR at the demodulator output can be written as

$$(\text{SNR})_D = E_{ff}\frac{P_T}{N_0W} \tag{7.38}$$

Recall that in Chapter 3 we defined the efficiency of an AM system as the ratio of sideband power to the total power in an AM signal. The preceding expression gives another, and perhaps better, way to view efficiency.

If the efficiency *could* be 1, AM would have the same postdetection SNR as the ideal DSB and SSB systems. Of course, as we saw in Chapter 3, the efficiency of AM is typically much less than 1 and the postdetection SNR is correspondingly lower. Note that an efficiency of 1 requires that the modulation index $a \to \infty$ so that the power in the ummodulated carrier is negligible compared to the total transmitted power. However, for $a > 1$ envelope demodulation cannot be used and AM loses its advantage.

EXAMPLE 7.1

An AM system operates with a modulation index of 0.5, and the power in the normalized message signal is 0.1W. The efficiency is

$$E_{ff} = \frac{(0.5)^2(0.1)}{1+(0.5)^2(0.1)} = 0.0244 \tag{7.39}$$

and the postdetection SNR is

$$(\mathrm{SNR})_D = 0.0244\frac{P_T}{N_0 W} \tag{7.40}$$

The detection gain is

$$\frac{(\mathrm{SNR})_D}{(\mathrm{SNR})_T} = 2E_{ff} = 0.0488 \tag{7.41}$$

This is more than 16 dB inferior to the ideal system requiring the same bandwidth. It should be remembered, however, that the motivation for using AM is not noise performance but rather that AM allows the use of simple envelope detectors for demodulation. The reason, of course, for the poor efficiency of AM is that a large fraction of the total transmitted power lies in the carrier component, which conveys no information since it is not a function of the message signal.

■

Envelope Demodulation of AM Signals

Since envelope detection is the usual method of demodulating an AM signal, it is important to understand how envelope demodulation differs from coherent demodulation in the presence of noise. The received signal at the input to the envelope demodulator is assumed to be $x_c(t)$ plus narrowband noise. Thus

$$\begin{aligned} x_r(t) = {} & A_c[1+am_n(t)]\cos(2\pi f_c t+\theta) \\ & +n_c(t)\cos(2\pi f_c t+\theta)-n_s(t)\sin(2\pi f_c t+\theta) \end{aligned} \tag{7.42}$$

where, as before, $\overline{n_c^2} = \overline{n_s^2} = 2N_0 W$. The signal $x_r(t)$ can be written in terms of envelope and phase as

$$x_r(t) = r(t)\cos[2\pi f_c t+\theta+\phi(t)] \tag{7.43}$$

where

$$r(t) = \sqrt{\{A_c[1+am_n(t)]+n_c(t)\}^2+n_s^2(t)} \tag{7.44}$$

and

$$\phi(t) = \tan^{-1}\left(\frac{n_s(t)}{A_c[1+am_n(t)]+n_c(t)}\right) \tag{7.45}$$

Since the output of an ideal envelope detector is independent of phase variations of the input, the expression for $\phi(t)$ is of no interest, and we will concentrate on $r(t)$. The envelope detector is assumed to be AC coupled so that

$$y_D(t) = r(t) - \overline{r(t)} \tag{7.46}$$

where $\overline{r(t)}$ is the average value of the envelope amplitude. Equation (7.46) will be evaluated for two cases. First, we consider the case in which $(\text{SNR})_T$ is large, and then we briefly consider the case in which the $(\text{SNR})_T$ is small.

***Envelope Demodulation: Large* $(\mathbf{SNR})_T$** For $(\text{SNR})_T$ sufficiently large, the solution is simple. From (7.44), we see that if

$$|A_c[1+am_n(t)]+n_c(t)| \gg |n_s(t)| \tag{7.47}$$

then *most of the time*

$$r(t) \cong A_c[1+am_n(t)]+n_c(t) \tag{7.48}$$

yielding, after removal of the DC component,

$$y_D(t) \cong A_c am_n(t)+n_c(t) \tag{7.49}$$

This is the final result for the case in which the SNR is large.

Comparing (7.49) and (7.29) illustrates that the output of the envelope detector is equivalent to the output of the coherent detector if $(SNR)_T$ is large. The detection gain for this case is therefore given by (7.34).

***Envelope Demodulation: Small* $(\mathbf{SNR})_T$** For the case in which $(\text{SNR})_T$ is small, the analysis is somewhat more complex. In order to analyze this case, we recall from Chapter 6 that $n_c(t)\cos(2\pi f_c t+\theta) - n_s(t)\sin(2\pi f_c t+\theta)$ can be written in terms of envelope and phase, so that the envelope detector input can be written as

$$\begin{aligned} e(t) &= A_c[1+am_n(t)]\cos(2\pi f_c t+\theta) \\ &\quad + r_n(t)\cos[2\pi f_c t+\theta+\phi_n(t)] \end{aligned} \tag{7.50}$$

For $(SNR)_T \ll 1$, the amplitude of $A_c[1+am_n(t)]$ will usually be much smaller than $r_n(t)$. Consider the phasor diagram illustrated in Figure 7.5, which is drawn for $r_n(t)$ greater than $A_c[1+am_n(t)]$. It can be seen that $r(t)$ is approximated by

$$r(t) \cong r_n(t) + A_c[l+am_n(t)]\cos[\phi_n(t)] \tag{7.51}$$

yielding

$$y_D(t) \cong r_n(t) + A_c[1+am_n(t)]\cos[\phi_n(t)] - \overline{r(t)} \tag{7.52}$$

The principal component of $y_D(t)$ is the Rayleigh-distributed noise envelope, and no component of $y_D(t)$ is proportional to the signal. Note that since $n_c(t)$ and $n_s(t)$ are random,

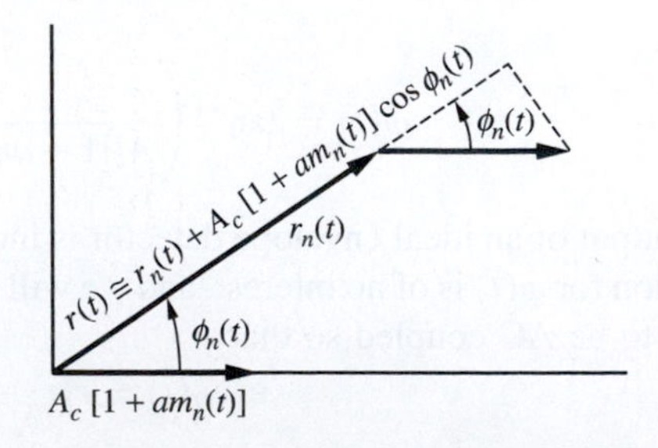

Figure 7.5
Phasor diagram for AM with $(\text{SNR})_T \ll 1$ (drawn for $\theta = 0$).

$\cos[\phi_n(t)]$ is also random. Thus the signal $m_n(t)$ is *multiplied* by a random quantity. This multiplication of the signal by a function of the noise has a significantly worse degrading effect than does additive noise.

This severe loss of signal at low-input SNR is known as the *threshold effect* and results from the nonlinear action of the envelope detector. In coherent detectors, which are linear, the signal and noise are additive at the detector output *if* they are additive at the detector input. The result is that the signal retains its identity even when the input SNR is low. For this reason, coherent detection is often desirable when the noise is large.

Square-Law Demodulation of AM Signals

The determination of the SNR at the output of a nonlinear system is often a very difficult task. The *square-law detector*, however, is one system for which this is not the case. In this section, we conduct a simplified analysis to illustrate the phenomenon of thresholding, which is characteristic of nonlinear systems.

In the analysis to follow, the postdetection bandwidth will be assumed twice the message bandwidth W. This is not a necessary assumption, but it does result in a simplification of the analysis without impacting the threshold effect. We will also see that harmonic and/or intermodulation distortion is a problem with square-law detectors, an effect that may preclude their use.

Square-law demodulators are implemented as a squaring device followed by a lowpass filter. The response of a square-law demodulator to an AM signal is $r^2(t)$, where $r(t)$ is defined by (7.44). Thus, the output of the square-law device can be written as

$$r^2(t) = \{A_c[1 + am_n(t)] + n_c(t)\}^2 + n_s^2(t) \tag{7.53}$$

We now determine the output SNR. Carrying out the indicated squaring operation gives

$$\begin{aligned} r^2(t) = {} & A_c^2 + 2A_c^2 am_n(t) + A_c^2 a^2 m_n^2(t) \\ & + 2A_c n_c(t) + 2A_c a n_c(t) m_n(t) + n_c^2(t) + n_s^2(t) \end{aligned} \tag{7.54}$$

First consider the first line of the preceding equation. The first term, A_c^2, is a DC term and is neglected. It is not a function of the signal and is not a function of noise. In addition, in most practical cases, the detector output is assumed AC coupled, so that DC terms are blocked. The second term is proportional to the message signal and represents the desired output. The third term is signal-induced distortion (harmonic and intermodulation) and will be considered separately. All four terms on the second line of (7.54) represent noise. We now consider the calculation of $(\text{SNR})_D$.

The signal and noise components of the output are written as

$$s_D(t) = 2A_c^2 am_n(t) \tag{7.55}$$

and

$$n_D(t) = 2A_c n_c(t) + 2A_c a n_c(t) m_n(t) + n_c^2(t) + n_s^2(t) \tag{7.56}$$

respectively. The power in the signal component is

$$S_D = 4A_c^4 a^2 \overline{m_n^2} \tag{7.57}$$

and the noise power is

$$N_D = 4A_c^2\overline{n_c^2} + 4A_c^2a^2\overline{n_c^2}\,\overline{m_n^2} + \sigma^2_{n_c^2+n_s^2} \tag{7.58}$$

The last term is given by

$$\sigma^2_{n_c^2+n_s^2} = E\left[\left[n_c^2(t)+n_s^2(t)\right]^2\right] - E^2\left[n_c^2(t)+n_s^2(t)\right] = 4\sigma_n^2 \tag{7.59}$$

where, as always, $\sigma_n^2 = \overline{n_c^2} = \overline{n_s^2}$. Thus,

$$N_D = 4A_c^2\sigma_n^2 + 4A_c^2a^2\overline{m_n^2(t)}\sigma_n^2 + 4\sigma_n^4 \tag{7.60}$$

This gives

$$(\mathrm{SNR})_D = \frac{a^2\overline{m_n^2}\,(A_c^2/\sigma_n^2)}{\left(1+a^2\overline{m_n^2}\right) + (\sigma_n^2/A_c^2)} \tag{7.61}$$

Recognizing that $P_T = \frac{1}{2}A_c^2\left(1+a^2\overline{m_n^2}\right)$ and $\sigma_n^2 = 2N_0W$, A_c^2/σ_n^2 can be written

$$\frac{A_c^2}{\sigma_n^2} = \frac{P_T}{\left[1+a^2\overline{m_n^2(t)}\right]N_0W} \tag{7.62}$$

Substitution into (7.61) gives

$$(\mathrm{SNR})_D = \frac{a^2\overline{m_n^2}}{\left(1+a^2\overline{m_n^2}\right)^2}\,\frac{P_T/N_0W}{1+N_0W/P_T} \tag{7.63}$$

For high SNR operation $P_T \gg N_0W$ and the last term in the denominator is negligible. For this case,

$$(\mathrm{SNR})_D = \frac{a^2\overline{m_n^2}}{\left(1+a^2\overline{m_n^2}\right)^2}\,\frac{P_T}{N_0W}, \quad P_T \gg N_0W \tag{7.64}$$

while for low SNR operation $N_0W \gg P_T$ and

$$(\mathrm{SNR})_D = \frac{a^2\overline{m_n^2}}{\left(1+a^2\overline{m_n^2}\right)^2}\left(\frac{P_T}{N_0W}\right)^2, \quad N_0W \gg P_T \tag{7.65}$$

Figure 7.6 illustrates (7.63) for several values of the modulation index a assuming sinusoidal modulation. We see that, on a log (decibel) scale, the slope of the detection gain characteristic below threshold is double the slope above threshold. The threshold effect is therefore obvious.

Recall that in deriving (7.63), from which (7.64) and (7.65) followed, we neglected the third term in (7.54), which represents signal-induced distortion. From (7.54) and (7.57) the distortion-to-signal-power ratio, denoted D_D/S_D, is

$$\frac{D_D}{S_D} = \frac{A_c^4a^4\overline{m_n^4}}{4A_c^4a^2\overline{m_n^2}} = \frac{a^2}{4}\frac{\overline{m_n^4}}{\overline{m_n^2}} \tag{7.66}$$

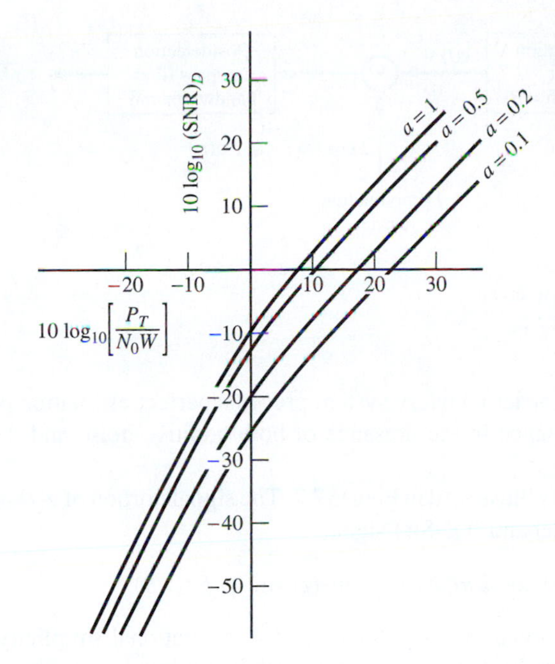

Figure 7.6
Performance of a square-law detector assuming sinusoidal modulation.

If the message signal is Gaussian with variance σ_m^2, the preceding becomes (see Problem 5.22.)

$$\frac{D_D}{S_D} = \frac{3}{4} a^2 \sigma_m^2 \tag{7.67}$$

We see that signal-induced distortion can be reduced by decreasing the modulation index. However, as illustrated in Figure 7.6, a reduction of the modulation index also results in a decrease in the output SNR.

The *linear envelope detector* defined by (7.44) is much more difficult to analyze over a wide range of SNRs because of the square root. However, to a first approximation, the performance of a linear envelope detector and a square-law envelope detector are the same. Harmonic distortion is also present in linear envelope detectors, but the amplitude of the distortion component is significantly less than that observed for square-law detectors. In addition, it can be shown that for high SNRs and a modulation index of unity, the performance of a linear envelope detector is better by approximately 1.8 dB than the performance of a square-law detector. (See Problem 7.16.)[1]

7.2 NOISE AND PHASE ERRORS IN COHERENT SYSTEMS

In the preceding section we investigated the performance of various types of demodulators. Our main interests were detection gain and calculation of the demodulated output SNR. Where coherent demodulation was used, the demodulation carrier was assumed to have *perfect* phase coherence with the carrier used for modulation. In a practical system, as we briefly discussed,

[1]For a detailed study of linear envelope detectores, see Bennett (1974).

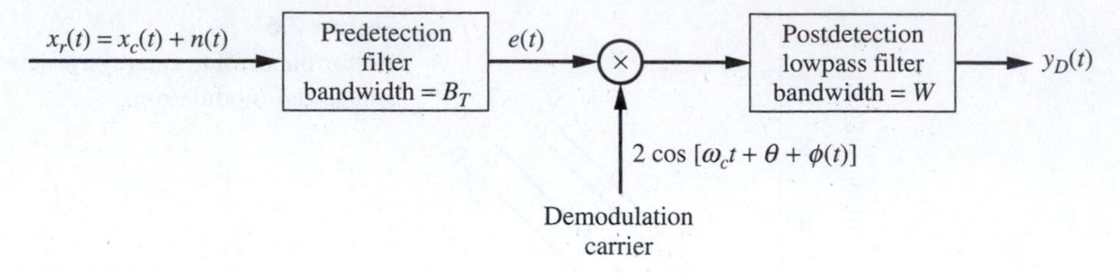

Figure 7.7
Coherent demodulator with phase error.

the presence of noise in the carrier recovery system prevents perfect estimation of the carrier phase. Thus, system performance in the presence of both additive noise and demodulation phase errors is of interest.

The demodulator model is illustrated in Figure 7.7. The signal portion of $e(t)$ is assumed to be the quadrature double-sideband (QDSB) signal

$$m_1(t)\cos(2\pi f_c t+\theta)+m_2(t)\sin(2\pi f_c t+\theta)$$

where any constant A_c is absorbed into $m_1(t)$ and $m_2(t)$ for notational simplicity. Using this model, a general representation for the error in the demodulated signal $y_D(t)$ is obtained. After the analysis is complete, the DSB result is obtained by letting $m_1(t)=m(t)$ and $m_2(t)=0$. The SSB result is obtained by letting $m_1(t)=m(t)$ and $m_2(t)=\pm\widehat{m}(t)$, depending upon the sideband of interest. For the QDSB system, $y_D(t)$ is the demodulated output for the direct channel. The quadrature channel can be demodulated using a demodulation carrier of the form $2\sin[2\pi f_c t+\theta+\phi(t)]$.

The noise portion of $e(t)$ is represented using the narrowband model

$$n_c(t)\cos(2\pi f_c t+\theta)-n_s(t)\sin(2\pi f_c t+\theta)$$

in which

$$\overline{n_c^2}=\overline{n_s^2}=N_0B_T=\overline{n^2}=\sigma_n^2 \tag{7.68}$$

where B_T is the bandwidth of the predetection filter, $\frac{1}{2}N_0$ is the double-sided power spectral density of the noise at the filter input, and σ_n^2 is the noise variance (power) at the output of the predetection filter. The phase error of the demodulation carrier is assumed to be a sample function of a zero mean Gaussian process of known variance σ_ϕ^2. As before, the message signals are assumed to have zero mean.

With the preliminaries of defining the model and stating the assumptions disposed of, we now proceed with the analysis. The assumed performance criterion is mean-square error in the demodulated output $y_D(t)$. Therefore, we will compute

$$\overline{\epsilon^2}=\overline{\{m_1(t)-y_D(t)\}^2} \tag{7.69}$$

for DSB, SSB, and QDSB. The multiplier input signal $e(t)$ in Figure 7.7 is

$$\begin{aligned}e(t) &= m_1(t)\cos(2\pi f_c t+\theta)+m_2(t)\sin(2\pi f_c t+\theta)\\ &\quad +n_c(t)\cos(2\pi f_c t+\theta)-n_s(t)\sin(2\pi f_c t+\theta)\end{aligned} \tag{7.70}$$

Multiplying by $2\cos[2\pi f_c t + \theta + \phi(t)]$ and lowpass filtering gives us the output

$$y_D(t) = [m_1(t) + n_c(t)]\ \cos\phi(t) - [m_2(t) - n_s(t)]\ \sin\phi(t) \tag{7.71}$$

The error $m_1(t) - y_D(t)$ can be written as

$$\epsilon = m_1 - (m_1 + n_c)\cos\phi + (m_2 - n_s)\sin\phi \tag{7.72}$$

where it is understood that $\epsilon, m_1, m_2, n_c, n_s$, and ϕ are all functions of time. The mean-square error can be written as

$$\begin{aligned}\overline{\epsilon^2} = {} & \overline{m_1^2} - \overline{2m_1(m_1 + n_c)\cos\phi} \\ & + \overline{2m_1(m_2 - n_s)\sin\phi} \\ & + \overline{(m_1 + n_c)^2\cos^2\phi} \\ & - \overline{2(m_1 + n_c)(m_2 - n_s)\sin\phi\cos\phi} \\ & + \overline{(m_2 - n_s)^2\sin^2\phi}\end{aligned} \tag{7.73}$$

The variables m_1, m_2, n_c, n_s, and ϕ are all assumed to be uncorrelated. It should be pointed out that for the SSB case, the power spectra of $n_c(t)$ and $n_s(t)$ will not be symmetrical about f_c. However, as pointed out in Section 6.5, $n_c(t)$ and $n_s(t)$ are still uncorrelated, since there is no time displacement. Thus, the mean-square error can be written as

$$\overline{\epsilon^2} = \overline{m_1^2} - \overline{2m_1^2\cos\phi} + \overline{m_1^2\cos^2\phi} + \overline{m_2^2\sin^2\phi} + \overline{n^2} \tag{7.74}$$

and we are in a position to consider specific cases.

First, let us assume the system of interest is QDSB with equal power in each modulating signal. Under this assumption, $\overline{m_1^2} = \overline{m_2^2} = \sigma_m^2$, and the mean-square error is

$$\overline{\epsilon_Q^2} = 2\sigma_m^2 - 2\sigma_m^2\overline{\cos\phi} + \sigma_n^2 \tag{7.75}$$

This expression can be easily evaluated for the case in which the maximum value of $|\phi(t)| \ll 1$ so that $\phi(t)$ can be represented by the first two terms in a power series expansion. Using the approximation

$$\overline{\cos\phi} \cong \overline{1 - \frac{1}{2}\phi^2} = 1 - \frac{1}{2}\sigma_\phi^2 \tag{7.76}$$

gives

$$\overline{\epsilon_Q^2} = \sigma_m^2\sigma_\phi^2 + \sigma_n^2 \tag{7.77}$$

In order to have an easily interpreted measure of system performance, the mean-square error is normalized by σ_m^2. This yields

$$\overline{\epsilon_{NQ}^2} = \sigma_\phi^2 + \frac{\sigma_n^2}{\sigma_m^2} \tag{7.78}$$

Note that the first term is the phase error variance and the second term is simply the reciprocal of the SNR. Note that for high SNR the important error source is the phase error.

The preceding expression is also valid for the SSB case, since an SSB signal is a QDSB signal with equal power in the direct and quadrature components. However, σ_n^2 may be different for the SSB and QDSB cases, since the SSB predetection filter bandwidth need only be half the bandwidth of the predetection filter for the QDSB case. Equation (7.78) is of such general interest that it is illustrated in Figure 7.8.

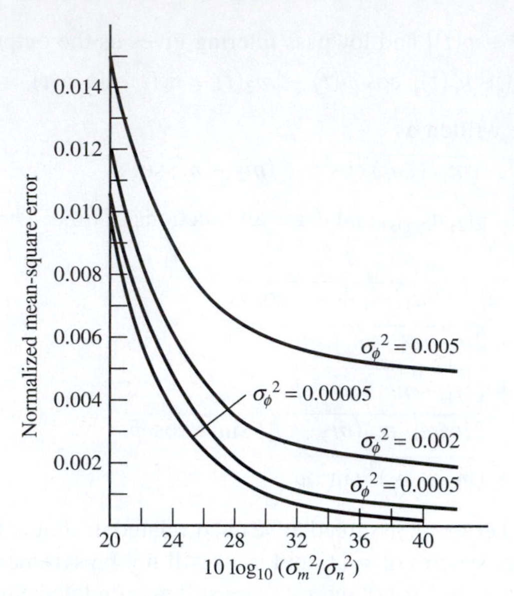

Figure 7.8
Mean-square error versus SNR for QDSB system.

In order to compute the mean-square error for a DSB system, we simply let $m_2 = 0$ and $m_1 = m$ in (7.74). This yields

$$\overline{\epsilon_D^2} = \overline{m^2} - 2\overline{m^2}\,\overline{\cos\phi} + \overline{m^2}\,\overline{\cos^2\phi} + \overline{n^2} \tag{7.79}$$

or

$$\overline{\epsilon_D^2} = \sigma_m^2\overline{(1 - \cos\phi)^2} + \overline{n^2} \tag{7.80}$$

which, for small ϕ, can be approximated as

$$\overline{\epsilon_D^2} \cong \sigma_m^2\left(\frac{1}{4}\right)\overline{\phi^4} + \overline{n^2} \tag{7.81}$$

If ϕ is zero-mean Gaussian with variance σ_ϕ^2 (see problem 5.22),

$$\overline{\phi^4} = \overline{\left(\phi^2\right)^2} = 3\sigma_\phi^4 \tag{7.82}$$

Thus

$$\overline{\epsilon_D^2} \cong \frac{3}{4}\sigma_m^2\sigma_\phi^4 + \sigma_n^2 \tag{7.83}$$

and the normalized mean-square error becomes

$$\overline{\epsilon_{ND}^2} = \frac{3}{4}\sigma_\phi^4 + \frac{\sigma_n^2}{\sigma_m^2} \tag{7.84}$$

Several items are of interest when comparing (7.84) and (7.78). First, equal output SNRs imply equal normalized mean-square errors for $\sigma_\phi^2 = 0$. This is easy to understand since the noise is additive at the output. The general expression for $y_D(t)$ is $y_D(t) = m(t) + n(t)$. The error is $n(t)$, and the normalized mean-square error is σ_n^2/σ_m^2. The analysis also illustrates that DSB systems are much less sensitive to demodulation phase errors than SSB or QDSB systems.

This follows from the fact that if $\phi \ll 1$, the basic assumption made in the analysis, then $\sigma_\phi^4 \ll \sigma_\phi^2$.

EXAMPLE 7.2

Assume that the demodulation phase-error variance of a coherent demodulator is described by $\sigma_\phi^2 = 0.01$. The SNR σ_m^2/σ_n^2 is 20 dB. If a DSB system is used, the normalized mean-square error is

$$\overline{\epsilon_{ND}^2} = \frac{3}{4}(0.01)^2 + 10^{-20/10} = 0.000075 \text{ (DSB)} \tag{7.85}$$

while for the SSB case the normalized mean-square error is

$$\overline{\epsilon_{ND}^2} = (0.01) + 10^{-20/10} = 0.02 \text{ (SSB)} \tag{7.86}$$

Note that for the DSB demodulator, the demodulation phase error can probably be neglected for most applications. For the SSB case the phase error contributes more significantly to the error in the demodulated output, and therefore, the phase error variance must clearly be considered. Recall that demodulation phase errors in a QDSB system result in crosstalk between the direct and quadrature message signals. Thus in SSB, demodulation phase errors result in a portion of $\widehat{m}(t)$ appearing in the demodulated output for $m(t)$. Since $m(t)$ and $\widehat{m}(t)$ are independent, this crosstalk can be a severely degrading effect unless the demodulation phase error is very small. ■

7.3 NOISE IN ANGLE MODULATION

Now that we have investigated the effect of noise on a linear modulation system, we turn our attention to angle modulation. We will find that there are significant differences between linear and angle modulation when noise effects are considered. We will also find significant differences between PM and FM. Finally, we will see that FM can offer greatly improved performance over both linear modulation and PM systems in noisy environments, but that this improvement comes at the cost of increased transmission bandwidth.

7.3.1 The Effect of Noise on the Receiver Input

Consider the system shown in Figure 7.9. The predetection filter bandwidth is B_T and is usually determined by Carson's rule. Recall from Chapter 3 that B_T is approximately $2(D+1)W$ Hz, where W is the bandwidth of the message signal and D is the deviation ratio, which is the peak frequency deviation divided by the bandwidth W. The input to the predetection filter is assumed to be the modulated carrier

$$x_c(t) = A_c \cos[2\pi f_c t + \theta + \phi(t)] \tag{7.87}$$

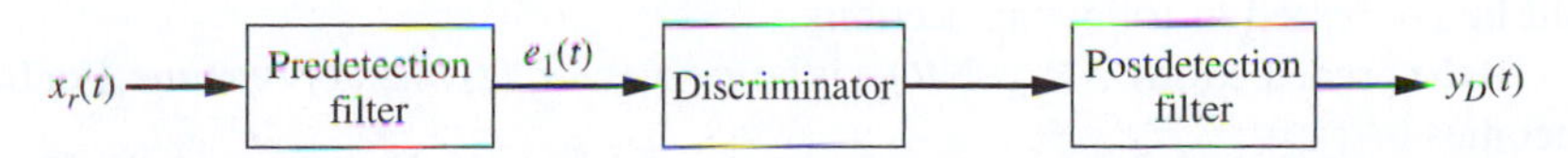

Figure 7.9
Angle demodulation system.

plus additive white noise that has the double-sided power spectral density $\frac{1}{2}N_0$ W/Hz. For angle modulation the phase deviation $\phi(t)$ is a function of the message signal $m(t)$.

The output of the predetection filter can be written as

$$\begin{aligned} e_1(t) = {} & A_c \cos[2\pi f_c t + \theta + \phi(t)] \\ & + n_c(t)\cos(2\pi f_c t + \theta) - n_s(t)\sin(2\pi f_c t + \theta) \end{aligned} \tag{7.88}$$

where

$$\overline{n_c^2} = \overline{n_s^2} = N_0 B_T \tag{7.89}$$

Equation (7.88) can be written as

$$e_1(t) = A_c \cos[2\pi f_c t + \theta + \phi(t)] + r_n(t)\cos[2\pi f_c t + \theta + \phi_n(t)] \tag{7.90}$$

where $r_n(t)$ is the Rayleigh-distributed noise envelope and $\phi_n(t)$ is the uniformly distributed phase. By replacing $2\pi f_c t + \phi_n(t)$ with $2\pi f_c t + \phi(t) + \phi_n(t) - \phi(t)$, we can write (7.90) as

$$\begin{aligned} e_1(t) = {} & A_c \cos[2\pi f_c t + \theta + \phi(t)] \\ & + r_n(t)\cos[\phi_n(t) - \phi(t)]\cos[2\pi f_c t + \theta + \phi(t)] \\ & - r_n(t)\sin[\phi_n(t) - \phi(t)]\sin[2\pi f_c t + \theta + \phi(t)] \end{aligned} \tag{7.91}$$

which is

$$\begin{aligned} e_1(t) = {} & \{A_c + r_n(t)\cos[\phi_n(t) - \phi(t)]\}\cos[2\pi f_c t + \theta + \phi(t)] \\ & - r_n(t)\sin[\phi_n(t) - \phi(t)]\sin[2\pi f_c t + \theta + \phi(t)] \end{aligned} \tag{7.92}$$

Since the purpose of the receiver is to recover the phase, we write the preceding expression as

$$e_1(t) = R(t)\cos[2\pi f_c t + \theta + \phi(t) + \phi_e(t)] \tag{7.93}$$

where $\phi_e(t)$ is the *phase deviation error due to noise* and is given by

$$\phi_e(t) = \tan^{-1}\left(\frac{r_n(t)\sin[\phi_n(t) - \phi(t)]}{A_c + r_n(t)\cos[\phi_n(t) - \phi(t)]}\right) \tag{7.94}$$

Since $\phi_e(t)$ adds to $\phi(t)$, which *conveys the message signal*, it is the noise component of interest.

If $e_1(t)$ is expressed as

$$e_1(t) = R(t)\cos[2\pi f_c t + \theta + \psi(t)] \tag{7.95}$$

the *phase deviation* of the discriminator input due to the combination of signal and noise is

$$\psi(t) = \phi(t) + \phi_e(t) \tag{7.96}$$

where $\phi_e(t)$ is the phase error due to noise. Since the demodulated output is proportional to $\psi(t)$ for PM, or $d\psi/dt$ for FM, we must determine $(\mathrm{SNR})_T$ for PM and for FM as separate cases. This will be addressed in following sections.

If the predetection SNR, $(\mathrm{SNR})_T$, is large, $A_c \gg r_n(t)$ *most of the time*. For this case (7.94) becomes

$$\phi_e(t) = \frac{r_n(t)}{A_c}\sin[\phi_n(t) - \phi(t)] \tag{7.97}$$

so that $\psi(t)$ is

$$\psi(t) = \phi(t) + \frac{r_n(t)}{A_c}\sin[\phi_n(t) - \phi(t)] \tag{7.98}$$

It is important to note that the effect of the noise $r_n(t)$ is suppressed if the transmitted signal amplitude A_c is increased. Thus the output noise is affected by the transmitted signal amplitude even for above-threshold operation.

In (7.98) note that $\phi_n(t)$, for a given value of t, is uniformly distributed over a 2π range. Also, for a given t, $\phi(t)$ is a constant that biases $\phi_n(t)$, and $\phi_n(t) - \phi(t)$ is in the same range mod(2π). We therefore neglect $\phi(t)$ in (7.98) and express $\psi(t)$ as

$$\psi(t) = \phi(t) + \frac{n_s(t)}{A_c} \tag{7.99}$$

where $n_s(t)$ is the quadrature noise component at the input to the receiver.

7.3.2 Demodulation of PM

Recall that for PM, the phase deviation is proportional to the message signal so that

$$\phi(t) = k_p m_n(t) \tag{7.100}$$

where k_p is the phase-deviation constant in radians per unit of $m_n(t)$ and $m_n(t)$ is the message signal normalized so that the peak value of $|m(t)|$ is unity. The demodulated output $y_D(t)$ for PM is given by

$$y_D(t) = K_D\psi(t) \tag{7.101}$$

where $\psi(t)$ represents the phase deviation of the receiver input due to the combined effects of signal and noise. Using (7.99) gives

$$y_{DP}(t) = K_D k_p m_n(t) + K_D\frac{n_s(t)}{A_c} \tag{7.102}$$

The output signal power for PM is

$$S_{DP} = K_D^2 k_p^2 \overline{m_n^2} \tag{7.103}$$

The power spectral density of the predetection noise is N_0, and the bandwidth of the predetection noise is B_T which, by Carson's rule, exceeds $2W$. We therefore remove the out-of-band noise by following the discriminator with a lowpass filter of bandwidth W. This filter has no effect on the signal but reduces the output noise power to

$$N_{DP} = \frac{K_D^2}{A_c^2}\int_{-W}^{W} N_0\, df = 2\frac{K_D^2}{A_c^2}N_0 W \tag{7.104}$$

Thus the SNR at the output of the phase discriminator is

$$(\text{SNR})_D = \frac{S_{DP}}{N_{DP}} = \frac{K_D^2 k_p^2 \overline{m_n^2}}{(2K_D^2/A_c^2)N_0 W} \tag{7.105}$$

Since the transmitted signal power P_T is $A_c^2/2$, we have

$$(\text{SNR})_D = k_p^2 \overline{m_n^2}\frac{P_T}{N_0 W} \tag{7.106}$$

The above expression shows that the improvement of PM over linear modulation depends on the phase-deviation constant and the power in the modulating signal. It should be remembered that if the phase deviation of a PM signal exceeds π radians, unique demodulation cannot be accomplished unless appropriate signal processing is used to ensure that the phase deviation due to $m(t)$ is continuous. If, however, we assume that the peak value of $|k_p m_n(t)|$ is π, the maximum value of $k_p^2 \overline{m_n^2}$ is π^2. This yields a *maximum* improvement of approximately 10 dB over baseband. In reality, the improvement is significantly less because $k_p^2 \overline{m_n^2}$ is typically much less than the maximum value of π^2. It should be pointed out that if the constraint that the output of the phase demodulator is continuous is imposed, it is possible for $|k_p m_n(t)|$ to exceed π.

7.3.3 Demodulation of FM: Above Threshold Operation

For the FM case,

$$\phi(t) = 2\pi f_d \int^t m_n(\alpha)\, d\alpha \tag{7.107}$$

where f_d is the deviation constant in Hz per unit amplitude of the message signal. If the maximum value of $|m(t)|$ is not unity, as is usually the case, the scaling constant K, defined by $m(t) = K m_n(t)$, is contained in k_p or f_d. The discriminator output $y_D(t)$ for FM is given by

$$y_D(t) = \frac{1}{2\pi} K_D \frac{d\psi}{dt} \tag{7.108}$$

where K_D is the discriminator constant. Substituting (7.99) into (7.108) and using (7.107) for $\phi(t)$ yields

$$y_{DF}(t) = K_D f_m m_n(t) + \frac{K_D}{2\pi A_c} \frac{dn_s(t)}{dt} \tag{7.109}$$

The output signal power at the output of the FM demodulator is

$$S_{DF} = K_D^2 f_d^2 \overline{m_n^2} \tag{7.110}$$

Before the noise power can be calculated, the power spectral density of the output noise must first be determined.

The noise component at the output of the FM demodulator is, from (7.109), given by

$$n_F(t) = \frac{K_D}{2\pi A_c} \frac{dn_s(t)}{dt} \tag{7.111}$$

It was shown in Chapter 6 that if $y(t) = dx/dt$, then $S_y(f) = (2\pi f)^2 S_x(f)$. Applying this result to (7.111) yields

$$S_{nF}(f) = \frac{K_D^2}{(2\pi)^2 A_c^2} (2\pi f)^2 N_0 = \frac{K_D^2}{A_c^2} N_0 f^2 \tag{7.112}$$

for $|f| < \frac{1}{2} B_T$ and zero otherwise. This spectrum is illustrated in Figure 7.10(b). The parabolic shape of the noise spectrum results from the differentiating action of the FM discriminator and has a profound effect on the performance of FM systems operating in the presence of noise. It is clear from Figure 7.10(b) that low-frequency message-signal components are subjected to

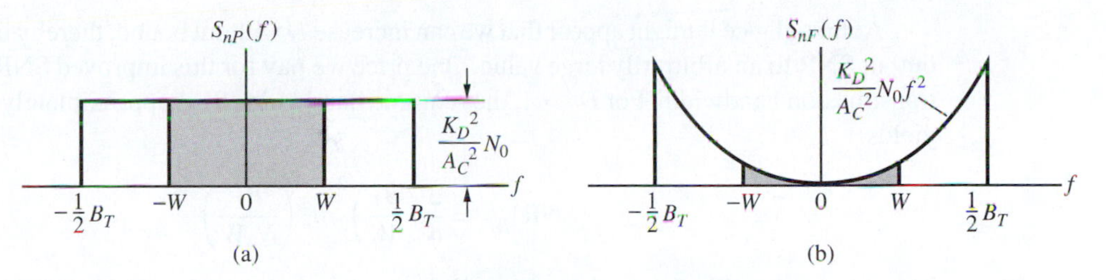

Figure 7.10
(a) Power spectral density for PM discriminator output, with portion for $|f| < W$ shaded. (b) Power spectral density for FM discriminator output, with portion for $|f| < W$ shaded.

lower noise levels than are high-frequency components. Once again, assuming that a lowpass filter having only sufficient bandwidth to pass the message follows the discriminator, the output noise power is

$$N_{DF} = \frac{K_D^2}{A_c^2} N_0 \int_{-W}^{W} f^2 \, df = \frac{2}{3}\frac{K_D^2}{A_c^2} N_0 W^3 \tag{7.113}$$

This quantity is indicated by the shaded area in Figure 7.10(b).

As usual, it is useful to write (7.113) in terms of P_T/N_0W. Since $P_T = A_c^2/2$ we have

$$\frac{P_T}{N_0 W} = \frac{A_c^2}{2N_0 W} \tag{7.114}$$

and from (7.113) the noise power at the output of the FM demodulator is

$$N_{DF} = \frac{1}{3} K_D^2 W^2 \left(\frac{P_T}{N_0 W}\right)^{-1} \tag{7.115}$$

Note that for both PM and FM the noise power at the discriminator output is inversely proportional to P_T/N_0W.

The SNR at the FM demodulator output is now easily determined. Dividing the signal power, defined by (7.110), by the noise power, defined by (7.115), gives

$$(\text{SNR})_{DF} = \frac{K_D^2 f_d^2 \overline{m_n^2}}{\frac{1}{3} K_D^2 W^2 \left(\frac{P_T}{N_0 W}\right)^{-1}} \tag{7.116}$$

which can be expressed as

$$(\text{SNR})_{DF} = 3\left(\frac{f_d}{W}\right)^2 \overline{m_n^2} \frac{P_T}{N_0 W} \tag{7.117}$$

where P_T is the transmitted signal power $\frac{1}{2}A_c^2$. Since the ratio of peak deviation to W is the deviation ratio D, the output SNR can be expressed

$$(\text{SNR})_{DF} = 3D^2 \overline{m_n^2} \frac{P_T}{N_0 W} \tag{7.118}$$

where, as always, the maximum value of $|m_n(t)|$ is unity. Note that the maximum value of $m(t)$, together with f_d and W, determines D.

At first glance it might appear that we can increase D without bound, thereby increasing the output SNR to an arbitrarily large value. One price we pay for this improved SNR is excessive transmission bandwidth. For $D \gg 1$, the required bandwidth B_T is approximately $2DW$, which yields

$$(\text{SNR})_{DF} = \frac{3}{4}\left(\frac{B_T}{W}\right)^2 \overline{m_n^2}\left(\frac{P_T}{N_0 W}\right) \tag{7.119}$$

This expression illustrates the trade-off that exists between bandwidth and output SNR. However, (7.119) is valid only if the discriminator input SNR is sufficiently large to result in operation *above threshold.* Thus the output SNR *cannot* be increased to any arbitrary value by simply increasing the deviation ratio and thus the transmission bandwidth. This effect will be studied in detail in a later section. First, however, we will study a simple technique for gaining additional improvement in the output SNR.

7.3.4 Performance Enhancement through the Use of De-emphasis

In Chapter 3 we saw that pre-emphasis and de-emphasis can be used to partially combat the effects of interference. These techniques can also be used to advantage when noise is present in angle modulation systems.

As we saw in Chapter 3, the de-emphasis filter is usually a first-order lowpass RC filter placed directly at the discriminator output. Prior to modulation, the signal is passed through a highpass pre-emphasis filter having a transfer function so that the combination of the pre-emphasis and de-emphasis filters has no net effect on the message signal. The de-emphasis filter is followed by a lowpass filter, assumed to be ideal with bandwidth W, which eliminates the out-of-band noise. Assume the de-emphasis filter to have the amplitude response

$$|H_{DE}(f)| = \frac{1}{\sqrt{1+(f/f_3)^2}} \tag{7.120}$$

where f_3 is the 3-dB frequency $1/(2\pi RC)$ Hz. The total noise power output with de-emphasis is

$$N_{DF} = \int_{-W}^{W} |H_{DE}(f)|^2 S_{nF}(f)\, df \tag{7.121}$$

Substituting $S_{nF}(f)$ from (7.112) and $|H_{DE}(f)|$ from (7.120) yields

$$N_{DF} = \frac{K_D^2}{A_c^2} N_0 f_3^2 \int_{-W}^{W} \frac{f^2}{f_3^2+f^2}\, df \tag{7.122}$$

or

$$N_{DF} = 2\frac{K_D^2}{A_c^2} N_0 f_3^3 \left(\frac{W}{f_3} - \tan^{-1}\frac{W}{f_3}\right) \tag{7.123}$$

In a typical situation, $f_3 \ll W$, so $\tan^{-1}(W/f_3) \cong \frac{1}{2}\pi$, which is small compared to W/f_3. For this case,

$$N_{DF} = 2\left(\frac{K_D^2}{A_c^2}\right) N_0 f_3^2 W \tag{7.124}$$

and the output SNR becomes

$$(\text{SNR})_{DF} = 3\left(\frac{f_d}{f_3}\right)^2 \overline{m_n^2} \frac{P_T}{N_0 W} \tag{7.125}$$

A comparison of (7.125) with (7.117) illustrates that for $f_3 \ll W$, the improvement gained through the use of pre-emphasis and de-emphasis is approximately $(W/f_3)^2$, which can be very significant in noisy environments.

EXAMPLE 7.3

Commercial FM operates with $f_d = 75$ kHz, $W = 15$ kHz, $D = 5$, and the standard value of 2.1 kHz for f_3. Assuming that $\overline{m_n^2} = 0.1$, we have, for FM without pre-emphasis and de-emphasis,

$$(\text{SNR})_{DF} = 7.5 \frac{P_T}{N_0 W} \tag{7.126}$$

With pre-emphasis and de-emphasis, the result is

$$(\text{SNR})_{DF,pe} = 128 \frac{P_T}{N_0 W} \tag{7.127}$$

With the chosen values, FM without de-emphasis is 8.75 dB superior to baseband, and FM with de-emphasis is 21 dB superior to baseband. The difference of 12.25 dB is approximately equivalent to a factor of 16. Thus, with the use of pre-emphasis and de-emphasis, the transmitter power can be reduced by a factor of 16. This improvement is obviously significant and more than justifies the use of pre-emphasis and de-emphasis.

■

As mentioned in Chapter 3, a price is paid for the SNR improvement gained by the use of pre-emphasis. The action of the pre-emphasis filter is to accentuate the high-frequency portion of the message signal relative to the low-frequency portion of the message signal. Thus pre-emphasis may increase the transmitter deviation and, consequently, the bandwidth required for signal transmission. Fortunately, many message signals of practical interest have relatively small energy in the high-frequency portion of their spectrum, so this effect is often of little or no importance.

7.4 THRESHOLD EFFECT IN FM DEMODULATION

Since angle modulation is a nonlinear process, demodulation of an angle-modulated signal exhibits a threshold effect. We now take a closer look at this threshold effect concentrating on FM demodulators or, equivalently, discriminators.

7.4.1 Threshold Effects in FM Demodulators

Significant insight into the mechanism by which threshold effects take place can be gained by performing a relatively simple laboratory experiment. We assume that the input to an FM discriminator consists of an unmodulated sinusoid plus additive bandlimited white noise having a power spectral density symmetrical about the frequency of the sinusoid. Starting out with a high SNR at the discriminator input, the noise power is gradually increased while continually observing the discriminator output on an oscilloscope. Initially, the discriminator output resembles bandlimited white noise. As the noise power spectral density is increased, thereby reducing the input SNR, a point is reached at which spikes or impulses appear in the discriminator output. The initial appearance of these spikes denotes that the discriminator is operating in the region of threshold.

The statistics for these spikes are examined in Appendix D. In this section we review the phenomenon of spike noise with specific application to FM demodulation. The system under consideration is that of Figure 7.9. For this case,

$$e_1(t) = A_c \cos(2\pi f_c t + \theta) + n_c(t)\cos(2\pi f_c t + \theta) - n_s(t)\sin(2\pi f_c t + \theta) \tag{7.128}$$

which is

$$e_1(t) = A_c \cos(2\pi f_c t + \theta) + r_n(t)\cos[2\pi f_c t + \theta + \phi_n(t)] \tag{7.129}$$

or

$$e_1(t) = R(t)\cos[2\pi f_c t + \theta + \psi(t)] \tag{7.130}$$

The phasor diagram of this signal is given in Figure 7.11. Like Figure D.2 in Appendix D, it illustrates the mechanism by which spikes occur. The signal amplitude is A_c and the angle is θ, since the carrier is assumed unmodulated. The noise amplitude is $r_n(t)$. The angle difference between signal and noise is $\phi_n(t)$. As threshold is approached, the noise amplitude grows until, at least part of the time, $|r_n(t)| > A_c$. Also, since $\phi_n(t)$ is uniformly distributed, the phase of the noise is sometimes in the region of $-\pi$. Thus the resultant phasor $R(t)$ can occasionally encircle the origin. When $R(t)$ is in the region of the origin, a relatively small change in the phase of the noise results in a rapid change in $\psi(t)$. Since the discriminator output is proportional to the time rate of change $\psi(t)$, the discriminator output is very large as the origin is encircled. This is essentially the same effect that was observed in Chapter 3 where the behavior of an FM discriminator operating in the presence of interference was studied.

The phase deviation $\psi(t)$ is illustrated in Figure 7.12 for the case in which the input SNR is -4.0 dB. The origin encirclements can be observed by the 2π jumps in $\psi(t)$. The output of an FM discriminator for several predetection SNRs is shown in Figure 7.13. The decrease in spike noise as the SNR is increased is clearly seen.

In Appendix D it is shown that the power spectral density of spike noise is given by

$$S_{d\psi/dt}(f) = (2\pi)^2(\nu + \overline{\delta\nu}) \tag{7.131}$$

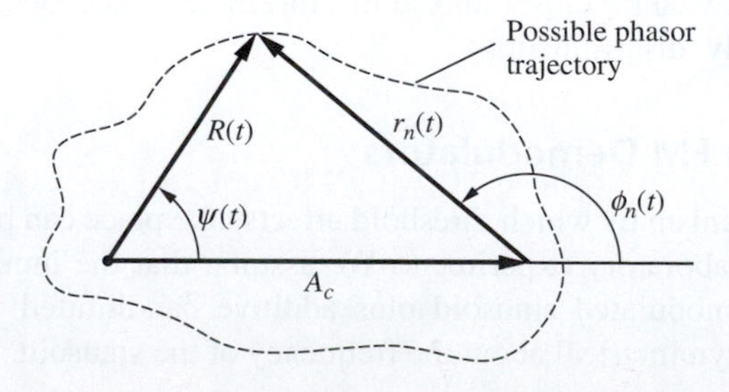

Figure 7.11
Phasor diagram near threshold (spike output) (drawn for $\theta = 0$).

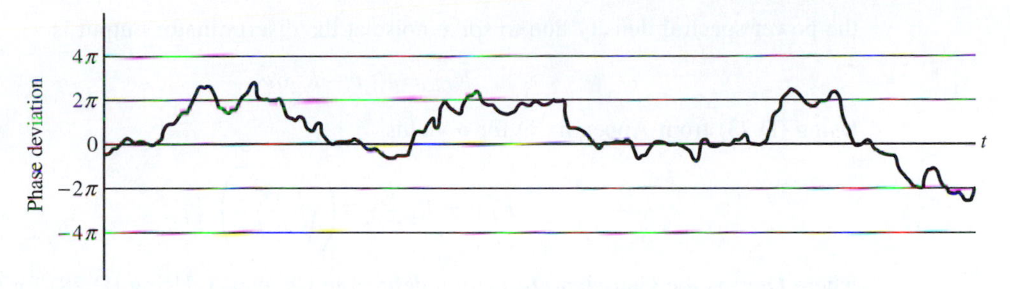

Figure 7.12
Phase deviation for a predetection SNR of −4.0 dB.

where ν is the average number of impulses per second resulting from an unmodulated carrier plus noise and $\overline{\delta\nu}$ is the net increase of the spike rate due to modulation. Since the discriminator output is given by

$$y_D(t) = \frac{1}{2\pi} K_D \frac{d\psi}{dt} \tag{7.132}$$

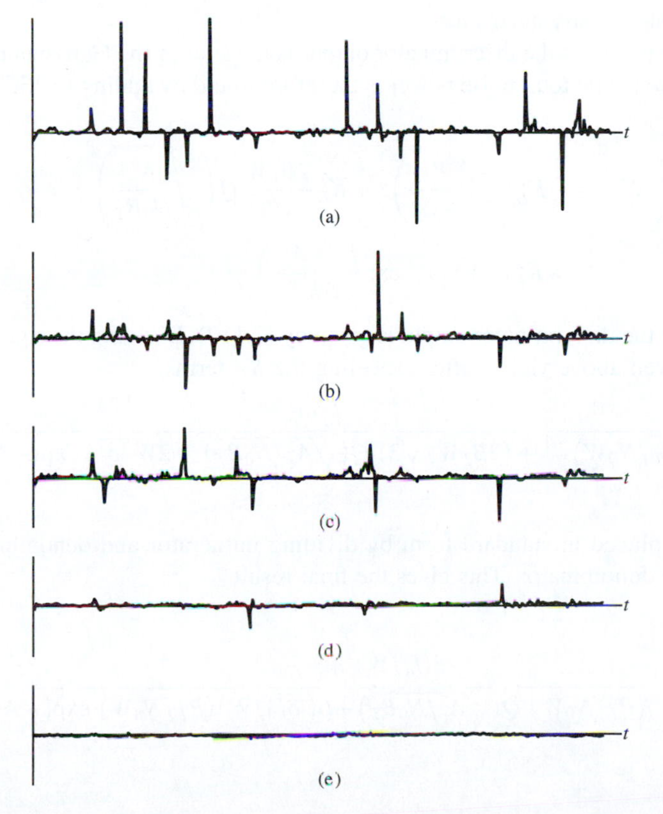

Figure 7.13
Output of FM discriminator due to input noise for various predetection SNRs.
(a) Predetection SNR = −10 dB.
(b) Predetection SNR = −4 dB.
(c) Predetection SNR = 0 dB.
(d) Predetection SNR = 6 dB.
(e) Predetection SNR = 10 dB.

the power spectral density due to spike noise at the discriminator output is

$$N_{D\delta} = K_D^2 \nu + K_D^2 \overline{\delta\nu} \tag{7.133}$$

Using (D.23) from Appendix D for ν yields

$$K_D^2 \nu = K_D^2 \frac{B_T}{\sqrt{3}} Q\left(\sqrt{\frac{A_c^2}{N_0 B_T}}\right) \tag{7.134}$$

where $Q(x)$ is the Gaussian Q-function defined in Chapter 5. Using (D.28) for $\overline{\delta\nu}$ yields

$$K_D^2 \overline{\delta\nu} = K_D^2 \, |\overline{\delta f}| \exp\left(\frac{-A_c^2}{2N_0 B_T}\right) \tag{7.135}$$

Since the spike noise at the discriminator output is white, the spike noise power at the discriminator output is found by multiplying the power spectral density by the two-sided postdetection bandwidth $2W$. Substituting (7.134) and (7.135) into (7.133) and multiplying by $2W$ yields

$$N_{D\delta} = K_D^2 \frac{2B_T W}{\sqrt{3}} Q\left(\sqrt{\frac{A_c^2}{N_0 B_T}}\right) + K_D^2 (2W) \, |\overline{\delta f}| \exp\left(\frac{-A_c^2}{2N_0 B_T}\right) \tag{7.136}$$

for the spike noise power. Now that the spike noise power is known, we can determine the total noise power at the discriminator output. After this is accomplished the output SNR at the discriminator output is easily determined.

The total noise power at the discriminator output is the sum of the Gaussian noise power and spike noise power. The total noise power is therefore found by adding (7.136) to (7.115). This gives

$$\begin{aligned} N_D = {} & \frac{1}{3} K_D^2 W^2 \left(\frac{P_T}{N_0 W}\right)^{-1} + K_D^2 \frac{2B_T W}{\sqrt{3}} Q\left(\sqrt{\frac{A_c^2}{N_0 B_T}}\right) \\ & + K_D^2 (2W) \, |\overline{\delta f}| \exp\left(\frac{-A_c^2}{2N_0 B_T}\right) \end{aligned} \tag{7.137}$$

The signal power at the discriminator output is given by (7.110). Dividing the signal power by the noise power given above yields, after canceling the K_D terms,

$$(\text{SNR})_D = \frac{f_d^2 \overline{m_n^2}}{\frac{1}{3} W^2 (P_T/N_0 W)^{-1} + (2B_T W/\sqrt{3}) \, Q\left(\sqrt{A_c^2/N_0 B_T}\right) + 2W \, |\overline{\delta f}| \exp\left(-A_c^2/2N_0 B_T\right)} \tag{7.138}$$

This result can be placed in standard form by dividing numerator and denominator by the leading term in the denominator. This gives the final result

$$(\text{SNR})_D = \frac{3 (f_d/W)^2 \overline{m_n^2} \frac{P_T}{N_0 W}}{1 + 2\sqrt{3} (B_T/W)(P_T/N_0 W) \, Q\left(\sqrt{A_c^2/N_0 B_T}\right) + 6 (|\overline{\delta f}|/W)(P_T/N_0 W) \exp\left(-A_c^2/2N_0 B_T\right)} \tag{7.139}$$

For operation above threshold, the region of input SNRs where spike noise is negligible, the last two terms in the denominator of the preceding expression are much less than one and may therefore be neglected. For this case the postdetection SNR is the above threshold result given by (7.117). It is worth noting that the quantity $A_c^2/(2N_0B_T)$ appearing in the spike noise terms is the predetection SNR. Note that the message signal explicitly affects two terms in the expression for the postdetection SNR through $\overline{m_n^2}$ and $|\overline{\delta f}|$. Thus, before $(\text{SNR})_D$ can be determined, a message signal must be assumed. This is the subject of the following example.

EXAMPLE 7.4

In this example the detection gain of an FM discriminator is determined assuming the sinusoidal message signal

$$m_n(t) = \sin(2\pi Wt) \tag{7.140}$$

The instantaneous frequency deviation is given by

$$f_d m_n(t) = f_d \sin(2\pi Wt) \tag{7.141}$$

and the average of the absolute value of the frequency deviation is therefore given by

$$|\overline{\delta f}| = 2W \int_0^{1/2W} f_d \sin(2\pi Wt)\,dt \tag{7.142}$$

Carrying out the integration yields

$$|\overline{\delta f}| = \frac{2}{\pi} f_d \tag{7.143}$$

Note that f_d is the peak frequency deviation, which by definition of the modulation index, β, is βW. (We use the modulation index β rather than the deviation ratio D since $m(t)$ is a sinusoidal signal.) Thus

$$|\overline{\delta f}| = \frac{2}{\pi}\beta W \tag{7.144}$$

From Carson's rule we have

$$\frac{B_T}{W} = 2(\beta + 1) \tag{7.145}$$

Since the message signal is sinusoidal $\beta = f_d/W$ and $\overline{m_n^2} = 1/2$. Thus

$$\left(\frac{f_d}{W}\right)^2 \overline{m_n^2} = \frac{1}{2}\beta^2 \tag{7.146}$$

Finally, the predetection SNR can be written

$$\frac{A_c^2}{2N_0B_T} = \frac{1}{2(\beta+1)}\frac{P_T}{N_0W} \tag{7.147}$$

Substituting (7.146) and (7.147) into (7.139) yields

$(SNR)_D$

$$= \frac{1.5\beta^2 P_T/N_0W}{1+4\sqrt{3}(\beta+1)(P_T/N_0W)Q\left(\sqrt{[1/(\beta+1)](P_T/N_0W)}+(12/\pi)\beta\exp\{[-1/2(\beta+1)](P_T/N_0W)\}\right.} \tag{7.148}$$

for the postdetection SNR.

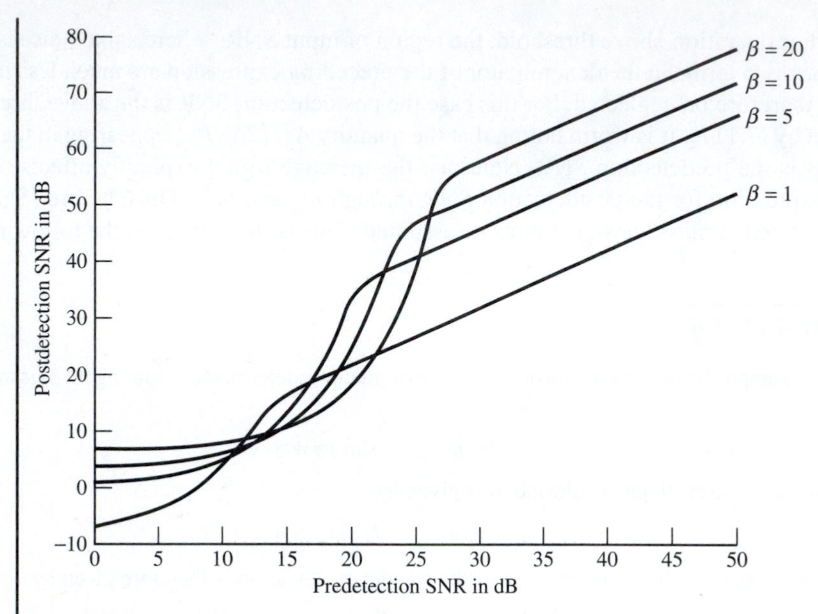

Figure 7.14
Frequency modulation system performance with sinusoidal modulation.

The postdetection SNR is illustrated in Figure 7.14 as a function of P_T/N_0W. The threshold value of P_T/N_0W is defined as the value of P_T/N_0W at which the postdetection SNR is 3 dB below the value of the postdetection SNR given by the above threshold analysis. In other words, the threshold value of P_T/N_0W is the value of P_T/N_0W for which the denominator of (7.148) is equal to 2. It should be noted from Figure 7.14 that the threshold value of P_T/N_0W increases as the modulation index β increases. The study of this effect is the subject of one of the computer exercises at the end of this chapter.

Satisfactory operation of FM systems requires that operation be maintained above threshold. Figure 7.14 shows the rapid convergence to the result of the above threshold analysis described by (7.117), with (7.146) used to allow (7.117) to be written in terms of the modulation index. Figure 7.14 also shows the rapid deterioration of system performance as the operating point moves into the below-threshold region.

■

COMPUTER EXAMPLE 7.1

The MATLAB program to generate the performance curves illustrated in Figure 7.14 follows.

```
% File:  c7ce1.m
zdB = 0:50; % predetection SNR in dB
z = 10.^(zdB/10); % predetection SNR
beta = [1 5 10 20]; % modulation index vector
hold on % hold for plots
for j=1:length(beta)
  bta = beta(j); % current index
  a1 = exp(-(0.5/(bta+1)*z)); % temporary constant
  a2 = q(sqrt(z/(bta+1))); % temporary constant
  num = (1.5*bta*bta)*z;
  den = 1+(4*sqrt(3)*(bta+1))*(z.*a2)+(12/pi)*bta*(z.*a1);
  result = num./den;
```

```
    resultdB = 10*log10(result);
    plot(zdB,resultdB,'k')
  end
  hold off
  xlabel('Predetection SNR in dB')
  ylabel('Postdetection SNR in dB')
  % End of script file.
```

■

EXAMPLE 7.5

Equation (7.148) gives the performance of an FM demodulator taking into account both modulation and additive noise. It is of interest to determine the relative effects of modulation and noise. In order to accomplish this, (7.148) can be written

$$(SNR)_D = \frac{1.5\beta^2 z}{1 + D_2(\beta, z) + D_3(\beta, z)} \tag{7.149}$$

where $z = P_T/N_0W$ and where $D_2(\beta, z)$ and $D_3(\beta, z)$ represent the second term (due to noise) and third term (due to modulation) in (7.148), respectively. The ratio of $D_3(\beta, z)$ to $D_2(\beta, z)$ is

$$\frac{D_3(\beta, z)}{D_2(\beta, z)} = \frac{\sqrt{3}}{\pi} \frac{\beta}{\beta + 1} \frac{\exp[-z/2(\beta+1)]}{Q[z/(\beta+1)]} \tag{7.150}$$

This ratio is plotted in Figure 7.15. It is clear that for $z > 10$, the effect of modulation on the denominator of (7.148) is considerably greater than the effect of noise. However, both $D_2(\beta, z)$ and $D_3(\beta, z)$ are

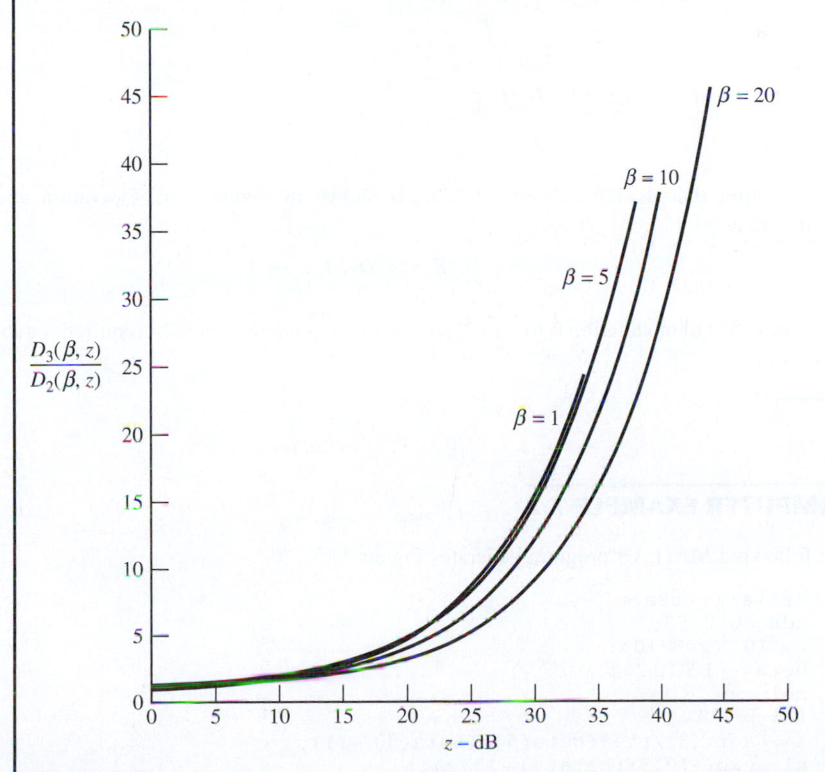

Figure 7.15
Ratio of $D_3(\beta, z)$ to $D_2(\beta, z)$.

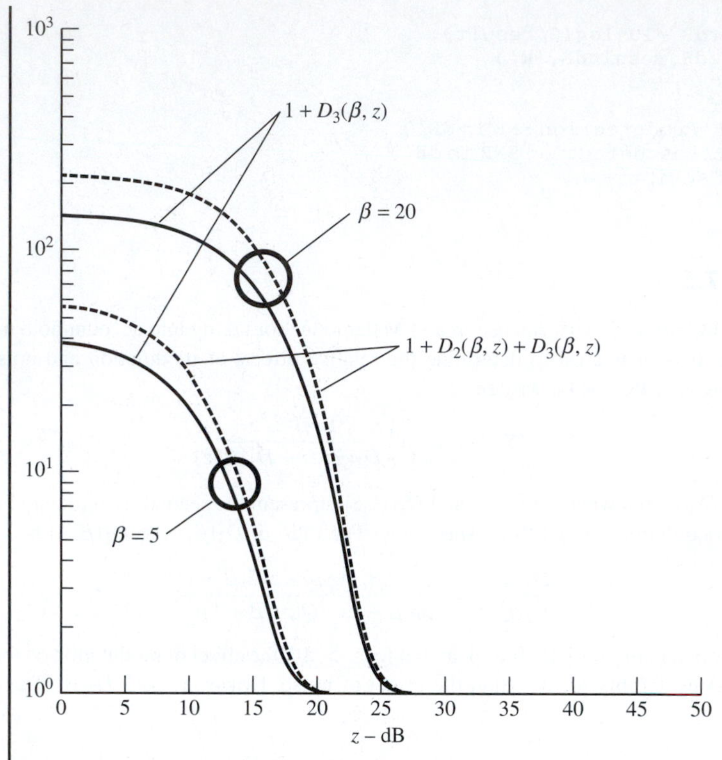

Figure 7.16
$1 + D_3(\beta, z)$ and $1 + D_2(\beta, z) + D_3(\beta, z)$.

much smaller than 1 above threshold. This is shown in Figure 7.16. Operation above threshold requires that

$$D_2(\beta, z) + D_3(\beta, z) \ll 1 \tag{7.151}$$

Thus, the effect of modulation is to raise the value of the predetection SNR required for above threshold operation.

■

COMPUTER EXAMPLE 7.2

The following MATLAB program generates Figure 7.15.

```
%File: c7ce2a.m
zdB = 0:2:50;
z = 10.^(zdB/10);
beta = [1 5 10 20];
hold on
for j=1:4
K = (sqrt(3)/pi)*(beta(j)/(beta(j)+1));
a1 = exp(-(0.5/(beta(j)+1)*z));
a2 = q(sqrt((1/(beta(j)+1))*z));
```

```
result = K*a1./a2;
plot(zdB,result)
end
hold off
xlabel('Predetection SNR in dB')
ylabel('D_3 / D_2')
% End of script file.
```

In addition, the following MATLAB program generates Figure 7.16.

```
File: c7ce2b.m
zdB = 0:0.1:40; % predetection SNR in dB
z = 10.^(zdB/10); % predetection SNR
beta = [5 20]; % modulation index vector
hold on % hold for plots
for j=1:2
  a2 = exp(-(0.5/(beta(j)+1)*z));
  a1 = q(sqrt((1/(beta(j)+1))*z));
  r1 = 1+(4*sqrt(3)*(beta(j)+1)*a2);
  r2 = r1+((12/pi)*beta(j)*a2);
  semilogy(zdB,r1,'k',zdB,r2,'k--')
end
hold off % release
xlabel('Predetection SNR in dB')
ylabel('1+D_3 and 1+D_2+D_3')
```

■

The threshold extension provided by a PLL is somewhat difficult to analyze, and many developments have been published.[2] Thus we will not cover it here. We state, however, that the threshold extension obtained with the PLL is typically on the order of 2 to 3 dB compared to the demodulator just considered. Even though this is a moderate extension, it is often important in high-noise environments.

■ 7.5 NOISE IN PULSE-CODE MODULATION

Pulse-code modulation was briefly discussed in Chapter 3, and we now consider a simplified performance analysis. There are two major error sources in PCM. The first of these results from quantizing the signal, and the other results from channel noise. As we saw in Chapter 3, quantizing involves representing each input sample by one of q quantizing levels. Each quantizing level is then transmitted using a sequence of symbols, usually binary, to uniquely represent each quantizing level.

7.5.1 Postdetection SNR

The sampled and quantized message waveform can be represented as

$$m_{\delta q}(t) = \sum m(t)\delta(t - iT_s) + \sum \epsilon(t)\delta(t - iT_s) \tag{7.152}$$

[2]See Taub and Schilling (1986), pp. 419–422, for an introductory treatment.

where the first term represents the sampling operation and the second term represents the quantizing operation. The ith sample of $m_{\delta q}(t)$ is represented by

$$m_{\delta q}(t_i) = m(t_i) + \epsilon_q(t_i) \tag{7.153}$$

where $t_i = iT_s$. Thus the SNR resulting from quantizing is

$$(\text{SNR})_Q = \frac{\overline{m^2(t_i)}}{\overline{\epsilon_q^2(t_i)}} = \frac{\overline{m^2}}{\overline{\epsilon_q^2}} \tag{7.154}$$

assuming stationarity. The quantizing error is easily evaluated for the case in which the quantizing levels have uniform spacing, S. For the uniform spacing case the quantizing error is bounded by $\pm\frac{1}{2}S$. Thus, assuming that $\epsilon_q(t)$ is uniformly distributed in the range

$$-\frac{1}{2}S \leq \epsilon_q \leq \frac{1}{2}S$$

the mean-square error due to quantizing is

$$\overline{\epsilon_q^2} = \frac{1}{S}\int_{-S/2}^{S/2} x^2\,dx = \frac{1}{12}S^2 \tag{7.155}$$

so that

$$(\text{SNR})_Q = 12\frac{\overline{m^2}}{S^2} \tag{7.156}$$

The next step is to express $\overline{m^2}$ in terms of q and S. If there are q quantizing levels, each of width S, it follows that the peak-to-peak value of $m(t)$, which is referred to as the *dynamic range* of the signal, is qS. Assuming that $m(t)$ is *uniformly* distributed in this range,

$$\overline{m^2} = \frac{1}{qS}\int_{-qS/2}^{qS/2} x^2 dx = \frac{1}{12}q^2S^2 \tag{7.157}$$

Substituting (7.157) into (7.156) yields

$$(\text{SNR})_Q = q^2 = 2^{2n} \tag{7.158}$$

where n is the number of binary symbols used to represent each quantizing level. We have made use of the fact that $q = 2^n$ for binary quantizing.

If the additive noise in the channel is sufficiently small, system performance is limited by quantizing noise. For this case (7.158) becomes the postdetection SNR and is independent of P_T/N_0W. If quantizing is not the only error source, the postdetection SNR depends on both P_T/N_0W and on quantizing noise. In turn, the quantizing noise is dependent on the signaling scheme.

An approximate analysis of PCM is easily carried out by assuming a specific signaling scheme and borrowing a result from Chapter 8. Each sample value is transmitted as a group of n pulses, and as a result of channel noise, any of these n pulses can be in error at the receiver output. The group of n pulses defines the quantizing level and is referred to as a *digital word.* Each individual pulse is a digital symbol, or bit assuming a binary system. We assume that the bit-error probability P_b is known, as it will be after the next chapter. Each of the n bits in the digital word representing a sample value is received correctly with probability $1 - P_b$. Assuming that errors occur independently, the probability that all n bits representing a digital word are received correctly is $(1 - P_b)^n$. The word-error probability P_w is therefore given by

$$P_w = 1 - (1 - P_b)^n \tag{7.159}$$

The effect of a word error depends on which bit of the digital word is in error. We assume that the bit error is the most significant bit (worst case). This results in an amplitude error of $\frac{1}{2}qS$. The effect of a word error is therefore an amplitude error in the range

$$-\frac{1}{2}qS \le \epsilon_w \le \frac{1}{2}qS$$

For simplicity we assume that ϵ_w is uniformly distributed in this range. The resulting mean-square word error is

$$\overline{\epsilon_w^2} = \frac{1}{12}q^2S^2 \tag{7.160}$$

which is equal to the signal power.

The total noise power at the output of a PCM system is given by

$$N_D = \overline{\epsilon_q^2}(1 - P_w) + \overline{\epsilon_w^2}P_w \tag{7.161}$$

The first term on the right-hand side of (7.161) is the contribution to N_D due to quantizing error, which is (7.155) weighted by the probability that all bits in a word are received correctly. The second term is the contribution to N_D due to word error weighted by the probability of word error. Using (7.161) for the noise power and (7.157) for signal power yields

$$(\text{SNR})_D = \frac{\frac{1}{12}q^2S^2}{\frac{1}{12}S^2(1 - P_w) + \frac{1}{12}q^2S^2P_w} \tag{7.162}$$

which can be written as

$$(\text{SNR})_D = \frac{1}{q^{-2}(1 - P_w) + P_w} \tag{7.163}$$

In terms of the wordlength n, using (7.158) the preceding result is

$$(\text{SNR})_D = \frac{1}{2^{-2n} + P_w(1 - 2^{-2n})} \tag{7.164}$$

The term 2^{-2n} is completely determined by the wordlength n, while the word-error probability P_w is a function of the SNR P_T/N_0W and the wordlength n.

If the word-error probability P_w is negligible, which is the case for a sufficiently high SNR at the receiver input,

$$(\text{SNR})_D = 2^{2n} \tag{7.165}$$

which, expressed in decibels, is

$$10\log_{10}(\text{SNR})_D = 6.02n \tag{7.166}$$

We therefore gain 6 dB in SNR for every bit added to the quantizer wordlength. The region of operation in which P_w is negligible and system performance is limited by quantization error is referred to as the *above-threshold region*.

Quantizing, and the effects of quantizing, play an important role in the design and implementation of digital communication systems. The subject of quantizing is covered in more detail in Appendix F. In Appendix F we will consider quantizing for the case in which the message signal is not uniformly distributed over the full dynamic range of the quantizer.

COMPUTER EXAMPLE 7.3

The purpose of this example is to examine the postdetection SNR for a PCM system. Before the postdetection SNR, $(\text{SNR})_D$, can be numerically evaluated, the word-error probability P_w must be known. As shown by (7.159) the word-error probability depends upon the bit-error probability. Borrowing a result from Chapter 8, however, will allow us to illustrate the threshold effect of PCM. If we assume frequency-shift keying (FSK), in which transmission using one frequency is used to represent a binary zero and a second frequency is used to represent a binary one, and a noncoherent receiver, the probability of bit error is

$$P_b = \frac{1}{2}\exp\left(-\frac{P_T}{2N_0B_T}\right) \tag{7.167}$$

In the preceding expression B_T is the bit-rate bandwidth, which is the reciprocal of the time required for transmission of a single bit in the n-symbol PCM digital word. The quantity P_T/N_0B_T is the predetection SNR. Substitution of (7.167) into (7.159) and substitution of the result into (7.164) yields the postdetection SNR, $(\text{SNR})_D$. This result is shown in Figure 7.17. The threshold effect can easily be seen. The following MATLAB program generates Figure 7.17.

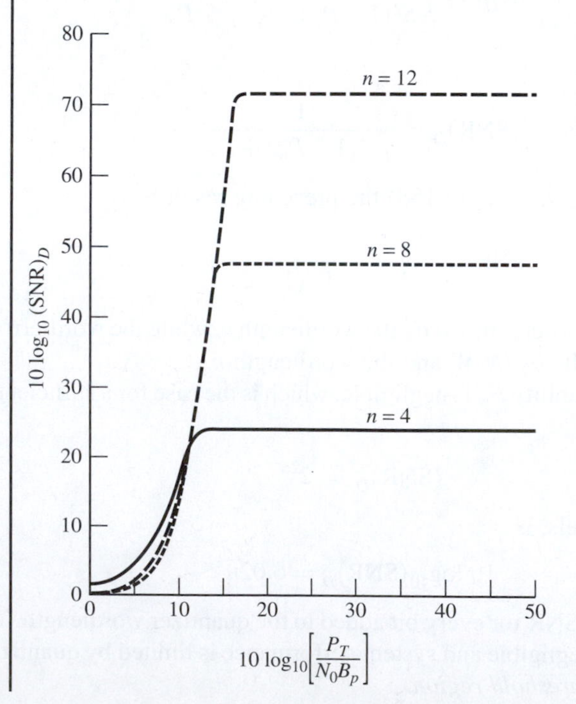

Figure 7.17
Signal-to-noise ratio at output of PCM system (FSK modulation used with noncoherent receiver).

```
% File c7ce3.m
n = [4 8 12]; % wordlengths
snrtdB = 0:0.1:30; % predetection snr in dB
snrt = 10.^(snrtdB/10); % predetection snr
Pb =0.5*exp(-snrt/2); % bit error probability
hold on % hold for multiple plots
for k=1:length(n)
 Pw = 1-(1-Pb).^n(k); % current value of Pw
 a = 2^(-2*n(k)); % temporary constant
 snrd = 1./(a+Pw*(1-a)); % postdetection snr
 snrddB = 10*log10(snrd); % postdetection snr in dB
 plot(snrtdB,snrddB)
end
hold off % release
xlabel('Predetection SNR in dB')
ylabel('Postdetection SNR in dB')
% End of script file.
```

Note that longer digital words give a higher value of $(\mathrm{SNR})_D$ above threshold due to reduced quantizing error. However, the longer digital word means that more bits must be transmitted for each sample of the original time-domain signal, $m(t)$. This increases the bandwidth requirements of the system. Thus, the improved SNR comes at the expense of a higher bit-rate or system bandwidth. We see again the threshold effect that occurs in nonlinear systems and the resulting trade-off between SNR and transmission bandwidth.

■

7.5.2 Companding

As we saw in Chapter 3, a PCM signal is formed by sampling, quantizing, and encoding an analog signal. These three operations are collectively referred to as *analog-to-digital conversion*. The inverse process of forming an analog signal from a digital signal is known as *digital-to-analog conversion*.

In the preceding section we saw that significant errors can result from the quantizing process if the wordlength n is chosen too small for a particular application. The result of these errors is described by the signal-to-quantizing-noise ratio expressed by (7.158). Keep in mind that (7.158) was developed for the case of a uniformly distributed signal.

The level of quantizing noise added to a given sample, (7.155), is independent of the signal amplitude, and small amplitude signals will therefore suffer more from quantizing effects than large amplitude signals. This can be seen from (7.156). There are essentially two ways to combat this problem. First, the quantizing steps can be made small for small amplitudes and large for large amplitude portions of the signal. This scheme is known as *nonuniform quantizing*. An example of a nonuniform quantizer is the Max quantizer, in which the quantizing steps are chosen so that the mean-square quantizing error is minimized. The Max quantizer is examined in detail in Appendix F.

The second technique, and the one of interest here, is to pass the analog signal through a nonlinear amplifier prior to the sampling process. The input–output characteristics of the amplifier are shown in Figure 7.18. For small values of the input x_{in}, the slope of the input–output characteristic is large. A change in a low-amplitude signal will therefore force the signal through more quantizing levels than the same change in a high-amplitude signal. This essentially yields smaller step sizes for small amplitude signals and therefore reduces the quantizing error for small amplitude signals. It can be seen from Figure 7.18 that the peaks of

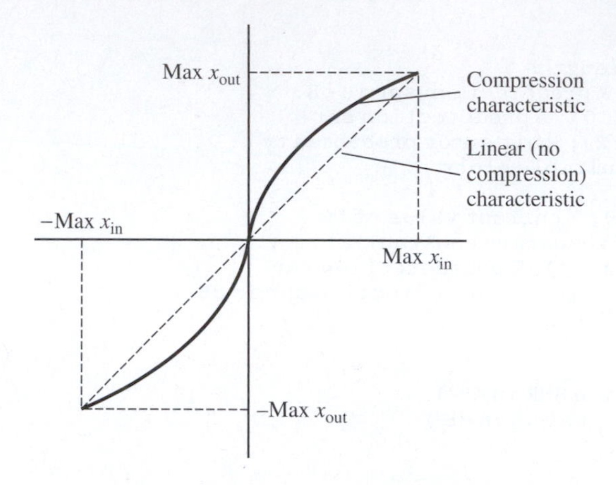

Figure 7.18
Input–output compression characteristic.

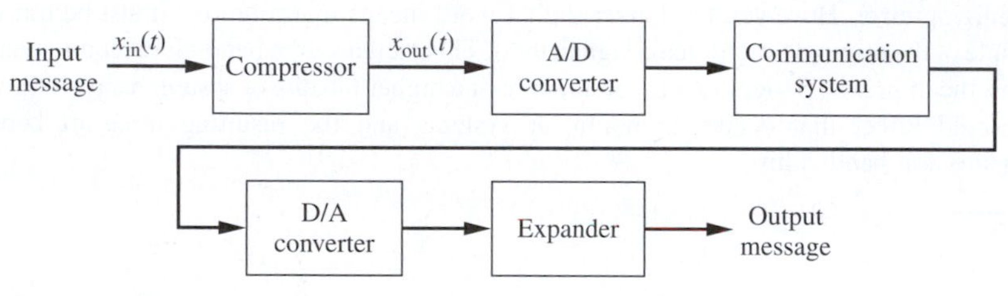

Figure 7.19
Example of companding.

the input signal are compressed. For this reason the characteristic shown in Figure 7.18 is known as a *compressor*.

The effect of the compressor must be compensated when the signal is returned to analog form. This is accomplished by placing a second nonlinear amplifier at the output of the DA converter. This second nonlinear amplifier is known as an *expander* and is chosen so that the cascade combination of the compressor and expander yields a linear characteristic, as shown by the dashed line in Figure 7.18. The combination of a compressor and an expander is known as a *compander*. A companding system is shown in Figure 7.19.

The concept of predistorting a message signal in order to achieve better performance in the presence of noise, and then removing the effect of the predistortion, should remind us of the use of pre-emphasis and de-emphasis filters in the implementation of FM systems.

Summary

1. The AWGN model is frequently used in the analysis of communications systems. However, the AWGN assumption is only valid over a certain bandwidth, and this bandwidth is a function of temperature. At a temperature of 3 K this bandwidth is approximately 10 GHz. If the temperature increases the bandwidth over which the white noise assumption is valid also increases. At standard temperature (290 K), the white noise assumption is valid to bandwidths exceeding 1000 GHz. Thermal noise results from the

combined effect of many charge carries. The Gaussian assumption follows from the central-limit theorem.

2. The SNR at the output of a baseband communication system operating in an additive Gaussian noise environment is P_T/N_0W, where P_T is the signal power, N_0 is the single-sided power spectral density of the noise ($\frac{1}{2}N_0$ is the two-sided power spectral density), and W is the signal bandwidth.
3. A DSB system has an output SNR of P_T/N_0W assuming perfect phase coherence of the demodulation carrier and a noise bandwidth of W.
4. A SSB system also has an output SNR of P_T/N_0W assuming perfect phase coherence of the demodulation carrier and a bandwidth of W. Thus, under ideal conditions, both SSB and DSB have performance equivalent to the baseband system.
5. An AM system with coherent demodulation achieves an output SNR of $E_{ff}P_T/N_0W$, where E_{ff} is the efficiency of the system. An AM system with envelope detection achieves the same output SNR as an AM system with coherent demodulation if the SNR is high. If the predetection SNR is small, the signal and noise at the demodulation output become multiplicative rather than additive. The output exhibits severe loss of signal for a small decrease in the input SNR. This is known as the *threshold effect*.
6. The square-law detector is a nonlinear system that can be analyzed for all values of P_T/N_0W. Since the square-law detector is nonlinear, a threshold effect is observed.
7. Using a quadrature double-sideband (QDSB) signal model, a generalized analysis is easily carried out to determine the combined effect of both additive noise and demodulation phase errors on a communication system. The result shows that SSB and QDSB are equally sensitive to demodulation phase errors if the power in the two QDSB signals are equal. Double-sideband is much less sensitive to demodulation phase errors than SSB or QDSB because SSB and QDSB both exhibit crosstalk between the quadrature channels for nonzero demodulation phase errors.
8. The analysis of angle modulation systems shows that the output noise is suppressed as the signal carrier amplitude is increased for system operation above threshold. Thus the demodulator noise power output is a function of the input signal power.
9. The demodulator output power spectral density is constant over the range $|f| < W$ for PM and is parabolic over the range if $|f| < W$ for FM. The parabolic power spectral density for an FM system is due to the fact that FM demodulation is essentially a differentiation process.
10. The demodulated output SNR is proportional to k_p^2 for PM, where k_p is the phase-deviation constant. The output SNR is proportional to D^2 for an FM system, where D is the deviation ratio. Since increasing the deviation ratio also increases the bandwidth of the transmitted signal, the use of angle modulation allows us to achieve improved system performance at the cost of increased bandwidth.

Table 7.1 Noise Performance Characteristics

System	Postdetection SNR	Transmission bandwidth
Baseband	$\frac{P_T}{N_0 W}$	W
DSB with coherent demodulation	$\frac{P_T}{N_0 W}$	$2W$
SSB with coherent demodulation	$\frac{P_T}{N_0 W}$	W
AM with envelope detection (above threshold) or AM with coherent demodulation. *Note*: E is efficiency	$\frac{EP_T}{N_0 W}$	$2W$
AM with square-law detection	$2\left(\frac{a^2}{2+a^2}\right)^2 \frac{P_T/N_0 W}{1+(N_0 W/P_T)}$	$2W$
PM above threshold	$k_p^2 \overline{m_n^2} \frac{P_T}{N_0 W}$	$2(D+1)W$
FM above threshold (without preemphasis)	$3D^2 \overline{m_n^2} \frac{P_T}{N_0 W}$	$2(D+1)W$
FM above threshold (with preemphasis)	$\left(\frac{f_d}{f_3}\right)^2 \overline{m_n^2} \frac{P_T}{N_0 W}$	$2(D+1)W$

11. The use of pre-emphasis and de-emphasis can significantly improve the noise performance of an FM system. Typical values result in a better than 10-dB improvement in the SNR of the demodulated output.

12. As the input SNR of an FM system is reduced, spike noise appears. The spikes are due to origin encirclements of the total noise phasor. The area of the spikes is constant at 2π, and the power spectral density is proportional to the spike frequency. Since the predetection bandwidth must be increased as the modulation index is increased, resulting in a decreased predetection SNR, the threshold value of P_T/N_0W increases as the modulation index increases.

13. An analysis of PCM, which is a nonlinear modulation process due to quantizing, shows that, like FM, a trade-off exists between bandwidth and output SNR. PCM system performance above threshold is dominated by the wordlength or, equivalently, the quantizing error. PCM performance below threshold is dominated by channel noise.

14. A most important result for this chapter is the postdetection SNRs for various modulation methods. A summary of these results is given in Table 7.1. Given in this table is the postdetection SNR for each technique as well as the required transmission bandwidth. The trade-off between postdetection SNR and transmission bandwidth is evident for nonlinear systems.

Further Reading

All the books cited at the end of Chapter 3 contain material about noise effects in the systems studied in this chapter. The books by Lathi (1998) and Haykin (2000) are especially recommended for their completeness. The book by Taub and Schilling (1986), although an older book, contains excellent sections on both PCM systems and threshold effects in FM systems. The book by Tranter et al. (2004) discusses quantizing in some depth.

Problems

Section 7.1

7.1. In discussing thermal noise at the beginning of this chapter, we stated that at standard temperature (290 K) the white noise assumption is valid to bandwidths exceeding 1000 GHz. If the temperature is reduced to 3 K, the variance of the noise is reduced, but the bandwidth over which the white noise assumption is valid is reduced to approximately 10 GHz. Express both of these reference temperatures (3 and 290 K) in degrees fahrenheit.

7.2. The waveform at the input of a baseband system has signal power P_T and white noise with single-sided power spectral density N_0. The signal bandwidth is W. In order to pass the signal without significant distortion, we assume that the input waveform is bandlimited to a bandwidth $B = 3W$ using a Butterworth filter with order n. Compute the SNR at the filter output for $n = 1$, 3, 5, and 10 as a function of P_T/N_0W. Also compute the SNR for the case in which $n \to \infty$. Discuss the results.

7.3. Derive the equation for $y_D(t)$ for an SSB system assuming that the noise is expanded about the frequency $f_x = f_c \pm \frac{1}{2}W$. Derive the detection gain and $(\text{SNR})_D$. Determine and plot $S_{n_c}(f)$ and $S_{n_s}(f)$.

7.4. Derive an expression for the detection gain of a DSB system for the case in which the bandwidth of the bandpass predetection filter is B_T and the bandwidth of the lowpass postdetection filter is B_D. Let $B_T > 2W$ and let $B_D > W$ simultaneously, where W is the bandwidth of the modulation. (There are two reasonable cases to consider.) Repeat for an AM signal.

7.5. A message signal is defined by

$$m(t) = A\cos(2\pi f_1 t + \theta_1) + B\cos(2\pi f_2 t + \theta_2)$$

where A and B are constants, $f_1 \neq f_2$, and θ_1 and θ_2 are random phases uniformly distributed in $[0, 2\pi)$. Compute $\widehat{m}(t)$ and show that the power in $m(t)$ and $\widehat{m}(t)$ are equal. Compute $E[m(t)\widehat{m}(t)]$, where $E[\cdot]$ denotes statistical expectation. Comment on the results.

7.6. In Section 7.1.3 we expanded the noise component about f_c. We observed, however, that the noise components for SSB could be expanded about $f_c \pm \frac{1}{2}W$, depending on the choice of sidebands. Plot the power spectral density for each of these two cases and for each case write the expressions corresponding to (7.16) and (7.17).

7.7. A message signal has the Fourier transform

$$M(f) = \begin{cases} A, & f_1 \leq |f| \leq f_2 \\ 0, & \text{otherwise} \end{cases}$$

Determine $m(t)$ and $\widehat{m}(t)$. Plot $m(t)$ and $\widehat{m}(t)$, and for f_2 fixed and $f_1 = 0, f_1 = -f_2/2$ and $f_1 = -f_2$. Comment on the results.

7.8. Assume that an AM system operates with an index of 0.6 and that the message signal is $12\cos(8\pi)$. Compute the efficiency, the detection gain in dB, and the output SNR in decibels relative to the baseband performance P_T/N_0W. Determine the improvement (in decibels) in the output SNR that results if the modulation index is increased from 0.6 to 0.9.

7.9. An AM system has a message signal that has a zero-mean Gaussian amplitude distribution. The peak value of $m(t)$ is taken as that value that $|m(t)|$ exceeds 0.5% of the time. If the index is 0.7, what is the detection gain?

7.10. The threshold level for an envelope detector is sometimes defined as that value of $(\text{SNR})_T$ for which $A_c > r_n$ with probability 0.99. Assuming that $a^2\overline{m_n^2} \cong 1$, derive the SNR at threshold, expressed in decibels.

7.11. An envelope detector operates above threshold. The modulating signal is a sinusoid. Plot $(\text{SNR})_D$ in decibels as a function of P_T/N_0W for the modulation index equal to 0.4, 0.5, 0.7, and 0.9.

7.12. A square-law demodulator for AM is illustrated in Figure 7.20. Assuming that $x_c(t) = A_c[1 + am_n(t)]\cos(2\pi f_c t)$ and $m(t) = \cos(2\pi f_m t) + \cos(4\pi f_m t)$, sketch the spectrum of each term that appears in $y_D(t)$. Do not neglect the noise that is assumed to be bandlimited white noise with bandwidth $2W$. In the spectral plot identify the desired component, the signal-induced distortion, and the noise.

7.13. Verify the correctness of (7.59).

7.14. Starting with (7.63) derive an expression for $(\text{SNR})_D$ assuming that the message is the sinusoid $m(t) = A\sin(2\pi f_m t)$. From this result verify the correct-

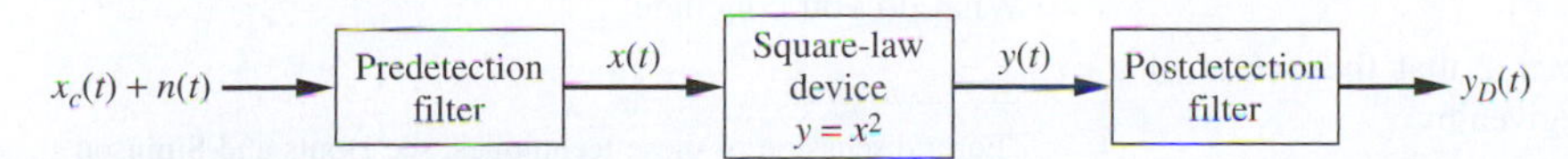

Figure 7.20

ness on Figure 7.6. Assuming this same signal for $m(t)$, plot D_D/S_D in decibels as a function of the index a. Finally, derive an expression for D_D/N_D as a function of P_T/N_0W with a as a parameter. Plot this last result for $a = 0.5$. What do you conclude?

7.15. Assume that a zero-mean message signal $m(t)$ has a Gaussian pdf and that in normalizing the message signal to form $m_n(t)$, the maximum value of $m(t)$ is assumed to be $k\sigma_m$, where k is a parameter and σ_m is the standard deviation of the message signal. Plot $(\text{SNR})_D$ as a function of P_T/N_0W with $a = 0.5$ and $k = 1, 3$, and 5. What do you conclude?

7.16. Compute $(\text{SNR})_D$ as a function of P_T/N_0W for a linear envelope detector assuming a high predetection SNR and a modulation index of unity. Compare this result to that for a square-law detector, and show that the square-law detector is inferior by approximately 1.8 dB. If necessary, you may assume sinusoidal modulation.

7.17. Consider the system shown in Figure 7.21, in which an RC highpass filter is followed by an ideal lowpass filter having bandwidth W. Assume that the input to the system is $A\cos(2\pi f_c t)$, where $f_c < W$, plus white noise with double-sided power spectral density $\frac{1}{2}N_0$. Determine the SNR at the output of the ideal lowpass filter in terms of N_0, A, R, C, W, and f_c. What is the SNR in the limit as $W \to \infty$?

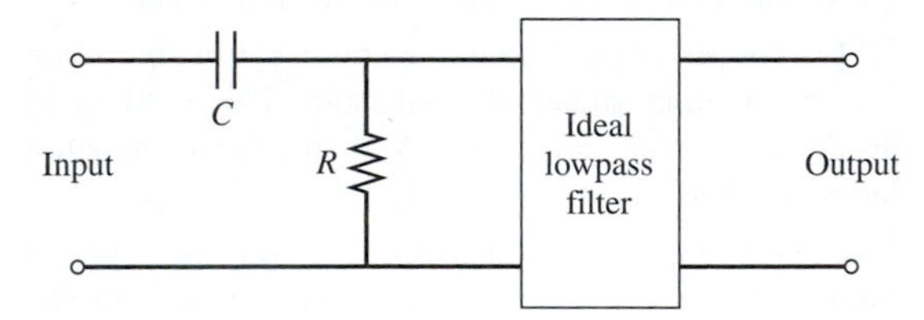

Figure 7.21

Section 7.2

7.18. An SSB system is to be operated with a normalized mean-square error of 0.05 or less. By making a plot of output SNR versus demodulation phase-error variance for the case in which normalized mean-square error is 0.4%, show the region of satisfactory system performance. Repeat for a DSB system. Plot both curves on the same set of axes.

7.19. It was shown in Chapter 2 that the output of a distortionless linear system is given by

$$y(t) = Ax(t - \tau)$$

where A is the gain of the system, τ is the system time delay, and $x(t)$ is the system input. It is often convenient to evaluate the performance of a linear system by comparing the system output with an amplitude-scaled and time-delayed version of the input. The mean-square error is then

$$\overline{\epsilon^2(A, \tau)} = \overline{[y(t) - Ax(t - \tau)]^2}$$

The values of A and τ that minimize this expression, denoted A_m and τ_m, respectively, are defined as the system gain and the system time delay. Show that with these definitions, the system gain is

$$A_m = \frac{R_{xy}(\tau_m)}{R_x(0)}$$

and the resulting system mean-square error is

$$\overline{\epsilon^2(A_m, \tau_m)} = R_y(0) - \frac{R_{xy}^2(\tau_m)}{R_x(0)}$$

Also show that the signal power at the system output is

$$S_D = A_m^2 R_x(0) = \frac{R_{xy}^2(\tau_m)}{R_x(0)}$$

and the output SNR is

$$\frac{S_D}{N_D} = \frac{R_{xy}^2(\tau_m)}{R_x(0)R_y(0) - R_{xy}^2(\tau_m)}$$

in which N_D is the mean-square error.[3]

Section 7.3

7.20. Draw a phasor diagram for an angle-modulated signal for $(\text{SNR})_T \gg 1$ illustrating the relationship between $R(t), A_c$, and $r_n(t)$. Show on this phasor diagram the relationship between $\psi(t)$, $\phi(t)$, and $\phi_n(t)$. Using the phasor diagram, justify that for $(\text{SNR})_T \gg 1$, the approximation

$$\psi(t) \approx \phi(t) + \frac{r_n(t)}{A_c}\sin[\phi_n(t) - \phi(t)]$$

is valid. Draw a second phasor diagram for the case in which $(\text{SNR})_T \ll 1$ and show that

$$\psi(t) \approx \phi_n(t) - \frac{A_c}{r_n(t)}\sin[\phi_n(t) - \phi(t)]$$

What do you conclude?

[3] For a discussion of these techniques, see Houts and Simpson (1968).

7.21. An FM demodulator operates above threshold, and therefore the output SNR is defined by (7.118). Using Carson's rule, write this expression in terms of B_T/W, as was done in (7.119). Plot $(\text{SNR})_T$ in decibels as a function of B_T/W with P_T/N_0W fixed at 30 dB. Determine the value of B_T/W that yields a value of $(\text{SNR})_T$ that is within 0.5 dB of the asymptotic value defined by (7.119).

7.22. The process of stereophonic broadcasting was illustrated in Chapter 3. By comparing the noise power in the $l(t) - r(t)$ channel to the noise power in the $l(t) + r(t)$ channel, explain why stereophonic broadcasting is more sensitive to noise than nonstereophonic broadcasting.

7.23. An FDM communication system uses DSB modulation to form the baseband and FM modulation for transmission of the baseband. Assume that there are eight channels and that all eight message signals have equal power P_0 and equal bandwidth W. One channel does *not* use subcarrier modulation. The other channels use subcarriers of the form

$$A_k \cos(2\pi k f_1 t), \quad 1 \leq k \leq 7$$

The width of the guardbands is $3W$. Sketch the power spectrum of the received *baseband* signal showing both the signal and noise components. Calculate the relationship between the values of A_k if the channels are to have equal SNRs.

7.24. Using (7.123), derive an expression for the ratio of the noise power in $y_D(t)$ with de-emphasis to the noise power in $y_D(t)$ without de-emphasis. Plot this ratio as a function of W/f_3. Evaluate the ratio for the standard values of $f_3 = 2.1$ kHz and $W = 15$ kHz, and use the result to determine the improvement, in decibels, that results through the use of de-emphasis. Compare the result with that found in Example 7.3.

7.25. White noise with two-sided power spectral density $\frac{1}{2}N_0$ is added to a signal having the power spectral density shown in Figure 7.22. The sum (signal plus noise) is filtered with an ideal lowpass filter with unity passband gain and bandwidth $B > W$. Determine the SNR at the filter output. By what factor will the SNR increase if B is reduced to W?

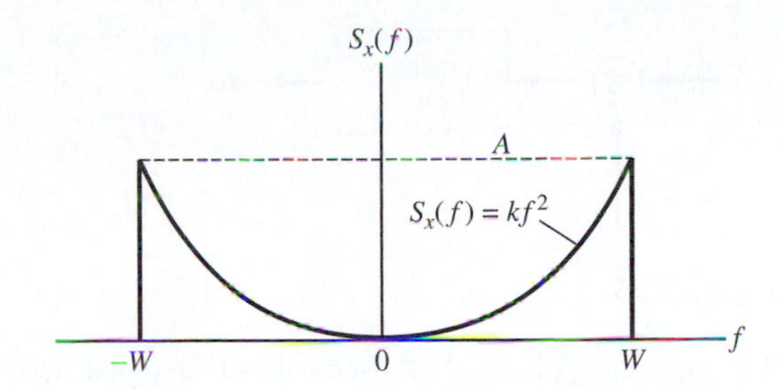

Figure 7.22

7.26. Consider the system shown in Figure 7.23. The signal $x(t)$ is defined by

$$x(t) = A\cos(2\pi f_c t)$$

The lowpass filter has unity gain in the passband and bandwidth W, where $f_c < W$. The noise $n(t)$ is white with two-sided power spectral density $\frac{1}{2}N_0$. The signal component of $y(t)$ is defined to be the component at frequency f_c. Determine the SNR of $y(t)$.

7.27. Repeat the preceding problem for the system shown in Figure 7.24.

7.28. Consider the system shown in Figure 7.25. The noise is white with two-sided power spectral density $\frac{1}{2}N_0$. The power spectral density of the signal is

$$S_x(f) = \frac{A}{1 + (f/f_3)^2}, \quad -\infty < f < \infty$$

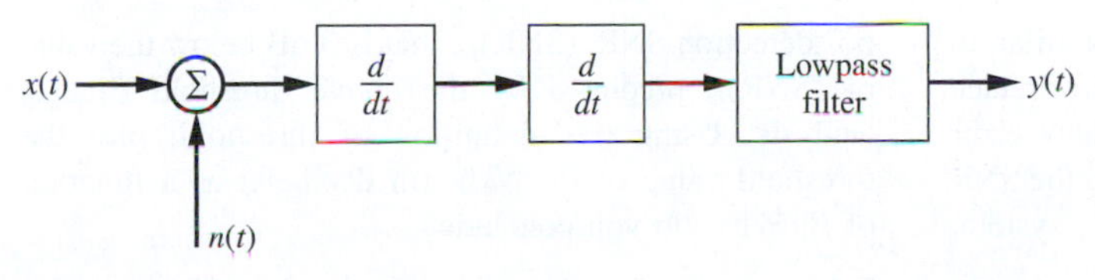

Figure 7.23

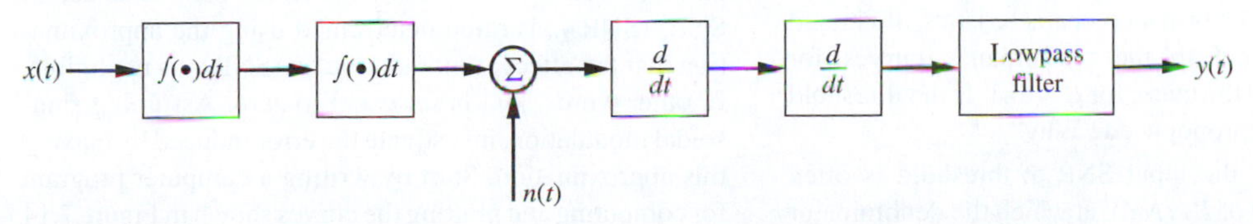

Figure 7.24

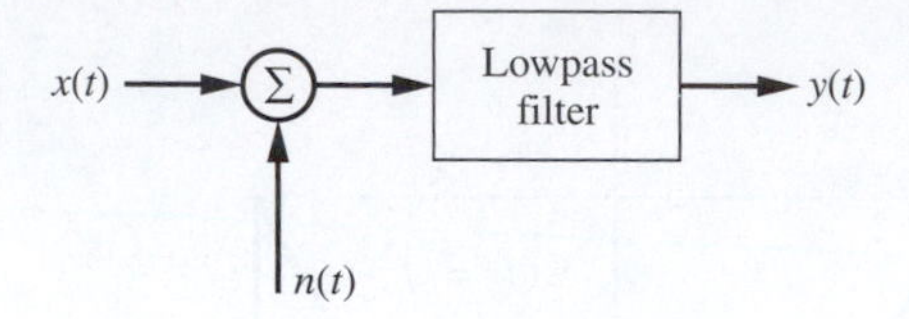

Figure 7.25

The parameter f_3 is the 3-dB bandwidth of the signal. The bandwidth of the ideal lowpass filter is W. Determine the SNR of $y(t)$. Plot the SNR as a function of W/f_3.

Section 7.4

7.29. Derive an expression, similar to (7.148), that gives the output SNR of an FM discriminator for the case in which the message signal is random with a Gaussian amplitude pdf. Assume that the message signal is zero mean and has variance σ_m^2.

7.30. In Example 7.4 we calculated the output SNR for an FM demodulator. We considered the effect of modulation on thresholding assuming that the message signal was a sinusoid. We now assume that the message signal is represented by the Fourier series

$$m(t) = \sum_{n=1}^{N} C_n \cos(2\pi n f_0 t + \theta_n)$$

Generalize (7.143) and (7.148) for this case.

7.31. Assume that the input to a perfect second-order PLL is an unmodulated sinusoid plus bandlimited AWGN. In other words, the PLL input is represented by

$$\begin{aligned} x_c(t) &= A_c \cos(2\pi f_c t + \theta) \\ &\quad + n_c(t) \cos(2\pi f_c t + \theta) \\ &\quad - n_s(t) \sin(2\pi f_c t + \theta) \end{aligned}$$

Also assume that the SNR at the loop input is large so that the phase jitter (error) is sufficiently small to justify use of the linear PLL model. Using the linear model derive an expression for the variance of the loop phase error due to noise in terms of the standard PLL parameters defined in Chapter 3. Show that the probability density function of the phase error is Gaussian and that the variance of the phase error is inversely proportional to the SNR at the loop input.

Section 7.5

7.32. Assume that a PPM system uses Nyquist rate sampling and that the minimum channel bandwidth is used for a given pulse duration. Show that the postdetection SNR can be written as

$$(\mathrm{SNR})_D = K \left(\frac{B_T}{W} \right)^2 \frac{P_T}{N_0 W}$$

and evaluate K.

7.33. The message signal on the input to an ADC is a sinusoid of 25 V peak to peak. Compute the signal-to-quantizing-noise power ratio as a function of the wordlength of the ADC. State any assumptions you make.

Computer Exercises

7.1. Develop a set of performance curves, similar to those shown in Figure 7.8, that illustrate the performance of a coherent demodulator as a function of the phase-error variance. Let the SNR be a parameter and express the SNR in decibels. As in Figure 7.8, assume a QDSB system. Repeat this exercise for a DSB system.

7.2. Execute the computer program used to generate the FM discriminator performance characteristics illustrated in Figure 7.14. Add to the performance curves for $\beta = 1, 5, 10$, and 20 the curve for $\beta = 0.1$. Is the threshold effect more or less pronounced? Why?

7.3. The value of the input SNR at threshold is often defined as the value of P_T/N_0W at which the denominator of (7.148) is equal to 2. Note that this value yields a postdetection SNR, $(\mathrm{SNR})_D$, that is 3 dB below the value of $(\mathrm{SNR})_D$ predicted by the above threshold (linear) analysis. Using this definition of threshold, plot the threshold value of P_T/N_0W (in decibels) as a function of β. What do you conclude?

7.4. In analyzing the performance of an FM discriminator, operating in the presence of noise, the postdetection SNR, $(\mathrm{SNR})_D$, is often determined using the approximation that the effect of modulation on $(\mathrm{SNR})_D$ is negligible. In other words, $|\overline{\delta f}|$ is set equal to zero. Assuming sinusoidal modulation, investigate the error induced by making this approximation. Start by writing a computer program for computing and plotting the curves shown in Figure 7.14 with the effect of modulation neglected.

7.5. In Chapter 3 we developed a MATLAB program that can be used to investigate the acquisition performance of a PLL. Using the same baseband model developed in Chapter 3, we now wish to examine acquisition performance in the presence of noise. Assume a perfect second-order PLL. Test the model by observing the number of cycles slipped in the acquisition process due to a step in the input frequency both with and without noise. It is your job to select the noise levels so that the impact of noise is satisfactory demonstrated.

7.6. The preceding computer exercise problem examined the behavior of a PLL in the acquisition mode. We now consider the performance in the tracking mode. Develop a computer simulation in which the PLL is tracking an unmodulated sinusoid plus noise. Let the predetection SNR be sufficiently high to ensure that the PLL does not lose lock. Using MATLAB and the histogram routine, plot the estimate of the pdf at the VCO output. Comment on the results.

7.7. Develop a computer program to verify the performance curves shown in Figure 7.17. Compare the performance of the noncoherent FSK system to the performance of both coherent FSK and coherent PSK with a modulation index of 1. We will show in the following chapter that the bit-error probability for coherent FSK is

$$P_b = Q\left(\sqrt{\frac{P_T}{N_0 B_T}}\right)$$

and that the bit-error probability for coherent BPSK with a unity modulation index is

$$P_b = Q\left(\sqrt{\frac{2P_T}{N_0 B_T}}\right)$$

where B_T is the system bit-rate bandwidth. Compare the results of the three systems studied in this example for $n = 8$ and $n = 16$.

7.8. In Problem 7.19 we described a technique for estimating the gain, delay, and the SNR at a point in a system given a reference signal. Develop a MATLAB program for implementing this technique. The delay τ_m is typically defined as the lag τ for which the cross-correlation $R_{xy}(\tau)$ is maximized. Develop and execute a testing strategy to illustrate that the technique is performing correctly. What is the main source of error in applying this technique? How can this error source be reduced, and what is the associated cost?

7.9. Assume a three-bit ADC (eight quantizing levels). We desire to design a companding system consisting of both a compressor and expander. Assuming that the input signal is a sinusoid, design the compressor such that the sinusoid falls into each quantizing level with equal probability. Implement the compressor using a MATLAB program, and verify the compressor design. Complete the compander by designing an expander such that the cascade combination of the compressor and expander has the desired linear characteristic. Using a MATLAB program, verify the overall design.

CHAPTER 8

PRINCIPLES OF DATA TRANSMISSION IN NOISE

In Chapter 7 we studied the effects of noise in analog communication systems. We now consider digital data modulation system performance in noise. Instead of being concerned with continuous-time, continuous-level message signals, we are concerned with the transmission of information from sources that produce discrete-valued symbols. That is, the input signal to the transmitter block of Figure 1.1 would be a signal that assumes only discrete values. Recall that we started the discussion of digital data transmission systems in Chapter 4, but without consideration of the effects of noise.

The purpose of this chapter is to consider various systems for the transmission of digital data and their relative performances. Before beginning, however, let us consider the block diagram of a digital data transmission system, shown in Figure 8.1, which is somewhat more detailed than Figure 1.1. The focus of our attention will be on the portion of the system between the optional blocks labeled *Encoder* (or simply *coder*) and *Decoder*. In order to gain a better perspective of the overall problem of digital data transmission, we will briefly discuss the operations performed by the blocks shown as dashed lines.

As discussed previously in Chapters 3 and 4, while many sources result in message signals that are inherently digital, such as from computers, it is often advantageous to represent analog signals in digital form (referred to as *analog-to-digital conversion*) for transmission and then convert them back to analog form upon reception (referred to as *digital-to-analog conversion*), as discussed in the preceding chapter. Pulse code modulation, introduced in Chapter 3, is an example of a modulation technique that can be employed to transmit analog messages in digital form. The SNR performance characteristics of a PCM system, which were presented in Chapter 7, show one advantage of this system to be the option of exchanging bandwidth for SNR improvement.[1]

Throughout most of this chapter we will make the assumption that source symbols occur with equal probability. Many discrete-time sources naturally produce symbols with equal probability. As an example, a binary computer file, which may be transmitted through a channel, frequently contains a nearly equal number of 1s and 0s. If source symbols do not occur with nearly equal probably, we will see in Chapter 11 that a process called *source coding* can be used to

[1]A device for converting voice signals from analog to digital and from digital to analog form is known as a *vocoder*.

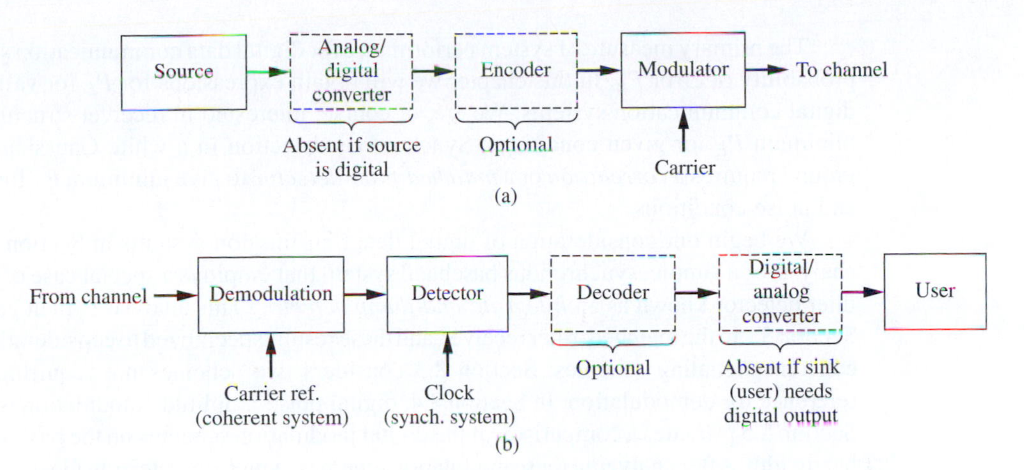

Figure 8.1
Block diagram of a digital data transmission system. (a) Transmitter. (b) Receiver.

create a new set of source symbols in which the binary states, 1 and 0, are equally likely. The mapping from the original set to the new set of source symbols is deterministic so that the original set of source symbols can be recovered from the data output at the receiver. The use of source coding is not restricted to binary sources. We will see Chapter 11 that the transmission of equally likely symbols ensures that the information transmitted with each source symbol is maximized, and therefore, the channel is used efficiently. In order to understand the process of source coding, we need a rigorous definition of information, which will be accomplished in Chapter 11.

Regardless of whether a source is purely digital or an analog source that has been converted to digital, it may be advantageous to add or remove redundant digits to the digital signal. Such procedures, referred to as *forward error-correction coding*, are performed by the encoder–decoder blocks of Figure 8.1 and also will be considered in Chapter 11.

We now consider the basic system in Figure 8.1, shown as the blocks with solid lines. If the digital signals at the modulator input take on one of only two possible values, the communication system is referred to as *binary*. If one of $M > 2$ possible values is available, the system is referred to as *M-ary*. For long-distance transmission, these digital baseband signals from the source may modulate a carrier before transmission, as briefly mentioned in Chapter 4. The result is referred to as *amplitude-shift keying* (ASK), *phase-shift keying* (PSK), or *frequency-shift keying* (FSK) if it is amplitude, phase, or frequency, respectively, that is varied in accordance with the baseband signal. An important M-ary modulation scheme, *quadriphase-shift keying* (QPSK), is often employed in situations in which bandwidth efficiency is a consideration. Other schemes related to QPSK include offset QPSK and minimum-shift keying (MSK). These schemes will be discussed in Chapter 9.

A digital communication system is referred to as *coherent* if a local reference is available for demodulation that is in phase with the transmitted carrier (accounting for fixed phase shifts due to transmission delays). Otherwise, it is referred to as *noncoherent*. Likewise, if a periodic signal is available at the receiver that is in synchronism with the transmitted sequence of digital signals (referred to as a *clock*), the system is referred to as *synchronous* (i.e., the data streams at transmitter and receiver are in lockstep); if a signaling technique is employed in which such a clock is unnecessary (e.g., timing markers might be built into the data blocks), the system is called *asynchronous*.

The primary measure of system performance for digital data communication systems is the probability of error P_E. In this chapter we will obtain expressions for P_E for various types of digital communication systems. We are, of course, interested in receiver structures that give minimum P_E for given conditions. Synchronous detection in a white Gaussian-noise background requires a *correlation* or a *matched-filter detector* to give minimum P_E for fixed signal and noise conditions.

We begin our consideration of digital data transmission systems in Section 8.1 with the analysis of a simple, synchronous baseband system that employs a special case of the matched filter detector known as an *integrate-and-dump detector*. This analysis is then generalized in Section 8.2 to the matched-filter receiver, and these results specialized to consideration of several coherent signaling schemes. Section 8.3 considers two schemes not requiring a coherent reference for demodulation. In Section 8.4, digital pulse-amplitude modulation is considered. Section 8.5 provides a comparison of the digital modulation schemes on the basis of power and bandwidth. After analyzing these modulation schemes, which operate in an ideal environment in the sense that infinite bandwidth is available, we look at zero-intersymbol interference signaling through bandlimited baseband channels in Section 8.6. In Sections 8.7 and 8.8, the effect of multipath interference and signal fading on data transmission is analyzed, and in Section 8.9, the use of equalizing filters to mitigate the effects of channel distortion is examined.

■ 8.1 BASEBAND DATA TRANSMISSION IN WHITE GAUSSIAN NOISE

Consider the binary digital data communication system illustrated in Figure 8.2(a), in which the transmitted signal consists of a sequence of constant-amplitude pulses of either A or $-A$ units in amplitude and T seconds in duration. A typical transmitted sequence is shown in Figure 8.2(b).

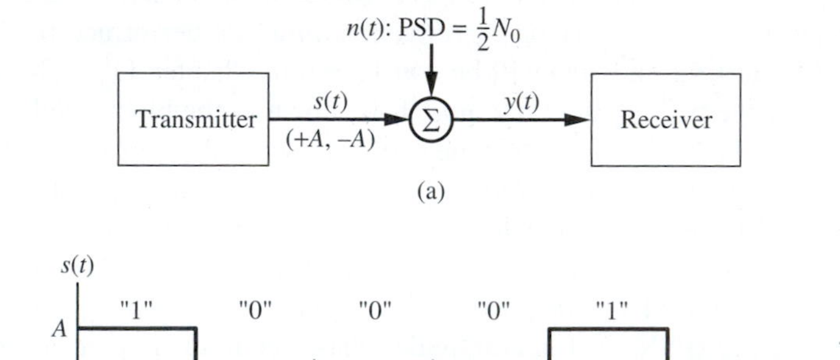

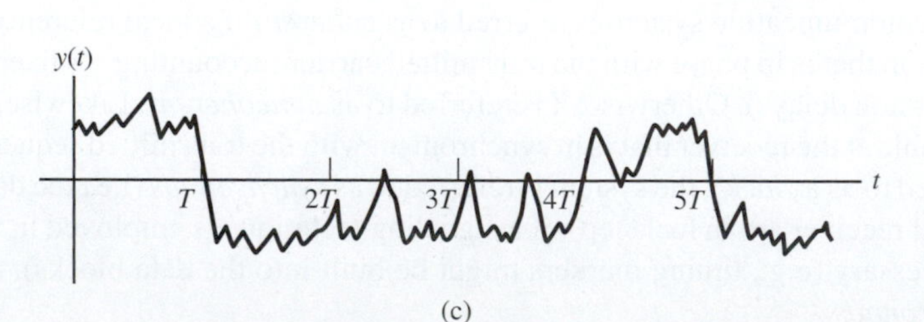

Figure 8.2
System model and waveforms for synchronous baseband digital data transmission. (a) Baseband digital data communication system. (b) Typical transmitted sequence. (c) Received sequence plus noise.

We may think of a positive pulse as representing a logic 1 and a negative pulse as representing a logic 0 from the data source. Each T-s pulse is called a *binit* for binary digit or, more simply, a *bit*. (In Chapter 11, the term *bit* will take on a new meaning.)

As in Chapter 7, the channel is idealized as simply adding white Gaussian noise with double-sided power spectral density $\frac{1}{2}N_0$ W/Hz to the signal. A typical sample function of the received signal plus noise is shown in Figure 8.2(c). For sketching purposes, it is assumed that the noise is bandlimited, although it is modeled as white noise later when the performance of the receiver is analyzed. It is assumed that the starting and ending times of each pulse are known by the receiver. The problem of acquiring this information, referred to as *synchronization*, briefly discussed in chapter 4, will not be considered at this time.

The function of the receiver is to decide whether the transmitted signal was A or $-A$ during each bit period. A straightforward way of accomplishing this is to pass the signal plus noise through a lowpass predetection filter, sample its output at some time within each T-s interval, and determine the sign of the sample. If the sample is greater than zero, the decision is made that $+A$ was transmitted. If the sample is less than zero, the decision is that $-A$ was transmitted. With such a receiver structure, however, we do not take advantage of everything known about the signal. Since the starting and ending times of the pulses are assumed known, a better procedure is to compare the area of the received signal-plus-noise waveform (data) with zero at the end of each signaling interval by integrating the received data over the T-s signaling interval. Of course, a noise component is present at the output of the integrator, but since the input noise has zero mean, it takes on positive and negative values with equal probability. Thus the output noise component has zero mean. The proposed receiver structure and a typical waveform at the output of the integrator are shown in Figure 8.3, where t_0 is the start of an arbitrary signaling interval. For obvious reasons, this receiver is referred to as an *integrate-and-dump detector*.

The question to be answered is the following: How well does this receiver perform, and on what parameters does its performance depend? As mentioned previously, a useful criterion of performance is probability of error, and it is this we now compute. The output of the integrator

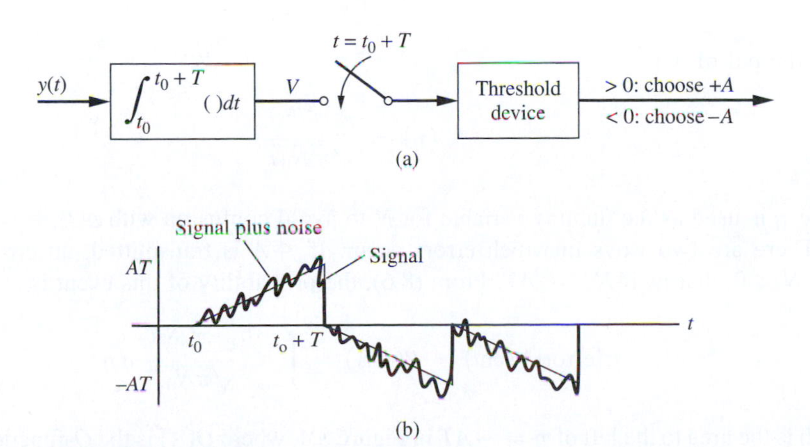

Figure 8.3
Receiver structure and integrator output. (a) Integrate-and-dump receiver. (b) Output from the integrator.

at the end of a signaling interval is

$$V = \int_{t_0}^{t_0+T} [s(t) + n(t)]\, dt = \begin{cases} +AT + N & \text{if } +A \text{ is sent} \\ -AT + N & \text{if } -A \text{ is sent} \end{cases} \tag{8.1}$$

where N is a random variable defined as

$$N = \int_{t_0}^{t_0+T} n(t)\, dt \tag{8.2}$$

Since N results from a linear operation on a sample function from a Gaussian process, it is a Gaussian random variable. It has mean

$$E[N] = E\left[\int_{t_0}^{t_0+T} n(t)\, dt\right] = \int_{t_0}^{t_0+T} E[n(t)]\, dt = 0 \tag{8.3}$$

since $n(t)$ has zero mean. Its variance is therefore

$$\begin{aligned} \text{var}[N] = E\left[N^2\right] &= E\left[\left(\int_{t_0}^{t_0+T} n(t)\, dt\right)^2\right] \\ &= \int_{t_0}^{t_0+T}\int_{t_0}^{t_0+T} E[n(t)n(\sigma)]\, dt\, d\sigma \\ &= \int_{t_0}^{t_0+T}\int_{t_0}^{t_0+T} \frac{1}{2} N_0 \delta(t-\sigma)\, dt\, d\sigma \end{aligned} \tag{8.4}$$

where we have made the substitution $E[n(t)n(\sigma)] = \frac{1}{2}N_0\delta(t-\sigma)$. Using the sifting property of the delta function, we obtain

$$\begin{aligned} \text{var}[N] &= \int_{t_0}^{t_0+T} \frac{1}{2} N_0\, d\sigma \\ &= \frac{1}{2} N_0 T \end{aligned} \tag{8.5}$$

Thus the pdf of N is

$$f_N(\eta) = \frac{e^{-\eta^2/N_0T}}{\sqrt{\pi N_0 T}} \tag{8.6}$$

where η is used as the dummy variable for N to avoid confusion with $n(t)$.

There are two ways in which errors occur. If $+A$ is transmitted, an error occurs if $AT + N < 0$, that is, if $N < -AT$. From (8.6), the probability of this event is

$$P(\text{error}|A \text{ sent}) = P(E|A) = \int_{-\infty}^{-AT} \frac{e^{-\eta^2/N_0T}}{\sqrt{\pi N_0 T}}\, d\eta \tag{8.7}$$

which is the area to the left of $\eta = -AT$ in Figure 8.4, where $Q(\cdot)$ is the Q-function.[2] Letting

[2]See Appendix G.1 for a discussion and tabulation of the Q-function.

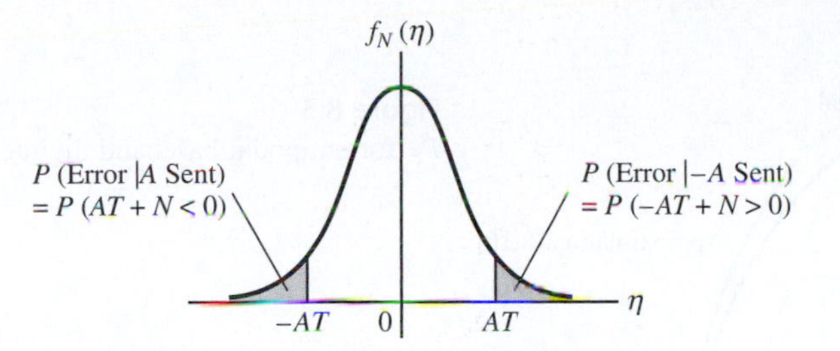

Figure 8.4
Illustration of error probabilities for binary signaling.

$u = \sqrt{2/N_0T}\eta$, we can write this as

$$P(E|A) = \int_{\sqrt{2A^2T/N_0}}^{\infty} \frac{e^{-u^2/2}}{\sqrt{2\pi}} du \triangleq Q\left(\sqrt{\frac{2A^2T}{N_0}}\right) \tag{8.8}$$

The other way in which an error can occur is if $-A$ is transmitted and $-AT+N>0$. The probability of this event is the same as the probability that $N > AT$, which can be written as

$$P(E\,|-A) = \int_{AT}^{\infty} \frac{e^{-\eta^2/N_0T}}{\sqrt{\pi N_0 T}} d\eta \triangleq Q\left(\sqrt{\frac{2A^2T}{N_0}}\right) \tag{8.9}$$

which is the area to the right of $\eta = AT$ in Figure 8.4. The average probability of error is

$$P_E = P(E|+A)P(+A) + P(E\,|-A)P(-A) \tag{8.10}$$

Substituting (8.8) and (8.9) into (8.10) and noting that $P(+A)+P(-A) = 1$, where $P(A)$ is the probability that $+A$ is transmitted, we obtain

$$P_E = Q\left(\sqrt{\frac{2A^2T}{N_0}}\right) \tag{8.11}$$

Thus the important parameter is A^2T/N_0. We can interpret this ratio in two ways. First, since the energy in each signal pulse is

$$E_b = \int_{to}^{t_0+T} A^2 dt = A^2T \tag{8.12}$$

and the ratio of signal energy per pulse to single-sided noise power spectral density is

$$z = \frac{A^2T}{N_0} = \frac{E_b}{N_0} \tag{8.13}$$

where E_b is called the *energy per bit*. Second, we recall that a rectangular pulse of duration T s has amplitude spectrum AT sinc Tf and that $B_p = 1/T$ is a rough measure of its bandwidth. Thus

$$z = \frac{A^2}{N_0(1/T)} = \frac{A^2}{N_0 B_p} \tag{8.14}$$

can be interpreted as the ratio of signal power to noise power in the signal bandwidth. The bandwidth B_p is sometimes referred to as the *bit-rate bandwidth*. We will refer to z as the

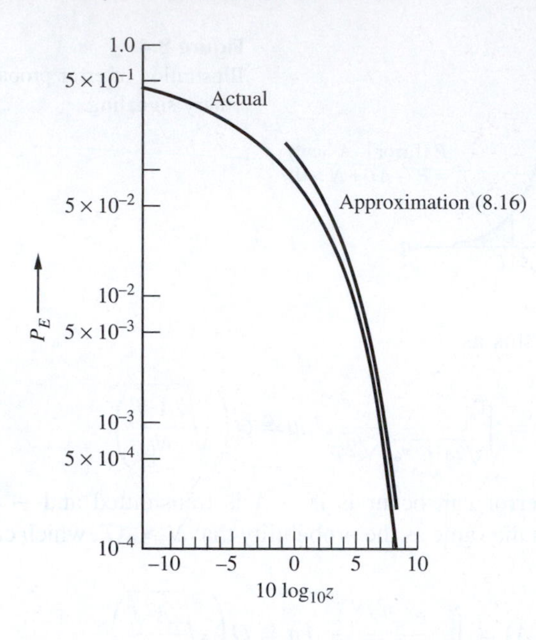

Figure 8.5
P_E for antipodal baseband digital signaling.

SNR. An often-used reference to this SNR in the digital communications industry is "E_b-over-N_0."[3]

A plot of P_E versus z is shown in Figure 8.5, where z is given in decibels. Also shown is an approximation for P_E using the asymptotic expansion for the Q-function:

$$Q(u) \cong \frac{e^{-u^2/2}}{u\sqrt{2\pi}}, \quad u \gg 1 \tag{8.15}$$

Using this approximation,

$$P_E \cong \frac{e^{-z}}{2\sqrt{\pi z}}, \quad z \gg 1 \tag{8.16}$$

which shows that P_E essentially decreases exponentially with increasing z. Figure 8.5 shows that the approximation of (8.16) is close to the true result of (8.11) for $z \gtrsim 3$ dB.

EXAMPLE 8.1

Digital data are to be transmitted through a baseband system with $N_0 = 10^{-7}$ W/Hz and the received signal amplitude $A = 20$ mV. (a) If 10^3 bps are transmitted, what is P_E? (b) If 10^4 bps are transmitted, to what value must A be adjusted in order to attain the same P_E as in part (a)?

Solution

To solve part (a), note that

$$z = \frac{A^2 T}{N_0} = \frac{(0.02)^2 (10^{-3})}{10^{-7}} = 4 \tag{8.17}$$

[3]A yet more distasteful term in use by some is *ebno*.

Using (8.16), $P_E \cong e^{-4}/2\sqrt{4\pi} = 2.58 \times 10^{-3}$. Part (b) is solved by finding A such that $A^2(10^{-4})/(10^{-7}) = 4$, which gives $A = 63.2$ mV.

■

EXAMPLE 8.2

The conditions are the same as in the preceding example, but a bandwidth of 5000 Hz is available. (a) What is the maximum data rate that can be supported by the channel? (b) Find the transmitter power required to give a probability of error of 10^{-6} at the data rate found in part (a).

Solution

(a) Since a rectangular pulse has Fourier transform

$$\Pi(t/T) \leftrightarrow T\,\mathrm{sinc}(fT)$$

we take the signal bandwidth to be that of the first null of the sinc function. Therefore, $1/T = 5000$ Hz, which implies a maximum data rate of $R = 5000$ bps. (b) To find the transmitter power to give $P_E = 10^{-6}$, we solve

$$10^{-6} = Q\left(\sqrt{\frac{2A^2T}{N_0}}\right) = Q(\sqrt{2z}) \tag{8.18}$$

Using the approximation (8.15) for the Q function, we need to solve

$$10^{-6} = \frac{e^{-z}}{2\sqrt{\pi z}}$$

iteratively. This gives the result

$$z \cong 10.53 \text{ dB} = 11.31 \text{ (ratio)}$$

Thus, $A^2T/N_0 = 11.31$, or

$$A^2 = 11.31\frac{N_0}{T} = 5.65 \times 10^{-3} \text{ V}^2$$

This corresponds to a signal amplitude of approximately 75.2 mV.

■

8.2 BINARY DATA TRANSMISSION WITH ARBITRARY SIGNAL SHAPES

In Section 8.1 we analyzed a simple baseband digital communication system. As in the case of analog transmission, it is often necessary to utilize modulation to condition a digital message signal so that it is suitable for transmission through a channel. Thus, instead of the constant-level signals considered in Section 8.1, we will let a logic 1 be represented by $s_1(t)$ and a logic 0 by $s_2(t)$. The only restriction on $s_1(t)$ and $s_2(t)$ is that they must have finite energy in a T-s interval. The energies of $s_1(t)$ and $s_2(t)$ are denoted by

$$E_1 \triangleq \int_{t_0}^{t_0+T} s_1^2(t)\,dt \tag{8.19}$$

Table 8.1 Possible Signal Choices for Binary Digital Signaling

Case	$s_1(t)$	$s_2(t)$	Type of signaling
1	0	$A\cos(\omega_c t)$	Amplitude-shift keying
2	$A\sin(\omega_c t+\cos^{-1}m)$	$A\sin(\omega_c t-\cos^{-1}m)$	Phase-shift keying with carrier ($\cos^{-1}m \triangleq$ modulation index)
3	$A\cos(\omega_c t)$	$A\cos(\omega_c+\Delta\omega)t$	Frequency-shift keying

and

$$E_2 \triangleq \int_{t_0}^{t_0+T} s_2^2(t)\,dt \tag{8.20}$$

respectively. In Table 8.1, three fundamental choices for $s_1(t)$ and $s_2(t)$ are given.

8.2.1 Receiver Structure and Error Probability

A possible receiver structure for detecting $s_1(t)$ or $s_2(t)$ in additive white Gaussian noise is shown in Figure 8.6. Since the signals chosen may have zero average value over a T-s interval (see the examples in Table 8.1), we can no longer employ an integrator followed by a threshold device as in the case of constant-amplitude signals. Instead of the integrator, we employ a filter with, as yet, unspecified impulse response $h(t)$ and corresponding frequency response function $H(f)$. The received signal plus noise is either

$$y(t) = s_1(t)+n(t), \quad t_0 \le t \le t_0+T \tag{8.21}$$

or

$$y(t) = s_2(t)+n(t), \quad t_0 \le t \le t_0+T \tag{8.22}$$

where the noise, as before, is assumed to be white with power spectral density $\frac{1}{2}N_0$. We can assume that $t_0 \doteq 0$ without loss of generality; that is, the signaling interval under consideration is $0 \le t \le T$.

To find P_E, we again note that an error can occur in either one of two ways. Assume that $s_1(t)$ and $s_2(t)$ were chosen such that $s_{01}(T) < s_{02}(T)$, where $s_{01}(t)$ and $s_{02}(t)$ are the outputs of the filter due to $s_1(t)$ and $s_2(t)$, respectively, at the input. If not, the roles of $s_1(t)$ and $s_2(t)$ at the input can be reversed to ensure this. Referring to Figure 8.6, if $v(T) > k$ where k is the threshold, we decide that $s_2(T)$ was sent; if $v(T) < k$, we decide that $s_1(t)$ was sent. Letting $n_0(t)$ be the noise component at the filter output, an error is made if $s_1(t)$ is sent and $v(T) = s_{01}(T)+n_0(T) > k$; if $s_2(t)$ is sent, an error occurs if $v(T) = s_{02}(T)+n_0(T) < k$. Since $n_0(t)$ is the result of passing white Gaussian noise through a fixed linear filter, it is a

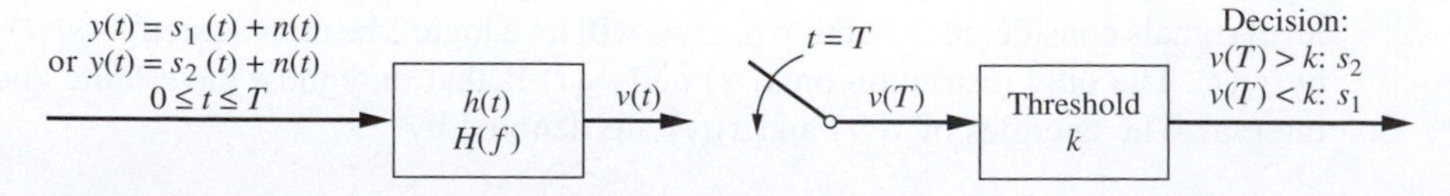

Figure 8.6
A possible receiver structure for detecting binary signals in white Gaussian noise.

Gaussian process. Its power spectral density is

$$S_{no}(f) = \frac{1}{2}N_0|H(f)|^2 \tag{8.23}$$

Because the filter is fixed, $n_0(t)$ is a stationary Gaussian random process with mean zero and variance

$$\sigma_0^2 = \int_{-\infty}^{\infty} \frac{1}{2}N_0|H(f)|^2 df \tag{8.24}$$

Since $n_0(t)$ is stationary, $N = n_0(T)$ is a random variable with mean zero and variance σ_0^2. Its pdf is

$$f_N(\eta) = \frac{e^{-\eta^2/2\sigma_0^2}}{\sqrt{2\pi\sigma_0^2}} \tag{8.25}$$

Given that $s_1(t)$ is transmitted, the sampler output is

$$V \triangleq v(T) = s_{01}(T) + N \tag{8.26}$$

and if $s_2(t)$ is transmitted, the sampler output is

$$V \triangleq v(T) = s_{02}(T) + N \tag{8.27}$$

These are also Gaussian random variables, since they result from linear operations on Gaussian random variables. They have means $s_{01}(T)$ and $s_{02}(T)$, respectively, and the same variance as N, that is, σ_0^2. Thus the conditional pdfs of V given $s_1(t)$ is transmitted, $f_V(v|s_1(t))$, and given $s_2(t)$ is transmitted, $f_V(v|s_2(t))$, are as shown in Figure 8.7. Also illustrated is a decision threshold k.

From Figure 8.7, we see that the probability of error, given $s_1(t)$ is transmitted, is

$$\begin{aligned} P(E|s_1(t)) &= \int_k^{\infty} f_V(v|s_1(t))\,dv \\ &= \int_k^{\infty} \frac{e^{-[v-s_{01}(T)]^2/2\sigma_0^2}}{\sqrt{2\pi\sigma_0^2}}\,dv \end{aligned} \tag{8.28}$$

which is the area under $f_V(v|\,s_1(t))$ to the right of $v = k$. Similarly, the probability of error, given $s_2(t)$ is transmitted, which is the area under $f_V(v\,|\,s_2(t))$ to the left of $v = k$, is given by

$$P(E|s_2(t)) = \int_{-\infty}^{k} \frac{e^{-[v-s_{02}(T)]^2/2\sigma_0^2}}{\sqrt{2\pi\sigma_0^2}}\,dv \tag{8.29}$$

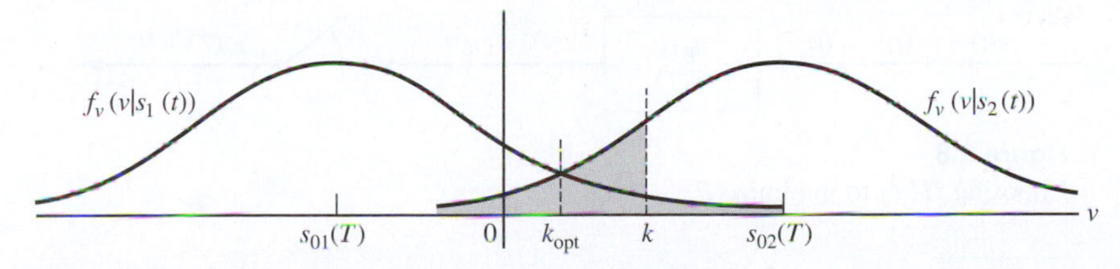

Figure 8.7
Conditional probability density functions of the filter output at time $t = T$.

Assuming that $s_1(t)$ and $s_2(t)$ are a priori equally probable,[4] the average probability of error is

$$P_E = \frac{1}{2}P[E\,|s_1(t)] + \frac{1}{2}P[E\,|s_2(t)] \tag{8.30}$$

The task now is to minimize this error probability by adjusting the threshold k and the impulse response $h(t)$.

Because of the equal a priori probabilities for $s_1(t)$ and $s_2(t)$ and the symmetrical shapes of $f_V(v|s_1(t))$ and $f_V(v|s_2(t))$, it is reasonable that the optimum choice for k is the intersection of the conditional pdfs, which is

$$k_{\text{opt}} = \frac{1}{2}[s_{01}(T) + s_{02}(T)] \tag{8.31}$$

The optimum threshold is illustrated in Figure 8.7 and can be derived by differentiating (8.30) with respect to k after substitution of (8.28) and (8.29). Because of the symmetry of the pdfs, the probabilities of either type of error, (8.28) or (8.29), are equal for this choice of k.

With this choice of k, the probability of error given by (8.30) reduces to

$$P_E = Q\left(\frac{s_{02}(T) - s_{01}(T)}{2\sigma_0}\right) \tag{8.32}$$

Thus we see that P_E is a function of the difference between the two output signals at $t = T$. Remembering that the Q-function decreases monotonically with increasing argument, we see that P_E decreases with increasing distance between the two output signals, a reasonable result. We will encounter this interpretation again in Chapters 9 and 10, where we discuss concepts of signal space.

We now consider the minimization of P_E by proper choice of $h(t)$. This will lead us to the matched filter.

8.2.2 The Matched Filter

For a given choice of $s_1(t)$ and $s_2(t)$, we wish to determine an $H(f)$, or equivalently, an $h(t)$ in (8.32), that maximizes

$$\zeta = \frac{s_{02}(T) - s_{01}(T)}{\sigma_0} \tag{8.33}$$

which follows because the Q-function is monotonically decreasing as its arguement increases. Letting $g(t) = s_2(t) - s_1(t)$, the problem is to find the $H(f)$ that maximizes $\zeta = g_0(T)/\sigma_0$, where $g_0(t)$ is the signal portion of the output due to the input $g(t)$.[5] This situation is illustrated in Figure 8.8.

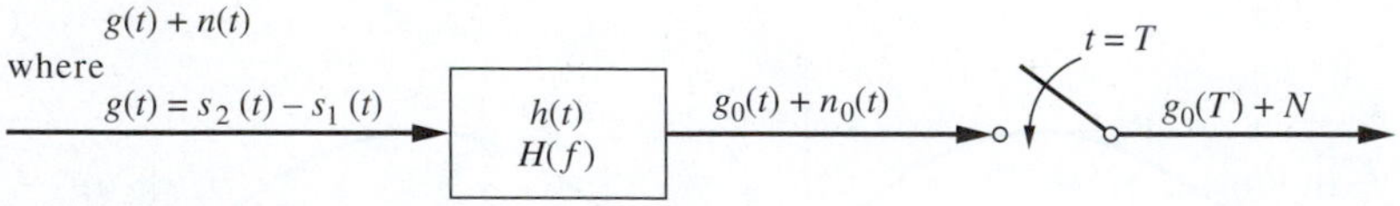

Figure 8.8
Choosing $H(f)$ to minimize P_E.

---

[4]See Problem 8.10 for the case of unequal a priori probabilities.

[5]Note that $g(t)$ is a fictitious signal. How it relates to the detection of digital signals will be apparent later.

We can equally well consider the maximization of

$$\zeta^2 = \frac{g_0^2(T)}{\sigma_0^2} = \frac{g_0^2(t)}{E\{n_0^2(t)\}}\bigg|_{t=T} \tag{8.34}$$

Since the input noise is stationary,

$$E\{n_0^2(t)\} = E\{n_0^2(T)\} = \frac{N_0}{2}\int_{-\infty}^{\infty} |H(f)|^2\, df \tag{8.35}$$

We can write $g_0(t)$ in terms of $H(f)$ and the Fourier transform of $g(t)$, $G(f)$, as

$$g_0(t) = \Im^{-1}[G(f)H(f)] = \int_{-\infty}^{\infty} H(f)\, G(f) e^{j2\pi ft}\, df \tag{8.36}$$

Setting $t = T$ in (8.36) and using this result along with (8.35) in (8.34), we obtain

$$\zeta^2 = \frac{\left|\int_{-\infty}^{\infty} H(f)G(f)e^{j2\pi fT}\, df\right|^2}{\frac{1}{2}N_0 \int_{-\infty}^{\infty} |H(f)|^2 df} \tag{8.37}$$

To maximize this equation with respect to $H(f)$, we employ *Schwarz's inequality.* Schwarz's inequality is a generalization of the inequality

$$|\mathbf{A}\cdot\mathbf{B}| = |AB\cos\theta| \le |\mathbf{A}||\mathbf{B}| \tag{8.38}$$

where **A** and **B** are ordinary vectors, with θ the angle between them, and $\mathbf{A}\cdot\mathbf{B}$ denotes their inner, or dot, product (A and B are their lengths). Since $|\cos\theta|$ equals unity if and only if θ equals zero or an integer multiple of π, equality holds if and only if **A** equals $k\mathbf{B}$, where k is a constant ($k > 0$ corresponds to $\theta = 0$ while $k < 0$ corresponds to $\theta = \pi$). Considering the case of two complex functions $X(f)$ and $Y(f)$, and defining the inner product as

$$\int_{-\infty}^{\infty} X(f)Y^*(f)\, df$$

Schwarz's inequality assumes the form[6]

$$\left|\int_{-\infty}^{\infty} X(f)Y^*(f) df\right| \le \sqrt{\int_{-\infty}^{\infty} |X(f)|^2 df}\sqrt{\int_{-\infty}^{\infty} |Y(f)|^2 df} \tag{8.39}$$

Equality holds if and only if $X(f) = kY(f)$, where k is, in general, complex. We will prove Schwarz's inequality in Chapter 10 with the aid of signal space notation.

We now return to our original problem, that of finding the $H(f)$ that maximizes (8.37). We replace $X(f)$ in (8.39) squared with $H(f)$ and $Y^*(f)$ with $G(f)e^{j2\pi Tf}$. Thus

$$\zeta^2 = \frac{2}{N_0}\frac{\left|\int_{-\infty}^{\infty} X(f)Y^*(f) df\right|^2}{\int_{-\infty}^{\infty} |H(f)|^2 df} \le \frac{2}{N_0}\frac{\int_{-\infty}^{\infty} |H(f)|^2 df \int_{-\infty}^{\infty} |G(f)|^2 df}{\int_{-\infty}^{\infty} |H(f)|^2 df} \tag{8.40}$$

Canceling the integral over $|H(f)|^2$ in the numerator and denominator, we find the maximum value of ζ^2 to be

$$\zeta^2_{\max} = \frac{2}{N_0}\int_{-\infty}^{\infty} |G(f)|^2 df = \frac{2E_g}{N_0} \tag{8.41}$$

[6]If more convenient for a given application, one could equally well work with the square of Schwarz's inequality.

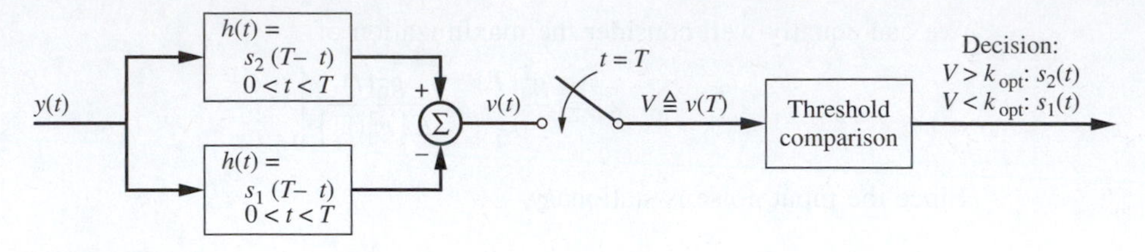

Figure 8.9
Matched-filter receiver for binary signaling in white Gaussian noise.

where $E_g = \int_{-\infty}^{\infty} |G(f)|^2 df$ is the energy contained in $g(t)$, which follows by Rayleigh's energy theorem. Equality holds in (8.40) if and only if

$$H(f) = k' G^*(f) e^{-j2\pi Tf} \tag{8.42}$$

where k' is an arbitrary constant. Since k' just fixes the gain of the filter (signal and noise are amplified the same), we can set it to unity. Thus the optimum choice for $H(f)$, $H_0(f)$, is

$$H_0(f) = G^*(f)\, e^{-j2\pi Tf} \tag{8.43}$$

The impulse response corresponding to this choice of $H_0(f)$ is

$$\begin{aligned} h_0(t) &= \Im^{-1}[H_0(f)] \\ &= \int_{-\infty}^{\infty} G^*(f)\, e^{-j2\pi Tf} e^{j2\pi fT} df \\ &= \int_{-\infty}^{\infty} G(-f)\, e^{-j2\pi f(T-t)} df \\ &= \int_{-\infty}^{\infty} G(f')\, e^{-j2\pi f'(T-t)} df' \end{aligned} \tag{8.44}$$

Recognizing this as the inverse Fourier transform of $g(t)$ with t replaced by $T - t$, we obtain

$$h_0(t) = g(T-t) = s_2(T-t) - s_1(T-t) \tag{8.45}$$

Thus, in terms of the original signals, the optimum receiver corresponds to passing the received signal plus noise through two parallel filters whose impulse responses are the time reverses of $s_1(t)$ and $s_2(t)$, respectively, and comparing the difference of their outputs at time T with the threshold given by (8.31). This operation is illustrated in Figure 8.9.

EXAMPLE 8.3

Consider the pulse signal

$$s(t) = \begin{cases} A, & 0 \le t \le T \\ 0, & \text{otherwise} \end{cases} \tag{8.46}$$

A filter matched to this signal has the impulse response

$$h_0(t) = s(t_0 - t) = \begin{cases} A, & t_0 - T \le t \le t_0 \\ 0, & \text{otherwise} \end{cases} \tag{8.47}$$

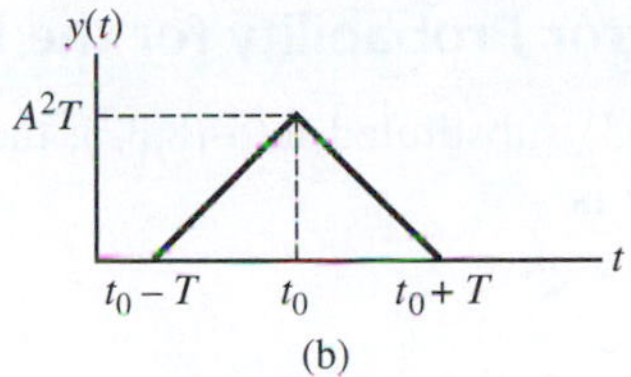

Figure 8.10
Signals pertinent to finding the matched-filter response of Example 8.3.

where the parameter t_0 will be fixed later. We note that if $t_0 < T$, the filter will be unrealizable, since it will have nonzero impulse response for $t < 0$. The response of the filter to $s(t)$ is

$$y(t) = h_0(t) * s(t) = \int_{-\infty}^{\infty} h_0(\tau)s(t-\tau)d\tau \tag{8.48}$$

The factors in the integrand are shown in Figure 8.10(a). The resulting integrations are familiar from our previous considerations of linear systems, and the filter output is easily found to be as shown in Figure 8.10(b). Note that the peak output signal occurs at $t = t_0$. This is also the time of peak-signal-to-rms-noise ratio, since the noise is stationary. Clearly, in digital signaling, we want $t_0 = T$. ∎

EXAMPLE 8.4

For a given value of N_0, consider the peak-signal-to-rms-noise ratio at the output of a matched filter for the two pulses

$$g_1(t) = A\Pi\left(\frac{t-t_0}{T}\right) \tag{8.49}$$

and

$$g_2(t) = B\cos\left(\frac{2\pi(t-t_0)}{T}\right)\Pi\left(\frac{t-t_0}{T}\right) \tag{8.50}$$

Relate A and B such that both pulses provide the same SNR at the matched filter output.

Solution

Since the SNR at the matched filter output by (8.41) is $2E_g/N_0$ and N_0 is the same for both cases, we can obtain equal SNR for both cases by computing the energy of each pulse and setting the two energies equal. The results are

$$E_{g_1} = \int_{t_0-T/2}^{t_0+T/2} A^2\,dt = A^2T \tag{8.51}$$

and

$$E_{g_2} = \int_{t_0-T/2}^{t_0+T/2} B^2\cos^2\left(\frac{2\pi(t-t_0)}{T}\right)dt = \frac{B^2T}{2} \tag{8.52}$$

Setting these equal, we have that $A = B/\sqrt{2}$ to give equal SNR. The peak signal-squared-to-mean-square-noise ratio is

$$\frac{2E_g}{N_0} = \frac{2A^2T}{N_0} = \frac{B^2T}{N_0} \tag{8.53}$$

∎

8.2.3 Error Probability for the Matched-Filter Receiver

From (8.33) substituted into (8.32), the error probability for the matched-filter receiver of Figure 8.9 is

$$P_E = Q\left(\frac{\zeta}{2}\right) \tag{8.54}$$

where ζ has the maximum value

$$\zeta_{\text{max}} = \left[\frac{2}{N_0}\int_{-\infty}^{\infty} |G(f)|^2 df\right]^{1/2} = \left[\frac{2}{N_0}\int_{-\infty}^{\infty} |S_2(f) - S_1(f)|^2 df\right]^{1/2} \tag{8.55}$$

given by (8.41). Using Parseval's theorem, we can write ζ^2_{max} in terms of $g(t) = s_2(t) - s_1(t)$ as

$$\begin{aligned} \zeta^2_{\text{max}} &= \frac{2}{N_0}\int_{-\infty}^{\infty} [s_2(t) - s_1(t)]^2\, dt \\ &= \frac{2}{N_0}\left[\int_{-\infty}^{\infty} s_2^2(t)\, dt + \int_{-\infty}^{\infty} s_1^2(t)\, dt - 2\int_{-\infty}^{\infty} s_1(t)s_2(t)\, dt\right] \end{aligned} \tag{8.56}$$

From (8.19) and (8.20), we see that the first two terms inside the braces are E_1 and E_2, respectively. We define

$$\rho_{12} = \frac{1}{\sqrt{E_1 E_2}}\int_{-\infty}^{\infty} s_1(t)\, s_2(t)\, dt \tag{8.57}$$

as the correlation coefficient of $s_1(t)$ and $s_2(t)$. Just as for random variables, ρ_{12} is a measure of the similarity between $s_1(t)$ and $s_2(t)$ and is normalized such that $-1 \leq \rho_{12} \leq 1$ (ρ_{12} achieves the end points for $s_1(t) = \pm k s_2(t)$, where k is a constant). Thus

$$\zeta^2_{\text{max}} = \frac{2}{N_0}\left(E_1 + E_2 - 2\sqrt{E_1 E_2}\,\rho_{12}\right) \tag{8.58}$$

and the error probability is

$$\begin{aligned} P_E &= Q\left[\left(\frac{E_1 + E_2 - 2\sqrt{E_1 E_2}\rho_{12}}{N_0}\right)^{1/2}\right] \\ &= Q\left[\left(2\frac{\frac{1}{2}(E_1 + E_2) - \sqrt{E_1 E_2}\rho_{12}}{N_0}\right)^{1/2}\right] \\ &= Q\left\{\left[\frac{2E}{N_0}\left(1 - \frac{\sqrt{E_1 E_2}}{E}\rho_{12}\right)\right]^{1/2}\right\} \end{aligned} \tag{8.59}$$

where $E = \frac{1}{2}(E_1 + E_2)$ is the average received signal energy per bit, since $s_1(t)$ and $s_2(t)$ are transmitted with equal a priori probability. It is apparent from (8.59) that in addition to depending on the signal energies, as in the constant-signal case, P_E also depends on the similarity between the signals through ρ_{12}. We note that (8.58) takes on its maximum value of

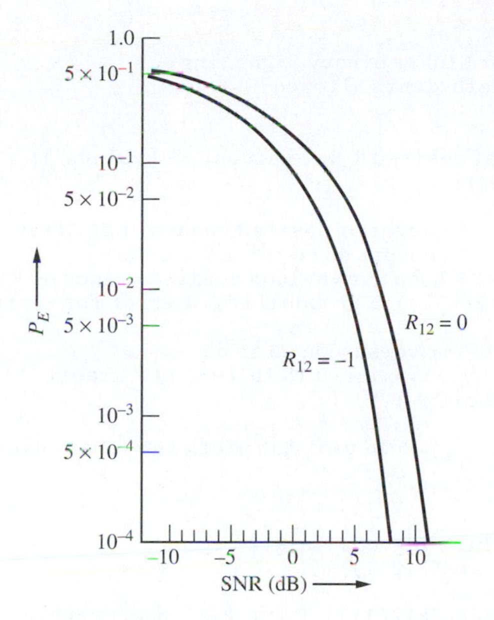

Figure 8.11
Probability of error for arbitrary waveshape case with $R_{12} = 0$ and $R_{12} = -1$.

$(2/N_0)\left(\sqrt{E_1} + \sqrt{E_2}\right)^2$ for $\rho_{12} = -1$, which gives the minimum value of P_E possible through choice of $s_1(t)$ and $s_2(t)$. This is reasonable, for then the transmitted signals are as dissimilar as possible. Finally, we can write (8.59) as

$$P_E = Q\left[\sqrt{z(1 - R_{12})}\right] \tag{8.60}$$

where $z = E/N_0$ is the average energy per bit divided by noise power spectral density as it was for the baseband system. The parameter R_{12} is defined as

$$R_{12} = \frac{2\sqrt{E_1 E_2}}{E_1 + E_2}\rho_{12} = \frac{\sqrt{E_1 E_2}}{E}\rho_{12} \tag{8.61}$$

and is a convenient parameter related to the correlation coefficient, but that should *not* be confused with a correlation function. The minimum value of R_{12} is -1, which is attained for $E_1 = E_2$ and $\rho_{12} = -1$. For this value of R_{12},

$$P_E = Q\left(\sqrt{2z}\right) \tag{8.62}$$

which is identical to (8.11), the result for baseband antipodal signals.

The probability of error versus the SNR is compared in Figure 8.11 for $R_{12} = 0$ (orthogonal signals) and $R_{12} = -1$ (antipodal signals).

COMPUTER EXAMPLE 8.1

A MATLAB program for computing the error probability for several values of correlation coefficient, R_{12}, is given below. Entering the vector [−1 0] in response to the first query reproduces the curves of Figure 8.11. Note that the user-defined function qfn(·) is used because MATLAB includes a function for erfc(u), but not $Q(u) = \frac{1}{2}\text{erfc}\left(u/\sqrt{2}\right)$.

```
% file:    c8ce1
% Bit error probability for binary binary signaling;
% vector of correlation coefficients allowed
%
clf
R12 = input('Enter vector of desired R_1_2 values; <= 3 values');
A = char('-','-.',':','--');
LR = length(R12);
z_dB = 0:.3:15;                  % Vector of desired values of Eb/N0 in dB
z = 10.^(z_dB/10);               % Convert dB to ratios
for k = 1:LR                     % Loop for various desired values of R12
   P_E=qfn(sqrt(z*(1-R12(k))));% Probability of error for vector
     of z-values
   % Plot probability of error versus Eb/N0 in dB
   semilogy(z_dB,P_E,A(k,:)),axis([0 15 10^(-6) 1]),xlabel
     ('E_b/N_0, dB'),ylabel('P_E'),...
   if k==1
     hold on; grid                 % Hold plot for plots for other values of
                                     R12
   end
end
if LR == 1                     % Plot legends for R12 values
   legend(['R_1_2 = ',num2str(R12(1))],1)
elseif LR == 2
   legend(['R_1_2 = ',num2str(R12(1))],['R_1_2 = ',num2str(R12
     (2))],1)
elseif LR == 3
    legend(['R_1_2 = ',num2str(R12(1))],['R_1_2 = ';,num2str(R12
      (2))],['R_1_2 = ',num2str(R12(3))],1)

% This function computes the Gaussian Q-function
%
function Q=qfn(x)
Q = 0.5*erfc(x/sqrt(2));
```

■

8.2.4 Correlator Implementation of the Matched-Filter Receiver

In Figure 8.9, the optimum receiver involves two filters with impulse responses equal to the time reverse of the respective signals being detected. An alternative receiver structure can be obtained by noting that the matched filter in Figure 8.12(a) can be replaced by a

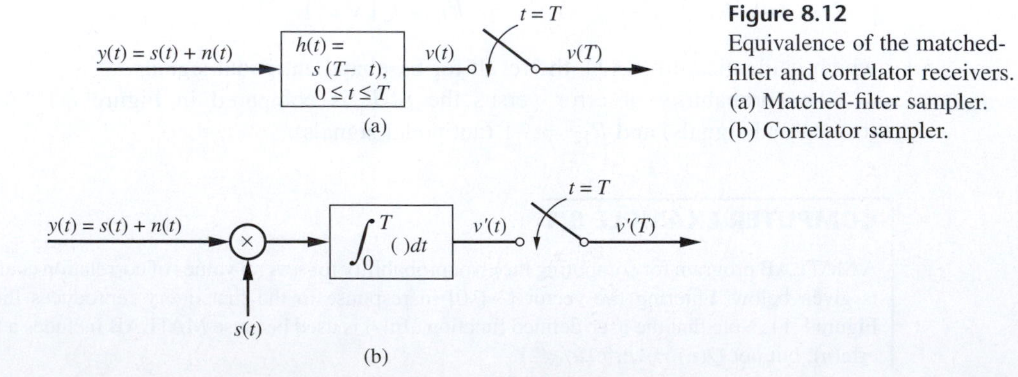

Figure 8.12
Equivalence of the matched-filter and correlator receivers. (a) Matched-filter sampler. (b) Correlator sampler.

multiplier–integrator cascade as shown in Figure 8.12(b). Such a series of operations is referred to as *correlation detection.*

To show that the operations given in Figure 8.12 are equivalent, we will show that $v(T)$ in Figure 8.12(a) is equal to $v'(T)$ in Figure 8.12(b). The output of the matched filter in Figure 8.12(a) is

$$v(t) = h(t) * y(t) = \int_0^T s(T - \tau) y(t - \tau)\, d\tau \tag{8.63}$$

which follows because $h(t) = s(T - t)$ for $0 \leq t < T$ and zero otherwise. Letting $t = T$ and changing variables in the integrand to $\alpha = T - \tau$, we obtain

$$v(T) = \int_0^T s(\alpha) y(\alpha)\, d\alpha \tag{8.64}$$

Considering next the output of the correlator configuration in Figure 8.12(b), we obtain

$$v'(T) = \int_0^T y(t) s(t)\, dt \tag{8.65}$$

which is identical to (8.64). Thus the matched filters for $s_1(t)$ and $s_2(t)$ in Figure 8.9 can be replaced by correlation operations with $s_1(t)$ and $s_2(t)$, respectively, and the receiver operation will not be changed. We note that the integrate-and-dump receiver for the constant signal case of Section 8.1 is actually a correlation or, equivalently, a matched-filter receiver.

8.2.5 Optimum Threshold

The optimum threshold for binary signal detection is given by (8.31), where $s_{01}(T)$ and $s_{02}(T)$ are the outputs of the detection filter in Figure 8.6 at time T due to the input signals $s_1(t)$ and $s_2(t)$, respectively. We now know that the optimum detection filter is a matched filter, matched to the difference of the input signals, and has the impulse response given by (8.45). From the superposition integral, we have

$$\begin{aligned} s_{01}(T) &= \int_{-\infty}^{\infty} h(\lambda) s_1(T - \lambda)\, d\lambda \\ &= \int_{-\infty}^{\infty} [s_2(T - \lambda) - s_1(T - \lambda)]\, s_1(T - \lambda)\, d\lambda \\ &= \int_{-\infty}^{\infty} s_2(u)\, s_1(u)\, du - \int_{-\infty}^{\infty} [s_1(u)]^2\, du \\ &= \sqrt{E_1 E_2} \rho_{12} - E_1 \end{aligned} \tag{8.66}$$

where the substitution $u = T - \lambda$ has been used to go from the second equation to the third, and the definition of the correlation coefficient (8.57) has been used to get the last equation along with the definition of energy of a signal. Similarly, it follows that

$$\begin{aligned} s_{02}(T) &= \int_{-\infty}^{\infty} [s_2(T - \lambda) - s_1(T - \lambda)]\, s_2(T - \lambda)\, d\lambda \\ &= \int_{-\infty}^{\infty} [s_2(u)]^2\, du - \int_{-\infty}^{\infty} s_2(u)\, s_1(u)\, du \\ &= E_2 - \sqrt{E_1 E_2} \rho_{12} \end{aligned} \tag{8.67}$$

Substituting (8.66) and (8.67) into (8.31), we find the optimum threshold to be

$$k_{\text{opt}} = \frac{1}{2}(E_2 - E_1) \tag{8.68}$$

Note that equal energy signals will always result in an optimum threshold of zero. Also note that the waveshape of the signals, as manifested through the correlation coefficient, has no effect on the optimum threshold value. Only the signal energies affect the threshold value.

8.2.6 Nonwhite (Colored) Noise Backgrounds

The question naturally arises about the optimum receiver for nonwhite noise backgrounds. Usually, the noise in a receiver system is generated primarily in the front-end stages and is due to thermal agitation of electrons in the electronic components (see Appendix A). This type of noise is well approximated as white. If a bandlimited channel precedes the introduction of the white noise, then we need only work with modified transmitted signals. If, for some reason, a bandlimiting filter follows the introduction of the white noise (for example, an IF amplifier following the RF amplifier and mixers where most of the noise is generated in a heterodyne receiver), we can use a simple artifice to approximate the matched-filter receiver. The colored noise plus signal is passed through a "whitening filter" with a frequency-response function that is the inverse square root of the noise spectral density. Thus, the output of this whitening filter is white noise plus a signal component that has been transformed by the whitening filter. We then build a matched-filter receiver with impulse response that is the difference of the time reverse of the "whitened" signals. The cascade of a whitening filter and matched filter (matched to the whitened signals) is called a *whitened matched filter*. This combination provides only an approximately optimum receiver for two reasons. Since the whitening filters will spread the received signals beyond the T-s signaling interval, two types of degradation will result:

1. The signal energy spread beyond the interval under consideration is not used by the matched filter in making a decision.
2. Previous signals spread out by the whitening filter will interfere with the matched filtering operation on the signal on which a decision is being made.

The latter is referred to as *intersymbol interference*, as first discussed in Chapter 4, and is explored further in Sections 8.7 and 8.9. It is apparent that degradation due to these effects is minimized if the signal *duration* is short compared with T, such as in a pulsed radar system. Finally, signal intervals adjacent to the interval being used in the decision process contain information that is relevant to making a decision on the basis of the correlation of the noise. In short, the whitened matched-filter receiver is nearly optimum if the signaling interval is large compared with the inverse bandwidth of the whitening filter. The question of bandlimited channels, and nonwhite background noise, is explored further in Section 8.6.

8.2.7 Receiver Implementation Imperfections

In the theory developed in this section, it is assumed that the signals are known *exactly* at the receiver. This is, of course, an idealized situation. Two possible deviations from this assumption are (1) the phase of the receiver's replica of the transmitted signal may be in error and (2) the exact arrival time of the received signal may be in error. These are called *synchronization* errors. The first case is explored in Section 8.3, and the latter is explored in the problems. Methods of synchronization are discussed in Chapter 9.

8.2.8 Error Probabilities for Coherent Binary Signaling

We now compare the performance of several fundamental coherent binary signaling schemes. Then we will examine noncoherent systems. To obtain the error probability for coherent systems, the results of Section 8.2 will be applied directly. The three types of coherent systems to be considered in this section are ASK, PSK, and FSK. Typical transmitted waveforms for these three types of digital modulation are shown in Figure 8.13. We also will consider the effect of an imperfect phase reference on the performance of a coherent PSK system. Such systems are often referred to as *partially coherent*.

Amplitude-Shift Keying

In Table 8.1, $s_1(t)$ and $s_2(t)$ for ASK are given as 0 and $A\cos(\omega_c t)$, where $f_c = \omega_c/2\pi$ is the carrier frequency. We note that the transmitter for such a system simply consists of an oscillator that is gated on and off; accordingly, ASK is often referred to as *on–off keying*. It is important to note that the oscillator runs continuously as the on–off gating is carried out.

The correlator realization for the optimum receiver consists of multiplication of the received signal plus noise by $A\cos(\omega_c t)$, integration over $(0, T)$ and comparison of the integrator output with the threshold $\frac{1}{4}A^2T$ as calculated from (8.68).

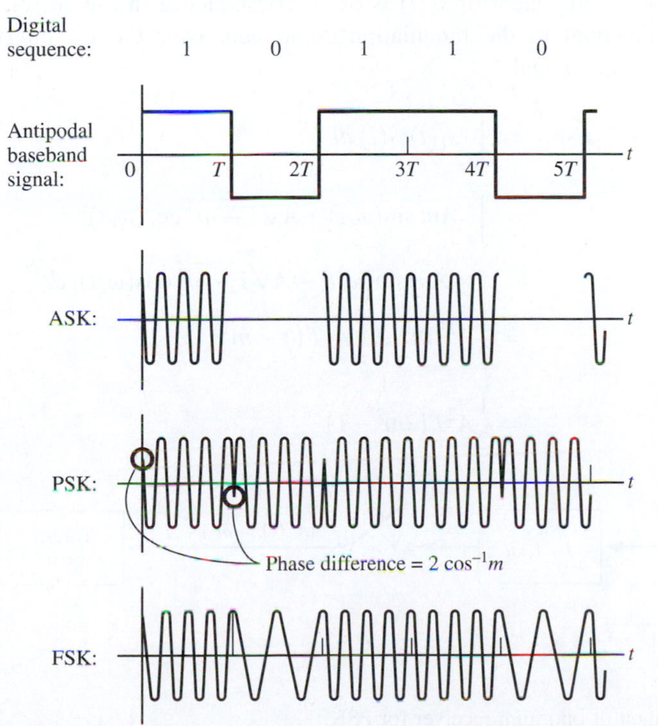

Figure 8.13
Waveforms for ASK, PSK, and FSK modulation.

From (8.57) and (8.61), $R_{12} = \rho_{12} = 0$ and the probability of error, from (8.60), is

$$P_E = Q(\sqrt{z}) \tag{8.69}$$

Because of the lack of a factor $\sqrt{2}$ in the argument of the Q-function, ASK is seen to be 3 dB worse in terms of SNR than antipodal baseband signaling. The probability of error versus SNR corresponds to the curve for $R_{12} = 0$ in Figure 8.11.

Phase-Shift Keying

From Table 8.1, the signals for PSK are

$$s_k(t) = A \sin\left[\omega_c t - (-1)^k \cos^{-1} m\right], \quad 0 \le t \le T,\ k = 1, 2 \tag{8.70}$$

where $\cos^{-1} m$, the modulation index, is written in this fashion for future convenience. For simplicity, we assume that $\omega_c = 2\pi n/T$, where n is an integer. Using $\sin(-x) = -\sin x$ and $\cos(-x) = \cos(x)$, we can write (8.70) as

$$s_k(t) = Am \sin(\omega_c t) - (-1)^k A\sqrt{1-m^2}\cos(\omega_c t), \quad 0 < t \le T, \quad k = 1, 2 \tag{8.71}$$

where we note that $\cos(\cos^{-1} m) = m$ and $\sin(\cos^{-1} m) = \sqrt{1-m^2}$.

The first term on the right-hand side of (8.71) represents a carrier component included in some systems for synchronization of the local carrier reference at the receiver to the transmitted carrier. The power in the carrier component is $\frac{1}{2}(Am)^2$, and the power in the modulation component is $\frac{1}{2}A^2(1-m^2)$. Thus m^2 is the fraction of the total power in the carrier component. The correlator receiver is shown in Figure 8.14, where, instead of two correlators, only a single correlation with $s_2(t) - s_1(t)$ is used. The threshold, calculated from (8.68), is zero. We note that the carrier component of $s_k(t)$ is of no consequence in the correlation operation because it is orthogonal to the modulation component over the bit interval. For PSK, $E_1 = E_2 = \frac{1}{2}A^2(1-m^2)T$ and

$$\begin{aligned}
\sqrt{E_1 E_2}\rho_{12} &= \int_0^T s_1(t)s_2(t)\,dt \\
&= \int_0^T \left[Am\sin(\omega_c t) + A\sqrt{1-m^2}\cos(\omega_c t)\right] \\
&\qquad \times \left[Am\sin(\omega_c t) - A\sqrt{1-m^2}\cos(\omega_c t)\right] dt \\
&= \frac{1}{2}A^2 T m^2 - \frac{1}{2}A^2 T(1-m^2) \\
&= \frac{1}{2}A^2 T(2m^2 - 1)
\end{aligned} \tag{8.72}$$

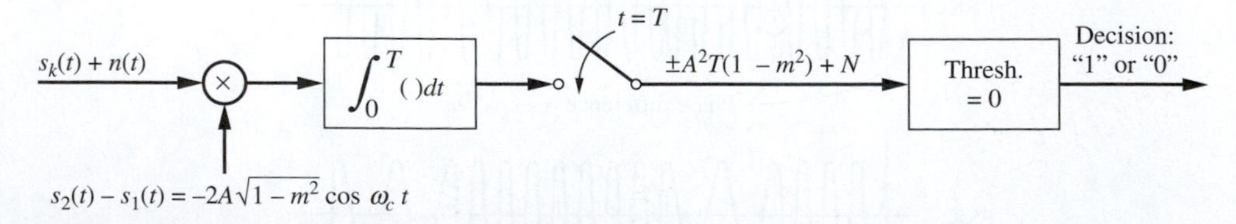

Figure 8.14
Correlator realization of optimum receiver for PSK.

Thus R_{12}, from (8.61), is

$$R_{12} = \frac{2\sqrt{E_1 E_2}}{E_1 + E_2}\rho_{12} = 2m^2 - 1 \tag{8.73}$$

and the probability of error for PSK is

$$P_E = Q\left[\sqrt{2(1-m^2)z}\right] \tag{8.74}$$

The effect of allocating a fraction m^2 of the total transmitted power to a carrier component is to degrade P_E by $10\log_{10}(1-m^2)$ dB from the ideal $R_{12} = -1$ curve of Figure 8.11.

For $m = 0$, the resultant error probability is 3 dB better than ASK and corresponds to the $R_{12} = -1$ curve in Figure 8.11. We will refer to the case for which $m = 0$ as *biphase-shift keying* (BPSK) to avoid confusion with the case for which $m \neq 0$.

EXAMPLE 8.5

Consider PSK with $m = 1/\sqrt{2}$. (a) By how many degrees does the modulated carrier shift in phase each time the binary data changes? (b) What percent of the total power is in the carrier, and what percent is in the modulation component? (c) What value of $z = E_b/N_0$ is required to give $P_E = 10^{-6}$?

Solution

(a) Since the change in phase is from $-\cos^{-1}m$ to $\cos^{-1}m$ whenever the phase switches, the phase change of the modulated carrier is

$$2\cos^{-1} m = 2\cos^{-1}\frac{1}{\sqrt{2}} = 2(45^\circ) = 90^\circ \tag{8.75}$$

(b) The carrier and modulation components are

$$\text{carrier} = Am\sin(\omega_c t) \tag{8.76}$$

and

$$\text{modulation} = \pm A\sqrt{1-m^2}\cos(\omega_c t) \tag{8.77}$$

respectively. Therefore, the power in the carrier component is

$$P_c = \frac{A^2 m^2}{2} \tag{8.78}$$

and the power in the modulation component is

$$P_m = \frac{A^2(1-m^2)}{2} \tag{8.79}$$

Since the total power is $A^2/2$, the percent power in each of these components is

$$\%P_c = m^2 \times 100 = 100\left(\frac{1}{\sqrt{2}}\right)^2 = 50\%$$

and

$$\%P_m = (1-m^2) \times 100 = 100\left(1 - \frac{1}{2}\right) = 50\%$$

respectively.

(c) We have, for the probability of error,

$$P_E = Q\left[\sqrt{2(1-m^2)z}\right] \cong \frac{e^{-(1-m^2)z}}{2\sqrt{\pi(1-m^2)z}} \tag{8.80}$$

Solving this iteratively, we obtain, for $m^2 = 0.5$, $z = 22.6$ or $E_b/N_0 = 13.54$ dB. Actually, we do not have to solve the error probability relationship iteratively again. From Example 8.2 we already know that $z = 10.53$ dB gives $P_E = 10^{-6}$ for BPSK (an antipodal signaling scheme). In this example we simply note that the required power is twice as much as for BPSK, which is equivalent to adding 3.01 dB on to the 10.53 dB required in Example 8.2.

■

Biphase-Shift Keying with Imperfect Phase Reference

The results obtained earlier for PSK are for the case of a perfect reference at the receiver. If $m = 0$, it is simple to consider the case of an imperfect reference at the receiver as represented by an input of the form $\pm A\cos(\omega_c t + \theta) + n(t)$ and the reference by $A\cos\left(\omega_c t + \hat{\theta}\right)$, where θ is an unknown carrier phase and $\hat{\theta}$ is the phase estimate at the receiver.

The correlator implementation for the receiver is shown in Figure 8.15. Using appropriate trigonometric identities, we find that the signal component of the correlator output at the sampling instant is $\pm AT\cos\phi$, where $\phi = \theta - \hat{\theta}$ is the phase error. It follows that the error probability *given* the phase error ϕ is

$$P_E(\phi) = Q\left(\sqrt{2z\cos^2\phi}\right) \tag{8.81}$$

We note that the performance is degraded by $20\log_{10}(\cos\phi)$ dB compared with the perfect reference case.

If we assume ϕ to be fixed at some maximum value, we may obtain an upper bound on P_E due to phase error in the reference. However, a more exact model is often provided by approximating ϕ as a Gaussian random variable with the pdf[7]

$$p(\phi) = \frac{e^{-\phi^2/2\sigma^2_\phi}}{\sqrt{2\pi\sigma^2_\phi}}, \quad |\phi| \le \pi \tag{8.82}$$

This is an especially appropriate model if the phase reference at the receiver is derived by means of a PLL operating with high SNR at its input. If this is the case, σ_ϕ^2 is related to the SNR at the input of the phase estimation device, whether it is a PLL or a bandpass-filter-limiter combination.

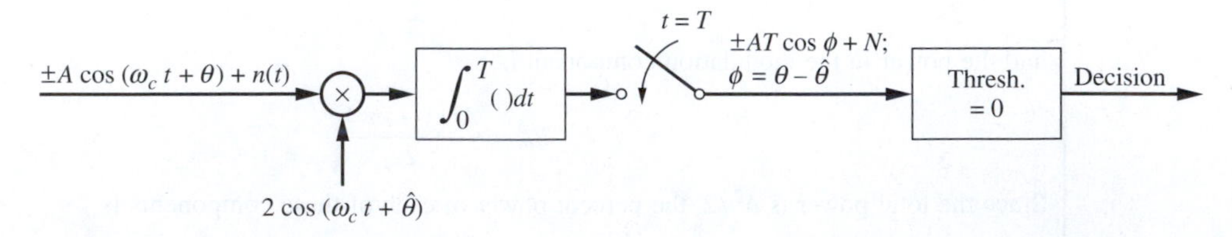

Figure 8.15
Effect of phase error in reference signal for correlation detection of BPSK.

---

[7]This is an approximation for the actual pdf for the phase error in a first-order PLL, which is known as Tikonov and is given by $p(\phi) = \exp\left(z_{\text{loop}}\cos\phi\right)/2\pi I_0\left(z_{\text{loop}}\right)$, $|\phi| \le \pi$, and 0 otherwise. z_{loop} is the SNR within the loop passband and $I_0(u)$ is the modified Bessel function of the first kind and order zero. Note that (8.82) should be renormalized so that its area is 1, but the error is small for σ_ϕ^2 small, which it is for z large.

Table 8.2 Effect of Gaussian Phase Reference Jitter on the Detection of BPSK

E/N_0, dB	$P_E, \sigma_\phi^2 = 0.01$ rad^2	$P_E, \sigma_\phi^2 = 0.05$ rad^2	$P_E, \sigma_\phi^2 = 0.1$ rad^2
9	3.68×10^{-5}	6.54×10^{-5}	2.42×10^{-4}
10	4.55×10^{-6}	1.08×10^{-5}	8.96×10^{-5}
11	3.18×10^{-7}	1.36×10^{-6}	3.76×10^{-5}
12	1.02×10^{-8}	1.61×10^{-7}	1.83×10^{-5}

To find the error probability averaged over all possible phase errors, we simply find the expectation of $P(E|\phi) = P_E(\phi)$, given by (8.81), with respect to the phase-error pdf, $p(\phi)$, that is,

$$P_E = \int_{-\pi}^{\pi} P_E(\phi)p(\phi)\,d\phi \tag{8.83}$$

The resulting integral must be evaluated numerically for typical phase error pdfs.[8] Typical results are given in Table 8.2 for $p(\phi)$ Gaussian.

Frequency-Shift Keying

In Table 8.1, the signals for FSK are given as

$$\begin{aligned} &s_1(t) = A\cos(\omega_c t) \\ \text{or} \quad &s_2(t) = A\cos(\omega_c + \Delta\omega)t \end{aligned} \quad 0 \le t \le T \tag{8.84}$$

For simplification, we assume that

$$\omega_c = \frac{2\pi n}{T} \tag{8.85}$$

and

$$\Delta\omega = \frac{2\pi m}{T} \tag{8.86}$$

where m and n are integers with $m \neq n$. This ensures that both $s_1(t)$ and $s_2(t)$ will go through an integer number of cycles in T s. As a result,

$$\begin{aligned} \sqrt{E_1E_2}\rho_{12} &= \int_0^T A^2\cos(\omega_c t)\cos(\omega_c + \Delta\omega)t\,dt \\ &= \frac{1}{2}A^2\int_0^T [\cos(\Delta\omega t) + \cos(2\omega_c + \Delta\omega)t]\,dt \\ &= 0 \end{aligned} \tag{8.87}$$

and $R_{12} = 0$. Thus

$$P_E = Q\left(\sqrt{z}\right) \tag{8.88}$$

which is the same as for ASK. The error probability versus SNR therefore corresponds to the curve $R_{12} = 0$ in Figure 8.11.

Note that the reason ASK and FSK have the same P_E versus SNR characteristics is that the comparison is being made on the basis of average signal power. If peak signal powers are constrained to be the same, ASK is 3 dB worse than FSK.

[8] See, for example, Van Trees (1968), Chapter 4.

We denote the three schemes just considered as coherent, binary ASK, PSK, and FSK to indicate the fact that they are binary. We consider M-ary ($M > 2$) schemes in Chapter 9.

EXAMPLE 8.6

Compare binary ASK, PSK, and FSK on the basis of E_b/N_0 required for $P_E = 10^{-6}$ and on the basis of transmission bandwidth for a constant data rate. Take the required bandwidth as the null-to-null bandwidth of the square-pulse modulated carrier. Assume the minimum bandwidth possible for FSK.

Solution

From before, we know that to give $P_{E,\text{BPSK}} = 10^{-6}$, the required E_b/N_0 is 10.53 dB. Amplitude-shift keying, on an average basis, and FSK require an SNR 3.01 dB above that of BPSK, or 13.54 dB, to give $P_E = 10^{-6}$. The Fourier transform of a square-pulse modulated carrier is

$$\Pi\left(\frac{t}{T}\right)\cos(2\pi f_c t) \leftrightarrow \left(\frac{T}{2}\right)\{\text{sinc}[T(f-f_c)]+\text{sinc}[T(f+f_c)]\}$$

The null-to-null bandwidth of the positive-frequency portion of this spectrum is

$$B_{\text{RF}} = \frac{2}{T}\text{ Hz} \tag{8.89}$$

For binary ASK and PSK, the required bandwidth is

$$B_{\text{PSK}} = B_{\text{ASK}} = \frac{2}{T} = 2R\text{ Hz} \tag{8.90}$$

where R is the data rate in bits per second. For FSK, the spectra for

$$s_1(t) = A\cos(\omega_c t), \quad 0 \le t \le T, \ \omega_c = 2\pi f_c$$

and

$$s_2(t) = A\cos(\omega_c t + \Delta\omega)t, \quad 0 \le t \le T, \ \Delta\omega = 2\pi\Delta f$$

are assumed to be separated by $1/2T$ Hz, which is the minimum spacing for orthogonality of the signals. Given that a cosinusoidal pulse has main-lobe half bandwidth of $1/T$ Hz, it can be roughly reasoned that the required bandwidth for FSK is therefore

$$B_{\text{CFSK}} = \underbrace{\frac{1}{T}+\frac{1}{2T}}_{f_c\text{ burst}} + \underbrace{\frac{1}{T}}_{f_c+\Delta f\text{ burst}} = \frac{2.5}{T} = 2.5R\text{ Hz} \tag{8.91}$$

We often specify bandwidth efficiency, R/B, in terms of bits per second per hertz. For binary ASK and PSK the bandwidth efficiency is 0.5 bps/Hz, while for binary coherent FSK it is 0.4 bps/Hz. ■

■ 8.3 MODULATION SCHEMES NOT REQUIRING COHERENT REFERENCES

We now consider two modulation schemes that do not require the acquisition of a local reference signal in phase coherence with the received carrier. The first scheme to be considered is referred to as *differentially coherent phase-shift keying* and may be thought of as the

Table 8.3 Differential Encoding Example

Message sequence:		1	0	0	1	1	1	0	0	0
Encoded sequence:	1	1	0	1	1	1	1	0	1	0
Reference digit:	↑									
Transmitted phase:	0	0	π	0	0	0	0	π	0	π

noncoherent version of BPSK considered in Section 8.2. Also considered in this section will be noncoherent, binary FSK (binary noncoherent ASK is considered in Problem 8.30).

8.3.1 Differential Phase-Shift Keying (DPSK)

One way of obtaining a phase reference for the demodulation of BPSK is to use the carrier phase of the preceding signaling interval. The implementation of such a scheme presupposes two things:

1. The mechanism causing the unknown phase perturbation on the signal varies so slowly that the phase is essentially constant from one signaling interval to the next.
2. The phase during a given signaling interval bears a known relationship to the phase during the preceding signaling interval.

The former is determined by the stability of the transmitter oscillator, time-varying changes in the channel, and so on. The latter requirement can be met by employing what is referred to as *differential encoding* of the message sequence at the transmitter.

Differential encoding of a message sequence is illustrated in Table 8.3. An arbitrary reference binary digit is assumed for the initial digit of the encoded sequence. In the example shown in Table 8.3, a 1 has been chosen. For each digit of the encoded sequence, the present digit is used as a reference for the following digit in the sequence. A 0 in the message sequence is encoded as a transition from the state of the reference digit to the opposite state in the encoded message sequence; a 1 is encoded as no change of state. In the example shown, the first digit in the message sequence is a 1, so no change in state is made in the encoded sequence, and a 1 appears as the next digit. This serves as the reference for the next digit to be encoded. Since the next digit appearing in the message sequence is a 0, the next encoded digit is the opposite of the reference digit, or a 0. The encoded message sequence then phase-shift keys a carrier with the phases 0 and π as shown in the table.

The block diagram in Figure 8.16 illustrates the generation of DPSK. The equivalence gate, which is the negation of an EXCLUSIVE-OR, is a logic circuit that performs the

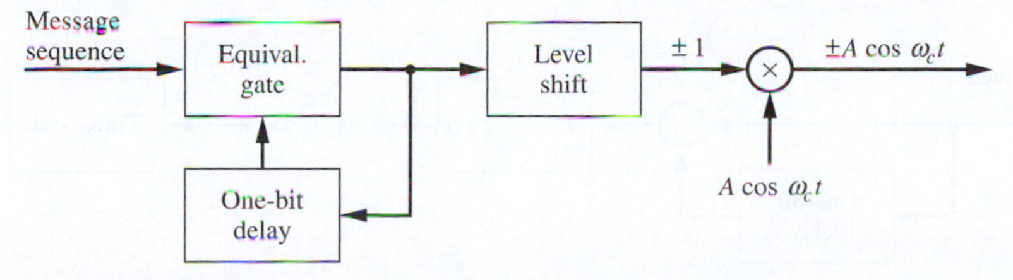

Figure 8.16
Block diagram of a DPSK modulator.

Table 8.4 Truth Table for the Equivalence Operation

Input 1 (message)	Input 2 (reference)	Output
0	0	1
0	1	0
1	0	0
1	1	1

operations listed in Table 8.4. By a simple level shift at the output of the logic circuit, so that the encoded message is bipolar, the DPSK signal is produced by multiplication by the carrier, or DSB modulation.

A possible implementation of a differentially coherent demodulator for DPSK is shown in Figure 8.17. The received signal plus noise is first passed through a bandpass filter centered on the carrier frequency and then correlated bit by bit with a one-bit delayed version of the signal plus noise. The output of the correlator is finally compared with a threshold set at zero, a decision being made in favor of a 1 or a 0, depending on whether the correlator output is positive or negative, respectively.

To illustrate that the received sequence will be correctly demodulated, consider the example given in Table 8.3, assuming no noise is present. After the first two bits have been received (the reference bit plus the first encoded bit), the signal input to the correlator is $S_1 = A\cos(\omega_c t)$, and the reference, or delayed, input is $R_1 = A\cos(\omega_c t)$. The output of the correlator is

$$v_1 = \int_0^T A^2 \cos^2(\omega_c t)\, dt = \frac{1}{2}A^2 T \tag{8.92}$$

and the decision is that a 1 was transmitted. For the next bit interval, the inputs are $R_2 = S_1 = A\cos(\omega_c t)$ and $S_2 = A\cos(\omega_c t + \pi) = -A\cos(\omega_c t)$, resulting in a correlator output of

$$v_2 = -\int_0^T A^2 \cos^2(\omega_c t)\, dt = -\frac{1}{2}A^2 T \tag{8.93}$$

and a decision that a 0 was transmitted is made. Continuing in this fashion, we see that the original message sequence is obtained if there is no noise at the input.

This detector, while simple to implement, is actually not optimum. The optimum detector for binary DPSK is shown in Figure 8.18. The test statistic for this detector is

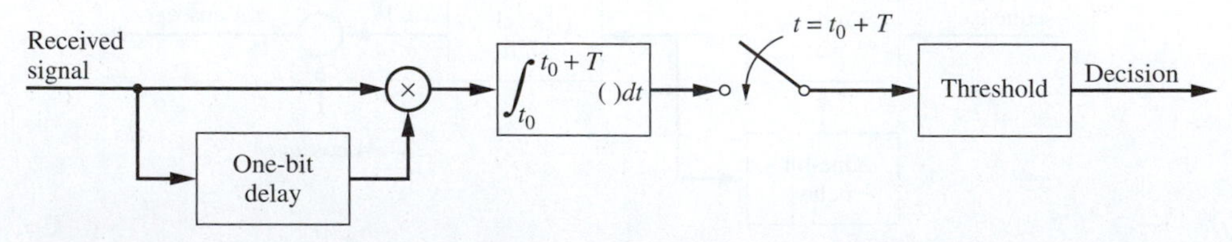

Figure 8.17
Demodulation of DPSK.

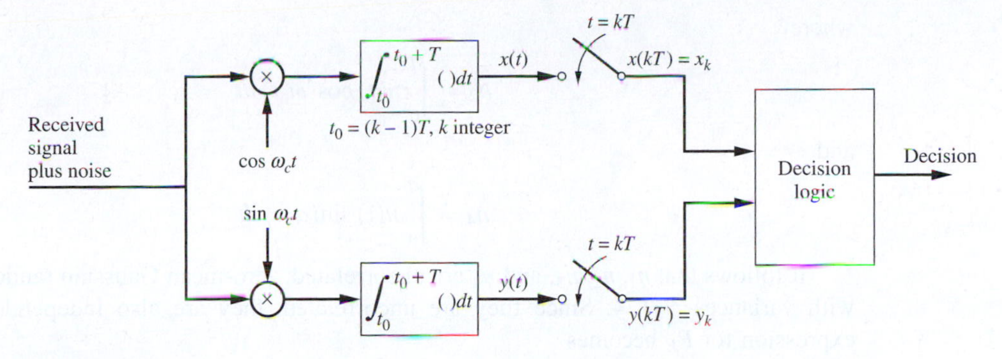

Figure 8.18
Optimum receiver for binary DPSK.

$$l = x_k x_{k-1} + y_k y_{k-1} \tag{8.94}$$

If $l > 0$, the receiver chooses the signal sequence

$$s_1(t) = \begin{cases} A\cos(\omega_c t + \theta), & -T \le t < 0 \\ A\cos(\omega_c t + \theta), & 0 \le t < T \end{cases} \tag{8.95}$$

as having been sent. If $l < 0$, the receiver chooses the signal sequence

$$s_2(t) = \begin{cases} A\cos(\omega_c t + \theta), & -T \le t < 0 \\ -A\cos(\omega_c t + \theta), & 0 \le t < T \end{cases} \tag{8.96}$$

as having been sent.

Without loss of generality, we can choose $\theta = 0$ (the noise and signal orientations with respect to the sine and cosine mixers in Figure 8.18 are completely random). The probability of error can then be computed from $P_E = \Pr\left[x_k x_{k-1} + y_k y_{k-1} < 0 \mid s_1 \text{ sent}, \theta = 0\right]$ (it is assumed that s_1 and s_2 are equally likely). Assuming that $\omega_c T$ is an integer multiple of 2π, we find the outputs of the integrators at time $t = 0$ to be

$$x_0 = \frac{AT}{2} + n_1 \quad \text{and} \quad y_0 = n_3 \tag{8.97}$$

where

$$n_1 = \int_{-T}^{0} n(t)\cos(\omega_c t)\, dt \tag{8.98}$$

and

$$n_3 = \int_{-T}^{0} n(t)\sin(\omega_c t)\, dt \tag{8.99}$$

Similarly, at time $t = T$, the outputs are

$$x_1 = \frac{AT}{2} + n_2 \quad \text{and} \quad y_1 = n_4 \tag{8.100}$$

where

$$n_2 = \int_0^T n(t)\cos(\omega_c t)\,dt \tag{8.101}$$

and

$$n_4 = \int_0^T n(t)\sin(\omega_c t)\,dt \tag{8.102}$$

It follows that n_1, n_2, n_3, and n_4 are uncorrelated, zero-mean Gaussian random variables with variances $N_0T/4$. Since they are uncorrelated, they are also independent, and the expression for P_E becomes

$$P_E = \Pr\left[\left(\frac{AT}{2} + n_1\right)\left(\frac{AT}{2} + n_2\right) + n_3 n_4 < 0\right] \tag{8.103}$$

This can be rewritten as

$$P_E = \Pr\left[\left(\frac{AT}{2} + \frac{n_1}{2} + \frac{n_2}{2}\right)^2 - \left(\frac{n_1}{2} - \frac{n_2}{2}\right)^2 + \left(\frac{n_3}{2} + \frac{n_4}{2}\right)^2 - \left(\frac{n_3}{2} - \frac{n_4}{2}\right)^2 < 0\right] \tag{8.104}$$

[To check this, simply square the separate terms in the argument of (8.104), collect like terms, and compare with the argument of (8.103).] Defining new Gaussian random variables as

$$\begin{aligned} w_1 &= \frac{n_1}{2} + \frac{n_2}{2} \\ w_2 &= \frac{n_1}{2} - \frac{n_2}{2} \\ w_3 &= \frac{n_3}{2} + \frac{n_4}{2} \\ w_4 &= \frac{n_3}{2} - \frac{n_4}{2} \end{aligned} \tag{8.105}$$

the probability of error can be written as

$$P_E = \Pr\left[\left(\frac{AT}{2} + w_1\right)^2 + w_3^2 < w_2^2 + w_4^2\right] \tag{8.106}$$

The positive square roots of the quantities on either side of the inequality sign inside the brackets can be compared just as well as the quantities themselves. From the definitions of w_1, w_2, w_3, and w_4, it can be shown that they are uncorrelated with each other and all are zero mean with variances $N_0T/8$. Since they are uncorrelated and Gaussian, they are also independent. It follows that

$$R_1 = \sqrt{\left(\frac{AT}{2} + w_1\right)^2 + w_3^2} \tag{8.107}$$

is a Ricean random variable (see Section 6.5.3). It is also true that

$$R_2 = \sqrt{w_2^2 + w_4^2} \tag{8.108}$$

is a Rayleigh random variable. It follows that the probability of error can be written as the double integral

$$P_E = \int_0^\infty \left[\int_{r_1}^\infty f_{R_2}(r_2)\, dr_2 \right] f_{R_1}(r_1)\, dr_1 \tag{8.109}$$

where $f_{R_1}(r_1)$ is a Ricean pdf and $f_{R_2}(r_2)$ is a Rayleigh pdf. Letting $\sigma^2 = N_0 T/8$ and $B = AT/2$ and using the Rayleigh and Ricean pdf forms given in Table 5.4 and by (6.149), respectively, this double integral becomes

$$\begin{aligned} P_E &= \int_0^\infty \left[\int_{r_1}^\infty \frac{r_2}{\sigma^2} \exp\left(-\frac{r_2^2}{2\sigma^2}\right) dr_2 \right] \frac{r_1}{\sigma^2} \exp\left(-\frac{r_1^2 + B^2}{2\sigma^2}\right) I_0\left(\frac{Br_1}{\sigma^2}\right) dr_1 \\ &= \int_0^\infty \left[\exp\left(-\frac{r_1^2}{2\sigma^2}\right) \right] \frac{r_1}{\sigma^2} \exp\left(-\frac{r_1^2 + B^2}{2\sigma^2}\right) I_0\left(\frac{Br_1}{\sigma^2}\right) dr_1 \\ &= \exp\left(-\frac{B^2}{2\sigma^2}\right) \int_0^\infty \frac{r_1}{\sigma^2} \exp\left(-\frac{r_1^2}{\sigma^2}\right) I_0\left(\frac{Br_1}{\sigma^2}\right) dr_1 \\ &= \frac{1}{2} \exp\left(-\frac{B^2}{2\sigma^2}\right) \exp\left(\frac{C^2}{2\sigma_0^2}\right) \int_0^\infty \frac{r_1}{\sigma_0^2} \exp\left(-\frac{r_1^2 + C^2}{2\sigma_0^2}\right) I_0\left(\frac{Cr_1}{2\sigma_0^2}\right) dr_1 \end{aligned} \tag{8.110}$$

where $C = B/2$ and $\sigma^2 = 2\sigma_0^2$. Since the integral is over a Ricean pdf, we have

$$\begin{aligned} P_E &= \frac{1}{2} \exp\left(-\frac{B^2}{2\sigma^2}\right) \exp\left(\frac{C^2}{2\sigma_0^2}\right) \\ &= \frac{1}{2} \exp\left(-\frac{B^2}{4\sigma^2}\right) = \frac{1}{2} \exp\left(-\frac{A^2 T}{2N_0}\right) \end{aligned} \tag{8.111}$$

Defining the bit energy E_b as $A^2T/2$ gives

$$P_E = \frac{1}{2} \exp\left(-\frac{E_b}{N_0}\right) \tag{8.112}$$

for the optimum DPSK receiver of Figure 8.18.

It has been shown in the literature that the suboptimum integrate-and-dump detector of Figure 8.17 with an input filter bandwidth of $B = 2/T$ gives an asymptotic probability of error at large E_b/N_0 values of

$$P_E \cong Q\left(\sqrt{E_b/N_0}\right) = Q(\sqrt{z}) \tag{8.113}$$

The result is about a 1.5-dB degradation in SNR for a specified probability of error from that of the optimum detector. Intuitively, the performance depends on the input filter bandwidth—a wide bandwidth results in excess degradation because more noise enters the detector (note that there is a multiplicative noise from the product of undelayed and delayed signals), and an excessively narrow bandwidth degrades the detector performance because of the intersymbol interference (ISI) introduced by the filtering.

Recalling the result for BPSK, (8.74) with $m = 0$, and using the asymptotic approximation $Q(u) \cong e^{-u^2/2}/(2\pi)^{1/2}u$, we obtain the following result for BPSK valid for large E_b/N_0:

$$P_E \cong \frac{e^{-E_b/N_0}}{2\sqrt{\pi E_b/N_0}} \qquad (\text{BPSK; } E_b/N_0 \gg 1) \tag{8.114}$$

For large E_b/N_0, DPSK and BPSK differ only by the factor $(\pi E_b/N_0)^{1/2}$, or roughly a 1-dB degradation in SNR of DPSK with respect to BPSK at low probability of error. This makes DPSK an extremely attractive solution to the problem of carrier reference acquisition required for demodulation of BPSK. The only significant disadvantages of DPSK are that the signaling rate is locked to the specific value dictated by the delay elements in the transmitter and receiver and that errors tend to occur in groups of two because of the correlation imposed between successive bits by the differential encoding process (the latter is the main reason for the 1-dB degradation in performance of DPSK over BPSK at high SNR).

COMPUTER EXAMPLE 8.2

A MATLAB Monte Carlo simulation of a delay-and-multiply DPSK detector is given below. A plot of estimated bit error probability may be made by fixing the desired E_b/N_0, simulating a long string of bits plus noise through the detector, comparing the output bits with the input bits, and counting the errors. Such a plot is shown in Figure 8.19 and is compared with the theoretical curves for the optimum detector, (8.110), as well as the asymptotic result, (8.113), for the suboptimum delay-and-multiply detector shown in Figure 8.17.

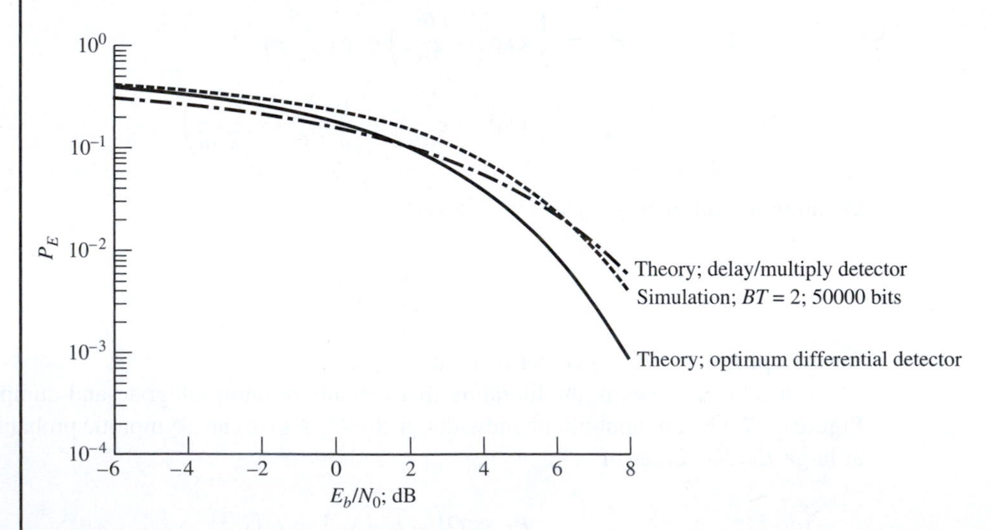

Figure 8.19
Simulated performance of a delay-and-multiply DPSK detector compared with theoretical results.

```
% file: c8ce2.m
% Simulation of suboptimum bandpass filter/delay-and-multiply demodulator
% with integrate-and-dump detection for DPSK; Butterworth filter at
  input.
```

```
%
clf
clear all
Eb_N0_dB_max = input('Enter maximum Eb/N0 in dB');
Eb_N0_dB_min = input('Enter minimum Eb/N0 in dB');
samp_bit = input('Enter number of samples per bit used in simulation');
n_order = input('Enter order of Butterworth prefilter');
BWT_bit = input('Enter filter bandwidth normalized by bit rate');
N_bits = input('Enter total number of bits in simulation');
ss = sign(rand(1,N_bits)-.5);  %Generate random +-1 sequence
                               (a digit/bit)
data = 0.5*(ss+1);              %Logical data is sequence of 1s and 0s
data_diff_enc = diff_enc(data); %Differentially encode data for DPSK
s = 2*data_diff_enc-1;          %Generate bipolar data for
                                modulation
T_bit = 1;                      %Arbitrarily take bit time as 1 second
BW = BWT_bit/T_bit;             %Compute filter bandwidth from
                               BW*T_bit
Eb_N0_dB = Eb_N0_dB_min:Eb_N0_dB_max;
Eb_N0 = 10.^(Eb_N0_dB/10);      %Convert desired Eb/N0 from dB to
                               ratio
Perror = zeros(size(Eb_N0_dB));
Eb = T_bit;                     %Bit energy is T_bit if ampl = 1
[num,den] = butter(n_order,
2*BW/samp_bit);                  %Obtain filter num/den coefficients
for k = 1:length(Eb_N0)         %Loop for each desired Eb/N0
Eb_N00 = Eb_N0(k);
  N0 = Eb/Eb_N00;                 %Compute noise PSD from Eb/N0
  del_t = T_bit/samp_bit;        %Compute sampling interval
  sigma_n = sqrt(N0/(2*del_t));  %Compute standard dev of noise
                                 samples
  sig = s(ones(samp_bit,1),:);   %Build array with columns samp_bit
                                 long
  sig = sig(:);                  %Convert bit sample matrix to vector
  bits_out = [];
  y_det = [];
  noise = sigma_n*randn (size(sig));
                                     %Form Gaussian noise sample
                                     sequence
  y = filter(num,den,sig+noise); %Filter signal + noise with chosen
                                     filter
  y_ref = delay1(y,samp_bit);    %Reference signal is 1-bit delayed
                                     S + N
  y_mult = y.*y_ref;             %MultiplyreceivedS + Nbyreference
  bits_out=int_and_dump(y_mult,samp_bit,N_bits);
  error_array=abs(bits_out-data);
                                  %Comparedetectedbitswithinput
                                  data
  error_array(1:5)=0;             %Excludefirst5bitsduetotransients
  ss=sum(error_array);            %Sumtogettotalnumberoferrors
  Perror(k)=ss/(N_bits-5);        %Subtract5;initial5bitsset=0
end
disp('E_b/N_0,dB;P_E')            %DisplaysimulatedPerrorwithEb/N0
disp([Eb_N0_dB'Perror'])
%Plot simulated bit error probabilities versus Eb/N0
semilogy(Eb_N0_dB,Perror,'-','LineWidth',1.5),grid,
  xlabel('E_b/N_0;dB'),...
     ylabel('P_E'),hold,...
     title('SimulationofBEPfordelay-and-multiplydetectorwith
        ButterworthprefilterforDPSK')
  %Plot theoretical bit error probability for optimum DPSK detector
  semilogy(Eb_N0_dB,0.5*exp(-10.^(Eb_N0_dB/10)),'-',
```

```
    'LineWidth',1.5)
    %Plot approximate theoretical result for suboptimum detector
    semilogy(Eb_N0_dB,qfn(sqrt(10.^(Eb_N0_dB/10))),'-.',
    'LineWidth,1.5)
    legend(['Simulation;BT=',num2str(BWT_bit),';',num2str(N_bits),
    'bits'],'Theory;optimumdifferentialdetector','Theory;delay/
    multiply(detector',3)
%diff_enc(input);functiontodifferentiallyencodebitstreamvector
%
output=diff_enc(input)
L_in=length(input);
output=[];
fork=1:L_in
    ifk==1
        output(k)=not(bitxor(input(k),1));
    else
        output(k)=not(bitxor(input(k),output(k-1)));
    end
end
%Shifts a vector by n_delay elements
%
function y_out=delay1(y_in,n_delay);
NN=length(y_in);
y_out=zeros(size(y_in));
y_out(n_delay+1:NN)=y_in(1:NN-n_delay);
%int_and_dump(input,samp_bit,N_bits);
%Function to integrate-and-dump detect
%
function bits_out=int_and_dump(input,samp_bit,N_bits)
%Reshape input vector with each bit occupying a column
samp_array=reshape(input,samp_bit,N_bits);
integrate=sum(samp_array);    %Integrate(sum) each bit (column)
bits_out=(sign(integrate)+1)/2;
```

A typical MATLAB command window interaction is given below:

```
>>comp_exam8_2
Enter maximum Eb/N0 in dB 8
Enter minimum Eb/N0 in dB -6
Enter number of samples per bit used in simulation 10
Enter order of Butterworth detection filter 2
Enter filter bandwidth normalized by bit rate 2
Enter total number of bits in simulation 50000
E_b/N_0, dB; P_E
-6.0000          0.4179
-5.0000          0.3999
-4.0000          0.3763
-3.0000          0.3465
-2.0000          0.3158
-1.0000           0.2798
0                 0.2411
1.0000            0.2000
2.0000            0.1535
3.0000            0.1142
4.0000            0.0784
5.0000            0.0463
6.0000            0.0243
7.0000            0.0115
8.0000            0.0039

Current plot held
```

■

8.3.2 Noncoherent FSK

The computation of error probabilities for noncoherent systems is somewhat more difficult than it is for coherent systems. Since more is known about the received signal in a coherent system than in a noncoherent system, it is not surprising that the performance of the latter is worse than the corresponding coherent system. Even with this loss in performance, noncoherent systems are often used when simplicity of implementation is a predominent consideration. Only noncoherent FSK will be discussed here.[9]

For noncoherent FSK, the transmitted signals are

$$s_1(t) = A\cos(\omega_c t + \theta), \quad 0 \le t \le T \tag{8.115}$$

and

$$s_2(t) = A\cos[(\omega_c + \Delta\omega)t + \theta], \quad 0 \le t \le T$$

where $\Delta\omega$ is sufficiently large that $s_1(t)$ and $s_2(t)$ occupy different spectral regions. The receiver for FSK is shown in Figure 8.20. Note that it consists of two receivers for noncoherent ASK in parallel. As such, calculation of the probability of error for FSK proceeds much the same way as for ASK, although we are not faced with the dilemma of a threshold that must change with SNR. Indeed, because of the symmetries involved, an exact result for P_E can be obtained. Assuming $s_1(t)$ has been transmitted, the output of the upper detector at time T, $R_1 \triangleq r_1(T)$ has the Ricean pdf

$$f_{R1}(r_1) = \frac{r_1}{N} e^{-(r_1^2 + A^2)/2N} I_0\left(\frac{Ar_1}{N}\right), \quad r_1 \ge 0 \tag{8.116}$$

where $I_0(\cdot)$ is the modified Bessel function of the first kind of order zero and we have made use of Section 6.5.3. The noise power is $N = N_0 B_T$. The output of the lower filter at time T, $R_2 \triangleq r_2(T)$, results from noise alone; its pdf is therefore Rayleigh:

$$f_{R_2}(r_2) = \frac{r_2}{N} e^{-r_2^2/2N}, \quad r_2 \ge 0 \tag{8.117}$$

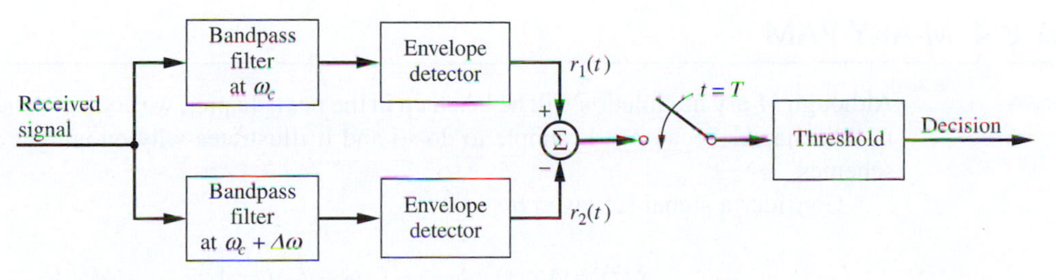

Figure 8.20
Receiver for noncoherent FSK.

---

[9]See Problem 8.30 for a sketch of the derivation of P_E for noncoherent ASK.

An error occurs if $R_2 > R_1$, which can be written as

$$P(E|s_1(t)) = \int_0^\infty f_{R_1}(r_1)\left[\int_{r_1}^\infty f_{R_2}(r_2)dr_2\right]dr_1 \tag{8.118}$$

By symmetry, it follows that $P(E|s_1(t)) = P(E|s_2(t))$, so that (8.118) is the average probability of error. The inner integral in (8.118) integrates to $\exp(-r_1^2/2N)$, which results in the expression

$$P_E = e^{-z}\int_0^\infty \frac{r_1}{N} I_0\left(\frac{A_{r_1}}{N}\right) e^{-r_1^2/N} dr_1 \tag{8.119}$$

where $z = A^2/2N$ as before. If we use a table of definite integrals (see Appendix G.4.2), we can reduce (8.119) to

$$P = \frac{1}{2}\exp\left(\frac{-z}{2}\right) \tag{8.120}$$

For coherent, binary FSK, the error probability for large SNR, using the asymptotic expansion for the Q-function, is

$$P_E \cong \frac{\exp(-z/2)}{\sqrt{2\pi z}} \quad \text{for } z \gg 1$$

Since these differ only by the multiplicative factor $\sqrt{2/\pi z}$, this indicates that *the power margin over noncoherent detection at large SNR is inconsequential*. Thus, because of the comparable performance and the added simplicity of noncoherent FSK, it is employed almost exclusively in practice instead of coherent FSK.

For bandwidth, we note that since the signaling bursts cannot be coherently orthogonal, as for coherent FSK, the minimum frequency separation between tones must be of the order of $2/T$ Hz for noncoherent FSK, giving a minimum null-to-null RF bandwidth of about

$$B_{\text{NCFSK}} = \frac{1}{T} + \frac{2}{T} + \frac{1}{T} = 4R \tag{8.121}$$

resulting in a bandwidth efficiency of 0.25 bps/Hz.

8.4 *M*-ARY PAM

Although M-ary modulation will be taken up in the next chapter, we consider one such scheme in this chapter because it is simple to do so and it illustrates why one might consider such schemes.

Consider a signal set given by

$$s_i(t) = A_i p(t), \quad t_0 \le t \le t_0 + T, \; i = 1, 2, \ldots, M \tag{8.122}$$

where $p(t)$ is the basic pulse shape that is 0 outside the interval $[t_0, t_0 + T]$ with energy

$$E_p = \int_{t_0}^{t_0+T} p^2(t)dt = 1 \tag{8.123}$$

and A_i is the amplitude of the ith possible transmitted signal with $A_1 < A_2 < \cdots < A_M$. Because of the assumption of unit energy for $p(t)$, the energy of $s_i(t)$ is A_1^2. Since we want to associate an integer number of bits with each pulse amplitude, we will restrict M to be an integer power of 2. For example, if $M = 8$, we can label the pulse amplitudes 000, 001, 010, 011, 100, 101, 110, and 111 thereby conveying three bits of information per transmitted pulse (an encoding technique called Gray encoding will be introduced later).

The received signal plus AWGN in the signaling interval $[t_0,\ t_0 + T]$ is given by

$$y(t) = s_i(t) + n(t) = A_i p(t) + n(t), \quad t_0 \leq t \leq t_0 + T \tag{8.124}$$

where for convenience, we set $t_0 = 0$. A reasonable receiver structure is to correlate the received signal plus noise with a replica of $p(t)$ and sample the output of the correlator at $t = T$, which produces

$$Y = \int_0^T [s_i(t) + n(t)]p(t)dt = A_i + N \tag{8.125}$$

where

$$N = \int_0^T n(t)p(t)dt \tag{8.126}$$

is a Gaussian random variable of zero mean and variance $\sigma_N^2 = N_0/2$ [the derivation is similar to (8.5)]. Following the correlation operation, the sample value is compared with a series of thresholds set at $(A_1 + A_2)/2,\ (A_2 + A_3)/2, \ldots,\ (A_{M-1} + A_M)/2$. The possible decisions are

$$\begin{aligned}
&\text{If } Y \leq \frac{A_1 + A_2}{2} \text{ decide that } A_1 p(t) \text{ was sent} \\
&\text{If } \frac{A_1 + A_2}{2} < Y \leq \frac{A_2 + A_3}{2} \text{ decide that } A_2 p(t) \text{ was sent} \\
&\text{If } \frac{A_2 + A_3}{2} < Y \leq \frac{A_3 + A_4}{2} \text{ decide that } A_3 p(t) \text{ was sent} \\
&\qquad \cdots \\
&\text{If } Y > \frac{A_{M-1} + A_M}{2} \text{ decide that } A_M p(t) \text{ was sent}
\end{aligned} \tag{8.127}$$

Recalling Section 2.3, we see that the correlation operation amounts to projecting the received signal plus noise into a generalized one-dimensional vector space with the result that the decision-making process can be illustrated as shown in Figure 8.21. The probability of making a decision error is the probability that a given pulse ampliltude was sent, say A_j, and a decision was made in favor of some other amplitude, averaged over all possible pulse amplitudes. Or, it can be alternatively computed as 1 minus the probability that A_j was sent

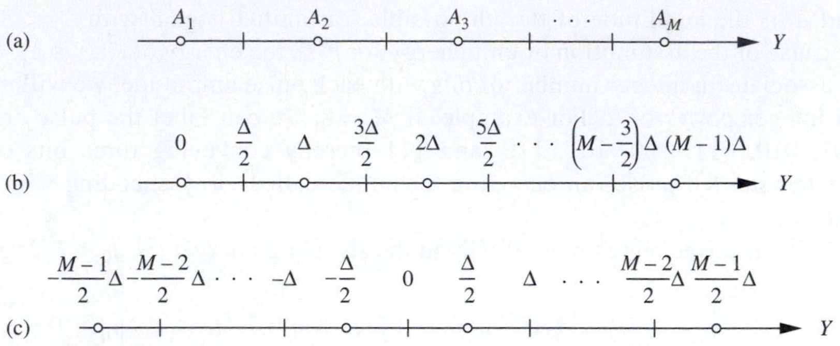

Figure 8.21
(a) Amplitudes and thresholds for PAM (b) Nonnegative-amplitude equally spaced case (c) Antipodal equally spaced case.

and a decision in favor of A_j was made, which is

$$P(E|A_j\text{ sent}) = \begin{cases} 1-\Pr\left[\dfrac{(A_{j-1}+A_j)}{2} < Y \leq \dfrac{A_j+A_{j+1}}{2}\right], & j=2,\ 3,\ \ldots,\ M-1 \\ 1-\Pr\left[Y \leq \dfrac{A_1+A_2}{2}\right], & j=1 \\ 1-\Pr\left[Y > \dfrac{A_{M-1}+A_M}{2}\right], & j=M \end{cases} \tag{8.128}$$

To simplify matters, we now make the assumption that for $A_j = (j-1)\Delta$ for $j = 1,\ 2,\ \ldots,\ M$. Thus,

$$\begin{aligned} P(E|A_j\text{ sent}) &= 1-\Pr\left[N < \frac{\Delta}{2}\right],\ j=1 \\ &= 1-\Pr\left[\frac{\Delta}{2} < \Delta+N \leq \frac{3\Delta}{2}\right] = 1-\Pr\left[-\frac{\Delta}{2} < N \leq \frac{\Delta}{2}\right], j=2 \\ &= 1-\Pr\left[\frac{3\Delta}{2} < 2\Delta+N \leq \frac{5\Delta}{2}\right] = 1-\Pr\left[-\frac{\Delta}{2} < N \leq \frac{\Delta}{2}\right], j=3 \\ &\cdots \\ &= 1-\Pr\left[\frac{(2M-3)\Delta}{2} \leq (M-1)\Delta+N\right] = 1-\Pr\left[N > -\frac{\Delta}{2}\right], j=M \end{aligned} \tag{8.129}$$

These reduce to

$$P(E|A_j \text{ sent}) = 1 - \int_{-\infty}^{\Delta/2} \frac{\exp(-\eta^2/N_0)}{\sqrt{\pi N_0}} d\eta$$
$$= \int_{\Delta/2}^{\infty} \frac{\exp(-\eta^2/N_0)}{\sqrt{\pi N_0}} d\eta = Q\left(\frac{\Delta}{\sqrt{2N_0}}\right), \quad j = 1, M \tag{8.130}$$

and

$$P(E|A_j \text{ sent}) = 1 - \int_{-\Delta/2}^{\Delta/2} \frac{\exp(-\eta^2/N_0)}{\sqrt{\pi N_0}} d\eta$$
$$= 2\int_{\Delta/2}^{\infty} \frac{\exp(-\eta^2/N_0)}{\sqrt{\pi N_0}} d\eta$$
$$= 2Q\left(\frac{\Delta}{\sqrt{2N_0}}\right), \quad j = 2, \ldots, M-1 \tag{8.131}$$

If all possible signals are equally likely, the average probability of error is

$$P_E = \frac{1}{M}\sum_{j=1}^{M} P(E|A_j \text{ sent})$$
$$= \frac{2(M-1)}{M} Q\left(\frac{\Delta}{\sqrt{2N_0}}\right) \tag{8.132}$$

Now the average signal energy is

$$E_{\text{ave}} = \frac{1}{M}\sum_{j=1}^{M} E_j = \frac{1}{M}\sum_{j=1}^{M} A_j^2 = \frac{1}{M}\sum_{j=1}^{M} (j-1)^2\Delta^2$$
$$= \frac{\Delta^2}{M}\sum_{k=1}^{M-1} k^2 = \frac{\Delta^2}{M}\frac{(M-1)M(2M-1)}{6} \tag{8.133}$$
$$= \frac{(M-1)(2M-1)\Delta^2}{6}$$

where the summation formula

$$\sum_{k=1}^{M-1} k^2 = \frac{(M-1)M(2M-1)}{6} \tag{8.134}$$

has been used. Thus

$$\Delta^2 = \frac{6E_{\text{ave}}}{(M-1)(2M-1)}, \quad M\text{-ary PAM} \tag{8.135}$$

so that

$$\begin{aligned} P_E &= \frac{2(M-1)}{M} Q\left(\sqrt{\frac{\Delta^2}{2N_0}}\right) \\ &= \frac{2(M-1)}{M} Q\left(\sqrt{\frac{3E_{\text{ave}}}{(M-1)(2M-1)N_0}}\right), \quad M\text{-ary PAM} \end{aligned} \tag{8.136}$$

If the signal amplitudes are symmetrically placed about 0 so that $A_j = (j-1)\Delta - \frac{(M-1)}{2}\Delta$ for $j = 1, 2, \ldots, M$, the average signal energy is

$$E_{\text{ave}} = \frac{(M^2-1)\Delta^2}{12}, \quad M\text{-ary antipodal PAM} \tag{8.137}$$

so that

$$\begin{aligned} P_E &= \frac{2(M-1)}{M} Q\left(\sqrt{\frac{\Delta^2}{2N_0}}\right) \\ &= \frac{2(M-1)}{M} Q\left(\sqrt{\frac{6E_{\text{ave}}}{(M^2-1)N_0}}\right), \quad M\text{-ary antipodal PAM} \end{aligned} \tag{8.138}$$

Note that binary antipodal PAM is 3 dB better than PAM. Also note that with $M = 2$, (8.138) for M-ary antipodal PAM reduces to the error probability for binary antipodal signaling.

In order to compare these M-ary modulation schemes with the other binary modulation schemes considered in this chapter, we need to do two things. The first is to express E_{ave} in terms of energy per bit. Since it was assumed that $M = 2^m$, where $m = \log_2 M$ is an integer number of bits, this is accomplished by setting $E_b = E_{\text{ave}}/m = E_{\text{ave}}/\log_2 M$ or $E_{\text{ave}} = E_b \log_2 M$. The second thing we need to do is convert the probabilities of error found above, which are symbol-error probabilities, to bit-error probabilities. This will be taken up in Chapter 9 where two cases will be discussed. The first is where mistaking the correct symbol in demodulation for any of the other possible symbols is equally likely. The second case, which is the case of interest here, is where adjacent symbol errors are more probable than nonadjacent symbol errors and encoding is used to ensure only one bit changes in going from a given symbol to an adjacent symbol (i.e., in PAM, going from a given amplitude to an adjacent amplitude). This can be ensured by using Gray encoding of the bits associated with the symbol amplitudes, which is demonstrated in Problem 8.32. If both of these conditions are satisfied, it then follows that $P_b \cong \frac{1}{(\log_2 M)} P_{\text{symbol}}$. Thus

$$P_{b,\,\text{PAM}} \cong \frac{2(M-1)}{M\log_2 M} Q\left(\sqrt{\frac{3(\log_2 M)E_b}{(M-1)(2M-1)N_0}}\right), \quad M\text{-ary PAM; Gray encoding} \tag{8.139}$$

and

$$P_{b,\,\text{antip. PAM}} = \frac{2(M-1)}{M\log_2 M} Q\left(\sqrt{\frac{6(\log_2 M)E_b}{(M^2-1)N_0}}\right), \quad M\text{-ary antipodal PAM; Gray encoding} \tag{8.140}$$

The bandwidth for PAM may be deduced by considering the pulses to be ideal rectangular of width $T = (\log_2 M)T_{\text{bit}}$. Their baseband spectra are therefore $S_k(f) = A_k\,\text{sinc}(Tf)$ for a 0 to first null bandwidth of

$$B_{\text{bb}} = \frac{1}{T} = \frac{1}{(\log_2 M)T_b} \text{ Hz}$$

If modulated on a carrier, the null-to-null bandwidth is twice the baseband value or

$$B_{\text{PAM}} = \frac{2}{(\log_2 M)T_b} = \frac{2R}{\log_2 M} \tag{8.141}$$

whereas BPSK, DPSK, and binary ASK have bandwidths of $B_{\text{RF}} = 2/T_b$ Hz. This illustrates that for a fixed bit rate, PAM requires less bandwidth the larger M. In fact the bandwidth efficiency for M-ary PAM is $0.5\log_2 M$ bps/Hz.

8.5 COMPARISON OF DIGITAL MODULATION SYSTEMS

Bit-error probabilities are compared in Figure 8.22 for the modulation schemes considered in this chapter. Note that the curve for antipodal PAM with $M = 2$ is identical to BPSK. Also note that the bit-error probability of antipodal PAM becomes worse the larger M. However, more bits are

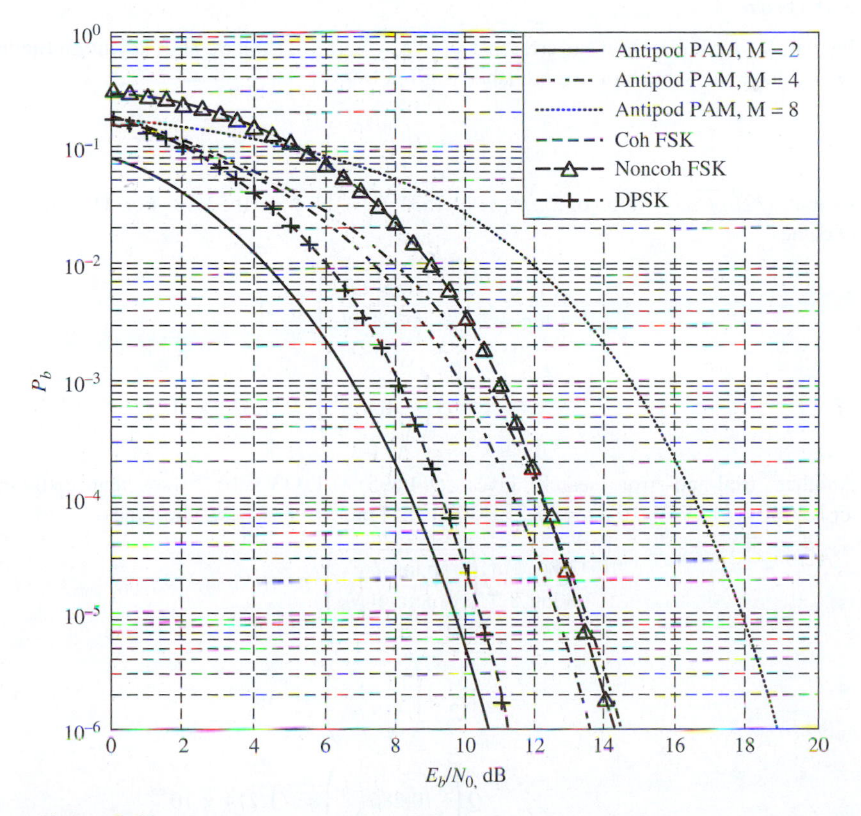

Figure 8.22
Error probabilities for several binary digital signaling schemes.

transmitted the larger M. In a bandlimited channel with sufficient signal power, it may desirable to send more bits per symbol. Noncoherent FSK and antipodal PAM with $M = 4$ have almost identical performance at large SNR. Note also the small difference in performance between BPSK and DPSK, with a slightly larger difference between coherent and noncoherent FSK.

In addition to cost and complexity of implementation, there are many other considerations in choosing one type of digital data system over another. For some channels, where the channel gain or phase characteristics (or both) are perturbed by randomly varying propagation conditions, use of a noncoherent system may be dictated because of the near impossibility of establishing a coherent reference at the receiver under such conditions. Such channels are referred to as *fading*. The effects of fading channels on data transmission will be taken up in Section 8.8.

The following example illustrates some typical SNR ratio and data rate calculations for the digital modulation schemes considered in this chapter.

EXAMPLE 8.7

Suppose $P_b = 10^{-6}$ is desired for a certain digital data transmission system. (a) Compare the necessary SNRs for BPSK, DPSK, antipodal PAM for $M = 2$, 4, 8, and noncoherent FSK. (b) Compare maximum bit rates for an RF bandwidth of 20 kHz.

Solution

For part (a), we find by trial and error that $Q(4.753) \approx 10^{-6}$. Biphase-shift keying and antipodal PAM for $M = 2$ have the same bit error probability, given by

$$P_b = Q\left(\sqrt{\frac{2E_b}{N_0}}\right) = 10^{-6}$$

so that $\sqrt{2E_b/N_0} = 4.753$ or $E_b/N_0 = (4.753)^2/2 = 11.3 = 10.53$ dB. For $M = 4$, Equation (8.140) becomes

$$\frac{2(4-1)}{4\log_2 4} Q\left(\sqrt{\frac{6\log_2 4}{4^2-1}\frac{E_b}{N_0}}\right) = 10^{-6}$$

$$Q\left(\sqrt{0.8\frac{E_b}{N_0}}\right) = 1.333 \times 10^{-6}$$

Another trial-and-error search gives $Q(4.695) \approx 1.333 \times 10^{-6}$ so that $\sqrt{0.8E_b/N_0} = 4.695$ or $E_b/N_0 = (4.695)^2/(0.8) = 27.55 = 14.4$ dB. For $M = 8$, (8.140) becomes

$$\frac{2(8-1)}{8\log_2 8} Q\left(\sqrt{\frac{6\log_2 8}{8^2-1}\frac{E_b}{N_0}}\right) = 10^{-6}$$

$$\frac{7}{12} Q\left(\sqrt{\frac{2}{7}\frac{E_b}{N_0}}\right) = 10^{-6}$$

$$Q\left(\sqrt{0.286\frac{E_b}{N_0}}\right) = 1.714 \times 10^{-6}$$

Yet another trial-and-error search gives $Q(4.643) \approx 1.714 \times 10^{-6}$ so that $\sqrt{0.286E_b/N_0} = 4.643$ or $E_b/N_0 = (4.643)^2/0.286 = 75.38 = 18.77$ dB.

For DPSK, we have

$$\frac{1}{2}\exp\left(\frac{-E_b}{N_0}\right) = 10^{-6}$$

$$\exp\left(\frac{-E_b}{N_0}\right) = 2 \times 10^{-6}$$

which gives

$$\frac{E_b}{N_0} = -\ln(2 \times 10^{-6}) = 13.12 = 11.18 \text{ dB}$$

For coherent FSK, we have

$$P_b = Q\left(\sqrt{\frac{E_b}{N_0}}\right) = 10^{-6}$$

so that

$$\sqrt{\frac{E_b}{N_0}} = 4.753 \text{ or } \frac{E_b}{N_0} = (4.753)^2 = 22.59 = 13.54 \text{ dB}$$

For noncoherent FSK, we have

$$\frac{1}{2}\exp\left(-0.5\frac{E_b}{N_0}\right) = 10^{-6}$$

$$\exp\left(-0.5\frac{E_b}{N_0}\right) = 2 \times 10^{-6}$$

which results in

$$\frac{E_b}{N_0} = -2\ln(2 \times 10^{-6}) = 26.24 = 14.18 \text{ dB}$$

For (b), we use the previously developed bandwidth expressions given by (8.90), (8.91), (8.121), and (8.141). Results are given in Table 8.5.

The results of Table 8.5 demonstrate that PAM is a modulation scheme that allows a trade-off between power efficiency (in terms of the E_b/N_0 required for a desired bit-error probability) and

Table 8.5 Comparison of Binary Modulation Schemes at $P_E = 10^{-6}$

Modulation method	Required SNR for $P_b = 10^{-6}$ (dB)	R for $B_{RF} = 20$ kHz (kbps)
BPSK	10.5	10
DPSK	11.2	10
Antipodal 4-PAM	14.4	20
Antipodal 8-PAM	18.8	30
Coherent FSK, ASK	13.5	8
Noncoherent FSK	14.2	5

bandwidth efficiency (in terms of maximum data rate for a fixed bandwidth channel). The power-bandwidth efficiency trade-off of other M-ary digital modulation schemes will be examined further in Chapter 9.

■

■ 8.6 PERFORMANCE OF ZERO-ISI DIGITAL DATA SYSTEMS

Although a fixed channel bandwidth was assumed in Example 8.7, the results of Chapter 4, Section 4.3, demonstrated that, in general, bandlimiting causes ISI and can result in severe degradation in performance. The use of pulse shaping to avoid ISI was also introduced in Chapter 4, where Nyquist's pulse shaping criterion was proved in Section 4.4.2. The frequency response characteristics of transmitter and receiver filters for implementing zero-ISI transmission were examined in Section 4.4.3, resulting in (4.54) and (4.55). In this section, we continue that discussion and derive an expression for the bit error probability of a zero-ISI data transmission system.

Consider the system of Figure 4.9, repeated in Fig. 8.23, where everything is the same except we now specify the noise as Gaussian and having a power spectral density of $G_n(f)$. The transmitted signal is

$$\begin{aligned} x(t) &= \sum_{k=-\infty}^{\infty} a_k \delta(t-kT) * h_T(t) \\ &= \sum_{k=-\infty}^{\infty} a_k h_T(t-kT) \end{aligned} \tag{8.142}$$

where $h_T(t)$ is the impulse response of the transmitter filter that has the lowpass frequency-response function $H_T(f) = \Im[h_T(t)]$. This signal passes through a bandlimited channel filter, after which Gaussian noise with power spectral density $G_n(f)$ is added to give the received signal

$$y(t) = x(t) * h_C(t) + n(t) \tag{8.143}$$

where $h_C(t) = \Im^{-1}[H_C(f)]$ is the impulse response of the channel. Detection at the receiver is accomplished by passing $y(t)$ through a filter with impulse response $h_R(t)$ and sampling its output at intervals of T. If we require that the cascade of transmitter, channel, and receiver filters satisfies Nyquist's pulse shaping criterion, it then follows that the output sample at time $t = t_d$,

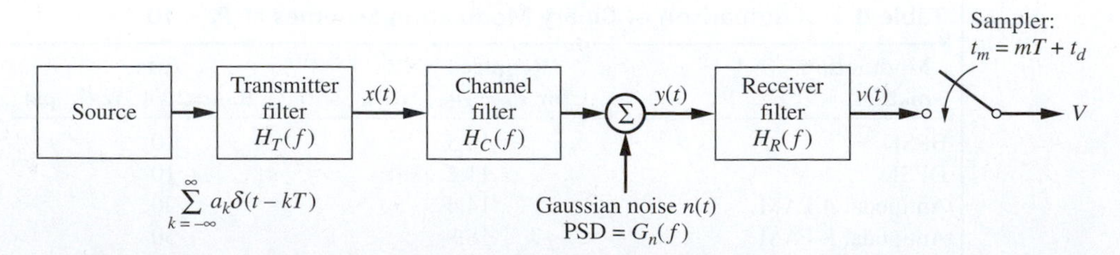

Figure 8.23
Baseband system for signaling through a bandlimited channel.

where t_d is the delay imposed by the channel and the receiver filters, is

$$\begin{aligned} V &= Aa_0 p(0) + N \\ &= Aa_0 + N \end{aligned} \tag{8.144}$$

where

$$AP(t - t_d) = h_T(t) * h_C(t) * h_R(t) \tag{8.145}$$

or, by Fourier transforming both sides, we have

$$AP(f)\exp(-j2\pi f t_d) = H_T(f)H_C(f)H_R(f) \tag{8.146}$$

In (8.145), A is a scale factor, t_d is a time delay accounting for all delays in the system, and

$$N = n(t) * h_R(t)|_{t=t_d} \tag{8.147}$$

is the Gaussian noise component at the output of the detection filter at time $t = t_d$.

For simplicity, we assume binary signaling ($a_m = +1$ or -1) so that the average probability of error is

$$\begin{aligned} P_E &= \Pr[a_m = 1]\Pr[Aa_m + N \le 0 \text{ given } a_m = 1] \\ &\quad + \Pr[a_m = -1]\Pr[Aa_m + N \ge 0 \text{ given } a_m = -1] \\ &= \Pr[Aa_m + N < 0 \text{ given } a_m = 1] \\ &= \Pr[Aa_m + N > 0 \text{ given } a_m = -1] \end{aligned} \tag{8.148}$$

where the latter two equations result by assuming $a_m = 1$ and $a_m = -1$ are equally likely and the symmetry of the noise pdf is invoked. Taking the last equation of (8.148), it follows that

$$P_E = \Pr[N \ge A] = \int_A^\infty \frac{\exp(-u^2/2\sigma^2)}{\sqrt{2\pi\sigma^2}}\,du = Q\left(\frac{A}{\sigma}\right) \tag{8.149}$$

where

$$\sigma^2 = \text{var}[N] = \int_{-\infty}^{\infty} G_n(f)|H_R(f)|^2\,df \tag{8.150}$$

Because the Q-function is a monotonically decreasing function of its argument, it follows that the average probability of error can be minimized through proper choice of $H_T(f)$ and $H_R(f)$ [$H_C(f)$ is assumed to be fixed], by maximizing A/σ or by minimizing σ^2/A^2. The minimization can be carried out, subject to the constraint in (8.146), by applying Schwarz's inequality. The result is

$$|H_R(f)|_{\text{opt}} = \frac{KP^{1/2}(f)}{G_n^{1/4}(f)|H_C(f)|^{1/2}} \tag{8.151}$$

and

$$|H_T(f)|_{\text{opt}} = \frac{AP^{1/2}(f)\,G_n^{1/4}(f)}{K\,|H_C(f)|^{1/2}} \tag{8.152}$$

where K is an arbitrary constant and any appropriate phase response can be used (recall that $G_n(f)$ is nonnegative since it is a power spectral density). $P(f)$ is assumed to have the zero-ISI property of (4.45) and to be nonnegative. Note that it is the cascade of transmitter, channel, and receiver filters that produces the overall zero-ISI pulse spectrum in accordance with (8.146).

The minimum value for the error probability is

$$P_{b,\,\min} = Q\left[\sqrt{E_T}\left(\int_{-\infty}^{\infty} \frac{G_n^{1/2}(f)\,P(f)}{|H_C(f)|}\,df\right)^{-1}\right] \tag{8.153}$$

where

$$E_T = E\{a_m^2\}\int_{-\infty}^{\infty} |h_T(t)|^2\,dt = \int_{-\infty}^{\infty} |H_T(f)|^2\,df \tag{8.154}$$

is the transmit signal energy and the last integral follows by Rayleigh's energy theorem. Also, $E\{a_m^2\} = 1$ since $a_m = 1$ or $a_m = -1$ with equal probability.

That (8.153) is the minimum error probability can be shown as follows. Taking the magnitude of (8.146), solving for $|H_T(f)|$, and substituting into (8.154), we may show that the transmitted signal energy is

$$E_T = A^2\int_{-\infty}^{\infty} \frac{P^2(f)\,df}{|H_C(f)|^2|H_R(f)|^2} \tag{8.155}$$

Solving (8.155) for $1/A^2$ and using (8.150) for var $(N) = \sigma^2$, it follows that

$$\frac{\sigma^2}{A^2} = \frac{1}{E_T}\int_{-\infty}^{\infty} G_n(f)|H_R(f)|^2\,df \int_{-\infty}^{\infty} \frac{P^2(f)\,df}{|H_C(f)|^2\,|H_R(f)|^2} \tag{8.156}$$

Schwarz's inequality (8.39) may now be applied to show that the minimum for σ^2/A^2 is

$$\left(\frac{\sigma}{A}\right)^2_{\min} = \frac{1}{E_T}\left(\int_{-\infty}^{\infty} \frac{G_n^{1/2}(f)P(f)}{|H_C(f)|}\,df\right)^2 \tag{8.157}$$

which is achieved for $|H_R(f)|_{\text{opt}}$ and $|H_T(f)|_{\text{opt}}$ given when (8.151) and (8.152) are used. The square root of the reciprocal of (8.157) is then the maximum A/σ that minimizes the error probability (8.149). In this case, Schwarz's inequality is applied in reverse with $|X(f)| = G_n^{1/2}(f)\,|H_R(f)|$ and $|Y(f)| = P(f)/[|H_C(f)||H_R(f)|]$. The condition for equality [i.e., achieving the minimum in (8.39)] is $X(f) = KY(f)$ or

$$G_n^{1/2}(f)\,|H_R(f)|_{\text{opt}} = K\frac{P(f)}{|H_C(f)||H_R(f)|_{\text{opt}}} \tag{8.158}$$

which can be solved for $|H_R(f)|_{\text{opt}}$, while $|H_T(f)|_{\text{opt}}$ is obtained by taking the magnitude of (8.146) and substituting $|H_R(f)|_{\text{opt}}$. (K is an arbitrary constant.)

A special case of interest occurs when

$$G_n(f) = \frac{N_0}{2}, \quad \text{all } f \text{ (white noise)} \tag{8.159}$$

and

$$H_C(f) = 1, \quad |f| \le \frac{1}{T} \tag{8.160}$$

In this case

$$|H_T(f)|_{\text{opt}} = |H_R(f)|_{\text{opt}} = K'P^{1/2}(f) \tag{8.161}$$

where K' is an arbitrary constant. If $P(f)$ is a raised cosine spectrum, then the transmit and receive filters are called *square-root raised-cosine filters* (in applications, the square-root raised cosine pulse shape is formed digitally by sampling). The minimum probability of error then simplifies to

$$P_{E,\,\min} = Q\left\{\sqrt{E_T}\left[\frac{N_0}{2}\int_{-1/T}^{1/T} P(f)\,df\right]^{-1}\right\} = Q\left(\sqrt{\frac{2E_T}{N_0}}\right) \tag{8.162}$$

where

$$p(0) = \int_{-1/T}^{1/T} P(f)\,df = 1 \tag{8.163}$$

follows because of the zero-ISI property expressed by (4.34). The result (8.162) is identical to that obtained previously for binary antipodal signaling in an infinite bandwidth baseband channel.

Note that the case of M-ary transmission can be solved with somewhat more complication in computing the average signal energy.

EXAMPLE 8.8

Show that (8.162) results from (8.153) if the noise power spectral density is given by

$$G_n(f) = \frac{N_0}{2}|H_C(f)|^2 \tag{8.164}$$

That is, the noise is colored with spectral shape given by the channel filter.

Solution

Direct substitution into the argument of (8.153) results in

$$\begin{aligned} \sqrt{E_T}\left(\int_{-\infty}^{\infty} \frac{G_n^{1/2}(f)P(f)}{|H_C(f)|}\,df\right)^{-1} &= \sqrt{E_T}\left(\int_{-\infty}^{\infty} \frac{\sqrt{N_0/2}|H_C(f)|P(f)}{|H_C(f)|}\,df\right)^{-1} \\ &= \sqrt{E_T}\left[\sqrt{\frac{N_0}{2}}\int_{-\infty}^{\infty} P(f)\,df\right]^{-1} \end{aligned} \tag{8.165}$$

$$= \sqrt{\frac{2E_T}{N_0}} \tag{8.166}$$

where (8.163) has been used. ■

EXAMPLE 8.9

Suppose that $G_n(f) = N_0/2$, and that the channel filter is fixed but unspecified. Find the degradation factor in E_T/N_0 over that for a infinite-bandwidth white-noise channel for the error probability of (8.153) due to pulse shaping and channel filtering.

Solution

The argument of (8.153) becomes

$$\begin{aligned}\sqrt{E_T}\left(\int_{-\infty}^{\infty}\frac{G_n^{1/2}(f)P(f)}{|H_C(f)|}\,df\right)^{-1} &= \sqrt{E_T}\left(\int_{-\infty}^{\infty}\frac{\sqrt{N_0/2}P(f)}{|H_C(f)|}\,df\right)^{-1}\\ &= \sqrt{\frac{2E_T}{N_0}}\left(\int_{-\infty}^{\infty}\frac{P(f)}{|H_C(f)|}\,df\right)^{-1}\\ &= \sqrt{2\left(\int_{-\infty}^{\infty}\frac{P(f)}{|H_C(f)|}\,df\right)^{-2}\frac{E_T}{N_0}}\\ &= \sqrt{\frac{2}{F}\frac{E_T}{N_0}}\end{aligned} \tag{8.167}$$

where

$$F = \left(\int_{-\infty}^{\infty}\frac{P(f)}{|H_C(f)|}\,df\right)^2 = \left(2\int_{0}^{\infty}\frac{P(f)}{|H_C(f)|}\,df\right)^2 \tag{8.168}$$

■

COMPUTER EXAMPLE 8.3

A MATLAB program to evaluate F of (8.168) assuming a raised cosine pulse spectrum and a Butterworth channel frequency response is given below. The degradation is plotted in decibels in Figure 8.24 versus

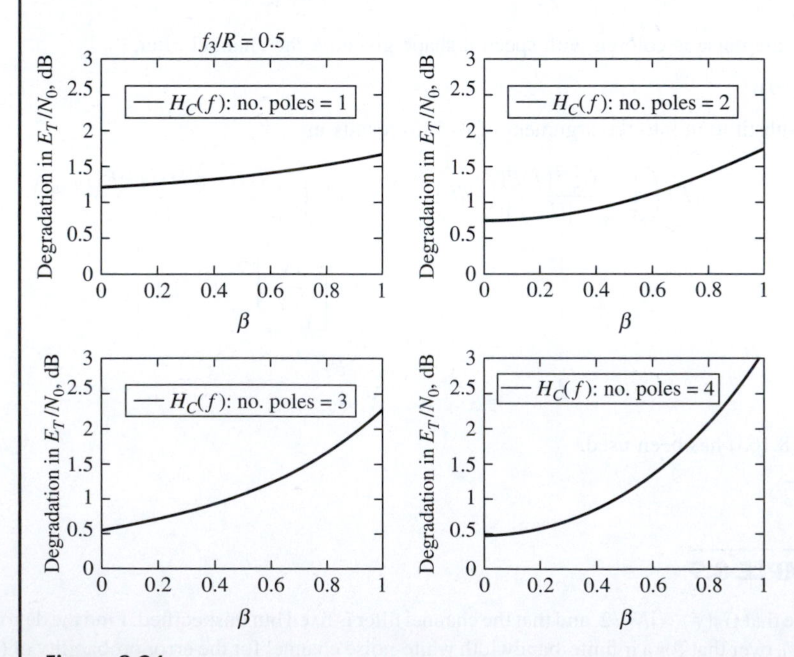

Figure 8.24
Degradations for raised cosine signaling through a Butterworth channel with additive Gaussian noise.

roll-off factor for a channel filter 3-dB cutoff frequency of 1/2 data rate. Note that the degradation, which is the decibel increase in E_T/N_0 needed to maintain the same bit-error probability as in a infinite bandwidth white Gaussian noise channel, ranges from less that 0.5 to 3 dB for the four-pole case as the raised cosine spectral width ranges from $f_3\,(\beta = 0)$ to $2f_3\,(\beta = 1)$

```
% file: c8ce3.m
% Computation of degradation for raised cosine signaling
% through a channel modeled as Butterworth
%
clf
T = 1
beta = 1;
f3 = 0.4/T;
for np = 1:4;
    beta = 0.001:.01:1;
    Lb = length(beta);
    for k = 1:Lb
        beta0 = beta(k);
        f1 = (1-beta0)/(2*T);
        f2 = (1+beta0)/(2*T);
        fmax = 1/T;
        f = 0:.001:fmax;
        I1 = find(f>=0 & f<f1);
        I2 = find(f>=f1 & f<f2);
        I3 = find(f>=f2 & f<=fmax);
        Prc = zeros(size(f));
        Prc(I1) = T;
        Prc(I2) = (T/2)*(1+cos((pi*T/beta0)*(f(I2)-(1-beta0)/(2*T))));
        Prc(I3) = 0;
        integrand = Prc.*sqrt(1+(f./f3).^(2*np));
        F(k) = (2*trapz(f, integrand)).^2;
    end
FdB = 10*log10(F);
subplot(2,2,np), plot(beta, FdB), xlabel('beta'), ylabel('Degr. in E_T/N_0,
dB'), ...
legend(['H_C(f): no. poles: ', num2str(np)]), axis([0 1 0 7])
if np == 1
    title(['f_3/R = ', num2str(f3*T)])
  end
end
```

■

■ 8.7 MULTIPATH INTERFERENCE

The channel models that we have assumed so far have been rather idealistic in that the only signal perturbation considered was additive Gaussian noise. Although realistic for many situations, additive Gaussian noise channel models do not accurately represent many transmission phenomena. Other important sources of degradation in many digital data systems are bandlimiting of the signal by the channel, as examined in the previous section; non-Gaussian noise, such as impulse noise due to lightning discharges or switches; RFI due to other transmitters; and multiple transmission paths, termed *multipath*, due to stratifications in the transmission medium or objects that reflect or scatter the propagating signal.

In this section we characterize the effects of multipath transmission because it is a fairly common transmission perturbation and its effects on digital data transmission can, in the simplest form, be analyzed in a straightforward fashion.

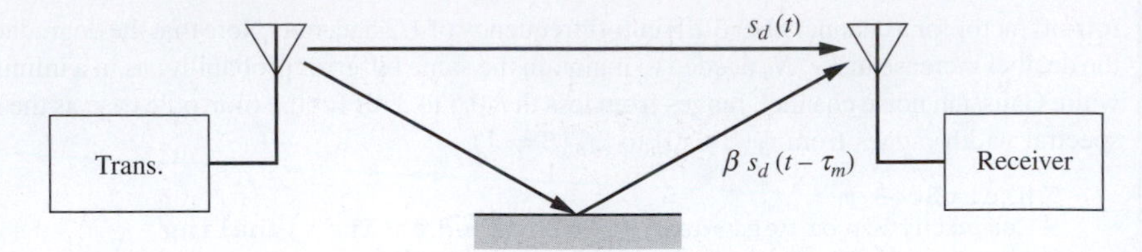

Figure 8.25
Channel model for multipath transmission.

Initially, we consider a two-ray multipath model as illustrated in Figure 8.25. In addition to the multiple transmission, the channel perturbs the signal with white Gaussian noise with double-sided power spectral density $\frac{1}{2}N_0$. Thus the received signal plus noise is given by

$$y(t) = s_d(t) + \beta s_d(t - \tau_m) + n(t) \tag{8.169}$$

where $s_d(t)$ is the received direct-path signal, β is the relative attenuation of the multipath component, and τ_m is its delay. For simplicity, consider the effects of this channel on binary BPSK. The direct-path signal can be represented as

$$s_d(t) = Ad(t)\cos(\omega_c t) \tag{8.170}$$

where $d(t)$, the data stream, is a continuous sequence of plus or minus 1-valued rectangular pulses, each of which is T in duration. Because of the multipath component, we must consider a sequence of bits at the receiver input. We will analyze the effect of the multipath component and noise on a correlation receiver as shown in Figure 8.26, which, we recall, detects the data in the presence of Gaussian noise alone with minimum probability of error. Writing the noise in terms of quadrature components $n_c(t)$ and $n_s(t)$, we find that the input to the integrator, ignoring double frequency terms, is

$$\begin{aligned} x(t) &= \mathrm{Lp}[2y(t)\cos(\omega_c t)] \\ &= Ad(t) + \beta A d(t - \tau_m)\cos(\omega_c \tau_m) + n_c(t) \end{aligned} \tag{8.171}$$

where $\mathrm{Lp}[\cdot]$ stands for the lowpass part of the bracketed quantity.

The second term in (8.171) represents interference due to the multipath. It is useful to consider two special cases:

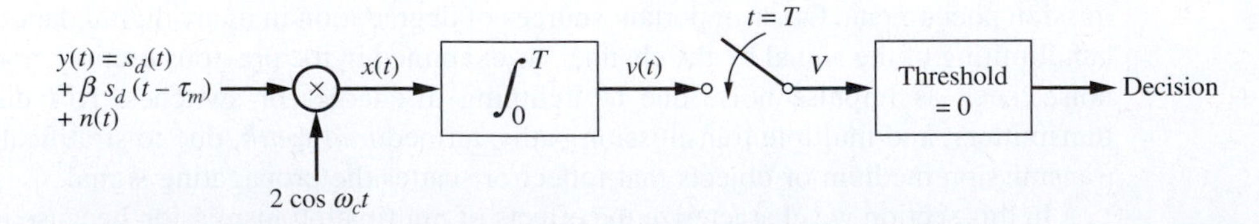

Figure 8.26
Correlation receiver for BPSK with signal plus multipath at its input.

1. $\tau_m/T \cong 0$, so that $d(t-\tau_m) \cong d(t)$. For this case, it is usually assumed that $\omega_0\tau_m$ is a uniformly distributed random variable in $(-\pi, \pi)$ and that there are many other multipath components of random amplitudes and phases. In the limit as the number of components becomes large, the sum process, composed of inphase and quadrature components, has Gaussian amplitudes. Thus, the envelope of the received signal is Rayleigh or Ricean (see Section 6.3.3), depending on whether there is a steady signal component present. The Rayleigh case will be analyzed in the next section.
2. $0 < \tau_m/T \leq 1$ so that successive bits of $d(t)$ and $d(t-t_m)$ overlap; in other words, there is ISI. For this case, we will let $\delta = \beta \cos\omega_c\tau_m$ be a parameter in the analysis.

We now analyze the receiver performance for case 2, for which the effect of ISI is nonnegligible. To simplify notation, let

$$\delta = \beta \cos(\omega_c\tau_m) \tag{8.172}$$

so that (8.171) becomes

$$x(t) = Ad(t) + A\delta d(t-\tau_m) + n_c(t) \tag{8.173}$$

If $\tau_m/T \leq 1$, only adjacent bits of $Ad(t)$ and $A\delta d(t-\tau_m)$ will overlap. Thus we can compute the signal component of the integrator output in Figure 8.26 by considering the four combinations shown in Figure 8.27. Assuming 1s and 0s are equally probable, the four combinations shown in Figure 8.27 will occur with equal probabilities of $\frac{1}{4}$. Thus the average probability of error is

$$P_E = \frac{1}{4}[P(E\,|\,{+}{+}) + P(E\,|\,{-}{+}) + P(E\,|\,{+}{-}) + P(E\,|\,{-}{-})] \tag{8.174}$$

where $P(E\,|\,{+}{+})$ is the probability of error given two 1's were sent, and so on. The noise component of the integrator output, namely

$$N = \int_0^T 2n(t)\cos(\omega_c t)\,dt \tag{8.175}$$

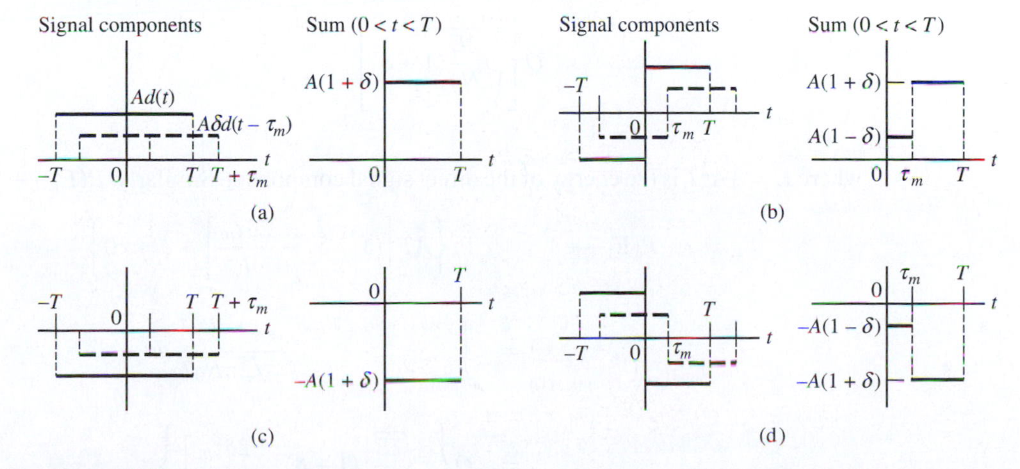

Figure 8.27
The various possible cases for ISI in multipath transmission.

is Gaussian with zero mean and variance

$$\begin{aligned}\sigma_n^2 &= E\left[4\int_0^T\int_0^T n(t)n(\sigma)\cos(\omega_c t)\cos(\omega_c\sigma)dt\, d\sigma\right] \\ &= 4\int_0^T\int_0^T \frac{N_0}{2}\delta(t-\sigma)\cos(\omega_c t)\cos(\omega_c\sigma)d\sigma\, dt \\ &= 2N_0\int_0^T \cos^2(\omega_c t)dt \\ &= N_0 T \quad (\omega_c T \text{ an integer multiple of } 2\pi)\end{aligned} \tag{8.176}$$

Because of the symmetry of the noise pdf and the symmetry of the signals shown in Figure 8.27, it follows that

$$P(E|++) = P(E|--) \text{ and } P(E|-+) = P(E|+-) \tag{8.177}$$

so that only two probabilities need be computed instead of four. From Figure 8.27, it follows that the signal component at the integrator output, given a 1, 1 was transmitted, is

$$V_{++} = AT(1+\delta) \tag{8.178}$$

and if a -1, 1 was transmitted, it is

$$\begin{aligned}V_{-+} &= AT(1+\delta) - 2A\delta\tau_m \\ &= AT\left[(1+\delta) - \frac{2\delta\tau_m}{T}\right]\end{aligned} \tag{8.179}$$

The conditional error probability $P(E\,|\,++)$ is therefore

$$\begin{aligned}P(E\,|\,++) &= \Pr[AT(1+\delta)+\mathrm{N}] < 0 = \int_{-\infty}^{-AT(1+\delta)} \frac{e^{-u^2/2N_0T}}{\sqrt{2\pi N_0 T}}du \\ &= Q\left[\sqrt{\frac{2E}{N_0}}(1+\delta)\right]\end{aligned} \tag{8.180}$$

where $E = \frac{1}{2}A^2T$ is the energy of the direct signal component. Similarly, $P(E|-+)$ is given by

$$\begin{aligned}P(E|-+) &= \Pr\left\{AT\left[(1+\delta) - \frac{2\delta\tau_m}{T}\right] + N < 0\right\} \\ &= \int_{-\infty}^{-AT(1+\delta)-2\delta\tau_m/T} \frac{e^{-u^2/2N_0T}}{\sqrt{2\pi N_0 T}}du \\ &= Q\left\{\sqrt{\frac{2E}{N_0}}\left[(1+\delta) - \frac{2\delta\tau_m}{T}\right]\right\}\end{aligned} \tag{8.181}$$

Substituting these results into (8.174) and using the symmetry properties for the other conditional probabilities, we have for the average probability of error

$$P_E = \frac{1}{2}Q\left[\sqrt{2z_0}(1+\delta)\right] + \frac{1}{2}Q\left\{\sqrt{2z_0}\left[(1+\delta) - \frac{2\delta\tau_m}{T}\right]\right\} \tag{8.182}$$

where $z_0 \triangleq E/N_0 = A^2T/2N_0$ as before.

A plot of P_E versus z_0 for various values of δ and τ_m/T, as shown in Figure 8.28, gives an indication of the effect of multipath on signal transmission. A question arises as to which curve in Figure 8.28 should be used as a basis of comparison. The one for $\delta = \tau_m/T = 0$ corresponds to the error probability for BPSK signaling in a nonfading channel. However, note that

$$z_m = \frac{E(1+\delta)^2}{N_0} = z_0(1+\delta)^2 \tag{8.183}$$

is the SNR that results if the total effective received signal energy, including that of the indirect component, is used. Indeed, from (8.182) it follows that this is the curve for $\tau_m/T = 0$ for a given value of δ. Thus, if we use this curve for P_E as a basis of comparison for P_E with τ_m/T

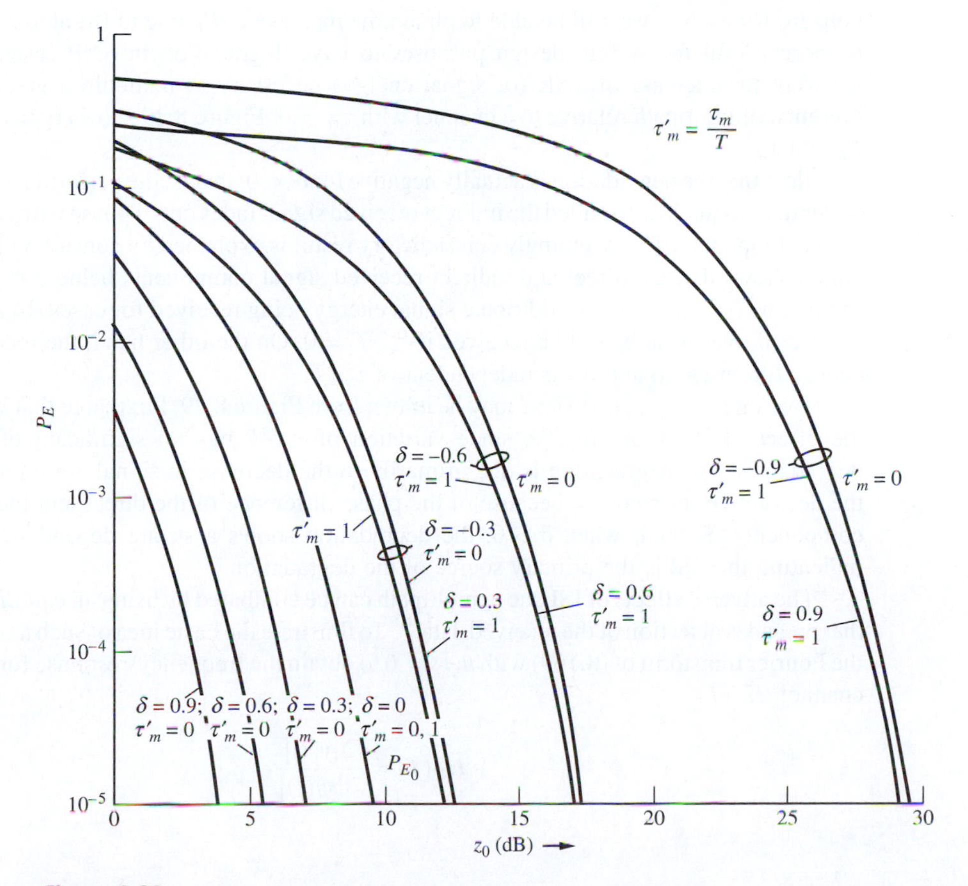

Figure 8.28
P_E versus z for various conditions of fading and ISI due to multipath.

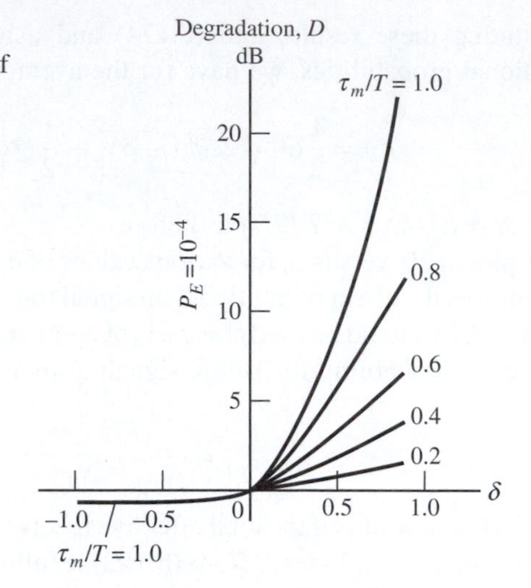

Figure 8.29
Degradation versus δ for correlation detection of BPSK in specular multipath for $P_E = 10^{-4}$.

nonzero for each δ, we will be able to obtain the increase in P_E due to ISI alone. However, it is more useful for system design purposes to have degradation in SNR instead. That is, we want the increase in SNR (or signal energy) necessary to maintain a given P_E in the presence of multipath relative to a channel with $\tau_m = 0$. Figure 8.29 shows typical results for $P_E = 10^{-4}$.

Note that the degradation is actually negative for $\delta < 0$; that is, the performance with ISI is better than for no ISI, provided the indirect received signal fades out of phase with respect to the direct component. This seemingly contradictory result is explained by consulting Figure 8.27, which shows that the direct and indirect received signal components being out of phase, as implied by $\delta < 0$, results in additional signal energy being received for cases (b) and (d) with $\tau_m/T > 0$ over what would be received if $\tau_m/T = 0$. On the other hand, the received signal energy for cases (a) and (c) is independent of τ_m/T.

Two interesting conclusions may be drawn from Figure 8.29. First, note that when $\delta < 0$, the effect of ISI is *negligible*, since variation of τ_m/T has no significant effect on the degradation. The degradation is due primarily to the decrease in signal amplitude owing to the destructive interference because of the phase difference of the direct and indirect signal components. Second, when $\delta > 0$, the degradation shows a strong dependence on τ_m/T, indicating that ISI is the primary source of the degradation.

The adverse effects of ISI due to multipath can be combated by using an *equalization filter* that precedes detection of the received data.[10] To illustrate the basic idea of such a filter, we take the Fourier transform of (8.169) with $n(t) = 0$ to obtain the frequency-response function of the channel, $H_C(f)$:

$$H_C(f) = \frac{\Im[y(t)]}{\Im[s_d(t)]} \tag{8.184}$$

[10]Equalization can be used to improve performance whenever intersymbol interference is a problem, for example, due to filtering as pointed out in Chapter 4.

If β and τ_m are known, the correlation receiver of Figure 8.26 can be preceded by a filter, referred to as an equalizer, with the frequency-response function

$$H_{\text{eq}}(t) = \frac{1}{H_C(f)} = \frac{1}{1+\beta e^{-j2\pi\tau_m f}} \tag{8.185}$$

to fully compensate for the signal distortion introduced by the multipath. Since β and τ_m will not be known exactly, or may even change with time, provision must be made for adjusting the parameters of the equalization filter. Noise, although important, is neglected for simplicity.

8.8 FLAT FADING CHANNELS

Returning to (8.169), we assume that there are several delayed multipath components with random amplitudes and phases.[11] Applying the central-limit theorem, it follows that the inphase and quadrature components of the received signal are Gaussian, the sum total of which we refer to as the diffuse component. In some cases, there may be one dominant component due to a direct line of sight from transmitter to receiver, which we refer to as the specular component. Applying the results of Section 6.5.3, it follows that the envelope of the received signal obeys a Ricean probability density function, given by

$$f_R(r) = \frac{r}{\sigma^2}\exp\left(-\frac{(r^2+A^2)}{2\sigma^2}\right)I_0\left(\frac{rA}{\sigma^2}\right), \quad r \geq 0 \tag{8.186}$$

where A is the amplitude of the specular component, σ^2 is the variance of each quadrature diffuse component, and $I_0(u)$ is the modified Bessel function of the first kind and order zero. Note that if $A = 0$, the Ricean pdf reduces to a Rayleigh pdf. We consider this special case because the general Ricean case is more difficult.

Implicit in this channel model as just discussed is that the envelope of the received signal varies slowly compared with the bit interval. This is known as a *slowly fading* channel. If the envelope (and phase) of the received signal envelope and/or phase varies nonnegligibly over the bit interval, the channel is said to be *fast fading*. This is a more difficult case to analyze than the slowly fading case and will not be considered here. A common model for the envelope of the received signal in the slowly fading case is a Rayleigh random variable, which is also the simplest case to analyze. Somewhat more general, but more difficult to analyze, is to model the envelope of the received signal as a Ricean random variable.

We illustrate a BPSK signal received from a Rayleigh slowly fading channel as follows. Let the demodulated signal be written in the simplified form

$$x(t) = Rd(t) + n_c(t) \tag{8.187}$$

where R is a Rayleigh random variable with pdf given by (8.186) with $A = 0$. If R were a constant, we know that the probability of error is given by (8.74) with $m = 0$. In other words, *given R*, we have for the probability of error

$$P_E(R) = Q(\sqrt{2Z}) \tag{8.188}$$

[11]For a prize-winning review of all aspects of fading channels, including statistical models, code design, and equalization, see the following paper: E. Biglieri, J. Proakis, and S. Shamai, Fading channels: Information-theoretic and communications aspects, *IEEE Transactions on Information Theory*, **44**: 2619–2692, October 1998.

where upper case Z is used because it is considered to be a random variable. In order to find the probability of error averaged over the envelope R, we a average (8.188) with respect to the pdf of R, which is assumed to be Rayleigh in this case. However, R is not explicitly present in (8.188) because it is buried in Z:

$$Z = \frac{R^2 T}{2N_0} \tag{8.189}$$

Now if R is Rayleigh distributed, it can be shown by transformation of random variables that R^2, and therefore Z, is exponentially distributed. Thus, the average of (8.188) is[12]

$$\overline{P}_E = \int_0^\infty Q\left(\sqrt{2z}\right) \frac{1}{\overline{Z}} e^{-z/\overline{Z}} dz \tag{8.190}$$

where $\overline{Z}$ is the average SNR. This integration can be carried out by parts with

$$u = Q\left(\sqrt{2z}\right) = \int_{\sqrt{2z}}^\infty \frac{\exp(-t^2/2)}{\sqrt{2\pi}} dt \quad \text{and} \quad dv = \frac{\exp(-z/\overline{Z})}{\overline{Z}} dz \tag{8.191}$$

Differentiation of the first expression and integration of the second expression gives

$$du = -\frac{\exp(-z)}{\sqrt{2\pi}} \frac{dz}{\sqrt{2z}} \quad \text{and} \quad v = -\exp\left(\frac{-z}{\overline{Z}}\right) \tag{8.192}$$

Putting this into the integration by parts formula, $\int u dv = uv - \int v du$, gives

$$\begin{aligned} \overline{P}_E &= -Q\left(\sqrt{2z}\right) \exp\left(\frac{-z}{\overline{Z}}\right)\Big|_0^\infty - \int_0^\infty \frac{\exp(-z)\exp(-z/\overline{Z})}{\sqrt{4\pi z}} dz \\ &= \frac{1}{2} - \frac{1}{2\sqrt{\pi}} \int_0^\infty \frac{\exp\left[-z(1+1/\overline{Z})\right]}{\sqrt{z}} dz \end{aligned} \tag{8.193}$$

In the last integral, let $w = \sqrt{z}$ and $dw = dz/2\sqrt{z}$, which gives

$$\overline{P}_E = \frac{1}{2} - \frac{1}{\sqrt{\pi}} \int_0^\infty \exp\left[-w^2\left(1 + \frac{1}{\overline{Z}}\right)\right] dw \tag{8.194}$$

We know that

$$\int_0^\infty \frac{\exp\left(-w^2/2\sigma_w^2\right)}{\sqrt{2\pi\sigma_w^2}} dw = \frac{1}{2} \tag{8.195}$$

[12]Note that there is somewhat of a disconnect here from reality—the Rayleigh model for the envelope corresponds to a uniformly distributed random phase in $(0, 2\pi)$ (a new phase and envelope random variable is assumed drawn each bit interval). Yet, a BPSK demodulator requires a coherent phase reference. One way to establish this coherent phase reference might be via a pilot signal sent along with the data-modulated signal. Experiment and simulation has shown that it is very difficult to establish a coherent phase reference directly from the Rayleigh fading signal itself, for example, by a Costas PLL.

because it is the integral over half of a Gaussian density function. Identifying $\sigma_w^2 = 1/2(1 + 1/\overline{Z})$ in (8.194) and using the integral (8.195) gives, finally, that

$$P_E = \frac{1}{2}\left(1 - \sqrt{\frac{\overline{Z}}{1+\overline{Z}}}\right), \quad \text{BPSK} \tag{8.196}$$

which is a well-known result.[13] A similar analysis for binary, coherent FSK results in the expression

$$\overline{P}_E = \frac{1}{2}\left(1 - \sqrt{\frac{\overline{Z}}{2+\overline{Z}}}\right), \quad \text{coherent FSK} \tag{8.197}$$

Other modulation techniques that can be considered in a similar fashion, and are more easily integrated than BPSK or coherent FSK are DPSK and noncoherent FSK. For these modulation schemes, the average error probability expressions are

$$\overline{P}_E = \int_0^\infty \frac{1}{2}e^{-z}\frac{1}{\overline{Z}}e^{-z/\overline{Z}}\,dz = \frac{1}{2(1+\overline{Z})}, \quad \text{DPSK} \tag{8.198}$$

and

$$\overline{P}_E = \int_0^\infty \frac{1}{2}e^{-z/2}\frac{1}{\overline{Z}}e^{-z/\overline{Z}}\,dz = \frac{1}{2+\overline{Z}}, \quad \text{noncoherent FSK} \tag{8.199}$$

respectively. The derivations are left to the problems. These results are plotted in Figure 8.30 and compared with the corresponding results for nonfading channels. Note that the penalty imposed by the fading is severe.

What can be done to combat the adverse effects of fading? We note that the degradation in performance due to fading results from the received signal envelope being much smaller on some bits than it would be for a nonfading channel, as reflected by the random envelope R. If the transmitted signal power is split between two or more subchannels that fade independently of each other, the degradation will most likely not be severe in all subchannels for a given binary digit. If the outputs of these subchannels are recombined in the proper fashion, it seems reasonable that better performance can be obtained than if a single transmission path is used. The use of such multiple transmission paths to combat fading is referred to as *diversity transmission.* There are various ways to obtain the independent transmission paths; chief ones are by transmitting over spatially different paths (space diversity), at different times (time diversity, often implemented by coding), with different carrier frequencies (frequency diversity), or with different polarizations of the propagating wave (polarization diversity).

In addition, the recombining may be accomplished in various fashions. First, it can take place either in the RF path of the receiver (predetection combining) or following the detector before making hard decisions (postdetection combining). The combining can be accomplished simply by adding the various subchannel outputs (equal-gain combining), weighting the various subchannel components proportionally to their respective SNR (maximal-ratio combining),

[13]See Proakis (2007), Chapter 14.

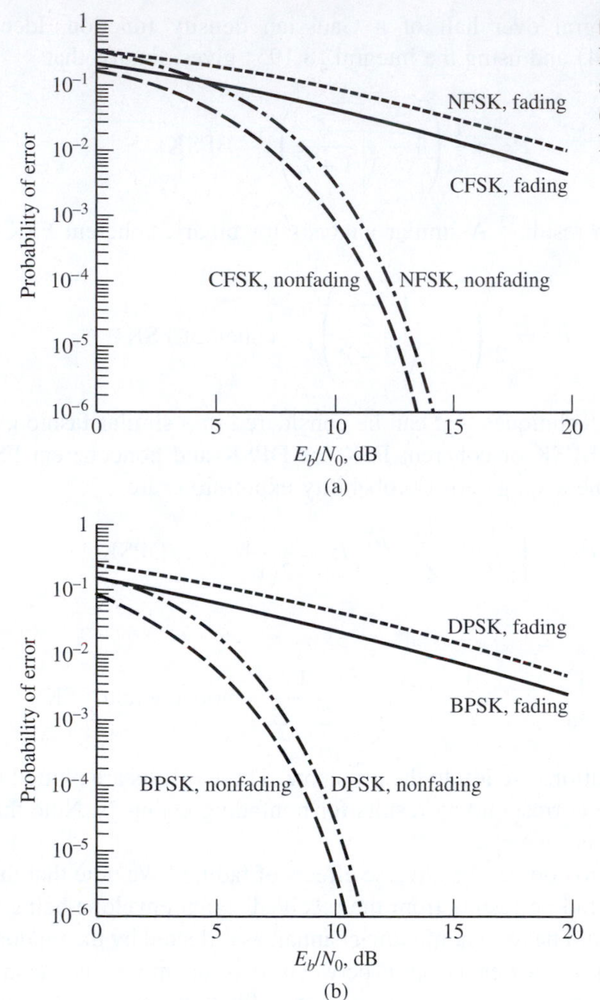

Figure 8.30
Error probabilities for various modulation schemes in flat fading Rayleigh channels. (a) Coherent and noncoherent FSK. (b) BPSK and DPSK.

or selecting the largest magnitude subchannel component and basing the decision only on it (selection combining).

In some cases, in particular, if the combining technique is nonlinear, such as in the case of selection combining, an optimum number of subpaths exist that give the maximum improvement. The number of subpaths L employed is referred to as the *order of diversity*.

That an optimum value of L exists in some cases may be reasoned as follows. Increasing L provides additional diversity and decreases the probability that most of the subchannel outputs are badly faded. On the other hand, as L increases with total signal energy held fixed, the average SNR per subchannel decreases, thereby resulting in a larger probability of error per subchannel. Clearly, therefore, a compromise between these two situations must be made. The problem of fading is again reexamined in Chapter 10 (Section 10.3), and the optimum selection of L is considered in Problem 10.27.

Finally, the reader is referred to Simon and Alouini (2005) for a generalized approach to performance analysis in fading channels.

COMPUTER EXAMPLE 8.4

A MATLAB program for computing the bit error probability of BPSK, coherent BFSK, DPSK, and noncoherent BFSK in nonfading and fading environments and providing a plot for comparison of nonfading and fading performance is given below.

```
% file: c8ce4.m
% Bit error probabilities for binary BPSK, CFSK, DPSK, NFSK in
  Rayleigh fading
% compared with same in nonfading
%
clf
mod_type = input('Enter mod. type: 1 = BPSK; 2 = DPSK; 3 = CFSK;
  4 = NFSK:');
z_dB = 0:.3:30;
z = 10.^(z_dB/10);
if mod_type == 1
    P_E_nf = qfn(sqrt(2*z));
    P_E_f = 0.5*(1-sqrt(z./(1+z)));
elseif mod_type == 2
    P_E_nf = 0.5*exp(-z);
    P_E_f = 0.5./(1+z);
elseif mod_type == 3
    P_E_nf = qfn(sqrt(z));
    P_E_f = 0.5*(1-sqrt(z./(2+z)));
elseif mod_type == 4
    P_E_nf = 0.5*exp(-z/2);
    P_E_f = 1./(2+z);
end
semilogy(z_dB,P_E_nf,'-'),axis([0 30 10^(-6) 1]),xlabel ('E_b/
  N_0, dB'),ylabel('P_E'),...
hold on
grid
semilogy(z_dB,P_E_f,'-')
if mod_type == 1
     title('BPSK')
elseif mod_type == 2
     title('DPSK')
elseif mod_type == 3
     title('Coherent BFSK')
elseif mod_type == 4
     title('Noncoherent BFSK')
end
legend ('No fading', 'Rayleigh Fading',1)
%
%            This function computes the Gaussian Q-function.
%
function Q=qfn(x)
Q = 0.5*erfc(x/sqrt(2));
```

The plot is the same as the noncoherent FSK curve of Figure 8.30. ■

■ 8.9 EQUALIZATION

As explained in Section 8.7, an equalization filter can be used to combat channel-induced distortion caused by perturbations such as multipath propagation or bandlimiting due to filters. According to (8.185), a simple approach to the idea of equalization leads to the concept of an inverse filter. As in Chapter 4, we specialize our considerations of an equalization filter to a particular form—a transversal or tapped-delay-line filter the block diagram of which is repeated in Figure 8.31.

We can take two approaches to determining the tap weights $\alpha_{-N}, \ldots, \alpha_0, \ldots \alpha_N$ in Figure 8.31 for given channel conditions. One is *zero-forcing*, and the other is *minimization of mean-square error*. We briefly review the first method, including a consideration of noise effects, and then consider the second.

8.9.1 Equalization by Zero Forcing

In Chapter 4 it was shown how the pulse response of the channel output, $p_c(t)$, could be forced to have a maximum value of 1 at the desired sampling time with N samples of 0 on either side of the maximum by properly choosing the tap weights of a $(2N+1)$-tap transversal filter. For a desired equalizer output at the sampling times of

$$\begin{aligned} p_{\text{eq}}(mT) &= \sum_{n=-N}^{N} \alpha_n p_c[(m-n)T] \\ &= \begin{cases} 1, & m=0 \\ 0, & m \neq 0, \end{cases} \quad m = 0, \pm 1, \pm 2, \ldots, \pm N \end{aligned} \tag{8.200}$$

the solution was to find the middle column of the inverse channel response matrix $[P_c]$:

$$[P_{\text{eq}}] = [P_c][A] \tag{8.201}$$

where the various matrices are defined as

$$[P_{\text{eq}}] = \begin{bmatrix} 0 \\ 0 \\ \vdots \\ 0 \\ 1 \\ 0 \\ 0 \\ \vdots \\ 0 \end{bmatrix} \begin{matrix} \left.\vphantom{\begin{matrix}0\\0\\\vdots\\0\end{matrix}}\right\} N \text{ zeros} \\ \\ \left.\vphantom{\begin{matrix}0\\0\\\vdots\\0\end{matrix}}\right\} N \text{ zeros} \end{matrix} \tag{8.202}$$

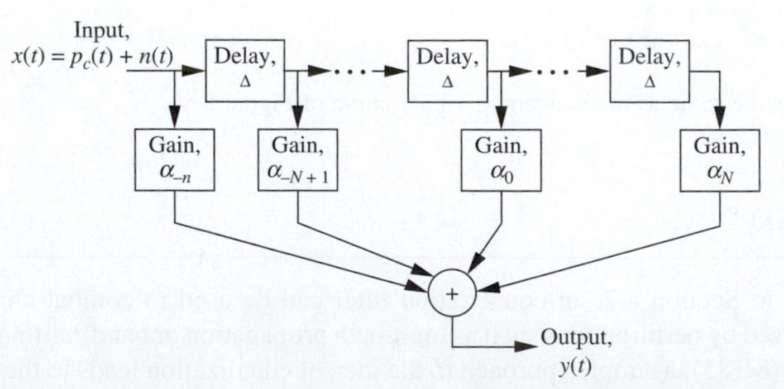

Figure 8.31
Transversal filter implementation for equalization of intersymbol interference.

$$[A] = \begin{bmatrix} \alpha_{-N} \\ \alpha_{-N+1} \\ \vdots \\ \alpha_N \end{bmatrix} \tag{8.203}$$

and

$$[P_c] = \begin{bmatrix} p_c(0) & p_c(-T) & \dots p_c(-2NT) \\ p_c(T) & p_c(0) & \dots p_c(-2N+1)T \\ \vdots & & \vdots \\ p_c(2NT) & \dots & p_c(0) \end{bmatrix} \tag{8.204}$$

That is the equalizer coefficient matrix is given by

$$[A]_{\text{opt}} = [P_c]^{-1}[P_{\text{eq}}] = \text{middle column of } [P_c]^{-1} \tag{8.205}$$

The equalizer response for delays less than $-NT$ or greater than NT are not necessarily zero. Since the zero-forcing equalization procedure only takes into account the received pulse sample values while ignoring the noise, it is not surprising that its noise performance may be poor in some channels. In fact, the noise spectrum is enhanced considerably at certain frequencies by a zero-forcing equalizer as a plot of its frequency response reveals:

$$H_{\text{eq}}(f) = \sum_{n=-N}^{N} \alpha_n \exp(-j2\pi nfT) \tag{8.206}$$

To assess the effects of noise, consider the input–output relation for the transversal filter with a signal pulse plus Gaussian noise of power spectral density $G_n(f) = (N_0/2)\Pi(f/2B)$ at its input, which can be written as

$$\begin{aligned} y(mT) &= \sum_{l=-N}^{N} \alpha_l \{p_c[(m-l)T] + n[(m-l)T]\} \\ &= \sum_{l=-N}^{N} \alpha_l p_c[(m-l)T] + \sum_{l=-N}^{N} \alpha_l n[(m-l)T] \\ &= p_{\text{eq}}(mT) + N_m,\ m = \dots,\ -2,\ -1,\ 0,\ 1,\ 2, \dots \end{aligned} \tag{8.207}$$

The random variables $\{N_m\}$ are zero-mean, Gaussian, and have variance

$$\begin{aligned} \sigma_N^2 &= E\{N_k^2\} \\ &= E\left\{\sum_{j=-N}^{N} \alpha_j n[(k-j)T] \sum_{l=-N}^{N} \alpha_l n[(k-l)T]\right\} \\ &= E\left\{\sum_{j=-N}^{N}\sum_{l=-N}^{N} \alpha_j \alpha_l n[(k-j)T]\, n[(k-l)T]\right\} \\ &= \sum_{j=-N}^{N}\sum_{l=-N}^{N} \alpha_j \alpha_l E\{n[(k-j)T]\, n[(k-l)T]\} \\ &= \sum_{j=-N}^{N}\sum_{l=-N}^{N} \alpha_j \alpha_l R_n[(j-l)T] \end{aligned} \tag{8.208}$$

where

$$R_n(\tau) = \Im^{-1}[G_n(f)] = N_0 B \,\mathrm{sinc}(2B\tau) \tag{8.209}$$

If it is assumed that $2BT = 1$ (consistent with the sampling theorem), then

$$R_n[(j-l)T] = N_0 B \,\mathrm{sinc}(j-l) = \frac{N_0}{2T}\,\mathrm{sinc}(j-l) = \begin{Bmatrix} \frac{N_0}{2T}, & j = l \\ 0 & j \neq l \end{Bmatrix} \tag{8.210}$$

and (8.208) becomes

$$\sigma_N^2 = \frac{N_0}{2T}\sum_{j=-N}^{N} \alpha_j^2 \tag{8.211}$$

For a sufficiently long equalizer, the signal component of the output, assuming binary transmission, can be taken as ± 1 equally likely. The probability of error is then

$$\begin{aligned} P_E &= \frac{1}{2}\Pr[-1 + N_m > 0] + \frac{1}{2}\Pr[1 + N_m < 0] \\ &= \Pr[N_m > 1] = \Pr[N_m < -1] \quad \text{(by symmetry of the noise pdf)} \\ &= \int_1^{\infty} \frac{\exp(-\eta^2/(2\sigma_N^2))}{\sqrt{2\pi\sigma_N^2}}\,d\eta = Q\left(\frac{1}{\sigma_N}\right) \\ &= Q\left(\frac{1}{\sqrt{\frac{N_0}{2T}\sum_j \alpha_j^2}}\right) = Q\left(\sqrt{\frac{2 \times 1^2 \times T}{N_0 \sum_j \alpha_j^2}}\right) = Q\left(\sqrt{\frac{1}{\sum_j \alpha_j^2}\frac{2E_b}{N_0}}\right) \end{aligned} \tag{8.212}$$

From (8.212) it is seen that performance is degraded in proportion to $\sum_{j=-N}^{N} \alpha_j^2$ which is a factor that directly enhances the output noise.

EXAMPLE 8.10

Consider the following pulse samples at a channel output:

$$\{p_c(n)\} = \{-0.01\ \ 0.05\ \ 0.004\ \ -0.1\ \ 0.2\ \ -0.5\ \ 1.0\ \ 0.3\ \ -0.4\ \ 0.04\ \ -0.02\ \ 0.01\ \ 0.001\}$$

Obtain the five-tap zero-forcing equalizer coefficients and plot the magnitude of its frequency response. By what factor is the SNR worsened due to noise enhancement?

Solution

The matrix $[P_c]$, from (8.204), is

$$[P_c] = \begin{bmatrix} 1 & -0.5 & 0.2 & -0.1 & 0.004 \\ 0.3 & 1 & -0.5 & 0.2 & -0.1 \\ -0.4 & 0.3 & 1 & -0.5 & 0.2 \\ 0.04 & -0.4 & 0.3 & 1 & -0.5 \\ -0.02 & 0.04 & -0.4 & 0.3 & 1 \end{bmatrix} \tag{8.213}$$

The equalizer coefficients are the middle column of $[P_c]^{-1}$, which is

$$[P_c]^{-1} = \begin{bmatrix} 0.889 & 0.435 & 0.050 & 0.016 & 0.038 \\ -0.081 & 0.843 & 0.433 & 0.035 & 0.016 \\ 0.308 & 0.067 & 0.862 & 0.433 & 0.050 \\ -0.077 & 0.261 & 0.067 & 0.843 & 0.435 \\ 0.167 & -0.077 & 0.308 & -0.081 & 0.890 \end{bmatrix} \tag{8.214}$$

Therefore, the coefficient vector is

$$[A]_{\text{opt}} = [P_c]^{-1}[P_{\text{eq}}] = \begin{bmatrix} 0.050 \\ 0.433 \\ 0.862 \\ 0.067 \\ 0.308 \end{bmatrix} \tag{8.215}$$

Plots of the input and output sequences are given in Figure 8.32(a) and (b), respectively, and a plot of the equalizer frequency response magnitude is shown in Figure 8.32(c). There is considerable enhancement of the output noise spectrum at low frequencies as is evident from the frequency response. Depending on the received pulse shape, the noise enhancement may be at higher frequencies in other cases. The noise enhancement, or degradation, factor in this example is

$$\sum_{j=-4}^{4} \alpha_j^2 = 1.0324 = 0.14\,\text{dB} \tag{8.216}$$

which is not severe in this case.

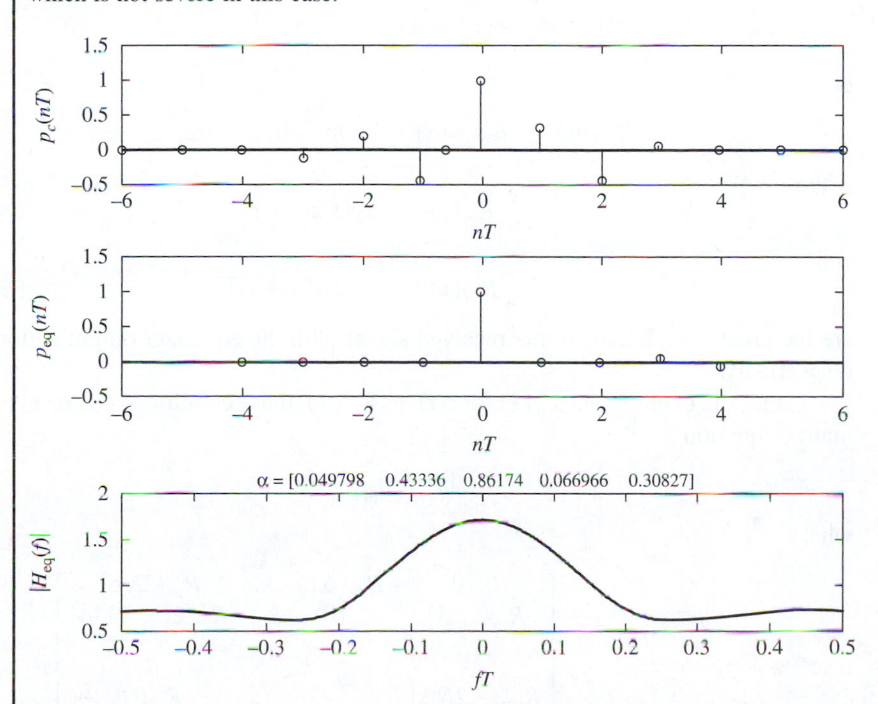

Figure 8.32
(a) Input and (b) output sample sequences for a five-tap zero-forcing equalizer. (c) Equalizer frequency response.

8.9.2 Equalization by Minimum Mean-Squared Error

Suppose that the desired output from the transversal filter equalizer of Figure 8.31 is $d(t)$. A minimum mean-squared error (MMSE) criterion then seeks the tap weights that minimize the mean-squared error between the desired output from the equalizer and its actual output. Since this output includes noise, we denote it by $z(t)$ to distinguish it from the pulse response of the equalizer. The MMSE criterion is therefore expressed as

$$\epsilon = E\left[[z(t) - d(t)]^2\right] = \text{minimum} \tag{8.217}$$

where if $y(t)$ is the equalizer input including noise, the equalizer output is

$$z(t) = \sum_{n=-N}^{N} \alpha_n y(t - n\Delta) \tag{8.218}$$

Since ϵ is a concave (bowl shaped) function of the tap weights, a set of sufficient conditions for minimizing the tap weights is

$$\frac{\partial \epsilon}{\partial \alpha_m} = 0 = 2E\left[[z(t) - d(t)]\frac{\partial z(t)}{\partial \alpha_m}\right], \quad m = 0, \pm 1, \ldots, \pm N \tag{8.219}$$

Substituting (8.218) in (8.219) and carrying out the differentiation, we obtain the conditions

$$E\left[[z(t) - d(t)]y(t - m\Delta)\right] = 0, \ \ m = 0, \pm 1, \pm 2, \ldots, \pm N \tag{8.220}$$

or

$$R_{yz}(m\Delta) = R_{yd}(m\Delta) = 0, \ \ m = 0, \pm 1, \pm 2, \ldots, \pm N \tag{8.221}$$

where

$$R_{yz}(\tau) = E\{y(t)z(t+\tau)\} \tag{8.222}$$

and

$$R_{yd}(\tau) = E\{y(t)d(t+\tau)\} \tag{8.223}$$

are the cross-correlations of the received signal with the equalizer output and with the data, respectively.

Using the expression (8.218) for $z(t)$ in (8.221), these conditions can be expressed as the matrix equation[14]

$$[R_{yy}][A]_{\text{opt}} = [R_{yd}] \tag{8.224}$$

where

$$[R_{yy}] = \begin{bmatrix} R_{yy}(0) & R_{yy}(\Delta) & \ldots & R_{yy}(2N\Delta) \\ R_{yy}(-\Delta) & R_{yy}(0) & \ldots & R_{yy}[2(N-1)\Delta] \\ \vdots & & & \vdots \\ R_{yy}(-2N\Delta) & & \ldots & R_{yy}(0) \end{bmatrix} \tag{8.225}$$

[14] These are known as the Wiener–Hopf equations. See Haykin (1996).

and

$$[R_{yd}] = \begin{bmatrix} R_{yd}(-N\Delta) \\ R_{yd}[-(N-1)\Delta] \\ \vdots \\ R_{yd}(N\Delta) \end{bmatrix} \tag{8.226}$$

and $[A]$ is defined by (8.203). Note that these conditions for the optimum tap weights using the MMSE criterion are similar to the conditions for the zero-forcing weights, except correlation-function samples are used instead of pulse-response samples.

The solution to (8.224) is

$$[A]_{\text{opt}} = [R_{yy}]^{-1}[R_{yd}] \tag{8.227}$$

which requires knowledge of the correlation matrices. The mean-squared error is

$$\begin{aligned} \epsilon &= E\left[\left[\sum_{n=-N}^{N} \alpha_n y(t-n\Delta) - d(t)\right]^2\right] \\ &= E\left\{d^2(t) - 2d(t)\sum_{n=-N}^{N} \alpha_n y(t-n\Delta) + \sum_{m=-N}^{N}\sum_{n=-N}^{N} \alpha_m\alpha_n y(t-m\Delta)y(t-n\Delta)\right\} \\ &= E\{d^2(t)\} - 2\sum_{n=-N}^{N} \alpha_n E\{d(t)y(t-n\Delta)\} + \sum_{m=-N}^{N}\sum_{n=-N}^{N} \alpha_m\alpha_n E\{y(t-m\Delta)y(t-n\Delta)\} \\ &= \sigma_d^2 - 2\sum_{n=-N}^{N} \alpha_n R_{yd}(n\Delta) + \sum_{m=-N}^{N}\sum_{n=-N}^{N} \alpha_m\alpha_n R_{yy}[(m-n)\Delta] \\ &= \sigma_d^2 - 2[A]^{\mathrm{T}}[R_{yd}] + [A]^{\mathrm{T}}[R_{yy}][A] \end{aligned} \tag{8.228}$$

where the superscript T denotes the matrix transpose and $\sigma_d^2 = E\{d^2(t)\}$. For the optimum weights (8.227), this becomes

$$\begin{aligned} \epsilon_{\min} &= \sigma_d^2 - 2\left\{[R_{yy}]^{-1}[R_{yd}]\right\}^{\mathrm{T}}[R_{yd}] + \left\{[R_{yy}]^{-1}[R_{yd}]\right\}^{\mathrm{T}}[R_{yy}]\left\{[R_{yy}]^{-1}[R_{yd}]\right\} \\ &= \sigma_d^2 - 2\left\{[R_{yd}]^{\mathrm{T}}[R_{yy}]^{-1}\right\}[R_{yd}] + [R_{yd}]^{\mathrm{T}}[R_{yy}]^{-1}[R_{yy}]\left\{[R_{yy}]^{-1}[R_{yd}]\right\} \\ &= \sigma_d^2 - 2[R_{yd}]^{\mathrm{T}}[A]_{\text{opt}} + [R_{yd}]^{\mathrm{T}}[A]_{\text{opt}} \\ &= \sigma_d^2 - [R_{yd}]^{\mathrm{T}}[A]_{\text{opt}} \end{aligned} \tag{8.229}$$

where the matrix relation $(\mathbf{AB})^{\mathrm{T}} = \boldsymbol{B}^{\mathrm{T}}\boldsymbol{A}^{\mathrm{T}}$ has been used along with the fact that the autocorrelation matrix is symmetric.

The question remains as to the choice for the time delay Δ between adjacent taps. If the channel distortion is due to multiple transmission paths (multipath) with the delay of a strong component equal to a fraction of a bit period, then it may be advantageous to set Δ equal to that expected fraction of a bit period (called a *fractionally spaced equalizer*).[15] On the other hand, if the shortest multipath delay is several bit periods then it would make sense to set $\Delta = T$.

[15]See J. R. Treichler, I. Fijalkow, and C. R. Johnson, Jr., Fractionally spaced equalizers. *IEEE Signal Processing Magazine*, 65–81, May 1996.

EXAMPLE 8.11

Consider a channel consisting of a direct path and a single indirect path plus additive Gaussian noise. Thus the channel output is

$$y(t) = Ad(t) + \beta Ad(t-\tau_m) + n(t) \tag{8.230}$$

where it is assumed that carrier demodulation has taken place so $d(t) = \pm 1$ in T-second bit periods is the data with assumed autocorrelation function $R_{dd}(\tau) = \Lambda(\tau/T)$ (i.e., a random coin-toss sequence). The strength of the multipath component relative to the direct component is β and its relative delay is τ_m. The noise $n(t)$ is assumed to be bandlimited with power spectral density $S_n(f) = (N_0/2)\Pi(f/2B)$ W/Hz so that its autocorrelation function is $R_{nn}(\tau) = N_0 B \operatorname{sinc}(2B\tau)$ where it is assumed that $2BT = 1$. Find the coefficients of a MMSE three-tap equalizer with tap spacing $\Delta = T$ assuming that $\tau_m = T$.

Solution

The autocorrelation function of $y(t)$ is

$$\begin{aligned} R_{yy}(\tau) &= E\{y(t)y(t+\tau)\} \\ &= E\{[Ad(t) + \beta Ad(t-\tau_m) + n(t)][Ad(t+\tau) + \beta Ad(t+\tau-\tau_m) + n(t+\tau)]\} \\ &= (1+\beta^2)A^2 R_{dd}(\tau) + R_{nn}(\tau) + \beta A^2[R_{dd}(\tau - T) + R_{dd}(\tau + T)] \end{aligned} \tag{8.231}$$

In a similar fashion, we find

$$\begin{aligned} R_{yd}(\tau) &= E\{y(t)d(t+\tau)\} \\ &= AR_{dd}(\tau) + \beta AR_{dd}(\tau + T) \end{aligned} \tag{8.232}$$

Using (8.225) with $N = 3$, $\Delta = T$, and $2BT = 1$ we find

$$[R_{yy}] = \begin{bmatrix} (1+\beta^2)A^2 + N_0 B & \beta A^2 & 0 \\ \beta A^2 & (1+\beta^2)A^2 + N_0 B & \beta A^2 \\ 0 & \beta A^2 & (1+\beta^2)A^2 + N_0 B \end{bmatrix} \tag{8.233}$$

and

$$[R_{yd}] = \begin{bmatrix} R_{yd}(-T) \\ R_{yd}(0) \\ R_{yd}(T) \end{bmatrix} = \begin{bmatrix} \beta A \\ A \\ 0 \end{bmatrix} \tag{8.234}$$

The condition (8.224) for the optimum weights becomes

$$\begin{bmatrix} (1+\beta^2)A^2 + N_0 B & \beta A^2 & 0 \\ \beta A^2 & (1+\beta^2)A^2 + N_0 B & \beta A^2 \\ 0 & \beta A^2 & (1+\beta^2)A^2 + N_0 B \end{bmatrix} \begin{bmatrix} \alpha_{-1} \\ \alpha_0 \\ \alpha_1 \end{bmatrix} = \begin{bmatrix} \beta A \\ A \\ 0 \end{bmatrix} \tag{8.235}$$

We may make these equations dimensionless by factoring out $N_0 B$ (recall that $2BT = 1$ by assumption) and defining new weights $c_i = A\alpha_i$, which gives

$$\begin{bmatrix} (1+\beta^2)\dfrac{2E_b}{N_0} + 1 & 2\beta\dfrac{E_b}{N_0} & 0 \\ 2\beta\dfrac{E_b}{N_0} & (1+\beta^2)\dfrac{2E_b}{N_0} + 1 & 2\beta\dfrac{E_b}{N_0} \\ 0 & 0 & (1+\beta^2)\dfrac{2E_b}{N_0} + 1 \end{bmatrix} \begin{bmatrix} c_{-1} \\ c_0 \\ c_1 \end{bmatrix} = \begin{bmatrix} 2\beta\dfrac{E_b}{N_0} \\ 2\dfrac{E_b}{N_0} \\ 0 \end{bmatrix} \tag{8.236}$$

where $E_b/N_0 \doteq A^2T/N_0$. For numerical values, we assume that $E_b/N_0 = 10$ and $\beta = 0.5$ which gives

$$\begin{bmatrix} 26 & 10 & 0 \\ 10 & 26 & 10 \\ 0 & 10 & 26 \end{bmatrix} \begin{bmatrix} c_{-1} \\ c_0 \\ c_1 \end{bmatrix} = \begin{bmatrix} 10 \\ 20 \\ 0 \end{bmatrix} \tag{8.237}$$

or, finding the inverse of the modified R_{yy} matrix using MATLAB, we get

$$\begin{bmatrix} c_{-1} \\ c_0 \\ c_1 \end{bmatrix} = \begin{bmatrix} 0.0465 & -0.0210 & 0.0081 \\ -0.0210 & 0.0546 & -0.0210 \\ 0.0081 & -0.0210 & 0.0465 \end{bmatrix} \begin{bmatrix} 10 \\ 20 \\ 0 \end{bmatrix} \tag{8.238}$$

giving finally that

$$\begin{bmatrix} c_{-1} \\ c_0 \\ c_1 \end{bmatrix} = \begin{bmatrix} 0.045 \\ 0.882 \\ -0.339 \end{bmatrix} \tag{8.239}$$

■

8.9.3 Tap Weight Adjustment

Two questions remain with regard to setting the tap weights. The first is what should be used for the desired response $d(t)$? In the case of digital signaling, one has two choices.

1. A known data sequence can be sent periodically and used for tap weight adjustment.
2. The detected data can be used if the modem performance is moderately good, since an error probability of only 10^{-2}, for example, still implies that $d(t)$ is correct for 99 out of 100 bits. Algorithms using the detected data as $d(t)$, the desired output, are called *decision directed*. Often, the equalizer tap weights will be initially adjusted using a known sequence, and after settling into nearly optimum operation, the adjustment algorithm will be switched over to a decision directed mode.

The second question is what procedure should be followed if the sample values of the pulse needed in the zero-forcing criterion or the samples of the correlation function required for the MMSE criterion are not available. Useful strategies to follow in such cases fall under the heading of *adaptive equalization*.

To see how one might implement such a procedure, we note that the mean-squared error (8.228) is a quadratic function of the tap weights with minimum value given by (8.229) for the optimum weights. Thus, the method of steepest descent may be applied. In this procedure, initial values for the weights, $[A]^{(0)}$, are chosen, and subsequent values are calculated according to[16]

$$[A]^{(k+1)} = [A]^{(k)} + \frac{1}{2}\mu\left[-\nabla\epsilon^{(k)}\right], \quad k = 0, 1, 2, \ldots \tag{8.240}$$

where the superscript k denotes the kth calculation time and $\nabla\epsilon$ is the gradient, or "slope," of the error surface. The idea is that starting with an initial guess of the weight vector, then the next closest guess is in the direction of the negative gradient. Clearly, the parameter $\mu/2$ is important in this stepwise approach to the minimum of ϵ, for one of two adverse things can happen:

1. A very small choice for μ means very slow convergence to the minimum of ϵ.
2. Too large of a choice for μ can mean overshoot of the minimum for ϵ with the result being damped oscillation about the minimum or even divergence from it.[17]

---

[16] See Haykin (1996), Section 8.2, for a full development.

[17] To guarantee convergence, the adjustment parameter μ should obey the relation $0 < \mu < 2/\lambda_{\max}$, where $\lambda_{\max}$ is the largest eigenvalue of the matrix $[Ryy]$ according to Haykin (1996).

Note that use of the steepest descent algorithm does not remove two disadvantages of the optimum weight computation: (1) It is dependent on knowing the correlation matrices $[R_{yd}]$ and $[R_{yy}]$; (2) It is computationally intensive in that matrix multiplications are still necessary (although no matrix inversions), for the gradient of ϵ can be shown to be

$$\begin{aligned}\nabla\epsilon &= \nabla\{\sigma_d^2 - 2[A]^T[R_{yd}] + [A]^T[R_{yy}][A]\} \\ &= -2[R_{yd}] + 2[R_{yy}][A]\end{aligned} \tag{8.241}$$

which must be recalculated for each new estimate of the weights. Substituting (8.241) into (8.240) gives

$$[A]^{(k+1)} = [A]^{(k)} + \mu\left[[R_{yd}] - [R_{yy}][A]^{(k)}\right], \quad k = 0, 1, 2, \ldots \tag{8.242}$$

An alternative approach, known as the *least-mean-square* (LMS) *algorithm*, that avoids both of these disadvantages, replaces the matrices $[R_{yd}]$ and $[R_{yy}]$ with instantaneous data based estimates. An initial guess for a_m is corrected from step k to step $k+1$ according to the recursive relationship

$$a_m^{(k+1)} = a_m^{(k)} - \mu y[(k-m)\Delta]\epsilon(k\Delta), \quad m = 0, \pm 1, \ldots, \pm N \tag{8.243}$$

where the error $\epsilon(k\Delta) = y(k\Delta) - d(k\Delta)$.

There are many more topics that could be covered on equalization, including decision feedback, maximum-likelihood sequence, and Kalman equalizers to name only a few.[18]

Summary

1. Binary baseband data transmission in AWGN with equally likely signals having constant amplitudes of $\pm A$ and of duration T results in an average error probability of

$$P_E = Q\left(\sqrt{\frac{2A^2T}{N_0}}\right)$$

where N_0 is the single-sided power spectral density of the noise. The hypothesized receiver was the integrate-and-dump receiver, which turns out to be the optimum receiver in terms of minimizing the probability of error.

2. An important parameter in binary data transmission is $z = E_b/N_0$, the energy per bit divided by the noise power spectral density (single sided). For binary baseband signaling, it can be expressed in the following equivalent forms:

$$z = \frac{E_b}{N_0} = \frac{A^2T}{N_0} = \frac{A^2}{N_0(1/T)} = \frac{A^2}{N_0 B_p}$$

[18]See Proakis (2007), Chapter 11.

where B_p is the "pulse" bandwidth, or roughly the bandwidth required to pass the baseband pulses. The latter expression then allows the interpretation that z is the signal power divided by the noise power in a pulse, or bit-rate, bandwidth.

3. For binary data transmission with arbitrary (finite energy) signal shapes, $s_1(t)$ and $s_2(t)$, the error probability for equally probable signals was found to be

$$P_E = Q(\sqrt{z})$$

where

$$z = \frac{1}{2N_0}\int_{-\infty}^{\infty} |S_1(f) - S_2(f)|^2 df$$
$$= \frac{1}{2N_0}\int_{-\infty}^{\infty} |s_1(t) - s_2(t)|^2 dt$$

in which $S_1(f)$ and $S_2(f)$ are the Fourier transforms of $s_1(t)$ and $s_2(t)$, respectively. This expression resulted from minimizing the average probability of error, assuming a linear-filter threshold-comparison type of receiver. The receiver involves the concept of a matched filter; such a filter is matched to a specific signal pulse and maximizes peak signal divided by rms noise ratio at its output. In a matched-filter receiver for binary signaling, two matched filters are used in parallel, each matched to one of the two signals, and their outputs are compared at the end of each signaling interval. The matched filters also can be realized as correlators.

4. The expression for the error probability of a matched-filter receiver can also be written as

$$P_E = Q\left\{[z(1 - R_{12})]^{1/2}\right\}$$

where $z = E/N_0$, with E being the *average* signal energy given by $E = \frac{1}{2}(E_1 + E_2)$. R_{12} is a parameter that is a measure of the similarity of the two signals; it is given by

$$R_{12} = \frac{2}{E_1 + E_2}\int_{-\infty}^{\infty} s_1(t)s_2(t)\, dt$$

If $R_{12} = -1$, the signaling is termed *antipodal*, while if $R_{12} = 0$, the signaling is termed *orthogonal*.

5. Examples of coherent (that is, the signal arrival time and carrier phase are known at the receiver) signaling techniques at a carrier frequency ω_c rad/s are the following:
PSK :

$$s_k(t) = A \sin\left[\omega_c t - (-1)^k \cos^{-1} m\right],\ nt_0 \leq t \leq nt_0 + T,\ k = 1, 2, \ldots$$
$$(\cos^{-1} m \text{ is called the } \textit{modulation index}),\ n = \textit{integer}$$

ASK:

$$s_1(t) = 0, \qquad nt_0 \leq t \leq nt_0 + T$$
$$s_2(t) = A\cos(\omega_c t), \quad nt_0 \leq t \leq nt_0 + T$$

FSK:

$$s_1(t) = A\cos(\omega_c t), \qquad nt_0 \leq t \leq nt_0 + T$$
$$s_2(t) = A\cos(\omega_c + \Delta\omega)t, \quad nt_0 \leq t \leq nt_0 + T$$

If $\Delta\omega = 2\pi l/T$ for FSK, where l is an integer, it is an example of an orthogonal signaling technique. If $m = 0$ for PSK, it is an example of an antipodal signaling scheme. A value of E_b/N_0 of approximately 10.53 dB is required to achieve an error probability of 10^{-6} for PSK with $m = 0$; 3 dB more than this is required to achieve the same error probability for ASK and FSK.

6. Examples of signaling schemes not requiring coherent carrier references at the receiver are DPSK and noncoherent FSK. Using ideal minimum-error-probability receivers, DPSK yields the error probability

$$P_E = \frac{1}{2}\exp\left(\frac{-E_b}{N_0}\right)$$

while noncoherent FSK gives the error probability

$$P_E = \frac{1}{2}\exp\left(\frac{-E_b}{2N_0}\right)$$

Noncoherent ASK is another possible signaling scheme with about the same error probability performance as noncoherent FSK.

7. In general, if a sequence of signals is transmitted through a bandlimited channel, adjacent signal pulses are smeared into each other by the transient response of the channel. Such interference between signals is termed *intersymbol interference* (ISI). By appropriately choosing transmitting and receiving filters, it is possible to signal through bandlimited channels while eliminating ISI. This signaling technique was examined by using Nyquist's pulse-shaping criterion and Schwarz's inequality. A useful family of pulse shapes for this type of signaling are those having raised cosine spectra.

8. One form of channel distortion is multipath interference. The effect of a simple two-ray multipath channel on binary data transmission was examined. Half of the time the received signal pulses interfere destructively, and the rest of the time they interfere constructively. The interference can be separated into ISI of the signaling pulses and cancelation due to the carriers of the direct and multipath components arriving out of phase.

9. Fading results from channel variations caused by propagation conditions. One of these conditions is multipath if the differential delay is short compared with the bit period but encompassing of many wavelengths. A commonly used model for a fading channel is one where the envelope of the received signal has a Rayleigh pdf. In this case, the signal power or energy has an exponential pdf, and the probability of error can be found by using the previously obtained error probability expressions for nonfading channels and averaging over the signal energy with respect to the assumed exponential pdf of the energy. Figure 8.30 compares the error probability for fading and nonfading cases for various modulation schemes. Fading results in severe degradation of the performance of a given modulation scheme. A way to combat fading is to use diversity.

10. Equalization can be used to remove a large part of the ISI introduced by channel filtering. Two techniques were briefly examined: zero-forcing and MMSE. Both can be realized by tapped delay-line filters. In the former technique, zero ISI is forced at sampling instants separated by multiples of a symbol period. If the tapped delay line is of length $(2N+1)$, then N zeros can be forced on either side of the desired pulse. In a MMSE equalizer, the tap weights are sought that give MMSE between the desired output from the equalizer and the actual output. The resulting weights for either case can be precalculated and preset, or adaptive circuitry can be implemented to automatically adjust the weights. The latter technique can make use of a training sequence periodically sent through the channel, or it can make use of the received data itself in order to carry out the minimizing adjustment.

Further Reading

A number of the books listed in Chapter 3 have chapters covering digital communications at roughly the same level as this chapter. For an authorative reference on digital communications, see Proakis (2007).

Problems

Section 8.1

8.1. A baseband digital transmission system that sends $\pm A$-valued rectangular pulses through a channel at a rate of 10,000 bps is to achieve an error probability of 10^{-6}. If the noise power spectral density is $N_0 = 10^{-7}$ W/Hz, what is the required value of A? What is a rough estimate of the bandwidth required?

8.2. Consider an antipodal baseband digital transmission system with a noise level of $N_0 = 10^{-5}$ W/Hz. The signal bandwidth is defined to be that required to pass the main lobe of the signal spectrum. Fill in the following table with the required signal power and bandwidth to achieve the error-probability and data rate combinations given.

Required Signal Powers A^2 and Bandwidth

R(bps)	$P_E = 10^{-3}$	$P_E = 10^{-4}$	$P_E = 10^{-5}$	$P_E = 10^{-6}$
1000				
10,000				
100,000				

8.3. Suppose $N_0 = 10^{-6}$ W/Hz and the baseband data bandwidth is given by $B = R = 1/T$ Hz. For the following bandwidths, find the required signal powers, A^2, to give a bit error probability of 10^{-5} along with the allowed data rates: (a) 5 kHz, (b) 10 kHz, (c) 100 kHz, (d) 1 MHz.

8.4. A receiver for baseband digital data has a threshold set at ϵ instead of zero. Rederive (8.8), (8.9), and (8.11) taking this into account. If $P(+A) = P(-A) = \frac{1}{2}$, find E_b/N_0 in decibels as a function of ϵ for $0 \leq \epsilon/\sigma \leq 1$ to give $P_E = 10^{-6}$, where σ^2 is the variance of N.

8.5. With $N_0 = 10^{-6}$ W/Hz and $A = 40$ mV in a baseband data transmission system, what is the maximum data rate (use a bandwidth of 0 to first null of the pulse spectrum) that will allow a P_E of 10^{-4} or less? 10^{-5}? 10^{-6}?

8.6. In a practical implementation of a baseband data transmission system, the sampler at the output of the integrate-and-dump detector requires 1 μs to sample the output. How much additional E_b/N_0, in decibels, is required to achieve a given P_E for a practical system over an ideal system for the following data rates? (a) 10 kbps, (b) 100 kbps, (c) 200 kbps.

8.7. The received signal in a digital baseband system is either $+A$ or $-A$, equally likely, for T-s contiguous intervals. However, the timing is off at the receiver so that the integration starts ΔT s late (positive) or early (negative). Assume that the timing error is less than one signaling interval. By assuming a zero threshold and considering two successive intervals [i.e., $(+A, +A)$, $(+A, -A)$, $(-A, +A)$, and $(-A, -A)$], obtain an expression for the probability of error as a function of ΔT. Show that it is

$$P_E = \frac{1}{2}Q\left(\sqrt{\frac{2E_b}{N_0}}\right) + \frac{1}{2}Q\left[\sqrt{\frac{2E_b}{N_0}}\left(1 - \frac{2|\Delta T|}{T}\right)\right]$$

Plot curves of P_E versus E_b/N_0 in decibels for $|\Delta T|/T = 0, 0.1, 0.2$ and 0.3 (four curves). Estimate the degradation in E_b/N_0 in decibels at $P_E = 10^{-4}$ imposed by timing misalignment.

8.8. Redo the derivation of Section 8.1 for the case where the possible transmitted signals are either 0 or A for T seconds. Let the threshold be set at $AT/2$. Express your result in terms of signal energy averaged over both signal possibilities, which are assumed equally probable; i.e., $E_{\text{ave}} = \frac{1}{2}(0) + \frac{1}{2}A^2T = A^2T/2$.

Section 8.2

8.9. As an approximation to the integrate-and-dump detector in Figure 8.3(a), we replace the integrator with a lowpass RC filter with frequency-response function

$$H(f) = \frac{1}{1 + j(f/f_3)}$$

where f_3 is the 3-dB cutoff frequency.

a. Find $s_{02}(T)/E\left[n_0^2(t)\right]$, where $s_{02}(T)$ is the value of the output signal at $t = T$ due to $+A$ being applied at $t = 0$ and $n_0(t)$ is the output noise. (Assume that the filter initial conditions are zero.)

b. Find the relationship between T and f_3 such that the SNR found in part (a) is maximized. (Numerical solution required.)

8.10. Assume that the probabilities of sending the signals $s_1(t)$ and $s_2(t)$ are not equal, but are given by p and $q = 1 - p$, respectively. Derive an expression for P_E that replaces (8.32) that takes this into account. Show that the error probability is minimized by choosing the threshold to be

$$k_{\text{opt}} = \frac{\sigma_0^2}{s_{0_1}(T) - s_{0_2}(T)} \ln\left(\frac{p}{q}\right) \frac{s_{0_1}(T) + s_{0_2}(T)}{2}$$

8.11. The general definition of a matched filter is a filter that maximizes peak signal-to-rms noise at some prechosen instant of time t_0.

a. Assuming white noise at the input, use Schwarz's inequality to show that the frequency-response function of the matched filter is

$$H_m(f) = S^*(f)\exp(-j2\pi f t_0)$$

where $S(f) = \Im[s(t)]$ and $s(t)$ is the signal to which the filter is matched.

b. Show that the impulse response for the matched-filter frequency-response function found in part (a) is

$$h_m(t) = s(t_0 - t)$$

c. If $s(t)$ is not zero for $t > t_0$, the matched-filter impulse response is nonzero for $t < t_0$; that is, the filter is noncausal and cannot be physically realized because it responds before the signal is applied. If we want a realizable filter, we use

$$h_{mr}(t) = \begin{cases} s(t_0 - t), & t \geq 0 \\ 0, & t < 0 \end{cases}$$

Find the realizable matched-filter impulse response corresponding to the signal

$$s(t) = A\Pi[(t - T/2)/T]$$

and t_0 equal to 0, $T/2$, T and $2T$.

d. Find the peak output signal for all cases in part (c). Plot these versus t_0. What do you conclude about the relation between t_0 and the causality condition?

8.12. Referring to Problem 8.11 for the general definition of a matched filter, find the following in relation to the two signals shown in Figure 8.33.

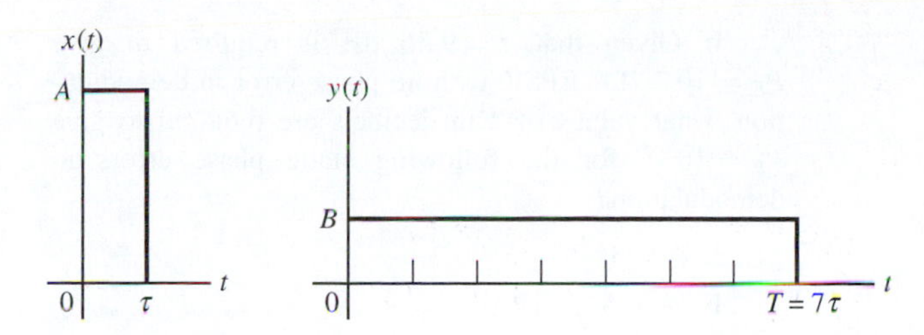

Figure 8.33

a. The causal matched-filter impulse responses. Sketch them.

b. Relate the constants A and B so that both cases give the same peak-signal-to-rms noise ratio at the matched-filter output.

c. Sketch the output of the matched filters as a function of time with signal only at the input.

d. Comment on the ability of the two matched filters for these signals to provide an accurate measurement of time delay. What do you estimate the maximum error to be in each case?

e. If *peak* transmitted power is a consideration, which waveform (and matched filter) is preferable?

8.13.

a. Find the optimum (matched-) filter impulse response $h_0(t)$, as given by (8.45) for $s_1(t)$ and $s_2(t)$, shown in Figure 8.34.

b. Find ζ^2 as given by (8.56). Plot ζ^2 versus t_0.

c. What is the best choice for t_0 such that the error probability is minimized?

d. What is the value of the threshold k as a function of t_0 to use according to (8.33)?

e. Sketch a correlator receiver structure for these signals.

8.14. Find the peak-signal-squared-to-mean-squared-noise ratio for the output of a matched filter for each of the following signals in terms of A and T. Take the noise spectral density (single sided) as N_0. Sketch each signal.

a. $s_1(t) = A\Pi[(t - T/2)/T]$.

b. $s_2(t) = (A/2)\{1 + \cos[2\pi(t - T/2)/T]\}$
$\Pi[(t - T/2)/T]$.

c. $s_3(t) = A\cos[\pi(t - T/2)/T]\Pi[(t - T/2)/T]$.

d. $s_4(t) = A\Lambda[(t - T/2)/T]$.

The signals $\Pi(t)$ and $\Lambda(t)$ are the unit-rectangular and unit-triangular functions defined in Chapter 2.

8.15. Given these signals:

$$s_A(t) = A\Pi\left(\frac{(t - T/2)}{T}\right)$$

$$s_B(t) = B\cos\left(\frac{\pi(t - T/2)}{T}\right)\Pi\left(\frac{t - T/2}{T}\right)$$

$$s_C(t) = \frac{C}{2}\left\{1 + \cos\left(\frac{2\pi(t - T/2)}{T}\right)\right\}\Pi\left(\frac{t - T/2}{T}\right)$$

Assume that they are used in a binary digital data transmission system in the following combinations. Express B and C in terms of A so that their energies are the same. Sketch each one and in each case, calculate R_{12} in (8.61) in terms of A and T. Write down an expression for P_E according to (8.60). What is the optimum threshold in each case?

a. $s_1(t) = s_A(t);\ s_2(t) = s_B(t)$.

b. $s_1(t) = s_A(t);\ s_2(t) = s_C(t)$.

c. $s_1(t) = s_B(t);\ s_2(t) = s_C(t)$.

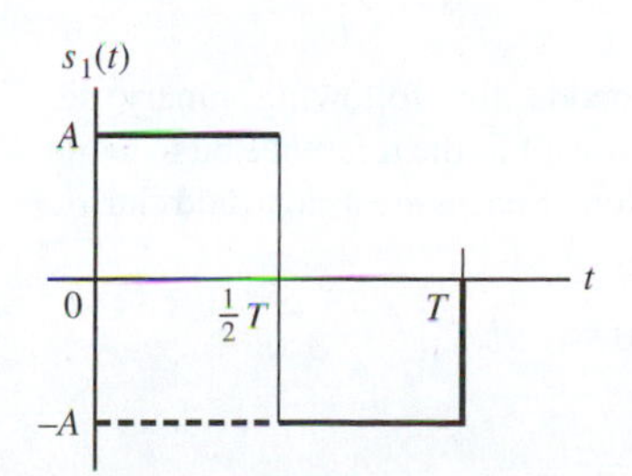

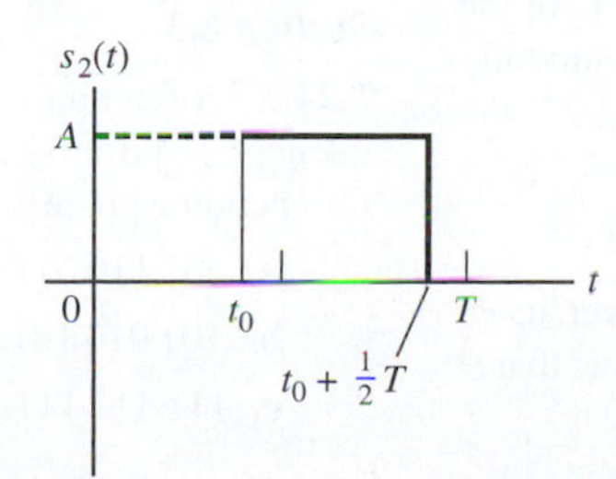

Figure 8.34

d. $s_1(t) = s_B(t); s_2(t) = -s_B(t)$.

e. $s_1(t) = s_C(t); s_2(t) = -s_C(t)$.

8.16. Given the three signals

$$s_A(t) = A\Pi\left(\frac{t-T/2}{T}\right)$$

$$s_B(t) = A\Pi\left(\frac{2(t-T/4)}{T}\right) - A\Pi\left(\frac{2(t-3T/4)}{T}\right)$$

$$s_C(t) = A\Pi\left(\frac{4(t-T/8)}{T}\right) - A\Pi\left(\frac{4(t-3T/8)}{T}\right) + A\Pi\left(\frac{4(t-5T/8)}{T}\right) - A\Pi\left(\frac{4(t-7T/8)}{T}\right)$$

a. Sketch each one and show that each has energy of A^2T.

b. Show that $R_{12} = 0$ for each of the combinations (A, B), (B, C), and (A, C). What is the optimum threshold for each of these signalling combinations?

c. What is P_E for each of the signaling combinations (A, B), (B, C), and (A, C)?

8.17. Consider PSK with $m = 1/2$.

a. By how many degrees does the modulated carrier shift in phase each time the binary data changes?

b. What percent of the total power is in the carrier, and what percent is in the modulation component?

c. What value of $z = E_b/N_0$ is required to give $P_E = 10^{-5}$?

8.18. Plot the results for P_E given in Table 8.2, page 407, versus $z = E_b/N_0$ in decibels with P_E plotted on a semilog axis. Estimate the additional E_b/N_0 at $P_E = 10^{-5}$ in decibels over the case for no phase error. Compare these results with that for constant phase error, as given by (8.81), of the same magnitude (ϕ for constant phase error equals σ_ϕ for the Gaussian phase-error case).

8.19. Find $z = E_b/N_0$ required to give $P_E = 10^{-6}$ for the following coherent digital modulation techniques: (a) binary ASK; (b) BPSK; (c) binary FSK; (d) BPSK with no carrier component but with a phase error of 5° in the demodulator; (e) PSK with no phase error in demodulation, but with $m = 1/\sqrt{2}$; (f) PSK with $m = 1/\sqrt{2}$ and with a phase error of 5° in the demodulator.

8.20.

a. Make a plot of degradation in decibels versus ϕ, the phase error in demodulation, for BPSK. Assume that ϕ is constant.

b. Given that $z = 9.56$ dB is required to give $P_E = 10^{-5}$ for BPSK with no phase error in demodulation, what values of z in decibels are required to give $P_E = 10^{-5}$ for the following static phase errors in demodulation?

i. $\phi = 3°$

ii. $\phi = 5°$

iii. $\phi = 10°$

iv. $\phi = 15°$

8.21. Plot the required SNR $z = E_b/N_0$, in decibels, to give(a) $P_E = 10^{-4}$, (b) $P_E = 10^{-5}$, and (c) $P_E = 10^{-6}$ versus m for PSK with a carrier component for $0 \le m \le 1$.

8.22.

a. Consider the transmission of digital data at a rate of $R = 50$ kbps and at an error probability of $P_E = 10^{-5}$. Using the bandwidth of the main lobe as a bandwidth measure, give an estimate of the required transmission bandwidth and E_b/N_0 in decibels required for the following coherent modulation schemes:

i. binary ASK

ii. BPSK

iii. binary coherent FSK (take the minimum spacing possible between the signal representing the logic 1 and that representing the logic 0).

b. Consider the same question as in part (a) but with $R = 500$ kbps and $P_E = 10^{-4}$.

8.23. Derive an expression for P_E for binary coherent FSK if the frequency separation of the two transmitted signals is chosen to give a minimum correlation coefficient between the two signals. That is, evaluate

$$\sqrt{E_1E_2}\rho_{12} = \int_0^T A^2 \cos(\omega_c t)\cos(\omega_c + \Delta\omega)t\,dt$$

as a function of $\Delta\omega$ and find the minimum value for R_{12}. How much improvement in E_b/N_0 in decibels over the orthogonal-signal case is obtained?
(*Hint:* Assume the sum-frequency term integrates to 0.)

Section 8.3

8.24. Differentially encode the following binary sequences. Arbitrarily choose a 1 as the reference bit to begin the encoding process. (Note: Spaces are used to add clarity.)

a. 101 110 011 100

b. 101 010 101 010

c. 111 111 111 111

d. 000 000 000 000

e. 111 111 000 000

f. 110 111 101 001

g. 111 000 111 000

h. 101 110 011 100

8.25.

a. Consider the sequence to 011 101 010 111. Differentially encode it, and assume that the differentially encoded sequence is used to biphase modulate a sinusoidal carrier of arbitrary phase. Prove that the demodulator of Figure 8.17 properly gives back the original sequence.

b. Now invert the sequence (i.e., 1s become 0s and vice versa). What does the demodulator of Figure 8.17 give now?

8.26.

a. In the analysis of the optimum detector for DPSK of Section 8.3.1, show that the random variables n_1, n_2, n_3, and n_4 have zero means and variances $N_0T/4$.

b. Show that w_1, w_2, w_3, and w_4 have zero means and variances $N_0T/8$.

8.27. Compare (8.110) and (8.113) to show that for large z, nonoptimum detection and optimum detection of DPSK differ by approximately 1.5 dB.

8.28.

a. Compute z in decibels required to give $P_E = 10^{-6}$ for noncoherent, binary FSK and for DPSK. For the latter, carry out the computation for both the optimum and suboptimum detectors.

b. Repeat part (a) for $P_E = 10^{-5}$.

c. Repeat part (a) for .$P_E = 10^{-4}$.

8.29. A channel of bandwidth 50 kHz is available. Using null-to-null RF bandwidths, what data rates may be supported by (a) BPSK, (b) coherent FSK (tone spacing = 1/2*T*), (c) DPSK, and (d) noncoherent FSK (tone spacing = 2/*T*).

8.30. Find the probability of error for noncoherent ASK, with signal set

$$s_i(t) = \begin{cases} 0, & 0 \le t \le T, \quad i = 1 \\ A\cos(2\pi f_c t + \theta), & 0 \le t \le T, \quad i = 2 \end{cases}$$

where θ is a uniformly distributed random variable in $[0, 2\pi)$. White Gaussian noise of two-sided power spectral density $N_0/2$ is added to this signal in the channel. The receiver is a bandpass filter of bandwidth $2/T$ Hz centered on f_c, followed by an envelope detector that is input to a sampler and threshold comparator. Assume that the signal, when present, is passed by the filter without distortion, and let the noise variance at the filter output be $\sigma_N^2 = N_0 B_T = 2N_0/T$.

Show that the envelope detector output with signal 1 present (i.e., zero signal) is Rayleigh distributed, and that the envelope detector output with signal 2 present is Ricean distributed. Assuming that the threshold is set at $A/2$, find an expression for the probability of error. You will not be able to integrate this expression. However, by making use of the approximation

$$I_0(\nu) \approx \frac{e^{\nu}}{\sqrt{2\pi\nu}}, \quad \nu \gg 1$$

you will be able to approximate the pdf of the sampler output for large SNR as Gaussian and express the probability of error in terms of a Q-function.
(*Hint:* Neglect the $\nu^{-1/2}$ in the above approximation.)

Show that the probability of error for SNR large is approximately

$$P_E = \frac{1}{2}P(E|S+N) + \frac{1}{2}P(E|N) \approx \frac{e^{-z}}{\sqrt{4\pi z}} + \frac{1}{2}e^{-z/2},$$

$$z = \frac{A^2}{4\sigma_N^2} \gg 1$$

Note that $z = A^2/4\sigma_N^2$ is the average signal-power (the signal is 0 half the time) -to-noise variance ratio. Plot the error probability versus the SNR and compare with that for DPSK and noncoherent FSK.

8.31. Integrate (8.119) by recasting the integrand into the form of a Ricean pdf, and therefore use the fact that it integrates to 1 [you will have to redefine some parameters and multiply and divide by $\exp(A^2/2N)$ similarly to the steps that led to (8.112)]. The result should be (8.120).

Section 8.4

8.32. Gray encoding of decimal numbers ensures that only one bit changes when the decimal number changes by one unit. Let $b_1 b_2 b_3 \cdots b_N$ represent an ordinary binary representation of a decimal number, with b_1 being the most significant bit. Let the corresponding Gray code bits be $g_1 g_2 g_3 \ldots g_N$. Then the Gray code representation is obtained by the algorithm

$$g_1 = b_1$$

$$g_n = b_n \oplus b_{n-1}$$

where $\oplus$ denotes modulo 2 addition (i.e., $0 \oplus 0 = 0$, $0 \oplus 1 = 1$, $1 \oplus 0 = 1$, and $1 \oplus 1 = 0$). Find the Gray code representation for the decimal numbers 0 through 31.

8.33. Show that (8.137) is the average energy in terms of Δ for M-ary antipodal PAM.

8.34. Consider a baseband antipodal PAM system with channel bandwidth of 10 kHz and a desired data rate of 20 kbps. (a) What is the required value for M? (b) What value of E_b/N_0 in decibels will give a bit-error probability of 10^{-6}? 10^{-5}? Find M as the nearest power of 2.

8.35. Reconsider Problem 8.34 but for a desired data rate of 25 kbps.

Section 8.5

8.36. Recompute the entries in Table 8.5 for a bit-error probability of 10^{-5} and an RF bandwidth of 200 kHz.

Section 8.6

8.37. Assume a raised cosine pulse with $\beta = 0.2$ [see (4.37)], additive noise with power spectral density

$$G_n(f) = \frac{\sigma_n^2/f_3}{1+(f/f_3)^2}$$

and a channel filter with transfer-function-squared magnitude given by

$$|H_C(f)|^2 = \frac{1}{1+(f/f_C)^2}$$

Find and plot the optimum transmitter and receiver filter amplitude responses for binary signaling for the following cases:

a. $f_3 = f_C = 1/2T$.

b. $f_C = 2f_3 = 1/T$.

c. $f_3 = 2f_C = 1/T$.

8.38.

a. Sketch the trapezoidal spectrum $P(f) = [b/(b-a)]\,\Lambda(f/b) - [a/(b-a)]\,\Lambda(f/a)$, $b > a > 0$, for $a = 1$ and $b = 2$.

b. By appropriate sketches, show that it satisfies Nyquist's pulse-shaping criterion.

8.39. Data are to be transmitted through a bandlimited channel at a rate $R = 1/T = 9600$ bps. The channel filter has frequency-response function

$$H_C(f) = \frac{1}{1+j(f/4800)}$$

The noise is white with power spectral density

$$\frac{N_0}{2} = 10^{-8} \text{ W/Hz}$$

Assume that a received pulse with raised cosine spectrum given by (4.37) with $\beta = 1$ is desired.

a. Find the magnitudes of the transmitter and receiver filter transfer functions that give zero ISI and optimum detection.

b. Using a table or the asymptotic approximation for the Q-function, find the value of A/σ required to give $P_{E,\min} = 10^{-4}$.

c. Find E_T to give this value of A/σ for the N_0, $G_n(f)$, $P(f)$, and $H_C(f)$ given above. (Numerical integration required.)

Section 8.7

8.40. Plot P_E from (8.182) versus z_0 for $\delta = 0.5$ and $\tau_m/T = 0.2$, 0.6, and 1.0. Develop a MATLAB program to plot the curves.

8.41. Redraw Figure 8.29 for $P_E = 10^{-5}$. Write a MATLAB program using the find function to obtain the degradation for various values of δ and τ_m/T.

Section 8.8

8.42. Fading margin can be defined as the incremental E_b/N_0, in decibels, required to provide a certain desired error probability in a fading channel as could be achieved with the same modulation technique in a nonfading channel. Assume that a bit-error probability of 10^{-3} is specified. Find the fading margin required for the following cases: (a) BPSK, (b) DPSK, (c) coherent FSK, and (d) noncoherent FSK.

8.43. Show the details in making the substitution $\sigma_w^2 = 1/2(1+1/\bar{Z})$ in (8.194) so that it gives (8.196) after integration.

8.44. Carry out the integrations leading to (8.197) [use (8.196) as a pattern], (8.198), and (8.199) given that the SNR pdf is given by $f_Z(z) = (1/\bar{Z})\exp(-z/\bar{Z})$, $z \geq 0$.

Section 8.9

8.45. Given the following channel pulse-response samples:

$$p_c(-3T) = 0.001 \qquad p_c(-2T) = -0.01$$
$$p_c(-T) = 0.1 \qquad p_c(0) = 1.0$$
$$p_c(T) = 0.2 \qquad p_c(2T) = -0.02$$
$$p_c(3T) = 0.005$$

a. Find the tap coefficients for a three-tap zero-forcing equalizer.

b. Find the output sample values for $mT = -2T$, T, 0, T, and $2T$.

c. Find the degradation in decibels due to noise enhancement.

8.46.

a. Consider the design of an MMSE equalizer for a multipath channel whose output is of the form

$$y(t) = Ad(t) + bAd(t - T_m) + n(t)$$

where the second term is a multipath component and the third term is noise independent of the data, $d(t)$. Assume $d(t)$ is a random (coin-toss) binary sequence with autocorrelation function $R_{dd}(\tau) = \Lambda(\tau/T)$. Let the noise have a lowpass-RC-filtered spectrum with 3-dB cutoff frequency $f_3 = 1/T$ so that the noise power spectral density is

$$S_{nn}(f) = \frac{N_0/2}{1 + (f/f_3)^2}$$

where $N_0/2$ is the two-sided power spectral density at the lowpass filter input. Let the tap spacing be $\Delta = T_m = T$. Express the matrix $[R_{yy}]$ in terms of the SNR $E_b/N_0 = A^2T/N_0$.

b. Obtain the optimum tap weights for a three-tap MMSE equalizer and at a SNR of 10 dB and $b = 0.5$.

c. Find an expression for the MMSE.

8.47. For the numerical auto- and cross-correlation matrices of Example 8.11, find explicit expressions (write out an equation for each weight) for the steepest-descent tap weight adjustment algorithm (8.242). Let $\mu = 0.01$. Justify this as an appropriate value using the criterion $0 < \mu < 2/\max(\lambda_i)$, where the λ_i are the eigenvalues of $[R_{yy}]$.

8.48. Using the criterion $0 < \mu < 2/\max(\lambda_i)$, where the λ_i are the eigenvalues of $[R_{yy}]$, find a suitable range of values for μ for Example 8.11.

8.49. Consider (8.237) with all elements of both $[R_{yy}]$ and $[R_{yd}]$ divided by 10. (a) Do the weights remain the same? (b) What is an acceptable range for μ for an adaptive MMSE weight adjustment algorithm (steepest descent) using the criterion $0 < \mu < 2/\max(\lambda_i)$, where the λ_i are the eigenvalues of $[R_{yy}]$.

8.50. Rework Example 8.11 for $E_b/N_0 = 20$ and $\beta = 0.1$. That is, recompute the matrices $[R_{yy}]$ and $[R_{yd}]$, and find the equalizer coefficients. Comment on the differences from Example 8.11.

Computer Exercises

8.1. Develop a computer simulation of an integrate-and-dump detector for antipodal baseband signaling based on (8.1). Generate AT or $-AT$ randomly by drawing a uniform random number in [0, 1] and comparing it with 1/2. Add to this a Gaussian random variable of zero mean and variance given by (8.5). Compare with a threshold of 0, and increment a counter if an error occurs. Repeat this many times, and estimate the error probability as the ratio of the number of errors to the total number of bits simulated. If you want to estimate a bit-error probability of 10^{-3}, for example, you will have to simulate at least $10 \times 1000 = 10000$ bits. Repeat for several SNR so that you can rough out a bit-error probablity curve versus E_b/N_0. Compare with theory given in Figure 8.5.

8.2. Write a computer program to evaluate the degradation imposed by bit timing error at a desired error probability as discussed in Problem 8.7.

8.3. Write a computer program to evaluate the degradation imposed by Gaussian phase jitter at a desired error probability as discussed in connection with the data presented in Table 8.2. This will require numerical integration.

8.4. Write a computer program to evaluate various digital modulation techniques:

a. For a specified data rate and error probability, find the required bandwidth and E_b/N_0 in decibels. Corresponding to the data rate and required E_b/N_0, find the required received signal power for $N_0 = 1$ W/Hz.

b. For a specified bandwidth and error probability find the allowed data rate and required E_b/N_0 in decibels. Corresponding to the data rate and required E_b/N_0, find the required received signal power for $N_0 = 1$ W/Hz.

8.5. Write a computer program to verify Figures 8.28 and 8.29.

8.6. Write a computer program to evaluate degradation due to flat Rayleigh fading at a specified error probability. Include PSK, FSK, DPSK, and noncoherent FSK.

8.7. Write a computer program to design equalizers for specified channel conditions for (a) the zero-forcing criterion and (b) the MMSE criterion.

CHAPTER 9

ADVANCED DATA COMMUNICATIONS TOPICS

In this chapter we consider some topics on data transmission that are more advanced than the fundamental ones considered in Chapter 8. The first topic considered is that of *M*-ary digital modulation systems, where $M > 2$. We will develop methods for comparing them on the basis of bit-error probability (power efficiency). We next examine bandwidth requirements for data transmission systems so that they may be compared on the basis of bandwidth efficiency. An important consideration in any communications system is synchronization including carrier, symbol, and word, which is considered next. Following this, modulation techniques that utilize bandwidths much larger than required for data modulation itself, called *spread spectrum*, are briefly considered. After spread spectrum modulation, an old concept called *multicarrier modulation* is reviewed (a special case of which is known as *orthogonal frequency division multiplexing*) and its application to delay spread channels is discussed. Application areas include wireless networks, digital subscriber lines, digital audio broadcasting, and digital video broadcasting. The next topic dealt with is satellite communications links. Finally, the basics of cellular wireless communications systems are briefly covered. The latter two topics provide specific examples of the application of some of the digital communications principles considered in Chapters 8 and 9.

■ 9.1 *M*-ARY DATA COMMUNICATIONS SYSTEMS

With the binary digital communications systems we have considered so far (with the exception of *M*-ary PAM in Chapter 8), one of only two possible signals can be transmitted during each signaling interval. In an *M*-ary system, one of M possible signals may be transmitted during each T_s-s signaling interval, where $M \geq 2$ (we now place a subscript s on the signaling interval T to denote "symbol"; we will place the subscript b on T to denote "bit" when $M = 2$). Thus binary data transmission is a special case of *M*-ary data transmission. We refer to each possible transmitted signal of an *M*-ary message sequence as a *symbol*.

9.1.1 *M*-ary Schemes Based on Quadrature Multiplexing

In Section 3.6 we demonstrated that two different messages can be sent through the same channel by means of quadrature multiplexing. In a quadrature-multiplexed system, the

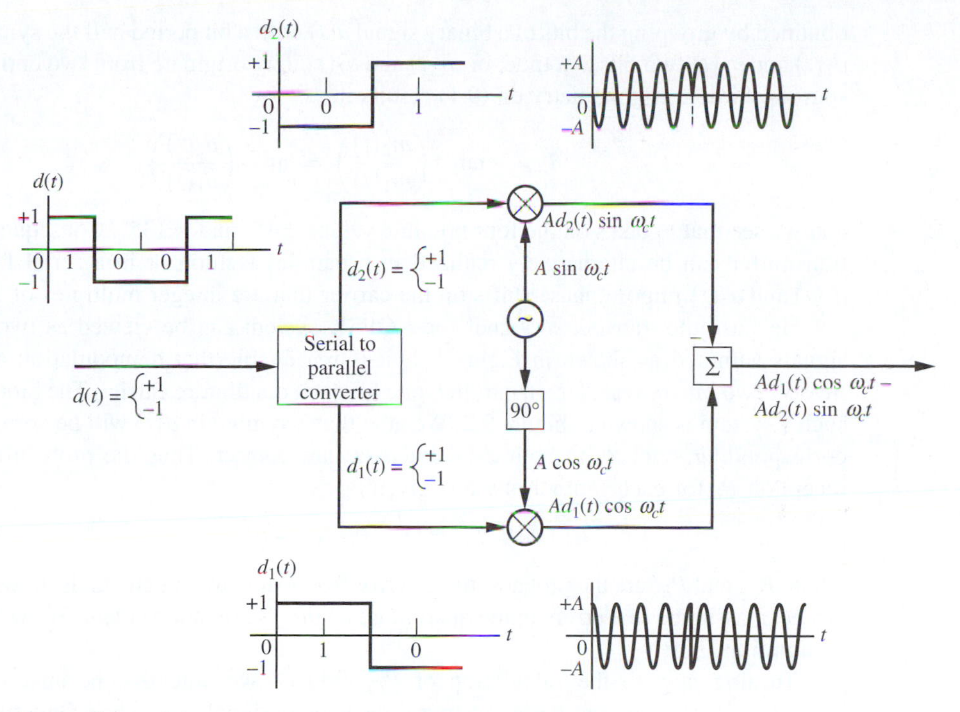

Figure 9.1
Modulator and typical waveforms for QPSK.

messages $m_1(t)$ and $m_2(t)$ are used to double-sideband modulate two carrier signals of frequency f_c Hz, which are in phase quadrature, to produce the modulated signal

$$\begin{aligned} x_c(t) &= A[m_1(t)\cos(2\pi f_c t) + m_2(t)\sin(2\pi f_c t)] \\ &\triangleq R(t)\cos[2\pi f_c t + \theta_i(t)] \end{aligned} \tag{9.1}$$

Demodulation at the receiver is accomplished by coherent demodulation with two reference sinusoids in phase quadrature that are ideally phase and frequency coherent with the quadrature carriers. This same principle can be applied to transmission of digital data and results in several modulation schemes, three of which will be described here: (1) quadriphase-shift keying (QPSK), (2) offset quadriphase-shift keying (OQPSK), and (3) minimum-shift keying (MSK).

In the analysis of these systems, we make use of the fact that coherent demodulation ideally results in the two messages $m_1(t)$ and $m_2(t)$ being separate at the outputs of the quadrature mixers. Thus these quadrature-multiplexed schemes can be viewed as two separate digital modulation schemes operating in parallel.

The block diagram of a parallel realization for a QPSK transmitter is shown in Figure 9.1, along with typical signal waveforms. In the case of QPSK, we set $m_1(t) = d_1(t)$ and $m_2(t) = -d_2(t)$, where d_1 and d_2 are ± 1-valued waveforms that have possible transitions each T_s s. Symbol transition instants are usually aligned for $d_1(t)$ and $d_2(t)$.[1] Note that we may think of $d_1(t)$ and $d_2(t)$, the symbol streams that modulate the quadrature carriers, as being

[1]The two data streams could be due to separate sources, not necessarily of the same data rate. At this point in the discussion, we assume that they are at the same rate.

obtained by grouping the bits of a binary signal $d(t)$ with a bit period half the symbol period of $d_1(t)$ and $d_2(t)$ two bits at a time, or $d_1(t)$ and $d_2(t)$ may originate from two entirely different sources. Simple trigonometry on (9.1) results in

$$\theta_i = -\tan^{-1}\left(\frac{m_2(t)}{m_1(t)}\right) = \tan^{-1}\left(\frac{d_2(t)}{d_1(t)}\right) \tag{9.2}$$

and we see that θ_i takes on the four possible values $\pm 45^\circ$ and $\pm 135^\circ$. Consequently, a QPSK transmitter can be alternatively realized in a parallel fashion or in a serial fashion where $d_1(t)$ and $d_2(t)$ impose phase shifts on the carrier that are integer multiples of 90°.

Because the transmitted signal for a QPSK system can be viewed as two binary PSK signals summed as shown in Figure 9.1, it is reasonable that demodulation and detection involve two binary receivers in parallel, one for each quadrature carrier. The block diagram of such a system is shown in Figure 9.2. We note that a symbol in $d(t)$ will be correct only if the corresponding symbols in both $d_1(t)$ and $d_2(t)$ are correct. Thus the probability of correct reception P_c for each symbol phase is given by

$$P_c = (1 - P_{E_1})(1 - P_{E_2}) \tag{9.3}$$

where P_{E_1} and P_{E_2} are the probabilities of error for the quadrature channels. In writing (9.3), it has been assumed that errors in the quadrature channels are independent. We will discuss this assumption shortly.

Turning now to the calculation of P_{E_1} and P_{E_2}, we note that because of symmetry, $P_{E_1} = P_{E_2}$. Assuming that the input to the receiver is signal plus white Gaussian noise with double-sided power spectral density $N_0/2$, that is,

$$\begin{aligned} y(t) &= x_c(t) + n(t) \\ &= Ad_1(t)\cos(2\pi f_c t) - Ad_2(t)\sin(2\pi f_c t) + n(t) \end{aligned} \tag{9.4}$$

we find that the output of the upper correlator in Figure 9.2 at the end of a signaling interval T_s is

$$V_1 = \int_0^{T_s} y(t)\cos(2\pi f_c t)\,dt = \pm\frac{1}{2}AT_s + N_1 \tag{9.5}$$

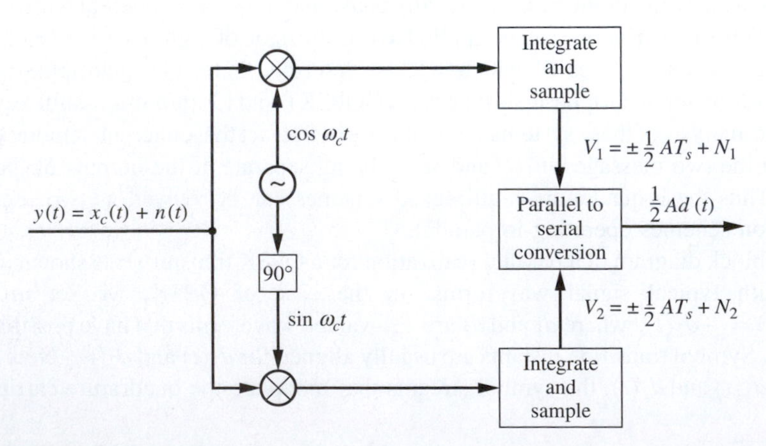

Figure 9.2
Demodulator for QPSK.

where

$$N_1 = \int_0^{T_s} n(t)\cos(2\pi f_c t)\,dt \tag{9.6}$$

Similarly, the output of the lower correlator at $t = T_s$ is

$$V_2 = \int_0^{T_s} y(t)\sin(2\pi f_c t)\,dt = \pm\frac{1}{2}AT_s + N_2 \tag{9.7}$$

where

$$N_2 = \int_0^{T_s} n(t)\sin(2\pi f_c t)\,dt \tag{9.8}$$

Errors at either correlator output will be independent if V_1 and V_2 are independent, which requires that N_1 and N_2 be independent. We can show that N_1 and N_2 are uncorrelated (Problem 9.4), and since they are Gaussian (why?), they are independent.

Returning to the calculation of P_{E_1}, we note that the problem is similar to the antipodal baseband case. The mean of N_1 is zero, and its variance is (the by now usual assumption is made that $f_c T_s$ is an integer)

$$\begin{aligned}
\sigma_1^2 = E\{N_1^2\} &= E\left[\left(\int_0^{T_s} n(t)\cos(2\pi f_c t)\,dt\right)^2\right] \\
&= \int_0^{T_s}\int_0^{T_s} E[n(t)n(\alpha)]\cos(2\pi f_c t)\cos(2\pi f_c \alpha)\,dt \\
&= \int_0^{T_s}\int_0^{T_s} \frac{N_0}{2}\delta\,(t-\alpha)\cos(2\pi f_c t)\cos(2\pi f_c \alpha)\,d\alpha\,dt \\
&= \frac{N_0}{2}\int_0^{T_s} \cos^2(2\pi f_c t)\,dt \\
&= \frac{N_0 T_s}{4}
\end{aligned} \tag{9.9}$$

Thus, following a series of steps similar to the case of binary antipodal signaling, we find that

$$\begin{aligned}
P_{E_1} &= \Pr[d_1 = +1]\Pr[E_1|d_1 = +1] + \Pr[d_1 = -1]\Pr[E_1|d_1 = -1] \\
&= \Pr[E_1|d_1 = +1] = \Pr[E_1|d_1 = -1]
\end{aligned} \tag{9.10}$$

where the latter equation follows by noting the symmetry of the pdf of V_1. But

$$\begin{aligned}
\Pr[E|d_1 = +1] &= \Pr\left[\frac{1}{2}AT_s + N_1 < 0\right] = \Pr\left[N_1 < -\frac{1}{2}AT_s\right] \\
&= \int_{-\infty}^{-AT_s/2} \frac{e^{-n_1^2/2\sigma_1^2}}{\sqrt{2\pi\sigma_1^2}}\,dn_1 = Q\left(\sqrt{\frac{A^2 T_s}{N_0}}\right)
\end{aligned} \tag{9.11}$$

Thus the probability of error for the upper channel in Figure 9.2 is

$$P_{E_1} = Q\left(\sqrt{\frac{A^2 T_s}{N_0}}\right) \tag{9.12}$$

with the same result for P_{E_2}. Noting that $\frac{1}{2}A^2T_s$ is the average energy for one quadrature channel, we see that (9.12) is identical to binary PSK. Thus, considered on a per channel basis, QPSK performs identically to binary PSK.

However, if we consider the probability of error for a single phase of a QPSK system, we obtain, from (9.3), the result

$$\begin{aligned} P_E &= 1 - P_c = 1 - (1 - P_{E_1})^2 \\ &\cong 2P_{E_1},\ P_{E_1} \ll 1 \end{aligned} \tag{9.13}$$

which is

$$P_E = 2Q\left(\sqrt{\frac{A^2 T_s}{N_0}}\right) \tag{9.14}$$

Noting that the energy per symbol is $A^2 T_s \triangleq E_s$ for the quadriphase signal, we may write (9.14) as

$$P_E \cong 2Q\left(\sqrt{\frac{E_s}{N_0}}\right) \tag{9.15}$$

A direct comparison of QPSK and BPSK on the basis of average symbol-energy-to-noise-spectral-density ratio shows that QPSK *is approximately 3 dB worse than binary PSK*. However, this is not a fair comparison since twice as many bits per signaling interval are being transmitted with the QPSK system as compared to the BPSK system, assuming T_s is the same. A comparison of QPSK and binary PSK on the basis of the systems transmitting equal numbers of bits per second (two bits per QPSK phase), shows that their performances are the same, as will be shown later. Binary PSK and QPSK are compared in Figure 9.3 on the basis of probability of error versus SNR $z = E_s/N_0$, where E_s is the average energy per symbol. Note that the curve for QPSK approaches $\frac{3}{4}$ as the SNR approaches zero ($-\infty$ dB). This is reasonable because the receiver will, on average, make only one correct decision for every four signaling intervals (one of four possible phases) if the input is noise alone.

9.1.2 OQPSK Systems

Because the quadrature data streams $d_1(t)$ and $d_2(t)$ can switch signs simultaneously in a QPSK system, it follows that the data-bearing phase θ_i of the modulated signal can occasionally change by 180°. This can have an undesirable effect in terms of envelope deviation if the modulated signal is filtered, which is invariably the case in a practical system. To avoid the possibility of 180° phase switching, the switching instants of the quadrature-channel data signals $d_1(t)$ and $d_2(t)$ of a quadriphase system can be offset by $T_s/2$ relative to each other, where T_s is the signaling interval in either channel. The resulting modulation scheme is referred to as *offset QPSK*, which is abbreviated OQPSK; it is also sometimes called *staggered QPSK*. With the offsetting or staggering of quadrature data streams by $T_s/2$, the maximum phase change due to data modulation of the transmitted carrier is 90°. Theoretically, the error probability performance of OQPSK and QPSK are identical. One limitation of an OQPSK system is that the data streams $d_1(t)$ and $d_2(t)$ *must have the same symbol durations*, whereas for QPSK they need not.

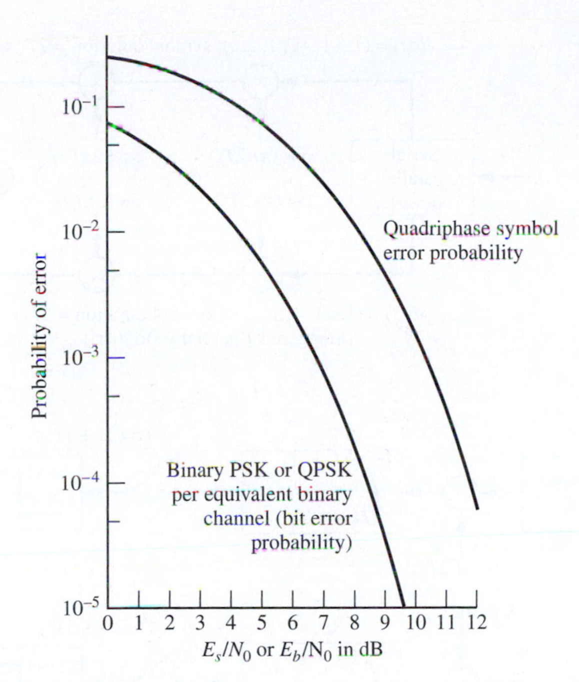

Figure 9.3
Symbol error probability for QPSK compared with that for BPSK.

9.1.3 MSK Systems

Type I and Type II MSK

In (9.1), suppose that message $m_1(t)$ is of the form

$$m_1(t) = d_1(t)\cos(2\pi f_1 t) \tag{9.16}$$

and message $m_2(t)$ is given by

$$m_2(t) = -d_2(t)\sin(2\pi f_1 t) \tag{9.17}$$

where $d_1(t)$ and $d_2(t)$ are binary data signals taking on the value $+1$ or -1 in symbol intervals of length $T_s = 2T_b$ s with switching times offset by T_b, and f_1 is the frequency in hertz of the weighting functions, $\cos(2\pi f_1 t)$ and $\sin(2\pi f_1 t)$, to be specified later. As in the case of QPSK, these data signals can be thought of as having been derived from a serial binary data stream whose bits occur each T_b s, with even-indexed bits producing $d_1(t)$ and odd-indexed bits producing $d_2(t)$, or vice versa. These binary data streams are weighted by a cosine or sine waveform as shown in Figure 9.4(a). If we substitute (9.16) and (9.17) into (9.1) and keep in mind that $d_1(t)$ and $d_2(t)$ are either $+1$ or -1, then, through the use of appropriate trigonometric identities, it follows that the modulated signal can be written as

$$x_c(t) = A\cos[2\pi f_c t + \theta_i(t)] \tag{9.18}$$

where

$$\theta_i(t) = \tan^{-1}\left[\frac{d_2(t)}{d_1(t)}\tan(2\pi f_1 t)\right] \tag{9.19}$$

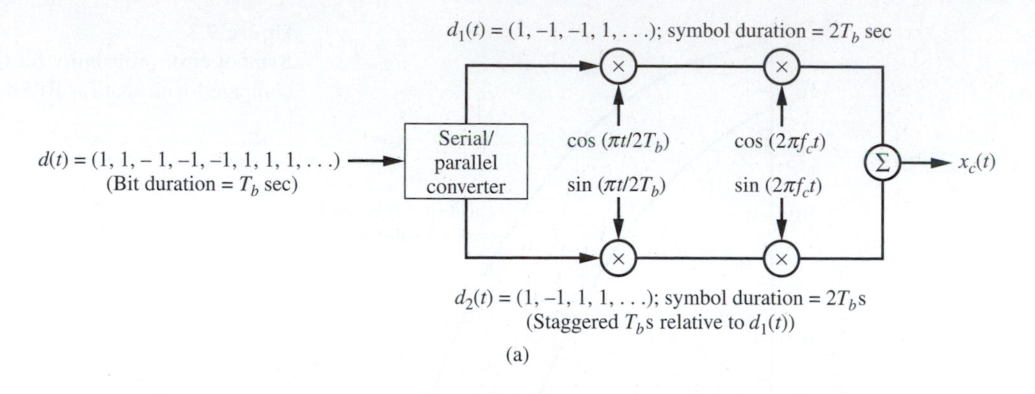

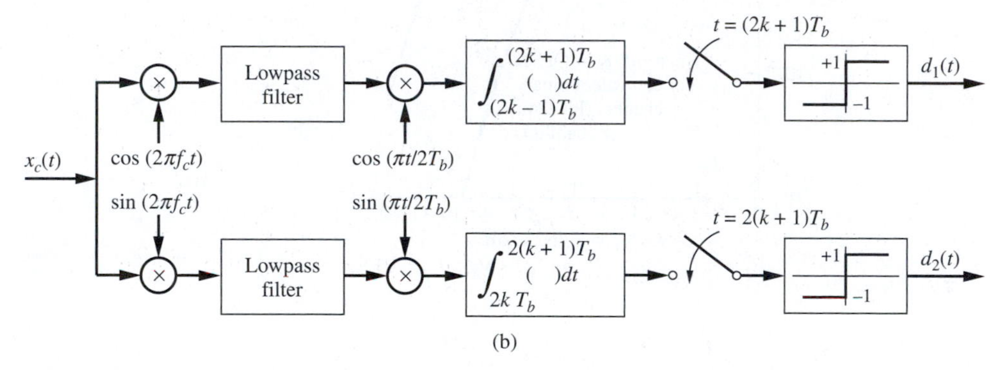

Figure 9.4
Block diagrams for parallel type I MSK modulator and demodulator. (a) Modulator. (b) Demodulator.

If $d_2(t) = d_1(t)$ (i.e., successive bits in the serial data stream are the same, either both 1 or both -1), then

$$\theta_i(t) = 2\pi f_1 t \tag{9.20}$$

whereas, if $d_2(t) = -d_1(t)$ (i.e., successive bits in the serial data stream are different), then

$$\theta_i(t) = -2\pi f_1 t \tag{9.21}$$

One form of MSK results if $f_1 = 1/2T_s = 1/4T_b$ Hz. In this case, each symbol of the data signal $d_1(t)$ is multiplied or weighted by one-half cycle of a cosine waveform, and each symbol of the data signal $d_2(t)$ is weighted by one-half cycle of a sine waveform, as shown in Figure 9.5(a). This form of MSK, wherein the weighting functions for each symbol are alternating half cycles of cosine or sine waveforms, is referred to as *MSK type I*. *Minimum-shift keying type II* modulation results if the weighting is always a positive half-cosinusoid or half-sinusoid, depending on whether it is the upper or lower arm in Figure 9.4 being referred to. This type of MSK modulation, which is illustrated in Figure 9.5(b), bears a closer relationship to OQPSK than to MSK type I.

Using $f_1 = 1/4T_b$ in (9.19) and substituting the result into (9.18) gives

$$x_c(t) = A\cos\left[2\pi\left(f_c \pm \frac{1}{4T_b}\right)t + u_k\right] \tag{9.22}$$

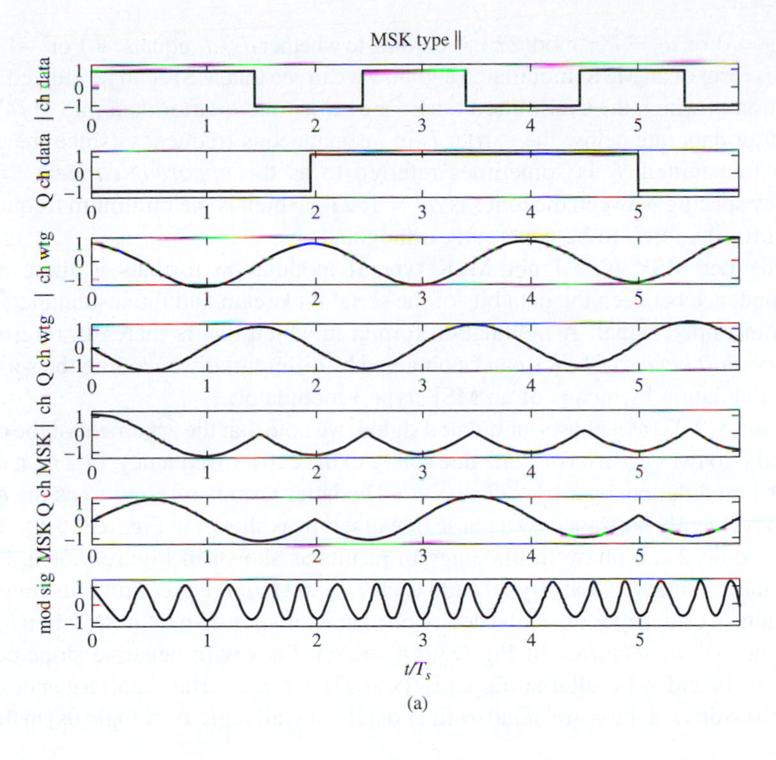

(a)

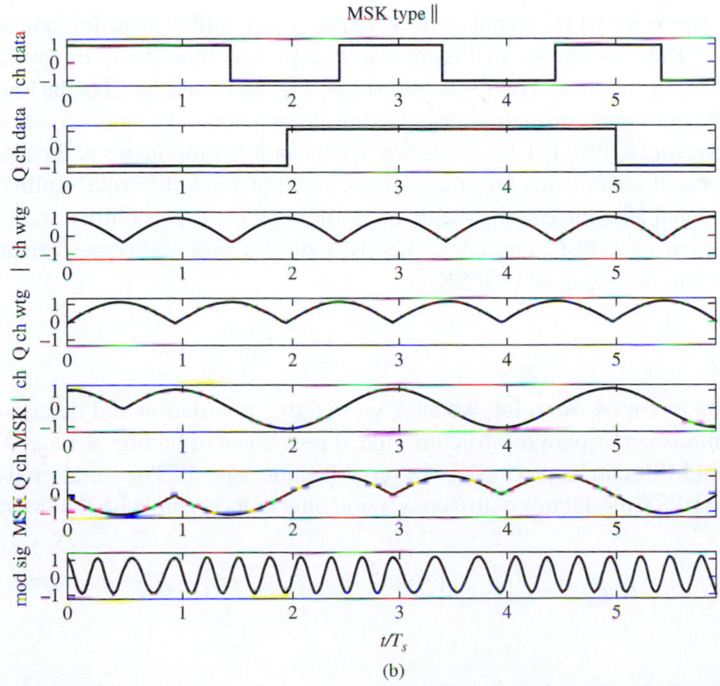

(b)

Figure 9.5
(a) Inphase and quadrature waveforms for MSK type I modulation. (b) MSK type II modulation.

where $u_k = 0$ or $u_k = k\pi \bmod(2\pi)$, according to whether d_2/d_1 equals $+1$ or -1, respectively. From this form of an MSK-modulated signal, we can see that MSK can be viewed as frequency modulation in which the transmitted tones[2] are either one-quarter data rate $(1/4T_b)$ above or one-quarter data rate below the carrier f_c in instantaneous frequency (since the carrier is not actually transmitted, f_c is sometimes referred to as the *apparent carrier*). Note that the frequency spacing between the tones is $\Delta f = 1/2T_b$, which is the minimum frequency spacing required for the tones to be coherently orthogonal.

In neither MSK type I nor MSK type II modulation formats is there a one-to-one correspondence between the data bits of the serial bit stream and the instantaneous frequency of the transmitted signal. A modulation format in which this is the case, referred to as *fast frequency-shift keying* (*FFSK*), can be obtained by differentially encoding the serial bit stream before modulation by means of an MSK type I modulator.

Viewing (9.22) as a phase-modulated signal, we note that the argument of the cosine can be separated into two phase terms, one due solely to the carrier frequency, or $2\pi f_c t$, and the other due to the modulation, or $\pm\pi(t/2T_b) + u_k$. The latter term is referred to as the *excess phase* and is conveniently portrayed by a phase tree diagram as shown in Figure 9.6(a). If the phase is shown modulo 2π, a phase trellis diagram results as shown in Figure 9.6(b). Note that the excess phase changes by exactly $\pi/2$ rad each T_b s and that it is a continuous function of time. This results in even better envelope deviation characteristics than OQPSK when filtered. In the excess-phase trellis diagram of Fig. 9.6(a), straight lines with negative slope correspond to alternating 1s and -1s (alternating logic 1s and 0s) in the serial-data sequence, and straight lines with positive slope correspond to all 1s or all -1s (all logic 1s or logic 0s) in the serial-data sequence.

The detector for MSK signals can be realized in parallel form in analogous fashion to QPSK or OQPSK, as shown in Figure 9.2, except that multiplication by $\cos(\pi t/2T_b)$ is required in the upper arm and multiplication by $\sin(\pi/2T_b)$ is required in the lower arm in order to realize the optimum correlation detector for the two data signals $d_1(t)$ and $d_2(t)$. As in the case of QPSK (or OQPSK), it can be shown that the noise components at the integrator outputs of the upper and lower arms are uncorrelated. Except for a different scaling factor (which affects the signal and noise components the same), the error probability analysis for MSK is identical to that for QPSK, and consequently, the error probability performance of MSK is identical to that of QPSK or OQPSK.

Serial MSK

In the discussion of MSK so far, we have viewed the modulation and detection processes as being accomplished by parallel structures like those shown in Figures 9.1 and 9.2 for QPSK. It turns out that MSK can be processed in a serial fashion as well. The serial modulator structure consists of a BPSK modulator with a conversion filter at its output with the frequency-response function

$$G(f) = \{\text{sinc}[(f - f_c]T_b - 0.25\} + \text{sinc}[(f + f_c)T_b + 0.25])e^{-j2\pi f t_0} \tag{9.23}$$

[2]One should not infer from this that the spectrum of the transmitted signal consists of impulses at frequencies $fc \pm 1/4T_b$.

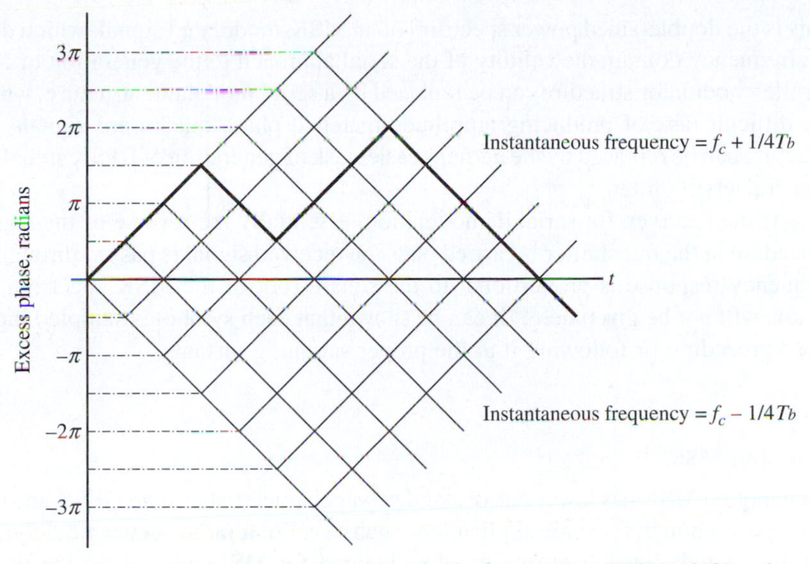

(a) Tree diagram showing the phase transitions for the data sequence 111011110101 as the heavy line.

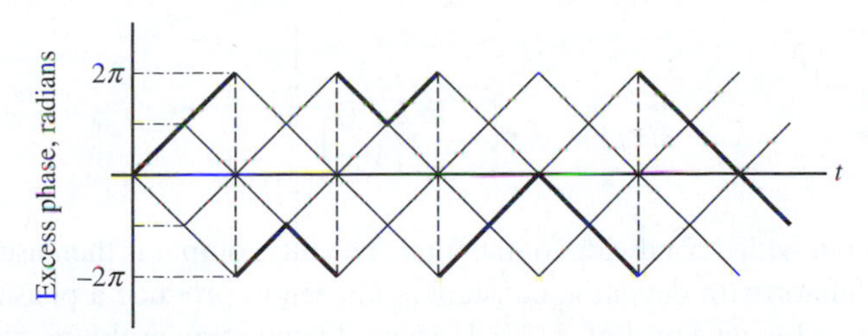

(b) Trellis diagram showing the same sequence as in (a) modulo 2π.

Figure 9.6
(a) Minimum-shift keying phase tree and (b) phase trellis diagrams.

where t_0 is an arbitrary filter delay and f_c is the apparent carrier frequency of the MSK signal. Note that the peak of the frequency response of the conversion filter is offset in frequency one-quarter data rate above the apparent carrier. The BPSK signal, on the other hand, is offset one-quarter data rate below the desired apparent carrier of the MSK signal. Its power spectrum can be written as

$$S_{\text{BPSK}}(f) = \frac{A^2 T_b}{2} \{\text{sinc}^2[(f - f_c)T_b + 0.25] + \text{sinc}^2[(f + f_c)T_b - 0.25]\} \tag{9.24}$$

The product of $|G(f)|^2$ and $S_{\text{BPSK}}(f)$ gives the power spectrum of the conversion filter output, which, after some simplification, can be shown to be

$$S_{\text{MSK}}(f) = \frac{32A^2T_b}{\pi^4}\left(\frac{\cos^2[2\pi T_b(f - f_c)]}{[1 - 16T_b^2(f - f_c)^2]^2} + \frac{\cos^2[2\pi T_b(f + f_c)]}{[1 - 16T_b^2(f + f_c)^2]^2}\right) \tag{9.25}$$

This is the double-sided power spectrum of an MSK-modulated signal, which demonstrates in the frequency domain the validity of the serial approach to the generation of MSK. Thus the parallel modulator structure can be replaced by a serial modulator structure, which means that the difficult task of producing amplitude-matched phase-quadrature signals in the parallel structure can be replaced by the perhaps easier task of generation of BPSK signals and synthesis of a conversion filter.

At the receiver, for serial demodulation, essentially the reverse of the signal-processing procedure at the transmitter is carried out. The received signal is passed through a filter whose frequency response is proportional to the square root of the MSK spectrum. Although the details will not be given here,[3] it can be shown that each symbol is sampled independently of those preceding or following it at the proper sampling instants.

Gaussian MSK

Even though MSK has lower out-of-band power characteristics than QPSK and OQPSK, it still is not good enough for some applications such as cellular radio. Better sidelobe suppression of the modulated signal spectrum can be obtained for MSK by making the phase transitions smoother than the straight-line characteristics shown in Figure 9.6. One means of doing this is to pass the NRZ-represented data through a lowpass filter with Gaussian frequency response given by[4]

$$H(f) = \exp\left[-\frac{\ln 2}{2}\left(\frac{f}{B}\right)^2\right] \tag{9.26}$$

where B is the 3-dB two-sided bandwidth of the filter. The filter output is then used as the input to a frequency modulator with deviation constant f_d chosen to produce a phase transition in going from a data bit -1 to data bit 1 of $\pi/2$ rad. An implementation problem is how to build a filter with frequency response given by (9.26), which corresponds to a filter with Gaussian impulse response (Table G.5)

$$h(t) = \sqrt{\frac{2\pi}{\ln 2}}B \exp\left(-\frac{2\pi^2 B^2}{\ln 2}t^2\right) \tag{9.27}$$

This is often done by digitally implementing a filter with Gaussian impulse response over a finite range of t. The step response of this filter is the integral of the impulse response,

$$y_s(t) = \int_{-\infty}^{t} h(\tau)\, d\tau \tag{9.28}$$

---

[3] See F. Amoroso and J. A. Kivett, Simplified MSK signaling technique. *IEEE Transactions on Communications*, **COM-25:** 433–441, April 1977.

[4] K. Morota and K. Haride, GMSK modulation for digital mobile radio telephony. *IEEE Transactions on Communications*, **COM-29:** 1044–1050, July, 1981.

Table 9.1 Ninety Percent Power Containment Bandwidths and Degradations in E_b/N_0 for GMSK

BT_b	90% containment BW (bit rates)*	Degradation from MSK (dB)
0.2	0.52	1.3
0.25	0.57	0.7
0.5	0.69	0.3
∞ (MSK)	0.78	0

*Double these for RF bandwidths.

so its response to a rectangular pulse, $\Pi(t/T_b)$, is

$$\begin{aligned} g(t) &= \int_{-\infty}^{t+T_b/2} h(\tau)d\tau - \int_{-\infty}^{t-T_b/2} h(\tau)\,d\tau \\ &= \frac{1}{2}\left\{\operatorname{erf}\left[\sqrt{\frac{2}{\ln 2}}\pi BT_b\left(\frac{t}{T_b}+\frac{1}{2}\right)\right] - \operatorname{erf}\left[\sqrt{\frac{2}{\ln 2}}\pi BT_b\left(\frac{t}{T_b}-\frac{1}{2}\right)\right]\right\} \\ &= \frac{1}{2}\left\{\operatorname{erf}\left[\sqrt{\frac{2}{\ln 2}}\pi BT_b\left(\frac{t}{T_b}+\frac{1}{2}\right)\right] + \operatorname{erf}\left[-\sqrt{\frac{2}{\ln 2}}\pi BT_b\left(\frac{t}{T_b}-\frac{1}{2}\right)\right]\right\} \end{aligned} \tag{9.29}$$

where T_b is the bit period and $\operatorname{erf}(u) = 2/\sqrt{\pi}\int_0^u \exp(-t^2)\,dt$ is the error function. The modulated waveform is produced by passing the entire NRZ-represented data stream through the Gaussian filter and then by using the filter output to frequency modulate the carrier. The excess phase of the resulting FM-modulated carrier is

$$\phi(t) = 2\pi f_d \sum_{n=-\infty}^{\infty} \alpha_n \int_{-\infty}^{t} g(\lambda - nT_b)\,d\lambda \tag{9.30}$$

where α_n is the sign of the nth bit and f_d is the deviation constant chosen to give phase transitions of $\pi/2$ rad. This modulation scheme, called *Gaussian MSK* (GMSK), can be shown to have a spectrum with very low sidelobes as determined by the product BT_b at the expense of more intersymbol interference the smaller BT_b. Gaussian MSK is used as the modulation scheme in the second-generation European cellular radio standard. Some results taken from Murota and Hirade giving 90% power containment bandwidth (i.e., the bandwidth within which 90% of the modulated signal power is contained) and degradation in E_b/N_0 from ideal MSK versus BT_b are given in Table 9.1.

9.1.4 *M*-ary Data Transmission in Terms of Signal Space

A convenient framework for discussing M-ary data transmission systems is that of signal space. The approach used here in terms of justifying the receiver structure is heuristic. It is placed on a firm theoretical basis in Chapter 10, where optimum signal detection principles are discussed.[5]

[5]Kotel'nikov (1947) was first to introduce the use of signal space into communication system characterization, followed later by Wozencraft and Jacobs (1965). For an analysis of several M-ary digital modulation schemes using signal space, see E. Arthurs and H. Dym, On the optimum detection of digital signals in the presence of white Gaussian noise—A geometric interpretation and a study of three basic data transmission systems. *IRE Transactions on Communications Systems*, **CS-10:** 336–372, December 1962.

We consider coherent communication systems with signal sets of the form

$$s_i(t) = \sum_{j=1}^{K} a_{ij}\phi_j(t), \ 0 \le t \le T_s, \ K \le M, \ i = 1, 2, \ldots, M \tag{9.31}$$

where the functions $\phi_j(t)$ are orthonormal over the symbol interval. That is,

$$\int_0^{T_s} \phi_m(t)\phi_n(t)\, dt = \begin{cases} 1, & m = n \\ 0, & m \neq n \end{cases} \tag{9.32}$$

Based on (9.31), we can visualize the possible transmitted signals as points in a space with coordinate axes $\phi_1(t), \phi_2(t), \phi_3(t), \ldots, \phi_K(t)$, much as illustrated in Figure 2.5.

At the output of the channel it is assumed that signal plus AWGN is received; that is,

$$y(t) = s_i(t) + n(t), \quad t_0 \le t \le t_0 + T_s, \ i = 1, 2, \ldots, M \tag{9.33}$$

where t_0 is an arbitrary starting time equal to an integer times T_s. As shown in Figure 9.7, the receiver consists of a bank of K correlators, one for each orthonormal function. The output of the jth correlator is

$$Z_j = a_{ij} + N_j, \quad j = 1, 2, \ldots, K, \quad i = 1, 2, \ldots, M \tag{9.34}$$

where the noise component N_j is given by ($t_0 = 0$ for notational ease)

$$N_j = \int_0^{T_s} n(t)\phi_j(t)\, dt \tag{9.35}$$

Since $n(t)$ is Gaussian and white, the random variables $N_1, N_2, \ldots, N_K$ can be shown to be independent, zero-mean, Gaussian random variables with variances $N_0/2$, which is the

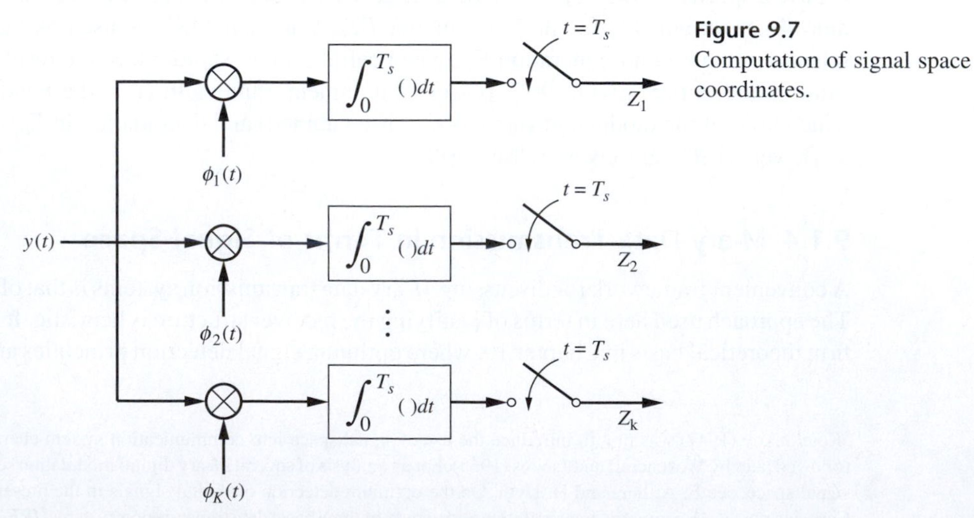

Figure 9.7 Computation of signal space coordinates.

Note: $y(t) = s_i(t) + n(t)$ where $n(t)$ is white Gaussian noise.

two-sided spectral density of the noise. That this is the case may be shown by considering the development

$$\begin{aligned}
E[N_j N_K] &= E\left[\int_0^{T_s} n(t)\phi_j(t)\,dt \int_0^{T_s} n(\lambda)\phi_k(\lambda)\,d\lambda\right] \\
&= E\left[\int_0^{T_s}\int_0^{T_s} n(t)n(\lambda)\phi_j(t)\phi_k(\lambda)\,d\lambda dt\right] \\
&= \int_0^{T_s}\int_0^{T_s} E[n(t)n(\lambda)]\phi_j(t)\phi_k(\lambda)\,d\lambda dt \\
&= \int_0^{T_s}\int_0^{T_s} \frac{N_0}{2}\delta(t-\lambda)\phi_j(t)\phi_k(\lambda)\,d\lambda dt \\
&= \frac{N_0}{2}\int_0^{T_s} \phi_j(t)\phi_k(t)\,dt \\
&= \begin{cases} N_0/2, & j = k \\ 0, & j \neq k \end{cases}
\end{aligned} \tag{9.36}$$

where the last line follows by virtue of the orthogonality of the $\phi_j(t)$s. Since $n(t)$ is zero mean, so are $N_1, N_2, \ldots, N_K$. The development leading to (9.36) shows that they are uncorrelated. Since they are Gaussian (each is a linear operation on a Gaussian random process), they are independent.

It can be shown that this signal space representation preserves all the information required to make a minimum error probability decision regarding which signal was transmitted. The next operation in the receiver is a decision box that performs the following function: Compare the received signal plus noise coordinates with the stored signal coordinates, a_{ij}. Choose as the transmitted signal that one closest to the received signal plus noise point with distance measured in the Euclidean sense; i.e., choose the transmitted signal as the one whose a_{ij} minimize

$$d_i^2 = \sum_{j=1}^{K} \left[Z_j - a_{ij}\right]^2 \tag{9.37}$$

This decision procedure will be shown in Chapter 10 to result in the minimum error probability possible with respect to the signal set.

EXAMPLE 9.1

Consider BPSK. Only one orthonormal function is required in this case, and it is

$$\phi(t) = \sqrt{\frac{2}{T_b}}\cos(2\pi f_c t), \quad 0 \leq t \leq T_b \tag{9.38}$$

The possible transmitted signals can be represented as

$$s_1(t) = \sqrt{E_b}\phi(t) \quad \text{and} \quad s_2(t) = -\sqrt{E_b}\phi(t) \tag{9.39}$$

where E_b is the bit energy; so $\alpha_{11} = \sqrt{E_b}$ and $\alpha_{21} = -\sqrt{E_b}$. For example, for a correlator output of $Z_1 = -1$ and with $E_b = 4$, Equation (9.37) becomes

$$d_1^2 = (-1 - \sqrt{4})^2 = 9$$
$$d_2^2 = (-1 + \sqrt{4})^2 = 1$$

so the decision would be made that $s_2(t)$ was sent.

9.1.5 QPSK in Terms of Signal Space

From Figures 9.7 and 9.2 we see that the receiver for QPSK consists of a bank of two correlators. Thus the received data can be represented in a two-dimensional signal space as shown in Figure 9.8. The transmitted signals can be represented in terms of two orthonormal functions $\phi_1(t)$ and $\phi_2(t)$ as

$$x_c(t) = s_i(t) = \sqrt{E_s}[d_1(t)\phi_1(t) - d_2(t)\phi_2(t)] = \sqrt{E_s}[\pm\phi_1(t) \pm \phi_2(t)] \tag{9.40}$$

where

$$\phi_1(t) = \sqrt{\frac{2}{T_s}}\cos(2\pi f_c t), \quad 0 \leq t \leq T_s \tag{9.41}$$

$$\phi_2(t) = \sqrt{\frac{2}{T_s}}\sin(2\pi f_c t), \quad 0 \leq t \leq T_s \tag{9.42}$$

E_s is the energy contained in $x_c(t)$ in one symbol interval. The resulting regions for associating a received data point with a possible signal point are also illustrated in Figure 9.8. It can be seen that the coordinate axes provide the boundaries of the regions that determine a given signal point to be associated with a received data point. For example, if the received datum point is in the first quadrant (region R_1), the decision is made that $d_1(t) = 1$ and $d_2(t) = 1$ (this will be denoted as signal point S_1 in the signal space). A simple bound on symbol-error probability can

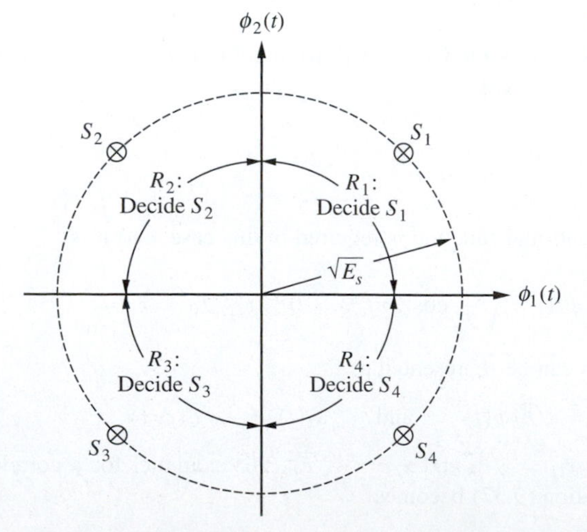

Figure 9.8
Signal space for QPSK.

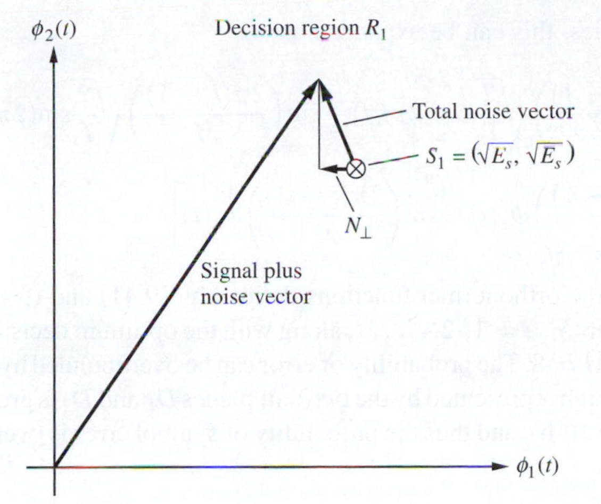

Figure 9.9
Representation of signal plus noise in signal space, showing $N_\perp$, the noise component that can cause the received data vector to land in R_2.

be obtained by recalling that the circular symmetry makes the conditional probability of error independent of the signal point chosen and noting that

$$\begin{aligned} P_E &= \Pr[Z \in R_2 \text{ or } R_3 \text{ or } R_4 | S_1 \text{ sent}] \\ &< \Pr[Z \in R_2 \text{ or } R_3 | S_1 \text{ sent}] + \Pr[Z \in R_3 \text{ or } R_4 | S_1 \text{ sent}] \end{aligned} \tag{9.43}$$

The two probabilities on the right-hand side of (9.43) can be shown to be equal. Thus,

$$\begin{aligned} P_E &< 2\Pr[Z \in R_2 \text{ or } R_3] = 2\Pr[\sqrt{E_s/2} + N_\perp < 0] \\ &= 2\Pr\left[N_\perp < -\sqrt{\frac{E_s}{2}}\right] \end{aligned} \tag{9.44}$$

where $N_\perp$, as shown in Figure 9.9, is the noise component perpendicular to the decision boundary between R_1 and R_2. It can be shown that it has zero mean and variance $N_0/2$. Thus,

$$P_E < 2\int_{-\infty}^{-\sqrt{E_s/2}} \frac{e^{-u^2/N_0}}{\sqrt{\pi N_0}}\, du = 2\int_{\sqrt{E_s/2}}^{\infty} \frac{e^{-u^2/N_0}}{\sqrt{\pi N_0}}\, du \tag{9.45}$$

Making the change of variables $u = v/\sqrt{N_0/2}$, we can reduce this to the form

$$P_E < 2Q\left(\sqrt{\frac{E_s}{N_0}}\right) \tag{9.46}$$

This is identical to (9.15), which resulted in neglecting the square of P_{E_1} in (9.13).

9.1.6 *M*-ary Phase-Shift Keying

The signal set for QPSK can be generalized to an arbitrary number of phases. The modulated signal takes the form

$$s_i(t) = \sqrt{\frac{2E_s}{T_s}} \cos\left(2\pi f_c t + \frac{2\pi(i-1)}{M}\right), \quad 0 \le t \le T_s,\ i = 1, 2, \ldots, M \tag{9.47}$$

Using trigonometric identities, this can be expanded as

$$s_i(t) = \sqrt{E_s}\left[\cos\left(\frac{2\pi(i-1)}{M}\right)\sqrt{\frac{2}{T_s}}\cos(2\pi f_c t) - \sin\left(\frac{2\pi(i-1)}{M}\right)\sqrt{\frac{2}{T_s}}\sin(2\pi f_c t)\right]$$

$$= \sqrt{E_s}\left[\cos\left(\frac{2\pi(i-1)}{M}\right)\phi_1(t) - \sin\left(\frac{2\pi(i-1)}{M}\right)\phi_2(t)\right] \quad (9.48)$$

where $\phi_1(t)$ and $\phi_2(t)$ are the orthonormal functions defined by (9.41) and (9.42).

A plot of the signal points S_i, $i = 1, 2, \ldots, M$, along with the optimum decision regions is shown in Figure 9.10(a) for $M = 8$. The probability of error can be overbounded by noting from Figure 9.10(b) that the total area represented by the two half planes D_1 and D_2 is greater than the total shaded area in Figure 9.10(b), and thus the probability of symbol error is overbounded by

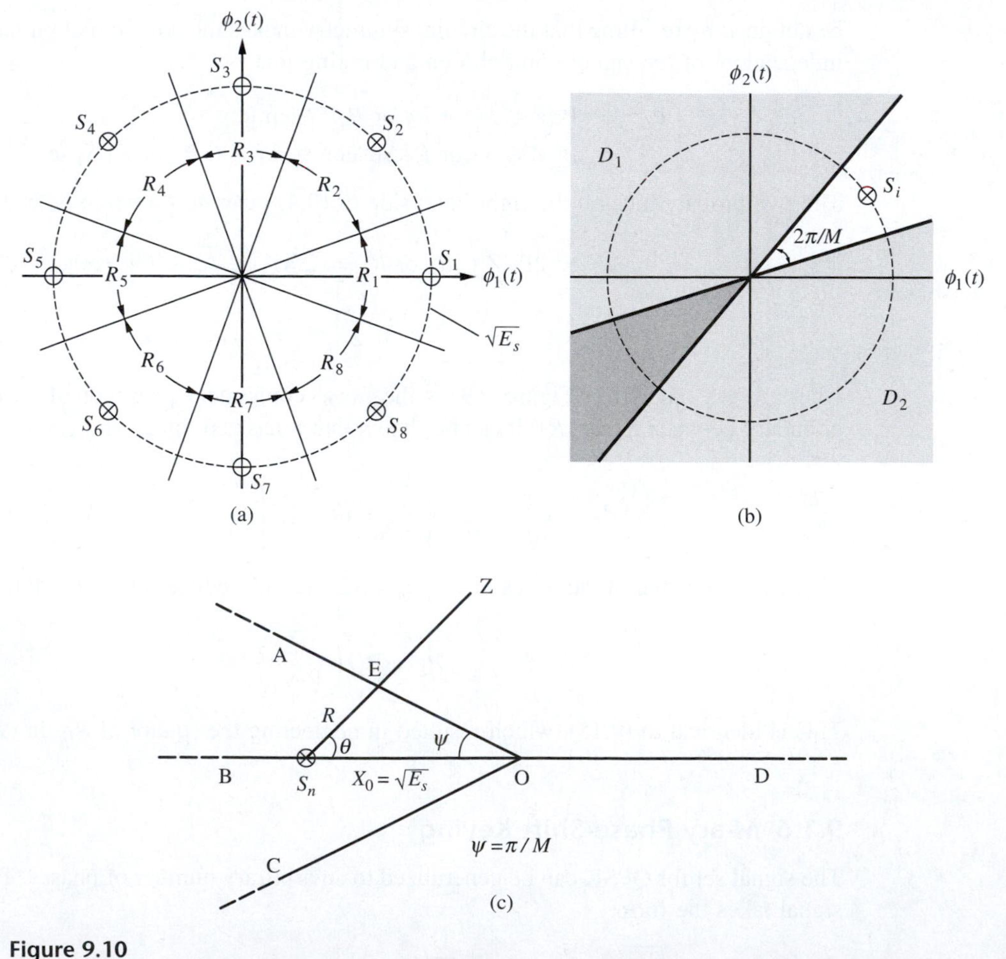

Figure 9.10
(a) Signal space for M-ary PSK with $M = 8$. (b) Signal space for M-ary PSK showing two half-planes that can be used to overbound P_E. (c) Coordinate setup for deriving Craig's exact integral for P_E.

the probability that the received data point Z_j lies in either half plane. Because of the circular symmetry of the noise distribution, both probabilities are equal. Consider a single half plane along with a single signal point, which is at a minimum distance of

$$d = \sqrt{E_s}\sin\left(\frac{\pi}{M}\right) \tag{9.49}$$

away from the boundary of the half plane. As in Figure 9.9, consider the noise component $N_\perp$, which is perpendicular to the boundary of the half plane. It is the only noise component that can possibly put the received datum point on the wrong side of the decision boundary; it has zero mean and a variance $N_0/2$. From this discussion and referring to Figure 9.10(b), it follows that the probability of error is overbounded by

$$\begin{aligned} P_E &< \Pr[Z \in D_1 \text{ or } D_2] = 2\Pr[Z \in D_1] \\ &= 2\Pr[d + N_\perp < 0] = 2\Pr[N_\perp < -d] \\ &= 2\int_{-\infty}^{-d} \frac{e^{-u^2/N_0}}{\sqrt{\pi N_0}}\, du = 2Q\left(\sqrt{\frac{2E_s}{N_0}}\sin\frac{\pi}{M}\right) \end{aligned} \tag{9.50}$$

From Figure 9.10(b) it can be seen that the bound becomes tighter as M gets larger (because the overlap of D_1 and D_2 becomes smaller with increasing M).

An exact expression for the symbol-error probability is[6]

$$P_E = \frac{1}{\pi}\int_0^{\pi - \pi/M} \exp\left(-\frac{(E_s/N_0)\,\sin^2(\pi/M)}{\sin^2\phi}d\phi\right) \tag{9.51}$$

The derivation, with the aid of Figure 9.10(c), is given below and follows that given in Craig's paper. Figure 9.10(c) shows the nth decision region for signal point S_n (recall that due to the circular symmetry, we can rotate this decision region to any convenient location). The probability of symbol error is the probability that the noise causes the received data point to land outside the wedge-shaped region bounded by the lines AO and CO, for example, the point Z, and is seen to be twice the probability that Z lies above the boundary AOD. It can be expressed as

$$P_E = 2\int_0^{\pi - \pi/M}\int_R^{\infty} f_{R\Theta}(r,\theta)\,dr\,d\theta \tag{9.52}$$

where R is the distance from the signal point to the boundary and $f_{R\Theta}(r,\ \theta)$ is the joint pdf of the noise components expressed in polar coordinates, which is

$$f_{R\Theta}(r,\ \theta) = \frac{r}{\pi N_0}\exp\left(-\frac{r^2}{N_0}\right), \quad r \geq 0,\ -\pi < \phi \leq \pi \tag{9.53}$$

[6]J. W. Craig, A new, simple and exact result for calculating the probability of error for two-dimensional signal constellations. *IEEE Milcom '91 Proceedings*, 571–575, October 1991.

(recall that the variance of the noise components is $N_0/2$). Substituting (9.53) into (9.52) and carrying out the integration over r, we get

$$P_E = \frac{1}{\pi}\int_0^{\pi - \pi/M} \exp\left(-\frac{R^2}{N_0}\right) d\theta \tag{9.54}$$

Now by the law of sines from Figure 9.10(c), we have

$$\frac{R}{\sin\psi} = \frac{X_0}{\sin(\pi - \theta - \psi)} = \frac{X_0}{\sin(\theta + \psi)}$$

or

$$R = \frac{X_0 \sin\psi}{\sin(\theta + \psi)} = \frac{\sqrt{E_s}\sin(\pi/M)}{\sin(\theta + \pi/M)} \tag{9.55}$$

Substitution of this expression for R into (9.54) gives

$$P_E = \frac{1}{\pi}\int_0^{\pi - \pi/M} \exp\left(-\frac{E_s \sin^2(\pi/M)}{N_0 \sin^2(\theta + \pi/M)}\right) d\theta \tag{9.56}$$

which, after the substitution $\phi = \pi - (\theta + \pi/M)$ gives (9.51). Performance curves computed from (9.51) will be presented later after conversion from symbol- to bit-error probabilities is discussed.

9.1.7 Quadrature-Amplitude Modulation

Another signaling scheme that allows multiple signals to be transmitted using quadrature carriers is *quadrature-amplitude modulation* (QAM), and the transmitted signal is represented as

$$s_i(t) = \sqrt{\frac{2}{T_s}}[A_i \cos(2\pi f_c t) + B_i \sin(2\pi f_c t)], \quad 0 \le t \le T_s \tag{9.57}$$

where A_i and B_i take on the possible values $\pm a, \ \pm 3a, \ldots, \ \pm(\sqrt{M} - 1)a$ with equal probability, where M is an integer power of 4. The parameter a can be related to the average energy of a symbol, E_s, as (see Problem 9.16)

$$a = \sqrt{\frac{3E_s}{2(M-1)}} \tag{9.58}$$

A signal space representation for 16-QAM is shown in Figure 9.11(a), and the receiver structure is shown in Figure 9.11(b). The probability of symbol error for M-QAM can be shown to be

$$P_E = 1 - \frac{1}{M}[(\sqrt{M} - 2)^2 P(C|\mathrm{I}) + 4(\sqrt{M} - 2)P(C|\mathrm{II}) + 4P(C|\mathrm{III})] \tag{9.59}$$

where the conditional probabilities $P(C|\mathrm{I})$, $P(C|\mathrm{II})$, and $P(C|\mathrm{III})$ are given by

$$P(C|\mathrm{I}) = \left[\int_{-a}^{a} \frac{\exp(-u^2/N_0)}{\sqrt{\pi N_0}} du\right]^2 = \left[1 - 2Q\left(\sqrt{\frac{2a^2}{N_0}}\right)\right]^2 \tag{9.60}$$

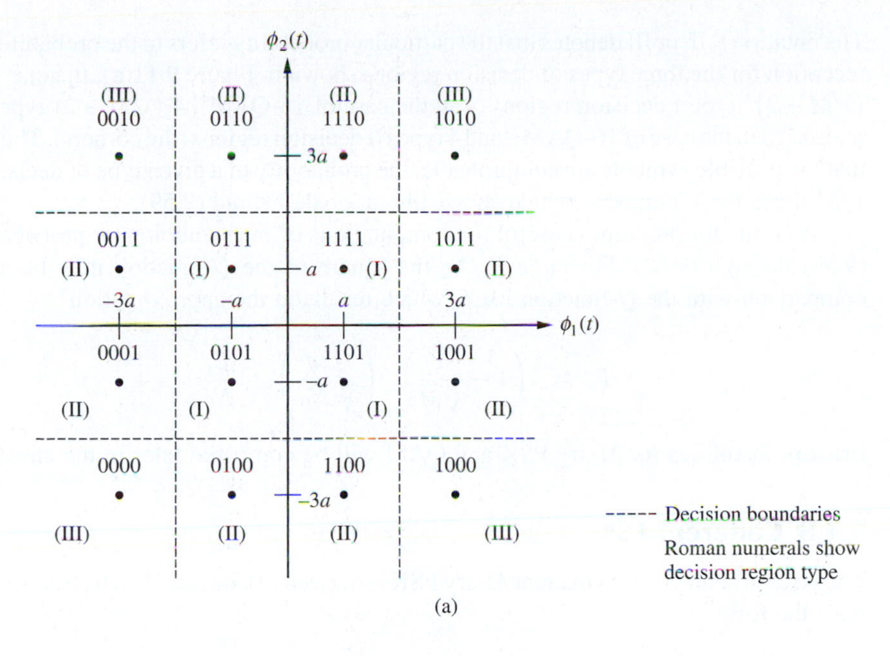

(a)

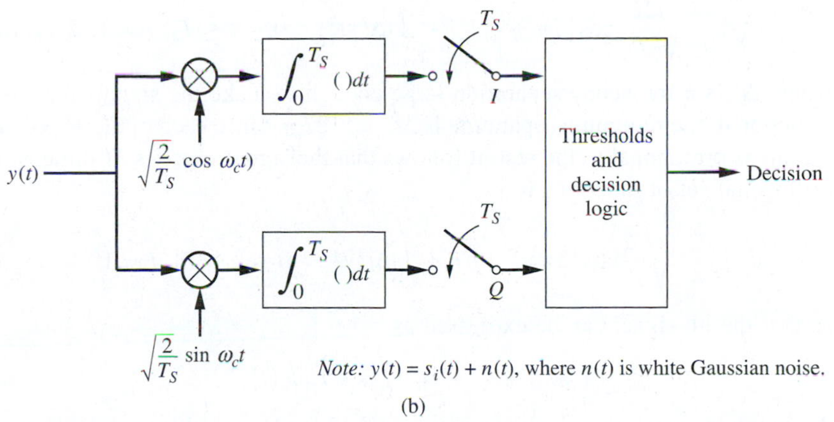

(b)

Figure 9.11
Signal space and detector structure for 16-QAM. (a) Signal constellation and decision regions for 16-QAM. (b) Detector structure for M-ary QAM. (Binary representations for signal points are Gray encoded.)

$$
\begin{aligned}
P(C|\text{II}) &= \int_{-a}^{a} \frac{\exp\left(-u^2/N_0\right)}{\sqrt{\pi N_0}}\, du \int_{-a}^{\infty} \frac{\exp(-u^2/N_0)}{\sqrt{\pi N_0}}\, du \\
&= \left[1 - 2Q\left(\sqrt{\frac{2a^2}{N_0}}\right)\right]\left[1 - Q\left(\sqrt{\frac{2a^2}{N_0}}\right)\right]
\end{aligned} \tag{9.61}
$$

$$
P(C|\text{III}) = \left(\int_{-\text{a}}^{\infty} \frac{\exp(-u^2/N_0)}{\sqrt{\pi N_0}}\, du\right)^2 = \left[1 - Q\left(\sqrt{\frac{2a^2}{N_0}}\right)\right]^2 \tag{9.62}
$$

The notation I, II, or III denotes that the particular probability refers to the probability of correct reception for the three types of decision regions shown in Figure 9.11(a). In general, there are $\left(\sqrt{M}-2\right)^2$ type I decision regions (4 in the case of 16-QAM), 4 $\left(\sqrt{M}-2\right)$ type II decision regions (8 in the case of 16-QAM), and 4 type III decision regions (the corners). Thus, assuming that the possible symbols are equiprobable, the probability of a given type of decision region is $1/M$ times these numbers, which shows the rationale behind (9.59).

A computer program is useful for computations of the symbol-error probabability using (9.59) through (9.62). For large E_s/N_0 the square of the Q-function may be neglected in comparison with the Q-function itself, which results in the approximation

$$P_s \cong 4\left(1-\frac{1}{\sqrt{M}}\right)Q\left(\sqrt{\frac{2a^2}{N_0}}\right), \quad \frac{E_s}{N_0} >> 1 \tag{9.63}$$

Error probabilities for M-ary PSK and QAM will be compared later in the chapter.

9.1.8 Coherent FSK

The error probability for coherent M-ary FSK is derived in Chapter 10. The transmitted signals have the form

$$s_i(t) = \sqrt{\frac{2E_s}{T_s}}\cos\{2\pi[f_c + (i-1)\Delta f]\,t\}, \quad 0 \le t \le T_s,\ i = 1, 2, \ldots, M \tag{9.64}$$

where Δf is a frequency separation large enough to make the signals represented by (9.64) orthogonal (the minimum separation is $\Delta f = 1/2T_s$). Since each of the M possible transmitted signals is orthogonal to the rest, it follows that the signal space is M-dimensional, where the orthogonal set of functions is

$$\phi_i(t) = \sqrt{\frac{2}{T_s}}\cos\{2\pi[f_c + (i-1)\Delta f]\,t\}, \quad 0 \le t \le T_s,\ i = 1, 2, \ldots, M \tag{9.65}$$

so that the ith signal can be expressed as

$$s_i(t) = \sqrt{E_s}\phi_i(t) \tag{9.66}$$

An example signal space is shown in Figure 9.12 for $M = 3$ (this unrealistic example is chosen for ease of drawing). An upper bound for the probability of error that becomes tighter as M gets larger is given by[7]

$$P_E \le (M-1)\,Q\left(\sqrt{\frac{E_s}{N_0}}\right) \tag{9.67}$$

which follows because, for an error to occur, the received data vector must be closer to any one of the $M-1$ incorrect signal points rather than the correct one. The probability of any one of these incorrect events is $Q\left(\sqrt{E_s/N_0}\right)$.

[7]This is derived by using the union bound of probability, which states that, for any set of K events that may be disjoint, $Pr\left[A_1 \cup A_2 \cup \cdots \cup A_k\right] \le \Pr\left[A_1\right] + \Pr\left[A_2\right] + \cdots + \Pr\left[A_k\right]$.

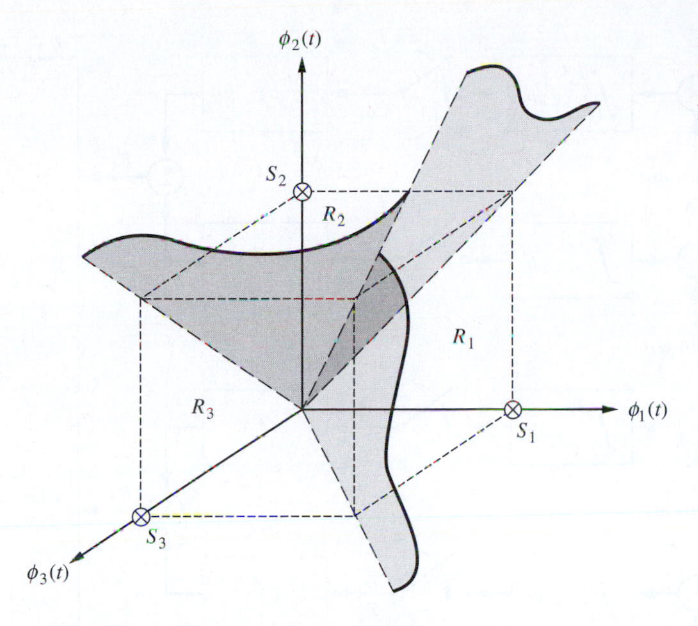

Figure 9.12
Signal space showing decision regions for tertiary coherent FSK.

9.1.9 Noncoherent FSK

Noncoherent M-ary FSK employs the same signal set as coherent FSK; however, a receiver structure is used that does not require the acquisition of a coherent carrier reference. A block diagram of a suitable receiver structure is shown in Figure 9.13. The symbol-error probability can be shown to be

$$P_E = \sum_{k=1}^{M-1} \binom{M-1}{k} \frac{(-1)^{k+1}}{k+1} \exp\left(-\frac{k}{k+1}\frac{E_s}{N_0}\right) \tag{9.68}$$

The derivation of the symbol-error probability may be sketched as follows. Referring to Figure 9.13, consider a received signal of the form

$$y(t) = \sqrt{\frac{2E_s}{T_s}}\cos(2\pi f_i t + \alpha),\ 0 \le t \le T_s,\ i = 1, 2, \ldots, M \tag{9.69}$$

where $|f_{i\pm 1} - f_i| \ge 1/T_s$ and α is an unknown phase angle. The orthognal basis functions for the jth correlator pair are

$$\begin{aligned} \phi_{2j-1}(t) &= \sqrt{\frac{2}{T_s}}\cos(2\pi f_j t), \quad 0 \le t \le T_s \\ \phi_{2j}(t) &= \sqrt{\frac{2}{T_s}}\sin(2\pi f_j t), \quad 0 \le t \le T_s, j = 1, 2, \ldots, M \end{aligned} \tag{9.70}$$

Given that that $s_i(t)$ was sent, the coordinates of the received data vector, denoted as $\mathbf{Z} = (Z_1, Z_2, Z_3, \ldots, Z_{2M-1}, Z_{2M})$, are

$$Z_{2j-1} = \begin{cases} N_{2j-1}, & j \ne i \\ \sqrt{E_s}\cos\alpha + N_{2i-1}, & i = j \end{cases} \tag{9.71}$$

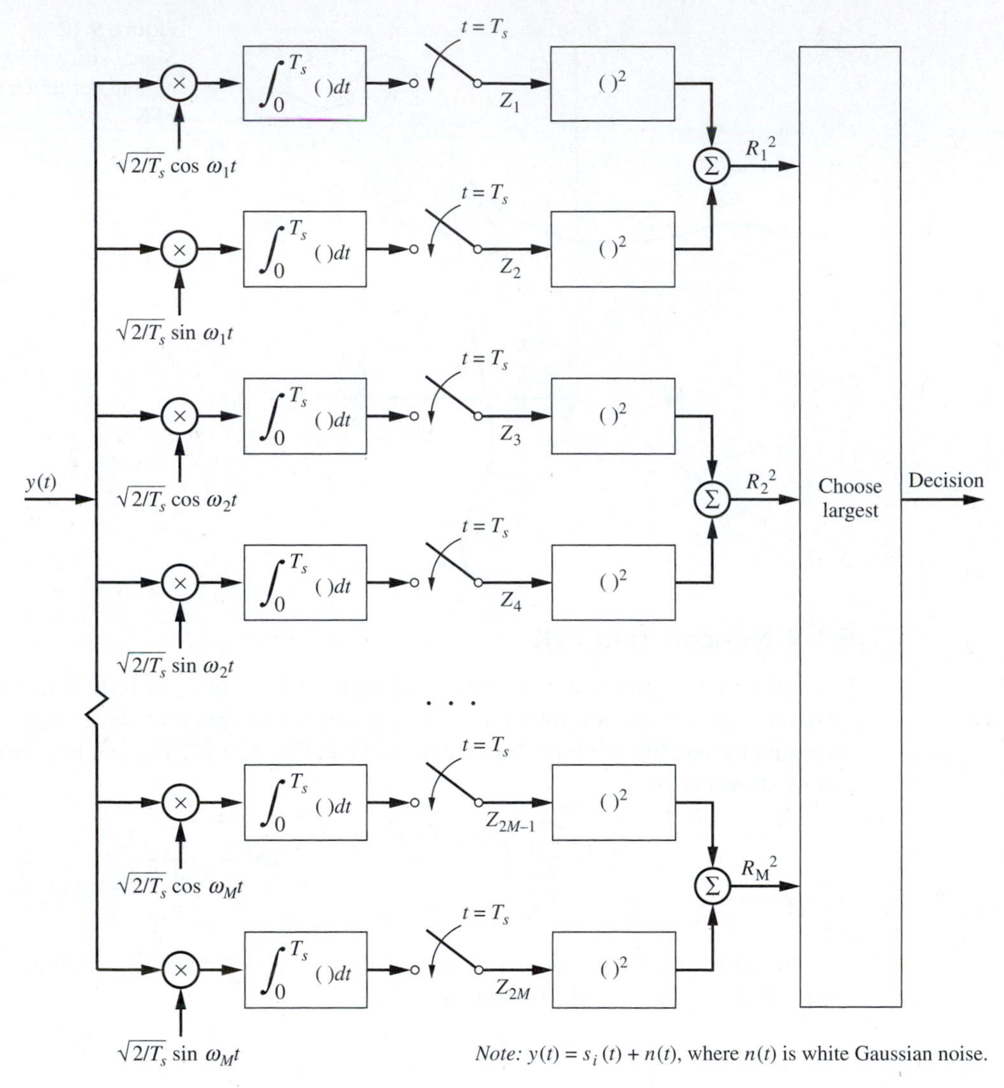

Figure 9.13
Receiver structure for noncoherent FSK.

and

$$Z_{2j} = \begin{cases} N_{2j}, & j \neq i \\ -\sqrt{E_s}\sin\alpha + N_{2i}, & i = j \end{cases} \tag{9.72}$$

where $j = 1, 2, \ldots, M$. The noise components are given by

$$\begin{aligned} N_{2j-1} &= \sqrt{\frac{2}{T_s}}\int_0^{T_s} n(t)\cos(2\pi f_j t)\,dt \\ N_{2j} &= \sqrt{\frac{2}{T_s}}\int_0^{T_s} n(t)\sin(2\pi f_j t)\,dt \end{aligned} \tag{9.73}$$

and are uncorrelated Gaussian random variables with zero means and variances $N_0/2$. Given that $s_i(t)$ was sent, a correct reception is made if

$$Z_{2j-1}^2 + Z_{2j}^2 < Z_{2i-1}^2 + Z_{2i}^2, \quad \text{all } j \neq i$$

or, equivalently, if

$$\sqrt{Z_{2j-1}^2 + Z_{2j}^2} < \sqrt{Z_{2i-1}^2 + Z_{2i}^2}, \quad \text{all } j \neq i \tag{9.74}$$

Evaluation of the symbol-error probability requires the joint pdf of the random variables $R_j = \sqrt{Z_{2j-1}^2 + Z_{2j}^2}$, $j = 1, 2, \ldots, M$. For $j = i$ and given α, Z_{2j-1} is a Gaussian random variable with mean $\sqrt{E_s}\cos\alpha$ and variance $N_0/2$, which follows from (9.71). Similarly, for $j = i$ and given α, Z_{2j} is a Gaussian random variable with mean $-\sqrt{E_s}\sin\alpha$ and variance $N_0/2$, which follows from (9.72). For $j \neq i$, both have zero means and variances $N_0/2$. Thus, the joint pdf of Z_{2j} and Z_{2j-1} given α is (x and y are the dummy variables for the pdf)

$$f_{Z_{2j-1},Z_{2j}}(x, y|\alpha) = \begin{cases} \dfrac{1}{\pi N_0}\exp\left\{-\dfrac{1}{N_0}\left[\left(x - \sqrt{E_s}\cos\alpha\right)^2 + \left(y + \sqrt{E_s}\sin\alpha\right)^2\right]\right\}, & j = i \\ \dfrac{1}{\pi N_0}\exp\left[-\dfrac{1}{N_0}(x^2 + y^2)\right] & j \neq i \end{cases} \tag{9.75}$$

To proceed, it is convenient to change to polar coordinates, defined by

$$x = \sqrt{\frac{N_0}{2}}\, r\,\sin\phi, \qquad y = \sqrt{\frac{N_0}{2}}\, r\,\cos\phi, \qquad r \geq 0,\ 0 \leq \phi < 2\pi \tag{9.76}$$

With this change of variables, minus the exponent in the first equation of (9.75) becomes

$$\begin{aligned} &\frac{1}{N_0}\left[\left(\sqrt{\frac{N_0}{2}}r\sin\phi - \sqrt{E_s}\cos\alpha\right)^2 + \left(\sqrt{\frac{N_0}{2}}r\cos\phi + \sqrt{E_s}\sin\alpha\right)^2\right] \\ &= \frac{1}{N_0}\left(\frac{N_0 r^2}{2}\sin^2\phi - \sqrt{2E_s N_0}\, r\sin\phi\cos\alpha + E_s\cos^2\alpha + \frac{N_0 r^2}{2}\cos^2\phi \right. \\ &\qquad \left. + \sqrt{2E_s N_0}\, r\cos\phi\sin\alpha + E_s\sin^2\alpha\right) \\ &= \frac{r^2}{2} - \sqrt{\frac{2E_s}{N_0}}\, r(\sin\phi\cos\alpha - \cos\phi\sin\alpha) + \frac{E_s}{N_0} \\ &= \frac{r^2}{2} + \frac{E_s}{N_0} - \sqrt{\frac{2E_s}{N_0}}\, r\sin(\phi - \alpha) \end{aligned} \tag{9.77}$$

When this is substituted into (9.75) we get (note that $(dx\,dy \rightarrow (N_0/2)r\,dr\,d\phi)$

$$f_{Rj\Phi j|a}(r,\phi|\alpha)=\frac{r}{2\pi}\exp\left\{-\left[\frac{r^2}{2}+\frac{E_s}{N_0}-\sqrt{\frac{2E_s}{N_0}}r\sin(\phi-\alpha)\right]\right\},\ j=i,\ r\geq 0,\ 0\leq\phi<2\pi$$

$$=\frac{r}{2\pi}\exp\left[-\left(\frac{r^2}{2}+\frac{E_s}{N_0}\right)\right]\exp\left[\sqrt{\frac{2E_s}{N_0}}r\sin(\phi-\alpha)\right] \tag{9.78}$$

The result for $j\neq i$ can be obtained by setting $E_s=0$ in (9.78). The unconditional pdf is found by averaging with respect to the pdf of α, which is uniform in any 2π range. Thus

$$f_{Rj\Phi j}(r,\phi)=\frac{r}{2\pi}\exp\left[-\frac{1}{2}\left(r^2+\frac{2E_s}{N_0}\right)\right]\int_{\phi}^{2\pi-\phi}\exp\left[\sqrt{\frac{2E_s}{N_0}}\,r\sin(\phi-\alpha)\right]\frac{d\alpha}{2\pi}$$

$$=\frac{r}{2\pi}\exp\left[-\frac{1}{2}\left(r^2+\frac{2E_s}{N_0}\right)\right]I_0\left(\sqrt{\frac{2E_s}{N_0}}\,r\right),\quad j=i,\ r\geq 0,\ 0\leq\phi<2\pi \tag{9.79}$$

where $I_0(\cdot)$ is the modified Bessel function of the first kind and order zero. The marginal pdf for R_j is obtained by integrating over ϕ, which gives

$$f_{Rj}(r)=r\exp\left[-\frac{1}{2}\left(r^2+\frac{2E_s}{N_0}\right)\right]I_0\left(\sqrt{\frac{2E_s}{N_0}}r\right),\quad j=i,\ r\geq 0 \tag{9.80}$$

which is a Ricean pdf. We get the result for $j\neq i$ by setting $E_s=0$, which gives

$$f_{Rj}(r)=r\exp\left(-\frac{r^2}{2}\right),\quad j\neq i, r\geq 0 \tag{9.81}$$

which is recognized as a Rayleigh pdf.

In terms of the random variables R_j, $j=1, 2, \ldots, M$, the detection criterion is

$$R_j<R_i,\ \text{all}\ j\neq i \tag{9.82}$$

Since the R_js are statistically independent random variables, the probability of this compound event is

$$\Pr\left[R_j<R_i,\ \text{all}\ j\neq i|R_i\right]=\prod_{j=1,\,j\neq i}^{M}\Pr\left[R_j<R_i|R_i\right] \tag{9.83}$$

But

$$\Pr\left[R_j<R_i|R_i\right]=\int_0^{R_i}r\exp\left(\frac{-r^2}{2}\right)dr=1-\exp\left(\frac{-R_i^2}{2}\right) \tag{9.84}$$

The probability of correct reception, given $s_i(t)$ was sent, is (9.84) averaged over R_i, where R_i has the Ricean pdf given by (9.80). This may be written, using (9.84) and (9.80), as the integral

$$P_s(C|s_i\ \text{sent})=\int_0^{\infty}\left[1-\exp\left(\frac{-r^2}{2}\right)\right]^{M-1}r\exp\left[-\frac{1}{2}\left(r^2+\frac{2E_s}{N_0}\right)\right]I_0\left(\sqrt{\frac{2E_s}{N_0}}r\right)dr \tag{9.85}$$

Now, by the binomial theorem

$$\left[1-\exp\left(\frac{-r^2}{2}\right)\right]^{M-1}=\sum_{k=0}^{M-1}\binom{M-1}{k}(-1)^k\exp\left(\frac{-kr^2}{2}\right) \tag{9.86}$$

Thus, interchanging the order of integration and summation, (9.85) may be written as

$$\begin{aligned}P_s(C|s_i\,\text{sent}) &= \sum_{k=0}^{M-1}\binom{M-1}{k}(-1)^k\int_0^\infty r\exp\left\{-\frac{1}{2}\left[(k+1)r^2+\frac{2E_s}{N_0}\right]\right\}I_0\left(\sqrt{\frac{2E_s}{N_0}}r\right)dr \\ &= \exp\left(\frac{-E_s}{N_0}\right)\sum_{k=0}^{M-1}\binom{M-1}{k}\frac{(-1)^k}{k+1}\exp\left(\frac{E_s}{(k+1)N_0}\right)\end{aligned} \tag{9.87}$$

where the definite integral

$$\int_0^\infty x\exp(-ax^2)I_0(bx)\,dx=\frac{1}{2a}\exp\left(\frac{b^2}{4a}\right),\quad a,b>0 \tag{9.88}$$

has been used.

Since this result is independent of the signal sent, it holds for any signal and therefore is the probability of correct reception independent of the particular $s_i(t)$ assumed. Hence, the probability of symbol error is given by

$$\begin{aligned}P_E &= 1-P_s(C|s_i\,\text{sent}) \\ &= 1-\exp\left(\frac{-E_s}{N_0}\right)\sum_{k=0}^{M-1}\binom{M-1}{k}\frac{(-1)^k}{k+1}\exp\left(\frac{E_s}{(k+1)N_0}\right)\end{aligned} \tag{9.89}$$

which can be shown to be equivalent to (9.68).

9.1.10 Differentially Coherent Phase-Shift Keying

Binary DPSK was introduced in Chapter 8 as a phase-shift-keyed modulation scheme where the previous bit interval is used as a reference for the current bit interval with the transmitted information conveyed in the phase difference by means of differential encoding. Recall that the loss in E_b/N_0 relative to coherent binary PSK is approximately 0.8 dB at low bit-error probabilities. The idea underlying binary DPSK is readily extended to the M-ary case, where the information is transmitted via the phase difference from one symbol interval to the next. The receiver then compares successive received signal phases to estimate the relative phase shift. That is, if successive transmitted signals are represented as

$$\begin{aligned}s_1(t) &= \sqrt{\frac{2E_s}{T_s}}\cos(2\pi f_c t),\quad 0\le t<T_s \\ s_i(t) &= \sqrt{\frac{2E_s}{T_s}}\cos\left(2\pi f_c t+\frac{2\pi(i-1)}{M}\right),\quad T_s\le t<2T_s\end{aligned} \tag{9.90}$$

then assuming the channel-induced phase shift α is constant over two successive signaling intervals, the received signal plus noise can be represented as

$$
\begin{aligned}
y_1(t) &= \sqrt{\frac{2E_s}{T_s}}\cos(2\pi f_c t + \alpha) + n(t), \quad 0 \le t < T_s \\
y_i(t) &= \sqrt{\frac{2E_s}{T_s}}\cos\left(2\pi f_c t + \alpha + \frac{2\pi(i-1)}{M}\right) + n(t), \quad T_s \le t < 2T_s
\end{aligned}
\tag{9.91}
$$

and the receiver's decision rule is then one of determining the amount of phase shift in $2\pi/M$ steps from one signaling interval to the next.

Over the years, several approximations and bounds have been derived for the symbol-error probability of M-ary DPSK (M-DPSK).[8] Just as for M-PSK, an exact expression for the symbol-error probability for M-DPSK has been published that utilizes the Craig expression for the Q-function given in Appendix G.[9] The result is

$$
P_E = \frac{1}{\pi}\int_0^{\pi - \pi/M} \exp\left(-\frac{(E_s/N_0)\sin^2(\pi/M)}{1 + \cos(\pi/M)\cos\phi}\right) d\phi \tag{9.92}
$$

Results for bit-error probabilities computed with the aid of (9.92) will be presented after the conversion of symbol to bit-error probabilities is discussed.

9.1.11 Bit-Error Probability from Symbol-Error Probability

If one of M possible symbols is transmitted, the number of bits required to specify this symbol is $\log_2 M$. It is possible to number the signal points using a binary code such that only one bit changes in going from a signal point to an adjacent signal point. Such a code is a *Gray code*, as introduced in Chapter 8, with the case for $M = 8$ given in Table 9.2.

Since mistaking an adjacent signal point for a given signal point is the most probable error, we assume that nonadjacent errors may be neglected and that Gray encoding has been used so that a symbol error corresponds to a single bit error (as would occur, for example, with M-ary

Table 9.2 Gray Code for $M = 8$

Digit	Binary code	Gray code
0	000	000
1	001	001
2	010	011
3	011	010
4	100	110
5	101	111
6	110	101
7	111	100

Note: The encoding algorithm is given in Problem 8.32.

[8] V. K. Prabhu, Error rate performance for differential PSK. *IEEE Trans. on Commun.*, **COM-30:** 2547–2550, December 1982. R. Pawula, Asymptotics and error rate bounds for M-ary DPSK. *IEEE Trans. on Commun.*, **COM-32:** 93–94, January 1984.

[9] R. F. Pawula, A New Formula for MDPSK Symbol Error Probability. *IEEE Commun. Letters*, **2:** 271–272, October 1998.

Table 9.3 Pertinent to the Computation of Bit-Error Probability for Orthogonal Signaling

M-ary signal	Binary representation	
1 (0)	0 0	0
2 (1)	0 0	1
3 (2)	0 1	0
4 (3)	0 1	1
5 (4)	1 0	0
6 (5)	1 0	1
7 (6)	1 1	0
8 (7)	1 1	1

PSK). We may then write the bit error probability in terms of the symbol error probability for an M-ary communications system for which these assumptions are valid as

$$P_{E,\,\text{bit}} = \frac{P_{E,\,\text{symbol}}}{\log_2 M} \tag{9.93}$$

Because we neglect probabilities of symbol errors for nonadjacent symbols, (9.93) gives a *lower bound* for the bit error probability.

A second way of relating bit error probability to symbol error probability is as follows. Consider an M-ary modulation scheme for which $M = 2^n$, n an integer. Then each symbol (M-ary signal) can be represented by an n-bit binary number, for example, the binary representation of the signal's index minus one. Such a representation is given in Table 9.3 for $M = 8$.

Take any column, say the last, which is enclosed by a box. In this column, there are $M/2$ zeros and $M/2$ ones. If a symbol (M-ary signal) is received in error, then for any given bit position of the binary representation (the rightmost bit in this example), there are $M/2$ of a possible $M - 1$ ways that the chosen bit can be in error (one of the M possibilities is correct). Therefore, the probability of a given data bit being in error, given that a signal (symbol) was received in error, is

$$P(B|S) = \frac{M/2}{M-1} \tag{9.94}$$

Since a symbol is in error if a bit in the binary representation of it is in error, it follows that the probability $P(S\,|\,B)$ of a symbol error given a bit error is unity. Employing Bayes' rule, we find the equivalent bit-error probability of an M-ary system can be approximated by

$$P_{E,\,\text{bit}} = \frac{P(B|S)P_{E,\,\text{symbol}}}{P(S|B)} = \frac{M}{2(M-1)}P_{E,\,\text{symbol}} \tag{9.95}$$

This result is especially useful for orthogonal signaling schemes such as FSK, where it is equally probable that any of the $M - 1$ incorrect signal points may be mistaken for the correct one.

Finally, in order to compare two communications systems using different numbers of symbols on an equivalent basis, the energies must be put on an equivalent basis. This is done by expressing the energy per symbol E_s in terms of the energy per bit E_b in each system by means of the relationship

$$E_s = (\log_2 M)E_b \tag{9.96}$$

which follows since there are $\log_2 M$ bits per symbol.

9.1.12 Comparison of M-ary Communications Systems on the Basis of Bit Error Probability

Figure 9.14 compares coherent and differentially coherent M-ary PSK systems on the basis of bit-error probability versus E_b/N_0 along with QAM. Figure 9.14 shows that the bit-error probability for these systems gets worse as M gets larger. This can be attributed to the signal points being crowded closer together in the two-dimensional signal space with increasing M. In addition, M-ary DPSK performs a few decibels worse than coherent PSK, which can be attributed to the noisy phase at the receiver for the former. Quadrature-amplitude modulation performs considerably better than PSK because it makes more efficient use of the signal space (since it varies in amplitude in addition to phase, the transmitted waveform has a nonconstant envelope that is disadvantageous from the standpoint of efficient power amplification).

Not all M-ary digital modulation schemes exhibit the undesirable behavior of increasing bit-error probability with increasing M. We have seen that M-ary FSK is a signaling scheme in which the number of dimensions in the signal space grows directly with M. This means that the bit-error probabilities for coherent and noncoherent M-ary FSK *decrease* as M increases because the increasing dimensionality means that the signal points are not crowded together as with M-ary PSK, for example, for which the signal space is two-dimensional regardless of the value of M (except for $M=2$). This is illustrated in Figure 9.15, which compares bit-error probabilities for coherent and noncoherent FSK for various values of M. Unfortunately, the bandwidth required for M-ary FSK (coherent or noncoherent) grows with M, whereas this is not the case for M-ary PSK. Thus, to be completely fair, one must compare M-ary communications systems on the basis of both their bit-error probability characteristics and their relative

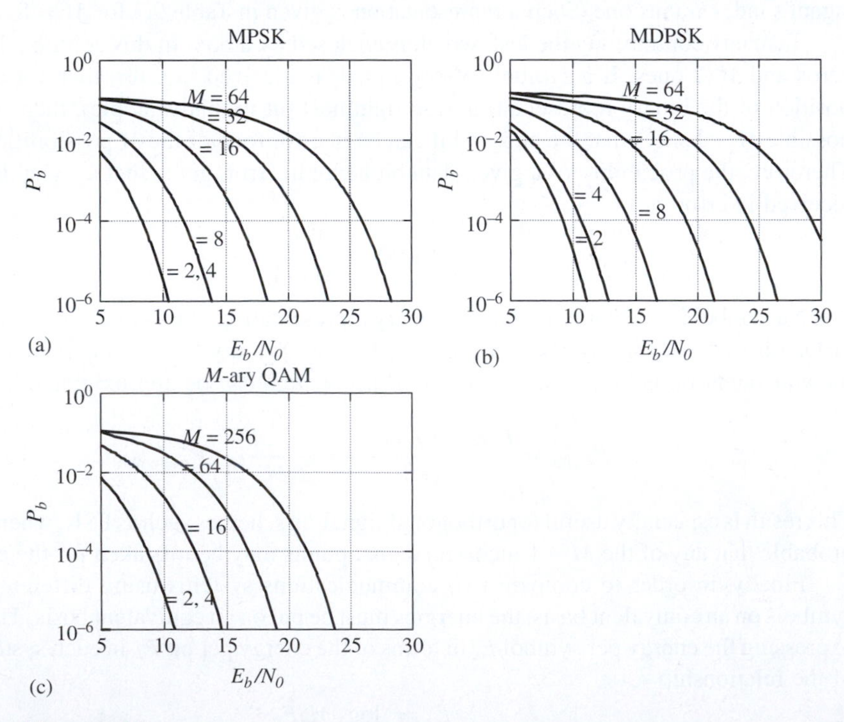

Figure 9.14
Bit-error probability versus E_b/N_0 for M-ary (a) PSK, (b) differentially coherent PSK, and (c) QAM.

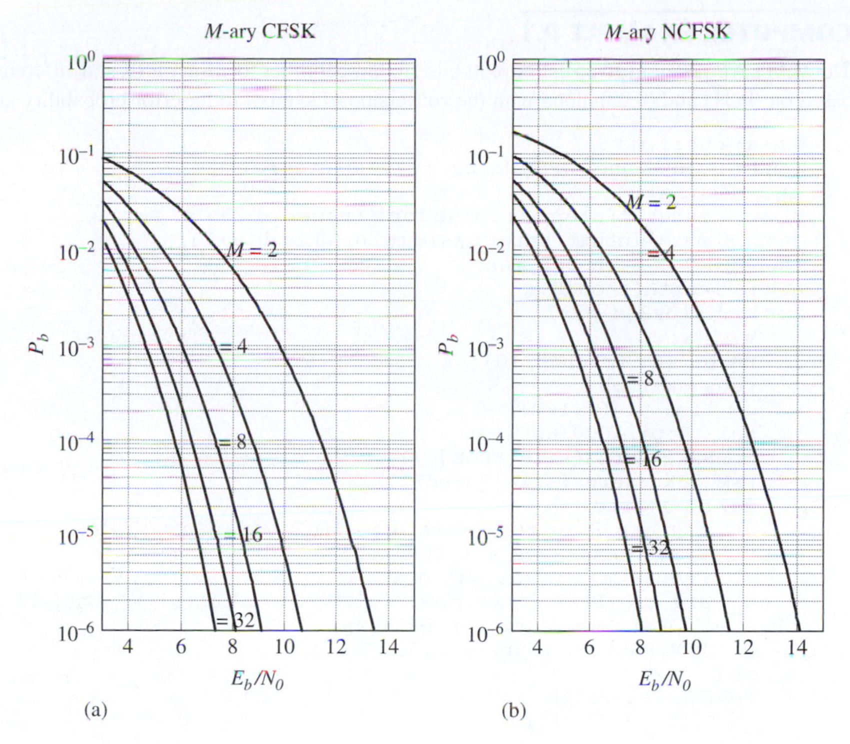

Figure 9.15
Bit-error probability versus E_b/N_0 for (a) coherent and (b) noncoherent M-ary FSK.

bandwidths. Note that the performance degradation of noncoherent over coherent FSK is not as severe as one might expect.

EXAMPLE 9.2

Compare the performances of noncoherent and coherent FSK on the basis of E_b/N_0 required to provide a bit-error probability of 10^{-6} for various values of M.

Solution

Using (9.67), (9.68), (9.95), and (9.96), the results in Table 9.4 can be obtained with the aid of appropriate MATLAB routines. Note that the loss in performance due to noncoherence is surprisingly small.

Table 9.4 Power Efficiencies for Noncoherent and Coherent FSK

	E_b/N_0 (dB) for $P_{E,bit} = 10^{-6}$	
M	Noncoherent	Coherent
2	14.20	13.54
4	11.40	10.78
8	9.86	9.26
16	8.80	8.22
32	8.02	7.48

■

COMPUTER EXAMPLE 9.1

The MATLAB program given below plots bit-error probabilities for M-ary PSK and differential M-ary PSK based on (9.51) and (9.92) along with the conversion of symbol to bit-error probability given by (9.93).

```
% file: c9ce1.m
% BEP for MPSK and MDPSK using Craig's integral
clf; clear all
M_max = input('Enter max value for M (power of 2) => ');
rhobdB_max = input('Enter maximum Eb/N0 in dB =>');
rhobdB = 5:0.5:rhobdB_max;
Lrho = length(rhobdB);
for k = 1:log2(M_max)
    M = 2^k;
    rhob = 10.^(rhobdB/10);
    rhos = k*rhob;
    up_lim = pi*(1-1/M);
    phi = 0:pi/1000:up_lim;
    PsMPSK = zeros(size(rhobdB));
    PsMDPSK = zeros(size(rhobdB));
    for m = 1:Lrho
        arg_exp_PSK = rhos(m)*sin(pi/M)^2./(sin(phi)).^2;
        Y_PSK = exp(-arg_exp_PSK)/pi;
        PsMPSK(m) = trapz(phi, Y_PSK);
        arg_exp_DPSK = rhos(m)*sin(pi/M)^2./(1+cos(pi/M)*cos(phi));
        Y_DPSK = exp(-arg_exp_DPSK)/pi;
        PsMDPSK(m) = trapz(phi, Y_DPSK);
    end
    PbMPSK = PsMPSK/k;
    PbMDPSK = PsMDPSK/k;
    if k ==1
       I = 4;
    elseif k == 2
       I = 5;
    elseif k == 3
       I = 10;
    elseif k == 4
       I = 19;
    elseif k == 5
       I = 28;
    end
    subplot(1,2,1), semilogy(rhobdB, PbMPSK), ...
       axis([min(rhobdB) max(rhobdB) 1e-6 1]), ...
        title('MPSK'), ylabel('{\itP_b}'), xlabel('{\itE_b/N}_0'), ...
        text(rhobdB(I)+.3, PbMPSK(I), ['{\itM} = ', num2str(M)])
    if k == 1
        hold on
        grid on
    end
    subplot(1,2,2), semilogy(rhobdB, PbMDPSK), ...
       axis([min(rhobdB) max(rhobdB) 1e-6 1]), ...
       title('MDPSK'), ylabel('{\itP_b}'), xlabel('{\itE_b/N}_0'),
        ...
       text(rhobdB(I+2)+.3, PbMPSK(I+2), ['{\itM} = ', num2str(M)])
    if k == 1
       hold on
       grid on
    end
end
```

Results computed using this program match those shown in Figure 9.14. ■

9.1.13 Comparison of M-ary Communications Systems on the Basis of Bandwidth Efficiency

If one considers the bandwidth required by an M-ary modulation scheme to be that required to pass the main lobe of the signal spectrum (null to null), it follows that the *bandwidth efficiencies* of the various M-ary schemes that we have just considered are as given in Table 9.5. These follow by extension of the arguements used in Chapter 8 for the binary cases. For example, analogous to (8.91) for coherent binary FSK, we have $1/T_s$ Hz on either end to the spectral null with $M-1$ spaces of $1/2T_s$ Hz inbetween for the remaining $M-2$ tone burst spectra ($M-1$ spaces $1/2T_s$ Hz wide), giving a total bandwidth of

$$\begin{aligned} B &= \frac{1}{T_s} + \frac{M-1}{2T_s} + \frac{1}{T_s} = \frac{M+3}{2T_s} \\ &= \frac{M+3}{2(\log_2 M)T_b} = \frac{(M+3)R_b}{2\log_2 M} \text{ Hz} \end{aligned} \tag{9.97}$$

from which the result for R_b/B given in Table 9.5 follows.

The reasoning for noncoherent FSK is similar except that tone burst spectra are assumed to be spaced by $2/T_s$ Hz[10] for a total bandwidth of

$$\begin{aligned} B &= \frac{1}{T_s} + \frac{2(M-1)}{T_s} + \frac{1}{T_s} = \frac{2M}{T_s} \\ &= \frac{2M}{(\log_2 M)T_b} = \frac{2MR_b}{\log_2 M} \text{ Hz} \end{aligned} \tag{9.98}$$

Phase-shift keying (including differentially coherent) and QAM have a single tone burst spectrum (of varying phase for PSK and phase/amplitude for QAM) for a total null-to-null bandwidth of

$$B = \frac{2}{T_s} = \frac{2}{(\log_2 M)T_b} = \frac{2R_b}{\log_2 M} \text{ Hz} \tag{9.99}$$

Table 9.5 Bandwidth Efficiencies of Various M-ary Digital Modulation Schemes

M-ary scheme	Bandwidth efficiency (bits/s/Hz)
PSK, QAM	$\frac{1}{2}\log_2 M$
Coherent FSK	$\frac{2\log_2 M}{M+3}$ (tone spacing of $1/2T_s$ Hz)
Noncoherent FSK	$\frac{\log_2 M}{2M}$ (tone spacing of $2/T_s$ Hz)

[10] This increased tone spacing as compared with coherent FSK is made under the assumption that frequency is not estimated in a noncoherent system to the degree of accuracy as would be necessary in a coherent system, where detection is implemented by correlation with the possible transmitted frequencies.

EXAMPLE 9.3

Compare bandwidth efficiencies on a main-lobe spectrum basis for PSK, QAM, and FSK for various M.

Solution

Bandwidth efficiencies in bits per second per hertz for various values of M are as given in Table 9.6. Note that for QAM, M is assumed to be a power of 4. Also note that the bandwidth efficiency of M-ary PSK *increases* with increasing M while that for FSK *decreases*.

Table 9.6 Bandwidth Efficiencies for Example 9.3 bps/Hz

M	QAM	PSK	Coherent FSK	Noncoherent FSK
2		0.5	0.4	0.25
4	1	1	0.57	0.25
8		1.5	0.55	0.19
16	2	2	0.42	0.13
32		2.5	0.29	0.08
64	3	3	0.18	0.05

■

■ 9.2 POWER SPECTRA FOR QUADRATURE MODULATION TECHNIQUES

The measures of performance for the various modulation schemes considered so far have been probability of error and bandwidth occupancy. For the latter, we used rough estimates of bandwidth based on null-to-null points of the modulated signal spectrum. In this section, we derive an expression for the power spectrum of quadrature modulated signals. This can be used to obtain more precise measures of the bandwidth requirements of quadrature modulation schemes such as QPSK, OQPSK, MSK, and QAM. One might ask why not do this for other signal sets, such as M-ary FSK. The answer is that such derivations are complex and difficult to apply (recall the difficulty of deriving spectra for analog FM). The literature on this problem is extensive, an example of which is given here.[11]

Analytical expressions for the power spectra of digitally modulated signals allow a definition of bandwidth that is based on the criterion of fractional power of the signal within a specified bandwidth. That is, if $S(f)$ is the double-sided power spectrum of a given modulation format, the fraction of total power in a bandwidth B is given by

$$\Delta P_{\mathrm{IB}} = \frac{2}{P_T} \int_{f_c - B/2}^{f_c + B/2} S(f)\, df \tag{9.100}$$

where the factor of 2 is used since we are only integrating over positive frequencies,

$$P_T = \int_{-\infty}^{\infty} S(f)\, df \tag{9.101}$$

[11] H. E. Rowe and V. K. Prabhu, Power spectrum of a digital, frequency-modulation signal. *The Bell System Technical Journal*, **54:** 1095–1125, July–August 1975.

is the total power, and f_c is the "center" frequency of the spectrum (usually the carrier frequency, apparent or otherwise). The percent out-of-band power ΔP_{OB} is defined as

$$\Delta P_{\mathrm{OB}} = (1 - \Delta P_{\mathrm{IB}}) \times 100\% \tag{9.102}$$

The definition of modulated signal bandwidth is conveniently given by setting ΔP_{OB} equal to some acceptable value, say 0.01 or 1%, and solving for the corresponding bandwidth. A curve showing ΔP_{OB} in decibels versus bandwidth is a convenient tool for carrying out this procedure, since the 1% out-of-band power criterion for bandwidth corresponds to the bandwidth at which the out-of-band power curve has a value of −20 dB. Later we will present several examples to illustrate this procedure.

As pointed out in Chapter 4, the spectrum of a digitally modulated signal is influenced both by the particular baseband data format used to represent the digital data and by the type of modulation scheme used to prepare the signal for transmission. We will assume nonreturn-to-zero (NRZ) data formatting in the following.

In order to obtain the spectrum of a quadrature-modulated signal using any of these data formats, the appropriate spectrum shown in Figure 9.16 is simply shifted up in frequency and centered around the carrier (assuming a single-sided spectrum).

To proceed, we consider a quadrature-modulated waveform of the form given by (9.1), where $m_1(t) = d_1(t)$ and $m_2(t) = -d_2(t)$ are random (coin toss) waveforms represented as

$$d_1(t) = \sum_{k=-\infty}^{\infty} a_k p(t - kT_s - \Delta_1) \tag{9.103}$$

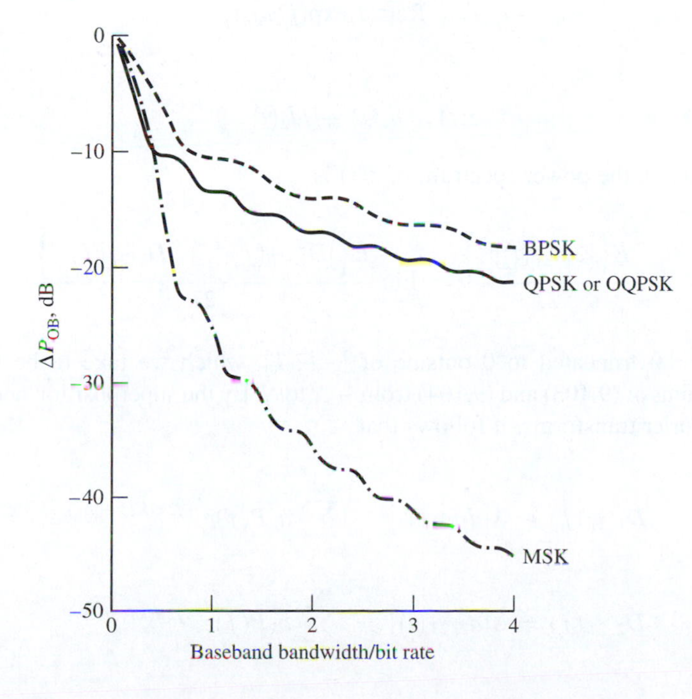

Figure 9.16
Fractional out-of-band power for BPSK, QPSK or OQPSK, and MSK.

and

$$d_2(t) = \sum_{k=-\infty}^{\infty} b_k p(t - kT_s - \Delta_2) \tag{9.104}$$

where $\{a_k\}$ and $\{b_k\}$ are independent, identically distributed (iid) sequences with

$$E\{a_k\} = E\{b_k\} = 0, \quad E\{a_k a_l\} = A^2\delta_{kl}, \quad E\{b_k b_l\} = B^2\delta_{kl} \tag{9.105}$$

in which $\delta_{kl} = 1$ for $k = l$ and 0 otherwise, is called the *Kronecker delta*.

The pulse shape functions $p(t)$ and $q(t)$ in (9.103) and (9.104) may be the same, or one of them may be zero. We now show that the double-sided spectrum of (9.1), with (9.103) and (9.104) substituted, is

$$S(f) = G(f - f_c) + G(f + f_c) \tag{9.106}$$

where

$$G(f) = \frac{A^2|P(f)|^2 + B^2|Q(f)|^2}{T_s} \tag{9.107}$$

in which $P(f)$ and $Q(f)$ are the Fourier transforms of $p(t)$ and $q(t)$, respectively. This result can be derived by applying (6.25). First, we may write the modulated signal in terms of its complex envelope as

$$x_c(t) = \text{Re}[z(t)\exp(j2\pi f_c t)] \tag{9.108}$$

where

$$z(t) = d_1(t) + jd_2(t) \tag{9.109}$$

According to (6.25), the power spectrum of $z(t)$ is

$$G(f) = \lim_{T\to\infty} \frac{E\left\{|\Im[z_{2T}(t)]|^2\right\}}{2T} = \lim_{T\to\infty} \frac{E\left\{|D_{1,\,2T}(f)|^2 + |D_{2,\,2T}(f)|^2\right\}}{2T} \tag{9.110}$$

where $z_{2T}(t)$ is $z(t)$ truncated to 0 outside of $[-T, T]$, which we take to be the same as truncating the sums of (9.103) and (9.104) from $-N$ to N. By the superposition and time-delay theorems of Fourier transforms, it follows that

$$\begin{aligned} D_{1,\,2T}(f) &= \Im[d_{1,\,2T}(t)] = \sum_{k=-N}^{N} a_k P(f) e^{-j2\pi(kT_s + \Delta_1)} \\ D_{2,\,2T}(f) &= \Im[d_{2,\,2T}(t)] = \sum_{k=-N}^{N} b_k P(f) e^{-j2\pi(kT_s + \Delta_2)} \end{aligned} \tag{9.111}$$

which gives

$$
\begin{aligned}
E\left\{\left|D_{1,2T}(f)\right|^2\right\} &= E\left\{\sum_{k=-N}^{N} a_k P(f) e^{-j2\pi(kT_s+\Delta_1)} \sum_{l=-N}^{N} a_l P^*(f) e^{j2\pi(lT_s+\Delta_1)}\right\} \\
&= |P(f)|^2 E\left\{\sum_{k=-N}^{N}\sum_{l=-N}^{N} a_k a_l e^{-j2\pi(k-l)T_s}\right\} \\
&= |P(f)|^2 \sum_{k=-N}^{N}\sum_{l=-N}^{N} E(a_k a_l) e^{-j2\pi(k-l)T_s} \\
&= |P(f)|^2 \sum_{k=-N}^{N}\sum_{l=-N}^{N} A^2 \delta_{kl} e^{-j2\pi(k-l)T_s} \\
&= |P(f)|^2 \sum_{k=-N}^{N} A^2 = (2N+1)|P(f)|^2 A^2
\end{aligned}
\tag{9.112}
$$

Similarly, it follows that

$$E\left\{\left|D_{2,2T}(f)\right|^2\right\} = (2N+1)|P(f)|^2 B^2 \tag{9.113}$$

Let $2T = (2N+1)T_s + \Delta t$, where $\Delta t < T_s$ accounts for end effects, and substitute (9.112) and (9.113) into (9.110), which becomes (9.107) in the limit.

This result can be applied to BPSK, for example, by letting $q(t) = 0$ and $p(t) = \Pi(t/T_b)$. The resulting baseband spectrum is

$$G_{\text{BPSK}}(f) = A^2 T_b \,\text{sinc}^2(T_b f) \tag{9.114}$$

The spectrum for QPSK follows by letting $A^2 = B^2$, $T_s = 2T_b$, and

$$p(t) = q(t) = \frac{1}{\sqrt{2}}\Pi\left(\frac{t}{2T_b}\right) \tag{9.115}$$

to get $P(f) = Q(f) = \sqrt{2}T_b\,\text{sinc}(2T_b f)$. This results in the baseband spectrum

$$G_{\text{QPSK}}(f) = \frac{2A^2|P(f)|^2}{2T_b} = 2A^2 T_b \,\text{sinc}^2(2T_b f) \tag{9.116}$$

This result also holds for OQPSK because the pulse shape function $q(t)$ differs from $p(t)$ only by a time shift that results in a factor of $\exp(-j2\pi T_b f)$ (magnitude of unity) in the amplitude spectrum $|Q(f)|$.

For M-ary QAM we use $A^2 = B^2$ (these are the mean-squared values of the amplitudes on the I and Q channels), $T_s = (\log_2 M)T_b$, and

$$p(t) = q(t) = \frac{1}{\sqrt{\log_2 M}}\Pi\left(\frac{t}{(\log_2 M)T_b}\right) \tag{9.117}$$

to get $P(f) = Q(f) = \sqrt{\log_2 M T_b}\,\text{sinc}[(\log_2 M)T_b f]$. This gives the baseband spectrum

$$G_{\text{MQAM}}(f) = \frac{2A^2|P(f)|^2}{(\log_2 M)T_b} = 2A^2 T_b \,\text{sinc}^2[(\log_2 M)T_b f] \tag{9.118}$$

The baseband spectrum for MSK is found by choosing the pulse shape functions

$$p(t) = q(t - T_b) = \cos\left(\frac{\pi t}{2T_b}\right)\Pi\left(\frac{t}{2T_b}\right) \tag{9.119}$$

and by letting $A^2 = B^2$. It can be shown (see Problem 9.25) that

$$\Im\left[\cos\left(\frac{\pi t}{2T_b}\right)\Pi\left(\frac{t}{2T_b}\right)\right] = \frac{4T_b\cos(2\pi T_b f)}{\pi\left[1-(4T_b f)^2\right]} \tag{9.120}$$

which results in the following baseband spectrum for MSK:

$$G_{\mathrm{MSK}}(f) = \frac{16A^2 T_b \cos^2(2\pi T_b f)}{\pi^2\left[1-(4T_b f)^2\right]^2} \tag{9.121}$$

Using these results for the baseband spectra of BPSK, QPSK (or OQPSK), and MSK in the definition of percent out-of-band power (9.102) results in the set of plots for fractional out-of-band power shown in Figure 9.16. These curves were obtained by numerical integration of the power spectra of (9.114), (9.116), and (9.121). From Figure 9.16, it follows that the RF bandwidths containing 90% of the power for these modulation formats are approximately

$$\begin{aligned} B_{90\%} &\cong \frac{1}{T_b}\ \text{Hz}\ \ (\text{QPSK, OQPSK, MSK}) \\ B_{90\%} &\cong \frac{2}{T_b}\ \text{Hz}\ \ (\text{BPSK}) \end{aligned} \tag{9.122}$$

These are obtained by noting the bandwidths corresponding to $\Delta P_{OB} = -10$ dB and doubling these values, since the plots are for baseband bandwidths.

Because the MSK out-of-band power curve rolls off at a much faster rate than do the curves for BPSK or QPSK, a more stringent in-band power specification, such as 99%, results in a much smaller containment bandwidth for MSK than for BPSK or QPSK. The bandwidths containing 99% of the power are

$$\begin{aligned} B_{99\%} &\cong \frac{1.2}{T_b} \quad (\text{MSK}) \\ B_{99\%} &\cong \frac{8}{T_b} \quad (\text{QPSK or OQPSK; BPSK off the plot}) \end{aligned} \tag{9.123}$$

For binary FSK, the following formula can be used to compute the power spectrum if the phase is continuous:[12]

$$G_f = G_+(f) + G_-(f) \tag{9.124}$$

where

$$G_\pm(f) = \frac{A^2\sin^2\left[\pi(f\pm f_1)T_b\right]\sin^2\left[\pi(f\pm f_2)T_b\right]}{2\pi^2 T_b\{1-2\cos[2\pi(f\pm\alpha)T_b]\cos(2\pi\beta T_b)+\cos^2(2\pi\beta T_b)\}}\left(\frac{1}{f\pm f_1}-\frac{1}{f\pm f_2}\right)^2 \tag{9.125}$$

[12]W. R. Bennett and S. O. Rice, Spectral density and autocorrelation functions associated with binary frequency-shift keying. *Bell System Technical Journal*, **42:** 2355–2385, September 1963.

In (9.125), the following definitions are used:

f_1, f_2 = the signaling frequencies in hertz $\left(\text{that is, } f_c - \frac{\Delta f}{2} \text{ and } f_c + \frac{\Delta f}{2}\right)$

$$\alpha = \frac{1}{2}(f_1 + f_2)$$
$$\beta = \frac{1}{2}(f_2 - f_1) \tag{9.126}$$

Equation (9.124) is used to get the bandpass (modulated) signal spectrum. Several examples of spectra are shown in Figure 9.17 for frequency-modulated spectra computed

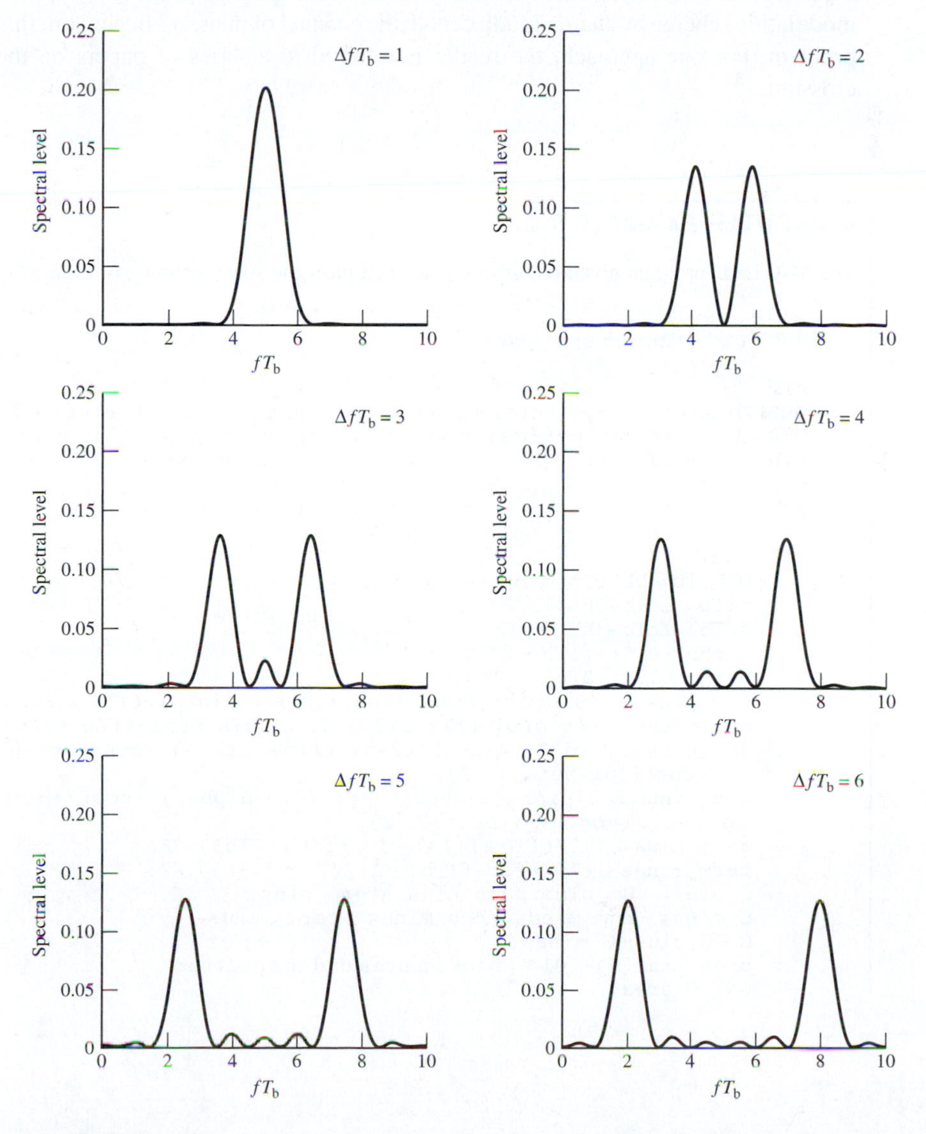

Figure 9.17
Power spectra for continuous-phase FSK computed from (9.125).

from (9.124) using (9.125) for a normalized carrier frequency of $f_c T_b = 5$ and normalized signaling frequency separations $(f_2 - f_1)T_b$ in steps from 1 to 6. Note that as the separation increases from 1 to 6, the spectrum goes from unimodal to bimodal with the bimodal components centered around the nominal signaling frequencies (e.g., $5 - 3 = 2$ and $5 + 3 = 8$ for the last case, which looks very much like the superposition of two BPSK spectra centered around the signaling frequencies).

The preceding approach to determining bandwidth occupancy of digitally modulated signals provides one criterion for selecting modulation schemes based on bandwidth considerations. It is not the only approach by any means. Another important criterion is adjacent channel interference. In other words, what is the degradation imposed on a given modulation scheme by channels adjacent to the channel of interest? In general, this is a difficult problem. For one approach, the reader is referred to a series of papers on the concept of crosstalk.[13]

COMPUTER EXAMPLE 9.2

The MATLAB program given below computes and plots the spectra shown in Figure 9.17.

```
% file: c9ce2.m
% Plot of FM power spectra
%
clf
DELfTb_min = input('Enter min freq spacing X bit period between tones =>');
DELfTb_0 = input('Enter step size in frequency spacing X bit period =>');
fTb = 0.009:0.01:10;            % Start fTb out at a value that avoids
                                zeros in denom
fcTb = 5;                       % Apparent carrier frequency,
                                normalized
for n=1:6
    DELfTb = DELfTb_min + (n-1)*DELfTb_0
    f1Tb = fcTb-DELfTb/2;
    f2Tb = fcTb+DELfTb/2;
    alpha = 0.5*(f1Tb + f2Tb);
    beta = 0.5*(f2Tb - f1Tb);
    num_plus = ((sin(pi*(fTb+f1Tb))).^2).*(sin(pi*(fTb+f2Tb))).^2;
    num_minus = ((sin(pi*(fTb-f1Tb))).^2).*(sin(pi*(fTb-f2Tb))).^2;
    den_plus = 2*pi^2*(1-2*cos(2*pi*(fTb+alpha)).*cos(2*pi*beta+eps)+
    ...(cos(2*pi*beta)).^2);
    den_minus = 2*pi^2*(1-2*cos(2*pi*(fTb-alpha)).*cos(2*pi*beta+
    eps)+... (cos(2*pi*beta)).^2);
    term_plus = (1./(fTb+f1Tb) - 1./(fTb+f2Tb)).^2;
    term_minus = (1./(fTb-f1Tb) - 1./(fTb-f2Tb)).^2;
    G_plus = num_plus./den_plus.*term_plus;
    G_minus = num_minus./den_minus.*term_minus;
    G = G_plus+G_minus;
    area = sum(G)*.01 % Check on area under spectrum
    GN = G/area;
```

[13]See I. Kalet, A look at crosstalk in quadrature-carrier modulation systems. *IEEE Transactions on Communications*, **COM-25:** 884–892, September 1977.

```
        subplot(3,2,n),xlabel('fT_b'),plot(fTb, GN), ...
            ylabel('Spectral level'), axis([0 10 0 max(GN)]),...
            legend(['DeltafT_b=' ,num2str(DELfTb)]),...
            if n == 5 n ==6
                xlabel('{itfT_b}')
            end
    end
```

9.3 SYNCHRONIZATION

We have seen that at least two levels of synchronization are necessary in a coherent communication system. For the known-signal-shape receiver considered in Section 8.2, the beginning and ending times of the signals must be known. When specialized to the case of coherent ASK, PSK, or coherent FSK, knowledge is required not only of the bit timing but of carrier phase as well. In addition, if the bits are grouped into blocks or words, the starting and ending times of the words are also required. In this section we will look at methods for achieving synchronization at these three levels. In order of consideration, we will look at methods for (1) carrier synchronization, (2) bit synchronization (already considered in Section 4.7 at a simple level), and (3) word synchronization. There are also other levels of synchronization in some communication systems that will not be considered here.

9.3.1 Carrier Synchronization

The main types of digital modulation methods considered were ASK, PSK, FSK, PAM, and QAM. Amplitude-shift keying and FSK can be noncoherently modulated, and PSK can be differentially modulated thus avoiding the requirement of a coherent carrier reference at the receiver (of course, we have seen that detection of noncoherently modulated signals entails some degradation in E_b/N_0 in data detection relative to the corresponding coherent modulation scheme). In the case of coherent ASK a discrete spectral component at the carrier frequency will be present in the received signal that can be tracked by a PLL to implement coherent demodulation (which is the first step in data detection). In the case of FSK discrete spectral components related to the FSK tones may be present in the received signal depending on the modulation parameters. For MPSK, assuming equally likely phases due to the modulation, a carrier component is not present in the received signal. If the carrier component is absent, one may sometimes be inserted along with the modulated signal (called a *pilot carrier*) to facilitate generation of a carrier reference at the receiver. Of course the inclusion of a pilot carrier robs power from the data-modulated part of the signal that will have to be accounted for in the power budget for the communications link.

We now focus attention on PSK. For BPSK, which really amounts to DSB modulation as considered in Chapter 3, two alternatives were illustrated in Chapter 3 for coherent demodulation of DSB. In particular these were a squaring PLL arrangement and a Costas loop. When used for digital data demodulation of BPSK, however, these loop mechanizations introduce a problem that was not present for demodulation of analog message signals. We note that either loop (squaring or Costas) will lock if $d(t)\cos(\omega_c t)$ or $-d(t)\cos(\omega_c t)$ is present at the loop input (i.e., we can't tell if the data-modulated carrier has been accidently inverted from our

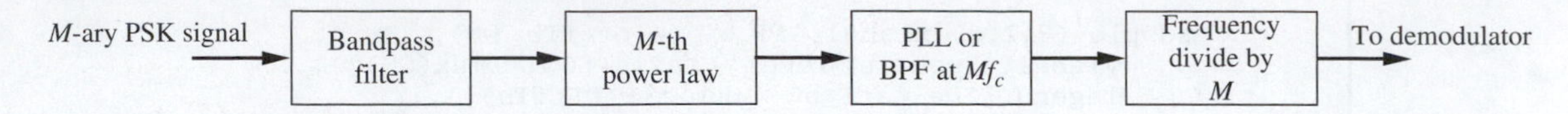

Figure 9.18
M-power law system for carrier synchronization of *M*-ary PSK.

perspective or not). Some method is usually required to resolve this sign ambiguity at the demodulator output. One method of doing so is to differentially encode the data stream before modulation and differentially decode it at the detector output with a resultant small loss in SNR. This is referred to as *coherent detection* of differentially encoded BPSK and is different from differentially coherent detection of DPSK.

Circuits similar to the Costas and squaring loops may be constructed for *M*-ary PSK. For example, the mechanism shown by the block diagram of Figure 9.18 will produce a coherent carrier reference from *M*-ary PSK, as the following development shows.[14] We take the Mth power of a PSK signal and get

$$
\begin{aligned}
y(t) &= [s_i(t)]^M = \left[\sqrt{\frac{2E_s}{T_s}}\cos\left(\omega_c t + \frac{2\pi(i-1)}{M}\right)\right]^M \\
&= A^M\left[\frac{1}{2}\exp\left(j\omega_c t + j\frac{2\pi(i-1)}{M}\right) + \frac{1}{2}\exp\left(-j\omega_c t - j\frac{2\pi(i-1)}{M}\right)\right]^M \\
&= \left(\frac{A}{2}\right)^M\left\{\sum_{m=0}^{M}\binom{M}{m}\exp\left[j(M-m)\omega_c t + j\frac{2\pi(M-m)(i-1)}{M}\right]\right. \\
&\qquad \left.\times\exp\left(-jm\omega_c t - j\frac{2\pi m(i-1)}{M}\right)\right\} \\
&= \left(\frac{A}{2}\right)^M\left\{\sum_{m=0}^{M}\binom{M}{m}\exp\left[j(M-2m)\omega_c t + j\frac{2\pi(M-2m)(i-1)}{M}\right]\right\} \\
&= \left(\frac{A}{2}\right)^M\{\exp[jM\omega_c t + j2\pi(i-1)] + \exp[-jM\omega_c t - j2\pi(i-1)] + \cdots\} \\
&= \left(\frac{A}{2}\right)^M\{2\cos[M\omega_c t + 2\pi(i-1)] + \cdots\} = \left(\frac{A}{2}\right)^M\{2\cos(M\omega_c t) + \cdots\}
\end{aligned}
\tag{9.127}
$$

where $A = \sqrt{2E_s/T_s}$ has been used for convenience and the binomial formula (see Appendix G.3) has been used to carry out the expansion of the *M*th power. Only the first and last terms of the sum in the fourth line are of interest (the remaining terms are indicated by

[14]Just as there is a binary phase ambiguity in Costas or squaring loop demodulation of BPSK, an *M*-phase ambiguity is present in the establishing of a coherent carrier reference in *M*-PSK by using the *M*-power technique illustrated here.

Table 9.7 Tracking Loop Error Variances

Type of modulation	Tracking loop error variance, σ_ϕ^2
None (PLL)	N_0B_L/P_c
BPSK (squaring or Costas loop)	$L^{-1}(1/z + 0.5/z^2)$
QPSK (quadrupling or data estimation loop)	$L^{-1}(1/z + 4.5/z^2 + 6/z^3 + 1.5/z^4)$

the three dots), for they make up the term $2\cos[M\omega_c t + 2\pi(i-1)] = 2\cos(2\pi Mf_c t)$, which can clearly be tracked by a PLL and a frequency divider used to produce a coherent reference at the carrier frequency. A possible disadvantage of this scheme is that M times the desired frequency must be tracked. Normally this would not be the carrier frequency itself but, rather, an IF. Costas-like carrier tracking loops for $M > 2$ have been presented and analyzed in the literature, but these will not be discussed here. We refer the reader to the literature on the subject[15], including the two-volume work by Meyr and Ascheid (1990).

The question naturally arises as to the effect of noise on these phase-tracking devices. The *phase error*, that is, the difference between the input signal phase and the VCO phase, can be shown to be approximately Gaussian with zero mean at high SNRs at the loop input. Table 9.7 summarizes the phase-error variance for these various cases.[16] When used with equations such as (8.83), these results provide a measure for the average performance degradation due to an imperfect phase reference. Note that in all cases, σ_ϕ^2 is inversely proportional to the SNR raised to integer powers and to the effective number L of symbols remembered by the loop in making the phase estimate. (See Problem 9.28.)

The terms used in Table 9.7 are defined as follows:

T_s = symbol duration.

B_L = single-sided loop bandwidth.

N_0 = single-sided noise spectral density.

L = effective number of symbols used in phase estimate.

P_c = signal power (tracked component only).

E_s = symbol energy.

$z = E_s/N_0$.

$L = 1/(B_LT_s)$.

EXAMPLE 9.4

Compare tracking error standard deviations of two BPSK systems: (1) One using a PLL tracking on a BPSK signal with 10% of the total transmit power in a carrier component and (2) the second using a Costas loop tracking a BPSK signal with no carrier component. The data rate is $R_b = 10$ kbps, and the received E_b/N_0 is 10 dB. The loop bandwidths of both the PLL and Costas loops are 50 Hz. (3) For the same parameter values what is the tracking error variance for a QPSK tracking loop?

[15] B. T. Kopp and W. P. Osborne, Phase jitter in MPSK carrier tracking loops: Analytical, simulation and laboratory results. *IEEE Transactions on Communications*, **COM-45:** 1385–1388, November 1997. S. Hinedi and W. C. Lindsey, On the self-noise in QASK decision-feedback carrier tracking loops. *IEEE Transactions on Communications*, **COM-37:** 387–392, April 1989.

[16] Stiffler (1971), Equation (8.3.13).

Solution

For (1), from Table 9.7, first row, the PLL tracking error variance and standard deviation are

$$\sigma^2_{\phi,\,\mathrm{PLL}} = \frac{N_0 B_L}{P_c} = \frac{N_0(T_b B_L)}{P_c T_b} = \frac{N_0}{0.1 E_b}\frac{B_L}{R_b}$$
$$= \frac{1}{0.1 \times 10}\frac{50}{10^4} = 5 \times 10^{-3}\ \mathrm{rad}^2$$
$$\sigma_{\phi,\,\mathrm{PLL}} = 0.0707\ \mathrm{rad}$$

For (2), from Table 9.7, second row, the Costas PLL tracking error variance and standard deviation are

$$\sigma^2_{\phi,\,\mathrm{Costas}} = B_L T_b\left(\frac{1}{z} + \frac{1}{2z^2}\right)$$
$$= \frac{50}{10^4}\left(\frac{1}{10} + \frac{1}{200}\right) = 5.25 \times 10^{-4}\ \mathrm{rad}^2$$
$$\sigma_{\phi,\,\mathrm{Costas}} = 0.0229\ \mathrm{rad}$$

The first case has the disadvantage that the loop tracks on only 10% of the received power. Not only is the PLL tracking on a lower power signal than the Costas loop, but either there is less power for signal detection (if total transmit powers are the same in both cases), or the transmit power for case 1 must be 10% higher than for case 2.

For (3), from Table 9.7, third row, the QPSK data tracking loop's tracking error variance and standard deviation are ($T_s = 2T_b$)

$$\sigma^2_{\phi,\,\mathrm{QPSK\ data\ est}} = 2B_L T_b\left(\frac{1}{z} + \frac{4.5}{z^2} + \frac{6}{z^3} + \frac{1.5}{z^4}\right)$$
$$= \frac{100}{10^4}\left(\frac{1}{10} + \frac{4.5}{100} + \frac{6}{1{,}000} + \frac{1.5}{10{,}000}\right)$$
$$= 1.5 \times 10^{-3}\ \mathrm{rad}^2$$
$$\sigma_{\phi,\,\mathrm{QPSK\ data\ est}} = 0.0389\ \mathrm{rad}$$

■

9.3.2 Symbol Synchronization

Three general methods by which symbol synchronization[17] can be obtained are

1. Derivation from a primary or secondary standard (for example, transmitter and receiver slaved to a master timing source with delay due to propagation accounted for)
2. Utilization of a separate synchronization signal (use of a pilot clock, or a line code with a spectral line at the symbol rate—for example, see the unipolar RZ spectrum of Figure 4.3)
3. Derivation from the modulation itself, referred to as *self-synchronization*, as explored in Chapter 4 (see Figure 4.16 and accompanying discussion).

---

[17] See Stiffler (1971) or Lindsey and Simon (1973) for a more extensive discussion.

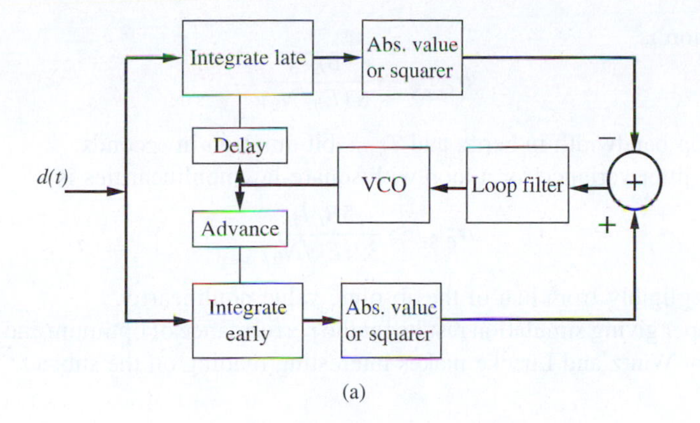

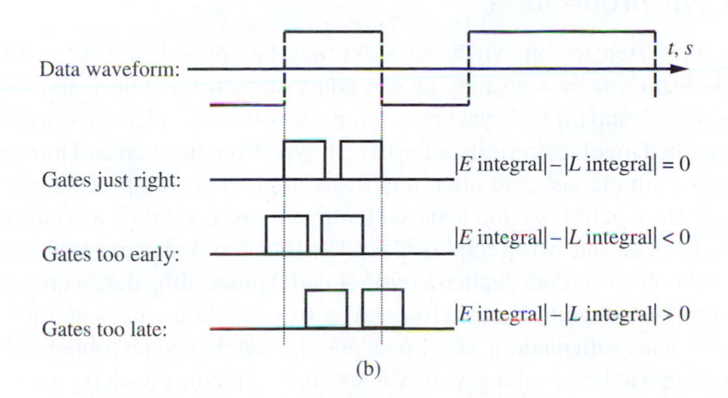

Figure 9.19
(a) Early-late gate type of bit synchronizer. (b) Waveforms pertinent to its operation.

Loop configurations for acquiring bit synchronization that are similar in form to the Costas loop are also possible.[18] One such configuration, called the *early-late gate synchronization loop*, is shown in Figure 9.19(a) in its simplest form. A binary NRZ data waveform is assumed as shown in Figure 9.19(b). Assuming that the integration gates' start and stop times are coincident with the leading and trailing edges, respectively, of a data bit 1 (or data bit -1), it is seen that the control voltage into the loop filter will be zero and the VCO will be allowed to put out timing pulses at the same frequency. On the other hand, if the VCO timing pulses are such that the gates are too early, the control voltage into the VCO will be negative, which will decrease the VCO frequency so that VCO timing pulses will delay the gate timing. Similarly, if the VCO timing pulses are such that the gates are too late, the control voltage into the VCO will be positive, which will increase the VCO frequency so that VCO timing pulses will advance the gate timing. The nonlinearity in the feedforward arms can be any even-order nonlinearity. It has been shown[19] that for an absolute value nonlinearity the variance of the timing jitter normalized

[18] See L. E. Franks, Carrier and bit synchronization in data communication—A tutorial review. *IEEE Transactions on Communications*, **Com-28:** 1107–1121, August 1980. Also see C. Georghiades and E. Serpedin, Synchronization, Chapter 19 in Gibson (2002).

[19] Simon, M. K., Nonlinear analysis of an absolute value type of an early-late gate bit synchronizer. *IEEE Transactions on Communication Technology*, **COM-18:** 589–596, October 1970.

by the bit duration is

$$\sigma^2_{\epsilon,\,\mathrm{AV}} \cong \frac{B_L T_b}{8\,(E_b/N_0)} \tag{9.128}$$

where $B_L =$ loop bandwidth in hertz, and $T_b =$ bit duration in seconds.

The timing jitter variance for a loop with square-law nonlinearities is

$$\sigma^2_{\epsilon,\,\mathrm{SL}} \cong \frac{5B_L T_b}{32(E_b/N_0)} \tag{9.129}$$

which differs negligibly from that of the absolute value nonlinearity.

An early paper giving simulation results for the performance of optimum and suboptimum synchronizers by Wintz and Luecke makes interesting reading on the subject.[20]

9.3.3 Word Synchronization

The same principles used for bit synchronization may be applied to word synchronization. These are (1) derivation from a primary or secondary standard, (2) utilization of a separate synchronization signal, and (3) self-synchronization. Only the second method will be discussed here. The third method involves the utilization of self-synchronizing codes. The construction of good codes is not a simple task and often requires computer search procedures.[21]

When a separate synchronization code is employed, this code may be transmitted over a channel separate from the one being employed for data transmission or over the data channel by inserting the synchronization code (called a *marker code*) preceeding data words. Such marker codes should have low-magnitude nonzero-delay autocorrelation values and low-magnitude cross-correlation values with random data. Some possible marker codes, obtained by computer search, are given in Table 9.8 along with values for their nonzero-delay peak correlation magnitudes.[22] Concatenation of the marker code and data sequence constitutes one *frame*.

Table 9.8 Marker Codes with Peak Nonzero-Delay Correlation Values

Code	Binary representation	Magnitude: peak correlation*
C7	1 0 1 1 0 0 0	1
C8	1 0 1 1 1 0 0 0	3
C9	1 0 1 1 1 0 0 0 0	2
C10	1 1 0 1 1 1 0 0 0 0	3
C11	1 0 1 1 0 1 1 1 0 0 0	1
C12	1 1 0 1 0 1 1 0 0 0 0 0	2
C13	1 1 1 0 1 0 1 1 0 0 0 0 0	3
C14	1 1 1 0 0 1 0 1 1 0 0 0 0 0	3
C15	1 1 1 1 1 0 0 1 1 0 1 0 1 1 0	3

*Zero-delay correlation = length of code

[20]P. A. Wintz and E. J. Luecke, Performance of optimum and suboptimum synchronizers. *IEEE Transactions on Communication Technology*, **Com-17:** 380–389, June 1969.

[21]See Stiffler (1971) or Lindsey and Simon (1973).

[22]R. A. Scholtz, Frame synchronization techniques. *IEEE Transactions on Communications*, **COM-28:** 1204–1213, August 1980.

Finally, it is important that correlation with the locally stored marker code be relatively immune to channel errors in the incoming marker code and in the received data frame. Scholtz gives a bound for the one-pass (i.e, on one marker sequence correlation) acquisition probability for frame synchronization. For a frame consisting of M marker bits and D data bits, it is

$$P_{\text{one-pass}} \geq [1 - (D + M - 1)P_{\text{FAD}}]P_{\text{TAM}} \tag{9.130}$$

where P_{FAD}, the probability of false acquisition on data alone, and P_{TAM}, the probability of true acquisition of the marker code, are given, respectively, by

$$P_{\text{FAD}} = \left(\frac{1}{2}\right)^M \sum_{k=0}^{h} \binom{M}{k} \tag{9.131}$$

and

$$P_{\text{TAM}} = \sum_{l=0}^{h} \binom{M}{l} (1 - P_e)^{M-l} P_e^l \tag{9.132}$$

in which h is the allowed disagreement between the marker sequence and the closest sequence in the received frame and P_e is the probability of a bit error due to channel noise.

To illustrate implemention of a search for the marker sequence in a received frame (with some errors due to noise), consider the received frame sequence

1 1 0 1 0 0 0 1 0 1 1 0 0 1 1 0 1 1 1 1

Suppose $h = 1$ and we want to find the closest match (to within one bit) of the 7-bit marker sequence 1 0 1 1 0 0 0. This amounts to counting the total number of disagreements, called the *Hamming distance*, between the marker sequence and a 7-bit block of the frame. This is illustrated by Table 9.9.

Table 9.9 Illustration of Word Synchronization with a Marker Code

1	1	0	1	0	0	0	1	0	1	1	0	0	1	1	0	1	1	1	1	(delay, Hamming distance)
1	0	1	1	0	0	0														(0, 2)
	1	0	1	1	0	0	0													(1, 2)
		1	0	1	1	0	0	0												(2, 5)
			1	0	1	1	0	0	0											(3, 4)
				1	0	1	1	0	0	0										(4, 4)
					1	0	1	1	0	0	0									(5, 4)
						1	0	1	1	0	0	0								(6, 4)
							1	0	1	1	0	0	0							(7, 1)
								1	0	1	1	0	0	0						(8, 5)
									1	0	1	1	0	0	0					(9, 5)
										1	0	1	1	0	0	0				(10, 3)
											1	0	1	1	0	0	0			(11, 3)
												1	0	1	1	0	0	0		(12, 6)
													1	0	1	1	0	0	0	(13, 5)

There is one match to within one bit; so the test has succeeded. In fact, one of four possibilities can occur each time we correlate a marker sequence with a frame: Let $\text{ham}(\mathbf{m}, \mathbf{d}_i)$ be the Hamming distance between the marker code $\mathbf{m}$ and the ith 7-bit (in this case) segment of the frame sequence $\mathbf{d}_i$. The possible outcomes are

1. We get $\text{ham}(\mathbf{m}, \mathbf{d}_i) \leq h$ for one, and only one, shift, and it is the correct one (sync detected correctly).
2. We get $\text{ham}(\mathbf{m}, \mathbf{d}_i) \leq h$ for one, and only one, shift, and it is the incorrect one (sync detected in error).
3. We get $\text{ham}(\mathbf{m}, \mathbf{d}_i) \leq h$ for two or more shifts (no sync detected).
4. We get no result for which $\text{ham}(\mathbf{m}, \mathbf{d}_i) \leq h$ (no sync detected).

If we do this experiment repeatedly, with each bit being in error with probability P_e, then $P_{\text{one-pass}}$ is approximately the ratio of correct syncs to the total number of trials. Of course, in an actual system, the test of whether the synchronization is successful is if the data can be decoded properly.

The number of marker bits to provide one-pass probabilities of 0.93, 0.95, 0.97, and 0.99, computed from (9.130), are plotted in Figure 9.20 versus the number of data bits for various bit-error probabilities. The disagreement tolerance is $h = 1$. Note that the number of marker bits required is surprisingly relatively insensitive to P_e. Also, as the data packet length increases, the number of marker bits required to maintain $P_{\text{one-pass}}$ at the chosen value increases, but not significantly. Finally, more marker bits are required on average the larger $P_{\text{one-pass}}$.

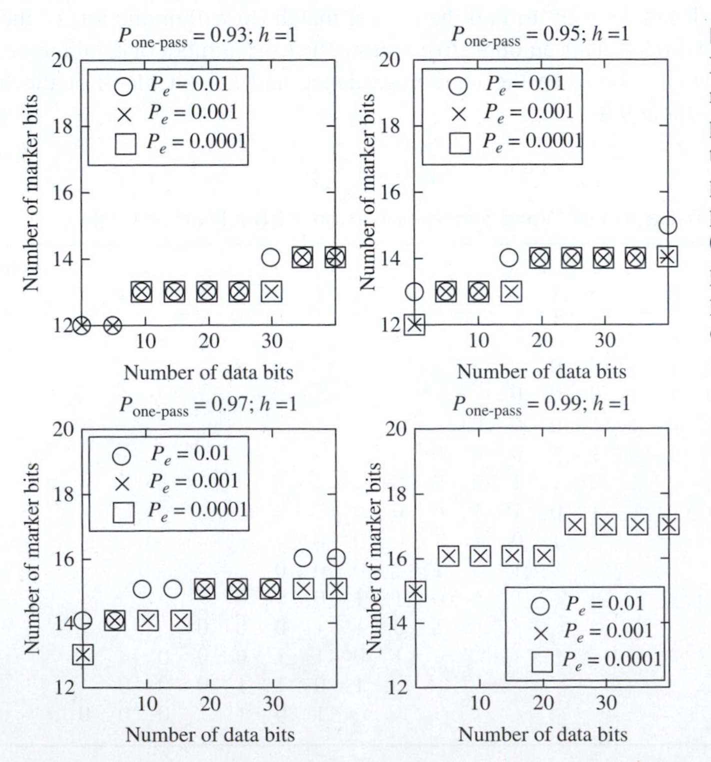

Figure 9.20 Number of marker bits required for various one-pass probabilities of word acquisition. (a) One-pass acquisition probability of 0.93. (b) One-pass acquisition probability of 0.95. (c) One-pass acquisition probability of 0.97. (d) One-pass acquisition probability of 0.99.

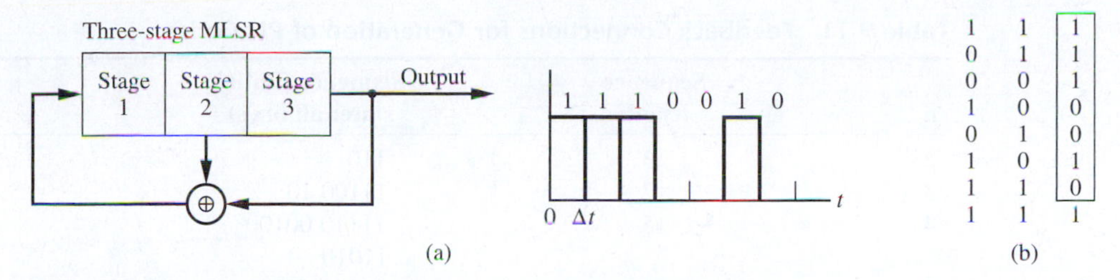

Figure 9.21
Generation of a 7-bit PN sequence. (a) Generation. (b) Shift register contents.

9.3.4 Pseudo-Noise Sequences

Pseudo-noise (PN) codes are binary-valued, noiselike sequences; they approximate a sequence of coin tossings for which a 1 represents a head and a 0 represents a tail. However, their primary advantages are that they are deterministic, being easily generated by feedback shift register circuits, and they have an autocorrelation function for a periodically extended version of the code that is highly peaked for zero delay and approximately zero for other delays. Thus they find application wherever waveforms at remote locations must be synchronized. These applications include not only word synchronization but also the determination of range between two points, the measurement of the impulse response of a system by cross-correlation of input with output, as discussed in Chapter 6 (Example 6.7), and in spread spectrum communications systems to be discussed in Section 9.4.

Figure 9.21 illustrates the generation of a PN code of length $2^3 - 1 = 7$, which is accomplished with the use of a shift register three stages in length. After each shift of the contents of the shift register to the right, the contents of the second and third stages are used to produce an input to the first stage through an EXCLUSIVE-OR (XOR) operation (that is, a binary add without carry). The logical operation performed by the XOR circuit is given in Table 9.10. Thus, if the initial contents (called the initial *state*) of the shift register are 1 1 1, as shown in the first row of Figure 9.21(b), the contents for seven more successive shifts are given by the remaining rows of this table. Therefore, the shift register again returns to the 1 1 1 state after $2^3 - 1 = 7$ more shifts, which is also the length of the output sequence taken at the third stage before repeating. By using an n-stage shift register with proper feedback connections, PN sequences of length $2^n - 1$ may be obtained. Note that $2^n - 1$ is the maximum possible length of the PN sequence because the total number of states of the shift register is 2^n, but one of these is the all-zeros state from which the feedback shift register will never recover if it were to end up in it. Hence, a proper feedback connection

Table 9.10 Truth Table for the XOR Operation

Input 1	Input 2	Output
1	1	0
1	0	1
0	1	1
0	0	0

Table 9.11 Feedback Connections for Generation of PN Codes

n	Sequence length	Sequence (initial state: all ones)	Feedback digit
2	3	110	$x_1 \oplus x_2$
3	7	11100 10	$x_2 \oplus x_3$
4	15	11110 00100 11010	$x_3 \oplus x_4$
5	31	11111 00011 01110 10100 00100 10110 0	$x_2 \oplus x_5$
6	63	11111 10000 01000 01100 01010 01111 01000 11100 10010 11011 10110 01101 010	$x_5 \oplus x_6$

will be one that cycles the shift register through all states except the all-zeros state; the total number of allowed states is therefore $2^n - 1$. Proper feedback connections for several values of n are given in Table 9.11.[23]

Considering next the autocorrelation function (normalized to a peak value of unity) of the periodic waveform obtained by letting the shift register in Figure 9.21(a) run indefinitely, we see that its values for integer multiples of the output pulse width $\Delta = n\Delta t$ are given by

$$R(\Delta) = \frac{N_A - N_U}{\text{sequence length}} \tag{9.133}$$

where N_A is the number of like digits of the sequence and a sequence shifted by n pulses and N_U is the number of unlike digits of the sequence and a sequence shifted by n pulses. This equation is a direct result of the definition of the autocorrelation function for a periodic waveform, given in Chapter 2, and the binary-valued nature of the shift register output. Thus the autocorrelation function for the sequence generated by the feedback shift register of Figure 9.21 (a) is as shown in Figure 9.22(a), as one may readily verify. Applying the definition of the autocorrelation function, we could also easily show that the shape for noninteger values of delay is as shown in Figure 9.22(a).

In general, for a sequence of length N, the minimum correlation is $-1/N$. One period of the autocorrelation function of a PN sequence of length $N = 2^n - 1$ can be written as

$$R_{\text{PN}}(\tau) = \left(1 + \frac{1}{N}\right)\Lambda\left(\frac{\tau}{\Delta t}\right) - \frac{1}{N}, \quad |\tau| \leq \frac{N\Delta t}{2} \tag{9.134}$$

where $\Lambda(x) = 1 - |x|$ for $|x| \leq 1$ and 0 otherwise is the unit-triangular function defined in Chapter 2.

[23]See R. E. Ziemer and R. L. Peterson (2001), Chapter 8, for additional sequences and proper feedback connections.

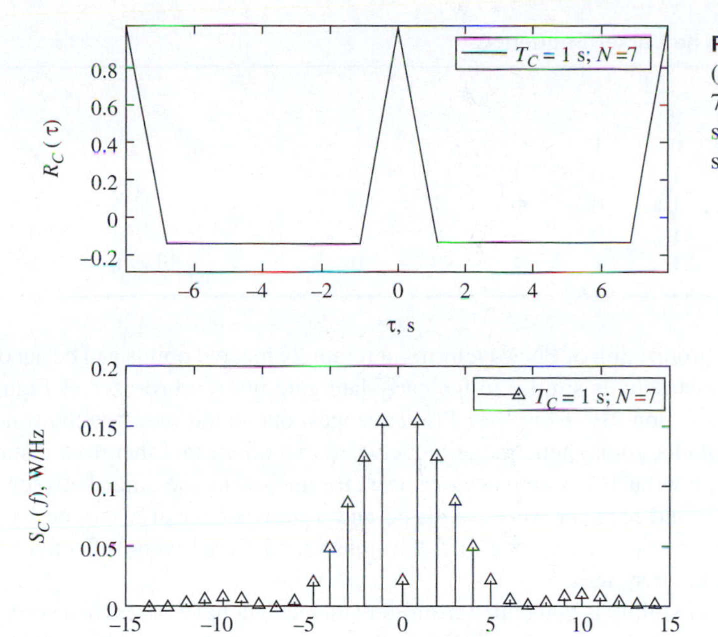

Figure 9.22
(a) Correlation function of a 7-chip PN code. (b) Power spectrum for the same sequence.

Its power spectrum is the Fourier transform of the autocorrelation function that can be obtained by applying (2.150). Consider only the first term of (9.134). The Fourier transform of it is

$$\Im\left[\left(1+\frac{1}{N}\right)\Lambda\left(\frac{\tau}{\Delta t}\right)\right]=\left(1+\frac{1}{N}\right)\Delta t\,\mathrm{sinc}^2(\Delta t f)$$

According to (2.150), this times $f_s = 1/(\mathrm{N}\ \Delta t)$ is the weight multiplier of the Fourier transform of the periodic correlation function (9.134), which is composed of impulses spaced by $f_s = 1/(N\Delta t)$, minus the contribution due to the $1/N$, so

$$\begin{aligned} S_{\mathrm{PN}}(f) &= \sum_{n=-\infty}^{\infty}\frac{1}{N}\left(1+\frac{1}{N}\right)\mathrm{sinc}^2\left[\Delta t\left(\frac{n}{N\Delta t}\right)\right]\delta\left(f-\frac{n}{N\Delta t}\right)-\frac{1}{N}\delta(f) \\ &= \sum_{n=-\infty,\,n\neq 0}^{\infty}\frac{N+1}{N^2}\,\mathrm{sinc}^2\left(\frac{n}{N}\right)\delta\left(f-\frac{n}{N\Delta t}\right)+\frac{1}{N^2}\delta(f) \end{aligned} \tag{9.135}$$

Thus, the impulses showing the spectral content of a PN sequence are spaced by $1/(N\Delta t)$ Hz and are weighted by $[(N+1)/N^2]\ \mathrm{sinc}^2(n/N)$ except for the one at $f=0$ which has weight $1/N^2$. Note that this checks with the DC level of the PN code, which is $-1/N$ corresponding to a DC power of $1/N^2$. The power spectrum for the 7-chip sequence generated by the circuit of Figure 9.21(a) is shown in Figure 9.22(b).

Because the correlation function of a PN sequence consists of a narrow triangle around zero delay and is essentially zero otherwise, it resembles that of white noise when used to drive any system whose bandwidth is small compared with the inverse pulse width. This is another manifestation of the reason for the name *pseudo-noise*.

Table 9.12 The Barker Sequences

1	0											
1	1	0										
1	1	0	1									
1	1	1	0	1								
1	1	1	0	0	1	0						
1	1	1	0	0	0	1	0	0	1	0		
1	1	1	1	1	0	0	1	1	0	1	0	1

The synchronization of PN waveforms at remotely located points can be accomplished by feedback loop structures similar to the early-late gate bit synchronizer of Figure 9.19 after carrier demodulation. By using long PN sequences, one could measure the time it takes for propagation of electromagnetic radiation between two points and therefore distance. It is not difficult to see how such a system could be used for measuring the range between two points if the transmitter and receiver were colocated and a transponder at a remote location simply retransmitted whatever it received or if the transmitted signal were reflected from a distant target as in a radar system.

Another possibility is that both transmitter and receiver have access to a very precise clock and that an epoch of the transmitted PN sequence is precisely known relative to the clock time. Then by noting the delay of the received code relative to the locally generated code, the receiver could determine the one-way delay of the transmission. This is, in fact, the technique used for the Global Positioning System (GPS), where delays of the transmissions from at least four satellites with accurately known positions are measured to determine the latitude, longitude, and altitude of a platform bearing a GPS receiver at any point in the vacinity of the earth. There are currently 24 such satellites in the GPS constellation, each at an altitude of about 12000 mi and making two orbits in less than a day, so it is highly probable that a receiver will be able to connect with at least four satellites no matter what its location. Modern GPS receivers are able to connect with up to 12 satellites and are accurate to within 15 m (one-way delay accuracy).

While the autocorrelation function of a PN sequence is very nearly ideal, sometimes the *aperiodic* autocorrelation function obtained by sliding the sequence past itself rather than past its periodic extension is important. Sequences with good aperiodic correlation properties, in the sense of low autocorrelation peaks at nonzero delays, are the Barker codes, which have aperiodic autocorrelation functions that are bounded by $(\text{sequence length})^{-1}$ for nonzero delays.[24] Unfortunately the longest known Barker code is of length 13. Table 9.12 lists all known Barker sequences (see Problem 9.32). Other digital sequences with good correlation properties can be constructed as combinations of appropriately chosen PN sequences (referred to as *Gold codes*).[25]

9.4 SPREAD-SPECTRUM COMMUNICATION SYSTEMS

We next consider a special class of modulation referred to as *spread-spectrum modulation*. In general, spread-spectrum modulation refers to any modulation technique in which the bandwidth of the modulated signal is spread well beyond the bandwidth of the modulating

[24] See Skolnik (1970), Chapter 20.

[25] See Peterson et al. (1995).

signal, independently of the modulating signal bandwidth. The following are reasons for employing spread-spectrum modulation:[26]

1. Provide resistance to intentional or unintentional jamming by another transmitter.
2. Provide a means for masking the transmitted signal in the background noise and prevent another party from eavesdropping.
3. Provide resistance to the degrading effects of multipath transmission.
4. Provide a means for more than one user to use the same transmission channel.
5. Provide range-measuring capability.

The two most common techniques for effecting spread-spectrum modulation are referred to as *direct sequence* (DS) and *frequency hopping* (FH). Figures 9.23 and 9.24 are block

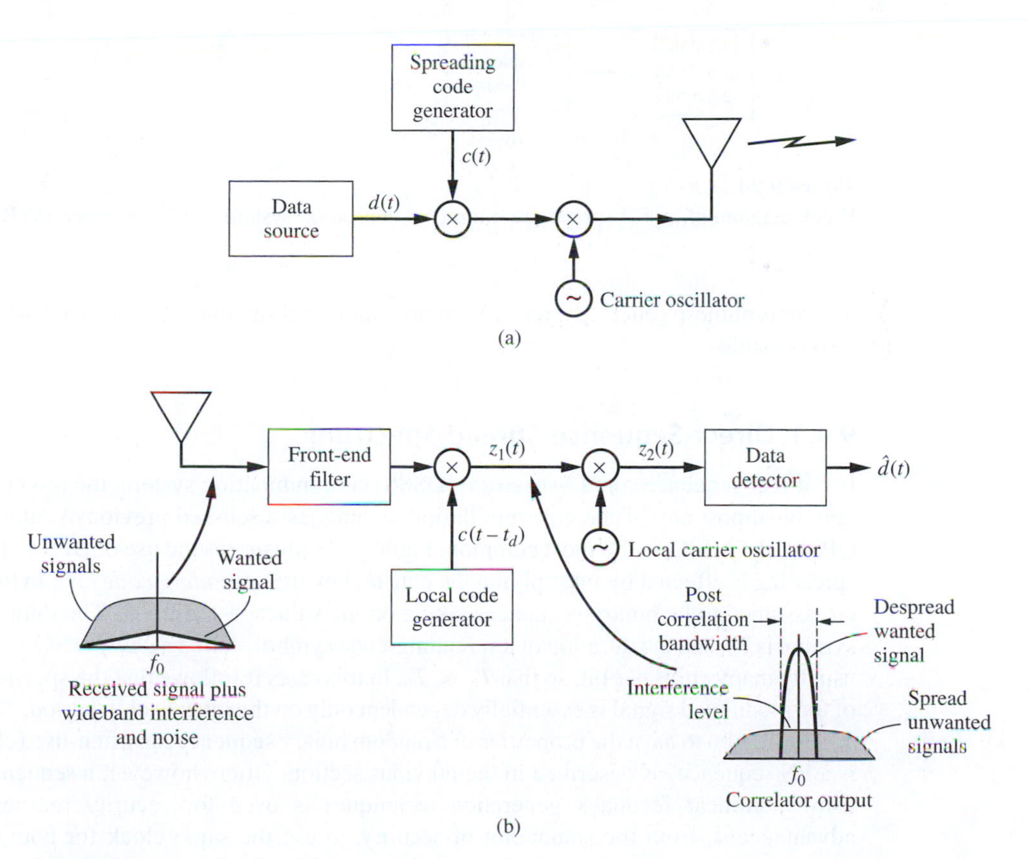

Figure 9.23
Block diagram of a DS spread-spectrum communication system. (a) Transmitter. (b) Receiver.

---

[26] A good survey paper on the early history of spread spectrum is Robert A. Scholtz, The origins of spread-spectrum communications. *IEEE Transaction on Communication*, **COM-30:** 822–854, May 1982.

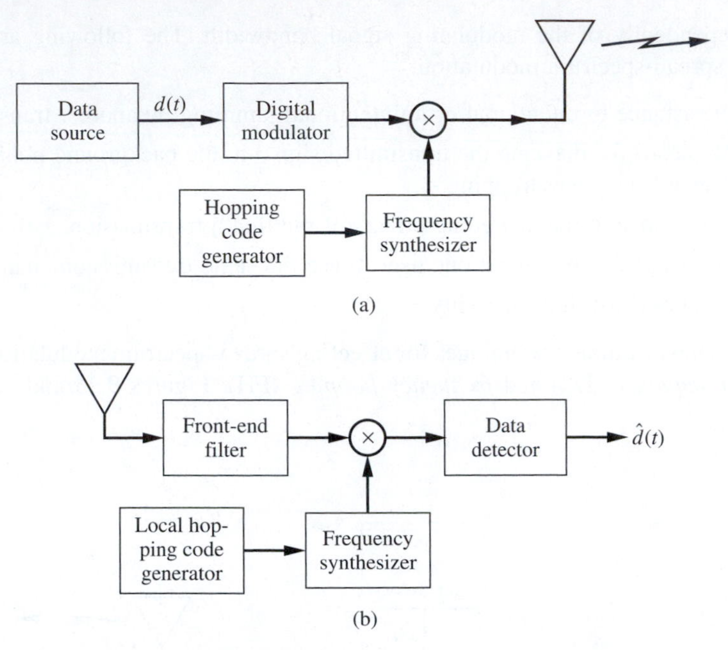

Figure 9.24
Block diagram of an FH spread-spectrum communication system. (a) Transmitter. (b) Receiver.

diagrams of these generic systems. Variations and combinations of these two basic systems are also possible.

9.4.1 Direct-Sequence Spread Spectrum

In a direct-sequence spread-spectrum (DSSS) communication system, the modulation format may be almost any of the coherent digital techniques discussed previously, although BPSK, QPSK, and MSK are the most common. Figure 9.23 illustrates the use of BPSK. The spectrum spreading is effected by multiplying the data $d(t)$ by the *spreading code* $c(t)$. In this case, both are assumed to be binary sequences taking on the values $+1$ and -1. The duration of a data symbol is T_b, and the duration of a spreading-code symbol, called a *chip period*, is T_c. There are usually many chips per bit, so that $T_c \ll T_b$. In this case, it follows that the spectral bandwidth of the modulated signal is essentially dependent only on the inverse chip period. The spreading code is chosen to have the properties of a random binary sequence; an often-used choice for $c(t)$ is a PN sequence, as described in the previous section. Often, however, a sequence generated using nonlinear feedback generation techniques is used for security reasons. It is also advantageous, from the standpoint of security, to use the same clock for both the data and spreading code so that the data changes sign coincident with a sign change for the spreading code. This is not necessary for proper operation of the system, however.

Typical spectra for the system illustrated in Figure 9.23 are shown directly below the corresponding blocks. At the receiver, it is assumed that a replica of the spreading code is available and is time synchronized with the incoming code used to multiply the BPSK-modulated carrier. This synchronization procedure is composed of two steps, called *acquisition* and *tracking*. A very brief discussion of methods for acquisition will be given later. For a fuller discussion and analyses of both procedures, the student is referred to Peterson et al. (1995).

A rough approximation to the spectrum of a DSSS signal employing BPSK data modulation can be obtained by representing the modulated, spread carrier as

$$x_c(t) = Ad(t)\, c(t) \cos(\omega_c t + \theta) \tag{9.136}$$

where it is assumed that θ is a random phase uniformly distributed in $[0,\ 2\pi]$ and $d(t)$ and $c(t)$ are independent, random binary sequences [if derived from a common clock, the independence assumption for $d(t)$ and $c(t)$ is not strictly valid]. With these assumptions, the autocorrelation function for $x_c(t)$ is

$$R_{x_c}(\tau) = \frac{A^2}{2} R_d(\tau) R_c(\tau) \cos(\omega_c \tau) \tag{9.137}$$

where $R_d(\tau)$ and $R_c(\tau)$ are the autocorrelation functions of the data and spreading code, respectively. If they are modeled as random "coin-toss" sequences as considered in Example 6.6 with spectrum illustrated in Figure 6.6(a), their autocorrelation functions are given by

$$R_d(\tau) = \Lambda\left(\frac{\tau}{T_b}\right) \tag{9.138}$$

and[27]

$$R_c(\tau) = \Lambda\left(\frac{\tau}{T_c}\right) \tag{9.139}$$

respectively. Their corresponding power spectral densities are

$$S_d(t) = T_b \operatorname{sinc}^2(T_b f) \tag{9.140}$$

and

$$S_c(t) = T_c \operatorname{sinc}^2(T_c f) \tag{9.141}$$

respectively, where the single-sided width of the main lobe of (9.140) is T_b^{-1} and that for (9.141) is T_c^{-1}.

The power spectral density of $x_c(t)$ can be obtained by taking the Fourier transform of (9.137):

$$S_{x_c}(f) = \frac{A^2}{2} S_d(f) * S_c(f) * \Im[\cos(\omega_c \tau)] \tag{9.142}$$

where the asterisk denotes convolution. Since the spectral width of $S_d(f)$ is much less than that for $S_c(f)$, the convolution of these two spectra is approximately $S_c(f)$.[28] Thus the spectrum of

[27]Note that since the spreading code is repeated, its autocorrelation function is periodic, and hence, its power spectrum is composed of discrete impulses whose weights follow a sinc-squared envelope. The analysis used here is a simplified one. See Peterson, et al. (1995) for a more complete treatment.

[28]Note that $\int_{-\infty}^{\infty} S_d(f)df = 1$ and, relative to $S_c(f)$, $S_d(f)$ appears to act more and more like a delta function as $1/T_b << 1/T_c$.

the DSSS modulated signal is very closely approximated by

$$\begin{aligned} S_{x_c}(f) &= \frac{A^2}{4}[S_c(f-f_c) + S_c(f+f_c)] \\ &= \frac{A^2 T_c}{4}\left\{\mathrm{sinc}^2[T_c(f-f_c)] + \mathrm{sinc}^2[T_c(f+f_c)]\right\} \end{aligned} \tag{9.143}$$

The spectrum, as stated above, is approximately independent of the data spectrum and has a null-to-null bandwidth around the carrier of $2/T_c$ Hz.

We next look at the error probability performance. First, assume a DSSS signal plus AWGN is present at the receiver. Ignoring propagation delays, the output of the local code multiplier at the receiver (see Figure 9.23) is

$$z_1(t) = Ad(t)c(t)c(t-\Delta)\cos(\omega_c t + \theta) + n(t)c(t-\Delta) \tag{9.144}$$

where Δ is the misalignment of the locally generated code at the receiver with the code on the received signal. Assuming perfect code synchronization ($\Delta = 0$), the output of the coherent demodulator is

$$z_2(t) = Ad(t) + n'(t) + \text{double frequency terms} \tag{9.145}$$

where the local mixing signal is assumed to be $2\cos(\omega_c t + \theta)$ for convenience, and

$$n'(t) = 2n(t)c(t)\cos(\omega_c t + \theta) \tag{9.146}$$

is a new Gaussian random process with zero mean. Passing $z_2(t)$ through an integrate-and-dump circuit, we have for the signal component at the output

$$V_0 = \pm AT_b \tag{9.147}$$

where the sign depends on the sign of the bit at the input. The noise component at the integrator output is

$$N_g = \int_0^{T_b} 2n(t)c(t)\cos(\omega_c t + \theta)\, dt \tag{9.148}$$

Since $n(t)$ has zero mean, N_g has zero mean. Its variance, which is the same as its second moment, can be found by squaring the integral, writing it as an iterated integral, and taking the expectation inside the double integral—a procedure that has been used several times before in this chapter and the previous one. The result is

$$\mathrm{var}(N_g) = E\left(N_g^2\right) = N_0 T_b \tag{9.149}$$

where N_0 is the single-sided power spectral density of the input noise. This, together with the signal component of the integrator output, allows us to write down an expression similar to the one obtained for the baseband receiver analysis carried out in Section 8.1 (the only difference is that the signal power is $A^2/2$ here, whereas it was A^2 for the baseband signal considered there). The result for the probability of error is

$$P_E = Q\left(\sqrt{\frac{A^2 T_b}{N_0}}\right) = Q\left(\sqrt{\frac{2E_b}{N_0}}\right) \tag{9.150}$$

With Gaussian noise alone, DSSS ideally performs the same as BPSK without the spread-spectrum modulation.

9.4.2 Performance in Continuous-Wave(CW) Interference Environments

Consider next a CW interference component of the form $x_I(t) = A_I \cos[(\omega_c + \Delta\omega)t + \phi]$. Now, the input to the integrate-and-dump detector, excluding double frequency terms, is

$$z'_2(t) = Ad(t) + n'(t) + A_I \cos(\Delta\omega t + \theta - \phi) \tag{9.151}$$

where A_I is the amplitude of the interference component, ϕ is its relative phase, and $\Delta\omega$ is its offset frequency from the carrier frequency in radians per second (rad/s). It is assumed that $\Delta\omega < 2\pi/T_c$. The output of the integrate-and-dump detector is

$$V'_0 = \pm AT_b + N_g + N_I \tag{9.152}$$

The first two terms are the same as obtained before. The last term is the result of interference and is given by

$$N_I = \int_0^{T_b} A_I c(t) \cos(\Delta\omega t + \theta - \phi)\, dt \tag{9.153}$$

Because of the multiplication by the wideband spreading code $c(t)$ and the subsequent integration, we approximate this term by an equivalent Gaussian random variable (the integral is a sum of a large number of random variables, with each term due to a spreading code chip). Its mean is zero, and for $\Delta\omega \ll 2\pi/T_c$, its variance can be shown to be

$$\text{var}\,(N_I) = \frac{T_c T_b A_I^2}{2} \tag{9.154}$$

With this Gaussian approximation for N_I, the probability of error can be shown to be

$$P_E = Q\left(\sqrt{\frac{A^2 T_b^2}{\sigma_T^2}}\right) \tag{9.155}$$

where

$$\sigma_T^2 = N_0 T_b + \frac{T_c T_b A_I^2}{2} \tag{9.156}$$

is the total variance of the noise plus interference components at the integrator output (permissible because noise and interference are statistically independent). The quantity under the square root can be further manipulated as

$$\begin{aligned} \frac{A^2 T_b^2}{2\sigma_T^2} &= \frac{A^2/2}{N_0/T_b + (T_c/T_b)(A_I^2/2)} \\ &= \frac{P_s}{P_n + P_I/G_P} \end{aligned} \tag{9.157}$$

where

$P_s = A^2/2$ is the signal power at the input.

$P_n = N_0/T_b$ is the Gaussian noise power in the bit-rate bandwidth.

$P_I = A_I^2/2$ is the power of the interfering component at the input.

$G_p = T_b/T_c$ is called the processing gain of the DSSS system.

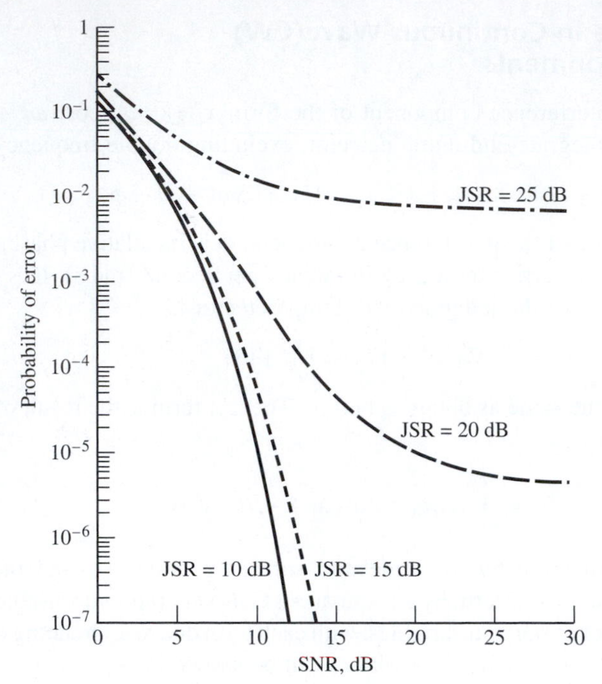

Figure 9.25
P_E versus SNR for DSSS with $G_p = 30$ dB for various jamming-to-signal ratios.

It is seen that the effect of the interference component is decreased by the processing gain G_p. Equation (9.157) can be rearranged as

$$\frac{A^2T_b^2}{2\sigma_T^2} = \frac{\text{SNR}}{1 + \text{SNR}(\text{JSR})/G_P} \tag{9.158}$$

where

SNR $= P_s/P_n = A^2T_b/(2N_0) = E_b/N_0$ is the signal-to-noise ratio.

JSR $= P_I/P_s$ is the jamming-to-signal power ratio.

Figure 9.25 shows P_E versus the SNR for several values of JSR where that the curves approach a horizontal asymptote for SNR sufficiently large, with the asymptote decreasing with decreasing JSR/G_p.

9.4.3 Performance in Multiple User Environments

An important application of spread-spectrum systems is multiple-access communications which means that several users may access a common communication resource to communicate with other users. If several users were at the same location communicating with a like number of users at another common location, the terminology used would be multiplexing (recall that frequency- and time-division multplexing were discussed in Chapter 3). Since the users are not assumed to be at the same location in the present context, the term *multiple access* is used. There are various ways to effect multiple-access communications including frequency, time, and code.

In frequency-division multiple access (FDMA), the channel resources are divided in frequency, and each active user is assigned a subband of the frequency resource. In time-division multiple access (TDMA) the communication resource is divided in time into contiguous frames which are composed of a series slots, and each active user is assigned a slot (see the discussion under Satellite Communications in Section 9.6). When all subbands or slots are assigned in FDMA and TDMA, respectively, no more users can be admitted to the system. In this sense, FDMA and TDMA are said to have *hard capacity limits*.

In the one remaining access system mentioned above, code-division multiple access (CDMA), each user is assigned a unique spreading code, and all active users can transmit simultaneously over the same band of frequencies. Another user who wants to receive information from a given user then correlates the sum total of all these receptions with the spreading code of the desired transmitting user and receives its transmissions assuming that the transmitter–receiver pair is properly synchronized. If the set of codes assigned to the users is not orthogonal or if they are orthogonal but multiple delayed components arrive at a given receiving user due to multipath, partial correlation with other users appears as noise in the detector of a particular receiving user of interest. These partial correlations will eventually limit the total number of users that can simultaneously access the system, but the maximum number is not fixed as in the cases of FDMA and TDMA. It will depend on various system and channel parameters, such as propagation conditions. In this sense, CDMA is said to have a *soft capacity limit*. (There is the possibility that all available codes are used before the soft capacity limit is reached.)

Several means for calculating the performance of a CDMA receivers have been published in the literature over the past few decades.[29] We take a fairly simplistic approach[30] in that the multiple-access interference is assumed sufficiently well represented by an equivalent Gaussian random process. In addition, we make the usual assumption that *power control* is used so that all users' transmissions arrive at the receiver of the user of interest with the same power. Under these conditions, it can be shown that the received bit-error probability can be approximated by

$$P_E = Q(\sqrt{\text{SNR}}) \tag{9.159}$$

where

$$\text{SNR} = \left(\frac{K-1}{3N} + \frac{N_0}{2E_b}\right)^{-1} \tag{9.160}$$

in which K is the number of active users and N is the number of chips per bit (i.e., the processing gain).

Figure 9.26 shows P_E versus E_b/N_0 for $N = 255$ and various numbers of users. It is seen that an error floor is approached as $E_b/N_0 \to \infty$ because of the interference from other users. For example, if 60 users are active and a P_E of 10^{-4} is desired, it cannot be achieved no matter what E_b/N_0 is used. This is one of the drawbacks of CDMA, and much research has gone into combating this problem, for example, multiuser detection, where the presence of multiple users is treated as a multihypothesis detection problem. Due to the overlap of signaling intervals, multiple symbols must be detected, and implementation of the true optimum receiver is

[29] See K. B. Letaief, Efficient evaluation of the error probabilities of spread-spectrum multiple-access communications. *IEEE Transactions on Communications*, 45, 239–246, February 1997.

[30] See M. B. Pursley, Performance evaluation of phase-coded spread-spectrum multiple-access communication—Part I: System analysis. *IEEE Transactions on Communications*, **COM-25:** 795–799, August 1977.

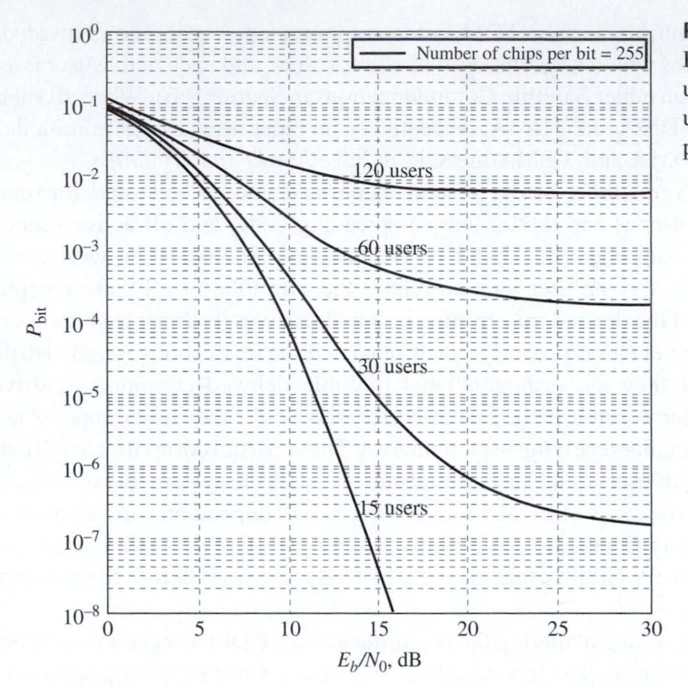

Figure 9.26 Bit-error probability for CDMA using DSSS with the number of users as a parameter; 255 chips per bit assumed.

computationally infeasible for moderate to large numbers of users. Various approximations to the optimum detector have been proposed and have been investigated.[31]

The situation is even worse if the received signals from the users have differing powers. In this case, the strongest user saturates the receiver, and the performances for the weaker users are unacceptable. This is known as the *near-far problem*.

A word about accuracy of the curves shown in Figure 9.26 is in order. The Gaussian approximation for multiple-access interference is almost always optimistic, with its accuracy becoming better the more users and the larger the processing gain (the conditions of the central-limit theorem are more nearly satisfied then).

COMPUTER EXAMPLE 9.3

The MATLAB program given below evaluates the bit-error probability for DSSS in a K-user environment. The program was used to plot Figure 9.26.

```
% file c9ce3.m
% Bit error probability for DSSS in multi-users
%
N = input('Enter processing gain (chips per bit) ');
K = input('Enter vector of number of users ');
clf
z_dB = 0:.1:30;
z = 10.^(z_dB/10);
LK = length(K);
for n = 1:LK
```

[31]See Verdu (1998).

```
    KK = K(n);
      SNR_1 = (KK-1)/(3*N) + 1./(2*z);
      SNR = 1./SNR_1;
      Pdsss=qfn(sqrt(SNR));
   semilogy(z_dB,Pdsss),axis([min(z_dB) max(z_dB) 10^(-8) 1]),...
      xlabel('{\itE_b\N}_0, dB'),ylabel('{\itP_E}'),...
       text(z_dB(170), 1.1*Pdsss(170), [num2str(KK), ' users'])
       if n == 1
          grid on
          hold on
       end
   end
   title(['Bit error probability for DSSS; number of chips per bit = ',num2str
   (N)])

   %       This function computes the Gaussian Q-function
   %
   function Q=qfn(x)
   Q = 0.5*erfc(x/sqrt(2));
```

9.4.4 Frequency-Hop Spread Spectrum

In the case of *frequency-hop spread spectrum* (FHSS), the modulated signal is hopped in a pseudorandom fashion among a set of frequencies so that a potential eavesdropper does not know in what band to listen or jam. Current FHSS systems may be classified as *fast hop* or *slow hop*, depending on whether one or several data bits are included in a hop, respectively. The data modulator for either is usually a noncoherent type such as FSK or DPSK, since frequency synthesizers are typically noncoherent from hop to hop. Even if one goes to the expense of building a coherent frequency synthesizer, the channel may not preserve the coherency property of the synthesizer output. At the receiver, as shown in Figure 9.24, a replica of the hopping code is produced and synchronized with the hopping pattern of the received signal and used to de-hop the received signal. Demodulation and detection of the de-hopped signal that is appropriate for the particular modulation used is then performed.

EXAMPLE 9.5

A binary data source has a data rate of 10 kbps, and a DSSS communication system spreads the data with a 127-chip short code system (i.e., a system where one code period is used per data bit). (1) What is the approximate bandwidth of the DSSS/BPSK transmitted signal? (2) A FHSS–BFSK (noncoherent) system is to be designed with the same transmit bandwidth as the DSSS–BPSK system. How many frequency-hop slots does it require?

Solution

(1) The bandwidth efficiency of BPSK is 0.5, which gives a modulated signal bandwidth for the unspread system of 20 kHz. The DSSS system has a transmit bandwidth of roughly 127 times this, or a total bandwidth of 2.54 MHz. (2) The bandwidth efficiency of coherent BFSK is 0.4, which gives a modulated signal bandwidth for the unspread system of 25 kHz. The number of frequency hops required to give the same spread bandwidth as the DSSS system is therefore 2,540,000/25,000 = 101.6. Since we can't have a partial hop slot, this is rounded up to 102 hop slots giving a total FHSS bandwidth of 2.55 MHz.

9.4.5 Code Synchronization

Only a brief discussion of code synchronization will be given here. For detailed discussions and analyses of such systems, the reader is referred to Peterson et al. (1995).[32]

Figure 9.27(a) shows a serial-search acquisition circuit for DSSS. A replica of the spreading code is generated at the receiver and multiplied by the incoming spread-spectrum signal (the carrier is assumed absent in Figure 9.27 for simplicity). Of course, the code epoch is unknown, so an arbitrary local code delay relative to the incoming code is tried. If it is within $\pm\frac{1}{2}$ chip of the correct code epoch, the output of the multiplier will be mostly despread data and its spectrum will pass through the bandpass filter whose bandwidth is of the order of the data bandwidth. If the code delay is not correct, the output of the multiplier remains spread and little power passes through the bandpass filter. The envelope of the bandpass filter output is

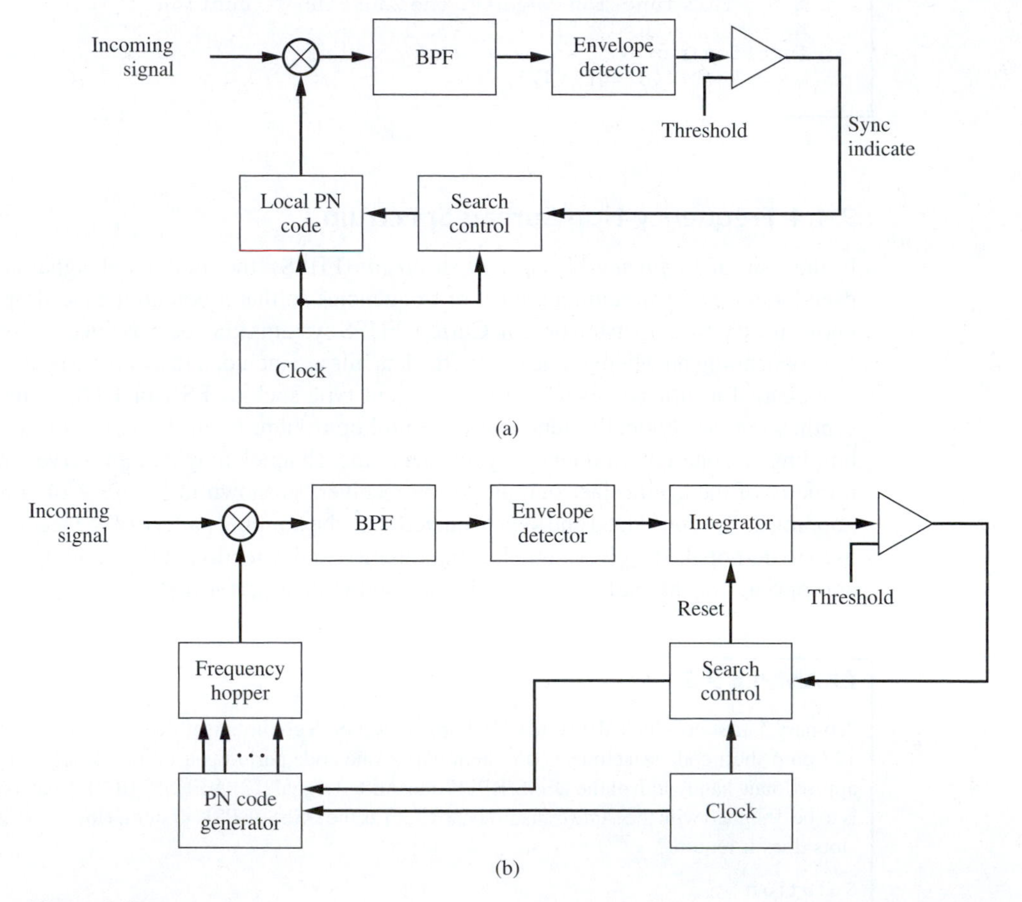

Figure 9.27
Code acquisition circuits for (a) DSSS and (b) FHSS using serial search.

[32]For an excellent tutorial paper on acquisition and tracking, see S. S. Rappaport and D. M. Grieco, Spread-spectrum signal acquisition: Methods and technology. *IEEE Communications Magazine*, **22** (6): 6–21 June 1984.

compared with a threshold—a value below threshold denotes an unspread condition at the multiplier output and, hence, a delay that does not match the delay of the spreading code at the receiver input, while a value above threshold indicates that the codes are approximately aligned. If the latter condition holds, the search control stops the code search and a tracking mode is entered. If the below-threshold condition holds, the codes are assumed to be not aligned, so the search control steps to the next code delay (usually a half chip) and the process is repeated. It is apparent that such a process can take a relatively long time to achieve lock. The mean time to acquisition is given by[33]

$$T_{\text{acq}} = (C-1)T_{\text{da}}\left(\frac{2-P_d}{2P_d}\right) + \frac{T_i}{T_d} \tag{9.161}$$

where

$C =$ code uncertainty region (the number of cells to be searched—usually the number of half chips.

$P_d =$ probability of detection.

$P_{\text{fa}} =$ probability of false alarm.

$T_i =$ integration time (time to evaluate one cell).

$T_{\text{da}} = T_i + T_{\text{fa}}P_{\text{fa}}$.

$T_{\text{fa}} =$ time required to reject an incorrect cell (typically several times T_i).

Other techniques are available that speed up the acquisition, but at the expense of more hardware or special code structures.

A synchronization scheme for FHSS is shown in Figure 9.27(b). The discussion of its operation would be similar to that for acquisition in DSSS except that the correct frequency pattern for despreading is sought.

EXAMPLE 9.6

Consider a DSSS system with code clock frequency of 3 MHz and a propagation delay uncertainty of ± 1.2 ms. Assume that $T_{\text{fa}} = 100T_i$ and that $T_i = 0.42$ ms. Compute the mean time to acquire for (a) $P_d = 0.82$ and $P_{\text{fa}} = 0.004$ (threshold of 41); (b) $P_d = 0.77$ and $P_{\text{fa}} = 0.002$ (threshold of 43); (c) $P_d = 0.72$ and $P_{\text{fa}} = 0.0011$ (threshold of 45);

Solution

The propagation delay uncertainty corresponds to a value for C of (one factor of 2 because of the ± 1.2 ms and the other factor of 2 because of the 1/2-chip steps)

$$C = 2 \times 2(1.2 \times 10^{-3}\ \text{s})(3 \times 10^6\ \text{chips/s}) = 14{,}400\ \text{half chips}$$

The result for the mean time to acquisition becomes

$$\begin{aligned} T_{\text{acq}} &= 14{,}399(T_i + 100T_iP_{\text{fa}})\left(\frac{2-P_d}{2P_d}\right) + \frac{T_i}{P_d} \\ &= \left[14{,}399(1 + 100P_{\text{fa}})\left(\frac{2-P_d}{2P_d}\right) + \frac{1}{P_d}\right]T_i \end{aligned}$$

[33]See Peterson, et al. (1995), Chapter 5.

With $T_i = 0.42$ ms and the values P_d of P_{fa} and given above we obtain the following for the mean time to acquire:

(a) $T_{acq} = 6.09$ s; (b) $T_{acq} = 5.80$ s; (c) $T_{acq} = 5.97$ s. There appears to be an optimum threshold setting.

■

9.4.6 Conclusion

From the preceding discussions and the block diagrams of the DS and FH spread-spectrum systems, it should be clear that *nothing is gained by using a spread-spectrum system in terms of performance in an additive white Gaussian noise channel.* Indeed, using such a system may result in slightly more degradation than by using a conventional system, owing to the additional operations required. The advantages of spread-spectrum systems accrue in environments that are hostile to digital communications—environments such as those in which multipath transmission or jamming of channels exist. In addition, since the signal power is spread over a much wider bandwidth than it is in an ordinary system, it follows that the average power density of the transmitted spread-spectrum signal is much lower than the power density when the spectrum is not spread. This lower power density gives the sender of the signal a chance to mask the transmitted signal by the background noise and thereby lower the probability that anyone may intercept the signal.

One last point is perhaps worth making: It is knowledge of the structure of the signal that allows the intended receiver to pull the received signal out of the noise. The use of correlation techniques is indeed powerful.

■ 9.5 MULTICARRIER MODULATION AND ORTHOGONAL FREQUENCY DIVISION MULTIPLEXING

One approach to combatting ISI, say, due to filtering or multipath imposed by the channel, and adapting the modulation scheme to the signal-to-noise characteristics of the channel is termed *multicarrier modulation* (MCM). Multicarrier modulation is actually a very old idea that has enjoyed a resurgence of attention in recent years because of the intense interest in maximizing transmission rates through twisted pair telephone circuits as one solution to the "last mile problem" mentioned in Chapter 1.[34] For a easy-to-read overview on its application to so-called digital subscriber lines (DSL), several references are available.[35] Another area that MCM has been applied with mixed succses is to digital audio broadcasting, particularly in Europe.[36] An extensive tutorial article directed toward wireless communications has been authored by Wang and Giannakis.[37]

The basic idea is the following for a channel that introduces ISI, e.g., a multipath channel or a severely bandlimited one such as local data distribution in a telephone channel, which is

[34]See, for example, R. W. Chang, and R. A. Gibby, A theoretical study of performance of an orthogonal multiplexing data transmission scheme. *IEEE Transactions on Communication Technology*, **COM-16:** 529–540, August 1968.

[35]See, for example, J. A. C. Bingham, Multicarrier modulation for data transmission: an idea whose time has come. *IEEE Communications Magazine*, **28:** 5–14, May 1990.

[36]http://en.wikipedia.org/wiki/Digital_audio_broadcasting.

[37]Z. Wang and G. B. Giannakis, Wireless multicarrier communications. *IEEE Signal Processing Magazine*, **17:** 29–48, May 2000.

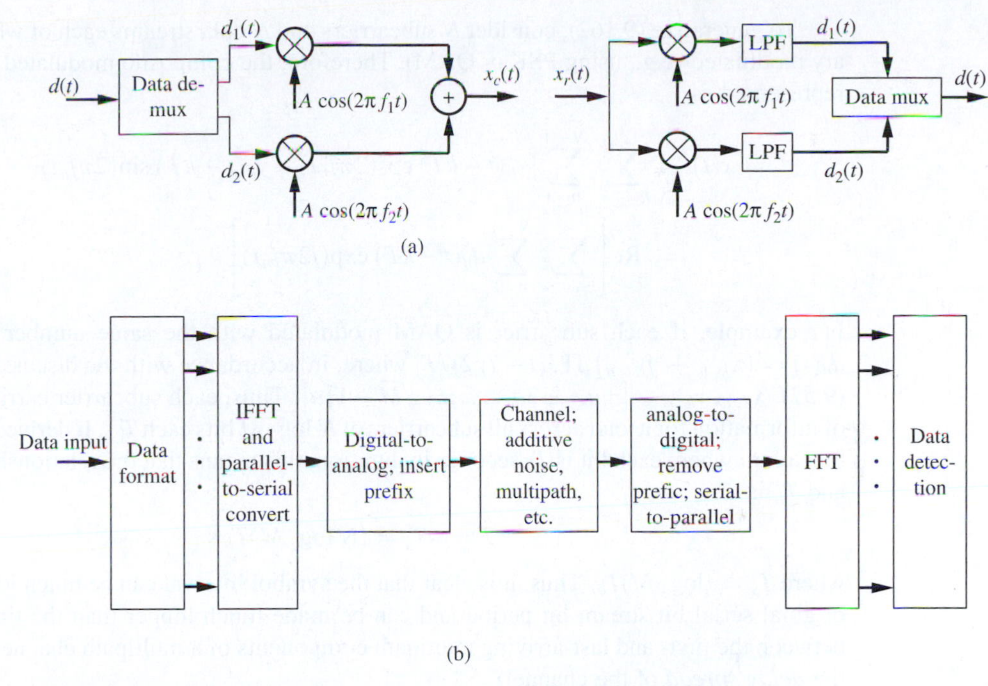

Figure 9.28
Basic concepts of MCM. (a) A simple two-tone MCM system. (b) A specialization of MCM to OFDM with FFT processing.

typically implemented by means of twisted pair wireline circuits. For simplicity of illustration, consider a digital data transmission scheme that employs two subcarriers of frequencies f_1 and f_2 each of which is BPSK modulated by bits from a single serial bit stream as shown in Figure 9.28(a). For example, the even-indexed bits from the serial bit stream, denoted d_1 in bipolar format, could modulate subcarrier 1 and the odd-indexed bits, denoted d_2, could modulate subcarrier 2, giving a transmitted signal in the nth transmission interval of

$$x(t) = A[d_1(t)\cos(2\pi f_1 t) + d_2(t)\cos(2\pi f_2 t)],\quad 2(n-1)T_b \le t \le 2nT_b \qquad (9.162)$$

Note that since every other bit is assigned to a given carrier, the symbol duration for the transmitted signal through the channel is twice the bit period of the original serial bit stream. The frequency spacing between subcarriers is assumed to be $f_2 - f_1 \ge 1/2T$, where $T = 2T_b$ in this case.[38] This is the minimum that the frequency separation can be in order for the subcarriers to be orthogonal; i.e., their product when integrated over an interval of $2T$ gives zero. The received signal is mixed with $\cos(2\pi f_1 t)$ and $\cos(2\pi f_2 t)$ in separate parallel branches at the receiver and each BPSK bit stream is detected separately. The separate parallel detected bit streams are then reassembled into a single serial bit stream. Because the durations of the symbols sent through the channel are twice the original bit durations of the serial bit stream at the input, this system should be more resistant to any ISI introduced by the channel than if the original serial bit stream were used to BPSK modulate a single carrier.

[38]With a frequency separation of $1/T$, MCM is usually referred to as *orthogonal frequency division multiplexing* (OFDM).

To generalize (9.162), consider N subcarriers and N data streams each of which could M-ary modulated (e.g., using PSK or QAM). Therefore, the composite modulated signal can be represented as

$$\begin{aligned} x(t) &= \sum_{k=-\infty}^{\infty} \sum_{n=0}^{N-1} [x_n(t-kT)\cos(2\pi f_n t) - y_n(t-kT)\sin(2\pi f_n t)] \\ &= \text{Re}\left[\sum_{k=-\infty}^{\infty} \sum_{n=0}^{N-1} d_n(t-kT)\exp(j2\pi f_n t)\right] \end{aligned} \tag{9.163}$$

For example, if each subcarrier is QAM modulated with the same number of bits, then $d_n(t) = (x_{k,n} + jy_{k,n})\, \Pi\, [(t - T/2)/T]$ where, in accordance with the discussion following (9.57), $x_{k,n}, y_{k,n} \in \left[\pm a,\ \pm 3a, \ldots, \pm(\sqrt{M} - 1)a\right]$. Thus, each subcarrier carries $\log_2 M$ bits of information for a total across all subcarriers of $N \log_2 M$ bits each T s. If derived from a serial bit stream where each bit is T_b seconds in duration, this means that the relationship between T and T_b is

$$T = NT_s = (N \log_2 M)T_b \text{ s} \tag{9.164}$$

where $T_s = (\log_2 M)T_b$. Thus, it is clear that the symbol interval can be much longer than the original serial bit stream bit period and can be made much longer than the time difference between the first- and last-arriving multipath components of a multipath channel (this defines the *delay spread* of the channel).

EXAMPLE 9.7

Consider a multipath channel with a delay spread of 10 μs through which it is desired to transmit data at a bit rate of 1 Mbps. Clearly this presents a severe intersymbol interference situation if the transmission takes place serially. Design an MCM system that has a symbol period that is at least a factor of 10 greater than the delay spread, thus resulting in multipath components spreading into succeeding symbols intervals by only 10%.

Solution

Using (9.164) with $T = 10 \times 10\ \mu$s and $T_b = 1/R_b = 1/10^6 = 10^{-6}$ s, we have

$$10 \times 10 \times 10^{-6} = (N \log_2 M) \times 10^{-6}$$

or

$$N \log_2 M = 100$$

Several values of M with the corresponding values for N, the number of subcarriers, are given below:

M	N
2	100
4	50
8	34
16	25
32	20

Note that since we can't have a fraction of a subcarrier, in the case of $M = 8$, N has been rounded up. Usually a coherent modulation scheme such as M-ary PSK or M-ary QAM would be used.

The synchronization required for the subcarriers would most likely be implemented by inserting pilot signals spaced in frequency and periodically in time.

■

Note that the powers of the individual subcarriers can be adjusted to fit the noise level characteristics of the channel. At frequencies where the SNR of the channel is low, we want a correspondingly low subcarrier power to be used, and at frequencies where the noise level of the channel is high, we want a correspondingly high subcarrier power to be used; i.e., the preferred transmission band is where the SNR is largest.[39]

An advantage of MCM is that it can be implemented by means of the DFT or its fast version, the FFT as introduced in Chapter 2. Consider (9.163) with just the data block at $k = 0$ and a subcarrier frequency spacing of $1/T = 1/NT_s$ Hz. The baseband complex modulated signal is then

$$\tilde{x}(t) = \sum_{n=0}^{N-1} d_n(t) \exp\left(\frac{j2\pi nt}{NT_s}\right) \tag{9.165}$$

If this is sampled at epochs $t = kT_s$, then (9.165) becomes

$$\tilde{x}(kT_s) = \sum_{n=0}^{N-1} d_n \exp\left(\frac{j2\pi nk}{N}\right), \quad k = 0, 1, \ldots, N-1 \tag{9.166}$$

which is recognized as the inverse DFT given in Chapter 2 (there is a factor $1/N$ missing, but this can be accommodated in the direct DFT).[40] In the form of (9.165) or (9.166), MCM is referred to as *orthogonal frequency division multiplexing* (OFDM) and is illustrated in Figure 9.28(b). The processing at the transmitter consists of the following steps:

1. Parse the incoming bit stream (assumed binary) into N blocks of $\log_2 M$ bits each.
2. Form the complex modulating samples, $d_n = x_n + jy_n$, $n = 0, 1, \ldots, N-1$.
3. Use these N blocks of symbols as the input to an inverse DFT or FFT algorithm.
4. Serially read out the inverse DFT output, interpolate, and use as the modulating signal on the carrier (not shown).

At the receiver, the inverse set of steps is performed. Note that the DFT at the receiver ideally produces $d_0, d_1, \ldots, d_{N-1}$. Since noise and ISI are present with practical channels, there will inevitably be errors. To combat the ISI, one of two things can be done:

1. A blank time interval can be inserted following each OFDM symbol, allowing a space to protect against the ISI.
2. An OFDM signal with a lengthened duration (greater than or equal to the channel memory) in which an added prefix repeats the signal from the end of the current symbol interval can be used (referred to as a cyclic prefix).

It can be shown that the latter procedure completely eliminates the ISI in OFDM.

[39]See G. David Forney, Jr., Modulation and coding for linear Gaussian channels. *IEEE Transactions on Information Theory*, **44:** 2384–2415, October 1998 for more explanation on this "water pouring" procedure, as it is known.

[40]This concept was reported in the paper S. B. Weinstein and Paul M. Ebert,"Data Transmission for Frequency Division Multiplexing Using the Discrete Fourier Transform." *IEEE Trans. on Commun. Technol.*, vol. 19, pp. 628–634, Oct. 1971.

As might be expected, the true state of affairs for MCM or OFDM is not quite so simple or desirable as outlined here. Some oversimplified features or disadvantages of MCM or OFDM are the following:

1. To achieve full protection against ISI as hinted at above, coding is necessary. With coding, it has been demonstrated that MCM affords about the same performance as a well-designed serial data transmission system with equalization and coding.[41]
2. The addition of several parallel subcarriers results in a transmitted signal with a highly varying envelope, even if the separate subcarriers employ constant-envelope modulation such as BPSK. This has implications regarding final power amplifier implementation at the transmitter. Such amplifiers operate most efficiently in a nonlinear mode (class B or C operation). Either the final power amplifier must operate linearly for MCM, with a penalty of lower efficiency, or distortion of the transmitted signal and subsequent signal degradation will take place.
3. The synchronization necessary for N subcarriers may be more complex than for a single-carrier system.
4. Clearly, using MCM adds complexity in the data transmission process; whether this complexity is outweighed by the faster processing speeds required of a serial transmission scheme employing equalization is not clear (with the overall data rates the same, of course).

■ 9.6 SATELLITE COMMUNICATIONS

In this section we look at the special application area of satellite communications to illustrate the use of some of the error probability results derived in Chapters 8 and 9.

Satellite communications were first conceived in the 1950s. The first satellite equipped with onboard radio transmitters was Sputnik 1, a Soviet satellite launched in the fall of 1957. The first U.S. communications satellite was Echo I, a passive reflecting sphere, which was launched in May of 1960. The first active U.S. satellite, Courier, where *active* refers to the satellite's ability to receive, amplify, and retransmit signals, was launched in 1960. It had only two transmitters and had a launch mass of only 500 lb. (Score was launched in 1958, but transmitted a prerecorded message.) In contrast, Intelsat VI, launched in 1986, had 77 transmitters and had a launch mass of 3600 lb. By comparison, Intelsat X has a launch mass of over 12,000 lb, 45 C-band transponders, and 16 Ku-band transponders. The first geostationary satellite over the Pacific Ocean was Syncom 3, launched in 1964 and was used to relay television coverage on the 1964 Summer Olympics in Tokyo to the United States. Over the Atlantic Ocean the first geostationary satellite was Intelesat I, launched in 1965.

Figure 9.29(a) shows a typical satellite repeater link, and Figure 9.29(b) shows a frequency-translating "bent-pipe" satellite communications system. Frequency translation is necessary to separate the receive and transmit frequencies and thus prevent "ring-around." Another type of satellite communication system, known as a demod–remod system [also referred to as *onboard processing* (OBP)], is shown in Figure 9.29(c). In such a satellite repeater, the data are actually demodulated and subsequently remodulated onto the downlink carriers. In addition to the relay communications system on board the satellite, other

[41]See H. Sari, G. Karam, and I. Jeanclaude, "Transmission Techniques for Digital Terrestrial TV Broadcasting," *IEEE Communications Magazine*, Vol. 33, pp. 100–109, Feb. 1995.

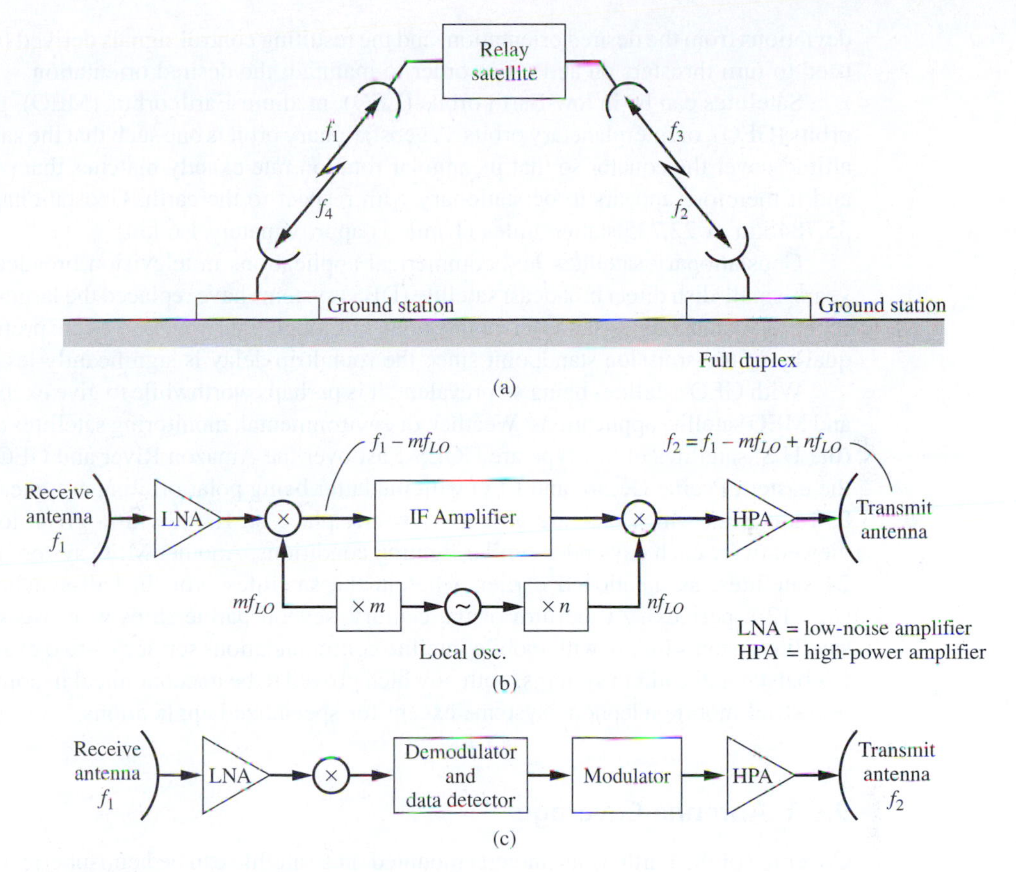

Figure 9.29
Various satellite relay link communications configurations. (a) Satellite repeater link. (b) Frequency-translation satellite communications relay. (c) Demod and/or remod, or on-board processing, satellite communications relay.

communications systems include ranging (to provide a range measurement to the satellite), command (to receive commands from an earth station to control the satellite), and telemetry (to relay data about the satellite's condition back to the earth).

Early satellite transmissions took place in the UHF, C, or X bands. Because of the subsequent crowding of these bands, new satellite frequency allocations were added at K, V, and Q bands (Table 1.2). Services are classified as *fixed-point* (communications between a satellite and fixed ground station), *broadcast* (transmission to many receivers), and *mobile* (e.g., communications to aircraft, ships, and land vehicles). *Intersatellite* refers to communications between satellites.

It is important that satellites be stabilized so that the antennas can be pointed to predetermined points on the earth's surface. Early satellites were *spin-stabilized*, which means that the satellites were physically spun about an axis that kept them oriented in a particular relationship to the earth as a result of the gyroscopic effect. Because of the difficulty in despinning ever-more-complicated antenna systems, present-day satellites are almost all *three-axis stabilized*. This means that a three-axis gyroscope system is on board to sense deviations from the desired orientation, and the resulting control signals derived from them are used to turn thrusters on and off in order to maintain the desired orientation.

Satellites can be in low-Earth orbits (LEO), medium-Earth orbits (MEO), geostationary orbits (GEO), or interplanetary orbits. A geostationary orbit is one such that the satellite is at an altitude over the equator so that its angular rotation rate exactly matches that of the Earth's, and it therefore appears to be stationary with respect to the earth. Geostationary altitude is 35,784 km or 22,235 statute miles (1 mile is approximately 1.6 km).

Geostationary satellites find commerical applications in television broadcast [in recent years, small-dish direct broadcast satellite (DBS) systems have replaced the large-dish systems of the past] and long-distance telephone relay (although lightwave cables are preferable from a quality of transmission standpoint since the roundtrip delay is significantly less).

With GEO satellites being so prevalent, it is perhaps worthwhile to give examples of LEO and MEO satellite applications. Weather, or environmental, monitoring satellites are both GEO (the U.S. satellites of this type are GOES-East over the Amazon River and GEOS-West over the eastern Pacific Ocean) and LEO with the latter being polar orbiting at typical altitudes of 850 km from which they are able to view any place on Earth with a given location being viewed twice each day under similar lighting conditions. Another MEO system is the GPS of 24 satellites, as mentioned earlier, wherein the satellites orbit in half-synchronous orbits (i.e., 12-h periods). At the turn of the century, several partnerships were working on LEO satellite systems for use with mobile satellite communications services—two examples are the Globalstar and Iridium systems, both of which proved to be uneconomical in comparison with terrestrial mobile telephone systems except for specialized applications.

9.6.1 Antenna Coverage

Coverage of the Earth by an antenna mounted on a satellite can be hemispherical, continental, or zonal depending on the antenna design. Antenna designs are now possible that cover several zones or spots simultaneously on the Earth's surface. Such designs allow *frequency reuse*, in that the same band of frequencies can be reused in separate beams, which effectively multiplies the bandwidth of the satellite transponder available for communications by the reuse factor. Figure 9.30 shows a typical antenna gain pattern in polar coordinates. The maximum gain can

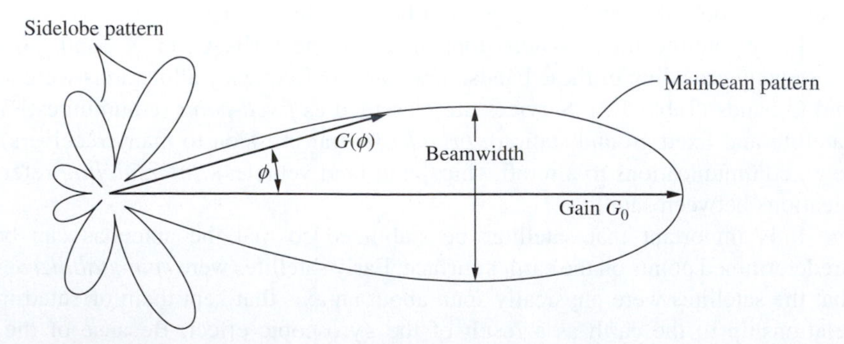

Figure 9.30
Polar representation of a general antenna gain function.

be roughly calculated from

$$G_0 = \rho_a \left(\frac{4\pi}{\lambda^2} \right) A \tag{9.167}$$

where

ρ_a = antenna efficiency (≤ 1)
λ = wavelength, m
A = aperture area, m^2

For a circular aperture of diameter d, $A = \pi d^2/4$ and (9.167) becomes

$$G_0 = \rho_a \left(\frac{\pi d}{\lambda} \right)^2 \tag{9.168}$$

The half-power beamwidth of an antenna can be approximated as

$$\phi_{3\text{dB}} = \frac{\lambda}{d\sqrt{\rho_a}} \text{ rad} \tag{9.169}$$

A convenient approximation for the antenna pattern of a parabolic reflector antenna for small angles off boresight (such that the gain is within 6 dB of the maximum value) is

$$g(\phi) = \rho_a \left(\frac{\pi d}{\lambda} \right)^2 \exp\left[-2.76 \left(\frac{\phi}{\phi_{3\text{dB}}} \right)^2 \right] \tag{9.170}$$

EXAMPLE 9.8

Find the aperture diameter and maximum gain for a transmit frequency of 10 GHz and $\rho_a = 0.8$ if from geosynchronous altitude, the following coverages are desired: (a) hemispherical, (b) continental United States (CONUS), and (c) a 150-mi-diameter spot.

Solution

The wavelength at 10 GHz is

$$\lambda = \frac{3 \times 10^8 \text{ m/s}}{10 \times 10^9 \text{Hz}} = 0.03 \text{ m}$$

$$= \frac{0.03 \text{ m}}{0.3048 \text{ m/ft}} = 0.0984 \text{ ft}$$

a. Geosynchronous altitude is 22,235 statute miles, and the earth's radius is 3963 mi. The angle subtended by the earth from geosynchronous altitude is

$$\phi_{\text{hemis}} = \frac{2(3963)}{22,235} = 0.356 \text{ rad}$$

Equating this to $\phi_{3\text{dB}}$ in (9.169) and solving for d, we have

$$d = \frac{0.0984}{0.356\sqrt{0.8}} = 0.31 \text{ ft}$$

b. The angle subtended by CONUS from geosynchronous altitude is

$$\phi_{\text{CONUS}} = \frac{4000}{22,235} = 0.18 \text{ rad}$$

Thus, from (9.169),

$$d = \frac{0.0984}{(0.18)\sqrt{0.8}} = 0.61 \text{ ft}$$

c. A 150-mi-diameter spot on the earth's surface directly below the satellite subtends an angle of

$$\phi_{150} = \frac{150}{22,235} = 0.0067 \text{ rad}$$

from geosynchronous orbit. The diameter of an antenna with this beamwidth is

$$d = \frac{0.0984}{0.0067\sqrt{0.8}} = 16.3 \text{ ft}$$

Note that doubling the frequency to 20 GHz would halve these diameters. ■

9.6.2 Earth Stations and Transmission Methods

Figure 9.31 shows a block diagram of the transmitter and receiving end of an Earth station. Signals from several sources enter the Earth station (e.g., telephone, television, etc.), whereupon two transmission options are available. First, the information from a single source can be placed on a single carrier. This is referred to as *single-channel-per-carrier* (SCPC). Second, information from several sources can be multiplexed together and placed onto contiguous intermediate frequency carriers, the sum translated to RF, power amplified, and transmitted. At the receiving end, the reverse process takes place.

At this point, it is useful to draw a distinction between multiplexing and multiple access. Multiple access (MA), like multiplexing, involves sharing of a common communications

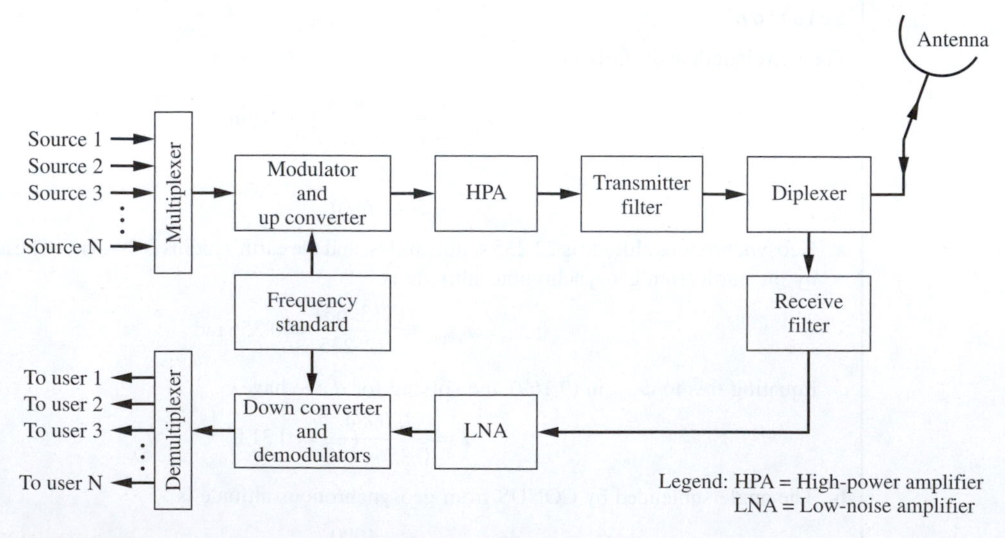

Figure 9.31
Satellite ground station receiver–transmitter configuration.

resource among several users. However, whereas multiplexing involves a fixed assignment of this resource at a local level, MA involves the remote sharing of a communication resource, and this sharing may under certain circumstances change dynamically under the control of a system controller. As mentioned before, there are three main techniques in use for jointly utilizing the communication resources of a remote resource, such as a relay satellite. These are

1. Frequency-division multiple access, wherein the communication resource is divided up in frequency
2. Time-division multiple access, wherein the resource is divided up in time
3. Code-division multiple access, wherein a unique code is assigned to each intended user and the separate transmissions are separated by correlation with the code of the desired transmitting party

Figure 9.32 illustrates these three accessing schemes. In FDMA, signals from various users are stacked up in frequency, just as for frequency-division multiplexing, as shown in Figure 9.32 (a). Guard bands are maintained between adjacent signal spectra to minimize crosstalk between channels. If frequency slots are assigned permanently to the users, the system is referred to as *fixed-assigned multiple access* (FAMA). If some type of dynamic allocation scheme is used to assign frequency slots, it is referred to as a *demand-assigned multiple access* (DAMA) system.

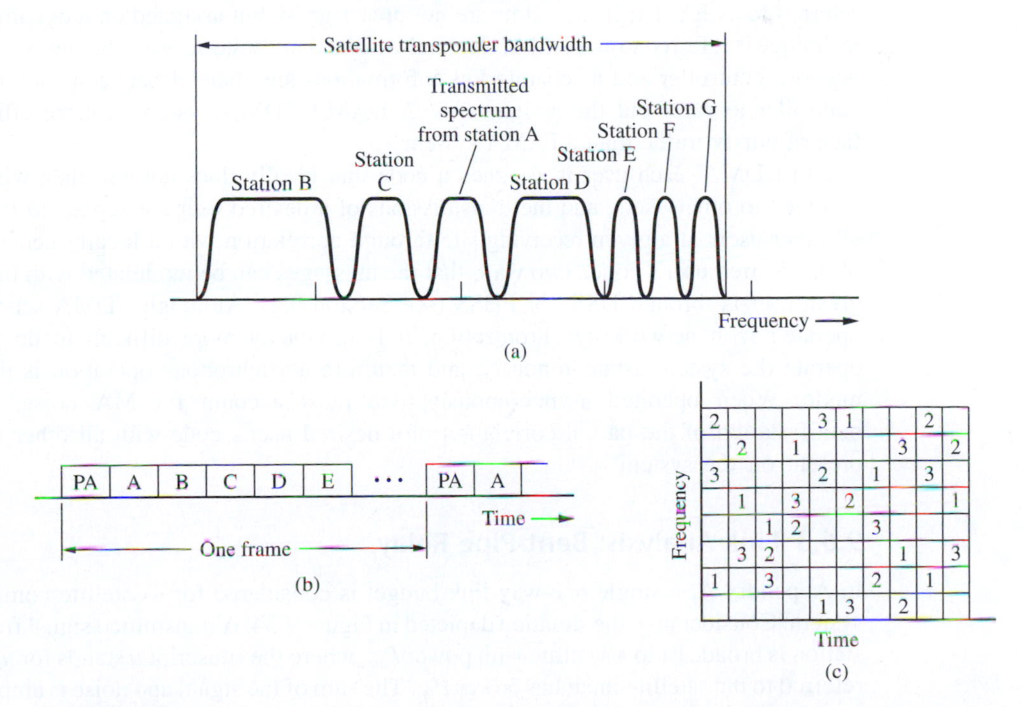

Figure 9.32
Illustration of MA techniques. (a) FDMA. (b) TDMA. (c) CDMA using frequency-hop modulation (numbers denote hopping sequences for channels 1, 2, and 3).

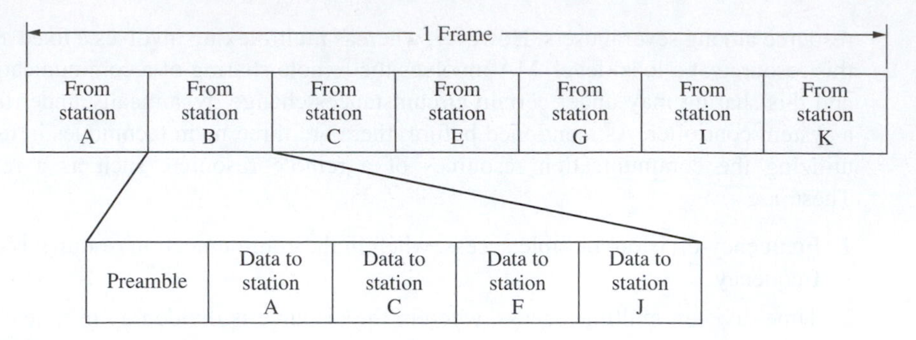

Figure 9.33
Details of a TDMA frame format.

In TDMA, the messages from various users are interlaced in time, just as for TDM, as shown in Figure 9.32 (b). As illustrated in Figure 9.33, the data from each user are conveyed in time intervals called *slots*. A number of slots make up a frame. Each slot is made up of a preamble plus information bits. The functions of the preamble are to provide identification and allow synchronization of the slot at the intended receiver. Guard times are utilized between each user's transmission to minimize crosstalk between channels.

It is necessary to maintain overall network synchronization in TDMA, unlike FDMA. If, in a TDMA system, the time slots that make up each frame are preassigned to specific sources, it is referred to as FAMA; if time slots are not preassigned, but assigned on a dynamic basis, the technique is referred to as DAMA. Demand-assigned multiple access schemes require a central network controller and a separate low-information-rate channel between each user and the controller to carry out the assignments. A DAMA TDMA system is more efficient in the face of bursty traffic than a FAMA system.

In CDMA, each user is assigned a code that ideally does not correlate with the codes assigned to other users, and the transmissions of a desired user are separated from those of all other users at a given receiving site through correlation with a locally generated replica of the desired user's code. Two ways that the messages can be modulated with the code for a given user is through DSSS or FHSS (see Section 9.4). Although CDMA schemes can be operated with network synchronization, it is obviously more difficult to do this than to operate the system asynchronously, and therefore asynchronous operation is the preferred mode. When operated asynchronously, one must account for MA noise, which is a manifestation of the partial correlation of a desired user's code with all other users' codes present on the system.

9.6.3 Link Analysis: Bent-Pipe Relay

In Appendix A, a single one-way link budget is considered for a satellite communications system. Consider now the situation depicted in Figure 9.34. A transmitted signal from a ground station is broadcast to a satellite with power P_{us}, where the subscript u stands for *uplink*. Noise referred to the satellite input has power P_{un}. The sum of the signal and noise is amplified by the satellite repeater to give a transmitted power from the satellite of

$$P_T = G(P_{us} + P_{un}) \tag{9.171}$$

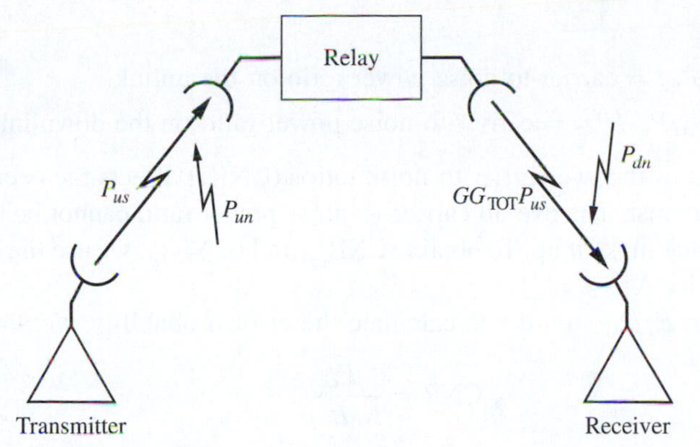

Figure 9.34
Signal and noise powers in the uplink and downlink portions of a bent-pipe satellite relay system.

where G is the power gain of the satellite repeater. The received signal power from the satellite at the receiving ground station is

$$P_{rs} = GG_{\mathrm{TOT}}P_{us} \tag{9.172}$$

where G_{TOT} represents total system losses and gains on the *downlink*. It can be expressed as

$$G_{\mathrm{TOT}} = \frac{G_t G_r}{L_a L_p} \tag{9.173}$$

where

G_t = gain of the satellite transmitter antenna.
L_a = atmospheric losses on the downlink.
L_p = propagation losses on the downlink.
G_r = gain of the ground station receive antenna.

The uplink noise power transmitted by the satellite repeater and appearing at the ground station input is

$$P_{run} = GG_{\mathrm{TOT}}P_{un} \tag{9.174}$$

Additional noise generated by the ground station itself is added to this noise at the ground station. The ratio of P_{rs} to total noise is the downlink carrier-to-noise power ratio. It is given by

$$(\mathrm{CNR})_r = \frac{P_{rs}}{P_{run} + P_{dn}} \tag{9.175}$$

Substituting previously derived expressions for each of the powers appearing on the right side of (9.175), we obtain

$$\begin{aligned}(\mathrm{CNR})_r &= \frac{GG_{\mathrm{TOT}}P_{us}}{GG_{\mathrm{TOT}}P_{un} + P_{dn}} \\ &= \frac{1}{P_{un}/P_{us} + P_{dn}/GG_{\mathrm{TOT}}P_{us}} \\ &= \frac{1}{(\mathrm{CNR})_u^{-1} + (\mathrm{CNR})_d^{-1}}\end{aligned} \tag{9.176}$$

where

$(\text{CNR})_u = P_{us}/P_{un}$ = carrier-to-noise power ratio on the uplink.

$(\text{CNR})_d = GG_{\text{TOT}}P_{us}/P_{dn}$ = carrier-to-noise power ratio on the downlink.

Note that the weakest of the two carrier-to-noise ratios (CNRs) affects the overall carrier-to-noise power ratio the most. The overall carrier-to-noise power ratio cannot be better than the worse of two CNRs that make it up. To obtain $(\text{CNR})_u$ and $(\text{CNR})_d$, we use the link equations developed in Appendix A.

To relate CNR to E_s/N_0 in order to calculate the error probability, we note that

$$\text{CNR} = \frac{P_c}{N_0 B_{RF}} \tag{9.177}$$

where

P_c = average carrier power.

N_0 = noise power spectral density.

B_{RF} = modulated signal (RF) bandwidth.

Multiplying numerator and denominator by the symbol duration T_s, we note that $P_c T_s = E_s$ is the symbol energy and obtain

$$\text{CNR} = \frac{E_s}{N_0 B_{RF} T_s}$$

or, solving for E_s/N_0,

$$\frac{E_s}{N_0} = (\text{CNR})B_{RF}T_s \tag{9.178}$$

Given a modulation scheme, we can use a suitable bandwidth criterion to determine $B_{RF}T_s$. For example, using the null-to-null bandwidth for BPSK as B_{RF}, we have $B_{RF} = 2/T_b$ or $T_b B_{RF} = 2$, where $T_s = T_b$, since we are considering binary signaling.

Because the CNR is related to E_s/N_0 by the constant $B_{RF}T_s$, we can write (9.176) as

$$\left(\frac{E_s}{N_0}\right)_r = \frac{1}{(E_s/N_0)_u^{-1} + (E_s/N_0)_d^{-1}} \tag{9.179}$$

where

$(E_s/N_0)_u$ = symbol-energy-to-noise-spectral-density ratio on the uplink

$(E_s/N_0)_d$ = symbol-energy-to-noise-spectral-density ratio on the downlink

EXAMPLE 9.9

Compute the relationship between $(E_s/N_0)_u$ and $(E_s/N_0)_d$ required to yield an error probability of $P_E = 10^{-6}$ on a bent-pipe satellite relay communications link if BPSK modulation is used.

Solution

For BPSK, $(E_b/N_0)_r \cong 10.53$ dB gives $P_E = 10^{-6}$. Thus (9.179) becomes

$$\frac{1}{(E_b/N_0)_u^{-1} + (E_s/N_0)_d^{-1}} = 10^{1.053} \cong 11.298 \tag{9.180}$$

Table 9.13 Uplink and Downlink Values of E_b/N_0 Required for $P_E = 10^{-6}$

$(E_b/N_0)_u$ (dB)	$(E_b/N_0)_d$ (dB)
20.0	11.06
15.0	12.47
14.0	13.14
13.54	13.54
12.0	15.98
11.0	20.52

Solving for $(E_b/N_0)_d$ from (9.180) in terms of $(E_b/N_0)_u$, we have the relationship

$$\left(\frac{E_b}{N_0}\right)_d = \frac{1}{0.0885 - (E_b/N_0)_u^{-1}} \tag{9.181}$$

Several pairs of values for $(E_b/N_0)_d$ and $(E_b/N_0)_u$ are given in Table 9.13.

A curve showing the graphical relationship between the uplink and downlink values of E_b/N_0 will be shown later in conjunction with another example. Note that the received E_b/N_0 is never better than the uplink or downlink values of E_b/N_0. For $(E_b/N_0)_u = (E_b/N_0)_d \cong 13.54$ dB, the value of $(E_b/N_0)_r$ is 10.53 dB, which is that value required to give $P_E = 10^{-6}$. Note that as either $(E_b/N_0)_u$ or $(E_b/N_0)_d$ approaches infinity, the other energy-to-noise-spectral-density ratio approaches 10.53 dB. ■

9.6.4 Link Analysis: OBP Digital Transponder

Consider a satellite relay link in which the modulation is digital binary and detection takes place on board the satellite with subsequent remodulation of the detected bits on the downlink carrier and subsequent demodulation and detection at the receiving ground station. This situation can be illustrated in terms of bit errors as shown in Figure 9.35.

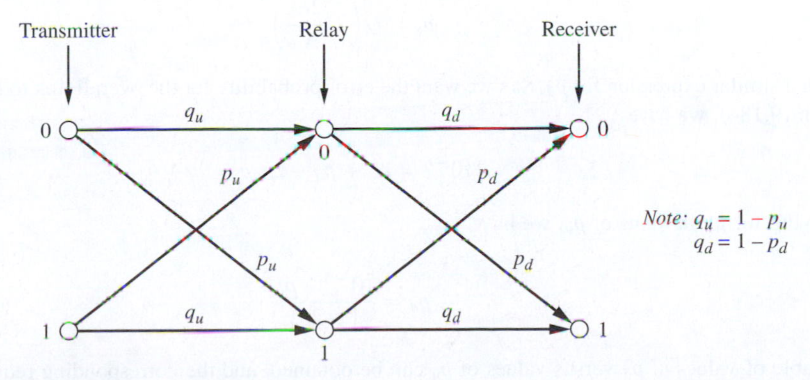

Figure 9.35
Transition probability diagram for uplink and downlink errors on a demod–remod satellite relay.

The channel is considered symmetrical in that errors for 1s and 0s are equally likely. It is also assumed that errors on the downlink are statistically independent of errors on the uplink and that errors in both links are independent of each other. From Figure 9.35, it follows that the overall probability of no error given that a 1 is transmitted is

$$P(C|1) = q_u q_d + p_u p_d \tag{9.182}$$

where

$q_u = 1 - p_u$ is the probability of no error on the uplink.

$q_d = 1 - p_d$ is the probability of no error on the downlink.

A similar expression holds for the probability $P(C|0)$ of correct transmission through the channel given that a 0 is transmitted, and it therefore follows that the probability of correct reception averaged over both 1s and 0s is

$$P(C) = \frac{1}{2}P(C|1) + \frac{1}{2}P(C|0) = P(C|1) = P(C|0) \tag{9.183}$$

The average probability of error is

$$\begin{aligned} P_E &= 1 - P(C) \\ &= 1 - (q_u q_d + p_u p_d) \\ &= 1 - (1 - p_u)(1 - p_d) - p_u p_d \\ &= p_u + p_d - 2p_u p_d \end{aligned} \tag{9.184}$$

The following example illustrates how to obtain the uplink and downlink signal energy-to-noise-spectral-density ratios required for an overall desired P_E for a given modulation technique.

EXAMPLE 9.10

Consider an OBP satellite communications link where BPSK is used on the uplink and the downlink. For this modulation technique,

$$p_u = Q\left(\sqrt{\frac{2E_b}{N_0}}\right) \tag{9.185}$$

with a similar expression for p_d. Say we want the error probability for the overall link to be 10^{-6}. Thus, from (9.184), we have

$$10^{-6} = p_u + p_d - 2p_u p_d \tag{9.186}$$

Solving for p_d in terms of p_u, we have

$$p_d = \frac{10^{-6} - p_u}{1 - 2p_u} \tag{9.187}$$

A table of values of p_d versus values of p_u can be obtained, and the corresponding required values of $(E_b/N_0)_u$ can then be calculated from (9.185) with a similar procedure followed for $(E_b/N_0)_d$. This is presented as Table 9.14.

Table 9.14 E_b/N_0 **Values for the Uplink and Downlink Required in an OBP Satellite Communications Link to Give** $P_E = 10^{-6}$

p_u	$(E_b/N_0)_u$ (dB)	p_d	$(E_b/N_0)_d$ (dB)
10^{-7}	11.30	9×10^{-7}	10.57
5×10^{-7}	10.78	5×10^{-7}	10.78
6×10^{-7}	10.71	4×10^{-7}	10.86
7×10^{-7}	10.66	3×10^{-7}	10.95

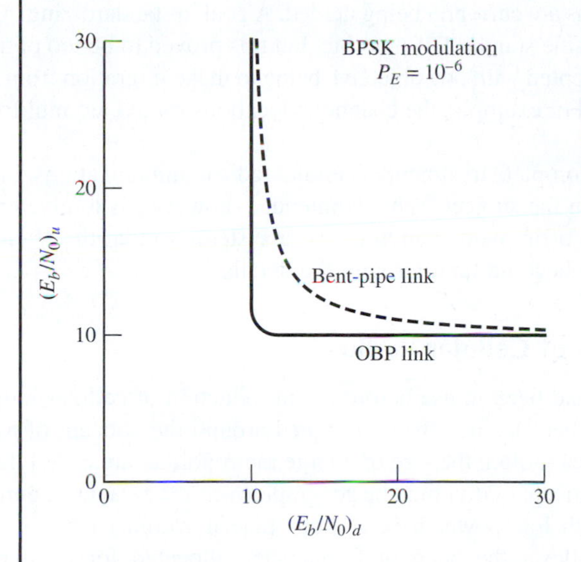

Figure 9.36
Comparison of bent-pipe and OBP relay characteristics for BPSK modulation and $P_E = 10^{-6}$.

Figure 9.36 shows $(E_b/N_0)_u$ versus $(E_b/N_0)_d$ for both the bent-pipe and OBP satellite relays for an overall bit error probability of 10^{-6}. Curves for other values of P_E or other digital modulation schemes can be obtained in a similar manner. This is left to the problems.

■

■ 9.7 CELLULAR RADIO COMMUNICATION SYSTEMS

Cellular radio communications systems were developed in the United States by Bell Laboratories, Motorola, and other companies in the 1970s and in Europe and Japan at about the same time. Test systems were installed in the United States in Washington, D.C. and Chicago in the late 1970s, and the first commerical cellular systems became operational in Japan in 1979, in Europe in 1981, and in the United States in 1983. The first system in the United States was designated AMPS for Advanced Mobile Phone System and proved to be very successful. The AMPS system used analog frequency modulation and a channel spacing of 33 kHz. Other standards used in Japan and Europe employed similar technology. In the early 1990s, there was more demand for cellular telephones than available capacity allowed so development of so-called second-generation (2G) systems began with the first 2G systems being fielded in the mid-1990s. All 2G systems use digital transmission, but with differing modulation and accessing schemes. The 2G European standard, called *Global System for Mobile* (GSM)

Communications, the Japanese system, and one U.S. standard [U.S. Digital Cellular (USDC) system] all employ TDMA, but with differing bandwidths and number of users per frame (see the discussion in Section 9.6 Satellite Communications for definitions of these terms). Another U.S. 2G standard, IS-95 (now cdmaOne) uses CDMA. A goal of 2G system development in the United States was backward compatibility because of the large AMPS infrastructure that had been installed with the first generation. Europe, on the other hand, had several first-generation standards, depending on the country, and their goal with 2G was to have a common standard across all countries. As a result, GSM has been widely adopted, not only in Europe but in much of the rest of the world. From the mid- to late 1990s work began on third-generation (3G) standards, and these systems are currently being fielded. A goal in standardizing 3G systems is to have a common world wide standard, if possible, but this proved to be too optimistic, so a family of standards was adopted with one objective being to make migration from 2G systems as convenient as possible. For example, the channel allocations for 3G are multiples of those used for 2G.

We will not provide a complete treatment of cellular radio communications. Indeed, entire books have been written on the subject. What is intended, however, is to give enough of an overview of the principles of implementation of these systems so that the reader may then consult other references to become familiar with the details.[42]

9.7.1 Basic Principles of Cellular Radio

Radio telephone systems had been in use before the introduction of cellular radio, but their capacity was very limited because they were designed around the concept of a single base station servicing a large area—often the size of a large metropolitan area. Cellular telephone systems are based on the concept of dividing the geographic service area into a number of cells and servicing the area with low-power base stations placed within each cell, usually the geographic center. This allows the band of frequencies allocated for cellular radio use (currently there are two bands in the 900 and 1800 MHz regions of the radio spectrum) to be reused over again a certain cell separation away, which depends on the accessing scheme used. For example, with AMPS, the reuse distance is three while for CDMA it is one. Another characteristic that the successful implementation of cellular radio depends on is the attenuation of transmitted power with frequency. Recall that for free space, power density decreases as the inverse square of the distance from the transmitter. Because of the propagation characteristics of terrestrial radio propagation, the decrease of power with distance is faster than an inverse square law, typically between the inverse third and fourth power of the distance. Were this not the case, it can be shown that the cellular concept would not work. Of course, because of the tessallation of the geographic area of interest into cells, it is necessary for the mobile user to be transfered from one base station to another as the mobile moves. This procedure is called *handoff* or *handover*. Also note that it is necessary to have some way of initializing a call to a given mobile and keeping track of it as it moves from cell to cell. This is the function of a *Mobile Switching Center* (MSC). MSCs also interface with the Public Switched Telephone Network (PSTN).

Consider Figure 9.37 which shows a typical cellular tessellation using hexagons. It is emphasized that real cells are never hexagonal; indeed, some cells may have very irregular

---

[42]Textbooks dealing with cellular communcations are: Stuber (2001), Rappaport (2002), Mark and Zhuang (2003), Goldsmith (2005), Tse and Viswanath (2005). Also recommended as an overview is Gibson (2002).

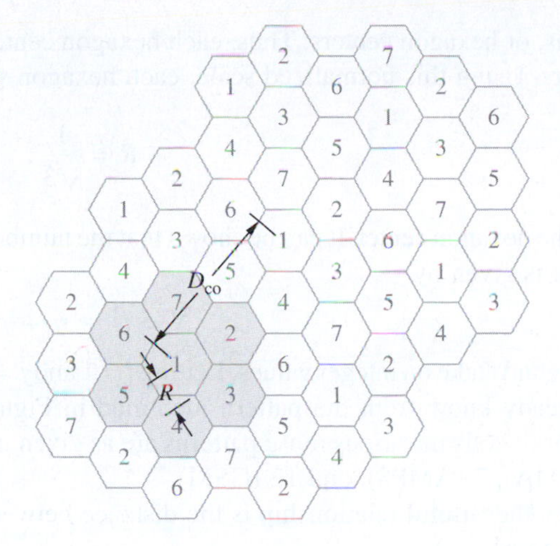

Figure 9.37
Hexagonal grid system representing cells in a cellular radio system; a reuse pattern of seven is illustrated.

shapes because of geographic features and illumination patterns by the transmit antenna. However, hexagons are typically used in theoretical discussions of cellular radio because a hexagon is one geometric shape that tessellates a plane and very closely approximates a circle which is what we surmise the contours of equal transmit power consist of in a relatively flat environment. Note that a seven-cell reuse pattern is indicated in Figure 9.37 via the integers given in each cell. Obviously there are only certain integers that work for reuse patterns, e.g., 1, 7, 12, . . . A convenient way to describe the frequency reuse pattern of an ideal hexagonal tessellation is to use a nonorthogonal set of axes, **U** and **V**, intersecting at 60° as shown in Figure 9.38. The normalized grid spacing of one unit represents distance between adjacent base

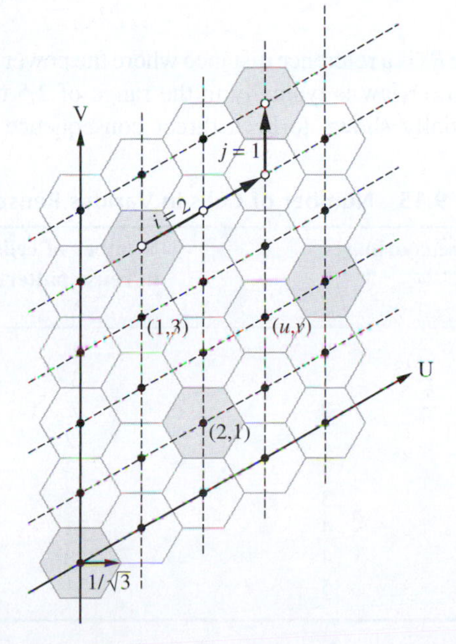

Figure 9.38
Hexagonal grid geometry showing coordinate directions; a reuse pattern of seven is illustrated.

stations, or hexagon centers. Thus, each hexagon center is at a point (u, v), where u and v are integers. Using this normalized scale, each hexagon vertex is

$$R = \frac{1}{\sqrt{3}} \tag{9.188}$$

from the hexagon center. It can be shown that the number of cells in an allowed frequency reuse pattern is given by

$$N = i^2 + ij + j^2 \tag{9.189}$$

where i and j take on integer values. Letting $i = 1$ and $j = 2$ (or vice versa), it is seen that $N = 7$ as we already know from the pattern identified in Figure 9.37. Putting in other integers, the number of cells in various reuse patterns are as given in Table 9.15. Typical reuse patterns are 1 (CDMA), 7 (AMPS), and 12 (GSM).

Another useful relationship is the distance between like-cell centers, D_{co}, which can be shown to be

$$D_{\text{co}} = \sqrt{3N}R = \sqrt{N} \tag{9.190}$$

which is an important consideration in computing *cochannel interference*, i.e., the inteference from a second user that is using the same frequency assignment as a user of interest. Clearly, if a reuse pattern has N cells in it, this interference could be a factor of N larger than that due to a single interfering user (not all cells at distance $\sqrt{N}$ from a user of interest may have an active call on that particular frequency). Note that there is a second ring of cells at $2\sqrt{N}$ that can intefere with a user of interest, but these are usually considered to be negligible compared with those within the first ring of intefering cells (and a third ring, etc.).

Assume a decrease in power with distance R of the form

$$P_r(R) = K\left(\frac{R_0}{R}\right)^{\alpha} \text{ W} \tag{9.191}$$

where R_0 is a reference distance where the power is known to be K W. As mentioned previously, the power law is typically in the range of 2.5 to 4 for terrestrial propagation, which can be analytially shown to be a direct consequence of the Earth's surface acting as a partially

Table 9.15 Number of Cells in Various Reuse Patterns

Reuse coorinates		Number of cells in reuse pattern	Normalized distance between repeat cells
i	j	N	$\sqrt{N}$
1	0	1	1
1	1	3	1.732
1	2	7	2.646
2	2	12	3.464
1	3	13	3.606
2	3	19	4.359
1	4	21	4.583
2	4	28	5.292
1	5	31	5.568

conducting reflector (other factors such as scattering from buildings and other large objects also come into play which accounts for the variation in α). In logarithmic terms, (9.191) becomes

$$P_{r,\text{dBW}}(R) = K_{\text{dB}} + 10\alpha \log_{10} R_0 - 10\alpha \log_{10} R \,\text{dBW} \tag{9.192}$$

Now consider reception by a mobile from a base station of interest, A, at distance d_A while at the same time being interfered with from a cochannel base station, B, at distance D_{co} from A. We assume for simplicity that the mobile is on a line connecting A and B. Thus, using (9.192), the signal-to-interference ratio (SIR) in decibels is

$$\begin{aligned} \text{SIR}_{\text{dB}} &= K_{\text{dB}} + 10\alpha \log_{10} R_0 - 10\alpha \log_{10} d_A \\ &\quad - [K_{\text{dB}} + 10\alpha \log_{10} R_0 - 10\alpha \log_{10} (D_{\text{co}} - d_A)] \\ &= 10\alpha \log_{10}\left(\frac{D_{\text{co}} - d_A}{d_A}\right) \\ &= 10\alpha \log_{10}\left(\frac{D_{\text{co}}}{d_A} - 1\right) \text{dB} \end{aligned} \tag{9.193}$$

Clearly, as $d_A \to D_{\text{co}}/2$, the argument of the logarithm approaches 1 and the SIR_{dB} approaches 0. At $d_A = D_{\text{co}}/2$ the mobile should ideally switch from A and begin using B as its base station.

We can also compute a worst-case SIR for a mobile of interest by using (9.193). If the mobile is using base station A as its source, the interference from the other cochannel base stations in the reuse pattern is no worse than that from B (the mobile was assumed to be on a line connecting A and B). Thus, the SIR_{dB} is underbounded by

$$\begin{aligned} \text{SIR}_{\text{dP,min}} &= 10\alpha \log_{10}\left(\frac{D_{\text{co}}}{d_A} - 1\right) - 10 \log_{10}(7 - 1) \text{ dB} \\ &= 10\alpha \log_{10}\left(\frac{D_{\text{co}}}{d_A} - 1\right) - 7.7815 \text{ dB} \end{aligned} \tag{9.194}$$

because the interference is increased by at worst a factor of $7 - 1$ (one station in the reuse pattern is the communicating base station to the mobile) due to the hexagonal tessellation.

EXAMPLE 9.11

Suppose that a cellular system uses a modulation scheme that requires a channel spacing of 25 kHz and an $\text{SIR}_{\text{dB,min}} = 20$ dB for each channel. Assume a total bandwidth of 6 MHz for both base-to-mobile (forward link) and mobile-to-base (reverse link) communications. Assume that the channel provides a propagation power law of $\alpha = 3.5$. Find the following: (a) the total number of users that can be accommodated within the reuse pattern, (b) The minimum reuse factor N, (c) the maximum number of users per cell, and (d) the efficiency in terms of voice circuits per base station per megahertz of bandwidth.

Solution

(a) The total bandwidth divided by the user channel bandwidth gives $6 \times 10^6/25 \times 10^3 = 240$ channels. Half of these are reserved for the downlink and half for the uplink, giving $240/2 = 120$ total users in the reuse pattern. (b) The $\text{SIR}_{\text{dB,min}}$ condition (9.194) gives

$$20 = 10(3.5)\log_{10}\left(\frac{D_{\text{co}}}{d_A} - 1\right) - 7.7815 \text{ dB}$$

which gives

$$\frac{D_{\text{co}}}{d_A} = 7.2201 = \sqrt{3N}$$

or

$$N = 17.38$$

Checking Table 9.14, we take the next largest allowed value of $N = 19$ ($i = 2$ and $j = 3$). (c) Dividing the total number of users by the number of cells in the reuse pattern, we obtain $\lfloor 120/19 \rfloor = 6$ users per cell, where the notation $\lfloor\ \rfloor$ means the largest interger not exceeding the bracketed quantity. The efficiency is

$$\eta_v = \frac{6 \text{ circuits}}{6 \text{ MHz}} = 1 \text{ voice circuit per base station per MHz}$$

■

EXAMPLE 9.12

Repeat Example 9.11 if $\text{SIR}_{\text{dB,min}} = 14$ dB is allowed.

Solution

Part (a) remains the same. (b) Part becomes

$$14 = 10(3.5)\log_{10}\left(\frac{D_{\text{co}}}{d_A} - 1\right) - 7.7815 \text{ dB}$$

which gives

$$\frac{D_{\text{co}}}{d_A} = 5.1908 = \sqrt{3N}$$

or

$$N = 8.98$$

which, from Table 9.15, translates to an allowed value of $N = 12$ ($i = 2$ and $j = 2$). (c) The number of users per cell is $\lfloor 120/12 \rfloor = 10$. (d) The efficiency is

$$\eta_v = \frac{10 \text{ circuits}}{6 \text{ MHz}} = 1.67 \text{ voice circuits per base station per megahertz}$$

■

9.7.2 Channel Perturbations in Cellular Radio

In addition to the Gaussian noise present in every communication link and the cochannel interference crudely analyzed above, another important source of degradation is fading. As the mobile moves the signal strength varies drastically because of multiple transmission paths. This fading can be characterized in terms of a Doppler spectrum, which is determined by the motion of the mobile (and to some small degree, the motion of the surroundings such a wind blowing trees or motion of reflecting vehicles). Another characteristic of the received fading signal is delay spread due to the differing propagation distances of the multipath components. As signaling rates increase, this becomes a more serious source of degradation. Equalization, as discussed in Chapter 7, can be used to compensate for it to some degree. Diversity can also be used to combat signal fading signal. In GSM and USDC, this takes the form of coding. For CDMA, diversity can be added in the form of simultaneous reception from two different base stations near cell boundaries. Other combinations of simultaneous transmissions[43] and receptions in a rich multipath environment (so-called MIMO techniques for *multiple-input, multiple-output*) are being proposed for future generation systems to significantly increase capacity. Also used in CDMA is a method called RAKE, which essentally detects the separate multipath components and puts them back together in a constructive fashion.

As progress is made in research, other means of combating detrimental channel effects are being considered for future cellular systems. These include multiuser detection to combat the cross-correlation noise due to other users in CDMA systems. As previously stated, what this does is to treat other users as sources that are detected and subtracted from the signal before the user of interest is detected.[44] Currently, multiuser detection is not included in the 3G standard, but is a distinct possibility for future cellular systems.

Another scheme that is being intensely researched currently as a means to extend the capacity of future cellular systems is smart antennas. This area entails any scheme where directivity of the antenna is used to increase the capacity of the system.[45]

A somewhat related area to that of smart antennas is space–time coding. These are codes that provide redundancy in both space and time. Space–time codes thereby exploit the channel redundancy in two dimensions and achieve more capacity than if the memory implicit in the channel is not made use of at all or if only one dimension is used.[46]

9.7.3 Characteristics of 1G and 2G Cellular Systems

Space does not allow much more than a cursory glance at the technical characteristics of first- and second-generation (1G and 2G) cellular radio systems - in particular, AMPS, GSM, and CDMA (referred to as IS-95 in the past, where the IS stands for Interim Standard, but now officially designated as cdmaOne). Second-generation cellular radio provides one of the most successful practical applications of many aspects of communications theory, including speech coding, modulation, channel coding, diversity techniques, and equalization. With the digital format used for 2G cellular, both voice and some data (limited to about 20 kbps) may be handled. Note that, while the accessing technique for GSM is said to be TDMA and that for cdmaOne is CDMA, both use FDMA in addition with 200-kHz spacing used for the former and 1.25-MHz spacing used for the latter.

---

[43] S. M. Alamouti, A simple transmit diversity technique for wireless communications. *IEEE Journal on Sel. Areas in Communication*, **16:** 1451–1458, October 1998. Also see the books by Paulraj, et al. (2003) and Tse and Viswanath (2005).

[44] See Verdu (1998).

[45] See Liberti and Rappaport (1999). For papers on smart antennas the *IEEE Transactions on Communications, The IEEE Journal on Selected Areas in Communications*, and *The IEEE Transactions on Wireless Communications* are recommended.

[46] A. F. Naguib, V. Tarokh, N. Seshadri, and A. R. Calderbank, A space-time coding modem for high-data-rate wireless communications. *IEEE Journal on Sel. Areas in Communication*, **16:** 1459–1478, October 1998. V. Tarokh, H. Jafarkhani, and A. R. Calderbank, Space-time block coding for wireless communications: performance results. *IEEE Journal on Sel. Areas in Communication*, **17:** 451–460, March 1999.

Table 9.16 Characteristics of First- and Second-Generation Cellular Radio Standards

	AMPS	GSM	CDMA
Carrier separation	30 kHz	200 kHz	1.25 MHz
No. channels/carrier	1	8	61 (64 Walsh codes; 3 sync, etc.)
Accessing technique	FDMA	TDMA–FDMA	CDMA-FDMA
Frame duration		4.6 ms with 0.58 ms slots	20 ms
User modulation	FM	GMSK, $BT = 0.3$ Binary, diff. encoded	BPSK, DL 64-ary orthog UL
DL–UL pairing	2 channels	2 slots	2 codes
Cell reuse pattern	7	12	1
Cochannel interference protection	≤ 15 dB	≤ 12 dB	NA
Error correction coding	NA	Rate–$\frac{1}{2}$ convolution Constraint length 5	Rate–$\frac{1}{2}$ convolution, Rate–$\frac{1}{3}$ convolution, UL Both constr. length 9
Diversity methods	NA	FH, 216.7 hops/s Equalization	Wideband signal Interleaving RAKE
Speech representation	Analog	Residual pulse excited, linear prediction coder	Code-excited vocoder
Speech coder rate	NA	13 kbps	9.6 kbps max

For complete details, the standard for each may be consulted. Before doing so, however, the reader is warned that these amount to thousands of pages in each case. Table 9.16 summarizes some of the most pertinent features of these three systems. For further details see some of the books referred to previously.

9.7.4 Characteristics of W-CDMA and cdma2000

As previously stated, in the late 1990s, work was begun by various standards bodies on third-generation (3G) cellular radio. The implementation of 3G cellular provides more capacity than 2G for voice in addition to much higher data capacity. At present, within a family of standards, there are two main competing standards for 3G, both using CDMA accessing. These are wideband CDMA (W-CDMA) promoted by Europe and Japan (harmonized with GSM characteristics), and cdma2000 which is based on IS-95 principles.

cdma2000

The most basic version of this wireless interface standard is referred to as $1 \times$ RTT for "1 times Radio Standard." Channelization still utilizes 1.25 MHz frequency as with cdmaOne, but increased capacity is achieved by increasing the number of user codes from 64 to 128 Walsh codes and changing the data modulation to QPSK on the forward link (BPSK in cdmaOne) and BPSK on the reverse link (64-ary orthogonal in cdmaOne). Spreading modulation is QPSK (balanced on the downlink and dual channel on the uplink). Accommodation of data is facilitated through media and link access control protocols and quality-of-service (QoS)

Table 9.17 Characteristics of Third-Generation Cellular Radio Standards

	W-CDMA	cdma2000
Channel BW (MHz)	5, 10, 20	1.25, 5, 10, 15, 20
FW link RF channel structure	DS	DS or MC
Chip rate, Mchips/s	4.096/8.192/16.384	1.2288/3.6864/7.3728/11.0593-14.7456, DS $n \times 1.2288$ $(n = 1, 3, \ldots)$, MC
Roll-off factor	0.22	Similar to TIA/EIA-95B
Frame length, ms	10, 20 (optional)	20 for data and control 5 for control on fundamental and dedicated control channels
Spreading modulation	Balanced QPSK (FW link) Dual channel QPSK (RV link) Complex spreading circuit	Balanced QPSK (FW link) Dual channel QPSK (RV link) Complex spreading circuit
Data modulation	QPSK (forward link) BPSK (reverse link)	QPSK (forward link) BPSK (reverse link)
Coherent detection	User dedicated time multiplexed pilot (forward link and reverse link), common pilot in forward link	Pilot time multiplexed with PC and EIB (reverse link) Common continuous pilot channel and auxilary pilot (forward link)
Channel multliplexing in reverse link	Control and pilot channel time multiplexed	Control, pilot fundamental, and supplemental code multiplexed
Multirate	Variable spreading and multicode	Variable spreading and multicode
Spreading factors	4-256 (4.096 Mcps)	4-256 (3.6864 Mcps)
Power control	Open and fast closed loop (1.6 kHz)	Open and fast closed loop (800 Hz)
Spreading (forward link)	Variable length orthogonal sequences for channel separation Gold sequences for user and cell separation	Variable length Walsh sequences for channel separation. m-sequence: 3×2^{15} (same sequence with time shift utilized in different cells; different sequences in I and Q channels)
Spreading (reverse link)	Variable length orthogonal sequences for channel separation. Gold sequence 2^{41} for user separation (different time shifts in I and Q channel, cycle 2^{16} radio frames)	Variable length orthogonal sequences—channel separation. m-sequence: 2^{15} (Same for all users; different sequences in I and Q channels). m-sequence: user separation (different time shifts for different users).
Handover	Soft handover. Interfrequency handover	Soft handover. Interfrequency handover

control, whereas no special provisions for data are present in cdmaOne. Data rates from 1.8 to 1036.8 kbps can be accommodated through varying cyclic redundancy check (CRC) bits, repetition, and deletions (at the highest data rate). Synchronization with the long code (common to each cell) in cdma2000 is facilitated by timing derived from GPS to localize where the long code epoch is within a given cell.

A higher data rate variation will use three $1 \times$ RTT channels on three carriers which may, but do not have to, use contiguous frequency slots. This is in a sense multicarrier modulation except that each carrier is spread in addition to the data modulation (in MCM as described in Section 9.5 the subcarriers were assumed to only have data modulation).

Wideband Code-Division Multiple Access

Wideband Code-Division Multiple Access (W-CDMA), as its name implies, is also based on CDMA access. It is the transmission protocol used by the Japanese NTT DoComo to provide high-speed wireless transmission (termed *Freedom of Mobile Multimedia Access*, or FOMO), and the most common wideband wireless transmission technology offered under the European Universal Mobile Telecommunications System (UMTS). Radio channels are 5 MHz wide, and they use QPSK spreading on both forward and reverse links in a slotted frame format (16 slots per frame for FOMA and 15 slots per frame for UMTS). Unlike cdma2000, it supports intercell asynchronous operation, with cell-to-cell handover being facilitated by a two-step synchronization process. Data rates from 7.5 to 5740 kbps can be accommodated by varying the spreading factor and assigning multiple codes (for the highest data rate).

Various system parameters and characteristics are summarized in Table 9.17[47] for both cdma2000 and W-CDMA.

Summary

1. When dealing with M-ary digital communications systems, with $M \geq 2$ it is important to distinguish between a bit and a symbol or character. A symbol conveys $\log_2 M$ bits. We must also distinguish between bit-error probability and symbol-error probability.
2. M-ary schemes based on quadrature multiplexing include QPSK, OQPSK, and MSK. All have a bit-error rate performance that is essentially the same as binary BPSK if precoding is used to ensure that only one bit error results from mistaking a given phase for an adjacent phase.
3. Minimum-shift keying can be produced by quadrature modulation or by serial modulation. In the latter case, MSK is produced by filtering BPSK with a properly designed conversion filter. At the receiver, serial MSK can be recovered by first filtering it with a bandpass matched filter and performing coherent demodulation with a carrier at $f_c + 1/4T_b$ (i.e., at the carrier plus a quarter data rate). Serial MSK performs identically to quadrature-modulated MSK and has advantageous implementation features at high data rates.
4. Gaussian MSK is produced by passing the ± 1-valued data stream (NRZ format) through a filter with Gaussian frequency response (and Gaussian impulse response), scaled by $2\pi f_d$, where f_d is the deviation constant in hertz

[47]From T. Ojanpera and S. Gray, An Overview of cdma2000, WCDMA, and EDGE, in Gibson (2002).

per volt, to produce the excess phase of an FM-modulated carrier. A GMSK spectrum has lower sidelobes than ordinary MSK at the expense of degradation in bit error probability due to the intersymbol interference introduced by the filtering of the data signal. Gaussian MSK was used in one of the second generation standards in the United States for cellular radio.

5. It is convenient to view M-ary data modulation in terms of signal space. Examples of data formats that can be considered in this way are M-ary PSK, QAM, and M-ary FSK. For the former two modulation schemes, the dimensionality of the signal space stays constant as more signals are added; for the latter, it increases directly as the number of signals added. A constant-dimensional signal space means signal points are packed closer as the number of signal points is increased, thus degrading the error probability; the bandwidth remains essentially constant. In the case of FSK, with increasing dimensionality as more signals are added, the signal points are not compacted, and the error probability decreases for a constant SNR; the bandwidth increases with an increasing number of signals, however.

6. Communication systems may be compared on the basis of power and bandwidth efficiencies. A rough measure of bandwidth is null-to-null of the main lobe of the transmitted signal spectrum. For M-ary PSK, QAM, and DPSK power efficiency decreases with increasing M (i.e., as M increases a larger value of E_b/N_o is required to provide a given value of bit-error probability) and bandwidth efficiency increases (i.e, the larger M, the smaller the required bandwidth for a given bit rate). For M-ary FSK (both coherent and noncoherent) the opposite is true. This behavior may be explained with the aid of signal space concepts—the signal space for M-ary PSK, QAM, and DPSK remains constant at two dimensions versus M (one-dimensional for $M = 2$), whereas for M-ary FSK it increases linearly with M. Thus, from a power efficiency standpoint the signal points are crowded together more as M increases in the former cases, whereas they are not in the latter case.

7. A convenient measure of bandwidth occupancy for digital modulation is in terms of out-of-band power or power-containment bandwidth. An ideal brick-wall containment bandwidth that passes 90% of the signal power is approximately $1/T_b$ Hz for QPSK, OQPSK, and MSK and about $2/T_b$ Hz for BPSK.

8. The different types of synchronization that may be necessary in a digital modulation system are carrier (only for coherent systems), symbol or bit, and possibly word. Carrier and symbol synchronization can be carried out by an appropriate nonlinearity followed by a narrowband filter or PLL. Alternatively, appropriate feedback structures may be used.

9. A PN sequence resembles a random "coin-toss" sequence but can be generated easily with linear feedback shift-register circuits. A PN sequence has a narrow correlation peak for zero delay and low sidelobes for nonzero delay, a property that makes it ideal for synchronization of words or measurement of range.

10. Spread-spectrum communications systems are useful for providing resistance to jamming, to provide a means for masking the transmitted signal from unwanted interceptors, to provide resistance to multipath, to provide a way for more than one user to use the same time–frequency allocation, and to provide range-measuring capability.
11. The two major types of spread-spectrum systems are DSSS and FHSS. In the former, a spreading code with rate much higher than the data rate multiplies the data sequence, thus spreading the spectrum, while for FHSS, a synthesizer driven by a pseudorandom code generator provides a carrier that hops around in a pseudorandom fashion. A combination of these two schemes, referred to as *hybrid spread spectrum*, is also another possibility.
12. Spread spectrum performs identically to whatever data-modulation scheme is employed without the spectrum spreading as long as the background is additive white Gaussian noise and synchronization is perfect.
13. The performance of a spread-spectrum system in interference is determined in part by its processing gain, which can be defined as the ratio of bandwidth of the spread system to that for an ordinary system employing the same type of data modulation as the spread-spectrum system. For DSSS the processing gain is the ratio of the data bit duration to the spreading code bit (or chip) duration.
14. An additional level of synchronization, referred to as *code synchronization*, is required in a spread-spectrum system. The serial search method is perhaps the simplest in terms of hardware and to explain, but it is relatively slow in achieving synchronization.
15. Multicarrier modulation is a modulation scheme where the data to be transmitted is multiplexed on several subcarriers that are summed before transmission. Each transmitted symbol is thereby longer by a factor of the number of subcarriers used than would be the case if the data were transmitted serially on a single carrier. This makes MCM more resistant to multipath than a serial transmission system, assuming both to be operating with the same data rate.
16. A special case of MCM wherein the subcarriers are spaced by $1/T$, where T is the symbol duration is called OFDM. Orthogonal frequency division multiplexing is often implemented by means of the inverse DFT at the transmitter and by a DFT at the receiver.
17. Satellite communications systems provide a specific example to which the digital modulation schemes considered in this chapter can be applied. Two general types of relay satellite configurations were considered in the last section of this chapter: bent-pipe and OBP (or demod–remod). In the OBP system, the data on the uplink are demodulated and detected and then used to remodulate the downlink carrier. In the bent-pipe relay, the uplink transmissions are translated in frequency, amplified, and retransmitted on the downlink. Performance characteristics of both types of links were considered by example.

18. Cellular radio provides an example of a communications technology that has been accepted faster and more widely by the public then first anticipated. First-generation systems were fielded in the early 1980s and used analog modulation. Second-generation systems were fielded in the mid-1990s. The introduction of 3G systems started around the year 2000. All 2G and 3G systems utilize digital modulation, with many based on CDMA.

Further Reading

In addition to the references given in Chapter 8, for a fuller discussion and in-depth treatment of the digital modulation techniques presented here, see Ziemer and Peterson (2001), and Peterson et al. (1995) for further discussion on spread spectrum and cellular communications. Another comprehensive reference is Proakis (2007).

Problems

Section 9.1

9.1. An M-ary communication system transmits at a rate of 4000 symbols per second. What is the equivalent bit rate in bits per second for $M = 4$? $M = 8$? $M = 16$? $M = 32$? $M = 64$? Generate a plot of bit rate versus $\log_2 M$.

9.2. A serial bit stream, proceeding at a rate of 10 kbps from a source, is given as

101110 000111 010011 (spacing for clarity)

Number the bits from left to right starting with 1 and going through 18 for the right most bit. Associate the odd-indexed bits with $d_1(t)$ and the even-indexed bits with $d_2(t)$ in Figure 9.1.

a. What is the symbol rate for d_1 or d_2?

b. What are the successive values of θ_i given by (9.2) assuming QPSK modulation? At what time intervals may θ_i switch?

c. What are the successive values of θ_i given by (9.2) assuming OQPSK modulation? At what time intervals may θ_i switch values?

9.3. Quadriphase-shift keying is used to transmit data through a channel that adds Gaussian noise with power spectral density $N_0 = 10^{-11}\ \mathrm{V}^2/\mathrm{Hz}$. What are the values of the quadrature-modulated carrier amplitudes required to give $P_{E,\text{ symbol}} = 10^{-5}$ for the following data rates?

a. 5 kbps

b. 10 kbps

c. 50 kbps

d. 100 kbps

e. 0.5 Mbps

f. 1 Mbps

9.4. Show that the noise components N_1 and N_2 for QPSK, given by (9.6) and (9.8), are uncorrelated; that is, show that $E[N_1 N_2] = 0$. (Explain why N_1 and N_2 are zero mean.)

9.5. A QPSK modulator produces a phase imbalanced signal of the form

$$x_c(t) = Ad_1(t)\cos\left(2\pi f_c t + \frac{\beta}{2}\right) - Ad_2(t)\sin\left(2\pi f_c t - \frac{\beta}{2}\right)$$

a. Show that the integrator outputs of Figure 9.2, instead of (9.5) and (9.7), are now given by

$$V'_1 = \frac{1}{2}AT_s\left(\pm\cos\frac{\beta}{2} \pm \sin\frac{\beta}{2}\right)$$

$$V'_2 = \frac{1}{2}AT_s\left(\pm\sin\frac{\beta}{2} \pm \cos\frac{\beta}{2}\right)$$

where the $\pm$ signs depend on whether the data bits $d_1(t)$ and $d_2(t)$ are $+1$ or -1.

b. Show that the probability of error per quadrature channel is

$$P'_{E,\text{ quad chan}} = \frac{1}{2}Q\left[\sqrt{\frac{2E_b}{N_0}}\left(\cos\frac{\beta}{2} + \sin\frac{\beta}{2}\right)\right]$$
$$+ \frac{1}{2}Q\left[\sqrt{\frac{2E_b}{N_0}}\left(\cos\frac{\beta}{2} - \sin\frac{\beta}{2}\right)\right]$$

Hint: For no phase imbalance, the correlator outputs were $V_1,\ V_2 = \pm\frac{1}{2}AT_s = \pm AT_b$ giving $E_b = V_1^2/T_b = V_2^2/T_b$ and

$$P_{E,\text{ quad chan}} = Q\left[\sqrt{\frac{2E_b}{N_0}}\right]$$

With phase imbalance, the best- and worst-case values for E'_b are

$$E'_b = E_b\left(\cos\frac{\beta}{2} + \sin\frac{\beta}{2}\right)^2$$

and

$$E'_b = E_b\left(\cos\frac{\beta}{2} - \sin\frac{\beta}{2}\right)^2$$

These occur with equal probability.

c. Plot P_E given by (9.12) and the above result for $P'_{E,\text{ quad chan}}$ on the same plot for $\beta = 0$, 2.5, 5, 7.5, and 10 degrees. Estimate and plot the degradation in E_b/N_o, expressed in decibels, due to phase imbalance at an error probability of 10^{-4} and 10^{-6} from these curves.

9.6.

a. A BPSK system and a QPSK system are designed to transmit at equal rates; that is, 2 bits are transmitted with the BPSK system for each symbol (phase) in the QPSK system. Compare their symbol-error probabilities versus E_s/N_0 (note that E_s for the BPSK system is $2E_b$).

b. A BPSK system and a QPSK system are designed to have equal transmission bandwidths. Compare their symbol-error probabilities versus SNR (note that for this to be the case, the symbol durations of both must be the same; i.e., $T_{s,\text{ BPSK}} = 2T_b = T_{s,\text{ QPSK}}$).

c. On the basis of parts (a) and (b), what do you conclude about the deciding factor(s) in choosing BPSK versus QPSK?

9.7. Given the serial data sequence

101011 010010 100110 110011

associate every other bit with the upper and lower data streams of the block diagrams of Figures 9.2 and 9.4. Draw on the same time scale (one below the other) the quadrature waveforms for the following data modulation schemes: QPSK, OQPSK, MSK type I, and MSK type II.

9.8. Sketch excess phase tree and phase trellis diagrams for each of the cases of Problem 9.7. Show as a heavy line the actual path through the tree and trellis diagrams represented by the data sequence given.

9.9. Derive (9.25) for the spectrum of an MSK signal by multiplying $|G(f)|^2$, given by (9.23), times $S_{\text{BPSK}}(f)$, given by (9.24). That is, show that serial modulation of MSK works from the standpoint of spectral arguments. (*Hint*: Work only with the positive-frequency portions of (9.23) and (9.24) to produce the first term of (9.25); similarly work with the negative-frequency portions to produce the second term of (9.25). In so doing you are assuming negligible overlap between positive- and negative-frequency portions.)

9.10. An MSK system has a carrier frequency of 10 MHz and transmits data at a rate of 50 kbps.

a. For the data sequence 1010101010..., what is the instantaneous frequency?

b. For the data sequence 0000000000..., what is the instantaneous frequency?

9.11. Show that (9.26) and (9.27) are Fourier transform pairs.

9.12.

a. Sketch the signal space with decision regions for 16-ary PSK [see (9.47)].

b. Use the bound (9.50) to write down and plot the symbol error probability versus E_b/N_0.

c. On the same axes, compute and plot the bit-error probability.

9.13.

a. Using (9.93) and appropriate bounds for $P_{E,\text{ symbol}}$, obtain E_b/N_0 required for achieving $P_{E,\text{ bit}} = 10^{-4}$ for M-ary PSK with $M = 8$, 16, 32.

b. Repeat for QAM for $M = 16$ and 64 using (9.63).

9.14. Derive the three equations numbered (9.60) through (9.62) for M-QAM.

9.15. By substituting (9.60) to (9.62) into (9.59), collecting all like-argument terms in the Q-function, and neglecting squared Q-function terms, show that the symbol-error probability for 16-QAM reduces to (9.63).

9.16. Show that for M-ary QAM

$$a = \sqrt{\frac{3E_s}{2(M-1)}}$$

where E_s is the symbol energy averaged over the constellation of M signals, which is (9.58). The summation formulas

$$\sum_{i=1}^{m} i = \frac{m(m+1)}{2} \quad \text{and} \quad \sum_{i=1}^{m} i^2 = \frac{m(m+1)(2m+1)}{6}$$

will prove useful.

9.17. Using (9.95), (9.96), and (9.67), obtain E_b/N_0 required for achieving $P_{E,\,\text{bit}} = 10^{-3}$ for M-ary coherent FSK for $M = 2, 4, 8, 16, 32, 64$. Program you calculator to do an iterative solution using MATLAB.

9.18. Using (9.95), (9.96), and (9.68), repeat Problem 9.17 for noncoherent M-ary FSK for $M = 2, 4, 8, 16, 32$.

9.19. Based on signal space arguements order the modulation schemes 16-PSK, 16-QAM, and coherent 16-CFSK from best to worst on the basis of:

a. Bandwidth efficiency

b. Communication efficiency (probability of bit error)

9.20. On the basis of null-to-null bandwidths, give the required transmission bandwidth to achieve a bit rate of 10 kbps for the following:

a. 16-QAM, 16-PSK, or 16-DPSK

b. 32-PSK

c. 64-PSK

d. 8-FSK, coherent

e. 16-FSK, coherent

f. 32-FSK, coherent

g. 8-FSK, noncoherent

h. 16-FSK, noncoherent

i. 32-FSK, noncoherent

Section 9.2

9.21. On the basis of 90% power-containment bandwidth, give the required transmission bandwidth to achieve a bit rate of 10 kbps for

a. BPSK

b. QPSK or OQPSK

c. MSK

d. 16-QAM

9.22. Generalize the results for power-containment bandwidth for quadrature-modulation schemes given in Section 9.2 to M-ary PSK. (Is it any different than the result for QAM?) With appropriate reinterpretation of the abscissa of Figure 9.16 and using the 90% power containment bandwidth, obtain the required transmission bandwidth to support a bit rate of 100 kbps for

a. 8-PSK

b. 16-PSK

c. 32-PSK

9.23. A binary PSK-modulated signal with carrier component can be written as

$$s_{\text{PSK}}(t) = A \sin\left[\omega_c t + \cos^{-1} md(t) + \theta\right]$$

where $m \leq 1$ is a constant which will be called the *modulation index* and $d(t) = \pm 1$ in T_b-s bit intervals.

a. Show that it can be expanded as

$$s_{\text{PSK}}(t) = Am \sin(\omega_c t + \theta) + A\sqrt{1 - m^2} d(t) \cos(\omega_c t + \theta)$$

Hints: Use the trigonometric identity for $\sin(\alpha + \beta)$ and the facts that $\cos(\cos^{-1} m) = m$ and $\sin(\cos^{-1} m) = \sqrt{1 - m^2}$ (justify these).

b. From part (a), note that the first term is an unmodulated carrier component and the second term is the modulated component. Find their average powers and show that

$$\frac{P_{\text{carrier}}}{P_{\text{modulation}}} = \frac{m^2}{1 - m^2}$$

9.24. Assume that a data stream $d(t)$ consists of a random (coin toss) sequence of $+1$ and -1 that is T s in duration. The autocorrelation function for such a sequence is

$$R_d(\tau) = \begin{cases} 1 - \dfrac{|\tau|}{T}, & \dfrac{|\tau|}{T} \leq 1 \\ 0, & \text{otherwise} \end{cases}$$

a. Find and sketch the power spectral density for an ASK-modulated signal given by

$$s_{\text{ASK}}(t) = \frac{1}{2} A[1 + d(t)] \cos(\omega_c t + \theta)$$

where θ is a uniform random variable in $(0, 2\pi]$.

b. Use the result of Problem 9.23(a) to compute and sketch the power spectral density of a PSK-modulated signal given by

$$s_{\text{PSK}}(t) = A \sin\{\omega_c t + \cos^{-1}[md(t)] + \theta\}$$

for the three cases $m = 0, 0.5$, and 1.

9.25. Derive the Fourier transform pair given by (9.120).

Section 9.3

9.26. Draw the block diagram of an M-power law circuit for synchronizing a local carrier for 8-PSK. Assume that $f_c = 10$ MHz and $T_s = 0.1$ ms. Carefully label all blocks, and give critical frequencies and bandwidths.

9.27. Plot σ_ϕ^2 versus z for the various cases given in Table 9.7. Assume 10% of the signal power is in the carrier for the PLL and all signal power is in the modulation for the Costas and data estimation loops. Assume values of $L =$ 100, 10, 5.

9.28. Find the difference in decibels between (9.128) and (9.129). That is, find the ratio $\sigma_{\epsilon,\,\mathrm{SL}}^2/\sigma_{\epsilon,\,\mathrm{AV}}^2$ expressed in decibels.

9.29. Consider the marker code C8 of Table 9.8. Find the Hamming distance between all possible shifts of it and the received sequence 10110 10110 00011 10101 (spaces for clarity). Is there a unique match to within $h = 1$ and this received sequence? If so, at what delay does it occur?

9.30. Fill in all the steps in going from (9.134) to (9.135).

9.31. An m-sequence is generated by a continuously running feedback shift resister with a clock rate of 10 kHz. Assume that the shift register has six stages and that the feedback connection is the proper one to generate a maximal length sequence. Answer the following questions:

a. How long is the sequence before it repeats?

b. What is the period of the generated sequence in milliseconds?

c. Provide a sketch of the autocorrelation function of the generated sequence. Provide critical dimensions.

d. What is the spacing between spectral lines in the power spectrum of this sequence?

e. What is the height of the spectral line at zero frequency? How is this related to the DC level of the m-sequence?

f. At what frequency is the first null in the *envelope* of the power spectrum?

9.32. Consider a 15-bit, maximal-length PN code. It is generated by feeding back the last two stages of a four-stage shift register. Assuming a 1 1 1 1 initial state, find all the other possible states of the shift register (show details). What is the sequence? Find and plot its periodic autocorrelation function, providing critical dimensions.

9.33. The aperiodic autocorrelation function of a binary code is of interest in some synchronization applications. In computing it, the code is not assumed to periodically repeat itself, but (9.133) is applied only to the overlapping part. For example, with the 3-chip Barker code of Table 9.12 the computation is as follows:

		$N_A - N_U$	$\dfrac{N_A - N_U}{N}$
Barker code	1 1 0		
Delay = 0	1 1 0	3	1
Delay = 1	1 1 0	0	0
Delay = 2	1 1 0	−1	− 1/3

For negative delays, we need not perform the calculation because autocorrelation functions are even.

a. Find the aperiodic autocorrelation functions of all the Barker sequences given in Table 9.12. What are the magnitudes of their maximum nonzero-delay autocorrelation values?

b. Compute the aperiodic autocorrelation function of the 15-chip PN sequence found in Problem 9.32. What is the magnitude of its maximum nonzero-delay autocorrelation values? Note from Table 9.5 that this is *not* a Barker sequence.

Section 9.4

9.34. Show that the variance of N_g as given by (9.148) is $N_0 T_b$.

9.35. Show that the variance of N_I as given by (9.153) is approximated by the result given by (9.154). (*Hint*: You will have to make use of the fact that $T_c^{-1}\Lambda(\tau/T_c)$ is approximately a delta function for small T_c).

9.36. A DSSS system employing BPSK data modulation operates with a data rate of 10 kbps. A processing gain of 1000 (30 dB) is desired.

a. Find the required chip rate.

b. What is the RF transmission bandwidth required (null to null.?

c. An SNR of 10 dB is employed. What is P_E for the following JSRs? 5 dB, 10 dB, 15 dB, and 30 dB.

9.37. Consider a DSSS system employing BPSK data modulation. $P_E = 10^{-5}$ is desired with $E_b/N_0 \to \infty$. For the following JSRs tell what processing gain, G_p, will give the desired P_E. If none, so state.

a. JSR = 30 dB.

b. JSR = 25 dB.

c. JSR = 20 dB.

9.38. Compute the number of users that can be supported at a maximum bit-error probability of 10^{-3} in a multiuser DSSS system with a code length of $n = 255$. [Hint: Take the limit as $E_b/N_0 \to \infty$ in (9.160), and set the resulting expression for $P_E = 10^{-4}$; then solve for N.]

9.39. Repeat Example 9.6 with everything the same except for a propagation delay uncertainty of ± 1.5 ms and a false alarm penalty of $T_{\text{fa}} = 1000\,T_i$.

Section 9.5

9.40.

a. Consider a multipath channel with a delay spread of 5 μs through which it is desired to transmit data at 500 kbps. Design an MCM system that has a symbol period at least a factor of 10 greater than the delay spread if the modulation to be used on each subcarrier is QPSK.

b. If an inverse FFT is to be used to implement this as an OFDM system, what size inverse FFT is necessary assuming that the FFT size is to be an integer power of 2?

Section 9.6

9.41.

a. Given a circular aperture transmit antenna with efficiency of 0.7 operating at 10 GHz and having a diameter of 1.5 m. Find its maximum gain.

b. Find its 3-dB beamwidth.

c. Plot its antenna gain pattern in decibels versus angle off boresight in degrees.

9.42. Rederive the curves shown in Figure 9.36, assuming BPSK modulation, for an overall P_E of (a) 10^{-3} and (b) 10^{-4}.

9.43. Rederive the curves shown in Figure 9.36, assuming $P_E = 10^{-5}$ if the modulation technique used on the uplink and downlink is (a) binary coherent FSK, (b) binary noncoherent FSK, and (c) binary DPSK.

9.44.

a. Find the diameter of an antenna aperture mounted on a geosynchronous altitude satellite that will provide a 100-mi spot on the earth's surface within its 3-dB beamwidth if the operating frequency on the downlink is 21 GHz and the antenna efficiency is 0.8

b. Find the maximum gain, G_0, of the antenna.

c. Plot the antenna gain pattern in decibels, $10 \log_{10} g(\phi)$, versus angle off boresight assuming the pattern of (9.170).

Section 9.7

9.45. Rework Examples 9.11 and 9.12 for an attenuation exponent of $\alpha = 4$.

9.46. Rework Example 9.11 with everything the same except assume $\text{SIR}_{\text{dB, min}} = 10$ dB.

Computer Exercises

9.1. Use MATLAB to plot curves of P_b versus E_b/N_0, $M = 2, 4, 8, 16,$ and 32 for

a. M-ary coherent FSK (Use the upper-bound expression as an approximation to the actual error probability.)

b. M-ary noncoherent FSK

Compare your results with Figure 9.15(a) and (b).

9.2. Use MATLAB to plot out-of-band power for M-ary PSK, QPSK (or OQPSK), and MSK. Compare with Figure 9.16. Use `trapz` to do the required numerical integration.

9.3. Approximate the power spectrum of coherent M-ary FSK by adding voltage spectra of sinusoidal bursts of duration T_b and of the appropriate frequency coherently, and then plotting the magnitude squared. What is the minimum spacing of the "tones" in order to maintain them coherently orthogonal?

9.4. Use MATLAB to plot curves like those shown in Figure 9.25. Use MATLAB to find the processing gain required to give a desired probability of bit error for a given JSR and SNR. Note that your program should check to see if the desired bit-error probability is possible for the given JSR and SNR.

9.5. Given a satellite altitude and desired illumination spot diameter on the earth's surface, use MATLAB to determine the antenna aperture diameter and maximum gain to give the desired spot diameter.

9.6. Develop a MATLAB program to plot Figure 9.36 for a given probability of bit error and

a. Binary PSK

b. Coherent binary FSK

c. Binary DPSK

d. Noncoherent binary FSK

9.7. Write a MATLAB simulation of GMSK that will simulate the modulated waveform. From this, compute and plot the power spectral density of the modulated waveform. Include the special case of ordinary MSK in your simulation so that you can compare the spectra of GMSK and MSK for several BT_B products. *Hint*: Do a "help psd" to find out how to use the power spectral density estimator in MATLAB to estimate and plot the power spectra of the simulated GMSK and MSK waveforms.

CHAPTER 10

OPTIMUM RECEIVERS AND SIGNAL SPACE CONCEPTS

For the most part, this book has been concerned with the *analysis* of communication systems. An exception occurred in Chapter 8, where we sought the best receiver in terms of minimum probability of error for binary digital signals of known shape. In this chapter we deal with the *optimization* problem; that is, we wish to find the communication system for a given task that performs the *best*, within a certain class, of all possible systems. In taking this approach, we are faced with three basic problems:

1. **What is the optimization criterion to be used?**
2. **What is the optimum structure for a given problem under this optimization criterion?**
3. **What is the performance of the optimum receiver?**

We will consider the simplest type of problem of this nature possible, that of fixed transmitter and channel structure with only the receiver to be optimized.

We have two purposes for including this subject in our study of information transmission systems. First, in Chapter 1 we stated that the application of probabilistic systems analysis techniques coupled with statistical optimization procedures has led to communication systems distinctly different in character from those of the early days of communications. The material in this chapter will, we hope, give you an indication of the truth of this statement, particularly when you see that some of the optimum structures considered here are building blocks of systems analyzed in earlier chapters. Additionally, the signal space techniques to be further developed later in this chapter provide a unification of the performance results for the analog and digital communication systems that we have obtained so far.

10.1 BAYES OPTIMIZATION

10.1.1 Signal Detection Versus Estimation

Based on our considerations in Chapters 8 and 9, we see that it is perhaps advantageous to separate the signal-reception problem into two domains. The first of these we shall refer to as *detection*, for we are interested merely in detecting the presence of a particular signal, among other candidate signals, in a noisy background. The second is referred to as *estimation*, in which we are interested in estimating some characteristic of a signal that is assumed to be present in a noisy environment. The signal characteristic of interest may be a time-independent parameter such as a constant (random or nonrandom) amplitude or phase or an estimate (past, present, or future value) of the waveform itself (or a functional of the waveform). The former problem is usually referred to as *parameter estimation*. The latter is referred to as *filtering*. We see that demodulation of analog signals (AM, DSB, and so on), if approached in this fashion, would be a signal-filtering problem.[1]

While it is often advantageous to categorize signal-reception problems as either detection or estimation, both are usually present in practical cases of interest. For example, in the detection of phase-shift-keyed signals, it is necessary to have an estimate of the signal phase available to perform coherent demodulation. In some cases, we may be able to ignore one of these aspects, as in the case of noncoherent digital signaling, in which signal phase was of no consequence. In other cases, the detection and estimation operations may be inseparable. However, we will look at signal detection and estimation as separate problems in this chapter.

10.1.2 Optimization Criteria

In Chapter 8, the optimization criterion that was employed to find the matched-filter receiver for binary signals was *minimum average probability of error*. In this chapter we will generalize this idea somewhat and seek signal detectors or estimators that *minimize average cost*. Such devices will be referred to as *Bayes* receivers for reasons that will become apparent later.

10.1.3 Bayes Detectors

To illustrate the use of minimum average cost optimization criteria to find optimum receiver structures, we will first consider detection. For example, suppose we are faced with a situation in which the presence or absence of a constant signal of value $k > 0$ is to be detected in the presence of an additive Gaussian noise component N (for example, as would result by taking a single sample of a signal plus noise waveform). Thus we may hypothesize two situations for the observed data Z:

Hypothesis 1 (H_1): $Z = N$ (noise alone); $P(H_1 \text{ true}) = p_0$.
Hypothesis 2 (H_2): $Z = k + N$ (signal plus noise); $P(H_2 \text{ true}) = 1 - p_0$.

Assuming the noise to have zero mean and variance σ_n^2, we may write down the pdfs of Z given hypotheses H_1 and H_2, respectively. Under hypothesis H_1, Z is Gaussian with mean zero and variance σ_n^2. Thus

$$f_Z(z|H_1) = \frac{e^{-z^2/2\sigma_n^2}}{\sqrt{2\pi\sigma_n^2}} \tag{10.1}$$

Under hypothesis H_2, since the mean is k,

$$f_Z(z|H_2) = \frac{e^{-(z-k)^2/2\sigma_n^2}}{\sqrt{2\pi\sigma_n^2}} \tag{10.2}$$

These conditional pdfs are illustrated in Figure 10.1. We note in this example that Z, the observed data, can range over the real line $-\infty < Z < \infty$. Our objective is to partition this

---

[1]See Van Trees (1968), Vol. I, for a consideration of filtering theory applied to optimal demodulation.

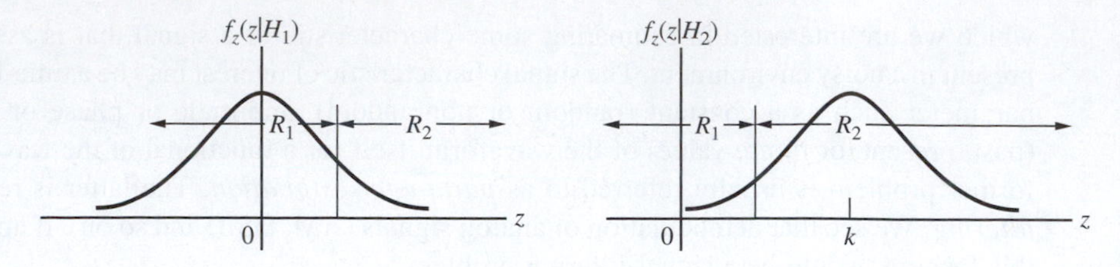

Figure 10.1
Conditional pdfs for a two-hypothesis detection problem.

one-dimensional observation space into two regions R_1 and R_2 such that if Z falls into R_1, we decide hypothesis H_1 is true, while if Z is in R_2, we decide H_2 is true. We wish to accomplish this in such a manner that the average cost of making a decision is minimized. It may happen, in some cases, that R_1 or R_2 or both will consist of multiple segments of the real line. (See Problem 10.2.)

Taking a general approach to the problem, we note that four a priori costs are required, since there are four types of decisions that we can make. These costs are

c_{11} = cost of deciding in favor of H_1 when H_1 is actually true.
c_{12} = cost of deciding in favor of H_1 when H_2 is actually true.
c_{21} = cost of deciding in favor of H_2 when H_1 is actually true.
c_{22} = cost of deciding in favor of H_2 when H_2 is actually true.

Given that H_1 was actually true, the conditional average cost of making a decision, $C(D|H_1)$, is

$$C(D|H_1) = c_{11}P[\text{decide } H_1|H_1 \text{ true}] + c_{21}P[\text{decide } H_2|H_1 \text{ true}] \tag{10.3}$$

In terms of the conditional pdf of Z given H_1, we may write

$$P[\text{decide } H_1|H_1 \text{ true}] = \int_{R_1} f_Z(z|H_1)\, dz \tag{10.4}$$

and

$$P[\text{decide } H_2|H_1 \text{ true}] = \int_{R_2} f_Z(z|H_1)\, dz \tag{10.5}$$

where the one-dimensional regions of integration are as yet unspecified.

We note that Z must lie in either R_1 or R_2, since we are forced to make a decision. Thus

$$P[\text{decide } H_1|H_1 \text{ true}] + P[\text{decide } H_2|H_1 \text{ true}] = 1 \tag{10.6}$$

or if expressed in terms of the conditional pdf $f_Z(z|H_1)$, we obtain

$$\int_{R_2} f_Z(z|H_1)\, dz = 1 - \int_{R_1} f_Z(z|H_1)\, dz \tag{10.7}$$

Thus, combining (10.3) through (10.6), the conditional average cost given H_1, $C(D|H_1)$, becomes

$$C(D|H_1) = c_{11}\int_{R_1} f_Z(z|H_1)\, dz + c_{21}\left[1 - \int_{R_1} f_Z(z|H_1)\, dz\right] \tag{10.8}$$

In a similar manner, the average cost of making a decision given that H_2 is true, $C(D|H_2)$, can be written as

$$\begin{aligned} C(D|H_2) &= c_{12}P[\text{decide } H_1|H_2 \text{ true}] + c_{22}P[\text{decide } H_2|H_2 \text{ true}] \\ &= c_{12}\int_{R_1} f_Z(z|H_2)\,dz + c_{22}\int_{R_2} f_Z(z|H_2)\,dz \\ &= c_{12}\int_{R_1} f_Z(z|H_2)\,dz + c_{22}\left[1 - \int_{R_1} f_Z(z|H_2)\,dz\right] \end{aligned} \tag{10.9}$$

To find the average cost without regard to which hypothesis is actually true, we must average (10.8) and (10.9) with respect to the prior probabilities of hypotheses H_1 and H_2, $p_0 = P[H_1 \text{ true}]$ and $q_0 = 1 - p_0 = P[H_2 \text{ true}]$. The average cost of making a decision is then

$$C(D) = p_0 C(D|H_1) + q_0 C(D|H_2) \tag{10.10}$$

Substituting (10.8) and (10.9) into (10.10) and collecting terms, we obtain

$$\begin{aligned} C(D) = {} & p_0\left\{c_{11}\int_{R_1} f_Z(z|H_1)\,dz + c_{21}\left[1 - \int_{R_1} f_Z(z|H_1)\,dz\right]\right\} \\ & + q_0\left\{c_{12}\int_{R_1} f_Z(z|H_2)\,dz + c_{22}\left[1 - \int_{R_1} f_Z(z|H_2)\,dz\right]\right\} \end{aligned} \tag{10.11}$$

for the average cost, or *risk*, in making a decision. Collection of all terms under a common integral that involves integration over R_1 results in

$$C(D) = [p_0 c_{21} + q_0 c_{22}] + \int_{R_1} \{[q_0(c_{12} - c_{22}) f_Z(z|H_2)] - [p_0(c_{21} - c_{11}) f_Z(z|H_1)]\}\,dz \tag{10.12}$$

The first term in brackets represents a fixed cost once p_0, q_0, c_{21}, and c_{22} are specified. The value of the integral is determined by those points which are assigned to R_1. Since wrong decisions should be more costly than right decisions, it is reasonable to assume that $c_{12} > c_{22}$ and $c_{21} > c_{11}$. Thus the two bracketed terms within the integral are positive because $q_0, p_0, f_Z(z|H_2)$, and $f_Z(z|H_1)$ are probabilities. Hence all values of z that give a larger value for the second term in brackets within the integral than for the first term in brackets should be assigned to R_1 because they contribute a negative amount to the integral. Values of z that give a larger value for the first bracketed term than for the second should be assigned to R_2. In this manner, $C(D)$ will be minimized. Mathematically, the preceding discussion can be summarized by the pair of inequalities

$$q_0(c_{12} - c_{22}) f_Z(Z|H_2) \underset{H_1}{\overset{H_2}{\gtrless}} p_0(c_{21} - c_{11}) f_Z(Z|H_1)$$

or

$$\frac{f_Z(Z|H_2)}{f_Z(Z|H_1)} \underset{H_1}{\overset{H_2}{\gtrless}} \frac{p_0(c_{21} - c_{11})}{q_0(c_{12} - c_{22})} \tag{10.13}$$

which are interpreted as follows: If an observed value for Z results in the left-hand ratio of pdfs being greater than the right-hand ratio of constants, choose H_2; if not, choose H_1. The left-hand side of (10.13), denoted by $\Lambda(Z)$,

$$\Lambda(Z) \triangleq \frac{f_Z(Z|H_2)}{f_Z(Z|H_1)} \tag{10.14}$$

is called the *likelihood ratio*. The right-hand side of (10.13)

$$\eta \triangleq \frac{p_0(c_{21}-c_{11})}{q_0(c_{12}-c_{22})} \tag{10.15}$$

is called the *threshold* of the test. Thus, the Bayes criterion of minimum average cost has resulted in a test of the likelihood ratio, which is a random variable, against the threshold value η. Note that the development has been general, in that no reference has been made to the particular form of the conditional pdfs in obtaining (10.13). We now return to the specific example that resulted in the conditional pdfs of (10.1) and (10.2).

EXAMPLE 10.1

Consider the pdfs of (10.1) and (10.2). Let the costs for a Bayes test be $c_{11} = c_{22} = 0$ and $c_{21} = c_{12}$.

a. Find $\Lambda(Z)$.

b. Write down the likelihood ratio test for $p_0 = q_0 = \frac{1}{2}$.

c. Compare the result of part (b) with the case $p_0 = \frac{1}{4}$ and $q_0 = \frac{3}{4}$.

Solution

a. The likelihood ratio is given by

$$\Lambda(Z) = \frac{\exp\left[-(Z-k)^2/2\sigma_n^2\right]}{\exp(-Z^2/2\sigma_n^2)} = \exp\left[\frac{2kZ-k^2}{2\sigma_n}\right] \tag{10.16}$$

b. For this case $\eta = 1$, which results in the test

$$\exp\left(\frac{2kZ-k^2}{2\sigma_n^2}\right) \underset{H_1}{\overset{H_2}{\gtrless}} 1 \tag{10.17}$$

Taking the natural logarithm of both sides [this is permissible because $\ln(x)$ is a monotonic function of x] and simplifying, we obtain

$$Z \underset{H_1}{\overset{H_2}{\gtrless}} \frac{k}{2} \tag{10.18}$$

which states that if the noisy received data are less than half the signal amplitude, the decision that minimizes risk is that the signal was absent, which is reasonable.

c. For this situation, $\eta = \frac{1}{3}$, and the likelihood ratio test is

$$\exp\left(\frac{2kZ-k^2}{2\sigma_n^2}\right) \underset{H_1}{\overset{H_2}{\gtrless}} \frac{1}{3} \tag{10.19}$$

or, simplifying,

$$Z \underset{H_1}{\overset{H_2}{\gtrless}} \frac{k}{2} - \frac{\sigma_n^2}{k}\ln 3 \tag{10.20}$$

Since is was assumed that $k > 0$, the second term on the right-hand side is positive and the optimum threshold is clearly reduced from the value resulting from signals having equal prior probabilities.

Thus, if the prior probability of a signal being present in the noise is increased, the optimum threshold is decreased so that the signal-present hypothesis (H_2) will be chosen with higher probability.

■

10.1.4 Performance of Bayes Detectors

Since the likelihood ratio is a function of a random variable, it is itself a random variable. Thus, whether we compare the likelihood ratio $\Lambda(Z)$ with the threshold η or we simplify the test to a comparison of Z with a modified threshold as in Example 10.1, we are faced with the prospect of making wrong decisions. The average cost of making a decision, given by (10.12), can be written in terms of the conditional probabilities of making wrong decisions, of which there are two.[2] These are given by

$$P_F = \int_{R_2} f_Z(z|H_1)\, dz \tag{10.21}$$

and

$$\begin{aligned} P_M &= \int_{R_1} f_Z(z|H_2)\, dz \\ &= 1 - \int_{R_2} f_Z(z|H_2)\, dz = 1 - P_D \end{aligned} \tag{10.22}$$

The subscripts F, M, and D stand for "false alarm," "missed detection," and "correct detection," respectively, a terminology that grew out of the application of detection theory to radar. (It is implicitly assumed that hypothesis H_2 corresponds to the signal-present hypothesis and that hypothesis H_1 corresponds to noise alone, or signal absent, when this terminology is used.) When (10.21) and (10.22) are substituted into (10.12), the risk per decision becomes

$$C(D) = p_0 c_{21} + q_0 c_{22} + q_0(c_{12} - c_{22})P_M - p_0(c_{21} - c_{11})(1 - P_F) \tag{10.23}$$

Thus, it is seen that if the probabilities P_F and P_M (or P_D) are available, the Bayes risk can be computed.

Alternative expressions for P_F and P_M can be written in terms of the conditional pdfs of the likelihood ratio given H_1 and H_2 as follows: Given that H_2 is true, an erroneous decision is made if

$$\Lambda(Z) < \eta \tag{10.24}$$

for the decision, according to (10.13), is in favor of H_1. The probability of inequality (10.24) being satisfied, given H_2 is true, is

$$P_M = \int_0^{\eta} f_\Lambda(\lambda|H_2)\, d\lambda \tag{10.25}$$

where $f_\Lambda(\lambda|H_2)$ is the conditional pdf of $\Lambda(Z)$ given that H_2 is true. The lower limit of the integral in (10.25) is $\eta = 0$ since $\Lambda(Z)$ is nonnegative, being the ratio of pdfs.

[2] As will be apparent soon, the probability of error introduced in Chapter 8 can be expressed in terms of P_M and P_F. Thus these conditional probabilities provide a complete performance characterization of the detector.

Similarly,

$$P_F = \int_{\eta}^{\infty} f_{\Lambda}(\lambda|H_1)\, d\lambda \tag{10.26}$$

because, given H_1, an error occurs if

$$\Lambda(Z) > \eta \tag{10.27}$$

[The decision is in favor of H_2 according to (10.13).] The conditional probabilities $f_{\Lambda}(\lambda|H_2)$ and $f_{\Lambda}(\lambda|H_1)$ can be found, in principle at least, by transforming the pdfs $f_Z(z|H_2)$ and $f_Z(z|H_1)$ in accordance with the transformation of random variables defined by (10.14). Thus two ways of computing P_M and P_F are given by using either (10.21) and (10.22) or (10.25) and (10.26). Often, however, P_M and P_F are computed by using a monotonic function of $\Lambda(Z)$ that is convenient for the particular situation being considered, as in Example 10.2.

A plot of $P_D = 1 - P_M$ versus P_F is called the *operating characteristic* of the likelihood ratio test, or the *receiver operating characteristic* (ROC). It provides all the information necessary to evaluate the risk through (10.23), provided the costs c_{11}, c_{12}, c_{21}, and c_{22} are known. To illustrate the calculation of an ROC, we return to the example involving detection of a constant in Gaussian noise.

EXAMPLE 10.2

Consider the conditional pdf's of (10.1) and (10.2). For an arbitrary threshold η, the likelihood ratio test of (10.13), after taking the natural logarithm of both sides, reduces to

$$\frac{2kZ - k^2}{2\sigma_n^2} \underset{H_1}{\overset{H_2}{\gtrless}} \ln \eta \quad \text{or} \quad \frac{Z}{\sigma_n} \underset{H_1}{\overset{H_2}{\gtrless}} \left(\frac{\sigma_n}{k}\right) \ln \eta + \frac{k}{2\sigma_n} \tag{10.28}$$

Defining the new random variable $X \triangleq Z/\sigma_n$ and the parameter $d \triangleq k/\sigma_n$, we can further simplify the likelihood ratio test to

$$X \underset{H_1}{\overset{H_2}{\gtrless}} d^{-1} \ln \eta + \frac{1}{2} d \tag{10.29}$$

Expressions for P_F and P_M can be found once $f_X(x|H_1)$ and $f_X(x|H_2)$ are known. Because X is obtained from Z by scaling by σ_n, we see from (10.1) and (10.2) that

$$f_X(x|H_1) = \frac{e^{-x^2/2}}{\sqrt{2\pi}} \quad \text{and} \quad f_X(x|H_2) = \frac{e^{-(x-d)^2/2}}{\sqrt{2\pi}} \tag{10.30}$$

That is, under either hypothesis H_1 or hypothesis H_2, X is a unity variance Gaussian random variable. These two conditional pdfs are shown in Figure 10.2. A false alarm occurs if, given H_1,

$$X > d^{-1} \ln \eta + \frac{1}{2} d \tag{10.31}$$

The probability of this happening is

$$\begin{aligned} P_F &= \int_{d^{-1}\ln\eta + d/2}^{\infty} f_X(x|H_1)\, dx \\ &= \int_{d^{-1}\ln\eta + d/2}^{\infty} \frac{e^{-x^2/2}}{\sqrt{2\pi}}\, dx = Q\left(d^{-1} \ln \eta + \frac{d}{2}\right) \end{aligned} \tag{10.32}$$

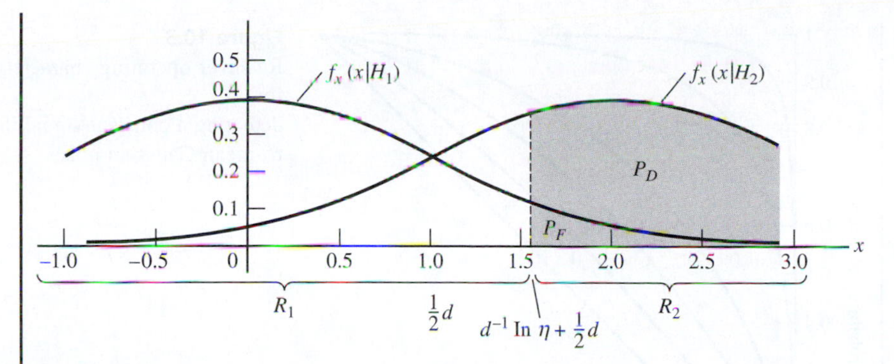

Figure 10.2
Conditional and decision regions for the problem of detecting a constant signal in zero-mean Gaussian noise.

which is the area under $f_X(x|H_1)$ to the right of $d^{-1}\ln\eta + \frac{1}{2}d$ in Figure 10.2. Detection occurs if, given H_2,

$$X > d^{-1}\ln\eta + \frac{1}{2}d \tag{10.33}$$

The probability of this happening is

$$\begin{aligned} P_D &= \int_{d^{-1}\ln\eta + d/2}^{\infty} f_X(x|H_2)\,dx \\ &= \int_{d^{-1}\ln\eta + d/2}^{\infty} \frac{e^{-(x-d)^2/2}}{\sqrt{2\pi}}\,dx = Q\left(d^{-1}\ln\eta - \frac{d}{2}\right) \end{aligned} \tag{10.34}$$

Thus P_D is the area under $f_X(x|H_2)$ to the right of $d^{-1}\ln\eta + d/2$ in Figure 10.2. Note that for $\eta = 0$, $\ln\eta = -\infty$ and the detector always chooses H_2 ($P_F = 1$). For $\eta = \infty$, $\ln\eta = \infty$ and the detector always chooses H_1 ($P_D = P_F = 0$). ■

COMPUTER EXERCISE 10.1

The ROC is obtained by plotting P_D versus P_F for various values of d, as shown in Figure 10.3. The curves are obtained by varying η from 0 to ∞. This is easily accomplished using the simple MATLAB code that follows.

```
% file: c10ce1
clear all;
d = [0 0.3 0.6 1 2 3];                  % vector of d values
eta = logspace(-2,2);                   % values of eta
lend = length(d);                       % number of d values
hold on                                 % hold for multiple plots
for j=1:lend                            % begin loop
     dj = d(j);                         % select jth value of d
     af = log(eta)/dj + dj/2;           % argument of Q for Pf
     ad = log(eta)/dj - dj/2;           % argument of Q for Pd
     pf = qfn(af);                      % compute Pf
     pd = qfn(ad);                      % compute Pd
     plot(pf,pd)                        % plot curve
end
```

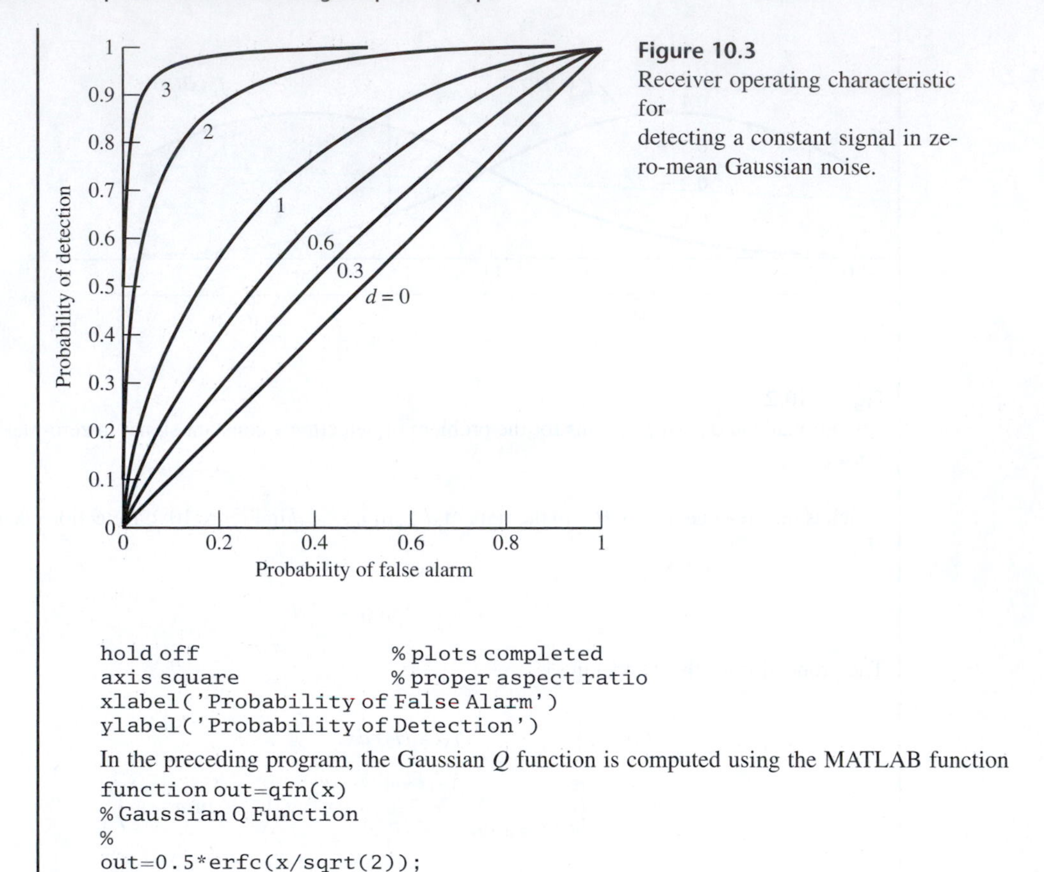

Figure 10.3
Receiver operating characteristic for
detecting a constant signal in zero-mean Gaussian noise.

```
hold off                        % plots completed
axis square                     % proper aspect ratio
xlabel('Probability of False Alarm')
ylabel('Probability of Detection')
```

In the preceding program, the Gaussian Q function is computed using the MATLAB function

```
function out=qfn(x)
% Gaussian Q Function
%
out=0.5*erfc(x/sqrt(2));
```

■

10.1.5 The Neyman–Pearson Detector

The design of a Bayes detector requires knowledge of the costs and a priori probabilities. If these are unavailable, a simple optimization procedure is to fix P_F at some tolerable level, say α, and maximize P_D (or minimize P_M) subject to the constraint $P_F \leq \alpha$. The resulting detector is known as the *Neyman–Pearson detector*. It can be shown that the Neyman–Pearson criterion leads to a likelihood ratio test identical to that of (10.13), except that the threshold η is determined by the allowed value of probability of false alarm α. This value of η can be obtained from the ROC for a given value of P_F, for it can be shown that the slope of a curve of an ROC at a particular point is equal to the value of the threshold η required to achieve the P_D and P_F of that point.[3]

10.1.6 Minimum-Probability-of-Error Detectors

From (10.12) it follows that if $c_{11} = c_{22} = 0$ (zero cost for making right decision) and $c_{12} = c_{21} = 1$ (equal cost for making either type of wrong decision), then the risk reduces to

[3]Van Trees (1968), Vol. 1.

$$
\begin{aligned}
C(D) &= p_0\left[1 - \int_{R_1} f_Z(z|H_1)\,dz\right] + q_0\int_{R_1} f_Z(z|H_2)\,dz \\
&= p_0\int_{R_2} f_Z(z|H_1)\,dz + q_0\int_{R_1} f_Z(z|H_2)\,dz \\
&= p_0P_F + q_0P_M
\end{aligned}
\tag{10.35}
$$

where we have used (10.7), (10.21), and (10.22). However, (10.35) is the probability of making a wrong decision, averaged over both hypotheses, which is the same as the probability of error used as the optimization criterion in Chapter 8. Thus Bayes receivers with this special cost assignment are minimum-probability-of-error receivers.

10.1.7 The Maximum a Posteriori Detector

Letting $c_{11} = c_{22} = 0$ and $c_{21} = c_{12}$ in (10.13), we can rearrange the equation in the form

$$
\frac{f_Z(Z|H_2)P(H_2)}{f_Z(Z)} \underset{H_1}{\overset{H_2}{\gtrless}} \frac{f_Z(Z|H_1)P(H_1)}{f_Z(Z)}
\tag{10.36}
$$

where the definitions of p_0 and q_0 have been substituted, both sides of (10.13) have been multiplied by $P(H_2)$, and both sides have been divided by

$$
f_Z(Z) \triangleq f_Z(Z|H_1)P(H_1) + f_Z(Z|H_2)P(H_2)
\tag{10.37}
$$

Using Bayes rule, as given by (5.10), (10.36) becomes

$$
P(H_2|Z) \underset{H_1}{\overset{H_2}{\gtrless}} P(H_1|Z) \quad (c_{11} = c_{22} = 0; \quad c_{12} = c_{21})
\tag{10.38}
$$

Equation (10.38) states that the most probable hypothesis, given a particular observation Z, is to be chosen in order to minimize the risk, which, for the special cost assignment assumed, is equal to the probability of error. The probabilities $P(H_1|Z)$ and $P(H_2|Z)$ are called *a posteriori probabilities*, for they give the probability of a particular hypothesis *after* the observation of Z, in contrast to $P(H_1)$ and $P(H_2)$, which give us the probabilities of the same events *before* observation of Z. Because the hypothesis corresponding to the maximum a posteriori probability is chosen, such detectors are referred to as *maximum a posteriori* (MAP) detectors. Minimum-probability-of-error detectors and MAP detectors are therefore equivalent.

10.1.8 Minimax Detectors

The minimax decision rule corresponds to the Bayes decision rule, where the a priori probabilities have been chosen to make the Bayes risk a maximum. For further discussion of this decision rule, see Van Trees (1968).

10.1.9 The *M*-ary Hypothesis Case

The generalization of the Bayes decision criterion to M hypotheses, where $M > 2$, is straightforward but unwieldy. For the M-ary case, M^2 costs and M a priori probabilities must be given. In effect, M likelihood ratio tests must be carried out in making a decision. If attention is restricted to the special cost assignment used to obtain the MAP detector for the binary case (that is, right decisions cost zero and wrong decisions are all equally costly), then a MAP decision rule results that is easy to visualize for the M-hypothesis case. Generalizing from (10.38), we have the MAP decision rule for the M-hypothesis case: Compute the M posterior probabilities $P(H_i|Z)$, $i = 1, 2, \ldots, M$, and choose as the correct hypothesis the one corresponding to the largest posterior probability. This decision criterion will be used when M-ary signal detection is considered.

10.1.10 Decisions Based on Vector Observations

If, instead of a single observation Z, we have N observations $Z \triangleq (Z_1, Z_2, \ldots, Z_N)$, all of the preceding results hold with the exception that the N-fold joint pdfs of Z, given H_1 and H_2, are to be used. If $Z_1, Z_2, \ldots, Z_N$ are conditionally independent, these joint pdfs are easily written since they are simply the N-fold products of the marginal pdfs of $Z_1, Z_2, \ldots, Z_N$, given H_1 and H_2. We will make use of this generalization when the detection of arbitrary finite energy signals in white Gaussian noise is discussed. We will find the optimum Bayes detectors for such problems by resolving the possible transmitted signals into a finite-dimensional *signal space*. In the next section, therefore, we continue the discussion of vector space representation of signals begun in Section 2.3.

10.2 VECTOR SPACE REPRESENTATION OF SIGNALS

We recalled, in Section 2.3, that any vector in three-dimensional space can be expressed as a linear combination of any three linearly independent vectors. Recall that such a set of three linearly independent vectors is said to *span* three-dimensional vector space and is referred to as a *basis-vector set* for the space. A basis set of unit-magnitude, mutually perpendicular vectors is called an *orthonormal basis set*.

Two geometrical concepts associated with vectors are magnitude of a vector and angle between two vectors. Both are described by the scalar (or dot) product of any two vectors **A** and **B** having magnitudes A and B, defined as

$$\mathbf{A} \cdot \mathbf{B} = AB \cos\theta \tag{10.39}$$

where θ is the angle between **A** and **B**. Thus

$$A = \sqrt{\mathbf{A} \cdot \mathbf{A}} \quad \text{and} \quad \cos\theta = \frac{\mathbf{A} \cdot \mathbf{B}}{AB} \tag{10.40}$$

Generalizing these concepts to signals in Section 2.3, we expressed a signal $x(t)$ with finite energy in an interval $(t_0,\ t_0 + T)$ in terms of a complete set of orthonormal basis functions $\phi_1(t), \phi_2(t), \ldots,$ as the series

$$x(t) = \sum_{n=1}^{\infty} X_n \phi_n(t) \tag{10.41}$$

where

$$X_n = \int_{t_0}^{t_0 + T} x(t)\phi_n^*(t)\, dt \tag{10.42}$$

which is a special case of (2.35) with $c_n = 1$ because the $\phi_n(t)$ are assumed orthonormal on the interval $(t_0,\ t_0 + T)$. This provided the alternative representation for $x(t)$ as the infinite-dimensional vector $(X_1, X_2, \ldots)$.

To set up a geometric structure on such a vector space, which will be referred to as *signal space*, we must first establish the linearity of the space by listing a consistent set of properties involving the members of the space and the operations between them. Second, we must establish the geometric structure of the space by generalizing the concept of scalar product, thus providing generalizations for the concepts of magnitude and angle.

10.2.1 Structure of Signal Space

We begin with the first task. Specifically, a collection of signals composes a linear signal space $\mathcal{S}$ if, for any pair of signals $x(t)$ and $y(t)$ in $\mathcal{S}$, the operations of addition (commutative and associative) of two signals and multiplication of a signal by a scalar are defined and obey the following axioms:

Axiom 1. The signal $\alpha_1 x(t) + \alpha_2 y(t)$ is in the space for any two scalars α_1 and α_2 (establishes $\mathcal{S}$ as linear).

Axiom 2. $\alpha[x(t) + y(t)] = \alpha x(t) + \alpha y(t)$ for any scalar α.

Axiom 3. $\alpha_1[\alpha_2 x(t)] = (\alpha_1\alpha_2)x(t)$

Axiom 4. The product of $x(t)$ and the scalar 1 reproduces $x(t)$.

Axiom 5. The space contains a unique zero element such that

$$x(t) + 0 = x(t)$$

Axiom 6. To each $x(t)$ there corresponds a unique element $-x(t)$ such that

$$x(t) + [-x(t)] = 0$$

In writing relations such as the preceding, it is convenient to suppress the independent variable t, and this will be done from now on.

10.2.2 Scalar Product

The second task, that of establishing the geometric structure, is accomplished by defining the scalar product, denoted (x, y), as a scalar-valued function of two signals $x(t)$ and $y(t)$ (in general, complex functions), with the following properties:

Property 1. $(x, y) = (y, x)^*$.

Property 2. $(\alpha x, y) = \alpha(x, y)$.

Property 3. $(x + y, z) = (x, z) + (y, z)$.

Property 4. $(x, x) > 0$ unless $x \equiv 0$, in which case $(x, x) = 0$.

The particular definition used for the scalar product depends on the application and the type of signals involved. Because we wish to include both energy and power signals in our future considerations, at least two definitions of the scalar product are required. If $x(t)$ and $y(t)$

are both of the same signal type, a convenient choice is

$$(x, y) = \lim_{T' \to \infty} \int_{-T'}^{T'} x(y)y^*(t)\, dt \tag{10.43}$$

for energy signals and

$$(x, y) = \lim_{T' \to \infty} \frac{1}{2T'} \int_{-T'}^{T'} x(y)y^*(t)\, dt \tag{10.44}$$

for power signals. In (10.43) and (10.44) T' has been used to avoid confusion with the signal observation interval T. In particular, for $x(t) = y(t)$, we see that (10.43) is the total energy contained in $x(t)$ and (10.44) corresponds to the average power. We note that the coefficients in the series of (10.41) can be written as

$$X_n = (x, \phi_n) \tag{10.45}$$

If the scalar product of two signals $x(t)$ and $y(t)$ is zero, they are said to be *orthogonal*, just as two ordinary vectors are said to be orthogonal if their dot product is zero.

10.2.3 Norm

The next step in establishing the structure of a linear signal space is to define the length, or norm $\|x\|$, of a signal. A particularly suitable choice, in view of the preceding discussion, is

$$\|x\| = (x, x)^{1/2} \tag{10.46}$$

More generally, the norm of a signal is any nonnegative real number satisfying the following properties:

Property 1. $\|x\| = 0$ if and only if $x \equiv 0$.

Property 2. $\|x + y\| \leq \|x\| + \|y\|$ (known as the *triangle inequality*).

Property 3. $\|\alpha x\| = |\alpha|\, \|x\|$, where α is a scalar.

Clearly, the choice $\|x\| = (x, x)^{1/2}$ satisfies these properties, and we will employ it from now on. A measure of the distance between, or dissimilarity of, two signals x and y is provided by the norm of their difference $\|x - y\|$.

10.2.4 Schwarz's Inequality

An important relationship between the scalar product of two signals and their norms is Schwarz's inequality, which was used in Chapter 8 without proof. For two signals $x(t)$ and $y(t)$, it can be written as

$$|(x, y)| \leq \|x\|\, \|y\| \tag{10.47}$$

with equality if and only if x or y is zero or if $x(t) = \alpha y(t)$ where α is a scalar.

To prove (10.47), we consider the nonnegative quantity $\|x + \alpha y\|^2$ where α is as yet an unspecified constant. Expanding it by using the properties of the scalar product, we obtain

$$\begin{aligned} \|x + \alpha y\|^2 &= (x + \alpha y,\, x + \alpha y) \\ &= (x, x) + \alpha^*(x, y) + \alpha(x, y)^* + |\alpha|^2(y, y) \\ &= \|x\|^2 + \alpha^*(x, y) + \alpha(x, y)^* + |\alpha|^2\|y\|^2 \end{aligned} \tag{10.48}$$

Choosing $\alpha = -(x, y)/\|y\|^2$, which is permissible since α is arbitrary, we find that the last two terms of (10.48) cancel, yielding

$$\|x + \alpha y\|^2 = \|x\|^2 - \frac{|(x, y)|^2}{\|y\|^2} \tag{10.49}$$

Since $\|x + \alpha y\|^2$ is nonnegative, rearranging (10.49) gives Schwarz's inequality. Furthermore, noting that $\|x + \alpha y\| = 0$ if and only if $x + \alpha y = 0$, we establish a condition under which equality holds in (10.47). Equality also holds, of course, if one or both signals are identically zero.

EXAMPLE 10.3

A familiar example of a space that satisfies the preceding properties is ordinary two-dimensional vector space. Consider two vectors with real components,

$$\mathbf{A}_1 = a_1 \hat{i} + b_1 \hat{j} \quad \text{and} \quad \mathbf{A}_2 = a_2 \hat{i} + b_2 \hat{j} \tag{10.50}$$

where $\hat{i}$ and $\hat{j}$ are the usual orthogonal unit vectors. The scalar product is taken as the vector dot product

$$(\mathbf{A}_1, \mathbf{A}_2) = a_1 a_2 + b_1 b_2 = \mathbf{A}_1 \cdot \mathbf{A}_2 \tag{10.51}$$

and the norm is taken as

$$\|\mathbf{A}_1\| = (\mathbf{A}_1, \mathbf{A}_1)^{1/2} = \sqrt{a_1^2 + b_1^2} \tag{10.52}$$

which is just the length of the vector. Addition is defined as vector addition,

$$\mathbf{A}_1 + \mathbf{A}_2 = (a_1 + a_2)\hat{i} + (b_1 + b_2)\hat{j} \tag{10.53}$$

which is commutative and associative. The vector $\mathbf{C} \triangleq \alpha_1 \mathbf{A}_1 + \alpha_2 \mathbf{A}_2$, where α_1 and α_2 are real constants, is also a vector in two-space (Axiom 1). The remaining axioms follow as well, with the zero element being $0\hat{i} + 0\hat{j}$.

The properties of the scalar product are satisfied by the vector dot product. The properties of the norm also follow, with property 2 taking the form

$$\sqrt{(a_1 + a_2)^2 + (b_1 + b_2)^2} \leq \sqrt{a_1^2 + b_1^2} + \sqrt{a_2^2 + b_2^2} \tag{10.54}$$

which is simply a statement that the length of the hypotenuse of a triangle is shorter than the sum of the lengths of the other two sides—hence the name *triangle inequality*. Schwarz's inequality squared is

$$(a_1 a_2 + b_1 b_2)^2 \leq \left(a_1^2 + b_1^2\right)\left(a_2^2 + b_2^2\right) \tag{10.55}$$

which simply states that $|\mathbf{A}_1 \cdot \mathbf{A}_2|^2$ is less than or equal to the length squared of $\mathbf{A}_1$ times the length squared of $\mathbf{A}_2$.

■

10.2.5 Scalar Product of Two Signals in Terms of Fourier Coefficients

Expressing two energy or power signals $x(t)$ and $y(t)$ in the form given in (10.41), we may show that

$$(x, y) = \sum_{m=1}^{\infty} X_m Y_m^* \tag{10.56}$$

Letting $y = x$ results in Parseval's theorem, which is

$$\|x\|^2 = \sum_{n=1}^{\infty} |X_m|^2 \tag{10.57}$$

To indicate the usefulness of the shorthand vector notation just introduced, we will carry out the proof of (10.56) and (10.57) using it. Let $x(t)$ and $y(t)$ be written in terms of their respective orthonormal expansions:

$$x(t) = \sum_{m=1}^{\infty} X_m \phi_m(t) \quad \text{and} \quad y(t) = \sum_{n=1}^{\infty} Y_n \phi_n(t) \tag{10.58}$$

where, in terms of the scalar product,

$$X_m = (x, \phi_m) \quad \text{and} \quad Y_n = (y, \phi_n) \tag{10.59}$$

Thus

$$(x, y) = \left(\sum_m X_m \phi_m, \sum_n Y_n \phi_n \right) = \sum_m X_m \left(\phi_m, \sum_n Y_n \phi_n \right) \tag{10.60}$$

by virtue of properties 2 and 3 of the scalar product. Applying property 1, we obtain

$$(x, y) = \sum_m X_m \left(\sum_n Y_n \phi_n, \phi_m \right)^* = \sum_m X_m \left[\sum_n Y_n^* (\phi_n, \phi_m)^* \right] \tag{10.61}$$

the last step of which follows by virtue of another application of properties 2 and 3. But the ϕ_n are orthonormal; that is, $(\phi_n, \phi_m) = \delta_{nm}$, where δ_{nm} is the Kronecker delta. Thus

$$(x, y) = \sum_m X_m \left[\sum_n Y_n^* \delta_{nm} \right] = \sum_m X_m Y_m^* \tag{10.62}$$

which proves (10.56). We may set $x(t) = y(t)$ to prove (10.57).

EXAMPLE 10.4

Consider a signal $x(t)$ and the approximation to it $x_a(t)$ of Example 2.5. Both x and x_a are in the signal space consisting of all finite-energy signals. All the addition and multiplication properties for signal space hold for x and x_a. Because we are considering finite-energy signals, the scalar product defined by (10.43) applies. The scalar product of x and x_a is

$$\begin{aligned} (x, x_a) &= \int_0^2 \sin(\pi t) \left[\frac{2}{\pi} \phi_1(t) - \frac{2}{\pi} \phi_2(t) \right] dt \\ &= \left(\frac{2}{\pi} \right)^2 - \frac{2}{\pi} \left(-\frac{2}{\pi} \right) = 2 \left(\frac{2}{\pi} \right)^2 \end{aligned} \tag{10.63}$$

The norm of their difference squared is

$$\begin{aligned}
\|x - x_a\|^2 &= (x - x_a, x - x_a) \\
&= \int_0^2 \left[\sin(\pi t) - \frac{2}{\pi}\phi_1(t) + \frac{2}{\pi}\phi_2(t)\right]^2 dt \\
&= 1 - \frac{8}{\pi^2}
\end{aligned} \tag{10.64}$$

which is just the minimum integral-squared error between x and x_a.

The norm squared of x is

$$\|x\|^2 = \int_0^2 \sin^2(\pi t)\, dt = 1 \tag{10.65}$$

and the norm squared of x_a is

$$\|x_a\|^2 = \int_0^2 \left[\frac{2}{\pi}\phi_1(t) - \frac{2}{\pi}\phi_2(t)\right]^2 dt = 2\left(\frac{2}{\pi}\right)^2 \tag{10.66}$$

which follows since $\phi_1(t)$ and $\phi_2(t)$ are orthonormal over the period of integration. Thus Schwarz's inequality for this case is

$$2\left(\frac{2}{\pi}\right)^2 < 1 \times \sqrt{2}\left(\frac{2}{\pi}\right) \tag{10.67}$$

which is equivalent to

$$\sqrt{2} < \frac{1}{2}\pi \tag{10.68}$$

Since x is not a scalar multiple of x_a, we must have strict inequality.

■

10.2.6 Choice of Basis Function Sets: The Gram–Schmidt Procedure

The question naturally arises as to how we obtain suitable basis sets. For energy or power signals, with no further restrictions imposed, we require infinite sets of functions. Suffice it to say that many suitable choices exist, depending on the particular problem and the interval of interest. These include not only the sines and cosines, or complex exponential functions of harmonically related frequencies, but also the Legendre functions, Hermite functions, and Bessel functions, to name only a few. All these are complete sets of functions.

A technique referred to as the *Gram–Schmidt procedure* is often useful for obtaining basis sets, especially in the consideration of *M*-ary signal detection. This procedure will now be described.

Consider the situation in which we are given a finite set of signals $s_1(t), s_2(t), \ldots, s_M(t)$ defined on some interval $(t_0, t_0 + T)$, and our interest is in all signals that may be written as linear combinations of these signals:

$$x(t) = \sum_{n=1}^{M} X_n s_n(t), \quad t_0 \le t \le t_0 + T \tag{10.69}$$

The set of all such signals forms an M-dimensional signal space if the $s_n(t)$ are linearly independent [that is, no $s_n(t)$ can be written as a linear combination of the rest]. If the $s_n(t)$ are not linearly independent, the dimension of the space is less than M. An orthonormal basis for the space can be obtained by using the Gram–Schmidt procedure, which consists of the following steps:

1. Set $v_1(t) = s_1(t)$ and $\phi_1(t) = v_1(t)/\|v_1\|$.
2. Set $v_2(t) = s_2(t) - (s_2, \phi_1)\phi_1$ and $\phi_2(t) = v_2(t)/\|v_2\|$; $v_2(t)$ is the component of $s_2(t)$ that is linearly independent of $s_1(t)$.
3. Set $v_3(t) = s_3(t) - (s_3, \phi_2)\phi_2(t) - (s_3, \phi_1)\phi_1(t)$ and $\phi_3(t) = v_3(t)/\|v_3\|$; $v_3(t)$ is the component of $s_3(t)$ linearly independent of $s_1(t)$ and $s_2(t)$.
4. Continue until all the $s_n(t)$ have been used. If the $s_n(t)$ are not linearly independent, then one or more steps will yield $v_n(t)$ for which $\|v_n\| = 0$. These signals are omitted whenever they occur so that a set of K orthonormal functions is finally obtained where $K \leq M$.

The resulting set forms an orthonormal basis set for the space since, at each step of the procedure, we ensure that

$$(\phi_n, \phi_m) = \delta_{nm} \tag{10.70}$$

where δ_{nm} is the Kronecker delta defined in Chapter 2, and we use all signals in forming the orthonormal set.

EXAMPLE 10.5

Consider the set of three finite-energy signals

$$\begin{aligned} s_1(t) &= 1, & 0 \leq t \leq 1 \\ s_2(t) &= \cos(2\pi t), & 0 \leq t \leq 1 \\ s_3(t) &= \cos^2(\pi t), & 0 \leq t \leq 1 \end{aligned} \tag{10.71}$$

We desire an orthonormal basis for the signal space spanned by these three signals.

Solution

We let $v_1(t) = s_1(t)$ and compute

$$\phi_1(t) = \frac{v_1(t)}{\|v_1\|} = 1, \quad 0 \leq t \leq 1 \tag{10.72}$$

Next, we compute

$$(s_2, \phi_1) = \int_0^1 1 \cos(2\pi t)\, dt = 0 \tag{10.73}$$

and we set

$$v_2(t) = s_2(t) - (s_2, \phi_1)\phi_1 = \cos(2\pi t), \quad 0 \leq t \leq 1 \tag{10.74}$$

The second orthonormal function is found from

$$\phi_2(t) = \frac{v_2}{\|v_2\|} = \sqrt{2}\cos(2\pi t), \quad 0 \leq t \leq 1 \tag{10.75}$$

To check for another orthonormal function, we require the scalar products

$$(s_3, \phi_2) = \int_0^1 \sqrt{2}\cos(2\pi t)\cos^2(\pi t)\, dt = \frac{1}{4}\sqrt{2} \tag{10.76}$$

and

$$(s_3, \phi_1) = \int_0^1 \cos^2(\pi t)\, dt = \frac{1}{2} \tag{10.77}$$

Thus

$$\begin{aligned} v_3(t) &= s_3(t) - (s_3, \phi_2)\phi_2 - (s_3, \phi_1)\phi_1 \\ &= \cos^2(\pi t) - \left(\frac{1}{4}\sqrt{2}\right)\sqrt{2}\cos(2\pi t) - \frac{1}{2} = 0 \end{aligned} \tag{10.78}$$

so that the space is two-dimensional. ■

10.2.7 Signal Dimensionality as a Function of Signal Duration

The sampling theorem, proved in Chapter 2, provides a means of representing strictly bandlimited signals, with bandwidth W, in terms of the infinite basis function set $\text{sinc}(f_s t - n)$, $n = 0, \pm 1, \pm 2, \ldots$ Because $\text{sinc}(f_s t - n)$ is not time-limited, we suspect that a strictly bandlimited signal cannot also be of finite duration (that is, time-limited). However, practically speaking, a time-bandwidth dimensionality can be associated with a signal provided the definition of bandlimited is relaxed. The following theorem, given without proof, provides an upper bound for the dimensionality of time-limited and bandwidth-limited signals.[4]

Dimensionality Theorem

Let $\{\phi_k(t)\}$ denote a set of orthogonal waveforms, all of which satisfy the following requirements:

1. They are identically zero outside a time interval of duration T, for example, $|t| \le \frac{1}{2}T$.
2. None has more than $\frac{1}{12}$ of its energy outside the frequency interval $-W < f < W$.

Then the number of waveforms in the set $\{\phi_k(t)\}$ is conservatively overbounded by $2.4TW$ when TW is large.

EXAMPLE 10.6

Consider the orthogonal set of waveforms

$$\begin{aligned} \phi_k(t) &= \Pi\left(\frac{t - k\tau}{\tau}\right) \\ &= \begin{cases} 1, & \frac{1}{2}(2k-1)\tau \le t \le \frac{1}{2}(2k+1)\tau, \quad k = 0, \pm 1, \pm 2, \pm K \\ 0, & \text{otherwise} \end{cases} \end{aligned}$$

[4]This theorem is taken from Wozencraft and Jacobs (1965), p. 294, where it is also given without proof. However, a discussion of the equivalence of this theorem to the original ones due to Shannon and to Landau and Pollak is also given.

where $(2K+1)\tau = T$. The Fourier transform of $\phi_k(t)$ is

$$\Phi_k(f) = \tau \operatorname{sinc}(\tau f) e^{-j2\pi k\tau f} \tag{10.79}$$

The total energy in $\phi_k(t)$ is τ, and the energy for $|f| \leq W$ is

$$\begin{aligned} E_W &= \int_{-W}^{W} \tau^2 \operatorname{sinc}^2(\tau f)\, df \\ &= 2\tau \int_0^{\tau W} \operatorname{sinc}^2 v\, dv \end{aligned} \tag{10.80}$$

where the substitution $v = \tau f$ has been made in the integral and the integration is carried out only over positive values of v, owing to the evenness of the integrand. The total pulse energy is $E = \tau$ so the ratio of energy in a bandwidth W to total energy is

$$\frac{E_W}{E} = 2\int_0^{\tau W} \operatorname{sinc}^2 v\, dv \tag{10.81}$$

This integral cannot be integrated in closed form, so we integrate it numerically using the MATLAB program below:[5]

```
% ex10_6
%
for tau_W = 1:.1:1.5
    v = 0:0.01:tau_W;
    y = (sinc(v)).^2;
    EW_E = 2*trapz(v, y);
    disp([tau_W, EW_E])
end
```

The results for E_W/E versus τW are given below:

τW	E_W/E
1.0	0.9028
1.1	0.9034
1.2	0.9066
1.3	0.9130
1.4	0.9218
1.5	0.9311

We want to choose τW such that $E_W/E \geq \frac{11}{12} = 0.9167$. Thus, $\tau W = 1.4$ will ensure that none of the $\phi_k(t)$s has more than $\frac{1}{12}$ of its energy outside the frequency interval $-W < f < W$.

Now $N = T/\tau$ orthogonal waveforms occupy the interval $(-\frac{1}{2}T, \frac{1}{2}T)$, where $\lfloor\ \rfloor$ signifies the integer part of T/τ. Letting $\tau = 1.4W^{-1}$, we obtain

$$N = \frac{TW}{1.4} = 0.714TW \tag{10.82}$$

which satisfies the bound given by the theorem.

■

[5]The integral can be expressed in terms of the sine-integral function that is tabulated. See Abramowitz and Stegun (1972).

10.3 MAXIMUM A POSTERIORI RECEIVER FOR DIGITAL DATA TRANSMISSION

We now apply the detection theory and signal space concepts just developed to digital data transmission. We will consider examples of coherent and noncoherent systems.

10.3.1 Decision Criteria for Coherent Systems in Terms of Signal Space

In the analysis of QPSK systems in Chapter 9, the received signal plus noise was resolved into two components by the correlators comprising the receiver. This made simple the calculation of the probability of error. The QPSK receiver essentially computes the coordinates of the received signal plus noise in a signal space. The basis functions for this signal space are $\cos(\omega_c t)$ and $\sin(\omega_c t)$, $0 \leq t \leq T$, with the scalar product defined by

$$(x_1, x_2) = \int_0^T x_1(t)x_2(t)\, dt \tag{10.83}$$

which is a special case of (10.43). These basis functions are orthogonal if $\omega_c T$ is an integer multiple of 2π, but are not normalized.

Recalling the Gram–Schmidt procedure, we see how this viewpoint might be generalized to M signals $s_1(t), s_2(t), \ldots, s_M(t)$ that have finite energy but are otherwise arbitrary. Thus, consider an M-ary communication system, depicted in Figure 10.4, wherein one of M possible signals of known form $s_i(t)$ associated with a message m_i is transmitted each T seconds. The receiver is to be constructed such that the probability of error in deciding which message was transmitted is minimized; that is, it is a MAP receiver. For simplicity, we assume that the messages are produced by the information source with equal a priori probability.

Ignoring the noise for the moment, we note that the ith signal can be expressed as

$$s_i(t) = \sum_{j=1}^{K} A_{ij}\phi_j(t), \quad i = 1, 2, \ldots, M,\ K \leq M \tag{10.84}$$

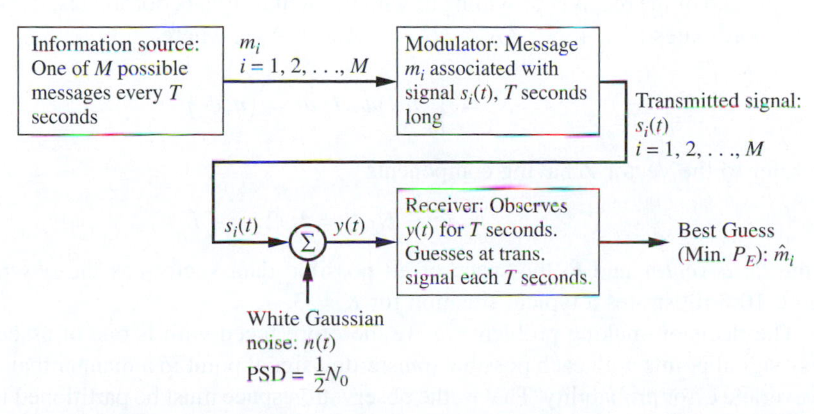

Figure 10.4
M-ary communication system.

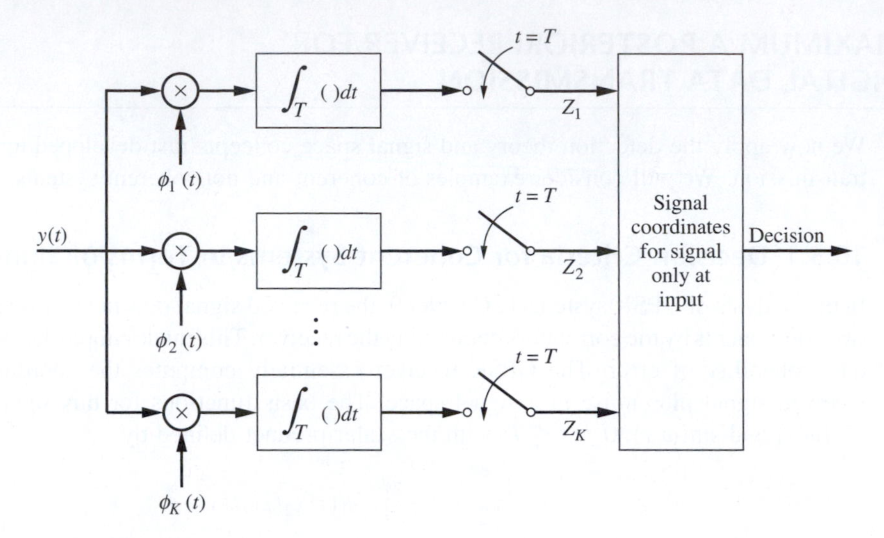

Figure 10.5
Receiver structure for resolving signals into K-dimensional signal space.

where the $\phi_j(t)$'s are orthonormal basis functions chosen according to the Gram–Schmidt procedure. Thus

$$A_{ij} = \int_0^T s_i(t)\phi_j(t)\,dt = (s_i, \phi_j) \tag{10.85}$$

and we see that the receiver structure shown in Figure 10.5, which consists of a bank of correlators, can be used to compute the generalized Fourier coefficients for $s_i(t)$. Thus we can represent each possible signal as a point in a K-dimensional signal space with coordinates $(A_{i1}, A_{i2}, \ldots, A_{iK})$, for $i = 1, 2, \ldots, M$.

Knowing the coordinates of $s_i(t)$ is as good as knowing $s_i(t)$, since it is uniquely specified through (10.84). The difficulty is, of course, that we receive the signals in the presence of noise. Thus, instead of the receiver providing us with the actual signal coordinates, it provides us with noisy coordinates $(A_{i1} + N_1, A_{i2} + N_2, \ldots, A_{iK} + N_K)$, where

$$N_j \triangleq \int_0^T n(t)\phi_j(t)\,dt = (n, \phi_j) \tag{10.86}$$

We refer to the vector $\mathbf{Z}$ having components

$$Z_j \triangleq A_{ij} + N_j,\ j = 1, 2, \ldots, K \tag{10.87}$$

as the *data vector* and to the space of all possible data vectors as the *observation space*. Figure 10.6 illustrates a typical situation for $K = 3$.

The decision-making problem we are therefore faced with is one of associating sets of noisy signal points with each possible transmitted signal point in a manner that will minimize the average error probability. That is, the observation space must be partitioned into M regions R_i, one associated with each transmitted signal, such that if a received data point falls into region R_ℓ, the decision "$s_\ell(t)$ transmitted" is made with minimum probability of error.

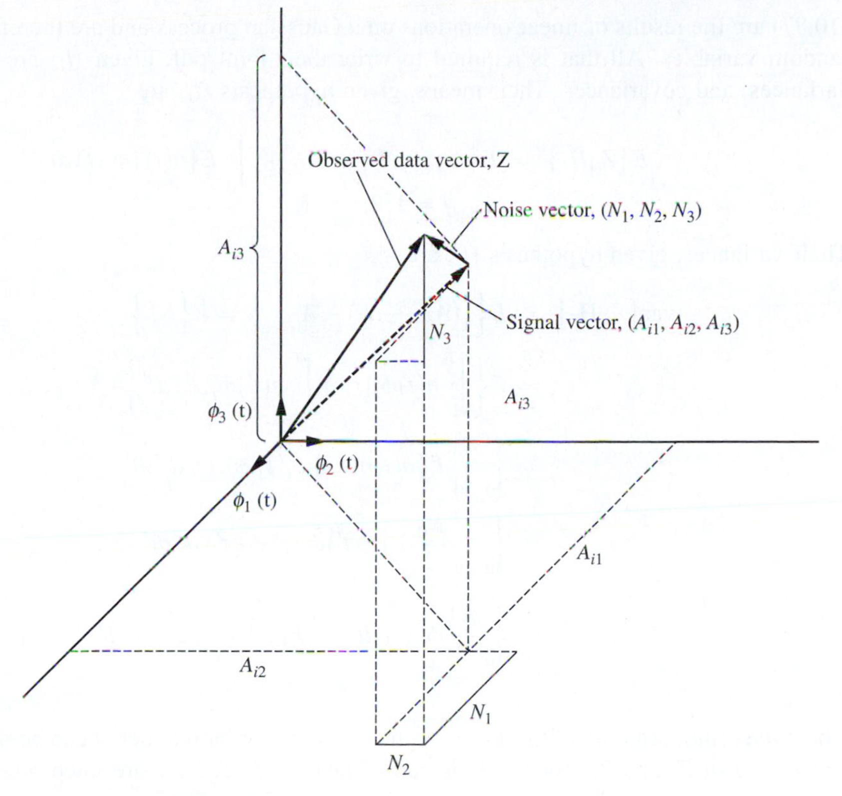

Figure 10.6
A three-dimensional observation space.

In Section 10.1, the minimum-probability-of-error detector was shown to correspond to a MAP decision rule. Thus, letting hypothesis H_ℓ be "signal $s_\ell(t)$ transmitted," we want to implement a receiver that computes

$$P(H_\ell|Z_1, Z_2, \ldots, Z_K), \quad \ell = 1, 2, \ldots, M, \quad K \le M \tag{10.88}$$

and chooses the largest.[6] To compute the posterior probabilities of (10.88), we use Bayes rule and assume that

$$P(H_1) = P(H_2) = \cdots = P(H_M) \tag{10.89}$$

Application of Bayes rule results in

$$P(H_\ell|z_1, \ldots, z_K) = \frac{f_Z(z_1, \ldots, z_K|H_\ell)P(H_\ell)}{f_Z(z_1, \ldots, z_K)} \tag{10.90}$$

However, since the factors $P(H_\ell)$ and $f_Z(z_1, \ldots, z_K)$ do not depend on ℓ, the detector can compute $f_Z(z_1, \ldots, z_K|H_\ell)$ and choose the H_ℓ corresponding to the largest. The Z_j given by

[6]Capital letters are used to denote components of data vectors because they represent coordinates of an observation that is *random*.

(10.87) are the results of linear operations on a Gaussian process and are therefore Gaussian random variables. All that is required to write their joint pdf, given H_ℓ, are their means, variances, and covariances. Their means, given hypothesis H_ℓ, are

$$\begin{aligned} E\{Z_j|H_\ell\} &= E\{A_{\ell j} + N_j\} = A_{\ell j} + \int_0^T E\{n(t)\}\phi_j(t)\,dt \\ &= A_{\ell j},\ j = 1, 2, \ldots, K \end{aligned} \tag{10.91}$$

Their variances, given hypothesis H_ℓ, are

$$\begin{aligned} \text{var}\{Z_j|H_\ell\} &= E\left\{\left[(A_{\ell j} + N_j) - A_{\ell j}\right]^2\right\} = E\left\{N_j^2\right\} \\ &= E\left\{\int_0^T n(t)\phi_j(t)dt \int_0^T n(t')\phi_j(t')\,dt'\right\} \\ &= \int_0^T\int_0^T E\{n(t)n(t')\}\phi_j(t)\,\phi_j(t')\,dt\,dt' \\ &= \int_0^T\int_0^T \frac{N_0}{2}\delta(t - t')\phi_j(t)\,\phi_j(t')\,dt\,dt' \\ &= \int_0^T \frac{N_0}{2}\phi_j^2(t)\,dt = \frac{1}{2}N_0,\ j = 1, 2, \ldots, K \end{aligned} \tag{10.92}$$

where the orthonormality of the ϕ_j has been used. In a similar manner, it can be shown that the covariance of Z_j and Z_k, for $j \neq k$, is zero. Thus $Z_1, Z_2, \ldots, Z_K$ are uncorrelated Gaussian random variables and, hence, are statistically independent. Thus

$$\begin{aligned} f_Z(z_1, \ldots, z_K|H_\ell) &= \prod_{j=1}^{K} \frac{\exp\left[-(z_j - A_{\ell j})^2/N_0\right]}{\sqrt{\pi N_0}} \\ &= \frac{1}{(\pi N_0)^{K/2}}\exp\left[-\sum_{j=1}^{K}(z_j - A_{\ell j})^2/N_0\right] \\ &= \frac{\exp\left\{-\|z - s_\ell\|^2/N_0\right\}}{(\pi N_0)^{K/2}} \end{aligned} \tag{10.93}$$

where

$$z = z(t) = \sum_{j=1}^{K} z_j\phi_j(t) \tag{10.94}$$

and

$$s_\ell = s_\ell(t) = \sum_{j=1}^{K} A_{\ell j}\phi_j(t) \tag{10.95}$$

Except for a factor independent of ℓ, (10.93), is the posterior probability $P(H_\ell|z_1, \ldots, z_K)$ as obtained by applying Bayes rule. Hence, choosing H_ℓ corresponding to the maximum posterior probability is the same as choosing the signal with coordinates $A_{\ell 1}, A_{\ell 2}, \ldots, A_{\ell K}$ so as to maximize (10.93) or, equivalently, so as to minimize the exponent. But $\|z - s_\ell\|$ is the distance between $z(t)$ and $s_\ell(t)$. Thus it has been shown that the decision criterion that minimizes the average probability of error is to choose as the transmitted signal the one whose signal point is closest to the received data point in observation space, distance being defined as the square root of the sum of the squares of the differences of the data and signal vector components. That is, choose H_ℓ such that[7]

$$\begin{aligned}(\text{Distance})^2 &= d^2 = \sum_{j=1}^{K}(Z_j - A_{\ell j})^2 \\ &= \|z - s_\ell\|^2 = \text{minimum}, \ \ell = 1, 2, \ldots, M\end{aligned} \tag{10.96}$$

which is exactly the operation to be performed by the receiver structure of Figure 10.5. We illustrate this procedure with the following example.

EXAMPLE 10.7

In this example we consider M-ary coherent FSK in terms of signal space. The transmitted signal set is

$$s_i(t) = A\cos\{2\pi[f_c + (i-1)\Delta f]t\}, \quad 0 \le t \le T_s \tag{10.97}$$

where

$$\Delta f = \frac{m}{2T_s}, \quad m \text{ an integer}$$

For mathematical simplicity, we assume that $f_c T_s$ is an integer. The orthonormal basis set can be obtained by applying the Gram–Schmidt procedure. Choosing

$$v_1(t) = s_1(t) = A\cos(2\pi f_c t), \quad 0 \le t \le T_s \tag{10.98}$$

we have

$$\|v_1\|^2 = \int_0^{T_s} A^2\cos^2(2\pi f_c t)\,dt = \frac{A^2 T_s}{2} \tag{10.99}$$

so that

$$\phi_1(t) = \frac{v_1}{\|v_1\|} = \sqrt{\frac{2}{T_s}}\cos(2\pi f_c t), \quad 0 \le t \le T_s \tag{10.100}$$

It can be shown in a straightforward fashion that $(s_2, \phi_1) = 0$ if $\Delta f = m/(2T_s)$, so that the second orthonormal function is

$$\phi_2(t) = \sqrt{\frac{2}{T_s}}\cos[2\pi(f_c + \Delta f)t], \quad 0 \le t \le T_s \tag{10.101}$$

and similarly for $M-2$ other orthonormal functions up to $\phi_M(t)$. Thus the number of orthonormal functions is the same as the number of possible signals; the ith signal can be written in terms of the ith

[7]Again, Z_j is the jth coordinate of an observation $z(t)$ that is random. Equation (10.96) is referred to as a *decision rule*.

orthonormal function as

$$s_i(t) = \sqrt{E_s}\phi_i(t) \tag{10.102}$$

We let the received signal plus noise waveform be represented as $y(t)$. When projected into the observation space, $y(t)$ has M coordinates, the ith one of which is given by

$$Z_i = \int_0^{T_s} y(t)\phi_i(t)\,dt \tag{10.103}$$

where $y(t) = s_i(t) + n(t)$. If $s_\ell(t)$ is transmitted, the decision rule (10.96) becomes

$$d^2 = \sum_{j=1}^{M} \left(Z_j - \sqrt{E_s}\delta_{\ell j}\right)^2 = \text{minimum over } \ell = 1, 2, \ldots, M \tag{10.104}$$

Taking the square root and writing the sum out, this can be expressed as

$$\begin{aligned} d &= \sqrt{Z_1^2 + Z_2^2 + \cdots + \left(Z_\ell - \sqrt{E_s}\right)^2 + \cdots + Z_M^2} \\ &= \text{minimum} \end{aligned} \tag{10.105}$$

For two dimensions (binary FSK), the signal points lie on the two orthogonal axes at a distance $\sqrt{E_s}$ out from the origin. The decision space consists of the first quadrant, and the optimum (minimum error probability) partition is a line at 45 degrees bisecting the right angle made by the two coordinate axes.

For M-ary FSK transmission, an alternative way of viewing the decision rule can be obtained by squaring the ℓth term in (10.104) so that we have

$$d^2 = \sum_{n=1}^{\infty} Z_j^2 + E_s - 2\sqrt{E_s}Z_\ell = \text{minimum} \tag{10.106}$$

Since the sums over j and E_s are independent of ℓ, d^2 can be minimized with respect to ℓ by choosing as the possible transmitted signal the one that will maximize the last term; that is, the decision rule becomes: Choose the possible transmitted signal $s_\ell(t)$ such that

$$\begin{aligned} \sqrt{E_s}Z_\ell &= \text{maximum or} \\ Z_\ell &= \int_0^T y(t)\phi_\ell(t)\,dt = \text{maximum with respect to } \ell \end{aligned} \tag{10.107}$$

In other words, we look at the output of the bank of correlators shown in Figure 10.5 at time $t = T_s$ and choose the one with the largest output as corresponding to the most probable transmitted signal. ■

10.3.2 Sufficient Statistics

To show that (10.96) is indeed the decision rule corresponding to a MAP criterion, we must clarify one point. In particular, the decision is based on the noisy signal

$$z(t) = \sum_{j=1}^{K} Z_j\phi_j(t) \tag{10.108}$$

Because of the noise component $n(t)$, this is not the same as $y(t)$, since an infinite set of basis functions would be required to represent all possible $y(t)$s. However, we may show that only K coordinates, where K is the signal space dimension, are required to provide all the information that is relevant to making a decision.

Assuming a complete orthonormal set of basis functions, $y(t)$ can be expressed as

$$y(t) = \sum_{j=1}^{\infty} Y_j \phi_j(t) \tag{10.109}$$

where the first K of the ϕ_j are chosen using the Gram–Schmidt procedure for the given signal set. Given that hypothesis H_ℓ is true, the Y_j are given by

$$Y_j = \begin{cases} Z_j = A_{\ell j} + N_j, & j = 1, 2, \ldots, K \\ N_j, & j = K+1, K+2, \ldots \end{cases} \tag{10.110}$$

where $Z_j, A_{\ell j}$, and N_j are as defined previously. Using a procedure identical to the one used in obtaining (10.91) and (10.92), we can show that

$$E\{Y_j\} = \begin{cases} A_{\ell j}, & j = 1, 2, \ldots, K \\ 0, & j > K \end{cases} \tag{10.111}$$

$$\text{var}\{Y_j\} = \frac{1}{2} N_0, \quad \text{all } j \tag{10.112}$$

with $\text{cov}\{Y_j Y_k\} = 0,\ j \neq k$. Thus the joint pdf of $Y_1, Y_2, \ldots$, given H_ℓ, is of the form

$$\begin{aligned} f_Y(y_1, y_2, \ldots, y_K, \ldots | H_\ell) &= C \exp\left\{ -\frac{1}{N_0} \left[\sum_{j=1}^{K} \left(y_j - A_{\ell j}\right)^2 + \sum_{j=K+1}^{\infty} y_j^2 \right] \right\} \\ &= C \exp\left(-\frac{1}{N_0} \sum_{j=K+1}^{\infty} y_j^2 \right) f_Z(y_1, \ldots, y_K, | H_\ell) \end{aligned} \tag{10.113}$$

where C is a constant. Since this pdf factors, $Y_{K+1}, Y_{K+2}, \ldots$ are independent of $Y_1, Y_2, \ldots, Y_K$ and the former provide no information for making a decision because they do not depend on $A_{\ell j}, j = 1, 2, \ldots, K$. Thus d^2 given by (10.106) is known as a sufficient statistic.

10.3.3 Detection of M-ary Orthogonal Signals

As a more complex example of the use of signal space techniques, let us consider an M-ary signaling scheme for which the signal waveforms have equal energies and are orthogonal over the signaling interval. Thus

$$\int_0^{T_s} s_i(t) s_j(t)\, dt = \begin{cases} E_s, & i = j \\ 0, & i \neq j,\ i = 1, 2, \ldots, M \end{cases} \tag{10.114}$$

where E_s is the energy of each signal in $(0, T_s)$.

A practical example of such a signaling scheme is the signal set for M-ary coherent FSK given by (10.97). The decision rule for this signaling scheme was considered in Example 10.7. It was found that $K = M$ orthonormal functions are required, and the receiver shown in Figure 10.5 involves M correlators. The output of the jth correlator at time T_s is given by (10.87). The decision criterion is to choose the signal point $i = 1, 2, \ldots, M$ such that d^2 given by (10.96) is minimized or, as shown in Example 10.7, such that

$$Z_\ell = \int_0^{T_s} y(t)\phi_\ell(t)\,dt = \text{maximum with respect to } \ell \tag{10.115}$$

That is, the signal is chosen that has the maximum correlation with the received signal plus noise. To compute the probability of symbol error, we note that

$$\begin{aligned} P_E &= \sum_{i=1}^{M} P[E|s_i(t) \text{ sent}]\,P[s_i(t) \text{ sent}] \\ &= \frac{1}{M}\sum_{i=1}^{M} P[E|s_i(t) \text{ sent}] \end{aligned} \tag{10.116}$$

where each signal is assumed a priori equally probable. We may write

$$P[E|s_i(t) \text{ sent}] = 1 - P_{ci} \tag{10.117}$$

where P_{ci} is the probability of a correct decision given that $s_i(t)$ was sent. Since a correct decision results only if

$$Z_j = \int_0^{T_s} y(t)s_j(t)\,dt < \int_0^{T_s} y(t)s_i(t)\,dt = Z_i \tag{10.118}$$

for all $j \neq i$, we may write P_{ci} as

$$P_{ci} = P\big(\text{all } Z_j < Z_i, \quad j \neq i\big) \tag{10.119}$$

If $s_i(t)$ is transmitted, then

$$\begin{aligned} Z_i &= \int_0^{T_s} \left[\sqrt{E_s}\phi_i(t) + n(t)\right]\phi_i(t)\,dt \\ &= \sqrt{E_s} + N_i \end{aligned} \tag{10.120}$$

where

$$N_i = \int_0^{T_s} n(t)\phi_i(t)\,dt \tag{10.121}$$

Since $Z_j = N_j$, $j \neq i$, given $s_i(t)$ was sent, it follows that (10.119) becomes

$$P_{ci} = P\big(\text{all } N_j < \sqrt{E_s} + N_i, \quad j \neq i\big) \tag{10.122}$$

Now N_i is a Gaussian random variable (a linear operation on a Gaussian process) with zero mean and variance

$$\text{var}\,[N_i] = E\left\{\left[\int_0^{T_s} n(t)\phi_j(t)\,dt\right]^2\right\} = \frac{N_0}{2} \tag{10.123}$$

Furthermore, N_i and N_j, for $i \neq j$, are independent, since

$$E\big[N_iN_j\big] = 0 \tag{10.124}$$

Given a particular value of N_i, (10.122) becomes

$$\begin{aligned} P_{ci}(N_i) &= \prod_{\substack{j=1 \\ j\neq 1}}^{M} P\left[N_j < \sqrt{E_s} + N_i\right] \\ &= \left(\int_{-\infty}^{\sqrt{E_s}+N_i} \frac{e^{-n_j^2/N_0}}{\sqrt{\pi N_0}} dn_j \right)^{M-1} \end{aligned} \tag{10.125}$$

which follows because the pdf of N_j is $n\left(0, \sqrt{N_0/2}\right)$. Averaged over all possible values of N_i, (10.125) gives

$$\begin{aligned} P_{ci} &= \int_{-\infty}^{\infty} \frac{e^{-n_i^2/N_0}}{\sqrt{\pi N_0}} \left(\int_{-\infty}^{\sqrt{E_s}+n_i} \frac{e^{-n_j^2/N_0}}{\sqrt{\pi N_0}} dn_j \right)^{M-1} dn_i \\ &= (\pi)^{-M/2} \int_{-\infty}^{\infty} e^{-y^2} \left(\int_{-\infty}^{\sqrt{E_s/N_0}+y} e^{-x^2} dx \right)^{M-1} dy \end{aligned} \tag{10.126}$$

where the substitutions $x = n_j/\sqrt{N_0}$ and $y = n_i/\sqrt{N_0}$ have been made. Since P_{ci} is independent of i, it follows that the probability of error is

$$P_E = 1 - P_{ci} \tag{10.127}$$

With (10.126) substituted into (10.127), a nonintegrable M-fold integral for P_E results, and one must resort to numerical integration to evaluate it.[8] Curves showing P_E versus $E_s/(N_0 \log_2 M)$ are given in Figure 10.7 for several values of M. We note a rather surprising behavior: As $M \to \infty$, error-free transmission can be achieved as long as $E_s/(N_0 \log_2 M) > \ln 2 = -1.59$ dB. This error-free transmission is achieved at the expense of infinite bandwidth, however, since $M \to \infty$ means that an infinite number of orthonormal functions are required. We will discuss this behavior further in Chapter 11.

10.3.4 A Noncoherent Case

To illustrate the application of signal space techniques to noncoherent digital signaling, let us consider the following binary hypothesis situation:

$$\begin{aligned} H_1: y(t) &= G\sqrt{2E/T}\cos(\omega_1 t + \theta) + n(t) \\ H_2: y(t) &= G\sqrt{2E/T}\cos(\omega_2 t + \theta) + n(t), \quad 0 \le t \le T \end{aligned} \tag{10.128}$$

where E is the energy of the transmitted signal in one bit period and $n(t)$ is white Gaussian noise with double-sided power spectral density $\frac{1}{2}N_0$. It is assumed that $|\omega_1 - \omega_2|/2\pi \gg T^{-1}$ so that the signals are orthogonal. Except for G and θ, which are assumed to be random variables, this problem would be a special case of the M-ary orthogonal signaling case just considered. (Recall also the consideration of coherent and noncoherent FSK in Chapter 8.)

The random variables G and θ represent random gain and phase perturbations introduced by a fading channel. The channel is modeled as introducing a random gain and phase shift during each bit interval. Because the gain and phase shift are assumed to remain constant

[8] See Lindsey and Simon (1973), pp. 199ff, for tables giving P_E.

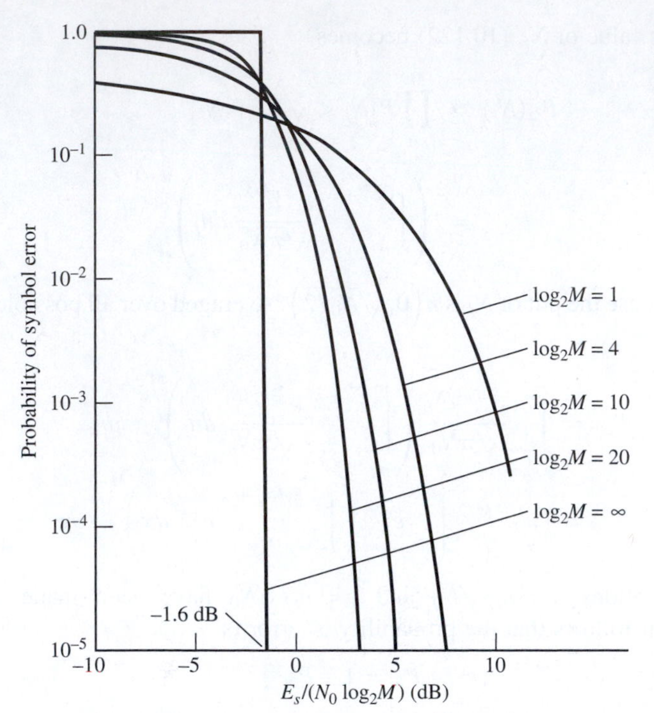

Figure 10.7
Probability of symbol error for coherent detection of M-ary orthogonal signals.

throughout a bit interval, this channel model is called *slowly fading*. We assume that G is Rayleigh and θ is uniform in $[0, 2\pi)$ and that G and θ are independent.

Expanding (10.128), we obtain

$$
\begin{aligned}
H_1 &: y(t) = \sqrt{\frac{2E}{T}}[G_1 \cos(\omega_1 t) + G_2 \sin(\omega_1 t)] + n(t) \\
H_2 &: y(t) = \sqrt{\frac{2E}{T}}[G_1 \cos(\omega_2 t) + G_2 \sin(\omega_2 t)] + n(t), \quad 0 \le t \le T
\end{aligned}
\tag{10.129}
$$

where $G_1 = G\cos\theta$ and $G_2 = -G\sin\theta$ are independent, zero-mean, Gaussian random variables (recall Example 5.15). We denote their variances by σ^2. Choosing the orthonormal basis set

$$
\left.
\begin{aligned}
\phi_1(t) &= \sqrt{\frac{2E}{T}}\cos(\omega_1 t) \\
\phi_2(t) &= \sqrt{\frac{2E}{T}}\sin(\omega_1 t) \\
\phi_3(t) &= \sqrt{\frac{2E}{T}}\cos(\omega_2 t) \\
\phi_4(t) &= \sqrt{\frac{2E}{T}}\sin(\omega_2 t)
\end{aligned}
\right\} \quad 0 \le t \le T
\tag{10.130}
$$

the term $y(t)$ can be resolved into a four-dimensional signal space, and decisions may be based on the data vector

$$\mathbf{Z} = (Z_1, Z_2, Z_3, Z_4) \quad (10.131)$$

where

$$Z_i = (y, \phi_i) = \int_0^T y(t)\phi_i(t)\, dt \quad (10.132)$$

Given hypothesis H_1, we obtain

$$Z_i = \begin{cases} \sqrt{E}G_i + N_i, & i = 1, 2 \\ N_i, & i = 3, 4 \end{cases} \quad (10.133)$$

and given hypothesis H_2, we obtain

$$Z_i = \begin{cases} N_i, & i = 1, 2 \\ \sqrt{E}G_{i-2} + N_i, & i = 3, 4 \end{cases} \quad (10.134)$$

where

$$N_i = (n, \phi_i) = \int_0^T n(t)\phi_i(t)\, dt, \quad i = 1, 2, 3, 4 \quad (10.135)$$

are independent Gaussian random variables with zero mean and variance $\frac{1}{2}N_0$. Since G_1 and G_2 are also independent Gaussian random variables with zero mean and variance σ^2, the joint conditional pdfs of Z, given H_1 and H_2, are the products of the respective marginal pdfs. It follows that

$$f_Z(z_1, z_2, z_3, z_4|H_1) = \frac{\exp\left[-\left(z_1^2 + z_2^2\right)/(2E\sigma^2 + N_0)\right]\exp\left[-\left(z_3^2 + z_4^2\right)/N_0\right]}{\pi^2(2E\sigma^2 + N_0)N_0} \quad (10.136)$$

and

$$f_Z(z_1, z_2, z_3, z_4|H_2) = \frac{\exp\left[-\left(z_1^2 + z_2^2\right)/N_0\right]\exp\left[-\left(z_3^2 + z_4^2\right)/(2E\sigma^2 + N_0)\right]}{\pi^2(2E\sigma^2 + N_0)N_0} \quad (10.137)$$

The decision rule that minimizes the probability of error is to choose the hypothesis H_ℓ corresponding to the largest posterior probability $P(H_\ell|z_1, z_2, z_3, z_4)$. Note that these probabilities differ from (10.136) and (10.137) only by a constant that is independent of H_1 or H_2. For a particular observation $\mathbf{Z} = (Z_1, Z_2, Z_3, Z_4)$, the decision rule is

$$f_Z(Z_1, Z_2, Z_3, Z_4|H_1) \underset{H_2}{\overset{H_1}{\gtrless}} f_Z(Z_1, Z_2, Z_3, Z_4|H_2) \quad (10.138)$$

which, after substitution from (10.136) and (10.137) and simplification, reduces to

$$R_2^2 \triangleq Z_3^2 + Z_4^2 \underset{H_2}{\overset{H_1}{\lessgtr}} Z_1^2 + Z_2^2 \triangleq R_1^2 \quad (10.139)$$

The optimum receiver corresponding to this decision rule is shown in Figure 10.8.

To find the probability of error, we note that both $R_1 \triangleq \sqrt{Z_1^2 + Z_2^2}$ and $R_2 \triangleq \sqrt{Z_3^2 + Z_4^2}$ are Rayleigh random variables under either hypothesis. Given H_1 is true, an error results if

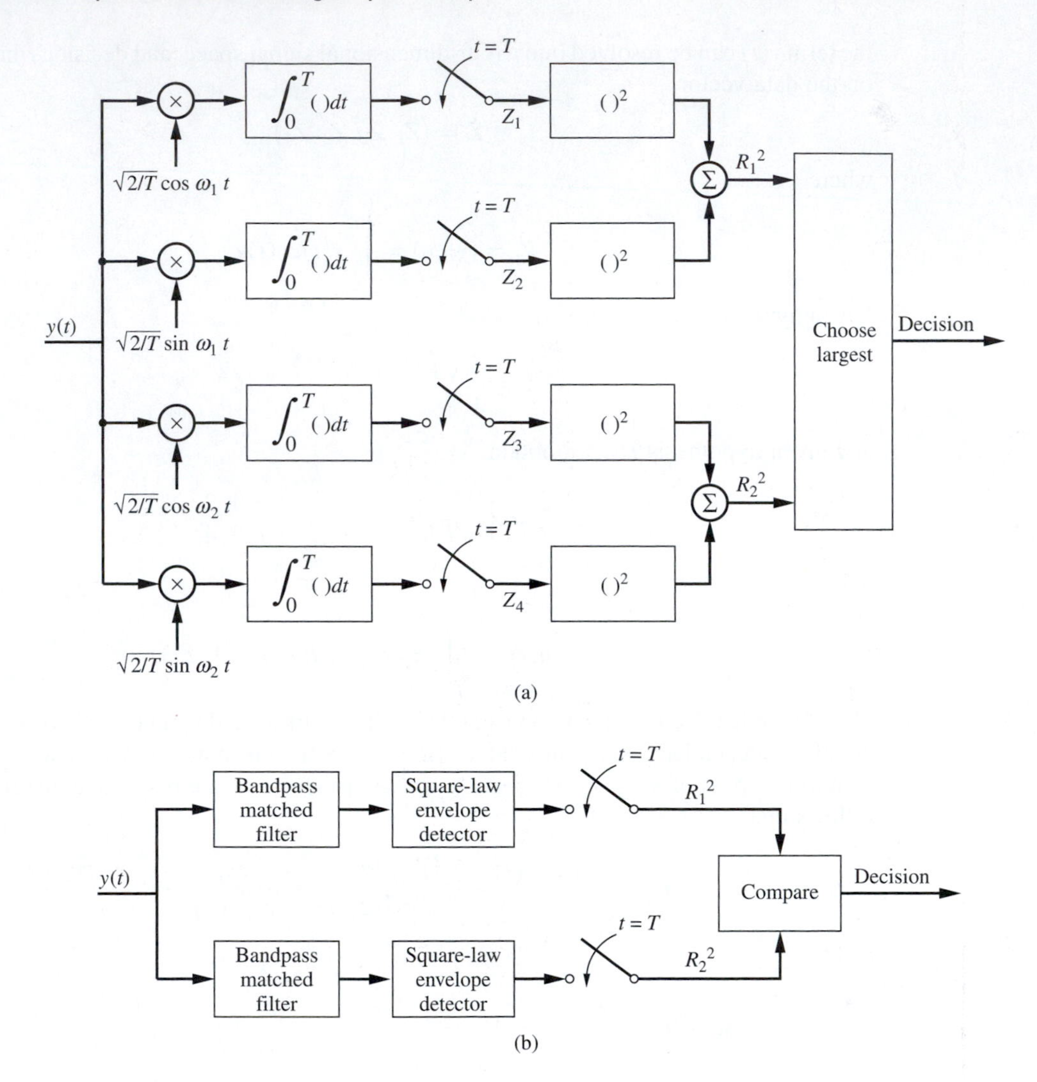

Figure 10.8
Optimum receiver structures for detection of binary orthogonal signals in Rayleigh fading. (a) Implementation by correlator and squarer. (b) Implementation by matched filter and envelope detector.

$R_2 > R_1$, where the positive square root of (10.139) has been taken. From Example 5.15, it follows that

$$f_{R_1}(r_1|H_1) = \frac{r_1 e^{-r_1^2/\left(2E\sigma^2 + N_0\right)}}{E\sigma^2 + \frac{1}{2}N_0}, \quad r_1 \geq 0 \tag{10.140}$$

and

$$f_{R_2}(r_2|H_1) = \frac{2r_2 e^{-r_2^2/N_0}}{N_0}, \quad r_2 \geq 0 \tag{10.141}$$

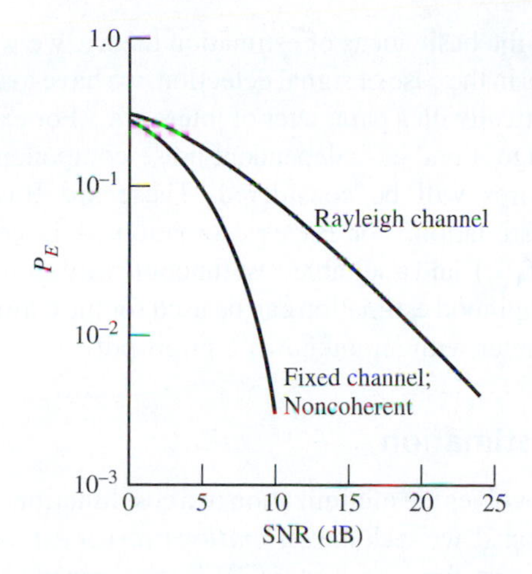

Figure 10.9
Comparison of P_E versus SNR for Rayleigh and fixed channels with noncoherent FSK signaling.

The probability that $R_2 > R_1$, averaged over R_1, is

$$\begin{aligned} P(E|H_1) &= \int_0^\infty \left[\int_{r_1}^\infty f_{R_2}(r_2|H_1)\, dr_2 \right] f_{R_1}(r_1|H_1)\, dr_1 \\ &= \frac{1}{2} \frac{1}{1 + 1/2(2\sigma^2 E/N_0)} \end{aligned} \tag{10.142}$$

where $2\sigma^2 E$ is the average received signal energy. Because of the symmetry involved, it follows that $P(E|H_1) = P(E|H_2)$ and that

$$P_E = P(E|H_1) = P(E|H_2) \tag{10.143}$$

The probability of error is plotted in Figure 10.9, along with the result from Chapter 8 for constant-amplitude noncoherent FSK (Figure 9.15). Whereas the error probability for non-fading, noncoherent FSK signaling decreases exponentially with the SNR, the fading channel results in an error probability that decreases only inversely with SNR.

One way to combat this degradation due to fading is to employ *diversity transmission*; that is, the transmitted signal power is divided among several independently fading transmission paths with the hope that not all of them will fade simultaneously. Several ways of achieving diversity were mentioned in Chapter 8. (See Problem 10.27).

10.4 ESTIMATION THEORY

We now consider the second type of optimization problem discussed in the introduction to this chapter—the estimation of parameters from random data. After introducing some background theory here, we will consider two applications of estimation theory to communication systems in Section 10.5.

In introducing the basic ideas of estimation theory, we will exploit several parallels with detection theory. As in the case of signal detection, we have available a noisy observation Z that depends probabilistically on a parameter of interest A.[9] For example, Z could be the sum of an unknown DC voltage A and an independent noise component $N: Z = A + N$. Two different estimation procedures will be considered. These are Bayes estimation and maximum-likelihood (ML) estimation. For Bayes estimation, A is considered to be random with a known a priori pdf $f_A(a)$, and a suitable cost function is minimized to find the optimum estimate of A. Maximum-likelihood estimation can be used for the estimation of nonrandom parameters or a random parameter with an unknown a priori pdf.

10.4.1 Bayes Estimation

Bayes estimation involves the minimization of a cost function, as in the case of Bayes detection. Given an observation Z, we seek the *estimation rule* (or *estimator*) $\hat{a}(Z)$ that assigns a value $\hat{A}$ to A such that the cost function $C[A, \hat{a}(Z)]$ is minimized. Note that C is a function of the unknown parameter A and the observation Z. Clearly, as the absolute error $|A - \hat{a}(Z)|$ increases, $C[A, \hat{a}(Z)]$ should increase, or at least not decrease; that is, large errors should be more costly than small errors. Two useful cost functions are the squared-error cost function, defined by

$$C[A, \hat{a}(Z)] = [A - \hat{a}(Z)]^2 \tag{10.144}$$

and the uniform cost function (square well), defined by

$$C[A, \hat{a}(Z)] = \begin{cases} 1, & |A - \hat{a}(Z)| > \Delta > 0 \\ 0, & \text{otherwise} \end{cases} \tag{10.145}$$

where Δ is a suitably chosen constant. For each of these cost functions, we wish to find the decision rule $\hat{a}(Z)$ that minimizes the average cost $E\{C[A, \hat{a}(Z)]\} = \overline{C[A, \hat{a}(Z)]}$. Because both A and Z are random variables, the average cost, or risk, is given by

$$\begin{aligned} \overline{C[A, \hat{a}(Z)]} &= \int_{-\infty}^{\infty}\int_{-\infty}^{\infty} C[A, \hat{a}(Z)] f_{AZ}(a, z)\, da\, dz \\ &= \int_{-\infty}^{\infty}\int_{-\infty}^{\infty} C[A, \hat{a}(Z)] f_{Z|A}(z|a) f_A(a)\, dz\, da \end{aligned} \tag{10.146}$$

where $f_{AZ}(a, z)$ is the joint pdf of A and Z and $f_{Z|A}(z|a)$ is the conditional pdf of Z given A. The latter can be found if the probabilistic mechanism that produces Z from A is known. For example, if $Z = A + N$, where N is a zero-mean Gaussian random variable with variance σ_n^2, then

$$f_{Z|A}(z|a) = \frac{\exp\left[-(z-a)^2/2\sigma_n^2\right]}{\sqrt{2\pi\sigma_n^2}} \tag{10.147}$$

[9]For simplicity, we consider the single-observation case first and generalize to vector observations later.

Returning to the minimization of the risk, we find it more advantageous to express (10.146) in terms of the conditional pdf $f_{A|Z}(a|z)$, which can be done by means of Bayes rule, to obtain

$$C[A, \hat{a}(Z)] = \int_{-\infty}^{\infty} f_Z(z)\left\{\int_{-\infty}^{\infty} C[a, \hat{a}(Z)]f_{A|Z}(a|z)\, da\right\}dz \qquad (10.148)$$

where

$$f_Z(z) = \int_{-\infty}^{\infty} f_{Z|A}(z|a)f_A(a)\, da \qquad (10.149)$$

is the pdf of Z. Since $f_Z(z)$ and the inner integral in (10.148) are nonnegative, the risk can be minimized by minimizing the inner integral for each z. The inner integral in (10.148) is called the *conditional risk*.

This minimization is accomplished for the squared-error cost function, (10.144), by differentiating the conditional risk with respect to $\hat{a}$ for a particular observation Z and setting the result equal to zero. The resulting differentiation yields

$$\begin{aligned}\frac{\partial}{\partial \hat{a}}\int_{-\infty}^{\infty} [a - \hat{a}(Z)]^2 f_{A|Z}(a|Z)\, da &= -2\int_{-\infty}^{\infty} a f_{A|Z}(a|Z)\, da \\ &\quad + 2\,\hat{a}(Z)\int_{-\infty}^{\infty} f_{A|Z}(a|Z)\, da\end{aligned} \qquad (10.150)$$

which, when set to zero, results in

$$\hat{a}_{se}(Z) = \int_{-\infty}^{\infty} a f_{A|Z}(a|Z) da \qquad (10.151)$$

where the fact that $\int_{-\infty}^{\infty} f_{A|Z}(a|Z)\, da = 1$ has been used. A second differentiation shows that this is a minimum. Note that $\hat{a}_{se}(Z)$, the estimator for a squared-error cost function, is the mean of the pdf of A given the observation Z, or the conditional mean. The values that $\hat{a}_{se}(Z)$ assume, $\hat{A}$, are random since the estimator is a function of the random variable Z.

In a similar manner, we can show that the uniform cost function results in the condition

$$f_{A|Z}(A|Z)|_{A=\hat{a}_{\mathrm{MAP}}(Z)} = \text{maximum} \qquad (10.152)$$

for Δ in (10.145) infinitesimally small. That is, the estimation rule, or estimator, that minimizes the uniform cost function is the maximum of the conditional pdf of A given Z, or the a posteriori pdf. Thus this estimator will be referred to as the MAP estimate. Necessary, but not sufficient, conditions that the MAP estimate must satisfy are

$$\frac{\partial}{\partial A} f_{A|Z}(A|Z)|_{A=\hat{a}_{\mathrm{MAP}}(Z)} = 0 \qquad (10.153)$$

or

$$\frac{\partial}{\partial A} \ln f_{A|Z}(A|Z)|_{A=\hat{a}_{\mathrm{MAP}}(Z)} = 0 \qquad (10.154)$$

where the latter condition is especially convenient for a posteriori pdfs of exponential type, such as Gaussian.

Often the MAP estimate is employed because it is easier to obtain than other estimates, even though the conditional-mean estimate, given by (10.151), is more general, as the following theorem indicates.

Theorem

If, as a function of a, the a posteriori pdf $f_{A|Z}(a|Z)$ has a single peak, about which it is symmetrical, and the cost function has the properties

$$C(A, \hat{a}) = C(A - \hat{a}) \tag{10.155}$$

$$C(x) = C(-x) \geq 0 \quad \text{(symmetrical)} \tag{10.156}$$

$$C(x_1) \geq C(x_2) \text{ for } |x_1| \geq |x_2| \quad \text{(convex)} \tag{10.157}$$

then the conditional-mean estimator is the Bayes estimate.[10]

10.4.2 Maximum-Likelihood Estimation

We now seek an estimation procedure that does not require a priori information about the parameter of interest. Such a procedure is ML estimation. To explain this procedure, consider the MAP estimation of a random parameter A about which little is known. This lack of information about A is expressed probabilistically by assuming the prior pdf of A, $f_A(a)$, to be broad compared with the posterior pdf, $f_{A|Z}(a|Z)$. If this were not the case, the observation Z would be of little use in estimating A. Since the joint pdf of A and Z is given by

$$f_{AZ}(a, z) = f_{A|Z}(a|z) f_Z(z) \tag{10.158}$$

the joint pdf, regarded as a function of a, must be peaked for at least one value of a. By the definition of conditional probability, we also may write (10.158) as

$$\begin{aligned} f_{ZA}(z, a) &= f_{Z|A}(z|a) f_A(a) \\ &\cong f_{Z|A}(z|a) \quad \text{(times a constant)} \end{aligned} \tag{10.159}$$

where the approximation follows by virtue of the assumption that little is known about A, thus implying that $f_A(a)$ is essentially constant. The ML estimate of A is defined as

$$f_{Z|A}(Z|A)|_{A=\hat{a}_{\text{ML}}(Z)} = \text{maximum} \tag{10.160}$$

From (10.160) and (10.160), the ML estimate of a parameter corresponds to the MAP estimate if little a priori information about the parameter is available. From (10.160), it follows that the ML estimate of a parameter A is that value of A which is most likely to have resulted in the observation Z; hence the name *maximum likelihood.* Since the prior pdf of A is not required to obtain an ML estimate, it is a suitable estimation procedure for random parameters whose prior pdf is unknown. If a deterministic parameter is to be estimated, $f_{Z|A}(z|A)$ is regarded as the pdf of Z with A as a parameter.

From (10.160) it follows that the ML estimate can be found from the necessary, but not sufficient, conditions

$$\left. \frac{\partial f_{Z|A}(Z|A)}{\partial A} \right|_{A=\hat{a}_{\text{ML}}(Z)} = 0 \tag{10.161}$$

[10] Van Trees (1968), pp. 60–61.

and

$$l(A) = \frac{\partial \ln f_{Z|A}(Z|A)}{\partial A}\bigg|_{A=\hat{a}_{\mathrm{ML}}(Z)} = 0 \tag{10.162}$$

When viewed as a function of A, $f_{Z|A}(Z|A)$ is referred to as the *likelihood function*. Both (10.161) and (10.162) will be referred to as *likelihood equations*.

From (10.154) and Bayes rule, it follows that the MAP estimate of a random parameter satisfies

$$\left[l(A) + \frac{\partial}{\partial A}\ln f_A(A)\right]\bigg|_{A=\hat{a}_{\mathrm{ML}}(Z)} = 0 \tag{10.163}$$

which is useful when finding both the ML and MAP estimates of a parameter.

10.4.3 Estimates Based on Multiple Observations

If a multiple number of observations are available, say $\mathbf{Z} \triangleq (Z_1, Z_2, \ldots, Z_K)$, on which to base the estimate of a parameter, we simply substitute the K-fold joint conditional pdf $f_{\mathbf{Z}|A}(\mathbf{z}|A)$ in (10.161) and (10.162) to find the ML estimate of A. If the observations are independent, when conditioned on A, then

$$f_{\mathbf{Z}|A}(\mathbf{z}|A) = \prod_{k=1}^{K} f_{Z_k|A}(z_k|A) \tag{10.164}$$

where $f_{Z_k|A}(z_k|A)$ is the pdf of the kth observation Z_k given the parameter A. To find $f_{A|\mathbf{Z}}(A|\mathbf{z})$ for MAP estimation, we use Bayes rule.

EXAMPLE 10.8

To illustrate the estimation concepts just discussed, let us consider the estimation of a constant-level random signal A embedded in Gaussian noise $n(t)$ with zero mean and variance σ_n^2:

$$z(t) = A + n(t) \tag{10.165}$$

We assume $z(t)$ is sampled at time intervals sufficiently spaced so that the samples are independent. Let these samples be represented as

$$Z_k = A + N_k,\ k = 1, 2, \ldots, K \tag{10.166}$$

Thus, given A, the Z_ks are independent, each having mean A and variance σ_n^2. Hence the conditional pdf of $\mathbf{Z} \triangleq (Z_1, Z_2, \ldots, Z_K)$ given A is

$$\begin{aligned} f_{\mathbf{Z}|A}(\mathbf{z}|A) &= \prod_{k=1}^{K} \frac{\exp\left[-(z_k - A)^2/2\sigma_n^2\right]}{\sqrt{2\pi\sigma_n^2}} \\ &= \frac{\exp\left[-\sum_{k=1}^{K}(z_k - A)^2/2_n^2\right]}{(2\pi\sigma_n^2)^{K/2}} \end{aligned} \tag{10.167}$$

We will assume two possibilities for A:

1. It is Gaussian with mean m_A and variance σ_A^2.
2. Its pdf is unknown.

In the first case, we will find the conditional-mean and the MAP estimates for A. In the second case, we will compute the ML estimate.

Case 1

If the pdf of A is

$$f_A(a) = \frac{\exp\left[-(a-m_A)^2/2\sigma_A^2\right]}{\sqrt{2\pi\sigma_A^2}} \tag{10.168}$$

its posterior pdf is, by Bayes' rule,

$$f_{A|\mathbf{Z}}(a|\mathbf{z}) = \frac{f_{\mathbf{Z}|A}(\mathbf{z}|a)f_A(a)}{f_{\mathbf{Z}}(z)} \tag{10.169}$$

After some algebra, it can be shown that

$$f_{A|\mathbf{Z}}(a|\mathbf{z}) = \left(2\pi\sigma_p^2\right)^{-1/2}\exp\left(\frac{-\left\{a-\sigma_p^2\left[\left(Km_s/\sigma_n^2\right)+\left(m_A/\sigma_A^2\right)\right]\right\}^2}{2\sigma_p^2}\right) \tag{10.170}$$

where

$$\frac{1}{\sigma_p^2} = \frac{K}{\sigma_n^2} + \frac{1}{\sigma_A^2} \tag{10.171}$$

and the sample mean is

$$m_s = \frac{1}{K}\sum_{k=1}^{K} Z_k \tag{10.172}$$

Clearly, $f_{A|\mathbf{Z}}(a|\mathbf{z})$ is a Gaussian pdf with variance σ_p^2 and mean

$$\begin{aligned} E\{A|\mathbf{Z}\} &= \sigma_p^2\left(\frac{Km_s}{\sigma_n^2} + \frac{m_A}{\sigma_A^2}\right) \\ &= \frac{K\sigma_A^2/\sigma_n^2}{1+K\sigma_A^2/\sigma_n^2}m_s + \frac{1}{1+K\sigma_A^2/\sigma_n^2}m_A \end{aligned} \tag{10.173}$$

Since the maximum value of a Gaussian pdf is at the mean, this is both the conditional-mean estimate (squared-error cost function, among other convex cost functions) and the MAP estimate (square-well cost function). The conditional variance var $\{A|\mathbf{Z}\}$ is σ_p^2. Because it is not a function of $\mathbf{Z}$, it follows that the average cost, or risk, which is

$$C[A, \hat{a}(Z)] = \int_{-\infty}^{\infty} \text{var}\{A|\mathbf{z}\}f_z(\mathbf{z})d\mathbf{z} \tag{10.174}$$

is just σ_p^2. From the expression for $E\{A|\mathbf{Z}\}$, we note an interesting behavior for the estimate of A, or $\hat{a}(Z)$. As $K\sigma_A^2/\sigma_n^2 \to \infty$,

$$\hat{a}(Z) \to m_s = \frac{1}{K}\sum_{k=1}^{K} Z_k \tag{10.175}$$

which says that as the ratio of signal variance to noise variance becomes large, the optimum estimate for A approaches the sample mean. On the other hand, as $K\sigma_A^2/\sigma_n^2 \to 0$ (small signal variance and/or large noise variance), $\hat{a}(\mathbf{Z}) \to m_A$, the a priori mean of A. In the first case, the estimate is weighted in favor of the observations; in the latter, it is weighted in favor of the known signal statistics. From the form of σ_p^2, we note that, in either case, the quality of the estimate increases as the number of independent samples of $z(t)$ increases.

Case 2

The ML estimate is found by differentiating $\ln f_{\mathbf{Z}|A}(\mathbf{z}|A)$ with respect to A and setting the result equal to zero. Performing the steps, the ML estimate is found to be

$$\hat{a}_{\mathrm{ML}}(\mathbf{Z}) = \frac{1}{K}\sum_{k=1}^{K} Z_k \tag{10.176}$$

We note that this corresponds to the MAP estimate if $K\sigma_A^2/\sigma_n^2 \to \infty$ (that is, if the a priori pdf of A is broad compared with the a posteriori pdf).

The variance of $\hat{a}_{\mathrm{ML}}(\mathbf{Z})$ is found by recalling that the variance of a sum of independent random variables is the sum of the variances. The result is

$$\sigma_{\mathrm{ML}}^2 = \frac{\sigma_n^2}{K} > \sigma_p^2 \tag{10.177}$$

Thus the prior knowledge about A, available through $f_A(a)$, manifests itself as a smaller variance for the Bayes estimates (conditional-mean and MAP) than for the ML estimate. ■

10.4.4 Other Properties of ML Estimates

Unbiased Estimates

An estimate $\hat{a}(\mathbf{Z})$ is said to be *unbiased* if

$$E\{\hat{a}(\mathbf{Z})|A\} = A \tag{10.178}$$

This is clearly a desirable property of any estimation rule. If $E\{\hat{a}(\mathbf{Z})|A\} - A = B \neq 0, B$ is referred to as the *bias of the estimate*.

The Cramer–Rao Inequality

In many cases it may be difficult to compute the variance of an estimate for a nonrandom parameter. A lower bound for the variance of an unbiased ML estimate is provided by the following inequality:

$$\mathrm{var}\{\hat{a}(\mathbf{Z})\} \geq \left(E\left\{\left[\frac{\partial\left[\ln f_{\mathbf{Z}|A}(\mathbf{Z}|a)\right]}{\partial a}\right]^2\right\}\right)^{-1} \tag{10.179}$$

or, equivalently,

$$\mathrm{var}\{\hat{a}(\mathbf{Z})\} \geq \left(-E\left\{\frac{\partial^2\left[\ln f_{\mathbf{Z}|A}(\mathbf{Z}|a)\right]}{\partial a^2}\right\}\right)^{-1} \tag{10.180}$$

where the expectation is only over $\mathbf{Z}$. These inequalities hold under the assumption that $\partial f_{\mathbf{Z}|A}/\partial a$ and $\partial^2 f_{\mathbf{Z}|A}/\partial a^2$ exist and are absolutely integrable. A proof is furnished by Van Trees (1968). Any estimate satisfying (10.179) or (10.180) with equality is said to be *efficient*.

A sufficient condition for equality in (10.179) or (10.80) is that

$$\frac{\partial\left[\ln f_{\mathbf{Z}|A}(\mathbf{Z}|a)\right]}{\partial a} = [\widehat{a}(\mathbf{Z}) - a]g(a) \tag{10.181}$$

where, $g(\cdot)$ is a function only of a. If an efficient estimate of a parameter exists, it is the ML estimate.

10.4.5 Asymptotic Qualities of ML Estimates

In the limit, as the number of independent observations becomes large, ML estimates can be shown to be *Gaussian*, *unbiased*, and *efficient*. In addition, the probability that the ML estimate for K observations differs by a fixed amount ϵ from the true value approaches zero as $K \rightarrow \infty$; an estimate with such behavior is referred to as *consistent*.

EXAMPLE 10.9

Returning to Example 10.8, we can show that $\widehat{a}_{\text{ML}}(\mathbf{Z})$ is an efficient estimate. We have already shown that $\sigma^2_{\text{ML}} = \sigma^2_n/K$. Using (10.180), we differentiate in $f_{\mathbf{Z}|A}$ once to obtain

$$\frac{\partial\left[\ln f_{\mathbf{Z}|A}\right]}{\partial a} = \frac{1}{\sigma_n^2}\sum_{k=1}^{K}(Z_k - a) \tag{10.182}$$

A second differentiation gives

$$\frac{\partial^2\left[\ln f_{\mathbf{Z}|A}\right]}{\partial a^2} = -\frac{K}{\sigma_n^2} \tag{10.183}$$

and (10.180) is seen to be satisfied with equality.

■

■ 10.5 APPLICATIONS OF ESTIMATION THEORY TO COMMUNICATIONS

We now consider two applications of estimation theory to the transmission of analog data. The sampling theorem introduced in Chapter 2 was applied in Chapter 3 in the discussion of several systems for the transmission of continuous-waveform messages via their sample values. One such technique is PAM, in which the sample values of the message are used to amplitude modulate a pulse-type carrier. We will apply the results of Example 10.8 to find the performance of the optimum demodulator for PAM. This is a *linear* estimator because the observations are linearly dependent on the message sample values. For such a system, the only way to decrease the effect of noise on the demodulator output is to increase the SNR of the received signal, since output and input SNR are linearly related.

Following the consideration of PAM, we will derive the optimum ML estimator for the phase of a signal in additive Gaussian noise. This will result in a PLL structure. The variance of the estimate in this case will be obtained for high input SNR by applying the Cramer–Rao inequality. For low SNRs, the variance is difficult to obtain because this is a problem in *nonlinear* estimation; that is, the observations are nonlinearly dependent on the parameter being estimated.

The transmission of analog samples by PPM or some other modulation scheme could also be considered. An approximate analysis of its performance for low input SNRs would show the threshold effect of nonlinear modulation schemes and the implications of the trade-off that is possible between bandwidth and output SNR. This effect was seen previously in Chapter 7 when the performance of PCM in noise was considered.

10.5.1 Pulse-Amplitude Modulation

In PAM, the message $m(t)$ of bandwidth W is sampled at T-s intervals, where $T \leq 1/2W$, and the sample values $m_k = m(t_k)$ are used to amplitude modulate a pulse train composed of time translates of the basic pulse shape $p(t)$, which is assumed zero for $t \leq 0$ and $t \geq T_0 < T$. The received signal plus noise is represented as

$$y(t) = \sum_{k=-\infty}^{\infty} m_k p(t - kT) + n(t) \tag{10.184}$$

where $n(t)$ is white Gaussian noise with double-sided power spectral density $\frac{1}{2}N_0$.

Considering the estimation of a single sample at the receiver, we observe

$$y(t) = m_0 p(t) + n(t), \quad 0 \leq t \leq T \tag{10.185}$$

For convenience, we assume that $\int_0^{T_0} p^2(t)dt = 1$. It follows that a sufficient statistic is

$$\begin{aligned} Z_0 &= \int_0^{T_0} y(t)p(t)dt \\ &= m_0 + N \end{aligned} \tag{10.186}$$

where the noise component is

$$N = \int_0^{T_0} n(t)p(t)dt \tag{10.187}$$

Having no prior information about m_0, we apply ML estimation. Following procedures used many times before, we can show that N is a zero-mean Gaussian random variable with variance $\frac{1}{2}N_0$. The ML estimation of m_0 is therefore identical to the single-observation case of Example 10.8, and the best estimate is simply Z_0. As in the case of digital data transmission, this estimator could be implemented by passing $y(t)$ through a filter matched to $p(t)$, observing the output amplitude prior to the next pulse, and then setting the filter initial conditions to zero. Note that the estimator is linearly dependent on $y(t)$.

The variance of the estimate is equal to the variance of N, or $\frac{1}{2}N_0$. Thus the SNR at the output of the estimator is

$$(\text{SNR})_0 = \frac{2m_0^2}{N_0} = \frac{2E}{N_0} \tag{10.188}$$

where $E = \int_0^{T_0} m_0^2 p^2(t)\,dt$ is the average energy of the received signal sample. Thus the only way to increase $(\text{SNR})_0$ is by increasing the energy per sample or by decreasing N_0.

10.5.2 Estimation of Signal Phase: The PLL Revisited

We now consider the problem of estimating the phase of a sinusoidal signal $A\cos(\omega_c t + \theta)$ in white Gaussian noise $n(t)$ of double-sided power spectral density $\frac{1}{2}N_0$. Thus the observed data are

$$y(t) = A\cos(\omega_c t + \theta) + n(t), \quad 0 \le t \le T \tag{10.189}$$

where T is the observation interval. Expanding $A\cos(\omega_c t + \theta)$ as

$$A\cos(\omega_c t)\cos\theta - A\sin(\omega_c t)\sin\theta$$

we see that a suitable set of orthonormal basis functions for representing the data is

$$\phi_1(t) = \sqrt{\frac{2}{T}}\cos(\omega_c t), \quad 0 \le t \le T \tag{10.190}$$

and

$$\phi_2(t) = \sqrt{\frac{2}{T}}\sin(\omega_c t), \quad 0 \le t \le T \tag{10.191}$$

Thus we base our decision on

$$z(t) = \sqrt{\frac{T}{2}}A\cos\theta\,\phi_1(t) - \sqrt{\frac{T}{2}}A\sin\theta\,\phi_2(t) + N_1\phi_1(t) + N_2\phi_2(t) \tag{10.192}$$

where

$$N_i = \int_0^T n(t)\phi_i(t)\,dt, \quad i = 1, 2 \tag{10.193}$$

Because $y(t) - z(t)$ involves only noise, which is independent of $z(t)$, it is not relevant to making the estimate. Thus we may base the estimate on the vector

$$\mathbf{Z} \triangleq (Z_1, Z_2) = \left(\sqrt{\frac{T}{2}}A\cos\theta + N_1,\ -\sqrt{\frac{T}{2}}A\sin\theta + N_2\right) \tag{10.194}$$

where

$$Z_i = (y(t), \phi_i(t)) = \int_0^T y(t)\phi_i(t)\,dt \tag{10.195}$$

The likelihood function $f_{\mathbf{z}|\theta}(z_1, z_2|\theta)$ is obtained by noting that the variance of Z_1 and Z_2 is simply $\frac{1}{2}N_0$, as in the PAM example. Thus the likelihood function is

$$f_{\mathbf{z}|\theta}(z_1, z_2|\theta) = \frac{1}{\pi N_0}\exp\left\{-\frac{1}{N_0}\left[\left(z_1 - \sqrt{\frac{T}{2}}A\cos\theta\right)^2 + \left(z_2 + \sqrt{\frac{T}{2}}A\sin\theta\right)^2\right]\right\} \tag{10.196}$$

which reduces to

$$f_{\mathbf{z}|\theta}(z_1, z_2|\theta) = C \exp\left[2\sqrt{\frac{T}{2}}\frac{A}{N_0}(z_1 \cos\theta - z_2 \sin\theta)\right] \tag{10.197}$$

where the coefficient C contains all factors that are independent of θ. The logarithm of the likelihood function is

$$\ln f_{\mathbf{z}|\theta}(z_1, z_2|\theta) = \ln C + \sqrt{2T}\frac{A}{N_0}(z_1 \cos\theta - z_2 \sin\theta) \tag{10.198}$$

which, when differentiated and set to zero, yields a necessary condition for the ML estimate of θ in accordance with (10.162). The result is

$$-Z_1 \sin\theta - Z_2 \cos\theta|_{\theta=\hat{\theta}_{\text{ML}}} = 0 \tag{10.199}$$

where Z_1 and Z_2 signify that we are considering a particular (random) observation. But

$$Z_1 = (y, \phi_1) = \sqrt{\frac{2}{T}}\int_0^T y(t)\cos(\omega_c t)\,dt \tag{10.200}$$

and

$$Z_2 = (y, \phi_2) = \sqrt{\frac{2}{T}}\int_0^T y(t)\sin(\omega_c t)\,dt \tag{10.201}$$

Therefore, (10.199) can be put in the form

$$-\sin\hat{\theta}_{\text{ML}}\int_0^T y(t)\cos(\omega_c t)\,dt - \cos\hat{\theta}_{\text{ML}}\int_0^T y(t)\sin(\omega_c t)\,dt = 0$$

or

$$\int_0^T y(t)\sin\left(\omega_c t + \hat{\theta}_{\text{ML}}\right)dt = 0 \tag{10.202}$$

This equation can be interpreted as the feedback structure shown in Figure 10.10. Except for the integrator replacing a loop filter, this is identical to the PLL discussed in Chapter 3.

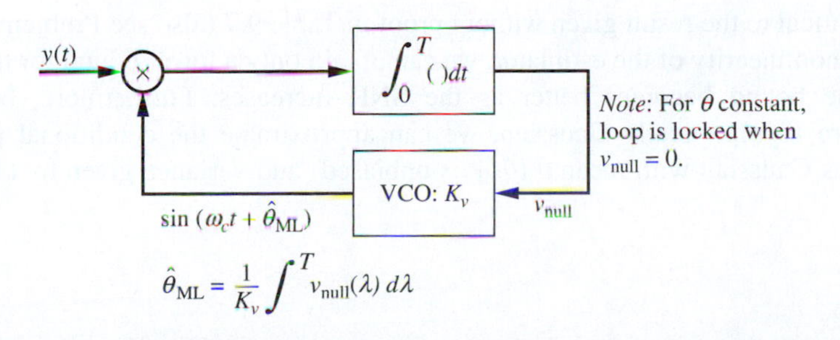

Figure 10.10
Maximum–Likelihood estimator for phase.

A lower bound for the variance of $\widehat{\theta}_{\text{ML}}$ is obtained from the Cramer–Rao inequality. Applying (10.180), we have for the first differentiation, from (10.198),

$$\frac{\partial\left(\ln f_{\mathbf{Z}|\theta}\right)}{\partial\theta} = \sqrt{2T}\frac{A}{N_0}\left(-Z_1 \sin\theta - Z_2 \cos\theta\right) \tag{10.203}$$

and for the second,

$$\frac{\partial^2\left(\ln f_{\mathbf{Z}|\theta}\right)}{\partial\theta^2} = \sqrt{2T}\frac{A}{N_0}\left(-Z_1 \cos\theta + Z_2 \sin\theta\right) \tag{10.204}$$

Substituting into (10.180), we have

$$\text{var}\left\{\widehat{\theta}_{\text{ML}}(Z)\right\} \geq \frac{1}{\sqrt{2T}}\frac{N_0}{A}\left(E\{Z_1\}\cos\theta - E\{Z_2\}\sin\theta\right)^{-1} \tag{10.205}$$

The expectations of Z_1 and Z_2 are

$$\begin{aligned} E\{Z_i\} &= \int_0^T E\{y(t)\}\phi_i(t)\,dt \\ &= \int_0^T \sqrt{\frac{T}{2}}A[(\cos\theta)\phi_1(t) - (\sin\theta)\phi_2(t)]\phi_i(t)\,dt \\ &= \begin{cases} \sqrt{\frac{T}{2}}A\cos\theta, & i = 1 \\ -\sqrt{\frac{T}{2}}A\sin\theta, & i = 2 \end{cases} \end{aligned} \tag{10.206}$$

where we used (10.192). Substitution of these results into (10.205) results in

$$\text{var}\left\{\widehat{\theta}_{\text{ML}}(Z)\right\} \geq \frac{1}{\sqrt{2T}}\frac{N_0}{A}\left[\sqrt{\frac{T}{2}}A\left(\cos^2\theta + \sin^2\theta\right)\right]^{-1} = \frac{N_0}{A^2 T} \tag{10.207}$$

Noting that the average signal power is $P_s = \frac{1}{2}A^2$ and defining $B_L = (2T)^{-1}$ as the equivalent noise bandwidth[11] of the estimator structure, we may write (10.207) as

$$\text{var}\left\{\widehat{\theta}_{\text{ML}}\right\} \geq \frac{N_0 B_L}{P_s} \tag{10.208}$$

which is identical to the result given without proof in Table 9.7 (also see Problem 9.27). As a result of the nonlinearity of the estimator, we can obtain only a lower bound for the variance. However, the bound becomes better as the SNR increases. Furthermore, because ML estimators are asymptotically Gaussian, we can approximate the conditional pdf of $\widehat{\theta}_{\text{ML}}$, $f_{\widehat{\theta}_{\text{ML}|\theta}}(\alpha|\theta)$, as Gaussian with mean θ ($\widehat{\theta}_{\text{ML}}$ is unbiased) and variance given by (10.207).

---

[11]The equivalent noise bandwidth of an ideal integrator of integration duration T is $(2T)^{-1}$ Hz.

Summary

1. Two general classes of optimization problems are signal detection and parameter estimation. Although both detection and estimation are often involved simultaneously in signal reception, from an analysis standpoint, it is easiest to consider them as separate problems.

2. Bayes detectors are designed to minimize the average cost of making a decision. They involve testing a likelihood ratio, which is the ratio of the a posteriori (posterior) probabilities of the observations, against a threshold, which depends on the a priori (prior) probabilities of the two possible hypotheses and costs of the various decision–hypothesis combinations. The performance of a Bayes detector is characterized by the average cost, or risk, of making a decision. More useful in many cases, however, are the probabilities of detection and false alarm P_D and P_F in terms of which the risk can be expressed, provided the a priori probabilities and costs are available. A plot of P_D versus P_F is referred to as the *receiver operating characteristic*.

3. If the costs and prior probabilities are not available, a useful decision strategy is the Neyman–Pearson detector, which maximizes P_D while holding P_F below some tolerable level. This type of receiver also can be reduced to a likelihood ratio test in which the threshold is determined by the allowed false-alarm level.

4. It was shown that a minimum-probability-of-error detector (that is, the type of detector considered in Chapter 8) is really a Bayes detector with zero costs for making right decisions and equal costs for making either type of wrong decision. Such a receiver is also referred to as an maximum a posteriori (MAP) detector, since the decision rule amounts to choosing as the correct hypothesis the one corresponding to the largest a posteriori probability for a given observation.

5. The introduction of signal space concepts allowed the MAP criterion to be expressed as a receiver structure that chooses as the transmitted signal the signal whose location in signal space is closest to the observed data point. Two examples considered were coherent detection of M-ary orthogonal signals and noncoherent detection of binary FSK in a Rayleigh fading channel.

6. For M-ary orthogonal signal detection, arbitrarily small probability of error can be achieved as $M \to \infty$ provided the ratio of energy per bit to noise spectral density is greater than -1.6 dB. This perfect performance is achieved at the expense of infinite transmission bandwidth, however.

7. For the Rayleigh fading channel, the probability of error decreases only inversely with the SNR rather than exponentially, as for the nonfading case. A way to improve performance is by using diversity.

8. Bayes estimation involves the minimization of a cost function, as for signal detection. The squared-error cost function results in the a posteriori conditional mean of the parameter as the optimum estimate, and a square-well cost function with infinitely narrow well results in the maximum of the a posteriori pdf of the data, given the parameter, as the optimum estimate

(MAP estimate). Because of its ease of implementation, the MAP estimate is often employed even though the conditional-mean estimate is more general, in that it minimizes any symmetrical, convex-upward cost function as long as the posterior pdf is symmetrical about a single peak.

9. An ML estimate of a parameter A is that value for the parameter, $\widehat{A}$, that is most likely to have resulted in the observed data A and is the value of A corresponding to the absolute maximum of the conditional pdf of Z given A. The ML and MAP estimates of a parameter are identical if the a priori pdf of A is uniform. Since the a priori pdf of A is not needed to obtain an ML estimate, this is a useful procedure for estimation of parameters whose prior statistics are unknown or for estimation of nonrandom parameters.
10. The Cramer–Rao inequality gives a lower bound for the variance of an ML estimate. In the limit, ML estimates have many useful asymptotic properties as the number of independent observations becomes large. In particular, they are asymptotically Gaussian, unbiased, and efficient (satisfy the Cramer–Rao inequality with equality).

Further Reading

Two classic textbooks on detection and estimation theory at the graduate level are Van Trees (1968) and Helstrom (1968). Both are excellent in their own way, Van Trees being somewhat wordier and containing more examples than Helstrom, which is closely written but nevertheless very readable. More recent treatments on detection and estimation theory are Poor (1994), Scharf (1990), and McDonough and Whalen (1995).

At about the same level as the above books is the book by Wozencraft and Jacobs (1965), which was the first book in the United States to use the signal space concepts exploited by Kotel'nikov (1959) in his doctoral dissertation in 1947 to treat digital signaling and optimal analog demodulation.

Two books by Kay (1993; 1998) cover estimation theory and detection theory in detail from a signal processing point of view. Algorithms are derived and discussed in detail, and in a number of cases, computer code is provided.

Problems

Section 10.1

10.1. Consider the hypotheses

$$H_1: Z = N \quad H_2: Z = S + N$$

where S and N are independent random variables with the pdfs

$$f_S(x) = 2e^{-2x}u(x) \quad \text{and} \quad f_N(x) = 10e^{-10x}u(x)$$

a. Show that

$$f_Z(z|H_1) = 10e^{-10z}u(z)$$

and

$$f_Z(z|H_2) = 2.5\left(e^{-2z} - e^{-10z}\right)u(z)$$

b. Find the likelihood ratio $\Lambda(Z)$.

c. If $P(H_1) = \frac{1}{3}, P(H_2) = \frac{2}{3}, c_{12} = c_{21} = 7$, and $c_{11} = c_{22} = 0$, find the threshold for a Bayes test.

d. Show that the likelihood ratio test for part (c) can be reduced to

$$Z \underset{H_1}{\overset{H_2}{\gtrless}} \gamma$$

Find the numerical value of γ for the Bayes test of part (c).

e. Find the risk for the Bayes test of part (c).

f. Find the threshold for a Neyman–Pearson test with P_F less than or equal to 10^{-3}. Find P_D for this threshold.

g. Reducing the Neyman-Pearson test of part (f) to the form

$$Z \underset{H_1}{\overset{H_2}{\gtrless}} \gamma$$

find P_F and P_D for arbitrary γ. Plot the ROC.

10.2. Consider a two-hypothesis decision problem where

$$f_Z(z|H_1) = \frac{\exp\left(-\frac{1}{2}z^2\right)}{\sqrt{2\pi}} \quad \text{and}$$

$$f_Z(z|H_2) = \frac{1}{2}\exp(-|z|)$$

a. Find the likelihood ratio $\Lambda(Z)$.

b. Letting the threshold η be arbitrary, find the decision regions R_1 and R_2 illustrated in Figure 10.1. Note that both R_1 and R_2 cannot be connected regions for this problem; that is, they will involve a multiplicity of line segments.

10.3. Assume that data of the form $Z = S + N$ are observed where S and N are independent, Gaussian random variables representing signal and noise, respectively, with zero means and variances σ_s^2 and σ_n^2. Design a likelihood ratio test for each of the following cases. Describe the decision regions in each case and explain your results.

a. $c_{11} = c_{22} = 0;\ \ c_{21} = c_{12};\ \ p_0 = q_0 = \frac{1}{2}$.

b. $c_{11} = c_{22} = 0;\ \ c_{21} = c_{12};\ \ p_0 = \frac{1}{4}; q_0 = \frac{3}{4}$.

c. $c_{11} = c_{22} = 0;\ \ c_{21} = \frac{1}{2}c_{12};\ \ p_0 = q_0 = \frac{1}{2}$.

d. $c_{11} = c_{22} = 0;\ \ c_{21} = 2c_{12};\ \ p_0 = q_0 = \frac{1}{2}$.

Hint: Note that under either hypothesis, Z is a zero-mean Gaussian random variable. Consider what the variances are under hypothesis H_1 and H_2, respectively.

10.4. Referring to Problem 10.3, find general expressions for the probabilities of false alarm and detection for each case. Assume that $c_{12} = 1$ in all cases. Numerically evaluate them for the cases where $\sigma_n^2 = 9$ and $\sigma_s^2 = 16$. Evaluate the risk.

Section 10.2

10.5. Show that ordinary three-dimensional vector space satisfies the properties listed in the subsection entitled Structure of Signal Space in Section 10.2, where $x(t)$ and $y(t)$ are replaced by vectors **A** and **B**.

10.6. For the following vectors in 3-space with x, y, z components as given, evaluate their magnitudes and the cosine of the angle between them ($\hat{i}$, $\hat{j}$, and $\hat{k}$ are the orthogonal unit vectors along the x, y, and z axes, respectively):

a. $\mathbf{A} = \hat{i} + 3\hat{j} + 2\hat{k}$; $\mathbf{B} = 5\hat{i} + \hat{j} + 3\hat{k}$;

b. $\mathbf{A} = 6\hat{i} + 2\hat{j} + 4\hat{k}$; $\mathbf{B} = 2\hat{i} + 2\hat{j} + 2\hat{k}$;

c. $\mathbf{A} = 4\hat{i} + 3\hat{j} + \hat{k}$; $\mathbf{B} = 3\hat{i} + 4\hat{j} + 5\hat{k}$;

d. $\mathbf{A} = 3\hat{i} + 3\hat{j} + 2\hat{k}$; $\mathbf{B} = -\hat{i} - 2\hat{j} + 3\hat{k}$.

10.7. Show that the scalar-product definitions given by (10.43) and (10.44) satisfy the properties listed in the subsection entitled Scalar Product in Section 10.2.

10.8. Using the appropriate definition, (10.43) or (10.44), calculate (x_1, x_2) for each of the following pairs of signals:

a. $e^{-|t|}, 2e^{-3t}u(t)$

b. $e^{-(4+j3)t}u(t), 2e^{-(3+j5)t}u(t)$

c. $\cos(2\pi t), \cos(4\pi t)$

d. $\cos(2\pi t), 5u(t)$

10.9. Let $x_1(t)$ and $x_2(t)$ be two real-valued signals. Show that the square of the norm of the signal $x_1(t) + x_2(t)$ is the sum of the square of the norm of $x_1(t)$ and the square of the norm of $x_2(t)$ if and only if $x_1(t)$ and $x_2(t)$ are orthogonal; that is, $\|x_1 + x_2\|^2 = \|x_1\|^2 + \|x_2\|^2$ if and only if $(x_1, x_2) = 0$. Note the analogy to vectors in three-dimensional space: the Pythagorean theorem applies only to vectors that are orthogonal or perpendicular (zero dot product).

10.10. Evaluate $\|x_1\|$, $\|x_2\|$, $\|x_3\|$, (x_2, x_1), and (x_3, x_1) for the signals in Figure 10.11. Use these numbers to construct a vector diagram and graphically verify that $x_3 = x_1 + x_2$.

10.11. Verify Schwarz's inequality for

$$x_1(t) = \sum_{n=1}^{N} a_n\phi_n(t) \quad \text{and} \quad x_2(t) = \sum_{n=1}^{N} b_n\phi_n(t)$$

where the $\phi_n(t)$s are orthonormal and the a_ns and b_ns are constants.

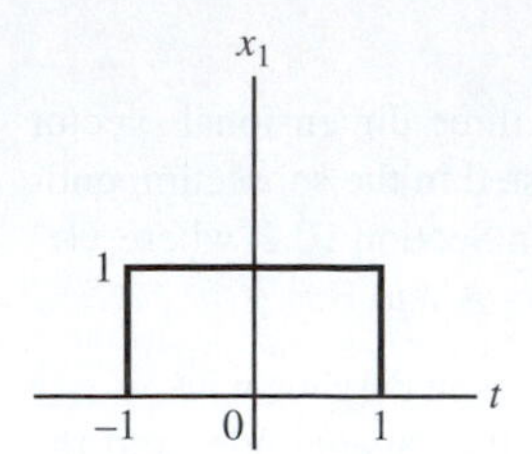

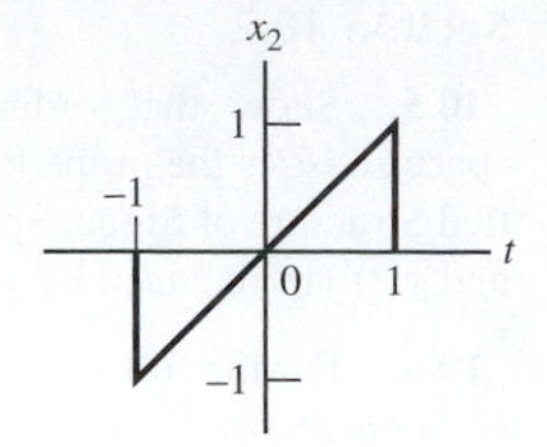

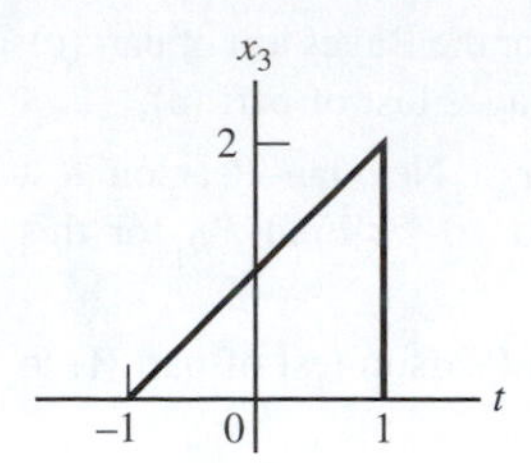

Figure 10.11

10.12. Verify Schwarz's inequality for the 3-space vectors of Problem 10.6.

10.13.

a. Use the Gram–Schmidt procedure to find a set of orthonormal basis functions corresponding to the signals given in Figure 10.12.

b. Express s_1, s_2, and s_3 in terms of the orthonormal basis set found in part (a).

10.14. Use the Gram–Schmidt procedure to find a set of orthonormal basis vectors corresponding the the vector space spanned by the vectors $x_1 = 3\hat{i} + 2\hat{j} - \hat{k}$, $x_2 = -2\hat{i} + 5\hat{j} + \hat{k}$, $\mathbf{x}_3 = 6\hat{i} - 2\hat{j} + 7\hat{k}$, and $x_4 = 3\hat{i} + 8\hat{j} - 3\hat{k}$.

10.15. Consider the set of signals

$$s_i(t) = \begin{cases} \sqrt{2}A\cos(2\pi f_c t + i\pi/4), & 0 \le f_c t \le N \\ 0, & \text{otherwise} \end{cases}$$

where N is an integer and $i = 0,1,2,3,4,5,6,7$.

a. Find an orthonormal basis set for the space spanned by this set of signals.

b. Draw a set of coordinate axes, and plot the locations of $s_i(t)$, $i = 0, 1, 2, \ldots, 7$, after expressing each one as a generalized Fourier series in terms of the basis set found in part (a).

10.16.

a. Using the Gram–Schmidt procedure, find an orthonormal basis set corresponding to the signals

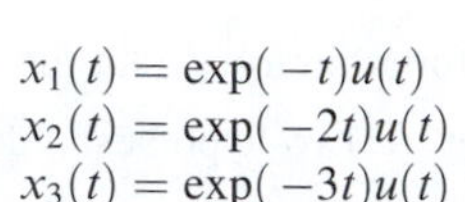

$$x_1(t) = \exp(-t)u(t)$$
$$x_2(t) = \exp(-2t)u(t)$$
$$x_3(t) = \exp(-3t)u(t)$$

b. See if you can find a general formula for the basis set for the signal set $x_1(t) = \exp(-t)u(t), \ldots, x_n(t) = \exp(-nt)u(t)$, where n is an arbitrary integer.

10.17.

a. Find a set of orthonormal basis functions for the signals given below that are defined on the interval $-1 \le t \le 1$:

$$x_1(t) = t$$
$$x_2(t) = t^2$$
$$x_3(t) = t^3$$
$$x_4(t) = t^4$$

b. Attempt to provide a general result for $x_n(t) = t^n$, $-1 \le t \le 1$.

10.18. Use the Gram–Schmidt procedure to find an orthonormal basis for the signal set given below. Express each signal in terms of the orthonormal basis set found.

$$s_1(t) = 1,\ 0 \le t \le 2$$
$$s_2(t) = \cos(\pi t),\ 0 \le t \le 2$$
$$s_3(t) = \sin(\pi t),\ 0 \le t \le 2$$
$$s_4(t) = \sin^2(\pi t),\ 0 \le t \le 2$$

10.19. Rework Example 10.6 for half-cosine pulses given by

$$\phi_k(t) = \Pi\left(\frac{t - k\tau}{\tau}\right)\cos\left[\pi\left(\frac{t - k\tau}{\tau}\right)\right],$$
$$k = 0, \pm 1, \pm 2, \ldots, \pm K$$

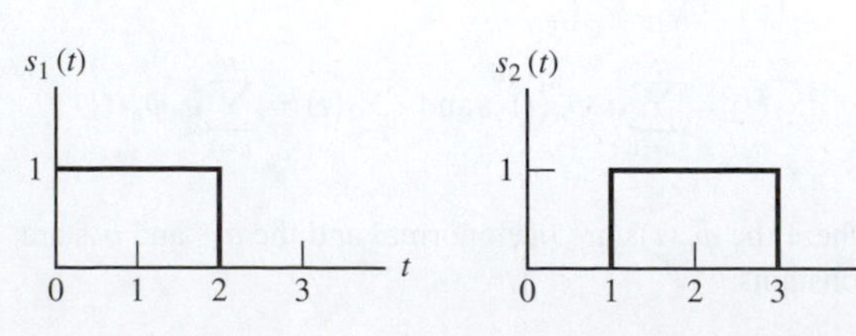

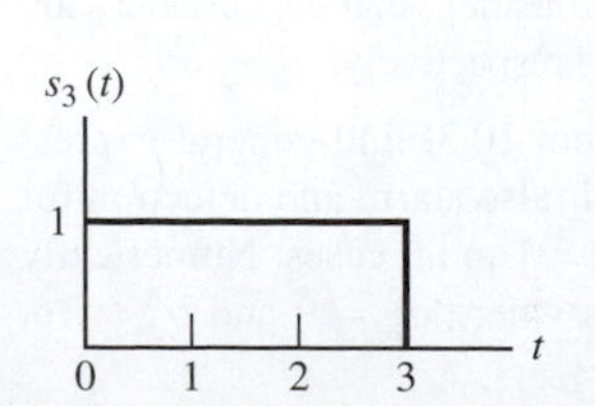

Figure 10.12

Section 10.3

10.20. For M-ary PSK/FSK, the transmitted signal is of the form

$$s_i(t) = A\cos\left(2\pi t + \frac{i\pi}{2}\right), \quad i = 0, 1, 2, 3, \text{ for } 0 \le t \le 1$$

$$s_i(t) = A\cos\left[4\pi t + (i-4)\frac{\pi}{2}\right], \quad i = 4, 5, 6, 7 \text{ for } 0 \le t \le 1$$

a. Find a set of basis functions for this signaling scheme. What is the dimension of the signal space? Express $s_i(t)$ in terms of these basis functions and the signal energy, $E = A^2/2$.

b. Sketch a block diagram of the optimum (minimum P_E) receiver.

c. Write down an expression for the probability of error. Do not attempt to integrate it.

10.21. Consider (10.126) for $M = 2$. Express P_E as a single Q-function. Show that the result is identical to binary, coherent FSK.

10.22. Consider *vertices-of-a-hypercube* signaling, for which the ith signal is of the form

$$s_i(t) = \sqrt{\frac{E_s}{n}} \sum_{k=1}^{n} \alpha_{ik}\phi_k(t), \quad 0 \le t \le T, n$$

in which the coefficients α_{ik} are permuted through the values $+1$ and -1, E_s is the signal energy, and the ϕ_ks are orthonormal. Thus $M = 2^n$, where $n = \log_2 M$ is an integer. For $M = 8$, $n = 3$, the signal points in signal space lie on the vertices of a cube in three-space.

a. Sketch the optimum partitioning of the observation space for $M = 8$.

b. Show that for $M = 8$ the symbol error probability is

$$P_E = 1 - P(C)$$

where

$$P(C) = \left[1 - Q\left(\sqrt{\frac{2E_s}{3N_0}}\right)\right]^3$$

c. Show that for n arbitrary the probability of symbol error is

$$P_E = 1 - P(C)$$

where

$$P(C) = \left[1 - Q\left(\sqrt{\frac{2E_s}{nN_0}}\right)\right]^n$$

d. Plot P_E versus E_s/N_0 for $n = 1, 2, 3, 4$. Compare with Figure 10.7.

Note that with the $\phi_k(t)$s chosen as cosinusoids of frequency spacing $1/T_s$ Hz vertices-of-a-hypercube modulation is the same as OFDM as described in Chapter 9 with BPSK modulation on the subcarriers.

10.23.

a. Referring to the signal set defined by (10.97), show that the minimum possible $\Delta f = \Delta\omega/2\pi$ such that $(s_i, s_j) = 0$ is $\Delta f = \frac{1}{2T_s}$

b. Using the result of part (a), show that for a given time-bandwidth product WT_s the maximum number of signals for M-ary FSK signaling is given by $M = 2WT_s$, where W is the transmission bandwidth and T_s is the signal duration. Use null-to-null bandwidth. Thus $W = \frac{M}{2T_s}$. (Note that this is smaller than the result justified in Chapter 9 because a wider tone spacing was used there.)

c. For vertices-of-a-hypercube signaling, described in Problem 10.22, show that the number of signals grows with WT_s as $M = 2^{2WT_s}$. Thus $W = (\log_2 M)/2T_s$ which grows slower with M than does FSK.

10.24. Go through the steps in deriving (10.142).

10.25. This problem develops the *simplex* signaling set.[12] Consider M orthogonal signals, $s_i(t)$, $i = 0, 1, 2, \ldots, M-1$, each with energy E_s. Compute the average of the signals

$$a(t) \triangleq \frac{1}{M}\sum_{i=0}^{M-1} s_i(t)$$

and define a new signal set

$$s_i'(t) = s_i(t) - a(t), \quad i = 0, 1, 2, \ldots, M-1$$

a. Show that the energy of each signal in the new set is

$$E_s' = E_s\left(1 - \frac{1}{M}\right)$$

b. Show that the correlation coefficient between each signal and another is

$$\rho_{ij} = -\frac{1}{M-1}, \; i, j = 0, 1, \ldots, M-1; \; i \neq j$$

c. Given that the probability of symbol error for an M-ary orthogonal signal set is

$$P_{s,\,\text{othog}} = 1 - \int_{-\infty}^{\infty} \left\{Q\left[-\left(v + \sqrt{\frac{2E_s}{N_0}}\right)\right]\right\}^{M-1} \frac{e^{-v^2/2}}{\sqrt{2\pi}} dv$$

[12]See Simon et al. (1995), pp. 204–205

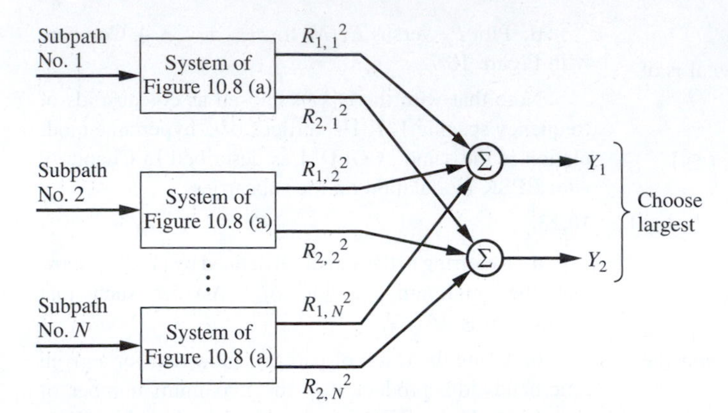

Figure 10.13

write down an expression for the symbol-error probability of the simplex signal set where, from (G.9), $Q(-x) = 1 - Q(x)$.

d. Simplify the expression found in part (c) using the union bound result for the probability of error for an orthogonal signaling set given by (9.67). Plot the symbol-error probability for $M = 2, 4, 8, 16$, and compare with that for coherent M-ary FSK.

10.26. Generalize the fading problem of binary noncoherent FSK signaling to the M-ary case. Let the ith hypothesis be of the form

$$H_i : y(t) = G_i\sqrt{\frac{2E_i}{T_s}}\cos(\omega_i t + \theta_i) + n(t),$$

$$i = 1, 2, \ldots, M;\ 0 \le t \le T_s$$

where G_i is Rayleigh, θ_i is uniform in $[0, 2\pi)$, E_i is the energy of the unperturbed ith signal of duration T_s, and $|\omega_i - \omega_j| \gg T_s^{-1}$, for $i \ne j$, so that the signals are orthogonal. Note that $G_i \cos\theta_i$ and $-G_i \sin\theta_i$ are Gaussian with mean zero; assume their variances to be σ^2.

a. Find the likelihood ratio test, and show that the optimum correlation receiver is identical to the one shown in Figure 10.8(a) with $2M$ correlators, $2M$ squarers, and M summers, where the summer with the largest output is chosen as the best guess (minimum P_E) for the transmitted signal if all E_i are equal. How is the receiver structure modified if the E_i are not equal?

b. Write down an expression for the probability of symbol error.

10.27. Investigate the use of diversity to improve the performance of binary noncoherent FSK signaling over the flat fading Rayleigh channel. Assume that the signal energy E_s is divided equally among N subpaths, all of which fade independently. For equal SNRs in all paths, the optimum receiver is shown in Figure 10.13.

a. Referring to Problem 5.37 of Chapter 5, show that Y_1 and Y_2 are chi-squared random variables under either hypothesis.

b. Show that the probability of error is of the form

$$P_E = \alpha^N \sum_{j=0}^{N-1} \binom{N+j-1}{j} (1-\alpha)^j$$

where

$$\alpha = \frac{\frac{1}{2}N_0}{\sigma^2 E' + N_0} = \frac{1}{2}\frac{1}{1 + \frac{1}{2}(2\sigma^2 E'/N_0)}, \quad E' = \frac{E_s}{N}$$

c. Plot P_E versus SNR $\triangleq 2\sigma^2 E_s/N_0$ for $N = 1, 2, 3, \ldots$, and show that an optimum value of N exists that minimizes P_E for a given SNR.

Section 10.4

10.28. Let an observed random variable Z depend on a parameter λ according to the conditional pdf

$$f_{Z|\Lambda}(z|\lambda) = \begin{cases} \lambda e^{-\lambda z}, & z \ge 0, \lambda > 0 \\ 0, & z < 0 \end{cases}$$

The a priori pdf of λ is

$$f_\Lambda(\lambda) = \begin{cases} \dfrac{\beta^m}{\Gamma(m)} e^{-\beta\lambda}\lambda^{m-1}, & \lambda \ge 0 \\ 0, & \lambda < 0 \end{cases}$$

where β and m are parameters and $\Gamma(m)$ is the gamma function. Assume that m is a positive integer.

a. Find $E\{\lambda\}$ and $\text{var}\{\lambda\}$ before any observations are made; that is, find the mean and variance of λ using $f_\Lambda(\lambda)$.

b. Assume one observation is made. Find $f_{\Lambda|Z}(\lambda|z_1)$ and hence the minimum mean-square error (conditional-mean) estimate of λ and the variance of the estimate. Compare with part (a). Comment on the similarity of $f_\Lambda(\lambda)$ and $f_{\Lambda|Z}(\lambda|z_1)$.

c. Making use of part (b), find the posterior pdf of λ given two observations $f_{\Lambda|Z}(\lambda|z_1, z_2)$. Find the minimum mean-square error estimate of λ based on two observations and its variance. Compare with parts (a) and (b), and comment.

d. Generalize the preceding to the case in which K observations are used to estimate λ.

e. Does the MAP estimate equal the minimum mean-square error estimate?

10.29. For which of the cost functions and posterior pdfs shown in Figure 10.14 will the conditional mean be the Bayes estimate? Tell why or why not in each case.

10.30. Show that the variance of $\widehat{a}_{\text{ML}}(\mathbf{Z})$ given by (10.176) is the result given by (10.177).

10.31. Given K independent measurements $(Z_1, Z_2, \ldots, Z_K)$ of a noise voltage $Z(t)$ at the RF filter output of a receiver:

a. If $Z(t)$ is Gaussian with mean zero and $\text{var}\{\sigma_n^2\}$, what is the ML estimate of the noise variance?

b. Calculate the expected value and variance of this estimate as functions of the true variance.

c. Is this an unbiased estimator?

d. Give a sufficient statistic for estimating the variance of Z.

10.32. Generalize the estimation of a sample of a PAM signal, expressed by (10.205), to the case where the sample value m_0 is a zero-mean Gaussian random variable with variance σ_m^2.

10.33. Consider the reception of a BPSK signal in noise with unknown phase, θ, to be estimated. The two hypotheses may be expressed as

$$H_1: y(t) = A\cos(\omega_c t + \theta) + n(t), 0 \le t \le T_s$$

$$H_2: y(t) = -A\cos(\omega_c t + \theta) + n(t), 0 \le t \le T_s$$

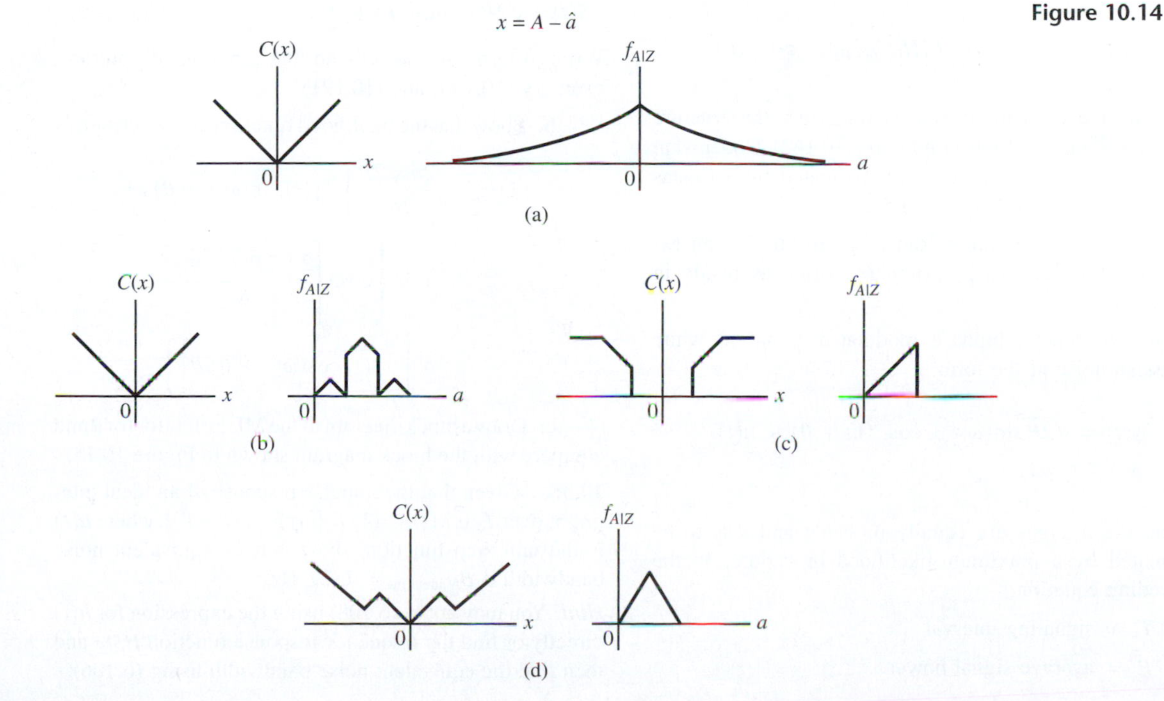

Figure 10.14

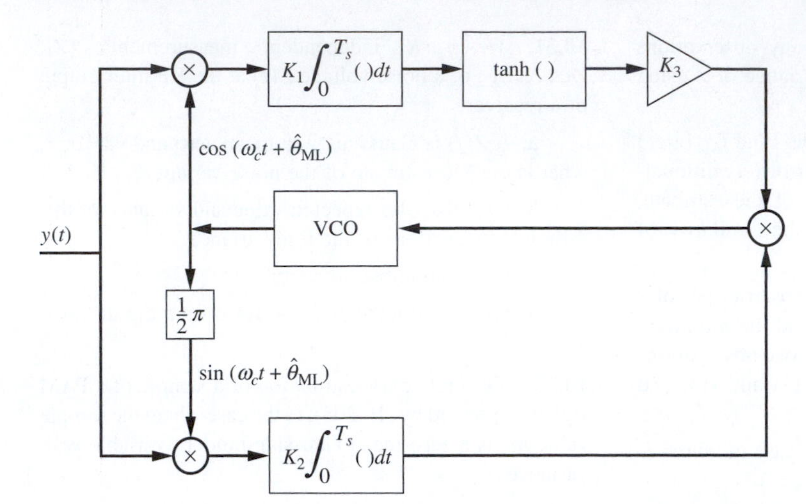

Figure 10.15

where A is a constant and $n(t)$ is white Gaussian noise with single-sided power spectral density N_0, and the hypotheses are equally probable $[P(H_1) = P(H_2)]$.

a. Using ϕ_1 and ϕ_2 as given by (10.190) and (10.191) as basis functions, write expressions for

$$f_{\mathbf{Z}|\theta,H_i}(z_1, z_2|\theta, H_i),\ i = 1,2$$

b. Noting that

$$f_{\mathbf{Z}|\theta,}(z_1, z_2|\theta) = \sum_{i=1}^{2} P(H_i) f_{\mathbf{Z}|\theta,H_i}(z_1, z_2|\theta, H_i)$$

show that the ML estimator can be realized as the structure shown in Figure 10.15 by employing (10.162). Under what condition(s) is this structure approximated by a Costas loop? (See Chapter 3, Figure 3.57.)

c. Apply the Cramer–Rao inequality to find an expression for $\text{var}\{\hat{\theta}_{ML}\}$. Compare with the result in Table 8.1.

10.34. Assume a biphase modulated signal in white Gaussian noise of the form

$$y(t) = \sqrt{2P}\sin\left(\omega_c t \pm \cos^{-1} m + \theta\right) + n(t),$$
$$0 \le t \le T_s$$

where the $\pm$ signs are equally probable and θ is to be estimated by a maximum-likelihood procedure. In the preceding equation,

$T_s =$ signaling interval
$P =$ average signal power
$\omega_c =$ carrier frequency(rad/s)
$m =$ modulation index
$\theta =$ RF phase (rad)

Let the double-sided power spectral density of $n(t)$ be $\frac{1}{2}N_0$.

a. Show that the signal portion of $y(t)$ can be written as

$$S(t) = \sqrt{2P}m\sin(\omega_c t + \theta) \pm \sqrt{2P}\sqrt{1-m^2}\cos(\omega_c t + \theta)$$

Write in terms of the orthonormal functions ϕ_1 and ϕ_2, given by (10.190) and (10.191).

b. Show that the likelihood function can be written as

$$L(\theta) = \frac{2m\sqrt{2P}}{N_0}\int_0^{T_s} y(t)\sin(\omega_c t + \theta)\,dt + \ln\left\{\cosh\left[\frac{2\sqrt{2P(1-m^2)}}{N_0} \times \int_0^{T_s} y(t)\cos(\omega_c t + \theta)dt\right]\right\}$$

c. Draw a block diagram of the ML estimator for θ and compare with the block diagram shown in Figure 10.15.

10.35. Given that the impulse response of an ideal integrator over T_s is $h(t) = (1/T)[u(t) - u(t-T)]$, where $u(t)$ is the unit step function, show that its equivalent noise bandwidth is $B_{N,\text{ideal int}} = 1/2T$ Hz.

Hint: You may apply (6.108) using the expression for $h(t)$ directly or find the frequency response function $H(f)$ and then find the equivalent noise bandwidth using (6.106).

Computer Exercises

10.1. In practical communications systems and radar systems we desire that the system operate with a probability of detection that is nearly one and a probability of false alarm that is only slightly greater than zero. For this case we have interest in a very small portion of the total receiver operating characteristic. With this in mind, make the necessary changes in the in the MATLAB program of Computer Example 10.1 so that the region of interest for practical operation is displayed. This region of interest is defined as $P_D \geq 0.95$ and $P_F \leq 0.01$. Determine the values of the parameter d that give operation in this region.

10.2. Write a computer program to make plots of σ_p^2 versus K, the number of observations, for fixed ratios of σ_A^2/σ_n^2, thus verifying the conclusions drawn at the end of Example 10.8.

10.3. Write a computer simulation of the PLL estimation problem. Do this by generating two independent Gaussian random variables to form Z_1 and Z_2 given by (10.194). Thus for a given θ, form the left-hand side of (10.199). Call the first value θ_0. Estimate the next value of θ, call it θ_1, from the algorithm.

$$\theta_1 = \theta_0 + \epsilon \tan^{-1}\left(\frac{Z_{2,0}}{Z_{1,0}}\right)$$

where $Z_{1,0}$ and $Z_{2,0}$ are the first values of Z_1 and Z_2 generated and ϵ is a parameter to be varied (choose the first value to be 0.01). Generate two new values of Z_1 and Z_2 (call them $Z_{1,1}$ and $Z_{2,1}$) and form the next estimate according to

$$\theta_2 = \theta_1 + \epsilon \tan^{-1}\left(\frac{Z_{2,1}}{Z_{1,1}}\right)$$

Continue in this fashion, generating several values of θ_i. Plot the θ_is versus i, the sequence index, to determine if they seem to converge toward zero phase. Increase the value of ϵ by a factor of 10 and repeat. Can you relate the parameter ϵ to a PLL parameter (see Chapter 3)? This is an example of Monte Carlo simulation.

CHAPTER 11

INFORMATION THEORY AND CODING

Information theory provides a different perspective for evaluating the performance of a communication system in that the performance can be compared with the theoretically best system for a given bandwidth and SNR. Significant insight into the performance characteristics of a communication system can often be gained through the study of information theory. More explicitly, information theory provides a quantitative measure of the information contained in message signals and allows us to determine the capability of a system to transfer this information from source to destination. Coding, a major application area of information theory, will be briefly presented in this chapter. We make no attempt in this chapter to be complete or rigorous. Rather we present an overview of basic ideas and illustrate these ideas through simple examples. We hope that students who study this chapter will be motivated to study these topics in more detail.

Information theory provides us with the performance characteristics of an *ideal*, or optimum, communication system. The performance of an ideal system provides a meaningful basis against which to compare the performance of the realizable systems studied in previous chapters. Performance characteristics of ideal systems illustrate the gain in performance that can be obtained by implementing more complicated transmission and detection schemes.

Motivation for the study of information theory is provided by *Shannon's coding theorem*, which can be stated as follows: If a source has an information rate less than the channel capacity, there exists a coding procedure such that the source output can be transmitted over the channel with an arbitrarily small probability of error. This is a powerful result. Shannon tells us that transmission and reception can be accomplished with *negligible* error, even in the presence of noise. An understanding of this process called *coding* and an understanding of its impact on the design and performance of communication systems require an understanding of several basic concepts of information theory.

We will see that there are two basic applications of coding. The first of these is referred to as source coding. Through the use of source coding, redundancy can be removed from message signals so that each transmitted symbol carries maximum information. In addition, through the use of channel, or error-correcting, coding, systematic redundancy can be induced into the transmitted signal so that errors caused by imperfect practical channels can be corrected.

11.1 BASIC CONCEPTS

Consider a hypothetical classroom situation occurring early in a course at the end of a class period. The professor makes one of the following statements to the class:

A. I shall see you next period.

B. My colleague will lecture next period.

C. Everyone gets an A in the course, and there will be no more class meetings.

What is the relative information conveyed to the students by each of these statements, assuming that there had been no previous discussion on the subject? Obviously, there is little information conveyed by statement (A), since the class would normally assume that their regular professor would lecture; that is, the probability $P(A)$ of the regular professor lecturing is nearly unity. Intuitively, we know that statement (B) contains more information, and the probability of a colleague lecturing $P(B)$ is relatively low. Statement (C) contains a vast amount of information for the entire class, and most would agree that such a statement has a very low probability of occurrence in a typical classroom situation. It appears that the lower the probability of a statement, or event, the greater is the information conveyed by that statement. Stated another way, the students' surprise on hearing a statement appears to be a good measure of the information contained in that statement. Information is defined consistent with this intuitive example.

11.1.1 Information

Let x_j be an event that occurs with probability $p(x_j)$. If we are told that event x_j has occurred, we say that we have received

$$I(x_j) = \log_a\left(\frac{1}{p(x_j)}\right) = -\log_a p(x_j) \tag{11.1}$$

units of information. This definition is consistent with the previous example since $I(x_j)$ increases as $p(x_j)$ decreases. Note that $I(x_j)$ is nonnegative since $0 \leq p(x_j) \leq 1$. The base of the logarithm in (11.1) is arbitrary and determines the units by which information is measured. R. V. Hartley,[1] who first suggested the logarithmic measure of information in 1928, used logarithms to the base 10 since tables of base 10 logarithms were widely available, and the resulting measure of information was the *hartley*. Today it is standard to use logarithms to the base 2, and the unit of information is the binary unit, or *bit*. If logarithms to the base e are used, the corresponding unit is the *nat*, or natural unit.

There are several reasons for us to adopt the base 2 logarithm to measure information. The simplest random experiment that one can imagine is an experiment with two equally likely outcomes. Flipping an unbiased coin is a common example. Knowledge of each outcome has associated with it one bit of information since the logarithm base is 2 and the probability of each outcome is 0.5. Since the digital computer is a binary machine, each logical 0 and each logical 1 has associated with it one bit of information, assuming that each of these logical states are equally likely.

[1]Hartley (1928)

EXAMPLE 11.1

Consider a random experiment with 16 equally likely outcomes. The information associated with each outcome is

$$I(x_j) = -\log_2 \frac{1}{16} = \log_2 16 = 4 \text{ bits} \tag{11.2}$$

where j ranges from 1 to 16. The information is associated with each outcome is greater than one bit, since the random experiment generates more than two equally likely outcomes, and therefore, the probability of each outcome is less than one-half.

■

11.1.2 Entropy

In general, the average information associated with the outcomes of an experiment is of interest rather than the information associated with a particular output. The average information associated with a discrete random variable X is defined as the entropy $H(X)$. Thus

$$H(X) = E\{I(x_j)\} = -\sum_{j=1}^{n} p(x_j) \log_2 p(x_j) \tag{11.3}$$

where n is the total number of possible outcomes. Entropy can be regarded as average uncertainty and therefore achieves a maximum when all outcomes are equally likely.

EXAMPLE 11.2

For a binary source let $p(1) = \alpha$ and $p(0) = 1 - \alpha = \beta$. From (11.3), the entropy is

$$H(\alpha) = -\alpha \log_2 \alpha - (1 - \alpha) \log_2 (1 - \alpha) \tag{11.4}$$

This is sketched in Figure 11.1. We note that if $\alpha = \frac{1}{2}$, each symbol is equally likely, and our uncertainty, and therefore the entropy, is a maximum. If $\alpha \neq \frac{1}{2}$, one of the two symbols becomes more likely than the other. Therefore uncertainty, and consequently the entropy, decreases. If α is equal to zero or one, our uncertainty is zero, since we know exactly which symbol will occur.

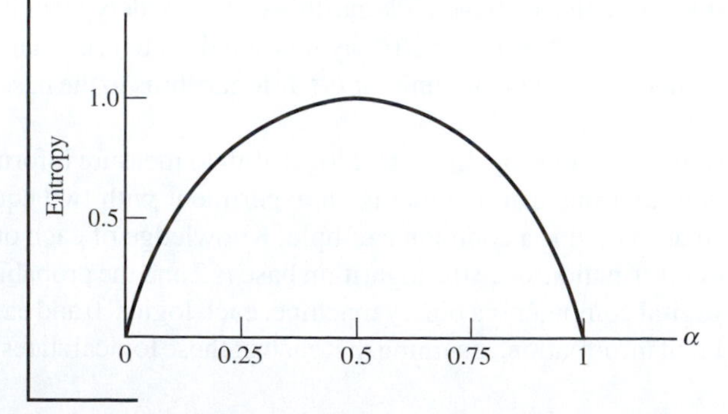

Figure 11.1
Entropy of a binary source.

■

From Example 11.2 we conclude, at least for the special case illustrated in Figure 11.1, that the entropy function has a maximum, which occurs when all probabilities are equal. This fact is of sufficient importance to warrant a more complete derivation. Assume that a chance experiment has n possible outcomes and that p_n is a dependent variable depending on the other probabilities. Thus

$$p_n = 1 - (p_1 + p_2 + \cdots + p_k + \cdots + p_{n-1}) \tag{11.5}$$

where p_j is concise notation for $p(x_j)$. The entropy associated with the chance experiment is

$$H = -\sum_{i=1}^{n} p_i \log_2 p_i \tag{11.6}$$

In order to find the maximum value of entropy, the entropy is differentiated with respect to p_k, holding all probabilities constant except p_k and p_n. This gives a relationship between p_k and p_n that yields the maximum value of H. Since all derivatives are zero except the ones involving p_k and p_n,

$$\frac{dH}{dp_k} = \frac{d}{dp_k}(-p_k \log_2 p_k - p_n \log_2 p_n) \tag{11.7}$$

Using (11.5) and

$$\frac{d}{dx}\log_a u = \frac{1}{u}\log_a e \frac{du}{dx} \tag{11.8}$$

gives

$$\frac{dH}{dp_k} = -p_k \frac{1}{p_k}\log_2 e - \log_2 p_k + p_n \frac{1}{p_n}\log_2 e + \log_2 p_n \tag{11.9}$$

or

$$\frac{dH}{dp_k} = \log_2 \frac{p_n}{p_k} \tag{11.10}$$

which is zero if $p_k = p_n$. Since p_k is arbitrary,

$$p_1 = p_2 = \cdots = p_n = \frac{1}{n} \tag{11.11}$$

To show that the preceding condition yields a maximum and not a minimum, note that when $p_1 = 1$ and all other probabilities are zero, the entropy is zero. From (11.6), the case where all probabilities are equal yields $H = \log_2 n$.

11.1.3 Discrete Channel Models

Throughout most of this chapter we will assume the communications channel to be *memoryless*. For such channels, the channel output at a given time is a function of the channel input *at that time* and is not a function of previous channel inputs. Discrete memoryless channels are completely specified by the set of conditional probabilities that relate the probability of each output state to the input probabilities. An example illustrates the technique. A diagram of a channel with two inputs and three outputs is illustrated in Figure 11.2. Each possible input-to-output path is indicated along with a conditional probability p_{ij}, which is concise notation for $p(y_j|x_i)$. Thus p_{ij} is the conditional probability of output y_j given input x_i and is called a *channel transition probability*. The complete set of transition probabilities defines the channel. In this

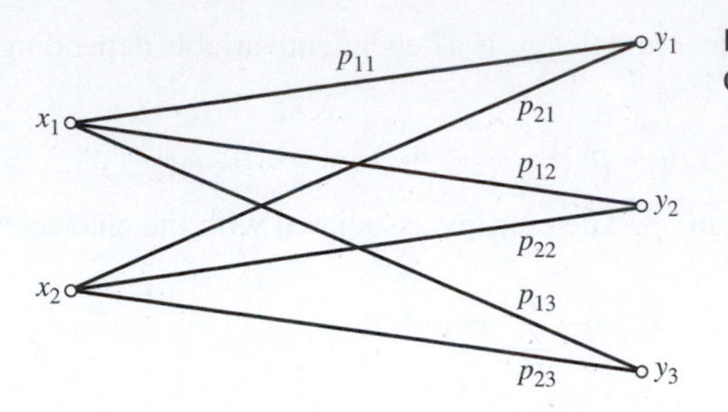

Figure 11.2
Channel diagram.

chapter, the transition probabilities are assumed constant. However, in many commonly encountered situations, the transition probabilities are time varying. An example is the wireless mobile channel in which the transmitter–receiver distance is changing with time.

We can see from Figure 11.2 that the channel is completely specified by the set of transition probabilities. Accordingly, the memoryless channel illustrated in Figure 11.2 can be defined by the matrix of transition probabilities $[P(Y|X)]$, where

$$[P(Y|X)] = \begin{bmatrix} p(y_1|\,x_1) & p(y_2|\,x_1) & p(y_3|\,x_1) \\ p(y_1|\,x_2) & p(y_2|\,x_2) & p(y_3|\,x_2) \end{bmatrix} \tag{11.12}$$

Since each channel input results in some output, each row of $[P(Y|X)]$ must sum to unity. We refer to the matrix of transition probabilities as the *channel matrix*.

The channel matrix is useful in deriving the output probabilities given the input probabilities. For example, if the input probabilities $P(X)$ are represented by the row matrix

$$[P(X)] = [p(x_1) \quad p(x_2)] \tag{11.13}$$

then

$$[P(Y)] = [p(y_1) \quad p(y_2) \quad p(y_3)] \tag{11.14}$$

which is computed by

$$[P(Y)] = [P(X)][P(Y|X)] \tag{11.15}$$

If $[P(X)]$ is written as a diagonal matrix, (11.15) yields a matrix $[P(X, Y)]$. Each element in the matrix has the form $p(x_i)p(y_j|x_i)$ or $p(x_j, y_j)$. This matrix is known as the *joint probability matrix*, and the term $p(x_i, y_j)$ is the joint probability of transmitting x_i and receiving y_j.

EXAMPLE 11.3

Consider the binary input–output channel shown in Figure 11.3. The matrix of transition probabilities is

$$[P(Y|X)] = \begin{bmatrix} 0.7 & 0.3 \\ 0.4 & 0.6 \end{bmatrix} \tag{11.16}$$

If the input probabilities are $P(x_1) = 0.5$ and $P(x_2) = 0.5$, the output probabilities are

$$[P(Y)] = [0.5 \quad 0.5] \begin{bmatrix} 0.7 & 0.3 \\ 0.4 & 0.6 \end{bmatrix} = [0.55 \quad 0.45] \tag{11.17}$$

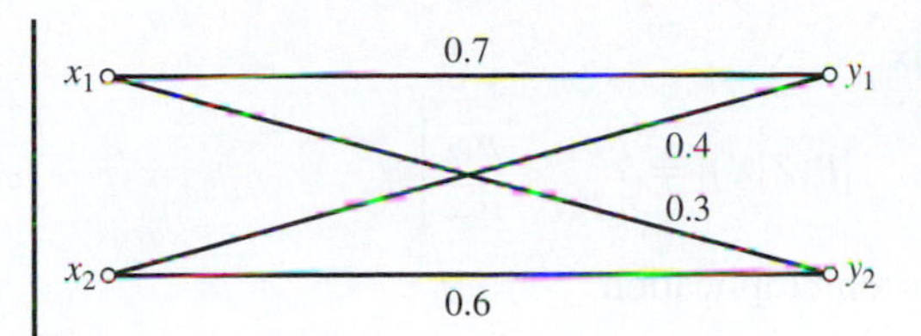

Figure 11.3
Binary channel.

and the joint probability matrix for the channel is

$$[P(X,Y)] = \begin{bmatrix} 0.5 & 0 \\ 0 & 0.5 \end{bmatrix} \begin{bmatrix} 0.7 & 0.3 \\ 0.4 & 0.6 \end{bmatrix} = \begin{bmatrix} 0.35 & 0.15 \\ 0.2 & 0.3 \end{bmatrix} \tag{11.18}$$

■

As we first observed in Chapter 9, a binary satellite communication system can often be represented by the cascade combination of two binary channels. This is illustrated in Figure 11.4(a), in which the first binary channel represents the uplink and the second binary channel represents the downlink. These channels can be combined as shown in Figure 11.4(b).

By determining all possible paths from x_i to z_j, it is clear that the following probabilities define the overall channel illustrated in Figure 11.4(b):

$$p_{11} = \alpha_1\beta_1 + \alpha_2\beta_3 \tag{11.19}$$

$$p_{12} = \alpha_1\beta_2 + \alpha_2\beta_4 \tag{11.20}$$

$$p_{21} = \alpha_3\beta_1 + \alpha_4\beta_3 \tag{11.21}$$

$$p_{22} = \alpha_3\beta_2 + \alpha_4\beta_4 \tag{11.22}$$

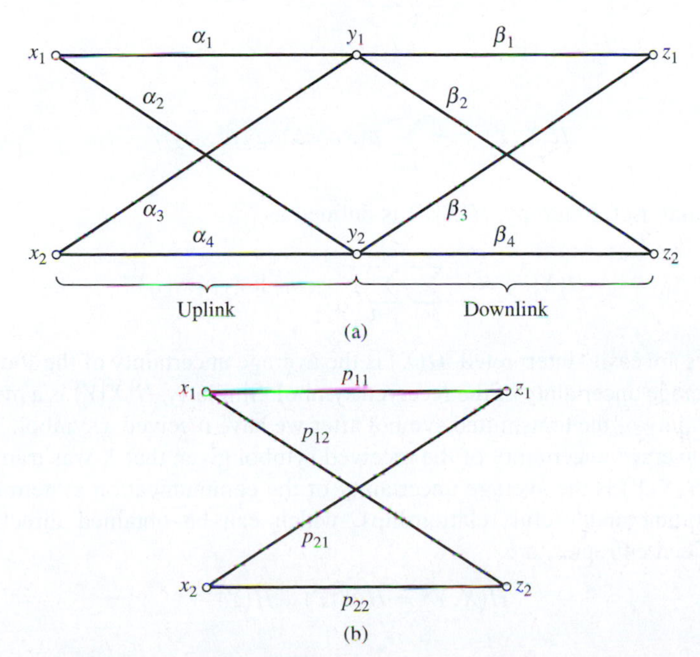

Figure 11.4
Two-hop satellite system.
(a) Binary satellite channel.
(b) Composite satellite channel.

Thus the overall channel matrix

$$[P(Z|X)] = \begin{bmatrix} p_{11} & p_{12} \\ p_{21} & p_{22} \end{bmatrix} \tag{11.23}$$

can be represented by the matrix multiplication

$$[P(Z|X)] = \begin{bmatrix} \alpha_1 & \alpha_2 \\ \alpha_3 & \alpha_4 \end{bmatrix} \begin{bmatrix} \beta_1 & \beta_2 \\ \beta_3 & \beta_4 \end{bmatrix} \tag{11.24}$$

For a two-hop communications system, the right-hand side of the preceding expression is simply the uplink channel matrix multiplied by the downlink channel matrix.

11.1.4 Joint and Conditional Entropy

Using the input probabilities $p(x_i)$, the output probabilities $p(y_j)$, the transition probabilities $p(y_j|x_i)$, and the joint probabilities $p(x_i, y_j)$, we can define several different entropy functions for a channel with n inputs and m outputs. These are

$$H(X) = -\sum_{i=1}^{n} p(x_i) \log_2 p(x_i) \tag{11.25}$$

$$H(Y) = -\sum_{j=1}^{m} p(y_j) \log_2 p(y_j) \tag{11.26}$$

$$H(Y|X) = -\sum_{i=1}^{n}\sum_{j=1}^{m} p(x_i, y_j) \log_2 p(y_j|x_i) \tag{11.27}$$

and

$$H(X, Y) = -\sum_{j=1}^{m} p(x_i, y_j) \log_2 p(x_i, y_j) \tag{11.28}$$

An important and useful entropy, $H(X|Y)$ is defined as

$$H(X|Y) = -\sum_{i=1}^{n}\sum_{j=1}^{m} p(x_i, y_j) \log_2 p(x_i|y_j) \tag{11.29}$$

These entropies are easily interpreted. $H(X)$ is the average uncertainty of the source, whereas $H(Y)$ is the average uncertainty of the received symbol. Similarly, $H(X|Y)$ is a measure of our average uncertainty of the transmitted symbol after we have received a symbol. The function $H(Y|X)$ is the average uncertainty of the received symbol given that X was transmitted. The joint entropy $H(X, Y)$ is the average uncertainty of the communication system as a whole.

Two important and useful relationships, which can be obtained directly from the previously defined entropies, are

$$H(X, Y) = H(X|Y) + H(Y) \tag{11.30}$$

and

$$H(X,Y) = H(Y|X) + H(X) \tag{11.31}$$

These are developed in Problem 11.13.

11.1.5 Channel Capacity

Consider for a moment an observer at the channel output. The observer's average uncertainty concerning the channel input will have value $H(X)$ before the reception of an output, and this average uncertainty of the input will typically decrease when the output is received. In other words, $H(X|Y) \leq H(X)$. The decrease in the average uncertainty of the transmitted signal when the output is received is a measure of the average information transmitted through the channel. This is defined as *mutual information* $I(X;Y)$. Thus

$$I(X;Y) = H(X) - H(X|Y) \tag{11.32}$$

It follows from (11.30) and (11.31) that we can also write (11.32) as

$$I(X;Y) = H(Y) - H(Y|X) \tag{11.33}$$

It should be observed that mutual information is a function of the source probabilities as well as of the channel transition probabilities.

It is easy to show mathematically that

$$H(X) \geq H(X|Y) \tag{11.34}$$

by showing that

$$H(X|Y) - H(X) = -I(X;Y) \leq 0 \tag{11.35}$$

Substitution of (11.29) for $H(X|Y)$ and (11.25) for $H(X)$ allows us to write $-I(X;Y)$ as

$$-I(X;Y) = -\sum_{i=1}^{n}\sum_{j=1}^{m} p(x_i, y_j) \log_2\left(\frac{p(x_i)}{p(x_i|y_j)}\right) \tag{11.36}$$

Since

$$\log_2 x = \frac{\ln x}{\ln 2} \tag{11.37}$$

and

$$\frac{p(x_i)}{p(x_i|y_j)} = \frac{p(x_i)p(y_j)}{p(x_i, y_j)} \tag{11.38}$$

we can write $-I(X;Y)$ as

$$-I(X;Y) = \frac{1}{\ln 2}\sum_{i=1}^{n}\sum_{j=1}^{m} p(x_i, y_j) \ln\left(\frac{p(x_i)p(y_j)}{p(x_i, y_j)}\right) \tag{11.39}$$

In order to carry the derivation further, we need the often used inequality

$$\ln x \leq x - 1 \tag{11.40}$$

which we can easily derive by considering the function

$$f(x) = \ln x - (x - 1) \tag{11.41}$$

The derivative of $f(x)$

$$\frac{df}{dx} = \frac{1}{x} - 1 \tag{11.42}$$

is equal to zero at $x = 1$. It follows that $f(1) = 0$ is the maximum value of $f(x)$, since we can make $f(x)$ less than zero by choosing x sufficiently large (>1). Using the inequality (11.40) in (11.39) results in

$$-I(X;Y) \le \frac{1}{\ln 2} \sum_{i=1}^{n} \sum_{j=1}^{m} p(x_i, y_j) \left(\frac{p(x_i)p(y_j)}{p(x_i, y_j)} - 1 \right) \tag{11.43}$$

which yields

$$-I(X;Y) \le \frac{1}{\ln 2} \left[\sum_{i=1}^{n} \sum_{j=1}^{m} p(x_i)p(y_j) - \sum_{i=1}^{n} \sum_{j=1}^{m} p(x_i, y_j) \right] \tag{11.44}$$

Since both the double sums equal 1, we have the desired result

$$-I(X;Y) \le 0 \qquad \text{or} \qquad I(X;Y) \ge 0 \tag{11.45}$$

Thus we have shown that mutual information is always nonnegative and, consequently, $H(X) \ge H(X|Y)$.

The *channel capacity* C is defined as the maximum value of mutual information, which is the maximum average information per symbol that can be transmitted through the channel for each channel use. Thus

$$C = \max[I(X;Y)] \tag{11.46}$$

The maximization is with respect to the source probabilities, since the transition probabilities are fixed by the channel. However, the channel capacity is a function of only the channel transition probabilities, since the maximization process eliminates the dependence on the source probabilities. The following examples illustrate the method.

EXAMPLE 11.4

The channel capacity of the discrete noiseless channel illustrated in Figure 11.5 is easily determined. We start with

$$I(X;Y) = H(X) - H(X|Y)$$

and write

$$H(X|Y) = -\sum_{i=1}^{n} \sum_{j=1}^{m} p(x_i, y_j) \log_2 p\,(x_i | y_j) \tag{11.47}$$

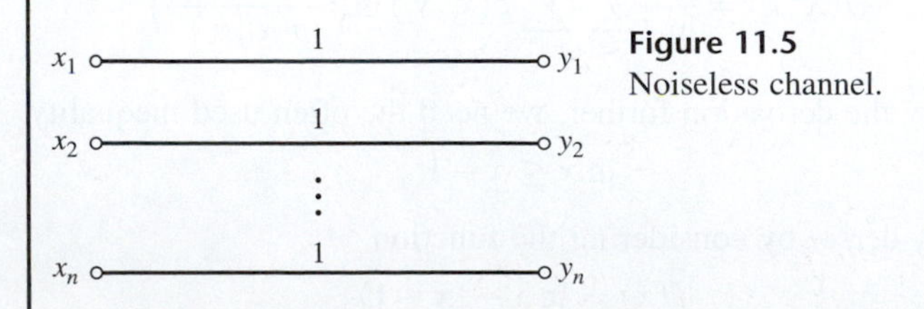

Figure 11.5
Noiseless channel.

For the noiseless channel, all $p(x_i, y_j)$ and $p(x_i|y_j)$ are zero unless $i = j$. For $i = j$, $p(x_i|y_j)$ is unity. Thus $H(X|Y)$ is zero for the noiseless channel, and

$$I(X;Y) = H(X) \tag{11.48}$$

We have seen that the entropy of a source is maximum if all source symbols are equally likely. Thus

$$C = \sum_{i=1}^{n} \frac{1}{n} \log_2 n = \log_2 n \tag{11.49}$$

■

EXAMPLE 11.5

An important and useful channel model is the *binary symmetric channel* (BSC) illustrated in Figure 11.6. We determine the capacity by maximizing

$$I(X;Y) = H(Y) - H(Y|X)$$

where

$$H(Y|X) = -\sum_{i=1}^{2}\sum_{j=1}^{2} p(x_i, y_j) \log_2 p(x_i|y_j) \tag{11.50}$$

Using the probabilities defined in Figure 11.6, we obtain

$$\begin{aligned} H(Y|X) = & -\alpha p \log_2 p - (1-\alpha)p \log_2 p \\ & -\alpha q \log_2 q - (1-\alpha)q \log_2 q \end{aligned} \tag{11.51}$$

or

$$H(Y|X) = -p \log_2 p - q \log_2 q \tag{11.52}$$

Thus

$$I(X;Y) = H(Y) + p\ \log_2 p + q \log_2 q \tag{11.53}$$

which is maximum when $H(Y)$ is maximum. Since the system output is binary, $H(Y)$ is a maximum when each output has a probability of $\frac{1}{2}$. Note that for a BSC equally likely outputs are for equally likely inputs. Since the maximum value of $H(Y)$ for a binary channel is unity, the channel capacity is

$$C = 1 + p \log_2 p + q \log_2 q = 1 - H(p) \tag{11.54}$$

where $H(p)$ is defined in (11.4).

The capacity of a BSC is sketched in Figure 11.7. As expected, if $p = 0$ or 1, the channel output is completely determined by the channel input, and the capacity is 1 bit per symbol. If p is equal to 0.5, an input symbol yields either output symbol with equal probability, and the capacity is zero.

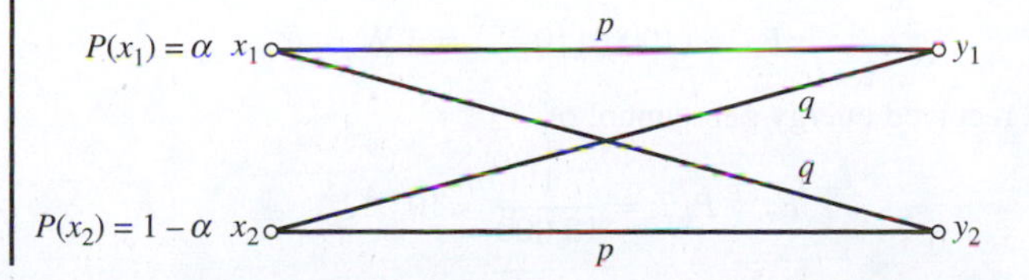

Figure 11.6
Binary symmetric channel.

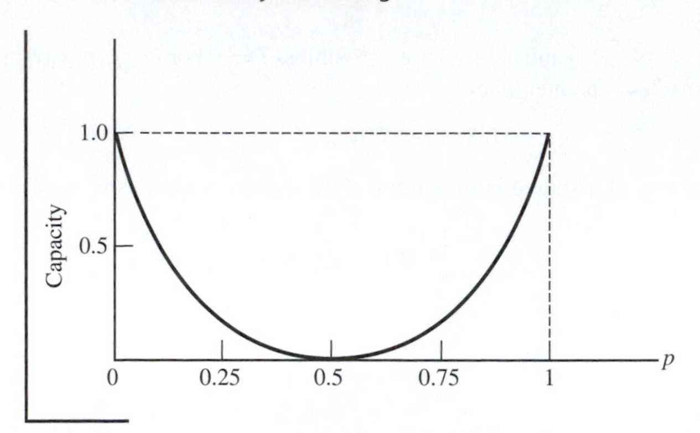

Figure 11.7
Capacity of a binary symmetric channel.

It is worth noting that the capacity of the channel illustrated in Figure 11.6 is most easily found by starting with (11.32), while the capacity of the channel illustrated in Figure 11.6 is most easily found starting with (11.33). Choosing the appropriate expression for $I(X;Y)$ can often save considerable effort. It sometimes takes insight and careful study of a problem to choose the expression for $I(X;Y)$ that yields the capacity with minimum computational effort.

The error probability P_E of a binary symmetric channel is easily computed. From

$$P_E = \sum_{i=1}^{2} p(e|x_i)p(x_i) \tag{11.55}$$

where $p(e|x_i)$ is the error probability given input x_i, we have

$$P_E = qp(x_1) + qp(x_2) = q[p(x_1) + p(x_2)] \tag{11.56}$$

Thus

$$P_E = q$$

which states that the *unconditional error probability* P_E is equal to the conditional error probability $p(y_j|x_i)$, $i \neq j$.

In Chapter 8 we showed that P_E is a decreasing function of the energy of the received symbols. Since the symbol energy is the received power multiplied by the symbol period, it follows that *if the transmitter power is fixed, the error probability can be reduced by decreasing the source rate*. This can be accomplished by removing the redundancy at the source through a process called *source coding*.

EXAMPLE 11.6

In Chapter 8 we showed that for binary coherent FSK systems, the probability of symbol error is the same for each transmitted symbol. Thus, a BSC model is a suitable model for FSK transmission. In this example we determine the channel matrix assuming that the transmitter power is 1000 W, the attenuation in the channel from transmitter to receiver input is 30 dB, the source rate r is 10,000 symbols per second, and that the noise power spectral density N_0 is 2×10^{-5} W/Hz.

Since the channel attenuation is 30 dB, the signal power P_R at the input to the receiver is

$$P_R = (1000)(10^{-3}) = 1 \text{ W} \tag{11.57}$$

This corresponds to a received energy per symbol of

$$E_s = P_R T = \frac{1}{10{,}000} = 10^{-4} \text{ J} \tag{11.58}$$

In Chapter 8 we saw that the error probability for a coherent FSK receiver is

$$P_E = Q\left(\sqrt{\frac{E_s}{N_0}}\right) \tag{11.59}$$

which, with the given values, yields $P_E = 0.0127$. Thus, the channel matrix is

$$[P(Y|X)] = \begin{bmatrix} 0.9873 & 0.0127 \\ 0.0127 & 0.9873 \end{bmatrix} \tag{11.60}$$

It is interesting to compute the change in the channel matrix resulting from a moderate reduction in source symbol rate with all other parameters held constant. If the source symbol rate is reduced 25% to 7500 symbols per second, the received energy per symbol becomes

$$E_s = \frac{1}{7500} = 1.333 \times 10^{-4}\ \text{J} \tag{11.61}$$

With the other given parameters, the symbol-error probability becomes $P_E = 0.0049$, which yields the channel matrix

$$[P(Y|X)] = \begin{bmatrix} 0.9951 & 0.0049 \\ 0.0049 & 0.9951 \end{bmatrix} \tag{11.62}$$

Thus the 25% reduction in source symbol rate results in an improvement of the system symbol-error probability by a factor of almost 3. In Section 11.2 we will investigate a technique that sometimes allows the source symbol rate to be reduced without reducing the source information rate. ■

■ 11.2 SOURCE CODING

We determined in the preceding section that the information from a source producing symbols according to some probability scheme could be described by the entropy $H(X)$. Since entropy has units of bits per symbol, we also must know the symbol rate in order to specify the source information rate in bits per second. In other words, the source information rate R_s is given by

$$R_s = rH(X)\ \text{bps} \tag{11.63}$$

where $H(X)$ is the source entropy in bits per symbol and r is the symbol rate in symbols per second.

Let us assume that this source is the input to a channel with capacity C bits per symbol or SC bits per second, where S is the available symbol rate for the channel. An important theorem of information theory, Shannon's *noiseless coding theorem*, as is stated as follows: *Given a channel and a source that generates information at a rate less than the channel capacity, it is possible to code the source output in such a manner that it can be transmitted through the channel.* A proof of this theorem is beyond the scope of this introductory treatment of information theory and can be found in any of the standard information theory textbooks.[2] However, we demonstrate the theorem by a simple example.

[2]See for example, Gallagher (1968).

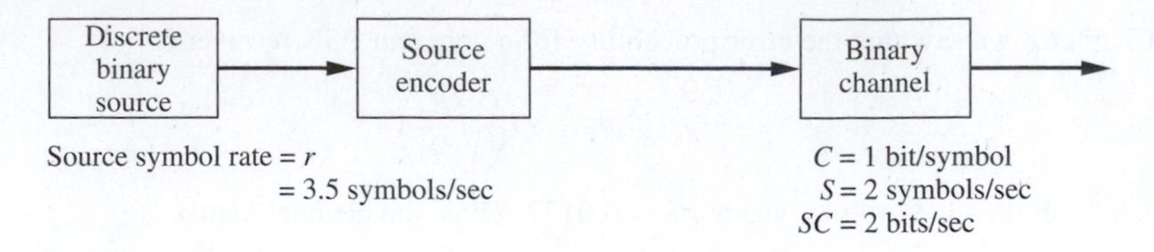

Figure 11.8
Transmission scheme.

11.2.1 An Example of Source Coding

Let us consider a discrete binary source that has two possible outputs A and B that have probabilities 0.9 and 0.1, respectively. Assume also that the source rate r is 3.5 symbols per second. The source output is input to a binary channel that can transmit a binary 0 or 1 at a rate of 2 symbols per second with negligible error, as shown in Figure 11.8. Thus, from Example 11.5 with $p = 1$, the channel capacity is 1 bit per symbol, which, in this case, is an information rate of 2 bits per second.

It is clear that the source *symbol* rate is greater than the channel capacity, so the source symbols cannot be placed directly into the channel. However, the source entropy is

$$H(X) = -0.1 \log_2 0.1 - 0.9 \log_2 0.9 = 0.469 \text{ bits/symbol} \tag{11.64}$$

which corresponds to a source *information* rate of

$$rH(X) = 3.5(0.469) = 1.642 \text{ bps} \tag{11.65}$$

Thus, the *information* rate is less than the channel capacity, so transmission is possible.

Transmission is accomplished by the process called *source coding*, whereby code words are assigned to n-symbol groups of source symbols. The shortest code word is assigned to the most probable group of source symbols, and the longest code word is assigned to the least probable group of source symbols. Thus source coding decreases the average symbol rate, which allows the source to be matched to the channel. The n-symbol groups of source symbols are known as the order n extension of the original source.

Table 11.1 illustrates the first-order extension of the original source. Clearly, the symbol rate at the coder output is equal to the symbol rate of the source. Thus the symbol rate at the channel input is still larger than the channel can accommodate.

The second-order extension of the original source is formed by taking the source symbols $n = 2$ at a time, as illustrated in Table 11.2. The average word length $\overline{L}$ is

$$\overline{L} = \sum_{i=1}^{2^n} p(x_i) l_i = 1.29 \tag{11.66}$$

Table 11.1 First-Order Extension

Source symbol	Symbol probability $P(\cdot)$	Code word	l_i	$P(\cdot)l_i$
A	0.9	0	1	0.9
B	0.1	1	1	0.1
				$\overline{L} = 1.0$

Table 11.2 Second-Order Source Extension

Source symbol	Symbol probability $P(\cdot)$	Code word	l_i	$P(\cdot)l_i$
AA	0.81	0	1	0.81
AB	0.09	10	2	0.18
BA	0.09	110	3	0.27
BB	0.01	111	3	0.03

where $p(x_i)$ is the probability of the ith symbol of the extended source and l_i is the length of the code word corresponding to the ith symbol. Since the source is binary, there are 2^n symbols in the extended source output, each of length n. Thus, for the second-order extension

$$\frac{\overline{L}}{n} = \frac{1}{n}\sum P(\cdot)l_i = \frac{1.29}{2} = 0.645 \text{ code symbols/source symbol} \tag{11.67}$$

and the symbol rate at the coder output is

$$r\frac{\overline{L}}{n} = 3.5(0.645) = 2.258 \text{ code symbols/second} \tag{11.68}$$

which is still greater than the 2 symbols per second that the channel can accept. It is clear that the symbol rate has been reduced, and this provides motivation to try again.

Table 11.3 shows the third-order source extension. For this case, the source symbols are grouped three at a time. The average word length $\overline{L}$ is 1.598, and

$$\frac{\overline{L}}{n} = \frac{1}{n}\sum P(\cdot)l_i = \frac{1.598}{3} = 0.533 \text{ code symbols/source symbol} \tag{11.69}$$

The symbol rate at the coder output is

$$r\frac{\overline{L}}{n} = 3.5(0.533) = 1.864 \text{ code symbols/second} \tag{11.70}$$

This rate can be accepted by the channel, and therefore transmission is possible using the third-order source extension.

It is worth noting in passing that if the source symbols appear at a constant rate, the code symbols at the coder output do not appear at a constant rate. As is apparent in Table 11.3, the source output AAA results in a single symbol at the coder output, whereas the source output

Table 11.3 Third-Order Source Extension

Source symbol	Symbol probability $P(\cdot)$	Code word	l_i	$P(\cdot)l_i$
AAA	0.729	0	1	0.729
AAB	0.081	100	3	0.243
ABA	0.081	101	3	0.243
BAA	0.081	110	3	0.243
ABB	0.009	11100	5	0.045
BAB	0.009	11101	5	0.045
BBA	0.009	11110	5	0.045
BBB	0.001	11111	5	0.005

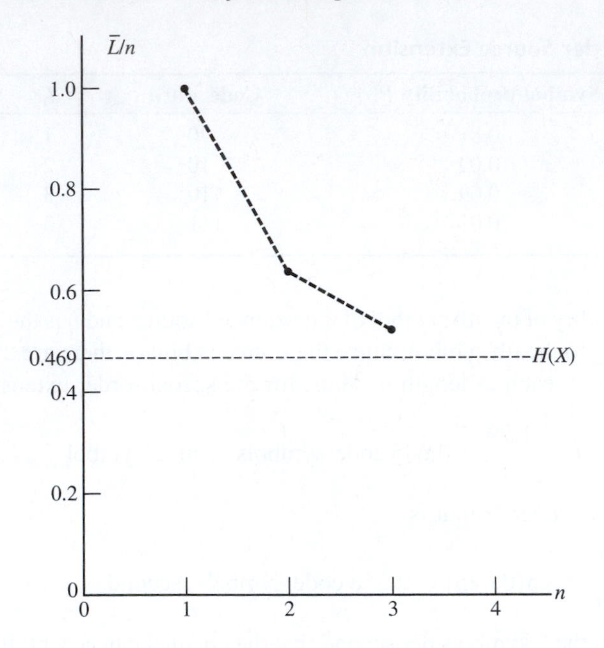

Figure 11.9
Behavior of $\overline{L}/n$.

BBB results in five symbols at the coder output. Thus symbol buffering must be provided at the coder output if the symbol rate into the channel is to be constant.

Figure 11.9 shows the behavior of $\overline{L}/n$ as a function of n. We see that $\overline{L}/n$ always exceeds the source entropy and converges to the source entropy for large n. This is a fundamental result of information theory.

To illustrate the method used to select the code words in this example, we consider the general problem of source coding.

11.2.2 Several Definitions

Before we discuss in detail the method of deriving code words, we pause to make a few definitions that will clarify our work.

Each code word is constructed from an *alphabet* that is a collection of symbols used for communication through a channel. For example, a binary code word is constructed from a two-symbol alphabet, wherein the two symbols are usually taken as 0 and 1. The *word length* of a code word is the number of symbols in the code word.

There are several major subdivisions of codes. For example, a code can be either *block* or *nonblock*. A block code is one in which each block of source symbols is coded into a fixed-length sequence of code symbols. A *uniquely decipherable* code is a block code in which the code words may be deciphered without using spaces. These codes can be further classified as *instantaneous* or *noninstantaneous*, according to whether it is possible to decode each word in sequence without reference to succeeding code symbols. Alternatively, noninstantaneous codes require reference to succeeding code symbols, as illustrated in Table 11.4. It should be remembered that a noninstantaneous code can be uniquely decipherable.

A useful measure of goodness of a source code is the *efficiency*, which is defined as the ratio of the minimum average word length of the code words $\overline{L}_{\min}$ to the average word length of the

Table 11.4 Instantaneous and Noninstantaneous Codes

Source symbols	Code 1 noninstantaneous	Code 2 instantaneous
x_1	0	0
x_2	01	10
x_3	011	110
x_4	0111	1110

code word $\overline{L}$. Thus

$$\text{Efficiency} = \frac{\overline{L}_{\min}}{\overline{L}} = \frac{\overline{L}_{\min}}{\sum_{i=1}^{n} p(x_i) l_i} \tag{11.71}$$

where $p(x_i)$ is the probability of the ith source symbol and l_i is the length of the code word corresponding to the ith source symbol. It can be shown that the minimum average word length is given by

$$\overline{L}_{\min} = \frac{H(X)}{\log_2 D} \tag{11.72}$$

where $H(X)$ is the entropy of the message ensemble being coded and D is the number of symbols in the code alphabet. This yields.

$$\text{Efficiency} = \frac{H(X)}{\overline{L} \log_2 D} \tag{11.73}$$

or

$$\text{Efficiency} = \frac{H(X)}{\overline{L}} \tag{11.74}$$

for a *binary* alphabet. Note that if the efficiency of a code is 100%, the average word length $\overline{L}$ is equal to the entropy, $H(X)$, as implied by Figure 11.9.

11.2.3 Entropy of an Extended Binary Source

In many problems of practical interest, the efficiency is improved by coding the order n source extension. This is exactly the scheme used in the preceding example of source coding. Computation of the efficiency of each of the three schemes used involves calculating the efficiency of the extended source. The efficiency can, of course, be calculated directly, using the symbol probabilities of the extended source, but there is an easier method.

The entropy of the order n extension of a discrete memoryless source, denoted $H(X^n)$, is given by

$$H(X^n) = nH(X) \tag{11.75}$$

This is easily shown by representing a message sequence from the output of the order n source extension as $(i_1, i_2, \ldots, i_n)$, where i_k can take on one of two states with probability p_{i_k}. The entropy of the order n extension of the source is

$$H(X^n) = -\sum_{i_1=1}^{2} \sum_{i_2=1}^{2} \cdots \sum_{i_n=1}^{2} (p_{i_1} p_{i_2} \cdots p_{i_n}) \log_2 (p_{i_1} p_{i_2} \cdots p_{i_n}) \tag{11.76}$$

or

$$H(X^n) = -\sum_{i_1=1}^{2}\sum_{i_2=1}^{2}\cdots\sum_{i_n=1}^{2}(p_{i_1}p_{i_2}\cdots p_{i_n})(\log_2 p_{i_1} + \log_2 p_{i_2} + \cdots + \log_2 p_{i_n}) \quad (11.77)$$

We can write the preceding expression as

$$\begin{aligned} H(X^n) = & -\sum_{i_1=1}^{2} p_{i_1}\log_2 p_{i_1}\left(\sum_{i_2=1}^{2} p_{i_2}\sum_{i_3=1}^{2} p_{i_3}\cdots\sum_{i_n=1}^{2} p_{i_n}\right) \\ & -\left(\sum_{i_1=1}^{2} p_{i_1}\right)\sum_{i_2=1}^{2} p_{i_2}\log_2 p_{i_2}\left(\sum_{i_3=1}^{2} p_{i_3}\cdots\sum_{i_n=1}^{2} p_{i_n}\right)\cdots \\ & -\left(\sum_{i_1=1}^{2} p_{i_1}\sum_{i_2=1}^{2} p_{i_2}\cdots\sum_{i_{n-1}=1}^{2} p_{i_{n-1}}\sum_{i_n=1}^{2} p_{i_n}\right)\sum_{i_n=1}^{2} p_{i_n}\log_2 p_{i_n} \end{aligned} \quad (11.78)$$

Each term in parentheses is equal to 1. Thus

$$H(X^n) = -\sum_{k=1}^{2}\sum_{i_k=1}^{2} p_{i_k}\log_2 p_{i_k} = \sum_{k=1}^{n} H(X) \quad (11.79)$$

which yields

$$H(X^n) = nH(X)$$

The efficiency of the extended source is therefore given by

$$\text{Efficiency} = \frac{nH(X)}{\overline{L}} \quad (11.80)$$

If efficiency tends to 100% as n approaches infinity, it follows that $\overline{L}/n$ tends to the entropy of the extended source. This is exactly the observation made from Figure 11.9.

11.2.4 Shannon–Fano Source Coding

There are several methods of coding a source output so that an instantaneous code results. We consider two such methods here. First, we consider the Shannon–Fano method, which is very easy to apply and usually yields source codes having reasonably high efficiency. In the next subsection we consider the Huffman source coding technique, which yields the source code having the shortest average word length for a given source entropy.

Assume that we are given a set of source outputs that are to be represented in binary form. These source outputs are first ranked in order of nonincreasing probability of occurrence, as illustrated in Figure 11.10. The set is then partitioned into two sets (indicated by line A-A') that are as close to equiprobable as possible, and 0s are assigned to the upper set and ls to the lower set, as seen in the first column of the codewords. This process is continued, each time partitioning the sets with as nearly equal probabilities as possible, until further partitioning is not possible. This scheme will give a 100% efficient code if the partitioning always results in equiprobable sets; otherwise, the code will have an efficiency less than 100%. For this particular example

$$\text{Efficiency} = \frac{H(X)}{\overline{L}} = \frac{2.75}{2.75} = 1 \quad (11.81)$$

since equiprobable partitioning is possible.

Source words	Probability	Code word	(Length)·(Probability)	
X_1	0.2500	00	2 (0.25)	= 0.50
X_2	0.2500	01	2 (0.25)	= 0.50
	A --------- A'			
X_3	0.1250	100	3 (0.125)	= 0.375
X_4	0.1250	101	3 (0.125)	= 0.375
X_5	0.0625	1100	4 (0.0625)	= 0.25
X_6	0.0625	1101	4 (0.0625)	= 0.25
X_7	0.0625	1110	4 (0.0625)	= 0.25
X_8	0.0625	1111	4 (0.0625)	= 0.25
			Average word length = 2.75	

Figure 11.10
Shannon–Fano source coding.

11.2.5 Huffman Source Coding

Huffman coding results in an optimum code in the sense that the Huffman code has the minimum average word length for a source of given entropy. The Huffman technique therefore yields the code having the highest efficiency. We shall illustrate the Huffman coding procedure using the same source output of eight messages previously used to illustrate the Shannon–Fano coding procedure.

Figure 11.11 illustrates the Huffman coding procedure. The source output consists of messages X_1, X_2, X_3, X_4, X_5, X_6, X_7, and X_8. They are listed in order of nonincreasing probability, as was done for Shannon–Fano coding. The first step of the Huffman procedure is to combine the two source messages having the lowest probability, X_7 and X_8.

The upper message, X_7, is assigned a binary 0 as the last symbol in the code word, and the lower message, X_8, is assigned a binary 1 as the last symbol in the code word. The combination of X_7 and X_8 can be viewed as a composite message having a probability equal to the sum of the probabilities of X_7 and X_8, which in this case is 0.1250, as shown. This composite message is

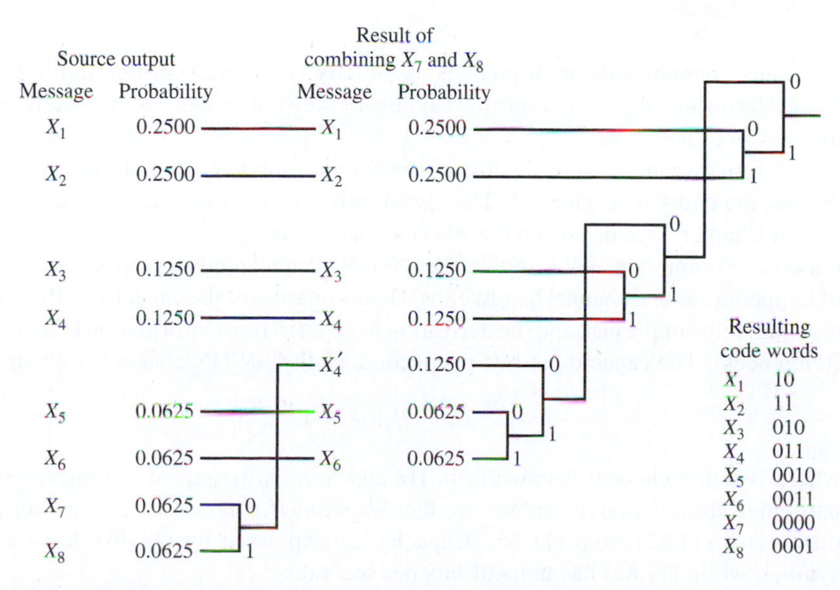

Figure 11.11
Example of Huffman source coding.

denoted X_4'. After this initial step, the new set of messages, denoted $X_1, X_2, X_3, X_4', X_5, X_6$, and X_4' are arranged in order of nonincreasing probability. Note that X_4' could be placed at any point between X_2 and X_5, although it was given the name X_4' because it was placed after X_4. The same procedure is then applied once again. The messages X_5 and X_6 are combined. The resulting composite message is combined with X_4'. This procedure is continued as far as possible. The resulting tree structure is then traced in reverse to determine the code words. The resulting code words are shown in Figure 10.11.

The code words resulting from the Huffman procedure are different from the code words resulting from the Shannon–Fano procedure because at several points the placement of composite messages resulting from previous combinations was arbitrary. The assignment of binary 0s or binary ls to the upper or lower messages was also arbitrary. Note, however, that the average word length is the same for both procedures. This must be the case for the example chosen because the Shannon–Fano procedure yielded 100% efficiency and the Huffman procedure can be no worse. There are cases in which the two procedures do not result in equal average word lengths.

11.3 COMMUNICATION IN NOISY ENVIRONMENTS: BASIC IDEAS

We now turn our attention to methods for achieving reliable communication in the presence of noise by combating the effects of that noise. We undertake our study with a promise from Claude Shannon of considerable success.

Shannon's Theorem (Fundamental theorem of Information Theory)
Given a discrete memoryless channel (each symbol is perturbed by noise independently of all other symbols) with capacity C and a source with positive rate R, where R < C, there exists a code such that the output of the source can be transmitted over the channel with an arbitrarily small probability of error.

Thus Shannon's theorem predicts essentially error-free transmission in the presence of noise. Unfortunately, the theorem tells us only of the existence of codes and tells nothing of how to construct these codes.

Before we start our study of constructing codes for noisy channels, we will take a minute to discuss the continuous channel. This detour will yield insight that will prove useful.

In Chapter 7 we discussed the AWGN channel and observed that, assuming that thermal noise is the dominating noise source, the AWGN channel model is applicable over a wide range of temperatures and channel bandwidths. Determination of the capacity of the AWGN channel is a relatively simple task and the derivation is given in most information theory textbooks (see References). The capacity, in bits per second, of the AWGN channel is given by

$$C_c = B \log_2 \left(1 + \frac{S}{N}\right) \tag{11.82}$$

where B is the channel bandwidth in Hz and S/N is the signal-to-noise power ratio. This particular formulation is known as the *Shannon-Hartley law*. The subscript is used to distinguish (11.82) from (11.46). Capacity, as expressed by (11.46), has units of bits per symbol, while (11.82) has units of bits per second.

The trade-off between bandwidth and SNR can be seen from the *Shannon–Hartley law*. For infinite SNR, which is the noiseless case, the capacity is infinite for any nonzero bandwidth.

We will show, however, that the capacity cannot be made arbitrarily large by increasing bandwidth if noise is present.

In order to understand the behavior of the Shannon–Hartley law for the large-bandwidth case, it is desirable to place (11.82) in a slightly different form. The energy per bit E_b is equal to the bit time T_b multiplied by the signal power S. At capacity, the bit rate R_b is equal to the capacity. Thus $T_b = 1/C_c$ s/bit. This yields, at capacity,

$$E_b = ST_b = \frac{S}{C_c} \tag{11.83}$$

The total noise power in bandwidth B is given by

$$N = N_0 B \tag{11.84}$$

where N_0 is the single-sided noise power spectral density in watts per hertz. The SNR can therefore be expressed as

$$\frac{S}{N} = \frac{E_b}{N_0}\frac{C_c}{B} \tag{11.85}$$

This allows the Shannon–Hartley law to be written in the equivalent form

$$\frac{C_c}{B} = \log_2\left(1 + \frac{E_b}{N_0}\frac{C_c}{B}\right) \tag{11.86}$$

Solving for E_b/N_0 yields

$$\frac{E_b}{N_0} = \frac{B}{C_c}\left(2^{C_c/B} - 1\right) \tag{11.87}$$

This expression establishes performance of the ideal system. For the case in which $B \gg C_c$

$$2^{C_c/B} = e^{(C_c/B)\ln 2} \cong 1 + \frac{C_c}{B}\ln 2 \tag{11.88}$$

where the approximation $e^x \cong 1 + x$, $|x| \ll 1$, has been used. Substitution of (11.88) into (11.87) gives

$$\frac{E_b}{N_0} \cong \ln 2 = -1.6\text{ dB} \qquad B \gg C_c \tag{11.89}$$

Thus, for the ideal system, in which $R_b = C_c$, E_b/N_0 approaches the limiting value of -1.6 dB as the bandwidth grows without bound.

A plot of E_b/N_0, expressed in decibels, as a function of R_b/B is illustrated in Figure 11.12. The ideal system is defined by $R_b = C_c$ and corresponds to (11.87). There are two regions of interest. The first region, for which $R_b < C_c$, is the region in which arbitrarily small error probabilities can be obtained. Clearly this is the region in which we wish to operate. The other region, for which $R_b > C_c$, does not allow the error probability to be made arbitrarily small.

An important trade-off can be deduced from Figure 11.12. If the bandwidth factor R_b/B is large so that the bit rate is much greater than the bandwidth, then a significantly larger value of E_b/N_0 is necessary to ensure operation in the $R_b < C_c$ region than is the case if R_b/B is small. Stated another way, assume that the source bit rate is fixed at R_b bits per second and the available bandwidth is large so that $B \gg R_b$. For this case, operation in the $R_b < C_c$ region requires only that E_b/N_0 is slightly greater than -1.6 dB. The required signal power is

$$S \cong R_b(\ln 2)N_0 W \tag{11.90}$$

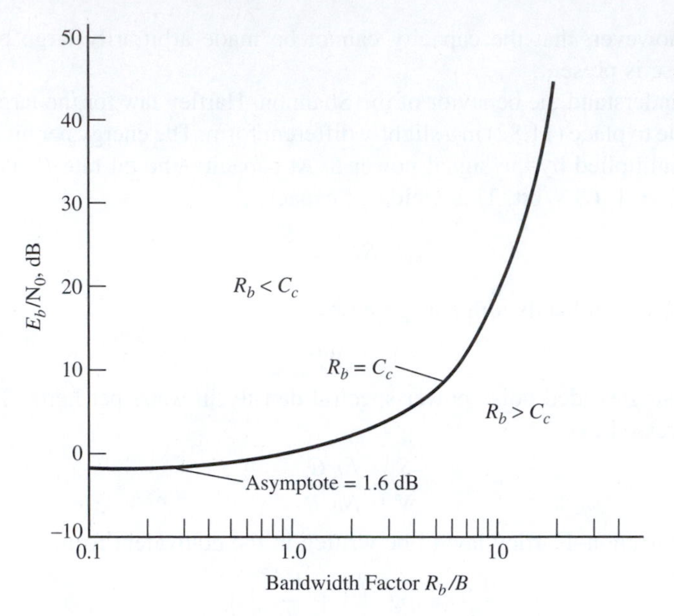

Figure 11.12
$R_b = C_c$ relationship for AWGN channel.

This is the minimum signal power for operation in the $R_b < C_c$ region. Therefore, operation in this region is desired for *power-limited operation.*

Now assume that bandwidth is limited so that $R_b \gg B$. Figure 11.12 shows that a much larger value of E_b/N_0 is necessary for operation in the $R_b < C_c$ region. Thus the required signal power is much greater than that given by (11.90). This is referred to as *bandwidth-limited operation.*

The preceding paragraphs illustrate that, at least in the AWGN channel–where the Shannon–Hartley law applies, a trade-off exists between power and bandwidth. This trade-off is of fundamental importance in the design of communication systems.

Realizing that we can theoretically achieve perfect system performance, even in the presence of noise, we start our search for system configurations that yield the performance promised by Shannon's theorem. Actually one such system was analyzed in Chapter 10. Orthogonal signals were chosen for transmission through the channel, and a correlation receiver structure was chosen for demodulation. The system performance is illustrated in Figure 10.7. Shannon's bound is clearly illustrated.

While are a number of techniques that can be used for combating the effects of noise, so that performance more closer to Shannon's limit is achieved, the most commonly used technique is *forward error correction.* The two major classifications of codes for forward error correction are block codes and convolutional codes. The following two sections treat these techniques.

■ 11.4 COMMUNICATION IN NOISY CHANNELS: BLOCK CODES

Consider a source that produces a serial stream of binary symbols at a rate of R_s symbols per second. Assume that these symbols are grouped into blocks T seconds long, so that each block contains $R_sT = k$ source or information symbols. To each of these k-symbol blocks is added redundant check symbols to produce a code word n symbols long. In a properly designed block code the $n-k$ check symbols provide sufficient information to the decoder to allow for the correction (or detection) of one or more errors that may occur in the transmission of the n symbol code word through the noisy channel. A coder that operates in this manner is said to produce an (n, k) block code. An important parameter of block codes is the code rate, which is defined as

$$R_s = \frac{k}{n} \tag{11.91}$$

since k bits of information are transmitted with each block of n symbols. A design goal is to achieve the required error-correcting capability with the highest possible rate.

Codes can either correct or merely detect errors, depending on the amount of redundancy contained in the check symbols. Codes that can correct errors are known as *error-correcting codes*. Codes that can only detect errors are also useful. As an example, when an error is detected but not corrected, a feedback channel can be used to request a retransmission of the code word found to be in error. We will discuss error-detection and feedback channels in a later section. If errors are more serious than a lost code word, the code word found to be in error can simply be discarded without requesting retransmission.

11.4.1 Hamming Distances and Error Correction

An understanding of how codes can detect and correct errors can be gained from a geometric point of view. A binary code word is a sequence of 1s and 0s that is n symbols in length. The *Hamming weight* $w(s_j)$ of code word s_j is defined as the number of 1s in that code word. The *Hamming distance* $d(s_i, s_j)$ or d_{ij} between code words s_i and s_j is defined as the number of positions in which s_i and s_j differ. It follows that Hamming distance can be written in terms of Hamming weight as

$$d_{ij} = w(s_i \oplus s_j) \tag{11.92}$$

where the symbol $\oplus$ denotes modulo-2 addition, which is binary addition without a carry.

EXAMPLE 11.7

Compute the Hamming distance between $s_1 = 101101$ and $s_2 = 001100$.

Solution

Since

$$101101 \oplus 001100 = 100001$$

we have

$$d_{12} = w(100001) = 2$$

which simply means that s_1 and s_2 differ in 2 positions.

■

A geometric representation of two code words is shown in Figure 11.13. The Cs represent two code words that are distance 5 apart. The code word on the left is the reference code word. The first "x" to the right of the reference represents a binary sequence distance 1 from the reference code word, where distance is understood to denote the Hamming distance.

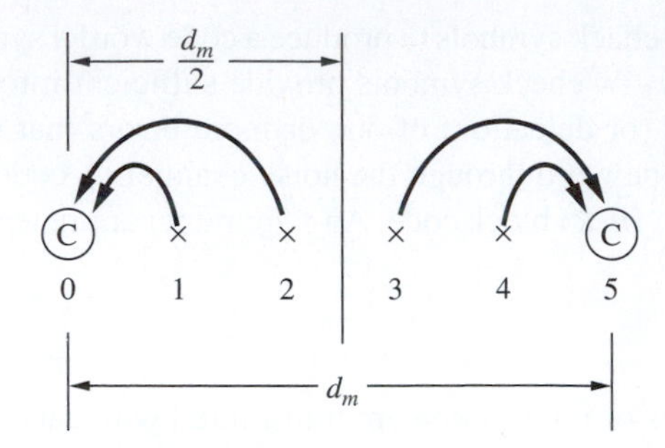

Figure 11.13 Geometric representation of two code words.

The second "x" to the right of the reference code word is distance 2 from the reference, and so on. Assuming that the two code words shown are the closest in Hamming distance of all the code words for a given code, the code is then a distance 5 code. Figure 11.13 illustrates the concept of a minimum-distance decoding, in which a given received sequence is assigned to the code word closest, in Hamming distance, to the received sequence. A minimum distance decoder will therefore assign the received sequences to the left of the vertical line to the code word on the left and the received sequences to the right of the vertical line to the code word on the right, as shown.

We deduce that a minimum-distance decoder can always correct as many as e errors, where e is the largest integer not to exceed

$$\frac{1}{2}(d_m - 1)$$

where d_m is the minimum distance between code words. It follows that if d_m is odd, all received words can be assigned to a code word. However, if d_m is even, a received sequence can lie halfway between two code words. For this case, errors are detected that cannot be corrected.

EXAMPLE 11.8

A code consists of eight code words [0001011, 1110000, 1000110, 1111011, 0110110, 1001101, 0111101, 0000000]. If 1101011 is received, what is the decoded code word?

Solution

The decoded code word is the code word closest in Hamming distance to 1101011. The calculations are

$$\begin{array}{ll} w(0001011 \oplus 1101011) = 2 & w(0110110 \oplus 1101011) = 5 \\ w(1110000 \oplus 1101011) = 4 & w(1001101 \oplus 1101011) = 3 \\ w(1000110 \oplus 1101011) = 4 & w(0111101 \oplus 1101011) = 4 \\ w(1111011 \oplus 1101011) = 1 & w(0000000 \oplus 1101011) = 5 \end{array}$$

The the decoded code word is therefore 1111011. ■

11.4.2 Single-Parity-Check Codes

A simple code capable of detecting, but not capable of correcting, single errors is formed by adding one check symbol to each block of k information symbols. This yields a $(k + 1, k)$ code.

Thus the rate is $k/(k+1)$. The added symbol is called a *parity-check symbol*, and it is added so that the Hamming weight of all code words is either odd or even. If the received word contains an *even* number of errors, the decoder will not detect the errors. If the number of errors is *odd*, the decoder will detect that an odd number of errors, most likely one, has been made.

11.4.3 Repetition Codes

The simplest code that allows for correction of errors consists of transmitting each symbol n times, which results in $n-1$ check symbols. This technique produces an $(n, 1)$ code having two code words; one of all 0s and one of all 1s. A received word is decoded as a 0 if the majority of the received symbols are 0s and as a 1 if the majority are ls. This is equivalent to minimum-distance decoding, wherein $\frac{1}{2}(n-1)$ errors can be corrected. Repetition codes have great error-correcting capability if the symbol error probability is low but have the disadvantage of having low rate. For example, if the information rate of the source is R bits per symbol, the rate R_c out of the coder is

$$R_c = \frac{k}{n}R = \frac{1}{n}R \quad \text{bits/symbol} \tag{11.93}$$

The process of repetition coding for a rate $\frac{1}{3}$ repetition code is illustrated in detail in Figure 11.14. The encoder maps the data symbols 0 and 1 into the corresponding code words 000 and 111. There are eight possible received sequences, as shown. The mapping from the transmitted sequence to the received sequence is random, and the statistics of the mapping are determined by the channel characteristics derived in Chapters 8 and 9. The decoder maps the received sequence into one of the two code words by a minimum Hamming distance decoding rule. Each decoded code word corresponds to a data symbol, as shown.

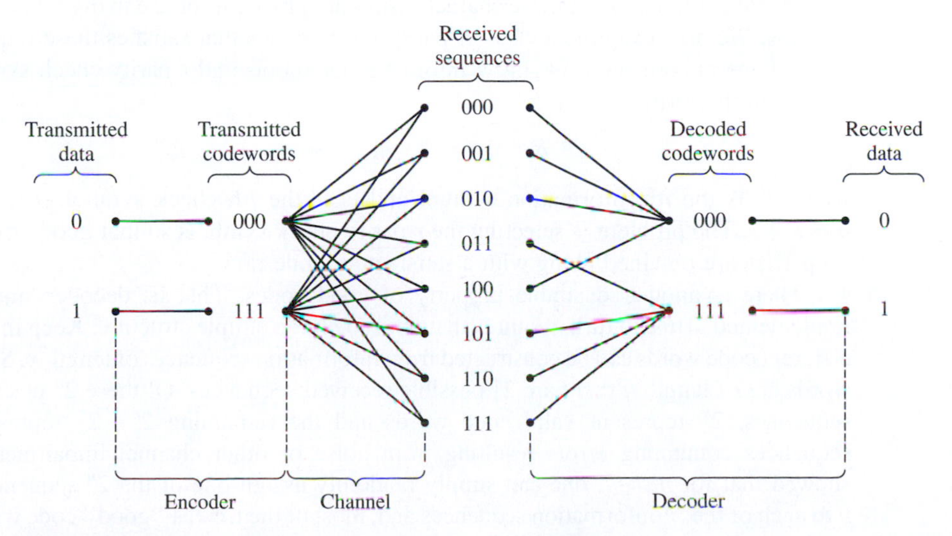

Figure 11.14
Example of rate $\frac{1}{3}$ repetition code.

EXAMPLE 11.9

Investigate the error-correcting capability of a repetition code having a code rate of $\frac{1}{3}$.

Solution

Assume that the code is used with a BSC with a conditional error probability equal to $(1-p)$, that is,

$$P(y_j|x_i) = 1-p, \quad i \neq j \tag{11.94}$$

Each source 0 is encoded as 000, and each source 1 is encoded as 111. An error is made if two or three symbols undergo a change in passing through the channel. Assuming that the source outputs are equally likely, the error probability P_e becomes

$$P_e = 3(1-p)^2 p + (1-p)^3 \tag{11.95}$$

For $1-p = 0.1, P_e = 0.028$, implying an improvement factor of slightly less than 4. For $1-p = 0.01$, the improvement factor is approximately 33. Thus the code performs best when $1-p$ is small.

We will see later that this simple example can be misleading since the error probability, p, with coding is not equal to the error probability, p, without coding. The example implies that performance increases as n, the Hamming distance between the code words, becomes larger. However, as n increases, the code rate decreases. In most cases of practical interest, the *information rate* must be maintained constant, which, for this example, requires that three code symbols be transmitted for each bit of information. An increase in redundancy results in an increase in symbol rate for a given information rate. Thus, coded symbols are transmitted with less energy than uncoded symbols. This changes the channel matrix so that p with coding is greater than p without coding. We will consider this effect in more detail in Computer Examples 11.1 and 11.2.

■

11.4.4 Parity-Check Codes for Single Error Correction

Repetition codes and single-parity-check codes are examples of codes that have either high error-correction capability or high information rate, but not both. Only codes that have a reasonable combination of these characteristics are practical for use in digital communication systems. We now examine a class of parity-check codes that satisfies these requirements.

A general code word having k information symbols and r parity check symbols can be written in the form

$$a_1 \quad a_2 \quad \cdots \quad a_k \quad c_1 \quad c_2 \quad \cdots \quad c_r$$

where a_i is the ith information symbol and c_j is the jth check symbol. The word length $n = k + r$. The problem is selecting the r parity check symbols so that good error-correcting properties are obtained along with a satisfactory code rate.

There is another desirable property of good codes. That is, decoders must be easily implemented. This, in turn, requires that the code has a simple structure. Keep in mind that 2^k different code words can be constructed from information sequences of length k. Since the code words are of length n there are 2^n possible received sequences. Of these 2^n possible received sequences, 2^k represent valid code words and the remaining $2^n - 2^k$ represent received sequences containing errors resulting from noise or other channel impairments. Shannon showed that for $n \gg k$, one can simply randomly assign one of the 2^n sequences of length n to each of the 2^k information sequences and, most of the time, a "good" code will result. The coder then consists of a table with these assignments. The difficulty with this strategy is that the code words lack structure and therefore table lookup is required for decoding. Table lookup is not desirable for most applications since it is slow and usually requires excessive memory. We now examine a structured technique for assigning information sequences to n-symbol code words.

Codes for which the first k symbols of the code word are the information symbols are called *systematic codes*. The $r = n - k$ parity check symbols are chosen to satisfy the r linear equations

$$\begin{aligned} 0 &= h_{11}a_1 \oplus h_{12}a_2 \oplus \cdots \oplus h_{1k}a_k \oplus c_1 \\ 0 &= h_{21}a_1 \oplus h_{22}a_2 \oplus \cdots \oplus h_{2k}a_k \oplus c_2 \\ \vdots \quad &\quad \vdots \qquad\qquad\qquad \vdots \\ 0 &= h_{r1}a_1 \oplus h_{r2}a_2 \oplus \cdots \oplus h_{rk}a_k \oplus c_r \end{aligned} \tag{11.96}$$

Equation (11.96) can be written as

$$[H][T] = [0] \tag{11.97}$$

where $[H]$ is called the parity-check matrix

$$[H] = \begin{bmatrix} h_{11} & h_{12} & \cdots & h_{1k} & 1 & 0 & \cdots & 0 \\ h_{21} & h_{22} & \cdots & h_{2k} & 0 & 1 & \cdots & 0 \\ \vdots & \vdots & \ddots & \vdots & \vdots & \vdots & \ddots & \vdots \\ h_{r1} & h_{r2} & \cdots & h_{rk} & 0 & 0 & \cdots & 1 \end{bmatrix} \tag{11.98}$$

and $[T]$ is the code-word vector

$$[T] = \begin{bmatrix} a_1 \\ a_2 \\ \vdots \\ a_k \\ c_1 \\ \vdots \\ c_r \end{bmatrix} \tag{11.99}$$

Now let the received sequence of length n be denoted $[R]$. If

$$[H][R] \neq [0] \tag{11.100}$$

we know that $[R]$ is not a code word, i.e., $[R] \neq [T]$, and at least one error has been made in the transmission of n symbols through the channel. If

$$[H][R] = [0] \tag{11.101}$$

we know that $[R]$ is a valid code word and, since the probability of symbol error on the channel is assumed small, the received sequence is *most likely* the transmitted code word.

The first step in the coding is to write $[R]$ in the form

$$[R] = [T] \oplus [E] \tag{11.102}$$

where $[E]$ represents the error pattern of length n induced by the channel. The decoding problem essentially reduces to determining $[E]$, since the code word can be reconstructed from $[R]$ and $[E]$. The structure induced by (11.96) defines the decoder.

As the first step in computing $[E]$, we multiply the received word $[R]$ by the parity-check matrix $[H]$. The product is denoted $[S]$. This yields

$$[S] = [H][R] = [H][T] \oplus [H][E] \tag{11.103}$$

Since $[H][T] = [0]$ we have

$$[S] = [H][E] \tag{11.104}$$

The matrix $[S]$ is known as the *syndrome*. Note that we cannot solve (11.104) directly since $[H]$ is not a square matrix and, therefore, the inverse of $[H]$ does not exist.

Assuming that a single error has taken place, the error vector will be of the form

$$[E] = \begin{bmatrix} 0 \\ 0 \\ \vdots \\ 1 \\ \vdots \\ 0 \end{bmatrix}$$

Multiplying $[E]$ by $[H]$ on the left-hand side shows that the syndrome is the ith column of the matrix $[H]$, where the error is in the ith position. The following example illustrates this method. Note that since the probability of symbol error on the channel is assumed small, the error vector having the smallest Hamming weight is the most likely error vector. Error patterns containing single errors are therefore the most likely.

EXAMPLE 11.10

A code has the parity-check matrix

$$[H] = \begin{bmatrix} 1 & 1 & 0 & 1 & 0 & 0 \\ 0 & 1 & 1 & 0 & 1 & 0 \\ 1 & 0 & 1 & 0 & 0 & 1 \end{bmatrix} \tag{11.105}$$

Assuming that 111011 is received, determine if an error has been made, and if so, determine the decoded code word.

Solution

First, we compute the syndrome, remembering that all operations are modulo 2. This gives

$$[S] = [H][R] = \begin{bmatrix} 1 & 1 & 0 & 1 & 0 & 0 \\ 0 & 1 & 1 & 0 & 1 & 0 \\ 1 & 0 & 1 & 0 & 0 & 1 \end{bmatrix} \begin{bmatrix} 1 \\ 1 \\ 1 \\ 0 \\ 1 \\ 1 \end{bmatrix} = \begin{bmatrix} 0 \\ 1 \\ 1 \end{bmatrix} \tag{11.106}$$

Since the syndrome is the third column of the parity-check matrix, the third symbol of the received word is assumed to be in error. Thus the decoded code word is 110011. This can be proved by showing that 110011 has a zero syndrome.

■

We now pause to examine the parity-check code in more detail. It follows from (11.96) and (11.98) that the parity checks can be written as

$$\begin{bmatrix} c_1 \\ c_2 \\ \vdots \\ c_r \end{bmatrix} = \begin{bmatrix} h_{11} & h_{12} & \cdots & h_{1k} \\ h_{21} & h_{22} & \cdots & h_{2k} \\ \vdots & \vdots & \ddots & \vdots \\ h_{r1} & h_{r2} & \cdots & h_{rk} \end{bmatrix} \begin{bmatrix} a_1 \\ a_2 \\ \vdots \\ a_k \end{bmatrix} \tag{11.107}$$

Thus the code-word vector $[T]$ can be written

$$[T] = \begin{bmatrix} a_1 \\ a_2 \\ \vdots \\ a_k \\ c_1 \\ \vdots \\ c_r \end{bmatrix} = \begin{bmatrix} 1 & 0 & \cdots & 0 \\ 0 & 1 & \cdots & 0 \\ \vdots & \vdots & \ddots & \vdots \\ 0 & 0 & \cdots & 1 \\ h_{11} & h_{12} & \cdots & h_{1k} \\ \vdots & \vdots & \ddots & \vdots \\ h_{r1} & h_{r2} & \cdots & h_{rk} \end{bmatrix} \begin{bmatrix} a_1 \\ a_2 \\ \vdots \\ a_k \end{bmatrix} \tag{11.108}$$

or

$$[T] = [G][A] \tag{11.109}$$

where $[A]$ is the vector of k information symbols,

$$[A] = \begin{bmatrix} a_1 \\ a_2 \\ \vdots \\ a_k \end{bmatrix} \tag{11.110}$$

and $[G]$, which is called the *generator matrix*, is

$$[G] = \begin{bmatrix} 1 & 0 & \cdots & 0 \\ 0 & 1 & \cdots & 0 \\ \vdots & \vdots & \vdots & \vdots \\ 0 & 0 & \cdots & 1 \\ h_{11} & h_{12} & \cdots & h_{1k} \\ \vdots & \vdots & \vdots & \vdots \\ h_{r1} & h_{r2} & \cdots & h_{rk} \end{bmatrix} \tag{11.111}$$

The relationship between the generator matrix $[G]$ and the parity-check matrix $[H]$ is apparent if we compare (11.98) and (11.111). If the m by m identity matrix is identified by $[I_m]$ and the matrix $[H_p]$ is defined by

$$[H_p] = \begin{bmatrix} h_{11} & h_{12} & \cdots & h_{1k} \\ h_{21} & h_{22} & \cdots & h_{2k} \\ \vdots & \vdots & \ddots & \vdots \\ h_{r1} & h_{r2} & \cdots & h_{rk} \end{bmatrix} \tag{11.112}$$

it follows that the generator matrix is given by

$$[G] = \begin{bmatrix} I_k \\ \cdots \\ H_p \end{bmatrix} \tag{11.113}$$

and that the parity-check matrix is given by

$$[H] = \left[H_p \vdots I_r \right] \tag{11.114}$$

which establishes the relationship between the generator and parity-check matrices for systematic codes.

Codes defined by (11.111) are referred to as *linear codes*, since the $k + r$ code word symbols are formed as a linear combination of the k information symbols. It is also worthwhile to note that if two different information sequences are summed to give a third sequence, then the code word for the third sequence is the sum of the two code words corresponding to the original two information sequences. This is easily shown. If two information sequences are summed, the resulting vector of information symbols is

$$[A_3] = [A_1] \oplus [A_2] \tag{11.115}$$

The code-word corresponding to $[A_3]$ is

$$[T_3] = [G][A_3] = [G]\{[A_1] \oplus [A_2]\} = [G][A_1] \oplus [G][A_2] \tag{11.116}$$

Since

$$[T_1] = [G][A_1] \tag{11.117}$$

and

$$[T_2] = [G][A_2] \tag{11.118}$$

it follows that

$$[T_3] = [T_1] \oplus [T_2] \tag{11.119}$$

Codes that satisfy this property are known as *group codes.*

11.4.5 Hamming Codes

A Hamming code is a particular parity-check code having distance 3. Since the code has distance 3, all single errors can be corrected. The parity-check matrix for the code has dimensions $2^{n-k} - 1$ by $n - k$ and is very easy to construct. If the i-th column of the matrix $[H]$ is the binary representation of the number i, this code has the interesting property in that, for a single error, the syndrome is the binary representation of the position in error.

EXAMPLE 11.11

Determine the parity-check matrix for a $(7, 4)$ code and the decoded code word if the received word is 1110001.

Solution

Since the ith column of the matrix $[H]$ is the binary representation of i, we have

$$[H] = \begin{bmatrix} 0 & 0 & 0 & 1 & 1 & 1 & 1 \\ 0 & 1 & 1 & 0 & 0 & 1 & 1 \\ 1 & 0 & 1 & 0 & 1 & 0 & 1 \end{bmatrix} \tag{11.120}$$

(Note that this is not a systematic code.) For the received word 1110001, the syndrome is

$$[S] = [H][R] = \begin{bmatrix} 0 & 0 & 0 & 1 & 1 & 1 & 1 \\ 0 & 1 & 1 & 0 & 0 & 1 & 1 \\ 1 & 0 & 1 & 0 & 1 & 0 & 1 \end{bmatrix} \begin{bmatrix} 1 \\ 1 \\ 1 \\ 0 \\ 0 \\ 0 \\ 1 \end{bmatrix} = \begin{bmatrix} 1 \\ 1 \\ 1 \end{bmatrix} \tag{11.121}$$

Thus the error is in the seventh position, and the decoded code word is 1110000.

We note in passing that for the $(7, 4)$ Hamming code, the parity checks are in the first, second, and fourth positions in the code words, since these are the only columns of the parity-check matrix containing only one nonzero element. The columns of the parity-check matrix can be permuted without changing the distance properties of the code. Therefore, the systematic code equivalent to (11.120) is obtained by interchanging columns 1 and 7, columns 2 and 6, and columns 4 and 5. ■

11.4.6 Cyclic Codes

The preceding subsections dealt primarily with the mathematical properties of parity-check codes, and the implementation of parity-check coders and decoders was not discussed. Indeed, if we were to examine the implementation of these devices, we would find that, in general, fairly complex hardware configurations are required. However, there is a class of parity-check codes, known as *cyclic codes*, that are easily implemented using feedback shift registers. A cyclic code derives its name from the fact that a cyclic permutation of any code word produces another code word. For example, if $x_1 x_2 \cdots x_{n-1} x_n$ is a code word, so is $x_n x_1 x_2 \cdots x_{n-1}$. In this section we examine not the underlying theory of cyclic codes but the implementation of coders and decoders. We will accomplish this by means of an example.

An (n, k) cyclic code can easily be generated with an $n - k$ stage shift register with appropriate feedback. The register illustrated in Figure 11.15 produces a $(7, 4)$ cyclic code. The switch is initially in position A, and the shift register stages initially contain all zeros. The $k = 4$ information symbols are then shifted into the coder. As each information symbol arrives, it is routed to the output and added to the value of $S_2 \oplus S_3$. The resulting sum is then placed into the first stage of the shift register. Simultaneously, the contents of S_1 and S_2 are shifted to S_2 and S_3, respectively. After all information symbols have arrived, the switch is moved to position B, and the shift register is shifted $n - k = 3$ times to clear it. On each shift, the sum of S_2 and S_3 appears at the output. This sum added to itself produces a 0 which is fed into S_1. After $n - k$ shifts, a code word has been generated that contains $k = 4$ information symbols and $n - k = 3$ parity-check symbols. It also should be noted that the register contains all 0s so that the coder is ready to receive the next $k = 4$ information symbols.

All $2^k = 16$ code words that can be generated with the example coder are also illustrated in Figure 11.15. The $k = 4$ information symbols, which are the first four symbols of each code word, were shifted into the coder beginning with the left-hand symbol. Also shown in Figure 11.15 are the contents of the register and the output symbol after each shift for the code word 1101.

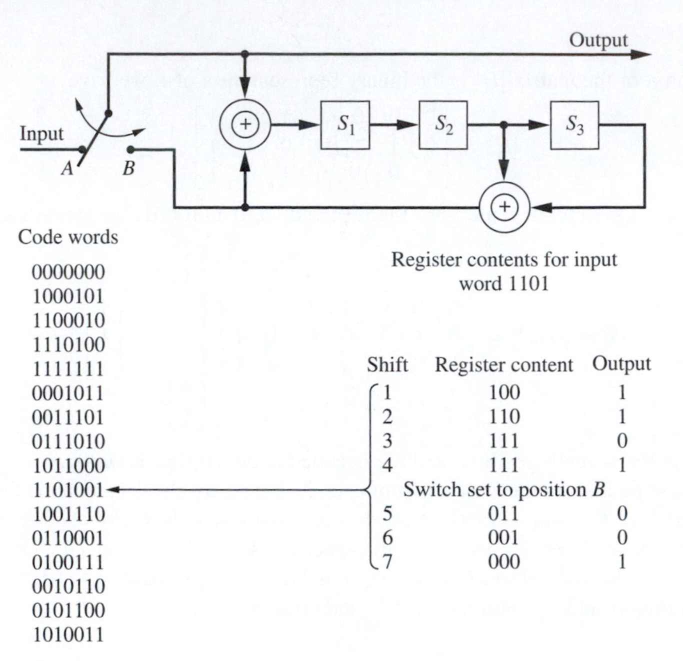

Figure 11.15 Coder for (7, 4) cyclic code.

The decoder for the (7, 4) cyclic code is illustrated in Figure 11.16. The upper register is used for storage, and the lower register and feedback arrangement are identical to the feedback shift register used in the coder. Initially, switch A is closed and switch B is open. The n received symbols are shifted into the two registers. If there are no errors, the lower register will contain all 0s when the upper register is full. The switch positions are then reversed, and the code word that is stored in the upper register is shifted out. This operation is illustrated in Figure 11.16 for the received word 1101001.

If, after the received word is shifted into the decoder the lower register does not contain all 0s, an error has been made. The error is corrected automatically by the decoder, since, when the incorrect symbol appears at the output of the shift register, a 1 appears at the output of the AND gate. This 1 inverts the upper register output and is produced by the sequence 100 in the lower register. The operation is illustrated in Figure 11.16.

Golay Code

The (23,12) Golay code has distance 7 and is therefore capable of correcting three errors in a block of 23 symbols. The rate is close to, but slightly greater than, $\frac{1}{2}$. Adding an additional parity symbol to the (23, 12) Golay code yields the (24, 12) extended Golay code which has distance 8. This allows correction of some, but not all, received sequences having four errors with a slight reduction in rate. The slight reduction in rate, however, has advantages. Since the rate of the extended Golay code is exactly $\frac{1}{2}$, the symbol rate through the channel is precisely twice the information rate. This factor of two difference between symbol rate and information rate frequently simplifies the design of timing circuits. The design of codes capable of correcting multiple errors is beyond the scope of this text. We will, however, consider the performance of the (23, 12) Golay code in an AWGN environment to the performance of a Hamming code in an example to follow.

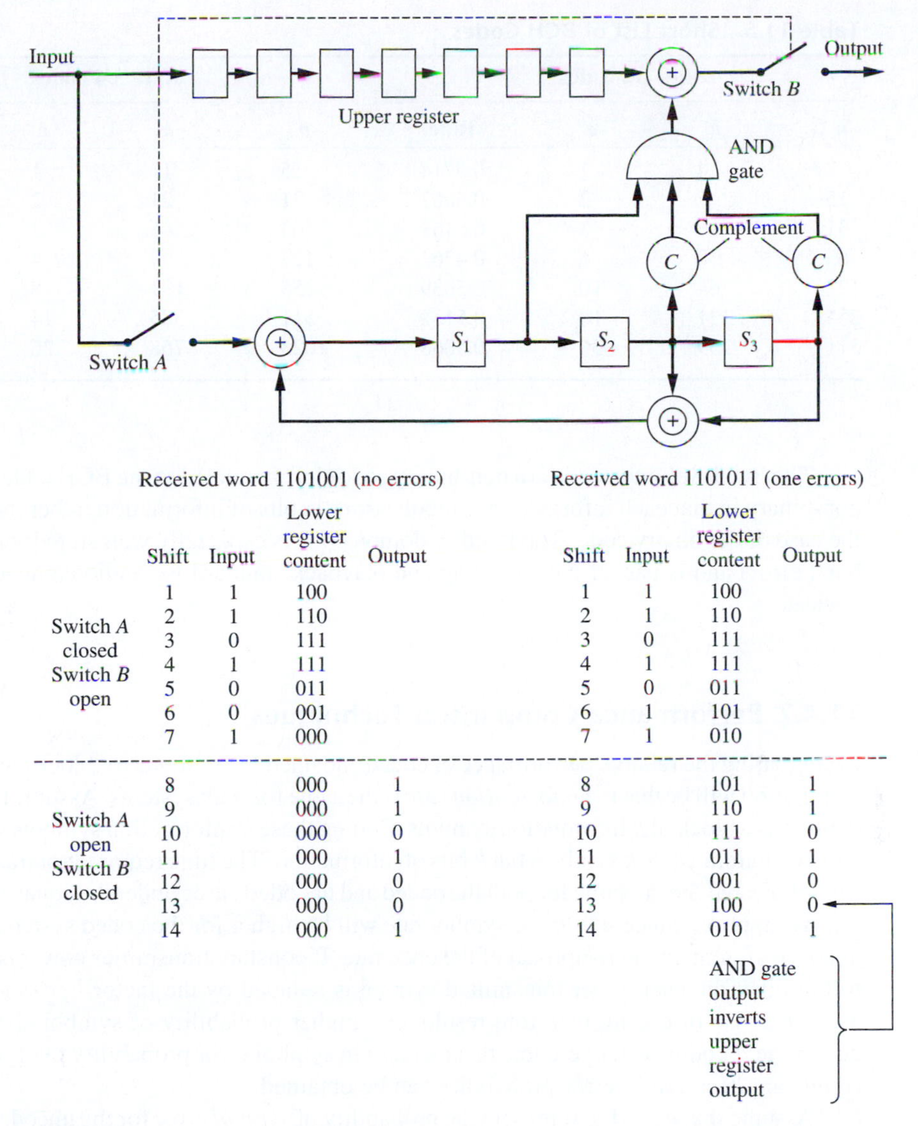

	Received word 1101001 (no errors)				Received word 1101011 (one errors)			
	Shift	Input	Lower register content	Output	Shift	Input	Lower register content	Output
Switch A closed, Switch B open	1	1	100		1	1	100	
	2	1	110		2	1	110	
	3	0	111		3	0	111	
	4	1	111		4	1	111	
	5	0	011		5	0	011	
	6	0	001		6	1	101	
	7	1	000		7	1	010	
Switch A open, Switch B closed	8		000	1	8		101	1
	9		000	1	9		110	1
	10		000	0	10		111	0
	11		000	1	11		011	1
	12		000	0	12		001	0
	13		000	0	13		100	0
	14		000	1	14		010	1

Figure 11.16
Decoder for $(7, 4)$ cyclic code.

Bose–Chaudhuri–Hocquenghem (BCH) Codes and Reed Solomon Codes

The binary codes are very flexible in that they can provide a variety of code rates with a given block length. This is illustrated in Table 11.5 which is a very brief list of a few BCH codes having code rates of approximately $\frac{1}{2}$ and $\frac{3}{4}$.[3] These codes are cyclic codes and therefore both coding and decoding can be accomplished using simple shift-register configurations as described previously.

[3]Tables giving acceptable values of n, k, and e for BCH codes are widely available. An extensive table for $n \leq 1023$ can be found in Lin and Costello (2004).

Table 11.5 Short List of BCH Codes

Rate 1/2 codes				Rate 3/4 codes			
n	k	e	Rate	n	k	e	Rate
7	4	1	0.5714	15	11	1	0.7333
15	7	2	0.4667	31	21	2	0.6774
31	16	3	0.5161	63	45	3	0.7143
63	30	6	0.4762	127	99	4	0.7795
127	64	10	0.5039	255	191	8	0.7490
255	131	18	0.5137	511	385	14	0.7534
511	259	30	0.5068	1023	768	26	0.7507

The Reed-Solomon code is a non-binary code closely related to the BCH code. The code is non-binary in that each information symbol carries m bits of information rather than 1 bit as in the case of the binary code. The Reed–Solomon code is especially well suited for controlling burst errors and is part of the recording and playback standard for audio compact disk (CD) devices.

11.4.7 Performance Comparison Techniques

In comparing the relative performance of coded and uncoded systems for block codes the basic assumption will be that the *information rate* is the same for both systems. Assume that a word is defined as a block of k information symbols. Coding these k information symbols yields a code word containing $n > k$ symbols but k bits of information. The time required for transmission of a word, T_w, will be the same for both the coded and uncoded cases under the equal-information-rate assumption. Since $n > k$ the symbol rate will be higher for the coded system than for the uncoded system by the reciprocal of the code rate. If constant transmitter power is assumed, it follows that the energy per transmitted symbol is reduced by the factor k/n when coding is used. The use of coding therefore results in a higher probability of symbol error. We must determine if coding can overcome this increase in symbol error probability to the extent that a significant decrease in error probability can be obtained.

Assume that q_u and q_c represent the probability of *symbol* error for the uncoded and coded systems, respectively. Also assume that P_{eu} and P_{ec} are the *word*-error probabilities for the uncoded and coded systems. The word error probability for the uncoded system is computed by observing that an uncoded word is in error if any of the k symbols in that word are in error. The probability that a symbol will be received correctly is $(1 - q_u)$, and since all symbol errors are assumed independent, the probability that all k symbols in a word are received correctly is $(1 - q_u)^k$. Thus the uncoded word-error probability is therefore given by

$$P_{eu} = 1 - (1 - q_u)^k \tag{11.122}$$

For the system using forward error correction, one or more symbol errors can possibly be corrected by the decoder, depending upon the code used. If the code is capable of correcting up to e errors, the probability of word error P_{ec} is equal to the probability that more than e errors are present in the received code word. Thus

$$P_{ec} = \sum_{i=e+1}^{n} \binom{n}{i} (1-q_c)^{n-i} q_c^i \tag{11.123}$$

where, as always,

$$\binom{n}{i} = \frac{n!}{i!(n-i)!} \tag{11.124}$$

The preceding equation for P_{ec}, (11.123), assumes that the code is a *perfect* code. A perfect code is a code in which e or fewer errors in an n-symbol code word are always corrected and a decoding failure always occurs if more than e errors are made in the transmission of an n-symbol code word. The only known perfect binary codes are the Hamming codes, for which $e = 1$, and the (23,12) Golay code, for which $e = 3$ as previously discussed. If the code is not a perfect code, one or more received sequences for which more than e errors occur can be corrected. In this case (11.123) is a worst-case performance bound. This bound is often tight, especially for high SNR.

Comparing word-error probabilities is only useful for those cases in which the n-symbol words, uncoded and coded, each carry an equal number of information bits. Comparing codes having different numbers of information bits per code word, or comparing codes having different error correcting capabilities, require that we compare codes on the basis of bit-error probability. Exact calculation of the bit-error probability from the channel symbol-error probability is often a difficult task and is dependent on the code generator matrix. However, Torrieri[4] derived both lower and upper bounds for the bit-error probability of block codes. These bounds are quite tight over most ranges of the channel SNR. The Torreri result expresses the bit-error probability as

$$P_b = \frac{q}{2(q-1)} \left[\sum_{i=e+1}^{d} \frac{d}{n} \binom{n}{i} P_s^i (1-P_s)^{n-i} + \sum_{i=d+1}^{n} i \binom{n}{i} P_s^i (1-P_s)^{n-i} \right] \tag{11.125}$$

where P_s is the channel symbol-error probability, e is the number of correctable errors per code word, d is the distance ($d = 2e+1$), and q is the size of the code alphabet. For binary codes $q = 2$ and for nonbinary codes, such as the Reed Solomon codes, $q = 2^m$.

In the coding examples to follow in the following section, we make use of (11.125). A MATLAB program is therefore developed to carry out the calculations required to map the symbol-error probabilities to bit-error probabilities as follows:

```
% File: ser2ber.m
function [ber] = ser2ber(q,n,d,t,ps)
lnps = length(ps);                 % length of error vector
ber = zeros(1,lnps);               % initialize output vector
for k=1:lnps                       % iterate error vector
    ser = ps(k);                   % channel symbol error rate
    sum1 = 0; sum2 = 0;            % initialize sums
```

[4]See D. J. Torreri, *Principles of Secure Communication Systems* (2nd ed.), Artech House, 1992, or D. J. Torreri, The information-bit error rate for block codes. *IEEE Transactions on Communications*, **COM-32** (4), Norwood, MA, April 1984.

```
    for i=(t+1):d
        term = nchoosek(n,i)*(ser^i)*((1-ser))^(n-i);
        sum1 = sum1+term;
    end
    for i=(d+1):n
        term = i*nchoosek(n,i)*(ser^i)*((1-ser)^(n-i));
        sum2 = sum2+term;
    end
    ber(k) = (q/(2*(q-1)))*((d/n)*sum1+(1/n)*sum2);
end
% End of function file.
```

11.4.8 Block Code Examples

The performance of a number of the coding techniques discussed in the preceding section are now considered.

COMPUTER EXAMPLE 11.1

In this example we investigate the effectiveness of a (7,4) single error-correcting code by comparing the word-error probabilities for the coded and uncoded systems. The symbol-error probabilities will also be determined. Assume that the code is used with a BPSK transmission system. As shown in Chapter 8, the symbol-error probability for BPSK in an AWGN environment is

$$q = Q\left(\sqrt{2z}\right) \tag{11.126}$$

where z is the SNR E_s/N_0. The symbol energy E_s is the transmitter power S times the word time T_w divided by k, since the total energy in each word is divided by k. Thus, the symbol-error probability without coding is given by

$$q_u = Q\left(\sqrt{\frac{2ST_w}{kN_0}}\right) \tag{11.127}$$

Assuming equal *word rates* for both the coded and uncoded system gives

$$q_c = Q\left(\sqrt{\frac{2ST_w}{nN_0}}\right) \tag{11.128}$$

for the coded symbol-error probability, since the energy available for k information symbols must be spread over $n > k$ symbols when coding is used. It follows that the *symbol*-error probability is increased by the use of coding as previously discussed. However, we shall show that the error-correcting capability of the code can overcome the increased symbol-error probability and indeed yield a net gain in word-error probability for certain ranges of the SNR. The uncoded word-error probability for the (7,4) code is given by (11.122) with $k = 4$. Thus

$$P_{eu} = 1 - (1 - q_u)^4 \tag{11.129}$$

Since $e = 1$, the word-error probability for the coded case, from (11.123), is

$$P_{ec} = \sum_{i=2}^{7} \binom{7}{i} (1 - q_c)^{7-i} q_c^i \tag{11.130}$$

The MATLAB program for performing the calculations outlined in the preceding two expressions follow.

```
% File: c11ce1.m
n = 7; k = 4; t = 1;                  % code parameters
zdB = 0:0.1:14;                       % set STw/No in dB
z = 10.^(zdB/10);                     % STw/No
lenz = length(z);                     % length of vector
qc = Q(sqrt(2*z/n));                  % coded symbol error prob.
qu = Q(sqrt(2*z/k));                  % uncoded symbol error prob.
peu = 1-((1-qu).^k);                  % uncoded word error prob.
pec = zeros(1,lenz);                  % initialize
for j=1:lenz
    pc = qc(j);                       % jth symbol error prob.
    s = 0;                            % initialize
    for i=(t+1):n
        termi = (pc^i)*((1-pc)^(n-i));
        s = s+nchoosek(n,i)*termi;
        pec(1,j) = s;                 % coded word error probability
    end
end
qq = [qc',qu',peu',pec'];
semilogy(zdB',qq)
xlabel('STw/No in dB')                % label x axis
ylabel('Probability')                 % label y
```

The word-error probabilities for the coded and uncoded systems are illustrated in Figure 11.17. The curves are plotted as a function of ST_W/N_0, which is word energy divided by the noise power spectral density.

Note that coding has little effect on system performance unless the value of ST_W/N_0 is in the neighborhood of 11 dB or above. Also, the improvement afforded by a $(7, 4)$ code is quite modest unless

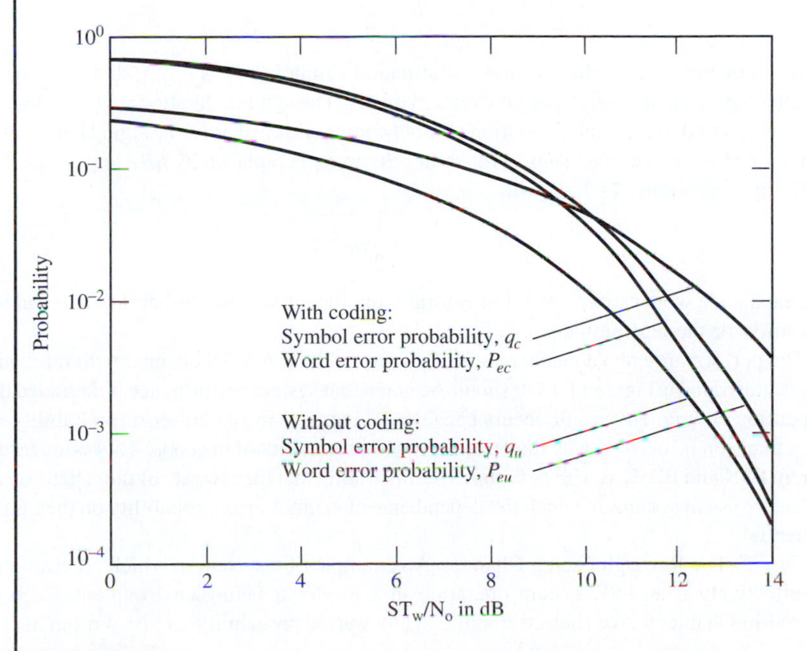

Figure 11.17
Comparison of uncoded and coded systems assuming a $(7, 4)$ code.

ST_W/N_0 is large, in which case system performance may be satisfactory without coding. However, in many systems, even small performance improvements are very important. Also, illustrated in Figure 11.17 are the uncoded and coded symbol-error probabilities q_u and q_c, respectively. The effect of spreading the available energy per word over a larger number of symbols is evident.

■

COMPUTER EXAMPLE 11.2

In this example we examine the performance of repetition codes in two different channels. Both cases utilize FSK modulation and a noncoherent receiver structure. In the first case, an AWGN channel is assumed. The second case assumes a Rayleigh fading channel. Distinctly different results will be obtained.

Case 1. The AWGN Channel: As was shown in Chapter 8 the error probability for a noncoherent FSK system in an AWGN channel is given by

$$q_u = \frac{1}{2} e^{-z/2} \tag{11.131}$$

where z is the ratio of signal power to noise power at the output of the receiver bandpass filter having bandwidth B_T. Thus z is given by

$$z = \frac{A^2}{2N_0 B_T} \tag{11.132}$$

where $N_0 B_T$ is the noise power in the signal bandwidth B_T. The performance of the system is illustrated by the $n = 1$ curve in Figure 11.18. When an n-symbol repetition code is used with this system, the symbol-error probability is given by

$$q_c = \frac{1}{2} e^{-z/2n} \tag{11.133}$$

This result occurs since coding a single information symbol (bit) as n repeated code symbols requires spreading the available energy per bit over n symbols. The symbol duration with coding is reduced by a factor n compared to the symbol duration without coding. Equivalently, the signal bandwidth is increased by a factor of n with coding. Thus, with coding B_T in q_u is replaced by nB_T to give q_c. The word-error probability is given by (11.123) with

$$e = \frac{1}{2}(n-1) \tag{11.134}$$

Since each code word carries one bit of information, the word-error probability is equal to the bit-error probability for the repetition code.

The performance of a noncoherent FSK system with an AWGN channel with rate $\frac{1}{3}$ and $\frac{1}{7}$ repetition codes is illustrated in Figure 11.18. It should be noted that system performance is degraded through the use of repetition coding. This result occurs because the increase in symbol-error probability with coding is greater than can be overcome by the error-correcting capability of the code. This same result occurs with coherent FSK and BPSK as well as with ASK, illustrating that the low rate of the repetition code prohibits its effective use in systems in which the dependence of symbol-error probability on the SNR is essentially exponential.

Case 2. The Rayleigh Fading Channel: An example of a system in which repetition coding can be used effectively is an FSK system operating in a Rayleigh fading environment. Such a system was analyzed in Chapter 9. We showed that the symbol-error probability can be written as

$$q_u = \frac{1}{2} \frac{1}{1 + E_a/2N_0} \tag{11.135}$$

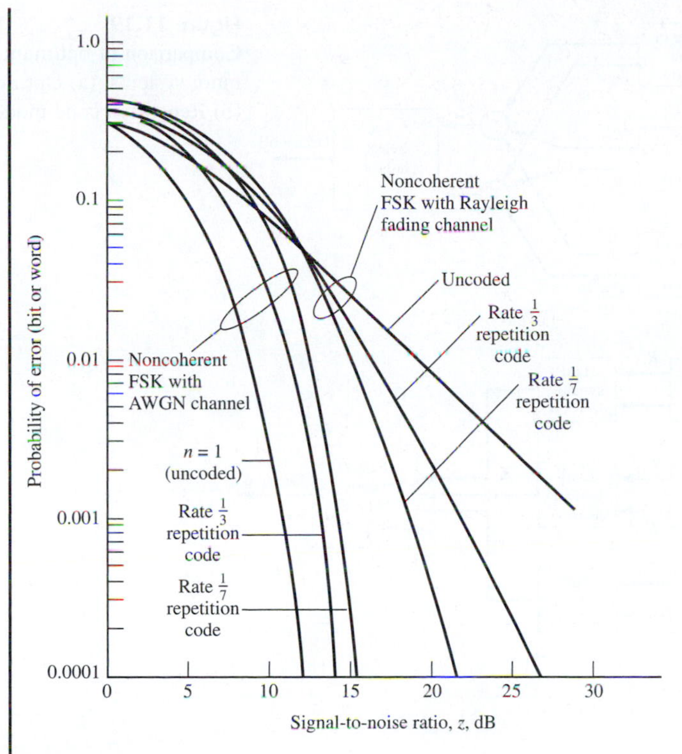

Figure 11.18
Performance of repetition codes on AWGN and Rayleigh fading channels.

in which E_a is the average received energy per symbol (or bit). The use of a repetition code spreads the energy E_a over the n symbols in a code word. Thus, with coding,

$$q_c = \frac{1}{2}\frac{1}{1 + E_a/2nN_0} \tag{11.136}$$

As in Case 1, the decoded bit-error probability is given by (11.123) with e given by (11.134). The Rayleigh fading results are also shown in Figure 11.18 for rate 1, $\frac{1}{3}$, and $\frac{1}{7}$ repetition codes, where, for this case, the SNR z is E_a/N_0. We see that the repetition code improves performance in a Rayleigh fading environment if E_a/N_0 is sufficiently large even though repetition coding does not result in a performance improvement in an AWGN environment.

Repetition coding can be viewed as time-diversity transmission since the n repeated symbols are transmitted in n different time slots or subpaths. We assume that energy per bit is held constant so that the available signal energy is divided equally among n subpaths. In Problem 10.26, it was shown that the optimal combining of the receiver outputs prior to making a decision on the transmitted information bit is as shown in Figure 11.19(a). The model for the repetition code considered in this example is shown in Figure 11.19(b). The essential difference is that a "hard decision" on each symbol of the n-symbol code word is made at the output of each of the n receivers. The decoded information bit is then in favor of the majority of the decisions made on each of the n symbols of the received code word.

When a hard decision is made at the receiver output, information is clearly lost, and the result is a degradation of performance. This can be seen in Figure 11.20, which illustrates the performance of the $n = 7$ optimal system of Figure 11.19(a) and that of the rate $\frac{1}{7}$ repetition code of Figure 11.19(b). Also shown for reference is the performance of the uncoded system.

Subpath 1
Receiver of Figure 10.8 (a)
R_{11}^2
R_{21}^2
Subpath 2
R_{12}^2
R_{22}^2
Subpath n
R_{1n}^2
R_{2n}^2
Σ
Y_1
Y_2
Select largest
Decision
(a)
Hard symbol decision
Select majority of hard decisions
(b)

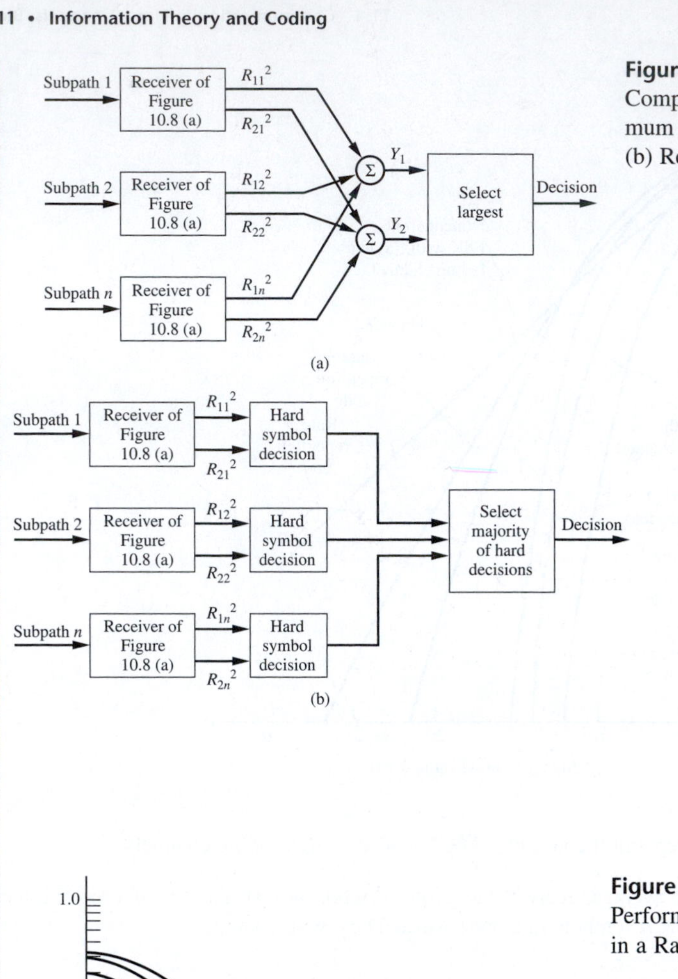

Figure 11.19
Comparison of optimum and suboptimum systems. (a) Optimum system. (b) Repetition code model.

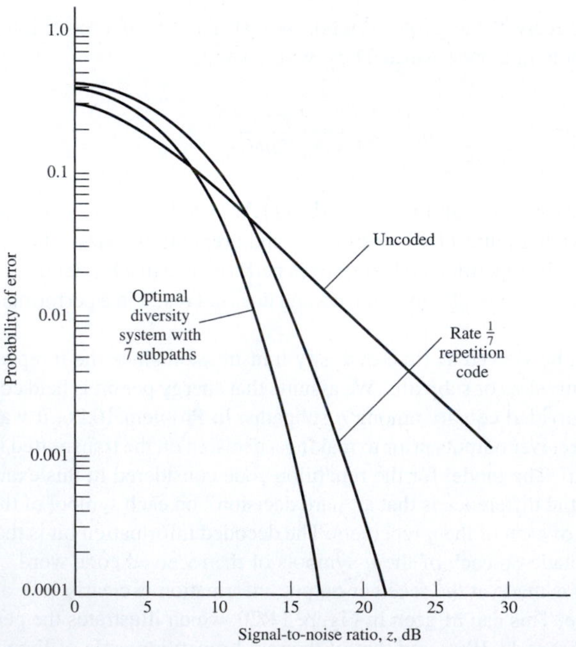

1.0
0.1
0.01
0.001
0.0001
Probability of error
0
5
10
15
20
25
30
Signal-to-noise ratio, z, dB
Uncoded
Optimal diversity system with 7 subpaths
Rate $\frac{1}{7}$ repetition code

Figure 11.20
Performance of noncoherent FSK in a Rayleigh fading channel.

COMPUTER EXAMPLE 11.3

In this example we compare the performance of a (15, 11) Hamming code and a (23, 12) Golay code with an uncoded system. A system using PSK modulation operating in an AWGN environment is assumed. Since the code words carry different numbers of information bits, comparisons based on the word-error probability cannot be used. We therefore use the Torrieri approximation given in (11.125). Since both codes are binary $q = 2$ for both cases. The MATLAB code follows and the results are illustrated in Figure 11.21.

```
% File: c11ce3.m
zdB = 0:0.1:10;                          % set Eb/No axis in dB
z = 10.^(zdB/10);                        % convert to linear scale
ber1 = q(sqrt(2*z)); % PSK result
ber2 = q(sqrt(12*2*z/23));               % CSER for (23,12) Golay code
ber3 = q(sqrt(11*z*2/15));               % CSER for (15,11) Hamming code
berg = ser2ber(2,23,7,3,ber2);           % BER for Golay code
berh = ser2ber(2,15,3,1,ber3);           % BER for Hamming code
semilogy(zdB,ber1,'k-',zdB,berg,'k-',zdB,berh,'k-.')
xlabel('E_b/N_o in dB')                  % label x axis
ylabel('Bit Error Probability')          % label y axis
legend('Uncoded','Golay code','Hamming code')
% End of script file.
```

The advantage of the Golay code is clear, especially for high E_b/N_0.

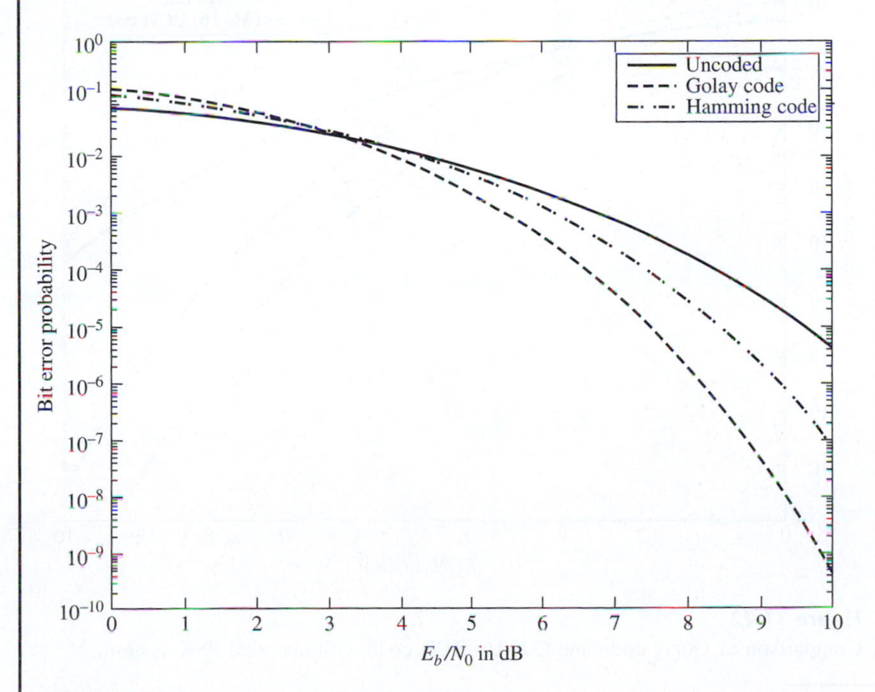

Figure 11.21
Performance comparisons for Golay code and Hamming code with uncoded system.

■

COMPUTER EXAMPLE 11.4

In this example we compare the performance of a (23, 12) Golay code and a (31, 16) BCH code with an uncoded system. Phase-shift keying modulation and operation in an AWGN environment are assumed. Note that both codes have rates of approximately $1/2$ and both codes are capable of correcting up to 3 errors per code word. The MATLAB code follows and the performance results are illustrated in Figure 11.22. Note that the BCH code provides improved performance.

```
% File: c11_ce4.m
zdB = 0:0.1:10;                          % set Eb/No in dB
z = 10.^(zdB/10);                        % convert to linear scale
ber1 = q(sqrt(2*z));                     % PSK result
ber2 = q(sqrt(12*2*z/23));               % SER for (23,12) Golay code
ber3 = q(sqrt(16*z*2/31));               % SER for (16,31) BCH code
berg = ser2ber(2,23,7,3,ber2);           % BER for (23,12) Golay code
berbch = ser2ber(2,23,7,4,ber3);         % BER for (16,31) BCH code
semilogy(zdB,ber1,'k-',zdB,berg,'k-',zdB,berbch,'k-.')
xlabel('E_b/N_o in dB')                  % label x axis
ylabel('Bit Error Probability')          % label y axis
legend('Uncoded','Gola y code','(31,16) BCH code')
% End of script file.
```

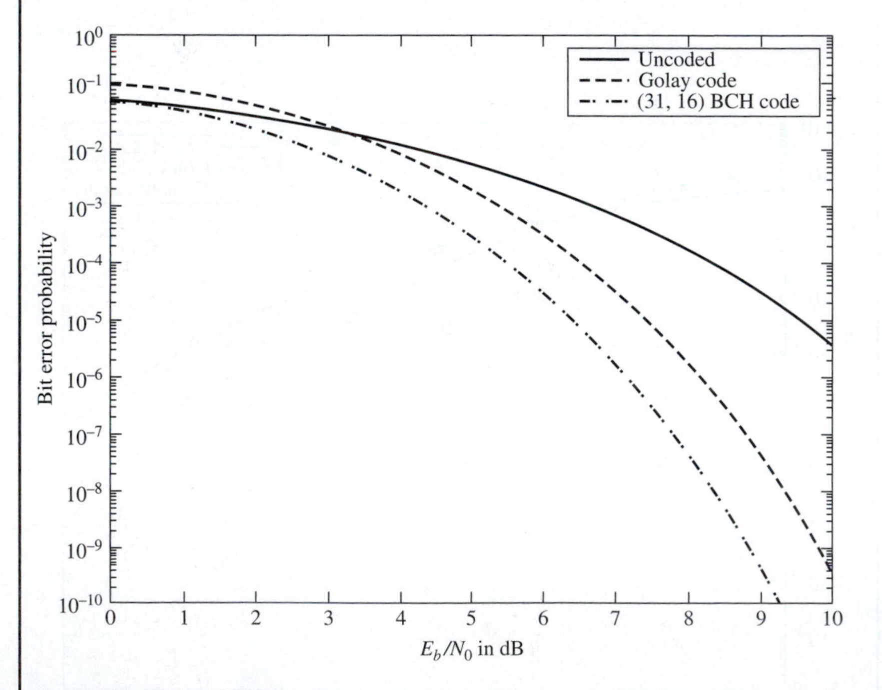

Figure 11.22
Comparison of Golay code and (31, 16) BCH code with uncoded PSK system.

■

11.5 COMMUNICATION IN NOISY CHANNELS: CONVOLUTIONAL CODES

The convolutional code is an example of a nonblock code. Rather than the parity-check symbols being calculated for a block of information symbols, the parity checks are calculated over a span of information symbols. This span, which is referred to as the *constraint span*, is shifted one information symbol each time an information symbol is input to the encoder.

A general convolutional coder is illustrated in Figure 11.23. The coder is rather simple and consists of three component parts. The heart of the coder is a shift register that holds k information symbols, where k is the constraint span of the code. The shift register stages are connected to v modulo-2 adders as indicated. Not all stages are connected to all adders. In fact, the connections are "somewhat random" and these connections have considerable impact on the performance of the resulting code. Each time a new information symbol is shifted into the coder, the adder outputs are sampled by the commutator. Thus v output symbols are generated for each input symbol yielding a code of rate $1/v$.[5]

A rate $\frac{1}{3}$ convolutional coder is illustrated in Figure 11.24. For each input, the output of the coder is the sequence $v_1 v_2 v_3$. For the coder of Figure 11.24

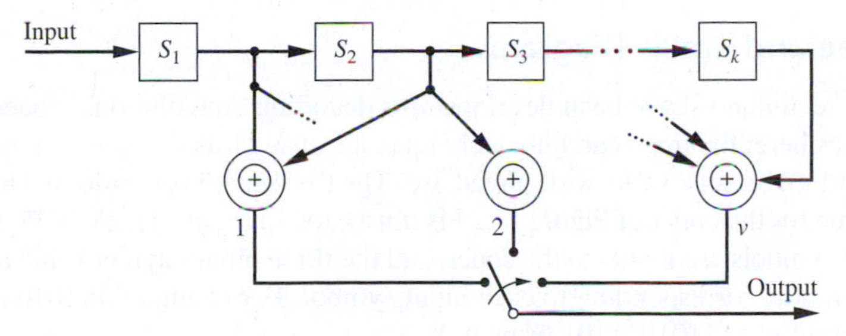

Figure 11.23
General convolutional coder.

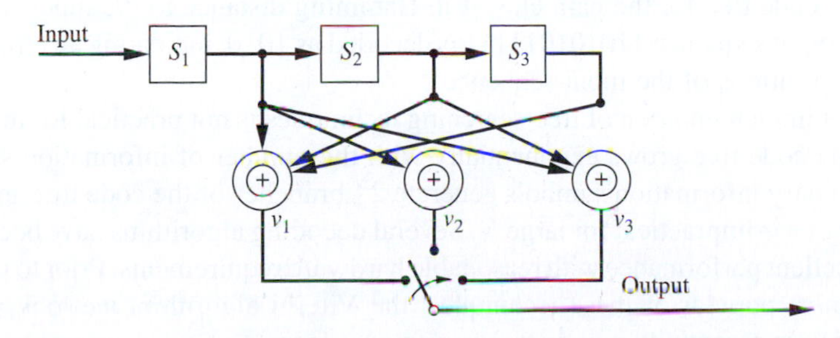

Figure 11.24
A rate $\frac{1}{3}$ convolutional coder.

[5] In this chapter we only consider convolutional coders having rate $1/v$. It is, of course, often desirable to generate convolutional codes having higher rates. If symbols are shifted into the coder k symbols at a time, rather than 1 symbol at a time, a rate k/v convolutional code results. These codes are more complex and beyond the scope of this introductory treatment. The motivated student should consult one of the standard textbooks on coding theory cited in the references.

$$v_1 = S_1 \oplus S_2 \oplus S_3 \tag{11.137}$$

$$v_2 = S_1 \tag{11.138}$$

$$v_3 = S_1 \oplus S_2 \tag{11.139}$$

We will see later that a well-performing code will have the property that, for S_2 and S_3 (the two previous inputs) fixed, $S_1 = 0$ and $S_1 = 1$ will result in outputs $v_1 v_2 v_3$ that are complements. The sequence $S_2 S_3$ will be referred to as the current state of the coder so that the current state, together with the current input, determine the output. Thus we see that the input sequence

$$101001 \cdots$$

results, assuming an initial state of 00, in the output sequence

$$111101011101100111 \cdots$$

At some point the sequence is terminated in a way that allows for unique decoding. This is accomplished by returning the coder to the initial 00 state and will be illustrated when we consider the Viterbi algorithm.

11.5.1 Tree and Trellis Diagrams

A number of techniques have been developed for decoding convolutional codes. We discuss two techniques here; the tree-searching technique, because of its fundamental nature, and the Viterbi algorithm, because of its widespread use. The tree search is considered first. A portion of the code tree for the coder of Figure 11.24 is illustrated in Figure 11.25. In Figure 11.25, the single binary symbols are inputs to the coder, and the three binary symbols in parentheses are the output symbols corresponding to each input symbol. For example, if 1010 is fed into the coder, the output is 111101011101 or path A.

The decoding procedure also follows from Figure 11.25. To decode a received sequence, we search the code tree for the path closest in Hamming distance to the input sequence. For example, the input sequence 110101011111 is decoded as 1010, indicating an error in the third and eleventh positions of the input sequence.

The exact implementation of tree-searching techniques is not practical for many applications since the code tree grows exponentially with the number of information symbols. For example, N binary information symbols generate 2^N branches of the code tree and storage of the complete tree is impractical for large N. Several decoding algorithms have been developed that yield excellent performance with reasonable hardware requirements. Prior to taking a brief look at the most popular of these techniques, the Viterbi algorithm, we look at the trellis diagram, which is essentially a code tree in compact form.

The key to construction of the trellis diagram is recognition that the code tree is repetitive after k branches, where k is the constraint span of the coder. This is easily recognized from the code tree shown in Figure 11.25. After the fourth input of an information symbol, 16 branches have been generated in the code tree. The coder outputs for the first eight branches match exactly the coder outputs for the second eight branches, except for the first symbol. After a little thought, you should see that this is obvious. The coder output depends only on the latest k inputs. In this case, the constraint span k is 3. Thus the output corresponding to the fourth

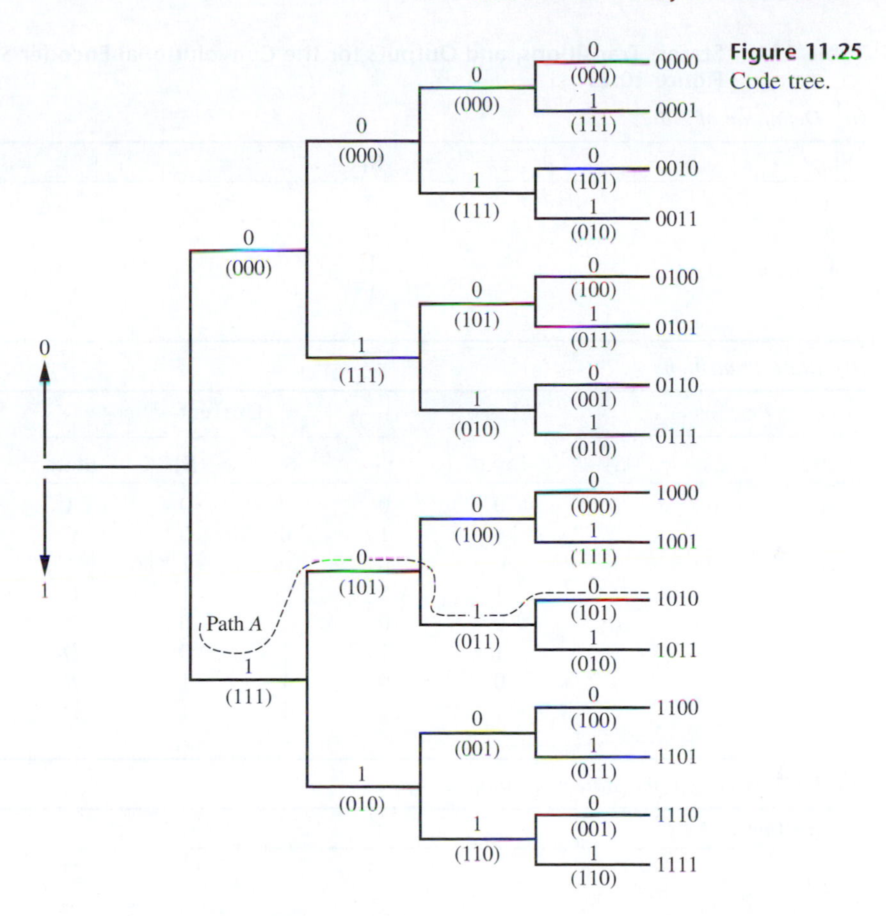

Figure 11.25
Code tree.

information symbol depends only on the second, third, and fourth coder inputs. It makes no difference whether the first information symbol was a binary 0 or a binary 1. (This should clarify the meaning of a constraint span.)

When the current information symbol is input to the coder, S_1 is shifted to S_2 and S_2 is shifted to S_3. The new state, S_2S_3 and the current input S_1 then determine the shift register contents $S_1S_2S_3$, which in turn determine the output $v_1v_2v_3$. This information is summarized in Table 11.6. The outputs corresponding to given state transitions are shown in parentheses, consistent with Figure 11.25.

It should be noted that states A and C can only be reached from states A and B. Also, states B and D can only be reached from states C and D. The information in Table 11.6 is often shown in a state diagram, as in Figure 11.26. In the state diagram, an input of binary 0 results in the transition denoted by a dashed line, and an input of binary 1 results in the transition designated by a solid line. The resulting coder output is denoted by the three symbols in parentheses. For any given sequence of inputs, the resulting state transitions and coder outputs can be traced on the state diagram. This is a very convenient method for determining the coder output resulting from a given sequence of inputs.

The trellis diagram illustrated in Figure 11.27 results directly from the state diagram. Initially, the coder is assumed to be in state A (all contents are 0s). A binary 0 input results in the coder remaining in state A, as indicated by the dashed line, and a binary 1 input results in a

Table 11.6 States, Transitions, and Outputs for the Convolutional Encoder Shown in Figure 10.23

(a) Definition of States

State	S_1	S_2
A	0	0
B	0	1
C	1	0
D	1	1

(b) State Transitions

Previous				Current				
State	S_1	S_2	**Input**	S_1	S_2	S_3	**State**	**Output**
A	0	0	0	0	0	0	*A*	(000)
			1	1	0	0	*C*	(111)
B	0	1	0	0	0	1	*A*	(100)
			1	1	0	1	C	(011)
C	1	0	0	0	1	0	*B*	(101)
			1	1	1	0	*D*	(010)
D	1	1	0	0	1	1	*B*	(001)
			1	1	1	1	D	(110)

(c) Encoder Output for State Transition $x \rightarrow y$

Transition	Output
$A \rightarrow A$	(000)
$A \rightarrow C$	(111)
$B \rightarrow A$	(100)
$B \rightarrow C$	(011)
$C \rightarrow B$	(101)
$C \rightarrow D$	(010)
$D \rightarrow B$	(001)
$D \rightarrow D$	(110)

transition to state *C*, as indicated by the solid line. Any of the four states can be reached by a sequence of two inputs. The third input results in the possible transitions shown. The fourth input results in exactly the same set of possible transitions. Therefore, after the second input, the trellis becomes completely repetitive, and the possible transitions are those labeled steady-state transitions. The coder can always be returned to state *A* by inputting two binary 0s as shown in Figure 11.27. As before, the output sequence resulting from any transition is shown by the sequence in parentheses.

11.5.2 The Viterbi Algorithm

In order to illustrate the Viterbi algorithm, we consider the received sequence that we previously considered to illustrate decoding using a code tree—namely, the sequence

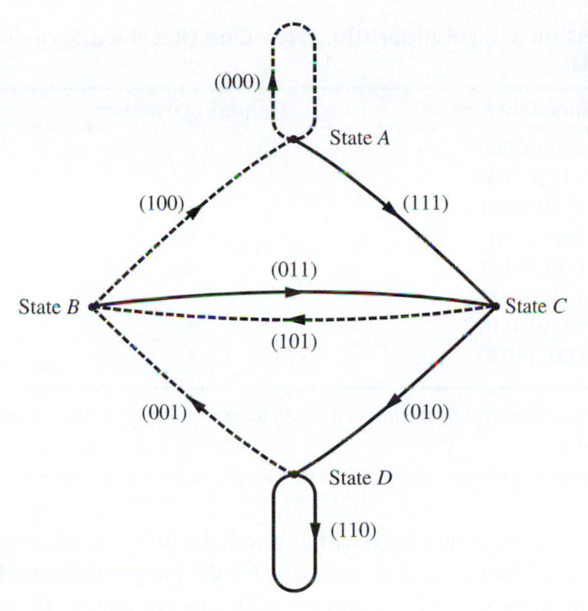

Figure 11.26
State diagram for the example convolutional coder.

110101011111. The first step is to compute the Hamming distances between the initial node (state A) and each of the four states three levels deep into the trellis. We must look three levels deep into the trellis because the constraint span of the example coder is 3. Since each of the four nodes can be reached from only two preceding nodes, eight paths must be identified, and the Hamming distance must be computed for each path. We therefore initially look three levels

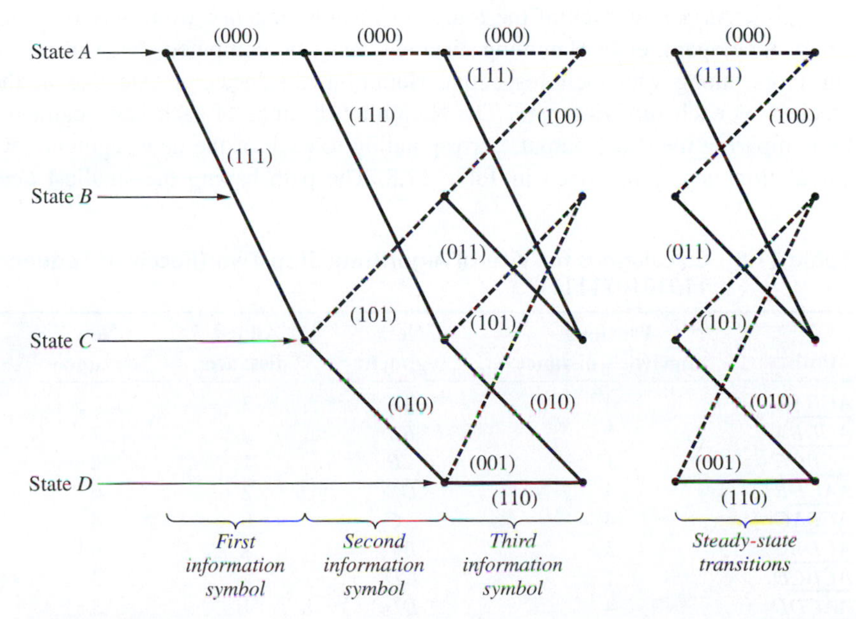

Figure 11.27
Trellis diagram.

Table 11.7 Calculations for Viterbi Algorithm: Step One (Received Sequence = 110101011)

Path[1]	Corresponding symbols	Hamming distance	Survivor?
AAAA	000000000	6	No
ACBA	111101100	4	Yes
ACDB	111010001	5	Yes[2]
AACB	000111101	5	No[2]
AAAC	000000111	5	No
ACBC	111101011	1	Yes
ACDD	111010110	6	No
AACD	000111010	4	Yes

[1]The initial and terminal states are identifted by the first and fourth letters, respectively. The second and third letters correspond to intermediate states.

[2]if two or more paths have the same Hamming distance, it makes no difference which is retained as the survivor.

deep into the trellis, and since the example coder has rate $\frac{1}{3}$, the first nine received symbols are initially considered. Thus the Hamming distances between the input sequence 110101011 and the eight paths terminating three levels deep into the trellis are computed. These calculations are summarized in Table 11.7. After the eight Hamming distances are computed, the path having the minimum Hamming distance to *each* of the four nodes is retained. These four retained paths are known as *survivors*. The other four paths are discarded from further consideration. The four survivors are identified in Table 11.7.

The next step in the application of the Viterbi algorithm is to consider the next three received symbols, which are 111 in the example being considered. The scheme is to compute once again the Hamming distance to the four states, this time four levels deep in the trellis. As before, each of the four states can be reached from only two previous states. Thus, once again, eight Hamming distances must be computed. Each of the four previous survivors, along with their respective Hamming distances, is extended to the two states reached by each surviving path. The Hamming distance of each new segment is computed by comparing the coder output, corresponding to each of the new segments, with 111. The calculations are summarized in Table 11.8. The path having the smallest new distance is

Table 11.8 Calculations for Viterbi Algorithm: Step Two (Received Sequence = 110101011111)

Path[1]	Previous survivor's distance	New segment	Added distance	New distance	Survivor?
ACBAA	4	*AA*	3	7	Yes
ACDBA	5	*BA*	2	7	No
ACBCB	1	*CB*	1	2	Yes
AACDB	4	*DB*	2	6	No
ACBAC	4	*AC*	0	4	Yes
ACDBC	5	*BC*	1	6	No
ACBCD	1	*CD*	2	3	Yes
AACDD	4	*DD*	1	5	No

[1]An underscore indicates the previous survivor.

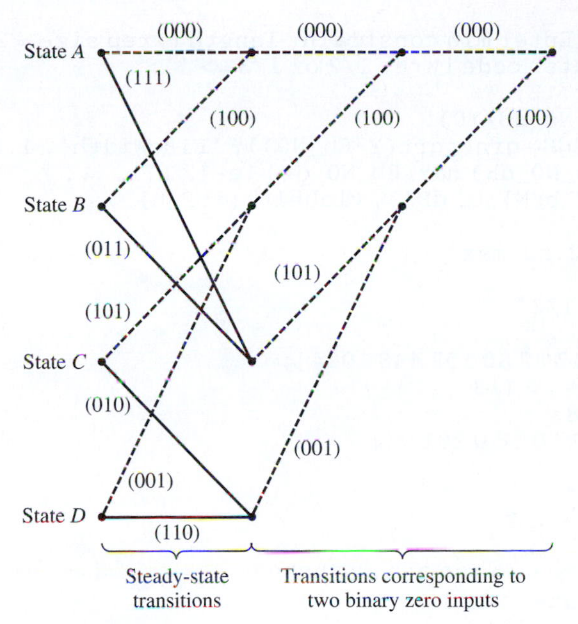

Figure 11.28 Termination of the trellis diagram.

path *ACBCB*. This corresponds to information sequence 1010 and is in agreement with the previous tree search.

For a general received sequence, the process identified in Table 11.8 is continued. After each new set of calculations, involving the next three received symbols, only the four surviving paths and the accumulated Hamming distances need be retained. At the end of the process, it is necessary to reduce the number of surviving paths from four to one. This is accomplished by inserting two dummy 0s at the end of the information sequence, corresponding to the transmission of six code symbols. As shown in Figure 11.28, this forces the trellis to terminate at state *A*.

The Viterbi algorithm has found widespread application in practice. It can be shown that the Viterbi algorithm is a maximum-likelihood decoder, and in that sense, it is optimal. Viterbi and Omura (1979) give an excellent analysis of the Viterbi algorithm. A paper by Heller and Jacobs (1971) summarizes a number of performance characteristics of the Viterbi algorithm.

11.5.3 Performance Comparisons for Convolutional Codes

As was done with block codes, a MATLAB program was developed that allows us to compare the bit error probabilities for convolutional codes having various parameters. The MATLAB program follows:

```
% File: c11_convcode.m
% BEP for convolutional coding in Gauss noise
% Rate 1/3 or 1/2
% Hard decisions
%
clf
nu_max = input('Enter max constraint length: 3-9, rate 1/2; 3-8, rate
1/3 => ');
```

```
nu_min = input('Enter min constraint length (step size = 2) => ');
rate = input('Enter code rate: 1/2 or 1/3 => ');
Eb_N0_dB = 0:0.1:12;
Eb_N0 = 10.^(Eb_N0_dB/10);
semilogy(Eb_N0_dB, qfn(sqrt(2*Eb_N0)), 'LineWidth', 1.5), ...
   axis([min(Eb_N0_dB) max(Eb_N0_dB) 1e-12 1]), ...
   xlabel('{itE_b/N}_0, dB'), ylabel('{itP_b}'), ...
hold on
for nu = nu_min:2:nu_max
    if nu == 3
       if rate == 1/2
          dfree = 5;
          c = [1 4 12 32 80 192 448 1024];
       elseif rate == 1/3
          dfree = 8;
          c = [3 0 15 0 58 0 201 0];
       end
    elseif nu == 4
       if rate == 1/2
          dfree = 6;
          c = [2 7 18 49 130 333 836 2069];
       elseif rate == 1/3
          dfree = 10;
          c = [6 0 6 0 58 0 118 0];
       end
    elseif nu == 5
       if rate == 1/2
          dfree = 7;
          c = [4 12 20 72 225 500 1324 3680];
       elseif rate == 1/3
          dfree = 12;
          c = [12 0 12 0 56 0 320 0];
       end
    elseif nu == 6
       if rate == 1/2
          dfree = 8;
          c = [2 36 32 62 332 701 2342 5503];
       elseif rate == 1/3
          dfree = 13;
          c = [1 8 26 20 19 62 86 204];
       end
    elseif nu == 7
       if rate == 1/2
          dfree = 10;
          c = [36 0 211 0 1404 0 11633 0];
       elseif rate == 1/3
          dfree = 14;
          c = [1 0 20 0 53 0 184 0];
       end
    elseif nu == 8
       if rate == 1/2
          dfree = 10;
          c = [2 22 60 148 340 1008 2642 6748];
       elseif rate == 1/3
          dfree = 16;
```

```
            c = [1 0 24 0 113 0 287 0];
        end
    elseif nu == 9
        if rate == 1/2
            dfree = 12;
            c = [33 0 281 0 2179 0 15035 0];
        elseif rate == 1/3
            disp('Error: there are no weights for nu = 9 and rate = 1/3')
        end
    end
    Pd = [];
    p = qfn(sqrt(2*rate*Eb_N0));
    kk = 1;
    for k = dfree:1:dfree+7;
        sum = 0;
        if mod(k,2) == 0
            for e = k/2+1:k
                sum = sum + nchoosek(k,e)*(p.^e).*((1-p).^(k-e));
            end
            sum = sum + 0.5*nchoosek(k,k/2)*(p.^(k/2)).*((1-p).^
            (k/2));
        elseif mod(k,2) == 1
            for e = (k+1)/2:k
                sum = sum + nchoosek(k, e)*(p.^e).*((1-p).^(k-e));
            end
        end
        Pd(kk, :) = sum;
        kk = kk+1;
    end
    Pbc = c*Pd;
    semilogy(Eb_N0_dB, Pbc, '--', 'LineWidth', 1.5), ...
        text(Eb_N0_dB(78)+.1, Pbc(78), ['nu = ', num2str(nu)])
    end
    legend(['BPSK uncoded'], ['Convol. coded; HD; rate = ', num2str
    (rate, 3)])
    hold off
    % End of script file.
```

The MATLAB code is based on the linearity of convolutional codes, which allows us to assume the all-zeros path through the trellis as being the correct path. A decoding error event then corresponds to a path that deviates from the all-zeros path at some point in the trellis and remerges with the all-zeros path a number of steps later. Since the all-zeros path is assumed to be the correct path, the number of information bit errors corresponds to the number of information ones associated with an error event path of a given length. The bit-error probability can then be upper bounded by

$$P_b < \sum_{k=d_{\text{free}}}^{\infty} c_k P_k \tag{11.140}$$

where d_{free} is the free distance of the code (the Hamming distance of the minimum-length error event path from the all-zeros path, or simply the Hamming weight of the minimum-length error event path), P_k is the probability of an error event path of length k occurring, and c_k is the weighting coefficient giving the number of information bit errors associated with all error event paths of length k in the trellis. The latter, called the *weight structure* of the code, can be found from the generating function of the code, which is a function that enumerates all nonzero paths through the trellis and gives the number of information ones associated with all paths of a given length. The partial (partial because the upper limit of the sum in (11.140) must be set to some finite number for computational purposes) weight structures of "good" convolutional codes have been found and published in the literature (the weights in the program above are given by the vectors labeled c).[6] The error event probabilities are given by[7]

$$P_k = \sum_{e=(k/2)+1}^{k} \binom{k}{e} p^e (1-p)^{k-e} + \frac{1}{2}\binom{k}{k/2} p^{k/2}(1-p)^{k/2}, \quad k \text{ even} \tag{11.141}$$

and

$$P_k = \sum_{e=(k+1)/2}^{k} \binom{k}{e} p^e (1-p)^{k-e}, \quad k \text{ odd} \tag{11.142}$$

where, for an AWGN channel,

$$p = Q\left(\sqrt{\frac{2kRE_b}{N_0}}\right) \tag{11.143}$$

in which R is the code rate.

Strictly speaking, when the upper limit of (11.140) is truncated to a finite integer, the upper bound may no longer be true. However, if carried out to a reasonable number of terms, the finite sum result of (11.140) is a sufficiently good approximation to the bit-error probability for moderate to low values of p as computer simulations have shown.[8]

COMPUTER EXAMPLE 11.5

As an example of the improvement one can expect from a convolutional code, estimates for the bit-error probability for rate $\frac{1}{2}$ and $\frac{1}{3}$ convolutional codes are plotted in Figures 11.29 and 11.30, respectively, as computed with the above MATLAB program. These results show that for codes of constraint length 7, a rate $\frac{1}{2}$ code gives about 3.5 dB improvement at a bit-error probability of 10^{-6} whereas a rate $\frac{1}{3}$ code gives almost 4 dB of improvement. For soft decisions (where the output of the detector is quantized into several levels before being input to the Viterbi decoder), the improvement would be significantly more (about 5.8 dB and 6.2 dB, respectively[9]).

[6] Odenwalder, J. P., Error Control, in *Data Communications, Networks, and Systems*, Thomas Bartree (ed.), Indianapolis: Howard W. Sams, 1985.

[7] See Ziemer and Peterson (2001), pp. 504–505.

[8] Heller, J. A. and I. M. Jacobs, Viterbi decoding for satellite and space communications. *IEEE Transactions on Communications Technology*, **COM-19:** 835–848, October 1971.

[9] See Ziemer and Peterson (2001), pp. 511 and 513.

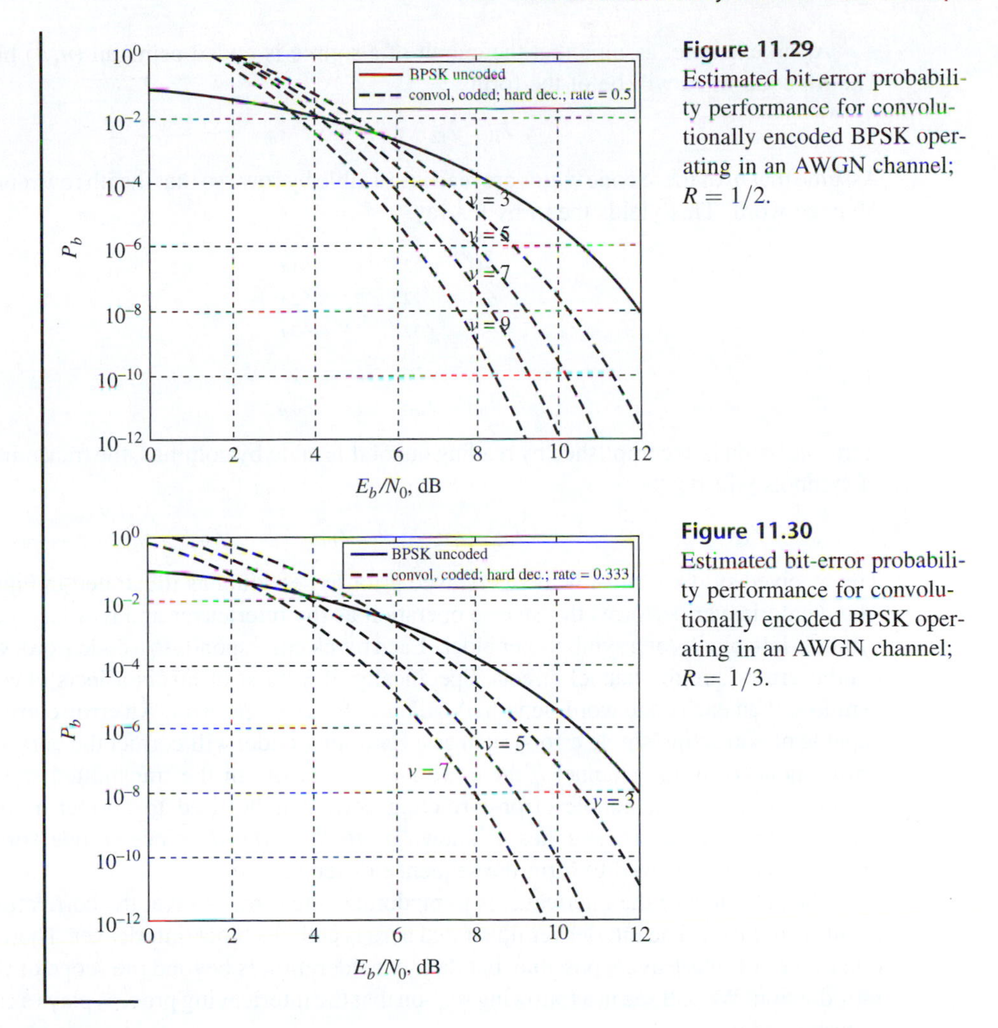

Figure 11.29
Estimated bit-error probability performance for convolutionally encoded BPSK operating in an AWGN channel; $R = 1/2$.

Figure 11.30
Estimated bit-error probability performance for convolutionally encoded BPSK operating in an AWGN channel; $R = 1/3$.

■

■ 11.6 COMMUNICATION IN NOISY CHANNELS: OTHER TECHNIQUES

For completeness we now very briefly consider a few other techniques.

11.6.1 Burst-Error-Correcting Codes

Many practical communication channels, such as those encountered in mobile communication systems, exhibit fading in which errors tend to group together in bursts. Thus, errors are no longer independent. Much attention has been devoted to code development for improving the performance of systems exhibiting burst-error characteristics. Most of these codes tend to be more complex than the simple codes previously considered. A code for correction of a single burst, however, is rather simple to understand and leads to a technique known as interleaving, which is useful in a number of situations.

As an example, assume that the output of a source is coded using an (n, k) block code. The ith code word will be of the form

$$\lambda_{i1} \quad \lambda_{i2} \quad \lambda_{i3} \quad \cdots \quad \lambda_{in}$$

Assume that m of these code words are read into a table by rows so that the ith row represents the ith code word. This yields the m by n array

$$\begin{array}{cccc} \lambda_{11} & \lambda_{12} & \cdots & \lambda_{1n} \\ \lambda_{21} & \lambda_{22} & \cdots & \lambda_{2n} \\ \lambda_{31} & \lambda_{32} & \cdots & \lambda_{3n} \\ \vdots & \vdots & \ddots & \vdots \\ \lambda_{m1} & \lambda_{m2} & \cdots & \lambda_{mn} \end{array}$$

If transmission is accomplished by reading out of this table by columns, the transmitted stream of symbols will be

$$\lambda_{11} \quad \lambda_{21} \quad \cdots \quad \lambda_{m1} \quad \lambda_{12} \quad \lambda_{22} \quad \cdots \quad \lambda_{m2} \quad \cdots \quad \lambda_{1n} \quad \lambda_{2n} \quad \cdots \quad \lambda_{mn}$$

The received symbols must be deinterleaved prior to decoding as illustrated in Figure 11.31. The deinterleaver performs the inverse operation as the interleaver and reorders the received symbols into blocks of n symbols per block. Each block corresponds to a code word, which may exhibit errors due to channel effects. Specifically, if a burst of errors affects m consecutive symbols, then each code word (length n) will have exactly one error. An error-correcting code capable of correcting single errors, such as a Hamming code, will correct the burst of channel errors induced by the channel *if* there are no other errors in the transmitted stream of mn symbols. Likewise, a double error-correcting code can be used to correct a single burst spanning $2m$ *symbols*. These codes are known as *interleaved codes* since m code words, each of length m, are interleaved to form the sequence of length mn.

The net effect of the interleaver is to randomize the errors so that the correlation of error events is reduced. The interleaver illustrated here is called a block interleaver. There are many other types of interleavers possible, but their consideration is beyond the scope of this simple introduction. We will see in a following section that the interleaving process plays a critical role in turbo coding.

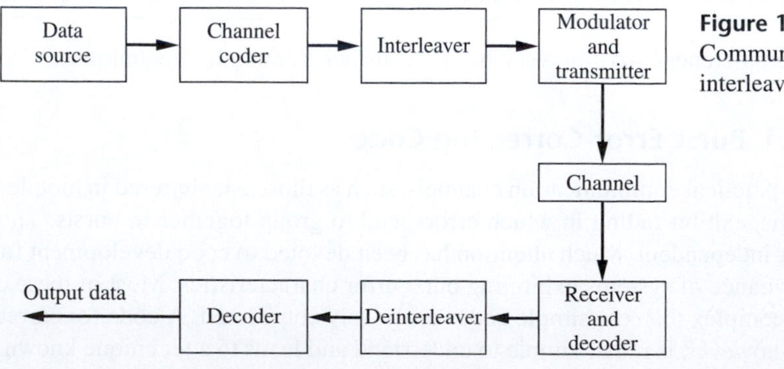

Figure 11.31 Communication system with interleaving.

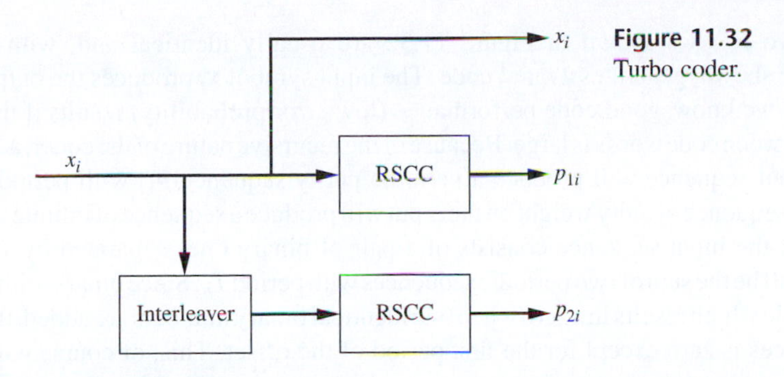

Figure 11.32
Turbo coder.

11.6.2 Turbo Coding

The study of coding theory has been a search for the coding scheme that yields a communications system having the performance closely approaching the Shannon bound. For the most part progress has been incremental. A large step in this quest for nearly ideal performance in the presence of noise was revealed in 1993 with the publication of a paper by Berrou, Glavieux, and Thitimajshima. It is remarkable that this paper was not the result of a search for a more powerful coding scheme, but was a result of a study of efficient clocking techniques for concatenated circuits. Their discovery, however, has revolutionized coding theory.

Turbo coding, and especially decoding, are complex tasks and a study of even simple implementations are well beyond the scope of this text. We will present, however, a few important concepts as motivation for further study.

The basic architecture of a turbo coder is illustrated in Figure 11.32. Note that the turbo coder consists of an interleaver, such as we studied previously and a pair of recursive systematic convolutional coders (RSCCs). An RSCC is shown in Figure 11.33. Note that the RSCC is much like the convolutional coders previously studied with one important difference. That difference lies in the feedback path from the delay elements back to the input. The conventional convolutional coder does not have this feedback path and therefore it behaves as a finite impulse response (FIR) digital filter. With the feedback path the filter becomes an infinite impulse response (IIR), or recursive, filter and here lies one of the attributes of the turbo code. The RSCC shown in Figure 11.33 a is a rate $\frac{1}{2}$ convolutional coder for which the input x_i generates an output sequence $x_i p_i$. Since the first symbol in the output sequence is the information symbol, the coder is systematic.

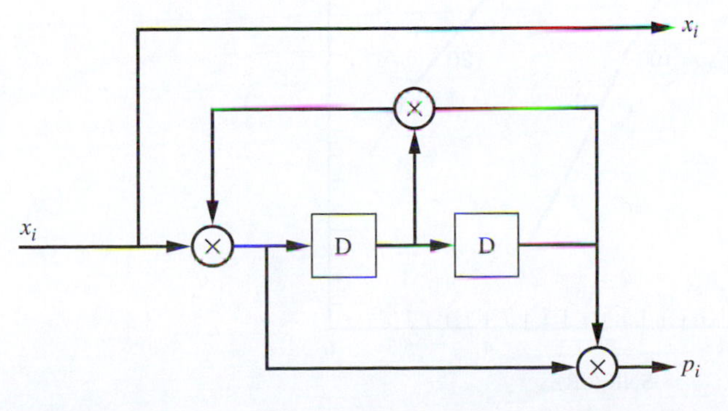

Figure 11.33
Recursive, systematic convolutional coder.

The two RSCCs shown in Figure 11.32 are usually identical and, with the parallel architecture shown, generates a rate $\frac{1}{3}$ code. The input symbol x_i produces the output sequence $x_i p_{1i} p_{2i}$. As we know, good code performance (low error probability) results if the Hamming distance between code words is large. Because of the recursive nature of the coder, a single binary 1 in the input sequence will produce a periodic parity sequence p_1, with period T_p. Strictly speaking, a sequence of unity weight on the input will produce a sequence of infinite weight for p_1. However, if the input sequence consists of a pair of binary ones separated by T_p, the parity sequence will be the sum of two periodic sequences with period T_p. Since binary arithmetic has an addition table which results in a zero when two identical binary numbers are added, the sum of the two sequences is zero except for the first period of the offset. This, of course will reduce the Hamming weight of the first parity sequence, which is an undesirable effect.

This is where the interleaver comes into play. The interleaver will change the separation between the two binary ones and therefore cancellation will, with high probability, not occur. It therefore follows that if one of the parity sequences has large Hamming weight, the other one will not.

Figure 11.34 illustrates the performance of a turbo code for two different interleaver sizes. The larger interleaver produces better performance results since it can do a better job of "randomizing" the interleaver output.

Most turbo decoding algorithms are based on the MAP estimation principle studied in the previous chapter. Of perhaps more importance is the fact that turbo decoding algorithms, unlike other decoding tools, are iterative in nature so that a given sequence passes through the decoder a number of times with the error probability decreasing with each pass. As a result, a trade-off exists between performance and decoding time. This attribute allows one the freedom to develop application specific decoding algorithms. This freedom is not available in other techniques. For example one can target various decoders for a given QoS by adjusting the number of iterations used in the decoding process. Decoders can also be customized to take

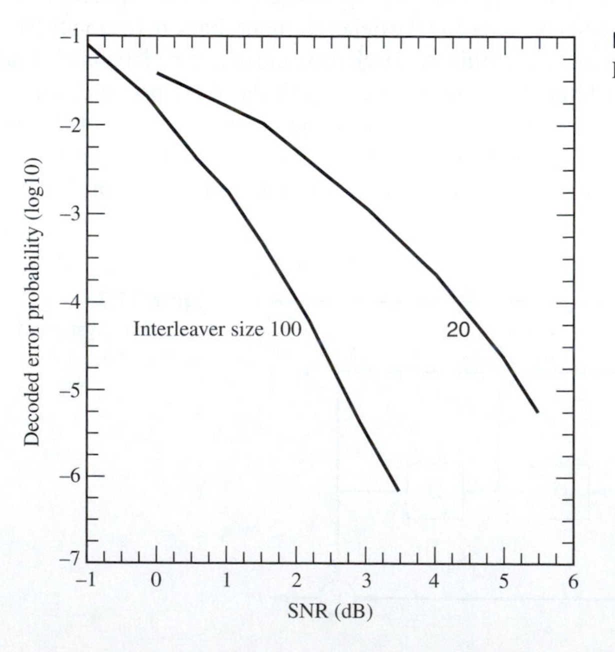

Figure 11.34
Performance curves for turbo code.

advantage of latency and/or performance trade-offs. As an example, data communications requires low bit-error probabilities but latency is not often a problem. Voice communications, however requires low latency but higher error probabilities can be tolerated.

11.6.3 Feedback Channels

In many practical systems, a feedback channel is available from receiver to transmitter. When available, this channel can be utilized to achieve a specified performance with decreased complexity of the coding scheme. Many such schemes are possible: decision feedback, error-detection feedback, and information feedback. In a decision-feedback scheme, a null-zone receiver is used, and the feedback channel is utilized to inform the transmitter either that no decision was possible on the previous symbol and to retransmit or that a decision was made and to transmit the next symbol. The null-zone receiver is usually modeled as a binary-erasure channel.

Error-detection feedback involves the combination of coding and a feedback channel. With this scheme, retransmission of code words is requested when errors are detected.

In general, feedback schemes tend to be rather difficult to analyze. Thus only the simplest scheme, the decision-feedback channel with perfect feedback assumed, will be treated here. Assume a binary transmission scheme with matched-filter detection. The signaling waveforms are $s_1(t)$ and $s_2(t)$. The conditional pdfs of the matched filter output at time T, conditioned on $s_1(t)$ and $s_2(t)$, were derived in Chapter 8 and are illustrated for our application in Figure 11.35. We shall assume that both $s_1(t)$ and $s_2(t)$ have equal a priori probabilities. For the null-zone receiver, two thresholds, a_1 and a_2, are established. If the sampled matched-filter output, denoted V, lies between a_1 and a_2, no decision is made, and the feedback channel is used to request a retransmission. This event is denoted an erasure and occurs with probability P_2. Assuming $s_1(t)$ transmitted, an error is made if $V > a_2$. The probability of this event is denoted P_1. By symmetry, these probabilities are the same for $s_2(t)$ transmitted.

Assuming independence, the probability of $j-1$ erasures followed by an error is

$$P(j-1 \text{ transmissions, error}) = P_2^{j-1}P_1 \tag{11.144}$$

The overall probability of error is the summation of this probability over all j. This is (note that $j=0$ is not included since $j=0$ corresponds to a correct decision resulting from a single transmission)

$$P_E = \sum_{j=1}^{\infty} P_2^{j-1}P_1 \tag{11.145}$$

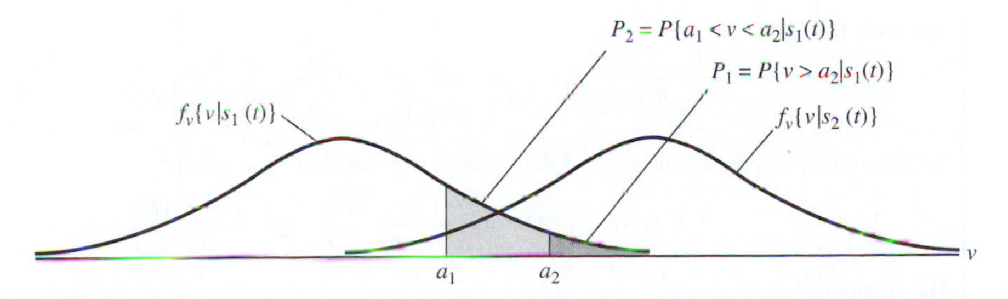

Figure 11.35
Decision regions for a null-zone receiver.

which is

$$P_E = \frac{P_1}{1 - P_2} \tag{11.146}$$

The expected number of transmissions N is also easily derived. The result is

$$N = \frac{1}{(1 - P_2)^2} \tag{11.147}$$

which is typically only slightly greater than one.

It follows from these results that the error probability can be reduced considerably without significantly increasing N. Thus performance is improved without a great sacrifice in information rate.

COMPUTER EXAMPLE 11.6

We now consider a baseband communications system with an integrate-and-dump detector. The output of the integrate-and-dump detector is given by

$$V = \begin{cases} +AT + N, & \text{if } +A \text{ is sent} \\ -AT + N, & \text{if } -A \text{ is sent} \end{cases}$$

where N is a random variable representing the noise at the detector output at the sampling instant. The detector uses two thresholds, a_1 and a_2, where $a_1 = -\gamma AT$ and $a_2 = \gamma AT$. A retransmission occurs if $a_1 < V < a_2$. Here we let $\gamma = 0.2$. The goal of this exercise is to compute and plot both the probability of error (Figure 11.36) and the expected number of transmissions (Figure 11.37) as a function of $z = A^2T/N_0$.

The probability density function of the sampled matched-filter output, conditioned on the transmission of $-A$, is

$$f_V(v|-A) = \frac{1}{\sqrt{2\pi}\sigma_n} \exp\left(-\frac{(v + AT)^2}{2\sigma_n^2}\right) \tag{11.148}$$

The probability of erasure is

$$P(\text{erasure}|-A) = \frac{1}{\sqrt{2\pi}\sigma_n} \int_{a_1}^{a_2} \exp\left(-\frac{(v + AT)^2}{2\sigma_n^2}\right) dv \tag{11.149}$$

With

$$y = \frac{v + AT}{\sigma_n} \tag{11.150}$$

(11.149) becomes

$$P(\text{erasure}|-A) = \frac{1}{\sqrt{2\pi}} \int_{(1-\gamma)AT/\sigma_n}^{(1+\gamma)AT/\sigma_n} \exp\left(-\frac{y^2}{2}\right) dy \tag{11.151}$$

which may be expressed in terms of the Gaussian Q-function. The result is

$$P(\text{erasure}|-A) = Q\left(\frac{(1-\gamma)AT}{\sigma_n}\right) - Q\left(\frac{(1+\gamma)AT}{\sigma_n}\right) \tag{11.152}$$

By symmetry

$$P(\text{erasure}|-A) = P(\text{erasure}|+A) \tag{11.153}$$

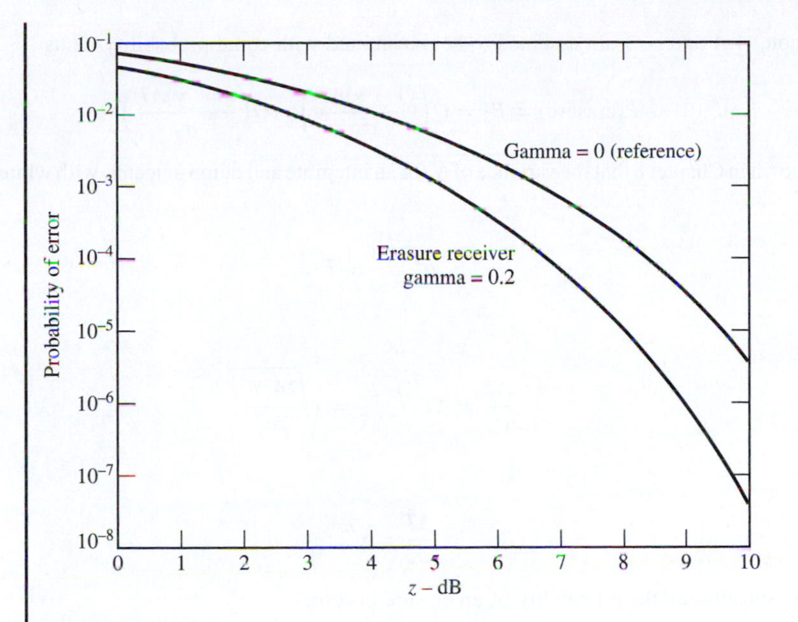

Figure 11.36
Probability of error.

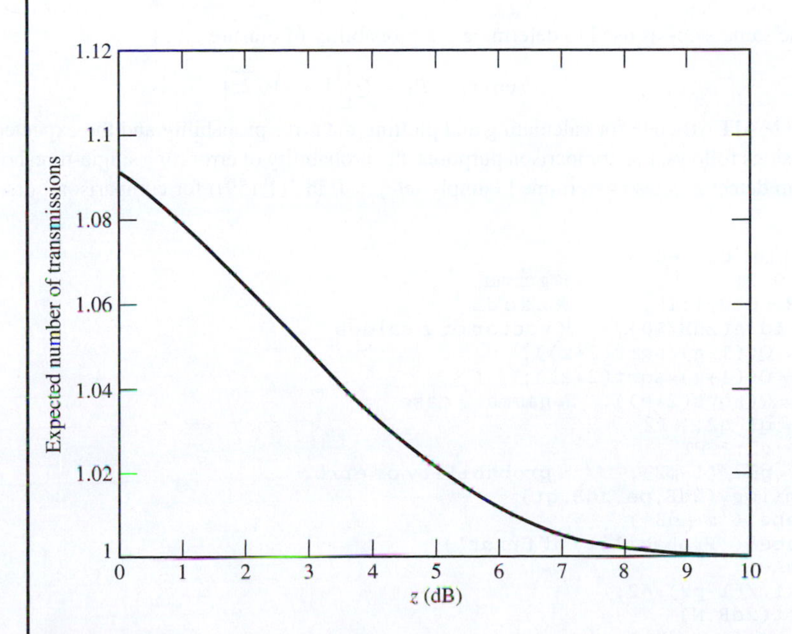

Figure 11.37
Expected number of transmissions.

In addition, $+A$ and $-A$ are assumed to be transmitted with equal probability. Thus

$$P(\text{erasure}) = P_2 = Q\left(\frac{(1-\gamma)AT}{\sigma_n}\right) - Q\left(\frac{(1+\gamma)AT}{\sigma_n}\right) \tag{11.154}$$

It was shown in Chapter 8 that the variance of N for an integrate and dump detector, with white noise input, is

$$\sigma_n^2 = \frac{1}{2}N_0T \tag{11.155}$$

Thus

$$\frac{AT}{\sigma_n} = AT\sqrt{\frac{2}{N_0T}} = \sqrt{\frac{2A^2T}{N_0}} \tag{11.156}$$

which is

$$\frac{AT}{\sigma_n} = \sqrt{2z} \tag{11.157}$$

With this substitution the probability of an erasure becomes

$$P_2 = Q\left[(1-\gamma)\sqrt{2z}\right] - Q\left[(1+\gamma)\sqrt{2z}\right] \tag{11.158}$$

The probability of error, conditioned on the transmission of $-A$ is

$$P(\text{error}|-A) = \frac{1}{\sqrt{2\pi}\sigma_n}\int_{a_2=\gamma AT}^{\infty} \exp\left(-\frac{(v+AT)^2}{2\sigma_n^2}\right)dv \tag{11.159}$$

Using the same steps as used to determine the probability of erasure gives

$$P(\text{error}) = P_1 = Q\left[(1+\gamma)\sqrt{2z}\right]$$

The MATLAB code for calculating and plotting the error probability and the expected number of transmissions follows. For comparison purposes, the probability of error for a single-threshold integrate-and-dump detector is also determined (simply let $\gamma = 0$ in (11.159)) for comparison purposes.

```
% File: c11ce6.m
g = 0.2;                % gamma
zdB = 0:0.1:10;         % z in dB
z = 10.^(zdB/10);       % vector of z values
q1 = Q((1-g)*sqrt(2*z));
q2 = Q((1+g)*sqrt(2*z));
qt = Q(sqrt(2*z));      % gamma=0 case
p2 = q1-q2; % P2
p1 = q2; % P1
pe = p1./(1-p2);        % probability of error
semilogy(zdB,pe,zdB,qt)
xlabel('z - dB')
ylabel('Probability of Error')
pause
N = 1./(1-p2).^2;
plot(zdB,N)
xlabel('z - dB')
ylabel('Expected Number of Transmissions')
% End of script file.
```

In the preceding program the Gaussian Q-function is calculated using the MATLAB routine

```
function out=Q(x)
out=0.5*erfc(x/sqrt(2));
```

■

■ 11.7 MODULATION AND BANDWIDTH EFFICIENCY

In Chapter 7, SNRs were computed at various points in a communication system. Of particular interest were the SNR at the input to the demodulator and the SNR of the demodulated output. These were referred to as the *predetection SNR*, $(\text{SNR})_T$, and the *postdetection SNR*, $(\text{SNR})_D$, respectively. The ratio of these parameters, the detection gain, has been widely used as a figure of merit for various systems. In this section we will compare the behavior of $(\text{SNR})_D$ as a function of $(\text{SNR})_T$ for several systems. First, however, we investigate the behavior of an *optimum*, but *unrealizable* system. This study will provide a basis for comparison and also provide additional insight into the concept of the trade-off of bandwidth for SNR.

11.7.1 Bandwidth and SNR

The block diagram of a communication system is illustrated in Figure 11.38. We will focus on the receiver portion of the system. The SNR at the output of the predetection filter, $(\text{SNR})_T$, yields the maximum rate at which information may arrive at the receiver. From the Shannon–Hartley law, this rate, C_T is

$$C_T = B_T \log_2\left[1 + (\text{SNR})_T\right] \tag{11.160}$$

where B_T, the predetection bandwidth, is typically the bandwidth of the modulated signal. Since (11.160) is based on the Shannon–Hartley law, it is valid only for AWGN cases. The SNR of the demodulated output, $(\text{SNR})_D$, yields the maximum rate at which information may leave the receiver. This rate, denoted C_D is given by

$$C_D = W \log_2\left[1 + (\text{SNR})_D\right] \tag{11.161}$$

where W is the bandwidth of the message signal.

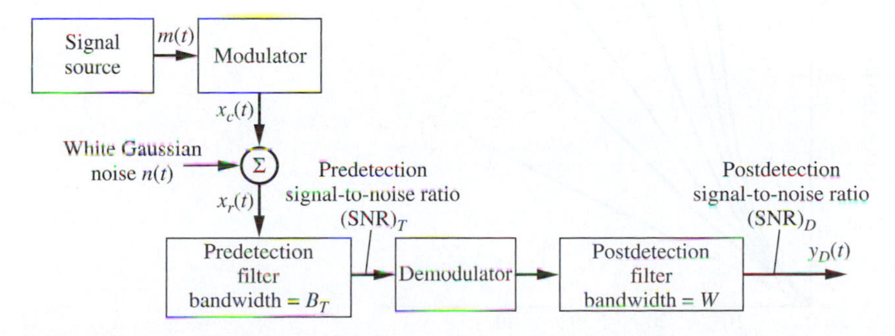

Figure 11.38
Block diagram of a communication system.

Optimal modulation is defined $C_D = C_T$. For this system, demodulation is accomplished, in the presence of noise, without loss of information. Equating C_D to C_T yields

$$(\text{SNR})_D = \left[1 + (SNR)_T\right]^{B_T/W} - 1 \tag{11.162}$$

which shows that the optimum exchange of bandwidth for SNR is *exponential*. Recall that we first encountered the trade-off between bandwidth and system performance, in terms of the SNR at the output of the demodulator in Chapter 7 when the performance of FM modulation in the presence of noise was studied.

The ratio of transmission bandwidth B_T to the message bandwidth W is referred to as the *bandwidth expansion factor* γ. To fully understand the role of this parameter, we write the predetection SNR as

$$(\text{SNR})_T = \frac{P_T}{N_0 B_T} = \frac{W}{B_T}\frac{P_T}{N_0 W} = \frac{1}{\gamma}\frac{P_T}{N_0 W} \tag{11.163}$$

Thus (11.162) can be expressed as

$$(\text{SNR})_D = \left[1 + \frac{1}{\gamma}\left(\frac{P_T}{N_0 W}\right)\right]^{\gamma} - 1 \tag{11.164}$$

The relationship between $(\text{SNR})_D$ and $P_T/N_0 W$ is illustrated in Figure 11.39.

11.7.2 Comparison of Modulation Systems

The concept of an optimal modulation system provides a basis for comparing system performance. For example, an ideal SSB system has a bandwidth expansion factor of one, since the transmission bandwidth is ideally equal to the message bandwidth. Thus the

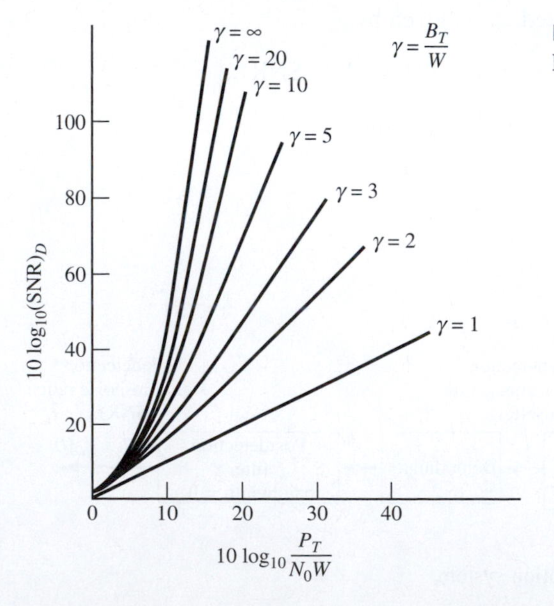

Figure 11.39
Performance of an optimum modulation system.

postdetection SNR of the optimal modulation system is, from (11.164) with γ equal to 1,

$$(\mathrm{SNR})_D = \frac{P_T}{N_0 W} \tag{11.165}$$

This is exactly the same result as obtained in Chapter 7 for an SSB system using coherent demodulation with a perfect phase reference. Therefore, if the transmission bandwidth B_T of an SSB system is *exactly* equal to the message bandwidth W, SSB is optimal, assuming that there are no other error sources. Of course, this can never be achieved in practice, since *ideal* filters would be required in addition to *perfect* phase coherence of the demodulation carrier.

The story is quite different with DSB, AM, and QDSB. For these systems, $\gamma = 2$. In Chapter 7 we saw that the postdetection SNR for DSB and QDSB, assuming perfect coherent demodulation, is

$$(\mathrm{SNR})_D = \frac{P_T}{N_0 W} \tag{11.166}$$

whereas for the optimal system it is given by (11.164) with $\gamma = 2$.

These results are shown in Figure 11.40 along with the result for AM with square-law demodulation. It can be seen that these systems are far from optimal, especially for large values of $P_T/N_0 W$.

Also shown in Figure 11.40 is the result for FM without preemphasis, with sinusoidal modulation, assuming a modulation index of 10. With this modulation index, the bandwidth expansion factor is

$$\gamma = \frac{2(\beta+1)W}{W} = 22 \tag{11.167}$$

The realizable performance of the FM system is taken from Figure 7.18. It can be seen that realizable systems fall far short of optimal if γ and $P_T/N_0 W$ are large.

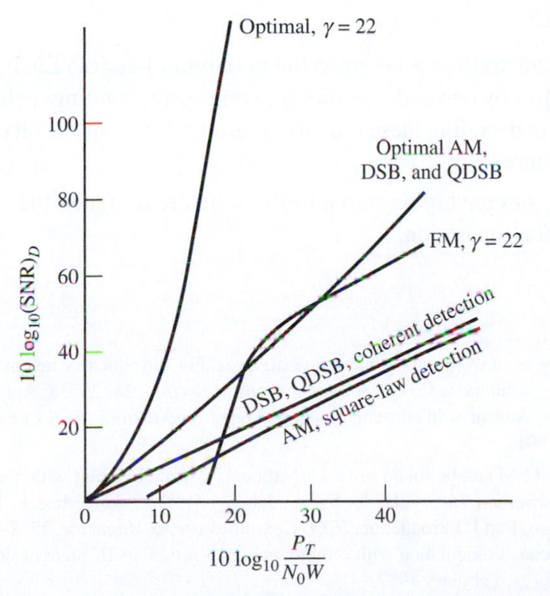

Figure 11.40
Performance comparison of analog systems.

11.8 BANDWIDTH AND POWER EFFICIENT MODULATION (TCM)

A desirable characteristic of any modulation scheme is the simultaneous conservation of bandwidth and power. Since the late 1970s, the approach to this challenge has been to combine coding and modulation. There have been two approaches: (1) continuous phase modulation (CPM)[10] with memory extended over several modulation symbols by cyclical use of a set of modulation indices; and (2) combining coding with an M-ary modulation scheme, referred to as *trellis-coded modulation* (TCM).[11] We briefly explore the latter approach in this section. For an introductory discussion of the former approach, see Ziemer and Peterson (1985), Chapter 4. Sklar (1988) is a well-written reference with more examples on TCM than given in this short section.

In Chapter 9 it was illustrated through the use of signal space diagrams that the most probable errors in an M-ary modulation scheme result from mistaking a signal point closest in Euclidian distance to the transmitted signal point as corresponding to the actual transmitted signal. Ungerboeck's solution to this problem was to use coding in conjunction with M-ary modulation to increase the minimum Euclidian distance between those signal points most likely to be confused without increasing the average power or bandwidth over an uncoded scheme transmitting the same number of bits per second. We illustrate the procedure with a specific example.

We wish to compare a TCM system and a QPSK system operating at the same data rates. Since the QPSK system transmits 2 bits per signal phase (signal space point), we can keep that same data rate with the TCM system by employing an 8-PSK modulator, which carries 3 bits per signal phase, in conjunction with a convolutional coder that produces three encoded symbols for every two input data bits, i.e., a rate $\frac{2}{3}$ coder. Figure 11.37(a) shows an coder for accomplishing this, and Figure 11.41(b) shows the corresponding trellis diagram. The coder operates by taking the first data bit as the input to a rate $\frac{1}{2}$ convolutional coder that produces the first and second encoded symbols, and the second data bit directly as the third encoded symbol. These are then used to select the particular signal phase to be transmitted according to the following rules:

1. All parallel transitions in the trellis are assigned the maximum possible Euclidian distance. Since these transitions differ by one code symbol (the one corresponding to the uncoded bit in this example), an error in decoding these transitions amounts to a single bit error, which is minimized by this procedure.
2. All transitions emanating or converging into a trellis state are assigned the next to largest possible Euclidian distance separation.

[10]Continuous phase Modulation has been explored by many investigators. For introductory treatments see C.-E. Sundberg, Continuous phase modulation. *IEEE Communications Magazine*, **24**: 25–38, April 1986, and J. B. Anderson and C.-E. Sundberg, Advances in constant envelope coded modulation. *IEEE Communications Magazine,* **29**: 36–45, December 1991.

[11]Three introductory treatments of TCM can be found in G. Ungerboeck, Channel coding with multilevel/phase signals. *IEEE Transactions on Information Theory*, IT-28: 55–66, January 1982; G. Ungerboeck, Trellis-coded modulation with redundant signal sets, Part I: Introduction. *IEEE Communications Magazine*, **25**: 5–11, February 1987; and G. Ungerboeck, Trellis-coded modulation with redundant signal sets, Part II: State of the art. *IEEE Communications Magazine*, **25**: 12–21, February 1987.

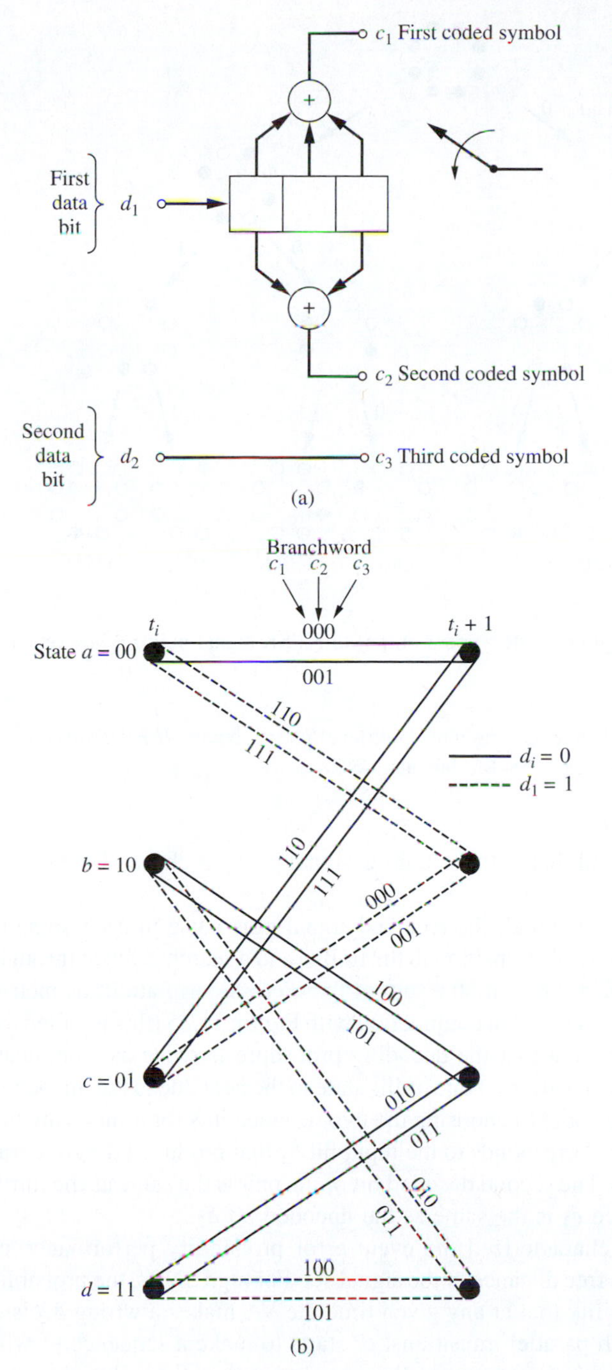

Figure 11.41
(a) Convolutional coder and (b) trellis diagram corresponding to 4-state, 8-PSK TCM.

The application of these rules to assigning the encoded symbols to a signal phase in an 8-PSK system can be done with a technique known as set partitioning, which is illustrated in Figure 11.42. If the coded symbol c_1 is a 0, the left branch is chosen in the first tier of the tree, whereas if c_1 is a 1, the right branch is chosen. A similar procedure is followed for tiers 2 and 3

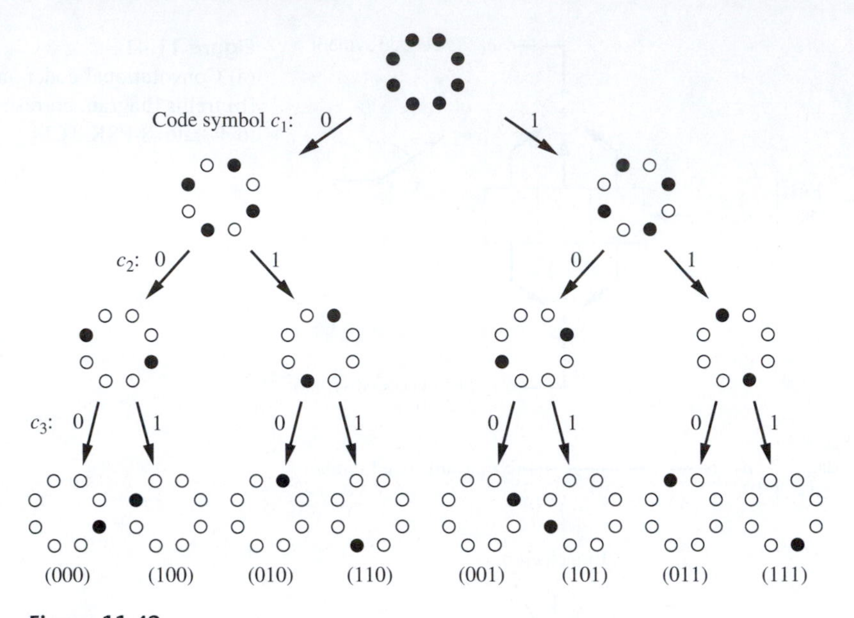

Figure 11.42
Set partitioning for assigning a rate $\frac{2}{3}$ coder output to 8-PSK signal points while obeying the rules for maximizing free distance.

From G. Ungerboeck, Channel coding with multilevel/phase signals. *IEEE Transactions on Information Theory*, **IT-28**: pp. 55–66, January 1982.

of the tree, with the result being that a unique signal phase is chosen for each possible coded output.

To decode the TCM signal, the received signal plus noise in each signaling interval is correlated with each possible transition in the trellis, and a search is made through the trellis by means of a Viterbi algorithm using the sum of these cross-correlations as metrics rather than Hamming distance as discussed in conjunction with Figure 11.25 (this is called the use of a *soft decision metric*). Also note that the decoding procedure is twice as complicated since two branches correspond to a path from one trellis state to the next due to the uncoded bit becoming the third symbol in the code. In choosing the two decoded bits for a surviving branch, the first decoded bit of the pair corresponds to the input bit b_1 that produced the state transition of the branch being decoded. The second decoded bit of the pair is the same as the third symbol c_3 of that branch word, since c_3 is the same as the uncoded bit b_2.

Ungerboeck has characterized the event error probability performance of a signaling method in terms of the free distance of the signal set. For high SNRs, the probability of an error event (i.e., the probability that at any given time the VA makes a wrong decision among the signals associated with parallel transitions, or starts to make a sequence of wrong decisions along some path diverging from more than one transition from the correct path) is well approximated by

$$P(\text{error event}) = N_{\text{free}} Q\left(\frac{d_{\text{free}}}{2\sigma}\right) \tag{11.168}$$

where N_{free} denotes the number of nearest-neighbor signal sequences with distance d_{free} that diverge at any state from a transmitted signal sequence, and reemerge with it after one or more transitions. (The free distance is often calculated by assuming the signal energy has been normalized to unity and that the noise standard deviation σ accounts for this normalization.)

For uncoded QPSK, we have $d_{\text{free}} = 2^{1/2}$ and $N_{\text{free}} = 2$ (there are two adjacent signal points at distance $d_{\text{free}} = 2^{1/2}$), whereas for 4-state-coded 8- PSK we have $d_{\text{free}} = 2$ and $N_{\text{free}} = 1$. Ignoring the factor N_{free}, we have an asymptotic gain due to TCM over uncoded QPSK of $2^2/(2^{1/2})^2 = 2 = 3$ dB. Figure 11.43, also from Ungerboeck, compares the asymptotic lower bound for the error event probability with simulation results.

It should be clear that the TCM coding–modulation procedure can be generalized to higher–level M-ary schemes. Ungerboeck shows that this observation can be generalized as follows:

1. Of the m bits to be transmitted per coder–modulator operation, $k \leq m$ bits are expanded to $k+1$ coded symbols by a binary rate $k/(k+1)$ convolutional coder.
2. The $k+1$ coded symbols select one of 2^{k+1} subsets of a redundant 2^{m+1}-ary signal set.
3. The remaining $m-k$ symbols determine one of 2^{m-k} signals within the selected subset.

It should also be stated that one may use block codes or other modulation schemes, such as M-ary ASK or QASK, to implement a TCM system.

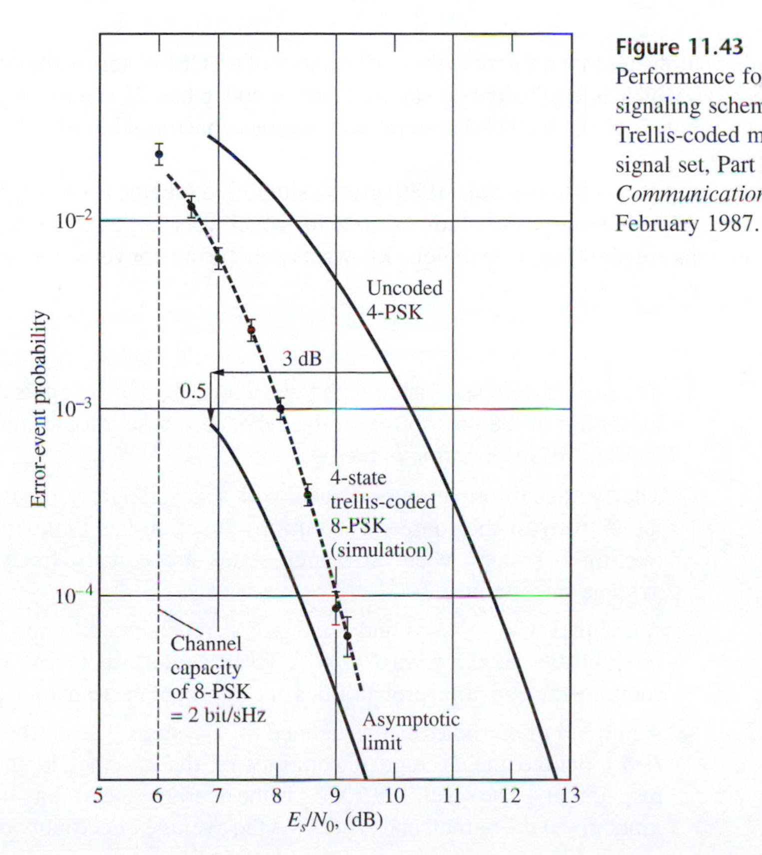

Figure 11.43
Performance for a 4-state, 8-PSK TCM signaling scheme. From G. Ungerboeck, Trellis-coded modulation with redundant signal set, Part l: Introduction. *IEEE Communications Magazine*, **25**: 5–11, February 1987.

Table 11.9 Asymptotic Coding Gains for TCM Systems

No. of States, 2^r	k	Asymtotic coding gain (dB) $G_{8PSK/QPSK}$ $m = 2$	$G_{16PSK/8PSK}$ $m = 3$
4	1	3.01	—
8	2	3.60	—
16	2	4.13	—
32	2	4.59	—
64	2	5.01	—
128	2	5.17	—
256	2	5.75	—
4	1	—	3.54
8	1	—	4.01
16	1	—	4.44
32	1	—	5.13
64	1	—	5.33
128	1	—	5.33
256	2	—	5.51

Source: Adapted from G. Ungerboeck, "Trellis-Coded Modulation with Redundant Signal Sets, Part II: States of the Art," *IEEE Communications Magazine.* Vol. 25. Feb. 1987, pp. 12–21.

Another parameter that influences the performance of a TCM system is the constraint span of the code, v, which is equivalent to saying that the coder has 2^v states. Ungerboeck has published asymptotic gains for TCM systems with various constraint lengths. These are given in Table 11.9.

Finally, the paper by Viterbi et al. (1989) gives a simplified scheme for M-ary PSK that uses a single rate $\frac{1}{2}$, 64-state binary convolutional code for which very large scale integrated circuit implementations are plentiful. A technique known as puncturing converts it to rate $(n-1)/n$.

Summary

1. The information associated with the occurrence of an event is defined as the logarithm of the probability of the event. If a base 2 logarithm is used, the measure of information is the bit.
2. The average information associated with a set of source outputs is known as the entropy of the source. The entropy function has a maximum, and the maximum occurs when all source states are equally likely. Entropy is average uncertainty.
3. A channel with n inputs and m outputs is represented by the nm transition probabilities of the form $P(y_j|x_i)$. The channel model can be a diagram showing the transition probabilities or a matrix of the transition probabilities.
4. A number of entropies can be defined for a system. The entropies $H(X)$ and $H(Y)$ denote the average uncertainty of the channel input and output, respectively. The quantity $H(X|Y)$ is the average uncertainty of the channel input given the output, and $H(Y|X)$ is the average uncertainty of the channel

output given the input. The quantity $H(X, Y)$ is the average uncertainty of the communication system as a whole.

5. The mutual information between the input and output of a channel is given by

$$I(X;Y) = H(X) - H(X|Y)$$

or

$$I(X;Y) = H(Y) - H(Y|X)$$

The maximum value of mutual information, where the maximization is with respect to the source probabilities, is known as the *channel capacity*.

6. Source coding is used to remove redundancy from a source output so that the information per transmitted symbol can be maximized. If the source rate is less than the channel capacity, it is possible to code the source output so that the information can be transmitted through the channel. This is accomplished by forming source extensions and coding the symbols of the extended source into code words having minimum average word length. The minimum average word length $\overline{L}$ approaches $H(X^n) = nH(X)$, where $H(X^n)$ is the entropy of the nth-order extension of a source having entropy $H(X)$, as n increases.

7. Two techniques for source coding were illustrated in this chapter. These were the Shannon–Fano technique and the Huffman technique. The Huffman technique yields an optimum source code, which is the source code having minimum average word length.

8. Error-free transmission on a noisy channel can be accomplished if the source rate is less than the channel capacity. This is accomplished using channel codes.

9. The capacity of an AWGN channel is

$$C_c = B \log_2 \left(1 + \frac{S}{N}\right)$$

where B is the channel bandwidth and S/N is the SNR. This is known as the Shannon–Hartley law.

10. An (n, k) block code is generated by appending $r = n - k$ parity symbols to a k-symbol source sequence. This yields an n-symbol code word.

11. Decoding is typically accomplished by computing the Hamming distance from the received n-symbol sequence to each of the possible transmitted code words. The code word closest in Hamming distance to the received sequence is the most likely transmitted code word. The two code words closest in Hamming distance determine the minimum distance of the code d_m. The code can correct $\frac{1}{2}(d_m - 1)$ errors.

12. A single-parity-check code is formed by adding a single-parity symbol to the information sequence. This $(k+1, k)$ code can detect single errors but provides no error-correcting capability.

13. The rate of a block code is k/n. The best codes provide powerful error-correction capabilities in combination with high rate.
14. Repetition codes are formed by transmitting each source symbol an odd number of times and therefore have rate $1/n$. Repetition codes do not provide improved performance in an AWGN environment but do provide improved performance in a Rayleigh fading environment. This simple example illustrated the importance of selecting an appropriate coding scheme for a given channel.
15. The parity-check matrix $[H]$ is defined such that $[H][T] = [0]$, where $[T]$ is the transmitted code word written as a column vector. If the received sequence is denoted by the column vector $[R]$, the syndrome $[S]$ is determined from $[S] = [H][R]$. This can be shown to be equivalent to $[S] = [H][E]$, where $[E]$ is the error sequence. If a single error occurs in the transmission of a code word, the syndrome is the column of $[H]$ corresponding to the error position.
16. The generator matrix $[G]$ of a parity-check code is determined such that $[T] = [G][A]$, where $[T]$ is the n-symbol transmitted sequence and $[A]$ is the k-symbol information sequence. Both $[T]$ and $[A]$ are written as column vectors.
17. For a group code, the modulo 2 sum of any two code words is another code word.
18. A Hamming code is a single error-correcting code such that the columns of the parity-check matrix correspond to the binary representation of the column index.
19. Cyclic codes are a class of block codes in which a cyclic shift of code-word symbols always yields another code word. These codes are very useful because implementation of both the coder and decoder is easily accomplished using shift registers and basic logic components.
20. The channel symbol-error probability of a coded system is greater than the symbol-error probability of an uncoded system since the available energy for transmission of k information symbols must be spread over the $n > k$ symbol code word rather than just the k information symbols. The error-correcting capability of the code often allows a net performance gain to be realized. The performance gain depends on the choice of code and the channel characteristics.
21. Convolutional codes are easily generated using simple shift registers and modulo 2 adders. Decoding is accomplished using a tree-search technique, which is often implemented using the Viterbi algorithm. The constraint span is the code parameter having the most significant impact on performance.
22. Interleaved codes are useful for burst-noise environments.
23. The feedback channel system makes use of a null-zone receiver, and a retransmission is requested if the receiver decision falls within the null zone. If a feedback channel is available, the error probability can be significantly reduced with only a slight increase in the required number of transmissions.
24. Use of the Shannon–Hartley law yields the concept of optimum modulation for a system operating in an AWGN environment. The result is the

performance of an optimum system in terms of predetection and postdetection bandwidth. The trade-off between bandwidth and SNR is easily seen.

25. Trellis-coded modulation is a scheme for combining M-ary modulation with coding in a way that increases the Euclidian distance between those signal points for which errors are most likely without increasing the average power or bandwidth over an uncoded scheme having the same bit rate. Decoding is accomplished using a Viterbi decoder that accumulates decision metrics (soft decisions) rather than Hamming distances (hard decisions).

Further Reading

An exposition of information theory and coding that was anywhere near complete would, of necessity, be presented at a level far beyond that intended for this text. The purpose in the present chapter is to present some of the basic ideas of information theory at a level consistent with the rest of this book. You should be motivated by this to further study.

The original paper by Shannon (1948) is stimulating reading at about the same level as this chapter. This paper is available as a paperback with an interesting postscript by W. Weaver (Shannon and Weaver, 1963).

A variety of textbooks on information theory are available. The book by Blahut (1987) is recommended. A current standard that is used in many graduate programs was authored by Cover and Thomas (2006).

There are also a number of textbooks available that cover coding theory at the graduate level. The book by Lin and Costello (2004) is a standard textbook. The book by Clark and Cain (1981) contains a wealth of practical information concerning coder and decoder design, in addition to the usual theoretical background material.

As mentioned in the last section of this chapter, the subject of bandwidth and power-efficient communications is very important to the implementation of modern systems. Continuous phase modulation is treated in the text by Ziemer and Peterson (1985). An introductory treatment of TCM, including a discussion of coding gain, is contained in the book by Sklar (2001). The book by Biglieri et al. (1991) is a complete treatment of TCM theory, performance, and implementation. A book on Turbo Codes is (Heegard and Wicker, 1999).

Problems

Section 11.1

11.1. A message occurs with a probability of 0.95. Determine the information associated with the message in bits, nats, and hartleys.

11.2. Assume that you have a standard deck of 52 cards (jokers have been removed).

a. What is the information associated with the drawing of a single card from the deck?

b. What is the information associated with the drawing of a pair of cards, assuming that the first card drawn is replaced in the deck prior to drawing the second card?

c. What is the information associated with the drawing of a pair of cards assuming that the first card drawn is not replaced in the deck prior to drawing the second card?

11.3. A source has five outputs denoted $[m_1, m_2, m_3, m_4, m_5]$ with respective probabilities [0.35, 0.25, 0.20, 0.15, 0.05]. Determine the entropy of the source. What is the maximum entropy of a source with five outputs?

11.4. A source consists of six outputs denoted $[A, B, C, D, E, F]$ with respective probabilities $[0.25, 0.25, 0.2, 0.1, 0.1, 0.1]$. Determine the entropy of the source.

11.5. A channel has the following transition matrix:

$$\begin{bmatrix} 0.7 & 0.2 & 0.1 \\ 0.2 & 0.5 & 0.3 \\ 0.1 & 0.2 & 0.7 \end{bmatrix}$$

a. Sketch the channel diagram showing all transition probabilities.

b. Determine the channel output probabilities assuming that the input probabilities are equal.

c. Determine the channel input probabilities that result in equally likely channel outputs.

d. Determine the joint probability matrix using part (c).

11.6. Describe the channel transition probability matrix and joint probability matrix for a noiseless channel.

11.7. Show that the cascade of N different binary symmetric channels yields a binary symmetric channel.

11.8. A binary symmetric channel has an error probability of 0.005. How many of these channels can be cascaded before the overall error probability exceeds 0.1?

11.9. A channel is described by the transition probability matrix

$$[P(Y|X)] = \begin{bmatrix} 3/4 & 1/4 & 0 \\ 0 & 0 & 1 \end{bmatrix}$$

Determine the channel capacity and the source probabilities that yield capacity.

11.10. A channel has two inputs, $(0, 1)$, and three outputs, $(0, e, 1)$, where e indicates an erasure; that is, there is no output for the corresponding input. The channel matrix is

$$\begin{bmatrix} 1-p & p & 0 \\ 0 & p & 1-p \end{bmatrix}$$

Compute the channel capacity.

11.11. A binary symmetric channel with error probability p_1 is followed by an erasure channel with erasure probability p_2. Describe the channel matrix that results from this cascade combination of channels. Comment on the results.

11.12. Determine the capacity of the channel described by the channel matrix shown below. Sketch your result as a function of p and give an intuitive argument that supports your sketch. (Note: $q = 1 - p$.). Generalize to N parallel binary symmetric channels.

$$\begin{bmatrix} p & q & 0 & 0 \\ q & p & 0 & 0 \\ 0 & 0 & p & q \\ 0 & 0 & q & p \end{bmatrix}$$

11.13. From the entropy definitions given in (11.25) through (11.29), derive (11.30) and (11.31).

11.14. The input to a quantizer is a random signal having an amplitude probability density function

$$f_X(x) = \begin{cases} ae^{-ax}, & x \geq 0 \\ 0, & x < 0 \end{cases}$$

The signal is to be quantized using four quantizing levels x_i as shown in Figure 11.44. Determine the values $x_i, i = 1, 2, 3$, in terms of a so that the entropy at the quantizer output is maximized.

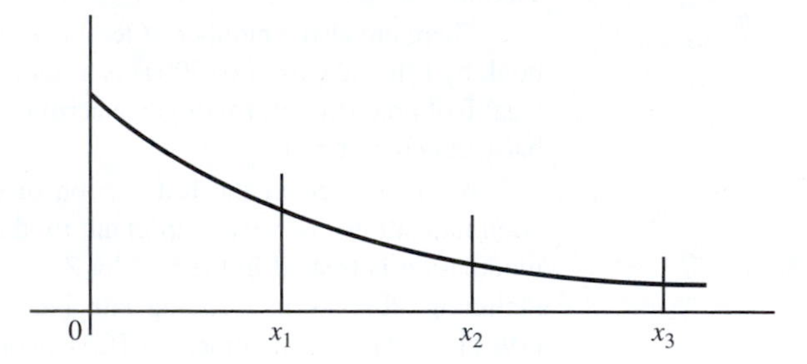

Figure 11.44

11.15. Repeat the preceding problem assuming that the input to the quantizer has the Rayleigh probability density function

$$f_X(x) = \begin{cases} \dfrac{x}{a^2} e^{-x^2/2a^2}, & x \geq 0 \\ 0, & x < 0 \end{cases}$$

11.16. A signal has a Gaussian amplitude-density function with zero mean and variance σ^2. The signal is sampled at 500 samples per second. The samples are quantized according to the following table. Determine the entropy at the quantizer output and the information rate in samples per second.

Quantizer input	Quantizer output
$-\infty < x_i < -2\sigma$	m_0
$-2\sigma < x_i < -\sigma$	m_1
$-\sigma < x_i < 0$	m_2
$0 < x_i < \sigma$	m_3
$\sigma < x_i < 2\sigma$	m_4
$2\sigma < x_i < \infty$	m_5

11.17. Determine the quantizing levels, in terms of σ, so that the entropy at the output of a quantizer is maximized. Assume that there are six quantizing levels and that the quantizer is a zero-mean Gaussian process as in the previous problem.

11.18. Two binary symmetrical channels are in cascade, as shown in Figure 11.45. Determine the capacity of each channel. The overall system with inputs x_1 and x_2 and outputs z_1 and z_2 can be represented as shown with p_{11}, p_{12}, p_{21}, and p_{22} properly chosen. Determine these four probabilities and the capacity of the overall system. Comment on the results.

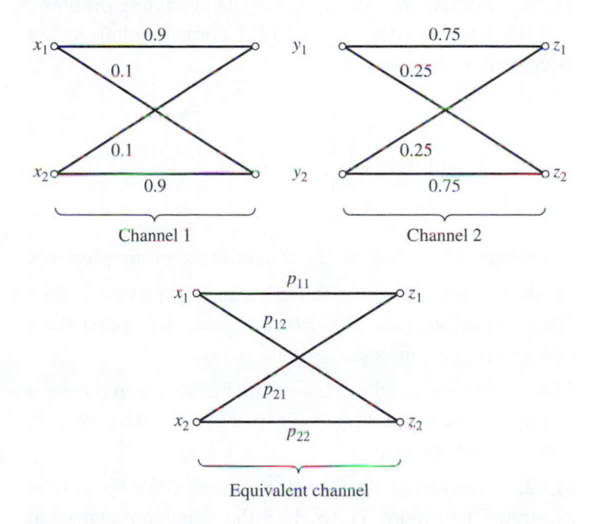

Figure 11.45

11.19. A two-hop satellite communications channel uses BPSK signaling. The uplink SNR is 8 dB, and the downlink SNR is 5 dB, where the SNR is the signal power divided by the noise power in the bit-rate bandwidth. Determine the overall error probability.

Section 11.2

11.20. A source has two outputs $[A, B]$ with respective probabilities $\left[\frac{3}{4}, \frac{1}{4}\right]$. Determine the entropy of the fourth-order extension of this source using two different methods.

11.21. Calculate the entropy of the fourth-order extension of the source defined in Table 11.1. Determine $\overline{L}/n$ for $n = 4$, and add this result to those shown in Figure 10.9. Determine the efficiency of the resulting codes for $n = 1, 2, 3$, and 4.

11.22. A source has five equally likely output messages. Determine a Shannon–Fano code for the source, and determine the efficiency of the resulting code. Repeat for the Huffman code, and compare the results.

11.23. A source has five outputs denoted $[m_1, m_2, m_3, m_4, m_5]$ with respective probabilities [0.41, 0.19, 0.16, 0.15, 0.9]. Determine the code words to represent the source outputs using both the Shannon–Fano and the Huffman techniques.

11.24. A binary source has output probabilities [0.85, 0.15]. The channel can transmit 350 binary symbols per second at the capacity of 1 bit/symbol. Determine the maximum source symbol rate if transmission is to be accomplished.

11.25. A source output consists of nine equally likely messages. Encode the source output using both binary Shannon–Fano and Huffman codes. Compute the efficiency of both of the resulting codes and compare the results.

11.26. Repeat the preceding problem assuming that the source has 12 equally likely outputs.

11.27. An analog source has an output described by the probability density function

$$f_X(x) = \begin{cases} 2x, & 0 \le x \le 1 \\ 0, & \text{otherwise} \end{cases}$$

The output of the source is quantized into 10 messages using the nine quantizing levels

$$x_i = 0.1k,\ k = 0, 1, \ldots, 10$$

The resulting messages are encoded using a binary Huffman code. Assuming that 250 samples of the source are transmitted each second, determine the resulting binary symbol rate in symbols per second. Also determine the information rate in bits per second.

11.28. A source output consists of four messages $[m_1, m_2, m_3, m_4]$ with respective probabilities [0.35, 0.3, 0.2, 0.15]. Determine the binary code words for the second-order source extension using the Shannon–Fano and Huffman coding techniques. Determine the efficiency of the resulting codes and comment on the results.

11.29. A source output consists of four messages $[m_1, m_2, m_3, m_4]$ with respective probabilities [0.35, 0.3,

0.2, 0.15]. The second-order extension of the source is to be encoded using a code with a three-symbol alphabet using the Shannon–Fano and coding technique Determine the efficiency of the resulting code.

11.30. It can be shown that a necessary and sufficient condition for the existence of an instantaneous binary code with word lengths l_i, $1 \leq i \leq N$, is that

$$\sum_{i=1}^{N} 2^{-l_i} \leq 1$$

This is known as the *Kraft inequality*. Show that the Kraft inequality is satisfied by the code words given in Table 11.3. (Note: The inequality given above must also be satisfied for uniquely decipherable codes.)

Section 11.3

11.31. A continuous bandpass channel can be modeled as illustrated in Figure 11.46. Assuming a signal power of 50 W and a noise power spectral density of 10^{-5} W/Hz, plot the capacity of the channel as a function of the channel bandwidth, and compute the capacity in the limit as $B \rightarrow \infty$.

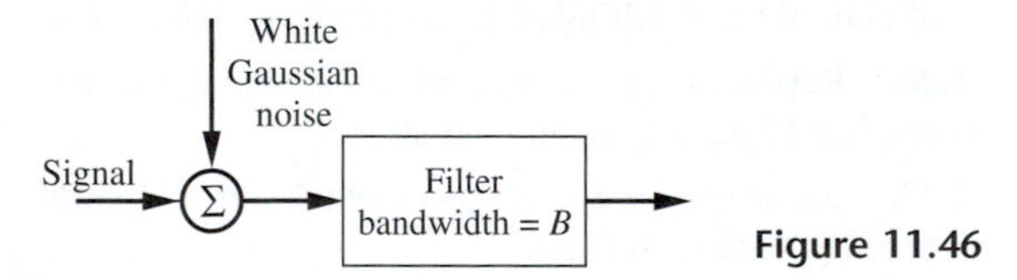

Figure 11.46

11.32. Consider again the bandpass channel illustrated in Figure 11.46. The noise power spectral density is 10^{-5} and the bandwidth is 10 kHz. Plot the capacity of the channel as a function of signal power P_T, and compute the capacity in the limit as $P_T \rightarrow \infty$. Contrast the result of the problem with the result of preceding problem.

Section 11.4

11.33. A (8,7) parity-check code is used on a channel having symbol-error probability p. Determine the probability of one or more undetected errors in an 8-symbol code word.

11.34. Derive an equation, similar to (11.95), that gives the error probability for a rate $\frac{1}{7}$ repetition code. Plot together, on the same set of axes, the error probability of both a rate of $\frac{1}{3}$ and rate $\frac{1}{7}$ repetition code as a function of $q = 1 - p$.

11.35. Develop an analysis that shows that increasing n for a rate $1/n$ always degrades system performance in an AWGN. In order to obtain specific results, assume PSK modulation.

11.36. Show that a (15, 11) Hamming code is a distance 3 code.

Hint: It is not necessary to find all code words.

11.37. Write the parity-check matrix and the generator matrix for a (15, 11) single error-correcting code. Assume that the code is systematic. Calculate the code word corresponding to the all-ones information sequence. Calculate the syndrome corresponding to an error in the third position assuming the code word corresponding to the all-ones input sequence.

11.38. A parity-check code has the parity-check matrix

$$[H] = \begin{bmatrix} 0 & 1 & 1 & 1 & 1 & 0 & 0 \\ 1 & 0 & 1 & 1 & 0 & 1 & 0 \\ 1 & 1 & 0 & 1 & 0 & 0 & 1 \end{bmatrix}$$

Determine the generator matrix and find all possible code words.

11.39. For the code described in the preceding problem, find the code words $[T_1]$ and $[T_2]$ corresponding to the information sequences

$$[A_1] = \begin{bmatrix} 0 \\ 1 \\ 1 \\ 1 \end{bmatrix} \qquad [A_2] = \begin{bmatrix} 1 \\ 0 \\ 1 \\ 0 \end{bmatrix}$$

Using these two code words, illustrate the group property.

11.40. Determine the generator matrix for a rate $\frac{1}{3}$ and a rate $\frac{1}{5}$ repetition code. Describe the generator matrix for a rate $1/n$ repetition code.

11.41. Determine the relationship between n and k for a Hamming code. Use this result to show that the code rate approaches 1 for large n.

11.42. Determine the generator matrix for the coder illustrated in Figure 11.15. Use the generator matrix to generate the complete set of code words and use the results to check the code words shown in Figure 11.15. Show that these code words constitute a cyclic code.

11.43. Use the result of the preceding problem to determine the parity-check matrix for the coder shown in Figure 11.15. Use the parity-check matrix to decode the received sequences 1101001 and 1101011. Compare your result with that shown in Figure 10.16.

11.44. Consider the coded system examined in Computer Example 11.1. Show that the probability of three symbol errors in a code word is negligible compared to the probability of two symbol errors in a code word for SNRs above a certain level.

11.45. The Hamming code was defined as a code for which the ith column of the parity-check matrix is the binary representation of the number i. With a little thought it becomes clear that the columns of the parity-check matrix can be permuted without changing the distance properties, and therefore the error-correcting capabilities, of the code. Using this fact, determine a generator matrix and the corresponding parity-check matrix of the systematic code equivalent to a (7, 4) Hamming code. How many different generator matrices can be found?

Section 11.5

11.46. Consider the convolutional coder shown in Figure 11.24. The shift register contents are $S_1S_2S_3$, where S_1 represents the most recent input. Compute the output sequence $v_1v_2v_3$ for $S_1 = 0$ and for $S_1 = 1$. Show that the two output sequences generated are complements.

11.47. Repeat the preceding problem for the convolutional coder illustrated in Figure 11.47. For the coder shown in Figure 11.47 the shift register contents are $S_1S_2S_3S_4$, where S_1 represents the most recent input.

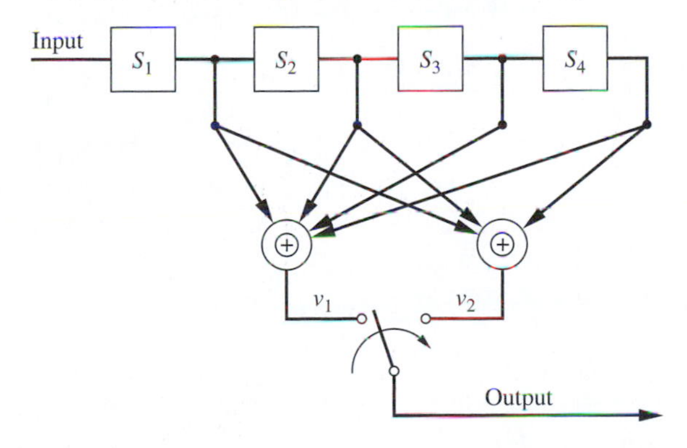

Figure 11.47

11.48. What is the constraint span of the convolutional coder shown in Figure 11.47? How many states are required to define the state diagram and the trellis diagram? Draw the state diagram, giving the output for each state transition.

11.49. Determine the state diagram for the convolutional coder shown in Figure 11.48. Draw the trellis diagram through the first set of steady-state transitions. On a second trellis diagram, show the termination of the trellis to the all-zero state.

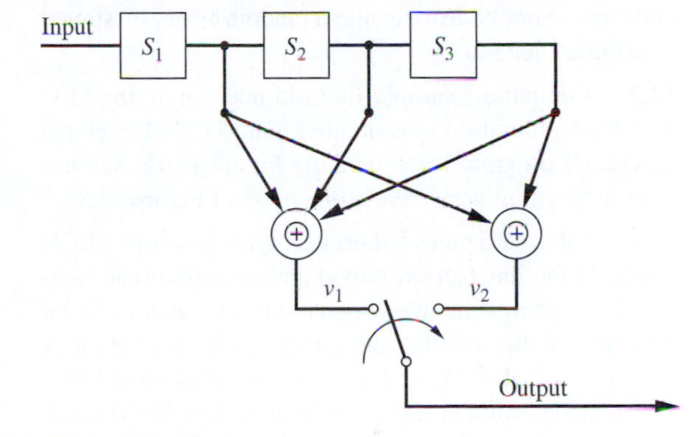

Figure 11.48

Section 11.6

11.50. A source produces binary symbols at a rate of 5000 symbols per second. The channel is subjected to error bursts lasting 0.2 s Devise an encoding scheme using an interleaved (n, k) Hamming code, which allows full correction of the error burst. Assume that the information rate out of the coder is equal to the information rate into the coder. What is the minimum time between bursts if your system is to operate properly?

11.51. Repeat the preceding problem assuming a (23,12) Golay code.

11.52. Develop the appropriate analysis to verify the correctness of (11.147).

Section 11.7

11.53. Compare FM with preemphasis to an optimal modulation system for $\beta = 1, 5,$ and 10. Consider only operation above threshold, and assume 20 dB as the value of P_T/N_0W at threshold.

Computer Exercises

11.1. Develop a computer program that allows you to plot the entropy of a source with variable output probabilities. We wish to observe that the maximum source entropy does indeed occur when the source outputs are equally likely. Start with a simple two-output source $[m_1, m_2]$ with respective probabilities $[a, 1 - a]$, and plot the entropy as a function of the parameter a. Then consider more complex cases such as a three-output source $[m_1, m_2, m_3]$ with respective probabilities $[a, b, 1-a-b]$. Be creative with the manner in which the results are displayed.

11.2. Develop a MATLAB program that generates the Huffman source code for an input random binary bit stream of arbitrary length.

11.3. Computer Example 11.2 did not contain the MATLAB program used to generate Figure 11.18. Develop a MATLAB program for generating Figure 11.18, and use your program to verify the correctness of Figure 11.18.

11.4. Table 11.5 gives a short list of rate $\frac{1}{2}$ and rate $\frac{3}{4}$ BCH codes. Using the Torrieri bound and an appropriate MATLAB program, plot together on a single set of axes the bit error probability for the rate $\frac{1}{2}$ BCH codes having block length $n = 7$, 15, 31, and 63. Assume PSK modulation with matched-filter detection. Repeat for rate $\frac{3}{4}$ BCH codes having block length $n = 15$, 31, 63, and 127. What conclusions can you draw from this exercise?

11.5. In implementing the Torrieri technique for comparing codes on the basis of information bit-error probability, the MATLAB function `nchoosek` was used. Using this function for large values of n and k can give rise to numerical precision difficulties that result from the factorial function. In order to illustrate this problem, execute the MATLAB function `nchoosek` with $n = 1000$ and $k = 500$. Develop an alternative technique for calculating `nchoosek` that mitigates some of these problems. Using your technique develop a performance comparison for (511,385) and (1023,768) BCH codes. Assume FSK modulation with coherent demodulation.

11.6. Develop a MATLAB program for generating the tree diagram illustrated in Figure 11.25.

11.7. Repeat Computer Example 11.6 for $\gamma = 0.1$ and $\gamma = 0.3$. What do you conclude from these results combined with the results of Computer Example 11.6, which were generated for $\gamma = 0.2$?

APPENDIX A

PHYSICAL NOISE SOURCES

As discussed in Chapter 1, noise originates in a communication system from two broad classes of sources: those external to the system, such as atmospheric, solar, cosmic, or man-made sources, and those internal to the system. The degree to which external noise sources influence system performance depends heavily upon system location and configuration. Consequently, the reliable analysis of their effect on system performance is difficult and depends largely on empirical formulas and on-site measurements. Their importance in the analysis and design of communication systems depends on their intensity relative to the internal noise sources. In this appendix, we are concerned with techniques of characterization and analysis of internal noise sources.

Noise internal to the subsystems that compose a communication system arises as a result of the random motion of charge carriers within the devices composing those subsystems. We now discuss several mechanisms that give rise to internal noise and suitable models for these mechanisms.

A.1 PHYSICAL NOISE SOURCES

A.1.1 Thermal Noise

Thermal noise is the noise arising from the random motion of charge carriers in a conducting or semiconducting medium. Such random agitation at the atomic level is a universal characteristic of matter at temperatures above absolute zero. Nyquist was one of the first to have studied thermal noise. *Nyquist's theorem* states that the mean-square noise voltage appearing across the terminals of a resistor of R Ω at temperature T K in a frequency band B Hz is given by

$$v_{\text{rms}}^2 = \langle v_n^2(t) \rangle = 4kTRB \text{ V}^2 \tag{A.1}$$

where $k =$ Boltzmann's constant $= 1.38 \times 10^{-23}$ J/K.

Thus a *noisy* resistor can be represented by an equivalent circuit consisting of a *noiseless* resistor in series with a noise generator of rms voltage v_{rms} as shown in Figure A.1(a). Short circuiting the terminals of Figure A.1(a) results in a short-circuit noise current of mean-square value

$$i_{\text{rms}}^2 = \langle i_n^2(t) \rangle = \frac{\langle v_n^2(t) \rangle}{R^2} = \frac{4kTB}{R} = 4kTGB \text{ A}^2 \tag{A.2}$$

where $G = 1/R$ is the conductance of the resistor. The Thevenin equivalent of Figure A.1(a) can therefore be transformed to the Norton equivalent shown in Figure A.1(b).

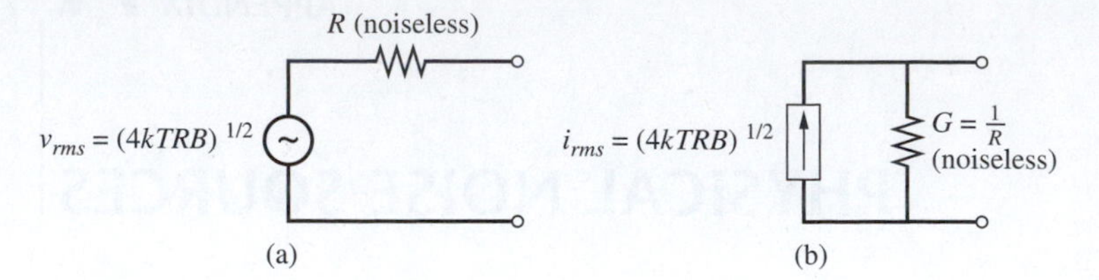

Figure A.1
Equivalent circuits for a noisy resistor. (a) Thevenin. (b) Norton.

EXAMPLE A.1

Consider the resistor network shown in Figure A.2. Assuming room temperature of $T = 290$ K, find the rms noise voltage appearing at the output terminals in a 100 kHz bandwidth.

Solution

We use voltage division to find the noise voltage due to each resistor across the output terminals. Then, since powers due to independent sources add, we find the rms output voltage v_0 by summing the square of the voltages due to each resistor (proportional to power), which gives the total mean-square voltage, and take the square root to give the rms voltage. The calculation yields

$$v_0^2 = v_{01}^2 + v_{02}^2 + v_{03}^2$$

where

$$v_{01} = \sqrt{4kTR_1B}\left(\frac{R_3}{R_1 + R_2 + R_3}\right) \tag{A.3}$$

$$v_{02} = \sqrt{4kTR_2B}\left(\frac{R_3}{R_1 + R_2 + R_3}\right) \tag{A.4}$$

and

$$v_{03} = \sqrt{4kTR_3B}\left(\frac{R_1 + R_2}{R_1 + R_2 + R_3}\right) \tag{A.5}$$

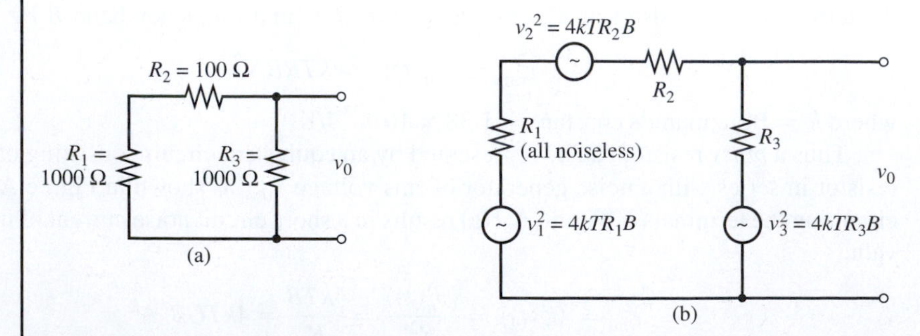

Figure A.2
Circuits for noise calculation. (a) Resistor network. (b) Noise equivalent circuit.

In the above expressions, $\sqrt{4kTR_iB}$ represents the rms voltage across resistor R_i. Thus

$$\begin{aligned} v_0^2 &= 4kTB\left(\frac{(R_1+R_2)R_3^2}{(R_1+R_2+R_3)^2} + \frac{(R_1+R_2)^2R_3}{(R_1+R_2+R_3)^2}\right) \\ &= (4\times 1.38\times 10^{-23}\times 290\times 10^5)\times\left(\frac{(1100)(1000)^2}{(2100)^2} + \frac{(1100)^2(1000)}{(2100)^2}\right) \\ &\cong 8.39\times 10^{-13}\ \mathrm{V}^2 \end{aligned} \tag{A.6}$$

Therefore,

$$v_0 = 9.16\times 10^{-7}\ \mathrm{V(rms)} \tag{A.7}$$

■

A.1.2 Nyquist's Formula

Although Example A.1 is instructive from the standpoint of illustrating noise computations involving several noisy resistors, it also illustrates that such computations can be exceedingly long if many resistors are involved. *Nyquist's formula*, which can be proven from thermodynamic arguments, simplifies such computations considerably. It states that the mean-square noise voltage produced at the output terminals of any one-port network containing only resistors, capacitors, and inductors is given by

$$\langle v_n^2(t)\rangle = 2kT\int_{-\infty}^{\infty} R(f)\,df \tag{A.8}$$

where $R(f)$ is real part of the complex impedance seen looking back into the terminals (in terms of frequency in hertz, $f=\omega/2\pi$). If the network contains only resistors, the mean-square noise voltage in a bandwidth B is

$$\langle v_n^2\rangle = 4kTR_{\mathrm{eq}}B\ \mathrm{V}^2 \tag{A.9}$$

where R_{eq} is the Thevenin equivalent resistance of the network.

EXAMPLE A.2

If we look back into the terminals of the network shown in Figure A.2, the equivalent resistance is

$$R_{\mathrm{eq}} = R_3\|(R_1+R_2) = \frac{R_3(R_1+R_2)}{R_1+R_2+R_3} \tag{A.10}$$

Thus

$$v_0^2 = \frac{4kTBR_3(R_1+R_2)}{R_1+R_2+R_3} \tag{A.11}$$

which can be shown to be equivalent to the result obtained previously.

■

A.1.3 Shot Noise

Shot noise arises from the discrete nature of current flow in electronic devices. For example, the *electron* flow in a saturated thermionic diode is due to the sum total of electrons emitted from the cathode which arrive randomly at the anode, thus providing an average *current flow* I_d (from anode to cathode when taken as positive) plus a randomly fluctuating component of mean-square value

$$i^2_{\text{rms}} = \langle i^2_n(t) \rangle = 2eI_d B \ \text{A}^2 \tag{A.12}$$

where $e =$ charge of the electron $= 1.6 \times 10^{-19}$ C. Equation (A.12) is known as *Schottky's theorem*.

Since powers from independent sources add, it follows that the squares of noise voltages or noise currents from independent sources, such as two resistors or two currents originating from independent sources, add. Thus, when applying Schottky's theorem to a p-n junction, the current flowing in a p-n junction diode is

$$I = I_s\left[\exp\left(\frac{eV}{kT}\right) - 1\right] \tag{A.13}$$

where V is the voltage across the diode and I_s is the reverse saturation current, can be considered as being caused by two independent currents $-I_s$ and $I_s \exp(eV/kT)$. Both currents fluctuate independently, producing a mean-square shot noise current given by

$$\begin{aligned} i^2_{\text{rms, tot}} &= \left[2eI_s\exp\left(\frac{eV}{kT}\right) + 2eI_s\right]B \\ &= 2e(I + 2I_s)\,B \end{aligned} \tag{A.14}$$

For normal operation, $I \gg I_s$ and the differential conductance is $g_0 = dI/dV = eI/kT$, so that (A.14) may be approximated as

$$i^2_{\text{rms, tot}} \cong 2eIB = 2kT\left(\frac{eI}{kT}\right)B = 2kTg_0B \tag{A.15}$$

which can be viewed as *half-thermal noise* of the differential conductance g_0 since there is a factor of 2 rather than a factor of 4 as in (A.2).

A.1.4 Other Noise Sources

In addition to thermal and shot noise, there are three other noise mechanisms that contribute to internally generated noise in electronic devices. We summarize them briefly here. A fuller treatment of their inclusion in the noise analysis of electronic devices is given by Van der Ziel (1970).

Generation–Recombination Noise

Generation–recombination noise is the result of free carriers being generated and recombining in semiconductor material. One can consider these generation and recombination events to be random. Therefore, this noise process can be treated as a shot noise process.

Temperature-Fluctuation Noise

Temperature-fluctuation noise is the result of the fluctuating heat exchange between a small body, such as a transistor, and its environment due to the fluctuations in the radiation and heat-conduction processes. If a liquid or gas is flowing past the small body, fluctuation in heat convection also occurs.

Flicker Noise

Flicker noise is due to various causes. It is characterized by a spectral density that increases with decreasing frequency. The dependence of the spectral density on frequency is often found to be proportional to the inverse first power of the frequency. Therefore, flicker noise is sometimes referred to as *one-over-f noise*. More generally, flicker noise phenomena are characterized by power spectra that are of the form constant/f^α, where α is close to unity. The physical mechanism that gives rise to flicker noise is not well understood.

A.1.5 Available Power

Since calculations involving noise involve transfer of power, the concept of maximum power available from a source of fixed internal resistance is useful. Figure A.3 illustrates the familiar theorem regarding maximum power transfer, which states that a source of internal resistance R delivers maximum power to a resistive load R_L if $R = R_L$ and that under these conditions, the power P produced by the source is evenly split between source and load resistances. If $R = R_L$, the load is said to be *matched* to the source, and the power delivered to the load is referred to as the *available power* P_a. Thus $P_a = \frac{1}{2}P$, which is delivered to the load only if $R = R_L$. Consulting Figure A.3(a), in which v_{rms} is the rms voltage of the source, we see that the voltage across $R_L = R$ is $\frac{1}{2}v_{\text{rms}}$. This gives

$$P_a = \frac{1}{R}\left(\frac{1}{2}v_{\text{rms}}\right)^2 = \frac{v_{\text{rms}}^2}{4R} \tag{A.16}$$

Similarly, when dealing with a Norton equivalent circuit as shown in Figure A.3(b), we can write the available power as

$$P_a = \left(\frac{1}{2}i_{\text{rms}}\right)^2 R = \frac{i_{\text{rms}}^2}{4G} \tag{A.17}$$

where $i_{\text{rms}} = v_{\text{rms}}/R$ is the rms noise current.

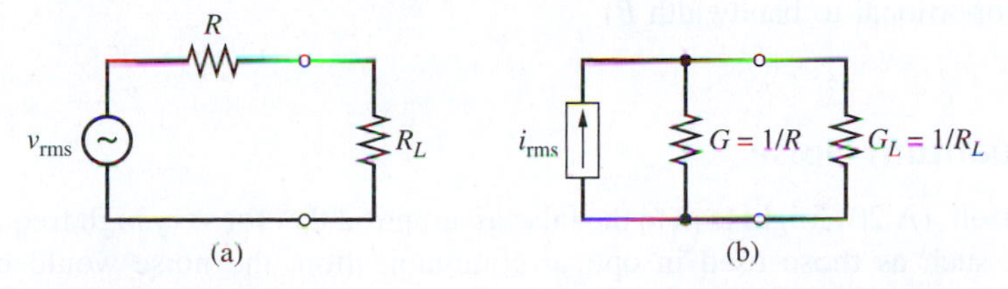

Figure A.3
Circuits pertinent to maximum power transfer theorem. (a) Thevenin equavalent for a source with load resistance R_L. (b) Norton equivalent for a source with load conductance G_L.

Returning to (A.1) or (A.2) and using (A.16) or (A.17), we see that a noisy resistor produces the available power

$$P_{a,R} = \frac{4kTRB}{4R} = kTB \text{ W} \tag{A.18}$$

Similarly, from (A.15), a diode with load resistance matched to its differential conductance produces the available power

$$P_{a,\,D} = \frac{1}{2}kTB \text{ W}, \quad I \gg I_s \tag{A.19}$$

EXAMPLE A.3

Calculate the available power per hertz of bandwidth for a resistance at room temperature, taken to be $T_0 = 290$ K. Express in decibels referenced to 1 W (dBW) and decibels referenced to 1 mW (dBm).

Solution

Power/hertz $= P_{a,R}/B = (1.38 \times 10^{-23})(290) = 4.002 \times 10^{-21}$ W/Hz.
Power/hertz in dBW $= 10 \log_{10}(4.002 \times 10^{-21}/1) \cong -204$ dBW/Hz.
Power/hertz in dBm $= 10 \log_{10}(4.002 \times 10^{-21}/10^{-3}) \cong -174$ dBm/Hz.

■

A.1.6 Frequency Dependence

In Example A.3, available power per hertz for a noisy resistor at $T_0 = 290$ K was computed and found, to good approximation, to be –174 dBm/Hz, independent of the frequency of interest. Actually, Nyquist's theorem, as stated by (A.1), is a simplification of a more general result. The proper quantum mechanical expression for available power per hertz, or available *power spectral density* $S_a(f)$, is

$$S_a(f) \triangleq \frac{P_a}{B} = \frac{hf}{\exp(hf/kT) - 1} \text{ W/Hz} \tag{A.20}$$

where $h =$ Planck's constant $= 6.6254 \times 10^{-34}$ Js.

This expression is plotted in Figure A.4, where it is seen that for all but very low temperatures and very high frequencies, the approximation is good that $S_a(f)$ is constant (that is, P_a is proportional to bandwidth B).

A.1.7 Quantum Noise

Taken by itself, (A.20) might lead to the false assumption that for very high frequencies where $hf \gg kT$, such as those used in optical communication, the noise would be negligible. However, it can be shown that a quantum noise term equal to hf must be added to (A.20) in order to account for the discrete nature of the electron energy. This is shown in Figure A.4 as the straight line, which permits the transition frequency between the thermal noise and quantum

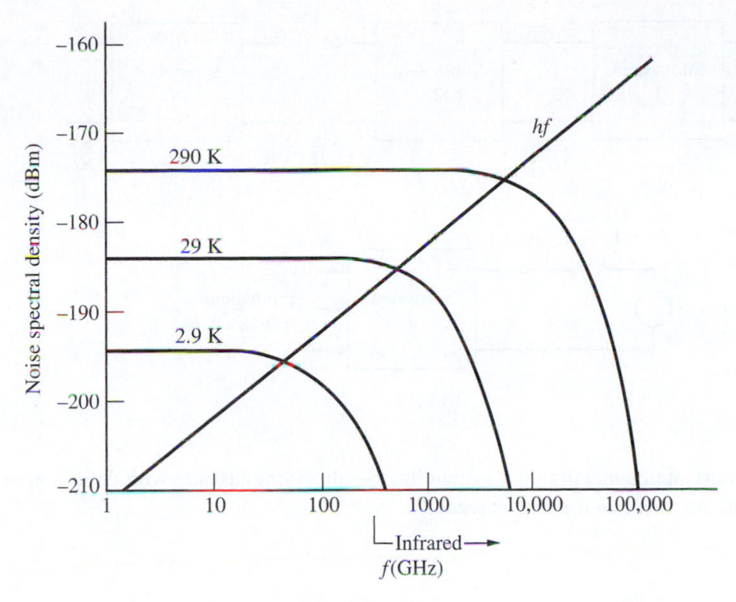

Figure A.4
Noise power spectral density versus frequency for thermal resistors.

noise regions to be estimated. This transition frequency is seen to be above 20 GHz even for $T = 2.9$ K.

A.2 CHARACTERIZATION OF NOISE IN SYSTEMS

Having considered several possible sources of internal noise in communication systems, we now wish to discuss convenient methods for characterization of the noisiness of the subsystems that make up a system, as well as overall noisiness of the system. Figure A.5 illustrates a cascade of N stages or subsystems that make up a system. For example, if this block diagram represents a superheterodyne receiver, subsystem 1 would be the RF amplifier, subsystem 2 the mixer, subsystem 3 the IF amplifier, and subsystem 4 the detector. At the output of each stage, we wish to be able to relate the signal-to-noise power ratio to that at the input. This will allow us to pinpoint those subsystems that contribute significantly to the output noise of the overall system, thereby enabling us to implement designs that minimize the noise.

A.2.1 Noise Figure of a System

One useful measure of system noisiness is the so-called *noise figure* F, defined as the ratio of the SNR at the system input to the SNR at the system output. In particular, for the lth subsystem in Figure A.5, the noise figure F_l is defined by the relation

$$\left(\frac{S}{N}\right)_l = \frac{1}{F_l}\left(\frac{S}{N}\right)_{l-1} \tag{A.21}$$

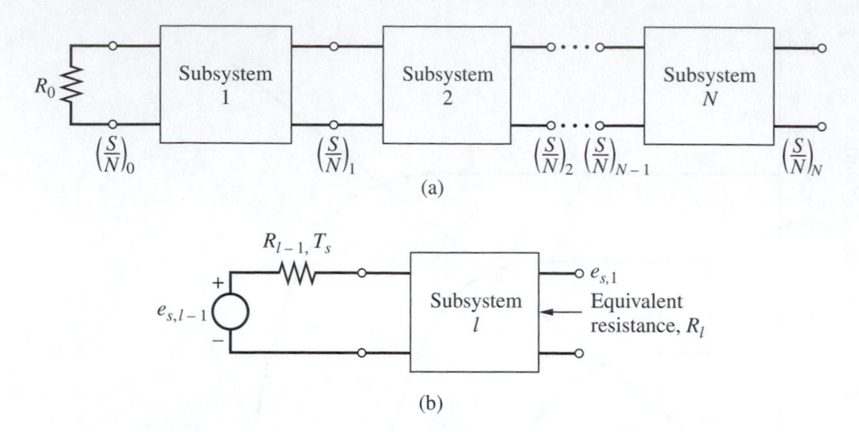

Figure A.5
Cascade of subsystems making up a system. (a) N-subsystem cascade with definition of SNRs at each point. (b) The lth subsystem in the cascade.

For an ideal, noiseless subsystem, $F_l = 1$; that is, the subsystem introduces no additional noise. For physical devices, $F_l > 1$.

Noise figures for devices and systems are often stated in terms of decibels. Specifically

$$F_{\text{dB}} = 10 \log_{10} F_{\text{ratio}} \tag{A.22}$$

Typical noise figures are 2 to 4.5 dB for a traveling wave tube amplifier (power gain of 20 to 30 dB) and 5 to 8 dB for mixers (a passive mixer has a loss of at least 3 dB due to the use of only one of the sidebands at its output). Further information is contained in Mumford and Schiebe (1968) or device manufacturer's data sheets.

The definition of noise figure given by (A.21) requires the calculation of both signal and noise powers at each point of the system. An alternative definition, equivalent to (A.21), involves the calculation of noise powers only. Although signal and noise powers at any point in the system depend on the loading of a subsystem on the preceding one, SNRs are independent of load, since both signal and noise appear across the same load. Hence any convenient load impedance may be used in making signal and noise calculations. In particular, we will use load impedances matched to the output impedance, thereby working with available signal and noise powers.

Consider the lth subsystem in the cascade of the system shown in Figure A.5. If we represent its input by a Thevenin equivalent circuit with rms signal voltage $e_{s,l-1}$ and equivalent resistant R_{l-1}, the available signal power is

$$P_{sa,l-1} = \frac{e_{s,l-1}^2}{4R_{l-1}} \tag{A.23}$$

If we assume that only thermal noise is present, the available noise power for a source temperature of T_s is

$$P_{na,l-1} = kT_sB \tag{A.24}$$

given an input SNR of

$$\left(\frac{S}{N}\right)_{l-1} = \frac{e^2_{s,l-1}}{4kT_sR_{l-1}B} \tag{A.25}$$

The available output signal power, from Figure A.5(b), is

$$P_{sa,l} = \frac{e^2_{s,l}}{4R_l} \tag{A.26}$$

We can relate $P_{sa,l}$ to $P_{sa,l-1}$ by the *available power gain* G_a of subsystem l, defined to be

$$P_{sa,l} = G_aP_{sa,l-1} \tag{A.27}$$

which is obtained if all resistances are matched. The output SNR is

$$\left(\frac{S}{N}\right)_l = \frac{P_{sa,l}}{P_{na,l}} = \frac{1}{F_l}\frac{P_{sa,l-1}}{P_{na,l-1}} \tag{A.28}$$

or

$$\begin{aligned} F_l &= \frac{P_{sa,l-1}}{P_{sa,l}}\frac{P_{na,l}}{P_{na,l-1}} = \frac{P_{sa,l-1}}{G_aP_{sa,l-1}}\frac{P_{na,l}}{P_{na,l-1}} \\ &= \frac{P_{na,l}}{G_aP_{na,l-1}} \end{aligned} \tag{A.29}$$

where any mismatches may be ignored, since they affect signal and noise the same.[1] Thus the noise figure is the ratio of the output noise power to the noise power that would have resulted had the system been noiseless. Noting that $P_{na,l} = G_aP_{na,l-1} + P_{\text{int},l}$, where $P_{\text{int},l}$ is the available internally generated noise power of subsystem l, and that $P_{na,l-1} = kT_sB$, we may write (A.29) as

$$F_l = 1 + \frac{P_{\text{int},l}}{G_akT_sB} \tag{A.30}$$

or, setting $T_s = T_0 = 290$ K to standardize the noise figure,[2] we obtain

$$F_l = 1 + \frac{P_{\text{int},l}}{G_akT_0B} \tag{A.31}$$

Thus, for $G_a \gg 1$, $F_l \cong 1$, which shows that the effect of internally generated noise becomes inconsequential for a system with a large gain. Conversely, a system with low gain enhances the importance of internal noise.

A.2.2 Measurement of Noise Figure

Using (A.29), with the available noise power at the output $P_{na,\text{out}}$ referred to the device input and representing this noise by a current generator $\overline{i_n^2}$ in parallel with the source resistance R_s or a

[1]This assumes that the gains for noise power and signal power are the same. If gain varies with frequency, then a spot noise figure can be defined, where signal power and noise power are measured in a small bandwidth Δf.

[2]If this were not done, the manufacturer of a receiver could claim superior noise performance of its product over that of a competitor simply by choosing T_s larger than the competitor. See Mumford and Scheibe (1968), pp. 53–56, for a summary of the various definitions of noise figure used in the past.

voltage generator $\overline{e_n^2}$ in series with it, we can determine the noise figure by changing the input noise a known amount and measuring the change in noise power at the device output. In particular, if we assume a current source represented by a saturated thermionic diode, so that

$$\overline{i_n^2} = 2eI_dB \text{ A}^2 \tag{A.32}$$

and sufficient current is passed through the diode so the noise power at the output is double the amount that appeared without the diode, then the noise figure is

$$F = \frac{eI_dR_s}{2kT_0} \tag{A.33}$$

where e is the charge of the electron in coulombs, I_d is the diode current in amperes, R_s is the input resistance, k is Boltzmann's constant, and T_0 is the standard temperature in kelvin.

A variation of the preceding method is the Y-factor method, which is illustrated in Figure A.6. Assume that two calibrated noise sources are available, one at effective temperature T_{hot} and the other at T_{cold}. With the first at the input of the unknown system with unknown temperature T_e, the available output noise power from (A.18) is

$$P_h = k(T_{\text{hot}} + T_e)\,BG \tag{A.34}$$

where B is the noise bandwidth of the device under test and G is its available power gain. With the cold noise source present, the available output noise power is

$$P_c = k(T_{\text{cold}} + T_e)\,BG \tag{A.35}$$

The two unknowns in these two equations are T_c and BG. Dividing the first by the second, we obtain

$$\frac{P_h}{P_c} = Y = \frac{T_{\text{hot}} + T_e}{T_{\text{cold}} + T_e} \tag{A.36}$$

When solved for T_e, this equation becomes [see (A.43) for the definition of T_e]

$$T_e = \frac{T_{\text{hot}} - YT_{\text{cold}}}{Y - 1} \tag{A.37}$$

which involves the two known source temperatures and the measured Y factor. The Y factor can be measured with the aid of the precision attenuator shown in Figure A.6 as follows:

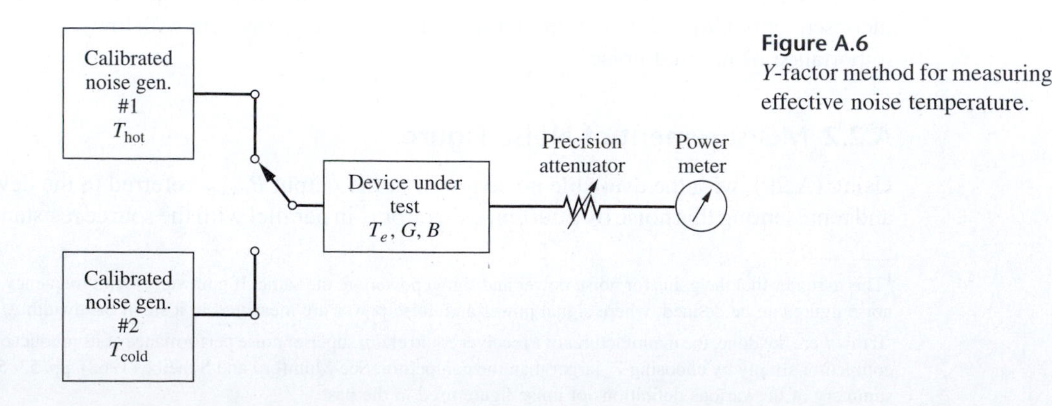

Figure A.6
Y-factor method for measuring effective noise temperature.

1. Connect the hot noise source to the system under test and adjust the attenuator for a convenient meter reading.
2. Switch to the cold noise source and adjust the attenuator for the same meter reading as before.
3. Noting the change in attenuator setting ΔA in decibels, calculate $Y = 10^{\Delta A/10}$.
4. Calculate the effective noise temperature using (A.37).

A.2.3 Noise Temperature

Equation (A.18) states that the available noise power of a resistor at temperature T is kTB W, independent of the value of R. We may use this result to define the *equivalent noise temperature* T_n of any noise source:

$$T_n = \frac{P_{n,\max}}{kB} \tag{A.38}$$

where $P_{n,\max}$ is the maximum noise power the source can deliver in bandwidth B.

EXAMPLE A.4

Two resistors R_1 and R_2 at temperatures T_1 and T_2, respectively, are connected in series to form a white-noise source. Find the equivalent noise temperature of the combination.

Solution

The mean-square voltage generated by the combination is

$$\langle v_n^2 \rangle = 4kBR_1T_1 + 4kBR_2T_2 \tag{A.39}$$

Since the equivalent resistance is $R_1 + R_2$, the available noise power is

$$P_{na} = \frac{\langle v_n^2 \rangle}{4(R_1 + R_2)} = \frac{4k(T_1R_1 + T_2R_2)B}{4(R_1 + R_2)} \tag{A.40}$$

The equivalent noise temperature is therefore

$$T_n = \frac{P_{na}}{kB}\frac{R_1T_1 + R_2T_2}{R_1 + R_2} \tag{A.41}$$

Note that T_n is not a physical temperature unless both resistors are at the same temperature. ■

A.2.4 Effective Noise Temperature

Returning to (A.30), we note that the second term, $P_{\text{int},l}/G_akT_0B$, which is dimensionless, is due solely to the internal noise of the system. Noting that $P_{\text{int},l}/G_akB$ has the dimensions of temperature, we may write the noise figure as

$$F_l = 1 + \frac{T_e}{T_0} \tag{A.42}$$

where

$$T_e = \frac{P_{\text{int},l}}{G_akB} \tag{A.43}$$

Thus,

$$T_e = (F_l - 1)T_0 \tag{A.44}$$

T_e is the *effective noise temperature* of the system and depends only on the parameters of the system. It is a measure of noisiness of the system referred to the input, since it is the temperature required of a thermal resistance, placed at the input of a noiseless system, in order to produce the same available noise power at the output as is produced by the internal noise sources of the system. Recalling that $P_{na,l} = G_a P_{na,l-1} + P_{\text{int},l}$ and that $P_{na,l-1} = kT_s B$, we may write the available noise power at the subsystem output as

$$\begin{aligned} P_{na,l} &= G_a k T_s B + G_a k T_e B \\ &= G_a k (T_s + T_e) B \end{aligned} \tag{A.45}$$

where the actual temperature of the source T_s is used. Thus the available noise power at the output of a system can be found by adding the effective noise temperature of the system to the temperature of the source and multiplying by $G_a kB$, where the term G_a appears because the noise power is referred to the system input.

A.2.5 Cascade of Subsystems

Considering the first two stages in Figure A.5, we see that noise appears at the output due to the following sources:

1. Amplified source noise, $G_{a_1} G_{a_2} k T_s B$.
2. Internal noise from the first stage amplified by the second stage, $G_{a_2} P_{a,\text{int}_1} = G_{a_2}(G_{a_1} k T_{e_1} B)$.
3. Internal noise from the second stage, $P_{a,\text{int}_2} = G_{a_2} k T_{e_2} B$.

Thus the total available noise power at the output of the cascade is

$$P_{na,2} = G_{a_1} G_{a_2} k \left(T_s + T_{e_1} + \frac{T_{e_2}}{G_{a_1}} \right) B \tag{A.46}$$

Noting that the available gain for the cascade is $G_{a_1} G_{a_2}$ and comparing with (A.45), we see that the effective temperature of the cascade is

$$T_e = T_{e_1} + \frac{T_{e_2}}{G_{a_1}} \tag{A.47}$$

From (A.42), the overall noise figure is

$$\begin{aligned} F = 1 + \frac{T_e}{T_0} &= 1 + \frac{T_{e_1}}{T_0} + \frac{1}{G_{a_1}} \frac{T_{e_2}}{T_0} \\ &= F_1 + \frac{F_2 - 1}{G_{a_1}} \end{aligned} \tag{A.48}$$

where F_1 is the noise figure of stage 1 and F_2 is the noise figure of stage 2. The generalization of this result to an arbitrary number of stages is known as *Friis's formula* and is given by

$$F = F_1 + \frac{F_2 - 1}{G_{a_1}} + \frac{F_3 - 1}{G_{a_1} G_{a_2}} + \cdots \tag{A.49}$$

whereas the generalization of (A.47) is

$$T_e = T_{e_1} + \frac{T_{e_2}}{G_{a_1}} + \frac{T_{e_3}}{G_{a_1}G_{a_2}} + \cdots \tag{A.50}$$

EXAMPLE A.5

A parabolic dish antenna is pointed up into the night sky. Noise due to atmospheric radiation is equivalent to a source temperature of 70 K. A low-noise preamplifier with noise figure of 2 dB and an available power gain of 20 dB over a bandwidth of 20 MHz is mounted at the antenna feed (focus of the parabolic reflector).

a. Find the effective noise temperature of the preamplifier.
b. Find the available noise power at the preamplifier output.

Solution

a. From (A.45), we have

$$T_{\text{eff, in}} = T_s + T_{e,\text{ preamp}} \tag{A.51}$$

but (A.44) gives

$$\begin{aligned} T_{e,\text{ preamp}} &= T_0(F_{\text{preamp}} - 1) \\ &= 290(10^{2/10} - 1) \\ &= 169.6\text{ K} \end{aligned} \tag{A.52}$$

b. From (A.45), the available output noise power is

$$\begin{aligned} P_{na,\text{ out}} &= G_a k(T_s + T_e)B \\ &= 10^{20/10}(1.38 \times 10^{-23})(169.6 + 70)(20 \times 10^6) \\ &= 6.61 \times 10^{-12}\text{ W} \end{aligned} \tag{A.53}$$

■

EXAMPLE A.6

A preamplifier with power gain to be found and a noise figure of 2.5 dB is cascaded with a mixer with a gain of 5 dB and a noise figure of 8 dB. Find the preamplifier gain such that the overall noise figure of the cascade is at most 4 dB.

Solution

Friis's formula specializes to

$$F = F_1 + \frac{F_2 - 1}{G_1} \tag{A.54}$$

Solving for G_1, we get

$$G_1 = \frac{F_2 - 1}{F - F_1} = \frac{10^{8/10} - 1}{10^{4/10} - 10^{2.5/10}} = 7.24\text{ (ratio)} = 8.6\text{ dB} \tag{A.55}$$

Note that the gain of the mixer is immaterial.

■

A.2.6 Attenuator Noise Temperature and Noise Figure

Consider a purely resistive attenuator that imposes a loss of a factor of L in available power between input and output; thus the available power at its output $P_{a,\text{out}}$ is related to the available power at its input $P_{a,\text{in}}$ by

$$P_{a,\text{ out}} = \frac{1}{L} P_{a,\text{ in}} = G_a P_{a,\text{ in}} \tag{A.56}$$

However, since the attenuator is resistive and assumed to be at the *same temperature* T_s as the equivalent resistance at its input, the available output power is

$$P_{na,\text{ out}} = kT_s B \tag{A.57}$$

Characterizing the attenuator by an effective temperature T_e and employing (A.45), we may also write $P_{na,\text{out}}$ as

$$\begin{aligned} P_{na,\text{ out}} &= G_a k(T_s + T_e)B \\ &= \frac{1}{L} k(T_s + T_e)B \end{aligned} \tag{A.58}$$

Equating (A.57) and (A.58) and solving for T_e, we obtain

$$T_e = (L - 1)T_s \tag{A.59}$$

for the effective noise temperature of a noise resistance of temperature T_s followed by an attenuator. From (A.42), the noise figure of the cascade of source resistance and attenuator is

$$F = 1 + \frac{(L-1)T_s}{T_0} \tag{A.60}$$

or

$$F = 1 + \frac{(L-1)T_0}{T_0} = L \tag{A.61}$$

for an attenuator at *room temperature*, T_0.

EXAMPLE A.7

Consider a receiver system consisting of an antenna with lead-in cable having a loss factor of $L = 1.5$ dB (gain of -1.5 dB), which at room temperature is also its noise figure F_1, and RF preamplifier with a noise figure of $F_2 = 7$ dB and a gain of 20 dB, followed by a mixer with a noise figure of $F_3 = 10$ dB and a conversion gain of 8 dB, and finally an integrated-circuit IF amplifier with a noise figure of $F_4 = 6$ dB and a gain of 60 dB.

a. Find the overall noise figure and noise temperature of the system

b. Find the noise figure and noise temperature of the system with preamplifier and cable interchanged (i.e., the preamplifier is mounted right at the antenna terminal).

Solution

a. Converting decibel values to ratios and employing (A.46), we obtain

$$\begin{aligned} F &= 1.41 + \frac{5.01 - 1}{1/1.41} + \frac{10 - 1}{100/1.41} + \frac{3.98 - 1}{100(6.3)/1.41} \\ &= 1.41 + 5.65 + 0.13 + 6.7 \times 10^{-3} = 7.19 = 8.57 \text{ dB} \end{aligned} \tag{A.62}$$

Note that the cable and RF amplifier essentially determine the noise figure of the system and that the noise figure of the system is enhanced because of the loss of the cable. If we solve (A.44) for T_e, we have an effective noise temperature of

$$T_e = T_0(F - 1) = 290(7.19 - 1) = 1796 \text{ K} \tag{A.63}$$

b. Interchanging the cable and RF preamplifier, we obtain the noise figure

$$\begin{aligned} F &= 5.01 + \frac{1.41 - 1}{100} + \frac{10 - 1}{100/1.41} + \frac{3.98 - 1}{100(6.3)/1.41} \\ &= 5.01 + (4.1 \times 10^{-3}) + 0.127 + (6.67 \times 10^{-3}) \\ &= 5.15 = 7.12 \text{ dB} \end{aligned} \tag{A.64}$$

The noise temperature is

$$T_e = 290(4.15) = 1203 \text{ K} \tag{A.65}$$

Now the noise figure and noise temperature are essentially determined by the noise level of the RF preamplifier.

■

We have omitted one possibly important source of noise which is the antenna. If the antenna is directive and pointed at source of significant thermal noise, such as the daytime sky (typical noise temperature of 300° F), its equivalent temperature may also be of importance in the calculation. This is particularly true when a low-noise preamplifier is employed.

■ A.3 FREE-SPACE PROPAGATION EXAMPLE

As a final example of noise calculation, we consider a free-space electromagnetic-wave propagation channel. For the sake of illustration, suppose the communication link of interest is between a synchronous-orbit relay satellite and a low-orbit satellite or aircraft, as shown in Figure A.7.

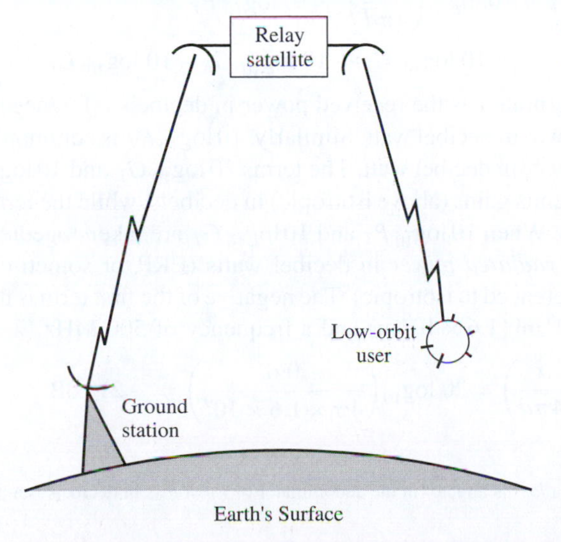

Figure A.7
A satellite-relay communication link.

This might represent part of a relay link between a ground station and a small scientific satellite or an aircraft. Since the ground station is high power, we assume the ground-station-to-relay-satellite link is noiseless and focus our attention on the link between the two satellites.

Assume a relay satellite transmitted signal power of P_T W. If radiated isotropically, the power density at a distance d from the satellite is given by

$$p_t = \frac{P_T}{4\pi d^2} \text{ W/m}^2 \tag{A.66}$$

If the satellite antenna has directivity, with the radiated power being directed toward the low-orbit vehicle, the antenna can be described by an antenna power gain G_T over the isotropic radiation level. For aperture type antennas with aperture area A_T large compared with the square of the transmitted wavelength λ^2, it can be shown that the maximum gain is given by $G_T = 4\pi A_T/\lambda^2$. The power P_R intercepted by the receiving antenna is given by the product of the receiving aperture area A_R and the power density at the aperture. This gives

$$P_R = \frac{P_T G_T}{4\pi d^2} A_R \tag{A.67}$$

However, we may relate the receiving aperture antenna to its maximum gain by the expression $G_R = 4\pi A_R/\lambda^2$, giving

$$P_R = \frac{P_T G_T G_R \lambda^2}{(4\pi d)^2} \tag{A.68}$$

Equation (A.68) includes only the loss in power from isotropic spreading of the transmitted wave. If other losses such as atmospheric absorption are important, they may be included as a loss factor L_0 in (A.68) to yield

$$P_R = \left(\frac{\lambda}{4\pi d}\right)^2 \frac{P_T G_T G_R}{L_0} \tag{A.69}$$

The factor $(4\pi d/\lambda)^2$ is sometimes referred to as the *free-space loss*.[3]

In the calculation of receiver power, it is convenient to work in terms of decibels. Taking $10 \log_{10} P_R$, we obtain

$$\begin{aligned} 10 \log_{10} P_R = {} & 20 \log_{10}\left(\frac{\lambda}{4\pi d}\right) + 10 \log_{10} P_T \\ & + 10 \log_{10} G_T + 10 \log_{10} G_R - 10 \log_{10} L_0 \end{aligned} \tag{A.70}$$

Now $10 \log_{10} P_R$ can be interpreted as the received power in decibels referenced to 1 W; it is commonly referred to as power in decibel watt. Similarly, $10 \log_{10} P_T$ is commonly referred to as the transmitted signal power in decibel watt. The terms $10 \log_{10} G_T$ and $10 \log_{10} G_R$ are the transmitter and receiver antenna gains (above isotropic) in decibels, while the term $10 \log_{10} L_0$ is the loss factor in decibels. When $10 \log_{10} P_T$ and $10 \log_{10} G_T$ are taken together, this sum is referred to as the *effective radiated power* in decibel watts (ERP, or sometimes EIRP, for effective radiated power referenced to isotropic). The negative of the first term is the free-space loss in decibels. For $d = 10^6$ mi (1.6×10^9 m) and a frequency of 500 MHz ($\lambda = 0.6$ m),

$$20 \log_{10}\left(\frac{\lambda}{4\pi d}\right) = 20 \log_{10}\left(\frac{0.6}{4\pi \times 1.6 \times 10^9}\right) = -210 \text{ dB} \tag{A.71}$$

[3]We take the convention here that a *loss* is a factor in the denominator of P_R; a *loss* in decibels is a positive quantity (a negative gain).

If λ or d change by a factor of 10, this value changes by 20 dB. We now make use of (A.70) and the results obtained for noise figure and temperature to compute the SNR for a typical satellite link.

EXAMPLE A.8

We are given the following parameters for a relay-satellite-to-user link:

Relay satellite effective radiated power ($G_T = 30$ dB; $P_T = 100$ W): 50 dBW

Transmit frequency: 2 GHz ($\lambda = 0.15$ m)

Receiver noise temperature of user (includes noise figure of receiver and background temperature of antenna): 700 K

User satellite antenna gain: 0 dB

Total system losses: 3 dB

Relay–user separation: 41,000 km

Find the signal-to-noise power ratio in a 50 kHz bandwidth at the user satellite receiver IF amplifier output.

Solution

The received signal power is computed using (A.69) as follows (+ and − signs in parentheses indicate whether the quantity is added or subtracted):

Free-space loss: $-20\log_{10}(0.15/4\pi \times 41 \times 10^6)$: 190.7 dB (−)

Effective radiated power: 50 dBW (+)

Receiver antenna gain: 0 dB (+)

System losses: 3 dB (−)

Received Signal Power: −143.7 dBW

The noise power level, calculated from (A.43), is

$$P_{\text{int}} = G_a k T_e B \tag{A.72}$$

where P_{int} is the receiver output noise power due to internal sources. Since we are calculating the SNR, the available gain of the receiver does not enter the calculation because both signal and noise are multiplied by the same gain. Hence, we may set G_a to unity, and the noise level is

$$\begin{aligned} P_{\text{int, dBW}} &= 10\log_{10}\left[kT_0\left(\frac{T_e}{T_0}\right)B\right] \\ &= 10\log_{10}(kT_0) + 10\log_{10}\left(\frac{T_e}{T_0}\right) + 10\log_{10} B \qquad \text{(A.73)} \\ &= -204 + 10\log_{10}\left(\frac{700}{290}\right) + 10\log_{10} 50{,}000 \\ &= -153.2\ \text{dBW} \end{aligned}$$

Hence, the SNR at the receiver output is

$$SNR_0 = -143.7 + 153.2 = 9.5\ \text{dB} \tag{A.74}$$

■

EXAMPLE A.9

To interpret the result obtained in the previous example in terms of the performance of a digital communication system, we must convert the SNR obtained to energy-per-bit-to-noise-spectral density ratio E_b/N_0 (see Chapter 8). By definition of SNR_0, we have

$$\text{SNR}_0 = \frac{P_R}{kT_eB} \tag{A.75}$$

Multiplying numerator and denominator by the duration of a data bit T_b, we obtain

$$\text{SNR}_0 = \frac{P_RT_b}{kT_eBT_b} = \frac{E_b}{N_0BT_b} \tag{A.76}$$

where $P_RT_b = E_b$ and $kT_e = N_0$ are the signal energy per bit and the noise power spectral density, respectively. Thus, to obtain E_b/N_0 from SNR_0, we calculate

$$\left.\frac{E_b}{N_0}\right|_{\text{dB}} = (\text{SNR}_0)_{\text{dB}} + 10\log_{10}(BT_b) \tag{A.77}$$

For example, from Chapter 8 we recall that the null-to-null bandwidth of a phase-shift keyed carrier is $2/T_b$ Hz. Therefore, BT_b for BPSK is 2 (3 dB) and

$$\left.\frac{E_b}{N_0}\right|_{\text{dB}} = 9.5 + 3 = 12.5 \text{ dB} \tag{A.78}$$

The probability of error for a binary BPSK digital communication system was derived in Chapter 8 as

$$\begin{aligned} P_E &= Q\left(\sqrt{\frac{2E_b}{N_0}}\right) \cong Q\left(\sqrt{2\times 10^{1.25}}\right) \\ &\cong 1.23\times 10^{-9} \quad \text{for} \quad \frac{E_b}{N_0} = 12.5 \text{ dB} \end{aligned} \tag{A.79}$$

which is a fairly small probability of error (anything less than 10^{-6} would probably be considered adequate). It appears that the system may have been overdesigned. However, no margin has been included as a safety factor. Components degrade or the system may be operated in an environment for which it was not intended. With only 3 dB allowed for margin, the performance in terms of error probability becomes 1.21×10^{-5}.

■

Further Reading

Treatments of internal noise sources and calculations oriented toward communication systems comparable to the scope and level of the presentation here may be found in most of the books on communications referenced in Chapters 2 and 3. A concise, but thorough, treatment at an elementary level is available in Mumford and Scheibe (1968). An in-depth treatment of noise in solid-state devices is available in Van der Ziel (1970). Another useful reference on noise is Ott (1988). For discussion of satellite-link power budgets, see Ziemer and Peterson (2001).

Problems

Section A.1

A.1. A true rms voltmeter (assumed noiseless) with an effective noise bandwidth of 30 MHz is used to measure the noise voltage produced by the following devices. Calculate the meter reading in each case.

a. A 10 kΩ resistor at room temperature, $T_0 = 290$ K.

b. A 10 k Ω resistor at 29 K.

c. A 10 k Ω resistor at 2.9 K.

d. What happens to all of the above results if the bandwidth is decreased by a factor of 4, a factor of 10, or a factor of 100?

A.2. Given a junction diode with reverse saturation current $I_s = 15\ \mu$A.

a. At room temperature (290 K), find V such that $I > 20I_s$, thus allowing (A.14) to be approximated by (A.15). Find the rms noise current.

b. Repeat part (a) for $T = 90$ K.

A.3. Consider the circuit shown in Figure A.8.

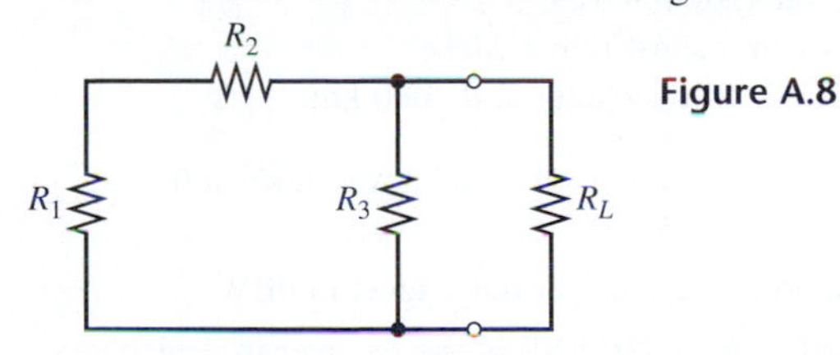

Figure A.8

a. Obtain an expression for the mean-square noise voltage appearing across R_L.

b. If $R_1 = 2000\ \Omega$, $R_2 = R_L = 300\ \Omega$, and $R_3 = 500\ \Omega$ find the mean-square noise voltage per hertz.

A.4. Referring to the circuit of Figure A.8, consider R_L to be a load resistance, and find it in terms of R_1, R_2, and R_3 so that the maximum available noise power available from R_1, R_2, and R_3 is delivered to it.

A.5. Assuming a bandwidth of 2 MHz, find the rms noise voltage across the output terminals of the circuit shown in Figure A.9 if it is at a temperature of 400 K.

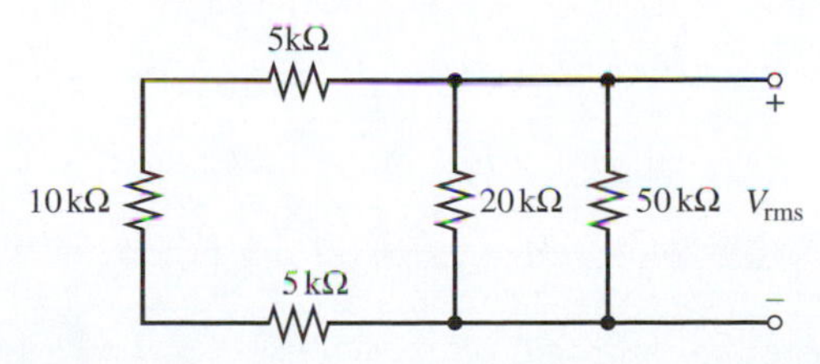

Figure A.9

Section A.2

A.6. Obtain an expression for F and T_e for the two-port resistive matching network shown in Figure A.10, assuming a source at $T_0 = 290$ K.

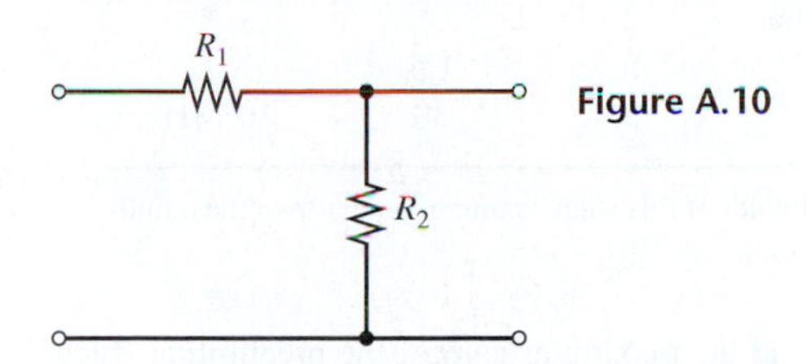

Figure A.10

A.7. A source with equivalent noise temperature $T_s = 1000$ K is followed by a cascade of three amplifiers having the specifications shown in Table A.1. Assume a bandwidth of 50 kHz.

Table A.1

Amplifier no.	F	T_e	Gain
1		300 K	10 dB
2	6 dB		30 dB
3	11 dB		30 dB

a. Find the noise figure of the cascade.

b. Suppose amplifiers 1 and 2 are interchanged. Find the noise figure of the cascade.

c. Find the noise temperature of the systems of parts (a) and (b).

d. Assuming the configuration of part (a), find the required input signal power to give an output SNR of 40 dB. Perform the same calculation for the system of part (b).

A.8. An attenuator with loss $L \gg 1$ is followed by an amplifier with noise figure F and gain $G_a = L$.

a. Find the noise figure of the cascade at temperature T_0.

b. Consider the cascade of two identical attenuator–amplifier stages as in part (a). Determine the noise figure of the cascade at temperature T_0.

c. Generalize these results to N identical attenuators and amplifiers at temperature T_0. How many decibels does the noise figure increase as a result of doubling the number of attenuators and amplifiers?

A.9. Given a cascade of a preamplifier, mixer, and amplifier with the specifications shown in Table A.2,

Table A.2

	Noise figure, dB	Gain, dB	Bandwidth
Preamplifier	2	7	*
Mixer	8	1.5	*
Amplifier	5	30	10 MHz

*The bandwidth of this stage is much greater than the amplifier bandwidth.

a. Find the maximum gain of the preamplifier such that the overall noise figure of the cascade is 5 dB or greater.

b. The preamplifier is fed by an antenna with noise temperature of 300 K (this is the temperature of Earth viewed from space). Find the temperature of the overall system using a preamplifier gain of 15 dB and also for the preamplifier gain found in part (a).

c. Find the noise power at the amplifier output for the two cases of part (b).

d. Repeat part (b) except now assume that a transmission line with loss of 2 dB connects the antenna to the preamplifier.

A.10. An antenna with a temperature of 300 K is fed into a receiver with a total gain of 80 dB, $T_e = 1500$ K, and a bandwidth of 3 MHz.

a. Find the available noise power at the output of the receiver.

b. Find the necessary signal power P_r in dBm at the antenna terminals such that the output SNR is 50 dB.

A.11. Referring to (A.37) and the accompanying discussion, suppose that two calibrated noise sources have effective temperatures of 600 K and 300 K.

a. Obtain the noise temperature of an amplifier with these two noise sources used as inputs if the difference in attenuator settings to get the same power meter reading at the amplifier's output is 1 dB, 1.5 dB, or 2 dB.

b. Obtain the corresponding noise figures.

Section A.3

A.12. Given a relay–user link as described in Section A.3 with the following parameters:

Average transmit power of relay satellite: 35 dBW
Transmit frequency: 7.7 GHz
Effective antenna aperture of relay satellite: 1 m^2
Noise temperature of user receiver (including antenna): 1000 K
Antenna gain of user: 6 dB
Total system losses: 5 dB
System bandwidth: 1 MHz
Relay–user separation: 41,000 km

a. Find the received signal power level at the user in dBW.

b. Find the receiver noise level in dBW.

c. Compute the SNR at the receiver in decibels.

d. Find the average probability of error for the following digital signaling methods: (1) BPSK, (2) binary DPSK, (3) binary noncoherent FSK, (4) QPSK.[4]

[4]This part of the problem requires results from Chapters 8 and 9.

APPENDIX B

JOINTLY GAUSSIAN RANDOM VARIABLES

In this appendix, we examine the joint pdf and the characteristic function for a set of Gaussian random variables $X_1, X_2, \ldots, X_N$. In Chapter 5 the joint pdf for $N=2$ was given as

$$f_{X_1X_2}(x_1, x_2) = \frac{\exp(-[1/2(1-\rho^2)]\{[(x_1-m_1)/\sigma_{x_1})]^2 - 2\rho[(x_1-m_1)/\sigma_{x_1}][(x_2-m_2)/\sigma_{x_2}] + [(x_2-m_2)/\sigma_{x_2}]^2\})}{2\pi\sigma_{x_1}\sigma_{x_2}\sqrt{1-\rho^2}} \tag{B.1}$$

where $m_i = E\{X_i\}$, $\sigma_{x_i}^2 = E\{[X_i - m_i]^2\}$, $i = 1, 2$, and $\rho = E\{(X_1-m_1)(X_2-m_2)/\sigma_{x_1}\sigma_{x_2}\}$. This important result is now generalized.

■ B.1 THE PROBABILITY DENSITY FUNCTION

The joint pdf of N jointly Gaussian random variables is

$$f_X(\mathbf{x}) = (2\pi)^{-N/2}|\det \mathbf{C}|^{-1/2}\exp\left[-\frac{1}{2}(\mathbf{x}-\mathbf{m})^T\mathbf{C}^{-1}(\mathbf{x}-\mathbf{m})\right] \tag{B.2}$$

where $\mathbf{x}$ and $\mathbf{m}$ are column matrices whose transposes are

$$\mathbf{x}^T = [x_1\ x_2\ \cdots\ x_N] \tag{B.3}$$

and

$$\mathbf{m}^T = [m_1\ m_2\ \cdots\ m_N] \tag{B.4}$$

respectively, and $\mathbf{C}$ is the positive definite matrix of correlation coefficients with elements

$$C_{ij} = E[(X_i - m_i)(X_j - m_j)] \tag{B.5}$$

Note that in (B.2) $\mathbf{x}^T$ and $\mathbf{m}^T$ are 1 by N row matrices and that $\mathbf{C}$ is an N by N square matrix.

■ B.2 THE CHARACTERISTIC FUNCTION

The joint characteristic function of the Gaussian random variables $X_1, X_2, \ldots, X_N$ is

$$M_{\mathbf{X}}(\mathbf{v}) = \exp\left[j\mathbf{m}^T\mathbf{v} - \frac{1}{2}\mathbf{v}^T\mathbf{C}\mathbf{v}\right] \tag{B.6}$$

where $\mathbf{v}^T = [v_1\ v_2\ \cdots\ v_N]$. From the power-series expansion of (B.6), it follows that for any four zero-mean Gaussian random variables,

$$E[X_1X_2X_3X_4] = E[X_1X_2]E[X_3X_4] + E[X_1X_3]E[X_2X_4] + E[X_1X_4]E[X_2X_3] \tag{B.7}$$

This is a rule that is useful enough to be worth memorizing.

B.3 LINEAR TRANSFORMATIONS

If a set of jointly Gaussian random variables is transformed to a new set of random variables by a linear transformation, the resulting random variables are jointly Gaussian. To show this, consider the linear transformation

$$\mathbf{y} = \mathbf{A}\mathbf{x} \tag{B.8}$$

where $\mathbf{y}$ and $\mathbf{x}$ are column matrices of dimension N and $\mathbf{A}$ is a nonsingular N by N square matrix with elements $[a_{ij}]$. From (B.8), the Jacobian is

$$J\begin{pmatrix} x_1, x_2, \ldots, x_N \\ y_1, y_2, \ldots, y_N \end{pmatrix} = \det(\mathbf{A}^{-1}) \tag{B.9}$$

where $\mathbf{A}^{-1}$ is the inverse matrix of $\mathbf{A}$. However, $\det(\mathbf{A}^{-1}) = 1/\det(\mathbf{A})$. Using this in (B.1), along with

$$\mathbf{x} = \mathbf{A}^{-1}\mathbf{y} \tag{B.10}$$

gives

$$f_{\mathbf{Y}}(\mathbf{y}) = (2\pi)^{-N/2}|\det\mathbf{C}|^{-1/2}|\det\mathbf{A}|^{-1} \times \exp\left[-\frac{1}{2}(\mathbf{A}^{-1}\mathbf{y}-\mathbf{m})^T\mathbf{C}^{-1}(\mathbf{A}^{-1}\mathbf{y}-\mathbf{m})\right] \tag{B.11}$$

Now $\det\mathbf{A} = \det\mathbf{A}^T$ and $\mathbf{A}\mathbf{A}^{-1} = \mathbf{I}$, the identity matrix. Therefore (B.11) can be written as

$$f_{\mathbf{Y}}(\mathbf{y}) = (2\pi)^{-N/2}|\det\mathbf{A}\mathbf{C}\mathbf{A}^T|^{-1/2} \times \exp\left\{-\frac{1}{2}[\mathbf{A}^{-1}(\mathbf{y}-\mathbf{A}\mathbf{m})]^T\mathbf{C}^{-1}[\mathbf{A}^{-1}(\mathbf{y}-\mathbf{A}\mathbf{m})]\right\} \tag{B.12}$$

But the equalities $(\mathbf{A}\mathbf{B})^T = \mathbf{B}^T\mathbf{A}^T$ and $(\mathbf{A}^{-1})^T = (\mathbf{A}^T)^{-1}$ allow the term inside the braces in (B.12) to be written as

$$-\frac{1}{2}[(\mathbf{y}-\mathbf{A}\mathbf{m})^T(\mathbf{A}^T)^{-1}\mathbf{C}^{-1}\mathbf{A}^{-1}(\mathbf{y}-\mathbf{A}\mathbf{m})]$$

Finally, the equality $(\mathbf{A}\mathbf{B})^{-1} = \mathbf{B}^{-1}\mathbf{A}^{-1}$ allows the above term to be rearranged to

$$-\frac{1}{2}[(\mathbf{y}-\mathbf{A}\mathbf{m})^T(\mathbf{A}\mathbf{C}\mathbf{A}^T)^{-1}(\mathbf{y}-\mathbf{A}\mathbf{m})]$$

Thus (B.12) becomes

$$f_{\mathbf{Y}}(\mathbf{y}) = (2\pi)^{-N/2}|\det\mathbf{A}\mathbf{C}\mathbf{A}^T|\ \exp\left\{-\frac{1}{2}(\mathbf{y}-\mathbf{A}\mathbf{m})^T(\mathbf{A}\mathbf{C}\mathbf{A}^T)^{-1}(\mathbf{y}-\mathbf{A}\mathbf{m})\right\} \tag{B.13}$$

We recognize this as a joint Gaussian density function for a random vector $\mathbf{Y}$ with mean vector $E[\mathbf{Y}] = \mathbf{A}\mathbf{m}$ and covariance matrix $\mathbf{A}\mathbf{C}\mathbf{A}^T$.

APPENDIX C

PROOF OF THE NARROWBAND NOISE MODEL

We now show that the narrowband noise model, introduced in Chapter 6, holds. To simplify notation, let

$$\widehat{n}(t) = n_c(t)\cos(\omega_0 t + \theta) - n_s(t)\sin(\omega_0 t + \theta) \tag{C.1}$$

where $\widehat{n}(t)$ is the noise representation defined by (C.1) and is not to be confused with the Hilbert transform. Thus, we must show that

$$E\{[n(t) - \widehat{n}(t)]\} = 0 \tag{C.2}$$

Expanding and taking the expectation term by term, we obtain

$$E\{(n - \widehat{n})^2\} = \overline{n^2} - \overline{2n\widehat{n}} + \overline{\widehat{n}^2} \tag{C.3}$$

where the argument, t, has been dropped to simplify notation.

Let us consider the last term in (C.3) first. By the definition of $\widehat{n}(t)$,

$$\begin{aligned} \overline{\widehat{n}^2} &= E\left\{[n_c(t)\cos(\omega_0 t + \theta) - n_s(t)\sin(\omega_0 t + \theta)]^2\right\} \\ &= \overline{n_c^2}\,\overline{\cos^2(\omega_0 t + \theta)} + \overline{n_s^2}\,\overline{\sin^2(\omega_0 t + \theta)} \\ &\quad - 2\overline{n_c n_s}\,\overline{\cos(\omega_0 t + \theta)\sin(\omega_0 t + \theta)} \\ &= \frac{1}{2}\overline{n_c^2} + \frac{1}{2}\overline{n_s^2} = \overline{n^2} \end{aligned} \tag{C.4}$$

where we have employed the fact that

$$\overline{n_c^2} = \overline{n_s^2} = \overline{n^2} \tag{C.5}$$

along with the averages

$$\overline{\cos^2(\omega_0 t + \theta)} = \frac{1}{2} + \frac{1}{2}\overline{\cos 2(\omega_0 t + \theta)} = \frac{1}{2} \tag{C.6}$$

$$\overline{\sin^2(\omega_0 t + \theta)} = \frac{1}{2} - \frac{1}{2}\overline{\cos 2(\omega_0 t + \theta)} = \frac{1}{2} \tag{C.7}$$

and

$$\overline{\cos(\omega_0 t + \theta)\sin(\omega_0 t + \theta)} = \frac{1}{2}\overline{\sin 2(\omega_0 t + \theta)} = 0 \tag{C.8}$$

Next, we consider $\overline{n\widehat{n}}$. By definition of $\widehat{n}(t)$, it can be written as

$$\overline{n\widehat{n}} = E\{n(t)[n_c(t)\cos(\omega_0 t + \theta) - n_s(t)\sin(\omega_0 t + \theta)]\} \tag{C.9}$$

From Figure 6.12,

$$n_c(t) = h(t') * [2n(t')\cos(\omega_0 t' + \theta)] \tag{C.10}$$

and

$$n_s(t) = -h(t') * [2n(t')\sin(\omega_0 t' + \theta)] \tag{C.11}$$

where $h(t')$ is the impulse response of the lowpass filter in Figure 6.12. The argument t' has been used in (C.10) and (C.11) to remind us that the variable of integration in the convolution is different from the variable t in (C.9). Substituting (C.10) and (C.11) into (C.9), we obtain

$$\begin{aligned}
\overline{n\widehat{n}} &= E\{n(t)h(t')*[2n(t')\cos(\omega_0 t'+\theta)]\cos(\omega_0 t+\theta) \\
&\quad + h(t')*[2n(t')\sin(\omega_0 t'+\theta)]\sin(\omega_0 t+\theta)]\} \\
&= E\{2n(t)h(t')*n(t')[\cos(\omega_0 t'+\theta)\cos(\omega_0 t+\theta) \\
&\quad + \sin(\omega_0 t'+\theta)\sin(\omega_0 t+\theta)]\} \\
&= E\{2n(t)h(t')*[n(t')\cos\omega_0(t-t')]\} \\
&= 2h(t')*[E\{n(t)n(t')\}\cos\omega_0(t-t')] \\
&= 2h(t')*[R_n(t-t')\cos\omega_0(t-t')] \\
&\triangleq 2\int_{-\infty}^{\infty} h(t-t')R_n(t-t')\cos\omega_0(t-t')\,dt'
\end{aligned} \tag{C.12}$$

Letting $u = t - t'$, gives

$$\overline{n\widehat{n}} = 2\int_{-\infty}^{\infty} h(u)\cos(\omega_0 u)R_n(u)\,du \tag{C.13}$$

Now, a general case of Parseval's theorem is

$$\int_{-\infty}^{\infty} x(t)y(t)\,dt = \int_{-\infty}^{\infty} X(f)Y^*(f)\,df \tag{C.14}$$

where $x(t) \leftrightarrow X(f)$ and $y(t) \leftrightarrow Y(f)$. In (C.13) we note that

$$h(u)\cos(\omega_0 u) \leftrightarrow \frac{1}{2}H(f-f_0) + \frac{1}{2}H(f+f_0) \tag{C.15}$$

and

$$R_n(u) \leftrightarrow S_n(f) \tag{C.16}$$

Thus, using (C.14), we may write (C.13) as

$$\overline{n\widehat{n}} = \int_{-\infty}^{\infty} [H(f - f_0) + H(f + f_0)] S_n(f)\, df \tag{C.17}$$

which follows because $S_n(f)$ is real. However, $S_n(f)$ is nonzero only where $H(f - f_0) + H(f + f_0) = 1$ because it was assumed narrowband. Thus (C.13) reduces to

$$\overline{n\widehat{n}} = \int_{-\infty}^{\infty} S_n(f)\, df = \overline{n^2(t)} \tag{C.18}$$

Substituting (C.18) and (C.4) into (C.3), we obtain

$$E\{(n - \widehat{n})^2\} = \overline{n^2} - 2\overline{n^2} + \overline{n^2} \equiv 0 \tag{C.19}$$

which shows that the mean-square error between $n(t)$ and $\widehat{n}(t)$ is zero.

APPENDIX D

ZERO-CROSSING AND ORIGIN ENCIRCLEMENT STATISTICS

In this appendix we consider a couple of problems frequently encountered in the study of FM demodulation of signals in additive Gaussian noise. Specifically, expressions are derived for the probability of a zero crossing of a bandlimited Gaussian process and for the average rate of origin encirclement of a constant-amplitude sinusoid plus narrowband Gaussian noise.

D.1 THE ZERO-CROSSING PROBLEM

Consider a sample function of a lowpass, zero-mean Gaussian process $n(t)$, as illustrated in Figure D.1. Denote the effective noise bandwidth by W, the power spectral density by $S_n(f)$, and the autocorrelation function by $R_n(\tau)$.

Consider the probability of a zero crossing in a small time interval Δ s in duration. For Δ sufficiently small, so that more than one zero crossing is unlikely, the probability $P_{\Delta-}$ of a minus-to-plus zero crossing in a time interval $\Delta \ll 1/2W$ is the probability that $n_0 < 0$ and $n_0 + \dot{n}_0\Delta > 0$. That is,

$$\begin{aligned} P_{\Delta-} &= \Pr[n_0 < 0 \text{ and } n_0 + \dot{n}_0\Delta > 0] \\ &= \Pr[n_0 < 0 \text{ and } n_0 > -\dot{n}_0\Delta, \text{all } \dot{n}_0 \geq 0] \\ &= \Pr[-\dot{n}_0\Delta < n_0 < 0, \text{all } \dot{n}_0 \geq 0] \end{aligned} \quad \text{(D.1)}$$

This can be written in terms of the joint pdf of n_0 and $\dot{n}_0$, $f_{n_0\dot{n}_0}(y, z)$, as

$$P_{\Delta-} = \int_0^\infty \left[\int_{-z\Delta}^0 f_{n_0\dot{n}_0}(y, z)\, dy \right] dz \quad \text{(D.2)}$$

where y and z are running variables for n_0 and $\dot{n}_0$, respectively. Now $\dot{n}_0$ is a Gaussian random variable, since it involves a linear operation on $n(t)$, which is Gaussian by assumption. It can be shown that

$$E\{n_0\dot{n}_0\} = \left.\frac{dR_n(\tau)}{d\tau}\right|_{\tau=0} \quad \text{(D.3)}$$

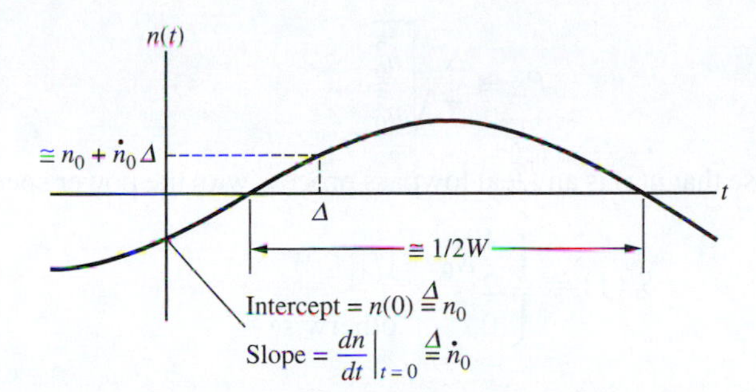

Figure D.1
Sample function of a lowpass Gaussian process of bandwidth W.

Thus, if the derivative of $R_n(\tau)$ exists at $\tau = 0$, it is zero because $R_n(\tau)$ must be even. It follows that

$$E\{n_0 \dot{n}_0\} = 0 \tag{D.4}$$

Therefore, n_0 and $\dot{n}_0$, which are samples of $n_0(t)$ and $dn_0(t)/dt$, respectively, are statistically independent, since uncorrelated Gaussian processes are independent. Thus, letting $\text{var}\{n_0\} = \overline{n_0^2}$ and $\text{var}\{\dot{n}_0\} = \overline{\dot{n}_0^2}$, the joint pdf of n_0 and $\dot{n}_0$ is

$$f_{n_0 \dot{n}_0}(y, z) = \frac{\exp(-y^2/2\overline{n_0^2})}{\sqrt{2\pi \overline{n_0^2}}} \frac{\exp(-z^2/2\overline{\dot{n}_0^2})}{\sqrt{2\pi \overline{\dot{n}_0^2}}} \tag{D.5}$$

which, when substituted into (D.2), yields

$$P_{\Delta -} = \int_0^\infty \frac{\exp(-z^2/2\overline{\dot{n}_0^2})}{\sqrt{2\pi \overline{\dot{n}_0^2}}} \left[\int_{-z\Delta}^0 \frac{\exp(-y^2/2\overline{n_0^2})}{\sqrt{2\pi \overline{n_0^2}}}\, dy \right] dz \tag{D.6}$$

For Δ small, the inner integral of (D.6) can be approximated as $z\Delta/\sqrt{2\pi \overline{n_0^2}}$, which allows (D.6) to be simplified to

$$P_{\Delta -} \cong \frac{\Delta}{\sqrt{2\pi \overline{n_0^2}}} \int_0^\infty z \frac{\exp(-z^2/2\overline{\dot{n}_0^2})}{\sqrt{2\pi \overline{\dot{n}_0^2}}}\, dz \tag{D.7}$$

Letting $\zeta = z^2/2\overline{\dot{n}_0^2}$ yields

$$\begin{aligned} P_{\Delta -} &\cong \frac{\Delta}{2\pi \sqrt{\overline{n_0^2}\, \overline{\dot{n}_0^2}}} \int_0^\infty \overline{\dot{n}_0^2}\, e^{-\zeta}\, d\zeta \\ &= \frac{\Delta}{2\pi} \sqrt{\frac{\overline{\dot{n}_0^2}}{\overline{n_0^2}}} \end{aligned} \tag{D.8}$$

for the probability of a minus-to-plus zero crossing in Δ s. By symmetry, the probability of a plus-to-minus zero crossing is the same. Thus, the probability of a zero crossing in Δ s, plus or minus, is

$$P_\Delta \cong \frac{\Delta}{\pi}\sqrt{\frac{\overline{\dot{n}_0^2}}{\overline{n_0^2}}} \tag{D.9}$$

For example, suppose that $n(t)$ is an ideal lowpass process with the power spectral density

$$S_n(f) = \begin{cases} \frac{1}{2}N_0, & |f| \le W \\ 0, & \text{otherwise} \end{cases} \tag{D.10}$$

Thus

$$R_n(\tau) = N_0 W \operatorname{sinc} 2W\tau$$

which possesses a derivative at $\tau = 0$. Therefore, n_0 and $\dot{n}_0$ are independent. It follows that

$$\overline{n_0^2} = \operatorname{var}\{n_0\} = \int_{-\infty}^{\infty} S_n(f)\,df = R_n(0) = N_0 W \tag{D.11}$$

and, since the transfer function of a differentiator is $H_d(f) = j2\pi f$, that

$$\begin{aligned} \overline{\dot{n}_0^2} &= \operatorname{var}\{\dot{n}_0\} = \int_{-\infty}^{\infty} |H_d(f)|^2 S_n(f)\,df = \int_{-W}^{W} (2\pi f)^2 \frac{1}{2}N_0\,df \\ &= \frac{1}{3}(2\pi W)^2 (N_0 W) \end{aligned} \tag{D.12}$$

Substitution of these results into (D.9) gives

$$2P_{\Delta -} = 2P_{\Delta +} = P_\Delta = \frac{\Delta}{\pi}\frac{2\pi W}{\sqrt{3}} = \frac{2W\Delta}{\sqrt{3}} \tag{D.13}$$

for the probability of a zero crossing in a small time interval Δ s in duration for a random process with an ideal rectangular lowpass spectrum.

D.2 AVERAGE RATE OF ZERO CROSSINGS

Consider next the sum of a sinusoid plus narrowband Gaussian noise:

$$\begin{aligned} z(t) &= A\cos(\omega_0 t) + n(t) \\ &= A\cos(\omega_0 t) + n_c(t)\cos(\omega_0 t) - n_s(t)\sin(\omega_0 t) \end{aligned} \tag{D.14}$$

where $n_c(t)$ and $n_s(t)$ are lowpass processes with statistical properties as described in Section 6.5. We may write $z(t)$ in terms of envelope $R(t)$ and phase $\theta(t)$ as

$$z(t) = R(t)\cos[\omega_0 t + \theta(t)] \tag{D.15}$$

where

$$R(t) = \sqrt{[A + n_c(t)]^2 + n_s^2(t)} \tag{D.16}$$

and

$$\theta(t) = \tan^{-1}\left(\frac{n_s(t)}{A + n_c(t)}\right) \tag{D.17}$$

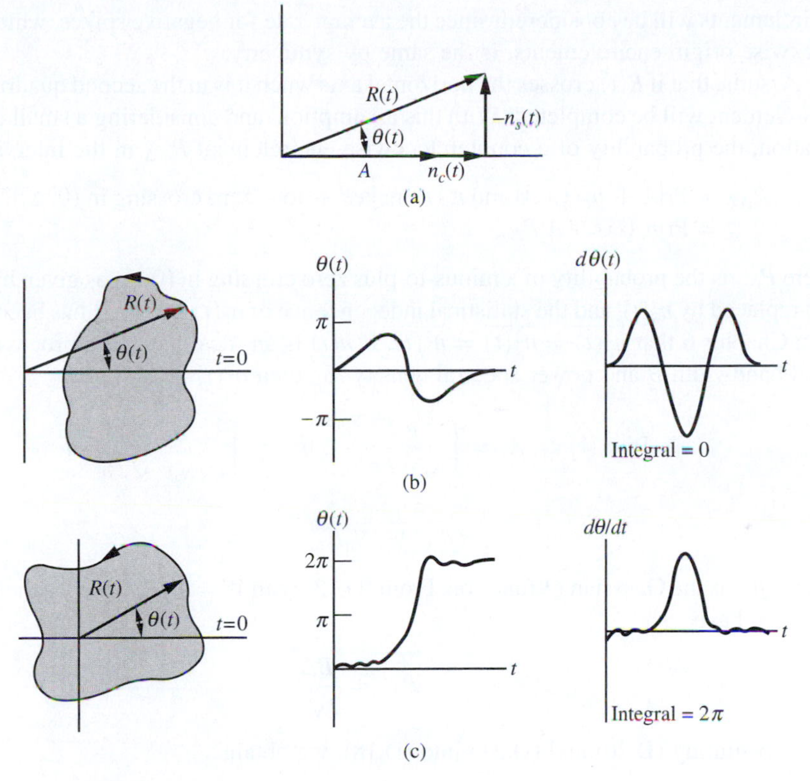

Figure D.2
Phasor diagrams showing possible trajectories for a sinusoid plus Gaussian noise. (a) Phasor representation for a sinusoid plus narrowband noise. (b) Trajectory that does not encircle origin. (c) Trajectory that does encircle origin.

A phasor representation for this process is shown in Figure D.2(a). In Figure D.2(b), a possible trajectory for the tip of $R(t)$ that does not encircle the origin is shown along with $\theta(t)$ and $d\theta(t)/dt$. In Figure D.2(c), a trajectory that encircles the origin is shown along with $\theta(t)$ and $d\theta(t)/dt$. For the case in which the origin is encircled the area under $d\theta/dt$ must be 2π rad. Recalling the definition of an ideal FM discriminator in Chapter 3, we see that the sketches for $d\theta/dt$ shown in Figure D.2 represent the output of a discriminator in response to input of an unmodulated signal plus noise or interference. For a high SNR, the phasor will randomly fluctuate near the horizontal axis. Occasionally, however, it will encircle the origin as shown in Figure D.2(c). Intuitively, these encirclements become more probable as the SNR decreases. Because of its nonzero area, the impulsive type of output illustrated in Figure D.2(c), caused by an encirclement of the origin, has a much more serious effect on the noise level of the discriminator output than does the noise excursion illustrated in Figure D.2(b), which has zero area.

We now derive an expression for the average number of noise spikes per second of the type illustrated in Figure D.2(c). Only positive spikes caused by counterclockwise origin encirclements will be considered, since the average rate for negative spikes, which result from clockwise origin encirclements, is the same by symmetry.

Assume that if $R(t)$ crosses the horizontal axis when it is in the second quadrant, the origin encirclement will be completed. With this assumption, and considering a small interval Δ s in duration, the probability of a counterclockwise encirclement $P_{cc\Delta}$ in the interval $(0, \Delta)$ is

$$\begin{aligned} P_{cc\Delta} &= \Pr[A + n_c(t) < 0 \text{ and } n_s(t) \text{ makes } + \text{ to } - \text{ zero crossing in } (0, \Delta)] \\ &= \Pr[n_c(t) < -A] P_{\Delta-} \end{aligned} \tag{D.18}$$

where $P_{\Delta-}$ is the probability of a minus-to-plus zero crossing in $(0, \Delta)$ as given by (D.13) with $n(t)$ replaced by $n_s(t)$, and the statistical independence of $n_c(t)$ and $n_s(t)$ has been used. Recall from Chapter 6 that $\overline{n_c^2(t)} = \overline{n_s^2(t)} = \overline{n^2(t)}$. If $n(t)$ is an ideal bandpass process with single-sided bandwidth B and power spectral density N_0, then $\overline{n^2(t)} = N_0 B$, and

$$\Pr[n_c(t) < -A] = \int_{-\infty}^{-A} \frac{e^{-n_c^2/2N_0B}}{\sqrt{2\pi N_0 B}} \, dn_c = \int_{A/\sqrt{N_0B}}^{\infty} \frac{e^{-u^2/2}}{\sqrt{2\pi}} \, du \tag{D.19}$$

$$= Q\left(\sqrt{A^2/N_0B}\right) \tag{D.20}$$

where $Q(\cdot)$ is the Gaussian Q-function. From (D.13) with $W = B/2$, which is the bandwidth of $n_s(t)$, we have

$$P_{\Delta-} = \frac{B\Delta}{2\sqrt{3}} \tag{D.21}$$

Substituting (D.20) and (D.21) into (D.18), we obtain

$$P_{cc\Delta} = \frac{B\Delta}{2\sqrt{3}} Q\left(\sqrt{\frac{A^2}{N_0 B}}\right) \tag{D.22}$$

The probability of a clockwise encirclement $P_{c\Delta}$ is the same by symmetry. Thus the expected number of encirclements per second, clockwise and counterclockwise, is

$$\begin{aligned} \nu &= \frac{1}{\Delta}(P_{c\Delta} + P_{cc\Delta}) \\ &= \frac{B}{2\sqrt{3}} Q\left(\sqrt{\frac{A^2}{N_0 B}}\right) \end{aligned} \tag{D.23}$$

We note that the average number of encirclements per second increases in direct proportion to the bandwidth and decreases essentially exponentially with increasing SNR $A^2/2N_0B$. We can see this in Figure D.3, which illustrates ν/B as a function of SNR. Figure D.3 also shows the asymptote as $A^2/2N_0B \to 0$ of $\nu/B = 1/2\sqrt{3} = 0.2887$.

The results derived above say nothing about the statistics of the number of impulses, N, in a time interval, T. However, it can be shown that the power spectral density of a periodic impulse noise process is given by

$$S_I(f) = \nu \overline{a^2} \tag{D.24}$$

where ν is the average number of impulses per second ($\nu = f_s$ for a periodic impulse train) and $\overline{a^2}$ is the mean-squared value of the impulse weights a_k. A similar result can be shown for

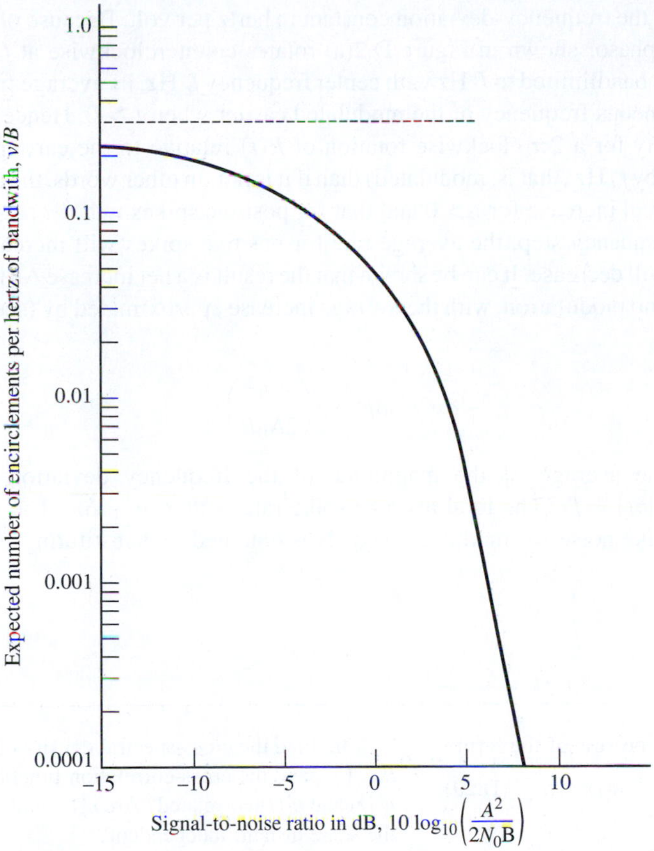

Figure D.3
Rate of origin encirclements as a function of SNR.

impulses which have exponentially distributed intervals between them (i.e., Poisson impulse noise). Approximating the impulse portion of $d\theta/dt$ as a Poisson impulse noise process with sample functions of the form

$$x(t) \triangleq \left.\frac{d\theta(t)}{dt}\right|_{\text{impulse}} = \sum_{k=-\infty}^{\infty} \pm 2\pi\delta(t-t_k) \tag{D.25}$$

where t_k is a Poisson point process with average rate ν given by (D.23), we may approximate the power spectral density of this impulse noise process as white with a spectral level given by

$$\begin{aligned} S_x(f) &= \nu(2\pi)^2 \\ &= \frac{4\pi^2 B}{\sqrt{3}} Q\left(\sqrt{\frac{A^2}{N_0 B}}\right), \ -\infty < f < \infty \end{aligned} \tag{D.26}$$

If the sinusoidal signal component in (D.14) is FM modulated, the average number of impulses per second is increased over that obtained for no modulation. Intuitively, the reason may be explained as follows. Consider a carrier that is FM modulated by a unit step. Thus

$$z(t) = A\cos\{2\pi[f_c + f_d u(t)]t\} + n(t) \tag{D.27}$$

where $f_d \leq \frac{1}{2}B$ is the frequency-deviation constant in hertz per volt. Because of this frequency step, the carrier phasor shown in Figure D.2(a) rotates counterclockwise at f_d Hz for $t > 0$. Since the noise is bandlimited to B Hz with center frequency f_c Hz, its average frequency is *less* than the instantaneous frequency of the modulated carrier when $t > 0$. Hence there will be a greater probability for a 2π clockwise rotation of $R(t)$ relative to the carrier phasor if it is frequency offset by f_d Hz (that is, modulated) than if it is not. In other words, the average rate for negative spikes will increase for $t > 0$ and that for positive spikes will decrease. Conversely, for a negative frequency step, the average rate for positive spikes will increase and that for negative spikes will decrease. It can be shown that the result is a net increase $\Delta\nu$ in the spike rate over the case for no modulation, with the average increase approximated by (see Problems D.1 and D.2)

$$\overline{\delta\nu} = \overline{|\delta f|}\exp\left(\frac{-A^2}{2N_0B}\right) \quad \text{(D.28)}$$

where $\overline{|\delta f|}$ is the average of the magnitude of the frequency deviation. For the case just considered, $\overline{|\delta f|} = f_d$. The total average spike rate is then $\nu + \overline{\delta\nu}$. The power spectral density of the spike noise for modulated signals is obtained by substituting $\nu + \overline{\delta\nu}$ for ν in (D.26).

Problems

D.1. Consider a signal-plus-noise process of the form

$$z(t) = A\cos[2\pi(f_0 + f_d)t] + n(t) \quad \text{(D.29)}$$

where $n(t)$ is given by

$$n(t) = n_c(t)\cos(2\pi f_0 t) - n_s(t)\sin(2\pi f_0 t) \quad \text{(D.30)}$$

Assume that $n(t)$ is an ideal bandlimited white-noise process with double-sided power spectral density equal to $\frac{1}{2}N_0$, for $-\frac{1}{2}B \leq f \pm f_0 \leq \frac{1}{2}B$, and zero otherwise. Write $z(t)$ as

$$z(t) = A\cos[2\pi(f_0 + f_d)t] + n_c'(t)\cos[2\pi(f_0 + f_d)t] - n_s'\sin[2\pi(f_0 + f_d)t]$$

a. Express $n_c'(t)$ and $n_s'(t)$ in terms of $n_c(t)$ and $n_s(t)$. Find the power spectral densities of $n_c'(t)$ and $n_s'(t)$, $S_{n_c'}(f)$ and $S_{n_s'}(f)$.

b. Find the cross-spectral density of $n_c'(t)$ and $n_s'(t)$, $S_{n_c'n_s'}(f)$, and the cross-correlation function, $R_{n_c'n_s'}(\tau)$. Are $n_c'(t)$ and $n_s'(t)$ correlated? Are $n_c'(t)$ and $n_s'(t)$, sampled at the same instant, independent?

D.2.

a. Using the results of Problem D.1, derive Equation (D.28) with $\overline{|\delta f|} = f_d$.

b. Compare equations (D.28) and (D.23) for a squarewave-modulated FM signal with deviation f_d by letting $\overline{|\delta f|} = f_d$ and $B = 2f_d$ for $f_d = 5$ and 10 for signal-to-noise ratios of $A^2/N_0B = 1, 10, 100, 1000$. Plot ν and $\overline{\delta\nu}$ versus A^2/N_0B.

APPENDIX **E**

CHI-SQUARE STATISTICS

Useful probability distributions result from sums of squares of independent Gaussian random variables of the form

$$Z = \sum_{i=1}^{n} X_i^2 \tag{E.1}$$

If each of the component random variables, X_i, is zero-mean and has variance σ^2, the probability density function of Z is

$$f_Z(z) = \frac{1}{\sigma^n 2^{n/2}\Gamma(n/2)} z^{(n-2)/2} \exp\left(\frac{-z}{2\sigma^2}\right), \quad z \geq 0 \tag{E.2}$$

The random variable Z is known as a *central chi-square*, or simply chi-square, random variable with n degrees of freedom. In (E.2), $\Gamma(x)$ is the Gamma function defined as

$$\Gamma(x) = \int_0^\infty t^{x-1} \exp(-t)\, dt, \quad x > 0 \tag{E.3}$$

The Gamma function has the properties

$$\Gamma(n) = (n-1)\Gamma(n-1) \tag{E.4}$$

and

$$\Gamma(1) = 1 \tag{E.5}$$

The two preceding equations give, for integer argument n,

$$\Gamma(n) = (n-1)! \text{ integer } n \tag{E.6}$$

Also

$$\Gamma\left(\frac{1}{2}\right) = \sqrt{\pi} \tag{E.7}$$

With the change of variables $z = y^2$, the central chi-square distribution with two degrees of freedom as obtained from (E.2) becomes the Rayleigh pdf, given by

$$f_Y(y) = \frac{y}{\sigma^2}\exp\left(\frac{-y^2}{2\sigma^2}\right), \quad y \geq 0 \tag{E.8}$$

If the component random variables in (E.1) are not zero mean but have means defined by $E(X_i) = m_i$, the resulting pdf of Z is

$$f_z(z) = \frac{1}{2\sigma^2}\left(\frac{z}{s^2}\right)^{(n-2)/4}\exp\left(-\frac{z+s^2}{2\sigma^2}\right) I_{n/2-1}\left(\frac{s\sqrt{z}}{\sigma^2}\right), \quad z \geq 0 \tag{E.9}$$

where

$$s^2 = \sum_{i=1}^{n} m_i^2 \tag{E.10}$$

and

$$I_m(x) = \sum_{k=0}^{\infty}\frac{(x/2)^{m+2k}}{k!\Gamma(m+k+1)}, \quad x \geq 0 \tag{E.11}$$

is the mth-order modified Bessel function of the first kind. The random variable defined by (E.9) is called a *noncentral chi-square* random variable. If we let $n = 2$ and make the change of variables $z = y^2$, (E.9) becomes

$$f_Y(y) = \frac{y}{\sigma^2}\exp\left(-\frac{y^2+s^2}{2\sigma^2}\right) I_0\left(\frac{sy}{\sigma^2}\right), \quad y \geq 0 \tag{E.12}$$

which is known as the Ricean pdf.

APPENDIX F

QUANTIZATION OF RANDOM PROCESSES

Quantization is important in any application where analog signals are sampled and converted to digital format for processing and transmission.[1]

Quantization of a random waveform can be accomplished by sampling at an appropriate rate, dictated by the sampling theorem, and processing each sample by a zero-memory nonlinear device defined as follows: Consider a set of $N+1$ decision levels $x_0, x_1, \ldots, x_N$ and a set of n output points $y_1, y_2, \ldots, y_N$. If the input sample X lies in the ith quantizing interval defined by

$$R_i = \{x_{i-1} < X \leq x_i\} \tag{F.1}$$

the quantizer produces the output y_i, where y_i is itself chosen to be some value in the interval R_i. The end values are chosen to be equal to the smallest and largest values, respectively, that the input samples may assume. Usually these values are $+\infty$ and $-\infty$, whereas the output values all have finite values. If $N = 2^n$, a unique n-bit binary word can be associated with each output value, and the quantizer is said to be an *n-bit quantizer.*[2]

The input–output characteristic $y = Q(x)$ of a quantizer has a staircase form. Figure F.1 shows two possible characteristics. Figure F.1(a) shows a midtread form with an output-quantization level located at $y = 0$. Figure F.1(b) shows a midriser form where $x = 0^+$ results in $y > 0$ and $x = 0^-$ results in $y < 0$.

The quantization process can be modeled as the addition to each input sample of a random-noise component $e = Q(X) - X$, which is dependent on the value of the input X. Figure F.1(c) shows a plot of the quantizing error e versus the input amplitude X. The quantizing error is conveniently described as *granular noise* when $x_1 < X < x_{N-1}$ and as *overload noise* when X is outside of this interval.

The performance of a quantizer can be characterized by the *mean-squared distortion*, defined by

$$D = \int_{-\infty}^{\infty} [Q(x) - x]^2 f_X(x)\, dx \tag{F.2}$$

[1] See Gibson (2002), Chapter 3 for general theory, Chapter 26 for application to PCM, and Chapter 81 for speech coding in cellular radio.

[2] For a well-written summary of quantization, see Allen Gersho, Quantization. *IEEE Communications Society Magazine*, **15**: 20–29, September 1977.

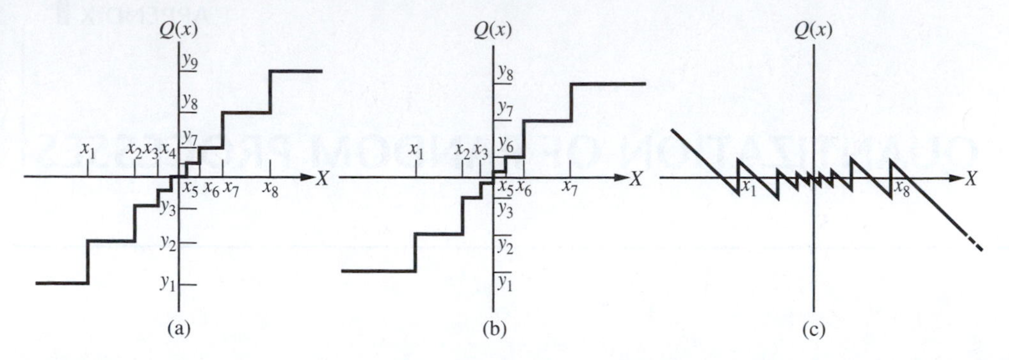

Figure F.1
Characteristics of quantizers (a) Characteristic of midtread quantizer (b) Characteristic of midriser quantizer (c) Typical plot of quantizing error versus input.

where $f_X(x)$ is the pdf of the input sample X. If the number of quantizing levels is very large, the distortion can be written as

$$D = \sum_{i=1}^{N} \int_{x_{i-1}}^{x_i} (y_i - x)^2 f_X(x)\, dx \tag{F.3}$$

which follows by breaking the region of integration into the separate intervals R_i noting that $Q(x) = y_i$ when x is in R_i. For large N, each interval R_i can be made small, and it is reasonable to approximate $f_X(x)$ as $f_X(x_i)$, a constant, within the interval R_i. In this case, the distortion becomes

$$D = \frac{1}{12} \sum_{i=1}^{N} f_X(x_i)\Delta_i^3 \tag{F.4}$$

where it is assumed that $y_i = (x_i + x_{i-1})/2$ and $\Delta_i = x_i - x_{i-1}$ is the length of R_i. Equation (F.4) implies that the overload points x_0 and x_N have been chosen so that the overload noise is negligible compared with the granular noise. If the quantizing intervals are equal in length, that is, if $\Delta_i = \Delta$, for all i, (F.4) becomes

$$D = \frac{\Delta^2}{12} \sum_{i=1}^{N} f_X(x_i)\Delta = \frac{\Delta^2}{12} \tag{F.5}$$

A convenient measure of performance in many cases is the *signal-to-noise ratio* (SNR), where the noise power is the mean-squared distortion D and the signal power is the variance, σ^2, of the input samples. A symmetrical uniform-interval quantizer is completely specified by giving the number of levels and either the step size or the overload level $V = x_N = -x_0$. The latter can be given in terms of the *loading factor* $y = V/\sigma$, which is commonly chosen to be $y = 4$ (four-sigma loading). If $y = 4$, the step size becomes $\Delta = 8\sigma/(N-2)$, which is found by employing a total amplitude range for the quantizing interval of 8σ, with $N - 2$ levels in that range. With $N = 2^n \gg 2$, it follows from the definition of the SNR that

$$\frac{\sigma^2}{D} = \frac{12\sigma^2}{(8\sigma/2^n)^2} = \frac{3}{16} 2^{2n} \tag{F.6}$$

which gives

$$\text{SNR} = 10\log_{10}\left(\frac{\sigma^2}{D}\right) = 6.02n - 7.3 \text{ dB} \tag{F.7}$$

Thus, the SNR increases by 6 dB per bit for a uniform quantizer. Varying the loading factor changes the constant term, 7.3, but not the multiplier of n.

For an input having a known pdf, one can optimally select the decision levels $x_0, x_1, \ldots, x_N$ and the output points such that the mean-squared error is minimized. In fact, Max[3] carried this out for a general kth absolute-value mean-error criterion and tabulated the optimum quantizer levels for a Gaussian input pdf for various values of N. The conditions for a minimum mean-squared error are

$$x_0 = -\infty, \qquad x_N = \infty, \qquad x_{N/2} = 0 \tag{F.8}$$

$$\int_{x_{i-1}}^{x_i} (x - y_i) f_x(x)\, dx = 0, \quad i = 1, 2, \ldots, N \tag{F.9}$$

and

$$x_i = \frac{y_i + y_{i+1}}{2}, \quad i = 1, 2, \ldots, N-1 \tag{F.10}$$

which is obtained by differentiating D with respect to the x_is and y_is. Equation (F.9) can be expressed as

$$y_i = \int_{x_{i-1}}^{x_i} x f_X(x)\, dx \tag{F.11}$$

which means that the output levels are the centroids under $f_X(x)$ between adjacent boundaries. These equations lead to closed-form solutions only in certain special cases.

Max suggested an iterative procedure to solve numerically for the values of x_i and y_i for Gaussian-distributed signals. Other input amplitude distributions of interest are Laplacian-distributed and gamma-distributed signals, which are often used to approximate the amplitude pdf of speech. Paez and Glisson[4] determined the optimum quantizer characteristics for Laplacian-distributed and gamma-distributed signals. For example, for Laplacian-distributed signals

$$f_X(x) = \left(\frac{a}{2}\right)\exp(-a|x|) \tag{F.12}$$

where the rms value is $\sigma = \sqrt{2}/a$. The optimum quantizer characteristics for various values of N are given in Table F.1.

[3] J. Max, Quantizing for minimum distortion, *IRE Transactions Information Theory*, **IT-6**: 7–12, March 1960.

[4] M. D. Paex and T. H. Glisson, Minimum mean-squared-error quantization in speech PCM and DPCM systems. *IEEE Transactions on Communications*, **COM-20**, 225–230, April 1972.

Table F.1 Optimum Quantizer Characteristics for Laplacian Inputs (σ=1)

i	$N=2$		$N=4$		$N=8$		$N=16$	
	x_i	y_i	x_i	y_i	x_i	y_i	x_i	y_i
0	$-\infty$	−0.707	$-\infty$	−1.810	$-\infty$	−2.994	$-\infty$	−4.316
1	∞	0.707	−1.102	−0.395	−2.286	−1.576	−3.605	−2.895
2			1.102	0.395	−1.181	−0.785	−2.499	−2.103
3			∞	1.810	−0.504	−0.222	−1.821	−1.504
4					0.504	0.222	−1.317	−1.095
5					1.181	0.785	−0.910	−0.726
6					2.286	1.576	−0.566	−0.407
7					∞	2.994	−0.266	−0.126
8							0.266	0.126
9							0.566	0.407
10							0.910	0.726
11							1.317	1.095
12							1.821	1.504
13							2.499	2.103
14							3.605	2.895
15							∞	4.316
MSE	0.5		0.1765		0.0548		0.0145	
SNR (dB)	3.01		7.53		12.16		18.12	

APPENDIX G

MATHEMATICAL AND NUMERICAL TABLES

This appendix contains several tables pertinent to the material contained in this book. The tables are

1. **The Gaussian Q-Function**
2. **Trigonometric Identities**
3. **Series Expansions**
4. **Integrals**
5. **Fourier Transform Pairs**
6. **Fourier Transform Theorems**

G.1 THE GAUSSIAN Q-FUNCTION

In this appendix we examine the Gaussian Q-function in more detail and discuss several approximations to the Q-function.[1] The Gaussian probability density function of unit variance and zero mean is

$$Z(x) = \frac{1}{\sqrt{2\pi}} e^{-x^2/2} \tag{G.1}$$

and the corresponding cumulative distribution function is

$$P(x) = \int_{-\infty}^{x} Z(t)\,dt \tag{G.2}$$

The Gaussian Q-function is defined as[2]

$$Q(x) = 1 - P(x) = \int_{x}^{\infty} Z(t)\,dt \tag{G.3}$$

An asymptotic expansion for $Q(x)$, valid for large x, is

$$Q(x) = \frac{Z(x)}{x}\left(1 - \frac{1}{x^2} + \frac{1(3)}{x^4} - \cdots + \frac{(-1)^n 1(3)\cdots(2n-1)}{x^{2n}}\right) + R_n \tag{G.4}$$

[1]The information given in this appendix is extracted from Abramowitz and Stegun, (1972) (originally published in 1964 as part of the National Bureau of Standards Applied Mathematics Series 55).

[2]For $x < 0$, $Q(x) = 1 - Q(|x|)$.

where the remainder is given by

$$R_n = (-1)^{n+1} 1(3) \cdots (2n+1) \int_x^\infty \frac{Z(t)}{t^{2n+2}}\, dt \tag{G.5}$$

which is less in absolute value than the first neglected term. For $x \geq 3$, less than 10% error results if only the first term in (G.4) is used to approximate the Gaussian Q-function.

A finite-limit integral for the Q-function, which is convenient for numerical integration, is[3]

$$Q(x) = \begin{cases} \dfrac{1}{\pi}\displaystyle\int_0^{\pi/2} \exp\left(-\dfrac{x^2}{2\sin^2\phi}\right) d\phi, & x \geq 0 \\ 1 - \dfrac{1}{\pi}\displaystyle\int_0^{\pi/2} \exp\left(-\dfrac{x^2}{2\sin^2\phi}\right) d\phi, & x < 0 \end{cases} \tag{G.6}$$

The error function can be related to the Gaussian Q-function by

$$\text{erf}(x) \triangleq \frac{2}{\sqrt{\pi}} \int_0^x e^{-t^2}\, dt = 1 - 2Q(\sqrt{2}x) \tag{G.7}$$

The complementary error function is defined as erfc $x = 1 -$ erf x so that

$$Q(x) = \frac{1}{2}\text{erfc}\left(\frac{x}{\sqrt{2}}\right) \tag{G.8}$$

which is convenient for computing values using MATLAB since erfc is a subprogram in MATLAB but the Q-function is not (unless you have a Communications Toolbox).

Table G.1 A Short Table of Q-Function Values

x	$Q(x)$	x	$Q(x)$	x	$Q(x)$
0	0.5	1.5	0.066807	3.0	0.0013499
0.1	0.46017	1.6	0.054799	3.1	0.00096760
0.2	0.42074	1.7	0.044565	3.2	0.00068714
0.3	0.38209	1.8	0.035930	3.3	0.00048342
0.4	0.34458	1.9	0.028717	3.4	0.00033693
0.5	0.30854	2.0	0.022750	3.5	0.00023263
0.6	0.27425	2.1	0.017864	3.6	0.00015911
0.7	0.24196	2.2	0.013903	3.7	0.00010780
0.8	0.21186	2.3	0.010724	3.8	7.2348×10^{-5}
0.9	0.18406	2.4	0.0081975	3.9	4.8096×10^{-5}
1.0	0.15866	2.5	0.0062097	4.0	3.1671×10^{-5}
1.1	0.13567	2.6	0.0046612	4.1	2.0658×10^{-5}
1.2	0.11507	2.7	0.0034670	4.2	1.3346×10^{-5}
1.3	0.096800	2.8	0.0025551	4.3	8.5399×10^{-6}
1.4	0.080757	2.9	0.0018658	4.4	5.4125×10^{-6}

[3]J. W. Craig, A new, simple and exact result for calculating the probability of error for two-dimensional signal constellations. *IEEE MILCOM'91 Conference Record.*, Boston, MA, 25.5.1–25.5.5, November 1991.

M. K. Simon and D. Divsalar, Some new twists to problems involving the Gaussian probability integral. *IEEE Transactions on Communication*, **46**: 200–210, February 1998.

A short table of values for $Q(x)$ is given in Table G.1. Note that values of $Q(x)$ for $x < 0$ can be found from the table by using the relationship

$$Q(-x) = 1 - Q(x) \tag{G.9}$$

For example, from Table G.1, $Q(-0.1) = 1 - Q(0.1) = 1 - 0.46017 = 0.53983$.

G.2 TRIGONOMETRIC IDENTITIES

$$\cos u = \frac{e^{ju} + e^{-ju}}{2}$$

$$\sin u = \frac{e^{ju} - e^{-ju}}{2j}$$

$$\cos^2 u + \sin^2 u = 1$$

$$\cos^2 u - \sin^2 u = \cos(2u)$$

$$2 \sin u \cos u = \sin(2u)$$

$$\cos u \cos v = \frac{1}{2}\cos(u - v) + \frac{1}{2}\cos(u + v)$$

$$\sin u \cos v = \frac{1}{2}\sin(u - v) + \frac{1}{2}\sin(u + v)$$

$$\sin u \sin v = \frac{1}{2}\cos(u - v) - \frac{1}{2}\cos(u + v)$$

$$\cos(u \pm v) = \cos u \cos v \mp \sin u \sin v$$

$$\sin(u \pm v) = \sin u \cos v \pm \cos u \sin v$$

$$\cos^2 u = \frac{1}{2} + \frac{1}{2}\cos(2u)$$

$$\cos^{2n} u = \frac{1}{2^{2n}}\left\{\sum_{k=0}^{n-1} 2\binom{2n}{k}\cos[2(n-k)u] + \binom{2n}{n}\right\}, \quad n \text{ a positive integer}$$

$$\cos^{2n-1} u = \frac{1}{2^{2n-2}}\left\{\sum_{k=0}^{n-1}\binom{2n-1}{k}\cos(2n-2k-1)u\right\}$$

$$\sin^2 u = \frac{1}{2} - \frac{1}{2}\cos(2u)$$

$$\sin^{2n} u = \frac{1}{2^{2n}}\left\{\sum_{k=0}^{n-1}(-1)^{n-k}\, 2\binom{2n}{k}\cos[2(n-k)u] + \binom{2n}{n}\right\}$$

$$\sin^{2n-1} u = \frac{1}{2^{2n-2}}\left[\sum_{k=0}^{n-1}(-1)^{n+k-1}\binom{2n-1}{k}\sin(2n-2k-1)u\right]$$

G.3 SERIES EXPANSIONS

$$(u+v)^n = \sum_{k=0}^{n} \binom{n}{k} u^{n-k} v^k, \quad \binom{n}{k} = \frac{n!}{(n-k)!k!}$$

Letting $u = 1$ and $v = x$, where $|x| \ll 1$ results in the approximations:

$$(1+x)^n \cong 1 + nx; \qquad (1-x)^n \cong 1 - nx; \qquad (1+x)^{1/2} \cong 1 + \frac{1}{2}x$$

$$\log_a u = \log_e u \log_a e; \quad \log_e u = \ln u = \log_e a \log_a u$$

$$e^u = \sum_{k=0}^{\infty} \frac{u^k}{k!} \cong 1 + u, \quad |u| \ll 1$$

$$\ln(1+u) \cong u, \quad |u| \ll 1$$

$$\sin u = \sum_{k=0}^{\infty} (-1)^k \frac{u^{2k+1}}{(2k+1)!} \cong u - \frac{u^3}{3!}, \quad |u| \ll 1$$

$$\cos u = \sum_{k=0}^{\infty} (-1)^k \frac{u^{2k}}{(2k)!} \cong 1 - \frac{u^2}{2!}, \quad |u| \ll 1$$

$$\tan u = u + \frac{1}{3}u^3 + \frac{2}{15}u^5 + \cdots$$

$$J_n(u) \cong \begin{cases} \dfrac{u^n}{2^n n!}\left(1 - \dfrac{u^2}{2^2(n+1)} + \dfrac{u^4}{2 \cdot 2^4 (n+1)(n+2)} - \cdots\right), |u| \ll 1 \\ \sqrt{\dfrac{2}{\pi u}} \cos\left(u - \dfrac{n\pi}{2} - \dfrac{\pi}{2}\right), |u| \gg 1 \end{cases}$$

$$I_0(u) \cong \begin{cases} 1 + \dfrac{u^2}{2^2} + \dfrac{u^4}{2^4} + \cdots \cong e^{u^2/4}, & 0 \le u \ll 1 \\ \dfrac{e^u}{\sqrt{2\pi u}}, & u \gg 1 \end{cases}$$

G.4 INTEGRALS

G.4.1 Indefinite

$$\int \sin(ax)\, dx = -\frac{1}{a}\cos(ax)$$

$$\int \cos(ax)\, dx = \frac{1}{a}\sin(ax)$$

$$\int \sin^2(ax)\, dx = \frac{x}{2} - \frac{1}{4a}\sin(2ax)$$

$$\int \cos^2(ax)\, dx = \frac{x}{2} + \frac{1}{4a}\sin(2ax)$$

$$\int x\sin(ax)\, dx = a^{-2}[\sin(ax) - ax\cos(ax)]$$

$$\int x\cos(ax)\, dx = a^{-2}[\cos(ax) + ax\sin(ax)]$$

$$\int x^m \sin x \, dx = -x^m \cos x + m \int x^{m-1} \cos x \, dx$$

$$\int x^m \cos x \, dx = x^m \sin x - m \int x^{m-1} \sin x \, dx$$

$$\int \exp(ax) \, dx = a^{-1} \exp(ax)$$

$$\int x^m \exp(ax) \, dx = a^{-1} x^m \exp(ax) - a^{-1} m \int x^{m-1} \exp(ax) \, dx$$

$$\int \exp(ax) \sin(bx) \, dx = (a^2 + b^2)^{-1} \exp(ax)[a \sin(bx) - b \cos(bx)]$$

$$\int \exp(ax) \cos(bx) \, dx = (a^2 + b^2)^{-1} \exp(ax)[a \cos(bx) + b \sin(bx)]$$

G.4.2 Definite

$$\int_0^\infty \frac{x^{m-1}}{1 + x^n} dx = \frac{\pi/n}{\sin(m\pi/n)}, \quad n > m > 0$$

$$\int_0^\pi \sin^2(nx) \, dx = \int_0^\pi \cos^2(nx) \, dx = \frac{\pi}{2}, \quad n \text{ an integer}$$

$$\int_0^\pi \sin(mx) \sin(nx) \, dx = \int_0^\pi \cos(mx) \cos(nx) \, dx = 0, \quad m \neq n, \ m \text{ and } n \text{ integer}$$

$$\int_0^\pi \sin(mx) \cos(nx) \, dx = \begin{cases} \dfrac{2m}{m^2 - n^2}, & m + n \text{ odd} \\ 0, & m + n \text{ even} \end{cases}$$

$$\int_0^\infty x^{a-1} \cos bx \, dx = \frac{\Gamma(a)}{b^a} \cos\left(\frac{\pi a}{2}\right), \quad 0 < |a| < 1, \quad b > 0$$

$$\int_0^\infty x^{a-1} \sin bx \, dx = \frac{\Gamma(a)}{b^a} \sin\left(\frac{\pi a}{2}\right), \quad 0 < |a| < 1, \quad b > 0$$

$$\int_0^\infty x^n \exp(-ax) \, dx = \frac{n!}{a^{n+1}}, \quad n \text{ an integer and } a > 0$$

$$\int_0^\infty \exp(-a^2 x^2) \, dx = \frac{\sqrt{\pi}}{2|a|}$$

$$\int_0^\infty x^{2n} \exp(-a^2 x^2) \, dx = \frac{1 \cdot (3) \cdot (5) \cdots (2n-1)\sqrt{\pi}}{2^{n+1} a^{2n+1}}, \quad a > 0$$

$$\int_0^\infty \exp(-ax) \cos(bx) \, dx = \frac{a}{a^2 + b^2}, \quad a > 0$$

$$\int_0^\infty \exp(-ax) \sin(bx) \, dx = \frac{b}{a^2 + b^2}, \quad a > 0$$

$$\int_0^\infty \exp(-a^2 x^2) \cos(bx) \, dx = \frac{\sqrt{\pi}}{2a} \exp\left(-\frac{b^2}{4a^2}\right)$$

$$\int_0^\infty x \exp(-ax^2) I_k(bx) \, dx = \frac{1}{2a} \exp\left(-\frac{b^2}{4a}\right), \quad a > 0$$

$$\int_0^\infty \frac{\cos(ax)}{b^2+x^2}\,dx = \frac{\pi}{2b}\exp(-ab), \quad a>0, \quad b>0$$

$$\int_0^\infty \frac{x\sin(ax)}{b^2+x^2}\,dx = \frac{\pi}{2}\exp(-ab), \quad a>0, \quad b>0$$

$$\int_0^\infty \operatorname{sinc}(x)\,dx = \int_0^\infty \operatorname{sinc}^2(x)\,dx = \frac{1}{2}$$

G.5 FOURIER TRANSFORM PAIRS

Signal	Fourier transform
$\Pi(t/\tau) = \begin{cases} 1, & \lvert t\rvert \le \frac{\tau}{2} \\ 0, & \text{otherwise} \end{cases}$	$\tau \operatorname{sinc}(f\tau) = \tau \frac{\sin(\pi f\tau)}{\pi f\tau}$
$2W \operatorname{sinc}(2Wt)$	$\Pi\left(\frac{f}{2W}\right)$
$\Lambda(t/\tau) = \begin{cases} 1-\frac{\lvert t\rvert}{\tau}, & \lvert t\rvert \le \tau \\ 0, & \text{otherwise} \end{cases}$	$\tau \operatorname{sinc}^2(f\tau)$
$W \operatorname{sinc}^2(Wt)$	$\Lambda\left(\frac{f}{W}\right)$
$\exp(-\alpha t)u(t), \quad \alpha > 0$	$\frac{1}{(\alpha + j2\pi f)}$
$t\exp(-\alpha t)u(t), \quad \alpha > 0$	$\frac{1}{(\alpha + j2\pi f)^2}$
$\exp(-\alpha\lvert t\rvert), \quad \alpha > 0$	$\frac{2\alpha}{\alpha^2 + (2\pi f)^2}$
$\exp\left[-\pi\left(\frac{t}{\tau}\right)^2\right]$	$\tau\exp[-\pi(\tau f)^2]$
$\delta(t)$	1
1	$\delta(f)$
$\cos(2\pi f_0 t)$	$\frac{1}{2}\delta(f-f_0) + \frac{1}{2}\delta(f+f_0)$
$\sin(2\pi f_0 t)$	$\frac{1}{2j}\delta(f-f_0) - \frac{1}{2j}\delta(f+f_0)$
$u(t)$	$\frac{1}{j2\pi f} + \frac{1}{2}\delta(f)$
$\frac{1}{(\pi t)}$	$-j\operatorname{sgn} f;\ \operatorname{sgn} f = \begin{cases} 1, & f>0 \\ -1, & f<0 \end{cases}$
$\sum_{m=-\infty}^{\infty}\delta(t-mT_s)$	$f_s\sum_{n=-\infty}^{\infty}\delta(f-nf_s); \quad f_s = \frac{1}{T_s}$

■ G.6 FOURIER TRANSFORM THEOREMS

Name	Time domain operation (signals assumed real)	Frequency domain operation
Superposition	$a_1x_1(t) + a_2x_2(t)$	$a_1X_1(f) + a_2X_2(f)$
Time delay	$x(t - t_0)$	$X(f)\exp(-j2\pi t_0 f)$
Scale change	$x(at)$	$\lvert a\rvert^{-1}X\left(\frac{f}{a}\right)$
Time reversal	$x(-t)$	$X(-f) = X^*(f)$
Duality	$X(t)$	$x(-f)$
Frequency translation	$x(t)\exp(j2\pi f_0 t)$	$X(f - f_0)$
Modulation	$x(t)\cos(2\pi f_0 t)$	$\frac{1}{2}X(f - f_0) + \frac{1}{2}X(f + f_0)$
Convolution*	$x_1(t) * x_2(t)$	$X_1(f)X_2(f)$
Multiplication	$x_1(t)x_2(t)$	$X_1(f) * X_2(f)$
Differentiation	$\frac{d^n x(t)}{dt^n}$	$(j2\pi f)^n X(f)$
Integration	$\int_{-\infty}^{t} x(\lambda)d\lambda$	$\frac{X(f)}{j2\pi f} + \frac{1}{2}X(0)\delta(f)$

*$x_1(t) * x_2(t) \triangleq \int_{-\infty}^{\infty} x_1(\lambda)x_2(t-\lambda)d\lambda$.

REFERENCES

HISTORICAL REFERENCES

1. Gallager, R. G. (1963) *Low Density Parity-Check Codes*, MIT Press, Cambridge, MA.

2. Hartley, R. V. L. (1928) Tranmission of information. *Bell System Technical Journal*, **7**: 535–563.

3. Kotel'nikov, V. A. (1959) *The Theory of Optimum Noise Immunity*, McGraw-Hill, New York. Doctoral dissertation presented in January 1947 before the Acadamic Council of the Molatov Energy Institute of Moscow.

4. Middleton, D. (1960) *An Introduction to Statistical Communication Theory*, McGraw Hill, New York. Reprinted by Peninsula Publishing, 1989.

5. North, D. O. (1943) An analysis of the factors which determine signal/noise discrimination in pulsed-carrier systems. *RCA Technical Report* PTR-6-C, June. Reprinted in the *Proceedings of the IEEE*, **51**: 1016–1027 (July 1963).

6. B. Oliver, J. Pierce, and C. Shannon (1948) The Philosophy of PCM, *Proc. IRE*, **16**: 1324–1331, November.

7. Rice, S. O. (1944) Mathematical analysis of random noise. *Bell System Technical Journal*, **23**: 282–332 (July 1944); **24**: 46–156 (January 1945).

8. Shannon, C. E. (1948) A mathematical theory of communications. *Bell System Technical Journal*, **27**: 379–423, 623–656 (July).

9. Wiener, N. (1949) *Extrapulation, Interpolation, and Smoothing of Stationary Time Series with Engineering Applications*, MIT Press, Cambridge, MA.

10. Woodward, P. M. (1953) *Probability and Information Theory with Applications to Radar*, Pergamon Press, New York.

11. Woodward, P. M. and J. L. Davies (1952) Information theory and inverse probability in telecommunication. *Proceedings of the Institute of Electrical Engineers*, **99**: Pt. III, 37–44 (March).

Books

12. Abramowitz, M. and I. Stegun (1972) *Handbook of Mathematical Functions*, Dover Publications, New York, (reprinting of the original National Bureau of Standards version).

13. Ash, C. (1992) *The Probability Turing Book*, IEEE Press, New York.

14. Bennett W. R. (1974) Envelope Detection of a Unit-Index Amplitude Modulated Carrier Accompanied by Noise. *IEEE Trans. Information Theory, **IT-20***, 723-728 (November, 1974).

15. Biglieri, B., D. Divsalar, P. J. McLane, and M. K. Simon (1991) *Introduction to Trellis-Coded Modulation with Applications*, Macmillan, New York.

16. Blahut, R. E. (1987) *Principles and Practice of Information Theory*, Addison-Wesley, Reading, MA.

17. Bracewell, R. (1986) *The Fourier Transform and Its Applications*, 2nd ed., McGraw-Hill, New York.

18. Carlson, A. B., P. B. Crilly, and J. Rutledge (2001) *Communication Systems*, 4th ed., McGraw-Hill, New York.

19. Clark, G. C., Jr. and J. B. Cain (1981) *Error-Correction Coding for Digital Communications*, Plenum Press, New York.

20. Couch, L. W., II (2007) *Digital and Analog Communication Systems*, 7th ed., Pearson Prentice Hall, Upper Saddle River, NJ.

21. Cover, T. M. and J. A. Thomas (2006) *Elements of Information Theory*, 2nd ed., Wiley, New York.

22. Gallager, Robert G. (1968) *Information Theory and Reliable Communication*, Wiley, New York.

23. Gardner, F. M. (1979) *Phaselock Techniques*, 2nd ed., Wiley, New York.

24. Gibson, J. D. (ed.) (2002) *The Communications Handbook*, 2nd ed., CRC Press, Boca Raton, FL.

25. Goldsmith, A. (2005) *Wireless Communications*, Cambridge University Press, Cambridge, UK.

26. Haykin, S. (1996) *Adaptive Filter Theory*, 3rd ed., Upper Saddle River, NJ: Prentice Hall.

27. Haykin, S. (2000) *Communication Systems*, 4th ed., Wiley, New York.

28. Heegard C. and S. B. Wicker (1999) *Turbo Coding*, Kluwer Academic Publishers, Boston, MA.

29. Helstrom, C. W. (1968) *Statistical Theory of Signal Detection*, 2nd ed., Pergamon Press, New York.

30. Holmes, J. K. (1987) *Coherent Spread Spectrum Systems*, Wiley, New York.

31. R. C. Houts and R. S. Simpson, "Analysis of waveform distortion in linear systems," IEEE Transactions on Eduction, vol. 12, pp. 122–125, June 1968.

32. Kamen, E. W. and B. S. Heck (2007) *Fundamentals of Signals and Systems*, 3rd ed., Prentice Hall, Upper Saddle River, NJ.

33. Kay, S. M. (1993) *Fundamentals of Statistical Signal Processing: Volume I, Estimation Theory*, Prentice Hall PTR, Upper Saddle River, NJ.

34. Kay, S. M. (1998) *Fundamentals of Statistical Signal Processing: Volume II, Detection Theory*, Prentice Hall PTR, Upper Saddle River, NJ.

35. Kobb, B. Z. (1996) *Spectrum Guide*, 3rd ed., New Signals Press, Falls Church, VA.

36. Kobb, B. Z. (2001) *Wireless Spectrum Finder*, McGraw-Hill, New York.

37. Lathi, B. P. (1998) *Modern Analog and Digital Communication Systems*, 3rd ed., Oxford University Press, Oxford.

38. Leon-Garcia, A. (1994) *Probability and Random Processes for Electrical Engineering*, Addison-Wesley, Reading, MA.

39. Liberti, J. C. and T. S. Rappaport (1999) *Smart Antennas for Wireless Communications: IS-95 and Third Generation CDMA Applications*, Prentice Hall PTR, Upper Saddle River, NJ.

40. Lin, S. and D. J. Costello, Jr. (2004) *Error Correcting Coding: Fundamentals and Applications*, 2nd ed., Prentice Hall, Upper Saddle River, NJ.

41. Lindsey, W. C. and M. K. Simon (1973), *Telecommunications Systems Engineering*, Prentice Hall, Upper Saddle River, NJ.

42. Mark, J. W. and W. Zhuang (2003) *Wireless Communications and Networking*, Prentice Hall, Upper Saddle River, NJ.

43. McDonough, R. N. and A. D. Whalen (1995) *Detection of Signals in Noise*, Academic Press, San Diego, CA.

44. Meyr, H. and G. Ascheid (1990) *Synchronization in Digital Communications*, Volumes I and II, Wiley, New York.

45. Mumford, W. W. and E. H. Scheibe (1968) *Noise Performance Factors in Communication Systems*, Horizen House-Microwave, Dedham, MA.

46. Ott, H. W. (1988) *Noise Reduction Techniques in Electronic Systems*, John Wiley and Sons, New York.

47. Paulraj, A., R. Nabar, and D. Gore (2003) *Introduction to Space-Time Wireless Communications*, Cambridge University Press, Cambridge, UK.

48. Papoulis, A. (1991) *Probability, Random Variables, and Stochastic Processes*, 3rd ed., McGraw-Hill, New York.

49. Peterson, R. L., R. E. Ziemer, and D. A. Borth (1995) *Introduction to Spread Spectrum Communications*, Prentice Hall, Upper Saddle River, NJ.

50. Poor, H. V. (1994) *An Introduction of Signal Detection and Estimation*, Springer, New York.

51. Proakis, J. G. (2007) *Digital Communications*, 6th ed., McGraw-Hill, New York.

52. Proakis, J. G. and M. Salehi (2005) *Fundamentals of Communication Systems*, Prentice Hall, Upper Saddle River, NJ.

53. Rappaport, T. S. (1996) *Wireless Communications: Principles and Practice*, Prentice Hall, Upper Saddle River, NJ.

54. Ross, S. (2002) *A First Course in Probability*, 6th ed., Prentice Hall, Upper Saddle River, NJ.

55. Scharf, L. (1990) *Statistical Signal Processing*, Addison-Wesley, Reading, MA.

56. Shannon, C. E. and W. Weaver (1963) *The Mathematical Theory of Communication*, University of Illinois Press, Urbana, IL.

57. Siebert, William (1986) *Circuits, Signals, and Systems*, McGraw-Hill, New York.

58. Simon, M. K. (2002) *Probability Distributions Involving Gaussian Random Variables*, Kluwer Academic Publishers, Boston.

59. Simon, M. K. and M. -S. Alouini (2005) *Digital Communication over Fading Channels: A Unified Approach to Performance Analysis*, Wiley, 2nd ed., New York.

60. Simon, M. K., S. M. Hinedi, and W. C. Lindsey (1995) *Digital Communication Techniques: Signal Design and Detection*, Prentice Hall, New York.

61. Sklar, B. (2001) *Digital Communications: Fundamentals and Applications*, 2nd ed., Prentice Hall, Upper Saddle River, NJ.

62. Skolnik, M. I. (ed.) (1970) *Radar Handbook*, 2nd ed., McGraw-Hill, New York.

63. Stiffler, J. J. (1971) *Theory of Synchronous Communications*, Prentice Hall, Upper Saddle River, NJ.

64. Stuber, G. L. (2001) *Principles of Mobile Communication*, 2nd ed., Kluwer Academic Publishers, Boston, MA.

65. Taub, H. and D. L. Schilling, (1986) *Principles of Communication Systems*, 2nd ed., McGraw-Hill, New York.

66. Tranter, W. H., K. S. Shanmugan, T. S. Rappaport, and K. L. Kosbar (2004) *Principles of Communication Systems Simulation with Wireless Applications*, Prentice Hall, Upper Saddle River, NJ.

67. Tse, D., and P. Viswanath (2005) *Fundamentals of Wireless Communication*, Cambridge University Press, Cambridge, UK.

68. Van der Ziel, A. (1970) *Noise: Sources, Characterization, Measurement*, Prentice Hall, Upper Saddle River, NJ.

69. Van Trees, H. L., (1968) *Detection, Estimation, and Modulation Theory*, Vol. I, Wiley, New York.

70. Van Trees, H. L., (1970) *Detection, Estimation, and Modulation Theory*, Vol. II, Wiley, New York.

71. Van Trees, H. L., (1971) *Detection, Estimation, and Modulation Theory*, Vol. III, Wiley, New York.

72. Verdu, S. (1998) *Multiuser Detection*, Cambridge University Press, Cambridge, UK.

73. Viterbi, A. J. (1966) *Principles of Coherent Communication*, McGraw-Hill, New York.

74. Viterbi, A. J., J. K. Wolf, E. Zehavi and R. Padovani. "A Pragmatic Approach to Trellis-Coded Modulation," *IEEE Communications Magazine*, **27**, 11–19 (July 1989).

75. Walpole, R. E., R. H. Meyers, S. L. Meyers, and K. Ye (2007) *Probability and Statistics for Engineers and Scientists*, 8th ed., Prentice Hall, Upper Saddle River, NJ.

76. Williams, A. B., and F. J. Taylor (1988) *Electronic Filter Design Handbook*, 2nd ed., McGraw-Hill, New York.

77. Wozencraft, J. M., and I. M. Jacobs, (1965) *Principles of Communication Engineering*, Wiley, New York.

78. Ziemer, R. E., and R. L. Peterson (2001) *Introduction to Digital Communication*, 2nd ed., Prentice Hall, Upper Saddle River, NJ.

79. Ziemer, R. E., W. H. Tranter, and D. R. Fannin (1998) *Signals and Systems: Continuous and Discrete*, 4th ed., Prentice Hall, Upper Saddle River, NJ.

AUTHOR INDEX

SUBJECT INDEX